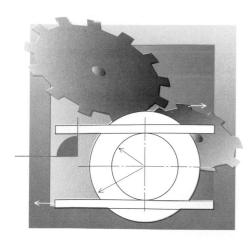

Technical Drawing

FIFTH EDITION

Technical Drawing

FIFTH EDITION

DAVID L. GOETSCH ■ **WILLIAM S. CHALK**
JOHN A. NELSON ■ **RAYMOND L. RICKMAN**

THOMSON

DELMAR LEARNING

Australia Canada Mexico Singapore Spain United Kingdom United States

THOMSON

DELMAR LEARNING

Technical Drawing, Fifth edition
David L. Goetsch, William S. Chalk, John A. Nelson, and Raymond L. Rickman

Vice President, Technology and Trades SBU:
Alar Elken

Editorial Director:
Sandy Clark

Senior Acquisitions Editor:
James DeVoe

Developmental Editor:
Christopher Shortt

Channel Manager:
Dennis Williams

Marketing Coordinator:
Casey Bruno

Production Director:
Mary Ellen Black

Production Manager:
Andrew Crouth

Production Editor:
Stacy Masucci

Art/Design Specialist:
Mary Beth Vought

Technology Project Manager:
Kevin Smith

Technology Project Specialist:
Linda Verde

Editorial Assistant:
Tom Best

Library of Congress Cataloging-in-Publication Data:
Card Number:

ISBN: 1-4018-5760-4

NOTICE TO THE READER

Brief Contents

Contents

Section 3

DESIGN DRAFTING APPLICATIONS

Section 4
RELATED TECHNOLOGIES, APPLICATIONS, AND PROCESSES

Section 5
EMPLOYABILITY SKILLS

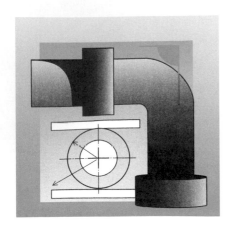

P r e f a c e

PURPOSES

Technical Drawing is intended for use in such courses as basic and advanced drafting, engineering graphics, descriptive geometry, mechanical drafting, machine drafting, tool and die design and drafting, and manufacturing drafting. It is appropriate for those courses offered in comprehensive high schools, area vocational schools, technical schools, community colleges, trade and technical schools, and at the freshman and sophomore levels in universities.

PREREQUISITES

There are no prerequisites. The text begins at the most basic level and moves step by step to the advanced levels. It is as well suited for students who have had no previous experience with technical drawing as it is for students with a great deal of prior experience.

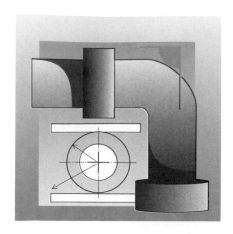

How To Use This Book

Technical Drawing is a comprehensive teaching and learning tool that contains several special features to promote the student's development and to make learning easier. Students and instructors can make use of the following features:

Career Profiles located in the Section Opener focus on the occupations of specific individuals. These profiles were chosen as representative of the types of jobs that students could acquire after completing an education in drafting. The career profiles also relate directly to the content covered in the section.

CAREER
PROFILE

Career Profile: Jon Whitney

In college, Jon Whitney studied English. On a whim, he applied to graduate school for architecture, and it ended up being the right decision.

After the three-and-one-half-year program, he began as a drafter at Goshow Architects in New York City. This meant he was responsible for turning architects' visions into dimensioned schematics. He would often go on-site to take measurements, then bring them back to insert into the AutoCAD program. He also would "pick up red marks"; that is, after a superior reviewed his drawings, Whitney served as a draftsman for three years, a typical amount of time for someone just out of an architecture master's program (although many drafters choose to keep drafting indefinitely).

While serving as a drafter, Whitney accumulated credits toward his IDP (internship development program); when you get enough credits, you can begin the year-long testing process to get your architects' license. Whitney is currently in the midst of that testing. Now that he is a junior project manager, he is still drafting, but he is also coordinating up to eight projects, each of which is in a different stage of the process.

The first stage is the schematic phase. At this pre-design stage, Whitney is likely to create a not-to-scale schematic by hand, making bubble diagrams to give a broad, general look at the project. For instance, how would a loft space look if it was adapted for office use? Often this stage is conceptual and artistic.

The next stage is design development, at which point the drawing starts to become more technical. This involves refining the schematic, starting to put down dimensions and materials. How big will the walls dividing offices be? Will they be made of masonry or steel? Do they

need windows? Whitney may also call in subcontractors (hazmat, mechanical engineers, landmarks conservancy consultants, etc.) if necessary.

Next, Whitney and his team will create construction documents, the technical drawings that are given to the contractor at the start of the project. These must be as clear as possible; in fact, Whitney says, this stage is never completed. One can always go one level deeper with detail. Therefore, you should do as much as possible in the allotted time. As a drafter, Whitney used to work primarily in this phase. Now, most mornings he will spend a few hours reviewing drawings (done either by himself or one of his drafters) with a red pen, looking for everything from spelling errors to technical problems. He must also look for clarity and any misleading information. In essence, the client is paying for the comprehensive and accurate packet of drawings and specifications developed here.

Next comes the bidding and contractor selection. Once a contractor is chosen, the architect hands over the construction documents and work on the project begins. In the construction administration phase, the architect and drafters may be called in to provide clarifying drawings on an unpredicted on-site issue (the floor plan may be 4" off, or perhaps there is a pipe behind a wall that wasn't noted in the drawings).

Being a project manager has been a big, but exciting, change for Whitney. He's gotten to step outside the tunnel vision required of drafters and has been able to get a broader look at each project. As a result, he sees each plan go from an artistic idea to a technical reality.

■

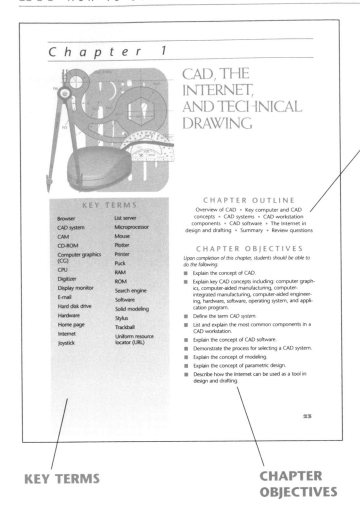

CHAPTER
OUTLINE

INDUSTRY
APPLICATION

KEY TERMS

CHAPTER
OBJECTIVES

Chapter Objectives at the beginning of each chapter identify the skills and knowledge that students will acquire from reading the material. When students have finished the chapter, they should review the objectives to ensure that they have met each one.

The **Chapter Outline** lists the title of each major topic covered in the chapter. This provides a preview of the content coverage.

Key Terms, listed at the beginning of each chapter, are important words and phrases that students will encounter as they study the chapter. Each key term is highlighted in **italics** at the first significant use in the text, and is included in the glossary or index at the back of the book.

Industry Application is a boxed article that explains how the skills and knowledge discussed in the chapter may be applied to a real-world job-related setting. A variety of skills will be covered, including math, science, communications, and computers.

A chapter **Summary** provides a recapitulation of the key topics covered in the chapter. This enables students to reassess their comprehension of the material before proceeding to the end-of-chapter questions and problems.

Review Questions are written to ensure that students have adequately read the chapter and that they understand the material. These questions prepare students for class tests given by the instructor and for the certification test offered by the American Design and Drafting Association. (Visit the American Design and Drafting Association web site at **www.adda.org** to learn about the certification programs and ADDA membership.) These questions do not require a computer or drafting materials.

Chapter Problems contain drawing projects that allow students to test their drafting skills. Advanced problems are marked by **icons**. These special icons have been placed adjacent to the advanced drawing projects to indicate the engineering and design field the drawing pertains to. These icons call out a specific drafting field or discipline.

Delmar's Drafting Web Site is located at **www.Delmar.com/drafting** This site is a tool to be used in conjunction with the material in this text. The site has links to industrial organizations, industry web sites, drafting certification sites, and the *Technical Drawing Online Companion*™. This Online Companion provides access to equivalents tables, fits tables, dimensions tables for screws and nuts, and much more.

CHAPTER PROBLEMS

SUMMARY

REVIEW QUESTIONS

ICONS

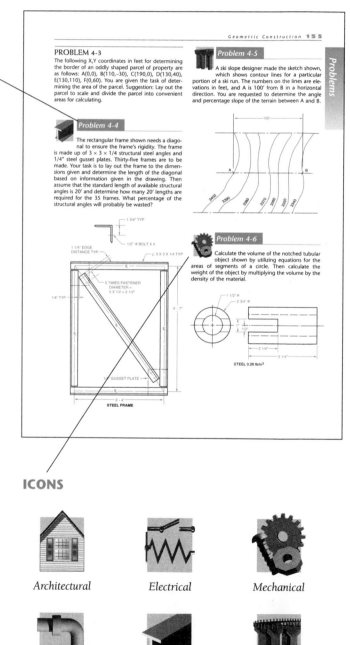

CONTENT OVERVIEW

Chapters 1–18 are to be used to help students develop the design and drafting skills that are fundamental to *all* drafting fields. Instructors are encouraged to use all of these chapters to build a solid footing of design and drafting knowledge for students.

Chapters 19–24 allow students to develop advanced knowledge and skills beyond the fundamentals. Instructors are encouraged to use these chapters to help their students develop an in-depth understanding of these discipline areas: Welding; Modern Manufacturing: Materials, Processes, and Automation; Drafting Applications: Pipe, Architectural, and Civil Engineering; Electronics and Printed Circuit Boards; Charts and Graphs; and The Design Process.

Chapter 24, The Design Process, is important in preparing the student for entry into the job force. The steps of the design process are defined and the reader is taught how to be creative in analyzing and solving problems. Modern design processes such as DFM, DFA, rapid prototyping, and reverse engineering are addressed.

Chapter 25 is provided to help students who have developed the technical skills needed to succeed on the job learn how to get a job. Instructors are encouraged to use this chapter to help students develop and refine their job-seeking skills.

Jobs in the modern workplace are done in teams. In order to succeed in today's workplace, design and drafting technicians must be good team players. Instructors are encouraged to use Chapter 26 to help students develop teamwork skills before completing their training.

INNOVATIONS

An advantage of the text is that it has evolved during a time when the world of technical drawing and design is going through a period of major transition from manual to automated techniques. Computer-aided drafting (CAD) has become the standard tool, but elements of the traditional methods and tools remain.

This transition created a need for a major text that deals with both CAD and traditional knowledge and skills. *Technical Drawing* fills this need. Even though the world of technical drawing and design has become automated, drafters and designers will still need to know the traditional basics and technical drawing fundamentals. Therefore, the fundamentals are treated in depth in this text.

What has changed, and will continue to change, is the way that drafters and designers prepare technical drawings. For this reason, CAD is treated in depth, and most of the drawings and illustrations were prepared on various CAD systems. Along with this treatment, *Technical Drawing* offers students and instructors a special blend of the manual and automated knowledge and techniques that will be needed into the twenty-first century.

Another advantage of the text is that it was written after the latest update of the most frequently used drafting standard—the ANSI series. Consequently, all dimensioning and tolerancing material in *Technical Drawing* is based on this most recent edition of the standard.

NEW FEATURES IN THE FIFTH EDITION

1. Chapter 1 was updated to reflect modern technological advances in such areas as CAD and solid modeling.

2. Chapter 5 was updated to include new material on three-dimensional visualization.

3. Chapter 11 was updated to strengthen the coverage of solid modeling.

4. Chapter 20 was updated to include a section on thermoplastics and thermosets.

5. Chapter 21 was updated to include coverage of geographic information systems (GIS), more civil engineering material, and coverage of preengineered/prefabricated metal buildings.

6. Chapter 26 was updated to include material on quality and competition in addition to teamwork.

7. All dimensional and drawing errors pointed out by reviewers and instructors who use the text were corrected.

8. CAD Instructions have been added for all drawing projects that are to be completed mechanically.

9. The supplemental package has been updated and expanded.

TESTED AND PROVEN FEATURES

- An enhanced supplement package contains a new Workbook, Instructor's Manual, e.resource CD-ROM with electronic instructional material, and access to World Class Learning™ distributed learning courseware.

- Chapter 24 (The Design Process) covers modern design practices and standards.

- Chapter 25 (Finding, Getting, and Keeping a Job) teaches students how to market themselves.

- Chapter 26 (Quality, Competition, and Teamwork on the Job) teaches students how to succeed in today's work environment.

- Discipline-specific icons highlight the drawing problems, enabling the reader to quickly choose problems pertaining to his or her drawing field.

- Chapter 1 (CAD, the Internet, and Technical Drawing) has been updated to contain Internet coverage.

- Review questions are formatted as multiple-choice and true/false for rapid testing and grading.

- New drawing problems are contained in the primary drawing chapters.

- Boxed articles covering real-world applications exist in all chapters. This reinforces the relevance of the chapter content to today's job environment.

- New math problems are contained in selected chapters for additional practice.

- Step-by-step explanations of drawing procedures and techniques.

- Written in language students will understand; technical terms are defined as they are used.

- Unique black and plum color format depicts isometric views more clearly than "flat" black-and-white drawings.

- Text and illustrations are located in *direct* relationship to each other wherever possible.

- Real-world techniques are highlighted in the Industry Application boxed articles.

- Although the emphasis is on mechanical drafting, other pertinent drafting subjects are included for a comprehensive, well-rounded approach to technical drawing.

- Contains in-depth drafting applications in architectural, structural, civil, piping, and electronics drafting.

THE LEARNING PACKAGE

The complete ancillary package was developed to achieve two goals:

1. To assist students in learning the essential information needed to prepare for the exciting field of drafting.

2. To assist instructors in planning and implementing their instructional programs for the most efficient use of time and other resources.

The *Technical Drawing* package was created as an integrated whole. Supplements are linked to and integrated with the text to create a comprehensive supplement package that supports students and instructors, beginning or veteran. The package includes:

Workbook—This newly revised and expanded workbook assists in developing the skills needed to create and read technical drawings. Advanced coverage includes dimensioning and tolerancing and commercial components such as fasteners, gears, cams, and springs. ISBN: 1-4018-5761-2.

Instructor's Manual—This ancillary contains general information, lesson plans, objective tests, and solutions to the text problems and review questions. Solutions are also provided for the *Workbook*, including the solutions to the CAD problems. ISBN: 1-4018-5763-9.

e.resource™—This is an educational resource that creates a truly electronic classroom. It is a CD-ROM containing tools and instructional resources that enrich your classroom and make your preparation time shorter. The elements of *e.resource* link directly to the text and tie together to provide a unified instructional system. With *e.resource* you can spend your time teaching, not preparing to teach. ISBN: 1-4018-5764-7.

Features contained in *e.resource* include:

- **Syllabus:** Lesson plans created by chapter. You have the option of using these lesson plans with your own course information.

- **Chapter Hints:** Objectives and teaching hints that provide the basis for a lecture outline that helps you to present concepts and material. Key points and concepts can be graphically highlighted for student retention.

- **Answers to Review Questions:** These solutions enable you to grade and evaluate end-of-chapter tests.

- **PowerPoint® Presentation:** These slides provide the basis for a lecture outline that helps you to present concepts and material. Key points and concepts can be graphically highlighted for student retention.

- **Computerized Test Bank:** Over 800 questions of varying levels of difficulty are provided in true/false and multiple-choice formats so you can assess student comprehension.

- **CAD Drawing Files:** Drawing files are provided that link to problems in the *Workbook*. Students download the files to complete the problems.

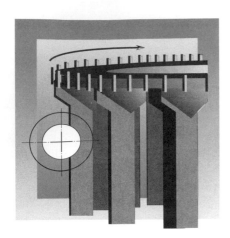

A c k n o w l e d g m e n t s

The authors would like to acknowledge the efforts of many people without whose assistance this project would not have been completed. Special acknowledgment is made to Evan Yares who revised and updated Jerry L. Roiter's Chapter 11, "Solid/3D Modeling: Computational Design and Analysis." We thank Deborah M. Goetsch for her assistance with photography and typing.

The following individuals reviewed the manuscript and made valuable suggestions to the authors. The authors and the publisher greatly appreciated their contributions to this textbook.

Robert Fox, Adirondack Community College

Michael Sinarski, General Motors Technical Academy

Art Leonard, Universal Technical Institute

George Bacza, Fairmont State College

Jim Hughes, Calhoun Community College

Jim Branciforte, Naugatuck Valley Technical
Community College

Ryan Brown, Illinois State University

Dan Walker, Southeast Career Center

John A. Warren, Mukwonago High School

Frank Katona, ITT Technical Institute

Timothy Riordan, ITT Technical Institute

Tom Bledsaw, ITT Technical Institute

Randy Blaschko, South Central Technical College

ABOUT THE AUTHORS

David L. Goetsch is Vice-President and Professor of Design, Drafting, Quality, and Safety at Okaloosa-Walton College in Niceville, Florida. His drafting and design program has won national acclaim for its pioneering efforts in the area of computer-aided drafting (CAD). In 1984, his school was selected as one of only 10 schools in the country to earn the distinguished Secretary's Award for an outstanding Vocational Program. Professor Goetsch is a widely acclaimed teacher, author, and lecturer on the subject of drafting and design. He won Outstanding Teacher of the Year honors in 1976, 1981, 1982, 1983, and 1984. In 1986 he won the Florida Vocational Association's Rex Gaugh Award for outstanding contributions to technical education in Florida. In 2003, Dr. Goetsch was selected as the University of West Florida's Distinguished Alumnus. He entered education full time after a successful career in design and drafting in the private sector, where he spent more than eight years as a Senior Drafter and Designer for a subsidiary of Westinghouse Corporation. He was elected Florida's Outstanding Economic Development person in 1992 and 1996. He is president of The Management Institute (TMI) and does consulting in the areas of design, drafting, quality, and safety. This is his 49th book.

Professor Emeritus **William S. Chalk** taught for 31 years in the College of Engineering at the University of Washington in the departments of General, Mechanical, and Nuclear engineering, and was Director of Nuclear Engineering Laboratories from 1976 to 1983. Professor Chalk had seven years of experience in industry with the Link-Belt Company and has worked with Boeing Airplane Company, The Puget Sound Naval Shipyard, and The Sandia Corporation.

Always interested in the practical applications of education, Professor Chalk has conducted a number of nationwide design workshops for industry. He is a recipient of Western Electric's Outstanding Instructor of the Year award. In addition, he is a former associate editor of *Design Graphics Journal.*

Professor Chalk holds a master's degree in mechanical engineering from the University of Washington and has participated in design training workshops at Dartmouth College and Stanford University. He is the coauthor of an engineering graphics text and workbook and is the author of a past ASEE paper.

Professor Chalk continues to be active in engineering by lecturing to engineers who plan to take engineering license examinations, and by lecturing in summer programs.

Jerry L. Roiter, author of Chapter 11, "Solid/3D Modeling: Computational Design and Analysis," has been teaching at the university level since 1972. He is currently a faculty member at the University of Wisconsin. Prior to beginning teaching he worked in industry as a designer and continues to work in industry on a regular basis as well as serving as a consultant. He was formerly Assistant Director of Visualization at the Cornell National Supercomputer Facility. He holds the bachelor of science, master of arts, and Doctor of Philosophy degrees.

Evan Yares, contributing author of Chapter 11, "Solid/3D Modeling: Computational Design and Analysis," is a strategic CAD consultant, working with vendors and users to help make CAD both easier and more productive. He is a contributor to several magazines, including CADalyst and Computer Graphics World, and is a well-known speaker on CAD subjects.

Raymond L. Rickman is chairman of the Manufacturing and Technology Department and Professor of Design and Drafting at Okaloosa-Walton College. Professor Rickman has extensive experience in the private sector and the classroom and is a consultant on the subject of geometric dimensioning and tolerancing. He is a member of the ANSI Y14.5 technical committee and the co-author of several career mathematics textbooks.

SECTION 1

Basics

Career Profile: Jon Whitney

In college, Jon Whitney studied English. On a whim, he applied to graduate school for architecture, and it ended up being the right decision.

After the three-and-one-half-year program, he began as a drafter at Goshow Architects in New York City. This meant he was responsible for turning architects' visions into dimensioned schematics. He would often go on-site to take measurements, then bring them back to insert into the AutoCAD program. He also would "pick up red marks"; that is, after a superior reviewed his drawings, Whitney served as a draftsman for three years, a typical amount of time for someone just out of an architecture master's program (although many drafters choose to keep drafting indefinitely).

While serving as a drafter, Whitney accumulated credits toward his IDP (internship development program); when you get enough credits, you can begin the year-long testing process to get your architects' license. Whitney is currently in the midst of that testing. Now that he is a junior project manager, he is still drafting, but he is also coordinating up to eight projects, each of which is in a different stage of the process.

The first stage is the schematic phase. At this pre-design stage, Whitney is likely to create a not-to-scale schematic by hand, making bubble diagrams to give a broad, general look at the project. For instance, how would a loft space look if it was adapted for office use? Often this stage is conceptual and artistic.

The next stage is design development, at which point the drawing starts to become more technical. This involves refining the schematic, starting to put down dimensions and materials. How big will the walls dividing offices be? Will they be made of masonry or steel? Do they need windows? Whitney may also call in subcontractors (hazmat, mechanical engineers, landmarks conservancy consultants, etc.) if necessary.

Next, Whitney and his team will create construction documents, the technical drawings that are given to the contractor at the start of the project. These must be as clear as possible; in fact, Whitney says, this stage is never completed. One can always go one level deeper with detail. Therefore, you should do as much as possible in the allotted time. As a drafter, Whitney used to work primarily in this phase. Now, most mornings he will spend a few hours reviewing drawings (done either by himself or one of his drafters) with a red pen, looking for everything from spelling errors to technical problems. He must also look for clarity and any misleading information. In essence, the client is paying for the comprehensive and accurate packet of drawings and specifications developed here.

Next comes the bidding and contractor selection. Once a contractor is chosen, the architect hands over the construction documents and work on the project begins. In the construction administration phase, the architect and drafters may be called in to provide clarifying drawings on an unpredicted on-site issue (the floor plan may be 4″ off, or perhaps there is a pipe behind a wall that wasn't noted in the drawings).

Being a project manager has been a big, but exciting, change for Whitney. He's gotten to step outside the tunnel vision required of drafters and has been able to get a broader look at each project. As a result, he sees each plan go from an artistic idea to a technical reality.

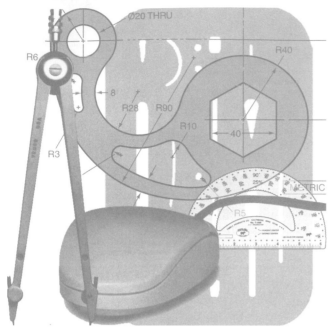

GRAPHIC COMMUNICATION AND TECHNICAL DRAWING

KEY TERMS

Axonometric projections

Cabinet

Cavalier

Design for manufacturability

Design process

Drafters

Drafting

Drafting technicians

Drawing

Graphic communication

Oblique projections

Orthographic projections

Parallel projection

Perspective projections

Projection

Projector

Technical drawing

CHAPTER OUTLINE

Graphic communication • Drawings described • Types of drawings • Types of technical drawings • Purpose of technical drawings • Applications of technical drawings • Regulation of technical drawings • What students of technical drawing, drafting, and CAD should learn • Technical drawing and quality/competitiveness • Summary • Review questions • Introduction problems

CHAPTER OBJECTIVES

Upon completion of this chapter, students should be able to do the following:

- Explain the concept of graphic communication.
- Define the term *drawing*.
- Differentiate between artistic and technical drawings.
- List and explain the types of technical drawings.
- Explain the purpose of technical drawings.
- Explain the different applications of technical drawings.
- Explain the concept of the regulation of technical drawings.
- Describe the role of design and drafting in promoting quality and competitiveness.

Graphic Communication

Graphic communication involves using visual material to relate ideas. Drawings, photographs, slides, transparencies, and sketches are all forms of graphic communication. Most children are able to draw before they are able to write. This is graphic communication. When one person sketches a rough map in giving another directions, this is graphic communication. Any medium that uses a graphic image to aid in conveying a message, instructions, or an idea is involved in graphic communication. One of the most widely used forms of graphic communication is the drawing.

Drawings Described

A *drawing* is a graphic representation of an idea, a concept, or an entity that actually or potentially exists in life. The drawing itself is: (1) a way of communicating all necessary information about an abstraction, such as an idea or a concept; or (2) a graphic representation of some real entity, such as a machine part, a house, or a tool, for example.

Drawing is one of the oldest forms of communication, dating back even farther than verbal communication. Cave dwellers painted drawings on the walls of their caves thousands of years before paper was invented. These crude drawings served as a means of communicating long before verbal communications had developed beyond the grunting stage. In later years, Egyptian hieroglyphics were a more advanced form of communicating through drawings.

The old adage "one picture is worth a thousand words" is still the basis of the need for technical drawings.

Types of Drawings

There are two basic types of drawings: artistic and technical. Some experts believe there are actually three types: the two mentioned and another type that combines these two. The third type is usually referred to as an illustration or rendering.

ARTISTIC DRAWINGS

Artistic drawings range in scope from the most simple line drawings to the most famous paintings. Regardless of their complexity or status, artistic drawings are used to express the feelings, beliefs, philosophies, or abstract ideas of the artist. This is why the lay person often finds it difficult to understand what is being communicated by a work of art.

In order to understand an artistic drawing, it is sometimes necessary to first understand the artist. Artists often take a subtle or abstract approach in communicating through their drawings. This gives rise to the various interpretations often associated with artistic drawings.

TECHNICAL DRAWINGS

The technical drawing, on the other hand, is not subtle or abstract. It does not require an understanding of its creator, only an understanding of technical drawings. A *technical drawing* is a means of clearly and concisely communicating all of the information necessary to transform an idea or a concept into reality. Therefore, a technical drawing often contains more than just a graphic representation of its subject. It also contains dimensions, notes, and specifications.

The mark of a good technical drawing is that it contains all of the information needed by individuals for converting the idea or concept into reality. The conversion process may involve manufacturing, assembly, construction, or fabrication. Regardless of the process involved, a good technical drawing allows the conversion process to proceed without having to ask designers or drafters for additional information or clarification.

Figure I-1 and *Figure I-2* contain samples of technical mechanical drawings that are used as guides by the people involved in various phases of manufacturing the represented parts. Notice that the drawings contain a graphic representation of the part, dimensions, material specifications, and notes.

ILLUSTRATIONS OR RENDERINGS

Illustrations or renderings are sometimes referred to as a third type of drawing because they are neither completely technical nor completely artistic; they combine elements of both, as shown in *Figures I-3, I-4, I-5,* and *I-6* (pages 5 to 6). They are technical in that they are drawn with mechanical instruments or on a computer-aided drafting system, and they contain some degree of technical information. However, they are also artistic in that they attempt to convey a mood, an attitude, a status, or other abstract, nontechnical feelings.

Types of Technical Drawings

Technical drawings are based on the fundamental principles of projection. A *projection* is a drawing or representation of an entity on an imaginary plane or planes. This projection plane serves the same purpose in technical drawing as is served by the movie screen in a theater.

As can be seen in *Figure I-7* (page 7), a projection involves four components: (1) the actual object that the drawing or projection represents, (2) the eye of the viewer looking at the object, (3) the imaginary projection

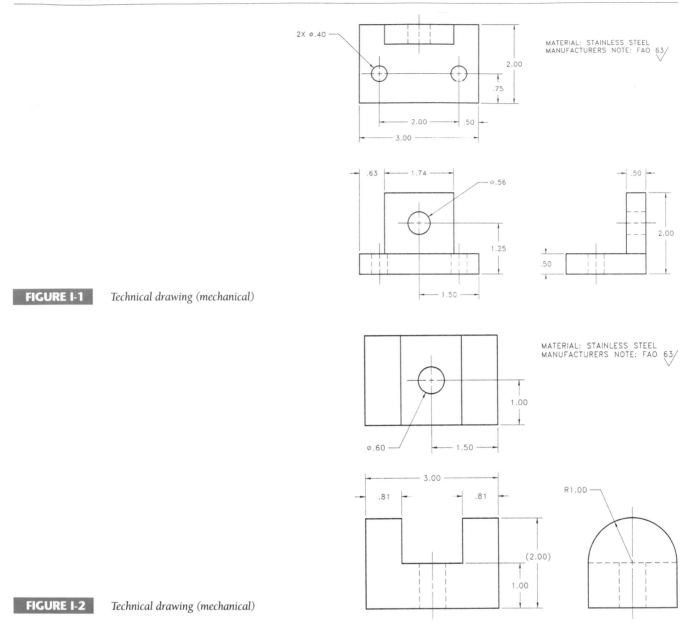

FIGURE I-1 *Technical drawing (mechanical)*

FIGURE I-2 *Technical drawing (mechanical)*

plane (the viewer's drawing paper or the graphics display in a computer-aided drafting system), and (4) imaginary lines of sight called *projectors*.

Two broad types of projection, both with several subclassifications, are parallel projection and perspective (converging) projection.

PARALLEL PROJECTION

Parallel projection is subdivided into the following three categories: orthographic, oblique, and axonometric projections.

Orthographic projections are drawn as multiview drawings that show flat representations of principal views of the subject, *Figure I-8* (page 7) *Oblique projections* actually show the full size of one view and are of

three varieties: *cabinet* (half-scale), *cavalier* (full-scale), and general (between half and full-scale). *Figures I-9* and *I-10* (page 7) shows cavalier and cabinet projections. *Axonometric projections* are three-dimensional drawings and are of three different varieties: isometric, dimetric, and trimetric, *Figures I-11, I-12,* and *I-13* (page 8).

PERSPECTIVE PROJECTION

Perspective projections are drawings that attempt to replicate what the human eye actually sees when it views an object. That is why the projectors in a perspective drawing converge. There are three types of perspective projections: one-point, two-point, and three-point projections, *Figures I-14, I-15,* and *I-16* (page 8).

FIGURE I-3   *Rendering*

FIGURE I-4   *Rendering*

Purpose of Technical Drawings

To appreciate the need for technical drawings, one must understand the design process. The design process is an orderly, systematic procedure used in accomplishing a needed design.

Any product that is to be manufactured, fabricated, assembled, constructed, built, or subjected to any other type of conversion process must first be designed. For example, a house must be designed before it can be built. An automobile must be designed before it can be

FIGURE I-5   *Mechanical illustration* (Courtesy Ken Elliott)

FIGURE I-6   *Mechanical illustration* ((Courtesy Ken Elliott)

manufactured. A printed circuit board must be designed before it can be fabricated.

THE DESIGN PROCESS

The *design process* is an organized, step-by-step procedure in which mathematical and scientific principles, coupled with experience, are brought to bear in order to solve a problem or meet a need. The design process has five steps. Traditionally, these steps have been (1) identification of the problem or a need, (2) development of initial ideas for solving the problem, (3) selection of a proposed solution, (4) development and testing of models or prototypes, and (5) developing working drawings, *Figure I-17* (page 9).

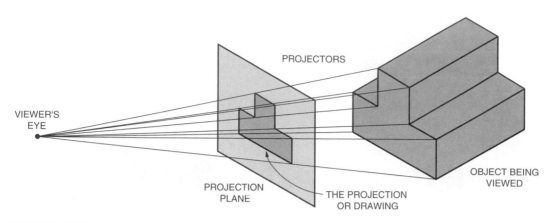

PROJECTORS

VIEWER'S
EYE

PROJECTION
PLANE

THE PROJECTION
OR DRAWING

OBJECT BEING
VIEWED

FIGURE I-7   *The projection plane*

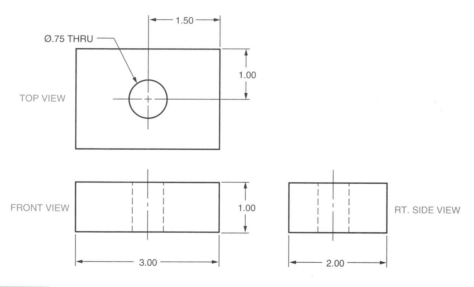

Ø.75 THRU

1.50

1.00

TOP VIEW

FRONT VIEW

1.00

3.00

RT. SIDE VIEW

2.00

FIGURE I-8   *Orthographic multiview drawing*

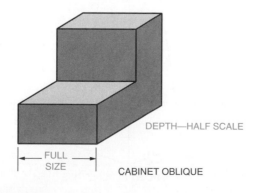

DEPTH—HALF SCALE

FULL
SIZE

CABINET OBLIQUE

FIGURE I-9   *Oblique projection (cabinet)*

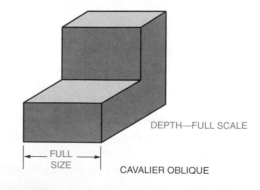

DEPTH—FULL SCALE

FULL
SIZE

CAVALIER OBLIQUE

FIGURE I-10   *Oblique projection (cavalier)*

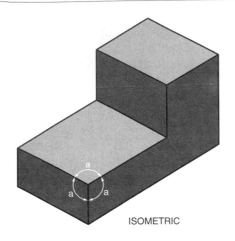

ISOMETRIC

FIGURE I-11 *Axonometric projection (isometric)*

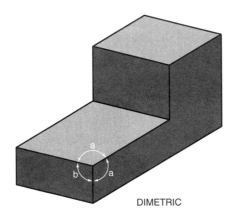

DIMETRIC

FIGURE I-12 *Axonometric projection (dimetric)*

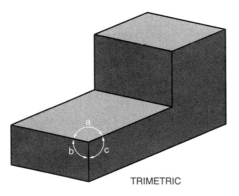

TRIMETRIC

FIGURE I-13 *Axonometric projection (trimetric)*

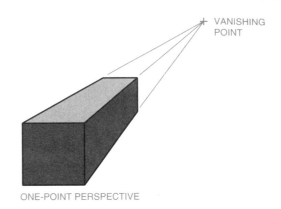

ONE-POINT PERSPECTIVE

FIGURE I-14 *One-point perspective projection*

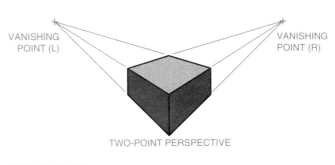

TWO-POINT PERSPECTIVE

FIGURE I-15 *Two-point perspective projection*

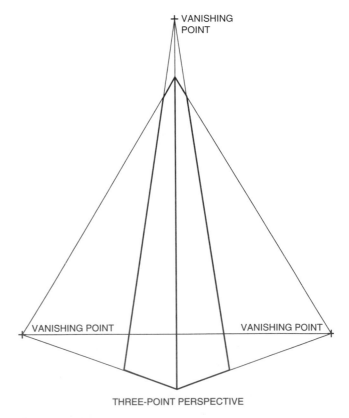

THREE-POINT PERSPECTIVE

FIGURE I-16 *Three-point perspective projection*

The age of computers has altered the design process slightly for those companies that have converted to computer-aided design and drafting. For these companies, the expensive, time-consuming fourth step in the design process—the making and testing of actual models or prototypes—has been substantially altered, *Figure I-18.* This fourth step has been replaced with three-dimensional computer models that can be quickly and easily produced on a CAD system using the database built up during the first three phases of the design process, *Figure I-19.*

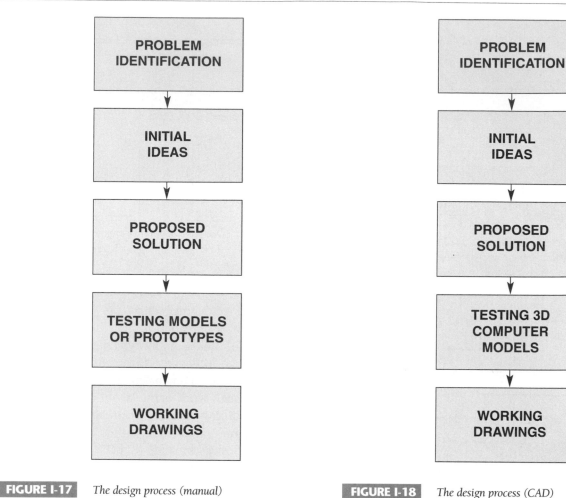

```
PROBLEM
IDENTIFICATION
      ↓
INITIAL
IDEAS
      ↓
PROPOSED
SOLUTION
      ↓
TESTING MODELS
OR PROTOTYPES
      ↓
WORKING
DRAWINGS
```

FIGURE I-17   *The design process (manual)*

```
PROBLEM
IDENTIFICATION
      ↓
INITIAL
IDEAS
      ↓
PROPOSED
SOLUTION
      ↓
TESTING 3D
COMPUTER
MODELS
      ↓
WORKING
DRAWINGS
```

FIGURE I-18   *The design process (CAD)*

Whether in the traditional design process or the more modern computer version, in either case, working drawings are an integral part of the design process from start to finish.

The purpose of technical drawings is to document the design process. Creating technical drawings to support the design process is called *drafting*. People who do drafting are known as *drafters* or *drafting technicians*. The words "draftsman" or "draughtsman" are no longer used.

In the first step of the design process, technical drawings are used to help clarify the problem or the need. The drawings may be old ones on file or they may be new ones created for the purpose of clarification. In the second step, technical drawings—often in the form of sketches or preliminary drawings—are used to document the various ideas and concepts formed. In the third step, technical drawings—again, usually preliminary drawings—are used to communicate the purposed solution.

If the traditional fourth step in the design process is being used, preliminary drawings and sketches from the first three steps will be used as guides in constructing models or prototypes for testing. If the more modern fourth step is being used, the database built up during documentation of the first three steps can be used in

FIGURE I-19   *Mountain bike helmet surface modeled using SURFCAM by Surfware* (Courtesy Surfware Inc.)

developing three-dimensional computer models. In both cases, the final step is the development of complete working drawings for guiding individuals involved in the conversion process. *Figure I-20* is a working drawing documenting the design of a simple mechanical part. The drawing was produced manually. *Figure I-21* is the same drawing produced on a CAD system.

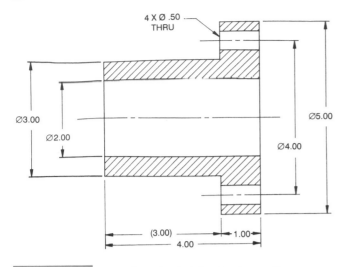

FIGURE I-20   *Simple mechanical drawing (manual)*

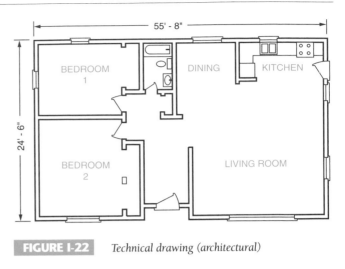

FIGURE I-22   *Technical drawing (architectural)*

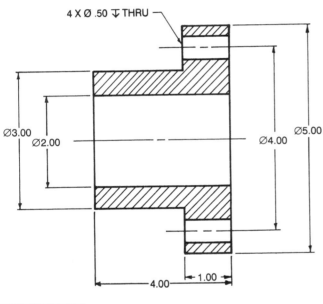

FIGURE I-21   *Simple mechanical drawing (CAD)*

Applications of Technical Drawings

Technical drawings are used in many different applications. They are needed in any setting that involves design and in any subsequent form of conversion process. The most common applications of technical drawings can be found in the fields of manufacturing, engineering, architecture, and construction and in all of their various related fields.

Architects use technical drawings to document their designs of residential, commercial, and industrial buildings, *Figures I-22* and *I-23*. Structural, electrical, and mechanical [heating, ventilating, air conditioning

(HVAC), and plumbing] engineers who work with architects also use technical drawings to document those aspects of the design for which they are responsible, *Figures I-24, I-25,* and *I-26*.

Surveyors and civil engineers use technical drawings to document such work as the layout of a new subdivision or the marking-off of the boundaries for a piece of property, *Figure I-27*. Contractors and construction personnel use technical drawings as their blueprints in converting architectural and engineering designs into reality, *Figures I-28* and *I-29* (pages 12–13).

Technical drawings are equally important to engineers, designers, and various other individuals working in the manufacturing industry. Manufacturing engineers use technical drawings to document their designs. Technical drawings guide the collective efforts of individuals who are concerned with the same common goal, *Figures I-30* and *I-31* (pages 14–15).

Regulation of Technical Drawings

Technical drawing practices must be regulated because of the diversity of their applications. Just as the English language must have certain standard rules of grammar, the graphic language must have certain rules of practice. This is the only way to ensure that all people attempting to communicate using the graphic language are speaking the same language.

STANDARDS OF PRACTICE

A number of different agencies have developed standards of practice for technical drawing. The most widely used standards of practice for technical drawing and drafting are those of the U.S. Department of Defense (DOD), the U.S. military (MIL), the American National

FIGURE I-23   *Technical drawing (architectural)*

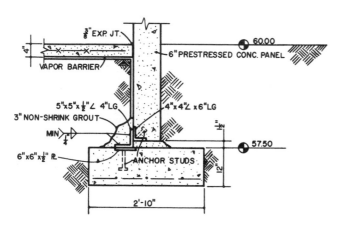

TYPICAL 6" WALL BASE CON FTG DETAIL

FIGURE I-24   *Technical drawing (structural)*

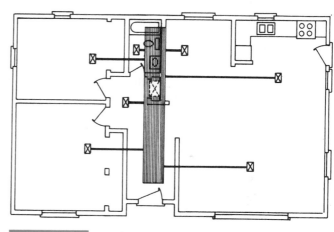

FIGURE I-26   *Technical drawing (HVAC)*

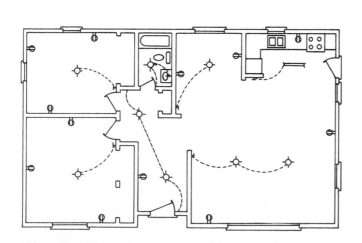

FIGURE I-25   *Technical drawing (electrical)*

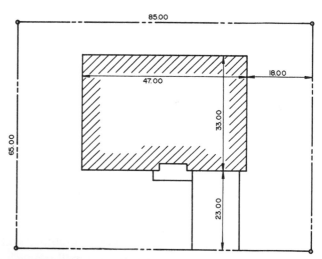

FIGURE I-27   *Technical drawing (civil)*

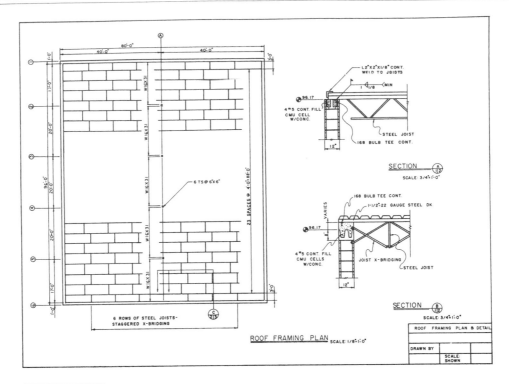

FIGURE I-28 *Architectural/engineering drawing*

Standards Institute (ANSI), and the American Society of Mechanical Engineers (ASME).

The American National Standards Institute does not limit its activities to the standardization of technical drawing and drafting practices. In fact, this is just one of the many fields for which ANSI maintains a continuously updated set of standards.

Standards of interest to drafters, designers, checkers, engineers, and architects are contained in the "Y" series of ANSI standards. *Figure I-32* (page 15) contains a list of ANSI standards frequently used in technical drawing and drafting specifications.

What Students of Technical Drawing, Drafting, and CAD Should Learn

Many people in the world of work use technical drawings in various forms. Engineers, designers, checkers, drafters, CAD technicians, and a long list of related occupations use technical drawings as an integral part of their jobs. Some of these people must be able to actually make drawings; others are only required to be able to read and interpret drawings; some must be able to do both.

What students of technical drawing, drafting, and CAD should learn depends on how they will use technical drawings in their jobs. Will they make them? Will

they read and interpret them? This textbook is written for students in the fields of engineering, design, drafting, and architecture, among others, who must be able to make, read, and interpret technical drawings. These students should develop a wide range of knowledge and skills, *Figure I-33* (page 15).

The learning required of technical drawing students can be divided into three categories: fundamental knowledge and skills, related knowledge, and advanced knowledge and skills.

In the fundamentals category, students of technical drawing, drafting, and CAD should develop knowledge and skills in the areas of drafting equipment; such fundamental drafting techniques as line work, lettering, scale use, and sketching; basic CAD system operation; geometric construction; multiview drawing; sectional views; descriptive geometry; auxiliary views; general dimensioning; and notation.

In the related knowledge category, students should develop a broad knowledge base in the areas of related math, welding, shop processes, and media and reproduction.

In the advanced category, students should develop knowledge and skills in the areas of development, geometric dimensioning and tolerancing, threads and fasteners, springs, cams, gears, machine design drawing, pictorial drawing, drafting shortcuts, and CAD/CAM technology and operations. The latter area represents a

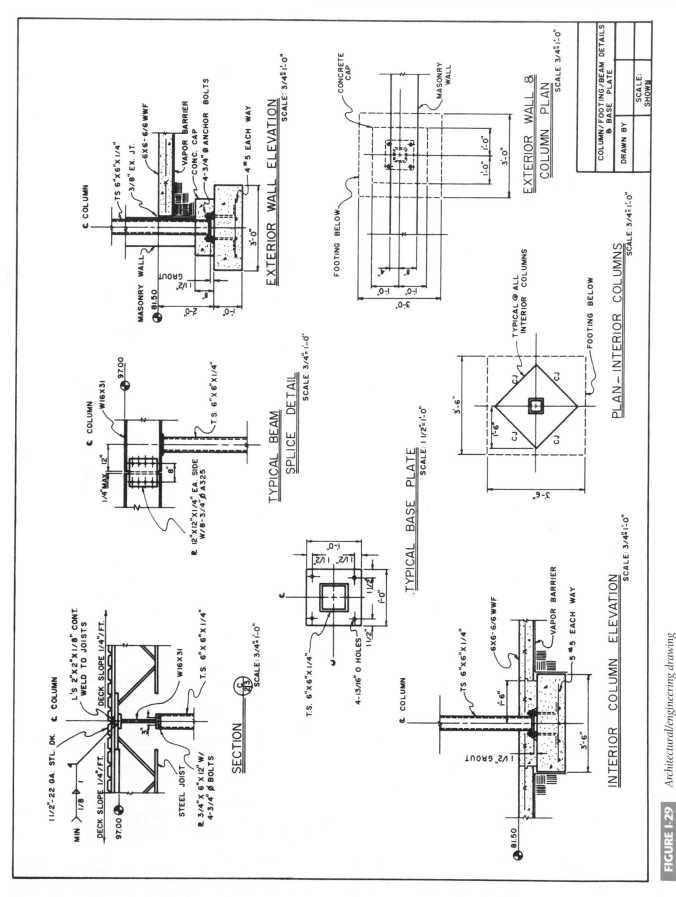

FIGURE I-29 *Architectural/engineering drawing*

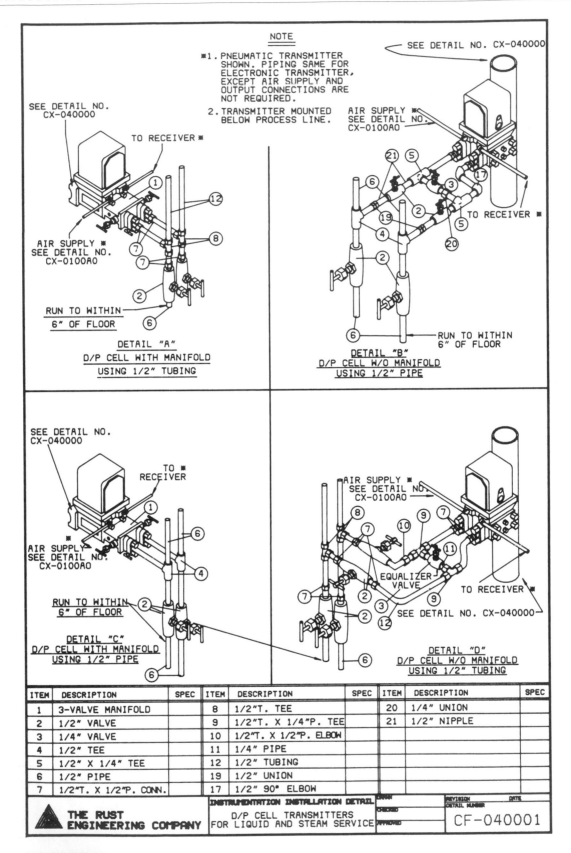

NOTE
* 1. PNEUMATIC TRANSMITTER SHOWN. PIPING SAME FOR ELECTRONIC TRANSMITTER, EXCEPT AIR SUPPLY AND OUTPUT CONNECTIONS ARE NOT REQUIRED.
2. TRANSMITTER MOUNTED BELOW PROCESS LINE.

DETAIL "A"
D/P CELL WITH MANIFOLD USING 1/2" TUBING

DETAIL "B"
D/P CELL W/O MANIFOLD USING 1/2" PIPE

DETAIL "C"
D/P CELL WITH MANIFOLD USING 1/2" PIPE

DETAIL "D"
D/P CELL W/O MANIFOLD USING 1/2" TUBING

ITEM	DESCRIPTION	SPEC	ITEM	DESCRIPTION	SPEC	ITEM	DESCRIPTION	SPEC
1	3-VALVE MANIFOLD		8	1/2"T. TEE		20	1/4" UNION	
2	1/2" VALVE		9	1/2"T. X 1/4"P. TEE		21	1/2" NIPPLE	
3	1/4" VALVE		10	1/2"T. X 1/2"P. ELBOW				
4	1/2" TEE		11	1/4" PIPE				
5	1/2" X 1/4" TEE		12	1/2" TUBING				
6	1/2" PIPE		19	1/2" UNION				
7	1/2"T. X 1/2"P. CONN.		17	1/2" 90° ELBOW				

THE RUST ENGINEERING COMPANY

INSTRUMENTATION INSTALLATION DETAIL
D/P CELL TRANSMITTERS
FOR LIQUID AND STEAM SERVICE

CF-040001

FIGURE I-30 *Isometric mechanical drawing* (Courtesy The Rust Engineering Company)

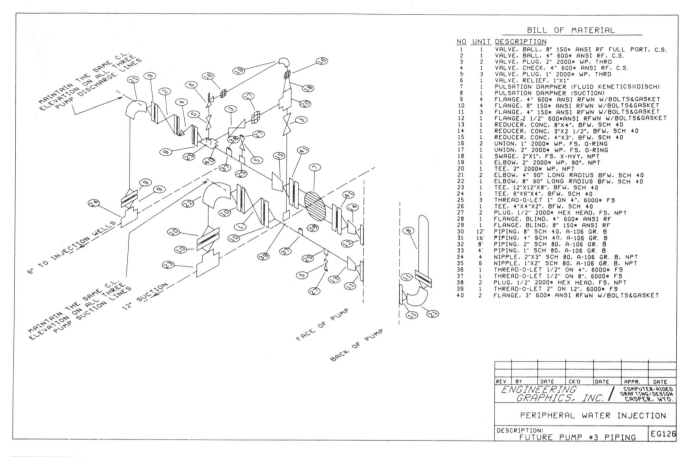

BILL OF MATERIAL

NO	UNIT	DESCRIPTION
1	1	VALVE, BALL, 8" 150# ANSI RF FULL PORT, C.S.
2	1	VALVE, BALL, 4" 600# ANSI RF, C.S.
3	2	VALVE, PLUG, 2" 2000# WP, THRD
4	1	VALVE, CHECK, 4" 600# ANSI RF, C.S.
5	3	VALVE, PLUG, 1" 2000# WP, THRD
6	1	VALVE, RELIEF, 1"X1"
7	1	PULSATION DAMPNER (FLUID KENETICS)(DISCH)
8	1	PULSATION DAMPNER (SUCTION)
9	4	FLANGE, 4" 600# ANSI RFWN W/BOLTS&GASKET
10	4	FLANGE, 8" 150# ANSI RFWN W/BOLTS&GASKET
11	3	FLANGE, 4" 150# ANSI RFWN W/BOLTS&GASKET
12	1	FLANGE,2 1/2" 600#ANSI RFWN W/BOLTS&GASKET
13	1	REDUCER, CONC, 8"X4", BFW, SCH 40
14	1	REDUCER, CONC, 3"X2 1/2", BFW, SCH 40
15	1	REDUCER, CONC, 4"X3", BFW, SCH 40
16	2	UNION, 1" 2000# WP, FS, O-RING
17	1	UNION, 2" 2000# WP, FS, O-RING
18	1	SWAGE, 2"X1", FS, X-HVY, NPT
19	1	ELBOW, 2" 2000# WP, 90#, NPT
20	1	TEE, 2" 2000# WP, NPT
21	2	ELBOW, 4" 90# LONG RADIUS BFW, SCH 40
22	1	ELBOW, 8" 90# LONG RADIUS BFW, SCH 40
23	1	TEE, 12"X12"X8", BFW, SCH 40
24	1	TEE, 6"X6"X4", BFW, SCH 40
25	3	THREAD-O-LET 1" ON 4", 6000# FS
26	1	TEE, 4"X4"X2", BFW, SCH 40
27	2	PLUG, 1/2" 2000# HEX HEAD, FS, NPT
28	1	FLANGE, BLIND, 4" 600# ANSI RF
29	1	FLANGE, BLIND, 8" 150# ANSI RF
30	12'	PIPING, 8" SCH 40, A-106 GR. B
31	16'	PIPING, 4" SCH 40, A-106 GR. B
32	8'	PIPING, 2" SCH 80, A-106 GR. B
33	4'	PIPING, 1" SCH 80, A-106 GR. B
34	4	NIPPLE, 2"X3" SCH 80, A-106 GR. B, NPT
35	6	NIPPLE, 1"X2" SCH 80, A-106 GR. B, NPT
36	1	THREAD-O-LET 1/2" ON 4", 6000# FS
37	1	THREAD-O-LET 1/2" ON 8", 6000# FS
38	2	PLUG, 1/2" 2000# HEX HEAD, FS, NPT
39	1	THREAD-O-LET 2" ON 12", 6000# FS
40	2	FLANGE, 3" 600# ANSI RFWN W/BOLTS&GASKET

REV	BY	DATE	CK'D	DATE	APPR.	DATE

ENGINEERING GRAPHICS, INC. / COMPUTER-AIDED DRAFTING/DESIGN CASPER, WYO.

PERIPHERAL WATER INJECTION

DESCRIPTION:
FUTURE PUMP #3 PIPING   EG126

FIGURE I-31   *Isometric piping schematic* (Courtesy Engineering Graphics Inc.)

DECIMAL INCH SIZE AND FORMAT _____ Y14.1M

LINE CONVENTIONS AND LETTERING ___ Y14.2M

MULTIVIEW AND SECTION _____ Y14.3M

PICTORIAL DRAWING _____ Y14.4M

DIMENSIONS AND TOLERANCING _____ Y14.5M

SCREW THREADS _____ Y14.6

GEARS, SPLINES, AND SERRATIONS ____ Y14.7.1

CASTINGS AND FORGINGS _____ Y14.8M

ENGINEERING DRAWING PRACTICES ___ Y14.100M

FIGURE I-32   *Sample list of drafting standards*

LEARNING CHECKLIST FOR STUDENTS OF TECHNICAL DRAWING

FUNDAMENTAL KNOWLEDGE AND SKILLS	RELATED KNOWLEDGE	ADVANCED KNOWLEDGE AND SKILLS
DRAFTING EQUIPMENT	MATH	DEVELOPMENT AND
FUNDAMENTAL DRAFTING	WELDING	INTERSECTIONS
TECHNIQUES	MEDIA AND	GEOMETRIC
SKETCHING	REPRODUCTION	DIMENSIONING AND
GEOMETRIC	MANUFACTURING	TOLERANCING
CONSTRUCTION	MATERIALS AND	THREADS AND
VISUALIZATION	PROCESSES	FASTENERS
MULTIVIEW DRAWING		SPRINGS
SECTION VIEWS		CAMS
GRAPHICAL DESCRIPTIVE		GEARS
GEOMETRY		MACHINE DESIGN
AUXILIARY VIEWS		DRAWING
GENERAL DIMENSIONING		PICTORIAL DRAWING
NOTATION		DRAFTING SHORTCUTS
BASIC CAD SYSTEM		CAD TECHNOLOGY
OPERATION		

FIGURE I-33   *Checklist for students of technical drawing*

significant change in techniques used to create, maintain, update, and store technical drawings, *Figures I-34 and I-35.*

Figures I-36 through *I-40* (pages 17–21) contain examples of several different kinds of technical drawings taken from the "real world" of drafting.

FIGURE I-34 *Typical CAD technology* (Courtesy Baystate Technologies Inc.)

FIGURE I-35 *Typical CAD system* (Courtesy Autodesk Inc.)

Technical Drawing and Quality/Competitiveness

Companies that design and manufacture products or design and build structures operate in a competitive environment. Surviving and succeeding in the intensely competitive modern marketplace requires companies to focus more attention than ever before on quality. Such concepts as flexible manufacturing, computer-integrated manufacturing (CIM), just-in-time manufacturing (JIT), ISO 9000, and statistical process control (SPC) are all attempts at improving quality and competitiveness.

As effective as these concepts may be at improving the quality of what takes place during manufacturing processes, they cannot overcome a poor design or misinterpretations that result from poorly prepared technical drawings. The best way to ensure quality and the competitive edge it can give a company is to do the following: (1) design in quality from the outset and, (2) prepare technical drawings that accurately and properly communicate the design to those who will manufacture or build it.

Quality can be defined as the extent to which a product conforms to or exceeds the customer's expectations. In design and manufacturing, customer expectations are typically communicated in the form of specifications. Consequently, in this field, quality is often viewed as the extent to which specifications are met or exceeded. Companies that consistently meet or exceed customer expectations are more competitive.

In addition to designing in quality, it is important to develop a design that is as easy as possible to manufacture. Often there will be several different design options that will meet or exceed customer expectations. When this is the case, the design chosen should be the one that is easiest to manufacture. This concept is known as *Design for manufacturability* (DFM) and is covered in Chapter 17.

Technical drawing plays a key role in determining the ultimate quality of a product. A good technical drawing is one that properly and conveniently communicates all of the information needed to transform a design into a product that meets or exceeds customer expectations. Anything less can detract from the quality of the product. The following examples illustrate this connection:

- An incorrect dimension on a technical drawing can cause the part to be produced at the wrong size. Each successive copy produced becomes waste.

- A dimension that is left off a technical drawing or just improperly placed can cause manufacturing personnel to interrupt their work to calculate the dimension or dimensions they need. The worst-case result is that they make a computation error and, consequently, produce a faulty part. The best-case result is that the interruption decreases productivity and, in turn, competitiveness.

- An entry left off a parts list can cause purchasing personnel to order an insufficient amount of material and, as a result, interrupt the flow of the manufacturing process.

- An improperly drawn view can cause a part to be manufactured in the wrong configuration.

These are only a few examples of how technical drawings can affect the ultimate quality of a product. By doing so they also affect the competitiveness of the

Industry Application

DESIGN FOR MANUFACTURABILITY

Metal Tech Manufacturing Company (MTM) had a reputation for quality products produced on time. Unfortunately, so did its principal competitor. Both companies kept their personnel and their processes up-to-date and working at peak performance levels. For years neither company could gain a sustainable competitive advantage over the other. Not until MTM decided to adopt *Design for manufacturability* (DFM) as a competitiveness strategy.

MTM created DFM teams for all of its products. These teams were comprised of design, drafting, and manufacturing personnel. Their charge was to develop designs that met the needs of customers, but could be manufactured as efficiently and in turn as economically as possible. In other words, with the DFM approach, designers had to consider not just functionality, but also manufacturability. A good product was no longer one that met the customer's needs in terms of function. It also had to be efficiently and economically manufactured.

The traditional attitude of "we designed it; making it is your problem" was replaced with "let's design it so that it works, and so that it can be efficiently produced." With manufacturing personnel involved in the design process, MTM's engineers and drafters found themselves being sent back to the drawing board frequently during the early stages of adopting DFM. Every time this happened the design was simplified, and the manufacturing process was streamlined as a result. Soon DFM became second nature, and MTM saw its production costs fall by 32 percent. The company was able to pass the savings along to customers and, as a result, within a year of adopting DFM had almost doubled its sales.

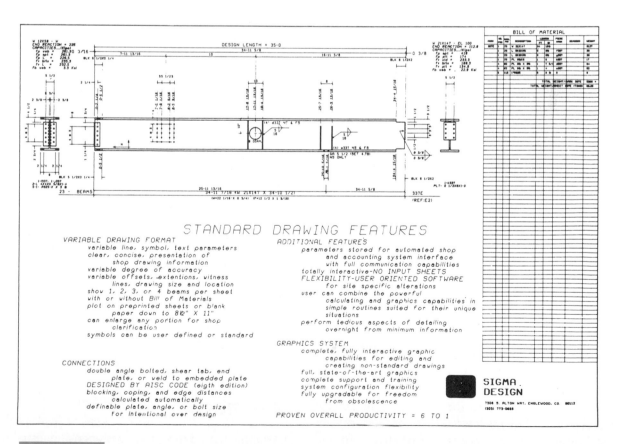

FIGURE I-36   *Structural steel drawing* (Courtesy Sigma Design)

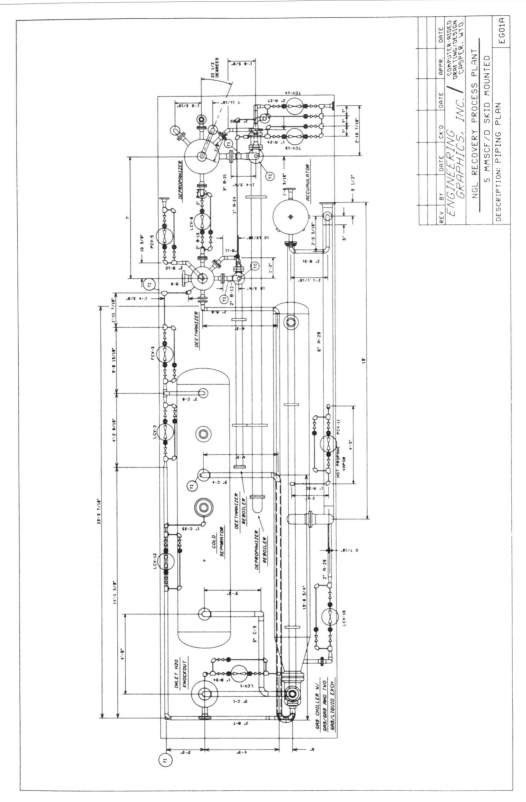

FIGURE I-37 *Piping drawing* (Courtesy Engineering Graphics Inc.)

company that manufactures the product. This book is designed to help students learn how to produce designs and technical drawings that make a positive contribution to quality and competitiveness.

Summary

- Graphic communication involves using visual material to relate ideas. Drawings, sketches, slides, photo-

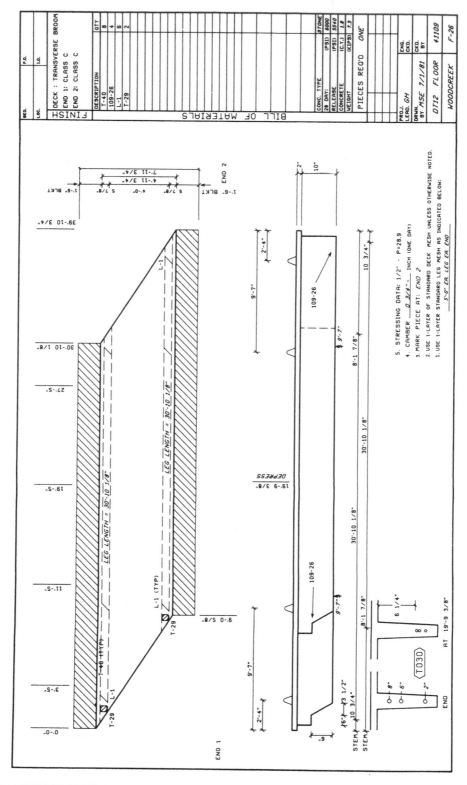

FIGURE I-38   *Prestressed concrete drawing* (Courtesy Sigma Design)

graphs, and transparencies are all forms of graphic communication.

- A drawing is a graphic representation of an idea, a concept, or an entity that actually or potentially exists in life.

- Artistic drawings are used to express the feelings, beliefs, philosophies, or abstract ideas of the artist. Technical drawings are used to clearly and concisely communicate all of the information necessary to transform an idea or concept into reality.

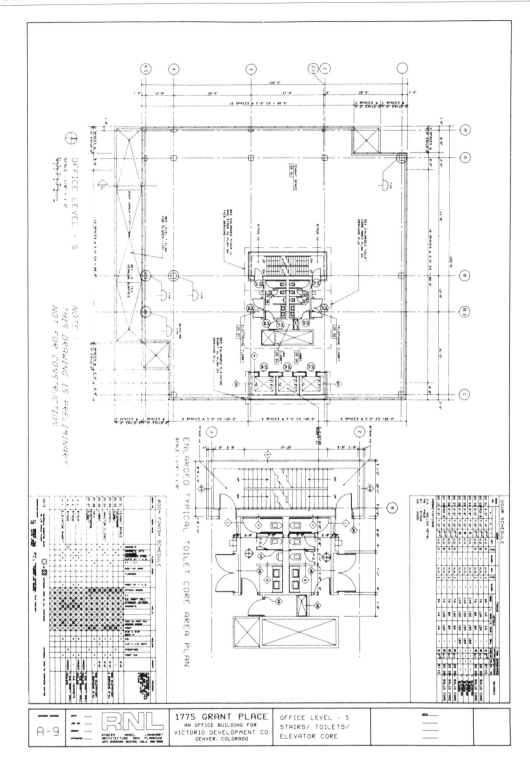

FIGURE I-39 *Architectural drawing* (Courtesy Sigma Design)

- Types of technical drawings include graphic, oblique, and axonometric projections (parallel projections), and one-, two-, and three-point perspective (perspective projections).

- The purpose of technical drawings is to document the design process.

- The principal applications of technical drawings are manufacturing, engineering, architecture, and construction.

- Technical drawings are regulated by such organizations as the U.S. Department of Defense, the U.S. military, the American National Standards Institute, and the American Society of Mechanical Engineers.

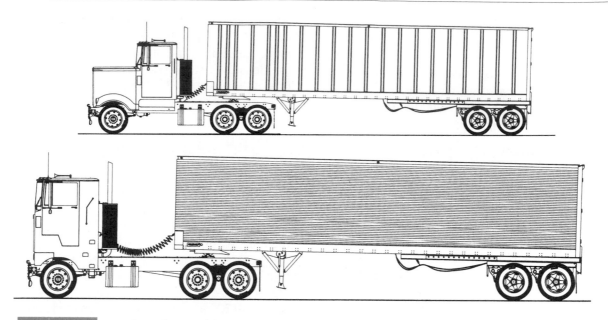

FIGURE I-40 *Mechanical pictorial drawing* (Courtesy Fruchaut Corporation)

• Design and drafting can promote quality competitiveness by ensuring that quality is designed in from the outset and ensuring that all drawings accurately communicate the design.

Review Questions

Answer the following questions either true or false.

1. A drawing is a graphic representation of an idea, a concept, or an entity that actually or potentially exists in life.

2. The old adage "a picture is worth a thousand words" is still the basis of the need for technical drawings.

3. The two basic types of drawings are practical and technical.

4. Technical drawings can be subtle or abstract.

5. Artistic drawings are commonly used in the manufacturing field.

6. A good technical drawing is one that properly and conveniently communicates all the information needed to transform a design into a product that meets or exceeds customer expectations.

Answer the following questions by selecting the best answer.

1. Which of the following is **not** a component of a projection?
   a. Imaginary lines of sight called projectors
   b. Three-dimensional view
   c. The eye of the viewer looking at the object
   d. The imaginary projection plane

2. Which of the following is a subdivision of a parallel projection?
   a. Trimetric projection
   b. Cavalier projection
   c. Cabinet projection
   d. Oblique projection

3. Which of the following is **not** a type of perspective projection?
   a. One-point
   b. Two-point
   c. Three-point
   d. Four-point

4. Which of the following is **not** a step in the design process?
   a. Identify a problem or a need.
   b. Identify ideas to solve the problem.
   c. Obtain examples of several different kinds of technical drawings taken from the real world of drafting.
   d. Select a proposed solution.

5. Which of the following is **not** a field that uses technical drawings extensively?
   a. Manufacturing
   b. Retail
   c. Engineering
   d. Architecture and construction

6. Which of the following is **not** a U.S. organization that regulates technical drawing practices?
   a. American Society of Mechanical Engineers (ASME)
   b. U.S. Department of Defense (DoD)
   c. International Standards Organization (ISO)
   d. U.S. military (MIL)

Introduction Problems

To the Student:

The majority of problems available for you in this text are related to technical drawing; however, additional problems are included that aim to do the following:

- Introduce you to the design process.
- Develop your critical thinking skills.
- Involve you in team activities.
- Exercise your imagination and creativity.
- Give you practice in using basic mathematics.
- Introduce you to some product design concepts.

You will need a scientific hand calculator with the usual functions; addition, subtraction, multiplying, dividing, trigonometric functions, logarithms, etc.

Some problems may have more information than you need to set up the problem for solution. Some problems may have more than one solution, which is typical of real-life problems in industry.

PROBLEM I-1

In a team of three or four students, discuss how the design of products might differ from present designs if everyone had only a thumb and an index finger on each hand. Summarize your ideas in freehand sketches and report orally to the rest of the class.

PROBLEM I-2

Figure I-7 illustrates the concept of a projection plane (for third-angle projection). In a team of three or four students, discuss how you would go about preparing a device to illustrate this concept to a class of junior high school students. As an aid to your discussions, follow the steps in the design process described in Figure I-17. Prepare a freehand sketch of your final solution. Include materials and approximate dimensions.

PROBLEM I-3

The weight of the flanged hub shown in Figures I-20 and I-21 needs to be calculated. The hub is steel with a density of 0.28 pounds per cubic inch. A suggestion for doing this type of calculation is to break up the object into convenient parts for doing the calcu-

lations. Separate the flanged hub into a hollow cylinder and a flange with one large hole and four small ones. Calculate the volume of metal and multiply by the density of steel. Use the format shown in the figure for Problem 4-30 (page 160) for your solution presentation.

PROBLEM I-4

Optimization in product design usually means satisfying the design requirements for the least cost. For a box design, least cost means using the least amount of material. Assume that a square box with an open top is to have a volume of 50 cubic inches. To visualize the optimum conditions for the box, prepare a plot of TOTAL AREA OF MATERIAL versus dimension D. That is, plot AREA on the vertical axis, dimension D on the horizontal axis.

Select a trial dimension H. Then calculate the total area of the material. The suggested axes and calculation steps are given in the following figure:

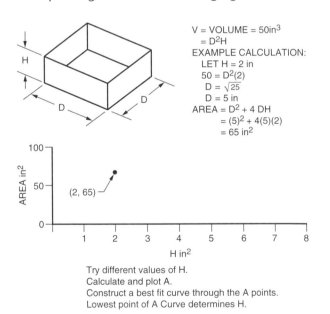

$$V = \text{VOLUME} = 50\text{in}^3$$
$$= D^2H$$

EXAMPLE CALCULATION:
LET H = 2 in
$$50 = D^2(2)$$
$$D = \sqrt{25}$$
$$D = 5 \text{ in}$$
$$\text{AREA} = D^2 + 4\,DH$$
$$= (5)^2 + 4(5)(2)$$
$$= 65 \text{ in}^2$$

Try different values of H.
Calculate and plot A.
Construct a best fit curve through the A points.
Lowest point of A Curve determines H.

PROBLEM I-5

A student was asked to study Figure I-33 (page 15) to compare what skills and knowledge the student already had to the topics listed in the checklist. One comment the student had was, "I believe that math should be on all three lists." What is your opinion regarding math and why? Discuss this question in a group of three or four and document your conclusions.

Chapter 1

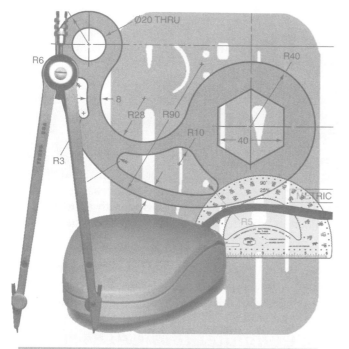

CAD, THE INTERNET, AND TECHNICAL DRAWING

KEY TERMS

Browser

CAD system

CAM

CD-ROM

Computer graphics (CG)

CPU

Digitizer

Display monitor

E-mail

Hard disk drive

Hardware

Home page

Internet

Joystick

List server

Microprocessor

Mouse

Plotter

Printer

Puck

RAM

ROM

Search engine

Software

Solid modeling

Stylus

Trackball

Uniform resource locator (URL)

CHAPTER OUTLINE

Overview of CAD • Key computer and CAD concepts • CAD systems • CAD workstation components • CAD software • The Internet in design and drafting • Summary • Review questions

CHAPTER OBJECTIVES

Upon completion of this chapter, students should be able to do the following:

- Explain the concept of CAD.
- Explain key CAD concepts including: computer graphics, computer-aided manufacturing, computer-integrated manufacturing, computer-aided engineering, hardware, software, operating system, and application program.
- Define the term *CAD system.*
- List and explain the most common components in a CAD workstation.
- Explain the concept of CAD software.
- Demonstrate the process for selecting a CAD system.
- Explain the concept of modeling.
- Explain the concept of parametric design.
- Describe how the Internet can be used as a tool in design and drafting.

This chapter provides a comprehensive treatment of modern computer-aided drafting as well as a section on using the Internet as a tool in design and drafting. These topics are covered conceptually rather than from the perspective of a specific vendor or Internet provider.

Overview of CAD

The dawning of the age of computers brought significant change to the fields of design, drafting, and engineering. The computer also brought corresponding changes to all of the various other fields related to design, drafting, and engineering. Although we think of the computer as a relatively recent tool, computers are actually not new. In fact, the first computer was developed in the 1830s by a British mathematician named Charles Babbage. Babbage's computer was a mechanical number-crunching machine that by today's standards was a dinosaur. However, many of the principles upon which Babbage based his Difference Engine are the same principles upon which today's microprocessor-based electronic computers are based.

Regardless of their relative speed and size, what makes computers such invaluable tools is their ability to store data, perform basic logic functions, and make mathematical computations. The ability to electronically link computers has allowed the design, drafting, and engineering functions in a company to be part of a fully integrated whole that includes marketing, accounting, manufacturing, and shipping.

Because of the computer's abilities, the design, drafting, and engineering functions of analysis, problem solving, and documentation (drawings, parts lists, specifications, etc.) have been significantly improved. So much so, in fact, that engineers have been able to explore new theories that were previously beyond their reach, and drafting technicians have been able to perfect new practices that have revolutionized the ways in which designs are documented.

Modern CAD workstations such as the one depicted in *Figure 1-1* had their origins in the 1950s when the United States Air Defense Command first used light pens for data entry. The next major development in the history of CAD occurred in 1963 at the Massachusetts Institute of Technology (MIT) in Boston, Massachusetts, when Professor Ivan Sutherland demonstrated his Sketchpad. The Sketchpad was a device that used a light pen to enter graphic data into a computer. Then, in 1964, IBM released the first commercially marketed CAD system. However, it was the development of the programmable integrated circuit in the 1960s that opened the door for the modern personal computer–based CAD sys-

FIGURE 1-1 *CAD workstation* (Courtesy Advanced Technologies Center)

tem. It is still the tiny but powerful microprocessor—a device that is based on integrated circuit technology—that is the fundamental enabling device in today's ever-smaller but more capable CAD systems.

Key Computer and CAD Concepts

There are a number of computer concepts with which today's design and drafting technician should be familiar. Building on these, there are several CAD concepts that should be part of the technician's knowledge base. All of these concepts represent knowledge that should become second nature to computer-literate design and drafting personnel.

The first two of these concepts are analog and digital computers. Of the two, it is the digital computer that people use most. For example, the personal, laptop, and notebook computers used in business, design, drafting, engineering, and manufacturing are all digital. Analog computers can be used to measure continuous physical properties such as heat, pressure, depth, and speed. Consequently, they are used most often to monitor and control mechanical, electronic, and hydraulic systems.

The most immediately noticeable difference between analog and digital computers is in how they display information for human use. Digital computers can display information in a readable form that requires no translation (i.e., using alphanumeric characters). Analog computers feed information back in ways that require interpretation or special devices for translation. Perhaps the best example of this difference is the clock. A clock with moving hands is an analog device. It measures the continuous passing of time. As time passes, the clock hands move accordingly. The clock face is inscribed with numbers arranged in a specified order that allows people to translate the positions of the hands into hours and minutes. A digital clock, on the other hand, simply displays the time using alphanumeric characters, *Figure 1-2*. CAD systems use digital computers.

Another important computer concept is microminiaturization. This concept accounts for the tendency of computers to become ever smaller, yet more powerful. The key enabling ingredient in the ongoing microminiaturization process is the programmable integrated circuit or IC chip, *Figure 1-3*. These microchips are placed on printed circuit boards that, because of the size of the chips, can now fit hundreds of components into a relatively small area. Integrated circuits allowed for the development of the microprocessor, the key technology in today's small but powerful computers, *Figure 1-4*.

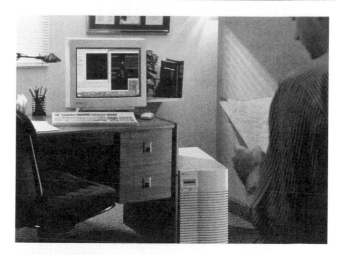

FIGURE 1-4 *Workstation* (Courtesy Hewlett-Packard Co.)

ANALOG CLOCK

DIGITAL CLOCK

FIGURE 1-2 *Analog and digital clock*

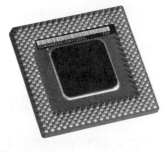

FIGURE 1-3 *Intel chip*

The rapid growth of the use of computers in the world of design and drafting brought forth a deluge of new, and often confusing, terms. Such terms as CAD, CADD, CG, CAE, AD, and hundreds of others spawned by the computer revolution began to be heard in drafting and design departments. As a result, drafting and design practitioners found themselves confused by it all.

The reader should be conversant in the language of computers as it applies to design and drafting. The first step is to develop an understanding of the term "computer-aided drafting" or CAD.

Before attempting to form a definition for CAD, the definition of "drafting" should be thoroughly understood. Most people think of drafting as simply the drawing of plans. True, drafters do draw plans. However, this is a very limited definition because drafters do much more than draw plans.

To understand drafting in its broadest sense, one must begin with the design process. The design process is an organized, systematic procedure used to accomplish a design that is needed to solve a problem or meet a need, *Figure 1-5*. Each step in the design process requires various types of documentation, such as sketches, preliminary drawings, working drawings, calculations, bills of material, parts lists, and schedules. Producing or "drafting" this documentation is the job of the drafter. Therefore, drafting means producing the documentation required in support of the design process.

With this understanding of drafting as a concept, one may now begin to develop an understanding of CAD. Computer-aided drafting or CAD means using the computer and peripheral devices in producing the documentation needed to support the design process. Put another way, CAD means using the computer and peripheral devices to do drafting.

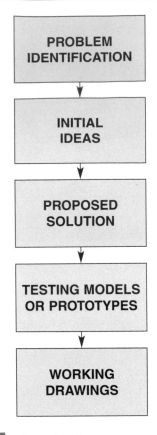

FIGURE 1-5 *The modern design process*

A well-informed drafter should be able to distinguish CAD from among the various other related terms frequently heard in modern drafting settings, such as AD, CADD, CAE, CG, and CAM.

CAD is usually taken to mean computer-aided drafting, but it can also be used to mean computer-aided design. It is up to the person using the term to make the distinction. AD means automated drafting, and it is synonymous with CAD or computer-aided drafting. Originally, AD was used when referring to the use of the computer to do drafting, and CAD was used when referring to the use of the computer as a design tool. However, AD did not catch on because it is a less distinctive term than CAD. Consequently, CAD began to be used to mean both computer-aided drafting and computer-aided design.

CADD is the acronym for computer-aided design and drafting. This is a broad concept in which a computer system is used for both design and drafting. CAE is the acronym for computer-aided engineering, which is sometimes used instead of computer-aided design.

CG is short for *computer graphics*. This is a term that is used and misused a great deal. It is sometimes used in place of CAD (drafting). However, in reality, it is a much broader term. CG means using the computer and peripheral devices for producing any type of graphic image, technical or artistic.

None of these terms has been formally standardized. However, there is a definite trend toward greater use of CAD when speaking of drafting and when speaking of design. We will follow this practice in this textbook.

The manufacturing equivalent of CAD is computer-aided manufacturing, or *CAM*. CAM refers to computer-controlled automation of the various manufacturing processes used in modern industry. CAD/CAM links together the processes of design, drafting, and manufacturing. The CAD/CAM process involves designing a product on the computer, using the computer to produce necessary documentation, and using the database stored in the computer from the design and drafting phases to issue the manufacturing instructions to automated machines and industrial robots.

CAD Systems

People who use the term *CAD system* are usually referring to a configuration of computer hardware, or the tools of the CAD technician. However, this is a misuse of the term. CAD hardware is just one of the three components that must be present in order to have a CAD system.

A CAD system actually consists of hardware, software, and users, *Figure 1-6*. A collection of CAD hardware is properly referred to as a hardware configuration. In order to turn a hardware configuration into a CAD system, software and well-trained users must be added.

CAD Workstation Components

Hardware is the term given to the computer's physical equipment or devices. Many different companies manufacture CAD hardware. Consequently, numerous configurations are on the market. However, a typical hardware configuration consists of a display monitor, input devices, output devices, storage devices, and a central processing unit (CPU), *Figure 1-7*.

DISPLAY MONITOR

The *display monitor* is an output device resembling a television that displays the information on which the CAD technician is working. As any type of documentation is

CAD SYSTEM =

HARDWARE + SOFTWARE + USERS

FIGURE 1-6 *Components of a CAD system*

FIGURE 1-9   *Any type of documentation created by the CAD technician can be displayed for viewing*

FIGURE 1-7   *Typical hardware configuration* (Courtesy Hewlett-Packard Co.)

created, it is displayed on the screen of the terminal, *Figures 1-8* and *1-9*.

There are three common types of terminal displays: (1) refresh displays, (2) raster displays, and (3) storage-tube displays. In a refresh display, the image is traced on the back of the screen by an electronic beam. This image must be constantly retraced or "refreshed." Raster displays create images by illuminating pixels (picture elements) on the screen. The more pixels available on the

FIGURE 1-8   *The display terminal screen is the CAD technicians drafting board*

screen for illumination, the higher the quality of the image. High resolution produces a high-quality image. Storage-tube displays create images by tracing them on the back of the display screen with an electron beam like refresh displays. However, unlike refresh displays, storage-tube images are stored and do not have to be refreshed. Graphics displays may be color or monochrome (amber or green) devices.

Most CAD configurations use a raster display monitor. These devices are similar to modern television sets in that the image is created by an electron beam that scans the back of the screen, illuminating pixels. The process is known as raster scanning.

The nature of the CAD technician's work demands that he or she have a large viewing area. Consequently, most CAD workstations use at least 17″ displays, and many use 21″ models.

Resolution is an issue when selecting a display monitor. It has to do with the level of detail that can be distinguished on a display. The more pixels per inch, the higher the resolution. The resolution of a display monitor and the image on the display are products of a video display card. This card determines the number of pixels per inch on the display monitor. A typical monitor will have 1600 × 1200 pixels or more.

Color is another important feature of display monitors. Like resolution, color is a function of the video display card. Colors on a display are produced in much the same way as an artist produces them on a canvas; by mixing different colors. Most displays can produce 256 different colors. A color display requires more video memory and processing power in the computer, and with higher resolution even more memory and processing power are needed. Because of this, video cards often have their own processor (known as a video accelerator) and their own memory.

CAD technicians need to understand the specifications that are typically given for display monitors.

Displays are rated according to their ability to refresh the screen—in other words, according to how fast the video display card can "repaint" the pixels, which it must do constantly in order to retain the image. The rating given for a display contains two specifications, one for vertical pixels and one for horizontal pixels. A low rating would be 60 Hz. The modern CAD display monitor should be in the higher ranges currently available, or eye fatigue will be a problem and productivity will suffer.

INPUT DEVICES

Input devices are those used by CAD technicians to interact with the CAD system. There are numerous different kinds of input devices, some that are used independently and some in conjunction with each other, *Figure 1-10*.

The keyboard is one of the most frequently used input devices in a CAD system. Keyboards are used for entering text, system commands, X-Y or X-Y-Z coordinates, and any other type of alphanumeric input.

Most CAD systems have keyboards that contain both the standard alphanumeric keys and special auxiliary keys, *Figure 1-11*. The special auxiliary keys may be part of the separate numeric pad or a group of display control keys.

Keying is the process used for entering text on drawings, entering certain system commands, entering dimensions not entered automatically, and for logging on to the system. Keys may also be used for cursor control.

The *mouse* is a widely used input device, *Figures 1-12* and *1-13*. The mouse is used for issuing commands and moving the screen cursor to different locations.

FIGURE 1-11 *HP9000 j-class workstation* (Courtesy Hewlett-Packard Co.)

FIGURE 1-12 *Mouse* (Courtesy IBM Corp.)

Checklist of Widely Used Input Devices
✓ Keyboard
✓ Mouse
✓ Digitizing Tablet
✓ Puck
✓ Stylus
✓ Trackball
✓ Joystick
✓ Light Pen
✓ Touch Panels
✓ Voice Recognition Devices

FIGURE 1-10 *Checklist of widely used input devices*

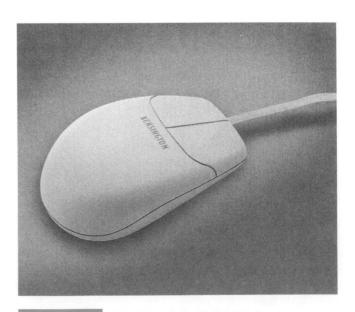

FIGURE 1-13 *Two-button mouse* (Courtesy Kensington)

There are two types of mice. One type (mechanical) has a roller on its underside that detects movement and transfers that movement to the screen cursor. The other type (optical) detects movement by reflecting a light beam off a special surface and transfers that movement to the cursor. Both have one to three buttons for command selection.

The *digitizer* is a special electromechanical input device that resembles an electronic tablet. In fact, the digitizer is frequently referred to as the tablet. Some CAD systems have digitizers that are as small as tablets, and some have digitizers that are as large as conventional drafting tables, *Figure 1-14*.

Digitizing is a process through which graphic data, such as a sketch or a drawing, is converted to digital data as it is input. A digitizer can be used for a number of different functions. It can be used in conjunction with a stylus or puck to control the screen cursor, and it can be used to create, locate, and manipulate graphic and alphanumeric data. A *stylus* resembles a ballpoint pen, while a *puck* resembles a mouse with a plastic window containing crosshairs, *Figure 1-15* and *1-16*.

Trackballs are less common in CAD configurations than they were in the earlier days of CAD, although they have re-emerged somewhat with the advent of laptop computers. A trackball consists of a ball mounted in a box that also has one to three buttons that are used like the buttons on a mouse, *Figures 1-17* and *1-18*. As the ball is rolled, sensors in the box detect the movement and transfer it to the cursor.

The *joystick* resembles a small version of what pilots of old airplanes used to control their aircraft, *Figure 1-19*. Most people who play video games are familiar with this device. The stick manipulates cursor movement, and

FIGURE 1-15   *Stylus* (Courtesy CalComp)

FIGURE 1-16   *Stylus and electronic pad* (Courtesy Gateway)

FIGURE 1-14   *Digitizer* (Courtesy CalComp)

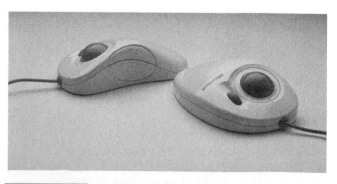

FIGURE 1-17   *Trackball* (Courtesy Microsoft)

FIGURE 1-18 *Trackball* (Courtesy Kensington)

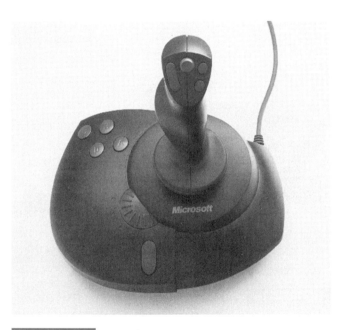

FIGURE 1-19 *Joystick* (Courtesy Microsoft)

buttons are typically available for command selection and entering coordinate data.

The light pen is rarely used in modern CAD configurations and is, in fact, one of the oldest of CAD input devices. A light pen looks like a pencil-sized flashlight. It is photosensitive and moves the cursor by touching the screen. A more contemporary device for doing the same thing is the touch panel, in which commands are given and options selected by touching the display screen with the finger. Touch-panel displays sense the pressure from the finger.

An emerging input technology is the voice recognition device. Commands are made into a microphone, and the system is trained to recognize the user's voice. Voice recognition technology still has too many drawbacks to make it a viable option on the modern CAD configuration. However, CAD technicians can expect to use this technology in the future as it continues to develop.

OUTPUT DEVICES

Hard copy output from a CAD system is produced by various types of *plotters* and *printers*, **Figure 1-20**. Pen plotters use pens of just one or a variety of colors to produce drawings on paper, vellum, or film. There are three broad categories of pen plotters: drum, flatbed, and microgrip.

A drum plotter consists of a long cylinder and a moveable pen carriage. The medium (paper, vellum, or film) moves back and forth over the cylinder while the pen carriage moves simultaneously over the medium.

With flatbed plotters the medium is mounted on a flat surface above which is a movable pen carriage. The medium remains static while the pen carriage moves over it along both an X and a Y axis. The surface area on a flatbed plotter corresponds to the standard sizes of drafting media (i.e., A, B, C, D, E).

Microgrip plotters resemble drum plotters and work in a similar manner. The term "microgrip" comes from the tiny rollers that grip the edge of the medium, moving it back and forth as the pen carriage moves simulta-

Checklist of Widely Used Output Devices
✓ Pen Plotters
▸ Drum
▸ Flatbed
▸ Microgrip
✓ Dot Matrix Plotter
✓ Electrostatic Plotter
✓ Printer/Plotters
▸ Inkjet Printer/Plotter
▸ Laser Printer/Plotter
▸ Dot Matrix Printer

FIGURE 1-20 *Checklist of widely used output devices*

neously. Microgrip plotters are popular because they are flexible, adaptable, and relatively inexpensive.

Key factors to consider when selecting pen plotters are accuracy, acceleration, repeatability, and speed, *Figure 1-21*. Accuracy is the difference between what should be drawn and what actually is expressed in thousandths of an inch. Acceleration is a measure of how long it takes the pen carriage to get up to plotting speed (like measuring how long it takes an automobile to go from 1 to 60 miles per hour). Acceleration is expressed in Gs (for gravitational force).

Pen speed is a measure of how fast the pen moves across the medium. A fast pen speed typically produces a higher-quality, more consistent line. A slower speed produces a darker line because the ink has more time to flow. However, a slow speed can also lead to ink bleeding and blotting. Repeatability is the ability of the plotter to redraw the same line over and over. With poor repeatability there will be double lines or progressively thickened lines.

The dot matrix plotter produces the image by printing a series of dots on the medium. The key evaluating component in a dot matrix plotter is a selenium drum that is electrostatically charged. The process of converting an image into dots is called rasterization.

Electrostatic plotters use a special type of medium that is coated so that toner or ink will adhere to it under an electrostatic charge. The process of rasterization converts the computerized image into dots and the dots are applied to the medium electrostatically, *Figure 1-22*.

FIGURE 1-22 *LED plotter* (Courtesy CalComp)

Resolution (the number of dots per inch) is an issue for both dot matrix and electrostatic plotters. The more dots per inch, the better the resolution.

Printer/plotters are output devices that can do double duty; they can be used as printers for alphanumeric output and as plotters for graphic output. An inkjet printer/plotter uses the rasterization process to convert the image or data into dots. Each dot then receives a tiny drop of ink, *Figure 1-23*. A laser printer/plotter also makes use of the rasterization process but uses a beam of light and an electrostatically charged medium to produce the image, *Figure 1-24*.

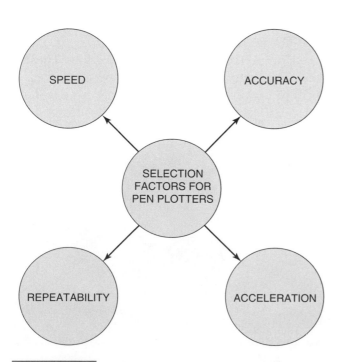

FIGURE 1-21 *Selection factors for pen plotters*

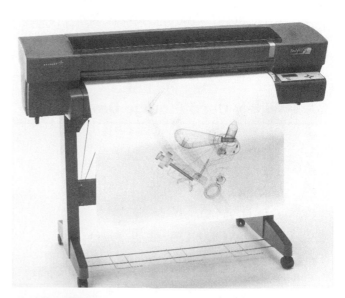

FIGURE 1-23 *Inkjet printer* (Courtesy CalComp)

FIGURE 1-24 *IBM network printer 12* (Courtesy IBM Corp.)

Dot matrix printers are used less and less, but they still may be found in some offices. With this device the dots are applied in one of two ways. The first way is a mechanical process in which a carbon or ink ribbon is struck by pins arranged in a print head (similar to an old mechanical typewriter). The other way is a thermal process in which heat is used to apply the dots.

STORAGE DEVICES

In a drafting and design department of the past, all drawings and other documentation were stored in filing cabinets, in hanging files, and in drawing cabinets. In modern drafting and design departments, documentation is stored electronically. *Figure 1-25* is a checklist of storage devices associated with modern CAD systems.

In order to understand the various devices used to store drafting and design documentation electronically,

it is necessary first to understand the terms RAM and ROM. *RAM* means "random access memory" or electronic memory that can be accessed in any order. *ROM* means "read only memory" or electronic memory that allows fast access to permanently stored data that can be retrieved by the user, but not altered.

The fixed disk drive is more commonly referred to as the hard disk. It is the most widely used type of electronic storage device in CAD systems and computers in general. *Hard disk drives* may be contained within the computer console or in an external cabinet that is attached to the processing unit. Key concerns relating to hard disks are storage capability (typically measured in megabytes and gigabytes of memory) and access time. Access time is the amount of time required to retrieve stored data. It is measured in milliseconds (ms). The lower the number of milliseconds, the less time it takes to retrieve data.

The floppy disk drive is so named because it accepts a removable plastic disk on which data are stored offline. Data that are not used continually may be stored on a floppy disk to avoid filling up the CAD system's hard drive. Floppy disks are a convenient way to store such data, but they do have drawbacks. Floppy disks are not as dense as hard disks. This means they are not able to hold as much data. In addition, data retrieval is slower with a floppy disk than with the hard disk.

Zip drives are an improvement over floppy disk drives in that they are much higher in density and the drives themselves are external. A zip drive can be used for downloading large amounts of information from one system and moving it to another, and for backup storage, *Figure 1-26.*

CD-ROM drives have become standard components in modern CAD configurations for two reasons. The first is that many high-volume standards, specifications, and vendor reference manuals can be obtained in a CD-ROM

Checklist of Widely Used Storage Devices

✓	Fixed Disk Drive
✓	Floppy Disk Drive
✓	Zip Drive
✓	CD-ROM Drive
✓	Optical Disk Drive
✓	Magnetic Tape Storage

FIGURE 1-25 *Checklist of widely used storage devices*

FIGURE 1-26 *Zip drive* (Courtesy Iomega Corp.)

format. Having a CD-ROM drive allows the CAD technician to use these reference materials online without having to spend time searching through a hard copy library. In addition, blank CDs are available in WORM (write-once-read-many) format. This means that large design and drafting projects can be downloaded onto CDs for permanent archiving. A CD can hold five to six times more data than a typical floppy disk, *Figure 1-27.*

Optical disk drives use a laser to place data on the disk and to change the data. An optical disk is a "write-many-read-many" device that can be erased and written over many times, *Figure 1-28.*

Magnetic tape storage is actually an old form of electronic storage. With magnetic tape (audio cassettes and cassette recorders/players), data are stored in the form of magnetic particles that attach themselves to plastic tape. A great deal of information can be stored on magnetic tape. Consequently, this medium is commonly used for

permanent backup storage. However, it has one major drawback. Data stored on magnetic tape can be accessed only sequentially. This is like having an audio cassette containing your favorite songs. If the one you want to hear now is in the middle of the tape, the only way to get to it is to fast forward through all of those that precede it.

CENTRAL PROCESSING UNIT

The central processing unit (*CPU*) in the computer consists of two major components—the control section and the arithmetic/logic section—as well as internal storage and interfacing mechanisms to tie it to peripherals (see *Figure 1-29*). It should be noted that although the main memory is often housed in the same console as the two components of the processing unit, it is not contained on the same printed circuit board (except in the case of the microcomputer). Consequently, the control section and the arithmetic/logic section taken together are called the central processing unit of the computer. When the control section and the arithmetic/logic section are both placed on a single silicon chip, you have a *microprocessor.*

The purpose of the control section is to direct all activities of the computer as set forth in stored programs. It controls the arithmetic/logic section, input and output operations, the transmittal of data to and from storage, and all other operational functions of the computer. The arithmetic/logic section performs all addition, subtraction, multiplication, division, and other mathematical operations and compares data so that logical decisions can be made by the computer.

FIGURE 1-27 *CD-Rom drive* (Courtesy IBM Corp.)

FIGURE 1-28 *Optical disk drive* (Courtesy Syquest Technology)

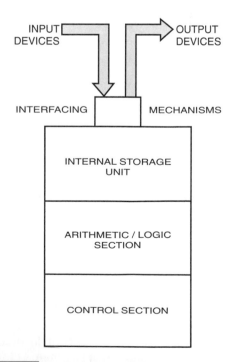

FIGURE 1-29 *Central processing unit*

One of the four characteristics that sets a computer apart from other machines is that it can modify program instructions by making logical decisions. A computer has this capability because the arithmetic/logic section of the central processing unit is able to make the types of comparisons needed for such logical decisions.

CAD Software

Software is the term for the nonmechanical components of a computer system. Such components include computer programs, documentation of those programs, and the various types of technical and reference manuals that go with the system. However, when most people use the term software, they are talking strictly about computer programs. A computer program is a specially coded set of instructions that directs the computer in performing all operations.

People who are trained to write programs are called computer programmers. You do not have to know how to program a computer to be an efficient and effective computer user. Most computer users do not know how to program a computer. However, it is helpful to understand what computer programs do.

Computers are capable of doing only what they are programmed to do. People write computer programs. Consequently, people control computers. Communicating with a computer by means of a computer program is similar to talking with a person from another country; one must use that person's language, or a language the person understands.

CAD software allows the computer to do what the user wants it to do when issuing commands and selecting options. There are many different types of CAD software produced by many different vendors. However, regardless of the type or the vendor, all CAD software has the following features: (1) commands for generating basic geometric shapes and figures, (2) viewing controls that determine what appears on the screen and how, and (3) modifying commands for editing, revising, and correcting.

CAD software is generally divided into two broad categories. Drafting software allows users to generate 2-, 2½-, and 3-dimensional drawings and other related documentation. Design software enhances the capabilities of designers.

DRAFTING SOFTWARE

Drafting software allows users to accomplish 2-, 2½-, and 3-dimensional drafting tasks. Two-dimensional (2D) tasks include orthographic drawings, bills of material, parts lists, and a variety of other types of documentation, such as tables, charts, and schedules. Two-and-one-half-dimensional (2 1/2D) drawings are oblique drawings.

Oblique drawings show the length, width, and depth of an object all on the same view. Three-dimensional (3D) drawings are isometric, dimetric, trimetric, or perspective drawings. Three-dimensional drawings are sometimes referred to as wireframe drawings. This is to differentiate them from the more complex solid models, which are sometimes developed in support of the design process. A 3D wireframe model defines only the exterior of the object. *Figures 1-30* and *1-31* are examples of 2D CAD drawings.

DESIGN SOFTWARE

Design software enables a number of specific design tasks to be undertaken on a CAD system. The most frequently used of these include solid modeling, finite-element analysis, kinematics, and simulation. *Solid modeling* involves constructing a mathematical, geometric model of a design part and displaying it on the monitor screen. A solid model is the most accurate, most realistic type of geometric model that CAD systems are currently capable of producing.

The most important difference between a solid model and the wireframe type of geometric model is that the wireframe model defines only the exterior of the object. A solid model, on the other hand, defines not only the exterior of the object, but the interior as well. Additional capabilities such as shading and coloring differentiate surfaces of the solid model, making it even more realistic and easy to understand, *Figures 1-32, 1-33,* and *1-34.*

Finite-element analysis is another frequently used design capability of CAD software. Most design products require some type of analysis involving stress, temperature, heat, and other types of calculations. The finite-element analysis capability allows the computer to divide the object into a series of small rectangular and triangular objects such that the surface of the geometric model appears to be a grid, *Figure 1-35.* In this way, the computer is able to analyze how each individual grid position will behave under certain types of stress and strain and other conditions, as well as how the entire object will behave under these conditions.

Kinematic software that can be run on CAD systems is popular among engineers and designers because it saves them a great deal of time and trouble. In the past, mechanical engineers and designers had to build cardboard models or elaborate drawings of the various types of mechanical linkages they designed. The cardboard models were used to determine if the required motion or linkage would actually work in the finished product as intended in the design. This is not a cost-effective way to design a linkage mechanism. With kinematic software, the engineers and designers can create a geometric model of the linkage, display it on the monitor, and

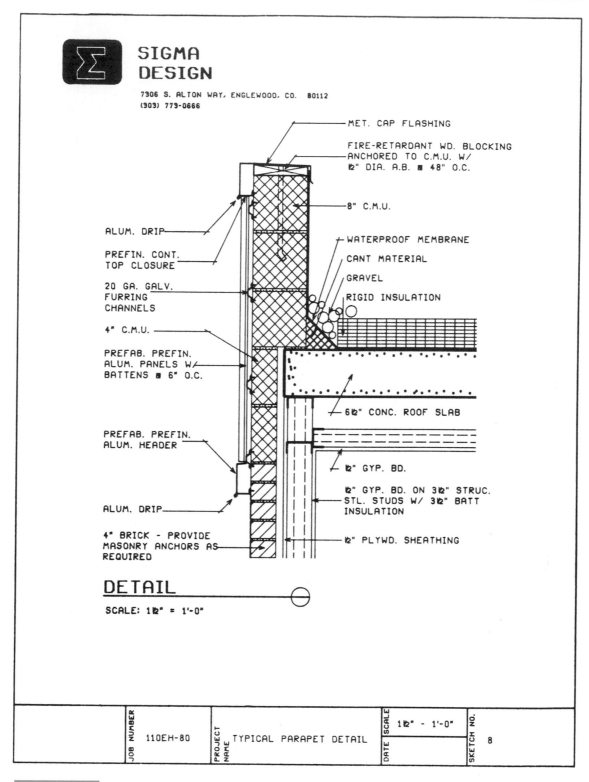

SIGMA DESIGN

7306 S. ALTON WAY, ENGLEWOOD, CO. 80112
(303) 779-0666

MET. CAP FLASHING

FIRE-RETARDANT WD. BLOCKING
ANCHORED TO C.M.U. W/
½" DIA. A.B. @ 48" O.C.

8" C.M.U.

WATERPROOF MEMBRANE

CANT MATERIAL

GRAVEL

RIGID INSULATION

ALUM. DRIP

PREFIN. CONT.
TOP CLOSURE

20 GA. GALV.
FURRING
CHANNELS

4" C.M.U.

PREFAB. PREFIN.
ALUM. PANELS W/
BATTENS @ 6" O.C.

6½" CONC. ROOF SLAB

PREFAB. PREFIN.
ALUM. HEADER

½" GYP. BD.

½" GYP. BD. ON 3½" STRUC.
STL. STUDS W/ 3½" BATT
INSULATION

ALUM. DRIP

½" PLYWD. SHEATHING

4" BRICK - PROVIDE
MASONRY ANCHORS AS
REQUIRED

DETAIL

SCALE: 1½" = 1'-0"

JOB NUMBER		PROJECT NAME		DATE	SCALE		SKETCH NO.
110EH-80		TYPICAL PARAPET DETAIL			1½" - 1'-0"		8

FIGURE 1-30 *Drawing produced on a CAD system* (Courtesy Sigma Design)

introduce motion to determine if the linkage works as designed. If the linkage does not work and revisions are required, designers and engineers can make the adjustments much faster and more easily using the various other capabilities of CAD systems.

Simulation software is available for CAD systems that allows electronic designers and engineers to test and verify the performance of a circuit before sending the design to production. In the past, electronic designers and engineers had to build live prototypes and test them

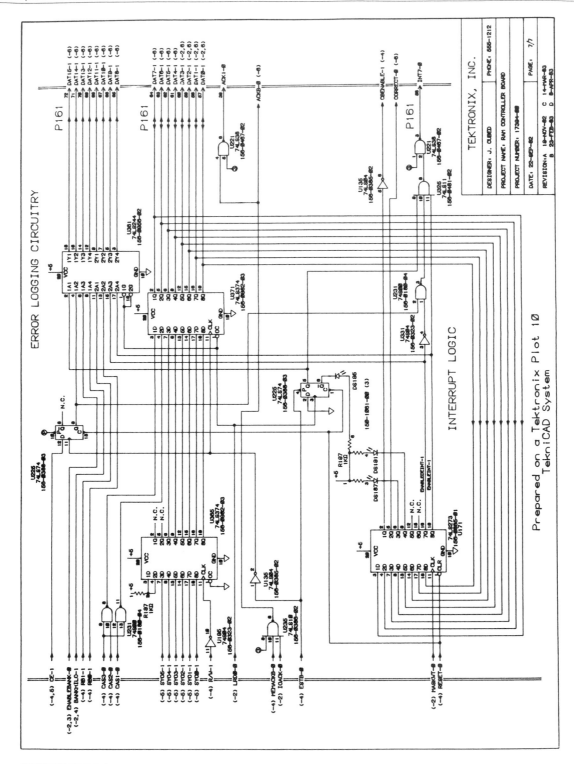

FIGURE 1-31 *Drawing produced on a CAD system* (Courtesy Tektronix Inc.)

to ensure that they worked properly before sending the circuits to production. This was a time-consuming, expensive process.

Simulation software is available for CAD systems that allows electronic designers and engineers to test and verify the production. By taking advantage of the many additional capabilities of CAD systems, designers and engineers can quickly and easily make corrections and revisions to circuits when simulation testing detects the need for them.

FIGURE 1-32   *Solid model* (Image courtesy of SolidWorks Corporation)

FIGURE 1-33   *Solid model of a machined and welded part* (Image courtesy of SolidWorks Corporation)

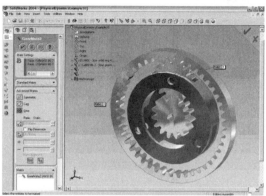

FIGURE 1-34   *Solid model of a machined gear* (Image courtesy of SolidWorks Corporation)

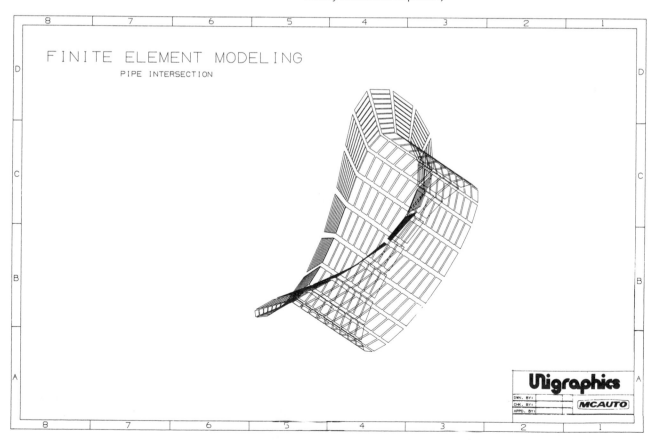

FIGURE 1-35   *Finite-element analysis* (Courtesy McAuto)

Industry Application

SELECTING THE RIGHT PLOTTER

Mark Sutton had been an excellent drafting and design student at Brookview Academy. He proved to be particularly adept at CAD. Consequently, it surprised no one when the chairman of Brookview's Drafting and Design Department asked Sutton to be a student representative on a faculty committee to purchase new plotters for the CAD lab.

At the first meeting of the committee it became apparent that no consensus existed concerning what type of plotters to purchase. One instructor favored flatbed pen plotters because they resembled traditional drafting boards. The resemblance helped him "relate better to the plotters." Another instructor recommended electrostatic plotters, but suggested that low-resolution models be purchased to save funds. She hoped to have funds left over to purchase some precision measurement tools for her mechanical drafting classes.

After two meetings and three hours of discussion and debate, the committee was no closer to achieving consensus than it was at the outset of its deliberations. In order to gain a better perspective, the committee chair asked the members to take a week to investigate, think, and research the issue. He asked the members to meet in a week. Before that meeting, each member was to hand him a sealed one-page recommendation with supportive rationale.

Mark Sutton, feeling somewhat intimidated by it all, decided to give the chair not his opinion but the opinion of practitioners. During the interim week he visited several companies and talked with five different drafting managers. Although their reasons differed somewhat, all five recommended microgrip plotters. Sutton, in preparation for the next meeting, prepared a one-page recommendation containing the reasons given by the various drafting managers.

The committee chair collected all of the sealed recommendations the day before the scheduled committee meeting and studied them overnight. The next day he opened the meeting by saying, "One of our members put a lot of effort into formulating a recommendation. I'd like to read it to you." Sutton slumped in his seat and his face flushed as the chair read his recommendation and reasons. The reasons Sutton had given were adaptability, flexibility, and relatively inexpensive pricing. Discussion began and within ten minutes the committee had formed a consensus around Sutton's recommendation. The other recommendations were put aside.

Once the decision to purchase microgrip plotters had been made, one of the committee members asked, "Whose recommendation did we adopt?" The chair smiled, nodded at Sutton, and said, "Our student member."

The Internet in Design and Drafting

The *Internet* is a worldwide network of loosely connected computer networks. To understand the Internet as a concept, think of the United States Senate. All 100 members of the Senate need to communicate with each other from time to time as they deliberate over issues and pass laws. To promote effective communication among its members, and to maintain order, the Senate operates according to an established set of rules. The Senate has a standing committee for establishing, maintaining, and updating the rules.

Think of the Internet as the rules committee. It simply defines the rules and languages by which they communicate. Just as the Senate Rules Committee has no power over how individual senators vote, the Internet has no control over what happens within the various networks and users who make up the World Wide Web. Consequently, the Internet has been described as the technological equivalent of the old American West before the days of law and order. There is currently an anything-goes aspect to the Internet.

Nothing is more important in the design process than information—particularly timely, accurate, up-to-date information. The Internet gives design and drafting personnel immediate access to a growing body of information; information that in the past was available only in manuals, catalogs, and other hard copy reference sources that were prone to being outdated. It also gives design and drafting personnel an effective vehicle for improving communication, another important aspect of the design process.

The primary uses of the Internet for design and drafting personnel are as follows: (1) person-to-person and person-to-group communication using e-mail, and (2) research.

E-MAIL COMMUNICATION

E-mail is short for electronic mail or person-to-person written communication sent electronically. E-mail is the electronic equivalent of the post office in that it is used to send written correspondence from one source to another. Only the method of delivery—and as a result, the speed of delivery—is fundamentally different.

Individuals and companies gain access to the Internet through Internet service providers. One of the functions of the service providers is to maintain a computer post office, which they do 24 hours a day, 365 days a year. All Internet users have an e-mail address, just as you have an address for deliveries from the U.S. Post Office. When you send e-mail correspondence to another Internet user, it first goes to the computer post office of your Internet service provider. Your service provider forwards it to the service provider of the recipient. That service provider, in turn, holds the correspondence until the recipient retrieves it. Of course, this all happens electronically in the blink of an eye, *Figure 1-36*.

E-mail senders and receivers have addresses. For example, the e-mail address jjones@mtc.com is for a user named John Jones (jjones) at a company named Manufacturing Technology Corporation (mtc). These two parts of the e-mail address are separated by the @ symbol, as they always are in an e-mail address.

In establishing an e-mail address in the United States, a special three-letter designation is used at the end of the address, *Figure 1-37*. This designation indicates the category of the user (e.g., business, educational institution, commercial Internet service provider, or any type of organization that is not a business). Outside the United

Organization Specifiers	
.com	A business
.edu	An educational institution
.net	A commercial Internet service provider
.org	An organization that is not a business

FIGURE 1-37   *Organization specifiers*

States, e-mail addresses end in two-letter designations that specify the country (e.g., *uk* for United Kingdom).

It is not uncommon for a project team to be spread out among several locations. E-mail is a convenient way to keep all members of the team informed with current information. E-mail can be forwarded to a group of people as easily as it can to an individual by using a *list server*—a special program that forwards e-mail to all individuals on a list.

List servers ensure that all individuals on a list see all messages that are sent to the list, including responses among individuals. This feature of e-mail can be especially beneficial when a project team includes members from different locations. For example, design and drafting personnel might be located in one location, manufacturing personnel in another, and marketing personnel in yet another. A list server can help ensure that all members of the team have the same information at the same time all the time.

THE INTERNET AS A RESEARCH TOOL

On any design and drafting project, research is fundamental. In the past, the necessary product and process information needed by drafting technicians has been housed in catalogs, reference manuals, and specification files. Many of the vendors, organizations, and agencies that have historically published their information in hard copy form now make it available via an Internet website. Information that is available on the World Wide Web has increased exponentially. As a result, there are few vendors a design and drafting team might need to interact with that do not have a home page on the Web.

A *home page* is the Internet version of an organization's actual location. To view the specifications and design information about a vendor's products, a design and drafting team can now visit the vendor's website. Immediate access to current information is the upside of the Internet as a research tool. The downside is a lack of overall organization.

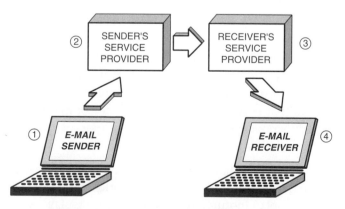

STEP 1: SENDER GENERATES MESSAGE
STEP 2: MESSAGE IS TRANSMITTED TO SENDER'S
SERVICE PROVIDER
STEP 3: MESSAGE IS TRANSMITTED TO
RECEIVER'S SERVICE PROVIDER
STEP 4: MESSAGE IS RETRIEVED BY RECEIVER

FIGURE 1-36   *How an e-mail message is sent*

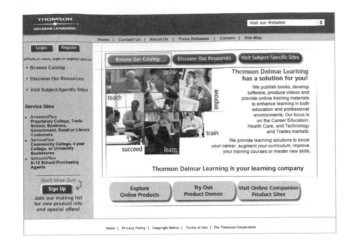

FIGURE 1-38 *Home page for Thomson Delmar Learning*

The World Wide Web is like a huge electronic library that has no card catalog. Picture a library wherein the only way to locate a specific book is to randomly browse through the shelves. To help identify specific websites, there are a number of search engines available through the Internet.

A *search engine* is an electronic enabler that can scan the Internet using keywords or phrases. For example, if a design and drafting team is interested in locating a specific type of valve, it can make a list of key terms related to the valve in question and enter them into a search engine. The search engine then guides the user to websites that have those keywords in them. It is still necessary to visit the actual website to determine if it contains the specific information needed.

Browsing is sometimes called surfing or navigating the Web. In order to surf the Web, users need a program called a *browser*. This program monitors where the user is in the Web at any given time, and it displays any information that is requested.

Every site on the web has an address called a *uniform resource locator* or *URL*. URLs begin with the following letter/symbol combination: **http://**

The colon and two slash characters following the http tell the browser program that the address that follows is a website. The browser will take users to the home page located in the Web at the specified address. An organization's home page is like the cover of a book in that it is usually attractive enough to make users want to look inside. It contains links that guide the user to more specific information about the organization and its products or services. For example, *Figure 1-38* is the home page for Thomson Delmar Learning. It is located at the following address: **http://www.delmar.com**

Summary

- Regardless of their relative speed and size, what makes computers such invaluable tools is their ability to store data, perform basic logic functions, and make mathematical computations. The ability to electronically link computers has allowed the design, drafting, and engineering functions of a company to be part of a fully integrated whole that includes marketing, accounting, manufacturing, and shipping.

- Modern CAD workstations had their origins in the 1950s when the United States Air Defense Command first used light pens for data entry. The next major development in the history of CAD occurred in 1963 at the Massachusetts Institute of Technology (MIT) when professor Ivan Sutherland demonstrated his Sketchpad. IBM released the first commercially marketed CAD system in 1964.

- All computers are either digital or analog. Analog computers are used to measure continuous physical properties such as heat, pressure, depth, and speed. They are used most often to monitor and control mechanical, electronic, and hydraulic systems. The digital computer is the type that most people recognize and use. Today's personal, laptop, and notebook computers are digital.

- Microminiaturization is the process that accounts for the fact that over time computers have gotten progressively smaller, but simultaneously more powerful. The enabling ingredient in the microminiaturization process is the programmable integrated circuit, which allowed the development of the microprocessor—the key component of today's small but powerful computers.

- Computer-aided design and computer-aided drafting are both referred to as CAD. Computer-aided manufacturing or CAM involves using the computer to enhance the capabilities of the production components of a company. Computer graphics is a broad term that encompasses the use of the computer and peripheral devices to produce any type of graphic image, either technical or artistic.

- CAD systems have three broad components: hardware, software, and users. CAD workstations consist of a display monitor (typically a raster display); various input devices such as a keyboard, digitizing tablet, stylus, trackball, puck, mouse, joystick, light pen, and/or voice recognition devices; various output devices such as pen plotters, dot matrix plotters, electrostatic plotters, printer/plotters, and printers (dot matrix, inkjet, or laser); various storage devices such as the fixed or hard disk drive, floppy disk drive, zip drive, CD-ROM drive, optical drive, and/or magnetic tape drive; and the central processing unit or CPU.

- CAD software has the following features: (1) commands for generating basic geometric shapes, figures, etc.; (2) viewing controls that determine what appears on the screen and how; and (3) modifying commands for editing, revising, and correcting. Drafting software allows users to produce 2-, 2 1/2-, and 3-dimensional drawings and other types of design documentation. Design software extends the designer's capabilities with such functions as solid modeling, finite-element analysis, kinematics, and simulation.

- The Internet is a worldwide network of loosely connected computer networks. The primary uses of the Internet for design and drafting personnel are person-to-person and person-to-group communication and research.

Review Questions

Answer the following questions either true or false.

1. The Internet is a worldwide network of loosely connected computer networks.

2. The modern CAD workstation originated in the late 1940s.

3. Other terms often used for *browsing* the Web are surfing and navigating.

4. All URLs on the Web begin with the letter/symbol combination of **www://**

5. The analog computer is the type that most people recognize and use.

6. CAD workstations usually use a *raster* type of display monitor.

Answer the following questions by selecting the best answer.

1. Which of the following is **not** true regarding what makes computers such valuable tools?
   a. Their ability to store data
   b. Their ability to perform basic logic functions
   c. Their ability to perform mathematical computations
   d. Their ability to provide linkages that guide the user

2. Which of the following is **not** one of the three broad components of a CAD system?
   a. Input device
   b. Hardware
   c. Software
   d. Users

3. Which of the following is **not** an input device?
   a. Trackball
   b. Joystick
   c. Light pen
   d. Zip drive

4. Which of the following is **not** one of the primary uses of the Internet for design and drafting personnel?
   a. Person-to-person communication
   b. Browsing
   c. Person-to-group communication
   d. Research

5. Which of the following is **not** a feature of CAD software?
   a. Commands for generating basic geometric shapes, figures, etc.
   b. Commands that control mechanical, electronic, and hydraulic systems
   c. Viewing controls that determine what appears on the screen and how
   d. Modifying commands for editing, revising, and correcting

6. Which of the following is **not** a storage device for a CAD workstation?
   a. CPU
   b. Zip drive
   c. CD-ROM drive
   d. Digitizing tablet

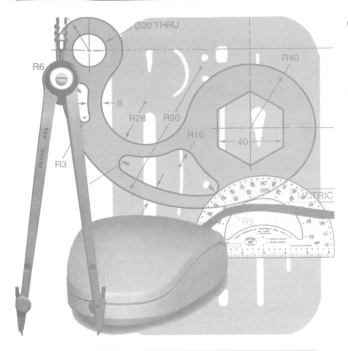

TECHNICAL DRAWING WITH INSTRUMENTS

CHAPTER OUTLINE

Conventional instruments • Conventional drafting requisites • Drawing sets • Scales • Measuring • Ink tools • Technical pens • Mechanical lettering sets • Butterfly scriber • Airbrush • Paper sizes • Whiteprinter • Files • Care of drafting equipment • Use of appliqués • Use of burnishing plates • Typewritten text • Overlay drafting • Scissors drafting techniques • Summary • Review questions • Chapter 2 problems

KEY TERMS

Adjustable curve

Appliqué

Bow compass

Burnishing

Compass

Decimal-inch scale

Drawing surface

Drop-bow compass

Dry transfer sheet

French curve

Full-divided scale

Hardware

Irregular curves

Kilo

Lettering machines

Milli

Open-carriage typewriter

Open-divided scale

Overlay drafting

Scissors drafting

Scriber templates

Slick

T-square

Transfer cards

Whiteprinter

CHAPTER OBJECTIVES

Upon completion of this chapter, students should be able to do the following:

■ Recognize the most frequently used conventional drafting instruments and their purposes.

■ Describe the contents of a typical drawing set and the purpose of each item.

■ Demonstrate the proper use of scales, including the mechanical engineer's, architect's, civil engineer's, and metric scale.

■ Demonstrate how to read the following precision measurement instruments: micrometer, vernier caliper, microfinish comparator, and the ellipses instrument.

■ Explain the two types of technical pens, their respective uses, and how to properly clean them.

■ Demonstrate the proper use of a mechanical lettering set.

■ Demonstrate the proper use of a butterfly scriber.

■ Explain the uses of the airbrush.

■ Describe the various standard sizes of paper used in drafting in both inches and millimeters.

■ Explain how the whiteprinting process works.

■ Explain the term *appliqué* and how the various kinds are used.

■ Explain the term *burnishing plate* and how one is used.

■ Describe the concept of overlay drafting.

■ Describe the concept of scissors drafting.

Note to Instructors and Students

A survey of users of this text indicates that there are three main approaches to teaching design and drafting: (1) begin students on the CAD system without instruction in conventional methods or instruments, (2) begin students with instruction in conventional methods and instruments before introducing them to CAD, (3) begin students on the CAD system, but provide instruction in conventional methods and instruments for historical purposes. To accommodate all of these instructional methodologies, the authors continue to provide this chapter on the use of conventional drafting and design instruments based on the assumption that instructors who prefer to begin students on the CAD system may simply skip this chapter.

The advent of computer-aided drafting or computer-aided design (CAD) has radically changed the instruments used to produce technical drawings. In fact, one rarely hears the term instruments used in a modern technical drawing setting. Instead one hears such terms as hardware, software, and systems. The tools used in a modern drafting department have changed continually over the years. Such changes are often associated with CAD, but in reality the tools used to produce technical drawings have always been in a state of continual evolution and always will be.

Engineers, designers, and drafters who are old enough can remember when technical drawings were prepared in ink on linen using T-squares. Technological changes eventually saw T-squares superseded by parallel bars that were, in turn, superseded by drafting machines. Linen was replaced by vellum. Templates replaced various mechanical instruments. Overlay and scissors drafting techniques eliminated a certain amount of duplication.

These nonautomated developments continued (and to a much lesser extent still continue) until the giant leaps in technical drawing technology brought about by CAD. With the advent of CAD, conventional drafting instruments began to be used less and less. However, they have not yet completely faded away, and it will be years before they do.

Therefore, this chapter covers first the conventional instruments and then the *hardware* and systems associated with CAD. In this way students who wish to learn about the evolution of technical drawing instruments, techniques, and media from the past to the present may do so. This approach is recommended to help today's technical drawing personnel gain a perspective on where they fit into the evolutionary process with regard to their profession.

Conventional Instruments

In drafting, no lines are made freehand. Each and every line is drawn using some kind of a drafting tool. It is up to the drafter to own a complete set of standard drafting tools in order to be fully functional.

When purchasing conventional drafting equipment, care must be taken to obtain quality equipment from a reliable dealer. It is advisable to consult with an experienced drafter, a drafting instructor, or a reputable dealer. The following is a list of the minimum required drafting equipment. Special templates and special equipment must be added to this list, depending upon the field of drafting and, in some cases, the actual product manufactured by the company. Each of the following pieces of equipment is illustrated in this chapter.

* Drawing board—24″ × 36″ (60 cm × 90 cm) minimum size
* T-square (parallel straightedge or drafting machine) to suit board
* 45° triangle—8″ (20 cm) size
* 30°–60° triangle—8″ (20 cm) size
* Triangular scale (depending upon the field of drafting)
* Center wheel bow compass—6″ (115 cm) with extension bar
* Drop bow compass (recommended)
* Irregular curves (two or three different configurations)
* Dividers—6″ (15 cm)
* Drafting brush
* Mechanical drafting pencils with lead
* Protractor or adjustable triangle
* Erasing shield
* Eraser
* Lead pointer with steel cutting wheel (pencil)
* Lead sandpaper or flat file (for compass lead)
* Circle template
* Ellipse template
* Drafting tape or drafting dots
* Calculator
* Dry cleaning pad (optional)

The following is a list of the required inking supplies.

* Technical pen #2 1/2 (for lettering scriber)
* Technical pen #1 (for lettering scriber)
* Technical pen #0 (for lettering scriber)
* India ink—black
* Pen cleaning solvent
* Lettering scriber

- Lettering template #100
- Lettering template #175
- Lettering template #290
- Compass adaptor for pen

Note that some specialized or expensive equipment is often furnished by the company.

Conventional Drafting Requisites

DRAWING TABLE

Drawing tables are available in a variety of styles. Most are adjustable up and down and can tilt to almost any angle from vertical 90° to horizontal, *Figures 2-1* and *2-2*.

DRAWING SURFACE

The *drawing surface,* whether it is a drawing tabletop or drawing board, must be flat, smooth and large enough to accommodate the drawing and some drafting equipment. If a *T-square* is used on a drawing board, at least one edge of the board must be absolutely true. Most quality drafting boards have a metal edge to ensure against warpage and against which to hold the T-square securely.

FIGURE 2-2   *Drafting table* (Courtesy Alvin & Co.)

Standard drafting boards range in size from small, 12″ × 17″ (30 cm × 43 cm), to large, 31″ × 42″ (78 cm × 105 cm). Standard drafting tabletops range in size from 31″ × 42″ (78 cm × 105 cm) to 37 1/2″ × 60″ (94 cm × 150 cm).

LIGHTING

It is important that the drawing surface be fully lighted without any shadows, *Figure 2-3*.

TOP COVER

The drafting board should have a top cover that protects the board surface and provides a perfect drawing surface. A good top cover actually seals over holes made by compass points, and it is easily cleaned, *Figure 2-4*.

FIGURE 2-1   *Drafting table* (Courtesy Alvin & Co.)

FIGURE 2-3   *Flourescent lamp* (Courtesy Waldman Lighting)

FIGURE 2-4 *Drafting board top cover* (Courtesy Modern School Supplies Inc.)

EFFICIENCY

To be fully efficient at drafting, one must keep all equipment clean, correctly adjusted and/or sharpened, and stored in a convenient location, ready for use at all times. It is good drafting practice to store each piece of equipment in a specific location and return it to its location after use. An organizer, such as the one shown in *Figure 2-5*, aids in keeping all equipment in its place.

LEFT-HANDED DRAFTERS

Most drafting equipment is designed for the right-handed drafter, although left-handed types of drafting machines can be purchased. The T-square is simply placed on the right side of the drafting board by left-handed drafters, but everything else is right handed. The left-handed drafter has to adapt. The lettering scriber is especially difficult to manipulate.

T-SQUARE

While T-squares are not used today in industry, they do provide a parallel straightedge for the beginning drafter. The T-square is used to draw horizontal lines. Draw lines only against the upper edge of the blade. Make sure the head is held securely against the left edge of the drawing board to guarantee parallel lines, *Figure 2-6*.

The T-square is composed of two parts: the head and the blade, *Figure 2-7*. The two parts are fastened together at an exact right angle. The blade must be straight and free of any nicks or imperfections. A transparent acrylic

FIGURE 2-6 *T-square and other traditional instruments* (Courtesy Teledyne Post)

FIGURE 2-5 *Equipment organizer* (Courtesy Stacor Corp.)

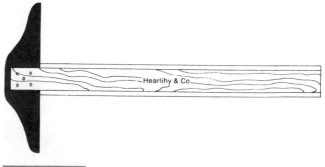

FIGURE 2-7 *Parts of the T-square* (Courtesy Hearlihy and Co.)

edge is recommended since this allows the drafter to see the drawing underneath the edge. T-squares can be purchased with adjustable heads for drawing specific angles.

PARALLEL STRAIGHTEDGE

The parallel straightedge is always parallel, regardless of where it is placed upon the drawing surface. Parallel control is accomplished by a system of cords and pulleys, *Figure 2-8*. The parallel straightedge replaced the T-square in industry and is still used somewhat today. Most straightedges come with a transparent acrylic edge, and some have rollers for a smooth gliding action. Some have a locking brake that permits the straightedge to be locked in any position.

DRAFTING MACHINES

A drafting machine is a device that attaches to the drafting table and replaces both the T-square and the parallel straightedge. There are two basic kinds of drafting machines. One is the arm type, *Figure 2-9,* and the other, the newer, is the track type, *Figure 2-10*. On both

FIGURE 2-10   *Track-type drafting machine* (Courtesy Alvin & Co.)

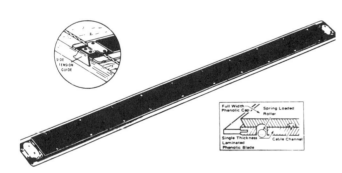

FIGURE 2-8   *Parallel straightedge* (Courtesy Modern School Supplies Inc.)

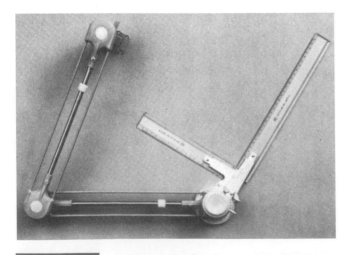

FIGURE 2-9   *Arm-type drafting machine* (Courtesy Teledyne Post)

types, a round head holds two straightedges at right angles to each other. The head can be rotated to set the straightedges at any angle. Most machines are available with interchangeable straightedges marked with different scales along their edges.

Drafting machines replace straightedges, scales, triangles, and protractors. They increase accuracy and greatly reduce drafting time. A drafting machine is one of the few tools that can be purchased either as a right-handed or a left-handed instrument.

Most drafting machines have a protractor and a vernier, which permit readings to 5 minutes of an arc, *Figure 2-11*. Notice that zero on the protractor is in line with the zero on the vernier. The vernier is graduated in 5-minute increments from zero to 60 minutes. To read the vernier, first read the protractor, *Figure 2-12*. In this example, the zero on the vernier points between 18° and 19°. On the vernier, notice that the only line that lines up with a line on the protractor is the 45; thus, this is read as 18°45′. Some drafting machine heads simplify this process by adding a digital readout, see *Figure 2-13*.

Drafting machine straightedges come in sizes of 9″ (23 cm), 12″ (30 cm), and 18″ (45 cm), graduated or ungraduated, in both transparent plastic and aluminum scale, *Figure 2-14*.

A drafting machine, although a precision instrument, should be checked for accuracy at least once a

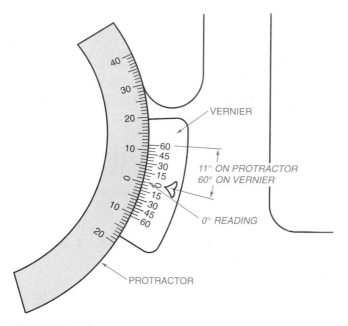

VERNIER

11° ON PROTRACTOR
60° ON VERNIER

0° READING

PROTRACTOR

FIGURE 2-11 *Protractor with vernier*

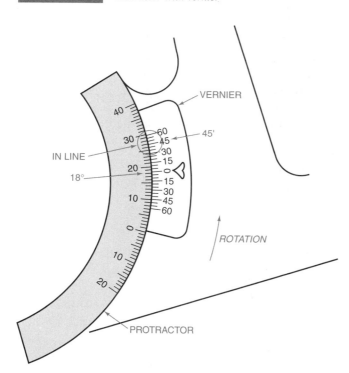

VERNIER

IN LINE

18°

45'

ROTATION

PROTRACTOR

FIGURE 2-12 *Reading the vernier*

FIGURE 2-13 *Drafting machine head with digital readout* (Courtesy Consul & Mutoh Ltd.)

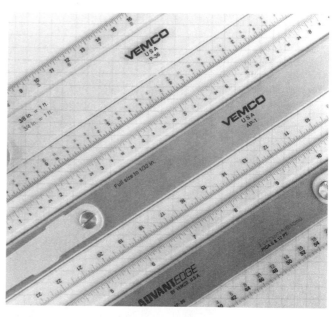

FIGURE 2-14 *Transparent and metal drafting machine straight-edges with scale (Courtesy Vemco Corp.)*

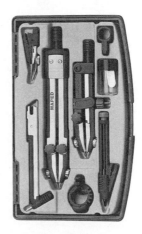

FIGURE 2-15 *Complete drafting set (Courtesy Alvin & Co.)*

week. The instructions for checking and adjusting a drafting machine are included with the manufacturer's information.

Drawing Sets

Typical drawing sets include *compasses*, dividers, and a ruling pen, *Figure 2-15.* Many sets include a variety of tools not normally used by the drafter. It is recommended that only those tools actually needed be purchased.

COMPASSES

There are two main types of compasses: the friction-joint type and the spring-bow type. The friction-joint type is still widely used for lightly laying out pencil drawings that will be inked. The disadvantage of this type of compass is that the setting may slip when strong pressure is applied to the lead.

The spring-bow type of compass, *Figure 2-16A,* is best for pencil drawings and tracings because it retains its setting even when strong pressure is applied to obtain dark lines. The spring, located at the top of the compass, holds the legs securely against the adjusting screw. The adjusting screw is used to make fine adjustments. Technical pen adapters allow ink attachments to be used, *Figure 2-16B.*

FIGURE 2-16A   *Spring-bow compass* (Courtesy Vemco Corp.)

FIGURE 2-16B   *Pen adapter* (Courtesy Vemco Corp.)

Compass leads should extend approximately ⅜″ (0.9 cm). The metal point of the compass extends slightly more than the lead to compensate for the distance the point penetrates the paper, *Figure 2-17.* The lead is sharpened with a sandpaper paddle to produce clean, sharp lines. The flat side of the lead faces outward in order to produce circles of very small diameter, *Figure 2-18.* Sharpen with a paddle in the direction of the arrow, as shown, in order to keep the lead sharp longer.

To draw a circle with the *bow compass,* the compass is revolved between the thumb and the index finger, *Figure 2-19.* Pressure is applied downward on the metal point to prevent the compass from jumping out of the center hole.

DROP-BOW COMPASS

The *drop-bow compass, Figure 2-20,* is used for circles of .03″ (0.08 cm) to .50″ (1.3 cm) diameter. The compass is adjusted to the required radius. The center point is located on the circle or arc swing point and held in place with the index finger. Rotate the knurled head of the compass between the thumb and second finger.

BOW COMPASS WITH LEAD CLUTCH

In order to eliminate the process of sharpening compass leads, some drafters use a compass with a lead clutch of

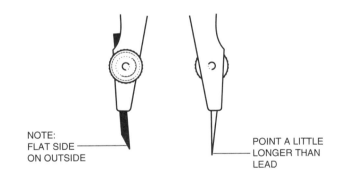

NOTE:
FLAT SIDE
ON OUTSIDE

POINT A LITTLE
LONGER THAN
LEAD

FIGURE 2-17   *Bow compass lead and metal points*

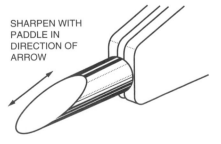

SHARPEN WITH
PADDLE IN
DIRECTION OF
ARROW

FIGURE 2-18   *Bow compass lead point shape* (Courtesy Drafting for Trades and Industry, Basic Skills, Nelson, Delmar Publishers Inc.)

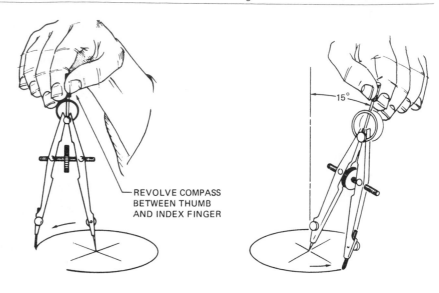

REVOLVE COMPASS
BETWEEN THUMB
AND INDEX FINGER

15°

FIGURE 2-19   *Drawing a circle with a bow compass.*

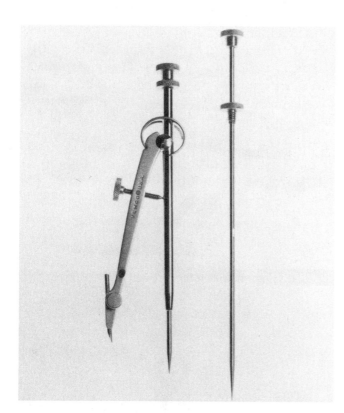

FIGURE 2-20   *Drop-bow compass* (Courtesy Vemco Corp.)

FIGURE 2-21   *Bow compass with special clutch* (Courtesy B. Carter Lykins)

0.5-mm lead, *Figure 2-21*. This compass saves time and ends messy lead sharpening. Special compasses are designed only for inking, *Figures 2-22* and *2-23*. An adaptor to attach to a standard compass to draw ink lines is illustrated in *Figure 2-24*.

BEAM COMPASS

A beam compass, *Figure 2-25*, is used to draw large circles or arcs. Fine line adjustments can be obtained and locked in place. Beam compasses come in sizes from 13″ (33 cm) bars and up.

FIGURE 2-22   *Inking bow compass* (Courtesy Koh-I-Noor Rapidograph)

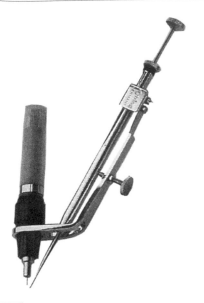

FIGURE 2-23 *Inking drop-bow compass* (Courtesy Koh-I-Noor Rapidograph)

FIGURE 2-24 *Standard compass ink adapter* (Courtesy Koh-I-Noor Rapidograph)

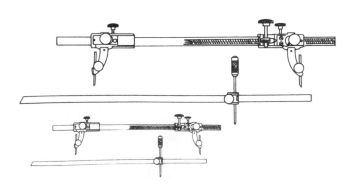

FIGURE 2-25 *Beam compass* (Courtesy Vemco Corp.)

ADJUSTABLE CURVE

An *adjustable curve, Figure 2-26*, has a locking knob and is used to draw any radius from 6.75″ to 200″ (17 cm to 500 cm). This tool takes over where the ordinary compass leaves off, and it eliminates the beam compass.

DIVIDER

A divider is similar to a compass except that it has a metal point on each leg. It is used to lay off distances and to transfer measurements, *Figure 2-27*.

PROPORTIONAL DIVIDERS

Proportional dividers are used to enlarge or reduce an object in scale. This tool has a sliding, adjustable pivot that varies the scale, *Figure 2-28*.

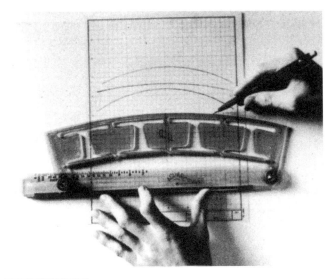

FIGURE 2-26 *Adjustable curve* (Courtesy Hoyle Products Inc.)

FIGURE 2-27 *Divider* (Courtesy Vemco Corp.)

FIGURE 2-28 *Proportional dividers* (Courtesy Modern School Supplies Inc.)

TRIANGLE

Two standard triangles are used by drafters. One is a 30–60-degree triangle, usually written as 30°–60° triangle. The other is a 45-degree triangle and the protractor, written as 45° triangle. The 45° triangle consists of two 45-degree angles, and one 90-degree angle, *Figure 2-29A*. The 30°–60° triangle contains a 30-degree angle, a 60-degree angle, and a 90-degree angle, *Figure 2-29B*.

Triangles are made of plastic and come in a variety of sizes other than those mentioned. When lines are laid out, triangles are placed firmly against the upper edge of the straightedge. Pencils are placed against the left edge of the triangles and lines are drawn upward, away from the straightedge. Parallel angular lines are made by mov-

ing the triangle to the right after each new line has been drawn, *Figure 2-30*.

ADJUSTABLE TRIANGLE

An adjustable triangle may take the place of both the 30°–60° and 45° triangles, *Figure 2-31*. It is recommended, however, that this tool be used only for drawing angles that cannot be made with the two standard triangles. The adjustable triangle is set by eye and thus is not as accurate as the solid triangle.

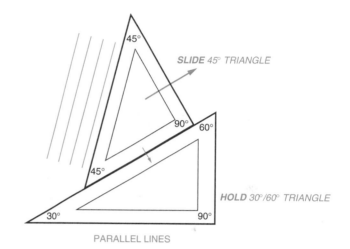

FIGURE 2-30   *Drawing parallel angular lines*

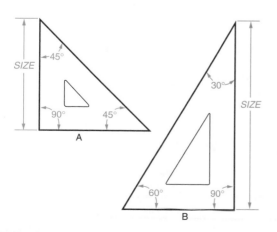

FIGURE 2-29   *(A) 45° angle, and (B) 30°–60° triangle*

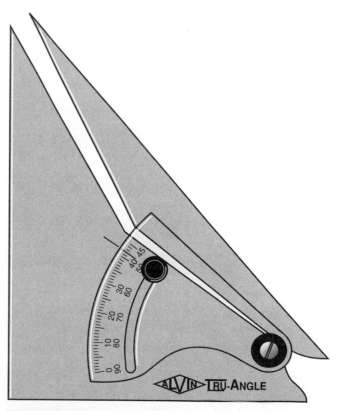

FIGURE 2-31   *Adjustable triangle* (Courtesy Modern School Supplies Inc.)

TEMPLATE

A template is a thin, flat piece of plastic containing various cutout shapes, *Figures 2-32, 2-33,* and *2-34.* It is designed to speed the work of the drafter and to make the finished drawing more accurate. Templates are available for drawing circles, ellipses, plumbing fixtures, bolts and nuts, screw threads, electronic symbols, springs, gears, and structural metals, to name just a few uses.

Templates come in many sizes to fit the scale being used on the drawing. A template should be used wher-ever possible to increase accuracy and speed. It is prefer-able to purchase templates that are stamped and not molded, because molded templates become brittle in time and break.

FRENCH CURVES

French curves are thin, plastic tools that come in an assort-ment of curved surfaces, *Figure 2-35.* They are used to pro-duce curved lines that cannot be made with a compass. Such lines are referred to as *irregular curves.* Most good

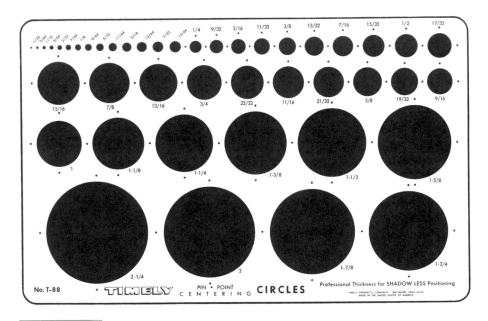

FIGURE 2-32   *Circle template* (Courtesy Timely Products Co.)

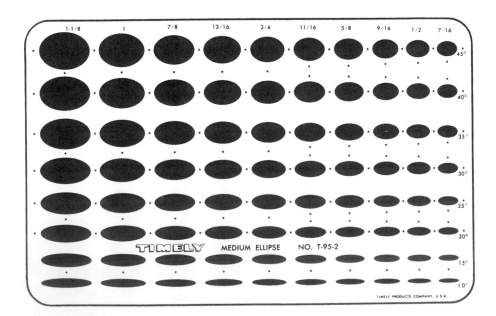

FIGURE 2-33   *Ellipse template* (Courtesy Timely Products Co.)

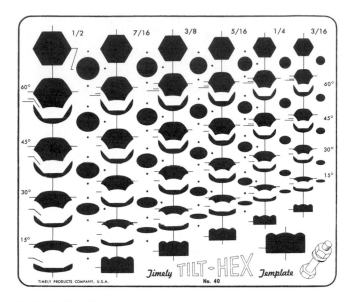

FIGURE 2-34 *Bolt and nut template* (Courtesy Timely Products Co.)

to get a general idea of where the curved line is going. If the line makes an abrupt turn, a line lightly sketched in place of the straight lines may be more useful. Starting from one side or the other, line up the French curve along as many points as possible and draw a dark line connecting these points, *Figure 2-36*. Readjust and align the French curve along all additional points, and continue drawing the curved line. Proceed in this manner until the line is completed.

ADJUSTABLE CURVE

Adjustable curves form smooth curves. *Figure 2-37* shows a flexible steel measuring tape that measures the perimeter of the curve to be drawn. The curve is held by friction between many layers of interlocking channels.

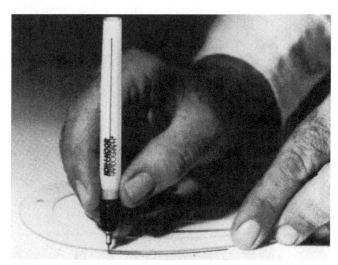

FIGURE 2-36 *Drawing an irregular curve* (Courtesy Koh-I-Noor Rapidograph)

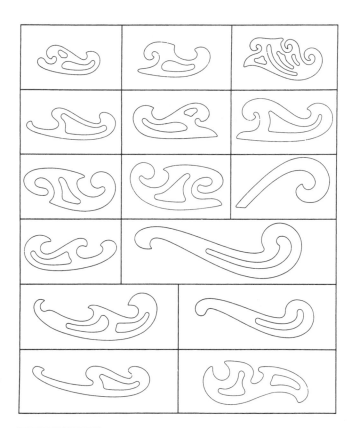

FIGURE 2-35 *Assortment of French curves* (Courtesy Modern School Supplies Inc.)

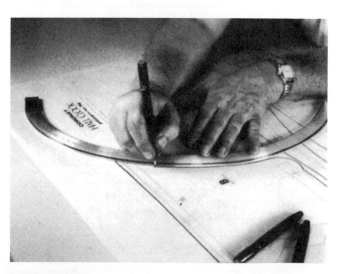

FIGURE 2-37 *Using an adjustable curve* (Courtesy Hoyle Products Inc.)

French curves are actually segments of such geometric curves as ellipses, parabolas, hyperbolas, and the like.

USING A FRENCH CURVE

To use a French curve, the irregular curve must be defined by a series of dots. Lightly connect straight lines

PROTRACTOR

A protractor is used to measure and lay out angles, *Figure 2-38*. It can be used in place of a drafting machine or an adjustable triangle.

To use the protractor, place the center point (located at the lower edge of the protractor) on the corner point of the angle. Align the base of the protractor along one side of the angle. The degrees are read along the semicircular edge.

PENCILS AND LEADS

Lead for a mechanical drafting pencil comes in 18 degrees of hardness, ranging from 9H, which is very hard, to 7B, which is very soft, *Figure 2-39*. For drafting purposes, the scale of hardness is as follows: 3H or 4H lead is recommended for layout work, and 2H, H, or HB leads are recommended for all other lines. Experiment with vari-

ous leads to determine which ones give the best line thickness. This varies depending upon the pressure applied to the point while drawing lines. *Figure 2-40* shows lead for a mechanical drafting pencil.

Regular pencils are sharpened with a pencil sharpener. It is important that enough wood is removed to ensure that the lead, not the wood, of the pencil comes in contact with the straightedge or triangle edges.

LEAD HOLDERS AND LEADS

Lead holders hold sticks of lead, *Figure 2-41*. The leads designed for lead holders come in the same range of hardness as those for regular mechanical pencils and are used for the same purposes. The main advantage is that they are more convenient to use. Leads are usually sharpened in a lead pointer. Electric lead pointers are fully automatic. A slight downward pressure of the lead starts the motor action, *Figure 2-42*. This machine produces a perfectly tapered point and eliminates all loose clinging graphite.

SANDPAPER PADDLE

A sandpaper paddle consists of several layers of sandpaper attached to a small wooden holder, *Figure 2-43*. The sandpaper is used to sharpen compass leads only. Do not sharpen leads over a drawing because the graphite will smear the drawing surface.

ERASERS

Various kinds of erasers are available to a drafter. One of the most commonly used is a soft, white block-type eraser, *Figure 2-44*. *Figure 2-45* shows a pencil-type

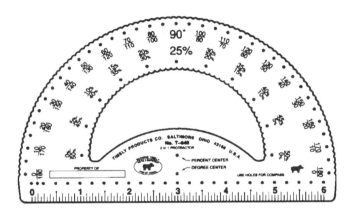

FIGURE 2-38 *Protractor* (Courtesy Alvin & Co.)

TASK	LEAD
CONSTRUCTION LINES	3H, 4H
GUIDE LINES	3H, 4H
LETTERING	H, F, HB
DIMENSION LINES	2H, H
LEADERLINES	2H, H
HIDDEN LINES	2H, H
CROSSHATCHING LINES	2H, H
CENTERLINES	2H, H
PHANTOM LINES	2H, H
STITCH LINES	2H, H
LONG BREAK LINES	2H, H
VISIBLE LINES	H, F, HB
CUTTING PLANE LINES	H, F, HB
EXTENSION LINES	2H, H
FREEHAND BREAK LINES	H, F, HB

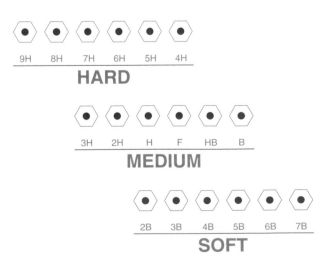

FIGURE 2-39 *Lead-lines chart (left) and grades of lead (right)*

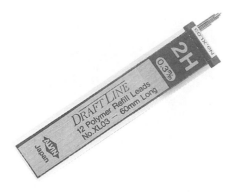

FIGURE 2-40 *Drafting lead* (Courtesy Alvin & Co.)

FIGURE 2-41 *Lead holders* (Courtesy Staedtler Inc.)

FIGURE 2-42 *Lead pointer*

FIGURE 2-43 *Sandpaper paddle* (Courtesy Keuffel & Esser Co.)

FIGURE 2-44 *Block-type eraser* (Courtesy Staedtler Mars)

FIGURE 2-45 *Pencil-type eraser with clutch* (Courtesy Staedtler Mars)

eraser with an adjustable clutch. If one develops good drawing habits, erasing can be kept to a minimum.

ELECTRIC ERASER

An electric eraser speeds up corrections. Some models take a 7″ (17.5 cm) long eraser strip. The model illustrated in *Figure 2-46* has a slip clutch to hold the eraser strip in place.

A cordless erasing machine can be used with or without the standard electric cord, and it uses rechargeable NiCad batteries, *Figure 2-47*. As the eraser is placed in the stand, the batteries are recharged.

Electric erasers do save time, but care must be taken not to burn through the drawing paper. This can be

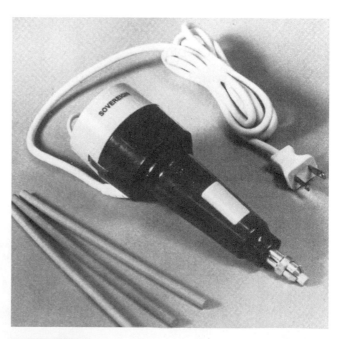

FIGURE 2-46 *Electric eraser with slip clutch* (Courtesy Rotex Co.)

FIGURE 2-47 *Rechargeable electric eraser* (Courtesy Rotex Co.)

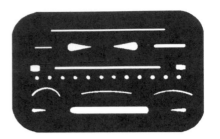

FIGURE 2-48 *Erasing shield* (Courtesy Alvin & Co.)

FIGURE 2-49 *Drafting brush* (Courtesy Alvin & Co.)

avoided by using an erasing shield and placing a thick sheet of paper beneath the drawing to cushion it.

ERASING SHIELD

An erasing shield restricts the erasing area so that correctly drawn lines will not be disturbed during the erasing procedure. It is made from a thin, flat piece of metal with variously sized cutouts, *Figure 2-48*. The shield is used by placing it over the line to be erased and erasing through the cutout.

DRAFTING BRUSH

The drafting brush is used to remove loose graphite and eraser crumbs from the drawing surface, *Figure 2-49*. Do not brush off a drawing surface by hand because this tends to smudge the drawing. Drafting brushes come in various sizes from 10½″ to 14″ (26 cm to 35 cm). The bristles can be either horsehair or nylon, and they can be cleaned with warm, soapy water.

DRY CLEANING POWDER

Cleaning powder is used to help keep drawings clean, to avoid smearing, and to speed up the drafting process.

Cleaning powder comes in a can or as pads, *Figure 2-50*, and is sprinkled over the drawing before starting. It is imperative that all cleaning powder is removed before placing the original drawing into a whiteprinter because the powder tends to stick to the roller. If good drafting habits are followed, the dry cleaning powder is not necessary.

Scales

Various kinds of scales are used by drafters, *Figure 2-51*. A number of different scales are included on each instrument. Scales save the drafter the work of computing new measurements every time a drawing is made larger or smaller than the original.

Scales come open divided and full divided. A *full-divided scale* is one in which the units of measurement are subdivided throughout the length of the scale. An

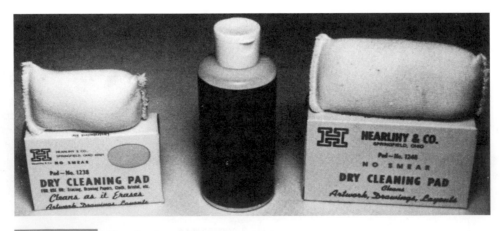

FIGURE 2-50 *Dry cleaning powder* (Courtesy Hearlihy & Co.)

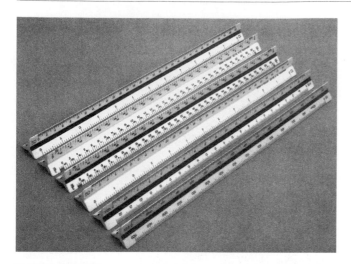

FIGURE 2-51 *A variety of drafting scales* (Courtesy Koh-I-Noor Rapidograph)

open-divided scale has its first unit of measurement subdivided, but the remaining units are open or free from subdivision.

MECHANICAL ENGINEER'S SCALE

Mechanical engineer's scales are divided into inches and parts to the inch. To lay out a full-size measurement, use the scale marked 16. This scale has each inch divided into 16 equal parts or divisions of 1/16 inch. It is used by placing the 0 on the point where measurement begins and stepping off the desired length, *Figure 2-52*.

To reduce a drawing 50 percent, use the scale marked 1/2. The large 0 at the end of the first subdivided measurement lines up with the other unit measurements that are part of the same scale. The large numbers crossed out in *Figure 2-53* go with the 1/4 scale starting at the

other end. These numbers are ignored while using the 1/2 scale. To lay out 1 3/4 inches at the 1/2 scale, read full inches to the right of 0 and fractions to the left of 0.

The 1/4 scale is used in the same manner as the 1/2 scale. However, measurements of full inches are made to the left of 0 and fractions to the right, because the 1/4 scale is located at the opposite end of the 1/2 scale, *Figure 2-54*.

ARCHITECT'S SCALE

The architect's scale is used primarily for drawing large buildings and structures. The full-size scale is used frequently for drawing smaller objects. Because of this, the architect's scale is generally used for all types of measurements. It is designed to measure in feet, inches, and fractions of an inch. Measure full feet to the right of 0, inches and fractions of an inch to the left of 0. The numbers crossed out in *Figure 2-55* correspond to the 1/2 scale. They can be used, however, as 6 inches because each falls halfway between full-foot divisions. Measurements from 0 are made in the opposite direction of the full scale because the 1/2 scale is located at the opposite end of the scale, *Figure 2-56*.

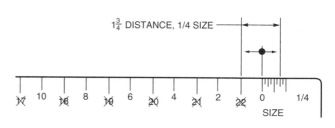

FIGURE 2-54 *Quarter-size scale*

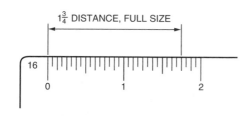

FIGURE 2-52 *Regular full-size scale*

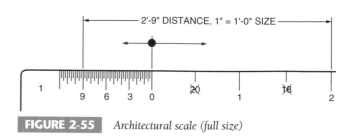

FIGURE 2-55 *Architectural scale (full size)*

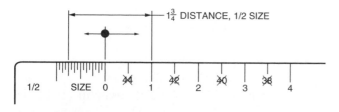

FIGURE 2-53 *Half-size scale*

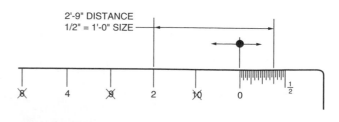

FIGURE 2-56 *Architectural scale (half size)*

With the 10 overlapping scales, a drawing may be made in various sizes from full size to 1/28 size. Since the architect's scale is showing divisions reciprocating feet and inches, a scale marked 1"=1'0", makes a drawing with this scale 1/12 size.

EXAMPLES

¾" = 1'0"	would be 1/16 size	(3/4 × 1/12)
¼" = 1'0"	would be 1/48 size	(1/4 × 1/12)
½" = 1'0"	would be 1/24 size	(1/2 × 1/12)
1" = 1'0"	would be 1/12 size	(1 × 1/12)

Full size is measured on the full-size side of the architect's scale where each division is 1/16.

Half size is measured by using the full-size side and dividing each measurement by 2.

Quarter size is measured by using the 3" = 1'0" scale. 3" = 1'0" would be 1/4 size (3 × 1/12 = 3/12 = 1/4)

Scale does not = size

incorrect   1/4 scale ≠ 1/4 size

correct   1/4 scale = 1/48 size

CIVIL ENGINEER'S SCALE

A civil engineer's scale is also called a *decimal-inch scale*. The number 10, located in the corner of the scale in *Figure 2-57*, indicates that each graduation is equal to 1/10 of an inch or .1". Measurements are read directly from the scale. The number 20, located in the corner of the scale shown in *Figure 2-58*, indicates that it is 1/20 of an inch.

Using the same scale for civil drafting, one inch equals two hundred feet, *Figure 2-59*, and one inch equals one hundred feet, *Figure 2-60*.

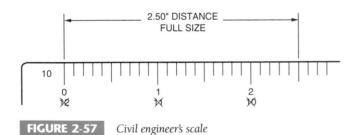

FIGURE 2-57   *Civil engineer's scale*

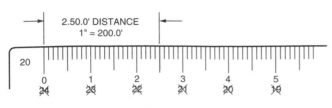

FIGURE 2-58   *Civil engineer's scale (half size)*

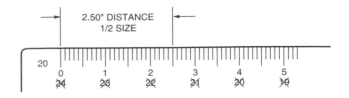

FIGURE 2-59   *Civil engineer's scale (half size)*

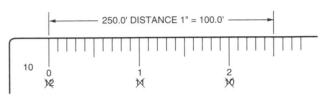

FIGURE 2-60   *Civil engineer's scale*

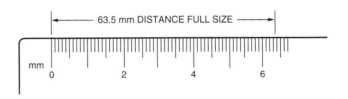

FIGURE 2-61   *Metric scale*

A metric scale is used if the millimeter is the unit of linear measurement. It is read the same as the decimal-inch scale except that it is in millimeters, *Figure 2-61*.

POCKET STEEL RULER

The drafter should make use of a pocket steel ruler. The pocket steel ruler is the easiest of all measuring tools to use. The inch scale, *Figure 2-62*, is six inches long, and is graduated in 10ths and 100ths of an inch on one side and 32nds and 64ths on the other side.

The metric scale is 150 millimeters long (approximately six inches) and is graduated in millimeters and half millimeters on one side, **Figure 2-63**. Sometimes metric pocket steel rulers are graduated in 64ths of an inch on the other side.

Measuring

The metric system uses the meter (m) as its basic dimension. A meter is 3.281 feet long or about 3 3/8 inches longer than a yardstick. Its multiples, or parts, are expressed by adding prefixes. These prefixes represent equal steps of 1000 parts. The prefix for a thousand (1000) is *kilo;* the prefix for a thousandth (1/1000) is *milli.* One thousand meters (1000 m), therefore, equals one kilometer (1.0 km). One thousandth of a meter

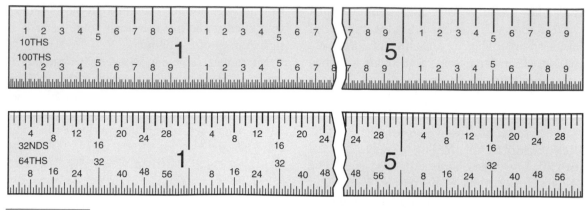

FIGURE 2-62 *Steel scale (inch)* (Courtesy L. S. Starrett Co.)

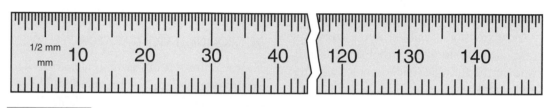

FIGURE 2-63 *Steel scale (metric)* (Courtesy L. S. Starrett Co.)

(1/1000 m) equals one millimeter (1.0 mm). Comparing metric to English then:

- One millimeter (1.0 mm) = 0.001 meter = .03937 inch
- One thousand millimeters (1000 mm) = 1.0 meter (1.0 m) = 3.281 feet
- One thousand meters (1000 m) = 1.0 kilometer (1.0 km) = 3281.0 feet

HOW TO READ A MICROMETER GRADUATED IN THOUSANDTHS OF AN INCH (.001″)

A micrometer consists of a highly accurate ground screw or spindle that is rotated in a fixed nut, thus opening or closing the distance between two measuring faces on the ends of the anvil and spindle, *Figure 2-64*. A piece of work is measured by placing it between the anvil and spindle faces and rotating the spindle by means of the thimble until the anvil and spindle both contact the work. The desired work dimension is then found from the micrometer reading indicated by the graduations on the sleeve and thimble, as described in the following paragraphs.

Since the pitch of the screw thread on the spindle is 1/40″ or 40 threads per inch in micrometers that are graduated to measure in inches, one complete revolution of the thimble advances the spindle face toward or away from the anvil face precisely 1/40 or .025 inch.

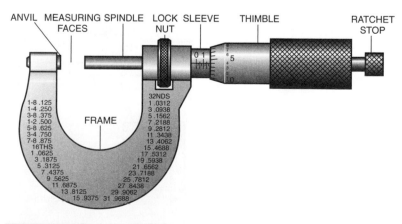

FIGURE 2-64 *Micrometer* (Courtesy L. S. Starrett Co.)

The reading line on the sleeve is divided into 40 equal parts by vertical lines that correspond to the number of threads on the spindle. Therefore, each vertical line designates 1/40 or .025 inch and every fourth line, which is longer than the others, designates hundreds of thousandths. For example: the line marked "1" represents .100", the line marked "2" represents .200", and the line marked "3" represents .300", and so forth.

The beveled edge of the thimble is divided into 25 equal parts, with each line representing .001" and every line numbered consecutively. Rotating the thimble from one of these lines to the next moves the spindle longitudinally 1/25 of .025" or .001 inch; rotating two divisions represents .002", and so forth. Twenty-five divisions indicate a complete revolution: .025 or 1/40 of an inch.

To read the micrometer in thousandths, multiply the number of vertical divisions visible on the sleeve by .025", and to this add the number of thousandths indicated by the line on the thimble that coincides with the reading line on the sleeve.

EXAMPLE (See *Figure 2-65*):

The 1" line on the sleeve is visible, representing .100"

Three additional lines are visible, each representing .025". Thus, 3 × .025" = .075"

The third line on the thimble coincides with the reading line on the sleeve, each line representing .001". Thus, 3 × .001" = .003"

The micrometer reading is 100" + .075" + .003" = .178"

An easy way to remember how to read a micrometer is to think of the various units as if you were making change from a ten-dollar bill. Count the figures on the sleeve as dollars, the vertical lines on the sleeve as quarters, and the divisions on the thimble as cents. Add up your change and put a decimal point instead of a dollar sign in front of the figures.

Micrometers come in both English and metric graduations. They are manufactured with an English size range of 1 inch through 60 inches and a metric size range of 25 millimeters to 1500 millimeters. The micrometer is a very sensitive device and must be treated with extreme care.

VERNIER CALIPER

Vernier calipers can measure both the outside and the inside of an object, *Figures 2-66* and *2-67*. Use the bottom scale when measuring an outside size. Use the top scale when measuring an inside size.

MICROFINISH COMPARATOR

The microfinish comparator is a handy tool for the drafter to approximate surface irregularities. Various kinds of microfinish comparators are available. *Figure 2-68* illustrates a comparator for cast surfaces.

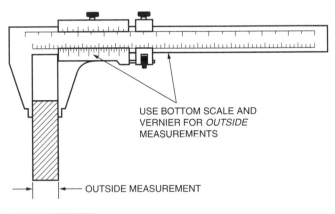

USE BOTTOM SCALE AND VERNIER FOR *OUTSIDE* MEASUREMENTS

OUTSIDE MEASUREMENT

FIGURE 2-66 *Caliper—outside measurement*

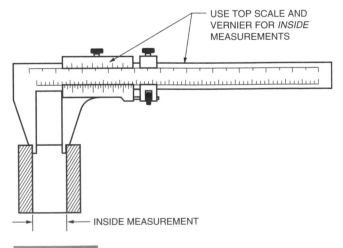

USE TOP SCALE AND VERNIER FOR *INSIDE* MEASUREMENTS

INSIDE MEASUREMENT

FIGURE 2-67 *Caliper—inside measurement*

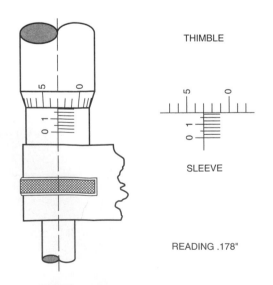

THIMBLE

SLEEVE

READING .178"

FIGURE 2-65 *Reading a micrometer* (Courtesy L. S. Starrett Co.)

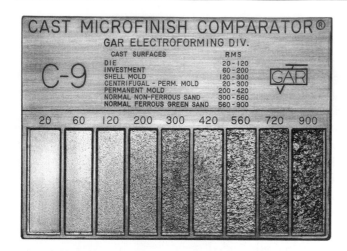

FIGURE 2-68 *Microfinish comparator* (Courtesy GAR Electroforming Div., Mite Corp.)

ELLIPSES INSTRUMENT

Two unique instruments are used to draw large ellipses. With these tools, the height and width of the ellipse are measured, locked in, and quickly drawn. A template is used to draw small ellipses.

Ink Tools

Some fields of drafting, such as civil (map) drafting, require that all drawings be done in ink. Some companies ink their drawings so that they can be reduced and filed on film. All artwork that is to be reproduced by camera, such as in the field of technical illustration, is done in ink. Ink drawing is no more difficult than pencil drawing, *Figure 2-69*.

Technical Pens

The key to successful inking is a good technical pen, *Figure 2-70*. Technical pens are produced in two styles. Notice the ends of the two pens in the Figure 2-70; one has a tapered end, the other a straight end. The tapered pen is used primarily for artwork; the straight-end style is used for drafting and mechanical lettering. Pens are available in various sizes and styles of pen-holder sets, *Figures 2-71* and *2-72*.

Technical pen points are manufactured of stainless steel, tungsten, or jewels. The stainless steel point is chromium plated for use on tracing paper or vellum. Tungsten points are long-wearing for use on abrasive, coated plotting film or triacetate. Jewel points are used on a plotter that has a controlled pen force.

Pen points are available in 13 standard sizes of varying widths, *Figure 2-73*. For general drafting inking, numbers .45/1 and .70/2½ are recommended.

CLEANING TECHNICAL PENS

Pens should be cleaned when they get sluggish or before storing them for long periods of time. The parts of most

FIGURE 2-70 *Drafting and art technical pens* (Courtesy Koh-I-Noor Rapidograph)

FIGURE 2-69 *Technical inking pen* (Courtesy Koh-I-Noor Rapidograph)

FIGURE 2-71 *Revolving pen holder* (Courtesy Koh-I-Noor Rapidograph)

technical inking pens are similar to those shown in *Figure 2-74*. When not in use, technical pens should be kept in a storage clamp or else capped to prevent ink from drying in the point. If a pen does get clogged, remove the point and hold it under warm tap water. This normally softens the ink. If the ink has dried, use an ultrasonic cleaner or a mild solvent. If the pens will not be used for a week or more, all ink should be removed and the pens stored empty and clean. Care must be taken when removing and replacing the cleaning needle. An ultrasonic cleaner is used quickly and efficiently to clean technical pens, *Figure 2-75*.

When pens are to be cleaned by hand, use the following recommended steps:

FIGURE 2-72 *Flat pack pen holder* (Courtesy Koh-I-Noor Rapidograph)

.13/5x0 .18/4x0 .25/3x0 .30/00 .35/0 .45/1 .50/2 .70/2½ .80/3 1.0/3½ 1.2/4 1.4/5 2.0/6

FIGURE 2-73 *Pen sizes* (Courtesy Staedtler Mars)

RESERVOIR PEN

INK CONTAINER — PEN BODY — CLEANING NEEDLE — POINT SECTION —

LOCK RING — SPACER RING — NEEDLE RETAINER — COVER OR CAP —

FIGURE 2-74 *Internal parts of a technical inking pen*

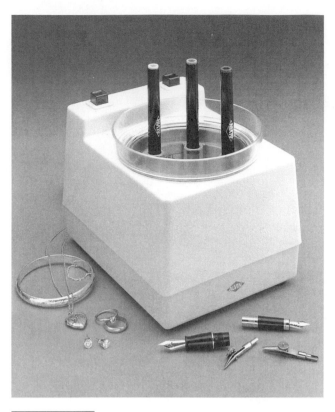

FIGURE 2-75 *Technical pen ultrasonic cleaner* (Courtesy Alvin & Co.)

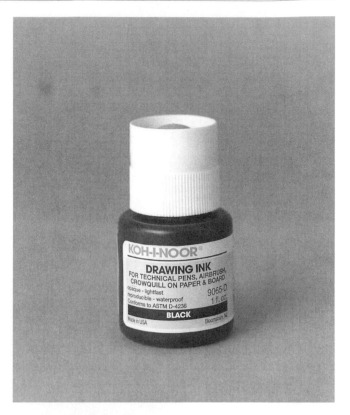

FIGURE 2-76 *Drawing ink* (Courtesy Alvin & Co.)

CLEANING

Pens can be ruined by improper cleaning. Study Steps 1 through 5 and follow them closely when cleaning pens. (Refer again to *Figure 2-74.*)

STEP 1 Remove the cap and the ink container.

STEP 2 Soak the body of the pen in hot water. The ink container should also be soaked if ink has dried in it.

STEP 3 After soaking, remove the pen body from the water. Hold the knurled part of the body with the top downward. Unscrew and remove the point section. Remove the end of the cleaning needle weight. Do not bend the cleaning needle or it will break.

STEP 4 Immerse all body parts in a good pen-cleaning fluid or hot water mixed half with ammonia.

STEP 5 Dry and clean.

FILLING

To fill the pen, follow Steps 1 through 5:

STEP 1 Unscrew and remove the knurled lock ring.

STEP 2 Remove the ink container. Leave the spacer ring in place.

STEP 3 Fill the ink container with lettering ink. Do not fill it more than 3/4″ from the top.

STEP 4 Hold the filled container upright and insert the pen body into the container.

STEP 5 Replace the knurled lock ring.

INK

A high-quality, fast-drying ink must be used in technical pens for the best results. The ink must be black and erasable, and it must not crack, chip, or peel, *Figure 2-76.* Keep inks out of extremely warm or cold temperatures. The bottles or jars should be kept airtight, and the excess ink should be cleaned from the neck of the container to keep it from drying in the cap. Inks in large containers should be transferred to smaller bottles or directly into pens, **away from working areas to avoid the possibility of spillage.**

Mechanical Lettering Sets

Lettering sets come in a variety of sizes and templates, *Figure 2-77.* All sets contain a scriber, various pen sizes, and templates.

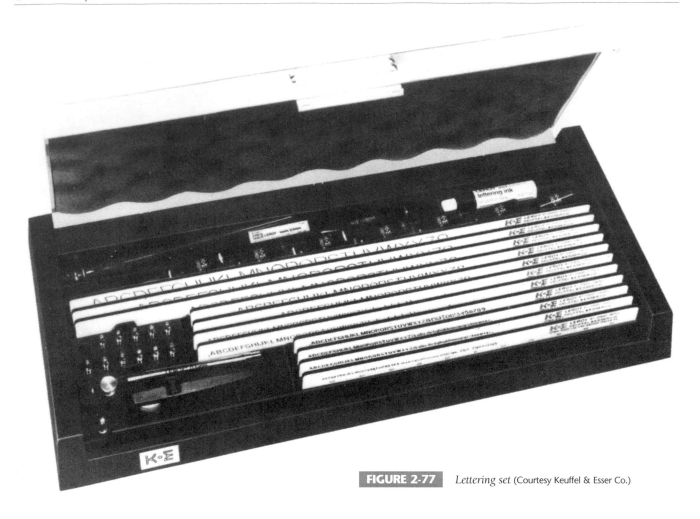

FIGURE 2-77 *Lettering set* (Courtesy Keuffel & Esser Co.)

Industry Application

DON'T THROW AWAY THOSE INSTRUMENTS

John McRae, director of the Design and Drafting Department at MicroTron Engineering, has a problem. He can remember a time when MicroTron had trouble finding drafting technicians with CADD skills. But over time, the computer has become a standard design tool. Now all new applicants have excellent CADD skills. Unfortunately, they have no knowledge of how to make technical drawings with instruments.

MicroTron, like many companies, still has occasion to use traditional drafting intruments—particularly when revising the drawings for older jobs that are still active. Sometimes it is more economical to revise drawings manually than to convert them to an electronic format. McRae's problem is that only a couple of the drafting technicians in his department know how to use traditional instruments, and many of them can't operate technical pens or mechanical lettering sets. McRae intends to call the local technical school and community college and ask why students are no longer learning to use instruments. In his words, ". . . even the best word processing technician still has to use a typewriter occasionally."

SCRIBER TEMPLATES

Scriber templates consist of laminated strips with engraved grooves that are used to form letters. A tracer pin moving in the grooves guides the scriber pen (or pencil) in forming the letters, *Figure 2-78*.

Guides for different sizes and kinds of letters are available for any of the lettering devices. Different point sizes are made for special pens so that fine lines can be used for small letters and wide lines for large letters. Scribers may be adjusted to form vertical or slanted let-

FIGURE 2-78 *Forming letters with a scriber* (Courtesy Kohl-I-Noor Rapidograph)

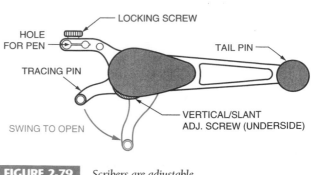

FIGURE 2-79 *Scribers are adjustable*

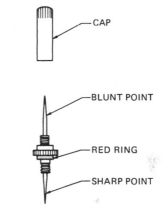

FIGURE 2-80 *Double tracing pen*

ters of several sizes from a single guide by simply unlocking the screw underneath the scriber and extending the arms, *Figure 2-79*.

One of the principal advantages of lettering guides is that they maintain uniform lettering. This is especially useful where many drafters are involved. Another important use is for the lettering of titles and note headings and numbers on drawings and reports.

Letters used to identify templates are:

U = UPPERCASE

L = LOWERCASE

N = NUMBERS

Thus, a template identified as 8-ULN means it is 8/16 inch high (½″) and has uppercase letters, lowercase letters, and numbers.

TRACING PIN

Better, more expensive scribers use a double tracing pin, *Figure 2-80*. The blunt end is used for single-stroke lettering templates or very large templates that have wide grooves. The sharp end is used for very small lettering templates, double-stroke letters, or script-type lettering using a fine groove. Most tracing pins have a sharp point, but some do not. Always screw the cap back on the unused end after turning the tracing pin to the desired tip. Be careful with the points because they will break if dropped and can cause a painful injury if mishandled.

STANDARD TEMPLATE

Learning to form mechanical letters requires a great deal of practice. *Figure 2-81* shows a template having three sets of uppercase and lowercase letters. Practice forming each size letter and number until they can be made rapidly and neatly. Use a very light, delicate touch so as not to damage the template, scriber, or pen.

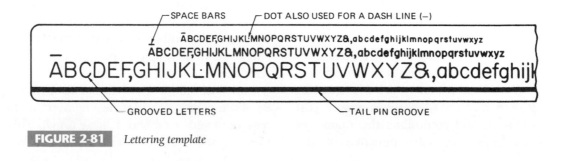

FIGURE 2-81 *Lettering template*

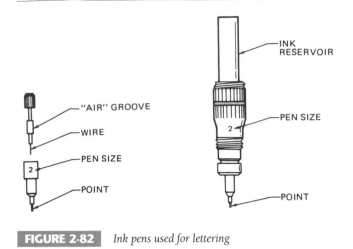

FIGURE 2-82 *Ink pens used for lettering*

FIGURE 2-83 *Butterfly scriber* (Courtesy Letterguide Inc.)

SIZE OF LETTERS

The size or height of the lettering on a template is called out by the number used to identify each set. Sizes are in thousandths of an inch. A #100 is .100 inch high, or slightly less than an eighth of an inch; a #240 is .240 inch high, or slightly less than a quarter of an inch.

Another system to determine template size uses simple numbers. These numbers are placed above the number 16 to indicate the fraction height of the letter. For instance, the number 3 placed above the number 16 would read as 3/16 inch in height.

PENS

There are two types of pens: the regular pen and the reservoir pen, *Figure 2-82*. The regular pen must be cleaned after each use. The reservoir pen should be cleaned when it gets sluggish or before being stored for long periods of time. This procedure is the same as it is for the cleaning of technical pens as described previously.

Butterfly Scriber

BASIC PARTS

The butterfly-type scriber shown in *Figure 2-83* is a delicate, precision tool that does its job without requiring any adjustments, repairs, or maintenance. The clear plastic base of the scriber bears the setting chart used in adjusting the pen arm for enlargements, reductions, verticals, and slants to be produced by tracing the engraved letters of a letter guide template.

The pen arm of the scriber holds the pen accessories for the various jobs to be performed. The pen and the arm have a thumb-tightening screw device for securing the pen being used and an adjustable pressure post screw with locking nut for controlling the amount of pressure at which the pen is set. The pressure post rides on the surface of the work when in use, and it is used only in conjunction with the swivel knife. The bull's-eye setting marker at the opposite end of the pen arm offers a concise, accurate means of setting the scriber for the various percentages and angles desired.

The tracing pin is the hardened tool steel point used in tracing the template letter. The tail pin serves as the pivot point for the triangular action of the scriber. This pin travels in the center groove of the template.

OPERATION OF THE BUTTERFLY SCRIBER

The butterfly scriber, a precision lettering tool, is the key to producing clean, sharp, controlled lettering. The setting chart, using the bull's-eye at the end of the pen arm for a marker, begins at the outer edge with a starting line marked "vertical." In this position, the scriber produces a vertical letter of normal size with the template being used. To enlarge this letter, set the bull's-eye at a position above the 100 percent intersection. At 120 percent, the scriber produces a letter 20 percent greater in height than it does at 100 percent. A reduction can be produced by setting the bull's-eye at a position below the 100 percent intersection. Variations in height range from 100 up to 140 percent and down to 60 percent. The extreme settings produce condensed letters, and the intermediate settings produce headings, subheadings, or large or small letters.

Slants in all sizes are easily produced by setting the bull's-eye on a line other than the vertical line. Normal slants or italics are produced in all height adjustments by setting the bull's-eye on either the 15° or 22½° line, and at the desired percent of height of the letter on the template.

Variations may be produced in slants ranging from 0° to 50° forward. Tracing the engraved template letter requires a very light and delicate touch. This results in more accurately traced letters and less wear on the equipment. Each lettering application requires its own specific pen and will place at the fingertips of the drafter the very best in standard typeface and hand-lettered alphabets for fast, easy rendering.

SPECIAL EFFECTS

By using one's imagination many special effects can be achieved, *Figures 2-84, 2-85,* and *2-86.*

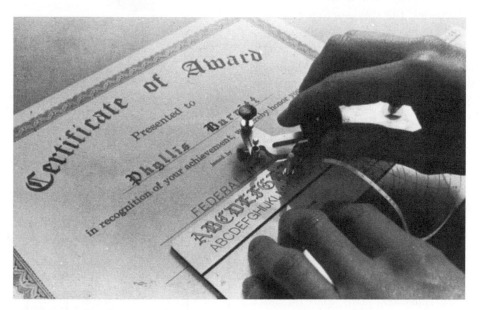

FIGURE 2-84 *Adjustable scriber creates special effects* (Courtesy Letterguide Inc.)

FIGURE 2-85 *Sample lettering styles* (Courtesy Letterguide Inc.)

FIGURE 2-86 *Additional special effects* (Courtesy Letterguide Inc.)

Airbrush

Airbrush guns are used for such purposes as production designing, pictorial rendering, portrait figure rendering, architectural rendering, and technical illustration. There are two kinds of airbrushes: the single-action type and the double-action type. In the single-action airbrush, the trigger controls the flow of air only. The fluid control is adjusted in front by the nozzle. In the double-action airbrush, the trigger controls both the flow of air and the amount of fluid to be sprayed, *Figure 2-87*.

Paper Sizes

Two basic standard paper sizes are 8½ × 11 inches and 9 × 12 inches. The basic standard metric size, A-4, is 210 × 297 millimeters. See *Figure 2-88*. Examples of paper folded to A-size are shown in *Figure 2-89*.

BORDERS

The location of the borders varies with each size sheet of paper, *Figure 2-90A* (page 69). This chart indicates the various standard borders used today. A standard horizontal border is shown in *Figure 2-90B* (page 69). A standard vertical border is shown in *Figure 2-90C* (page 70).

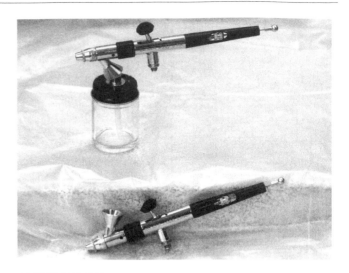

FIGURE 2-87   *Airbrush* (Courtesy Badger Airbrush Co.)

	INCHES			MILLIMETERS
SIZE	DIMENSIONS		SIZE	DIMENSIONS
A	8 1/2 x 11	9 x 12	A-4	210 x 297
B	11 x 17	12 x 18	A-3	297 x 420
C	17 x 22	18 x 24	A-2	420 x 594
D	22 x 34	24 x 36	A-1	594 x 841
E	34 x 44	36 x 48	A-0	841 x 1189

FIGURE 2-88   *Paper sizes*

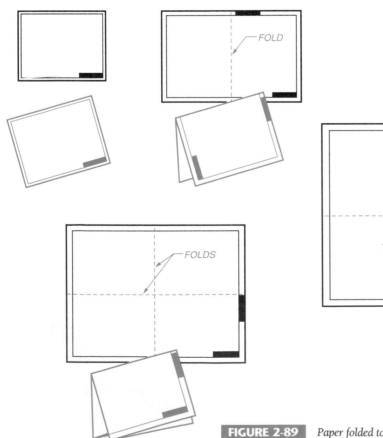

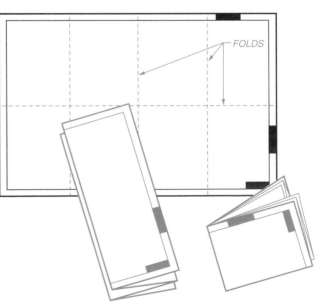

FIGURE 2-89   *Paper folded to A-size*

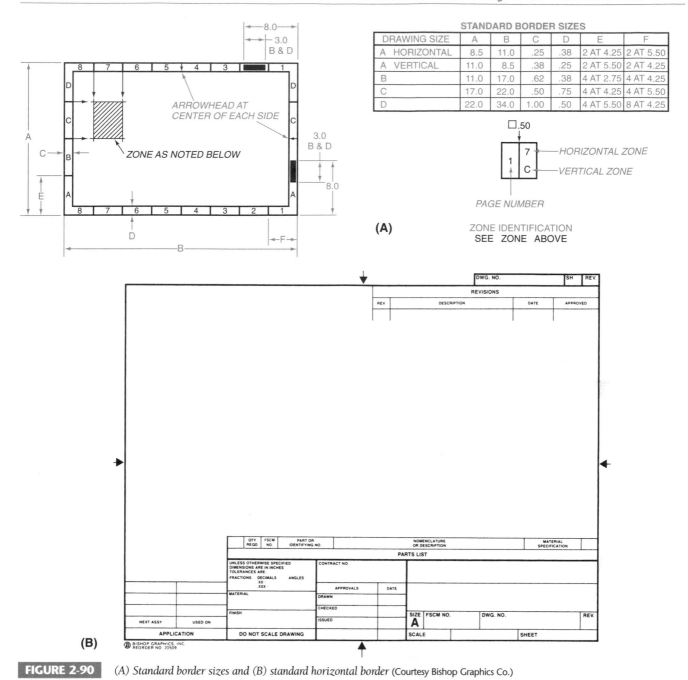

STANDARD BORDER SIZES						
DRAWING SIZE	A	B	C	D	E	F
A HORIZONTAL	8.5	11.0	.25	.38	2 AT 4.25	2 AT 5.50
A VERTICAL	11.0	8.5	.38	.25	2 AT 5.50	2 AT 4.25
B	11.0	17.0	.62	.38	4 AT 2.75	4 AT 4.25
C	17.0	22.0	.50	.75	4 AT 4.25	4 AT 5.50
D	22.0	34.0	1.00	.50	4 AT 5.50	8 AT 4.25

(A)

ZONE IDENTIFICATION
SEE ZONE ABOVE

FIGURE 2-90   *(A) Standard border sizes and (B) standard horizontal border* (Courtesy Bishop Graphics Co.)

ZONING

Zoning is used to pinpoint a particular detail on a drawing. The exact rectangular zone is located by the use of numbers running horizontally and letters running vertically in the margins. By extending these imaginary lines, the exact rectangular zone, Zone 7-C, is located as shown in *Figure 2-90A*. See the corresponding symbol below the chart. The number at the left (1) indicates the page number, and the number at the top right (7) indicates the corresponding number on the horizontal margin. The letter at the lower right (C) indicates the corresponding letter in the vertical margin.

Whiteprinter

Many types of whitepapers are available for use in drafting rooms. A *whiteprinter, Figure 2-91* (page 70), reproduces a drawing through a chemical process. Most of these machines work on the same basic principle, *Figure 2-92* (page 71). A bright light passes through the translucent original drawing and onto a coated whiteprint paper. The light breaks down the coating on the whiteprint paper, but wherever lines have been drawn on the original drawing, no light strikes the coated sheet. Then the whiteprint paper is passed through ammonia vapor for developing. This chemical developing

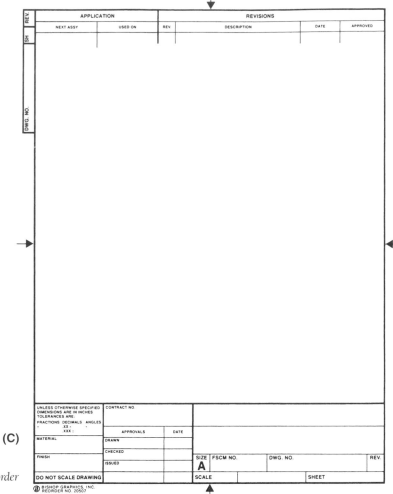

(C)

FIGURE 2-90 *(Continued) (C) Standard vertical border*
(Courtesy Bishop Graphics Co.)

FIGURE 2-91 *Multifunction copier, printer, scanner (Courtesy OCE)*

causes the unexposed areas—those that were shaded by lines on the original—to turn blue or black.

Most whiteprinters have controls to regulate the speed and flow of the developing chemical. Each type of machine requires different settings and has different controls. Before operating any whiteprinter, read all of the manufacturer's instructions.

Today, with the advent of new technology, copies are often made on a printer, *Figure 2-93.*

Files

A finished drawing represents a great deal of valuable drafting time and is, therefore, a costly investment. Drawings must be stored flat in a clean storage area, *Figure 2-94.* Vertical drawing storage is provided by hangers, *Figure 2-95* (page 71). Most engineering firms keep their files in fireproof and theftproof vaults.

Care of Drafting Equipment

Drafting tools are precision instruments, and the proper care will ensure that they last a lifetime.

PLASTIC TOOLS

Plastic drafting tools, such as T-squares, parallel straight-edges, templates, and triangles, should be wiped immediately after use with a damp cloth to remove ink or

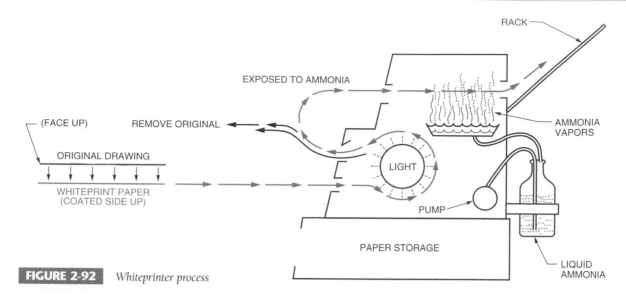

FIGURE 2-92 *Whiteprinter process*

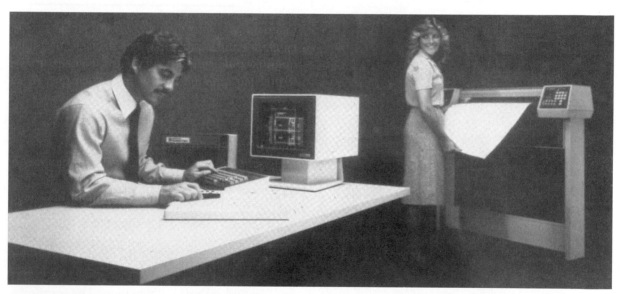

FIGURE 2-93 *Printer* (Courtesy J. S. Staedtler Inc.)

FIGURE 2-94 *Drawing file system* (Courtesy Safco Products)

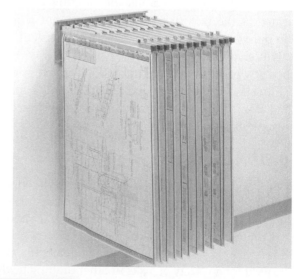

FIGURE 2-95 *Vertical drawing file* (Courtesy Safco Products)

graphite that may stain the tools or be carried to the next drawing. Once a plastic instrument is stained, a mild soap or ammonia solution will dissolve many water- and oil-based inks. Be careful *not* to use a solvent such as paint thinner, lacquer thinner, or alcohol.

Plastic drafting tools should be kept out of direct sunlight and away from warm surfaces to prevent them from becoming brittle, cracked, and warped. They should be stored in a flat position with cloth or paper between them to avoid scratching the surface.

A great number of plastics are used in drafting instruments. Most are made from either styrene or acrylic plastic. Styrene is a more flexible and softer plastic than acrylic. Although acrylic instruments are harder, they are more prone to chipping. Because both types of plastic are relatively soft, plastic drafting instruments should never be used for a cutting edge.

COMPASSES

Almost all compasses are made of brass that is chrome- or nickel-plated. To clean these instruments, use a mild solution of soap and water to remove residue and dirt.

Compasses should not normally need oiling, unless they are kept in a damp area that could cause rust. If a compass is oiled unnecessarily, there is a risk of soiling the next drawing on which it is used.

TABLES AND CHAIRS

Wooden drafting furniture is cared for in the same manner as any other wooden furniture. It may be polished or waxed with ordinary products. Do not polish the insides of drawers or cabinets. These areas may retain the wax, which can then be transferred to drawings.

Steel furniture can be cleaned with soap and water and then waxed.

The gears and joints on adjustable drafting tables are lubricated at the factory and generally do not require further oiling. Additional oiling increases the risk of getting oil or grease on a drawing.

The tops of most drafting tables are coated with a vinyl film such as melamine or a phenol-laminate material. A glass cleaner or mild ammonia solution is used to clean these surfaces.

Use of Appliqués

The word *appliqué* is a generic term used to describe a variety of shortcut products used in drafting. These products include such items as tapes, pads, and various other ready-made appliqués for creating printed circuit board artwork, *Figure 2-96*. These same materials may also be used for a variety of tasks in other drafting fields, *Figure 2-97*. For example, architects use tapes for making lines and walls on floor plans.

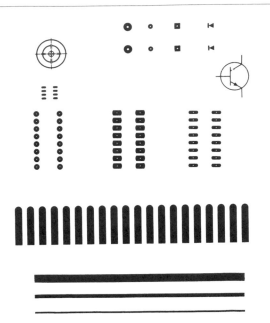

FIGURE 2-96 *Tapes and pads for printed circuit board drafting*

Transfer cards are used primarily as substitutes for mechanical lettering, but any type of symbol or frequently used piece of graphic data can be placed on a transfer card. Transfer cards are especially designed to fit against a parallel bar, drafting rule, or other straightedge for ease of alignment. Symbols are transferred from the card by rubbing them with a blunt point.

Dry transfer sheets are designed according to the same principles as transfer cards. The major differences are that transfer sheets are just that, sheets—not cards. Dry transfer sheets are used a great deal in architectural drafting and technical illustration. The transfer is made by rubbing the symbols on the sheet with a blunt, rounded point or a special burnisher.

Dry transfer materials do have some drawbacks. The heat of ammonia-developing print machines tends to lift dry transfer material from the sheet. In addition, the material may dry out and crack with age.

Use of Burnishing Plates

Burnishing is another shortcut for creating graphic symbology fast and easily. *Burnishing* involves placing a specially textured plate under the drafting medium (usually paper or vellum) and rubbing the drafting surface with a pencil. The pencil may be soft and dark or light and hard, depending on the amount of emphasis desired. Two of the most commonly used symbols on burnishing plates are bricks and stone, but any symbol could be made into a plate.

One weakness of burnishing plates is that the symbols they produce do not reproduce well.

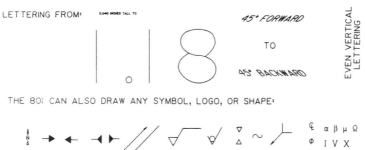

LINEX 80I SCRIBER

DIRECT INK LETTERING ON DRAFTING SURFACES

THE LINEX 80I SCRIBER DOES SCRIBER QUALITY LETTERING IN A
FRACTION OF THE TIME IT TAKES TO DO MANUALLY!

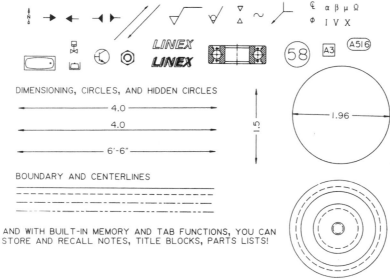

LETTERING FROM: 0.040 INCHES TALL TO

45° FORWARD

TO

45 BACKWARD

EVEN VERTICAL LETTERING

THE 80I CAN ALSO DRAW ANY SYMBOL, LOGO, OR SHAPE:

DIMENSIONING, CIRCLES, AND HIDDEN CIRCLES

|——————— 4.0 ———————|

|——————— 4.0 ———————|

|——————— 6'-6" ———————|

1.96

1.5

BOUNDARY AND CENTERLINES

AND WITH BUILT-IN MEMORY AND TAB FUNCTIONS, YOU CAN
STORE AND RECALL NOTES, TITLE BLOCKS, PARTS LISTS!

FIGURE 2-97 *Sample appliqué* (Courtesy A.D.S./Linex Inc.)

Typewritten Text

Text on drawings and other types of documentation consists of dimensions, notes, and callouts. Creating text, or lettering, is one of the slowest, least productive manually performed tasks in drafting. Typewritten text is a shortcut for improving on lettering in manual drafting situations.

A number of different methods for typing text are used in drafting. The most frequently used are the open-carriage typewriter and the lettering machine. The *open-carriage typewriter* is any brand of typewriter that has been specially designed to hold larger-than-normal media, such as drawings, bills of material, and parts lists. Once a drawing is completed and ready for annotation, the drafter or engineering secretary rolls it into an open-carriage typewriter and types the required text. Typewritten text is fast, neat, and consistent, *Figures 2-98* and *2-99*.

Lettering machines provide another means for accomplishing typewritten text. Lettering machines are used primarily for titles on drawings, but they may be used in any situation where Leroy lettering is used.

Lettering machines, such as the Kroy machine, output the letters on a clear tape that is pressed onto the drawing surface, *Figure 2-100.* Lettering machines are capable of producing text in a number of different sizes and styles. There are some drawbacks to this shortcut. The machine and its type fonts are expensive.

Overlay Drafting

Overlay drafting is a complete drafting process that uses advanced reproduction techniques and materials to reduce the amount of time spent in preparing drafting documentation. Actually more than a shortcut technique, the underlying principle of overlay drafting is

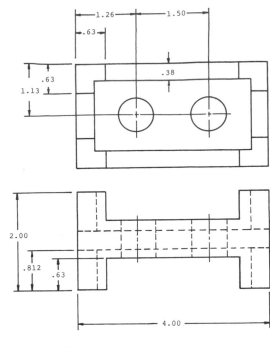

FIGURE 2-98 *Typewritten text on drawings*

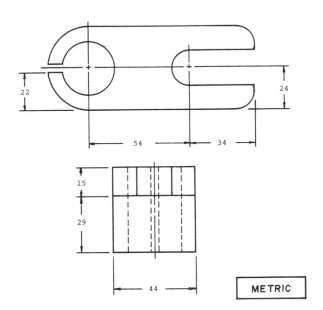

FIGURE 2-99 *Typewritten text on drawings*

Lettering Sample
Lettering Sample
𝕷𝖊𝖙𝖙𝖊𝖗𝖎𝖓𝖌 𝕾𝖆𝖒𝖕𝖑𝖊
Lettering Sample
LETTERING SAMPLE
LETTERING SAMPLE

FIGURE 2-100 *Kroy lettering samples*

sheet is automatically lined up (registered) by the pin bar. The simplest example of how overlay drafting works is in preparing a set of commercial architectural plans.

In manual drafting, this preparation requires drawing the floor plan four separate times: once for the floor plan, once for the electrical plan, once for the plumbing plan, and once for the HVAC plan. Overlay drafting eliminates this time-consuming repetition. In overlay drafting, the floor plan is drawn once. Then, before it is dimensioned, three additional originals are created from it, using a special sensitized polyester film, a flatbed vacuum frame printer, and a diazo print machine. In general terms, this is how the process works:

STEP 1 A sheet of punched polyester film is placed on the drafting board. The holes across the top of the film fit over the pin bar that is permanently attached to the top edge of the drafting board. The floor plan is drawn on this base sheet, but, for the moment, the dimensions and all other annotations are left off. These will be added later, after the base sheet is used to reproduce several other originals that do not require dimensions or annotation.

STEP 2 The base sheet is taken from the drafting board. A sheet of punched, sensitized polyester film is placed on top of it. The two sheets are fastened together with plastic registration pins. Together, they are placed in the flatbed vacuum frame exposure unit. The lights in the unit expose the sensitized film, "burning away" all of the special light-sensitive emulsion, except where the light is blocked out by lines from the base sheet. The exposed polyester film is then run through the ammonia developing section of a print machine, producing what is called a *slick*. A slick is a polyester reproduction of the base sheet original. It is not drawn on; rather, it is used as a base sheet over which electrical,

nonrepetition. Nonrepetition means that once any type of graphic data or symbology has been drawn the first time, it should never have to be drawn again.

The special tools and materials of overlay drafting include pin bars, registration tabs, prepunched polyester film or a punch to punch holes in standard film, a flatbed vacuum frame printer, and a diazo print machine.

Overlay drafting involves placing a punched base sheet on a drafting table over a pin bar and placing various overlay sheets on top of it. Each successive overlay

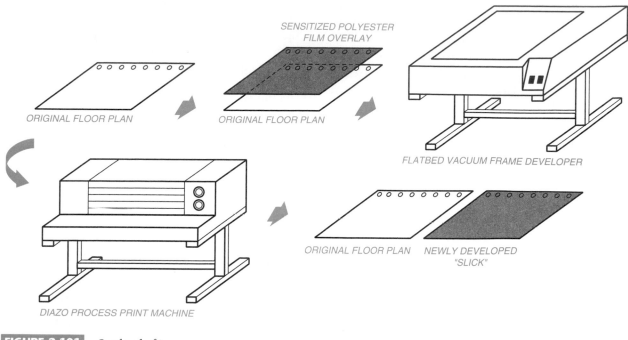

SENSITIZED POLYESTER
FILM OVERLAY

ORIGINAL FLOOR PLAN

ORIGINAL FLOOR PLAN

FLATBED VACUUM FRAME DEVELOPER

DIAZO PROCESS PRINT MACHINE

ORIGINAL FLOOR PLAN   NEWLY DEVELOPED
"SLICK"

FIGURE 2-101   *Overlay drafting process*

plumbing, and HVAC plans may be overlaid. The original base sheet of the floor plan may now be completed by adding dimensions and other annotation.

STEP 3   The new slick may now be placed over the pin bar on a drafting board, and a clear sheet of polyester film placed on top of it. The electrical symbols for the electrical plan are added on this new sheet. Only information for the electrical plan is entered on this overlay sheet. To get a print of the electrical plan superimposed on the floor plan, the slick base sheet containing the floor plan and the electrical plan overlay are placed in the flatbed vacuum frame printer, along with a sheet of ammonia developing print paper. All three sheets are secured together with plastic registration pins. When the exposure step is completed, the sheets are separated and the print paper is run through the ammonia developing section of a diazo print machine, thus producing a print of the electrical plan. The same process is repeated for the plumbing plan and the electrical plan. To save even more time, three slicks of the floor plan could have been made and given to three different drafters; one drafter would create the electrical plan overlay, one the plumbing plan overlay, and the other the HVAC plan overlay. This process is illustrated in *Figure 2-101*.

Scissors Drafting Techniques

Scissors drafting is an extension of overlay drafting. The two combined can bring substantial productivity benefits in manual drafting situations. In *scissors drafting*, if any part of a set of drawings has ever been drawn before, say a typical detail or sectional view, it need not be redrawn. Rather, the techniques outlined previously to create a slick are used, and the details and other data needed from the slick are cut out and taped to a carrier sheet, *Figure 2-102*. A new slick is created using the carrier sheet as the original.

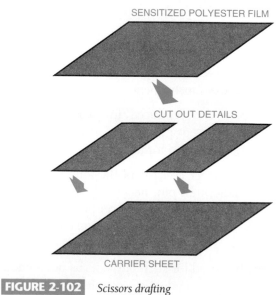

SENSITIZED POLYESTER FILM

CUT OUT DETAILS

CARRIER SHEET

FIGURE 2-102   *Scissors drafting*

The scissors drafting technique is best illustrated by example. A mechanical drafter must construct a sheet of typical details and sectional views for a documentation package. All of the needed details have been drawn before, but in different jobs. Some of the details needed are in one job, some in another, and so on. Rather than redrawing, the drafter decides to use scissors drafting techniques.

First, the drafter locates all of the details needed and pulls the required sheets from the filing drawers. A slick of each sheet is then made. After cutting out the details from the slicks, the drafter assembles them on a carrier sheet. The first slicks allow the original drawings to be kept in case they are ever needed again.

Using the carrier sheet as an original, the drafter creates a slick containing all of the details. The slick may then be copied onto polyester film that accepts plastic lead or ink, and this new medium is used like any other original. The entire process takes about 20 minutes, whereas completely redrawing each detail would take hours.

Summary

- The most frequently used conventional drafting instruments include the following: drawing board, triangles, scales, compasses, irregular curves, dividers, drafting brush, mechanical pencils, protractors, adjustable triangle, erasing shield, eraser, lead pointer, flat tile or lead sandpaper, circle template, ellipse template, drafting tape or dots, calculator, and dry cleaning pad.

- A typical drawing set will include dividers, various compasses, and ink attachments.

- The most frequently used scales are the mechanical engineer's, architect's, civil engineer's, and metric scales.

- Precision measuring instruments that technical drawing students should be able to use include the following: micrometer, vernier caliper, microfinish comparator, and the ellipses instrument.

- Technical pens are produced in two styles. The tapered pen is used primarily for artwork. The straight-end style is used for technical drawing.

- The butterfly scriber consists of the following parts: bull's-eye handle, trail pin, height adjustment screw, pen tightening screw, pen arm, tracer pin, tail pin slot, calibrations for italics, and calibrations for height variations.

- The airbrush is used for such purposes as production designing, pictorial rendering, portrait figure rendering, architectural rendering, and technical illustration.

- Standard paper sizes in inches are as follows: A = 8½ × 11, B = 11 × 17, C = 17 × 22, D = 22 × 34, and E = 34 × 44. Standard sizes in millimeters are as follows: A-4 = 210 × 297, A-3 = 297 × 420, A-2 = 420 × 594, A-1 = 594 × 841, and A-0 = 841 × 1189.

- In the whiteprinting process an original drawing is placed on top of coated whiteprint paper (coated side up). The two are exposed to intense light that burns off all of the coating except that covered by the pencil or ink lines of the original drawing. The original drawing is then removed, and the whiteprint paper is exposed to ammonia vapors. These vapors turn the remaining coating a deep purplish-blue color, creating a print of the original.

- Appliqués are various types of ready-made products that can be applied to an original drawing in the same way as tape is applied. They are time-saving devices.

- Burnishing plates are used to create graphic symbology quickly. A burnishing plate is a specially textured device that is placed under an original drawing. The paper surface is then rubbed to create the image in question.

- Overlay and scissors drafting are two methods used to reduce the time and work needed to produce a drawing. Details that have been drawn before are copied, cut out, and placed on a carrier sheet. The carrier sheet is then copied onto a special medium (polyester film) to create an original.

Review Questions

Answer the following questions either true or false.

1. Dry cleaning powder is used after the completion of a drawing to clean up any specks or smearing.

2. The transfer card and the dry transfer sheet are not used to perform the same drafting function.

3. For drafting purposes, a 4H lead is recommended for layout work, and 2H, H, or HB leads are recommended for all other lines.

4. The friction-joint type compass is recommended for drafting because the setting stays secure even when strong pressure is applied.

5. A drafting brush is used to remove loose graphite and eraser crumbs from the drawing surface.

6. In most cases, scissors drafting can save much time.

7. An easy way to remember how to read a micrometer is to compare it to making change for a $1 bill.

8. An open-divided scale is one in which the units of measurement are subdivided throughout the length of the scale.

Answer the following questions by selecting the best answer.

1. Which of the following is **not** a type of drafting triangle?
   a. 30°–60°
   b. 45°
   c. Beam triangle
   d. Adjustable triangle

2. Which of the following drafting tools is **not** commonly used to draw horizontal lines?
   a. T-square
   b. Drafting machine square
   c. Protractor
   d. Parallel bar

3. Which of the following paper sizes are commonly used in industry today to produce engineering drawings?
   a. 8.5 × 11
   b. 9 × 12
   c. 22 × 34
   d. Both b and c

4. Which of the following is **not** an important feature/use of the drawing surface or top cover?
   a. To provide a perfect drawing surface
   b. A metal edge that ensures against warpage
   c. Must be easily cleaned
   d. To cover over the holes made by the compass

5. Which of the following is **not** true regarding the vernier protractor?
   a. Permits readings to 5 minutes of an arc
   b. Is graduated in increments from 0 to 60 minutes
   c. Takes the place of the arm-type drafting machine
   d. The zero on the protractor is in line with the zero on the vernier.

6. Which of the following pieces of hardware would typically **not** be a part of a modern CAD configuration?
   a. Graphics display
   b. Copier
   c. Keyboard
   d. Digitizer

7. The word *appliqué* in drafting describes:
   a. A variety of shortcut products
   b. The variance from a specified dimension or design requirement
   c. A special device for holding the work in a machine tool
   d. None of the above

8. Which of the following is **true** regarding the Kroy lettering machine?
   a. It outputs the letters on a clear tape.
   b. It is an expensive piece of equipment.
   c. The fonts and sizes of letters are very limited.
   d. Both a and b

9. Which of the following is **not** true regarding the whiteprinter?
   a. Most of them work on different basic principles.
   b. It reproduces a drawing through a chemical process.
   c. The chemical used is ammonia.
   d. The chemical developing causes the exposed areas to turn black or blue.

10. Which of the following tools are necessary to do overlay drafting?
    a. Pin bars
    b. Prepunched polyester film
    c. Registration tabs
    d. All of the above

Chapter 2 Problems

Unless otherwise specified use the format shown in Problem 4-37 (page 161) to present your solutions to the following problems.

PROBLEM 2-1

On an A-size sheet of paper lay out lines to scale as indicated:
   Scale 16 (Figure 2-52)
      5-7/16"
      3-5/8"
   Scale 1/2 (Figure 2-53)
      9-1/2"
      11-3/8"
   Scale 1/4 (Figure 2-54)
      11-3/8"
      7-7/8"
   Scale 1 (Figure 2-55)
      3" – 8-1/2"
      4" – 1-3/4"
   Scale 1/2 (Figure 2-56)
      5" – 9-3/4"
      7" – 8-1/4"
   Scale 10 (Figure 2-57)
      3-3/8"
      5-7/16"
   Scale mm (Figure 2-61)
      42.5 mm
      37.7 mm

PROBLEM 2-2

Assume that a whiteprinter, Figure 2-93, can process paper at 25 feet per minute (fpm). Approximately how many seconds would the printer take to print a D-size drawing the long way? (Dimension B. See Figure 2-91.) Note that the paper passes through the printer two times while traveling a total of 26 inches.

PROBLEM 2-3

A rich uncle says to you that he will contribute to your education by giving you a check equal to a stack of one-dollar bills equal to your height. How large should the check be? A micrometer shown in Figure 2-64 would be helpful. What else could you do to determine the thickness of a one-dollar bill?

PROBLEM 2-4

At a certain time of day a drafter used a standard steel tape to measure the shadow of a 1.8 m tall person. The shadow measured approximately 0.9 m. Next, the drafter measured the shadow of a nearby water tower and found it to be approximately 7.7 m. Determine the approximate height of the water tower using two methods.
1. Use a proportional equation.
2. Use trigonometry.

PROBLEM 2-5

Read the micrometer shown to determine the outside dimension of a typical drafting pencil.

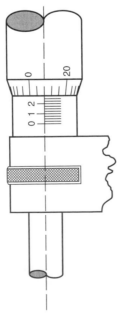

PROBLEM 2-6

Prepare sketches to show how you would use a 30°–60° triangle with a 45° triangle in conjunction with a T-square to construct lines at the following angles to the horizontal: 15°, 75°, 105°, and 165°. Refer to Figure 2-30.

PROBLEMS 2-7 THROUGH 2-12

Carefully measure each line in inches, or in millimeters if metric is indicated. Neatly enter your answers on a sheet of paper. For extra practice, measure each line full size as given, half size as given, quarter size as given, or ten-times scale as assigned by the instructor.

PROBLEM 2-7

FULL SIZE

1)
2)
3)
4)
5)
6)
7)
8)
9)
10)

PROBLEM 2-8

HALF SIZE

1)
2)
3)
4)
5)
6)
7)
8)
9)
10)

PROBLEM 2-9

3/4 SCALE

1)
2)
3)
4)
5)
6)
7)
8)
9)
10)

PROBLEM 2-10

METRIC (FULL SIZE) MILLIMETERS

1)
2)
3)
4)
5)
6)
7)
8)
9)
10)

PROBLEM 2-11

METRIC (HALF SIZE) MILLIMETERS

1)
2)
3)
4)
5)
6)
7)
8)
9)
10)

PROBLEM 2-12

METRIC (QUARTER SIZE) MILLIMETERS

1)
2)
3)
4)
5)
6)
7)
8)
9)
10)

PROBLEMS 2-13 THROUGH 2-28

Construct each object using the given dimensions. Use consistent thick, black object lines throughout. Keep all corners tight and sharp. Where indicated, draw thin, black center lines.

PROBLEM 2-13

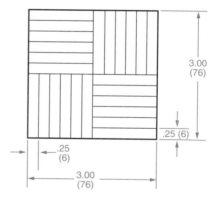

PROBLEM 2-14

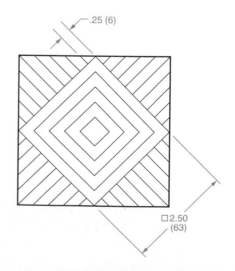

PROBLEM 2-15

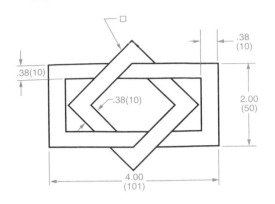

PROBLEM 2-18

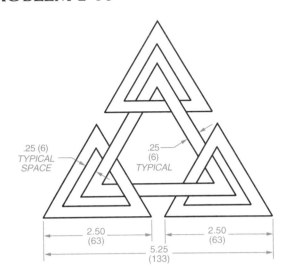

PROBLEM 2-16

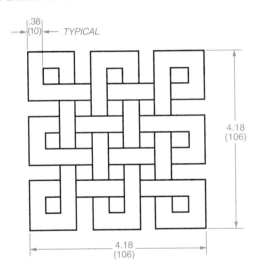

PROBLEM 2-19

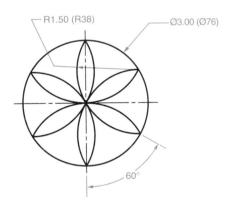

PROBLEM 2-17

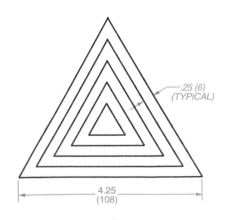

PROBLEM 2-20

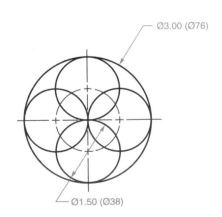

PROBLEM 2-21

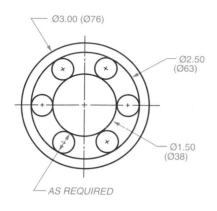

Ø3.00 (Ø76)

Ø2.50 (Ø63)

Ø1.50 (Ø38)

AS REQUIRED

PROBLEM 2-22

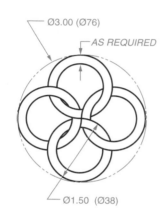

Ø3.00 (Ø76)

AS REQUIRED

Ø1.50 (Ø38)

PROBLEM 2-23

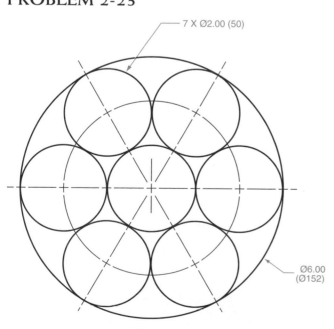

7 X Ø2.00 (50)

Ø6.00 (Ø152)

PROBLEM 2-24

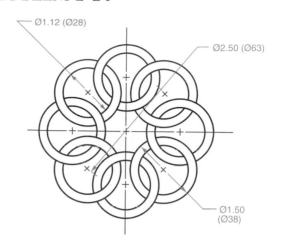

Ø1.12 (Ø28)

Ø2.50 (Ø63)

Ø1.50 (Ø38)

PROBLEM 2-25

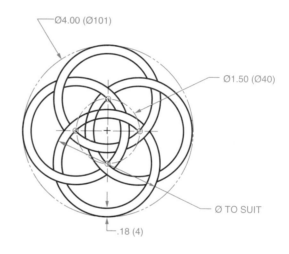

Ø4.00 (Ø101)

Ø1.50 (Ø40)

Ø TO SUIT

.18 (4)

PROBLEM 2-26

Ø5.50 (Ø139)

ALL SPACES = .25 (6)

PROBLEM 2-27

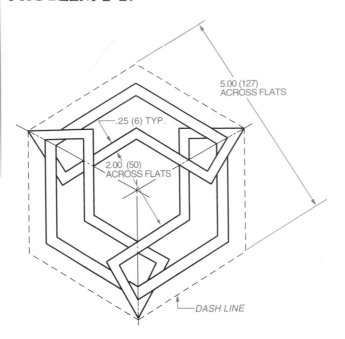

5.00 (127)
ACROSS FLATS

.25 (6) TYP.

2.00 (50)
ACROSS FLATS

—DASH LINE

PROBLEM 2-28

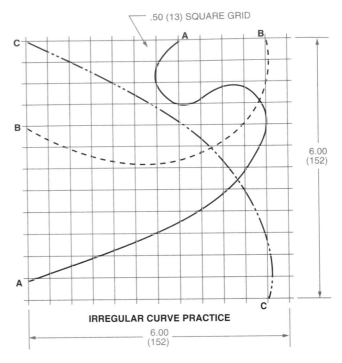

.50 (13) SQUARE GRID

6.00
(152)

IRREGULAR CURVE PRACTICE

6.00
(152)

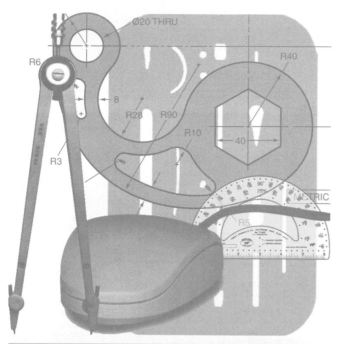

MANUAL LETTERING, SKETCHING, AND LINE TECHNIQUES

KEY TERMS

Axonometric sketching

Boardroom sketching

Center line

Conversational sketching

Cutting plane line

Dimension line

Extension line

Hidden line

Isometric (sketching)

Line work

Oblique sketching

Orthographic sketching

Perspective sketching

Phantom line

Text

Visible line

CHAPTER OUTLINE

Talking sketching • Freehand lettering • Freehand lettering techniques • Line work • Sketching • Types of sketches • Sketching materials • Sketching techniques • Summary • Review questions • Chapter 3 problems

CHAPTER OBJECTIVES

Upon completion of this chapter students should be able to do the following:

■ Explain the concept of talking sketching, including the two kinds.

■ List the various styles of freehand lettering and the characteristics of good lettering.

■ Explain the techniques one must know in order to do freehand lettering.

■ Illustrate the various types of lines used on technical drawings.

■ Illustrate the four types of sketches.

■ Explain what materials are needed in order to make sketches.

■ Demonstrate the most commonly used sketching techniques.

Note to Instructors and Students

Although modern CAD technologies have altered radically the way that design documentation is prepared, there is still a need for engineers, designers, and CAD technicians to develop their skills in the areas of sketching, hand lettering, and line work. Sketches remain an important form of communication between engineers, designers, and the CAD technicians who will convert their ideas into finished drawings and other types of documentation. Consequently, the authors continue to offer this chapter and recommend that instructors use it to help students develop the sketching, lettering, and line work skills they will need in the workplace.

Talking Sketching

Two types of sketching, which are often neglected in technical drawing texts but nevertheless complement the skills of a successful drafter, designer, or engineer, are conversational sketching and boardroom sketching.

CONVERSATIONAL SKETCHING

Conversational sketching usually occurs between two or more individuals huddled around a drafting board or a cafeteria table. The one doing the talking combines several types of sketches in one drawing as the idea is talked through. The first part of the sketch is most likely an orthographic view, the easiest projection to draw rapidly. Then, as the talking progresses, a pictorial is used to give the observers a feeling of depth. For example, a designer communicating an idea for an AM/FM receiver would begin with the orthographic view, *Figure 3-1*, and finish with a pictorial, *Figure 3-2*. A drafter, designer, or engineer sketches rapidly while talking and doesn't have time to develop a single specific kind of sketching, such as orthographic views, an isometric, an oblique, or a two-point perspective.

BOARDROOM SKETCHING

Communication with more individuals than a drafting board can accommodate requires a more visible surface for the larger group. The drawing surface for *boardroom sketching* sometimes is a chalkboard, but more often is a pastel-colored porcelainized enamel, a glass panel, or a dry-erase surface. All of these surfaces require a special marker and can be erased with a cloth.

Quality talking sketches, like quality sketching discussed earlier in this chapter, require practice to develop visible lines and especially to develop lettering large enough to be read by those sitting at the back of the room.

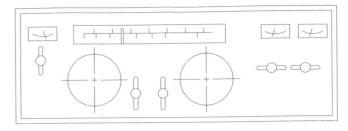

FIGURE 3-1 *Talking sketch (orthographic view)*

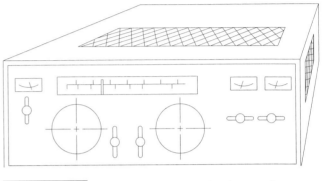

FIGURE 3-2 *Talking sketch completed with pictorial*

Note that sketching and lettering on overhead projectors is another medium for talking sketching. Practice to make sure that all those present can read all the lettering and see the lines. Additional comments on the use of overhead projectors are in Chapter 24.

Freehand Lettering

Text is an important part of a technical drawing. Not all information required on technical drawings can be communicated graphically, the most obvious being dimensions. *Text* on technical drawings consists of dimensions, notes, legends, and other data that are best conveyed using alphanumeric characters, *Figure 3-3*.

Several different ways are used to create text on technical drawings. The traditional method is by freehand lettering. Other methods include such mechanical lettering techniques as scriber templates, typewritten notation, and typed lettering generated by computer-aided drafting systems. This chapter focuses on freehand lettering. Other methods are described elsewhere in this text.

LETTERING STYLES

There are numerous different lettering styles or fonts, *Figure 3-4*. The standard style for freehand lettering on technical drawings, as established in American National Standards document Y14.2-1973, is single-stroke Gothic lettering. Vertical, single-stroke Gothic letters are

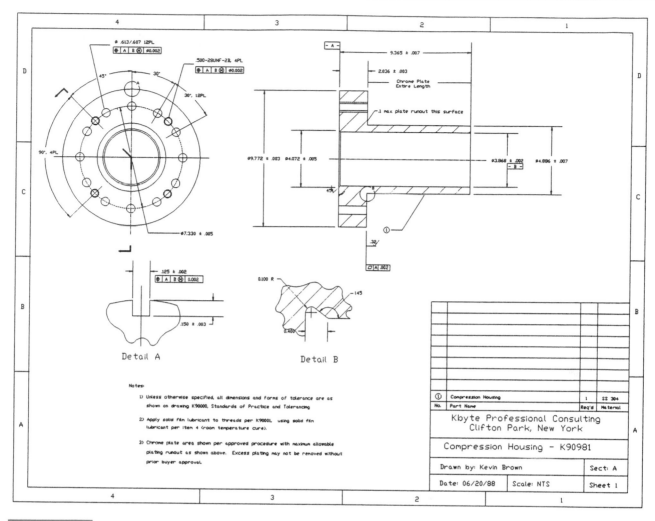

FIGURE 3-3   *Examples of text on a technical drawing*

THIS IS BLOCK FONT

THIS IS FAST FONT

THIS IS FUTURA FONT

THIS IS LEROY FONT

𝕿𝕳𝕴𝕾 𝕴𝕾 𝕺𝕷𝕯 𝕰𝕹𝕲𝕷𝕴𝕾𝕳 𝕱𝕺𝕹𝕿

THIS IS RIVERA FONT

THIS IS TIMES FONT

THIS IS HELVET FONT

BAUSCH & LOMB ▼ FONTS

FIGURE 3-4   *Sample lettering fonts (Courtesy Bausch & Lomb, Inc.®)*

SINGLE - STROKE GOTHIC LETTERING SAMPLE

FIGURE 3-5   *Single-stroke Gothic lettering*

the most universally used of the various styles available to drafters, *Figure 3-5*.

Some modifications of the standard Gothic configuration of letters are often made, without actually changing from the Gothic style of lettering. One way is through the use of uppercase and lowercase letters, *Figure 3-6*, but this is seldom acceptable on technical drawings. Another method is to condense or extend the letters, *Figure 3-7*.

The most common way of modifying Gothic letters is by inclining them slightly to the right, *Figure 3-8*. Inclined lettering is easier to make as it lends itself to a natural direction of wrist action. The correct angle of the inclined

UPPERCASE GOTHIC
lowercase Gothic

FIGURE 3-6 *Uppercase and lowercase Gothic lettering*

EXTENDED VARIATION
CONDENSED VARIATION

FIGURE 3-7 *Extended and condensed variations of Gothic lettering*

SAMPLE OF INCLINED LETTERING

FIGURE 3-8 *Inclined Gothic lettering*

SLOPPY LETTERING IS DIFFICULT TO READ

FIGURE 3-9 *Sloppy lettering is difficult to read.*

UNIFORMITY ◄═══ GUIDELINES

FIGURE 3-10 *Guidelines improve uniformity.*

B E F R 3 8 2   TOP HEAVY
B E F R 3 8 2   CORRECT FORM

FIGURE 3-11 *Top-heavy letters are not balanced.*

SPACING WITHIN WORDS SHOULD BE CLOSE.

SPACING BETWEEN WORDS SHOULD BE FAR.

FIGURE 3-12 *The proper spacing of letters and words*

THIS IS A PROPERLY SPACED SENTENCE.

THISISANIMPROPERLYSPACEDSENTENCE.

FIGURE 3-13 *Spacing between words is important.*

elements is a two-unit incline to the right for each five units of letter height. Errors are not as detectable with inclined letters as they are with vertical elements. Because the inclined elements are longer, they are easier to read. However, inclined lettering is not universally accepted, and caution must be exercised not to conflict with customary drafting styles. A backhanded or left-leaning inclination is never an acceptable modification.

CHARACTERISTICS OF GOOD LETTERING

Good freehand lettering, regardless of whether it is uppercase or lowercase, condensed or extended, or vertical or inclined, must have certain characteristics. These requisites include neatness, uniformity, stability, proper spacing, and speed.

Neat lettering is important so that the information being conveyed can be easily read. Few things detract from the appearance and quality of a technical drawing more than sloppy lettering, *Figure 3-9*.

For uniformity, all letters should be the same in height, proportion, and inclination. A necessary tactic for maintaining uniformity is the use of guidelines, *Figure 3-10*. The customary heights of characters in technical drawing are 1/8″ (3 mm) for regular text, and 3/16″ (4.5 mm) for headings and titles.

The proper stability or balance of letters is an important characteristic in freehand lettering. Each letter should appear balanced and firmly positioned to the human eye. Top-heavy letters are not balanced because they appear about to topple over, *Figure 3-11*.

The proper spacing of letters and words is important, and it takes a lot of practice to accomplish. A good rule of thumb to follow in terms of spacing is to use close spacing within words, and far spacing between words, *Figure 3-12*. The proper positions of letters relative to one another in words is accomplished by spacing the letters in the word equally in the area, not by trying to equalize the spacing between letters. This becomes automatic if the drafter concentrates on the word being lettered, not on each letter. Another rule of thumb for spacing is to allow the width of one round letter, such as O, C, Q, or G, between words. *Figure 3-13* illustrates how this type of spacing can make the lettering much easier to read.

In the modern drafting room, because time is money, speed in freehand lettering is critical. Typically, freehand lettering is one of the slowest, most time-consuming tasks drafters must perform. It takes many hours of practice to develop freehand lettering that is neat, uniform, balanced, properly spaced, and fast. Some drafters never reach this goal. Those who do, reach it through constant practice, coupled with continual efforts to improve.

Freehand Lettering Techniques

Freehand lettering techniques are learned by knowing what grades of lead to use, how to make the basic lettering strokes, and how to use guidelines, and by constantly practicing and trying to improve.

Lettering in ink has been greatly simplified in recent years. Old-fashioned tools, such as adjustable-nib ruling pens and speedball pens, have been replaced by the less cumbersome, easier-to-use technical pen, *Figure 3-14*.

When lettering in ink, drafters still use light guidelines made with pencil lead. The actual lettering is done with the desired pen point size. Commonly used pen points for lettering in ink are sizes 0, 1, and 2, which are standard American sizes. In metrics, these point sizes represent line widths of 0.35 mm, 0.50 mm, and 0.60 mm.

All letters and numbers are created using six basic strokes, *Figure 3-15*. The first stroke is a single stroke made downward and to the right at approximately 45°. The second stroke is made downward and to the left at approximately 45°. The third stroke is vertical and is made from top to bottom. Stroke number four is horizontal and is made from left to right. The fifth stroke is a half-circular stroke to the left, made from top to bottom. The sixth stroke is a half-circular stroke to the right, made from top to bottom. All alphanumeric characters can be created using combinations of these six strokes. *Figure 3-16* shows how these strokes are used for making selected characters.

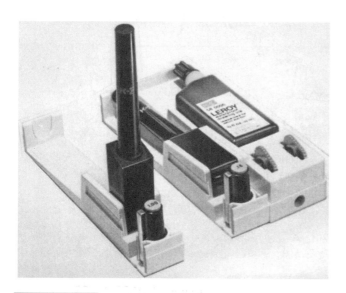

FIGURE 3-14   *Modern technical pens* (Courtesy Keuffel & Esser Co.)

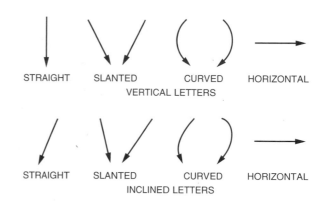

FIGURE 3-15   *Basic strokes used for lettering*

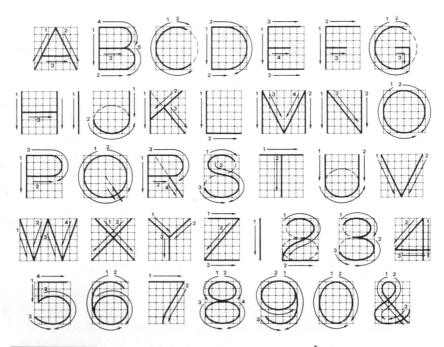

FIGURE 3-16   *Forming uppercase Gothic letters and numerals—vertical style*

LETTERING GUIDELINES

Guidelines are a critical part of freehand lettering. Uniformity, neatness, and stability cannot be achieved without using guidelines.

Line Work

Line work is the generic term given to all of the various techniques used in creating the graphic data on technical drawings. Mechanical line work is made using either mechanical pencils or technical pens. Such devices as parallel bars, drafting machines, triangles, scales, and numerous other tools are used to guide the line-making. Since inking is dealt with in the first chapter, this chapter focuses on pencil line work.

CHARACTERISTICS OF LINES

Twelve basic types of lines are used in manual drafting. Each has its own individual characteristics. The *visible line* is thick and dark. The *hidden line* is a series of short dashes separated by even shorter breaks. The hidden line is thinner than the visible line. *Dimension lines* and *extension lines* are solid, thin lines of approximately the same width as hidden lines. Dimension lines should be broken for dimensions and should have arrowheads for terminations.

The *center line* is broken with one short dash in its center. It is the same width as the hidden, dimension, and extension lines. The *phantom line* is just like the center line except that it has two dashes. The dashes are repeated approximately every two inches. The *cutting plane line* is thick like the visible line and consists of a series of long, equally spaced dashes. All lines used on technical drawings should closely match those in *Figure 3-17*.

HORIZONTAL AND VERTICAL LINES

Horizontal lines are formed by pressing the straightedge (T-square, parallel bar, drafting machine scale, and so forth) against the worksheet with one hand and moving the pencil with the other. Uniformity of line widths and weights can be achieved by holding the pencil at approximately 60° from the drawing surface, maintaining an even pressure downward, and slowly revolving the pencil axially as it moves across the drawing surface, *Figure 3-18*. This keeps the lead tip symmetrical.

Vertical lines are created according to the same principles, except that the drafter's hand moves upward rather than from left to right. The angle of inclination, the amount of pressure, and the rotating motion are the same as they are for horizontal lines.

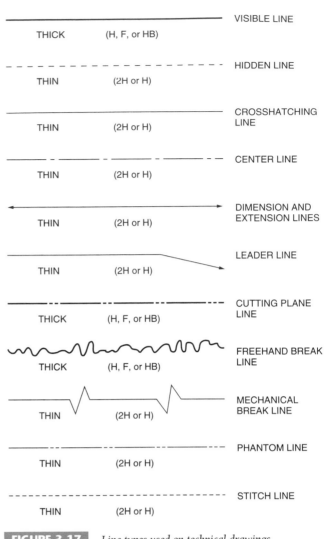

FIGURE 3-17   *Line types used on technical drawings*

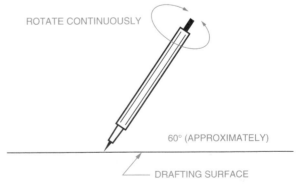

FIGURE 3-18   *Maintaining uniformity of lines*

ANGULAR LINES

Many modern devices are available to assist drafters in making angular lines. These include protractors, adjustable triangles, and adjustable arms on drafting machines. However, most angular lines can be created simply by using the standard 30°–60° and 45° triangles alone and in various combinations, *Figures 3-19, 3-20, 3-21,* and *3-22.* These standard tools create angles of 15°, 30°, 45°, 60°, and 75°.

PARALLEL LINES

Parallel lines can be created in a number of different ways. Vertical (and horizontal) parallel lines are made by simply moving the straightedge the required distance and making each successive line, *Figure 3-23.*

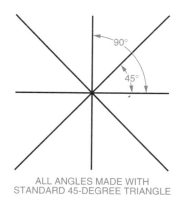

FIGURE 3-19   *Making angular lines with the 45° triangle*

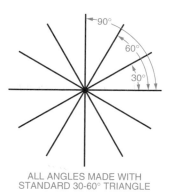

FIGURE 3-20   *Making angular lines with the 30°–60° triangle*

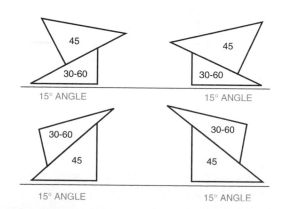

FIGURE 3-21   *Making 15° angles*

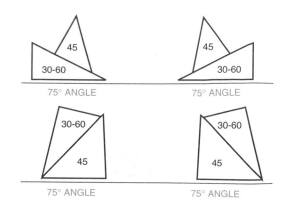

FIGURE 3-22   *Making 75° angles*

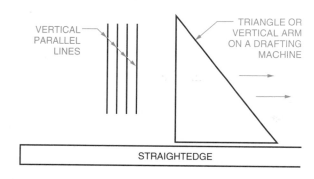

FIGURE 3-23   *Making vertical parallel lines*

Parallel lines at angles can be created by using the 30°–60° and 45° triangles in combination much the same as they are used for making angular lines. When one uses triangles to create angular lines, the first line is created at the desired angle. Aligning one edge of a triangle to the line, register any side of the second triangle against one of the nonaligned edges of the first triangle. Holding the second triangle to prevent it from moving and sliding the first triangle along the engaged edge of the second triangle will reposition the originally aligned edge to any desired parallel position. Successive parallel lines are created in the same way, *Figure 3-24.*

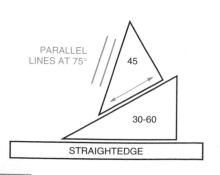

FIGURE 3-24   *Creating successive parallel lines*

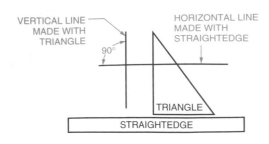

FIGURE 3-25 *Making perpendicular lines*

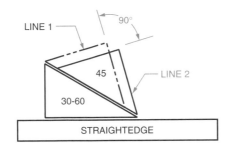

FIGURE 3-26 *Creating a line perpendicular to a nonvertical or nonhorizontal line*

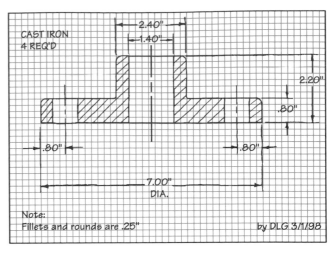

FIGURE 3-27 *Typical design sketch*

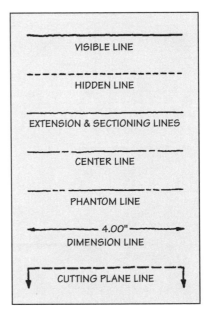

FIGURE 3-28 *Lines used in sketching*

PERPENDICULAR LINES

Drawing perpendicular lines can be accomplished in a manner similar to drawing parallel lines. Horizontal and vertical perpendicular lines can be created using a straightedge and a triangle, *Figure 3-25.*

Creating a line perpendicular to a nonhorizontal or nonvertical line is accomplished by using triangles in conjunction with a straightedge, *Figure 3-26.* Line 1 in this figure is drawn first. Then the 45° triangle is slid along the 30°–60° triangle, and the perpendicular line is created with the opposite side of the 45° triangle.

Sketching

Even in the world of high technology and computers, sketching is still one of the most important skills for drafters and designers. Sketching is one of the first steps in communicating ideas for a design, and it is used in every step thereafter. It is common practice for designers to prepare sketches that are turned over to drafters for conversion to finished working drawings. *Figure 3-27* is an example of a typical design sketch.

SKETCHING LINES

The lines used in creating sketches closely correspond to those used in creating technical drawings except, of course, that they are not as sharp and crisp. *Figure 3-28* illustrates the various types of lines used in making sketches.

The basic line types are: visible line, hidden line, center line, dimension line, sectioning line, extension line, and cutting plane line. These lines represent the various lines available for creating sketches. The character of each line, as illustrated in *Figure 3-29,* should be closely adhered to when making sketches.

Types of Sketches

The types of sketches correspond to the types of technical drawings. There are four types of sketching: orthographic, axonometric, oblique, and perspective, *Figures 3-30, 3-31, 3-32,* and *3-33.*

Orthographic sketching relates to flat, graphic facsimiles of a subject showing no depth. Six principal views of

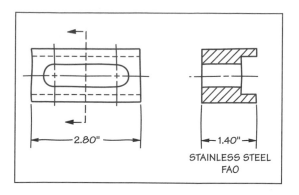

FIGURE 3-29 *Sample design sketch*

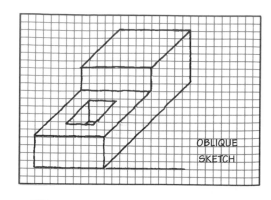

FIGURE 3-32 *Oblique sketch*

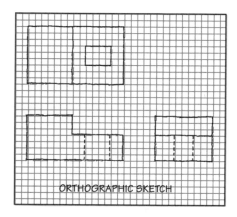

FIGURE 3-30 *Orthographic sketch*

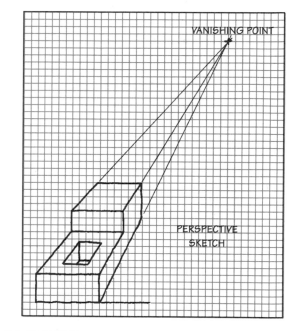

FIGURE 3-33 *Perspective sketch*

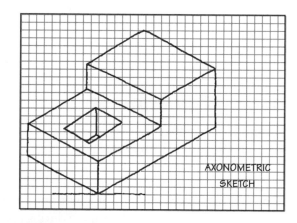

FIGURE 3-31 *Axonometric sketch*

a subject may be incorporated in an orthographic sketch: top, front, bottom, rear, right side, and left side, *Figure 3-34.* The views selected for use in a sketch depend on the nature of the subject and the judgment of the sketcher.

Axonometric sketching may be one of three types: isometric, dimetric, or trimetric, *Figure 3-35.* The type most frequently used is *isometric,* in which length and width lines recede at 30° to the horizontal and height

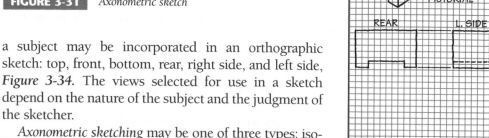

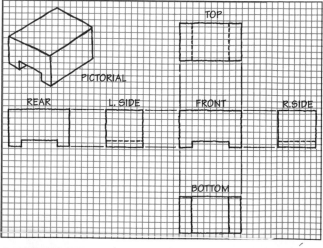

FIGURE 3-34 *Six principal views in orthographic sketching*

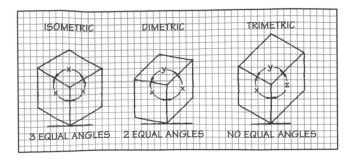

FIGURE 3-35 *Three types of axonometric sketches*

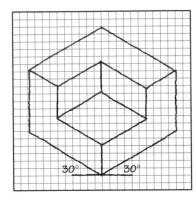

FIGURE 3-36 *Isometric sketch*

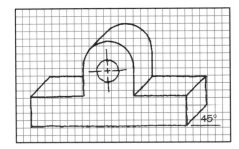

FIGURE 3-37 *Oblique sketch*

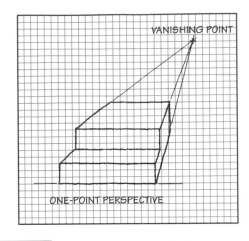

FIGURE 3-38 *One-point perspective sketch*

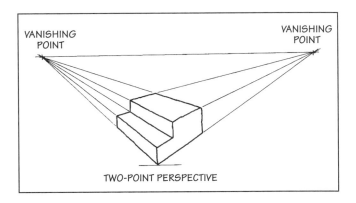

FIGURE 3-39 *Two-point perspective sketch*

lines are vertical, *Figure 3-36*. In sketching, the use of these terms is academic as they relate to proportional scales and angle positions of height, width, and depth, which are only estimated in sketching.

Oblique sketching involves a combination of a flat, orthographic front surface with depth lines receding at a selected angle of 30°, 60° or 45° (usually 45°), *Figure 3-37*.

Perspective sketching involves creating a graphic facsimile of the subject. Consequently, depth lines must recede to a hypothetical vanishing point (or points), *Figures 3-38* and *3-39*. In fact, all pictorial sketches naturally tend to assume characteristics of perspective sketches as a result of how the eye views the apparent relative proportions of objects. This is not necessarily undesirable.

Sketching Materials

An advantage of sketching is that it requires very few material aids. Whereas drafters must have a complete collection of tools, equipment, and materials in order to do working drawings, sketching requires nothing more than a pencil and a piece of paper. It is not uncommon for a sketch to be drawn on a paper napkin during a hurried luncheon meeting.

Sketching done in an office environment requires three basic materials: pencil, media (paper or graph paper), and an eraser. Graph paper simplifies the sketching process considerably, especially for students just learning, and it should be used freely.

Sketching Techniques

Sketches, as with drawings, consist of straight and curved lines. With practice, drafters can become skilled in creating neat, sharp, clear examples—straight or curved—of

Industry Application

SKETCHING ABILITY LEADS TO PROMOTION

"Don't they teach sketching in school any more? I can't find one drafting technician in this department who can make a decent sketch." Dan Johnson, chief drafter for Precision Machining, Inc. (PMI), had been frustrated when he made this statement. The problem he had faced at the time was simple. He needed to send a drafting technician into the field to make sketches of several parts for a warehouse full of machines that were to be refurbished and retrofitted. PMI had a contract to machine the parts. Unfortunately, the machines were old enough that technical drawings of the parts were no longer available.

Johnson's only option was to send a drafting technician to the warehouse armed with a sketch pad and micrometers. The parts would be measured, sketches would be made, and the drafting department would have the information needed to create a new technical drawing package. It was after seven technicians in a row had told Johnson that they either never learned sketching or had forgotten how that the chief drafter had vented his frustration. Fortunately for Johnson, Maria Sims overheard him. Sims was PMI's newest junior drafting technician, having graduated from Clark County College of Technology just a month earlier. Sims told Johnson that not only could she make comprehensive, accurate, readable sketches, but that she could read a micrometer, too. She was given the assignment on the spot and her sketching skills served her well.

Johnson was so impressed with his newest junior drafting technician that he not only promoted her, he asked her to give the entire department a seminar on sketching.

all the various line types introduced previously. When sketching, the following general rules apply.

1. Hold the pencil firmly, but not so tightly as to create tension or hand fatigue.

2. Grip the pencil approximately one inch to one and one-half inches up from the point.

3. Maintain a comfortable angle between the pencil and the sketching strokes.

4. Draw horizontal lines from left to right using short, slightly overlapping strokes.

5. Draw vertical lines from top to bottom using short, slightly overlapping strokes.

6. Draw curved lines using short, slightly overlapping strokes.

In addition to these general rules, some specific techniques are used in making the various line types for sketching.

SKETCHING STRAIGHT LINES AND CURVES

Making straight lines on graph paper is a simple process of guiding the pencil using the existing lines. If graph paper is not available, pencil dots can be positioned to plot the path of the line, *Figure 3-40*. In this figure, the sketcher enters a series of pencil dots on the paper that provide a basic outline of the shape of the object. Then, with a series of short, slightly overlapping strokes, the pencil dots are connected, *Figures 3-41, 3-42*, and *3-43*. This technique is also used for curved lines, *Figure 3-44*.

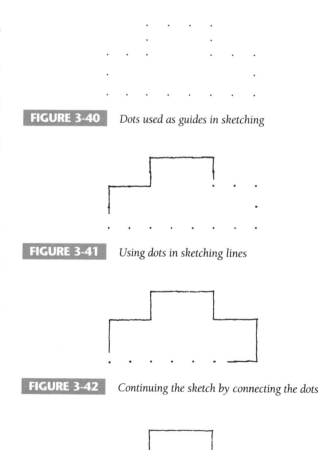

FIGURE 3-40   *Dots used as guides in sketching*

FIGURE 3-41   *Using dots in sketching lines*

FIGURE 3-42   *Continuing the sketch by connecting the dots*

FIGURE 3-43   *Completing the sketch made by using dots as guides*

FIGURE 3-44   *Using dots in sketching curved lines*

SKETCHING CIRCLES

Figure 3-45 illustrates a series of six steps that can be used for sketching a circle. Vertical and horizontal center lines are sketched, which positions the center of the circle (Step 1), and the radial distances of the desired circle size are marked on each of these lines, equidistant from the center (Step 2). A square is drawn symmetrically around the center, with the sides located at the radial line marks (Step 3). On the diagonals of the square (Step 4), the radial distances are again marked off from the center (Step 5). This provides four positions for the circumference to pass through at the sides of the square, and four more positions on the diagonals of the square. The right half of the circle is sketched in from top to bottom using short, slightly overlapping strokes, and then the left half is sketched in the same manner (Step 6).

SKETCHING ELLIPSES

A similar technique is used for sketching ellipses on an orthographic view, except that the square becomes a rectangle, *Figure 3-46*. Ellipses are oriented on the pictorial of the object being sketched as shown in the diagram in *Figure 3-47*.

PROPORTION IN SKETCHING

Sketches are not done to scale, but it is important that they be made proportionately accurate. All of the various components of a sketch should be kept in proportion to those of the actual object. This technique takes a great deal of practice to master.

Some methods for achieving proportion recommend using a pencil or a strip of paper as a simulated scale. These techniques are not only unrealistic in terms of the real world, they defeat the very purpose of sketching. A skilled sketcher must learn to maintain proportion without the use of tools and aids. The best device for accomplishing proportion in sketching is the human eye. With practice, the drafter can become proficient in maintaining proportion without the use of extraneous, time-consuming devices. The following general rules relating to proportion will also help.

STEP 1   In sketching, use graph paper whenever possible.

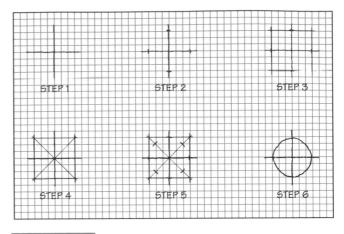

FIGURE 3-45   *Sketching a circle*

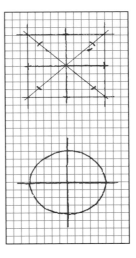

FIGURE 3-46   *Sketching an ellipse*

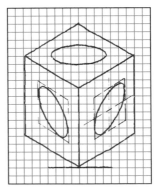

FIGURE 3-47   *Orienting an ellipse on an object*

STEP 2   Examine the object to be sketched and mentally break it into its component parts.

STEP 3   Beginning with the largest components (width and height), estimate the proportion, such as the width is 4/3 times the height or 5/2 times the height, and so on.

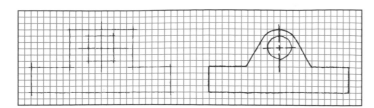

FIGURE 3-48 *Blocking in components*

STEP 4 Lay out the largest component according to the proportions decided upon in Step 3. Use construction line squares and rectangles to block in irregularly shaped components, *Figure 3-48*.

Repeat Steps 3 and 4 until the entire object is finished.

ORTHOGRAPHIC SKETCHING

Orthographic sketching may involve sketching any combination of the six principal views of the subject. The top, front, and right-side views are normally selected for representing an object in an orthographic sketch. However, these views are not always appropriate. The sketcher must learn to choose the most appropriate views. These are the views that show the most detail and the fewest hidden lines. A good rule of thumb to use in selecting views is to select the views that would give you all of the information you would need if you had to make the object yourself.

Once the views have been selected, the orthographic sketch may be laid out using the techniques set forth earlier in this chapter. To ensure that the sketched views align, the entire sketch should be blocked in before adding details, *Figure 3-49*. Once the layout is blocked in, the details can be added one view at a time, *Figures 3-50, 3-51,* and *3-52*.

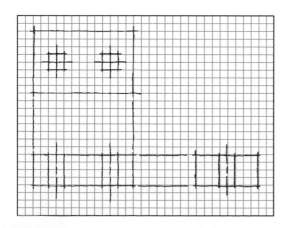

FIGURE 3-49 *Blocking in an orthographic sketch*

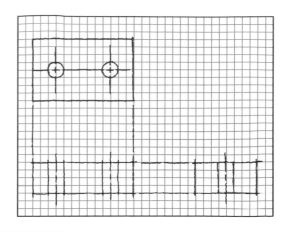

FIGURE 3-50 *Completing the top view*

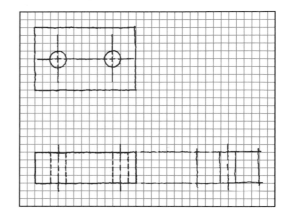

FIGURE 3-51 *Completing the front view*

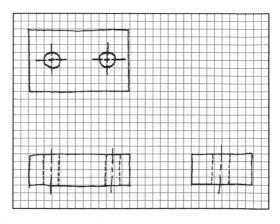

FIGURE 3-52 *Completing the sketch*

AXONOMETRIC SKETCHING

As was mentioned earlier, there are three types of axonometric projection: isometric, dimetric, and trimetric. Isometric projection is used in sketching. Dimetric and trimetric projection have little application in sketching, due to the difficulty in proportioning scale values of depth, width, and height. Isometric views have the same scaling value in all three directions, eliminating the need to vary proportions among the three directions. In an isometric sketch, height lines are vertical, and width and depth lines recede at approximately 30° and 150° (180°–30°) from the horizontal.

The first step in creating an isometric sketch is to lay out the isometric axis, *Figure 3-53*. All normal lines will be parallel to one of the axis lines. The next step is to block in the object using construction lines, *Figure 3-54*. Five steps in the development of an isometric sketch are shown in *Figures 3-55, 3-56, 3-57, 3-58,* and *3-59*.

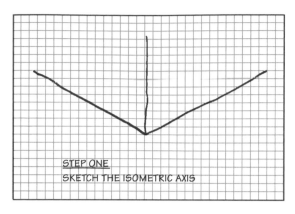

FIGURE 3-55   *Step 1 in making an isometric sketch*

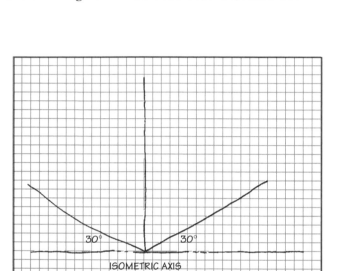

FIGURE 3-53   *The isometric axis in sketching*

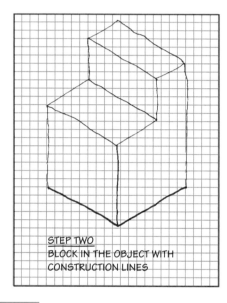

FIGURE 3-56   *Step 2 in making an isometric sketch*

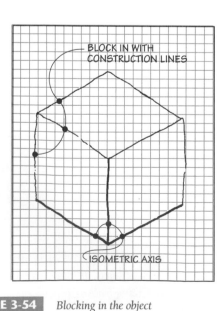

FIGURE 3-54   *Blocking in the object*

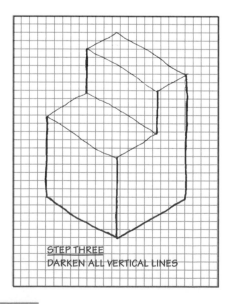

FIGURE 3-57   *Step 3 in making an isometric sketch*

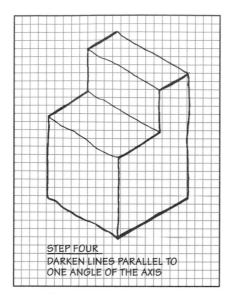

FIGURE 3-58 *Step 4 in making an isometric sketch*

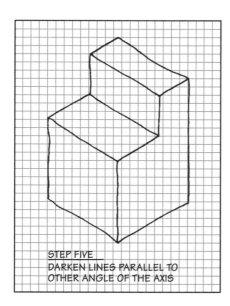

FIGURE 3-59 *Step 5 in making an isometric sketch*

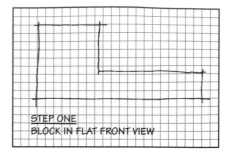

FIGURE 3-60 *Step 1 in making an oblique sketch*

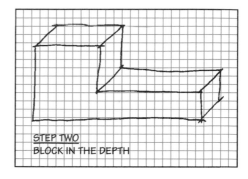

FIGURE 3-61 *Step 2 in making an oblique sketch*

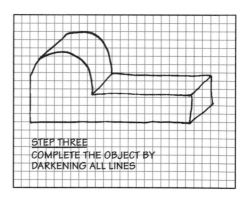

FIGURE 3-62 *Step 3 in making an oblique sketch*

OBLIQUE SKETCHING

Oblique sketching involves laying out the front view of an object and showing the depth lines receding at an angle (usually 45°) from the horizontal. Oblique sketching is particularly useful for dealing with an object having round components. Oblique sketching allows round components to be drawn round, rather than elliptical.

Using the blocking-in method, the flat front surface of the object is laid out. The depth is then blocked in using parallel lines, and the sketch is completed by outlining the exposed profile of the rear surface. *Figures 3-60, 3-61,* and *3-62* illustrate three steps in creating an oblique sketch.

PERSPECTIVE SKETCHING

Perspective sketching closely approximates how the human eye actually sees an object. Two common types of perspective sketches are one-point and two-point perspectives.

A one-point perspective sketch is similar to an oblique sketch, except that depth lines recede to a vanishing point instead of receding parallel to one another, *Figure 3-63.* In constructing a one-point perspective, the following procedures apply.

STEP 1 Lay out the flat front surface of the object using the blocking-in method, *Figure 3-64.*

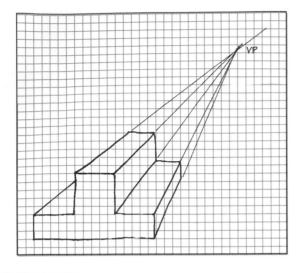

FIGURE 3-63 *One-point perspective sketch*

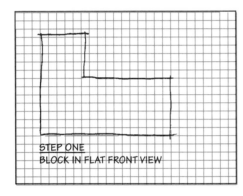

FIGURE 3-64 *Step 1 in making a one-point perspective sketch*

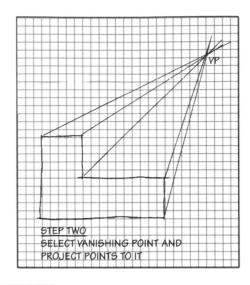

FIGURE 3-65 *Step 2 in making a one-point perspective sketch*

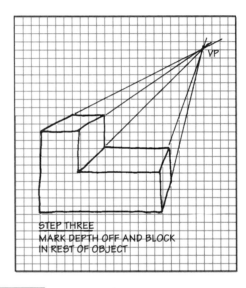

FIGURE 3-66 *Step 3 in making a one-point perspective sketch*

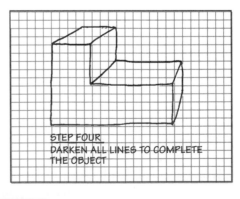

FIGURE 3-67 *Step 4 in making a one-point perspective sketch*

STEP 2 Select and mark a single vanishing point. Project all points on the front surface back to the vanishing point, *Figure 3-65.*

STEP 3 Estimate the depth of the object and mark it off on all line projectors, *Figure 3-66.*

STEP 4 Complete the sketch by outlining the exposed profile of the rear surface, *Figure 3-67.*

A two-point perspective resembles an isometric sketch, except that width and depth lines recede to the left and right vanishing points rather than receding in parallel, *Figure 3-68.*

In constructing a two-point perspective, the following procedures apply.

STEP 1 Lay out the two-point perspective frame, which consists of the vertical height line, the vanishing point left, the vanishing

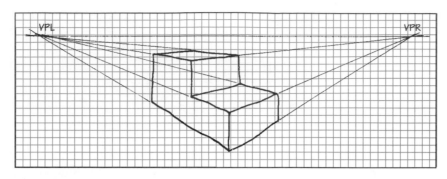

FIGURE 3-68   *Two-point perspective sketch*

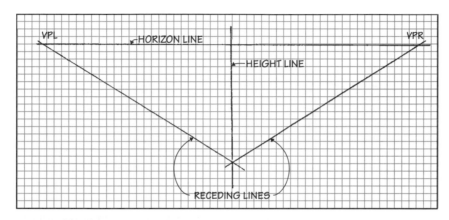

FIGURE 3-69   *Laying out the two-point perspective frame*

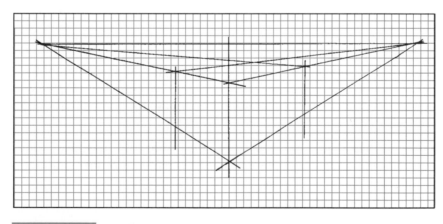

FIGURE 3-70   *Blocking in the object*

point right, and the receding lines (all estimated locations), *Figure 3-69*. The horizon line when positioned below the view provides a view of the bottom of the object, and when positioned above it shows the top. The vanishing points must be on the horizon.

STEP 2   Block in the object, estimating the length and width for proportion, *Figure 3-70*.

STEP 3   Lay out the details, lightly giving special attention to proportion, *Figure 3-71*.

STEP 4   Complete the two-point perspective sketch, *Figure 3-72*.

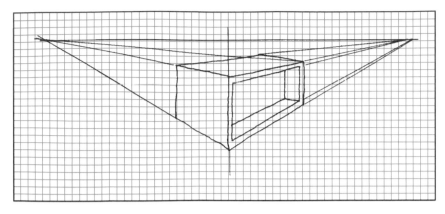

FIGURE 3-71 *Laying out the details*

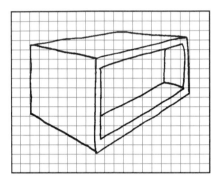

FIGURE 3-72 *Completing the two-point perspective sketch*

Summary

- Talking sketching is of two types; conversational and boardroom sketching. Conversational sketches are developed as two or three individuals discuss an idea. Boardroom sketches are developed on a marker board, chalkboard, or flip chart while a group discusses an idea.

- Freehand lettering may be done in several styles including Block, Fast, Futura, Leroy, Old English, Times, and Helvet. Single-stroke Gothic lettering may be done in upper- or lowercase, extended or condensed, and vertical or inclined styles. The characteristics of good lettering are neatness, uniformity, stability, proper spacing, and speed.

- Students should be able to make and demonstrate the proper use of the following types of lines: visible, hidden, crosshatching, center, dimension, extension, leader, cutting plane, freehand break, mechanical break, phantom, and stitch.

- Students should be able to illustrate the following types of sketches: orthographic, axonometric, oblique, and perspective.

- The materials needed for making sketches are pencils, paper or graph paper, and an eraser.

- Sketching techniques with which students should be proficient are making straight lines, curves, circles, and ellipses. Students should also be proficient in maintaining proper proportion when sketching.

Review Questions

Answer the following questions either true or false.

1. The standard style of freehand lettering on technical drawings is vertical, single-stroke Arial.

2. The correct slope of slanted lettering is a two-unit incline to the right for each five units of letter height.

3. Three strokes are required to make the letter B.

4. In regard to space left between words, it is a rule of thumb to leave the width of one round letter such as Q, C, or O.

5. Uniformity, neatness, and stability can all be achieved without using guidelines.

6. A setting of 8 on a lettering guide will produce letters 1/4″ high.

7. *Line work* is a term used to describe one specific technique used to create graphic data on technical drawings.

Answer the following questions by selecting the best answer.

1. Which of the following is **not** a characteristic of good freehand lettering?
   a. Neatness
   b. Speed
   c. Uniformity
   d. Incline

2. Which of the following is **not** one of the basic line types used on technical drawings?
   a. Angular, perpendicular, nonhorizontal, nonvertical
   b. Visible, hidden, crosshatch
   c. Cutting plane, freehand break, mechanical
   d. Center, dimension, extension, leader

3. Which of the following is an advantage of sketching compared to mechanical drawing?
   a. It requires few material aids
   b. It can be mastered without much practice
   c. It can be done with just a pencil and piece of paper
   d. Both a and c

4. Which axonometric projection is preferred for sketching?
   a. Isometric
   b. Dimetric
   c. Trimetric
   d. Quadmetric

5. Which kind of projection exhibits a circle as an elliptic shape?
   a. Mercator
   b. Isometric
   c. Multiview
   d. Auxiliary

6. How many principal views are there in an orthographic projection?
   a. Two
   b. Four
   c. Six
   d. Eight

7. Guidelines are a critical part of freehand lettering because:
   a. They improve uniformity.
   b. They improve neatness.
   c. They improve stability.
   d. All of the above

8. What is the customary height of characters in a technical drawing?
   a. 1/16″ for regular text and 3/16″ for headings and titles
   b. 1/8″ for regular text and 3/16″ for headings and titles
   c. 1/8″ for regular text and 1/2″ for headings and titles
   d. 1/4″ for regular text and 1/2″ for headings and titles

9. Which of the following is a type of axonometric sketch?
   a. Perspective
   b. Orthographic
   c. Oblique
   d. Trimetric

Chapter 3 Problems

Problems 3-1 through 3-9 are intended to give beginning drafters experiences in visualizing that objects are made of convenient shapes and features. Visualizing these shapes and features will help you to completely dimension an object. Another objective of these nine problems is to encourage you to use approximations, when appropriate, to obtain ballpark answers before diving into detailed calculations.

Problem 3-22 will be used here as an example to illustrate the requirements for the nine problems. Assume that you have been asked to calculate the weight of the object in Problem 3-22, which is made of aluminum with a density of 0.097 pounds per cubic inch. To make a ballpark calculation, consider the base to be $1″ \times 2″ \times 4″$ or a volume of approximately 8 cubic inches and the vertical portion $1/2″ \times 1\text{-}1/2″ \times 4″$ or 3 cubic inches. A total of $8 + 3 = 11$ cubic inches, which multiplied by the density equals approximately 1.0 pound. The correct volume is calculated by deducting the semicircular 1″ radius hole and deducting some volume for the rounded corners of 1/2″ radius. The correct volume is found to be 9.84 cubic inches. Multiply by the density to obtain 0.954 pounds for the correct weight.

Unless otherwise specified, use the format shown in Problem 4-37 to present your solutions for the following problems.

PROBLEM 3-1
Assume that the object in Problem 3-17 is aluminum. Calculate its weight.

PROBLEM 3-2
Assume that the object in Problem 3-18 is aluminum. Calculate its weight.

PROBLEM 3-3
Assume that the object in Problem 3-21 is aluminum. Calculate its weight.

PROBLEM 3-4

Assume that the object in Problem 3-23 is steel with a density of 0.28 pounds per cubic inch. Calculate its weight.

PROBLEM 3-5

Assume that the object in Problem 3-28 is brass with a density of 0.32 pounds per cubic inch. Calculate its weight.

PROBLEM 3-6

Assume that the object in Problem 3-29 is brass with a density of 0.32 pounds per cubic inch. Calculate its weight.

PROBLEM 3-7

Assume that the object in Problem 3-31 is steel with a density of 0.28 pounds per cubic inch. Calculate its weight.

PROBLEM 3-8

Assume that the object in Problem 3-33 is steel with a density of 7,765 kilograms per cubic meter. Calculate its weight in kilograms. Note that one kilogram equals 2.2 pounds. Also, the squares in Problem 3-33 are given as 10 mm or .010 m.

PROBLEM 3-9

Assume that the object in Problem 3-25 is cast iron with a density of 6,933 kilograms per cubic meter. Calculate its weight in kilograms.

PROBLEMS 3-10 THROUGH 3-13

The following problems are intended to give the beginning drafter practice in making neat uppercase, vertical, and Gothic-style letters and numbers. On an A-size sheet of vellum, carefully lay out light .12 spaces as shown. Use .50 borders all around and **be sure to skip a space** between lines of letters or numbers. Practice letters, numbers, words, sentences, and paragraphs, as given.

PROBLEM 3-10

.12
(3)

A
B
C
D
E
F
G
H
I
J
K
L
M
N
O
P
Q
R
S
T
U
V
W
X
Y
Z

.09
(2)

1
2
3
4
5
6
7
8
9
0

PROBLEM 3-11

.12
(3)

YOUR FIRST NAME

YOUR LAST NAME

.09
(2)

DATE (EXAMPLE, 8 AUG 88)

YOUR SCHOOL–COLLEGE NAME

YOUR TOWN

YOUR STATE

YOUR MAIL ZIP CODE

YOUR TELEPHONE NUMBER

PROBLEM 3-12

.12
(3)

GOOD LETTERING TAKES PRACTICE

.09
(2)

IMPROVEMENT IN LETTERING REQUIRES EFFORT

PATIENCE AND HARD WORK EQUALS GOOD LETTERING

SINGLE-STROKE, UPPERCASE, VERTICAL GOTHIC
LETTERING IS USED IN THE MECHANICAL DRAFTING FIELD

PROBLEM 3-13

LETTERING IS EXTREMELY IMPORTANT. IT MUST BE NEAT,
LEGIBLE, CLEAR, AND IN PROPORTION AND MUST BE DONE
IN A REASONABLE AMOUNT OF TIME

.12
(3)

.09
(2)

PROBLEMS 3-14 THROUGH 3-33

These problems are intended to give the beginning drafter practice in sketching. On an A-size sheet of vellum, neatly sketch the top, front, and right-side views of each pictorial figure shown. The grid is provided for proportion only. Try to keep the object in scale as you sketch; keep all parallel lines parallel, and sketch as neatly as possible. Do not use any drawing instruments. Add all lines, including hidden lines and center lines.

Next, sketch each object from your three-view drawings into isometric pictorial views.

PROBLEM 3-14

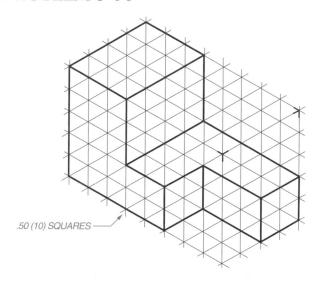

.50 (10) SQUARES

PROBLEM 3-15

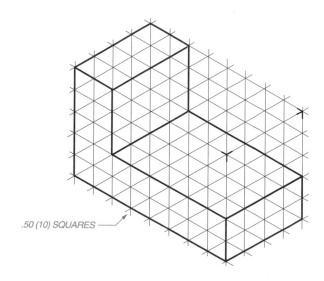

.50 (10) SQUARES

PROBLEM 3-16

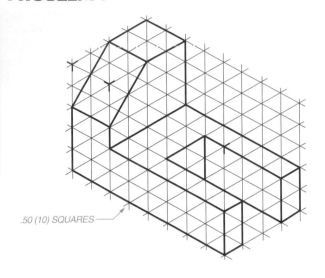

.50 (10) SQUARES

PROBLEM 3-19

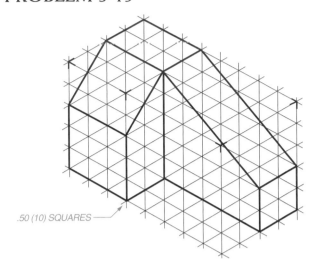

.50 (10) SQUARES

PROBLEM 3-17

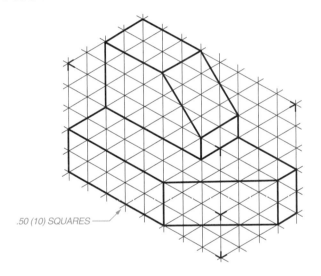

.50 (10) SQUARES

PROBLEM 3-20

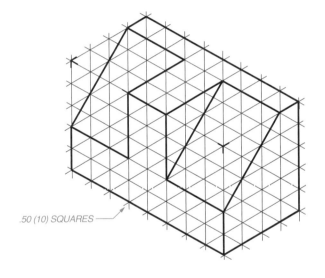

.50 (10) SQUARES

PROBLEM 3-18

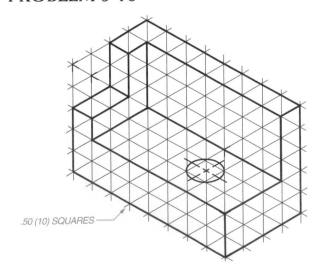

.50 (10) SQUARES

PROBLEM 3-21

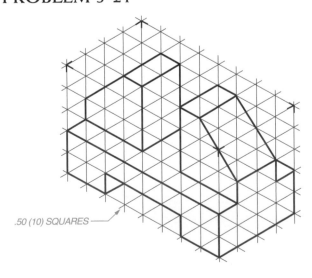

.50 (10) SQUARES

PROBLEM 3-22

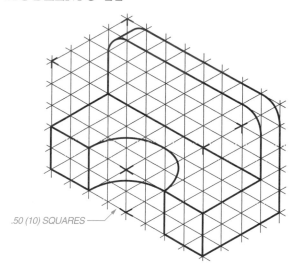

.50 (10) SQUARES

PROBLEM 3-25

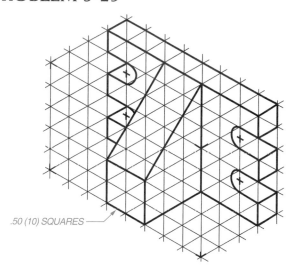

.50 (10) SQUARES

PROBLEM 3-23

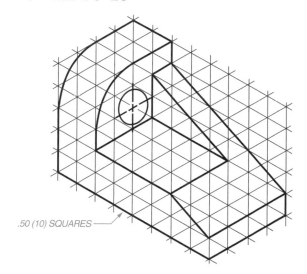

.50 (10) SQUARES

PROBLEM 3-26

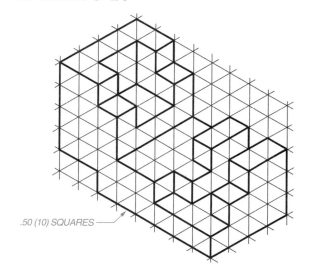

.50 (10) SQUARES

PROBLEM 3-24

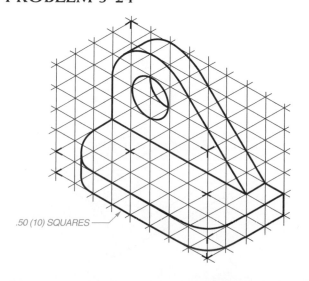

.50 (10) SQUARES

PROBLEM 3-27

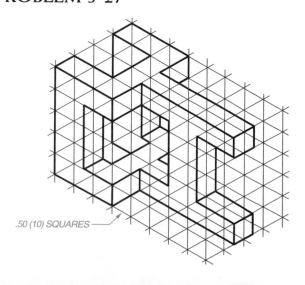

.50 (10) SQUARES

PROBLEM 3-28

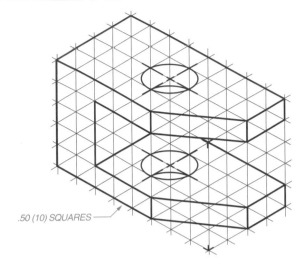

.50 (10) SQUARES

PROBLEM 3-31

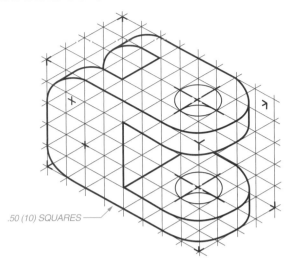

.50 (10) SQUARES

PROBLEM 3-29

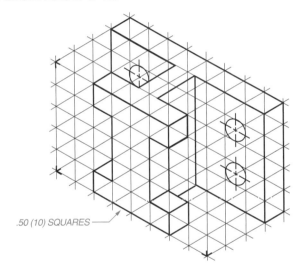

.50 (10) SQUARES

PROBLEM 3-32

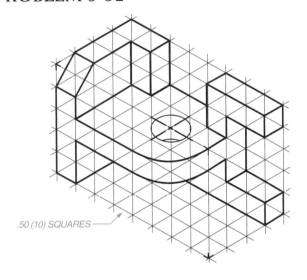

.50 (10) SQUARES

PROBLEM 3-30

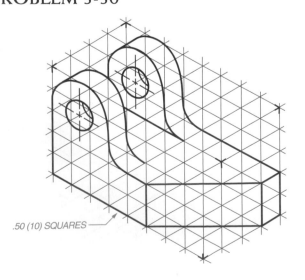

.50 (10) SQUARES

PROBLEM 3-33

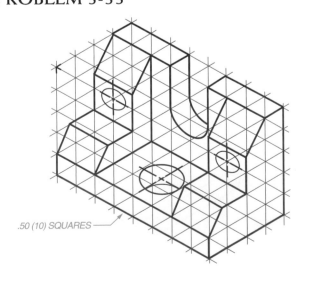

.50 (10) SQUARES

PROBLEMS 3-34 THROUGH 3-37

These few problems are intended to provide future drafters, designers, and engineers with experiences in developing an appreciation for sketching from memory, for recording preliminary design ideas, and for orally communicating technical ideas while sketching.

PROBLEM 3-34

Visit a facility and mentally (no notes or rough sketches) study an object, such as a machine component, an object or a tool in a shop, a part of an automobile, or a piece of sports equipment with a goal in mind that upon returning to class you will sketch the object(s) from memory. Then, prepare orthographic or isometric-type sketches.

The skill emphasized here is that if you observe and study an object as though you were required to sketch it later, you will **see more** than if you just casually look at it. Develop this skill collecting technical ideas for designing.

PROBLEM 3-35

Sketch orthographic views and an isometric of a representative object to be used as your nameplate for a desk, to match your career interests. For example; a bridge could represent a career in construction or civil engineering; a pressurized tank, for welding; an airplane, for flying; a computer, for computer science; and so on.

PROBLEM 3-36

Sketch orthographic views of your design of one of the following.
a. a modern chair
b. a storage cabinet for
   1. wardrobe
   2. drafting tools
   3. typical home use garden tools
c. a basic outline of a sports car
d. a basic outline of an electric car
e. a cabinet for trophies
f. a two-person skateboard
g. an outdoor logo for your school
h. a toolbox for
   1. car maintenance tools
   2. woodworking tools
   3. hobbies (fishing, models, etc.)

PROBLEM 3-37

Visit one of the locations listed and select a device, not too complicated, to sketch. Make a rough sketch (for your use only) for later use in a drafting class. Be prepared to sketch the device, while describing its attributes (shape, size, function, material, aesthetic properties, etc.) to either another student or at the board, to the class.
a. metal shop
b. wood shop
c. electronics shop
d. hardware store
e. garage
f. kitchen
g. computer store

Chapter 4

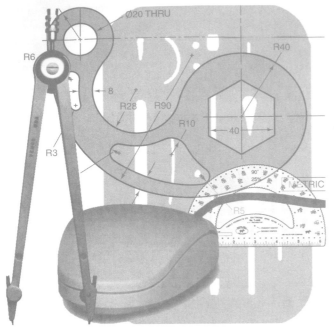

GEOMETRIC CONSTRUCTION

KEY TERMS

Acute angle	Obtuse angle
Altitude	Obtuse angle triangle
Angle	Ogee curve
Bisect	Parallelogram
Central angle	Point
Chord	Pole
Circle	Polygon
Circumference	Polyhedron
Concentric circle	Prism
Cone	Pyramid
Conic section	Quadrant
Cylindrical helix	Quadrilateral
Diameter	Radius
Eccentric circle	Right angle
Epicycloid	Right triangle
Equilateral triangle	Scalene triangle
Helix	Sector
Hemisphere	Segment
Hypocycloid	Semicircle
Involute	Sphere
Isosceles triangle	Tangent point
Lead	Triangle
Line	Truncated

CHAPTER OBJECTIVES

Upon completion of this chapter, students should be able to do the following:

■ Define the most frequently used terms in geometric nomenclature.

■ Properly apply the elemental principles of geometric construction.

■ Demonstrate the proper procedures for polygon construction.

■ Demonstrate the proper procedures for circular construction.

■ Demonstrate the proper procedures for supplementary construction including the following: spiral; helix; involute of a line, triangle, square, and circle; and cycloidal curve.

To be truly proficient in the layout of both simple and complex drawings, the drafter must know and fully understand the many geometric construction methods used. These methods are illustrated in this chapter and are basically simple principles of pure geometry. These simple principles are used to develop a drawing with complete accuracy, and in the fastest time possible, without wasted motion or any guesswork. Applying these geometric construction principles give drawings a finished, professional appearance.

In laying out the various geometric constructions, it is important to use a very sharp, 4H lead and to be extremely accurate at all times. Always draw light construction lines that can hardly be seen when held at arm's length. These light construction lines should not be erased as this takes up valuable drawing time, and they can also be reused to check layout work if necessary.

Geometric Nomenclature

POINTS IN SPACE

A *point* is an exact location in space or on a drawing surface, *Figure 4-1*. A point is actually represented on the drawing by a crisscross at its exact location. The exact point in (drawing) space is where the two lines of the crisscross intersect. When a point is located on an existing line, a light, short, dashed line or crossbar is placed on the line at the location of the exact point. Never represent a point on a drawing by a dot, except for sketching locations. This is not accurate enough, and it is considered to be poor drafting practice.

LINE

A straight line is the shortest distance between two points, *Figure 4-2*. It can be drawn in any direction. A *line* can be straight, an arc, a circle, or a free curve, as illustrated in *Figure 4-3*. If a line is indefinite, and the ends are not fixed in length, the actual length is a matter of convenience. If the end points of a line are important, they must be marked by means of small, mechanically drawn crossbars, as described by a point in space.

FIGURE 4-2 *A straight line is the shortest distance between two points.*

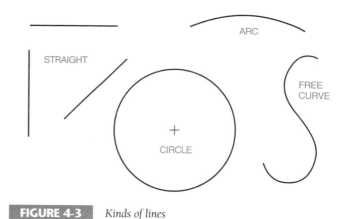

FIGURE 4-3 *Kinds of lines*

Straight lines and curved lines are considered to be parallel if the shortest distance between them remains constant. The symbol used for parallel lines is //. Lines that are tangent and at 90° are considered perpendicular. The symbol for perpendicular lines is ⊥ (singular), *Figure 4-4,* and ⊥'s (plural). The symbol for an angle is ∠ (singular) and ∠'s (plural). To draw an angle, use the drafting machine, a triangle, or a protractor. For extra accuracy, use the vernier on the drafting machine or a vernier protractor.

ANGLE

An *angle* is formed by the intersection of two lines. There are three major kinds of angles: right angles, acute angles, and obtuse angles, *Figure 4-5*. The *right angle* is an angle of 90°. An *acute angle* is an angle at less than 90°, and an *obtuse angle* is an angle at more than 90°. Note that a straight line is 180°. Figure 4-5 also illustrates the complementary and supplementary angles of a given angle.

There are 360 degrees (360°) in a full circle. Each degree is divided into 60 minutes (60′). Each minute is

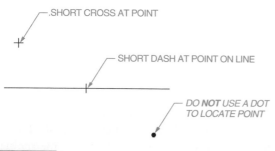

FIGURE 4-1 *Points in space or on a surface*

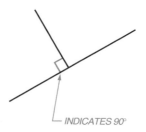

FIGURE 4-4 *Perpendicular lines*

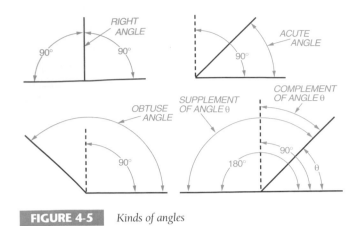

FIGURE 4-5 *Kinds of angles*

divided into 60 seconds (60″). Example: 48°28′38″ is read as 48 degrees, 28 minutes, and 38 seconds.

To convert minutes and seconds to decimal degrees, divide minutes by 60 and seconds by 3,600.

EXAMPLE:

21°18′27″ = 21° + ($\frac{18}{60}$)° + ($\frac{27}{3600}$)° = 21.3075°

To convert decimal degrees to degrees, minutes, and seconds, multiply the degree decimal by 60 to obtain minutes, and the minute decimal by 60 to obtain seconds.

EXAMPLE:

77.365° = 77° + (.365 × 60)′ = 77°21′.9′
77°21′.9′ = 77°21′ + (.9 × 60)″ = 77°21′54″

A vernier may be used to measure and read off minutes and seconds of a degree. The vernier scale is discussed in Chapter 1.

TRIANGLE

A *triangle* is a closed plane figure with three straight sides and three interior angles. The sum of the three internal angles is always exactly 180°, half of the 360° in a full circle. *Figure 4-6* shows the various kinds of triangles: a right triangle, an equilateral triangle, an isosceles triangle, a scalene triangle, and an obtuse angle triangle.

A *right triangle* is a triangle having a right angle or an angle of 90°. The two sides forming the right angle are called legs, and the third side (the longest) is the hypotenuse. Any triangle inscribed in a semicircle is a right triangle, *Figure 4-7*.

An *equilateral triangle*, as its name implies, is a triangle with all sides of equal length. All of its interior angles are also equal. An *isosceles triangle* has two sides of equal length and two equal interior angles. A *scalene triangle* has no equal sides or angles. An *obtuse angle triangle* is a triangle having an obtuse angle greater than 90°, with no equal sides.

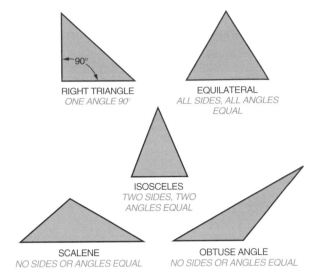

FIGURE 4-6 *Kinds of triangles*

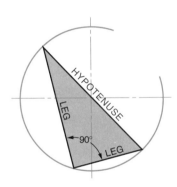

FIGURE 4-7 *Any triangle inscribed within a semicircle is a right triangle*

POLYGON

A *polygon* is a closed plane figure with three or more straight sides, *Figure 4-8*. More specifically, shown in the figure are regular polygons, meaning that all sides are equal in each of these examples. The most important of these polygons as they relate to drafting are probably the triangle with three sides, the square with four sides, the hexagon with six sides, and the octagon with eight sides.

QUADRILATERAL

A *quadrilateral* is a plane figure bounded by four straight sides. When opposite sides are parallel, the quadrilateral is also considered to be a *parallelogram*, *Figure 4-9*.

CIRCLE

A *circle* is a closed curve with all points on the circle at the same distance from the center point. The major components of a circle are the diameter, the radius, and the circumference, *Figure 4-10*.

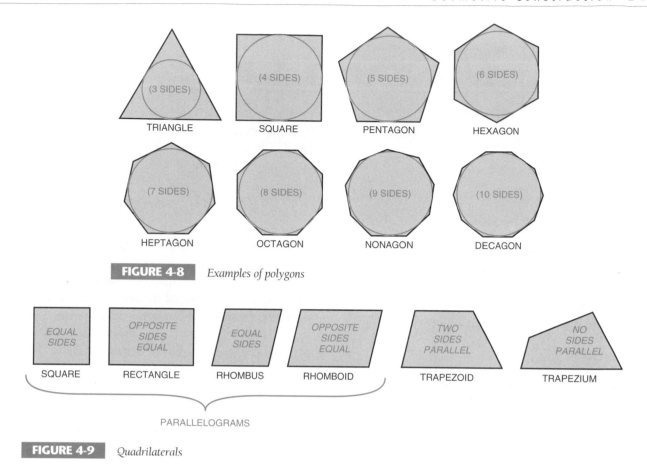

FIGURE 4-8   *Examples of polygons*

FIGURE 4-9   *Quadrilaterals*

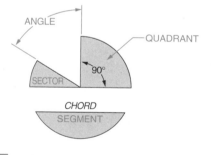

FIGURE 4-11   *Other parts of a circle*

FIGURE 4-10   *The major components of a circle*

The *diameter* of a circle is the straight distance from one outside curved surface through the center point to the opposite outside curved surface. The diameter of a circle is twice the size of the radius.

The *radius* of a circle is the distance from the center point to the outside curved surface. The radius is half the diameter and is used to set the compass when drawing a diameter.

The *circumference* of a circle is the distance around the outer surface of the circle. To calculate the circumference of a circle, multiply the value of π (use the approxima-

tion 3.1416) by the diameter. A chart similar to the one found in Appendix A may be used.

Other important parts of a circle are the central angle, the sector, the quadrant, the chord, and the segment, *Figure 4-11.*

A *central angle* is an angle formed by two radial lines from the center of the circle.

A *sector* is the area of a circle lying between two radial lines and the circumference.

A *quadrant* is a sector with a central angle of 90°, and usually with one of the radial lines oriented horizontally.

A *chord* is any straight line whose opposite ends terminate on the circumference of the circle. (A diameter is a chord passing through the center of the circle.)

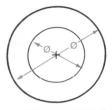 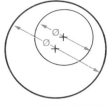

CONCENTRIC CIRCLE   ECCENTRIC CIRCLE

FIGURE 4-12   *Concentric and eccentric circles*

A *segment* is the smaller portion of a circle separated by a chord.

Concentric circles are two or more circles with a common center point, *Figure 4-12*.

Eccentric circles are two or more circles without a common center point, *Figure 4-12*.

A *semicircle* is half of a circle.

POLYHEDRON

A *polyhedron* is a solid object bounded by plane surfaces. Each surface is called a face. If the faces are equal, regular polygons, the solid figure is a regular polyhedron, *Figure 4-13*.

Prism

A *prism* is a solid having ends that are parallel, matched polygons, and sides that are parallelograms, *Figure 4-14*. This definition also applies to round or circular objects, such as a cylinder. When the polygon on one end of a prism is not parallel to the other end, it is said to be *truncated*. The *altitude* of a prism is the perpendicular distance between its end polygons (or bases).

Pyramid and Cone

A *pyramid* is a polyhedron having a polygon as its base. Three or more triangles form its lateral sides, which meet at a common vertex, *Figure 4-15*. A *cone* is a pyramid with a central axis and an infinite number of sides that form a continuous curved lateral surface. When the vertex of a pyramid or cone has been removed by a plane that intersects all the lateral sides (which forms a new polygon), the pyramid or cone is said to be truncated, *Figure 4-16*.

Sphere

A *sphere* is a closed surface, every point on which is equidistant from a common point or center, *Figure 4-17*. If a

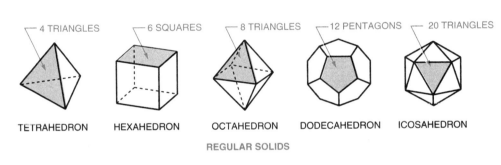

4 TRIANGLES   6 SQUARES   8 TRIANGLES   12 PENTAGONS   20 TRIANGLES

TETRAHEDRON   HEXAHEDRON   OCTAHEDRON   DODECAHEDRON   ICOSAHEDRON

REGULAR SOLIDS

FIGURE 4-13   *Polyhedrons (solids)*

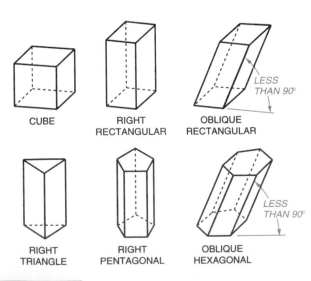

CUBE   RIGHT RECTANGULAR   OBLIQUE RECTANGULAR

LESS THAN 90°

RIGHT TRIANGLE   RIGHT PENTAGONAL   OBLIQUE HEXAGONAL

LESS THAN 90°

FIGURE 4-14   *Prisms*

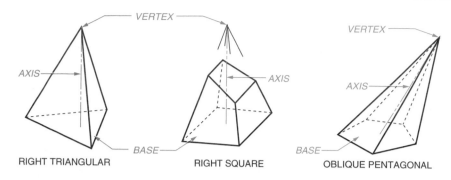

FIGURE 4-15 *Pyramids*

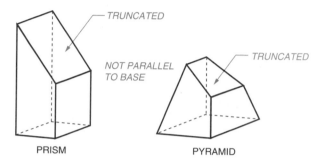

FIGURE 4-16 *Truncated prism and pyramid*

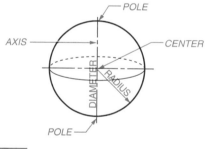

FIGURE 4-17 *Sphere*

sphere is cut into two equal parts, the parts are called *hemispheres*. *Poles* are two reference measuring positions on the surface of the sphere on opposite sides of its center.

Elemental Construction Principles

The remaining portion of this chapter is devoted to illustrating step-by-step the many geometric construction principles used by the drafter to develop various geometric forms. It is important that each step be fully understood and followed. As the beginning drafter uses these geometric construction principles, various shortcuts will become evident, thus reducing the drawing time and increasing accuracy even more. At the end of the chapter, each of these techniques is incorporated or used in some way to complete the problems.

HOW TO BISECT A LINE

To *bisect* a line means to divide it in half or to find its center point. In the given process, a line will also be constructed at the exact center point of exactly 90°.

GIVEN: Line A-B, *Figure 4-18A.*

STEP 1 Set the compass approximately two-thirds of the length of line A-B and swing an arc from point A, *Figure 4-18B.*

STEP 2 Using the same compass setting, swing an arc from point B, *Figure 4-18c.*

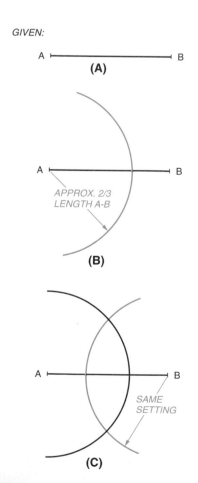

FIGURE 4-18 *(A) How to bisect a line; (B) Step 1; (C) Step 2*

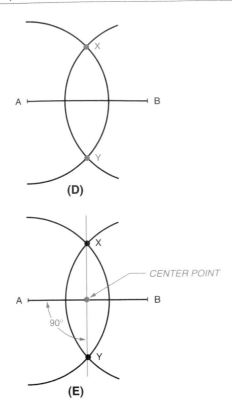

FIGURE 4-18 (Continued) (D) Step 3; (E) Step 4

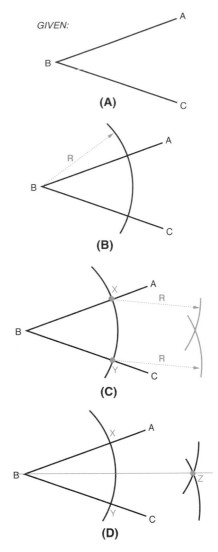

FIGURE 4-19 (A) How to bisect an angle; (B) Step 1; (C) Step 2; (D) Step 3

STEP 3 At the two intersections of these arcs, locate points X and Y, *Figure 4-18D.*

STEP 4 Draw a straight line connecting point X with point Y. Where this line intersects line A-B is the bisect of line A-B, *Figure 4-18E.* Line X-Y is also perpendicular to line A-B at the center point.

HOW TO BISECT AN ANGLE

To bisect an angle means to divide it in half or to cut it into two equal angles.

GIVEN: Angle ABC, *Figure 4-19A.*

STEP 1 Set the compass at any convenient radius and swing an arc from point B, *Figure 4-19B.*

STEP 2 Locate points X and Y on the legs of the angle, and swing two arcs of the same length from points X and Y, respectively, *Figure 4-19C.*

STEP 3 Where these arcs intersect, locate point Z. Draw a straight line from B to Z. This line will bisect angle ABC and establish two equal angles: ABZ and ZBC, *Figure 4-19D.*

HOW TO DRAW AN ARC OR CIRCLE (RADIUS) THROUGH THREE GIVEN POINTS

GIVEN: Three points in space at random: A, B, and C, *Figure 4-20A.*

STEP 1 With straight lines, lightly connect points A to B, and B to C, *Figure 4-20B.*

STEP 2 Using the method outlined for bisecting a line, bisect lines A-B and B-C, *Figure 4-20C.*

STEP 3 Locate point X where the two extended bisectors meet. Point X is the exact center of the arc or circle, *Figure 4-20D.*

STEP 4 Place the point of the compass on point X and adjust the lead to any of the points A, B, or C (they are the same distance), and swing the circle. If all work is done correctly, the arc

GIVEN:

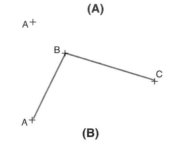

(A)

(B)

(C)

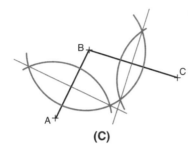

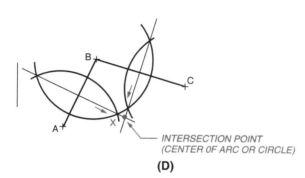

INTERSECTION POINT
(CENTER 0F ARC OR CIRCLE)

(D)

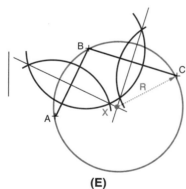

(E)

FIGURE 4-20 *(A) How to draw an arc or circle (radius) through three given points; (B) Step 1; (C) Step 2; (D) Step 3; (E) Step 4*

or circle should pass through each point, as shown in *Figure 4-20E.*

HOW TO DRAW A LINE PARALLEL TO A STRAIGHT LINE AT A GIVEN DISTANCE

GIVEN: Line A-B, and a required distance to the parallel line, *Figure 4-21A.*

GIVEN:

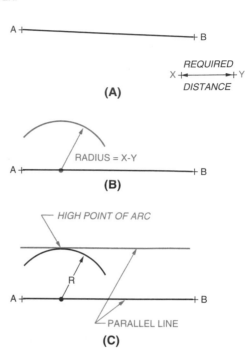

FIGURE 4-21 *(A) How to draw a line parallel to a straight line at a given distance; (B) Step 1; (C) Step 2*

STEP 1 Set the compass at the required distance to the parallel line. Place the point of the compass at any location on the given line, and swing a light arc whose radius is the required distance, *Figure 4-21B.*

STEP 2 Adjust the straightedge of either a drafting machine or an adjustable triangle so that it lines up with line A–B, slide the straightedge up or down to the extreme high point, which is the tangent point of the arc, then draw the parallel line, *Figure 4-21C.*

NOTE: *The distance between parallel lines is measured on any line that is perpendicular to both.*

HOW TO DRAW A LINE PARALLEL TO A CURVED LINE AT A GIVEN DISTANCE

GIVEN: Curved line A-B, and a required distance to the parallel line, *Figure 4-22A.*

STEP 1 Set the compass at the required distance to the parallel line. Starting from either end of the curved line, place the point of the compass on the given line, and swing a series of light arcs along the given line, *Figure 4-22B.*

STEP 2 Using an irregular curve, draw a line along the extreme high points of the arcs, *Figure 4-22C.*

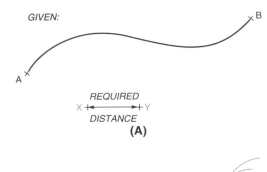

GIVEN:

REQUIRED
X ⟵——⟶ Y
DISTANCE

(A)

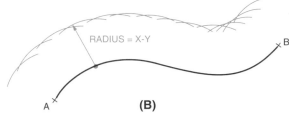

RADIUS = X-Y

B

A

(B)

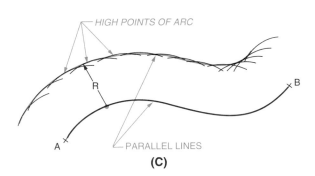

HIGH POINTS OF ARC

R

B

A

PARALLEL LINES

(C)

FIGURE 4-22 (A) How to draw a line parallel to a curved line at a given distance; (B) Step 1; (C) Step 2

HOW TO DRAW A PERPENDICULAR TO A LINE AT A POINT (METHOD 1)

GIVEN: Line A-B with point P on the same line, *Figure 1-23A*.

STEP 1 Using P as the center, make two arcs of equal radius or one continuous arc (R1) to intercept line A-B on either side of point P, at points S and T, *Figure 4-23B*.

STEP 2 Swing larger but equal arcs (R2) from each of points S and T to cross each other at point U, *Figure 4-23C*.

STEP 3 A line from U to P is perpendicular to line A-B at point P, *Figure 4-23D*.

HOW TO DRAW A PERPENDICULAR TO A LINE AT A POINT (METHOD 2)

GIVEN: Line A-B with point P on the line, *Figure 4-24A*.

STEP 1 Swing an arc of any convenient radius whose center O is at any convenient location not on line A-B, but positioned to make the arc cross line A-B at points P and Q, *Figure 4-24B*.

STEP 2 A line from point Q through center O intercepts the opposite side of the arc at point R, *Figure 4-24C*.

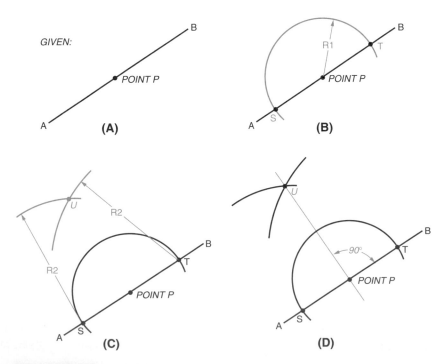

FIGURE 4-23 (A) How to draw a perpendicular to a line at a point (method 1); (B) Step 1; (C) Step 2; (D) Step 3

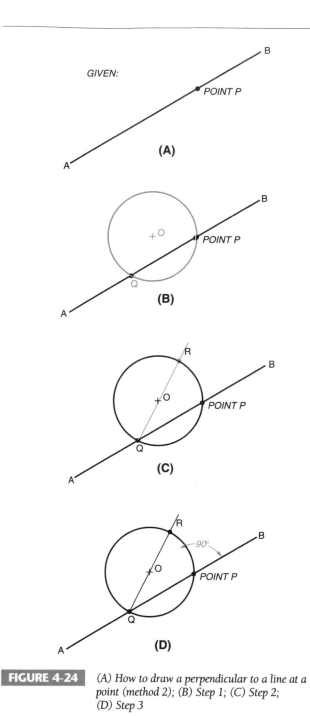

FIGURE 4-24 (A) How to draw a perpendicular to a line at a point (method 2); (B) Step 1; (C) Step 2; (D) Step 3

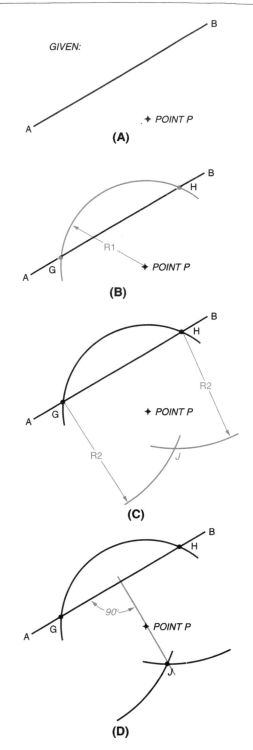

FIGURE 4-25 (A) How to draw a perpendicular to a line from a point not on the line; (B) Step 1; (C) Step 2; (D) Step 3

STEP 3 Line R-P is perpendicular to line A-B. (A right triangle has been inscribed in a semicircle.) See *Figure 4-24D*.

HOW TO DRAW A PERPENDICULAR TO A LINE FROM A POINT NOT ON THE LINE

GIVEN: Line A-B and point P, *Figure 4-25A*.

STEP 1 Using P as a center, swing an arc (R1) to intercept line A-B at points G and H, *Figure 4-25B*.

STEP 2 Swing larger, but equal length arcs (R2) from each of the points G and H to intercept each other at point J, *Figure 4-25C*.

STEP 3 Line P-J is perpendicular to line A-B, *Figure 4-25D*.

HOW TO DIVIDE A LINE INTO EQUAL PARTS

GIVEN: Line A-B, *Figure 4-26A.*

STEP 1 Draw a line 90° from either end of the given line. A 90° line from point B is illustrated in *Figure 4-26B,* but either end or either direction will work as well.

STEP 2 Place a scale with its zero at point A of the given line. If three equal parts are required, pivot the scale until the three-inch measurement (or any length representing three equal units of measure: 30, 60 and 90 mm, for example) is on the perpendicular line drawn in Step 1. Place a short dash at these points. The example in *Figure 4-26C* shows these dashes at the 1″ and 2″ marks.

STEP 3 Project lines downward from these points and add short hash marks where these projected lines cross the original given line A-B. This divides the line A-B into three equal parts. Check all work, comparing your final step with *Figure 4-26D.* An example of equal spacing, where equally spaced holes are required with a given length, is illustrated in *Figure 4-26E.*

HOW TO DIVIDE A LINE INTO PROPORTIONAL PARTS

GIVEN: Line A-B, *Figure 4-27A. Problem:* Locate point X at 2/3 of the distance from point A to point B.

STEP 1 Draw a line 90° from either end of the line. A 90° line from point B is illustrated in *Figure 4-27B.*

STEP 2 Place a scale with its zero on point A of the given line. Because a 2/3 proportion is required, pivot the scale until any multiple of three units of measure intersects the perpendicular line drawn in Step 1. In this example, 6 is used, representing three 2-unit increments. The 2/3 position of this length is two 2-unit increments, or the 4 position, where a hash

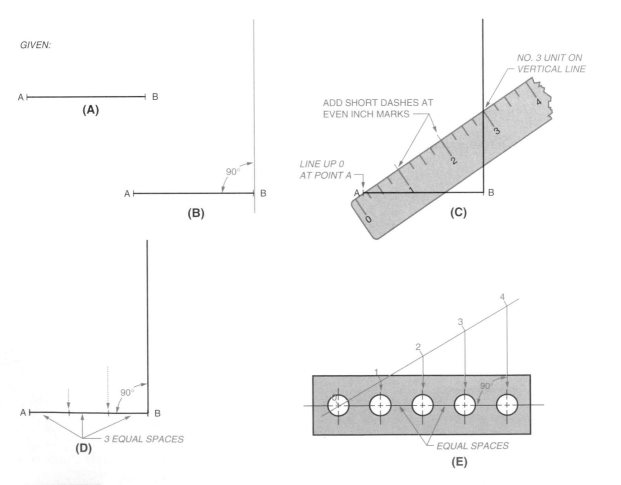

FIGURE 4-26 (A) How to divide a line into equal parts; (B) Step 1; (C) Step 2; (D) Step 3; (E) An example of equal spacing within a given length

GIVEN:

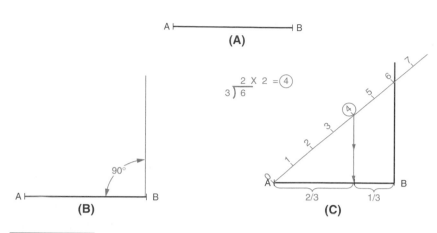

FIGURE 4-27 *(A) How to divide a line into proportional parts; (B) Step 1; (C) Step 2*

mark is made, as shown in *Figure 4-27C.* Projecting this point downward to line A-B, it becomes point X, which is 2/3 the overall distance from point A.

HOW TO TRANSFER AN ANGLE

GIVEN: An angle formed by two straight lines, 0A-0B, *Figure 4-28A,* and one location of where the transferred angle begins (point 0′), *Figure 4-28B.*

STEP 1 Refer to *Figure 4-28C.* Draw an arc through both legs of a given angle (R1) and then duplicate this radius at the transferred angle location, *Figure 4-28D.*

STEP 2 Transfer the chord length between the two angle legs at the intersection of the arc (R2) to the arc at the transfer angle location.

STEP 3 A line from the arc center to the intersection of arc and chord length forms the second line, forming an angle equal to the original, *Figures 4-28E* and *4-28F.*

HOW TO TRANSFER AN ODD SHAPE

GIVEN: Triangle ABC, *Figure 4-29A.*

STEP 1 Letter or number the various corners and point locations of the odd shape in counter-clockwise order around its perimeter. In this example, place the compass point at point A of the original shape and extend the lead to point B. Refer back to *Figure 4-29A.* Swing a light arc at the new desired location, *Figure 4-29B.* Letter the center point as A′

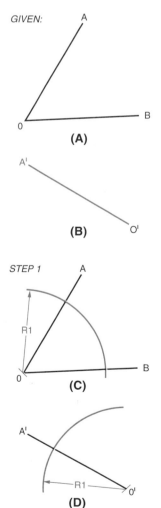

FIGURE 4-28 *(A) How to transfer an angle; (B) Given: Point 0′; (C) Step 1; (D) Transferred angle*

STEP 2

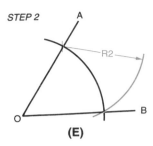

(E)

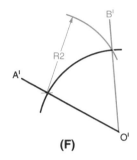

(F)

FIGURE 4-28 (Continued) (E) Step 2; (F) Step 3

and add letter B′ at any convenient location on the arc. It is a good habit to lightly letter each point as you proceed.

STEP 2 Place the compass point at letter B of the original shape, *Figure 4-29C,* and extend the compass lead to letter C of the original shape.

STEP 3 Transfer this distance, B-C, to the layout, *Figure 4-29D.*

STEP 4 Going back to the original object, place the compass point at letter A, *Figure 4-29E,* and extend the compass lead to letter C.

STEP 5 Transfer the distance A-C as illustrated in *Figure 4-29F.* Locate and letter each point.

STEP 6 Connect points A′, B′, and C′ with light, straight lines. This completes the transfer of the object, *Figure 4-29G.* Recheck all work and, if correct, darken lines to the correct line weight.

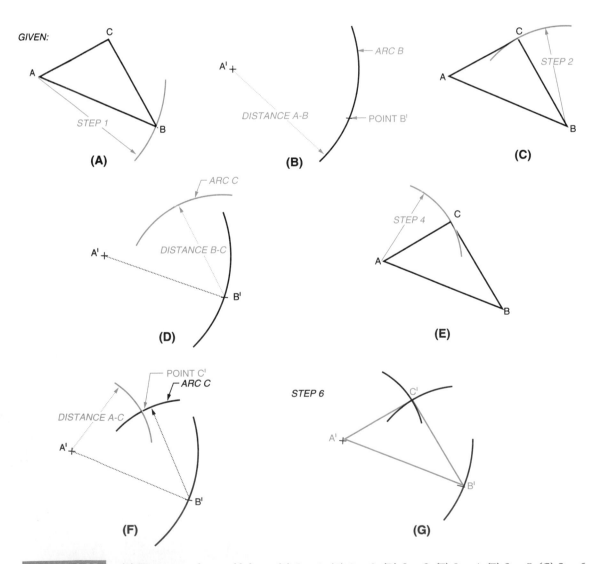

FIGURE 4-29 (A) How to transfer an odd shape; (B) Step 1; (C) Step 2; (D) Step 3; (E) Step 4; (F) Step 5; (G) Step 6

HOW TO TRANSFER COMPLEX SHAPES

A complex shape can be transferred in exactly the same way by reducing the shape into simple triangles and transferring each triangle using the foregoing method.

GIVEN: An odd shape, A, B, C, D, E, F, G, *Figure 4-30A*. Letter or number the various corners and point locations of the odd shape in clockwise order around the perimeter, *Figure 4-30A*. Use the longest line or any convenient line as a starting point. Line A-B is chosen here as the example.

STEP 1 Lightly divide the shape into triangle divisions, using the baseline if possible. Transfer each triangle in the manner described in Figures 4-29A through 4-29G. Suggested

triangles to be used in example Figure 4-30A are ABC, ABD, ABE, ABF, and ABG, *Figure 4-30B*.

STEP 2 This completes the transfer. Check all work and, if correct, darken in lines to correct line thickness. See *Figure 4-30C*.

HOW TO PROPORTIONATELY ENLARGE OR REDUCE A SHAPE

GIVEN: A rectangle, *Figure 4-31A*. *Problem:* To enlarge or reduce its size proportionately.

STEP 1 Draw a line from corner to corner diagonally, and extend it as shown in *Figure 4-31B*.

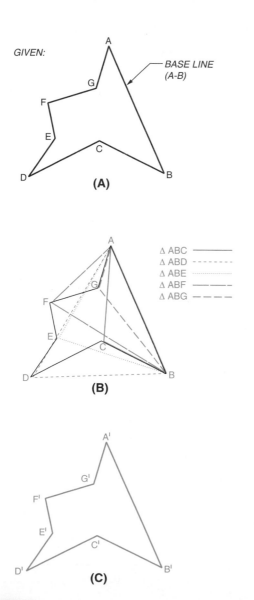

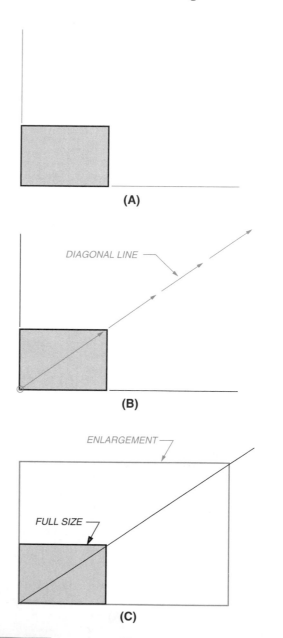

FIGURE 4-30 *(A) How to transfer complex shapes; (B) Step 1; (C) Step 2*

FIGURE 4-31 *(A) How to proportionately enlarge or reduce a shape; (B) Step 1; (C) Step 2*

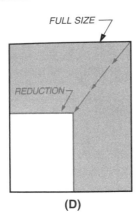

FULL SIZE

REDUCTION

(D)

FIGURE 4-31   *(Continued) (D) Step 3*

STEP 2   The rectangle is enlarged to any size proportionately if the vertical and horizontal sides are located from the extended diagonal line, *Figure 4-31C.*

STEP 3   The rectangle is reduced proportionately if the vertical and horizontal lines are located on the unextended diagonal line, *Figure 4-31D.*

Polygon Construction

HOW TO DRAW A TRIANGLE WITH KNOWN LENGTHS OF SIDES

GIVEN: Lengths 1, 2, and 3, Figure *4-32A.*

STEP 1   Draw the longest length line, in this example length 3, with endpoints A and B, *Figure 4-32B.* Swing an arc (R1) from point A whose radius is either length 1 or length 2; in this example, length 1.

STEP 2   Using the radius length not used in Step 1, swing an arc (R2) from point B to intercept the arc swung from point A at point C, *Figure 4-32C.*

STEP 3   Connect A to C and B to C to complete the triangle, *Figure 4-32D.*

HOW TO DRAW AN EQUILATERAL TRIANGLE

GIVEN: A baseline and the given length of each side, *Figure 4-33A.*

STEP 1   Either adjust the drafting machine angle to 60° or use a 30°–60° triangle and project from

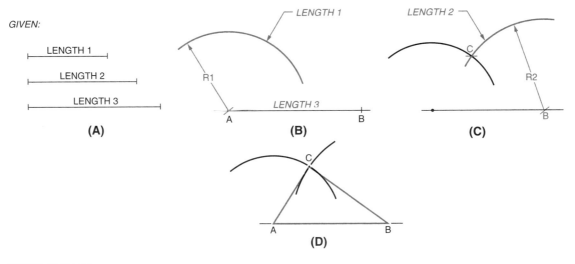

GIVEN:

LENGTH 1

LENGTH 2

LENGTH 3

(A)

LENGTH 1

R1

LENGTH 3

A   B

(B)

LENGTH 2

C

R2

B

(C)

C

A   B

(D)

FIGURE 4-32   *(A) How to draw a triangle with known lengths of sides; (B) Step 1; (C) Step 2; (D) Step 3*

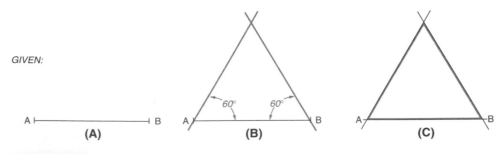

GIVEN:

A   B

(A)

A   60°   60°   B

(B)

A   B

(C)

FIGURE 4-33   *(A) How to draw an equilateral triangle; (B) Step 1; (C) Step 2*

Industry Application

A CLASSROOM DEBATE: WHY SHOULD WE LEARN THIS?*

The drafting class at Franklin Grove Community College was halfway through its lesson on geometric construction, and some of the students were becoming frustrated. "This is the most boring, tedious work I've ever done," said one student. "Why should we learn this when the computer will do it for us anyway?" Overhearing this comment, the instructor, John Diamond, asked the class to stop working and gather around. This is what he told the class.

"I know that geometric construction can be tedious and boring. I didn't enjoy learning how to make them either when I was a student, but I'm glad I did. I began teaching technical drawing around the time that CADD was in its infancy. Software developers were having problems back then trying to build programs that could create good technical drawings. One problem was that the personal computer hardware of the time had neither the memory nor the processing power to create complex technical drawings. Of course time and technology would solve this problem."

"Another problem was that computer programmers didn't understand geometric constructions or other technical drawing techniques well enough to write programs that would do what drafting technicians and engineers needed them to do. What they needed was someone who really understood geometric constructions to advise the computer programmers. That's where I became involved."

"Because I understood geometric constructions to a sufficient level of detail to explain them to computer programmers, I was hired as a consultant. Not only did I make extra money, but I was involved in the original development of one of today's most successful CADD programs."

"You see, its always better to understand the *why* behind the *how*. Not learning geometric constructions before learning CADD is like not learning mathematical computations before using a calculator. The student who depends on the computer instead of learning fundamentals such as geometric construction will be like the auto mechanic who can only change parts because he does not know how to diagnose the problem."

* The names in this article have been changed to protect privacy.

points A and B at 60° using light lines, *Figure 4-33B.* Allow light construction lines to cross as shown; do not erase them.

STEP 2 Check to see that there are three equal sides and, if so, darken in the actual triangle using correct line thickness, *Figure 4-33C.* Care should be exercised in constructing the sharp corners. Again, do not erase the light construction lines.

HOW TO DRAW A SQUARE

GIVEN: The location of the center and the required distance across the "flats" of a square, *Figure 4-34A.*

STEP 1 Lightly draw a circle with a diameter equal to the distance around the flats of the square. Set the compass at half the required diameter, *Figure 4-34B.*

STEP 2 Using a triangle, lightly complete the square by constructing tangent lines to the circle. Allow the light construction lines to project

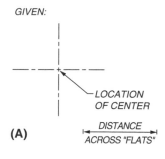

(A)

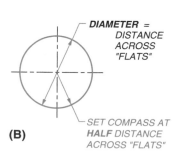

(B)

FIGURE 4-34   *(A) How to draw a square; (B) Step 1*

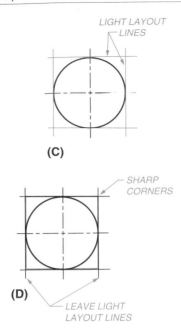

(C)

(D)

LIGHT LAYOUT LINES

SHARP CORNERS

LEAVE LIGHT LAYOUT LINES

FIGURE 4-34   *(Continued) (C) Step 2; (D) Step 3*

from the square, as shown, without erasing them, *Figure 4-34C*.

STEP 3 Check to see that there are four equal sides and, if so, darken in the actual square using the correct line thickness, *Figure 4-34D*. Care

should be exercised in constructing the sharp corners. Again, do **not** erase the light construction lines.

HOW TO DRAW A PENTAGON (5 SIDES)

GIVEN: The location of the pentagon center and the diameter that will circumscribe the pentagon, *Figure 4-35A*.

STEP 1 Locate point A at the top-center of the circle and, using a drafting machine, position an angle of 72° (360/5) from the horizontal (or 18° from the vertical) through point A to locate point B where the angle crosses the circumference of the circle, *Figure 4-35B*.

STEP 2 Draw a horizontal line from point B to locate point C on the circumference of the circle on the opposite side, *Figure 4-35C*.

STEP 3 Set the compass at the distance from point B to point C, and swing this distance from the points as illustrated in *Figure 4-35D* to locate points X and Y.

STEP 4 Lightly connect the points. Check to see that there are five equal sides and, if correct, darken in the actual pentagon taking care to construct five sharp corners, *Figure 4-35E*.

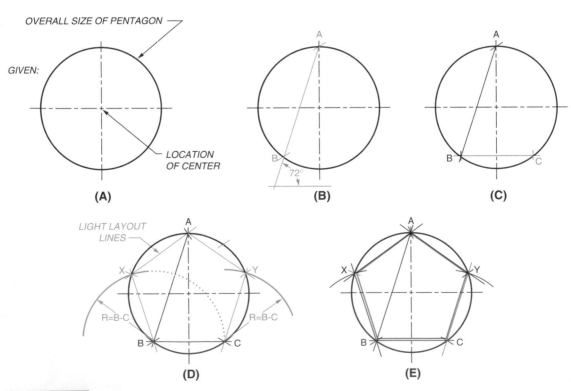

FIGURE 4-35   *(A) How to draw a pentagon; (B) Step 1; (C) Step 2; (D) Step 3; (E) Step 4*

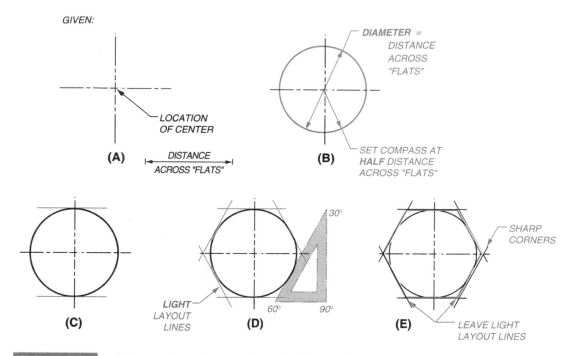

FIGURE 4-36 *(A) How to draw a hexagon; (B) Step 1; (C) Step 2; (D) Step 3; (E) Step 4*

HOW TO DRAW A HEXAGON (6 SIDES)

GIVEN: The location of the required center and the required distance across the flats of a hexagon, *Figure 4-36A*.

STEP 1 Lightly draw a circle with a diameter equal to the distance across the flats of the hexagon. Set the compass at half the required diameter, *Figure 4-36B*.

STEP 2 Draw two horizontal lines tangent to the curve, or two vertical lines if the hexagon is to be oriented at 90° to the illustrated position, *Figure 4-36C*.

STEP 3 Using a 30°–60° triangle, lightly complete the hexagon by constructing tangent lines to the circle, *Figure 4-36D*. Allow the light construction lines to extend as shown; do **not** erase them.

STEP 4 Check to see that there are six equal sides and, if so, darken in the actual hexagon using correct line thickness and taking care to construct six sharp corners, *Figure 4-36E*. Again, do **not** erase the light construction lines.

HOW TO DRAW AN OCTAGON (8 SIDES)

GIVEN: The location of the required center and the required distance across the flats of an octagon, *Figure 4-37A*.

STEP 1 Lightly draw a circle with a diameter equal to the distance across the flats of the octagon. Set the compass at half the required diameter, *Figure 4-37B*.

STEP 2 Lightly draw two horizontal lines and two vertical lines tangent to the circle, as illustrated in *Figure 4-37C*.

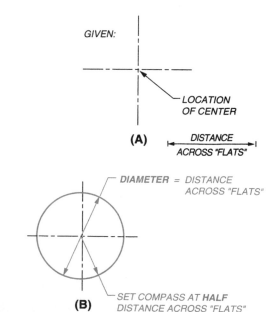

FIGURE 4-37 *(A) How to draw an octagon; (B) Step 1*

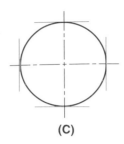

(C)

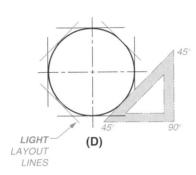

LIGHT
LAYOUT   **(D)**
LINES

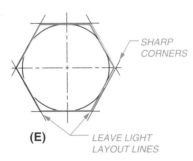

SHARP
CORNERS

(E)   LEAVE LIGHT
LAYOUT LINES

FIGURE 4-37   *(Continued) (C) Step 2; (D) Step 3; (E) Step 4*

STEP 3   Using a 45° triangle, lightly complete the octagon by constructing tangent lines to the circle, *Figure 4-37D*. Allow the light lines to extend.

STEP 4   Check that there are eight equal sides and, if so, darken in the actual octagon using correct line thickness and taking care to construct eight sharp corners, *Figure 4-37E*. Again, do **not** erase the light construction lines.

Circular Construction

HOW TO LOCATE THE CENTER OF A GIVEN CIRCLE

GIVEN: A circle without a center point, *Figure 4-38A*.

STEP 1   Using the drafting machine or T-square, draw a horizontal line across the circle approximately halfway between the **estimated** center of the given circle and the uppermost point on the circumference. Label the end points of the chord thus formed as A and B, *Figure 4-38B*.

STEP 2   Draw perpendicular lines (90°) downward from points A and B. Locate points C and D where these two lines pass through the circle, *Figure 4-38C*.

STEP 3   Carefully draw a straight line from point A to point D and from point C to point B. Where these lines cross is the exact center of the given circle. Place a compass point on the center point; adjust the lead to the edge of the circle and swing an arc to check that the center is accurate, *Figure 4-38D*.

Alternatively, the intersection of perpendicular bisectors of any two nonparallel chords serve to locate the center. This is a modification of the previous construction for passing a circle through any three non-aligned points.

GIVEN:

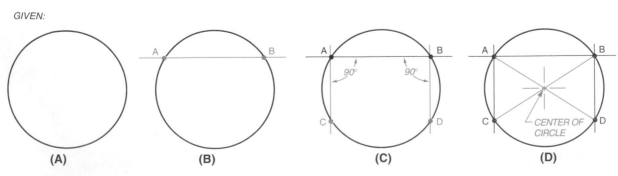

(A)   **(B)**   **(C)**   **(D)**

FIGURE 4-38   *(A) How to locate the center of a given circle; (B) Step 1; (C) Step 2; (D) Step 3*

TANGENT POINTS

A *tangent point* is the exact location or point where one line stops and another line begins. Tangent also means to "touch." As an example, a tangent point is the exact point where a curved line stops and a straight line begins, *Figure 4-39.*

HOW TO LOCATE TANGENT POINTS

The tangent point is a point 90° from the straight line to the swing point of the curved line. Place a light hash mark at each tangent point. It is a good habit to always find all tangent points on the object in all views before darkening in the drawing.

The tangent points for a right-angle bend are illustrated in *Figure 4-40A.* The tangent points for an acute angle bend are illustrated in *Figure 4-40B.* The tangent points for an obtuse angle bend are illustrated in *Figure 4-40C.*

In each preceding example, a light line is constructed 90° from the straight line to the exact swing point of the

radius. Find all the tangent points before darkening in the final work. Always darken in all compass work first, followed by the straight lines.

The tangent point between two arcs or circles is the exact point where one arc or circle ends and the next arc or circle begins. The tangent point could also be where one arc or circle touches another arc or circle, *Figure 4-40D.*

The tangent point is found by drawing a straight line from the swing point of the first arc or circle to the swing point of the second arc or circle, *Figure 4-40E.* Place a short, light dash at the exact tangent point.

HOW TO CONSTRUCT AN ARC TANGENT TO A RIGHT ANGLE (90°)

GIVEN: A right angle (90°), lines A and B, and a required radius, *Figure 4-41A.*

STEP 1 Set the compass at the required radius and, out of the way, swing a radius from line A and one from line B, as illustrated in *Figure 4-41B.*

STEP 2 From the extreme high points of each radius, construct a light line parallel to line A and another line parallel to line B, *Figure 4-41C.*

STEP 3 Where these lines intersect is the exact location of the required swing point. Set the compass point on the swing point and lightly construct the required radius, *Figure 4-41D.* Allow the radius swing to extend past the required area, as shown in the figure. It is

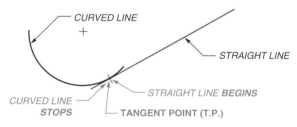

FIGURE 4-39 *Tangent points*

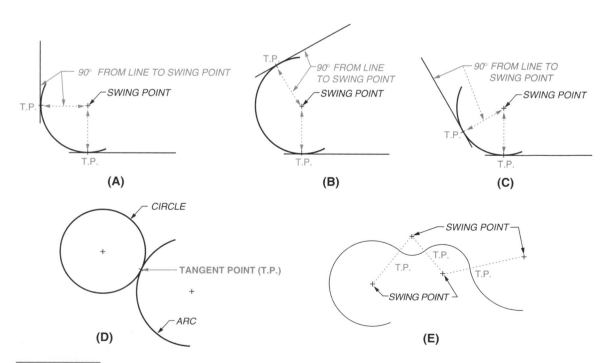

FIGURE 4-40 *How to locate tangent points (A) On a right angle; (B) On an acute angle; (C) On an obtuse angle; (D) Between arcs or circles; (E) Tangent points between arcs or circles*

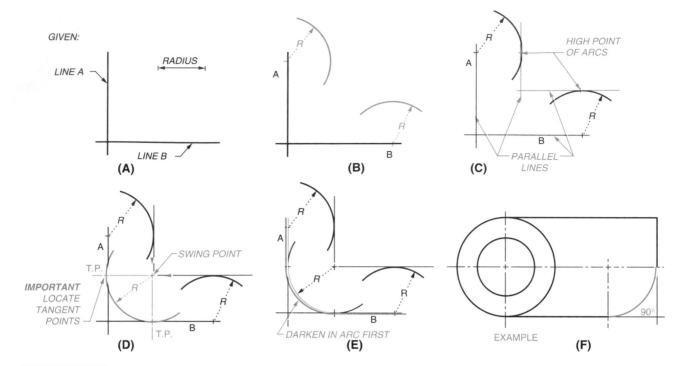

FIGURE 4-41 (A) How to construct an arc tangent to a right angle; (B) Step 1; (C) Step 2; (D) Step 3; (E) Step 4; (F) Example of an arc tangent to two lines at a right angle

important to locate all tangent points (T.P.) before darkening in.

STEP 4 Check all work and darken in the radius using the correct line thickness. Darken in connecting straight lines as required. Always construct compass work first, followed by straight lines. Leave all light construction lines, *Figure 4-41E*.

An example of an arc tangent to two lines at a right angle (90°) is illustrated in *Figure 4-41F*.

HOW TO CONSTRUCT AN ARC TANGENT TO AN ACUTE ANGLE (LESS THAN 90°)

GIVEN: An acute angle, lines A and B, and a required radius, *Figure 4-42A*. Follow the same procedure as outlined in the preceding example.

STEP 1 Set the compass at the required radius and, out of the way, swing a radius from line A and one from line B, as illustrated in *Figure 4-42B*.

STEP 2 From the extreme high points of each radius, construct a light line parallel to line A and another line parallel to line B, *Figure 4-42C*.

STEP 3 Where these lines intersect is the exact location of the required swing point. Set the compass point on the swing point and lightly construct the required radius, *Figure 4-42D*. Allow the radius swing to extend past the required area, as shown in Figure 4-42D.

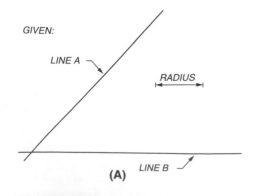

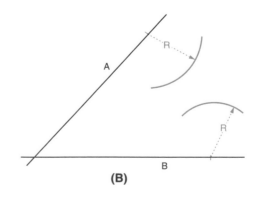

FIGURE 4-42 (A) How to construct an arc tangent to an acute angle; (B) Step 1

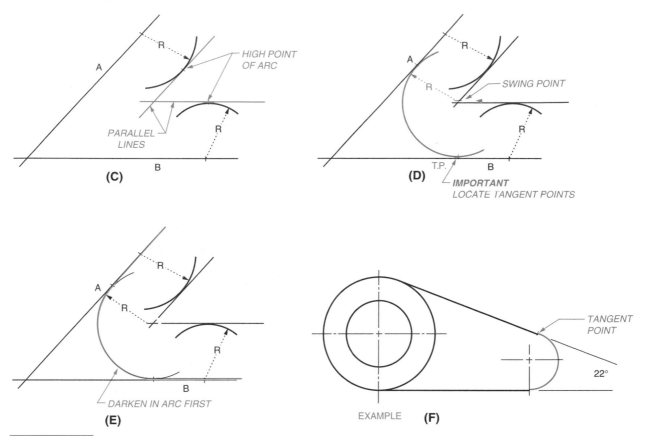

FIGURE 4-42   *(Continued) (C) Step 2; (D) Step 3; (E) Step 4; (F) Example of an arc tangent to two lines at an acute angle*

STEP 4 Check all work and darken in the radius using the correct line thickness. Darken in connecting straight lines as required. Always construct compass work first, followed by straight lines. Leave all light construction lines, *Figure 4-42E*.

An example of an arc tangent to two lines at an acute angle (22°) is illustrated in *Figure 4-42F*.

HOW TO CONSTRUCT AN ARC TANGENT TO AN OBTUSE ANGLE (MORE THAN 90°)

GIVEN: An obtuse angle between lines A and B, and a required radius, *Figure 4-43A*. Follow the same procedure as outlined in the two preceding examples.

STEP 1 Set the compass at the required radius and, out of the way, swing a radius from line A and one from line B, as illustrated in *Figure 4-43B*.

STEP 2 From the extreme high points of each radius, construct a light line parallel to line A and one parallel to line B, *Figure 4-43C*.

STEP 3 Where these lines intersect is the exact location of the required swing point. Set the compass point on the swing point and lightly

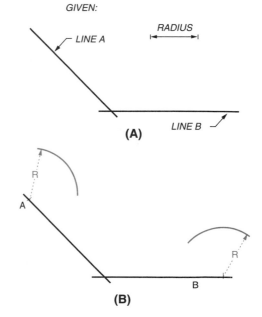

FIGURE 4-43   *(A) How to construct an arc tangent to an obtuse angle; (B) Step 1*

construct the required radius. Allow the radius swing to extend past the required area, as shown in *Figure 4-43D*.

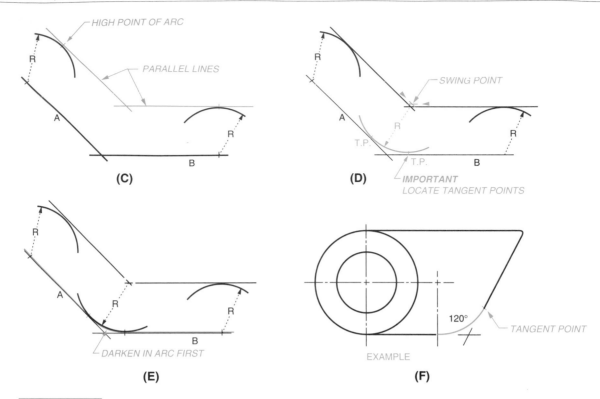

FIGURE 4-43 *(Continued) (C) Step 2; (D) Step 3; (E) Step 4; (F) Example of an arc tangent to two lines at an obtuse angle*

STEP 4 Check all work and darken in the radius using the correct line thickness. Darken in connecting straight lines as required. Always construct compass work first, followed by straight lines. Leave all light construction lines, see *Figure 4-43E*.

An example of an arc tangent to two lines at an obtuse angle (120°) is illustrated in *Figure 4-43F*.

HOW TO CONSTRUCT AN ARC TANGENT TO A STRAIGHT LINE AND A CURVE

GIVEN: Straight line A, an arc B with a center point, and a required radius, *Figure 4-44A*.

STEP 1 Set the compass at the required radius and, out of the way, swing a radius from the given arc B and one from the given straight line A, *Figure 4-44B*.

STEP 2 From the extreme high points of each radius, construct a light straight line parallel to line A, and construct a radius outside the given arc B equal to the required radius, as illustrated in **Figure 4-44C**.

STEP 3 Where these lines intersect is the exact location of the required swing point. Set the compass point on the swing point and lightly construct the required radius, *Figure 4-44D*.

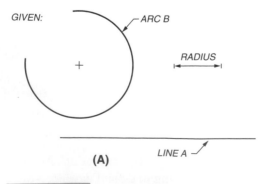

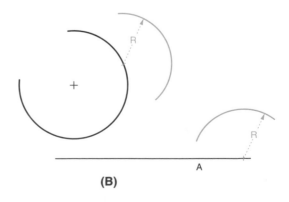

FIGURE 4-44 *(A) How to construct an arc tangent to a straight line and a curve; (B) Step 1*

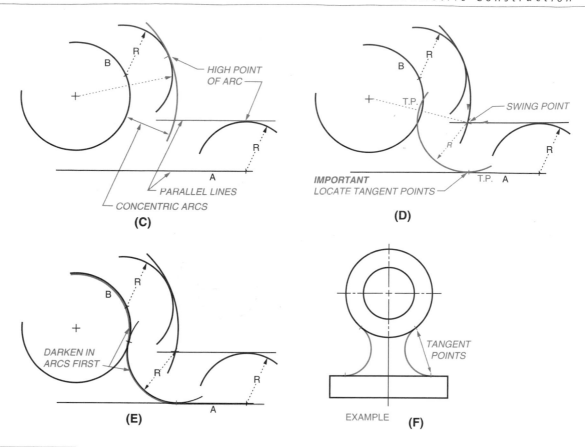

FIGURE 4-44 *(Continued) (C) Step 2; (D) Step 3; (E) Step 4; (F) Example of an arc tangent to a straight line and a curve*

Allow the radius swing to extend past the required area as shown. Locate all tangent points (T.P.) before darkening in.

STEP 4 Check all work, darken in the radius using the correct line thickness. Darken in the arcs first and the straight line last, *Figure 4-44E.*

An example of an arc tangent to a straight line and a curve is illustrated in *Figure 4-44F.*

HOW TO CONSTRUCT AN ARC TANGENT TO TWO RADII OR DIAMETERS

GIVEN: Diameter A and arc B with center points located, and the required radius, *Figure 4-45A.*

STEP 1 Set the compass at the required radius and, out of the way, swing a radius of the required length from a point on the circumference of given diameter A. Out of the way, swing a required radius from a point on the circumference of a given arc B, *Figure 4-45B.*

STEP 2 From the extreme high points of each radius, construct a light radius outside of the

given radii A and B, as illustrated in *Figure 4-45C.*

STEP 3 Where these arcs intersect is the exact location of the required swing point. Set the compass point on the swing point and lightly construct the required radius, *Figure 4-45D.* Allow the radius swing to extend past the required area as shown. Before darkening in, it is important to locate all tangent points (T.P.), *Figure 4-45D.*

STEP 4 Check all work, darken in the radii using the correct line thickness. Darken in the arcs or radii in consecutive order from left to right or from right to left, thus constructing a smooth connecting line having no apparent change in direction, *Figure 4-45E.*

An example of an arc tangent to two radii is illustrated in *Figure 4-45F.*

HOW TO DRAW AN OGEE CURVE

An *ogee curve* is used to join two parallel lines. It forms a gentle curve that reverses itself in a neat, symmetrical geometric form.

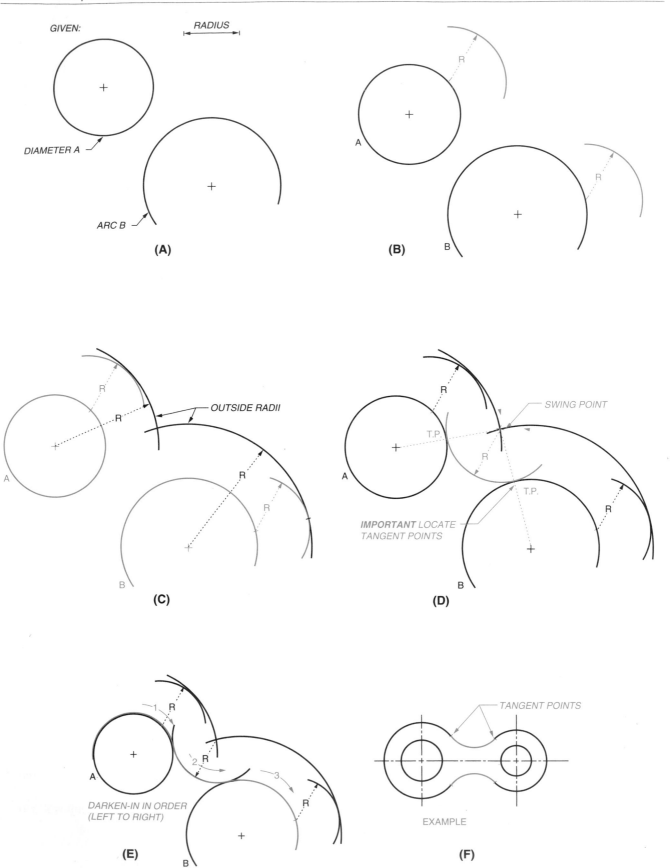

FIGURE 4-45 (A) How to construct an arc tangent to two radii or diameters; (B) Step 1; (C) Step 2; (D) Step 3; (E) Step 4; (F) Example of an arc tangent to two radii

GIVEN: Parallel lines A-B and C-D, *Figure 4-46A*.

STEP 1 Draw a straight line connecting the space between the parallel lines. In this example, from point B to point C, *Figure 4-46B*.

STEP 2 Make a perpendicular bisector to line B-C to establish point X, *Figure 4-46C*.

STEP 3 Make perpendicular bisectors to the lines B-X and X-C, *Figure 4-46D*.

STEP 4 Draw a perpendicular from line A-B at point B to intersect the perpendicular bisector of B-X, which locates the first required swing center. Draw a perpendicular from line C-D at

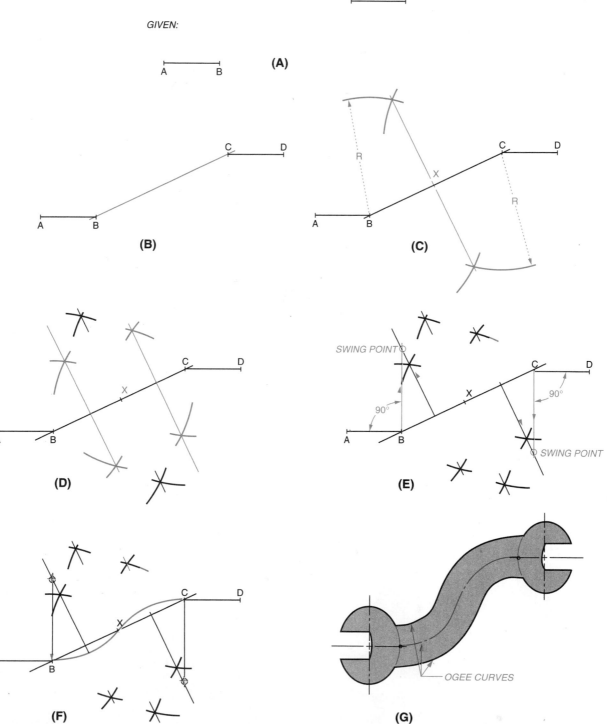

FIGURE 4-46 (A) How to draw an ogee curve; (B) Step 1; (C) Step 2; (D) Step 3; (E) Step 4; (F) Step 5; (G) Step 6

point C to intersect the perpendicular bisector of C-X, which locates the second required swing center, *Figure 4-46E.*

STEP 5 Place the compass point on the first swing point and adjust the compass lead to point B, and swing an arc from B to X. Place the compass point on the second swing point and swing an arc from X to C, *Figure 4-46F.* This completes the ogee curve.

NOTE: Point X is the tangent point between arcs. Check and, if correct, darken in all work.

An example of an ogee curve is illustrated in *Figure 4-46G.*

CONIC SECTIONS

A *conic section* is a section cut by a plane passing through a cone. These sections are bounded by various kinds of shapes. Depending upon where the section is cut, the shape can be a triangle, a circle, an ellipse, a parabola or a hyperbola, *Figures 4-47A* through *4-47E.*

TRIANGLE

Figure 4-47A. This shape results when a plane passes through the apex of the cone.

CIRCLE

Figure 4-47B. This shape results when a plane passes through the cone, parallel with the base and perpendicular to the axis.

ELLIPSE

Figure 4-47C. This shape results when a plane passes through the cone inclined to the axis.

PARABOLA

Figure 4-47D. This shape results when a plane passes through the cone parallel to one element.

HYPERBOLA

Figure 4-47E. This shape results when a plane passes through the cone parallel with the axis of the cone.

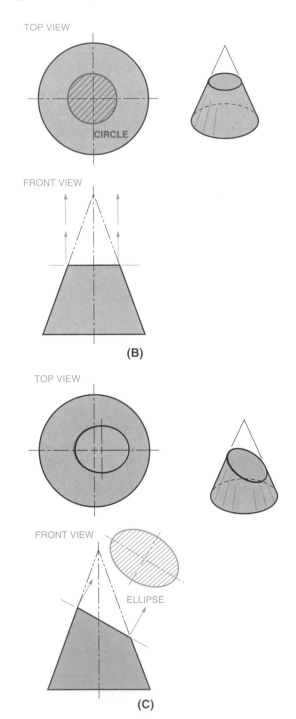

(B)

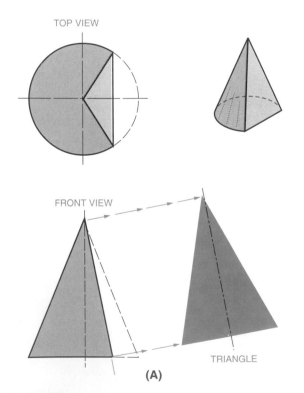

(A)

(C)

FIGURE 4-47 (A) Triangle section; (B) Circle section; (C) Ellipse section

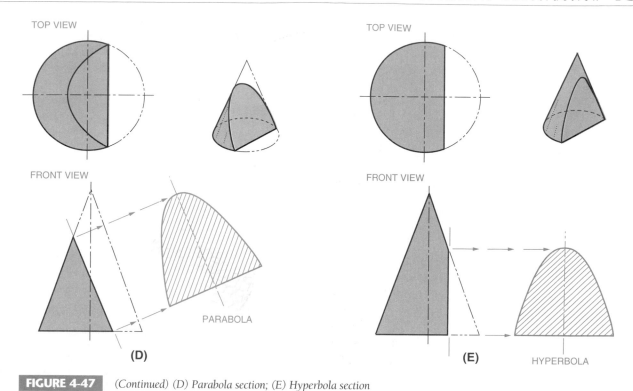

TOP VIEW

FRONT VIEW

PARABOLA

(D)

TOP VIEW

FRONT VIEW

HYPERBOLA

(E)

FIGURE 4-47 *(Continued) (D) Parabola section; (E) Hyperbola section*

HOW TO DRAW AN ELLIPSE, CONCENTRIC CIRCLE METHOD

GIVEN: The location of the center point, the major diameter, and the minor diameter, *Figure 4-48A.*

STEP 1 Lightly draw one circle equal to the major diameter and another circle equal to the minor diameter, *Figure 4-48B.*

STEP 2 Divide both circles into 12 equal divisions by passing lines through the center at every 30°. Number points 1 through 5 in clockwise consecutive order on the major diameter, positioning the 3 location on the right horizontal axis. Number points 7 through 11 in clockwise consecutive order, positioning the 9 location on the left horizontal axis. Point 6 is on the lower vertical axis at the minor diameter, and point 12 is on the upper vertical axis at the minor diameter, *Figure 4-48C.* If the ellipse were to be positioned at 90° from this example, the point locations would be rotated accordingly.

STEP 3 Project down from point 1 on the major diameter, and to the right from point 1 on the minor diameter. Where these lines intersect is the exact location where point 1 will be on the ellipse, *Figure 4-48D.*

STEP 4 Project down from point 2 on the major diameter, and to the right from point 2 on the minor diameter. Where these lines intersect is the exact location where point 2 will be on the ellipse, *Figure 4-48E.*

STEP 5 This completes the first quadrant of the ellipse. Continue around the circle to locate points 4, 5, 6, 7, 8, 9, 10, and 11 in the same manner except in reverse, *Figure 4-48F.*

STEP 6 Lightly draw straight lines connecting points 1 through 12 in order to get a general idea of the ellipse outline, *Figure 4-48G.*

STEP 7 Darken in the ellipse using an irregular curve. Carefully connect all the points with a smooth, continuous line of the correct thickness, *Figure 4-48H.* It is sometimes helpful to divide into 10° or 15° spaces the two ends where the ellipse curves the fastest in order to have more points around the extreme ends. In this example, this is done between points 2–4 and 8–10.

Examples of ellipses are illustrated in *Figure 4-48,* showing how a rotated circle would look at 0°, 15°, 45°, 60°, and 90°.

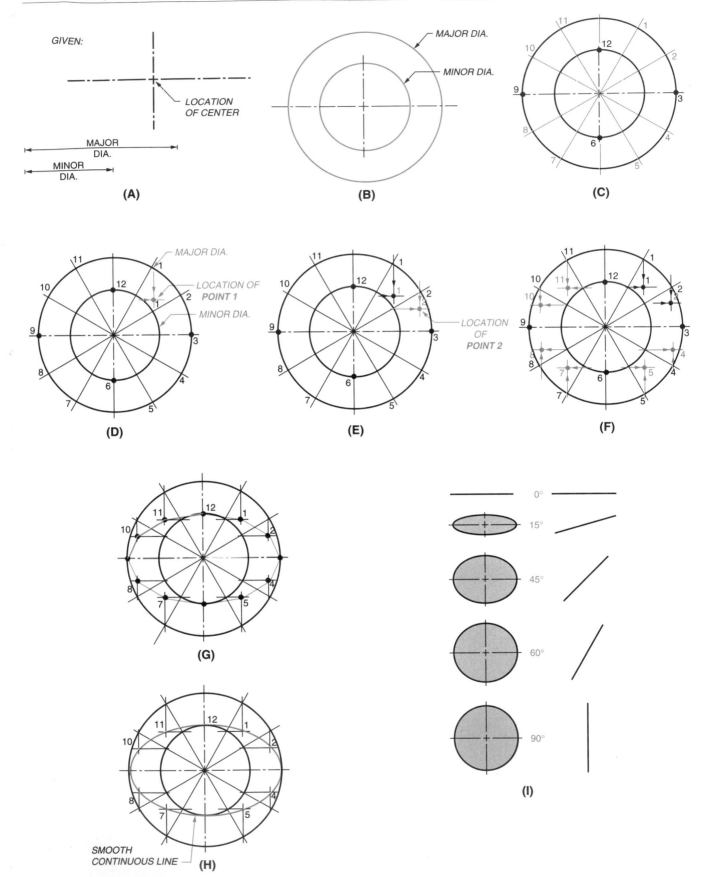

FIGURE 4-48 (A) How to draw an ellipse (concentric method); (B) Step 1; (C) Step 2; (D) Step 3; (E) Step 4; (F) Step 5; (G) Step 6; (H) Step 7; (I) Example of ellipses at various degrees

HOW TO DRAW A TRAMMEL ELLIPSE

GIVEN: A major or minor axis of a required ellipse, and any located point through which the ellipse must pass, *Figure 4-49A*. If a major axis is given, mark the end points A and B. If the minor axis is given, as is shown in the figure, mark the end points C and D. Mark as P the known location of where the ellipse must pass.

STEP 1 Draw a perpendicular bisector of the given axis to locate the unknown axis position, crossing the known axis at a point that will become 0, *Figure 4-49B*.

STEP 2 Mark an end corner of a separate strip of paper as point 0, and mark along the edge of the strip one-half of the distance A-B or C-D, whichever is known. Mark this point on the strip as A′ or C′ to correspond to the known axis, *Figure 4-49C*. Position the strip so that point 0 lies on the known point through which the ellipse must pass. Position the other strip mark on the axis that is not represented by the half-length on the strip, at the position (of two possible positions) that is nearest to the ellipse center.

STEP 3 With the strip positioned as specified in Step 2, mark the strip at the location of where it crosses the other axis. The length of end point 0 to this new position represents one-half of the unknown axis length, and it should correspondingly be labeled as A′ or C′, *Figure 4-49D*.

STEP 4 Using the two marked strip locations, reposition the strip so that the mark representing one-half of the major axis always lies on the minor axis, and the mark representing one-half of the minor axis always lies on the major axis. Repeat as needed to establish enough points for constructing the ellipse, *Figure 4-49E*.

STEP 5 Using a French curve, pass a smooth line through each of these established points, *Figure 4-49F*.

HOW TO DRAW A FOCI ELLIPSE

GIVEN: Major and minor axes of a required ellipse. The end points of the major axis are labeled A and B, the end points of the minor axis are labeled C and D, and the ellipse center is labeled 0, *Figure 4-50A*.

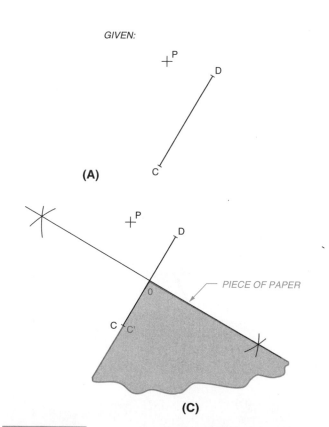

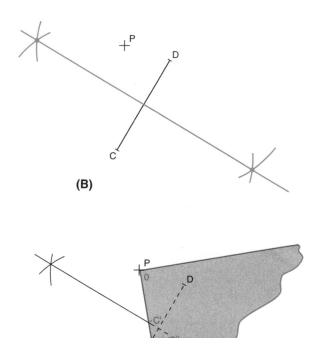

FIGURE 4-49   *(A) Trammel ellipse; (B) Step 1; (C) Step 2; (D) Step 3*

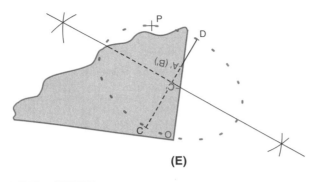

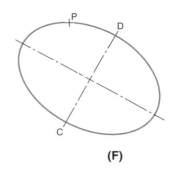

FIGURE 4-49 *(Continued)(E) Step 4; (F) Step 5*

STEP 1 Swing a radius of one-half of the major axis (0A) from the end point of the minor axis (C or D) to locate the focal points E and F at the points where the arc crosses the major axis, *Figure 4-50B.*

STEP 2 Select a random array of points on the major axis between one focal point and the ellipse center (space those at the ends more closely). Number these in consecutive order, *Figure 4-50C.*

STEP 3 Swing an arc whose radius is the distance from point B to number 1. Swing this distance from focal point F. Intersect this arc with an arc whose radius is the distance from point A to the same numbered point, number 1 in this example, but whose center is at focal point E, *Figure 4-50D.*

STEP 4 Repeat Step 3, but reverse the focal point positions of the arc centers, *Figure 4-50E.*

STEP 5 Repeat Steps 3 and 4 for each point selected in Step 2, *Figure 4-50F.*

STEP 6 Connect the points into a smooth curve, using a French curve, *Figure 4-50G.*

HOW TO LOCATE THE MAJOR AND MINOR AXES OF A GIVEN ELLIPSE WITH A LOCATED CENTER

GIVEN: An ellipse, with center 0 located, *Figure 4-51A.*

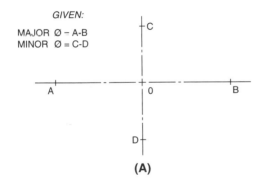

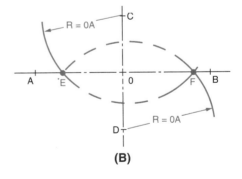

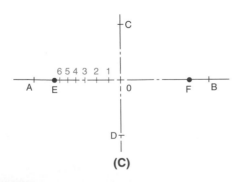

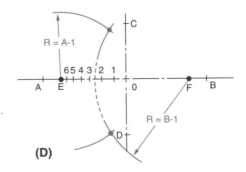

FIGURE 4-50 *(A) How to draw a foci ellipse; (B) Step 1; (C) Step 2; (D) Step 3*

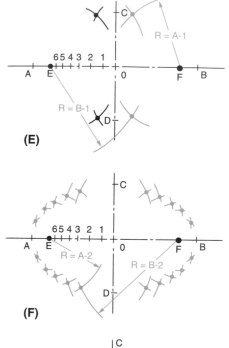

(E)

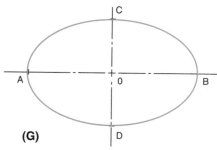

(F)

(G)

FIGURE 4-50 (Continued) (E) Step 4; (F) Step 5; (G) Step 6

STEP 1 From the ellipse center 0, draw a circle at a radius that allows intersecting the ellipse at any four locations. Label these points in successive order 1, 2, 3, and 4, moving clockwise around the ellipse center, *Figure 4-51B.*

STEP 2 Draw a rectangle by connecting each of the labeled points in successive order. As shown in *Figure 4-51C,* the major and minor axes are parallel to these sides, found by drawing perpendicular bisectors of two consecutive sides.

STEP 3 Draw the major and minor axes parallel to sides 1-2 and 2-3 through point 0, *Figure 4-51D.*

HOW TO DRAW A TANGENT TO AN ELLIPSE AT A POINT ON THE ELLIPSE

GIVEN: An ellipse, and a point P on its perimeter, *Figure 4-52A.* If not provided, locate the

GIVEN:

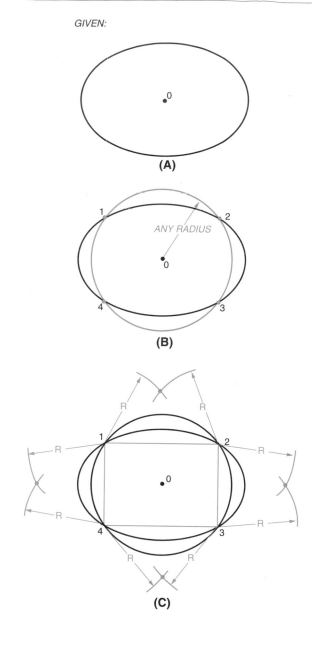

(A)

(B)

(C)

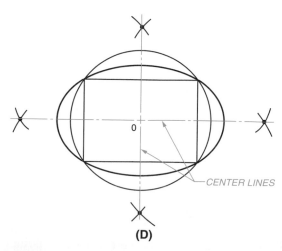

(D)

FIGURE 4-51 (A) How to locate the major and minor axes of a given ellipse; (B) Step 1; (C) Step 2; (D) Step 3

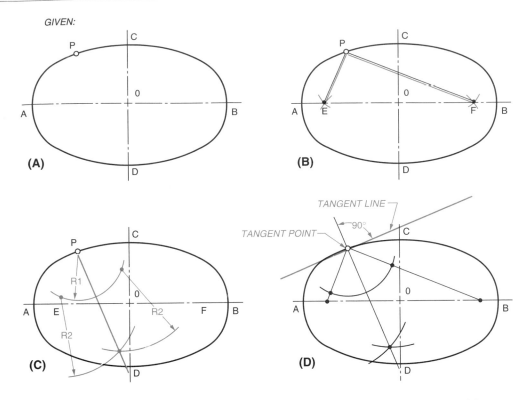

GIVEN:

(A)

(B)

TANGENT LINE

TANGENT POINT

90°

(C)

(D)

FIGURE 4-52 (A) How to draw a tangent to an ellipse at a point on the ellipse; (B) Step 1; (C) Step 2; (D) Step 3

major axis A-B and minor axis C-D and the focal points E and F.

STEP 1 Draw E-P and F-P, *Figure 4-52B*.

STEP 2 Bisect the angle EPF, *Figure 4-52C*.

STEP 3 Draw a perpendicular to the angle bisector of angle EPF at point P. This is the required tangent, *Figure 4-52D*.

HOW TO DRAW A TANGENT TO AN ELLIPSE FROM A DISTANT POINT

GIVEN: An ellipse and a distant point P, *Figure 4-53A*.

STEP 1 If not provided, locate the major axis A-B and focal points E and F, *Figure 4-53B*.

STEP 2 Swing an arc R1 from point P to pass through either focal point. F is selected for this example, *Figure 4-53C*.

STEP 3 Swing an arc whose radius is equal to the major axis from the other focal point (E in this example). Intersect the arc from Step 2 at points R and S, *Figure 4-53D*.

STEP 4 Extend the lines from E to each of points R and S, to cross the ellipse at U and V, *Figure 4-53E*.

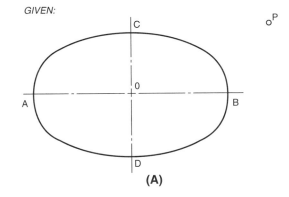

GIVEN:

(A)

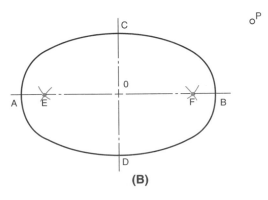

(B)

FIGURE 4-53 (A) How to draw a tangent to an ellipse from a distant point; (B) Step 1

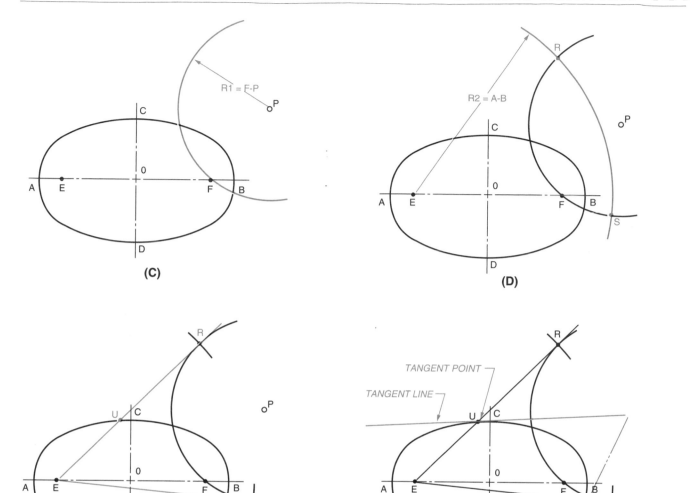

FIGURE 4-53   *(Continued) (C) Step 2; (D) Step 3; (E) Step 4; (F) Step 5*

STEP 5 The two possible points of tangency of the two possible lines from point P are U and V. Select and draw either, or both, *Figure 4-53F.*

HOW TO DRAW A PARABOLA (METHOD 1)

GIVEN: The front view and plan view of cone ABC with cutting plane X-Y, *Figure 4-54A.*

STEP 1 Locate points Y′ and Y″ in the plan view. In the front view, place the point of the compass on point Y and position the lead on point X. Swing point X to the extended baseline and down into the plan view, as illustrated in *Figure 4-54B.*

STEP 2 Locate a point along line X-Y (in this example, point D) and, using point Y as a swing point, swing point D to the extended baseline and down into the plan view. In the front view, reset the compass to a distance equal to the horizontal distance from the cone axis to the outer edge (identified as length 1), illustrated in *Figure 4-54C.* Transfer this distance to the plan view. Project point D down into the plan view and, where point D intersects the arc, locate points D′ and D″. Project these points to the right, intersecting with point D from above at the actual intersection points D′ and D″.

STEP 3 Choose other points along line X-Y in the front view, and project them over and down as described in Step 2, *Figure 4-54D.*

STEP 4 Using an irregular curve, connect all points. This completes the parabola, *Figure 4-54E.*

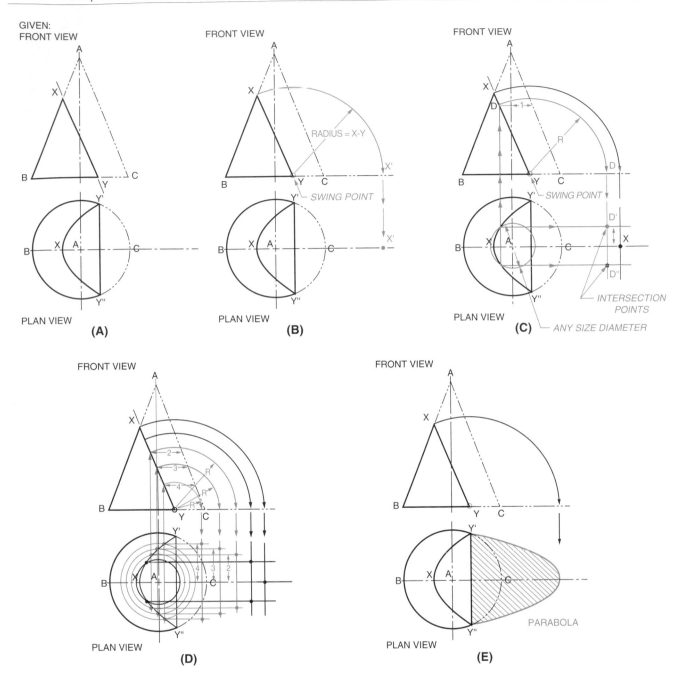

FIGURE 4-54 *(A) How to draw a parabola (method 1); (B) Step 1; (C) Step 2; (D) Step 3; (E) Step 4*

HOW TO DRAW A PARABOLA (METHOD 2)

GIVEN: Required rise A-B, and required span A-C.
Problem: Construct a parabola within the required rise and span, *Figure 4-55A*.

STEP 1 Divide half the span distance A-0 into any number of equal parts. (In this example, 4 equal parts are used.) See *Figure 4-55B*. Divide half the rise A-B into equal parts amounting to the square of the equal parts in Step 1 (in this example, $4^2 = 16$ equal parts).

STEP 2 From line A-0, each point on the parabola is offset by a number of units equal to the square of the numbers of units from point 0, *Figure 4-55C*. For example, point 1 projects 1 unit; point 2 projects 4 units; point 3 projects 9 units, and so forth.

STEP 3 Using an irregular curve, connect the points to form a smooth parabolic curve, *Figure 4-55D*.

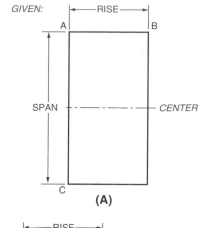

(A)

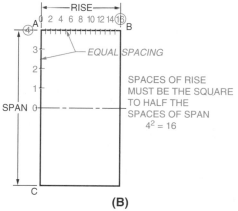

(B)

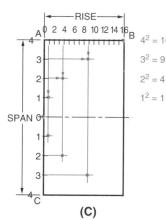

(C)

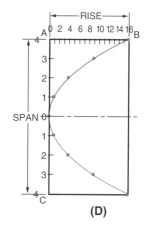

(D)

FIGURE 4-55 (A) How to draw a parabola (method 2); (B) Step 1; (C) Step 2; (D) Step 3

HOW TO FIND THE FOCUS POINT OF A PARABOLA

GIVEN: A parabolic curve with points A, 0, and B, *Figure 4-56A. Problem:* Find the focus point of the parabola A0B.

STEP 1 Draw a line from point A to point B. Continue the center line to a point equal to length X, distance Y, to find point C, *Figure 4-56B.*

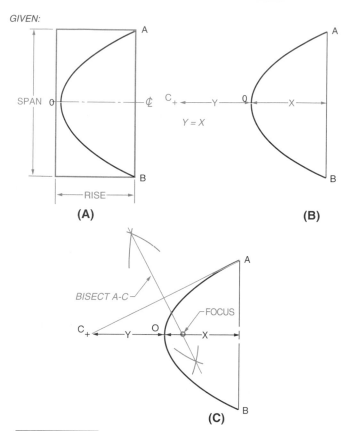

FIGURE 4-56 (A) How to find the focus point of a parabola; (B) Step 1; (C) Step 2

STEP 2 Draw a line from point C to point A and bisect it. The intersection of the bisect line and the axis is the focus of the parabola, *Figure 4-56C.*

HOW TO DRAW A HYPERBOLA (METHOD 1)

GIVEN: The front view and plan view of cone ABC with cutting plane X-Y, *Figure 4-57A.*

STEP 1 Locate points Y′ and Y″ in the plan view. In the front view, place the point of the compass on point Y and set the lead to point X. Swing point X to the extended baseline and down to the plan view, as illustrated in *Figure 4-57B.*

STEP 2 Locate a point along line X-Y (in this example, point D). Using Y as a swing point, swing point D to the extended baseline and down into the plan view. In the front view, reset the compass to a distance equal to the horizontal distance from the cone axis to the outer edge (length 1), as illustrated in *Figure 4-57C.* Transfer the arc to the plan view; where the

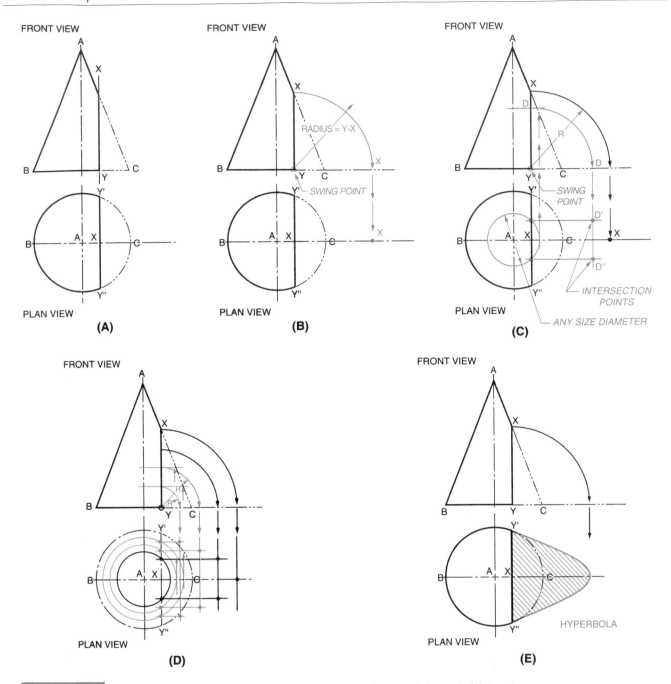

FIGURE 4-57 (A) How to draw a hyperbola (method 1); (B) Step 1; (C) Step 2; (D) Step 3; (E) Step 4

arc intersects with the cutting plane line X-Y, project to the right, and where these points intersect with point D from above is the actual location of points D′ and D″.

STEP 3 Choose other points along line X-Y in the front view and project them over and down as described in Step 2, *Figure 4-57D.*

STEP 4 Using an irregular curve, connect all points. This completes the hyperbola, *Figure 4-57E.*

HOW TO DRAW A HYPERBOLA (METHOD 2)

GIVEN: Coordinates, a square whose sides equal the transverse axis, and points A and B, *Figure 4-58A.*

STEP 1 With the center at the intersection of the diagonal lines from opposite corners of the square, place the compass lead at the corner of the given square and swing arcs to intersect the horizontal line through the center. This

GIVEN:
SQUARE EQUAL TO THE TRANSVERSE AXIS WITH
POINTS A-B AS SHOWN

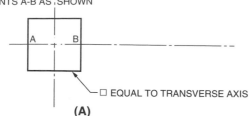

□ EQUAL TO TRANSVERSE AXIS

(A)

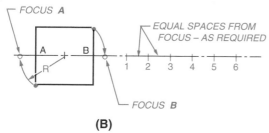

FOCUS A

EQUAL SPACES FROM
FOCUS – AS REQUIRED

FOCUS B

(B)

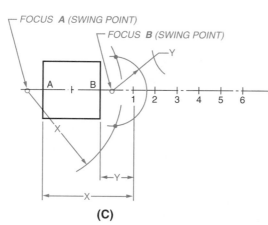

FOCUS A (SWING POINT)

FOCUS B (SWING POINT)

(C)

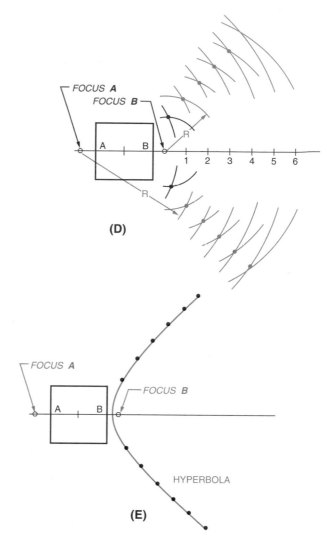

FOCUS A
FOCUS B

(D)

FOCUS A

FOCUS B

HYPERBOLA

(E)

FIGURE 4-58 *(A) How to draw a hyperbola (method 2); (B) Step 1; (C) Step 2; (D) Step 3; (E) Step 4*

locates focus A and focus B. Progressing outward along the horizontal line, mark off equal spaces of arbitrary length from the focus points, *Figure 4-58B.*

STEP 2 Set the compass at dimension X (A-1) and swing an arc from focus A. Set the compass at dimension Y (B-1) and swing an arc from focus B, *Figure 4-58C.*

STEP 3 Follow the same procedure for as many points as are required (in this example, six points). Set the compass at distance A-2 and swing from focus A. Set the compass at distance B-2 and swing from focus B, and so on, *Figure 4-58D.*

STEP 4 Using an irregular curve, carefully complete the hyperbola curve, *Figure 4-58E.*

HOW TO JOIN TWO POINTS BY A PARABOLIC CURVE

GIVEN: Points X and Y with 0 an assumed point of tangency, *Figures 4-59A, 4-59B, and 4-59C.*

STEP 1 Divide line X-0 into an equal number of parts; divide 0-Y into the same number of equal parts, *Figures 4-59D, 4-59E, and 4-59F.*

STEP 2 Connect corresponding points, *Figures 4-59G, 4-59H, and 4-59I.* Using an irregular curve, draw the parabolic curve as a smooth, flowing curve.

Supplementary Construction

HOW TO DRAW A SPIRAL

GIVEN: Crossed axes, *Figure 4-60A.*

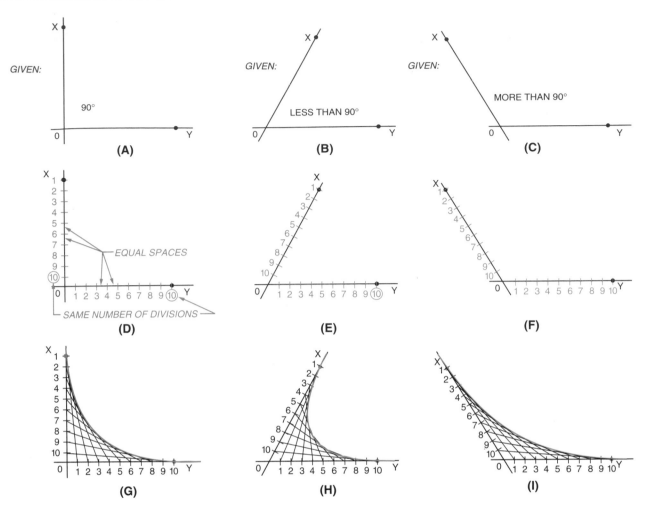

FIGURE 4-59 *(A) How to join two points by a parabolic curve (given: 90° angle); (B) Given: Less than 90° angle; (C) Given: More than 90° angle; (D) Step 1; (E) Step 1; (F) Step 1; (G) Step 2; (H) Step 2; (I) Step 2*

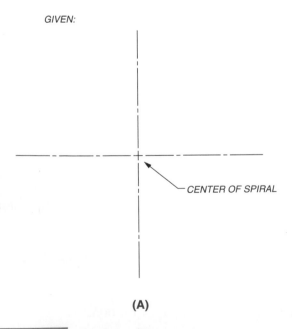

(A)

FIGURE 4-60 *(A) How to draw a spiral*

STEP 1 Divide the circle into equal angles (in this example, 30°). Set the compass at a required radius and draw various diameters to suit, evenly spaced, as illustrated in *Figure 4-60B*.

STEP 2 Starting any place along the angles, step over one angle and up one diameter until the required spiral is completed, *Figure 4-60C*.

HOW TO DRAW A HELIX

A *helix*, a form of spiral, is used in screw threads, worm gears, and spiral stairways, to mention but a few uses. A helix is generated by moving a point around and along the surface of a cylinder with uniform angular velocity about its axis. A *cylindrical helix* is simply known as a helix, and the distance measured parallel to the axis traversed by a point in one revolution is called the *lead*.

GIVEN: The top and front views of a cylinder with a required lead, *Figure 4-61A*.

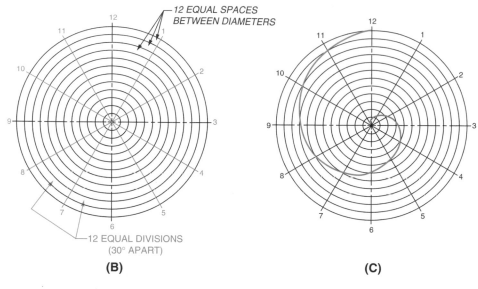

(B)   **(C)**

FIGURE 4-60   *(Continued) (B) Step 1; (C) Step 2*

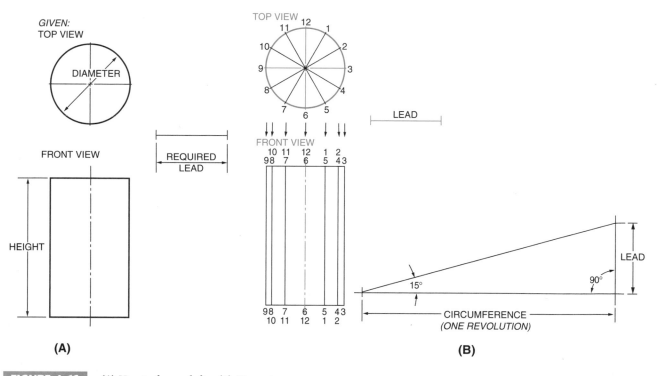

(A)   **(B)**

FIGURE 4-61   *(A) How to draw a helix; (B) Given views*

STEP 1   Divide the top view into an equal number of spaces (12 in this example). Draw a line equal to one revolution and/or circumference, project the lead distance from the end, and label all points per *Figure 4-61B*.

STEP 2   Divide the circumference into the same amount of equal spaces, and project each from the inclined line to the corresponding point projected from the front view, *Figure 4-61C*.

STEP 3   Connect all points of the helix, and draw as hidden lines the lines that would disappear from view on an actual helix form, *Figure 4-61D*.

Notice that a right-hand helix is illustrated. To construct a left-hand helix, simply project all the points in the opposite direction, *Figure 4-61E*.

HOW TO DRAW AN INVOLUTE OF A LINE

The curved path of a point on a string as it unwinds from a line, a triangle, a square, or a circle is an *involute*.

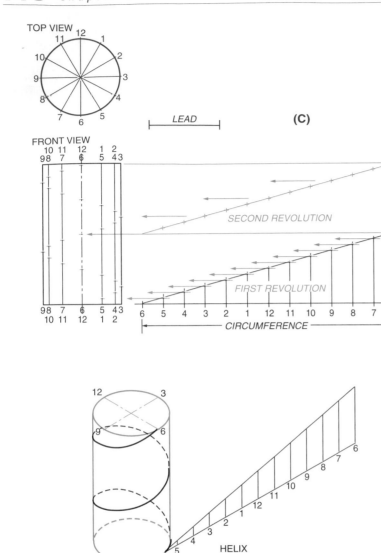

FIGURE 4-61 *(Continued)(C) Steps 1 and 2; (D) and (E)*

The involute is used to construct involute gears. The involute forms the face and part of the flank of the teeth of the gear.

GIVEN: Line A-B, *Figure 4-62A.*

STEP 1 Extend line A-B, as illustrated in *Figure 4-62B.* Set the compass at the length A-B, and, using point B as the swing point, swing semicircle A-C. With the compass set at length A-C, and, using point A as the swing point, swing semicircle A-D. Continue in this manner, alternating between points A and B until the required involute is completed, as shown in the figure.

HOW TO DRAW AN INVOLUTE OF A TRIANGLE

GIVEN: Triangle ABC, *Figure 4-63A.*

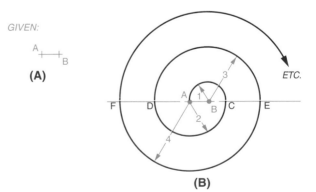

FIGURE 4-62 *(A) How to draw an involute of a line; (B) Step 1*

STEP 1 Extend straight lines from the triangle, as illustrated in *Figure 4-63B.* Set the compass at length A-C, and, using point A as the swing

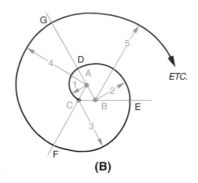

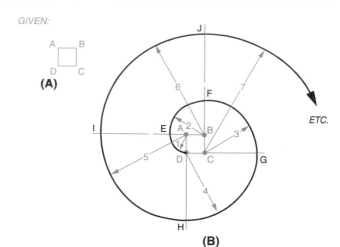

FIGURE 4-63 *(A) How to draw an involute of a triangle; (B) Step 1*

point, swing semicircle A-C. With the compass set at length B-D, and, using point B as the swing point, swing semicircle B-E. Continue similarly around the triangle until the required involute is completed, as shown in the figure.

FIGURE 4-64 *(A) How to draw an involute of a square; (B) Step 1*

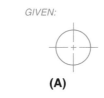

HOW TO DRAW AN INVOLUTE OF A SQUARE

GIVEN: Square ABCD, *Figure 4-64A*.

STEP 1 Extend straight lines from the square, as illustrated in *Figure 4-64B*. Set the compass at length A-B, and using point A as the swing point, swing semicircle A-E. With the compass set at length B-E, and, using point B as the swing point, swing semicircle B-F. Continue in the same manner around the square until the required involute is completed, as shown in the figure.

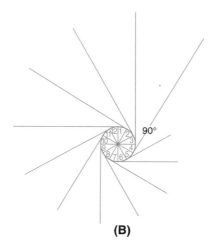

HOW TO DRAW AN INVOLUTE OF A CIRCLE

GIVEN: Circle A, *Figure 4-65A*.

STEP 1 Divide the circle into a number of equal parts and number each point clockwise around the circle (in this example, 12 equal parts). Project a line perpendicular to the radius at each point in a clockwise direction, *Figure 4-65B*.

STEP 2 Set the compass at length 1-2, and, using point 2 as the swing point, swing semicircle 1-2. With the compass set at length 1-2 plus 2-3, and, using point 3 as the swing point, swing semicircle 1-3. Continue in the same manner around the circle until the required involute is completed, as shown in *Figure 4-65C*.

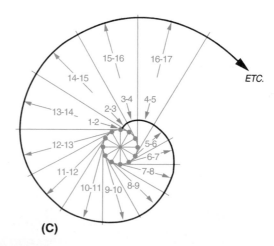

FIGURE 4-65 *(A) How to draw an involute of a circle; (B) Step 1; (C) Step 2*

HOW TO DRAW A CYCLOIDAL CURVE

GIVEN: A required span of a cycloid.

STEP 1 As the span represents the rolling distance of one revolution of a diameter, divide the span length by pi to find the required diameter. Divide the span into an equal number of divisions. Twelve is a convenient number, *Figure 4-66A.*

STEP 2 Draw the rolling diameter tangent to the given span at point 0/12. Divide this diameter into the same number of equal spaces as in the Step 1 span division. Consecutively number the radial end points of these divisions to make them meet their corresponding number of the span divisions as the diameter rolls along the span, *Figure 4-66B.*

STEP 3 At each span division draw the rolling diameter tangent to the span in the rolled position, with the division numbers of span and diameter matching at their contact point. Locate the point 0/12 position on the diameter at each of these locations, marked with a small cross, *Figure 4-66C.*

STEP 4 Connect all the point 1 diameter division positions with a smooth curve to complete the cycloidal curve, *Figure 4-66D.*

NOTE: If the given span is a concave arc, a hypocycloid will be drawn, *Figure 4-67.* If the span is convex, an epicycloid will be drawn, *Figure 4-68.*

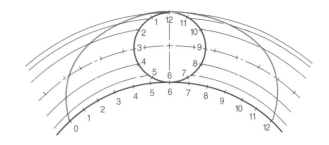

FIGURE 4-67 A hypocycloid

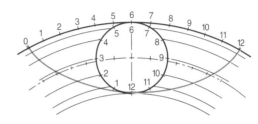

FIGURE 4-68 An epicycloid

Summary

• Geometric nomenclature primarily of the following concepts: points in space, line, angle, triangle, polygon, quadrilateral, circle, polyhedron, prism, pyramid, cone, and sphere.

• Elements and polygon construction principles consist primarily of how to do the following: bisect a line; bisect an angle; draw an arc or circle through three points; draw a line parallel to a straight line at a given distance; draw a line perpendicular to a line at a point; draw a line parallel to a curved line at a given distance; draw a per-

GIVEN:

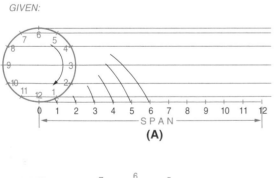

(A)

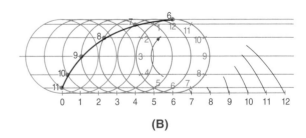

(B)

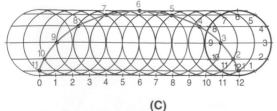

(C)

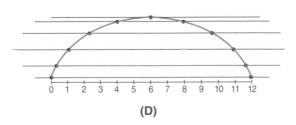

(D)

FIGURE 4-66 (A) How to draw a cycloidal curve Step 1; (B) Step 2; (C) Step 3; (D) Step 4

pendicular to a line from a point not on the line; divide a line into equal parts; divide a line into proportional parts; transfer an angle; transfer an odd shape; transfer complex shapes; proportionately enlarge or reduce a shape; draw a triangle with sides of known length; draw a square; draw a pentagon; and draw a hexagon.

- Circular construction consists primarily of knowing how to do the following: locate the center of a given circle; construct an arc tangent to a right angle, an acute angle, an obtuse angle, a straight line, a curve, and two radii or diameters; draw an ogee curve; draw an ellipse using the concentric circle method; draw a trammel ellipse; draw a foci ellipse; locate the major and minor axes of a given ellipse with a located center; draw a tangent to an ellipse at a point on the ellipse and from a distance point; draw a parabola; find the focus point of a parabola; draw a hyperbola; and join two points by a parabolic curve.

- Supplementary constructions require knowing how to do the following: draw a spiral; draw a helix; draw an involute of a line, triangle, square, and circle; and draw a cycloidal curve.

Review Questions

Answer the following questions either true or false.

1. A right angle is an angle of 90°.
2. An acute angle is an angle at more than 90°.
3. An obtuse angle is an angle at more than 90°.
4. There are six types of polygons.
5. Geometric construction procedures are not particularly important to the drafter.
6. Lines that are tangent and at 90° are considered perpendicular.

Answer the following questions by selecting the best answer.

1. Which of the following is not a part of a circle?
   a. Angle
   b. Center
   c. Sector
   d. Quadrant
2. Which of the following is true regarding a line?
   a. The shortest distance between two points
   b. Can be drawn in any direction
   c. Can be straight, an arc, a circle, or a free curve
   d. All of the above
3. Which of the following is true regarding the sum of three internal angles of a triangle?
   a. Always exactly 180 degrees
   b. Always half of the 360 degrees in a full circle

c. Equal to half the sum of the three straight sides
   d. Both a and b
4. The circumference of a circle can be calculated by:
   a. Multiplying pi by the diameter of the circle.
   b. Dividing the diameter of the circle by pi.
   c. Multiplying pi by 2 times the radius.
   d. Both a and c
5. Which of the following is not a kind of triangle?
   a. Equilateral
   b. Obtuse
   c. Quadrilateral
   d. Scalene
6. A point in space is represented on a drawing by:
   a. A crisscross at its exact location.
   b. A dot at its exact location.
   c. A crossbar on an existing line.
   d. Both a and c
7. Which of the following is **not** true regarding concentric and eccentric circles?
   a. Concentric circles have a common center point.
   b. Eccentric circles have no common center point.
   c. Concentric circles have no common center point.
   d. Two or more circles with a common center point are called concentric.
8. Which of the following is not true regarding an angle?
   a. There are three major types of angles.
   b. There are four major types of angles.
   c. An obtuse angle is one of the major types of angles.
   d. An angle is formed by two intersecting lines.

Chapter 4 Problems

CONVENTIONAL INSTRUCTIONS

The following problems are intended to give the beginning student practice in the many geometric construction techniques used to develop drawings. Accuracy, line work, neatness, speed, and centering are stressed. It is recommended that dimensions not be added to problems at this time. The beginning student should practice using drafting instruments and correct line thicknesses, and should concentrate on developing good drafting habits.

The steps to follow in laying out all drawings throughout this book are:

STEP 1 All geometric construction work should be made very accurately using a sharp 4H lead.

STEP 2 All geometric construction work must be laid out very lightly.

STEP 3 Do **not** erase construction lines. If constructed lightly, they will not be seen on the whiteprint copy.

STEP 4 Make a rough sketch of each problem before beginning to calculate the overall shape.

STEP 5 Lightly draw each problem completely first.

STEP 6 Locate all tangent points as you proceed. Make short, light dashes at each tangent point.

STEP 7 Try to center the problem in the work area.

STEP 8 Check each dimension for accuracy.

STEP 9 Darken in the drawing using the correct line thickness and the following steps:

- Locate and draw all center lines with a thin black line.

- Darken in all diameters using a compass.

- Darken in all radii using either a compass or a circle template.

- Darken in all horizontal lines, either from right to left or from left to right.

- Recheck all dimensions.

- Check all lines for correct thickness.

- Fill out the title block using light guidelines and neat lettering.

CAD INSTRUCTIONS

Students who plan to complete the following problems on a CAD system should begin with these general instructions. Before reading the specific instructions for each activity, go through each step in the following planning checklist. The checklist applies to any CAD system and will help ensure the optimum use of your time and resources.

1. Analyze the problem carefully. Decide exactly what you are being asked to do.

2. Determine what resources and references (if any) you will need in order to complete the problem. Collect those references and resources and have them readily available.

3. Decide if any particular standards apply to the project, and have those standards available.

4. Determine what types of views will be required and how many of each.

5. Determine what the final plotted scale of the drawing will need to be and select the appropriate paper size for plotting/printing.

6. Plan your drawing sequence. In what order will you develop the drawing (i.e., lines, features, dimension lines, leaders, dimensions)?

7. Review the various CAD commands you will have to use in order to develop the drawing.

8. Examine your CAD system to ensure that everything is in working order, then begin the project.

Problem 4-1

Triangular spaces are utilized for rigidity in frames and trusses. The scissors truss shown is typical of trusses premade for house construction. Using the dimensions and slopes given, lay out an approximate truss and determine approximate lengths of the 2 × 4s. Assume that a lumber store near you has 2 × 4s in stock in 8-, 10-, 12-, and 16-foot lengths. How many and what lengths of 2 × 4s are needed? (Note that the actual dimensions of standard 2 × 4s are 1-1/2" × 3-1/2".) Approximately what percentage of the lumber is wasted?

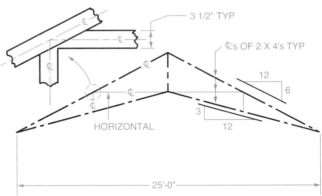

NOTE: SPECIAL GUSSET PLATES (NOT SHOWN) ARE USED AS JOINTS

Problem 4-2

The given drawing shows preliminary measurements to determine the distance across the river from line AB to the outcropping of rock on the other side. Using the information given, calculate the shortest distance from the 1,000-foot-long line to the rocks.

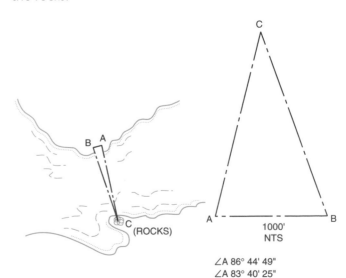

∠A 86° 44' 49"
∠A 83° 40' 25"

PROBLEM 4-3

The following X,Y coordinates in feet for determining the border of an oddly shaped parcel of property are as follows: A(0,0), B(110,–30), C(190,0), D(130,40), E(130,110), F(0,60). You are given the task of determining the area of the parcel. Suggestion: Lay out the parcel to scale and divide the parcel into convenient areas for calculating.

Problem 4-4

The rectangular frame shown needs a diagonal to ensure the frame's rigidity. The frame is made up of 3 × 3 × 1/4 structural steel angles and 1/4" steel gusset plates. Thirty-five frames are to be made. Your task is to lay out the frame to the dimensions given and determine the length of the diagonal based on information given in the drawing. Then assume that the standard length of available structural angles is 20' and determine how many 20' lengths are required for the 35 frames. What percentage of the structural angles will probably be wasted?

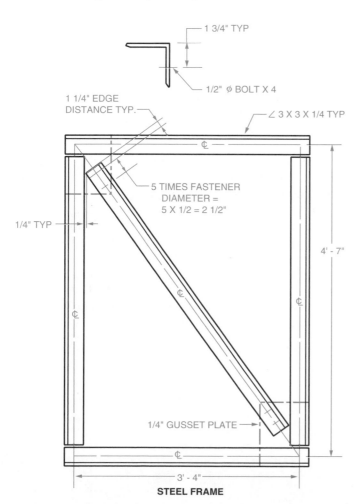

STEEL FRAME

Problem 4-5

A ski slope designer made the sketch shown, which shows contour lines for a particular portion of a ski run. The numbers on the lines are elevations in feet, and A is 100' from B in a horizontal direction. You are requested to determine the angle and percentage slope of the terrain between A and B.

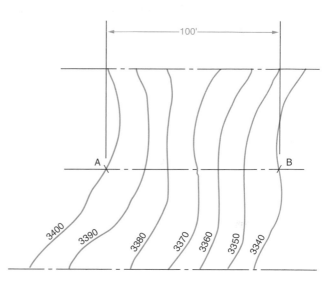

Problem 4-6

Calculate the volume of the notched tubular object shown by utilizing equations for the areas of segments of a circle. Then calculate the weight of the object by multiplying the volume by the density of the material.

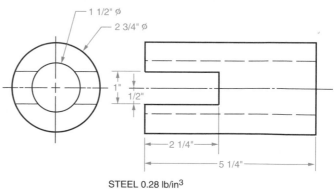

STEEL 0.28 lb/in³

PROBLEM 4-7

Construct an ellipse using the concentric circle method shown in Figure 4-48. Minor Diameter = 25 mm. Major Diameter = 40 mm.

PROBLEM 4-8

Divide the work area into four equal spaces. In the upper left-hand space, draw an inclined line 3.25 long. Bisect this line. Show all construction lines lightly.

In the upper right-hand space, draw an acute angle with intersecting lines, each approximately 2.75 long. Bisect this angle. Show all light construction lines.

In the lower left-hand space, draw an inclined line 2.88 long. Divide this line into five spaces. Show all light construction lines.

In the lower right-hand space, draw an inclined line 3.125 long. Divide this line into three proportional parts 2, 3, and 4. Show all light construction lines.

PROBLEM 4-9

Divide the work area into four equal spaces. In the center of the upper left-hand space, draw an equilateral triangle having sides of 2.0. Bisect the interior angles. Show all light construction lines.

In the center of the upper right-hand space, draw a square having sides of 2.0. Show all light construction lines.

In the center of the lower left-hand space, draw a hexagon having the distance of 2.0 across the flats. Show all light construction lines.

In the center of the lower right-hand space, draw an octagon having the distance of 2.0 across the flats. Show all light construction lines.

PROBLEM 4-10

Divide the work area into four equal spaces. In the upper left-hand space, draw two lines intersecting at 60° and draw an arc with a .88 radius tangent to the two lines. Show all light construction lines.

In the upper right-hand space, draw two lines intersecting at 120° and draw an arc with a .625 radius tangent to the two lines. Show all light construction lines.

In the lower left-hand space, draw two lines intersecting at 90° and draw an arc with 1.25 radius tangent to the two lines. Show all light construction lines.

In the lower right-hand space, draw an ellipse with a major diameter of 3.00 and a minor diameter of 1.75.

PROBLEM 4-11

Divide the work area into four equal spaces. In the upper left-hand space, center a line 10 mm long and label it line A-B. Construct a straight line involute with five arcs.

In the upper right-hand space, center a triangle with sides equal to .25 and label the triangle ABC. Construct a triangular involute with five arcs.

In the lower left-hand space, center a square with sides equal to 10 mm and label the square ABCD. Construct a square involute with five arcs.

In the lower right-hand space, center a circle with a diameter of 1.00. Construct a circle involute with as many arcs as space will allow.

PROBLEM 4-12

Divide the work area into four equal spaces. In the center of the upper left-hand space draw a spiral with 30° spaces, at 4 mm increments, for one 360° rotation.
In the center of the upper right-hand space, draw two lines at right angles intersecting a point 0. One line is to be 3.5 long, and the other 2.5 long. Draw a parabolic curve between these two lines.

In the center of the lower left-hand space, draw a rectangle 2.0 × 4.50 in size. Label the 2.0 side as the rise, and the 4.50 as the span. Construct a parabola (Method 2), and locate the focus point.

In the center of the lower right-hand space, draw a line 80 mm long, and label its end points A and B. Locate a point on line A-B at 5/8 of the line's length from point A. Label the point X.

PROBLEMS 4-13 THROUGH 4-26

Using the art for Problems 4-13 through 4-26, center each object within the work area using correct line thickness.

Do not erase light construction lines.

Locate and draw a light, short dash at all tangent points.

Do not dimension objects.

Problem 4-13

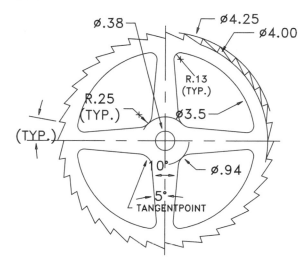

Problem 4-14

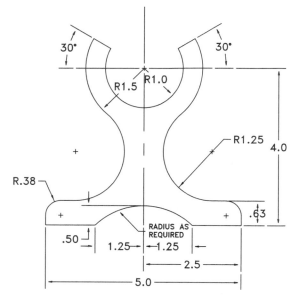

Problem 4-15

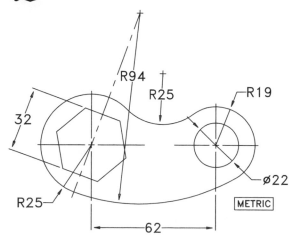

Problem 4-16

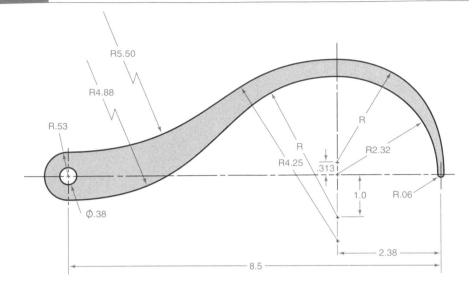

Problem 4-17

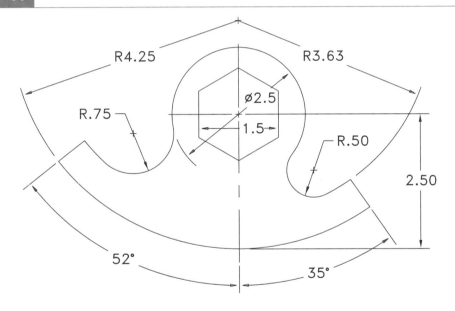

Problem 4-18

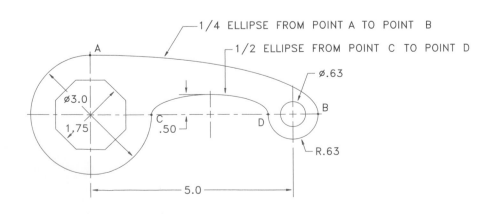

Problem 4-19

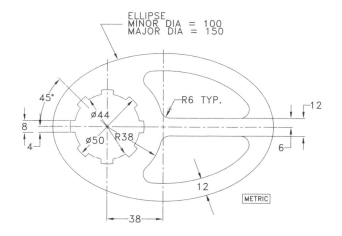

ELLIPSE
MINOR DIA = 100
MAJOR DIA = 150

R6 TYP.

45°

ø44

ø50 R38

8
4

12

6

12

38

METRIC

Problem 4-21

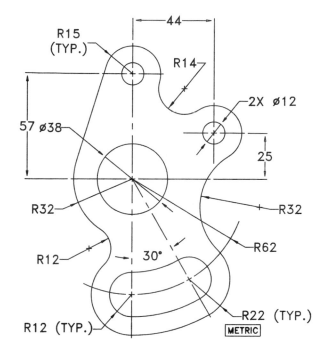

44

R15 (TYP.)

R14

2X ø12

57 ø38

25

R32

R32

R62

R12

30°

R22 (TYP.)

R12 (TYP.)

METRIC

Problem 4-20

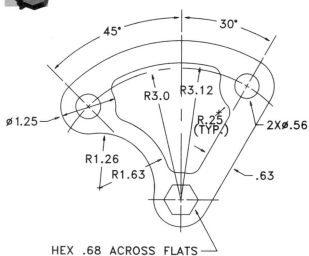

45°

30°

R3.0 R3.12

R.25 (TYP.)

ø1.25

2Xø.56

R1.26

R1.63

.63

HEX .68 ACROSS FLATS

Problem 4-22

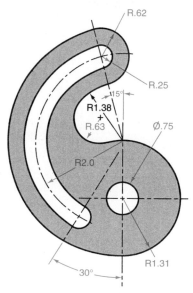

R.62

R.25

15°

R1.38

R.63

Ø.75

R2.0

30°

R1.31

Problem 4-23

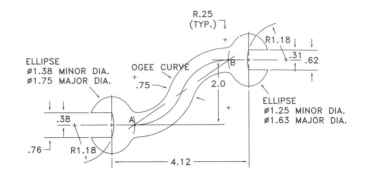

R.25
(TYP.)

R1.18

.31 .62

ELLIPSE
Ø1.38 MINOR DIA.
Ø1.75 MAJOR DIA.

OGEE CURVE

B

.75

2.0

ELLIPSE
Ø1.25 MINOR DIA.
Ø1.63 MAJOR DIA.

.38

A

.76

R1.18

4.12

Problem 4-24

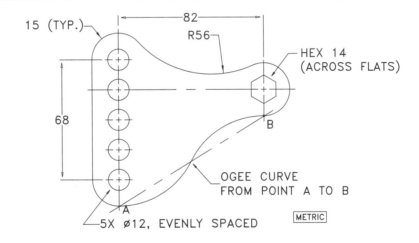

15 (TYP.)

82

R56

HEX 14
(ACROSS FLATS)

68

B

OGEE CURVE
FROM POINT A TO B

A

5X Ø12, EVENLY SPACED

METRIC

Problem 4-25

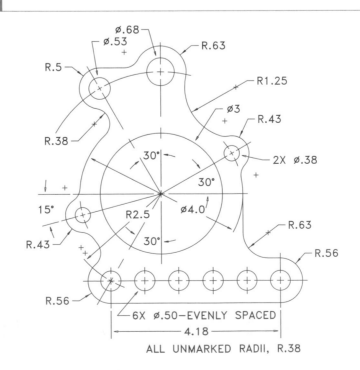

Ø.68
Ø.53

R.63

R.5

R1.25

Ø3

R.43

R.38

2X Ø.38

30°

30°

15°

R2.5

Ø4.0

R.63

R.43

30°

R.56

R.56

6X Ø.50—EVENLY SPACED

4.18

ALL UNMARKED RADII, R.38

Problem 4-26

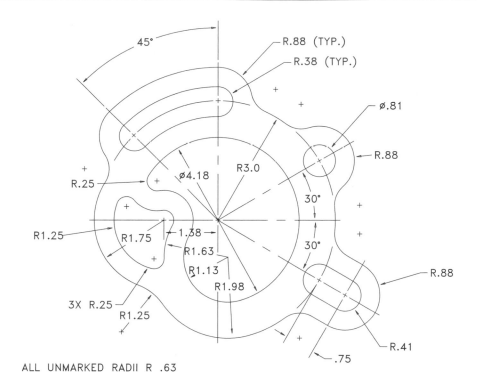

45°

R.88 (TYP.)

R.38 (TYP.)

ø.81

R.88

ø4.18

R3.0

30°

R.25

R1.25

R1.75

1.38

R1.63

30°

R1.13

R1.98

R.88

3X R.25

R1.25

R.41

.75

ALL UNMARKED RADII R .63

PROBLEMS 4-27 THROUGH 4-32

On an A-size sheet of vellum, carefully construct each of the assigned figures. Use thin, black center lines and thick, black object lines. Leave all light construction work. Do not add dimensions.

Problem 4-27

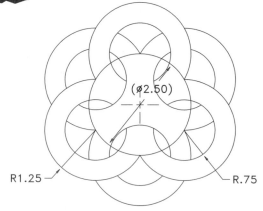

(ø2.50)

R1.25

R.75

Problem 4-28

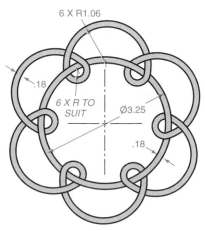

6 X R1.06

.18

6 X R TO SUIT

ø3.25

.18

Problem 4-29

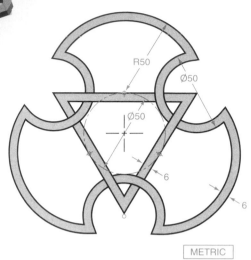

R50
Ø50
Ø50
6
6

METRIC

Problem 4-31

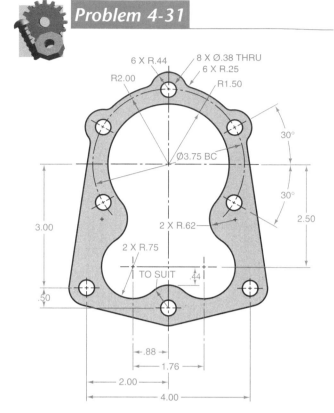

6 X R.44
8 X Ø.38 THRU
6 X R.25
R2.00
R1.50
30°
Ø3.75 BC
30°
2.50
2 X R.62
3.00
2 X R.75
TO SUIT | .44
.50
.88
1.76
2.00
4.00

Problem 4-30

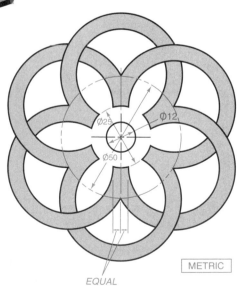

Ø25
Ø12
Ø50

METRIC

EQUAL

Problem 4-32

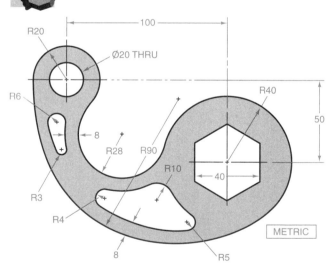

R20
100
Ø20 THRU
R6
R40
8
50
R28   R90
R3
R10
40
R4
8
R5

METRIC

PROBLEM 4-33

Design and lay out to scale, a clock face to match your interest(s) in technology.

PROBLEM 4-34

Find a logo for one of the following. Prepare a drawing of it.

- **a.** an automobile
- **b.** a computer company
- **c.** an airline
- **d.** a product of your choice
- **e.** an organization

PROBLEM 4-35

Refer to the figures indicated in this chapter and do the following:

- **a.** Figure 4-22—Draw a curve parallel to a curve similar to the lower edge of the object for Problem 4-9.
- **b.** Figure 4-30—Reproduce the drawing for Figure 4-30 and transfer it.
- **c.** Figure 4-46—Change the spacing of lines A-B and C-D and draw an ogee curve to join them.
- **d.** Figure 4-61—Draw the helix shown.
- **e.** Figure 4-65—Follow the steps and draw an involute curve.

NOTE: The involute curve is the basis for the profiles of spur gears and helical gears.

- **f.** Figure 4-68—Construct the epicycloid shown.

Problem 4-36

The golden section shown is attributed to the Greeks who used it as a guide to create pleasing proportions for architecture, sculpture, and pottery. Three nested golden sections can be formed by following the construction pattern shown. Note that the ratio of the sides of each golden rectangle is 1.618 to 1.

$(2.618)/(1.618) = 1.618$; and $(4.236)/(2.618) = 1.618$

(For interest, check the ratios of the sides of credit cards and some food packaging.) Refer to the following figure and reconstruct the three golden rectangles. Then lay out some shelves for a living room for books

and hi-fi equipment. Include vertical dividers that incorporate the golden section proportions.

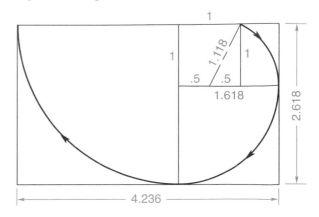

Problem 4-37

The freehand sketch given shows a format for a typical classroom engineering problem statement and solution. On a sheet of 8.5 × 11 (A-size) paper redo the statement (GIVEN and FIND) and the solution. Use a straightedge for the sketch and diagrams and good lettering style for the statement and solutions.

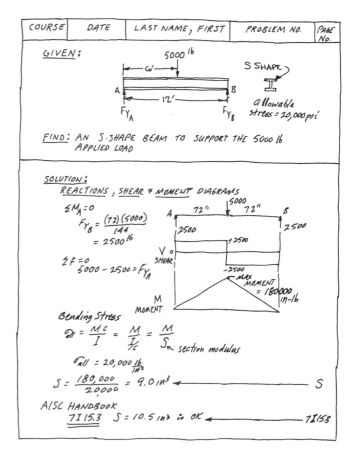

Problem 4-38

You have been asked to help an individual lay out a tentative plan for a service garage with parking facilities. The commercially zoned lot measures 25 m × 35 m. Sketch No. 1 shows a preferred arrangement of the service-pits building with parking. Lay out the lot using the given information. Use a scale to make the best use of an A-size drawing.

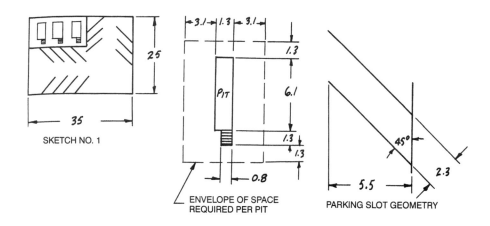

SKETCH NO. 1

ENVELOPE OF SPACE
REQUIRED PER PIT

PARKING SLOT GEOMETRY

SECTION 2

Technical Drawing Fundamentals

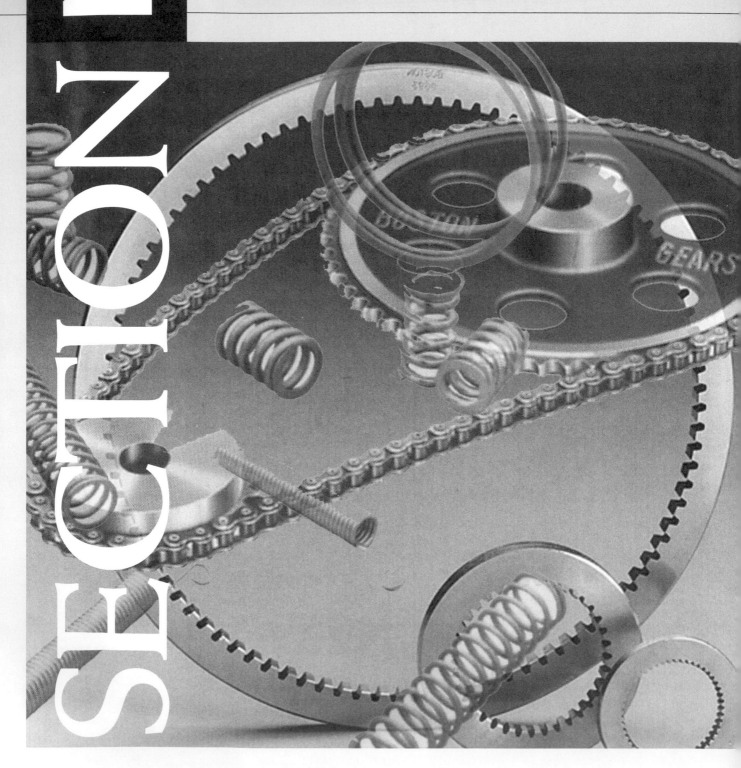

Career Profile:
Alexander Torres

Graduating with a master's degree in fine arts, Alexander Torres thought he might become a painter—but then reality (and bills!) set in. Having studied technical drawing in high school and then gone on to get a degree in engineering and drafting, he knew that he *did* have experience in a more lucrative field.

Using the skills learned in college and high school, he started doing illustrations and flowcharts for companies like Lockheed Martin. After a few years, he found his way into magazine work, which required both his drafting skills and his artistry.

Today, as a technical illustrator for TK magazine, he often has to create drawings that communicate the ideas explained in an article. The function of these schematics is part artistic (to illustrate the point: He recently used drafting software to draw a candy bar, to metaphorically show that when working with motherboards you need to break it into its components) and part functional (the art must often explain complex parts of computers in ways that mere text cannot).

On this day, he has just been handed an article about circuit boards. His first move in planning the drawing: Read the abstract. He will then go over the concept with the editors to find out what they are looking for—do they want a dynamic look (in which case he might present an orthographic projection) or true representation? If the latter, what focal point should he take? In drawing something like this, he tries to look at the real thing. If he doesn't have it accessible at home, he will search the Internet. These will help him to visualize the drawing that he is about to do. Then Torres does a pencil sketch, organizing the information so that it will work in a rectilinear format.

He will then plug in numbers into Infinity, a computer drafting system that allows him to build in 3D—a benefit because then he can move the point of view of the camera as needed. Sometimes he may decide that a fisheye view will help the object jump off the page; other times he must stick with the realistic focal point. Either way, the artist in him tries to ensure that the final product will not be merely technical and accurate, but also an element that draws the reader into the magazine.

Chapter 5

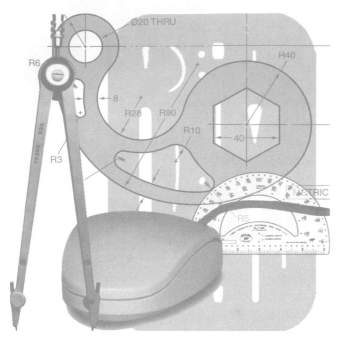

SPATIAL VISUALIZATION AND MULTIVIEW DRAWINGS

CHAPTER OBJECTIVES

Upon completion of this chapter, students should be able to do the following:

- Explain the concept of spatial visualization.
- Demonstrate proficiency in the use of spatial visualization.
- Demonstrate proficiency in the use of orthographic projection.
- Demonstrate proficiency in all five steps of planning a drawing.
- Demonstrate proficiency in centering a drawing.
- Demonstrate proficiency in making proper use of technical drawing conventions.
- Demonstrate proficiency in applying visualization techniques.
- Explain the difference between third- and first-angle projections.

KEY TERMS

Blind hole

Conventional break

Counterbored hole

Countersunk hole

Fillets

Fold line

Foreshortening

Front view

Hidden lines

Orthographic projection

Perspective projection

Principal viewing planes

Right-side view

Rounds

Runouts

Spatial visualization

Spotfaced hole

Tapered hole

Through hole

Top view

Drafters, designers, and engineers follow a design process to solve technical product problems and use technical drawings to document and develop their ideas. The steps in the process leading to the technical drawings, primarily working drawings, were noted in the general comments in the Introduction. In Chapter 2, the drawing tools used to prepare the drawings were discussed, followed by another chapter that emphasized the first format drafters, designers, and engineers use to record their ideas, sketching techniques. This chapter focuses on the ever-present challenge to a design person to prepare multiview drawings, to scale, of technical, 3D concepts on a 2D surface so that the users of the drawings will clearly understand, in 3D, what the design person had in mind, and have sufficient information to accurately produce the concept.

Providing accurate shape descriptions of objects by drawing methods requires that three-dimensional object information be presented in a flat, two-dimensional drawing space. This is usually done by drawing images of the object from multiple directions. Commonly shared methods and interpretations are essential for all who make or use such drawings. Technical drawings seldom use more than lines to outline an object's features. Visual qualities, such as color and texture, are more accurately specified by written requirements.

Viewing an object by eye creates a depth distortion. This phenomenon is recreated in a field of drawing called *perspective projection*, **Figure 5-1**. This distortion occurs because the visual rays used to create the image in the viewer's eye are not parallel. This distortion, while useful in providing the illusion of distance, results in a loss of image accuracy and fails to provide the information needed for most detailed technical communication. In order to develop multiview drawings, you must be proficient in the concept of spatial visualization.

Spatial Visualization

We see the world in three dimensions. When we look at an object, it has height, length, and depth. But when working in engineering or drafting, we sometimes have to communicate our ideas in a two-dimensional format. Making two-dimensional representations of three-dimensional objects or looking at a two-dimensional representation and visualizing the object in three dimensions requires a skill known as *spatial visualization*.

Spatial visualization is the act of (1) forming an accurate mental picture of how a three-dimensional object will look when represented in a two-dimensional format or (2) forming an accurate mental picture of how a three-dimensional object will look based on the information given in a two-dimensional representation. Some people are innately talented at spatial visualization and some are not. However, with practice most people can become proficient at spatial visualization.

TWO-DIMENSIONAL SPATIAL VISUALIZATION

To gain a better understanding of the concept of spatial visualization, try the following experiment:

1. Pick up your copy of this book and look at it as a three-dimensional object (see *Figure 5-2*).

2. Now orient the book so that your line of sight is perpendicular to the front cover. You should see no depth—just a flat surface. This is part of the concept of orthographic projection used when constructing multiple views of an object. Orthographic projection is covered later in this chapter.

3. Repeat Step 2 for all sides of the book. You should see the following sides: front, back, top, bottom, right side, and left side (see *Figure 5-3*).

4. At this point, do not concern yourself with the orientation of the six views of the book shown in Figure 5-3.

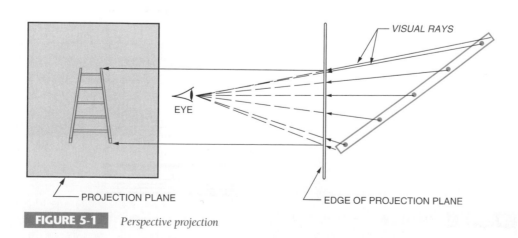

FIGURE 5-1 *Perspective projection*

VISUAL RAYS

EYE

PROJECTION PLANE

EDGE OF PROJECTION PLANE

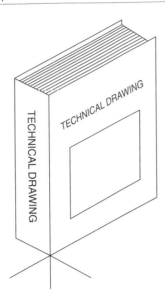

FIGURE 5-2   *Three-dimensional view of this book*

Orienting views is important, but it is covered later in this chapter. For now you need only concentrate on visualizing what the views should look like in a two-dimensional format, and you should understand that any three-dimensional object has these six views when applying the concept of orthographic projection. Even a plain globe has six views (all six would look the same).

Try this experiment with several other objects that are conveniently available, such as a chair, soda can, computer mouse, coin, or any other object that is handy. But this time, try to visualize each view without actually turning the object around and looking at it. Based on what you can see in three dimensions, what would the front, back, right side, left side, top, and bottom of the object look like in two dimensions? Make a sketch of

each view of each object. This is a convenient experiment that can be conducted almost anywhere. Continue the experiment for a week with various objects in your environment, and you will find yourself becoming skilled at two-dimensional spatial visualization.

THREE-DIMENSIONAL SPATIAL VISUALIZATION

Visualizing what a three-dimensional object will look like when drawn in two dimensions is half of the spatial visualization challenge. The other half involves using a two-dimensional representation of an object as the basis for visualizing what the actual object will look like in three dimensions. For example, look at the two-dimensional drawing in *Figure 5-4*. How would this object look in three dimensions?

Looking at the front and back views, you can see that the object in question is shaped like an L. Looking at the right and left sides, you can discern how wide the object is. The top and bottom views show in which directions the legs of the L point. With this information and a little practice, you can determine that the three-dimensional object in question would look like *Figure 5-5*.

The ability to apply spatial visualization is important to engineers and drafting technicians. For most people, developing this skill takes time, patience, and practice—lots of practice. Fortunately, spatial visualization is a concept that can be practiced anywhere. Until your spatial visualization skills are well developed, practice it on objects in your environment.

Solid modeling software (Chapter 11) is making spatial visualization easier for engineers and drafting technicians by doing much of the work for you. But you cannot develop a three-dimensional solid model every time you need to visualize an object. Consequently, it is important to develop your skills in applying spatial visualization manually.

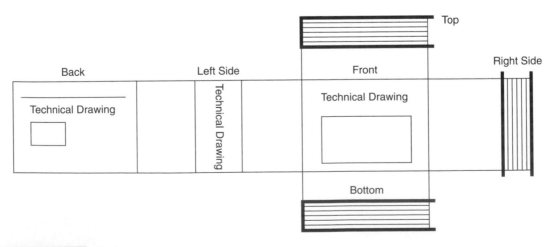

FIGURE 5-3   *Orthographic views of this book*

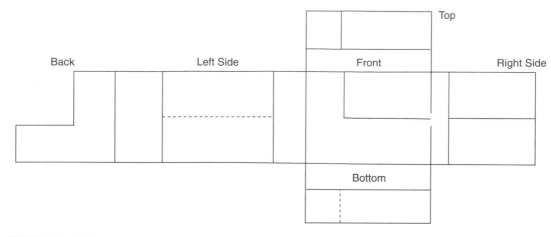

FIGURE 5-4   *Multiple views of an object*

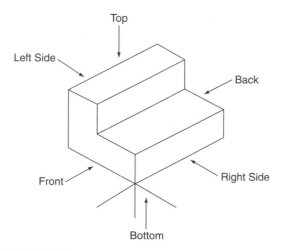

FIGURE 5-5   *Three-dimensional view of an object*

Orthographic Projection

Orthographic projection is the most accurate term for shape description wherein an undistorted image of the object appears in a flat, transparent, but imaginary projection plane. A projection plane may be thought of as an imaginary pane of glass, with an object underneath it or behind it, *Figure 5-6A*. Viewed from the side, the pane of glass appears as a line, *Figure 5-6B*. Assume that light beams emitting from the surfaces of the object are projected to the projection plane, and that each light beam from each exposed surface is directed toward the viewing plane, *Figures 5-6C* and *D*. The image shown in Figure 5-6A traces the path of each light beam as it intersects the viewing plane. This image is called a projection, and it illustrates the features of the exposed surfaces of the object.

NORMAL SURFACES

The surface images of the figures just described do not represent the size and shape of their corresponding real surfaces. Rectangular surfaces appear as parallelograms. The lengths of edges are shorter in the image than their actual lengths (a phenomenon called *foreshortening*). However, if the viewing plane were positioned parallel to several of the object's surfaces, as shown in *Figure 5-7A*, then the image appearing in that viewing plane would provide the actual size and shape of those surfaces, as shown in *Figure 5-7B*. A surface that is parallel to a viewing plane is said to be "normal" to the viewing plane.

A simple object, such as the cannonball in *Figure 5-8*, needs only one view to describe its true size and shape. An object such as the flat gasket in *Figure 5-9* needs only one view and a callout stating its required thickness. The callout can be noted on the drawing, as illustrated, or listed in the title block under "material." *Figure 5-10A* shows an object with a viewing plane placed in a normal position relative to some surfaces of the object, resulting in the orthographic views shown in *Figure 5-10B*. The sizes and shapes of the object's normal surfaces are shown in perfect outline in the orthographic view, but the viewer still cannot determine other features of the object. For example, *Figures 5-10C, 5-10D,* and *5-10E* are different objects that provide exactly the same projected view. Furthermore, not all the surfaces of the projected views of Figures 5-10C and 5-10E are shown in their real size and shape as they are not normal to the viewing plane. There is no way to fully distinguish these without additional information.

TWO ORTHOGRAPHIC VIEWS

In order to present the images of each viewing plane in a flat drawing area, one of the viewing planes must be repositioned to lie in the same plane of the drawing as the other. The procedure used to do this is to create a *fold line* where the imaginary perpendicular-orthographic

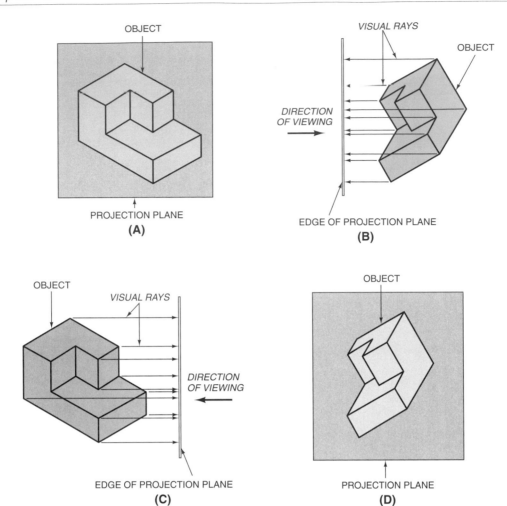

OBJECT

PROJECTION PLANE

(A)

VISUAL RAYS

OBJECT

DIRECTION OF VIEWING

EDGE OF PROJECTION PLANE

(B)

OBJECT

VISUAL RAYS

DIRECTION OF VIEWING

EDGE OF PROJECTION PLANE

(C)

OBJECT

PROJECTION PLANE

(D)

FIGURE 5-6 *(A) Imaginary windowpane in front of object; (B) Side view of windowpane; (C) and (D) Reversed projection*

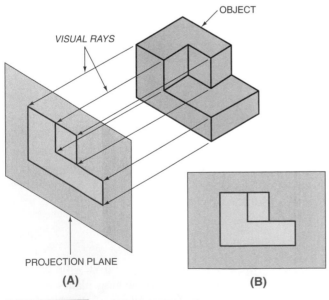

OBJECT

VISUAL RAYS

PROJECTION PLANE

(A)

(B)

FIGURE 5-7 *(A) Viewing plane parallel to surfaces; (B) viewing plane provides actual size and shape of surfaces*

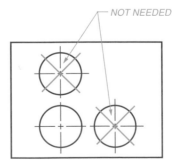

NOT NEEDED

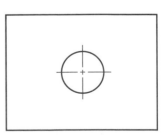

ONE-VIEW DRAWING

FIGURE 5-8 *Simple one-view drawing*

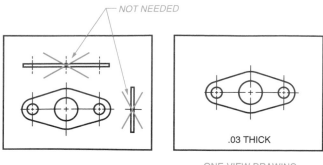

NOT NEEDED

.03 THICK

ONE-VIEW DRAWING

FIGURE 5-9 *Simple one-view drawing*

viewing planes meet each other along the straight edge of intersection, as shown in *Figure 5-11A*. Figure 5-11A shows the object in Figure 5-10A with a second viewing plane in position. The fold line acts as a hinge as the two intersecting viewing planes are swung into the same plane, *Figure 5-11B*. The resulting orthographic views are shown in *Figure 5-11C*. Each view contains the image of a 90° rotated view or projection of the other. In this procedure, the features and surfaces are always aligned to each other and thus can be located.

Conversely, the same two orthographic views are shown in *Figures 5-11D, 5-11E,* and *5-11F* in haphazard relationship to each other, putting them each in error and making it impossible to determine surface positions or the real shape of the object.

LABELING TWO VIEWS

Figures 5-12A and *B* show the positioning of two adjoining orthographic viewing planes to describe the object shown in Figure 5-10C. The orthographic view that presents the most characteristic shape of the object is usually selected and identified as the *front view.* Once this view is selected, the orthographic view that is projected to the right of the front view is identified as the *right-side view.* The front view is often referred to as the front elevation, and the right-side view is referred to as the right profile.

Figure 5-13A shows viewing planes with view images that accurately describe Figure 5-10D. But finished orthographic drawings should not include the borders of viewing planes. For example, *Figure 5-13B* shows the views without the borders. Viewing planes were earlier identified as imaginary; therefore, the fold line is

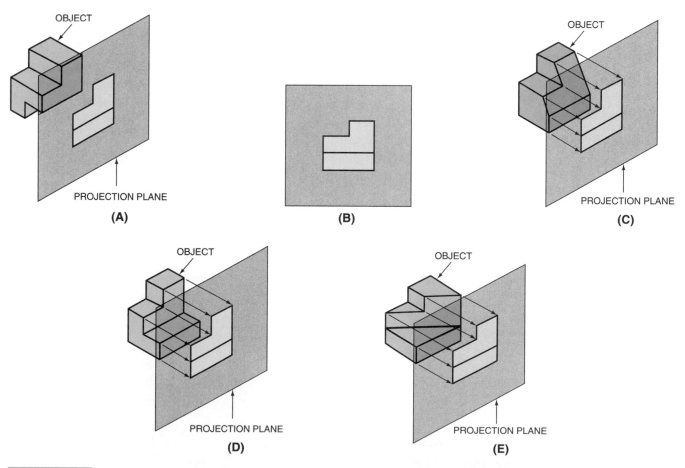

OBJECT

PROJECTION PLANE

(A)

(B)

OBJECT

PROJECTION PLANE

(C)

OBJECT

PROJECTION PLANE

(D)

OBJECT

PROJECTION PLANE

(E)

FIGURE 5-10 *(A) Viewing plane in a normal position; (B) orthographic view of the object; (C) different object with the same orthographic view; (D) another object with the same orthographic view; (E) another object with the same orthographic view*

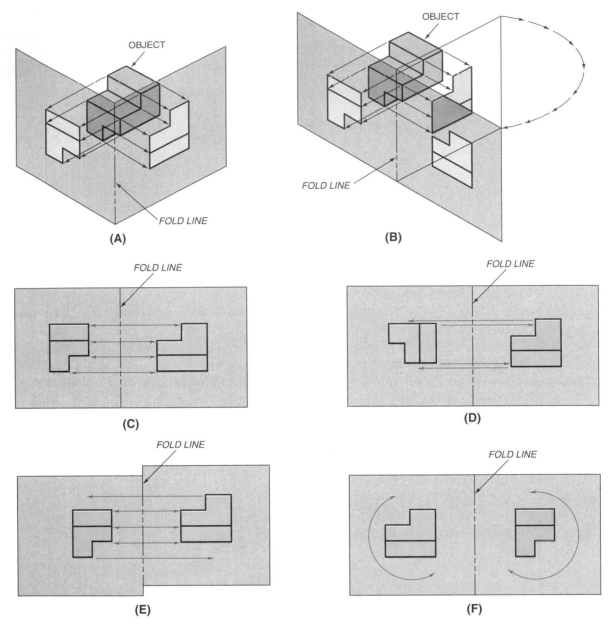

OBJECT

FOLD LINE

(A)

OBJECT

FOLD LINE

(B)

FOLD LINE

(C)

FOLD LINE

(D)

FOLD LINE

(E)

FOLD LINE

(F)

FIGURE 5-11 *(A) Fold line; (B) fold line acts as a hinge; (C) projecting features from one view to the next view; (D), (E), and (F) incorrect positioning of views*

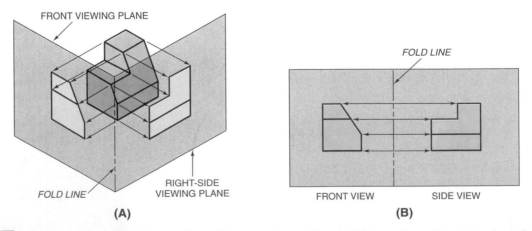

FRONT VIEWING PLANE

FOLD LINE

RIGHT-SIDE VIEWING PLANE

(A)

FOLD LINE

FRONT VIEW   SIDE VIEW

(B)

FIGURE 5-12 *(A) Positioning of two adjoining orthographic viewing planes; (B) two adjoining orthographic viewing planes flattened out*

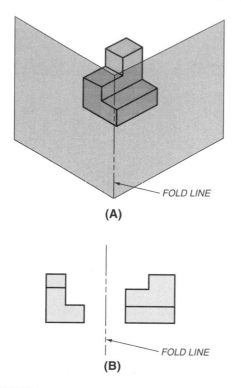

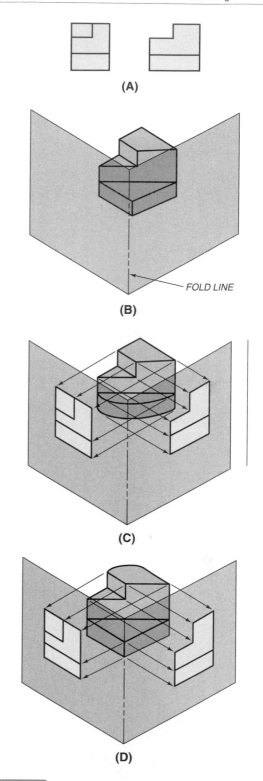

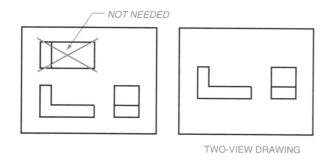

(A)

(B)

FIGURE 5-13 *(A) Viewing planes and images; (B) Orthographic drawing omitting viewing planes*

TWO-VIEW DRAWING

FIGURE 5-14 *Only two views are required*

imaginary also. Often, fold lines are needed for view constructions, as illustrated later in this chapter and future chapters (especially Chapter 8), but in general, fold lines are removed on technical drawings.

The distance of the image from the fold line identifies the object's location from the viewing plane. The drafter decides what this distance will be in order to make the best use of the space on the drawing paper and to provide room for dimensions and notes, see *Figure 5-14.*

MULTIPLE ORTHOGRAPHIC VIEWS

Figure 5-15A shows two orthographic views that do not completely describe the object. Multiple interpretations

FIGURE 5-15 *(A) Orthographic views of an object; (B) Correct interpretation; (C) and (D) Incorrect interpretations*

of the objects are possible, causing the viewer to be misled by inadequate information. Correctly interpreted, the object appears as shown in *Figure 5-15B.* Incorrect interpretations are shown in *Figures 5-15C, 5-15D,* and *5-15E.*

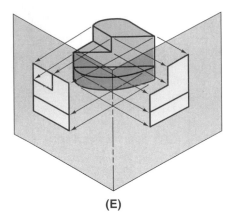

(E)

FIGURE 5-15 *(Continued) (E) Incorrect interpretation*

A third viewing plane is commonly used to ensure adequate definition of the represented object. The third viewing plane is usually above the object and perpendicular to the other two. It is called the *top view*. The top view is sometimes called the plan view in construction-related drawings. This third view is shown in *Figure 5-16.*

Select the views to best present the object's shape and make the most efficient use of the drawing area, *Figure 5-17A,* or to maintain the orientation of the object that is most easily understood by the reader, as illustrated in *Figure 5-17B.* Recall that the view selected should be an attempt to show the most characteristic shape of an object. For example, seeing a building drawn sideways makes reading the drawing very awkward.

The three viewing planes in Figure 5-16 are called *principal viewing planes,* as they are all perpendicular to one another. Orient the front view first. Three other principal planes complete the "box"; they make up the sides of a transparent box that completely surrounds the object, *Figure 5-18A. Figure 5-18B* shows the possible

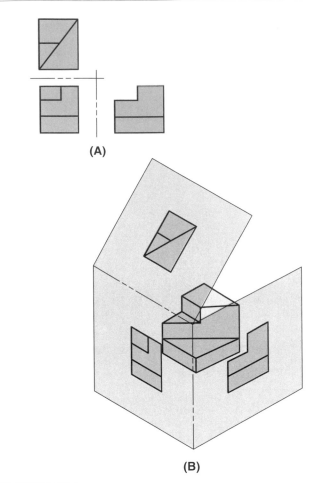

(B)

FIGURE 5-17 *(A) Orthographic view of an object; (B) Three principal planes*

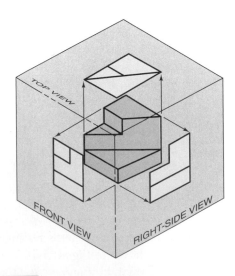

FIGURE 5-16 *Third viewing plane added*

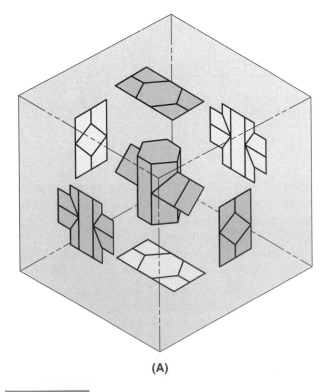

(A)

FIGURE 5-18 *(A) Transparent box*

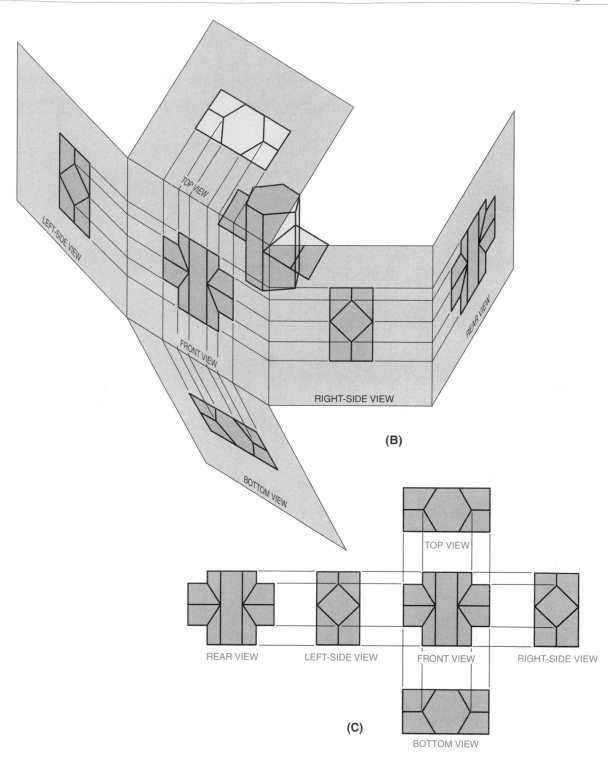

FIGURE 5-18 *(Continued) (B) Fold lines selection; (C) Resulting multiple views*

fold line selections that could be used with orthographic views on the principal planes. *Figure 5-18C* shows the six principal views.

OBJECT DESCRIPTION REQUIREMENTS

The six views of Figure 5-18C are not all necessary to provide sufficient information; three are sufficient for

this object. The common arrangement of top, front, and right-side views is shown in *Figure 5-19*. The top view aligns directly above the front view, and the right-side view aligns directly to the right of the front view. Both the top view and the right-side view images are the same distance behind the front view projection plane. See dimension b in *Figure 5-20*.

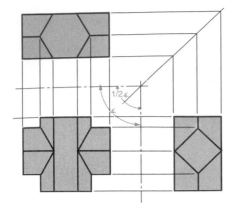

FIGURE 5-19 *Transferring dimensions—45° projection method*

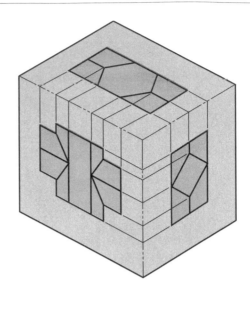

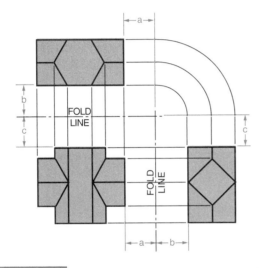

FIGURE 5-20 *Transferring dimensions—compass method*

DIMENSION TRANSFER METHODS

Figure 5-19 shows a method of transferring dimensions between the top and right-side views. This method, used only with principal views, employs a line at 45° to the fold lines of the principal planes.

Use a compass for a different method shown in Figure 5-20.

Figure 5-21 shows the transfer method, which is usually the most accurate. It is the method used in descriptive geometry discussed in Chapter 8. Draw the front and top views first and arbitrarily position a fold line between them. Recall that the fold line must be perpendicular to the projection lines between views. Construct a second fold line perpendicular to the projection lines that will go to the right to be used for the right-side view. This second fold line can be at any distance from the front view. Once the fold line is drawn, the distances from it to the right-side view must agree with distances to the top view, dimensions a, b, and c. Transfer these dimensions with a scale or dividers. If the front and

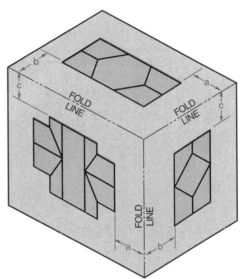

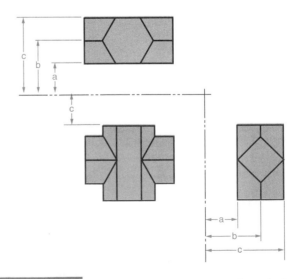

FIGURE 5-21 *Transferring dimensions—transfer method*

right-side views are drawn first, then construct a fold line for the top view in a similar manner.

HIDDEN LINES

Figure 5-22 is an isometric view of an object. The orthographic top, front, and right-side views are illustrated in *Figure 5-23*. *Hidden lines* represent feature outlines, for example, intersections of plane surfaces, real but invisible, and are used in each viewing plane where appropriate.

A hidden line may be covered by a visible (solid) object line in a particular view. This concept is shown in *Figures 5-24A* and *5-24B*. Figure 5-24B shows surface B aligned behind the solid line of intersection of plane A with the inclined plane. As expected in orthographic projection, there are many hidden lines covered by solid lines. Mentally perceiving where the hidden lines lie in orthographic views requires visualization.

Figure 5-25 shows typical hidden line length and spacing, approximately 1/8″ to 1/4″ long spaced approximately 1/32″ to 1/16″ (3.2 mm to 6.4 mm and .8 mm to 1.6 mm, respectively). The lengths and spaces can vary depending on the feature and the space available on the drawing, but consistency is encouraged. Hidden lines are narrower than solid lines. Use the accepted drafting practices of hidden line applications shown in *Figure 5-26*.

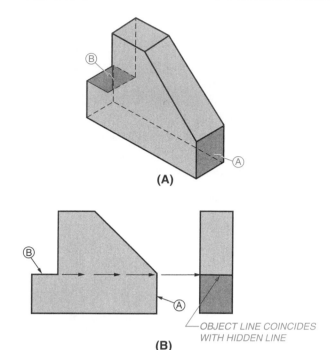

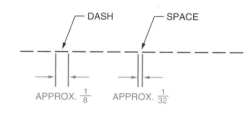

FIGURE 5-24 *(A) Isometric view of an object; (B) Object line takes precedence over hidden line*

FIGURE 5-25 *Average hidden line construction*

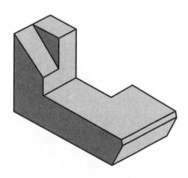

FIGURE 5-22 *Isometric view of an object*

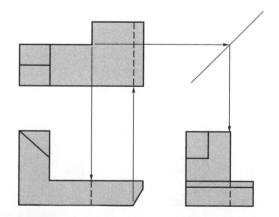

FIGURE 5-23 *Orthographic views of the same object*

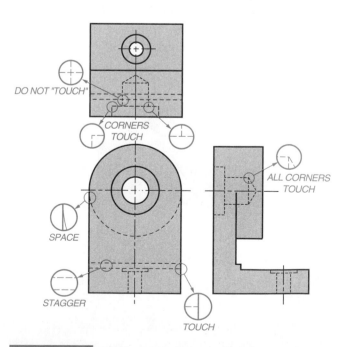

FIGURE 5-26 *Hidden line drafting practices*

CENTER LINES

Center lines are used primarily as locations of circular, cylindrical, or spherical features. They can also be used to specify locations of other principal symmetries as well, such as the center line of a ship or a rocket. Crossed perpendicular center lines indicate a two-coordinate location of where an axis exists. A single center line indicates the path of an axis. Identify all axes, on all views, of circular features except for noncritical radii, Figure 5-26.

SURFACE CATEGORIES

Recall that normal surfaces are parallel to one of the principal projection planes of the imaginary transparent box surrounding the object. In *Figure 5-27*, these surfaces are identified as the top, front, and right side. The viewing planes containing these images are identified as horizontal, frontal, and profile projection planes, respectively. A normal surface's true size and shape appear in one of the principal views, *Figure 5-28*, but the surface appears as an edge view, a line, in the other principal views.

Surfaces A, B, and C, Figures 5-27 and 5-28, are inclined surfaces. They are not parallel to any one of the principal viewing planes, but each inclined surface is perpendicular to one of them. For example, inclined plane A is perpendicular to the frontal projection plane. Moreover, a surface that is perpendicular to a viewing plane will appear as an edge or line when projected on that plane. Surface A appears smaller than its true size in both the top and right-side views, Figure 5-28, because of foreshortening. Inclined surfaces B and C are configured similarly to surface A, with B perpendicular to the profile projection plane, and C is perpendicular to the horizontal projection plane. Surface D is not perpendicular to any of the principal projection planes and is

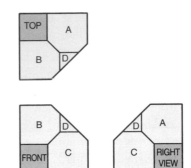

FIGURE 5-28   *Orthographic views*

therefore categorized as an oblique plane. It appears foreshortened in all of the principal projection planes.

The transparent box of principal projection planes that surround cylindrical surfaces uses the cylindrical axis as the line of orientation, *Figure 5-29*. If the cylindrical axis is made perpendicular to one of the principal viewing planes, then the cylindrical surface appears as a circle or an arc. Figure 5-29 shows a cylinder whose axis is perpendicular to the top horizontal principal projection plane, and the circular image that results in the top view is shown in *Figure 5-30*. The curved surface of the

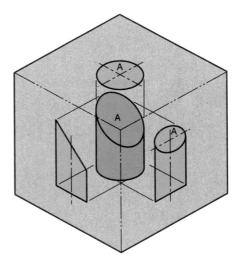

FIGURE 5-29   *Object inside transparent box*

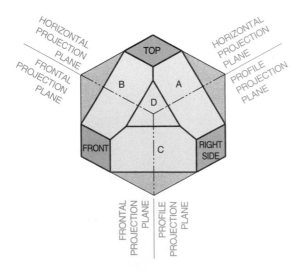

FIGURE 5-27   *Surface categories*

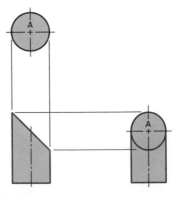

FIGURE 5-30   *Circular image results in this top view*

cylinder cannot be classified as normal, inclined, or oblique. Plane surface A in Figure 5-30 is inclined to the horizontal viewing plane and shows as an ellipse in the right-side view. If surface A were at 45° to the horizontal viewing plane, a circle would appear as surface A in the right-side view.

Planning the Drawing

Before beginning a three-view drawing of any object, make preliminary sketches of the object. Sketches of the object in *Figure 5-31A* are shown in *Figure 5-31B*. The sketches were used as an aid in selecting the best front view and its best position. Sketches are helpful in selecting the required views and their positions.

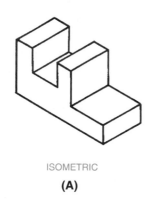

ISOMETRIC

(A)

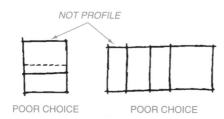

NOT PROFILE

POOR CHOICE   POOR CHOICE

GOOD CHOICE–POOR POSITION

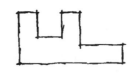

GOOD CHOICE–GOOD POSITION

(B)

FIGURE 5-31 (A) Isometric view of an object; (B) Sketches of object to determine front view

HOW TO PREPARE A THREE-VIEW DRAWING

STEP 1 Visualize the object to be sure you have a good mental picture of it.

STEP 2 Decide which view to use as the front view by sketching the object in several positions, Figure 5-31B. Keep in mind the following:

- The front view is the most important view; viewers see this one first.
- The front view should show the most basic profile; for example, use a side view of a vehicle with the wheels on the ground.
- The front view should appear stable; place the heavy part at the bottom.
- The front view should be positioned to minimize the number of hidden lines in all views; this enhances visualization.

STEP 3 Decide how many views are needed to completely illustrate the object.

STEP 4 Decide, keeping in mind the guidelines in Step 2, in which position to place the front view. Study *Figures 5-32, 5-33, 5-34,* and *5-35* to understand why 5-35 is the best position for the object from Figure 5-31A.

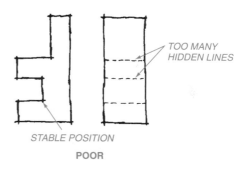

TOO MANY HIDDEN LINES

STABLE POSITION

POOR

FIGURE 5-32 *Poor front view—too many hidden lines*

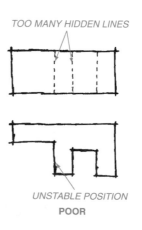

TOO MANY HIDDEN LINES

UNSTABLE POSITION

POOR

FIGURE 5-33 *Poor front view—unstable, too many hidden lines*

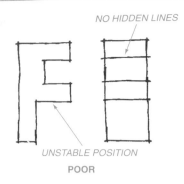

NO HIDDEN LINES

UNSTABLE POSITION

POOR

FIGURE 5-34   *Poor front view—unstable*

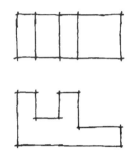

FIGURE 5-35   *Best position—front view*

STEP 5   Decide on the best placement of the views in the work area, usually centered. Do neat work.

POSITIONING THE VIEWS WITHIN THE WORK AREA

The example shown on the left of *Figure 5-36* is centered, but poor judgment was exercised in using the space on the drawing. Space is wasted on both sides, and little room is available at the top and bottom for dimensions and notes. An inch (25.4 mm) of space near the borders is desirable. The example shown on the right in Figure 5-36 is balanced and well centered. Try to keep open spaces evenly distributed around views. A systematic method for centering views is discussed next, after

some comments on a general sketching procedure for centering views.

Sketching Procedure

Sketch efficiently by doing it freehand, quickly, and only to approximate scale on a sheet of practice drawing paper. First, select the front view, using the guidelines previously discussed, and sketch a light outline of the object in approximately the best position. Next, sketch outlines of the top and right-side views. Locate center lines of important features. Finally, heavy in the outline and center lines.

Centering the Drawing

Figures 5-37 and *5-38* show a simple object and a procedure used to center a sketch of the object within a specified work area. The simple object in Figure 5-37 is shown in a dimensioned isometric drawing. The views of the object in both figures are to be sketched freehand. Use an 8 1/2 × 11 practice sheet, and follow the procedure given. There is no need to use a scale, but a drawing sheet with fade-out lines would be helpful.

Calculate the total horizontal distance needed for the front view plus the right-side view by adding the front view width of 4.0 plus 2.0 for the depth of the right-side view, for a total of 6.0 inches. To center these views, subtract 7.0 from 11.0 to get 4.0. Divide this in thirds for the spacing at each end and between views, D = 1.66 inches. (See Figure 5-37.)

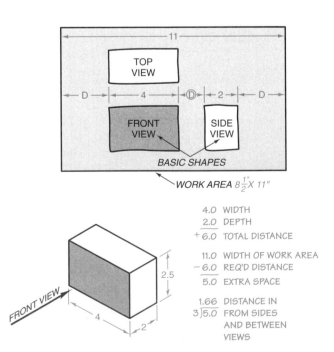

TOP VIEW

D → ← 4 → ← D → ← 2 → ← D

FRONT VIEW   SIDE VIEW

BASIC SHAPES

WORK AREA $8\frac{1}{2}"X\ 11"$

FRONT VIEW

2.5

4

2

4.0	WIDTH
2.0	DEPTH
+ 6.0	TOTAL DISTANCE
11.0	WIDTH OF WORK AREA
− 6.0	REQ'D DISTANCE
5.0	EXTRA SPACE

1.66   DISTANCE IN
3)5.0   FROM SIDES
AND BETWEEN
VIEWS

TOO MUCH WHITE SPACE

EVEN SPACING ALL AROUND

TOP VIEW

FRONT VIEW

SIDE VIEW

FRONT VIEW

TOO CLOSE TO EDGE
POOR

WORK AREA
BEST

FIGURE 5-36   *Positioning the views within the work area*

FIGURE 5-37   *Centering views horizontally*

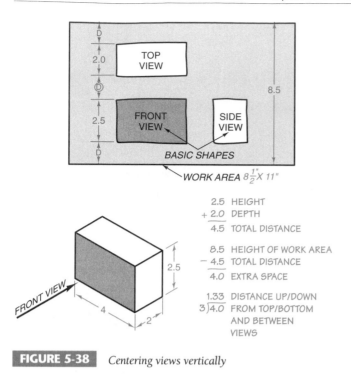

FIGURE 5-38 *Centering views vertically*

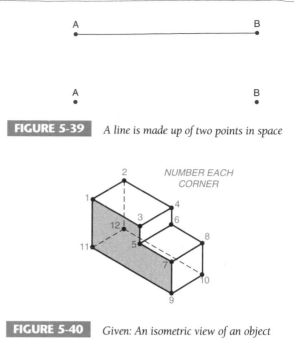

FIGURE 5-39 *A line is made up of two points in space*

FIGURE 5-40 *Given: An isometric view of an object*

Calculate the total vertical distance needed for the front view plus the top view by adding the front view height of 2.5, plus 2.0 for the depth of the object, for a total of 4.5 inches. To center these views, subtract from 8.5 to get 4.0, and divide this in thirds for the upper and lower and between-view spacing, D = 1.33 inches. (See Figure 5-38.)

Try an alternate approach for centering a drawing. Lay out to scale on a practice sheet, with drawing instruments, outlines of the front view and the right-side view with a 1.0-inch space between the two views. Next, lay out the top view with a 1.0-inch space between the front and top views. Next, place a good drawing sheet over the practice sheet and move it around until the drawing is basically centered. Finally, trace the drawing onto the good sheet.

NUMBERING FOR MAKING DRAWINGS

Sometimes a multiview drawing of an object, to be prepared from an isometric drawing, is so complicated that the person making the drawing is not quite sure how some part of a view will look. These more difficult drawings are easier to visualize if points of the various features are numbered, then systematically analyzed. Think of the corners of the features as lines connecting points. For instance, think of each line as connecting two points in space, *Figure 5-39*. If the ends of a line can be found, all that needs to be done to complete the line is to connect the ends. Once the ends have been found and numbered in one view, the same can be done in other views by simple projection. *Figure 5-40* is an isometric view of

a simple object, simple for discussion purposes. Each end of each line has been assigned a number.

Use the following steps for practice in numbering and analyzing drawings.

STEP 1 Lightly draw the basic shape outline of the required views and the 45° projection line, *Figure 5-41*.

STEP 2 Assign the numbers from the isometric sketch to each corner of the view.

STEP 3 Project these points up to the top view, and number them as shown in *Figure 5-42*. Start with the number seen first, followed by a comma or slash, and then the number hidden.

STEP 4 Project these points from the front view to the right-side view, and project the same numbered points from the top view via the

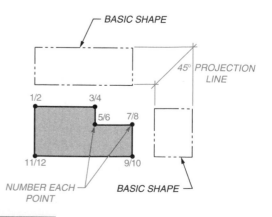

FIGURE 5-41 *Number one view*

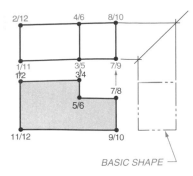

FIGURE 5-42 *Project numbers into the top view*

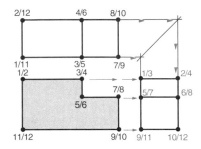

FIGURE 5-43 *Project numbers into the right-side view*

45° line. The exact position of each point is located where the projection lines cross; number them, *Figure 5-43.* Projecting and locating points can be done with alternative methods as shown in Figures 5-20 and 5-21.

NOTE: *If a point can be located in any two of the three views, it can be projected into the third view.*

STEP 5 Connect points together, for example, point 1 to point 2, 3 to 4, and so on, until all the line connections are made. Check the planes made by the lines. Refer to the isometric view, Figure 5-40, as needed.

COMMENT: *These steps describe how to make a complicated drawing. Read and understand a complicated drawing already made by first locating and numbering points on an object in at least three views. Connect the numbers and systematically study the object, line by line, plane by plane.*

Technical Drawing Conventions

Assume now that preliminary sketching has been done to locate the best positions for views of an object for a multiview drawing, and that the views are going to be drawn, to scale, with the drawing tools discussed in Chapter 2. The following topics are important for multiview, technical drawings.

- Rounds and Fillets
- Runouts

- Intersecting Surfaces
- Curve Plotting
- Cylindrical Intersections
- Incomplete Views
- Aligned Features
- How to Represent Holes
- Conventional Breaks

The following discussions on recommended techniques for preparing multiview technical drawings all emphasize the same message: effective communication of three-dimensional concepts, that is, to enhance the visualization process for the use of the drawings.

ROUNDS AND FILLETS

Metal casting processes require an object's material to flow into cavities and corners in a mold. Sharp corners are not only undesirable on a casting but are difficult to cast, so *rounds* are specified for outside corners and *fillets* for inside corners. Cast objects tend to crack due to the strain placed on the metal during cooling. Fillets distribute the strain and alleviate the tendency to crack. Rounds and fillets improve an object's ability to withstand cyclic fatigue loading, and also enhance the appearance of an object, *Figures 5-44* and *5-45.*

RUNOUTS

Runouts are curved surfaces formed where a flat and a curved surface meet, Figure 5-45. To find the exact

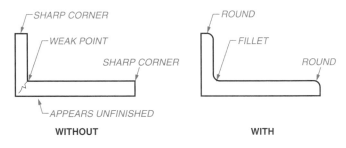

FIGURE 5-44 *Rounds and fillets*

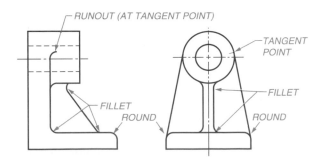

FIGURE 5-45 *Runouts*

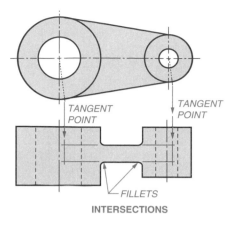

FIGURE 5-46   *Locate the tangent points.*

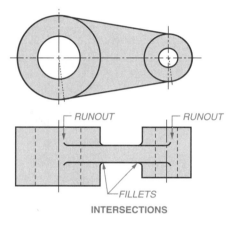

FIGURE 5-47   *Complete runouts at tangent points.*

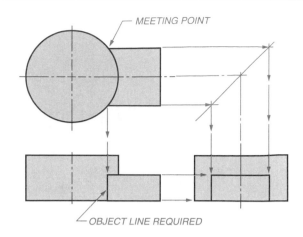

FIGURE 5-48   *Intersecting surfaces*

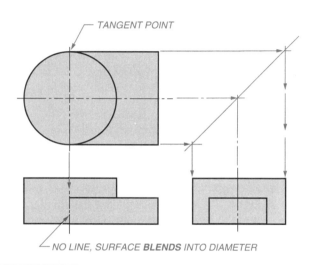

FIGURE 5-49   *Intersecting surfaces*

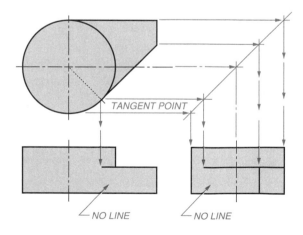

FIGURE 5-50   *Intersecting surfaces*

intersection where a runout occurs, locate the tangent points of the curved surfaces, *Figure 5-46.* Project these tangent points into the next appropriate view and add the runouts as shown in *Figure 5-47.* Note the direction of each runout.

TREATMENT OF INTERSECTING SURFACES

Figures 5-48 through *5-51* show how to illustrate various intersecting surfaces. In these examples, rounds, fillets, and runouts have been omitted to simplify the drawings. In actual practice, rounds, fillets, and runouts would have been added. In Figure 5-48, the flat surface meets the cylinder sharply (see top view), making a visible edge necessary in the front view. The meeting point in the top view is the exact location of the object line in the front view and right-side view.

In Figure 5-49, the flat surface blends into the cylinder, stopping at the point of tangency. In the top view, the tangent point is located at the center line and blends in at that location in the front view.

In Figure 5-50, the surface blends into the cylinder, stopping at the point of tangency, similar to Figure 5-49

except on an angle. The tangent point is projected into the front view and right-side view.

In Figure 5-51, the radius blends into the base, stopping at the point of tangency. In the top view a line is drawn extending to the tangent point of the arc in the front view.

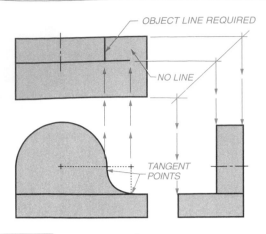

FIGURE 5-51 *Intersecting surfaces*

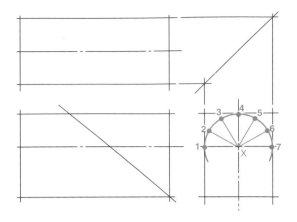

FIGURE 5-53 *Curve plotting (Steps 2 and 3)*

CURVE PLOTTING

Curved surfaces in multiview drawings need to be projected into other views. Follow the steps listed to plot a curved surface.

STEP 1 Lightly complete the basic outline of each view, *Figure 5-52*. Locate the center of an arc, point X.

STEP 2 Draw the arc and divide it into equal spaces, *Figure 5-53*. Use 30° spacing for smaller arcs, but 10° spacing for larger ones. If accuracy is not needed, use larger spacing to save time.

STEP 3 Number the points in order clockwise, Figure 5-53.

STEP 4 Project these points up to the 45° axis line, and to the left to the top view, *Figure 5-54*.

STEP 5 Project from the right-side view to the slanted surface in the front view and locate each point, *Figure 5-55*.

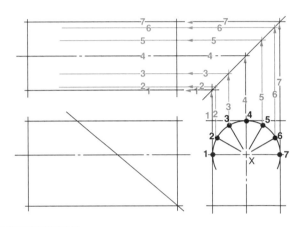

FIGURE 5-54 *Curve plotting (Step 4)*

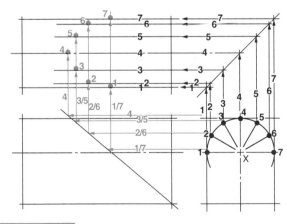

FIGURE 5-55 *Curve plotting (Step 5)*

STEP 6 Project these points up from the front view to the top view.

STEP 7 Number the points 1 through 7 in the top view, where the numbered lines from the right-side view intersect the numbered lines from the front view, *Figure 5-56*.

NOTE: *The method shown in Figure 5-21 could be used for projecting and locating curves.*

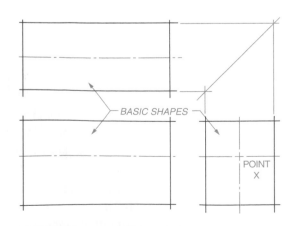

FIGURE 5-52 *Curve plotting (Step 1)*

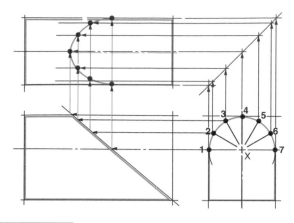

FIGURE 5-56 *Curve plotting (completed)*

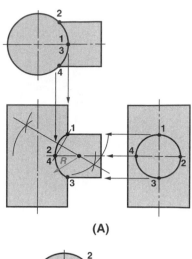

(A)

CYLINDRICAL INTERSECTIONS

If there is a considerable difference in the diameters of each of the intersecting cylinders, then use a straight line for the intersection, *Figures 5-57A* and *5-57B*. If there is little difference between the intersecting cylindrical diameters, then use a simple arc, constructed or approximated through three points, *Figures 5-58A* and *5-58B*.

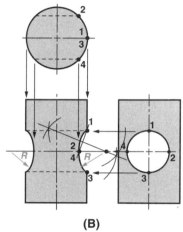

(B)

FIGURE 5-58 *(A) Cylindrical intersections (arc);*
(B) Cylindrical intersections (arc)

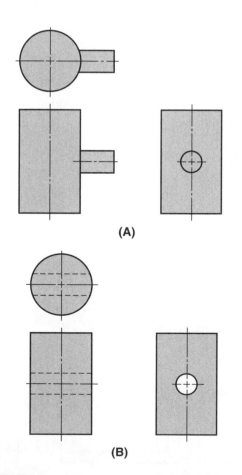

(A)

(B)

FIGURE 5-57 *(A) Cylindrical intersections (straight line);*
(B) Cylindrical intersections (straight line)

In sheet metal construction, locate the actual edge of intersection as shown in *Figures 5-59* and *5-60*. Figure 5-59 shows line elements on each of two intersecting cylindrical surfaces. The line elements are parallel to their respective cylindrical axes and to a common (frontal) viewing plane. For example, line elements number 1 on each cylinder are the same distance from the frontal viewing plane. Also, the line elements numbers 2, 3, and 4 are each the same distance from the frontal plane, respectively. Use this relationship of the elements to construct a curved line of intersection between the large cylinder and the smaller cylinder shown in Figure 5-59.

STEP 1 Refer to *Figure 5-60A* and locate line elements 1, 2, 3, and 4 on the smaller cylinder in the right-side view. Project them into the front view.

STEP 2 Locate a fold line between the front and right-side views, *Figure 5-60B,* at a distance A from the center line of the large cylinder. Locate a fold line at the distance A from the top view.

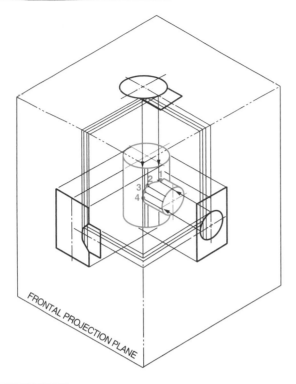

FIGURE 5-59 *Two intersecting cylindrical surfaces*

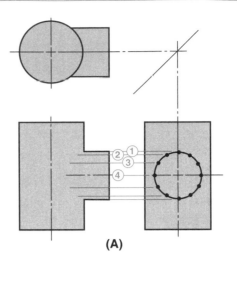

(A)

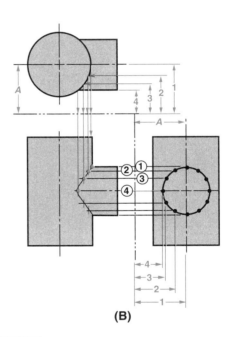

(B)

FIGURE 5-60 *(A) Step 1; (B) Step 2*

Transfer the distances labeled 1 through 4 in the right-side view to the top view as shown. Next, because each corresponding line element from each cylinder is parallel to the front viewing plane and the same distance from it, they must intersect. Fit a curve through the points of intersection, Figure 5-60B. In a like manner, complete the lower portion of the intersection.

INCOMPLETE VIEWS

A drawing can be absolutely correct but confusing, as in *Figure 5-61*. The left- and right-side views are correct, but they are difficult to read due to the overlapping of the details on both ends of the bracket. Omit some of the features in order to make a view clearer and easier to understand, *Figure 5-62*. For example, the right-side view shows only the right-side details. Omitting some of the details in a view for clarity is a standard practice used in technical drawing. Sometimes a note is added such as, "only right side shown."

ALIGNED FEATURES

Show features at their actual distance from an axis or center line, rather than in the projected or foreshortened position. *Figure 5-63A* is an isometric view of a symmetrical part. *Figure 5-63B* indicates how this symmet-

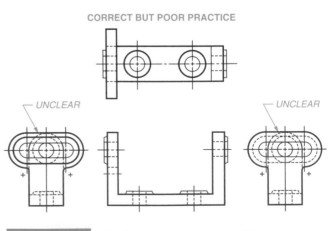

CORRECT BUT POOR PRACTICE

— UNCLEAR   — UNCLEAR

FIGURE 5-61 *Overlapping views—poor practice*

INCORRECT BUT BETTER PRACTICE

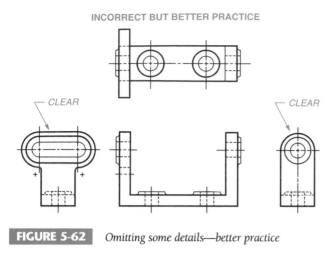

FIGURE 5-62 *Omitting some details—better practice*

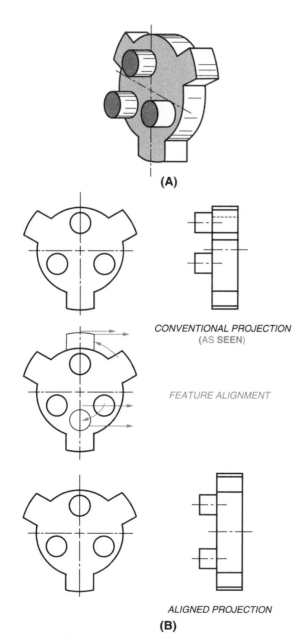

CONVENTIONAL PROJECTION
(AS SEEN)

FEATURE ALIGNMENT

ALIGNED PROJECTION

(B)

FIGURE 5-63 *(A) Isometric view of a symmetrical part;*
(B) Aligned projection

rical part appears confusing in conventional projection (top illustration), but gives a better indication of the part's symmetry when projected from the aligned position (lower illustration).

Chapter 6 has further information on aligning views for clarity.

HOW TO REPRESENT HOLES

Since products contain various kinds of holes, be able to draw each kind and, therefore, be able to identify different kinds of holes when reading drawings.

PLAIN HOLES

A *through hole* goes completely through an object; a *blind hole* is machined to a specific depth. Through holes and blind holes constitute the majority of plain holes in products, *Figure 5-64*. The drawing at the left in Figure 5-64 shows a top view and a front view of a through hole. Holes are usually drilled. A drilled hole is not as accurate as a reamed hole. For example, the tolerance for a 1″ diameter drilled hole might be ±.002″, but ±.0002″ for a 1″ diameter reamed hole. A hole, specified in a callout to be reamed, is first drilled to a size approximately .015″ smaller in diameter than the size specified, and then reamed to the callout specifications. The drafter, designer, or engineer specifies the reamed hole required; the machinist selects the proper drill.

A blind hole is drilled to a specific depth, Figure 5-64. Twist drills as shown in the figure have a conical point of approximately 118° for general-purpose drilling. The full depth of the hole is called out for the cylindrical portion of the hole only.

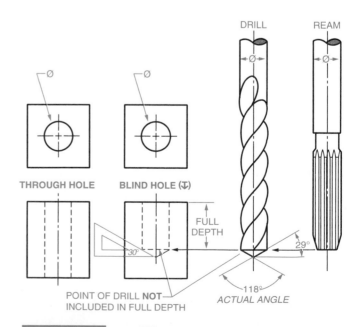

FIGURE 5-64 *Through hole and blind hole*

TAPERED HOLES

A *tapered hole, Figure 5-65,* must be drilled first .015″ smaller in diameter than the smallest diameter of the tapered hole, and then reamed.

COUNTERSUNK HOLES

A *countersunk hole* is usually a plain through hole with an upper portion enlarged conically, *Figure 5-66,* most often to 82° to accommodate standard, flat head machine screws. A machinist makes the countersink to a depth that produces a specified diameter slightly larger than the head of the flat head screw.

COUNTERBORED HOLES

A *counterbored hole* is usually a plain through hole with an upper portion enlarged cylindrically, *Figure 5-67,* to a specified diameter and depth. A callout would specify the

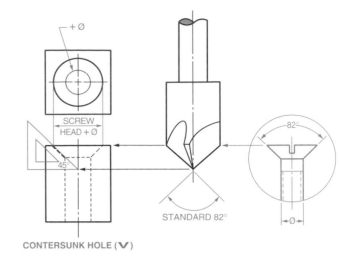

CONTERSUNK HOLE (∨)

FIGURE 5-66   *Countersunk hole*

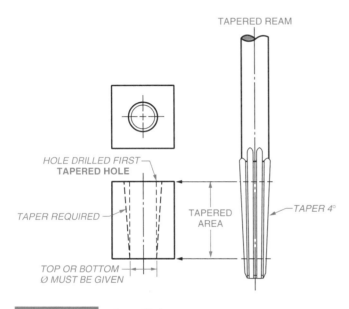

FIGURE 5-65   *Tapered hole*

through hole to be slightly larger than the fastener, and the counterbore to be slightly larger than the head of the fastener and slightly deeper than the height of the head.

SPOTFACED HOLES

A *spotfaced hole* is essentially a plain through hole with a shallow counterbore, *Figure 5-68,* deep enough to provide a flat surface for a head of a bolt, machine screw, or a nut for a threaded member.

CONVENTIONAL BREAKS

Use a *conventional break* for an effective way to draw long objects of constant cross section, such as a pipe, tubing, structural members, or shafting, on a relatively small sheet of drawing paper. A long pipe, drawn to 1/4 size, is shown in *Figure 5-69A,* but redrawn to full size, with

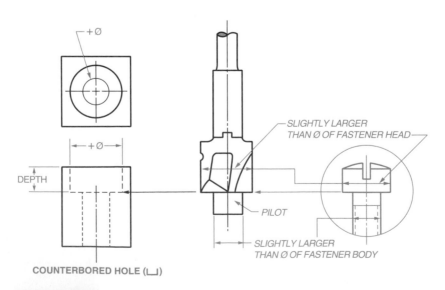

COUNTERBORED HOLE (⊔)

FIGURE 5-67   *Counterbored hole*

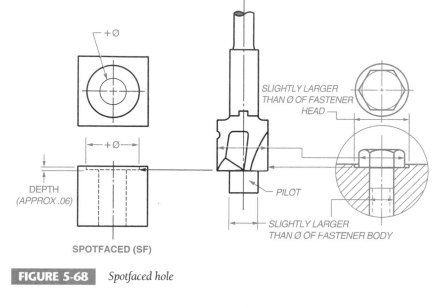

FIGURE 5-68 *Spotfaced hole*

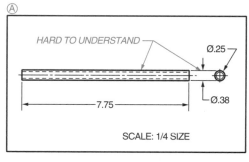

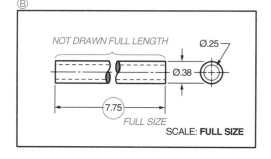

FIGURE 5-69 *Conventional breaks*

a conventional break, in *Figure 5-69B*. This presents a much clearer understanding of the pipe, or other member, especially if there are holes to be located and dimensioned at either end.

KINDS OF CONVENTIONAL BREAKS

Three kinds of breaks are commonly used in technical drawings: the S, the Z, and the freehand. Follow the steps in *Figure 5-70* for constructing an S break for a shaft or tube. The S break may be drawn with instruments or done freehand. Use *Figure 5-71A* as a guide for constructing a Z break, usually used for long, wide parts, of constant cross section. The Z part is usually done freehand. The freehand break, *Figure 5-71B*, is only one example of the use of the freehand break line. For example, broken-out sections will be discussed in Chapter 6.

Visualization

Visualization is the process of re-creating a three-dimensional image of an object in a person's mind, using the evidence and clues provided by orthographic drawings. This process has been the focus of much

investigation, often used for studying a person's ability to perceive (such as in intelligence tests). The goal of reading an orthographic drawing is to visualize accurately information about the relative positions of an object's surfaces and geometric features.

There is some evidence that an active what-if imagination is a key ingredient in the visualization process. For example, a solid line circle in a view causes a what-if visualizer to seek further evidence. If an adjoining view has a solid line rectangle, bisected by a center line (axis) that aligns with the circle's center, a solid cylinder is indicated, *Figure 5-72*. If the aligned rectangle has hidden lines, a hole is indicated, *Figure 5-73*. A single orthographic view seldom is capable of providing three-dimensional evidence, and the reader must consider two or more views simultaneously in seeking an understanding of feature outlines.

Engineering drawings all follow the same procedures, and none is any more or less difficult to read than another. However, the quantity of geometrical information varies considerably among drawings, and the time required to interpret the information varies accordingly. Visualization practice using simple drawings is required

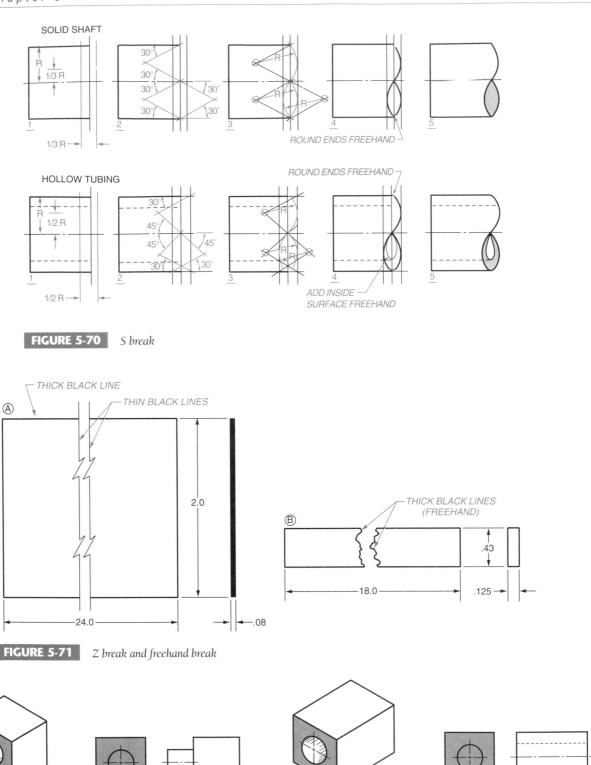

SOLID SHAFT

R
1/3 R
1/3 R

30°
30°
30°
30°
30°
30°

R
R
R

ROUND ENDS FREEHAND

HOLLOW TUBING

R
1/2 R
1/2 R

30°
45°
45°
30°
45°
30°

R
R
R
R

ROUND ENDS FREEHAND

ADD INSIDE
SURFACE FREEHAND

FIGURE 5-70   *S break*

THICK BLACK LINE

THIN BLACK LINES

Ⓐ

2.0

24.0

.08

THICK BLACK LINES
(FREEHAND)

Ⓑ

.43

18.0

.125

FIGURE 5-71   *Z break and freehand break*

FIGURE 5-72   *Visualization of an object*

FIGURE 5-73   *Visualization of an object*

by most individuals to gain experience in mental three-dimensional image construction. Even experienced designers often use modeling clay to form a three-dimensional object from orthographic views. Some build models out of cardboard and glue. Some do quasi-isometric drawings of the overall object, based on the orthographic views, and then systematically try to draw in lines, planes, and solids until all parts of the puzzle fit into a three-dimensional concept. Chapter 18 has excellent guidelines for creating pictorial drawings from orthographic views.

Industry Application

CONFUSION OVER THE ANGLE OF PROJECTION

Mary Anderson had never seen anything like it. Her company, Precision Metal Components (PMC), had recently teamed with a German company on a large contract to manufacture parts for automobile engines. The plan was for the German partner to provide a set of existing drawings that PMC personnel would revise as necessary depending on design changes.

Anderson had expected metric dimensioning, so that aspect of the drawings came as no surprise. What she hadn't expected was the arrangement of views on the drawings. "Look at these drawings!," said Anderson. "The top view is where the front view should be, and the left side is where the right side should be." In every respect the drawings were well drawn. In fact, they were some of the best drawings Anderson had ever seen, except, that is, for the arrangement of views. Anderson couldn't understand how someone could produce a set of drawings that could be so perfect in every way but one. It didn't make sense.

As she pondered the question, Anderson remembered a different angle of projection she had learned in school. Taking her old technical drawing book off the shelf, Anderson blew away the accumulated dust and turned to the chapter on orthographic projection. She soon had her answer. While the United States uses third-angle projection, the method Anderson and her team are accustomed to, Germany uses first-angle projection.

Anderson made copies of the pertinent pages for her team members. Smiling, Anderson commented to no one in particular, "We are going to need a little training before we revise this drawing package."

First-Angle Projection

While the United States, Britain, and Canada use the third-angle projection multiview drawing system, most of the rest of the industrial world uses first-angle projection. The main difference between the two systems is the position of the projection plane. Recall the following sentence from the beginning of this chapter: "A projection plane may be thought of as an imaginary pane of glass, with an object underneath it or behind it, Figure 5-6A." That is, it is observer first, then projection plane, and last, the object. For first-angle projection, it is observer first, then the object, and last, the projection plane.

First-angle projection, like third-angle projection, starts with the front view as the most important view. Refer to *Figure 5-74* and imagine that the observer is at the left of the front view and looking toward the object,

the letter E. What the observer sees, looking at the object, is on the projection plane behind the object, that is, observer-object-plane. Next, imagine that the observer is above the front view and looking toward the object. What the observer sees is the top view, which is placed below the object on the drawing.

Due to the increase in the international exchange of parts and drawings, the projection method used should be indicated on the drawing, with the standard symbol, or the name of the country issuing the drawing. The standard symbol is usually placed on a drawing in or near the title block. *Figure 5-75A* shows the standard symbol used to denote third-angle projection. *Figure 5-75B* shows first-angle projection. Note that in Figure 5-75A the observer is assumed to be at the right of the front view (F.V.) and what the observer sees is shown in the right-side view (R.S.V.). In Figure 5-75B the observer is also assumed to be at the right of the object and what the observer sees is shown in a right-side view, but the projection plane is behind the object.

FIGURE 5-74 | *First-angle projection*

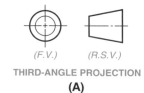

THIRD-ANGLE PROJECTION
(A)

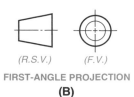

FIRST-ANGLE PROJECTION
(B)

FIGURE 5-75 | *(A) Third-angle projections symbol; (B) First-angle projection symbol*

Summary

- Spatial visualization is the act of forming an accurate mental picture of how a three-dimensional object will look when represented in a two-dimensional format or forming an accurate mental picture of how a three-dimensional object will look based on the information given in a two-dimensional representation.

- Students studying orthographic projection techniques should master the following: projecting normal surfaces; projecting two orthographic views; projecting multiple orthographic views; deciding how many views and which views are needed to adequately describe an object; how to transfer dimensions between views; proper use of hidden lines; proper use of center lines; and projection of inclined surfaces.

- Planning the development of a typical drawing requires five steps: (1) visualize the object; (2) select which view will be the front view; (3) decide how many views are needed; (4) decide where to place the front view; and (5) decide on the best placement of the other views.

- Properly centering a drawing within a specialized work area requires the use of simple mathematical calculations. The amount of space occupied by the various views to be drawn (vertically and horizontally) is subtracted from the space available. The remaining space is then divided proportionally.

- Students should learn to make proper use of the following technical drawing conventions: rounds and fillets; runouts; intersecting surfaces; curve plotting; cylindrical intersections; incomplete views; aligned features; holes; and conventional breaks.

- Students should learn how to mentally visualize an object in orthographic views and how to visualize from orthographic views to the object.

- First-angle projection is used in the industrialized countries of the world except the United States, Great Britain, and Canada. With first-angle projection you have the viewer first, then the object, and last the projection plane.

Review Questions

Answer the following questions either true or false.

1. Before beginning a three-view drawing, it is a good practice to make preliminary sketches of the object.

2. The work area on a drawing is the size of the piece of paper you are working with.

3. An orthographic projection is the most accurate description of shape wherein an undistorted image of the object appears in the flat, transparent, but imaginary projection plane.

4. Runout is the formation of a curved surface where a flat and a curved surface meet.

5. Fillets are interior rounded intersections between two surfaces.

6. The usual depth (call off) for a counterbored hole is **slightly** deeper than the height of the head of the fastener.

7. The lightly constructed 45° angle projection line is used to represent visual rays.

8. A micrometer is used in dimensioning a blind hole.

9. Three views should be used to describe an object.

10. All drawings should be lightly numbered.

Answer the following questions by selecting the best answer.

1. Which of the following is **not** a frequent use of center lines?
   a. The location of a circular or cylindrical feature
   b. The location of an intersection of inclined planes
   c. The location of a spherical feature
   d. To specify the location of other principal symmetries

2. The difference between an inclined plane and an oblique plane is:
   a. Position in the drawing.
   b. Different projection angle.
   c. Inclined planes are used in third-angle projection only.
   d. Inclined planes are not parallel to any normal surface, but are perpendicular to at least one. Oblique planes are neither parallel nor perpendicular.

3. Which of the following is **true** regarding ways to draw a cylindrical intersection?
   a. If there is a considerable difference in the dimensions, use a straight line for the intersection.
   b. If there is little difference in the dimensions, use a simple arc.
   c. Both a and b
   d. Neither a nor b

4. Where is the third-angle projection multiview system used?
   a. The United States
   b. Britain
   c. Canada
   d. All of the above

5. What is the main difference between the first-angle and the third-angle projection?
   a. The standard symbols
   b. The time required to interpret the information
   c. The position of the projection plane
   d. None of the above

6. To illustrate a countersunk hole, what angle is used to draw the countersunk portion?
   a. 81°
   b. 82°
   c. 83°
   d. 84°
   e. None of the above

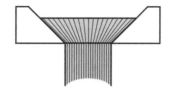

Chapter 5 Problems

The following problems are intended to give the beginning student experience in visualizing objects that are made of convenient shapes and features. Visualizing these shapes and features will help you to completely dimension an object.

Problem 3-22 (page 105) will be used here as an example to illustrate the requirements for the problems. Assume that you have been asked to calculate the weight of the object in Problems 3-22, which is made of aluminum with a density of 0.097 pounds per cubic inch. You calculate the volume to be 9.84 cubic inches, which when multiplied by 0.097 pounds per cubic inch equals 0.954 pounds. Unless otherwise specified, use the format shown in Problem 4-37 (page 167) to present your solutions to the following problems.

PROBLEM 5-1

Assume that the object in Problem 5-8 is cast iron with a density of 0.25 pounds per cubic inch. Calculate the weight of the object in pounds.

PROBLEM 5-2

Assume that the object in Problem 5-14 is steel with a density of 7,765 kilograms per cubic meter. Calculate the weight of the object in kilograms.

PROBLEM 5-3

Assume that the object in Problem 5-18 is aluminum with a density of 0.097 pounds per cubic inch. Calculate the weight of the object in pounds.

PROBLEM 5-4

Assume that the object in Problem 5-27 is aluminum with a density of 0.097 pounds per cubic inch. Calculate the weight of the object in pounds.

PROBLEM 5-5

Assume that the object in Problem 5-41 is steel with a density of 7,765 kilograms per cubic meter. Calculate the weight of the object in kilograms.

PROBLEM 5-6

Assume that the object shown here is steel with a density of 0.28 pounds per cubic inch. Calculate the weight of the object in pounds.

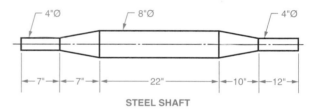

STEEL SHAFT

CONVENTIONAL INSTRUCTIONS

The following problems are intended to give the beginning student practice in visualizing the mutiview system, choosing and sketching the required view, using drafting instruments, and using correct line thickness. As these are beginning problems, no dimensions will be used at this time.

The steps to follow in laying out all drawings throughout this book are:

STEP 1 Study the problem carefully.

STEP 2 Choose the view with the most detail as the front view.

STEP 3 Position the front view so that there will be the least number of hidden lines in the other views.

STEP 4 Make a sketch of all required views.

STEP 5 Center the required views within the work area with a 1″ (25 mm) space between each view.

STEP 6 Use light projection lines. Do **not** erase them.

STEP 7 Lightly complete all views.

STEP 8 Check to see that all views are centered within the work area.

STEP 9 Check to see that there is a 1″ (25 mm) space between all views.

STEP 10 Carefully check all dimensions in all views.

STEP 11 Darken in all views using correct line thickness.

STEP 12 Recheck all work, and, if correct, neatly fill out the title block using light guidelines and neat lettering.

CAD INSTRUCTIONS

Students who plan to complete the following problems on a CAD system should begin with these general instructions. Before reading the specific instructions for each activity, go through each step in the following planning checklist. The checklist applies to any CAD system and will help ensure the optimum use of your time and resources.

1. Analyze the problem carefully. Decide exactly what you are being asked to do.

2. Determine what resources and references (if any) you will need in order to complete the problem. Collect those references and resources and have them readily available.

3. Decide if any particular standards apply to the project, and have those standards available.

4. Determine what types of views will be required and how many of each.

5. Determine what the final plotted scale of the drawing will need to be, and select the appropriate paper size for plotting/printing.

6. Plan your drawing sequence. In what order will you develop the drawing (i.e., lines, features, dimension lines, leaders, dimensions)?

7. Review the various CAD commands you will have to use in order to develop the drawing.

8. Examine your CAD system to ensure that everything is in working order, then begin the project.

PROBLEMS 5-7 THROUGH 5-17

Construct a 3-view drawing of each object, using the listed steps.

PROBLEM 5-7

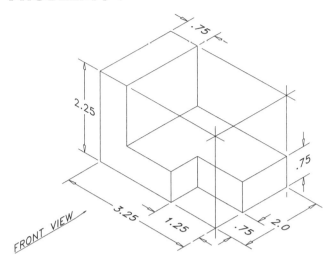

PROBLEM 5-8

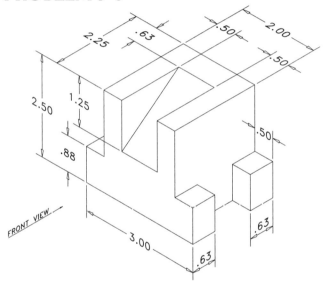

PROBLEM 5-9

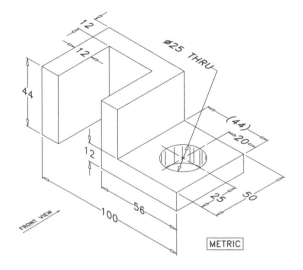

METRIC

PROBLEM 5-10

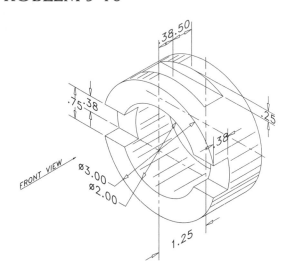

PROBLEM 5-13

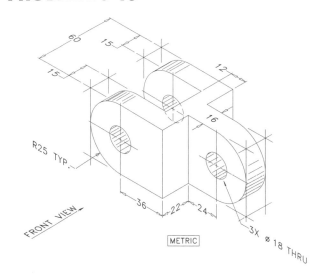

PROBLEM 5-11

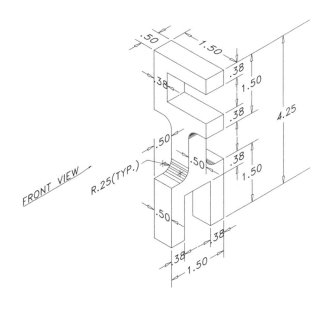

PROBLEM 5-14

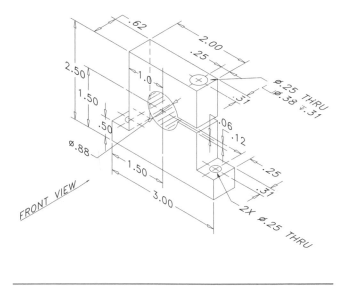

PROBLEM 5-12

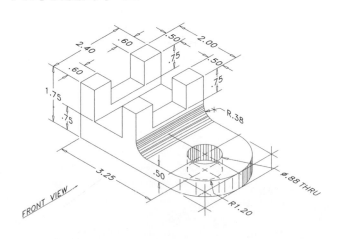

PROBLEM 5-15

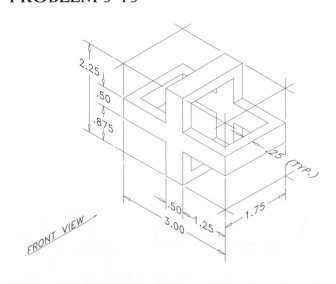

PROBLEM 5-16

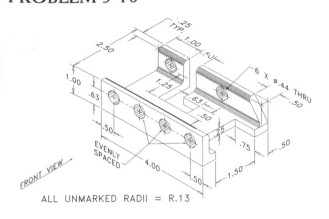

ALL UNMARKED RADII = R.13

PROBLEM 5-17

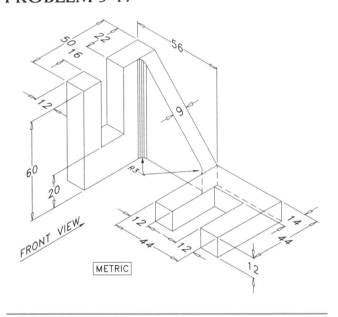

METRIC

PROBLEM 5-18

Construct a 3-view drawing of this object. Project points 1, 2, and 3 into the right-side view and into the top view. Complete all views using the listed steps.

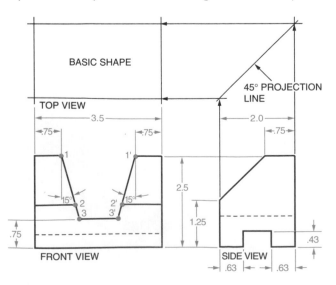

PROBLEM 5-19

Construct a 3-view drawing of this object. Project points 1 through 8 into the top view and into the front view. Complete all views using the listed steps.

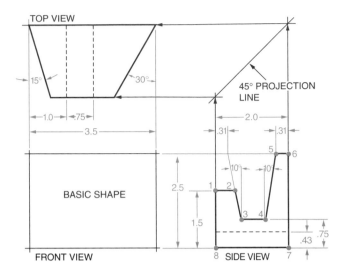

PROBLEM 5-20

Construct a 3-view drawing of this object. Project points 1 through 6 into the front view and into the right-side view. Complete all views using the listed steps.

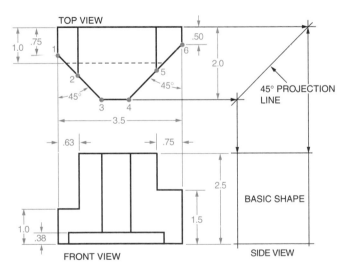

PROBLEM 5-21

Construct a 3-view drawing of this object. Project points 1 through 7 into the right-side view and into the front view. Complete all views using the listed steps.

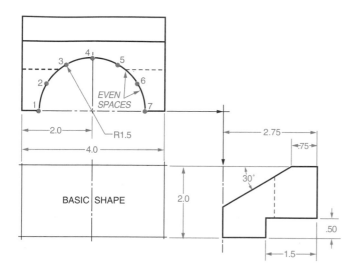

PROBLEM 5-23

Construct a 3-view drawing of this object. Locate and number 12 points around the 1.5 diameter hole. Project points 1 through 12 into the front view and into the right-side view. Complete all views using the listed steps.

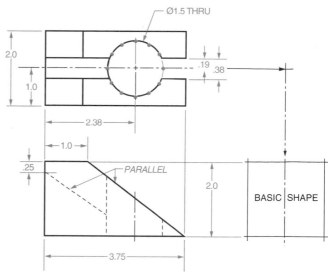

PROBLEM 5-22

Construct a 3-view drawing of this object. Project points 1 through 7 into the right-side view and into the top view. Complete all views using the listed steps.

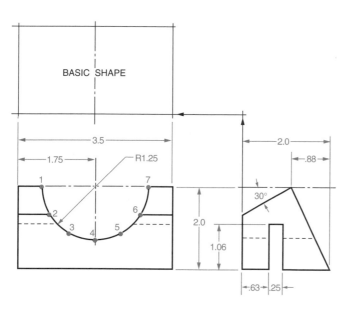

PROBLEMS 5-24 THROUGH 5-31

Construct a 3-view drawing of each object, using the listed steps.

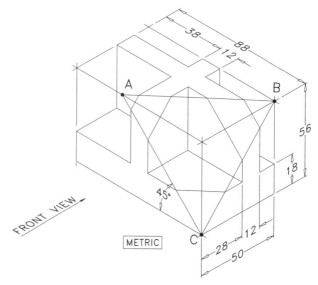

PROBLEM 5-25

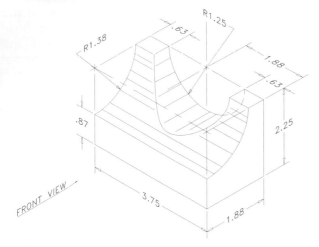

PROBLEM 5-28

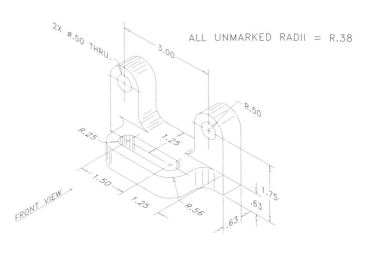

PROBLEM 5-26

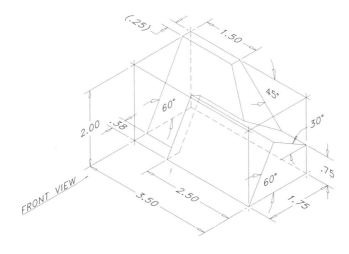

PROBLEM 5-29

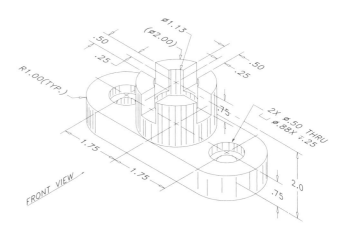

PROBLEM 5-27

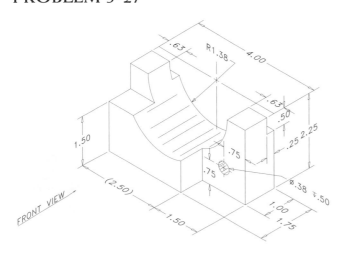

PROBLEM 5-30

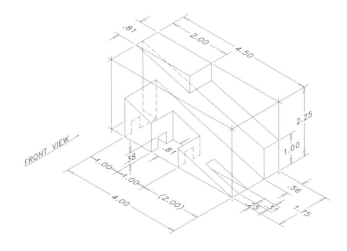

PROBLEM 5-31

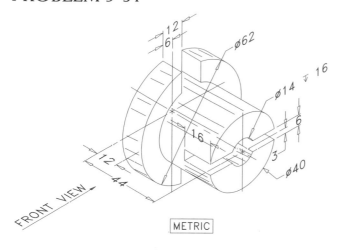

METRIC

PROBLEMS 5-33 THROUGH 5-42

Construct a 3-view drawing of each object, using the listed steps.

PROBLEM 5-33

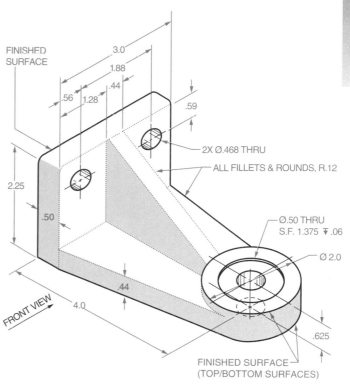

PROBLEM 5-32

Construct a 4-view drawing of this object (front, top, right, and left-side views). Use the listed steps.

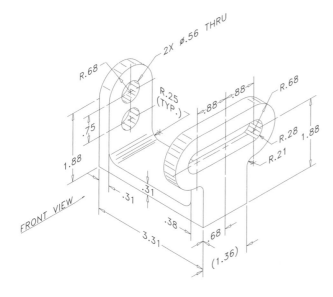

PROBLEM 5-34

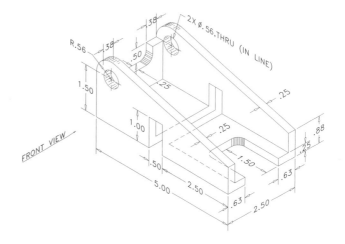

PROBLEM 5-35

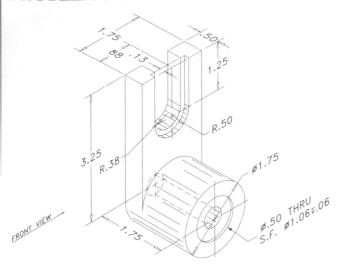

PROBLEM 5-36

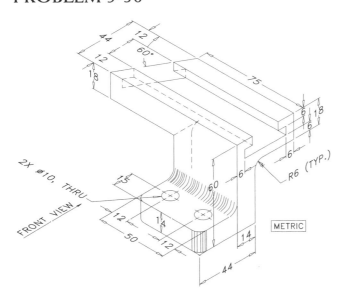

METRIC

PROBLEM 5-37

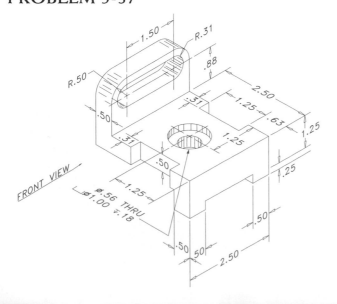

PROBLEM 5-38

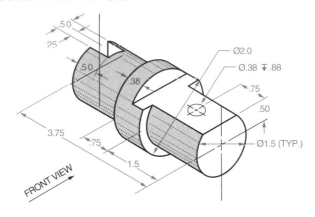

PROBLEM 5-39

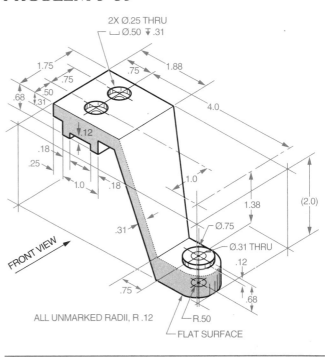

ALL UNMARKED RADII, R .12

PROBLEM 5-40

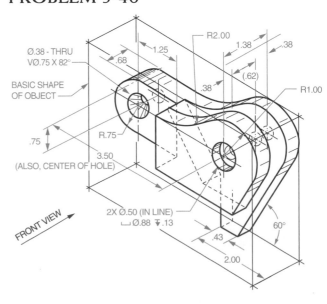

PROBLEM 5-41

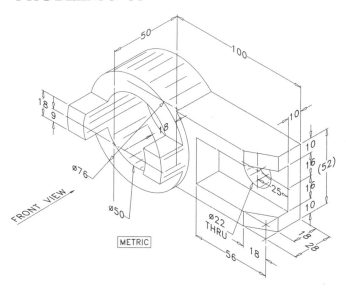

PROBLEM 5-42

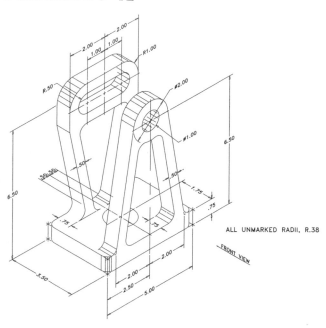

ALL UNMARKED RADII, R.38

PROBLEMS 5-43 THROUGH 5-56

Choose the front view and construct a 3-view drawing of the objects in Problems 5-43 through 5-56, using the listed steps.

PROBLEM 5-43

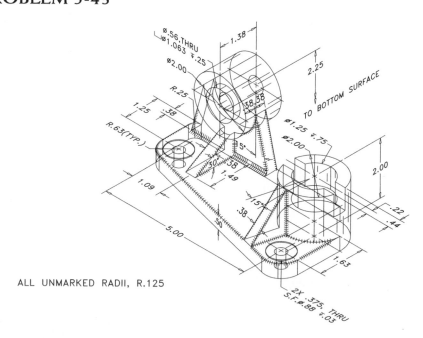

ALL UNMARKED RADII, R.125

PROBLEM 5-44

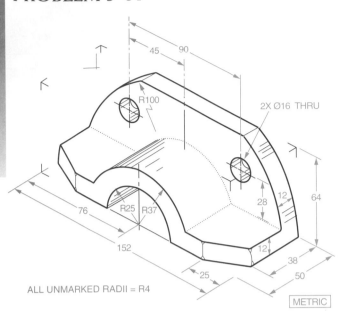

R100
2X Ø16 THRU
45
90
R25 R37
76
152
28
12
64
12
38
25
50
ALL UNMARKED RADII = R4
METRIC

PROBLEM 5-46

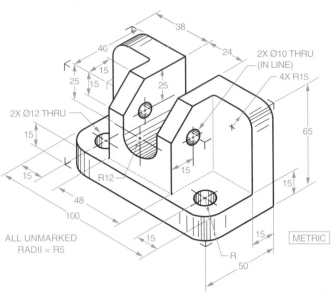

38
40
24
2X Ø10 THRU
(IN LINE)
4X R15
15
25
15
25
2X Ø12 THRU
65
15
R12
15
15
48
100
15
15
ALL UNMARKED
RADII = R5
R
50
METRIC

PROBLEM 5-45

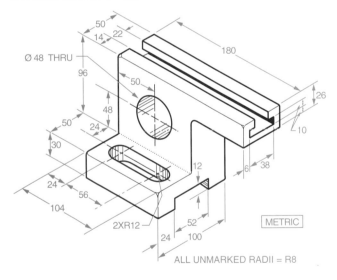

50
14 22
180
Ø 48 THRU
96
50
50
26
48
24
10
30
12
6 38
24
56
52
104
2XR12
24
100
METRIC
ALL UNMARKED RADII = R8

PROBLEM 5-47

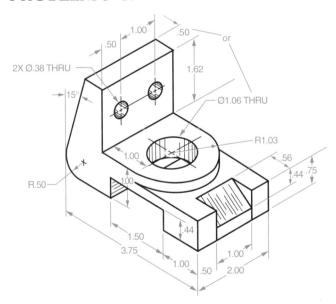

1.00
.50
or
.50
2X Ø.38 THRU
1.62
Ø1.06 THRU
15°
R1.03
.56
1.00
.44 .75
R.50
1.00
1.50
.44
3.75
1.00
1.00
.50 2.00

PROBLEM 5-48

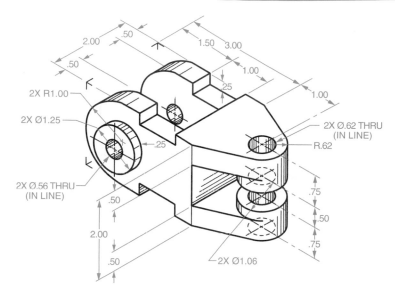

PROBLEM 5-49

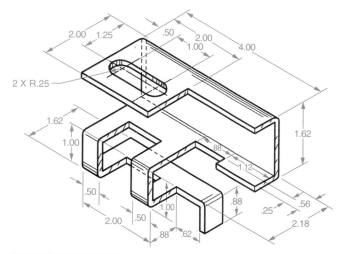

MATERIAL THICKNESS .125
INTERNAL RADII .0125

PROBLEM 5-50

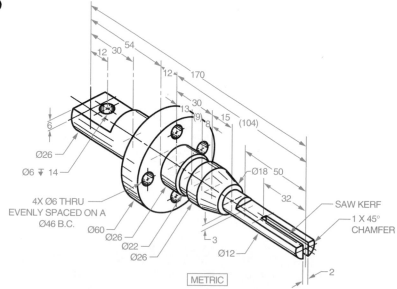

METRIC

PROBLEM 5-51

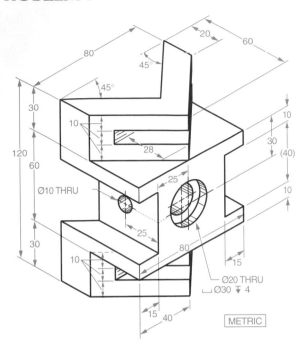

METRIC

PROBLEM 5-52

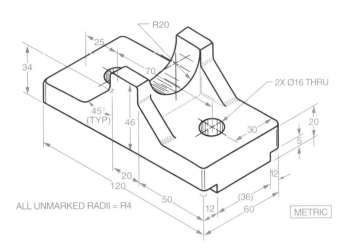

ALL UNMARKED RADII = R4

METRIC

PROBLEM 5-53

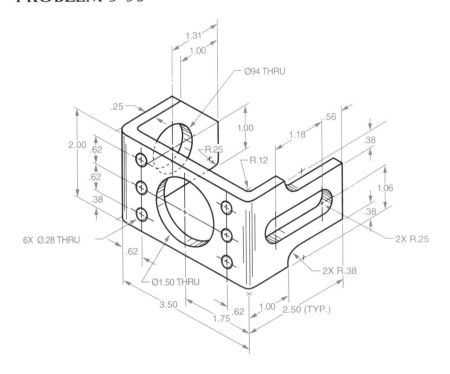

PROBLEM 5-54

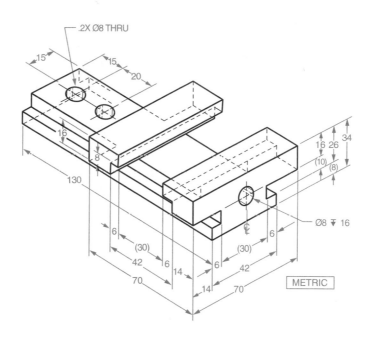

PROBLEM 5-55

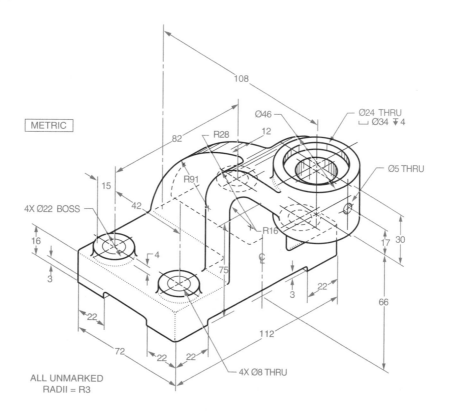

PROBLEM 5-56

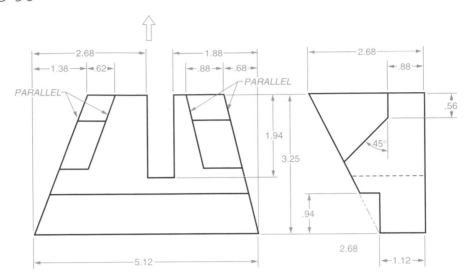

PROBLEM 5-57

For practice in reading multiview drawings that have already been drawn, study the pattern of the sample responses related to the following numbered multiview drawing, and then complete the surface analyses as indicated.

Sample Responses

1. Surface	25-26	in R is 5-6-10-9	in T
2. Surface	16-17	in F is 29-30	in R
3. Surface	16-17	in F is 2-3-6-10-11-7	in T

Complete the Responses

1. Surface	13-14-15-16	in F is _____	in T
2. Surface	24-25-26-27-28	in R is _____	in T
3. Surface	29-25-26-27-30	in R is _____	in T
4. Surface	8-7-12-11	in T is _____	in F
5. Surface	13-14	in F is _____	in T
6. Surface	29-25-24	in R is _____	in F
7. Surface	11-7-2	in T is _____	in F
8. Surface	11-7-2	in T is _____	in R
9. Surface	32-30-27-28	in R is _____	in T
10. Surface	6-5-9-10	in T is _____	in F

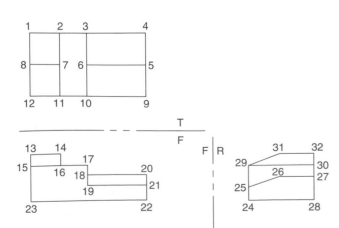

PROBLEM 5-58

Six possible T views, all correctly drawn, are shown for the one F view. Draw an A view for each one of the T views.

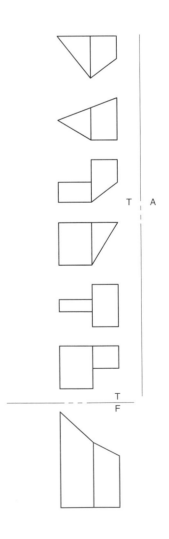

PROBLEM 5-59

The steel end-cover shown has eight through holes 9/16″ diameter for 1/2″ diameter fillister head cap screws. (See Appendix A.) Change the callout for the holes, to be counterbored. (Refer to Chapter 10 for callouts.) The heads of the cap screws will not protrude above the surface of the cover. (This change is for safety reasons; some counterbored holes are specified for aesthetic reasons.) Prepare a two-view drawing of the cover, with the new callout and the dimensions. Show one hole in cross section, in a separate view.

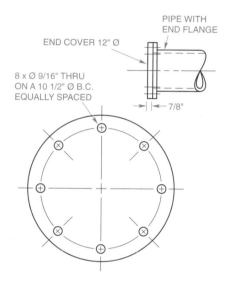

PROBLEM 5-60

Visit a lab or shop listed, and do a field sketch (rough sketch) of a piece of apparatus (not too complicated). Later, use the sketch to prepare a multiview drawing of the apparatus. Show approximate dimensions and note general materials.

    a. Chemistry lab
    b. Physics lab
    c. Metal shop
    d. Wood shop
    e. Electrical/Electronics shop
    f. Computer lab

PROBLEM 5-61

Visit a track and field area and a gymnasium and do a field sketch (rough sketch) of one of the following. Later, use the sketch to prepare a multiview drawing of the item sketched. Give approximate dimensions and note general materials.

    a. Hurdle
    b. Starting block
    c. Javelin
    d. High-jump post

    e. Horse
    f. Springboard (used with a horse)
    g. Parallel bars
    h. Other

PROBLEM 5-62

Visit a weight room and do a field sketch (rough sketch) of a portion of a complicated assembly or all of a relatively simple assembly. Later, use the sketch to prepare a multiview drawing of the item sketched. Give approximate dimensions and note general materials.

PROBLEM 5-63

Do a conceptual design of a contemporary desk lamp for a student's desk. That is, do the general outline of the lamp only, not the insides. Prepare a multiview drawing of the lamp and indicate approximate sizes.

PROBLEM 5-64

Design a ladder for use around a typical home. Use wood and fasten the rungs to the side rails with glue, unless otherwise directed. Prepare a multiview drawing of the ladder, to scale, with approximate dimensions. (A Z break may be useful here.)

PROBLEM 5-65

Utilize the following anthropometric data and design one of the objects listed. Use the dimensions given or use your own body dimensions as a guide. Prepare a multiview drawing of the design, with approximate dimensions and general materials.

    A door (standard height 6′8″ to 7′0″; doorknob or handle, 38″ above floor)
    A working counter for the center of a kitchen (standard height 36″ to 40″)
    A bar stool (standard height 27″ to 30″)
    A simple workbench (standard height 36″)
    A computer workplace (25″ to 26″ knee clearance)

PROBLEM 5-66

Design a simple metal grill for use over an open fire, collapsible and small enough to fit in a backpack. Prepare a multiview drawing of the grill. Include approximate dimensions.

PROBLEM 5-67

Design, for a junior high school science class, an inexpensive, easy to assemble, portable, demonstration apparatus for illustrating the change of supporting force, F, (in pounds) in the diagonal member BC as the angle θ is changed, while the weight W is constant. Refer to the frame outline shown. Prepare a multiview drawing of the design, with approximate dimensions, and note the materials used. (A scale may need to be designed; for example, use a spring with a pointer attached, calibrated for different forces pulling on the spring.)

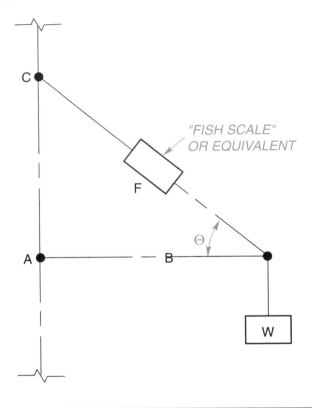

PROBLEM 5-68

Design an easy-to-assemble apparatus for a junior high school science class. It should be portable and inexpensive to demonstrate two facets of a pendulum. First, demonstrate how the period (a complete cycle back and forth) of a pendulum changes with a change in length. Second, demonstrate how the period doesn't change with a change in weight for any one length. Prepare a multiview drawing of the design, indicate approximate dimensions, and note the materials in general. The equation for the period T of a pendulum is T = 2π × square root of (length/g). The g in the equation is the gravity constant. If the length is in meters and the gravity constant equals 9.81 meters/second squared, then the period will be in seconds. For example, a swing (a type of pendulum) with a length of 12 meters and your best friend in the swing, would have a period of approximately 6.9 seconds. (This would be about a 40' long swing.)

PROBLEM 5-69

The drawing shown is for a special aluminum circular sealing plate, used on an experimental apparatus. The plate has a groove for an O-ring seal and threaded holes for a pressure transducer. A similar plate is needed for another project, but the outside diameter is to be changed to 200 mm, and the threaded holes on the 41 mm bolt circle are to be drilled and tapped for a metric fastener M8 × 1.25. The depth of the threaded portion of holes varies from 1 times the diameter of the fastener if the material is steel, to 2 times the fastener diameter for aluminum. Prepare a similar drawing for the circular plate and use your own judgment for the changes in the dimensions. Note that the title block has been omitted for convenience; however, the third-angle projection symbol is shown. Note: Problem 5-69 is a "real-world" drawing. It does not conform to standard drafting conventions in all cases.

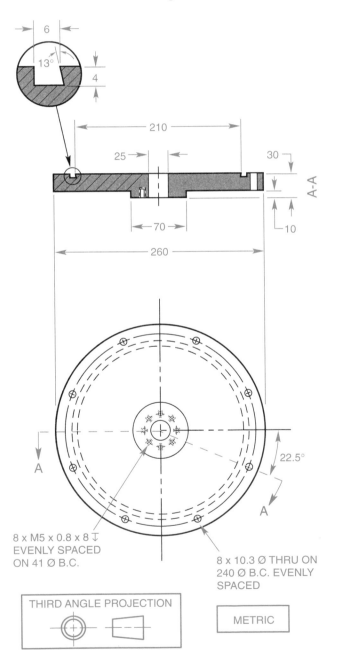

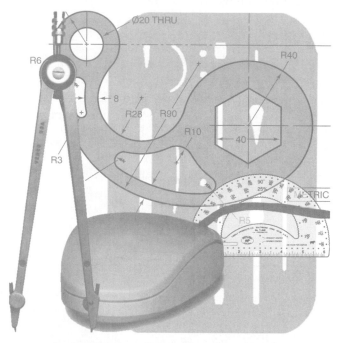

SECTIONAL VIEWS

CHAPTER OUTLINE

Sectional views • Cutting-plane line • Direction of sight • Section lining • Multisection views • Kinds of sections • Sections through ribs or webs • Holes, ribs and webs, spokes and keyways • Aligned sections • Fasteners and shafts in section • Intersections in section • Summary • Review questions • Chapter 6 problems

CHAPTER OBJECTIVES

Upon completion of this chapter, students should be able to do the following:

- Describe the concept of a sectional view.

- Define what is meant by a cutting plane.

- Illustrate section lining.

- Know how to prepare the following kinds of sections: full, offset, half, broken-out, revolved, removed, auxiliary, thinwall, and assembly.

- Know how to deal with the following in sections: ribs, webs, holes, spokes, fasteners, shafts, and keyways.

- Explain aligned sectioning.

- Know how to treat intersections in sections.

Sectional Views

A *sectional view* communicates more information about objects with complicated internal features than the conventional multiview drawing method can with hidden lines. The multiview drawing method, discussed in Chapter 5, is an excellent way to illustrate various external features of an object; however, complicated internal features are difficult to illustrate with hidden lines only. Sectional views solve the problem.

Sectional views are somewhat similar to x-ray images and very similar to magnetic resonance imaging in that each one shows internal details of an object not otherwise available to the observer.

A sectional view is a view of the surfaces that would be seen if an object could be cut open. The view is commonly referred to as a cross section, or simply section, *Figure 6-1.*

Hidden lines are omitted in sectional views, unless it is absolutely necessary to illustrate some unique feature.

This hidden-line convention and other sectioning conventions described in ASME Y14.3M-1994 are part of the language of technical drawing. Discussions of the conventions follow.

Cutting-Plane Line

The *cutting-plane line* indicates the path that an imaginary cutting plane follows to slice through an object. Think of the cutting-plane line as a saw blade that is used to cut through the object. The cutting-plane line is represented by a thick black dashed line, as shown in *Figure 6-2.* The

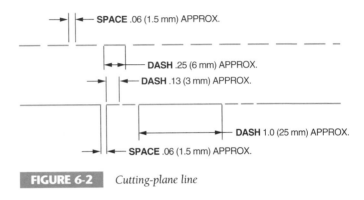

SPACE .06 (1.5 mm) APPROX.

DASH .25 (6 mm) APPROX.

DASH .13 (3 mm) APPROX.

DASH 1.0 (25 mm) APPROX.

SPACE .06 (1.5 mm) APPROX.

FIGURE 6-2   *Cutting-plane line*

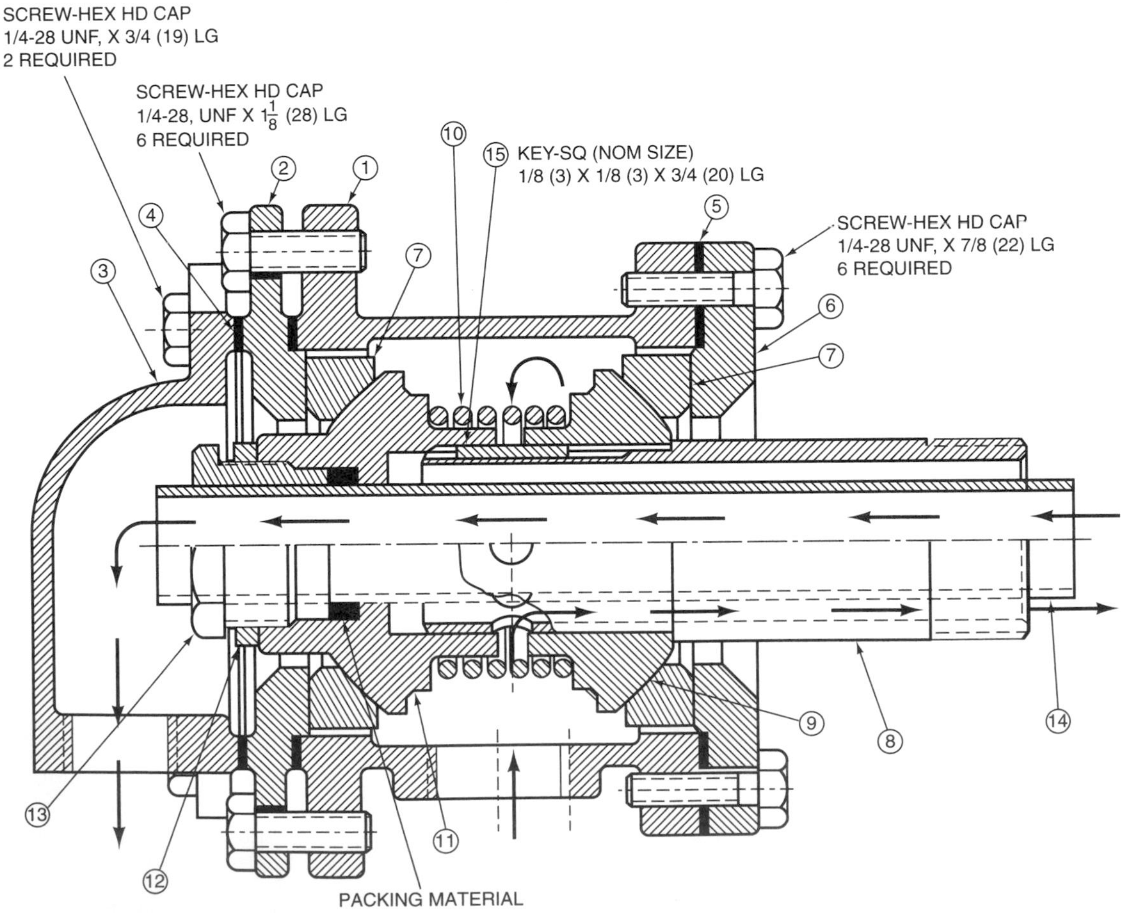

SCREW-HEX HD CAP
1/4-28 UNF, X 3/4 (19) LG
2 REQUIRED

SCREW-HEX HD CAP
1/4-28, UNF X $1\frac{1}{8}$ (28) LG
6 REQUIRED

KEY-SQ (NOM SIZE)
1/8 (3) X 1/8 (3) X 3/4 (20) LG

SCREW-HEX HD CAP
1/4-28 UNF, X 7/8 (22) LG
6 REQUIRED

PACKING MATERIAL

FIGURE 6-1   *Sectional view*

line on top is the newer cutting-plane line; sizes are approximated and spaced by eye. If the cutting-plane line passes through the center of the object, and it is very obvious where the cutting plane is located, it can be omitted. This is the drafter's decision.

Direction of Sight

The drafter must indicate the direction in which the object is to be viewed after it is sliced or cut through. This is accomplished by adding a short leader and arrowhead to the ends of the cutting-plane line, *Figure 6-3*. These arrowheads indicate the direction of sight. Letters are usually added to the ends of the cutting-plane line to indicate exactly what cutting plane is used. The cutting plane extends past the object by approximately 1/2″, as shown.

Section Lining

Section lining shows where the object is sliced or cut by the cutting-plane line. Section lining is represented by thin black lines drawn at 45° to the horizon, unless there is some specific reason for using a different angle. Section lining is spaced by eye from 1/16″ (1.5 mm) to 1/4″ (6 mm) apart, depending upon the overall size of the object. The average spacing used for most drawings is .13″ (3 mm), *Figure 6-4*. Section lines must be of uniform thickness (thin black) and evenly spaced by eye.

If a cutting plane passes through two parts, each part has section lines using a 45° angle or other principal

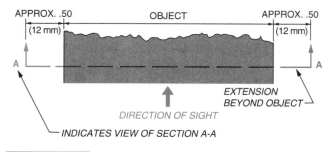

FIGURE 6-3   *The direction of sight*

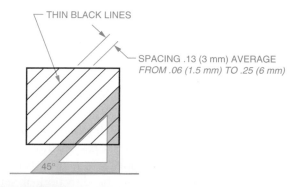

FIGURE 6-4   *Section lining*

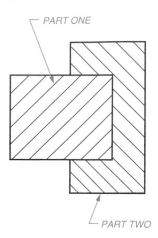

FIGURE 6-5   *Two parts with section lining*

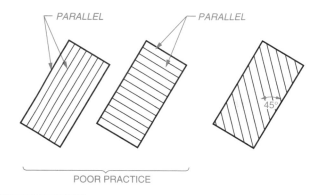

FIGURE 6-6   *Section lining angle*

angle. These section lines should not be aligned in the same direction and should not be joined together at the common visible edge, *Figure 6-5*. If the cutting plane passes through more than two parts, the section lining of each part must be drawn at a different angle. When an angle other than 45° is used, the angle should be 30° or 60°. Section lining should **not** be parallel with the sides of the object to be section lined, *Figure 6-6*.

In past years, section lining used various symbols to indicate the type of material used to make the object. These symbols used only such general type identifications as cast iron, steel, brass, aluminum, and so forth. (Refer ahead to Figure 6-33 (page 219) for examples). Today, because there are so many different kinds of material, section lining symbols have been eliminated, and a single, all-purpose section lining is used (as illustrated in Figure 6-4). Specific information as to the type of material is given in the title block under "material," Figure 6-34 (page 219).

Multisection Views

When an object is very complicated and cut in more than one place, each cutting-plane line must be labeled

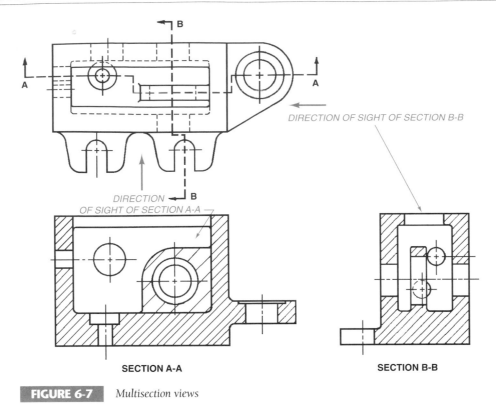

SECTION A-A

SECTION B-B

FIGURE 6-7 *Multisection views*

starting with section A-A, followed by B-B, and so forth, *Figure 6-7*.

Kinds of Sections

Nine kinds of sections are used today in industry.

- Full section
- Offset section
- Half section
- Broken-out section
- Revolved section (rotated section)
- Removed section
- Auxiliary section
- Thinwall section
- Assembly section

Some sections are made up of a combination of the nine kinds of sections. Each is explained in full detail in the following paragraphs.

FULL SECTION

A *full section* is simply a section of one of the regular multiviews that is sliced or cut completely in two. See the given problem, *Figure 6-8,* a regular three-view drawing of an object.

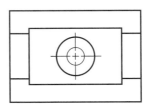

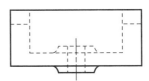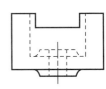

FIGURE 6-8 *Given: Regular three views of an object*

Determine which view contains many hard-to-understand hidden lines. In this example, it is the front view. Add a cutting plane to either the top view or the right-side view. In this example, the top view is chosen. Indicate how the front view is to be viewed, that is, the direction of sight. After determining where the object is to be sliced or cut and viewed, change the front view into section A-A, *Figure 6-9*.

Think of the object as a pictorial drawing, *Figure 6-10*. An imaginary cutting-plane line is passed through the object, *Figure 6-11*. The front portion is removed and the remaining section is viewed by the direction of sight, *Figure 6-12*.

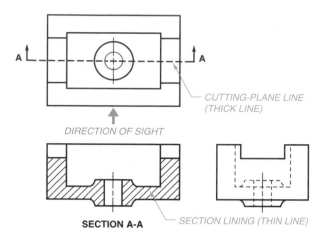

CUTTING-PLANE LINE
(THICK LINE)

DIRECTION OF SIGHT

SECTION A-A

SECTION LINING (THIN LINE)

FIGURE 6-9   *Section A-A added*

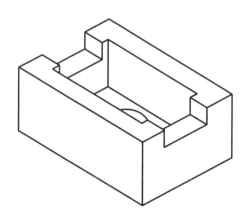

FIGURE 6-10   *Pictorial view of the object*

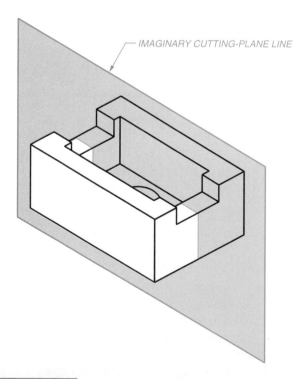

IMAGINARY CUTTING-PLANE LINE

FIGURE 6-11   *Imaginary cutting plane added*

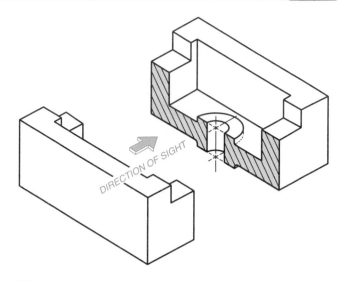

DIRECTION OF SIGHT

FIGURE 6-12   *Pictorial view of the full section*

Notice that section lining is applied only to the area the imaginary cutting plane passed through. The back side of the hole and the back sides of the notches are **not** section lined.

OFFSET SECTION

Many times, important features do not fall in a straight line as they do in a full section. These important features can be illustrated in an *offset section* by bending or off-setting the cutting-plane line. An offset section is very similar to a full section, except that the cutting-plane line is not straight, *Figure 6-13*.

Note that the features of the countersunk holes A, projection B with its counterbore, and groove C with a shoulder are not aligned with one another. The cutting-plane line is added, and changes of direction (staggers) are formed by right angles to pass through these features. An offset cutting-plane line A-A is added to the top view, and the material behind the cutting plane is

GIVEN:

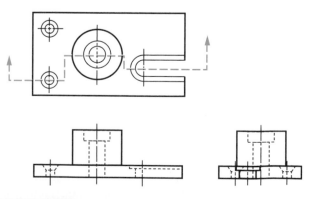

FIGURE 6-13   *Offset section*

Industry Application

WORKPLACE SKILLS

John Diesel, a CAD technician in an automotive company, was approached by his chief engineer, who described a fuel tank problem to John. Production needed the volume of an oddly shaped fuel tank, 30" long and partially elliptical-rectangular in cross section. The tank had been detailed but had not been built. Did John have any ideas?

One solution John proposed was to construct a prototype of the tank and measure the volume. John also noted that a less-expensive method would be to use the details and construct cross-sectional views of the tank at three-inch intervals, determine the area of each cross section, and finally calculate the volume by summing the products of each area multiplied by three inches.

The cross-sectional areas were determined and found to be 144, 146, 150, 152, 149, 145, 144, 142, 140, and 138 square inches. The total sum of each of these areas multiplied by three equals 4,350 cubic inches. When this figure was divided by 231 cubic inches per gallon, the volume of the tank was found to be 18.8 gallons.

viewed in section A-A, *Figure 6-14.* The front view is changed into an offset section, similar to a full-sectional view. The actual bends of the cutting-plane lines are omitted in the offset section, *Figure 6-15.* With the use of a sectional view, another view often may be omitted. In this example, the right-side view could have been omitted as it adds nothing to the drawing and takes extra time to draw.

HALF SECTION

In a *half section,* the object is cut only halfway through and a quarter section is removed. See the given problem, *Figure 6-16,* showing two views of an object. A cutting plane is added to the front view, with only one arrowhead to indicate the viewing direction. Also, a quarter section is removed and, in this example, the right side is sectioned accordingly, *Figure 6-17.* A pictorial view of this half section is illustrated in *Figure 6-18.* The visible half of the object that is not removed shows the exterior of the object, and the removed half shows the interior of

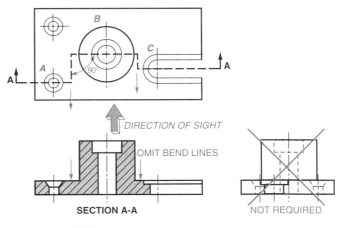

| **FIGURE 6-15** | *Bends omitted from section view* |

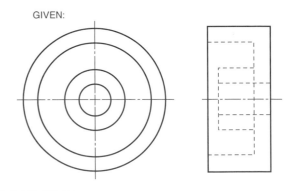

| **FIGURE 6-16** | *Given: Regular two views of an object* |

the object. The half of the object not sectioned can be drawn as it would normally be drawn, with the appropriate hidden lines.

Half sections are best used when the object is symmetrical, that is, the same shape and size on both sides

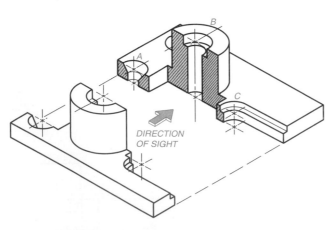

| **FIGURE 6-14** | *Pictorial view of the offset section* |

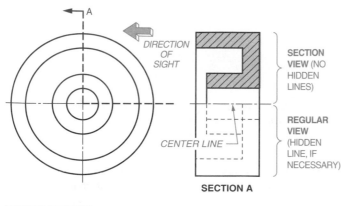

FIGURE 6-17   *Half section*

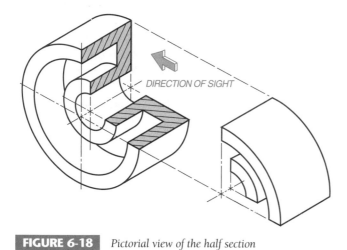

FIGURE 6-18   *Pictorial view of the half section*

of the cutting-plane line. A half-section view is capable of illustrating both the inside and the outside of an object in the same view. In this example, the top half of the right side illustrates the interior; the bottom half illustrates the exterior. A center line is used to separate the two halves of the half section (refer back to Figure 6-17). A solid line would indicate the presence of a real edge, which would be false information.

BROKEN-OUT SECTION

Sometimes, only a small area needs to be sectioned in order to make a particular feature or features easier to understand. In this case, a *broken-out section* is used, *Figure 6-19*. As drawn, the top section is somewhat confusing and could create a question. To clarify this area, a portion is removed, *Figure 6-20*. The finished drawing would be drawn as illustrated in *Figure 6-21*. The broken line is put in freehand and is drawn as a visible thick line. The actual cutting-plane line is usually omitted.

REVOLVED SECTION (ROTATED SECTION)

A *revolved section,* sometimes referred to as a *rotated section,* is used to illustrate the cross section of ribs, webs,

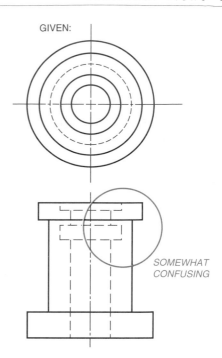

FIGURE 6-19   *Given: Regular two views of an object*

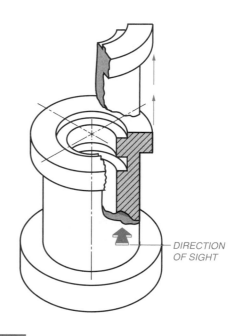

FIGURE 6-20   *Pictorial view of the broken-out section*

bars, arms, spokes, or other similar features of an object. *Figure 6-22* is a two-view drawing of an arm. The cross-sectional shape of the center portion of the arm is not defined. In drafting, no feature should remain questionable, and a section through the center portion of the arm would provide the complete information.

A revolved section is made by assuming a cutting plane perpendicular to the axis of the feature of the object to be described, *Figure 6-23*. Note that the rotation point occurs at the cutting-plane location and, theoretically,

Industry Application

WORKPLACE SKILLS

Betty Konrad was a CAD technician employed by a company that produced household appliances such as ovens, dishwashers, clothes washers, and dryers. The company planned to launch an advertising campaign to boost sales and expand their market base. They needed an effective means to communicate to the public just how their appliances worked.

Betty was assigned the task of organizing a group of two CAD technicians and two illustrators to prepare informational drawings to show how the appliances performed and what they were made of. To organize the team and to get them looking forward to working together, Betty called a meeting and tossed out several questions. "What do you think the users want to know about the appliances?" "What types of drawings would best show how they work and how they are assembled?" The group concluded that the most prevalent technical drawing showing the "insides" of appliances were pictorials with broken-out sections (Figure 6-20). Furthermore, the group felt that the users wanted to know how easy an appliance was to use and its capacity, such as what size turkey fit in an oven and how large loads could be in the washer and dryer.

Betty assigned two appliances to each CAD technician for the creation of isometric broken-out sections. The illustrators were to use this output to prepare colorful presentations. The illustrators would also include capacities.

As the team launched their efforts, Betty prepared a planning chart so each member would have a visual reminder of the team's schedule.

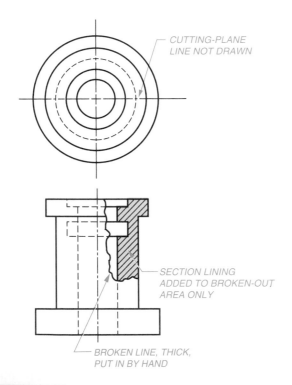

FIGURE 6-21 *Broken-out section*

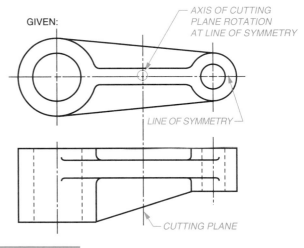

FIGURE 6-22 *Given: Regular two views of an object*

will be rotated 90°. Rotate the imaginary cutting-plane line about the rotation point of the object, *Figure 6-24.* Notice that dimension X is transferred from the top view to the sectional view of the feature—in this example, the front view. Dimension Y in the top view is also transferred to the front view. The section is now drawn in

place. The finished drawing is illustrated in *Figure 6-25.* Note that the break lines in the front view are on each side of the sectional view and are put in freehand.

The revolved section is not used as much today as it was in past years. Revolved sections tend to be confusing, and they often create problems for the people who must interpret the drawings. Today, we are encouraged to use a removed section instead of a revolved or rotated section.

REMOVED SECTION

A *removed section* is very similar to a rotated section except that, as the name implies, it is drawn removed or away from the regular views, *Figure 6-26.* The removed

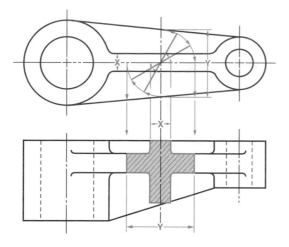

FIGURE 6-23 *Revolved section*

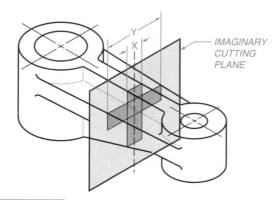

FIGURE 6-24 *Pictorial view of revolved section*

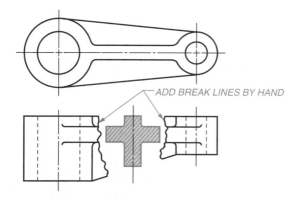

FIGURE 6-25 *Revolved section view*

section, as with the revolved section, is also used to illustrate the cross section of ribs, webs, bars, arms, spokes, or other similar features of an object.

A removed section is made by assuming that a cutting plane, perpendicular to the axis of the feature of the object, is added through the area that is to be sectioned. (Refer back to Figures 6-23 and 6-24.) Transfer dimensions X and Y to the removed views, exactly as was done in the

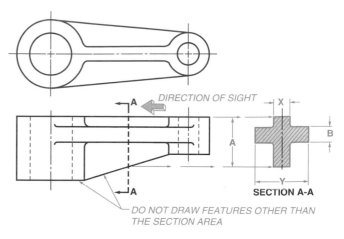

FIGURE 6-26 *Removed section*

rotated section. Height features, such as dimensions A and B, are transferred from the front view in this example.

Note that a removed section must identify the cutting-plane line from which it was taken. In the sectional view, do not draw features other than the actual section.

Removed sections are labeled section A-A, section B-B, and so forth, corresponding to the letters at the ends of the cutting-plane line. The sections are usually placed on the drawing in alphabetical order from left to right or from top to bottom, away from the regular views.

Sometimes a removed section is simply drawn on a center line that is extended from the object, *Figure 6-27*. A removed section can be drawn to an enlarged scale if necessary to illustrate and/or dimension a small feature. The scale of the removed section must be indicated directly below the sectional view, *Figure 6-28*.

In the field of mechanical drafting, the removed section should be drawn on the same page as the regular views. If there is not room enough on the same page and the removed section is drawn on another page, a page number cross reference must be given as to where the removed section may be found. The page where the removed section is located must refer back to the page from which the section is taken. For example, section A-A on sheet 2 of 4.

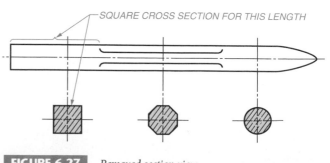

FIGURE 6-27 *Removed section view*

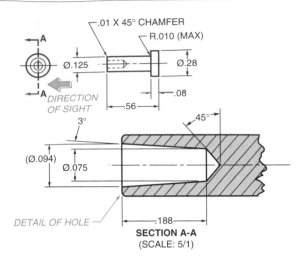

FIGURE 6-28 *Enlarged removed section*

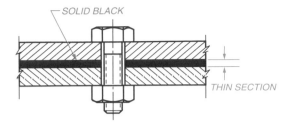

FIGURE 6-30 *Thinwall section*

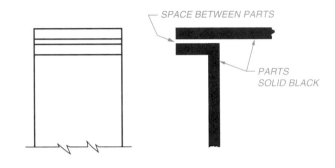

FIGURE 6-31 *Space between thinwall sections*

AUXILIARY SECTION

If a sectional view of an object is intended to illustrate the true size and shape of an object's boundary, the cutting-plane path must be perpendicular to the axis or surfaces of the object. An *auxiliary section* is projected in the same way as any normal auxiliary view, *Figure 6-29*.

THINWALL SECTION

Any very thin object that is drawn in section, such as sheet metal, a gasket, or a shim, should be filled-in solid black, as it is impossible to show the actual section lining. This is called a *thinwall section, Figure 6-30*. If several thin pieces that are filled-in solid black are touching one another, a small white space is left between the solid thinwall sections, *Figure 6-31*.

ASSEMBLY SECTION

When a sectional drawing is made up of two or more parts it is called an *assembly section, Figure 6-32*. An assembly section can be a full section (as it is in this example), an offset section, a half section, or a combination of the various kinds of sectional views. The assembly section shows how the various parts go together.

Each one of the section linings in Figure 6-32 is the accepted standard described in the beginning of the chapter. Nevertheless, sometimes you will be given drawings using lining symbols to represent different materials

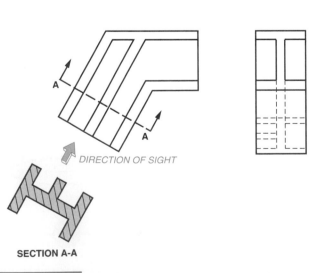

FIGURE 6-29 *Auxiliary section*

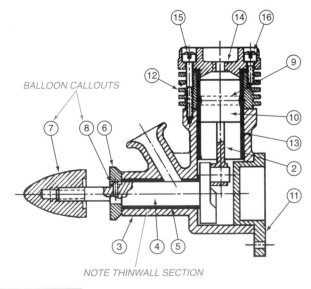

FIGURE 6-32 *Assembly section*

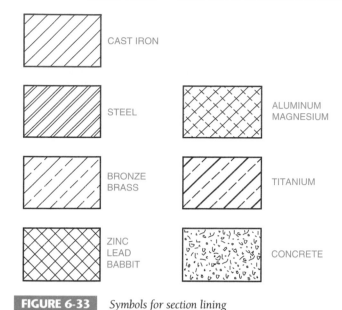

FIGURE 6-33   *Symbols for section lining*

as illustrated in *Figure 6-33*. The symbols shown represent the more commonly used engineering metals and concrete. Symbols are also available for rubber, plastic, cork, insulation, slate, earth, rock, sand, and wood.

Each part in the assembly must be labeled with a name, part or plan number, and the quantity required for one complete assembly. If the assembly section does not have many parts, this information is added by a note alongside each part. If the assembly has many parts, and there is not enough room to prevent the drawing from appearing cluttered, each individual part may be identified by a number within a circle called a *balloon*. The balloon callout system is used in this example. A table must be added to the drawing, listing the name, part or plan number, the quantity required for one complete assembly, and a cross reference to the corresponding balloon number. This is called a *parts list*. The exact form of the list varies from company to company. *Figure 6-34* is an example of a parts list used with the balloon system of callouts. Notice that entries are sometimes listed in reverse (bottom to top) order, as illustrated.

Occasionally drafters deviate from conventional standards for clarity, even though the deviation may violate a basic standard. The first part of this chapter covered conventional standards for sectioning; the latter part will focus on deviations for clarity.

Sections through Ribs or Webs

True projection of a sectioned view often produces incorrect impressions of the actual shape of the object. *Figure 6-35* has given a front view and a right-side view. Its pictorial view would look like *Figure 6-36*. A full section A-A would appear as it does in *Figure 6-37*. This is a true projection of section A-A, as the cutting-plane line passes through the rib.

CORRESPONDS TO BALLOON
NUMBERS IN FIGURE 6-32

PART NO.	NO. REQD.	DESCRIPTION	MATERIAL	NOTES
6	1	DRIVE WASHER	1018 STEEL	
5	1	BUSHING	LEADED BRONZE	
4	1	CRANKSHAFT	AISI 1040 STEEL	
3	1	CRANKCASE	ALUMINUM ALLOY	
2	1	CONNECTING ROD	ALUMINUM ALLOY 7075T6	
1				(NO. NOT USED)

USUAL CALLOUT INFORMATION

FIGURE 6-34   *Examples of a parts list*

GIVEN:

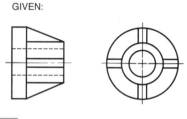

FIGURE 6-35   *Given: Two-view drawing of an object*

PICTORIAL VIEW

FIGURE 6-36   *Pictorial view of an object*

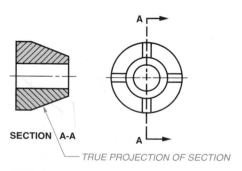

SECTION A-A

TRUE PROJECTION OF SECTION

FIGURE 6-37   *True projection of an object*

Industry Application

CLEARING UP THE CONFUSION

The drafting team was dealing with a complex, irregularly shaped machine part. To adequately provide all of the information needed to manufacture the part, the team had to draw all six orthographic views. This did not completely solve the problem. "This part just has too many hidden lines," said the team leader. "I don't think I could make this part from these drawings. With all of the hidden lines, it's hard to determine what is required. Any suggestions?"

"Why don't we cut some sections?" asked one of the team members. "I've been out of school too long," said the team leader. "Where would you cut them and how many?" "As I see it," said the team member, "we need to cut a full section to show the right side, and another for the left side. Then we need an offset section from the direction of the front view and one broken-out section."

Various team members began sketching as they talked. Soon the types and numbers of sections needed had been determined and the drafting team set to work revising the drawings. The final result was a set of drawings that gave manufacturing personnel the information they needed to make the part without difficulty.

However, such a sectional view gives an incorrect impression of the object's actual shape, and it is poor drawing practice. It misleads the viewer into thinking the object is actually shaped as it is shown in *Figure 6-38*. The conventional practice used to illustrate this section is to draw the section view as illustrated in *Figure 6-39*, which is not a true projection. Note that the web or rib is not section lined.

Some companies use another method to compensate for this problem. It is somewhat of a middle ground or a combination of true projection and correct representation, *Figure 6-40*. This is called alternate section lining. Section lining, over the rib or web section, is drawn

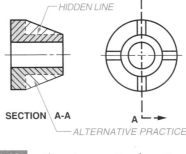

FIGURE 6-40   *Alternate conventional practice*

using every **other** section line, and the actual shape is indicated by hidden lines. However, most companies do not use alternate section lining.

Another example of a cutting plane passing through a rib or web is shown in *Figure 6-41*. This example is a true projection, but it is poor drafting practice because it gives the impression that the center portion is a thick,

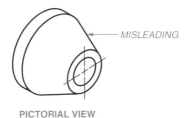

PICTORIAL VIEW

FIGURE 6-38   *True projection can be misleading*

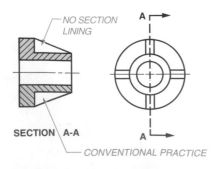

FIGURE 6-39   *Conventional practice—web or rib not sectioned*

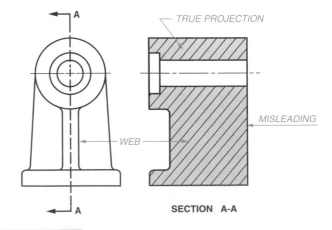

FIGURE 6-41   *Example of true projection*

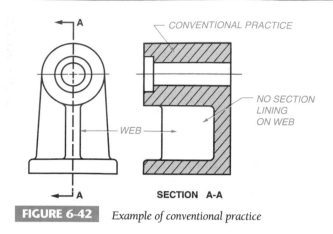

Example of conventional practice

solid mass. *Figure 6-42* is drawn incorrectly, but it does not give the false impression of the object's center portion. This is the conventional practice used.

Holes, Ribs and Webs, Spokes and Keyways

Holes located around a bolt circle are sometimes not aligned with the cutting-plane line, *Figure 6-43*. The cutting plane passes through only one hole. This is a true projection of the object, but poor drafting practice. In actual practice, the top hole is theoretically revolved to the cutting-plane line and projected to the sectional view, *Figure 6-44*. This practice is called *aligning of features*.

RIBS AND WEBS

Ribs or webs sometimes do not align with the cutting-plane line, *Figure 6-45*. The cutting plane passes through only one web and only one hole. This is a true projection of the object, but poor drafting practice. In actual practice, one of the webs is theoretically revolved up to the cutting-plane line and projected to the sectional view, *Figure 6-46*. Notice that the bottom hole is unaffected and is projected normally. This is another example of aligning of features.

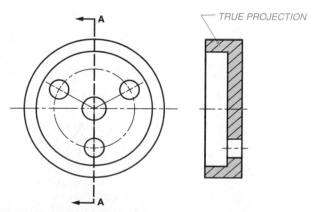

Holes using true projection

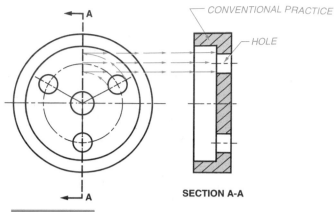

Holes using conventional practice (aligning of features)

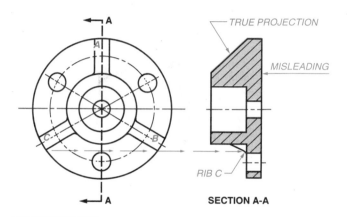

Rib or web using true projection

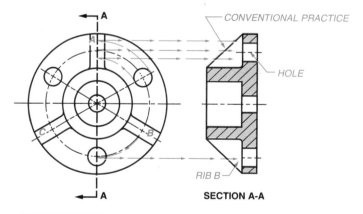

Rib or web using conventional practice

SPOKES AND KEYWAYS

Spokes and keyways and other important features sometimes do not align with the cutting-plane line, *Figure 6-47*. The cutting-plane line passes through only one spoke and misses the keyway completely. This also is a true projection of the object, but poor drafting practice. In conventional practice, spoke B is revolved to the cutting-plane line and projected to the sectional view, *Figure 6-48*. The keyway is also projected as illustrated. This is another example of aligning of features.

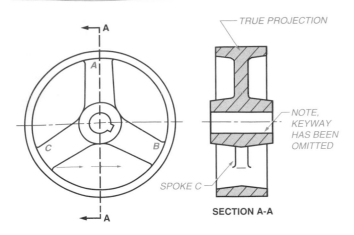

FIGURE 6-47 *Spokes and keyway using true projection*

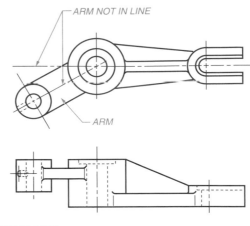

FIGURE 6-49 *Two-view drawing*

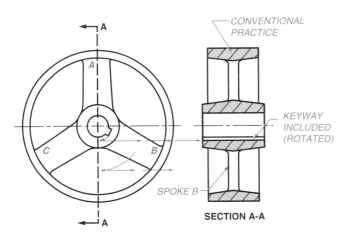

FIGURE 6-48 *Spokes and keyway using conventional practice*

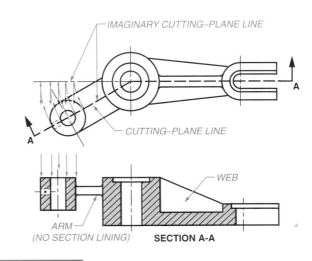

FIGURE 6-50 *Aligned section*

Aligned Sections

Arms and other similar features are revolved to alignment in the cutting plane, as were spokes in the preceding section. This procedure is used if the cutting-plane line cannot align completely with the object, as illustrated in *Figure 6-49*.

The arm or feature is now revolved to the imaginary cutting plane and projected down to the sectional view, *Figure 6-50*. The actual cutting-plane line is bent and drawn through the arm or feature and then revolved to a straight, aligned vertical position. Notice that section lining is **not** applied to the arm, and it is also omitted from the web area.

Fasteners and Shafts in Section

If a cutting plane passes lengthwise through any kind of fastener or shaft, the fastener or shaft is **not** sectioned, *Figure 6-51*. Section lining of a fastener or shaft would

have no interior detail, thus it would serve no purpose and only add confusion to the drawing. The round head machine screw, the hex head cap screw with nut, and the rivet are not sectioned. The other objects in the figure such as fasteners, ball bearing rollers, and so forth, are also not sectioned.

If a cutting-plane line passes perpendicularly through the axis of a fastener or shaft, section lining is added to the fastener or shaft, *Figure 6-52*. The end view has section lining added as shown.

Intersections in Section

Where an intersection of a small or relatively unimportant feature is cut by a cutting-plane line, it is not drawn as a true projection, *Figure 6-53*. Since a true projection takes drafting time, it is preferred that it be disregarded, and the feature drawn, using conventional practice, as shown in *Figure 6-54*. This procedure is much quicker and more easily understood.

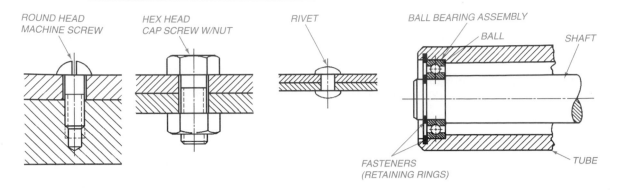

ROUND HEAD
MACHINE SCREW

HEX HEAD
CAP SCREW W/NUT

RIVET

BALL BEARING ASSEMBLY

BALL

SHAFT

FASTENERS
(RETAINING RINGS)

TUBE

FIGURE 6-51 *Parts not sectioned*

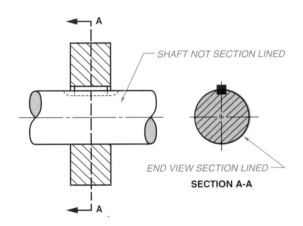

SHAFT NOT SECTION LINED

END VIEW SECTION LINED

SECTION A-A

FIGURE 6-52 *Shaft sectioned in end view only*

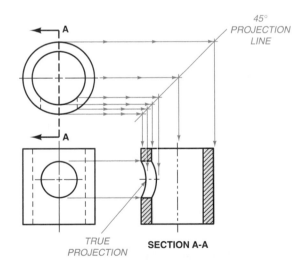

45°
PROJECTION
LINE

TRUE
PROJECTION

SECTION A-A

FIGURE 6-53 *Intersection using true projection*

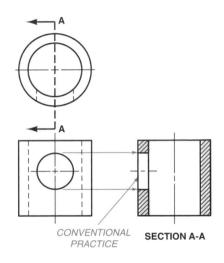

CONVENTIONAL
PRACTICE

SECTION A-A

FIGURE 6-54 *Intersection using conventional practice*

- The direction of sight is shown by arrows on the cutting-plane line. It indicates the direction of sight for viewing the completed sectional view.

- Section lining shows where the object is sliced or cut by the cutting-plane line. It indicates the surface that is touched by the cutting plane.

- When an object is complicated and cut in more than one place, each cutting-plane line must be labeled starting with section A-A, B-B, C-C, and so on.

- The kinds of sections students can expect to draw are as follows: full; offset; half; broken-out; revolved; removed; auxiliary; thinwall; and assembly.

- A section cut through a rib or a web can result in a misleading representation if true projection is applied. Consequently, in conventional practice section lines are not drawn in the rib or web.

- When cutting sections through objects that have holes, ribs, webs, spokes, and/or keyways that are not obliqued, a cutting plane can pass through one feature and miss another. This problem is solved by revolving the missed feature over to the cutting plane. This same

Summary

- A sectional view is a view of the surfaces of an object that would be seen if an object could be cut open.

- A cutting-plane line indicates the path that an imaginary cutting plane follows to slide through an object.

technique is used in creating obliqued sections of objects with arms or other features that do not line up with a cutting plane.

- If a cutting plane passes lengthwise through any type of fastener or shaft, the fastener or shaft is not section lined.

Review Questions

Answer the following questions either true or false.

1. True projection of a sectional view always produces the actual shape of the object.

2. Section lining should be parallel with the sides of the object to be section lined.

3. A revolved section is the same as a rotated section.

4. A removed section is very similar to a rotated section except that it is drawn away from the regular view.

5. Hidden lines are usually not used in a sectional view.

6. A removed section is sometimes drawn to larger scale to illustrate and/or dimension a small feature.

7. Half sections are best used when an object is **not** symmetrical.

8. Today, it is recommended to use the removed section rather than the revolved section.

9. If a cutting-plane line passes lengthwise through any kind of fastener or shaft, the fastener or shaft is not sectioned.

10. If a cutting-plane line passes perpendicularly through the axis of a fastener or shaft, the fastener or shaft is not sectioned.

Answer the following questions by selecting the best answer.

1. Which of the following is **not** a list of types of sectional views?
   a. Full, half, offset
   b. Front, back, profile
   c. Broken-out, revolved, removed
   d. Thinwall, auxiliary, assembly

2. Which of the following is **not true** regarding alternate section lining?
   a. It is a combination of true projection and correct representation.
   b. Over the rib or web section, every other section line is drawn.
   c. The rib or web section is not lined.
   d. The actual shape is indicated by hidden lines.

3. What kind of section view illustrates both the exterior and interior of the object?

   a. Half section
   b. Full section
   c. Broken-out
   d. Revolved

4. Which of the following is **not** true regarding an assembly section?
   a. When a sectional drawing is made up or two or more parts, it is called an assembly section.
   b. A table, called a parts list, may be added to the drawing.
   c. Each part number may be identified by a number within a circle called a balloon.
   d. The assembly section is always represented using a full section view.

5. What must be done if a removed section needs to be placed on a page other than the page on which the cutting-plane line is placed?
   a. The page where the removed section is located must refer back to the page from which it was taken.
   b. A page number cross-reference must be given to where the removed section may be found.
   c. Both a and b
   d. None of the above

6. What must be included for each part in an assembly section?
   a. Name of part
   b. Part number or plan number
   c. Quantity required to complete one assembly
   d. All of the above

Chapter 6 Problems

The following problems are intended to give the beginning drafter practice in using the various kinds of sectional views used in industry. As these are beginning problems, no dimensions will be used at this time.

CONVENTIONAL INSTRUCTIONS

The steps to follow in laying out all problems in this chapter are:

STEP 1 Study the problem carefully.

STEP 2 Choose the view with the most detail as the front view.

STEP 3 Position the front view so there will be the least amount of hidden lines in the other views.

STEP 4 Make a sketch of all required views.

STEP 5 Determine what should be drawn in section, what type of section should be used, and where to place the cutting-plane line.

STEP 6 Center the required views within the work area with a 1″ (25 mm) space between views.

STEP 7 Use light projection lines. Do **not** erase them.

STEP 8 Lightly complete all views.

STEP 9 Check to see that all views are centered within the work area.

STEP 10 Check to see that there is a 1″ (25 mm) space between all views.

STEP 11 Carefully check all dimensions in all views.

STEP 12 Darken in all views using correct line thickness.

STEP 13 Add a cutting-plane line and section lining as required.

STEP 14 Recheck all work and, if it is correct, neatly fill out the title block using light guidelines and neat lettering.

CAD Instructions

Students who plan to complete the following problems on a CAD system should begin with these general instructions. Before reading the specific instructions for each activity, go through each step in the following planning checklist. The checklist applies to any CAD system and will help ensure the optimum use of your time and resources.

1. Analyze the problem carefully. Decide exactly what you are being asked to do.

2. Determine what resources and references (if any) you will need in order to complete the problem. Collect those references and resources and have them readily available.

3. Decide if any particular standards apply to the project and have those standards available.

4. Determine what types of views will be required and how many of each.

5. Determine what the final plotted scale of the drawing will need to be and select the appropriate paper size for plotting/printing.

6. Plan your drawing sequence. In what order will you develop the drawing (i.e., lines, features, dimension lines, leaders, dimensions)?

7. Review the various CAD commands you will have to use in order to develop the drawing.

8. Examine your CAD system to ensure that everything is in working order, then begin the project.

Problem 6-1

Center three views within the work area, and make the front view a full section.

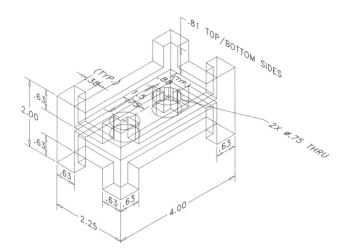

Problem 6-2

Center two views within the work area, and make one view a full section. Use correct drafting practices for the ribs.

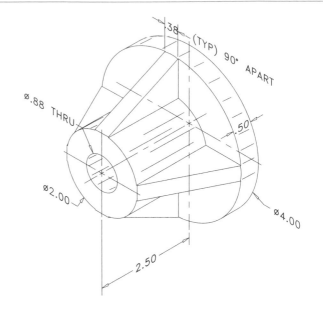

Problem 6-3

Center two views within the work area, and make one view a full section. Use correct drafting practices for the holes.

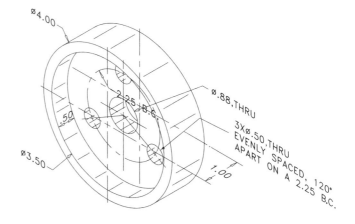

Problem 6-4

Center the front view and top view within the work area. Make one view a full section.

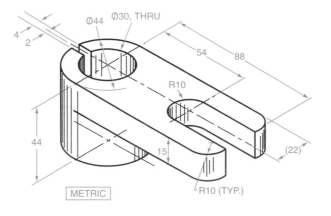

Problem 6-5

Center two views within the work area, and make one view a full section. Use correct drafting practices for the arms, horizontal hole, and keyway.

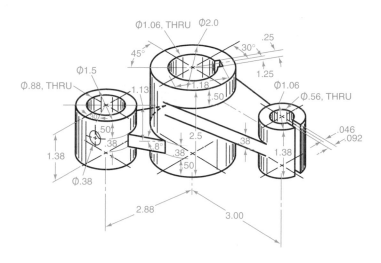

Problem 6-6 X

Center two views within the work area, and make one view a full section. Use correct drafting practices for the keyway, ribs, and holes.

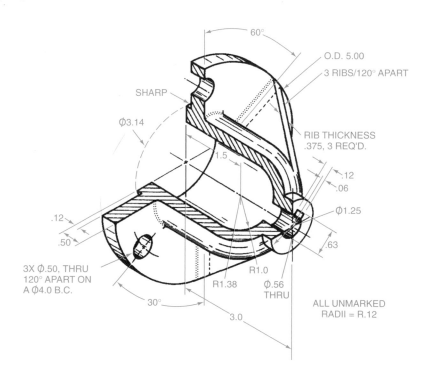

Problem 6-7 ✗

Center three views within the work area, and make one view an offset section. Be sure to include three major features.

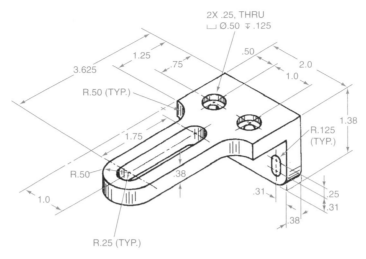

Problem 6-8

Center three views within the work area, and make one view an offset section. Be sure to include three major features.

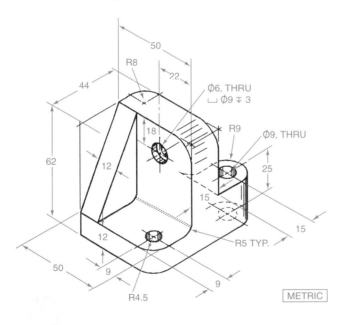

Problem 6-9

Center three views within the work area, and make one view an offset section. Be sure to include as many of the important features as possible.

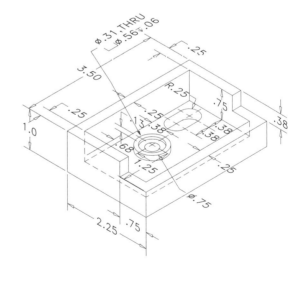

Problem 6-10

Center three views within the work area, and make one view an offset section. Be sure to include as many of the important features as possible.

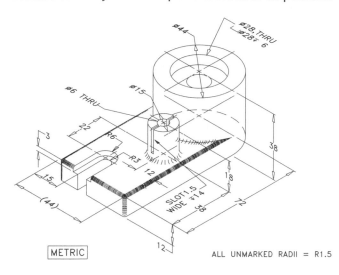

METRIC

ALL UNMARKED RADII = R1.5

Problem 6-12

Center the front view and top view within the work area. Make one view a half section.

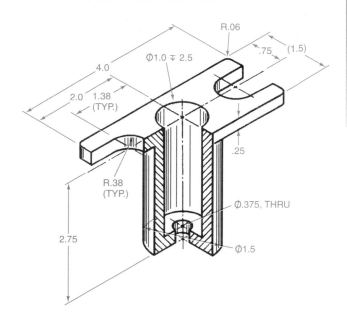

Problem 6-11

Center two views within the work area, and make one view an offset section. Be sure to include as many of the important features as possible.

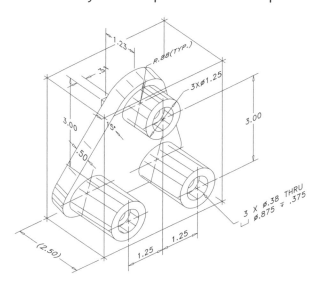

Problem 6-13

Center two views within the work area, and make one view a half section.

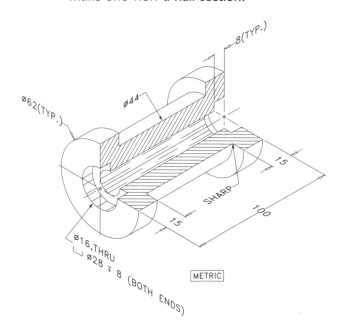

METRIC

Problem 6-14

Center the two views within the work area, and make one view a half section.

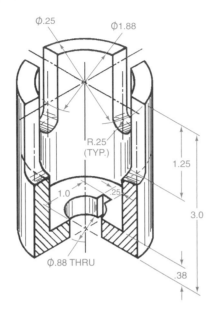

Ø.25 Ø1.88

R.25 (TYP.)

1.0 .25

1.25

3.0

.38

Ø.88 THRU

Problem 6-15

Center two views within the work area, and make one view a half section.

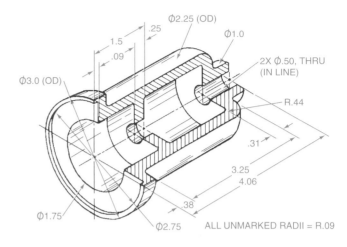

Ø2.25 (OD) Ø1.0

1.5 .25

.09

Ø3.0 (OD)

2X Ø.50, THRU (IN LINE)

R.44

.31

3.25
4.06

38

Ø1.75

Ø2.75

ALL UNMARKED RADII = R.09

Problem 6-16

Center two views within the work area, and make one view a half section.

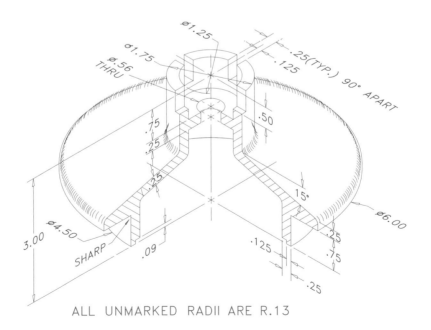

ø1.25

ø1.75

ø.56

THRU

.25 (TYP.) 90° APART

.125

.75

.25

.50

.25

15°

ø6.00

3.00

ø4.50

SHARP

.09

.125

.25

.75

.25

ALL UNMARKED RADII ARE R.13

Problem 6-17

Center the required views within the work area, and make one view a broken-out section to illustrate the complicated interior area.

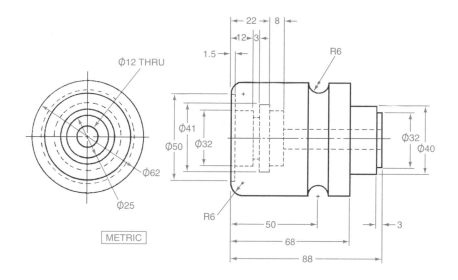

Problem 6-18

Center the required views within the work area, and make one view a broken-out section as required.

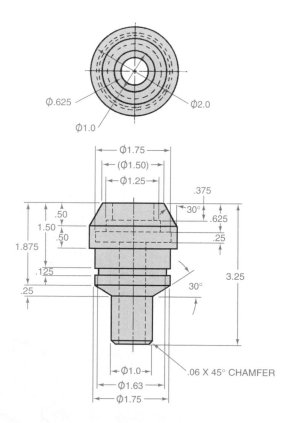

Problem 6-19

Center three views within the work area, and add removed sections A-A and B-B.

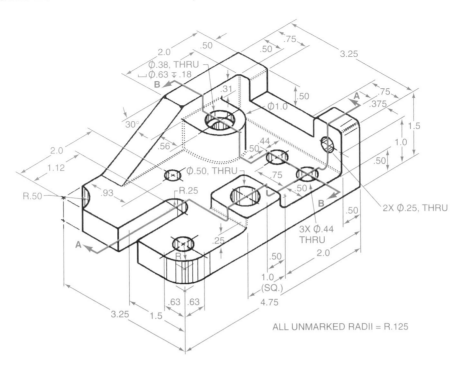

ALL UNMARKED RADII = R.125

Problem 6-20

Center two views within the work area, and add removed sections A-A, B-B, and C-C.

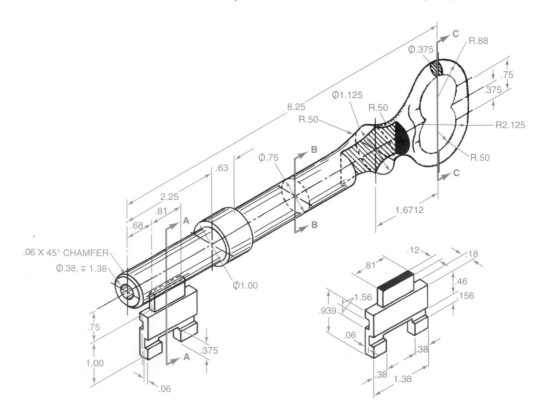

Problem 6-21 X

Center the required views within the work area, and add the removed section as required.

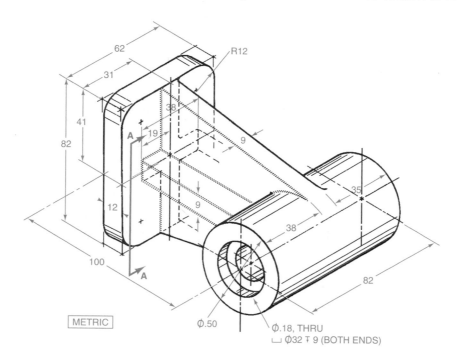

62
31
R12
41
38
82
A
19
9
12
9
35
100
A
38
82
METRIC
Ø.50
Ø.18, THRU
Ø32 ⊤ 9 (BOTH ENDS)

Problem 6-22

Center the required views within the work area, and add the removed section as required.

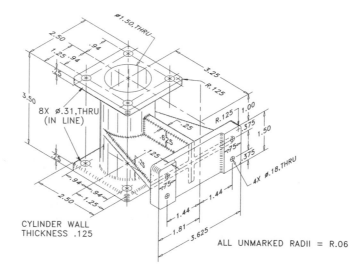

Ø1.50, THRU
2.50
.94
1.25 .94
.25
3.25
3.50
R.125
8X Ø.31,THRU
(IN LINE)
R.125 1.00
.25
.375
.625
1.50
.375
.125
.25
.94 .94
.75
2.50 1.25
4X Ø.18,THRU
.75
CYLINDER WALL
THICKNESS .125
1.44
1.44 1.44
1.81
3.625
ALL UNMARKED RADII = R.06

Problem 6-23

Center the required views within the work area, and add removed sections A-A and B-B.

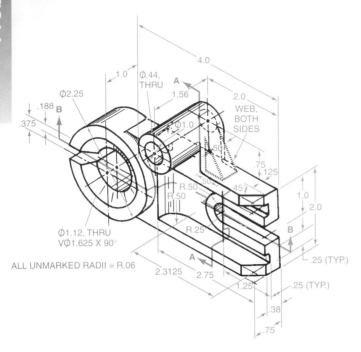

Problem 6-24

Center the required views within the work area, and add removed section A-A.

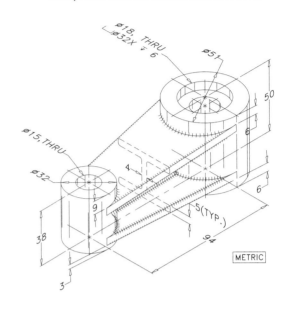

Problem 6-25

Center the four views within the work area. Make the top view section A-A and the right-side view section B-B.

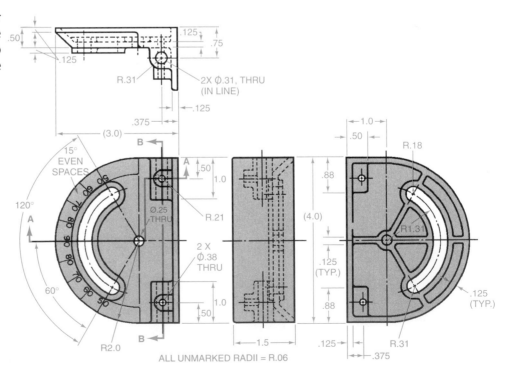

Problem 6-26

Center the required views within the work area, and add removed sections A-A and B-B.

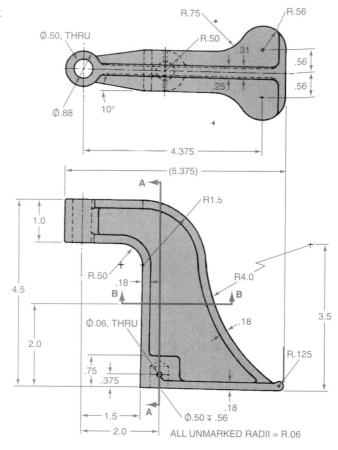

Problem 6-27

Center the required views within the work area, and add removed sections A-A and B-B.

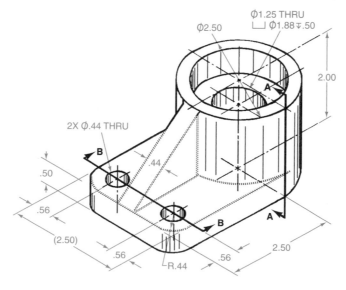

ALL UNMARKED RADII = R.06

Problem 6-28

Center the front view, side view, and removed sections A-A, B-B, and C-C within the work area.

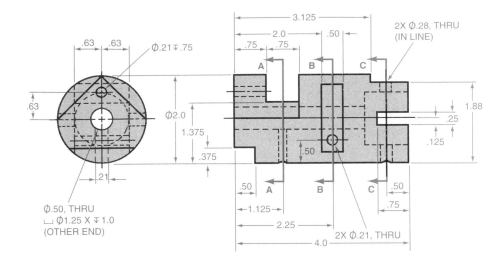

Problem 6-29

Make a two-view assembly drawing of parts 1, 2, 3, and 4. Make one view a full-section assembly. Use correct section lining and all conventional drafting practices.

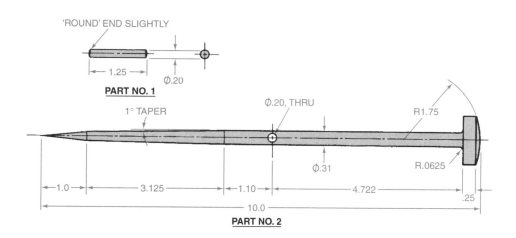

'ROUND' END SLIGHTLY

1.25 Ø.20

PART NO. 1

1° TAPER Ø.20, THRU R1.75

Ø.31 R.0625

1.0 3.125 1.10 4.722 .25

10.0

PART NO. 2

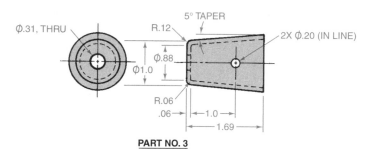

Ø.31, THRU R.12 5° TAPER 2X Ø.20 (IN LINE)

Ø.88 Ø1.0

R.06 .06 1.0 1.69

PART NO. 3

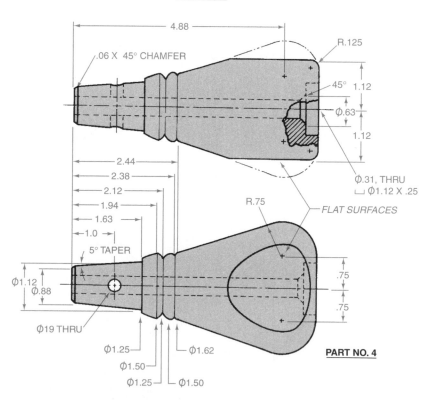

4.88 R.125

.06 X 45° CHAMFER 45° 1.12 Ø.63 1.12

2.44 Ø.31, THRU ⌴ Ø1.12 X .25

2.38 2.12 1.94 1.63 1.0 R.75 FLAT SURFACES

5° TAPER

Ø1.12 Ø.88 .75 .75

Ø19 THRU

Ø1.25 Ø1.62 Ø1.50 Ø1.25 Ø1.50

PART NO. 4

Problems

Problem 6-30

Make a two-view assembly drawing of parts 1, 2, and 3. Make one view a full-section assembly. Use correct section lining and all conventional drafting practices.

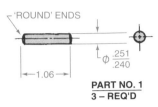

'ROUND' ENDS

$\phi \frac{.251}{.240}$

1.06

PART NO. 1
3 – REQ'D

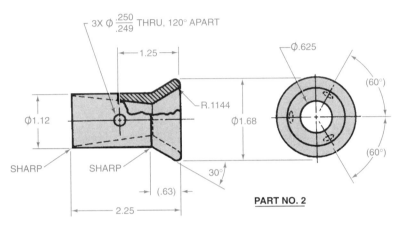

3X $\phi \frac{.250}{.249}$ THRU, 120° APART

1.25

ϕ.625

(60°)

R.1144

ϕ1.12

ϕ1.68

(60°)

SHARP   SHARP   30°

(.63)

2.25

PART NO. 2

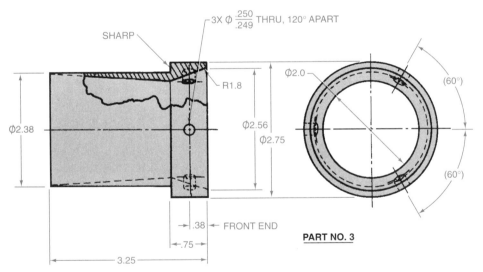

3X $\phi \frac{.250}{.249}$ THRU, 120° APART

SHARP

R1.8

ϕ2.0

(60°)

ϕ2.38

ϕ2.56

ϕ2.75

(60°)

.38 — FRONT END

.75

3.25

PART NO. 3

Problem 6-31

Make a two-view assembly drawing of parts 1, 2, and 3. Make one view a full-section assembly. Use correct section lining and all conventional drafting practices.

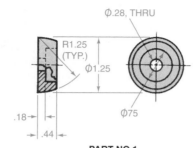

PART NO.1

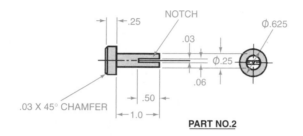

PART NO.2

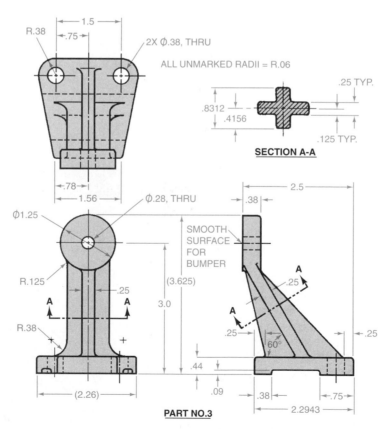

SECTION A-A

PART NO.3

PROBLEMS 6-32 THROUGH 6-37

Center required views within the work area. Leave a 1" or 25 mm space between views. Make one view into a section view to fully illustrate the object. Use either a full, half, offset, broken-out, revolved, or removed section. Do not add dimensions.

Problem 6-32

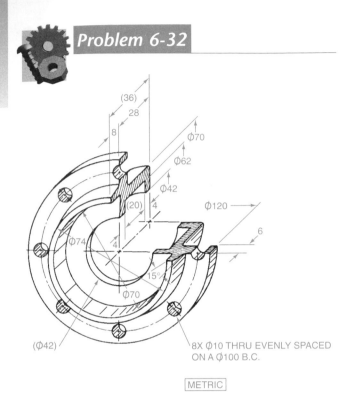

8X Ø10 THRU EVENLY SPACED
ON A Ø100 B.C.

METRIC

Problem 6-34

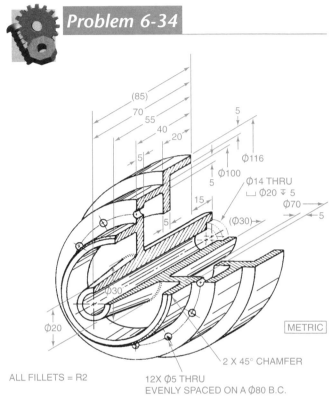

Ø14 THRU
⌴ Ø20 ⍱ 5

2 X 45° CHAMFER

12X Ø5 THRU
EVENLY SPACED ON A Ø80 B.C.

ALL FILLETS = R2

METRIC

Problem 6-33

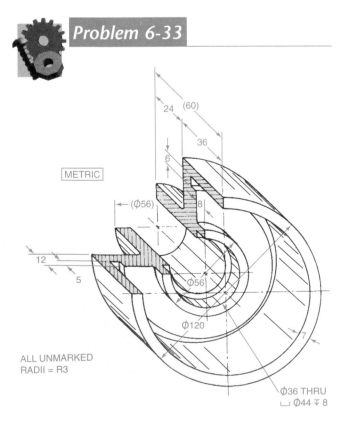

METRIC

ALL UNMARKED
RADII = R3

Ø36 THRU
⌴ Ø44 ⍱ 8

Problem 6-35

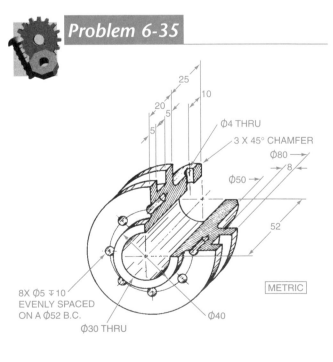

Ø4 THRU

3 X 45° CHAMFER

8X Ø5 ⍱ 10
EVENLY SPACED
ON A Ø52 B.C.

Ø30 THRU

Ø40

METRIC

Problem 6-36

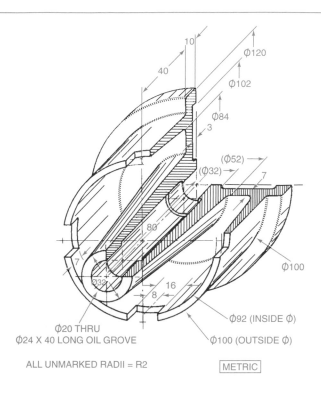

Ø120
Ø102
Ø84
3
10
40
(Ø52)
(Ø32)
7
80
Ø100
7
Ø32
16
8
Ø20 THRU
Ø24 X 40 LONG OIL GROVE
Ø92 (INSIDE Ø)
Ø100 (OUTSIDE Ø)

ALL UNMARKED RADII = R2

METRIC

Problem 6-37

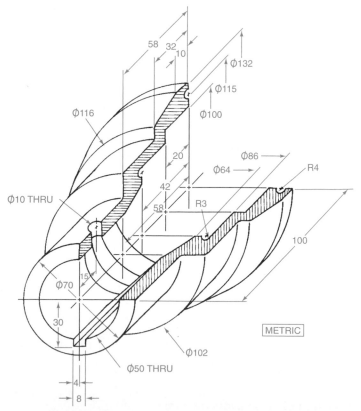

58 32
10
Ø132
Ø115
Ø100
Ø116
20
Ø86
Ø64
R4
Ø10 THRU
42
58
R3
100
Ø70
15
30
Ø102
Ø50 THRU
4
8

ALL UNMARKED RADII = R5

METRIC

PROBLEM 6-38

Refer to Problem 17-9 and use the full-section drawing for information on dimensions to determine the overall distance from the outside of part 1 on the left to the outside of part 1 on the right, along the center line of the pulley. Prepare a sketch with dimensions for your answer.

PROBLEM 6-39

Refer to Problem 18-52 (assembly) and prepare an assembly section (front-view orientation) of the vise. Indicate approximate dimensions only.

PROBLEM 6-40

Refer to Problem 18-50 and prepare a full-section view of the object.

PROBLEM 6-41

Refer to Problem 10-21 and prepare an offset section view (profile orientation) of the object through the center line. Include a threaded hole.

Problem 6-42

Select an object having a nonuniform cross section like the scissor half shown and prepare removed section views to illustrate detail of the object. Suggested objects: hammer, pliers, bolt cutters, boat oar, snow ski, water ski, propeller (marine, airplane, contemporary wind machine, etc.), nozzle, handles, or spokes.

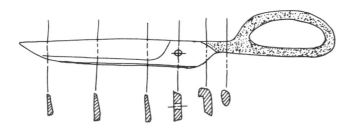

PROBLEM 6-43

Refer to Problem 10-12 and prepare an aligned section view of the object. Show the threaded holes and the keyway using conventional practices.

PROBLEM 6-44

Refer to Problem 10-37 and prepare an aligned section view of the object.

PROBLEM 6-45

Refer to Problem 10-43. Prepare a conventional section view, showing the ribs of the object.

PROBLEM 6-46

Refer to Problem 17-13 (top view). Prepare a section, drawn to scale.

Problem 6-47

Refer to the drawings shown. Prepare an offset section C-C, full size, of the distributor cap. Dimensions are not required.

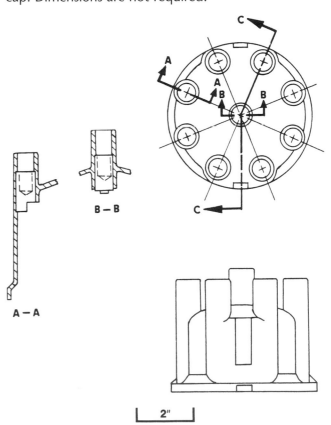

PROBLEM 6-48

Select one of the following common devices and prepare a section view(s) to best show inside details, and how the device works, for a nontechnical person. Simplify the details of the device for clarity. How-things-work books would be useful references. Indicate approximate sizes only.

   a. clock
   b. toaster
   c. internal-combustion-engine cylinder
   d. ski binding
   e. electric drill
   f. car window mechanism (manual or electrical)
   g. electric motor application in an average home, including an automobile.

NOTE: The average home with an automobile has approximately 35 electric motors.

PROBLEM 6-49

Find material on threaded inserts (usually a threaded collar made of steel, inserted in softer material such as aluminum). Prepare a section drawing showing an application. Indicate approximate dimensions.

PROBLEM 6-50

Select an object from one of the locations or categories listed and prepare a multiview drawing of your redesign ideas for the object. Use section views to replace traditional views where appropriate. (Refer to Figure 6-15.)

   a. classroom
   b. lab
   c. shop
   d. garage
   e. automobile
   f. hobby
   g. sports equipment

Problem 6-51

Use the partially completed drawings shown and do the following.

   a. Assume that the beam used has a depth of 152 mm and a flange width of 84 mm. Estimate dimensions of the parts that make up dimension X, and show in a freehand sketch dimension X and your assumptions.

   b. Make additional estimates: (1) of the thickness of the plates for the trolley wheels, and (2) of the distances from the trolley wheels to the lower bolted connection and to the hole. Prepare an offset section through the trolley wheels, through the bolted connection, and through the hole.

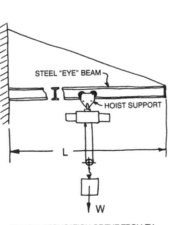

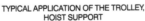

TYPICAL APPLICATION OF THE TROLLEY, HOIST SUPPORT

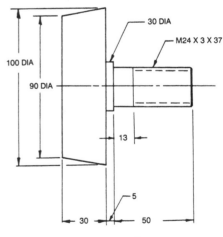

HALF SCALE

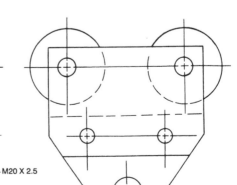

PROBLEM 6-52

Design a lever-type door handle for handicapped individuals who have prosthetics for hands and cannot use knob-type door openers. Use removed sections to show the details of the lever arm. Indicate approximate dimensions.

PROBLEM 6-53

Design a desk for a personal computer, keyboard, printer with track-fed paper, a mouse with pad, and a monitor. Prepare a multiview drawing of the desk and show a cross-sectional view (profile view orientation) of the desk. A 26″ knee standard is a typical standard for computer desks. Indicate approximate dimensions.

PROBLEM 6-54

Name or describe the type of sectioning used in the figures and problems listed.

Figure 10-105 (page 420), Figure 12-41 (page 501), Figure 13-14 (page 535), Figure 14-17 (page 583), Problem 17-9 (page 633), Figure 19-35 (page 725), Figure 20-13 (page 751), Figure 21-2 (page 789), and Figure 21-76 (page 838).

PROBLEM 6-55

Name or describe the type of sectioning used in the figures and problems listed.

Figure 10-127 (page 434), Figure 13-31 (page 540), Figure 16-30 (page 622), Problem 17-10 (page 634), Figure 20-3 (page 746), Figure 21-17 (page 800), Figure 21-69 (page 830), and Figure 21-77 (page 839).

PROBLEM 6-56

Name or describe the type of sectioning used in the figures and problems listed.

Problem 10-39 (page 457), Figure 12-5 (page 489), Figure 12-67 (page 511), Problem 12-25 (page 526), Figure 13-73 (page 564), Figure 17-17 (page 639), Problem 17-18 (page 640), Figure 20-8 (page 748), Figure 21-75 (page 837), and Figure 24-13 (page 949).

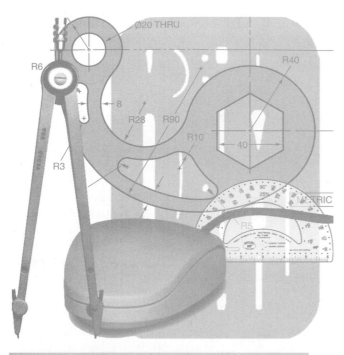

AUXILIARY VIEWS

CHAPTER OUTLINE

Auxiliary views defined • Secondary auxiliary views • Partial views • Auxiliary section • Half auxiliary views • Oblique surfaces, viewing planes, and auxiliary view enlargements • Summary • Review questions • Chapter 7 problems

CHAPTER OBJECTIVES

Upon completion of this chapter, students should be able to do the following:

- Define the concept of an auxiliary view and list its advantages.
- Describe and use the reference plane method to draw an auxiliary view.
- Describe and use the folding line method to draw an auxiliary view.
- Draw round and curved surfaces in auxiliary views.
- Illustrate secondary auxiliary views.
- Illustrate partial auxiliary views.
- Draw enlarged auxiliary views.
- Describe the use of a viewing plane for auxiliary views.

Auxiliary Views Defined

Many objects have inclined surfaces that are not always parallel to the regular planes of projection. For example, in *Figure 7-1*, the front view is correct as shown, but the top and right-side views do **not** correctly represent the inclined surface. To truly represent the inclined surface and to show its true shape, an auxiliary view must be drawn. An *auxiliary view* has a line of sight that is perpendicular to the inclined surface, as viewed looking directly at the inclined surface. Auxiliary views are always projected 90° from the inclined surface.

An auxiliary view serves three purposes:

- It illustrates the true size of a surface.
- It illustrates the true shape of a surface, including all true angles and/or arcs.
- It is used to project and complete other views.

An auxiliary view can be constructed from any of the regular views. An auxiliary view projected from the front view would appear as it does in Figure 7-1. This is referred to as a *front view auxiliary*. An auxiliary view projected from the top view would appear as it does in *Figure 7-2*. This is referred to as a *top view auxiliary*. An auxiliary view projected from the right-side view would appear as it does in *Figure 7-3*. This is referred to as a *side view auxiliary*. Note in each case that the auxiliary view is projected 90° from the inclined or slanted surface

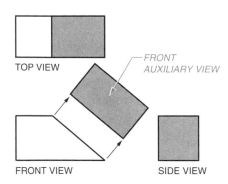

FIGURE 7-1 *Front view auxiliary view*

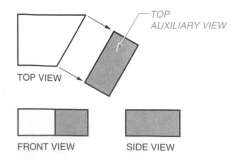

FIGURE 7-2 *Top view auxiliary view*

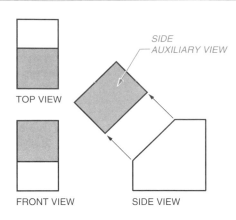

FIGURE 7-3 *Side view auxiliary view*

and is viewed from a line of sight 90° to the inclined or slanted surface, or as viewed looking directly down upon the inclined surface.

HIDDEN LINES IN AN AUXILIARY VIEW

Hidden lines should be omitted in an auxiliary view, unless they are needed for clarity. This is the drafter's prerogative or decision.

HOW TO DRAW AN AUXILIARY VIEW

GIVEN: The pictorial view of an object, *Figure 7-4*. Notice the inclined surface. Because the inclined surface is on the front view, this will be a front view auxiliary. The usual three views of an object, front view, top view and right-side view, are shown in *Figure 7-5*.

STEP 1 Label all important points of the auxiliary view, as illustrated in *Figure 7-6A*.

STEP 2 Construct a reference line, which is also the edge view of a reference plane, in the right-side view. Always construct a reference line so that it is vertical and passes through as many points as possible. In this example, it passes through points a–d, *Figure 7-6B*.

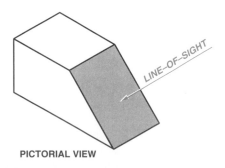

FIGURE 7-4 *Drawing an auxiliary view*

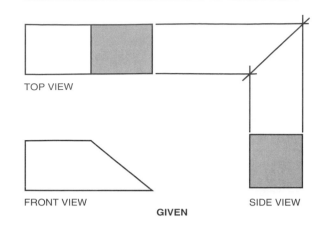

TOP VIEW

FRONT VIEW

SIDE VIEW

GIVEN

FIGURE 7-5 *Given: Three views of an object*

STEP 3 Draw light projection lines 90° from the inclined surface, and construct a reference line parallel to the inclined surface at any convenient distance, as shown in *Figure 7-6C*. Label all important points established thus far. Notice points a and d are **on** the reference line.

STEP 4 In the right-side view, measure the distance each point is **from** the reference line. Project these distances back to the inclined surface of the front view and up to 90° from the inclined surface to the reference line above. Transfer each distance, and lightly label each point as illustrated in *Figure 7-6D*.

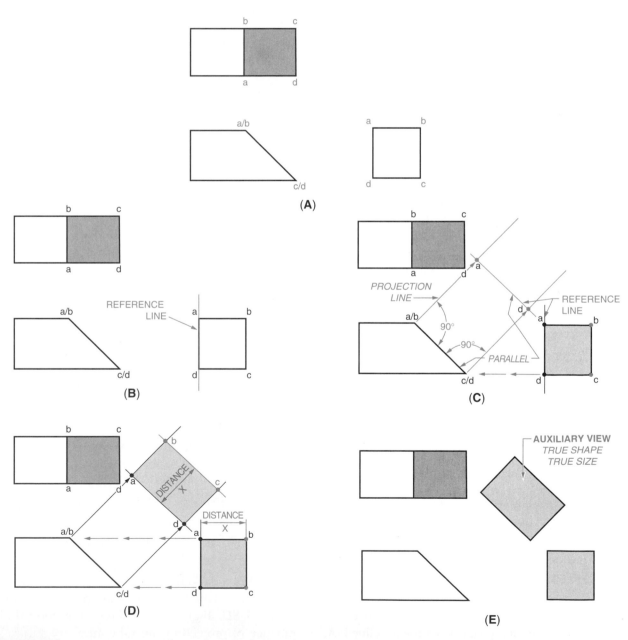

FIGURE 7-6 *(A) Step 1; (B) Step 2; (C) Step 3; (D) Step 4; (E) Step 5 (finished drawing)*

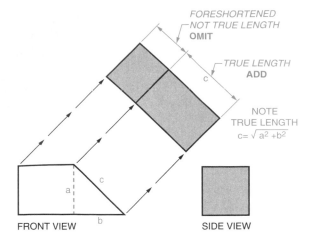

FRONT VIEW

SIDE VIEW

FIGURE 7-7 *Draw only the inclined surface in the auxiliary view.*

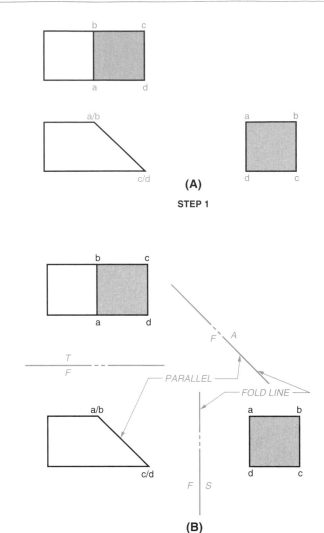

(A)

STEP 1

(B)

STEP 2

FIGURE 7-8 *(A) Step 1; (B) Step 2*

STEP 5 Recheck all work. Be sure:

- The projection is 90° from the inclined surface, in this example the front view.
- The reference line is parallel to the inclined surface of the front view.
- All distances have been transferred accurately.

If correct, carefully darken in and complete all views. The final finished drawing will appear as it does in *Figure 7-6E*. Notice that **only** the inclined surface is projected into the auxiliary view. Anything else would be foreshortened and, thus, not of true size or shape, and therefore of no use. Good drafting practice is to project **only** the surface of the inclined line, *Figure 7-7*.

HOW TO DRAW AN AUXILIARY VIEW USING A FOLD LINE

GIVEN: The pictorial view of an object, Figure 7-4, which is also shown in the usual three orthographic views, front, top, and side, Figure 7-5. Because the inclined surface is on the front view, we will use a front view auxiliary for this illustration.

STEP 1 Label all the important points of the auxiliary surface as shown in *Figure 7-8A*.

STEP 2 Construct and label *fold lines,* which also represent the lines of intersection of adjacent orthographic projection planes, *Figure 7-8B*. (Refer to Figure 5-7.)

STEP 3 Draw light projection lines 90° from the inclined surface, and locate points a, b, c, and d in the auxiliary view from the fold line F/A, *Figure 7-8C*. These distances are identical to

the distances from a, b, c, and d to the fold line F/S and fold line F/T. Lightly connect a, b, c, and d bordering the inclined surface.

STEP 4 Recheck all work to ensure the following:

- That the projection lines are 90° from the inclined surface, which appears as an edge in the front view.
- That the fold line F/A is parallel to the inclined surface in the front view.
- That all distances have been transferred accurately.

Carefully darken in and complete the auxiliary view. The finished drawing will appear as it does in *Figure 7-8D*. Figure 7-8D shows the good drafting practice of projecting only the auxiliary surface, the surface of interest.

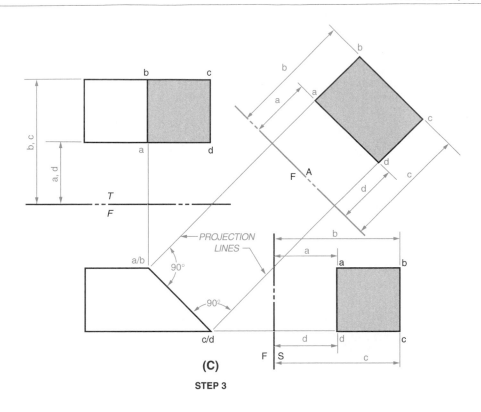

(C)

STEP 3

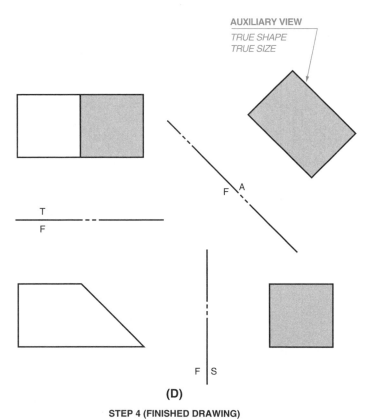

(D)

STEP 4 (FINISHED DRAWING)

FIGURE 7-8　　*(Continued) (C) Step 3;(D) Step 4 (finished drawing)*

HOW TO PROJECT A ROUND SURFACE FROM AN INCLINED OR SLANTED SURFACE

GIVEN: The usual three views of an object, front view, right-side view, and unfinished top view, are shown in *Figure 7-9*. (The usual 45° projection angle line will be omitted, as a slightly newer projection method will be used to complete the top view.) Refer also to the pictorial drawing of this object, *Figure 7-10*.

STEP 1 Divide the rounded view into equal spaces. In this example, using a 30° triangle, the right-side view is rounded and divided into 12 equal parts, *Figure 7-11A*. Letter each point clockwise, as shown in the figure.

STEP 2 Construct a vertical reference line in the right-side view so it passes through the center, in this example, through points a and g. (Always place the reference line through the center of any symmetrical object.) Construct a reference line in the top view that runs through the center, as illustrated in *Figure 7-11B*.

STEP 3 Draw light projection lines from the 12 points in the right-side view to the inclined edge in the front view. Project these same 12 points

directly up to the top view from the inclined edge in the front view. Notice points a and g are on the reference line, *Figure 7-11C*.

STEP 4 In the right-side view, measure the distance each point is from the reference line. Transfer each of these distances from the right-side view reference line to the top view reference line. Label each point lightly as each is found, *Figure 7-11D*.

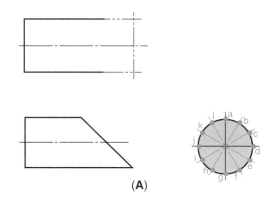

(A)

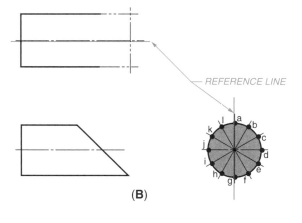

REFERENCE LINE

(B)

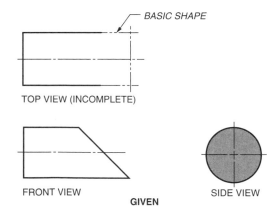

BASIC SHAPE

TOP VIEW (INCOMPLETE)

FRONT VIEW   SIDE VIEW

GIVEN

FIGURE 7-9   *Projecting a round surface from an inclined surface*

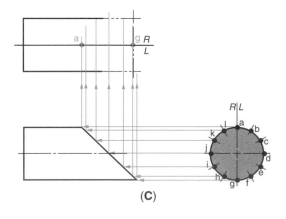

(C)

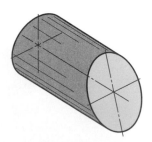

FIGURE 7-10   *Pictorial view of an object*

FIGURE 7-11   *(A) Step 1; (B) Step 2; (C) Step 3*

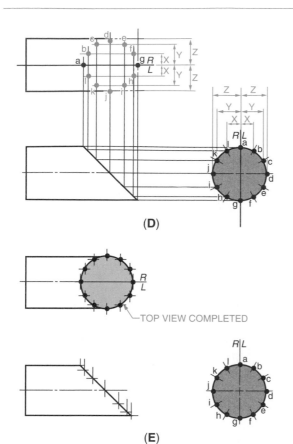

(D)

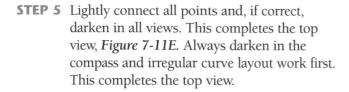

(E)

FIGURE 7-11   (Continued) (D) Step 4; (E) Step 5

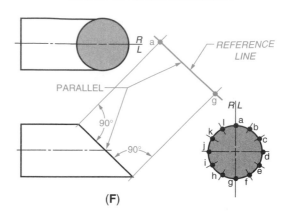

(F)

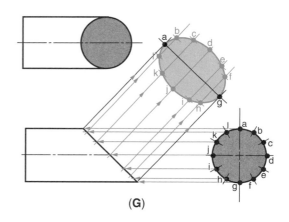

(G)

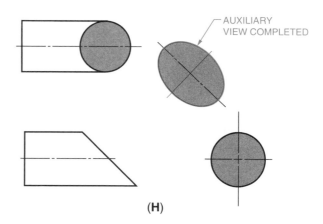

(H)

FIGURE 7-11   (Continued) (F) Step 6; (G) Step 7; (H) Step 8 (finished drawing)

STEP 5 Lightly connect all points and, if correct, darken in all views. This completes the top view, *Figure 7-11E*. Always darken in the compass and irregular curve layout work first. This completes the top view.

HOW TO DRAW AN AUXILIARY VIEW OF A ROUND SURFACE

STEP 6 Draw light projection lines 90° from the inclined surface of the front view. Construct the reference line parallel to the inclined surface at any convenient distance. Label the points that are on the reference line; in this example, a and g, *Figure 7-11F.*

STEP 7 Draw light projection lines from the 12 points in the right-side view to the inclined edge in the front view (Step 3). Project these same 12 points directly up to the auxiliary view from the inclined edge in the front view. Again, notice points a and g are on the reference line, *Figure 7-11G*. In the right-side view, measure the distance each point is from the reference line. Transfer each of these dis-

tances from the right-side view reference line to the auxiliary view reference line. Label each point lightly as each is found.

STEP 8 Lightly connect all points and, if correct, darken in the auxiliary view. This completes the auxiliary view, *Figure 7-11H*. Notice that only the inclined surface has been projected into the auxiliary view.

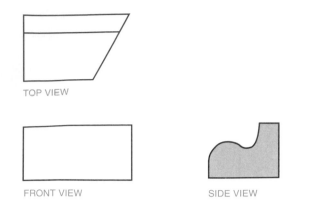

TOP VIEW

FRONT VIEW

SIDE VIEW

FIGURE 7-12 *Plotting an irregular curved surface*

HOW TO PLOT AN IRREGULAR CURVED SURFACE

GIVEN: The usual three views: front view (incomplete), top view, and the right-side view, *Figure 7-12*. This is an example of a top view auxiliary.

STEP 1 Add various points at random along the curved surface. Even spaces are not necessary, but try to choose points that pick up high and low points along the line, *Figure 7-13*. Always label the points in a clockwise direction.

STEP 2 From the right-side view, project these points up to the 45° projection line and over to the top view. Where these points intersect with the inclined surface, project directly down into the front view. Project these same points from the right-side view to the front view. Where the points intersect is where the point actually is. Lightly locate and connect all lines and complete the front view, *Figure 7-13B*.

STEP 3 Establish a reference line in the right-side view. Construct this reference line so it passes through as many points as possible; in this example, points a and j, *Figure 7-13C*.

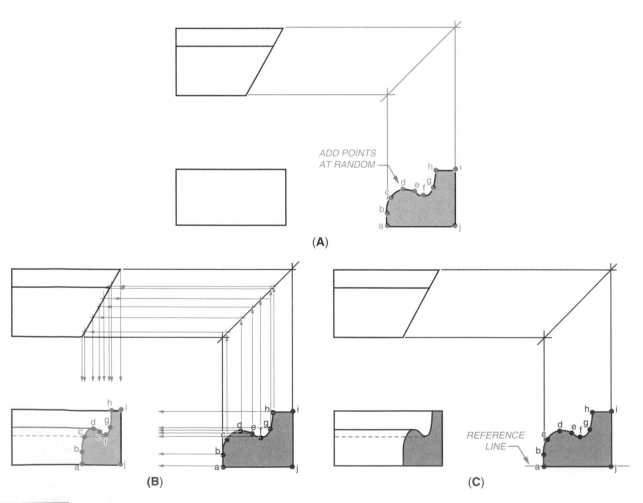

FIGURE 7-13 *(A) Step 1; (B) Step 2; (C) Step 3*

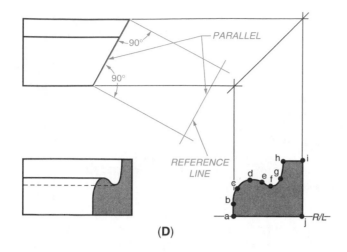

(D)

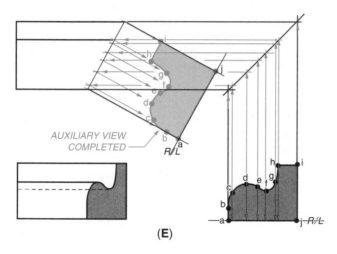

AUXILIARY VIEW
COMPLETED

(E)

FIGURE 7-13 *(Continued) (D) Step 4; (E) Step 5 (finished drawing)*

STEP 4 Add projection lines at 90° from the inclined surface. Add a reference line parallel to the inclined surface. Notice that points a and j again fall **on** the reference line, *Figure 7-13D*.

STEP 5 Project all points from the right-side view up and over to the inclined surface of the top view. Where the points intersect with the inclined surface, project 90° from the inclined surface. Transfer all distances from the reference line in the right-side view to the reference line in the auxiliary view. Lightly connect all points and, if correct, darken in all views, *Figure 7-13E*.

Secondary Auxiliary Views

Up to this point, primary auxiliary views have been dealt with. A *primary auxiliary view* can be projected from any of the regular views, as has been illustrated thus far.

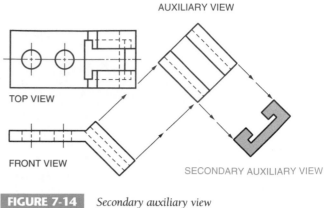

FIGURE 7-14 *Secondary auxiliary view*

Sometimes, a primary auxiliary view is not enough to fully illustrate an object; a secondary auxiliary view is needed. A *secondary auxiliary view* is projected directly from the auxiliary view. Many times, the auxiliary view and/or some of the other views cannot be fully completed until the secondary view has been completed first, *Figure 7-14*.

Partial Views

The use of a *partial auxiliary view* makes it possible to eliminate one or more of the regular views which, in turn, saves drafting time and cost. *Figure 7-15* is an example of a front view, top view, right-side view, and auxiliary view. Note the auxiliary view is always drawn partial; only the inclined surface is drawn on the auxiliary view. By using a partial top and right-side view in conjunction with the particular auxiliary view, the drawing can be simplified without detracting from its clarity; in fact, in most cases, these partial views make the drawing easier to read, *Figure 7-16*.

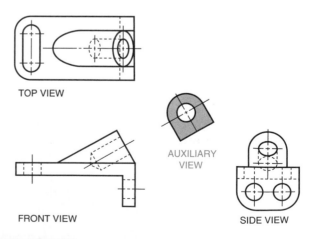

FIGURE 7-15 *Given: Regular three views and partial auxiliary view*

Industry Application

SAVING TIME AND CUTTING COSTS

"We've got to cut costs in this department," said Anna Valdez, CAD supervisor for Montgomery Engineering Company (MEC). MEC was undergoing a company-wide effort to cut costs, increase productivity, improve quality, and enhance competitiveness. The CAD department was feeling the pressure because MEC's latest contract consisted of more than 50 separate machine parts, all oddly shaped and complex. Valdez had called a departmental meeting to brainstorm ways to reduce drafting time on this contract. "We can save a lot of time by using partial auxiliary views where appropriate and by eliminating regular views when we do," suggested John Compton, a senior CAD technician. Compton laid a stack of drawings on the table and pointed out several places where full auxiliary views had been drawn when partials would have sufficed. "Not only could we be using partials," said Compton, pointing to the drawings for emphasis, "but the partial auxiliary views would eliminate the need for regular views in some cases." "Good point, John," said Valdez. "Just this one suggestion would reduce drafting time by at least 30 percent. What other suggestions are there?"

"Look at these objects," said Tram Ly. "They are symmetrical. Why do we waste time drawing full auxiliary views when half views would be sufficient?" "A good question," said Valdez. "Is anyone opposed to using half auxiliary views on symmetrical parts?" Hearing no comments, Valdez said, "Good. Let's do it." By the time the meeting finally ended, MEC's drafting department had identified several solid suggestions for reducing drafting time and cutting costs.

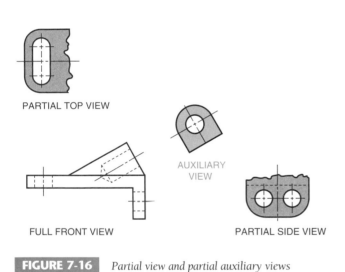

PARTIAL TOP VIEW

FULL FRONT VIEW

AUXILIARY VIEW

PARTIAL SIDE VIEW

FIGURE 7-16 *Partial view and partial auxiliary views*

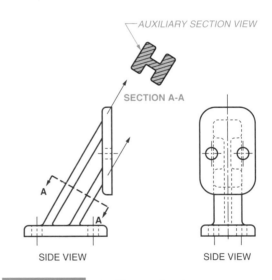

AUXILIARY SECTION VIEW

SECTION A-A

SIDE VIEW

SIDE VIEW

FIGURE 7-17 *Auxiliary section views*

Auxiliary Section

An *auxiliary section*, as its name implies, is an auxiliary view in section. An auxiliary section is drawn exactly as is any removed sectional view, and it is projected in exactly the same way as any auxiliary view, *Figure 7-17*. All the usual auxiliary view rules apply, and generally only the surface cut by the cutting-plane line is drawn.

Half Auxiliary Views

If an auxiliary view is symmetrical, and space is limited, it is permitted to draw only half of the auxiliary view, *Figure 7-18*. Use of the half auxiliary view saves some time, but it should only be used as a last resort, as it

could be confusing to those interpreting the drawing. Always draw the **near** half, as shown in the figure.

Oblique Surfaces, Viewing Planes, and Auxiliary View Enlargements

Oblique surfaces were described in Chapter 5 (Figure 5-23) as not being perpendicular to any one of the three principal projection planes. The Letter A Billboard in *Figure 7-19A* qualifies as an oblique surface. It is used in this discussion to illustrate three auxiliary views: finding an auxiliary view from an oblique surface, using a

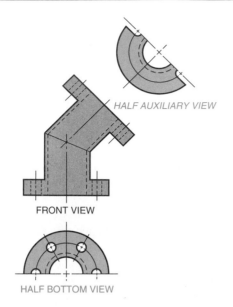

HALF AUXILIARY VIEW

FRONT VIEW

HALF BOTTOM VIEW

FIGURE 7-18   *Half auxiliary view*

viewing plane to establish an auxiliary view, and enlarging an auxiliary view to show more information.

Finding the true size of an oblique surface requires the descriptive geometry procedure for finding a plane in true size. As the descriptive geometry technique (Figure 8-17) is thoroughly discussed in the next chapter, only a brief summary will be offered here.

Any line in *true length* in the oblique surface may be used. Line b_1-b_2 in the Letter A Billboard shown in Figure 7-19A is used to illustrate the technique. Construct fold line T/F parallel to b_1-b_2 in the **front** view so b_1-b_2 will be in true length in the **top** view. Make fold line T/A_1 perpendicular to the direction of line b_1-b_2 in the top view so b_1-b_2 will appear as a point in the A_1 view and the oblique surface will appear as an edge. Draw fold line A_1/A_2 parallel to the edge view of the oblique surface so that the oblique surface will appear in *true size* in the A_2 view. (Note that line b_1-b_2 will appear in true length.)

To illustrate the use of an auxiliary viewing plane, align viewing plane V-V in *Figure 7-19B* parallel to the

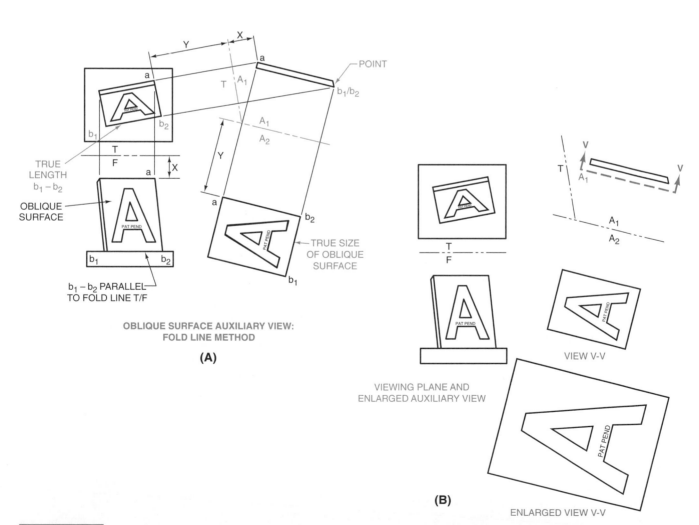

(A)

OBLIQUE SURFACE AUXILIARY VIEW:
FOLD LINE METHOD

(B)

VIEWING PLANE AND
ENLARGED AUXILIARY VIEW

VIEW V-V

ENLARGED VIEW V-V

FIGURE 7-19   *(A) Oblique surface auxiliary view: fold line method; (B) Viewing plane and enlarged auxiliary view*

edge view of the Letter A Billboard. View V-V will then appear in true size. Auxiliary view V-V could be placed at other locations on the drawing if more room is needed.

A two-to-one *enlargement of the auxiliary view* V-V is shown in Figure 7-19B. Construct enlargements when more detail is needed.

An additional type of enlargement drawing for better visualization is illustrated in Figure 3-3 and in a number of drawings in Chapter 21, "Drafting Applications: Pipe, Structural, Architectural, Civil Engineering, and GIS." The example in Figure 3-3 is typical. Refer to the isolated drawing labeled **Detail B.** Also notice the full-section drawing of the compression housing. The partial drawing in the detail is an enlargement of the portion within the small circle located where the flange joins the tubular portion. The letter B appears in the circumference of the small circle.

Summary

• An auxiliary view is a view with its line of sight perpendicular to an inclined surface of an object. An auxiliary view serves three purposes: (1) it illustrates the true size of a surface, (2) it illustrates the true shape of a surface, and (3) it can project and complete other views.

• A secondary auxiliary view is one that is projected from a primary auxiliary view to more fully illustrate an object.

• A partial auxiliary view is used to save drafting time and costs by eliminating one or more regular views.

• An auxiliary section is an auxiliary view in sections. It is drawn in the same way as a removed section.

• A half auxiliary view can be used when an auxiliary view is symmetrical. When this is used, the near half is drawn.

Review Questions

Answer the following questions either true or false.

1. A partial auxiliary view makes it possible to eliminate one or more of the regular views.

2. Hidden lines in an auxiliary view should be omitted unless they are needed for clarity.

3. A primary auxiliary view is always projected from the front view.

4. A reference line should be drawn horizontally and pass through as many points as possible.

5. Use of the half auxiliary view is very common because it saves time.

6. When laying out a drawing with an auxiliary view, always choose the view with the most detail as the front view.

Answer the following questions by selecting the best answer.

1. Which of the following is **not** a purpose that an auxiliary view serves?
   a. To avoid having to create a removed sectional view
   b. To illustrate the true size of a surface
   c. To illustrate the true shape of a surface
   d. To project and complete other views

2. Which of the following is **not** a major kind of auxiliary view?
   a. Front view
   b. Top view
   c. Side view
   d. Bottom view

3. When we draw an auxiliary view, projection lines must be drawn at what angle from the inclined surface?
   a. 60°
   b. 45°
   c. 90°
   d. None of the above

4. Which of the following is **not** true regarding a secondary auxiliary view?
   a. It is projected directly from the auxiliary view.
   b. Many times the secondary auxiliary view is completed first.
   c. A secondary auxiliary view is always created.
   d. None of the above

5. Which of the following is **not** true regarding an auxiliary sectional view?
   a. It is an auxiliary view in sections.
   b. It is drawn exactly as is any removed section view.
   c. Generally, only the surface cut by the cutting-plane line is drawn.
   d. Not all of the auxiliary view rules apply.

Chapter 7 Problems

The following problems are intended to give the beginning drafter practice in sketching and laying out multiviews with an auxiliary view.

CONVENTIONAL INSTRUCTIONS

The steps to follow in laying out any drawing with an auxiliary view are:

STEP 1 Study the problem carefully.

STEP 2 Choose the view with the most detail as the front view.

STEP 3 Position the front view so there will be the least number of hidden lines in the other views.

STEP 4 Determine from which view to project the auxiliary view.

STEP 5 Make a sketch of all views, including the auxiliary view.

STEP 6 Center the required views within the work area with approximately 1 ″ (25 mm) of space between the views. Adjust the regular views to accommodate the auxiliary view.

STEP 7 Use light projection lines. Do **not** erase them.

STEP 8 Lightly complete all views.

STEP 9 Check to see that all views are centered within the work area.

STEP 10 Carefully check all dimensions in all views.

STEP 11 Darken in all views using correct line thickness.

STEP 12 Recheck all work and, if it is correct, neatly fill out the title block using light guidelines and neat lettering.

CAD Instructions

Students who plan to complete the following problems on a CAD system should begin with these general instructions. Before reading the specific instructions for each activity, go through each step in the following planning checklist. The checklist applies to any CAD system and will help ensure the optimum use of your time and resources.

1. Analyze the problem carefully. Decide exactly what you are being asked to do.

2. Determine what resources and references (if any) you will need in order to complete the problem. Collect those references and resources and have them readily available.

3. Decide if any particular standards apply to the project and have those standards available.

4. Determine what types of views will be required and how many of each.

5. Determine what the final plotted scale of the drawing will need to be and select the appropriate paper size for plotting/printing.

6. Plan your drawing sequence. In what order will you develop the drawing (i.e., lines, features, dimension lines, leaders, dimensions)?

7. Review the various CAD commands you will have to use in order to develop the drawing.

8. Examine your CAD system to ensure that everything is in working order, then begin the project.

PROBLEMS 7-1 THROUGH 7-4

Draw the front view, top view, right-side view, and auxiliary view. Complete all views using the listed steps.

PROBLEM 7-1

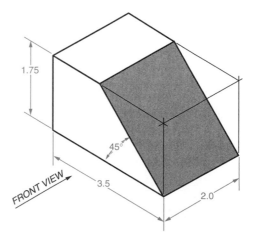

PROBLEM 7-2

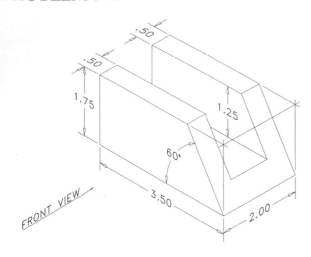

PROBLEMS 7-5 AND 7-6

Draw the front view, top view, right-side view, and *two* auxiliary views to illustrate the true size and shape of the slanted surfaces. Complete all views using the listed steps.

PROBLEM 7-5

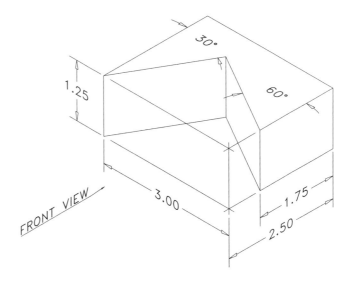

PROBLEM 7-3

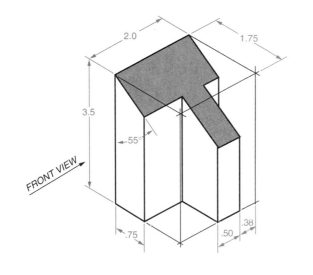

PROBLEM 7-6

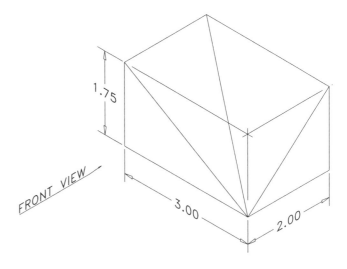

PROBLEM 7-4

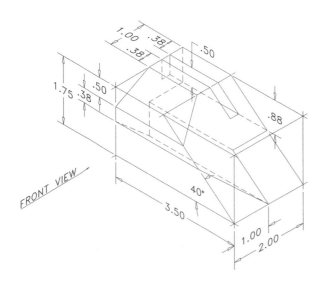

PROBLEMS 7-7 THROUGH 7-13

Draw the front view, top view, right-side view, and auxiliary view. Complete all views using the listed steps.

PROBLEM 7-7

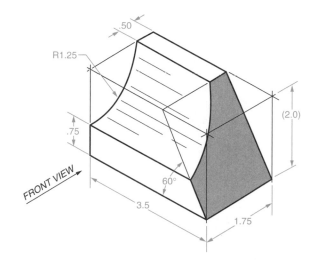

PROBLEM 7-8

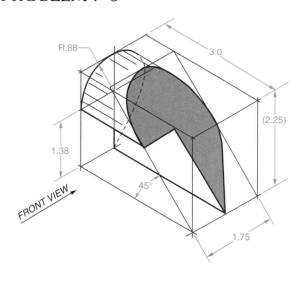

PROBLEM 7-9

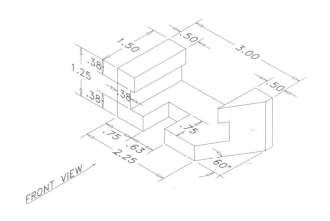

PROBLEM 7-10

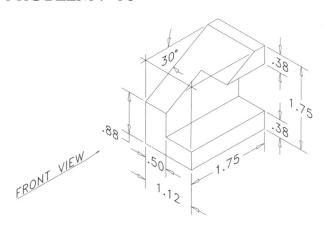

PROBLEM 7-11

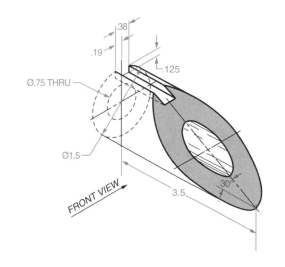

PROBLEM 7-12

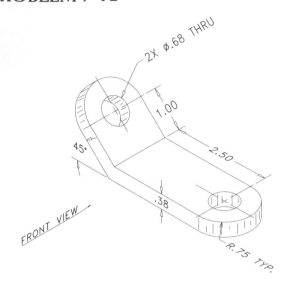

PROBLEM 7-14

Draw the front view, top view, right-side view, and *two* auxiliary views to illustrate the true size and shape of all surfaces. Complete all views using the listed steps.

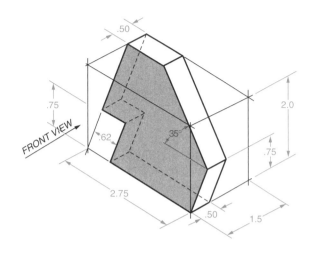

PROBLEM 7-13

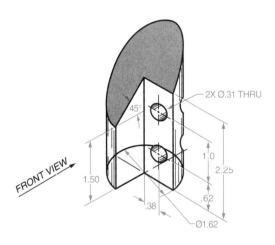

PROBLEMS 7-15 THROUGH 7-22

Draw the required views to fully illustrate each object. Complete all views using the listed steps.

PROBLEM 7-15

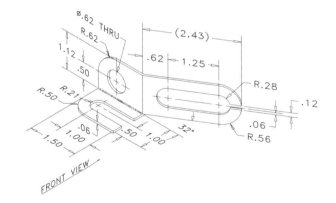

PROBLEM 7-16

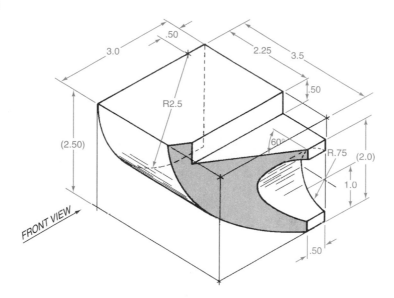

PROBLEM 7-17

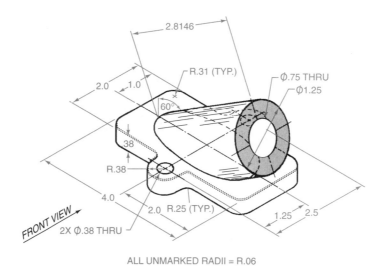

ALL UNMARKED RADII = R.06

PROBLEM 7-18

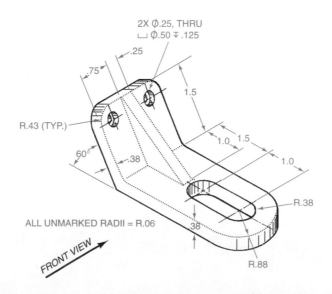

ALL UNMARKED RADII = R.06

PROBLEM 7-19

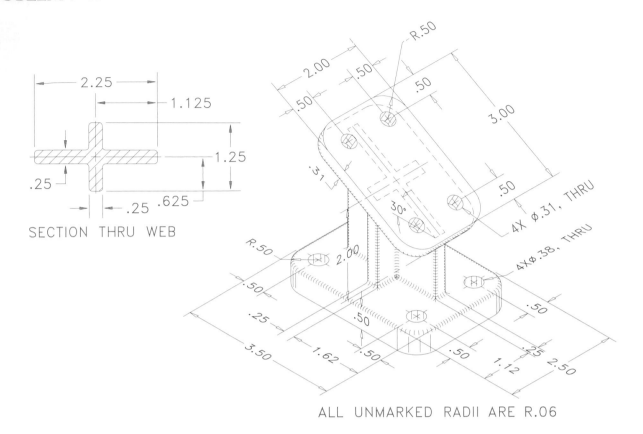

SECTION THRU WEB

ALL UNMARKED RADII ARE R.06

PROBLEM 7-20

PROBLEM 7-21

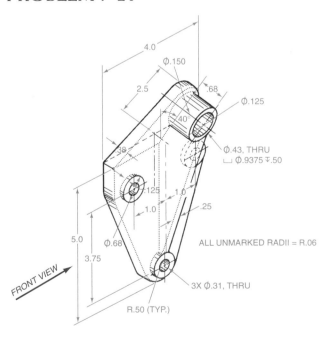

PROBLEM 7-22

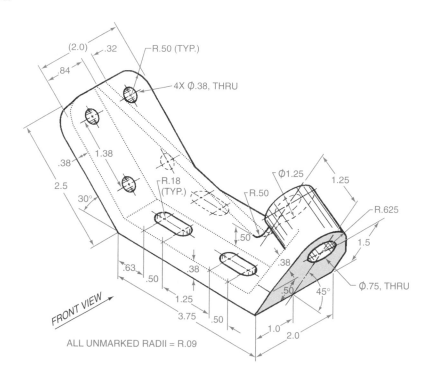

FRONT VIEW

ALL UNMARKED RADII = R.09

PROBLEMS 7-23 THROUGH 7-29

Make a finished drawing of selected problems as assigned by the instructor. Draw the given front and right-side views, add the top view with hidden lines if required, and add an auxiliary view. If assigned, design your own right-side view consistent with the given front view, and add a complete auxiliary view. Do *not* add dimensions.

PROBLEM 7-23

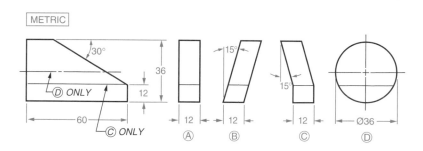

PROBLEM 7-24

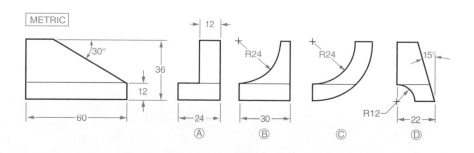

PROBLEM 7-25

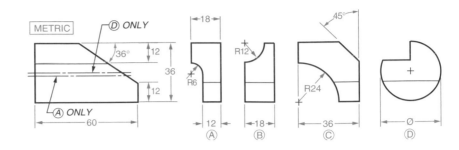

PROBLEM 7-26

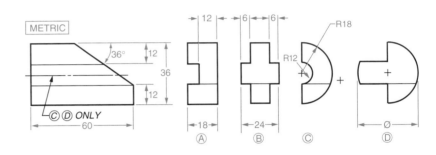

PROBLEM 7-27

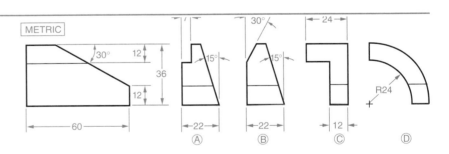

PROBLEM 7-28

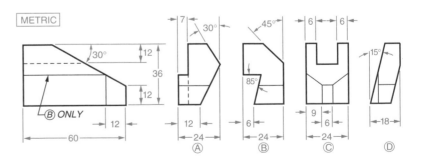

PROBLEM 7-29

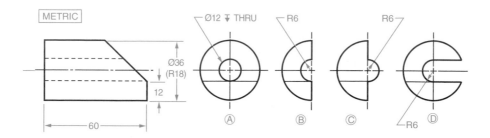

PROBLEMS 7-30 THROUGH 7-37

Draw the given views and add the required auxiliary view. Present the views within the work area so all views have a neat, centered appearance. Do *not* add dimensions.

PROBLEM 7-30

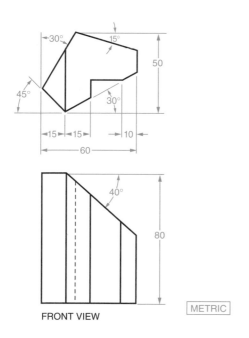

PROBLEM 7-31

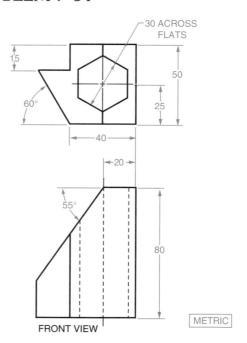

PROBLEM 7-32

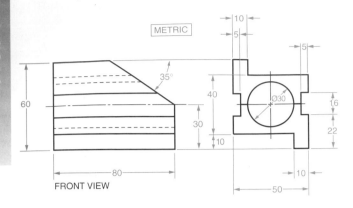

FRONT VIEW

PROBLEM 7-33

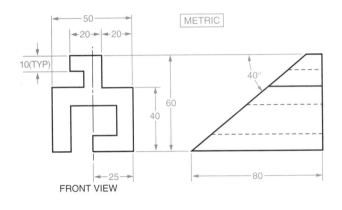

FRONT VIEW

PROBLEM 7-34

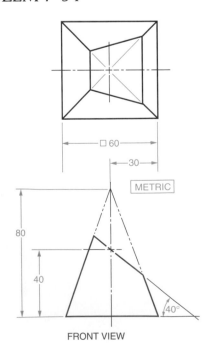

FRONT VIEW

PROBLEM 7-35

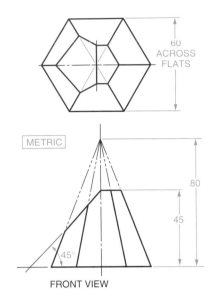

FRONT VIEW

PROBLEM 7-36

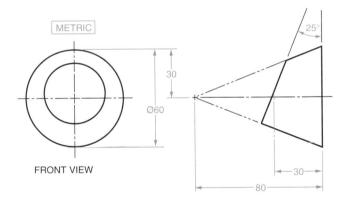

FRONT VIEW

PROBLEM 7-37

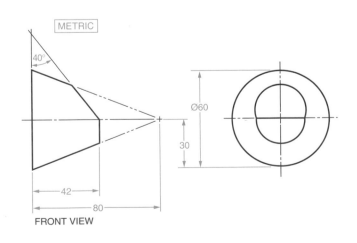

FRONT VIEW

PROBLEM 7-38

The partially completed building, in outline form, is to have the name of your city painted on one side of the roof, in letters to be read by individuals in aircraft as high as 5,000' above the roof sign. A rule of thumb for good readability is that the letters on a sign, in a classroom, in an auditorium, or on a roof, should be as high as approximately 1/250 of the distance the viewer is from the sign. Use an auxiliary view of the roof to lay out your city's name. Then specify the length of the building to suit your sign. Refer to Figure 3-4 for example fonts.

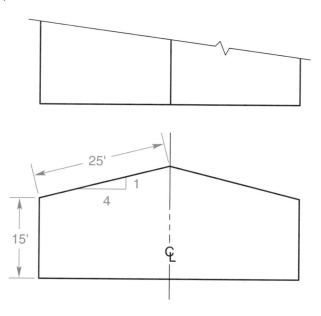

PROBLEM 7-39

The 750 mm outside diameter flange for the special 45° elbow shown is to have M24 × 3.0, in through holes on a 650 mm bolt circle. The recommended distance between fasteners (for strength and to provide space for wrenches) is 2.5 to 3.0 times the diameter of the fastener. Decide on the best, even number of bolts for the 650 mm bolt circle on this flange. Show your answer in an auxiliary view of the flange with the bolt holes.

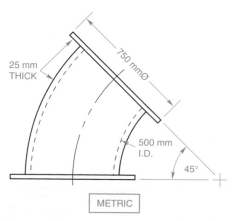

PROBLEM 7-40

Panels A, B, and C, in the control station shown, satisfy anthropometric requirements. They are 30" from the operator's head at location O, and the edge of the desk is 10" from the operator. Panel B needs to be replaced with a new panel that has the five new instruments shown, located in a reasonable arrangement in the panel. Prepare an auxiliary view of panel B and locate the five instruments, using your own judgment for the arrangement. Note approximate dimensions to their locations. Suggestion: Construct the auxiliary view of B, which will show the panel in true size. Cut out five rectangles of paper to represent the five instruments, and make trial arrangements of them on your auxiliary view. Select the best arrangement, in your opinion. Mark their locations and give approximate dimensions to their locations.

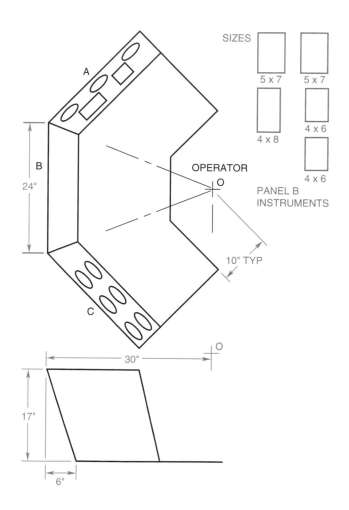

PROBLEM 7-41

The hopper shown needs a vibrator attached to the left side to shake down material to the conveyor below. Without the vibrator, some materials tend to hang up. Prepare an auxiliary view of the left side of the hopper to determine the approximate dimensions to locate a mounting plate that already has threaded holes, shown. (The plate is to be welded to the hopper.) The center of the plate should be approximately 1.9 m above the floor level so maintenance personnel can reach the vibrator. Also, locate the plate to either side of center so the maintenance person won't have to reach over the conveyor.

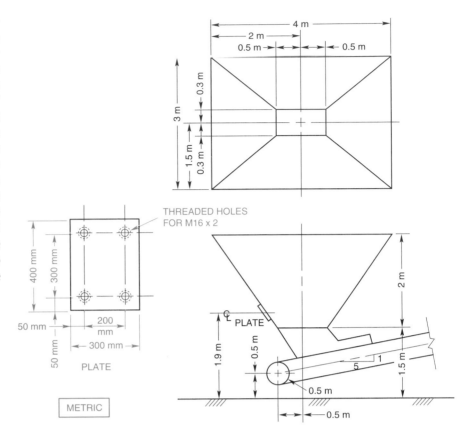

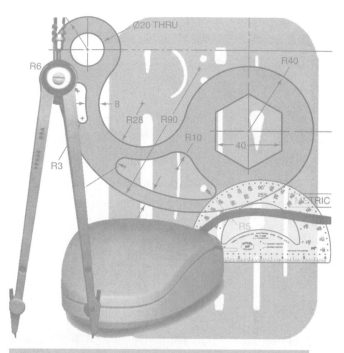

DESCRIPTIVE GEOMETRY

KEY TERMS

Azimuth bearing	North/south deviation
Batter	Notations
Bearing	Parallelogram law
Bevel	Piercing point
Clockwise	Pitch
Contour interval	Scalars
Contour line	Slope
Cuts and fills	Strike
Dip	Top view
Equilibrant	True length
Fold line	True slope
Grade	Trusses
Graphical method	Two-force member
Heading	Unlimited plane
Line	Uppercase letters
Lowercase end points	Vector
Lowercase letters	Viewing plane

CHAPTER OUTLINE

Descriptive geometry projection • Steps used in descriptive geometry projection • Notations • Fold lines • Bearings, slope, and grade • Applied descriptive geometry • Topographic topics • Vectors • Summary • Review questions • Chapter 8 problems

CHAPTER OBJECTIVES

Upon completion of this chapter, students should be able to do the following:

■ Explain the reasons for using the descriptive geometry method of projection.

■ List the steps used to solve descriptive geometry problems.

■ Demonstrate proficiency in the use of systems of notations.

■ Define the term *fold line.*

■ Demonstrate proficiency at using fold lines.

■ Demonstrate proficiency in determining the position of a line in space using its bearing, slope, and grade.

■ Demonstrate proficiency in plotting the boundaries of real estate.

■ Demonstrate proficiency in applying the concepts of cut and fill.

■ Demonstrate proficiency in applying the concepts of strike and dip.

■ Demonstrate proficiency in the use of vectors.

Descriptive Geometry Projection

Primary auxiliary views are views that have been projected perpendicular from an inclined surface and viewed looking directly at that surface. As studied in Chapter 7, the auxiliary view shows the true size and true shape of the inclined surface and is sometimes used to complete other views. If a surface is oblique, that is, slanting in more than one direction and not 90° from one of the other views, the true shape and true size of the successive surface is based on auxiliary views.

Using the descriptive geometry method of projection not only gives true size and shape, but it also can be used to find intersections, true distances of lines in space, true angles between surfaces, and exact piercing points. Descriptive geometry graphically shows the solution to problems dealing with points, lines and planes, and their relationship in space. In order to be able to apply descriptive geometry to various drafting problems, the drafter must know and understand the various basic steps involved. This chapter explains these basic steps. The basic theories covered in this chapter are:

- Projecting a line into other views
- Projecting points into other views
- Determining the true length of a line
- Determining the point view of a line
- Finding the true distance between a line and a point in space
- Determining the true distance between two lines in space
- Projecting a plane surface in space
- Developing an edge view of a plane surface in space
- Determining the true distance between a plane surface and a point in space
- Determining the true angle between plane surfaces in space

Steps Used in Descriptive Geometry Projection

All problems, regardless of their complexity, use the same procedures or steps. These steps ultimately must be done in the same order each time, although some intermediate steps may be selected or are optional. Like climbing a flight of stairs, each step must be executed, and only experience can give one the ability to perform multiple steps simultaneously. An example of the sequential steps needed in descriptive geometry problems is as follows: In order to find an edge view of a plane (flat surface), we must locate a true-length line in

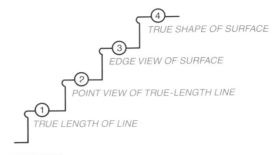

TRUE SHAPE OF SURFACE

EDGE VIEW OF SURFACE

POINT VIEW OF TRUE-LENGTH LINE

TRUE LENGTH OF LINE

FIGURE 8-1 *Steps used to solve descriptive geometric problems*

that plane (Step 1); a point view of the true-length line must be projected, which yields an edge view of the plane (Step 2); see *Figure 8-1*. By projective means, it would be impossible to find the edge view of an oblique surface from normal views without performing Steps 1 and 2.

Notations

As we progress through sequential steps it is important to keep track of all views and points in space. (Recall that a straight line is composed of two spatially located end points.) To do this, we should use a system of *notations* or labeling. Each view and each point should be labeled to provide an accurate identity of each at all times. For purposes of this text, the following notations are used in this chapter.

Each point in space is called out in *lowercase letters*.

EXAMPLE:

a, b, c, d
Each *line* in space is identified by the two *lowercase end points*.

EXAMPLE:

Line a-b
Each line in space is assumed to end at the indicated end points, unless it is specified as an extended line (continues without end).

EXAMPLE:

Line p-q and Extended Line j-k
Each plane in space, such as a triangle, is called out in lowercase letters.

EXAMPLE:

Triangle abc
Each plane in space is assumed to be limited (stops at the indicated boundaries) unless otherwise defined as an *unlimited plane* (limitless extension beyond the indicated boundaries).

EXAMPLE:

Plane defg and Unlimited Plane mno
Each *viewing plane* is called out in *uppercase letters*.

EXAMPLE:

F = Frontal viewing plane (a principal plane).
T = Top (horizontal) viewing plane (a principal plane).
R = Right-side (profile) viewing plane (principal plane).
L = Left-side (profile) viewing plane (principal plane).
A(digit) = an auxiliary viewing plane, with the number of successive projections from a principal viewing plane.
B(digit) = same as A(digit), but using a different series of successive projections.

A combination of lowercase and uppercase letters is used to fully describe each point in space and the view in which it is located.

EXAMPLE:

aF would be a location as seen in the frontal viewing plane.
aT-bT would be line a-b location as seen in the top (horizontal) viewing plane.
aR-bR-cR would be plane a-b-c as seen in the right-side (profile) viewing plane.

Fold Lines

A fold line is represented by a thin black line, similar to a phantom line. It indicates a 90° intersection of two viewing planes. Each fold line must be labeled using uppercase letters, *Figure 8-2*. In this example, the fold line is placed between the front view and the top view. This placement eliminates the usual 45° projection line, *Figure 8-3*. It is important to add these notations in order to keep track of exactly which viewing plane is being executed.

FIGURE 8-2   *Fold line*

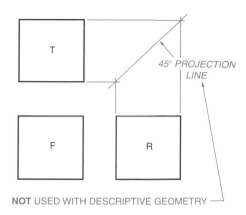

FIGURE 8-3   *Regular multiview drawing using 45° projection line*

HOW TO PROJECT A LINE INTO OTHER VIEWS

Fold lines are placed between successive orthographic views and labeled with uppercase letters, *Figure 8-4*. Any appropriate spacing is permissible, but the distance from the image to the frontal viewing plane fold line must be equal at both the top and right-side views. The spatial positioning of a line a-b (surrounded by an imaginary transparent box) is shown in *Figure 8-5*. Point a is closer to the frontal viewing plane than is point b, but it is farther below the top (horizontal) viewing plane. Represented on a normal layout drawing, it would be illustrated and labeled as shown in *Figure 8-6*.

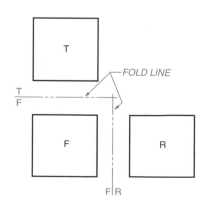

FIGURE 8-4   *Regular multiview drawing using fold lines instead of 45° projection line*

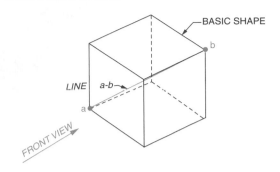

FIGURE 8-5   *Spatial positioning of line a-b*

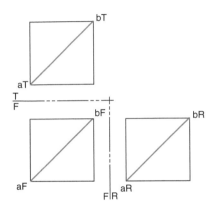

FIGURE 8-6   *Line a-b in all three regular views*

The customary 45° projection line is not used in descriptive geometry projection. All points are projected along light projection lines drawn at 90° to the fold lines. All measurements are taken along these light projection lines and measured from the fold line to the point in space, *Figure 8-7*.

An important rule to remember is to always skip a view between all measurements. In Figure 8-7, notice in the right-side view that dimension X (distance from F/R fold line to point aR) is projected into and through the

front view and up to the top view and transferred into the top view. (Distance from F/T fold line to point aT.) The front view was skipped.

Again, referring to the right-side view of Figure 8-7, dimension Y (distance from F/R fold line to point bR) is transferred into the top view (distance from F/T fold line to point bT), and the front view was skipped. In each of the instances, the distance is measured as the perpendicular distance from the fold line to the point.

EXAMPLE:

To project a top view of a line from a given front and right-side view.

GIVEN: Line a-b in the front view and line a-b in the right-side view, *Figure 8-8*. Extend projection lines into the top view from the end points in the front view, aF and bF. Find the distances X and Y from the line end points in the right-side view aR and bR to the frontal viewing plane at fold line F/R and transfer them into the top view. Label all points and fold lines in all views. Be sure that all projections are made at 90° from the relevant fold lines, *Figure 8-9*.

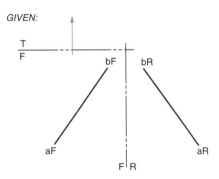

FIGURE 8-8   *Locating line a-b in top view*

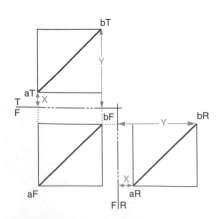

FIGURE 8-7   *Skip a view when transferring distances*

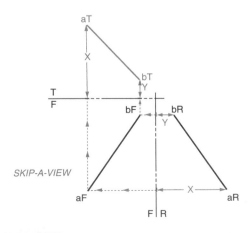

FIGURE 8-9   *Line a-b in the top view using the fold line*

HOW TO LOCATE A POINT IN SPACE (RIGHT VIEW)

A point in space is projected and measured in exactly the same way as a line in space, except that the point is a line with a single end point, *Figure 8-10A*.

EXAMPLE:

To project the right-side view of a point from a given top and front view.

GIVEN: Point a in the front view and top view (*Figure 8-10A*).

Project point aF from the front view into the right-side view. Point aR must lie on this projection line. Find X, which is the distance from fold line F/T to aT in the top view and project it into and through the front view, *Figure 8-10B*. Project it over into the right-side view. Transfer distance X to find point a. Be sure to always label all points and fold lines in all views.

HOW TO FIND THE TRUE LENGTH OF A LINE

Any line that is parallel to a fold line will appear in its true length in the next successive view adjoining that fold line.

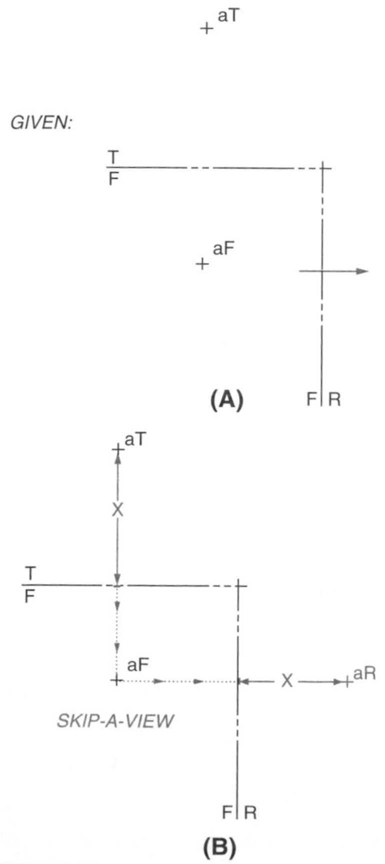

To find the true length of any line:

STEP 1 Draw a fold line parallel to the line of which the true length is required. This can be done at any convenient distance, such as approximately one-half inch.

STEP 2 Label the fold line A for auxiliary view.

STEP 3 Extend projection lines from the end points of the line being projected into the auxiliary view. These must always be at 90° to the fold line.

STEP 4 Transfer the end point distances from the fold line in the second preceding view from the one being drawn, to locate the corresponding end points in the view being drawn.

EXAMPLE:

Refer to *Figure 8-11A*.

GIVEN: Line a-b in the front view, side view, and top view. The problem is to find the true length of line a-b. A true length can be projected from any of the three principal views by placing a fold line parallel to the line in any view. The right-side view is selected in this example.

STEP 1 Draw a fold line parallel to line a-b, and label it as shown in *Figure 8-11B*. Extend light projection lines at 90° to the fold line from points a and b into the auxiliary view.

STEP 2 Determine distances X and Y from the front view to the near-fold line and transfer them into the auxiliary view, as shown in *Figure 8-11C*. Label all points and fold lines. The result will be the true length of line a-b.

Use these steps to find the true length of any line.

HOW TO CONSTRUCT A POINT VIEW OF A LINE

To construct a point view of a line:

STEP 1 Find the true length of the line.

STEP 2 Draw a fold line perpendicular to the true-length line at any convenient distance from either end of the true-length line.

STEP 3 Label the fold line A-B (B indicates a secondary auxiliary view).

STEP 4 Extend a light projection line from the true-length line into the secondary auxiliary view.

STEP 5 Transfer the distance of the line end points into the secondary auxiliary view (B) from the corresponding points in the second preceding view.

GIVEN:

(A)

(B)

FIGURE 8-10   (A) *Locating a point in space (right view);* (B) *Step 1*

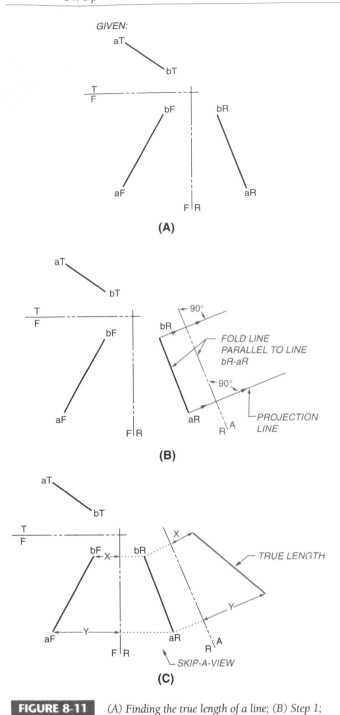

FIGURE 8-11 *(A) Finding the true length of a line; (B) Step 1; (C) Step 2*

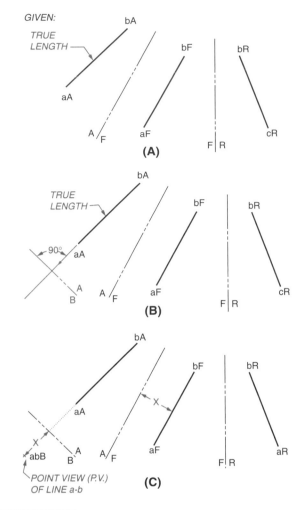

FIGURE 8-12 *(A) Constructing a point view of a line; (B) Step 1; (C) Step 2*

EXAMPLE:

Refer to *Figure 8-12A.*

GIVEN: Line a-b in the front view, side view, and auxiliary view. The true length is located in the auxiliary viewing plane (A), which is projected from the front view. (The true length could have been projected from any of the given views.)

STEP 1 At any convenient distance from either end, draw a fold line A/B perpendicular to the true length of line a-b. This will establish a viewing plane that is perpendicular to the direction of the line's path. Extend a projection line aligned with the true-length line into this new secondary auxiliary view. The projection lines from both points a and b appear to align in this common projection line. Therefore, both points a and b must be located on this projection line in the secondary auxiliary view. In *Figure 8-12B,* the fold line A/B was added to the left of the true-length line a-b.

STEP 2 The front view is the second preceding view to the secondary auxiliary view being constructed. Therefore, distance X in the front view from the fold line A/F to line a/b is transferred to the secondary auxiliary view from the fold line A/B, *Figure 8-12C.* As points a and b in the front view are both at the same distance from the fold line A/F, and they both lie on the same projection line in the secondary auxiliary view, they will appear to coincide at the same location. This provides evidence that the end view of the line has been achieved.

HOW TO FIND THE TRUE DISTANCE BETWEEN A LINE AND A POINT IN SPACE

The true distance between a line and a point in space will be evident in a view that shows the end point of the line and that point, simultaneously.

STEP 1 Project an auxiliary view (viewing plane A) to find the true length of the line, and project the point into the same view.

STEP 2 Project a secondary auxiliary view (B) to find the point view of the line, and project the point into the same view.

STEP 3 The observable distance between the end view of the line and the point will be the actual distance between them. The location in the line that is nearest to the point is on the path that is perpendicular to the line from the point, but this is not discernible in the secondary auxiliary view.

EXAMPLE:

Refer to *Figure 8-13A*.

GIVEN: Line a-b and point c in the front view and right-side view.

STEP 1 Project the true length of line a-b by placing a fold line parallel to a given view of the line, and project point c into the auxiliary view. Recall that point locations are transferred from the second preceding view. Label all fold lines and all points, *Figure 8-13B*.

STEP 2 Draw a fold line perpendicular to the true length of line a-b, and label the fold line. Project line a-b into the secondary auxiliary view (B) to find the point view of line a-b. Project point c along into the second auxiliary view (B). The actual distance between the point view of line a-b/B and point cB is evident, *Figure 8-13C*.

The location in the line a-b that is nearest to point c lies on a path that is perpendicular to line ab and passes through c. Any path that is perpendicular to a line will appear perpendicular in a view where the line is true length. Therefore, the path can be drawn in the preceding view to locate point p on the line. Point p can be projected to its correct location on the original views of line a-b.

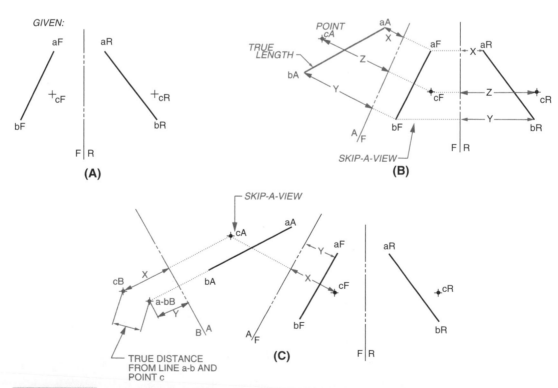

FIGURE 8-13   *(A) Finding the true distance between a line and a point in space; (B) Step 1; (C) Step 2*

HOW TO FIND THE TRUE DISTANCE BETWEEN TWO PARALLEL LINES

If two lines are actually parallel, they will appear parallel in all views. A view is needed to show the end view of both lines simultaneously, where the real distance between them will be apparent.

STEP 1 Find the true length of each of the two lines. Projecting a view that will provide the true length of a line will automatically provide the true length of any other parallel line that is projected into the same view.

STEP 2 Find the point view of each of the two lines. Projecting a view that will provide the end view of a line will automatically provide the end view of any other parallel line that is projected into the same view.

STEP 3 The true distance between the parallel lines is the straight-line path between their end points. There is no single location within the length of the lines where this occurs, as each

location in a line has a corresponding closest point location on the other line, each on the path connecting them and perpendicular to their respective lines.

EXAMPLE:

Refer to *Figure 8-14A*.

GIVEN: The parallel lines a-b and c-d, in a front view, right-side view, and a top view.

STEP 1 Find the true lengths of line a-b and line c-d in auxiliary view A. The two parallel lines will also be parallel in the auxiliary, if they are actually parallel, *Figure 8-14B*.

STEP 2 Draw a fold line perpendicular to the true-length lines, a-b and c-d. Project the lines into the secondary auxiliary view (B) to find the point view of the two lines a-b and c-d. Measure the true distance between the point views of lines a-b/B and c-d/B, *Figure 8-14C*.

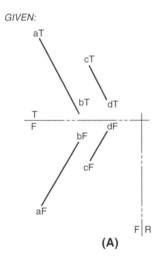

(A)

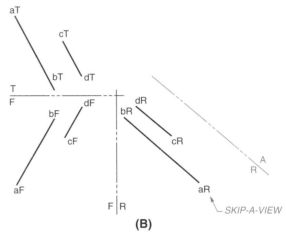

(B)

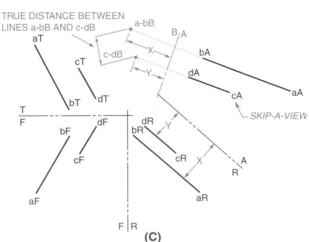

(C)

FIGURE 8-14 *(A) Finding the true distance between two parallel lines; (B) Step 1; (C) Step 2*

HOW TO FIND THE TRUE DISTANCE BETWEEN TWO NONPARALLEL (OR SKEWED) LINES

If two lines are not parallel, they will appear nonparallel in at least one view. It is wise to first check two nonparallel (or skewed) lines to determine if they actually intersect, which would make the distance between them zero. This can be verified by projecting the apparent point of intersection of a view to the adjoining view. If the apparent points of intersection align, then the lines actually intersect.

To determine the true distance between two nonparallel lines:

STEP 1 Project the true length of either one of the two lines, and project the other line into that auxiliary view (A).

STEP 2 Find the point view of the true-length line, and project the other line into that secondary auxiliary view (B).

STEP 3 Measure the true distance between the point view line at a location that is perpendicular to the other line.

EXAMPLE:

Refer to *Figure 8-15A.*

GIVEN: Nonparallel lines a-b and c-d in a front view and right-side view.

STEP 1 Project the true length of line a-b in an auxiliary view (A), and project line c-d along into that auxiliary view, *Figure 8-15B*. Notice that c-d is not the true length.

STEP 2 Project the point view of line a-b into the secondary view (B) and project line c-d into this view. Measure the true distance between the point view of line a-b/B, perpendicular from line c-d/B, *Figure 8-15C*.

HOW TO PROJECT A PLANE INTO ANOTHER VIEW

A limited plane is located by points in space joined by straight lines. In order to transfer a plane, find at least three points in the plane and project each point into the next view.

GIVEN:

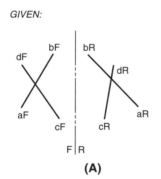

(A)

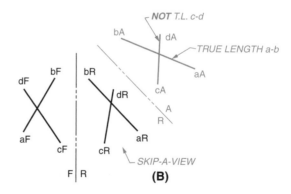

(B)

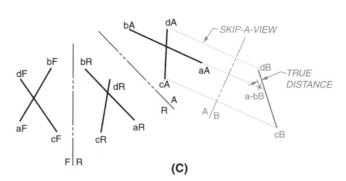

(C)

FIGURE 8-15   *(A) Finding the true distance between two nonparallel (or skewed) lines; (B) Step 1; (C) Step 2*

EXAMPLE:

Refer to *Figure 8-16A.*

GIVEN: Triangular plane abc shown in the top view and front view.

STEP 1 In the top view, find the distance from the fold line to points aT, bT, and cT, *Figure 8-16B.*

STEP 2 Project these points from the front view into the right-side view, and transfer the corresponding distances from the top view points to the top view fold line into the right-side view, *Figure 8-16C.* Construct straight lines between points a, b, and c. This transfers the plane surface.

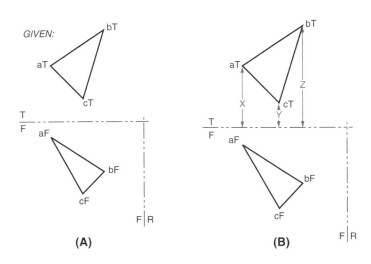

(A)   **(B)**

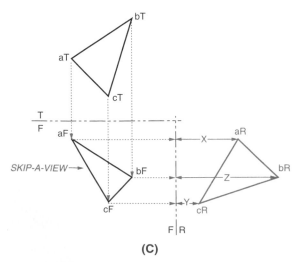

(C)

FIGURE 8-16   *(A) Projecting a plane into another view; (B) Step 1; (C) Step 2*

HOW TO CONSTRUCT AN EDGE VIEW OF A PLANE SURFACE

If the end view of a line that is in a plane is shown, then the edge view of that plane is shown also. Recall that to find the end view of a line, a true length must be found first. When the boundaries of a plane are provided, projecting a view to find the true length of a selected boundary is a simple procedure. It is necessary only to locate a fold line parallel to the boundary to ensure its true length in the next projected view.

A shorter method of securing a true-length line in a plane is also used, if a true-length line is not present already. A line may be added on the plane in any view, arranged to be parallel to an adjoining fold line, and to have its terminations at the plane boundaries. Projecting the added-line terminations to the corresponding boundaries in the adjoining view relocates the added line in that view. Moreover, it will be a true-length line, as it was made to be parallel to the fold line in the preceding view.

Whichever method is used to find a true-length line in the plane, all point locations within the plane should be projected to the view where the true-length line exists, if not there already. A fold line perpendicular to the true-length line will provide a direction for projecting the edge view of the plane. Projecting all the points in the plane will provide evidence of this, as they will align into a single, straight path.

EXAMPLE:

Refer to *Figure 8-17A.*

GIVEN: Plane surface abc in the top view and front view.

STEP 1 Construct a line parallel to fold line F/R and through one or more points, if possible. In this example, the line passes through point c. Label this newly constructed line cF/xF, *Figure 8-17B.*

STEP 2 Construct the true length of line cF/xF in the next view, *Figure 8-17C.*

STEP 3 Construct the point view of the true-length line cR/xR. Project the remaining points of the plane surface into this view. Also, label all points and fold lines. If done correctly, the result should be a straight line that passes through the point view. This straight line is the edge view of the plane, *Figure 8-17D.*

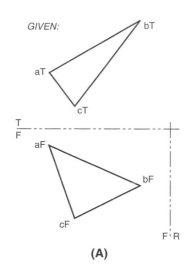

STEP 4 Construct a fold line A/B parallel to plane abc (shown as an edge in view A) and construct the plane surface abc in view B. Plane surface abc is now seen in true size, in view B, *Figure 8-17E*, because plane abc was seen as an edge in the adjacent view A.

HOW TO FIND THE TRUE DISTANCE BETWEEN A PLANE SURFACE AND A POINT IN SPACE

Finding the true distance between a given point and a plane in space requires a view that shows the edge view of the plane and the point in the same view. The distance from the point to the plane is the path that is perpendicular to the plane and passing through the point.

STEP 1 Add a line on the plane parallel to a fold line, if a true-length line is not already available in the plane.

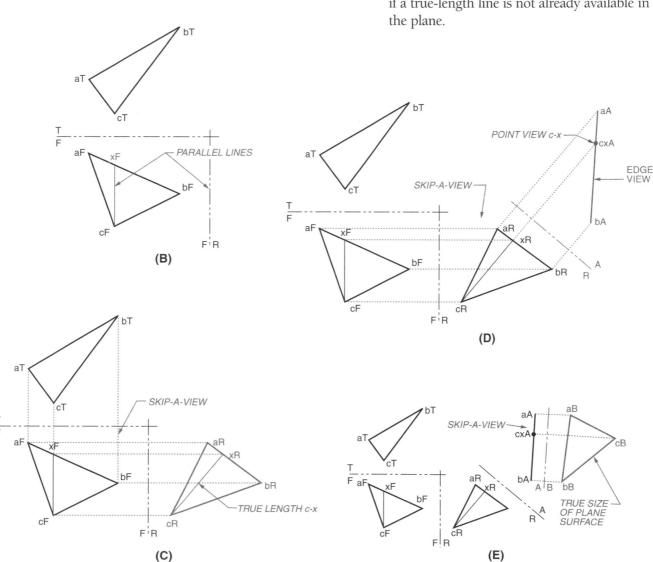

FIGURE 8-17 (A) Constructing an edge view of a plane surface; (B) Step 1; (C) Step 2; (D) Step 3; (E) Step 4

STEP 2 Project the added line to an adjoining view, where it will be seen in true length.

STEP 3 Project a point view of the added true-length line.

STEP 4 Project the points of the plane into this view. The result should be a straight line that passes through the added line end view, which is the edge view. Project the point in space into this view. Measure the perpendicular distance from the plane's edge view to the point in space. This is the true distance between the plane surface and the point in space.

EXAMPLE:

Refer to *Figure 8-18A.*

GIVEN: Triangular plane abc and point X in the front view and right-side view.

STEP 1 Construct a line parallel to the fold line F/R and through one or more points. Point c is a convenient line termination. Label this newly constructed line cR/zR, *Figure 8-18B.*

STEP 2 Construct the true length of line cR/zR in the front view, and label the line cF/zF, *Figure 8-18C.* Bring point X into this view, also. Label all points and fold lines.

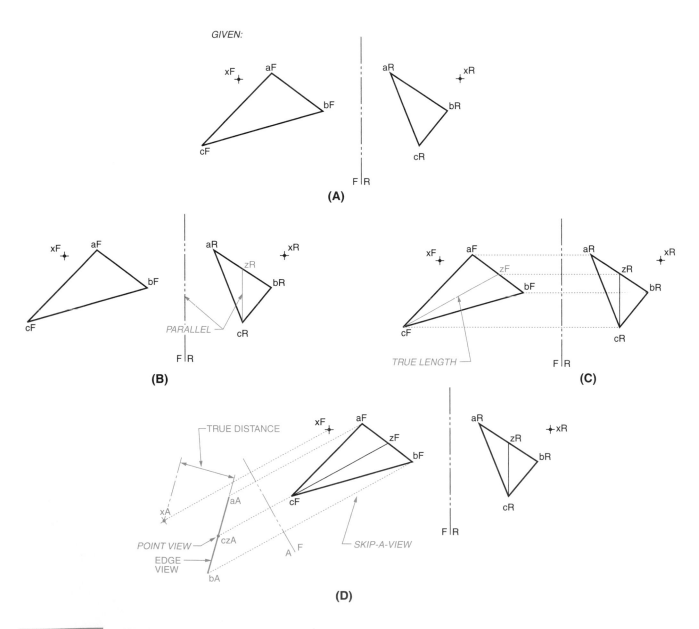

FIGURE 8-18   *(A) Finding the true distance between a plane surface and a point in space; (B) Step 1; (C) Step 2; (D) Step 3*

STEP 3 Project the end view of the true-length line cF/zF in the auxiliary view, *Figure 8-18D*. Project the points of the plane surface into this view, and also point X, projecting 90° from the fold line. Label all points and fold lines. Measure the perpendicular distance from the edge view to the point in space. This is the true distance between the plane surface and the point X in space.

HOW TO FIND THE TRUE ANGLE BETWEEN TWO PLANES (DIHEDRAL ANGLE)

The edge of intersection between two planes is a line that is common to both planes. The end view of that line will provide the edge view of both planes simultaneously, and the angle between them will be evident.

To find the true angle between two surfaces:

STEP 1 Construct the true length of the intersection between the two surfaces, and project all other points of both planes into that auxiliary view.

STEP 2 Project the end view of the true-length edge of intersection line and project the points of both planes into this secondary auxiliary view. Label all points and fold lines. Measure the true angle between the two surfaces.

EXAMPLE:

Refer to *Figure 8-19A*.

GIVEN: Plane abc and plane abd that intersect one edge, a-b, in the front view, side view, and top view.

STEP 1 Project the true length of the edge of intersection a-b, and project all other points into this auxiliary view (A), *Figure 8-19B*.

STEP 2 Project the point view of the true length of the edge of intersection a-b and project all other points into this secondary auxiliary view (B), *Figure 8-19C*. Measure the true angle between two surfaces.

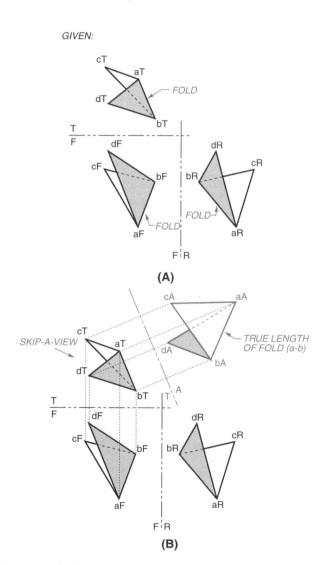

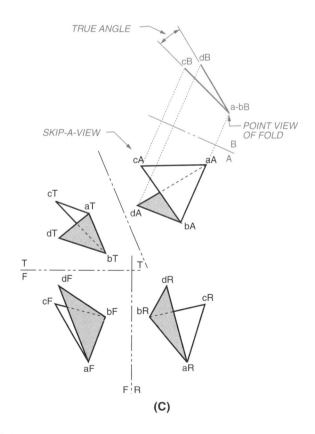

FIGURE 8-19 *(A) Finding the angle between two surfaces; (B) Step 1; (C) Step 2*

HOW TO DETERMINE THE VISIBILITY OF LINES

To determine which line of an apparent intersection of two lines is closer to the viewer, the following steps are used.

STEP 1 From the exact crossover point of the lines, project to an adjoining view.

STEP 2 In the adjoining view, determine which of the lines is closest to the fold line between the views, on the projection line from the first view (or the first line that the projection line encounters on its path from the first view).

STEP 3 Whichever line is closest to the fold line at that point only is the line that is in front of the other line in the first view.

EXAMPLE:

Refer to *Figure 8-20A.*

GIVEN: Lines a-b and c-d in the front view and top view.

STEP 1 At the exact crossover of lines a-b and c-d in the top view, project down through the fold line to the corresponding lines in the front view. At this exact point along the lines, line c-d is closer to the fold line and line a-b is farther away from the fold line; thus, in the top view, line c-d is in front of line a-b, *Figure 8-20B.*

STEP 2 See *Figure 8-20C.* At the exact crossover of lines a-b and c-d in the front view, project up through the fold line to the corresponding lines in the top view. At this exact point along the lines, line a-b is closer to the fold line and line c-d is farther away from the fold line; thus, in the front view, line a-b is in front of line c-d.

STEP 3 The end result is drawn to illustrate this crossover, *Figure 8-20D.*

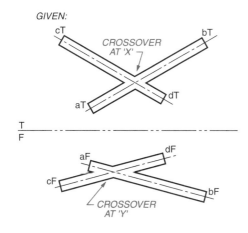

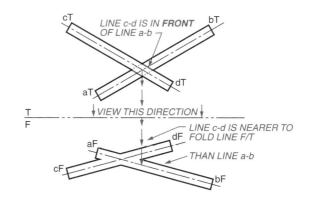

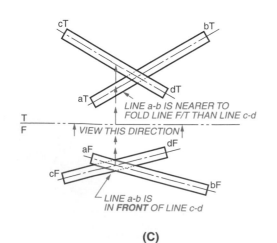

(C)

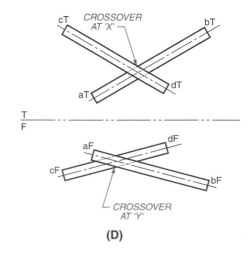

(D)

FIGURE 8-20 *(A) Determining the visibility of lines; (B) Step 1; (C) Step 2; (D) Step 3*

HOW TO DETERMINE THE PIERCING POINT BY INSPECTION

The *piercing point* is the exact location of the intersection of a surface and a line. The exact piercing point is determined from the view where the surface appears as an edge view. In *Figure 8-21A,* the top view illustrates the edge view of the surface. The piercing point is established in the top view and projected down into the front view, *Figure 8-21B.* In *Figure 8-22A,* the front view illustrates the edge view of the surface. The piercing point is established in the front view and projected up into the top view, *Figure 8-22B.*

HOW TO DETERMINE THE PIERCING POINT BY CONSTRUCTION

To determine the exact piercing point by construction:

STEP 1 Construct a true length of a line on the plane surface.

STEP 2 Project a point view of the true-length line and an edge view of the plane surface. The edge view of the plane lies in the path of the line in this view, indicating the piercing point location.

STEP 3 Transfer the piercing point back into the other views.

STEP 4 Determine the visibility of lines to find which part of the line is visible from the viewing direction. Use the fold line to determine visibility, if necessary.

EXAMPLE:

Refer to *Figure 8-23A.*

GIVEN: Plane surface abc and line x-y, in the top view and front view.

STEP 1 In the front view of plane abc, construct a line parallel to fold line bF-dF and find its true length in the top view, *Figure 8-23B.*

STEP 2 Project an auxiliary view to find the point view of the true-length line from the top view and the edge view of the plane. The line

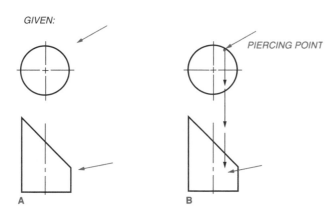

FIGURE 8-21   *Determining the piercing point by inspection*

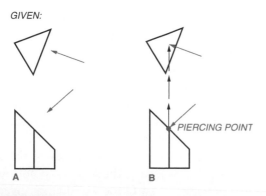

FIGURE 8-22   *Determining the piercing point by inspection*

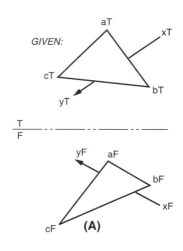

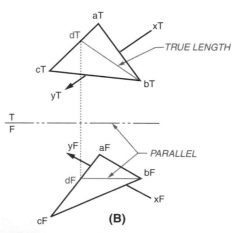

FIGURE 8-23   *(A) Determining the piercing point by construction; (B) Step 1*

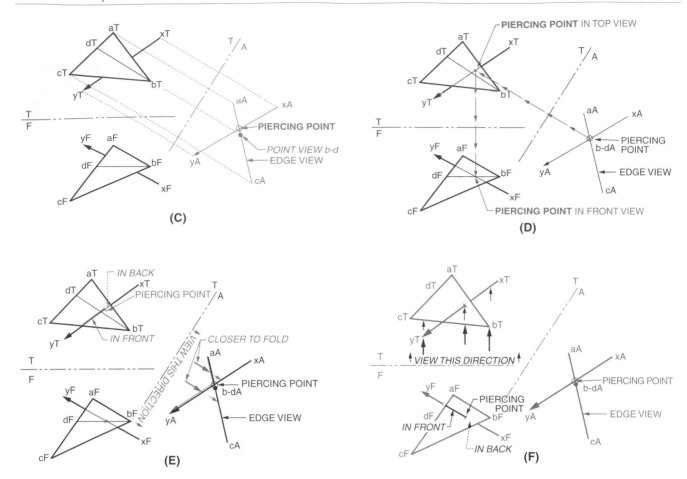

FIGURE 8-23 *(Continued) (C) Step 2; (D) Step 3; (E) Step 4; (F) Step 5*

intersection with the edge view of the plane indicates the piercing point location, *Figure 8-23C.*

STEP 3 Project the piercing point back into other views, *Figure 8-23D.*

STEP 4 In each view that the piercing point is being projected to, the portion of the line that is visible can be determined by checking its preceding view, *Figure 8-23E.*

Notice in the auxiliary view that line x-y is closer to the fold line from the piercing point to point yA than is the edge view of the plane surface from the piercing point to point cA. Therefore, in the top view, line x-y is seen only from the piercing point to yT.

STEP 5 By viewing back into the top view, determine the visibility of lines of the plane surface and line x-y in the front view, *Figure 8-23F.* Notice in the top view that line x-y is closer to the fold line from the piercing point to point yT than is the edge view of the plane surface cT/bT; therefore, in the front view, line x-y is seen only from the piercing point to yF.

When the edge view does not appear in the regular views, it must be constructed using the preceding steps, *Figure 8-24A.* Once the edge view is constructed, the piercing point can easily be seen. The piercing point is now projected down into the front view and up into the top view, *Figure 8-24B.*

HOW TO DETERMINE THE PIERCING POINT BY LINE PROJECTION

An alternate method of determining the piercing point simply locates the path of a cutting plane that passes through the line, leaving a "scar" on the surface of the given plane. Since the line lies in the cutting plane, an intersection of the line and the scar on the given plane locates the piercing point. The end points of the scar are found by aligning the cutting plane with the given line in one view, where it crosses the boundary of the given plane in two places. Projecting the cutting plane intersection to the corresponding boundaries of the given plane in the next view locates the scar end points in that view, and it provides a view of the scar that the given line can cross at the piercing point.

GIVEN:

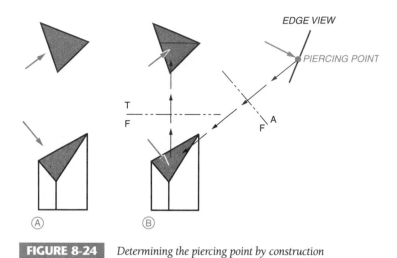

FIGURE 8-24   *Determining the piercing point by construction*

EXAMPLE:

See Figure 8-25A.

GIVEN: Skewed surface abc, and a line.

STEP 1   In the top view, lightly draw the piercing line and determine points x and y where they cross the boundary of plane abc. The piercing point is located somewhere between these points.

STEP 2   Project points x and y down into the front view to the corresponding edges of the plane abc, *Figure 8-25B.* Where line x-y crosses the given line is the exact piercing point.

STEP 3   Project the piercing point to the top view, *Figure 8-25C.* The visibility of the piercing line is determined by inspection.

Figure 8-26A illustrates a cylinder pierced by a line. Lightly draw the piercing line in the top view and find point x. Project point x down into the front view, *Figure 8-26B,* to find piercing point x. Because the piercing line exits the skewed surface, piercing point y is found on the edge view of the front view, *Figure 8-26C.* Project piercing point y up into the top view. Visibility of the piercing line is determined by inspection.

If the object is a cone, line segments must be located and drawn from the base of the cone up to the vertex of the cone.

GIVEN:

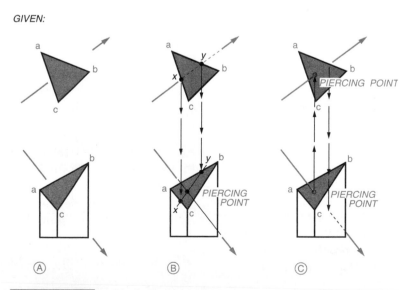

FIGURE 8-25   *Determining piercing points by line projection (skewed surface)*

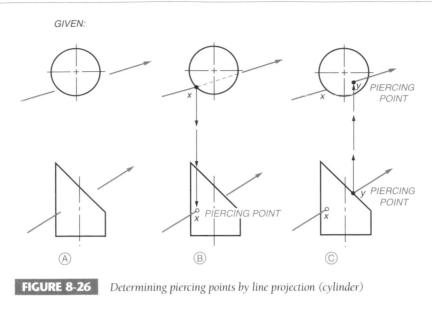

FIGURE 8-26 *Determining piercing points by line projection (cylinder)*

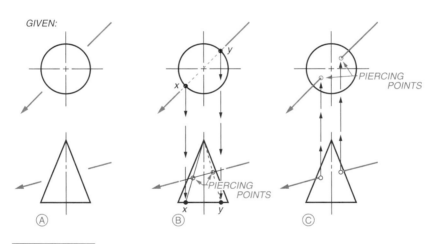

FIGURE 8-27 *Determining piercing points by line projection (cone)*

GIVEN: A cone with a piercing line, *Figure 8-27A.*

Lightly draw the piercing line in the top view and find points x and y, *Figure 8-27B*. Project points x and y down to the base of the cone. Project light line segments from points x and y on the base up to the vertex of the cone. Where these lines cross the piercing line is the location of the two piercing points. Visibility of the piercing line is determined by inspection, *Figure 8-27C*.

If the object is spherical, an imaginary flat surface on the sphere must be established along the piercing line. GIVEN: A sphere with a piercing line, *Figure 8-28A*. Lightly draw the piercing line in the top view. Where the piercing line intersects with the sphere is the location of the imaginary flat surface, *Figure 8-28B*. This also establishes the diameter of the flat surface. Transfer the imag-

inary flat surface to the front view. Where the piercing line intersects the imaginary flat surface is the exact piercing point locations. Once the piercing point locations are found they are projected up into the top view, *Figure 8-28C*. The visibility of the piercing line is determined by inspection.

HOW TO FIND THE INTERSECTION OF TWO PLANES BY LINE PROJECTION

The intersection of two planes can be determined by generating an edge view of one of the planes. Another method, somewhat simpler, is the line-projection method.

GIVEN: Two plane surfaces, *Figure 8-29A*.

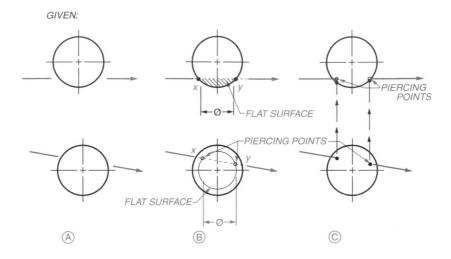

FIGURE 8-28 *Determining piercing points by line projection (sphere)*

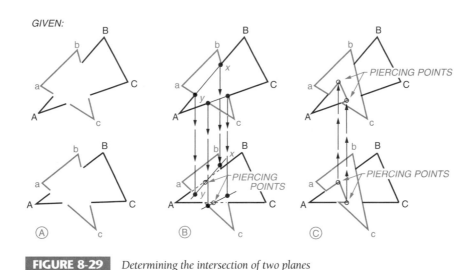

FIGURE 8-29 *Determining the intersection of two planes*

In the top view, locate points x and y on edge ab of one of the planes. Project points x and y down into the front view as illustrated in *Figure 8-29B,* and draw a light line from x-y in the front view. The piercing point is where line x-y crosses the edge ab. Project points from edge ac of the same plane. Project these points down into the front view. Draw a line connecting these points; where they cross edge ac is the exact piercing point. Complete the views as illustrated in *Figure 8-29C.*

HOW TO FIND THE INTERSECTION OF A CYLINDER AND A PLANE SURFACE BY LINE PROJECTION

The intersection of a cylinder and a plane surface can also be determined by generating an edge view of the plane surface. Another method is the line-projection method.

GIVEN: A cylinder and a plane surface, *Figure 8-30A.*

Divide the circle into equal parts; in this example, 12 equal parts. Extend these lines out to the edges of the plane surface, *Figure 8-30B.* The example illustrated uses points 1 o'clock and 7 o'clock to find points x and y on the edge of the plane surface. Project points x and y down into the front view. Draw a line from x-y in the front view. Points 1 o'clock and 7 o'clock are located somewhere on this line. To find their exact locations, project points 1 and 7 from the top view to the line in the front view. Continue around the various points to locate each of the 12 points. Complete the drawing as shown in *Figure 8-30C.*

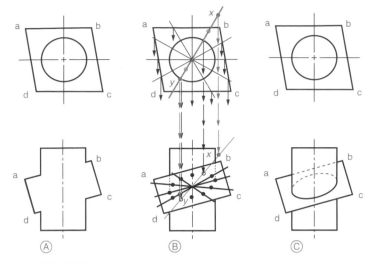

FIGURE 8-30 *Determining the intersection of a cylinder and a plane surface*

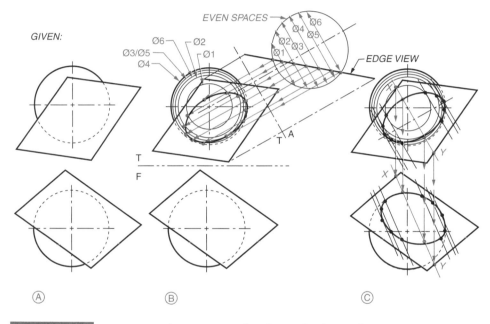

FIGURE 8-31 *Determining the intersection of a sphere and a plane surface*

HOW TO FIND THE INTERSECTION OF A SPHERE AND A PLANE SURFACE

GIVEN: A sphere and a plane, *Figure 8-31A*.

Construct an edge view of the plane, *Figure 8-31B*. Divide the sphere into even spaces as illustrated. Think of each evenly spaced line as an imaginary sliced-off portion of the sphere. Draw each imaginary slice in the top view, and where the projection from the corresponding point on the edge view crosses is the piercing point. Complete the top view as illustrated. Once these points have been found in the top view, the line-

projection method is used to transfer them to the front view, *Figure 8-31C*.

HOW TO FIND THE INTERSECTION OF TWO PRISMS (METHOD 1)

GIVEN: The top, front, and right-side views of two intersecting prisms, *Figure 8-32*.

Label the various points as illustrated. Project each point to the 45° projection and into the next view, and where they intersect is the location of each point, as

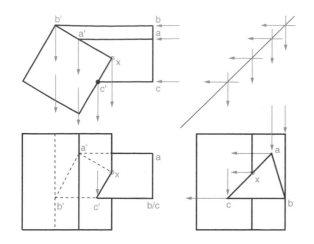

FIGURE 8-32 *Finding the intersection of two prisms (method 1)*

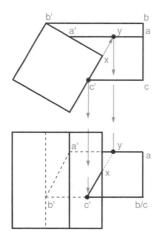

FIGURE 8-33 *Finding the intersection of two prisms (method 2)*

illustrated. The only point in question is point x; using the three views, it can be easily found.

HOW TO FIND THE INTERSECTION OF TWO PRISMS (METHOD 2)

With only two views, there is a problem locating point X in the front view. Refer to *Figure 8-33*. In the top view, extend line c'-x where it crosses line segment a-a'. Project point y down into the front view to line a-a'. Draw a line from c' to y to locate point x in the front view.

Bearings, Slope, and Grade

In some fields of drafting the exact position of a line in space is described by its *bearing* and *slope* or by its bearing and *grade*.

BEARING OF A LINE

The bearing or *heading* of a line is the direction of that line as it is drawn in the *top view* of a drawing on a map. Using a map of a given area on the earth's surface, a view looking directly down would be an example of how a bearing of a line is used.

There are two methods used to call off a bearing. It is called off by either its *north/south deviation* or by its *azimuth bearing*.

NORTH/SOUTH DEVIATION

A bearing of a line is measured in degrees with respect to the north or the south. It is customary to consider the north as being located from the top of the page and the south as being located from the bottom of the page, unless otherwise noted, *Figure 8-34*. Bearings are called off in angles less than 90° and are always given from the north or south, heading toward either the east or west, *Figure 8-35*. In this example a line, 1-2, is projected 45° from the north and toward the west. It has a bearing of, and is called off as, N 45° W. A line, 1-2, is projected 30° from the south toward the east, *Figure 8-36*. It has a bearing of, and is called off as, S 30° E. It is important to know where the point of origin or the beginning of the line is, and in which direction the line is heading, *Figure 8-37*. Line 2-2' could be called off as either N 45° W or as S 45° E, depending on its point of origin. If the line is projected from point 1 to point 2, the bearing would be N 45° W; if from 1' to point 2', the bearing would be S 45° E.

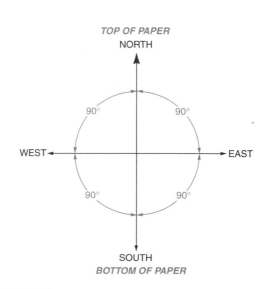

FIGURE 8-34 *North, south, east, west headings; north is usually located at the top of the page*

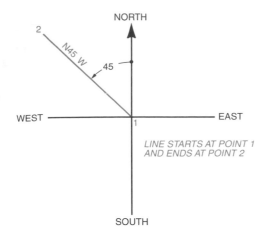

FIGURE 8-35 *A line with a heading of N 45° W; headings are always drawn in the top view*

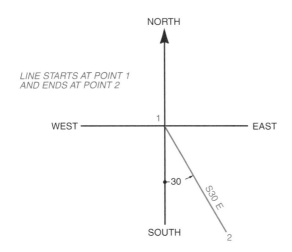

FIGURE 8-36 *A line with a heading of S 30° E*

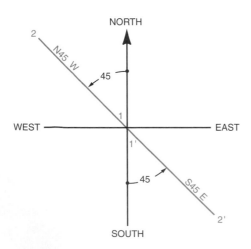

FIGURE 8-37 *This line could be either N 45° W or S 45° W, depending upon the starting point and the direction in which it is projected.*

AZIMUTH BEARING

In using the azimuth bearing method, the total angle is used, going *clockwise* from the north. A given line, 1-2, is drawn at 45° from the north and heading toward the west, *Figure 8-38*. Using the azimuth method and measuring clockwise from the north, it has a bearing of 315°, (90° plus 90° plus 90° plus 45°). It is usually understood that the north bearing is the starting point, thus the N is usually omitted in verbally describing the bearing; on the drawing, however, it is best to use the N or North so as to avoid any confusion. Note that bearing is sometimes noted as heading.

SLOPE ANGLE

Using the slope angle is a way to describe the inclination of a line. The slope of a line is the angle, in degrees, that the line makes with a horizontal (level) plane, *Figure 8-39*. The line is heading from point 1 toward point 2, and it is rising; therefore, it is a plus (+) slope angle.

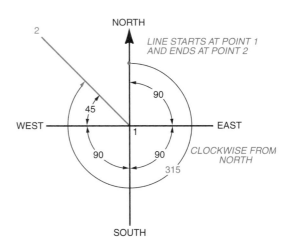

FIGURE 8-38 *Azimuth bearing is indicated from the north clockwise to the line.*

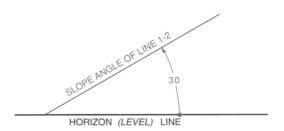

FIGURE 8-39 *Slope angle is the angle in degrees from the horizon.*

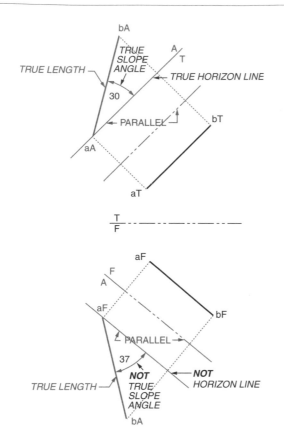

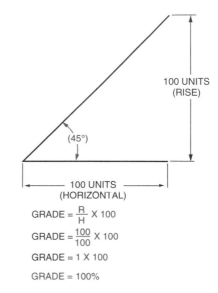

GRADE = $\frac{R}{H}$ X 100

GRADE = $\frac{100}{100}$ X 100

GRADE = 1 X 100

GRADE = 100%

FIGURE 8-41 *Grade percent is the inclination of a line in relation to the horizon, noted in percent.*

FIGURE 8-40 *True slope angle of a line can only be seen in a view that has a horizon line. It is noted in degrees.*

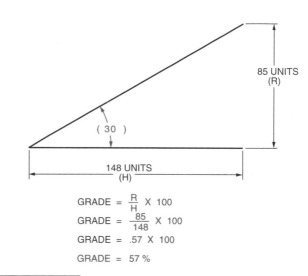

GRADE = $\frac{R}{H}$ X 100

GRADE = $\frac{85}{148}$ X 100

GRADE = .57 X 100

GRADE = 57 %

FIGURE 8-42 *Example of a line with a grade of 57%*

Note that the *true slope* of a line can only be seen in the view that has a horizon (level) line and in which it is drawn as a *true length, Figure 8-40.* In this example, the true length can be found from the front view and/or from the top view. However, projecting from the front view will **not** give a horizon (level) line—it is only found by projecting from the top view.

GRADE PERCENT

Another way to describe the inclination of a line in relation to the horizon is by the ratio of the vertical rise (R) to the horizontal (H). It is expressed in percent and calculated by the simple formula:

PERCENT OF GRADE = $\dfrac{\text{UNITS OF VERTICAL RISE (R)}}{\text{UNITS OF HORIZON (H)}}$ **X 100**

Figure 8-41 illustrates a simple 45° angle line with 100 horizontal units and 100 vertical units. 100 divided by 100 times 100 equals a 100% grade. If the angle were 30°, the grade would be 57%, *Figure 8-42.* The most common method of measuring the horizontal and vertical units is with an engineer's scale using multiples of 10.

Percent of grade is another way to express the slope of a line, but, as in laying out grade percents, **the view that has both the true horizon and true length must be used.** Refer back to Figure 8-40.

OTHER WAYS TO CALL OFF THE SLOPE OF A LINE

Various fields of drafting use slightly different methods to call off the specified inclination of lines. Each method is similar to that of calling off percent of slope and/or percent of grade in that each expresses the relationship

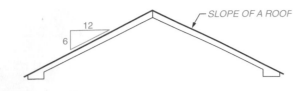

FIGURE 8-43   *Pitch is the slope of a roof and is a ratio of vertical rise for every 12 inches or 12 feet of horizontal span.*

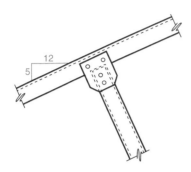

FIGURE 8-44   *Bevel is used in structural engineering to indicate the slope of a structural member.*

FIGURE 8-45   *Batter is used in the civil engineering field to indicate the slope of a grade.*

of the line to the horizontal (level) plane. Three common methods are pitch, bevel, and batter.

Pitch.   Pitch is usually used in the architectural field of drafting and is used to call off the slope of a roof. It is expressed as the ratio of vertical rise to 12 inches or 12 feet of horizontal span, *Figure 8-43*. In each horizontal 12 inches the roof rises 6 inches.

Bevel.   In structural engineering, where steel members are used in the construction of the building, the slope of a member is called off by the *bevel* of the beam, *Figure 8-44*. It is very similar to that of the pitch used in architectural drafting.

Batter.   In the field of civil engineering the slope of the grade (earth) is used and is referred to as *batter*. It is called off as illustrated by *Figure 8-45*. This means that for each unit of rise (R), there are four horizontal (H) units.

ANGLES BETWEEN BEARINGS OR HEADINGS

If you know the bearings of two lines, the angle between the lines can be determined, *Figure 8-46*. Given are lines 1-2 and 1-3, as illustrated. Both these lines fall within one 90° quadrant; therefore, to find the angle between the lines, subtract one bearing from the other, *Figure 8-47*. Fifteen degrees is subtracted from 80°, and the true angle between the bearings is 65°.

If the bearings fall within two or more quadrants, the actual angles in each quadrant must be added or subtracted, as illustrated in *Figures 8-48, 8-49, 8-50,* and *8-51*.

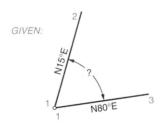

FIGURE 8-46   *Angle between two headings in one quadrant (90°)*

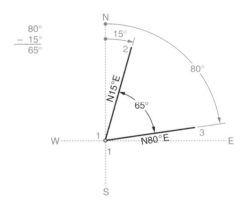

FIGURE 8-47   *How to calculate the angle between headings in one quadrant.*

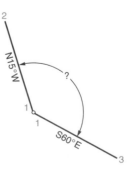

FIGURE 8-48   *Angle between two headings in three quadrants*

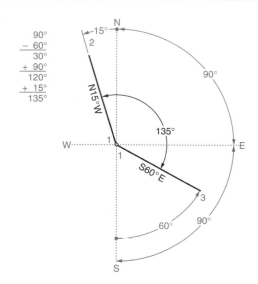

FIGURE 8-49 *How to calculate the angle between headings in three quadrants*

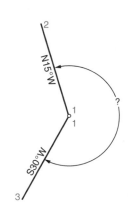

FIGURE 8-50 *Angle between two headings in four quadrants*

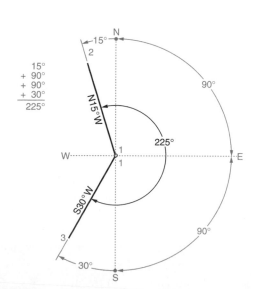

FIGURE 8-51 *How to calculate the angle between headings in four quadrants*

HOW TO CONSTRUCT A LINE WITH A SPECIFIED BEARING, SLOPE ANGLE, AND LENGTH

The bearing is always drawn in the top view, and the true length can be drawn if the line is to be exactly horizontal. Any line with an incline up (+) or down (−) from the horizon **must** be laid out using the following steps:

GIVEN: The starting point aT and aF of a line with a bearing of N 60° E, with a slope angle of +15°, and 230.0 feet long, *Figure 8-52A*.

Because the line is **not** horizontal and has a slope angle of plus (+) 15°, the true length must be constructed before the heading length can be drawn in the top view.

STEP 1 In the top view, draw a line starting from aT, at 60° from the north and toward the east, *Figure 8-52B*. Temporarily draw the line longer than the 230.0 feet required.

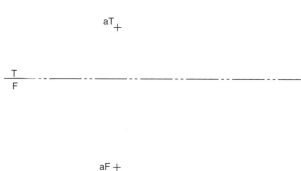

(A)

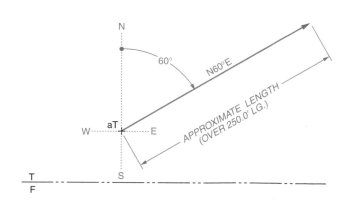

(B)

FIGURE 8-52 *(A) Constructing a line with a specified bearing, slope angle, and length; (B) Step 1*

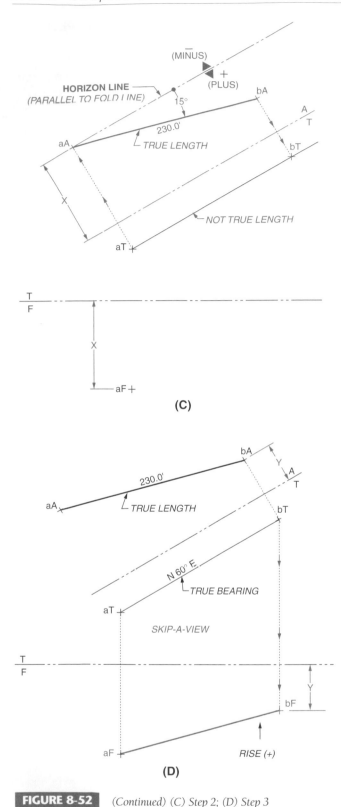

FIGURE 8-52 | (Continued) (C) Step 2; (D) Step 3

STEP 2 Find the true length of the line by adding a fold line, T/A, parallel to the bearing line aT. Project point a into the auxiliary view to find aA, *Figure 8-52C,* and refer to dimension X. Don't forget to skip a view.

From point aA in the auxiliary view, draw a light horizon line parallel to the fold line T/A. From point aA, construct a 15° angle toward the fold line as shown.

*NOTE: Draw the line toward the fold line for a rise (+) and away from the fold line for a drop (–) in elevation. Along this line measure the true length, 230.0 feet, to establish point bA. Project bA back into the top view to find point bT. Remember, this line is **not** the true length in the top view.*

STEP 3 Project point bT down into the front view to find point bF. Use the Y dimension to find point bF, but do not forget to skip a view, *Figure 8-52D.* Construct line aF/bF. Note that the line rises from a to b in the front view, which indicates a plus (+) rise.

The true bearing is found in the top view and the true length is found in the auxiliary view. Remember, true length must be found before the other views can be completed.

HOW TO CONSTRUCT A LINE WITH A SPECIFIED BEARING, PERCENT OF GRADE, AND LENGTH

Approximately the same steps are used in constructing a line with a specified bearing, percent of grade, and length as was used in constructing a line with a specified bearing, slope angle, and length. The bearing is always drawn in the top view, and the true length of the line can be drawn if the line is to be **exactly** horizontal. Any line with an incline up (+) or down (–) from the horizon **must** be laid out using the following steps:

GIVEN: The starting point aT and aF of a line with a bearing of N 45° W, with a grade of (–) 40%, and 190.0 feet long, *Figure 8-53A.* Because the line is **not** horizontal and has a grade of minus (–) 40%, the true length must be constructed before the bearing length can be drawn in the top view.

STEP 1 In the top view, draw a line starting from aT, at 45° from the north and toward the west, *Figure 8-53B.* Temporarily draw the line actually longer than the 190.0 feet required.

STEP 2 Find the true length of the line by adding a fold line, T/A, parallel to the bearing line aT. Project point a into the auxiliary view to find aA, *Figure 8-53C,* and refer to dimension X. Don't forget to skip a view.

From point aA in the auxiliary view, draw a light horizon line parallel to the fold line T/A. From point aA, and along the horizon

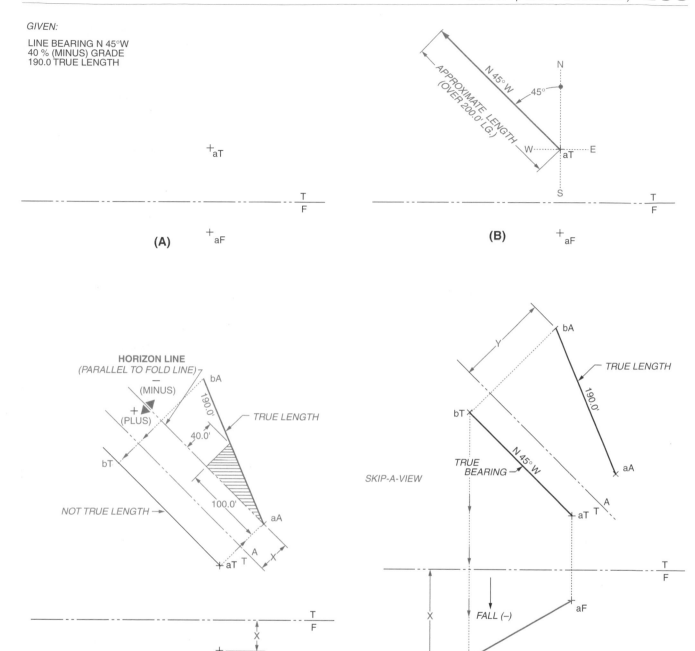

GIVEN:

LINE BEARING N 45°W
40 % (MINUS) GRADE
190.0 TRUE LENGTH

FIGURE 8-53 (A) Constructing a line with a specified bearing, percent of grade, and length; (B) Step 1; (C) Step 2; (D) Step 3

line, measure 100 units. From this point, turn 90° away from the fold line and measure 40 units, and put a point there. Construct a line from aA to this new point and extend it out to the 190.0 feet true length to find point bA.

NOTE: *Construct the triangle toward the fold line for a rise (+) and away from the fold line for a drop (–) in elevation. Project bA back into the top view to find point bT. Remember, this line is not the true length in the top view.*

STEP 3 Project point bT down into the front view to find point bF. Use the Y dimension to find point bF, but do not forget to skip a view, *Figure 8-53D*. Construct the line aF/bF. Note that the line falls from a to b in the front view, which indicates a minus (–) drop. The true bearing is found in the top view and the true length is found in the auxiliary view. Remember, true length must be found before the other views can be completed.

Applied Descriptive Geometry

The topics in the remainder of this chapter are essentially applications of the descriptive geometry basics already presented. Some of the topics are contours and topographic maps, vectors, forces in trusses, and revolution.

Topographic Topics

SURVEYS

Compass directions with distances describe real estate boundaries in cities and counties for commercial and domestic property owners. The descriptions usually include references to governmental surveys, such as being located in a government lot, section, township, and range and they have specific coordinates so a surveyor can locate the property and stake out the area. Architects and builders have drafters prepare scaled drawings of the areas for planning the locations of buildings and houses.

HOW TO PLOT THE BOUNDARY OF REAL ESTATE

EXAMPLE:

GIVEN: Compass bearings with distances in feet, outlining a real estate lot, shown in a rough sketch with North in the traditional direction toward the top of the sheet, *Figure 8-54A*.

STEP 1 Lay out the boundary to scale starting with the S17° W 59', then the S31° W 47', and next the S54°30'E 170' while utilizing the drawing area efficiently: for example, the direction North may not be toward the top of the sheet, *Figure 8-54B*. Note that the 71°30' angle shown is an internal angle, and that the sum of all the internal angles equals (number of sides N − 2) × 180°. For the layout shown, (N − 2) × 180° = (5 − 2) × 180° = 540°. The 85°30' angle is called a deflection angle; this is the smallest angle the surveyor must turn the transit to align with the next side. The sum of all the deflection angles must equal 360° for a complete traverse around the property.

STEP 2 Complete the boundary with N26°25'E 104' next and N54°30'W 176' last. The last segment should close on the starting point, *Figure 8-54C*.

For excavation and landscaping planning, an architect may ask a drafter to draw in contour lines. This is the topic of the next section.

CONTOUR LINES

A *contour line* on a plot or map is a line at a constant elevation on the surface of the earth, above sea level, *Figure 8-55A*. A *contour interval* is the difference in ele-

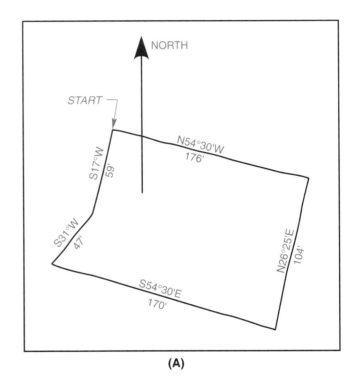

(A)

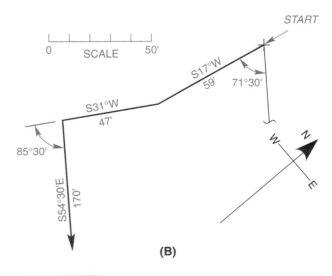

(B)

FIGURE 8-54 (A) Compass bearings and distances: Real estate lot; (B) Step 1

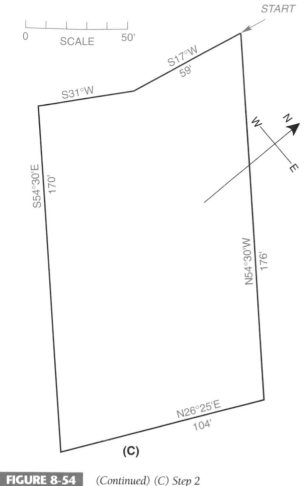

FIGURE 8-54 *(Continued) (C) Step 2*

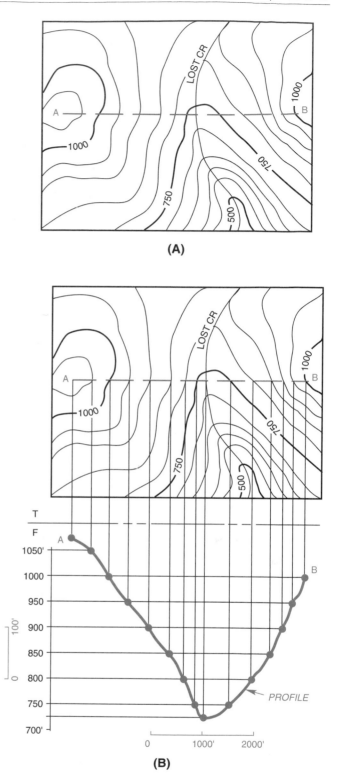

FIGURE 8-55 *(A) Contour map; (B) Profile*

vation between neighboring contour lines. It is the same value for any one drawing. For example, 50 feet is the interval in Figure 8-55A due to mountainous terrain. Ten-foot intervals are common; architectural plots may only have one-foot intervals on relatively level terrain for houses or buildings.

Usually, every fifth contour line is heavier and its elevation indicated. Lines close together represent steep terrain.

PROFILES

Imagine a vertical plane cutting through a contoured map and being able to see the cross section at the cut. The cross section would show a profile of the terrain in an elevation view, as illustrated in *Figure 8-55B* for a straight path from point A to point B. The fold line is added as a reminder that all topographic maps are T views, and elevations, of course, occur in views projected off a T view.

CUTS AND FILLS

Highways passing through hilly terrain require that the earth of some hills be excavated and the earth used to fill in valleys—*cuts and fills, Figure 8-56*. The volume of the

PROFILES FOR CUTS AND FILLS

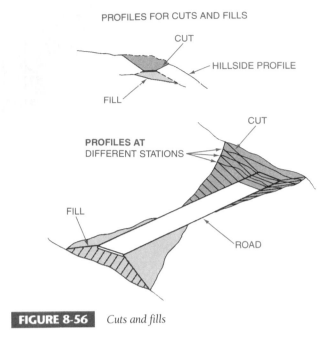

FIGURE 8-56 *Cuts and fills*

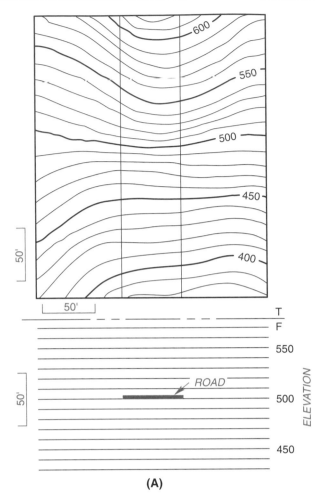

(A)

FIGURE 8-57 *(A) Proposed road*

earth to be cut and filled needs to be determined prior to having an army of construction workers and their equipment arrive at the job site, for efficient use of labor and equipment. The volume of the cut determines how much fill can be completed. Volumes of cuts and fills are determined from boundary maps.

HOW TO DETERMINE THE BOUNDARIES OF CUTS AND FILLS FROM TOPOGRAPHIC MAPS WITH CONTOUR LINES

EXAMPLE:

GIVEN: A portion of a topographic map with contours and the proposed location of a 60-foot wide, level road, at an elevation of 500', *Figure 8-57A*. The fold line T/F is perpendicular to the direction of the road, and the F view shows elevation planes at 10-foot intervals; the topographic map and the elevations are drawn to the same scale.

STEP 1 Construct cutting planes at 45° to the horizontal in the elevation view F, to each side of the proposed 60-foot-wide road, *Figure 8-57B*. Where the cutting planes intersect the elevation planes, mark the corresponding intersections on the map. Connect the intersections, freehand, to show the boundaries of the excavation cut, Figure 8-57B.

STEP 2 In a similar manner construct cutting planes at 45° to represent the proposed earth fill, and mark the boundaries as before, *Figure 8-57C*. Note that the angles of cuts and fills vary— for example, a commonly used fill has a slope of 1.5 horizontal to 1.0 vertical.

HOW TO DETERMINE THE VOLUMES OF CUTS AND FILLS (GRAPHICAL, MECHANICAL, AND ELECTRONIC)

EXAMPLE:

GIVEN: The cut and fill boundaries from the preceding example shown in a simplified T view, *Figure 8-58A*.

STEP 1 Start at the 500-foot elevation. Divide the cut and the fill with vertical planes at 25-foot intervals called stations (STA), *Figure 8-58B*.

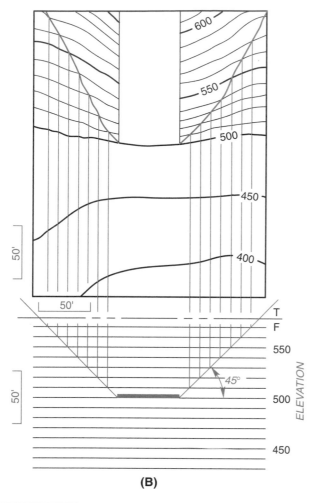

(B)

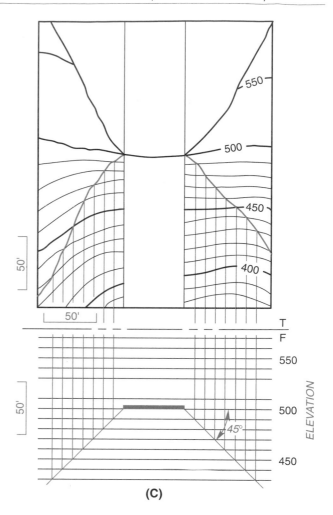

(C)

FIGURE 8-57 (Continued) (B) Step 1; (C) Step 2

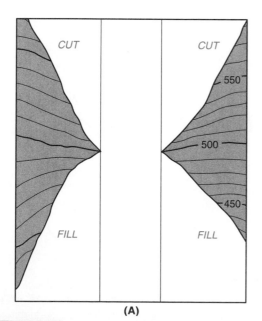

(A)

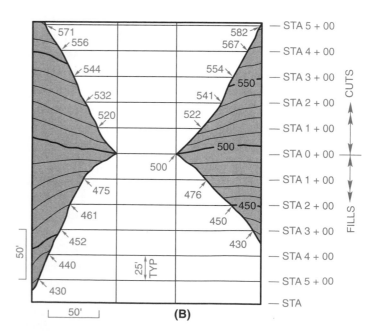

(B)

FIGURE 8-58 (A) Cut and fill boundaries; (B) Step 1

Estimate the elevations of the points where the vertical planes at each station intersect the boundaries by reading the contour lines.

STEP 2 Construct scaled drawings of the profiles at the stations, and determine the area of each profile by using a graphical method for calculating geometric areas, *Figure 8-58C*. Tabulate the results.

Three methods used to calculate geometric areas of cross sections are graphical, mechanical, and electronic. The three are applied to a variety of problems, such as cuts and fills, cross sections of ship and boat hulls, and cross sections of layers of odd shapes for volume calculations.

A *graphical method* for approximating areas was used in Figure 8-58C. The method is to divide the area into convenient shapes, such as trapezoids and triangles, to scale appropriate lengths, and to calculate the areas.

A mechanical method utilizes a planimeter shown in the conceptual sketch, *Figure 8-58D*. Pole arm OA is confined to swivel about point O. Point P is a pointer the operator uses to trace the complete boundary, drawn to scale, causing an indicating wheel to rotate and slide. The wheel's final position gives a reading of the area traversed.

An electronic method employs a computer and a digitizer (looks like a computer mouse). A cross-haired window at the bottom of the digitizer is positioned at small increments of distances around the boundary of an area, *Figure 8-58E*. At each increment along the boundary, a key on the digitizer is clicked. This causes X-Y coordinates to be stored in a computer. Finally, the computer uses the coordinates to calculate the digitized area.

STEP 3 Calculate the amount of earth to excavate (cut) between two stations (STA) by averaging the areas at the stations and multiplying by the 25-foot distance between the stations. For example, the area STA 2 + 00 of the cut = 3,502 square feet; the area at STA 3 + 00 of the cut = 5,316 square feet; the volume between STAs 2 and 3 of the cut is $(3,502 + 5,316)/2 \times (25) = 110,225$ cubic feet. The volume in cubic yards (called yards in construction) is $110,225/27 = 4,082$. Use this calculation procedure to

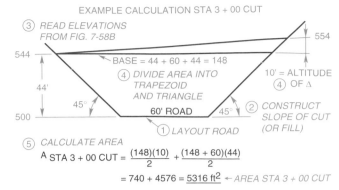

EXAMPLE CALCULATION STA 3 + 00 CUT

③ *READ ELEVATIONS FROM FIG. 7-58B*

④ *DIVIDE AREA INTO TRAPEZOID AND TRIANGLE*

② *CONSTRUCT SLOPE OF CUT (OR FILL)*

① *LAYOUT ROAD*

10' = ALTITUDE ④ OF Δ

⑤ *CALCULATE AREA*

$$A \text{ STA } 3 + 00 \text{ CUT} = \frac{(148)(10)}{2} + \frac{(148 + 60)(44)}{2}$$

$$= 740 + 4576 = \underline{5316 \text{ ft}^2} \leftarrow \text{AREA STA } 3 + 00 \text{ CUT}$$

	STA	ft²
CUTS	5 + 00	10,412
	4 + 00	7,442
	3 + 00	5,316
	2 + 00	3,502
	1 + 00	900
	0	0
FILLS	1 + 00	2,070
	2 + 00	4,620
	3 + 00	6,900

(C)

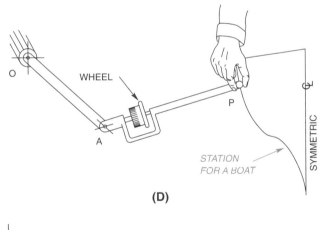

WHEEL

STATION FOR A BOAT

SYMMETRIC

(D)

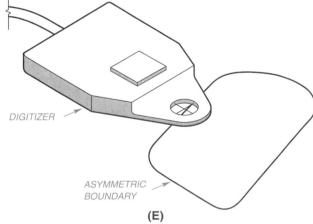

DIGITIZER

ASYMMETRIC BOUNDARY

(E)

FIGURE 8-58 *(Continued) (C) Step 2 (graphical method); (D) Mechanical method; (E) Electronic method*

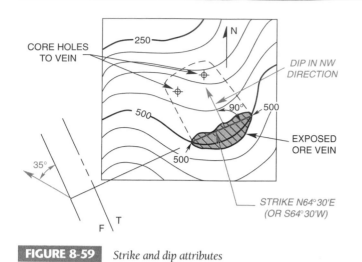

FIGURE 8-59 *Strike and dip attributes*

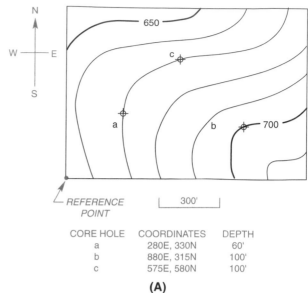

CORE HOLE	COORDINATES	DEPTH
a	280E, 330N	60'
b	880E, 315N	100'
c	575E, 580N	100'

(A)

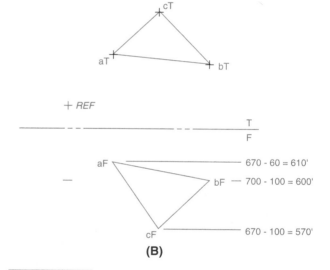

(B)

FIGURE 8-60 *(A) Core holes for vein location; (B) Step 1*

determine the volume of earth between STAs 0 + 00 and 5 + 00 of the cut, Figure 8-58B. Get 20,709 yards available for fills. The fill requirement (using the same calculation procedure) for the volume between STAs 0 + 00 and 3 + 00 FILLS is only 9,390 yards. The remaining earth from the cut would be used to fill to additional stations.

STRIKE AND DIP

Strike and dip are terms used in mining to describe three attributes of ore veins, *Figure 8-59*. *Strike* is a compass direction of a horizontal line lying in the plane of an ore vein. *Dip* is the downward angle of the vein with respect to a horizontal plane at the site of the ore vein. Dip also includes the general compass direction of the downward slope.

The vein in Figure 8-59 would be described as N64°30'E, 35° NW, that is, the compass direction of the strike and the 35° dip downward toward the NW.

HOW TO DETERMINE THE STRIKE AND DIP OF A VEIN OF ORE (OR STRATUM OF ROCK)

EXAMPLE:

GIVEN: The depths of three core-holes, a, b, and c, sunk over a discovered ore vein, *Figure 8-60A.*

STEP 1 Lay out the points a, b, and c in a T view to scale, Figure 8-60A. Since the points, which are at different depths, do not fall in a straight

line, they describe a triangular plane lying in the vein, *Figure 8-60B.*

STEP 2 Find plane abc as an edge in view A by using fold line T/A, *Figure 8-60C,* and the technique described in Figures 8-17A through 8-17D in this chapter.

STEP 3 In view A measure the dip angle, and in the T view measure the strike bearing, *Figure 8-60D.* Note the general direction of the dip to complete the attributes of the ore vein—N76° W, 52° NE.

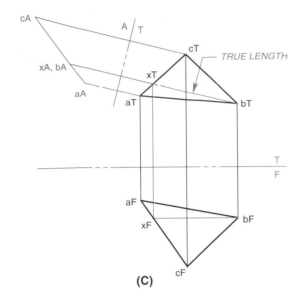

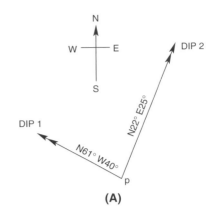

(A)

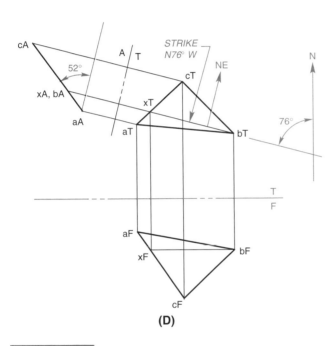

FIGURE 8-60 *(Continued) (C) Step 2; (D) Step 3*

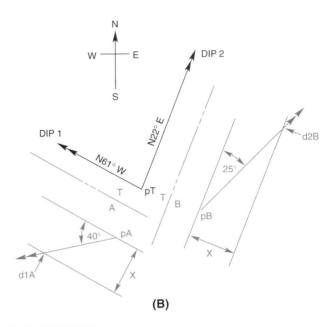

FIGURE 8-61 *(A) Given: Dip directions and angles; (B) Step 1*

STEP 1 Construct view A using fold line T/A parallel to DIP1 to obtain the true view of the dip. Construct view B using T/B parallel to DIP2 to obtain its true view, *Figure 8-61B*. From points pA and pB lay out the same distance, X, downward. Label the marks d1A and d2B, respectively.

STEP 2 Project the points d1A and d2B into the T view and note that the line d1T-d2T is horizontal and, therefore, the strike of the ore vein, *Figure 8-61C*.

STEP 3 Project the triangle (p-d1-d2)T into a view C using fold line T/C perpendicular to the strike line, and measure the true dip angle, *Figure 8-61D*. The true strike and dip are N52° E, 43° NW.

HOW TO DETERMINE THE STRIKE AND DIP OF A VEIN GIVEN TWO LINES IN THE PLANE OF THE VEIN AND INTERSECTING

EXAMPLE:

GIVEN: Two apparent dip directions (not strikes) and angles of the dips as follows: N61° W, 40° and N22° E, 25° both originating from the same point p, a bore-hole, *Figure 8-61A*.

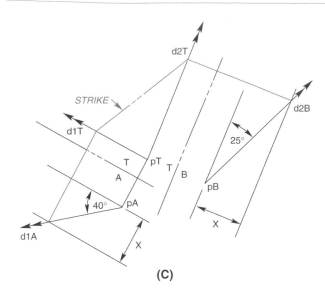

(C)

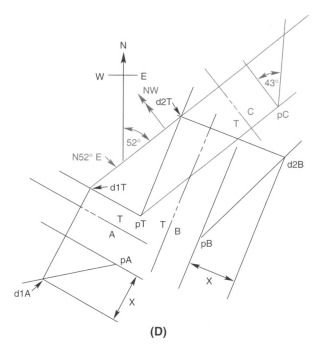

(D)

FIGURE 8-61 *(Continued) (C) Step 2; (D) Step 3*

Vectors

A *vector* as defined in engineering, technology, and mathematics must have both magnitude and direction. The vector shown in *Figure 8-62,* drawn to scale and having an arrowhead at one end, could represent any one of the following: displacement (length and direction); force (pounds and direction); velocity (distance per unit of time and direction); or acceleration (distance per unit of time squared and direction). As a compari-

M ————————► N

1.5 feet ————————	1.5'
15 pounds —————————	15 lbf
150 feet/second —————	150 fps
1.5 ft/sec/sec —————————	1.5 ft/sec²

FIGURE 8-62 *Vector*

son, direction only is a characteristic of the omnipresent straight line with an arrowhead at one end. A straight line with arrowheads at both ends in dimensioning indicates magnitude. These are not vectors, but they efficiently convey information in technical material, such as compass bearings, cursor positions, callouts, identifying features, and of course dimensions on orthographic drawings, *Figure 8-63.* Magnitudes without directions are called *scalars,* such as temperature and pressure.

Vector equivalents of physical systems are simplifications used in engineering so that the vectors can be manipulated efficiently in mathematical formulas. For example, bulky objects or portions of a bridge deck are assumed to be loads concentrated at a point. Internal loads, although acting over an area, are assumed to be point loads acting at the centroid of the area. All parts of an object traveling at the same velocity are represented by one velocity vector acting at the center of mass of the object *Figure 8-64.* Always keep in mind that these simplifications do not change the physical situation, and that point loadings are only approximate. Thus, graphical solutions employing these simplifications are as accurate as mathematical solutions. A number of the vector manipulations that can be done graphically are discussed

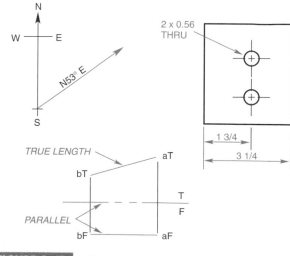

FIGURE 8-63 *Non-vectors*

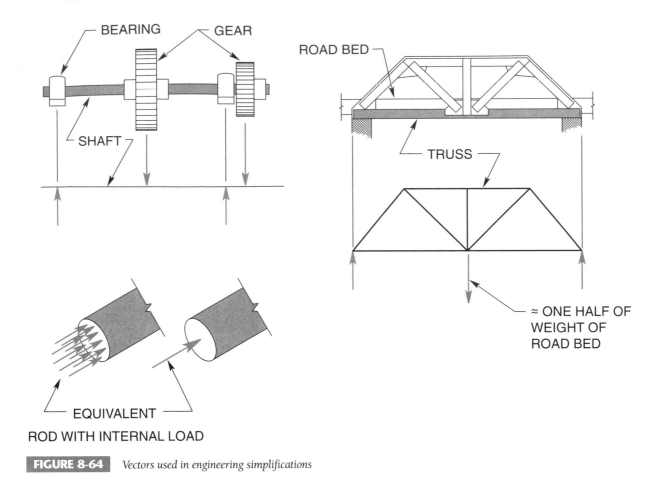

FIGURE 8-64   *Vectors used in engineering simplifications*

in this chapter, for two main reasons—to demonstrate the efficacy of graphics and to emphasize that when you can visualize what is going on, you will have a better understanding of the analysis and the answer.

Vectors will be labeled with capital letters, e.g., F(AB) is a force F in the direction of, or in, member AB. An applied load would be W or F. A velocity vector would be V.

VECTOR SYSTEMS

Six possibilities exist for analyses using vectors: coplanar (concurrent, parallel, and nonparallel) and noncoplanar (concurrent, parallel, and nonparallel). Coplanar-concurrent and noncoplanar-concurrent are the categories to be emphasized in this chapter.

COPLANAR-CONCURRENT FORCE SYSTEMS— THE PARALLELOGRAM LAW: TWO VECTORS

The *parallelogram law* basically states that two vectors, not parallel, drawn to scale, having the same units, and acting at the same point can be added or subtracted as shown in *Figure 8-65A*. Use vector A plus vector B to construct a parallelogram. The diagonal of the parallelogram vector (A + B) is called the resultant R, which is equivalent to the effect of the two vectors, *Figure 8-65B*. Vector A minus vector B can also be used to construct a parallelogram. The resultant vector (A–B) is the diagonal, *Figure 8-65C*.

If the vectors being added or subtracted are force vectors, the equal and opposite reaction to the resultant is defined as the *equilibrant*. The reaction must be present to satisfy a basic law of mechanics, which is that the sum of the forces acting on an object that is in equilibrium must add to zero in any one direction, that is, $\Sigma F = 0$. A force may be defined as the action of one body on another, and it tends to change the state of rest or motion of the body acted on.

A vector force, acting on a body can be considered to be moved along its line of action without changing its effect on the body.

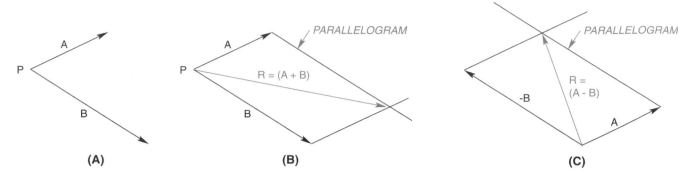

FIGURE 8-65   *(A) Coplanar concurrent vectors; (B) Adding vectors: Parallelogram law; (C) Subtracting vectors: Parallelogram law*

EXAMPLE:

Refer to *Figure 8-66A*.

GIVEN: Force vectors P and Q acting on body B. Also, S and T acting on W.

STEP 1 Move vectors P and Q along their respective lines of action until they meet and are tail to tail, *Figure 8-66B*. Also, move S and T.

STEP 2 The resultant R of vectors P and Q shows the combined effect of P and Q on body B,

Figure 8-66C. Also, the resultant U of vectors S and T shows the effect on body W.

HOW TO ADD MORE THAN TWO COPLANAR-CONCURRENT VECTORS

Two graphical methods can be used to add more than two coplanar-concurrent vectors. One requires adding any two vectors first, and then using the resultant to add to the third, etc. The second involves adding vectors "tail to head," in any order, to obtain a resultant.

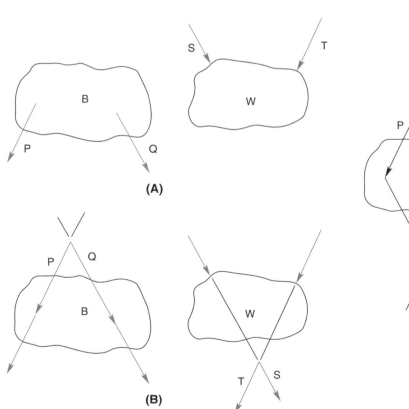

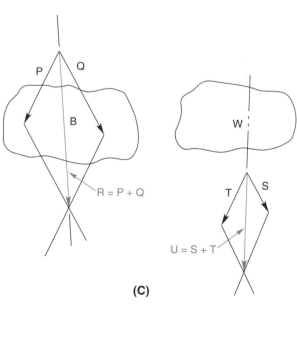

FIGURE 8-66   *(A) Vector line of action; (B) Step 1; (C) Step 2*

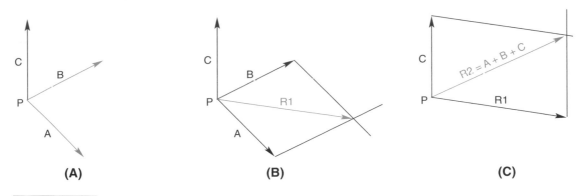

FIGURE 8-67   *(A) Three coplanar concurrent vectors (method 1); (B) Step 1; (C) Step 2*

EXAMPLE: METHOD 1

Refer to *Figure 8-67A*.

GIVEN: Vectors A, B, and C acting at point P.

STEP 1   Add vectors A and B using the parallelogram law to obtain resultant R1, *Figure 8-67B*.

STEP 2   Add vectors R1 and vector C using the parallelogram law to obtain resultant R2, *Figure 8-67C*. R2 is the resultant of the three vectors and would have the same effect as A, B, and C acting together at point P.

EXAMPLE: METHOD 2

Refer to *Figure 8-68A*.

GIVEN: Vectors A, B, and C acting at point P.

STEP 1   Move vector B parallel to its given position so that the tail of B is at the head of vector A, *Figure 8-68B*.

STEP 2   Locate vector C in a similar manner so that the tail of C is at the head of vector B, *Figure 8-68C*.

STEP 3   A vector from point P to the head of vector C is the resultant of the three vectors, that is, the resultant would have the same effect on point P as vectors A, B, and C, *Figure 8-68D*.

RESOLUTION OF VECTORS

A given vector can be resolved into two coplanar vector components that act along two axes that intersect at the tail of the given vector.

HOW TO RESOLVE A VECTOR ONTO PERPENDICULAR AXES

EXAMPLE 1:

Refer to *Figure 8-69A*.

GIVEN: Vector A and axes X and Y intersecting at O.

STEP 1   From the tip of vector A construct lines parallel to the two axes so that the lines intersect the axes X and Y at X(1) and Y(1), *Figure 8-69B*.

STEP 2   The two vector components that replace vector A are OX(1) and OY(1) with arrowheads at X(1) and Y(1), respectively, *Figure 8-69C*.

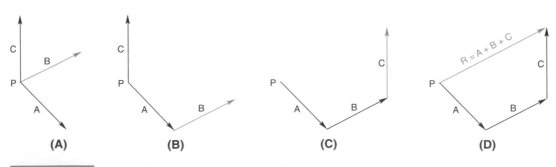

FIGURE 8-68   *Four coplanar concurrent vectors (method 2)*

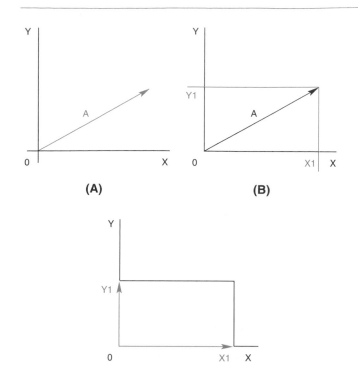

FIGURE 8-69 (A) Resolution on perpendicular axes; (B) Step 1; (C) Step 2

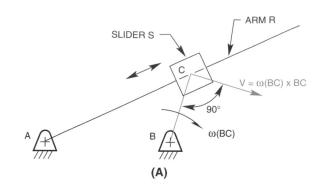

(A)

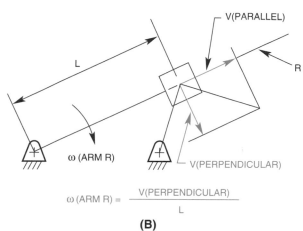

(B)

FIGURE 8-70 (A) Quick-return mechanism; (B) Steps 1 and 2

EXAMPLE 2:

A three-dimensional mechanism problem in dynamics, but simplified to a statics problem at the instant shown. A further simplification presents the components as straight lines. The simplified mechanism and the velocity vector are drawn to scale. Refer to *Figure 8-70A*.

GIVEN: Arm R confined to rotate about point A. Slider S pinned to C of crank BC. Crank BC confined to rotate about point B, 360 degrees. (The apparent interference of arm R with crank BC doesn't occur in the three-dimensional mechanism.) Slider C is free to slide along arm R.

Slider C at the instant shown has a velocity V(C) m/sec = ω(BC) radians/sec × BC m. The direction of the velocity of slider S at the instant shown is perpendicular to the crank BC because slider C is attached to point C.

TO FIND: The angular velocity ω(ARM)

STEP 1 Resolve vector V(C), drawn to scale, into two components, one parallel to arm R and one perpendicular, V(perpendicular), to arm R, *Figure 8-70B*.

STEP 2 Divide V(perpendicular) m/sec by L m to obtain ω(arm R) radians/sec. (Note that radians/sec = $2 \times \pi \times$ RPS. RPS = revolutions per second).

HOW TO RESOLVE A VECTOR ONTO TWO AXES NOT PERPENDICULAR

EXAMPLE:

GIVEN: Vector V originating at o and axes m and n intersecting at o, *Figure 8-71A*.

STEP 1 From the tip of vector V construct lines parallel to the two axes so that they intersect axes m and n at M(1) and N(1), *Figure 8-71B*.

STEP 2 The two components that replace vector V are OM(1) and ON(1) with arrowheads at M(1) and N(1), respectively, *Figure 8-71C*.

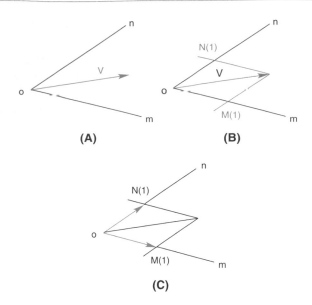

(A) **(B)**

(C)

FIGURE 8-71   *(A) Resolving a vector onto two axes not perpendicular; (B) Step 1; (C) Step 2*

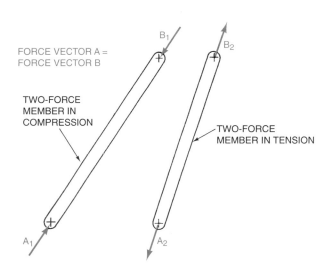

FORCE VECTOR A = FORCE VECTOR B

TWO-FORCE MEMBER IN COMPRESSION

TWO-FORCE MEMBER IN TENSION

(WEIGHTS OF THE MEMBERS ARE USUALLY NEGLIGIBLE COMPARED TO THE EFFECTS OF THE APPLIED FORCES.)

FIGURE 8-72   *Two-force members*

TWO-FORCE MEMBERS

Two-force members have forces at each end, equal in magnitude, in the opposite direction, and collinear. Therefore, they are in equilibrium. The weight of the member is considered to be negligible due to the usually much greater effect of the forces applied. See *Figure 8-72.*

HOW TO DETERMINE THE FORCES IN THE MEMBERS OF A PINNED JOINT

EXAMPLE:

GIVEN: A space diagram showing weight W supported by a frame in equilibrium, made up of two-force members AB and BC, drawn to scale, and pinned to a wall, *Figure 8-73A.*

STEP 1   Isolate pinned joint B by "cutting" the members AB and BC and redrawing the joint, parallel to the original drawing. An engineering expedient is to represent members of a frame (or truss) as straight lines. The lines of action of the internal forces F(AB) and F(BC) intersect at joint B. Their directions must be tentatively assumed by inspection. The direction of force F(AB) must keep the joint away from the wall. Force F(BC) keeps the joint from dropping. See *Figure 8-73B.*

STEP 2   Construct a vector force diagram of the forces at joint B. First, lay out the applied force W, from a point O, vertically downward, to scale. The two remaining forces F(AB) and F(BC) must add—tail to head—to W and end at the starting point O because joint B is in equilibrium. To determine the magnitudes of F(AB) and F(BC), construct a light line through the

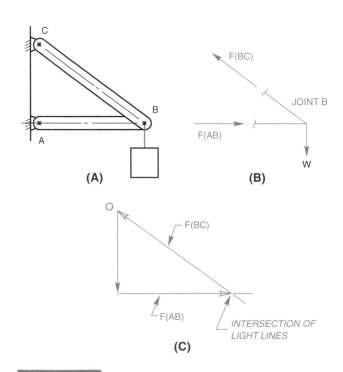

(A) **(B)**

(C)

FIGURE 8-73   *(A) Determining forces of a pinned joint; (B) Step 1; (C) Step 2*

head of vector W parallel to the direction F(AB) and a light line through the tail of vector W parallel to the direction of F(BC), *Figure 8-73C*. Where the two lines intersect determines the magnitudes of F(AB) and F(BC). The tail-to-head method of adding vectors determines their directions. The assumed directions are confirmed in the vector force diagram.

Note that only two unknown magnitudes of forces can be determined in these coplanar joint analyses. Their directions are determined in the space diagram.

TRUSSES

Trusses are frames that are usually two-dimensional and are made of structural shapes, such as angles, tubes, rounds, and channels. They are assembled in triangular patterns for rigidity. The members are considered to be two-force members, and the joints are assumed to be pinned even though welded, bolted, or riveted. Observe some features of the truss shown in *Figure 8-74A* and

the simplified drawing of the truss in *Figure 8-74B*. Both are drawn to scale.

Trusses seldom stand alone. They are used in pairs or in multiples. For example, some highway bridges and railroad bridges employ pairs. Multiples are used to support roofs, *Figure 8-75A*. In a coplanar analysis, each truss is assigned its share of the roadbed loading or the roof loads at the appropriate joints of the trusses, *Figure 8-75B*.

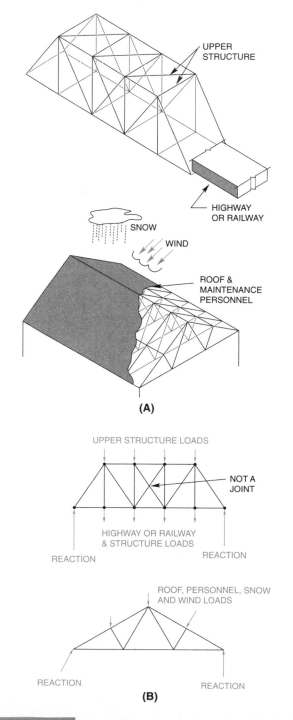

(A)

(B)

FIGURE 8-75 *(A) Truss applications; (B) Truss drawings for analyses*

TWO BOLTS OR
RIVETS FOR SAFETY

CENTER LINE

GUSSET
PLATES

BOLTS

APPLIED
LOAD

CONTINUOUS
MEMBER

(A)

TRIANGULAR
SPACES

PINNED
JOINTS
(TYPICAL)

REACTION   *REACTION*

APPLIED LOAD

(B)

FIGURE 8-74 *(A) Example truss; (B) Simplified drawing of truss*

HOW TO DETERMINE THE FORCES IN THE MEMBERS OF A TRUSS

EXAMPLE 1(A): METHOD OF JOINTS

GIVEN: Line drawing to scale of a symmetrical truss with one applied load and two equal reactions, *Figure 8-76A*.

STEP 1 Select a joint in the truss with only two unknowns. A and C meet this criterion. Use joint A and isolate it in a space diagram, *Figure 8-76B*. Assume the direction of F(AD) by inspection to "go into" the joint, that is, to be a compressive force. Assume F(AB) "goes away from" the joint and is thus a tensile force. Next, lay out a force diagram to scale, starting with the known force of 2500 N.

Follow the procedure described in Step 2 of the preceding discussion, Figure 8-73C, where the light lines indicate the lines of action of the two unknown forces. Their intersection point determines the magnitudes. Scale the magnitudes and put arrowheads on the vectors so that the vectors add head to tail back to the tail of the 2500 N force, the starting point. Finally, make a small freehand sketch of the truss and circle the known members, AD and AB.

STEP 2 Joints B or D have only two unknowns. Select either one and isolate it in an equilibrium sketch. Use joint B, *Figure 8-76C*. Construct a force diagram, starting with known information, the 5,000 N applied load, and the

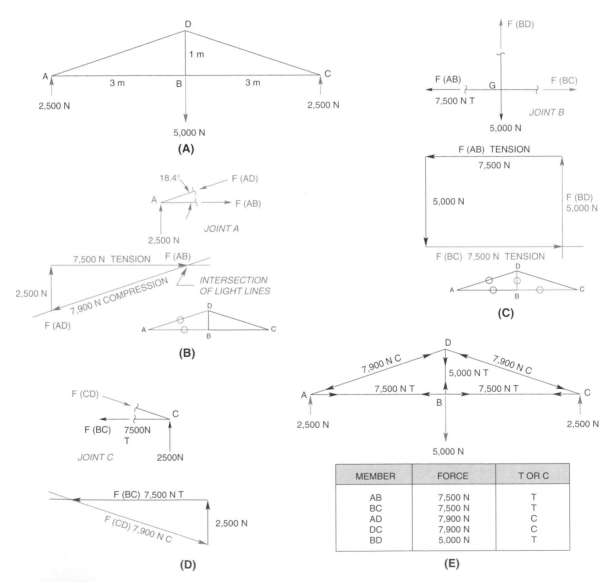

MEMBER	FORCE	T OR C
AB	7,500 N	T
BC	7,500 N	T
AD	7,900 N	C
DC	7,900 N	C
BD	5,000 N	T

FIGURE 8-76 *(A) Example 1 (a) method of joints; (B) Step 1; (C) Step 2; (D) Step 3; (E) Step 4*

7,500 N tensile force in AB. Construct the light lines for the forces in members BD and BC. Scale their magnitudes and add arrowheads so the forces add head to tail back to the tail of the 5,000 N force. On a new freehand sketch, circle the known members, AD, AB, BD, and BC.

STEP 3 The remaining member DC is known due to the symmetry of the truss, F(CD) = F(AD), and is a compressive force. If the truss or the loading were not symmetrical, joint C would be isolated and a force diagram prepared as shown, *Figure 8-76D*.

STEP 4 Prepare a summary of the forces on a sketch of the truss or in a table, *Figure 8-76E*.

EXAMPLE 1(B): METHOD: MAXWELL DIAGRAM WITH BOW'S NOTATION

GIVEN: The same truss used for the method of joints, but to be labeled with Bow's notation scheme, *Figure 8-77A*. Label the triangular spaces

within the truss with numbers or letters. Create enclosed spaces between the applied forces and reaction forces by drawing a light envelope line surrounding the truss and passing through each applied force and reaction. Label each "space" with a letter.

STEP 1 Construct a force diagram from reading the external spaces in the envelope in a clockwise direction, *Figure 8-77B*. Use a convenient scale to keep the force diagram on the drawing sheet and to use the space efficiently. Read spaces a to b for the left reaction, 2,500 N. Then read spaces b to c for 2,500 N. Finally, read c to a for the 5,000 N force, which brings the force diagram back to the starting point. Thus, the truss is in equilibrium.

STEP 2 Select any joint with only two unknown members and read spaces clockwise around the joint, *Figure 8-77C*. Construct lines parallel to the two members crossed. For example, start at joint A and go a to b, b to 1, and 1 to a. Point 1 is located where the construction lines cross.

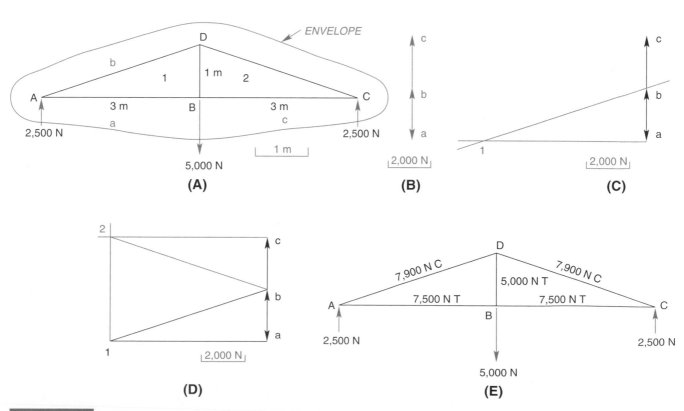

FIGURE 8-77 *(A) Example 1 (b) Maxwell diagram with Bow's notation; (B) Step 1; (C) Step 2; (D) Step 3; (E) Step 4*

STEP 3 Select the next joint with only two unknown members, joint B or D. Around B go a to 1, 1 to 2, 2 to c, and c to a, *Figure 8-77D.* The two light lines for 1 to 2 and 2 to c locate point 2. To finish the force diagram for the truss, go around joint C or joint D. Choose joint C and go b to c, c to 2, and 2 to b.

STEP 4 Prepare a summary of the forces on a sketch of the truss, *Figure 8-77E.* Go clockwise around each joint of the truss to determine tensile and compressive forces. For example, reading clockwise around joint A, 1 to a in the force diagram goes in a direction away from A. Therefore, AB is in tension, and the magnitude is scaled to be 7,500 N. Going b to 1 goes into joint A, thus producing a compressive force of 7,900 N. In a similar clockwise pattern, determine the type and magnitude of the force for each member.

The next example focuses on a more complicated truss with asymmetrical loading.

EXAMPLE 2(A): METHOD OF JOINTS

GIVEN: A freehand sketch and a scaled line drawing of a symmetrical roof truss, but with asymmetrical loading due to anticipated wind and snow conditions, *Figure 8-78A.* The reactions at A and D were determined analytically using the fundamental equations of statics, $\Sigma F = 0$ and $\Sigma M = 0$. At joint A, the small freehand sketch shows a typical symbol for a pinned support, that is, it provides support vertically and horizontally. The symbol at the right end indicates support in one direction only, normal to the surface of support.

STEP 1 Start at a joint with only two unknowns, A or D. Isolate joint A, *Figure 8-78B,* and construct a force diagram to find F(AB) and F(AG). (Use the technique discussed with respect to Figure 8-73C.) Then, to keep a handy record, circle the known members AB and AG in a miniature sketch of the truss. Circle additional known members as they are found.

STEP 2 Isolate joint G, the next joint with only two unknowns, *Figure 8-78C.* Construct a force diagram starting with known forces F(AG) and the 5,000 N applied force. Find F(GF) and F(GB) and circle those members in a miniature sketch.

STEP 3 Isolate joint B next, *Figure 8-78D,* and construct a force diagram starting with F(AB), then F(GB) to find F(BF) and F(BC). Note that the light line for F(BC) coincides with F(AB) because of the arbitrary clockwise selection of forces around joint B. However, any order of taking the forces around a joint can be used for the method of joints; the final magnitudes and directions would be identical. Circle the known members, in a miniature sketch.

STEP 4 Isolate joint F, *Figure 8-78E.* Construct a force diagram by using any order of forces. A clockwise order was used for the figure, starting with known F(FB), then F(FG) followed by the 7,000 N applied force to determine F(FE) and F(FC). Add these to a miniature sketch.

Determine F(CE) by inspection to be 10,000 N. A force acting normal to a joint that has members at 90° to each other can only transmit force to the truss member along the same line as the force. Circle this member in the sketch.

STEP 5 Isolate the last joint D, *Figure 8-78F,* to determine F(DC) and F(DE). Observe that force F(DE) should equal F(EF) due to the 90° angles between member DE and CE, and between DE and the 10,000 N force. Circle the last two members to complete the analysis.

STEP 6 Prepare a table of the forces in the truss members, *Figure 8-78G.*

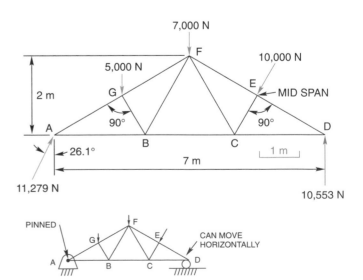

FIGURE 8-78   *(A) Example 2 (a) method of joints*

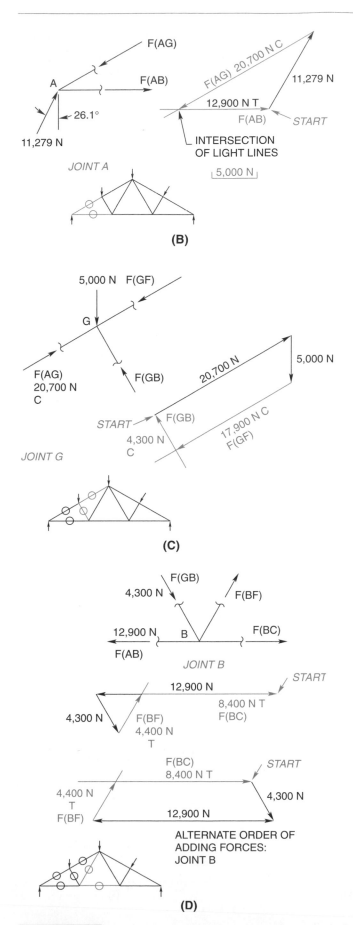

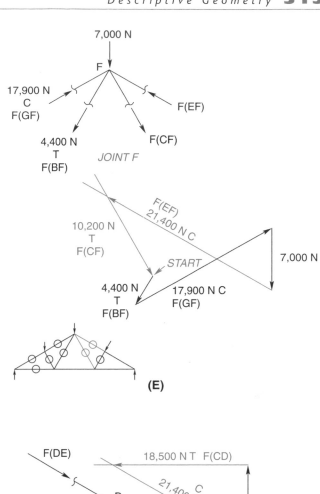

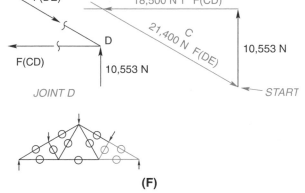

MEMBER	FORCE N	T OR C
AB	12900	T
BC	8400	T
CD	18500	T
DE	21400	C
EF	21400	C
FG	17900	C
GA	20700	C
BG	4300	C
BF	4400	T
CF	10200	T
CE	10000	C

(G)

FIGURE 8-78 *(Continued) (B) Step 1; (C) Step 2; (D) Step 3; (E) Step 4; (F) Step 5; (G) Step 6*

EXAMPLE 2(B): METHOD: MAXWELL DIAGRAM WITH BOW'S NOTATION

GIVEN: The same symmetrical truss, drawn to scale, with asymmetrical loads, used in the preceding method of joints example, *Figure 8-79A*. Label the triangular spaces within the truss with numbers. Draw a light line envelope surrounding the truss and passing through each applied force and reaction. Label these spaces with letters.

STEP 1 Start a force diagram (to a scale to keep the diagram on the drawing sheet) by reading the external spaces in a clockwise direction, a to

b, b to c, c to d, d to e, and e to a. Plot each applied force and reaction, *Figure 8-79B*.

Either clockwise or counterclockwise will work, but once a pattern is started it must be used throughout the complete problem both for the Maxwell diagram and for determining whether a member is in tension or compression.

STEP 2 Start at a joint with only two unknowns, A or D. At joint D, go clockwise a to 1, 1 to e, and e to a to locate point 1, *Figure 8-79C*.

STEP 3 Now joint E has only two unknowns, so continue d to e, e to 1, 1 to 2, and 2 to d to locate point 2, *Figure 8-79D*.

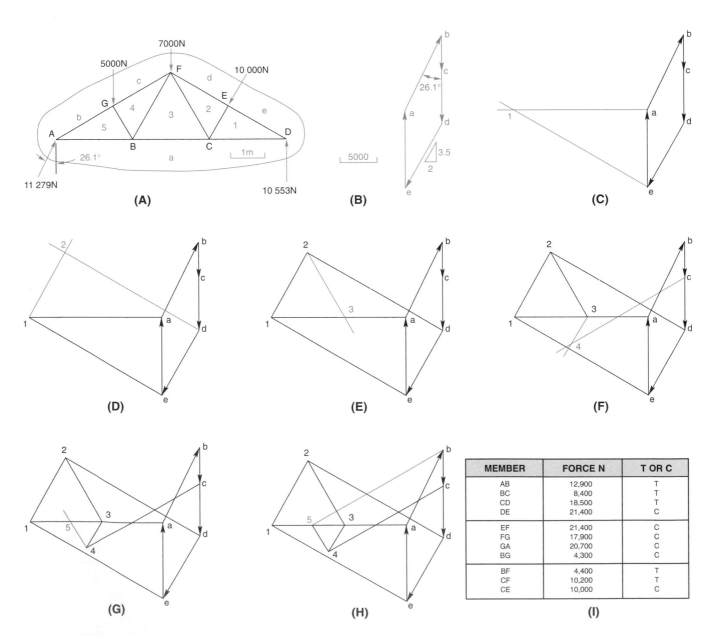

MEMBER	FORCE N	T OR C
AB	12,900	T
BC	8,400	T
CD	18,500	T
DE	21,400	C
EF	21,400	C
FG	17,900	C
GA	20,700	C
BG	4,300	C
BF	4,400	T
CF	10,200	T
CE	10,000	C

FIGURE 8-79 *(A) Example 2 (b) Maxwell diagram with Bow's notation; (B) Step 1; (C) Step 2; (D) Step 3; (E) Step 4; (F) Step 5; (G) Step 6; (H) Step 7; (I) Step 8*

STEP 4 For joint C go a-3-2-1-a to locate point 3, *Figure 8-79E*.

STEP 5 Joint F has only two unknowns, so go d-2-3-4-c-d to locate point 4, *Figure 8-79F*.

STEP 6 Around point B go a-5-4-3-a to locate point 5, *Figure 8-79G*.

STEP 7 For the last joint, A, go a-b-5-a, *Figure 8-79H*. The closing line in direction 5-b should intersect point b on the Maxwell diagram for a correct solution.

STEP 8 Prepare a table of the forces in each member; scale the diagram for magnitudes and use the clockwise pattern to determine whether the force goes into the joint (compression) or goes away (tension), *Figure 8-79I*. For exam-

ple, around joint C for member CF, 3-2 goes away from the joint; thus, CF is in tension. Scale the magnitude to be 10,000 N. Around joint F, 2–3 goes away from the joint F as expected because CF is a tension member.

NONCOPLANAR-CONCURRENT FORCE SYSTEMS

HOW TO DETERMINE THE FORCES IN THE MEMBERS OF A NONCOPLANAR-CONCURRENT FORCE SYSTEM

A noncoplanar-concurrent force system could be described as a "tripod" with a force or forces applied at the apex of the tripod, that is, three two-force members, usually rigid, meeting at a point and supporting a load, *Figure 8-80*. Only three unknown forces may be present

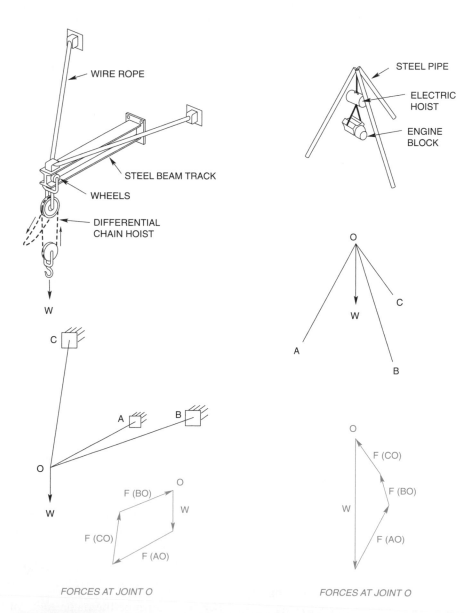

FORCES AT JOINT O   FORCES AT JOINT O

FIGURE 8-80 *Noncoplanar-concurrent force systems*

Industry Application

DESCRIPTIVE GEOMETRY PAYS OFF

Du Vo Minh could remember really disliking descriptive geometry during his years as a student. Projecting lines into other views, determining the true length of a line, identifying the true angle between surfaces, and constructing force diagrams were not what Minh had thought drafting would be about. He wanted to design expensive mansions and unique commercial buildings. He wanted to draw the plans for majestic skyscrapers and towering parking decks. Minh's teacher had reassured him at the time that he would probably do all of these things after graduation, and that he would do them even better if he understood the proper application of descriptive geometry. At the time, the instructor's advice had been a bitter pill to swallow, but Minh had taken it and learned descriptive geometry.

Now, six years later, Minh silently thanks the instructor for insisting that he learn force diagrams and all of the other applications of descriptive geometry. The company he worked for, Commercial Construction, Inc. (CCI), had been awarded a contract to renovate the historic district of downtown Williamsboro, Tennessee, a prestigious multi-million dollar contract. But things were not going well, or at least they hadn't been.

CCI was using trusses in the renovation of several colonial-style buildings, and the trusses were failing. Several had twisted under only partial loading, and truss members had actually snapped on one. The engineers who had designed the trusses were stumped; they couldn't find the problem. That's where Minh became a hero and earned a promotion.

Using the descriptive geometry concepts he had learned in school, Minh developed a force diagram for the trusses in question. He also created a member table showing the forces for each vector in the diagram. By examining Minh's diagram and table, CCI's engineers were able to see clearly what they had missed mathematically. The trusses were quickly redesigned, and CCI's construction crews resumed work. Du Vo Minh was promoted to senior drafter and assigned to the design team responsible for roof trusses.

---

in order to solve for their magnitudes using basic statics; again the governing equation is $\Sigma F = 0$. Furthermore, the vectors, representing the forces in the three members, plus the applied load(s) must add tail to head starting at the apex and ending at the apex because the system is in equilibrium, Figure 8-80.

Graphical solutions of three-member, noncoplanar-concurrent force systems incorporate two fundamental descriptive geometry techniques; one, finding the piercing point of a line in a plane and two, finding the true length of a line. In a space diagram, the direction line of a force vector of one member pierces the plane made by the other two vectors. In a force diagram the true lengths of the three vectors representing the forces in the members are found. The space diagram and the force diagram are drawn to scale and are usually superimposed for convenience.

EXAMPLE 1:

GIVEN: A track beam for a hoist, attached to a wall, held up by two wire ropes, and supporting a load at the end as illustrated in the simplified line drawing, *Figure 8-81A*. The line drawing of the beam and wire ropes is drawn to scale. The applied load of 40,000 N is drawn to scale.

STEP 1 Draw a direction line, originating at the tip of the applied load vector, and parallel to any one of the three members, *Figure 8-81B*. Find the piercing point made by this line in the plane made by the other two members.

Select this first direction line for convenience in arriving at a solution. For example, the plane made by (b-o)T and (c-o)T in view T shows as an edge in view F. Having the edge view expedites finding a piercing point. Therefore, construct a line parallel to the third member (a-o)F, starting the line at the tip of vector F, and note where the line pierces the (b-o-c)F plane, which actually extends beyond the apparent boundaries of (b-o-c)F. (Keep in mind that a willingness to "go beyond arbitrary boundaries" is a step toward being a creative individual.)

STEP 2 Find F(BO) and F(CO) now that F(AO) has been established. The force vectors parallel to members OB and OC, and lying in the plane of BOC extended, must add tail to head to F(AO) and end at the origin of the force vector F, *Figure 8-81C*.

STEP 3 Determine the magnitudes of the three unknown forces and prepare a summary sketch, *Figure 8-81D*. Use the principle of revolution (to be discussed in the next section of this chapter) to determine the magnitudes. The sketch of isolated joint O with

directions of the three forces found in the force diagram indicate whether each member is in tension or compression. Of course, cables supporting a load must be in tension.

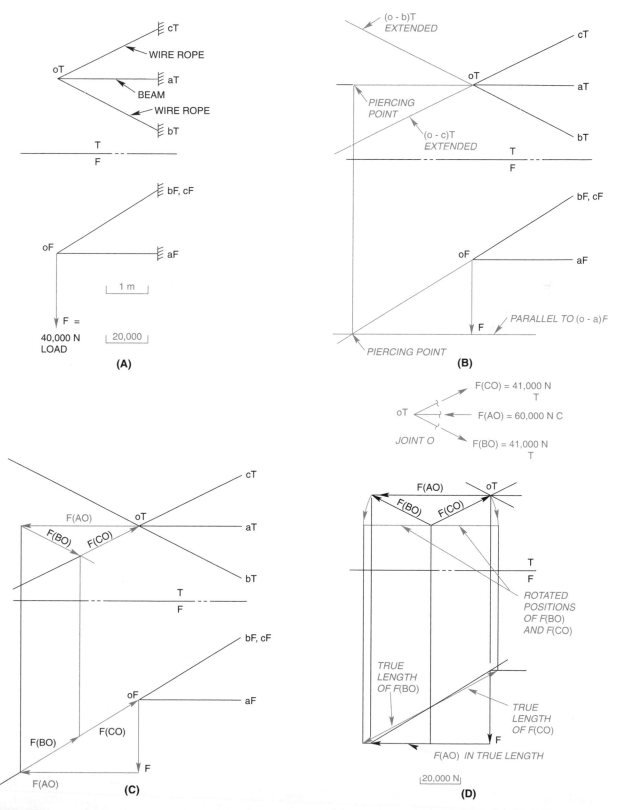

FIGURE 8-81 *(A) Determining the forces of a noncoplanar-concurrent force system—example 1; (B) Step 1; (C) Step 2; (D) Step 3*

EXAMPLE 2:

GIVEN: A tripod on a sloping terrain supporting a 12,000 N load, *Figure 8-82A*. The tripod and the applied load are drawn to scale.

STEP 1 Select a plane made by two of the members, to expedite the solution, and construct a view showing this plane as an edge, *Figure 8-82B*. Use members (a-o)F and (c-o)F to establish the plane using auxiliary line (a-x)F, and project the plane (a-o-c)T into view A. Use the technique shown in Figure 8-17, discussed earlier in this chapter, for finding an edge view of a plane surface.

STEP 2 Find the piercing point in plane (a-o-c)A of the vector representing the force in member (o-b)A by constructing a direction line in view A, originating at the tip of the 12,000 N force and parallel to member (o-b)A. Label the piercing point pA and show F(OB), *Figure 8-82C*. Project a line from point pA into view T and locate pT by noting where extended line (o-b)T crosses the projected line from pA in view A.

STEP 3 In the T view, construct a line from the piercing point pT, parallel to (o-a)T, until it intersects (o-c)T. This establishes F(OA) and F(OC), *Figure 8-82D*. Project the point of intersection of F(OA) with (o-c)T into view A and indicate both vectors F(OA) and F(OC) in view A.

STEP 4 Determine the magnitudes of the three unknowns—F(OB), F(OA), and F(OC), and list them in a table, *Figure 8-82E*. All three unknown forces could be determined by the true-length-of-a-line method. For example, F(OA) and F(OC) could be determined in an auxiliary view B by making a fold line A/B parallel to the edge view of plane (a-o-c)A. Then F(OA) and F(OC) would be in true size in view B. A different method, however, will be used for the three unknowns—the technique of revolution. For clarity, the technique of revolution will be shown for determining one force only, force F(OB). Revolution will be discussed in the next section.

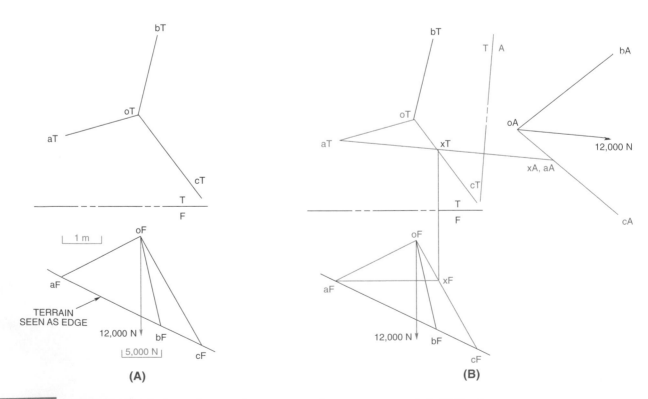

(A)   (B)

FIGURE 8-82 *(A) Determining the forces of a noncoplanar-concurrent force system—example 2; (B) Step 1*

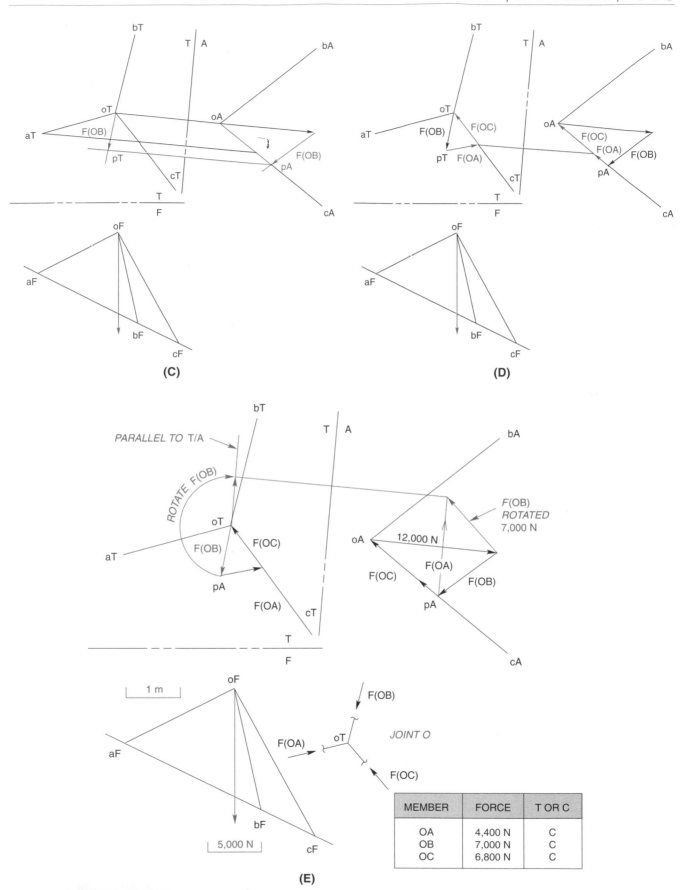

FIGURE 8-82 *(Continued) (C) Step 2; (D) Step 3; (E) Step 4*

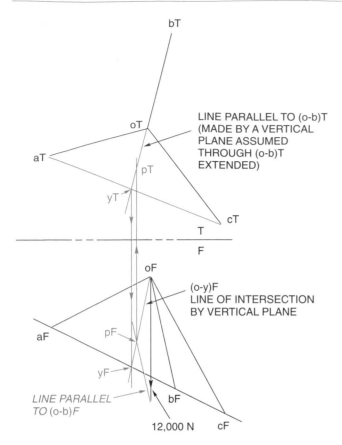

FIGURE 8-83 *Piercing point by line projection method*

COMMENT: An alternative method of finding the piercing point described in Step 2 would be to employ the line projection method, Figure 8-25, discussed earlier in this chapter. The method is shown in views F and T, Figure 8-83.

REVOLUTION

The revolution technique in graphics saves time and space. The technique can be used to solve many true-length and true-angle problems while requiring fewer views.

Consider the basic attributes of the technique by observing the straight member, a-b, rigidly attached to an axis m-n (which can rotate 360°), shown in the T and F views, *Figure 8-84A*. The true length of member a-b is seen twice in the F view as the axis makes one revolution, *Figure 8-84B*. The end bT of member a-b traces a circular path in the T view, and bF traces a straight line in the F view. The circle made by bT is in true size in the T view, and bF's path shows as an edge parallel to the fold line T/F in the F view.

The revolution technique works only if the axis, whether real or imaginary, shows as a point in one view and in true length in any adjacent view.

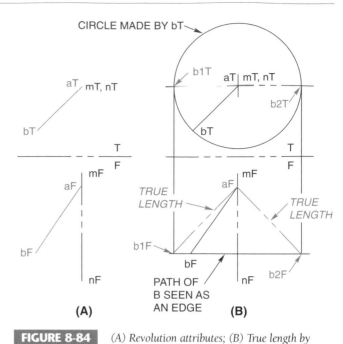

FIGURE 8-84 *(A) Revolution attributes; (B) True length by revolution*

HOW TO DETERMINE THE TRUE LENGTH OF A LINE

EXAMPLE:

GIVEN: The T and F views from Figure 8-11A are redrawn here, *Figure 8-85A*. In the T view, assume that an axis (m-n)T goes through bT; but it only shows as a point, mT,nT. Then, in the F view, axis mF-nF shows in true length.

STEP 1 In the T view, rotate member (a-b)T about axis mT,nT until (a-b)T is parallel to the fold line T/F, *Figure 8-85B*. Next, in the F view,

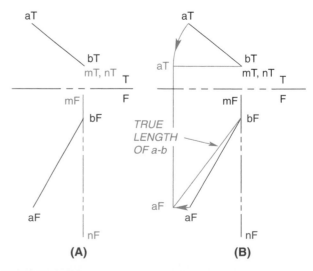

FIGURE 8-85 *(A) Determining the true length of a line—example 1; (B) Step 1*

move point aF parallel to the fold line T/F until it lines up with the projection line from aT's new position in the T view. Redraw (a-b)F in the revolved position where it is now in true length.

COMMENT: *Revolution techniques also save time in laying out developments, Chapter 9.*

HOW TO DETERMINE THE TRUE ANGLE A LINE MAKES WITH A PLANE

EXAMPLE 1:

GIVEN: Plane (a-b-c-d)T in true size and line (k-j)T piercing the plane at point pT, in the T view, *Figure 8-86A*. In the F view, the line (k-j)F pierces plane (a-b-c-d)F, which shows as an edge.

STEP 1 Construct an auxiliary axis m-n that shows as point mT,nT in the T view and goes through piercing point pT, *Figure 8-86B*. Construct the axis (m-n)F as a true-length line in the F view, perpendicular to the T/F fold line.

STEP 2 Rotate line (k-j)T about the axis, in the direction of the smallest angle, until line (k-j)T is parallel to the fold line T/F, *Figure 8-86C*. Construct projection lines from the new positions of kT and jT into the F view. Then locate the new positions of kF and jF by moving

them parallel to the fold line T/F. The point kF moves to the left to the projection line from kT, and point jF moves to the right to the projection from jT. Draw (k-j)F in its revolved position and observe that it is now in true length. Measure the true angle between the line (k-j)F and the plane (a-b-c-d)F.

In summary, the true angle a line makes with a plane must be seen in a view where the plane shows as an edge and the line is in true length.

EXAMPLE 2:

GIVEN: Views F, T and A from Figure 8-23F and redrawn here, *Figure 8-87A*.

This example shows that with only one more view, view B, the true angle between the piercing line (x-y)A and the plane (a-b-c)A can be found easily using revolution.

Construct a fold line A/B parallel to the edge view of plane (a-b-c)A shown in the A view, *Figure 8-87B*. Project the plane and the piercing line into the B view. The plane (a-b-c)B will be in true size, but the line (x-y)B is not.

Therefore, in view A, construct a true-length axis (m-n)A going through the piercing point and perpendicular to plane (a-b-c)A. The axis will show as a point

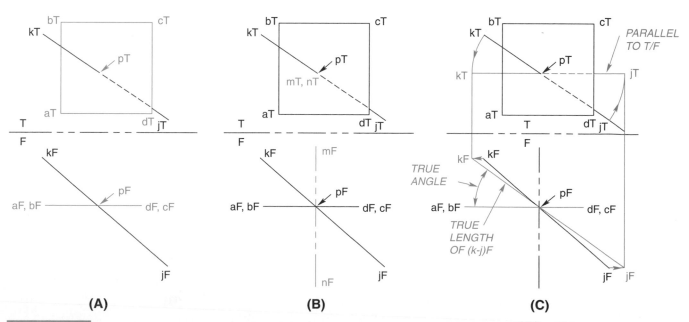

(A)   **(B)**   **(C)**

FIGURE 8-86   *(A) Determining the true angle; (B) Step 1; (C) Step 2*

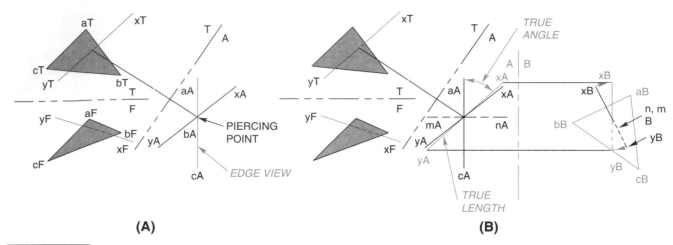

(A)

(B)

FIGURE 8-87 *(A) Determining the true angle—example 2 (Refer to Figure 8-23F); (B) True angle*

(m-n)B in view B, at the piercing point. Rotate the line (x-y)B about the axis in view B until the line is parallel to the fold line A/B. Then in view A, move point xA parallel to fold line A/B to coincide with the new position of xB in view B. In view A, draw the line (x-y)A in its rotated position, where it now shows in true length. In view A measure the true angle between the line (x-y)A and the edge view of plane (a-b-c)A.

HOW TO DETERMINE CLEARANCES USING REVOLUTION

EXAMPLE:

GIVEN: The top and front views of a proposed installation of a vertical wind turbine and turbine blade details, *Figure 8-88A*. Requirement:

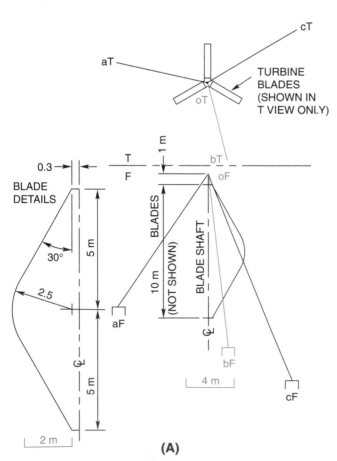

(A)

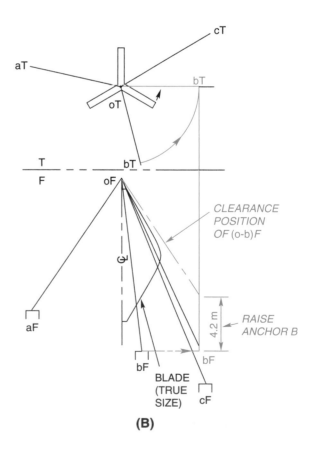

(B)

FIGURE 8-88 *(A) Vertical wind turbine; (B) Step 1*

Determine the vertical distance to raise anchor B so cable (o-b)F will clear any portion of the turbine blades by at least .3 meter. Cables (o-a)F and (o-c)F are assumed to meet this requirement, but they should be checked.

STEP 1 Rotate (o-b)T in the T view until it is parallel to the fold line T/F, *Figure 8-88B*. Note that the vertical axis of the turbine qualifies as the axis of revolution because it is in true length in the F view and shows as a point in the adjacent view, view T. Move point bF to the right in the F view as shown. Then, lay out the true size of the turbine blade in the F view.

Construct a straight line from oF, representing (o-b)F, which clears the turbine blade by at least .3 meter. Find the intersection point of this straight line (o-b)F with a vertical line directly above anchor B.

The vertical distance down from the intersection point to the original elevation of anchor B is the required distance to raise anchor B.

Summary

• Descriptive geometry shows graphically the solution to problems involving points, lines, and planes and their relationship in space.

• Basic theories of descriptive geometry include the following: projecting a line into other views; projecting points into other views; determining the true length of a line; determining the point view of a line; finding the true distance between a point and a line, a plane or two lines in space; projecting a plane surface in space; developing an edge view of a plane surface in space; and determining the true angle between plane surfaces in space.

• The bearing or heading of a line is the direction of the line as it is drawn on a map in top view. There are two methods used for calling out a bearing: north/south deviation and azimuth bearing.

• The slope of a line is the angle, in degrees, that the line makes with a horizontal plane.

• The grade percent of a line describes the inclination of the line in relation to the horizon by the ratio of the vertical rise to the horizontal.

• The pitch of a line or surface describes the slope of a roof and is expressed as the ratio of vertical rise to 12 inches or 12 feet in horizontal span.

• A contour line on a plot or map is a line at a constant elevation of the earth, above sea level. Contour interval is the difference in elevation between neighboring contour lines.

• Strike is a compass direction of a horizontal line lying in the plane of an ore vein. Dip is the downward angle of the vein with respect to a horizontal plane at the site of the ore vein.

• A vector is a graphic representation of force, displacement, velocity, or acceleration. A vector has both direction and magnitude.

Review Questions

Answer the following questions either true or false.

1. Descriptive geometry is used by drafters, designers, and engineers because it graphically shows the solutions to problems dealing with points, lines and planes, and their relationship in space.

2. Notations are like footnotes.

3. Notations are important because they are used to keep track of all views and points in space.

4. The point view of a line describes the inclination of a line in relation to the horizon.

5. In order to find an edge view of a plane, a true-length line must be located in that plane.

6. A fold line is represented by a thin black line and must be labeled using lowercase letters.

7. The first step to determining the visibility of two intersecting lines (which of the two lines is closer) is to project the exact crossover point of the lines to an adjoining view.

8. The bearing and grade, or bearing and slope, refers to the exact position of a fold line.

9. A contour line on a plot or a map is a line at a constant elevation on the surface of the earth, above sea level.

10. The significance of contour lines being close together on a map is steep terrain.

11. The term cuts and fills refers to earth being excavated for highways.

12. A vector is a directed line segment which, in computer graphics, is defined by its two end points.

13. Trusses are usually used in pairs or in multiples.

14. A truss is a frame that is usually three-dimensional and made of structural shapes.

15. Space diagrams and force diagrams are used in truss analysis to determine the forces in the members of a truss.

16. Methods of assembling truss joints include welding, riveting, and bolting.

17. When projecting a plane into another view, you must first find at least four points in the plane and project each point into the next view.

Answer the following questions by selecting the best answer.

1. Which of the following is **not** an acceptable method for determining the intersection of two planes?
   a. By inspection
   b. By generating an edge view of the plane surface
   c. By using the line projection method
   d. Both b and c

2. Which of the following is **not** a step involved in determining the true distance between two parallel lines in space?
   a. Find the true length of each of the two lines.
   b. Find the point view of each of the two lines.
   c. Project a point view of the added true-length line.
   d. The true distance between the parallel lines is the straight-line path between their end points.

3. Which of the following is **not** a step involved in determining the true distance between two nonparallel (or skewed) lines in space?
   a. Project the true length of either one of the two lines, and project the other line into that auxiliary view (A).
   b. Find the point view of the true-length line, and project the other line into that secondary auxiliary view (B).
   c. Measure the true distance between the point view line at a location that is perpendicular to the other line.
   d. Project the same points from the front view into the right-side view.

4. A point is called out in a given view by:
   a. Lowercase letters.
   b. Uppercase letters.
   c. Either upper- or lowercase letters.
   d. All of the above

5. Which of the following is an important rule to remember when projecting from one view to another?
   a. Always project the top view of a line from a given front.
   b. The customary 45° projection line is used.

   c. The fold line is always placed between the front V and the top V.
   d. Always skip a view between all measurements.

6. Which of the following is **not** one of the basic steps involved in finding the true length of any straight line?
   a. Draw a fold line parallel to the line for which the true length is required and label it A for auxiliary.
   b. Extend projection lines from the end points of the line being projected into the auxiliary view.
   c. Projection lines must always be parallel to the fold line.
   d. Transfer the end point distances from the fold line in the second preceding view from the one being drawn to locate the corresponding end points in the view being drawn.

7. Which of the following is **not** one of the basic steps used to find the point view of a line as projected from its true length?
   a. Find the true length of the line.
   b. Draw a fold line perpendicular to the true-length line at any convenient distance from either end of the true-length line and label it A-B.
   c. Extend a light projection line from the true-length line into the secondary auxiliary view.
   d. Measure the true distance between the point view line at a location perpendicular to the other line.

8. Which of the following is a use of vectors in engineering and technology?
   a. Displacement
   b. Force
   c. Velocity
   d. All of the above

9. Which of the following is **not** part of the parallelogram law?
   a. Two vectors
   b. Not parallel, drawn to scale
   c. Parallel, drawn to scale
   d. Having the same units, and acting at the same point can be added or subtracted using a specific formula

10. Which of the following is the correct formula for determining the percent of grade?
    a. Units of horizon ÷ units of vertical rise × 100
    b. Units of vertical rise ÷ units of horizon × 100
    c. Units of horizon × units of vertical rise ÷ 100

11. A piercing point is:
    a. The exact location of an intersection of two lines.
    b. The exact location of an intersection of two surfaces.
    c. The exact location of an intersection of a plane and a surface line.
    d. None of the above

12. Given the bearing, slope, and length of a line, which of the following is **not** true when constructing the line in a drawing?
    a. The bearing is always drawn in the top view.
    b. The true length can be drawn if the line is to be exactly horizontal.
    c. Any line with an incline up (+) or down (–) from the horizon must be laid out using specific steps.
    d. If the line is horizontal, the true length must be constructed before the heading length can be drawn in the top view.

13. Which of the following is **not** true regarding trusses?
    a. Trusses are seldom used in multiples.
    b. Trusses seldom stand alone.
    c. Trusses are frames that are usually two-dimensional.
    d. Joints are assumed to be pinned, although they can be welded, bolted, or riveted.

Chapter 8 Problems

The following problems are intended to give the beginning drafter practice in using the various principles of descriptive geometry. Problems 1 through 13 deal with each of the various principles used to solve actual design problems. Problems 14 through 32 apply these principles to develop required views of various objects.

CONVENTIONAL INSTRUCTIONS

The steps to follow in laying out problems 1 through 13 and 23 through 32 are:

1. On an 8½ × 11 sheet of paper with a sharp 4-H lead, locate all lines, points, and fold lines per the given dimensions.
2. Complete the problem per the given instructions. Project in the direction of the large arrow.

The steps to follow in laying out problems 14 through 27 are:

STEP 1   Study the problem carefully.

STEP 2   Using the given front view, make a sketch of all required views.

STEP 3   Center the required views within the work area with a 1-inch (25 mm) space.

STEP 4   Use light projection lines. Do **not** erase them.

STEP 5   Lightly complete all views.

STEP 6   Check to see that all views are centered within the work area.

STEP 7   Check that there is a 1-inch (25 mm) space between all views.

STEP 8   Carefully check all dimensions in all views.

STEP 9   Darken in all views using correct line thickness.

STEP 10 Recheck all work, and, if correct, neatly fill out the title block using light guidelines and neat lettering.

CAD INSTRUCTIONS

Students who plan to complete the following problems on a CAD system should begin with these general instructions. Before reading the specific instructions for each activity, go through each step in the following planning checklist. The checklist applies to any CAD system and will help ensure the optimum use of your time and resources.

1. Analyze the problem carefully. Decide exactly what you are being asked to do.
2. Determine what resources and references (if any) you will need in order to complete the problem. Collect those references and resources and have them readily available.
3. Decide if any particular standards apply to the project, and have those standards available.
4. Determine what types of views will be required and how many of each.
5. Determine what the final plotted scale of the drawing will need to be, and select the appropriate paper size for plotting/printing.
6. Plan your drawing sequence. In what order will you develop the drawing (i.e., lines, features, dimension lines, leaders, dimensions)?
7. Review the various CAD commands you will have to use in order to develop the drawing.
8. Examine your CAD system to ensure that everything is in working order, then begin the project.

PROBLEM 8-1

Locate line a-b and fold lines per the given dimensions. Locate line a-b in the right-side view. Label all points in all views.

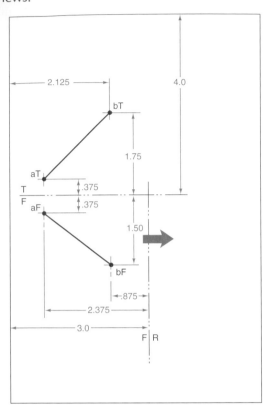

PROBLEM 8-2

Locate line a-b and fold lines per the given dimensions. Locate line a-b in the top view. Locate point c on line a-b in all views. Label all points in all views.

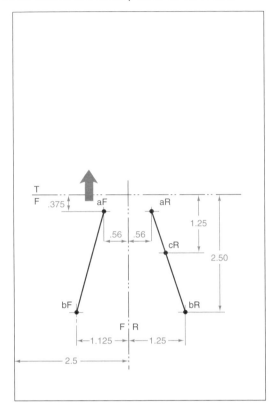

PROBLEM 8-3

Locate line a-b and fold lines per the given dimensions. Locate line a-b in the right view. Find the true length of line a-b, projecting from the right side. Label all points in all views.

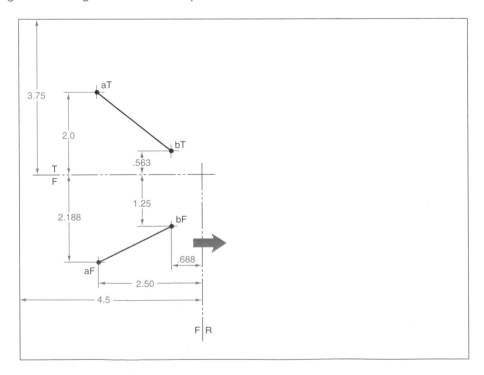

PROBLEM 8-4

Locate the line a-b and fold lines per the given dimensions. Locate line a-b in the top view. Find the point view of line a-b, projecting from the top view. Label all points in all views.

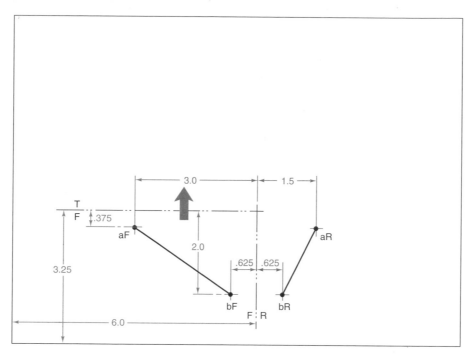

PROBLEM 8-5

Locate line a-b, point c, and fold lines per the given dimensions. Find the true distance from line a-b to point c. Project from the right view. Label all points in all views.

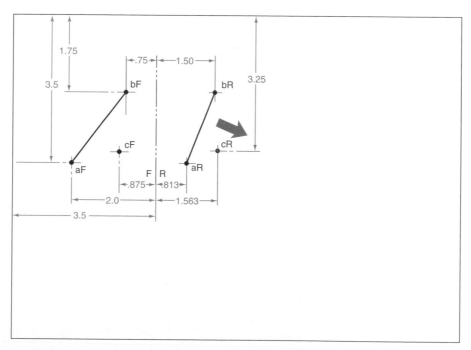

PROBLEM 8-6

Locate parallel lines a-b and c-d and fold lines per the given dimensions. Find the true distance between lines a-b and c-d. Project from the front view. Label all points in all views.

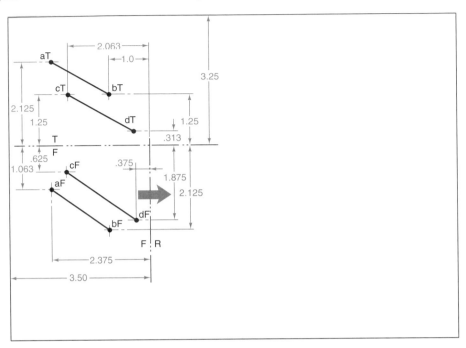

PROBLEM 8-7

Locate lines a-b and c-d and fold lines per the given dimensions. Find the true distance between lines a-b and c-d. Project from the top view. Label all points in all views.

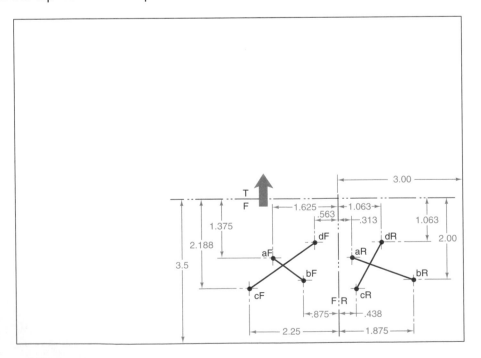

PROBLEM 8-8

Locate points a, b, and c and fold lines per the given dimensions. Connect points a, b, and c to form a plane. Project plane abc into the top view. Label all points in all views.

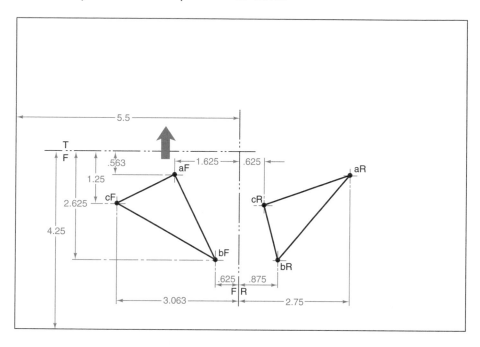

PROBLEM 8-9

Locate points a, b, c, and d, point x, and fold lines per the given dimensions. Connect points a, b, c, and d to form a plane. Project plane abcd into the right view; locate point x in all views. Label all points in all views.

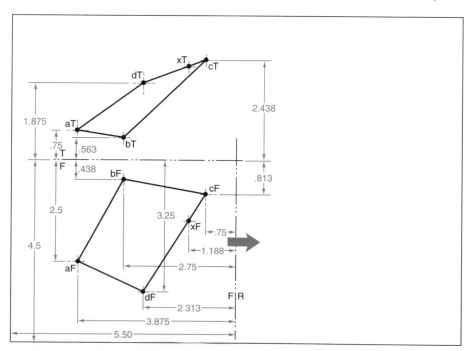

PROBLEM 8-10

Locate points a, b, c, and fold lines per the given dimensions. Connect points a, b, and c to form a plane. Draw a line from point cF, parallel to fold line F-R; label this line cF-xF. Find the edge view of plane surface abc. Project from the right view. Label all points in all views.

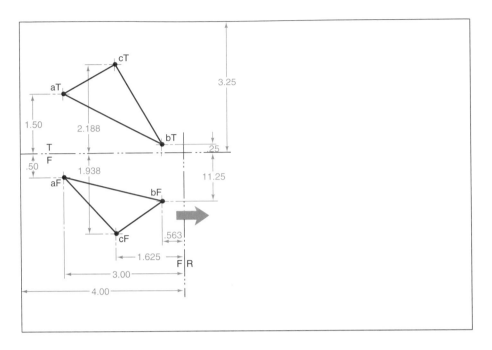

PROBLEM 8-11

Locate points a, b, and c and fold lines per the given dimensions. Connect points a, b, and c to form a plane. Draw a line from point bF, parallel to fold line F-T; label this line bF-xF. Find the edge view of the plane abc and project from the front view. Label all points in all views.

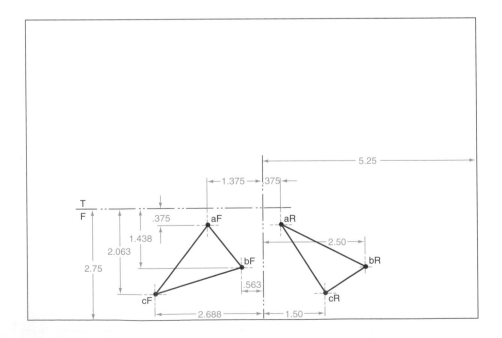

PROBLEM 8-12

Locate points a, b, and c, point x, and fold lines per the given dimensions. Connect points a, b, and c to form a plane. Find the true distance between plane surface abc and point x. Project from the front view. Label all points in all views.

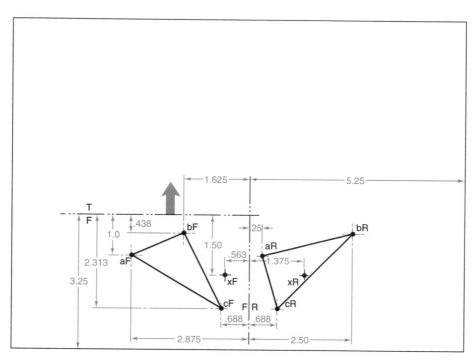

PROBLEM 8-13

Locate points a, b, c, and d and fold lines per the given dimensions. Connect points a, b, c, and d to form a folded object. Find the true angle between the plane surfaces as viewed directly along the fold, a-b. Project from the right view. Label all points in all views.

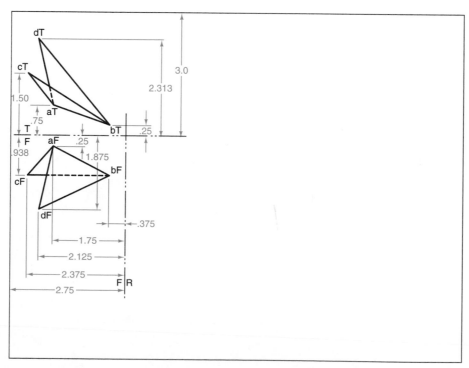

PROBLEMS 8-14 THROUGH 8-22

Draw the front view, top view, right view, and auxiliary view per the listed steps.

PROBLEM 8-16

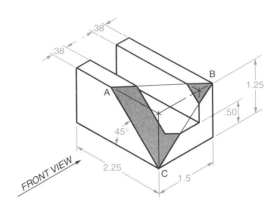

PROBLEM 8-14

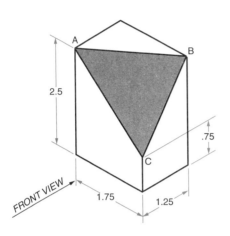

PROBLEM 8-17

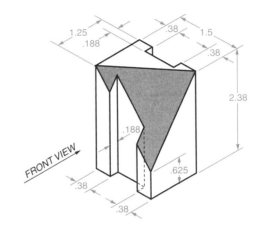

PROBLEM 8-15

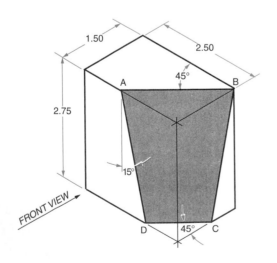

PROBLEM 8-18

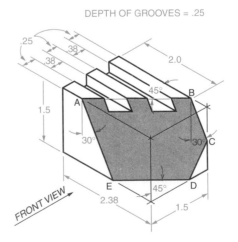

PROBLEM 8-19

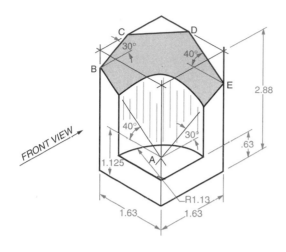

PROBLEM 8-22

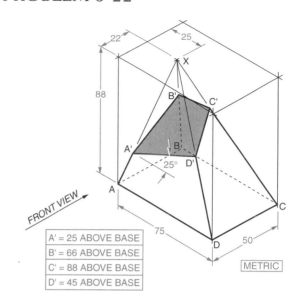

A' = 25 ABOVE BASE	
B' = 66 ABOVE BASE	
C' = 88 ABOVE BASE	
D' = 45 ABOVE BASE	

METRIC

PROBLEM 8-20

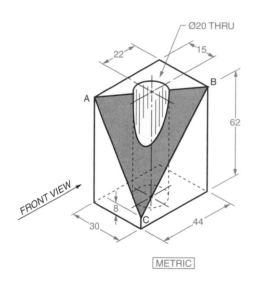

METRIC

PROBLEM 8-23

Draw the front view, top view, right view, and required auxiliary views per the listed steps. Find the true angle at abc, as viewed along the fold.

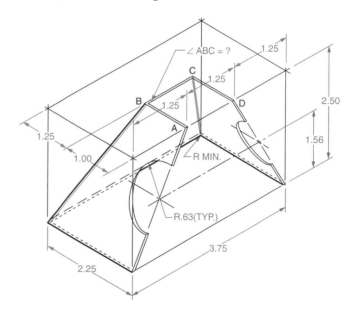

PROBLEM 8-21

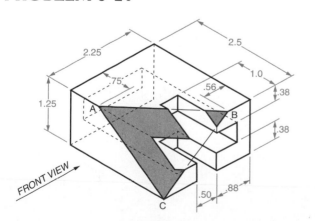

PROBLEMS 8-24 AND 8-25

Draw the front view, top view, right view, and auxiliary view per the listed steps.

PROBLEM 8-24

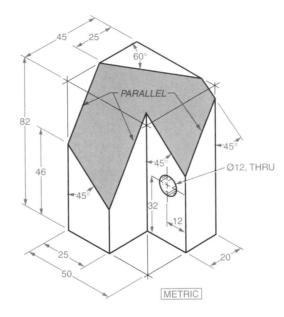

PROBLEM 8-25

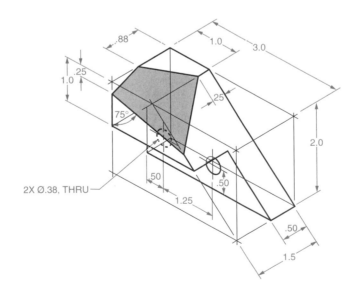

PROBLEMS 8-26 AND 8-27

Draw the front view, top view, right view, and required auxiliary views per the listed steps. Find the true angle between the various surfaces as viewed along the fold.

PROBLEM 8-26

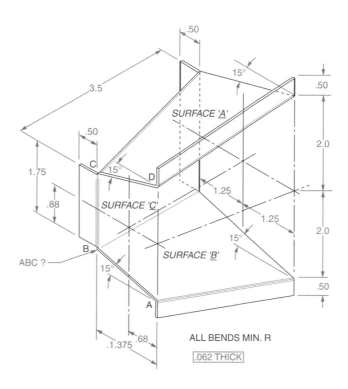

SURFACE 'A'

SURFACE 'C'

SURFACE 'B'

ABC ?

.50
3.5
.50
1.75
.88
.68
.1.375
15°
15°
15°
15°
15°
.50
.50
2.0
1.25
1.25
2.0
.50

ALL BENDS MIN. R

.062 THICK

PROBLEM 8-27

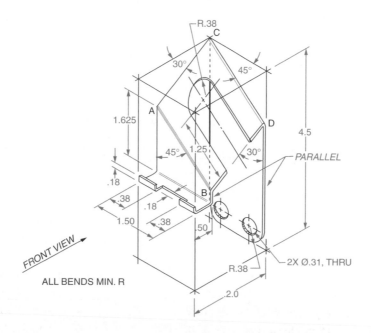

R.38

30°
45°
45°
30°
PARALLEL

1.625
.18
.38
.18
1.50
.38
.50
1.25

4.5

2X Ø.31, THRU
R.38
2.0

FRONT VIEW

ALL BENDS MIN. R

PROBLEM 8-28

Complete the front view using the line-projection method. Draw the correct visibility of all lines in the front view.

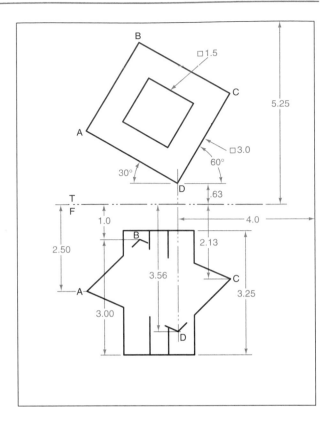

PROBLEM 8-29

Complete the top and front views using the line-projection method. Draw the correct visibility of all lines by inspection.

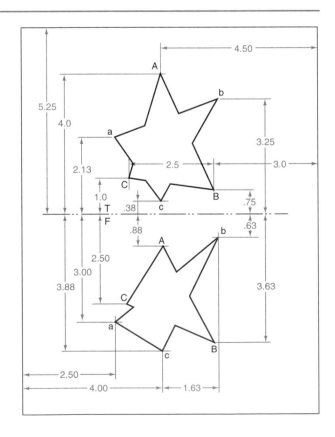

PROBLEM 8-30

Complete the front view by inspection and projection. Draw the correct visibility of all lines.

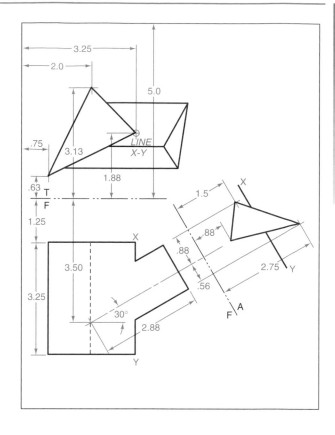

PROBLEM 8-31

Complete the front view using the line-projection method. Draw the correct visibility of all lines.

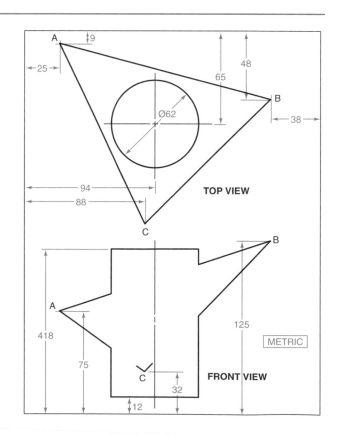

PROBLEM 8-32

Complete the top and front views using the line-projection method. Do not use the right-side view for solving the problem. Draw the correct visibility of all lines.

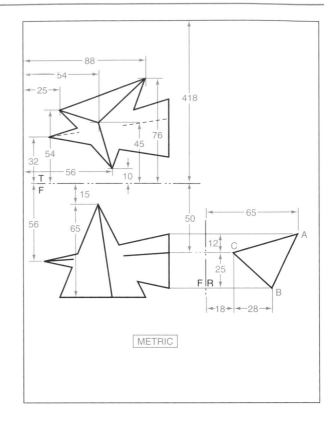

METRIC

PROBLEMS 8-33 THROUGH 8-42

Lightly lay out and draw the two views and the fold line on an A-size sheet of vellum using the given dimensions. Complete all views using correct line thickness—add all hidden lines.

PROBLEM 8-33

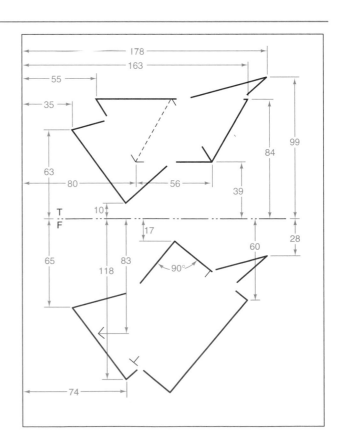

PROBLEM 8-34

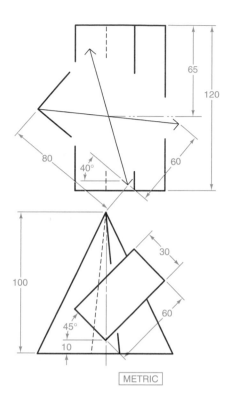

METRIC

PROBLEM 8-35

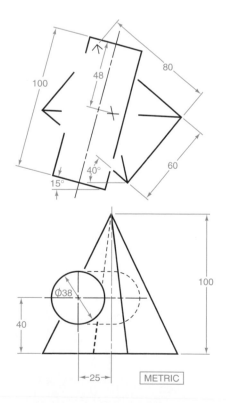

METRIC

PROBLEM 8-36

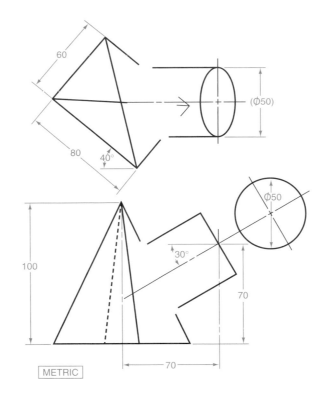

METRIC

PROBLEM 8-37

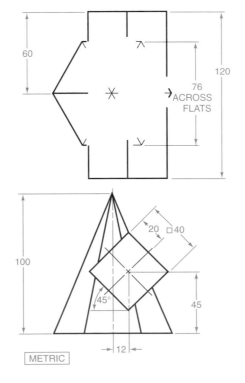

METRIC

PROBLEM 8-38

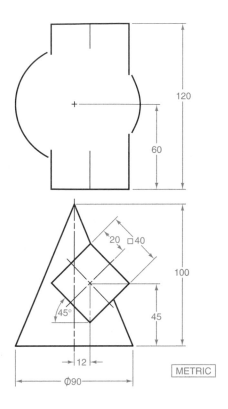

METRIC

PROBLEM 8-39

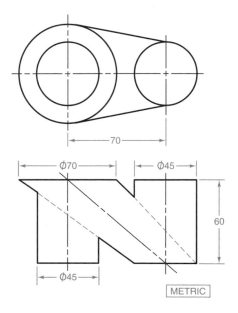

METRIC

PROBLEM 8-40

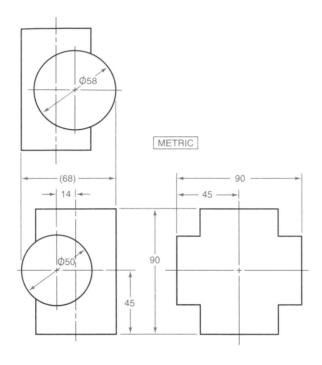

PROBLEM 8-41

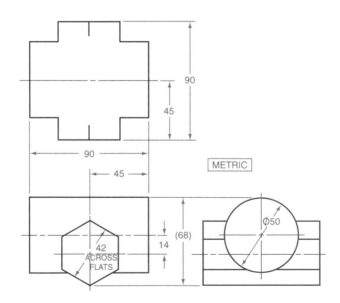

PROBLEM 8-42

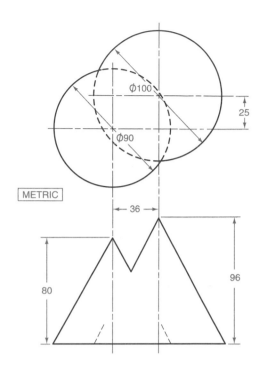

METRIC

PROBLEM 8-43

Calculate the angles using the given bearings for A through F. Show all math work.

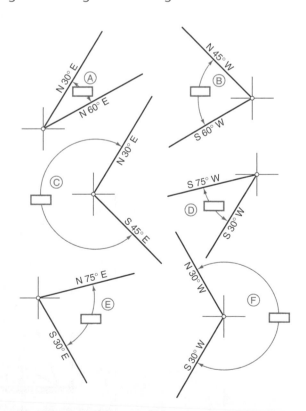

PROBLEM 8-44

On a B-size paper, draw the *top view* and *front view* of a straight roadway (a-b) 280.0 feet long, with an azimuth bearing of N 288°, and a grade of –20%. Use a 1″ = 50.0′ scale. Show all light construction work.

GIVEN: Location of fold line F/T and location of point aT and aF.

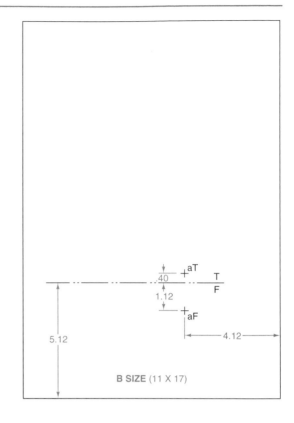

B SIZE (11 X 17)

PROBLEM 8-45

On a B-size paper, lay out the *top view* and *front view* of a radio tower. Use a 1″ = 50.0′ scale. Show all light construction work. Answer the following questions:
1. What is the grade percent for the guide wire a/b?
2. What is the slope angle of the guide wire?
3. What is the true length of the wire?

GIVEN: Location of the radio tower, guide wires, and fold line F/T.

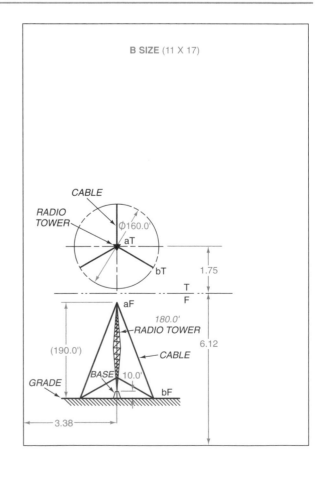

B SIZE (11 X 17)

PROBLEM 8-46

On a B-size paper, draw an existing conveyor belt system with the given specifications (see specifications in box). The conveyor belt system is to be redesigned to go from point a directly to point d. Use a 1" = 50.0' scale. Show all light construction work. Answer the following questions:

1. What is the new redesigned bearing from points a to d?
2. What is the new redesigned true length?
3. What is the new grade percent?
4. How much shorter is the new design than the old design?

GIVEN: Location of grade, starting point a, and fold line F/T.

BELT	BEARING	TRUE LG.	GRADE %
a – b	N 30° W	180.0'	40%
b – c	N 80° W	140.0'	30%
c – d	S 45° W	145.0'	15%
ANSWER			
a – d			

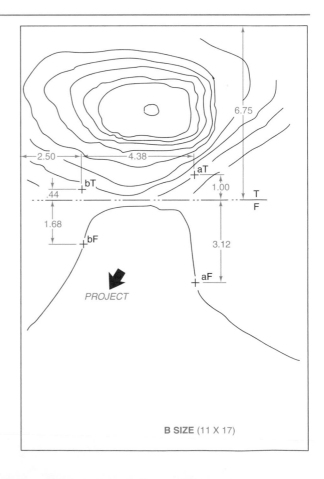

PROBLEM 8-47

On a B-size paper, construct two tunnels, a and b. Tunnel a starts at point a and has a bearing of N 70° W. It is 240.0 feet long and has a plus slope angle of 34°. Tunnel b starts at point b and has a bearing of N 60° E. It is 200.0 feet long and has a plus slope angle of 19°. Use a 1" = 50' scale. Show all light construction work. Project from the front view. Answer the following questions:

1. Do the tunnels ever intersect?
2. If not, how far apart are they?

GIVEN: Locations of tunnel entrances a and b and fold line F/T.

PROBLEM 8-48

On a B-size paper, construct two mining shafts into a mountain from two different directions and elevations. Both are cut straight into the mountain at different percents of grades. One shaft starts at point a and the other starts at point b. Shaft a has a bearing of S 60° W, with a grade of + 35%, and is 165.0 feet long. Shaft b has a bearing of S 45° E, with a grade of + 40%, and is 176.0 feet long. Use a 1″ = 50.0′ scale. Show all light construction work. Answer the following questions:

1. Do the shafts intersect?
2. What is the true distance from the entrance of shaft a to the entrance of shaft b?
3. What is the bearing and percent of grade from the two entrances?

GIVEN: Locations of the shaft entrances a and b, fold line F/T, and mountain.

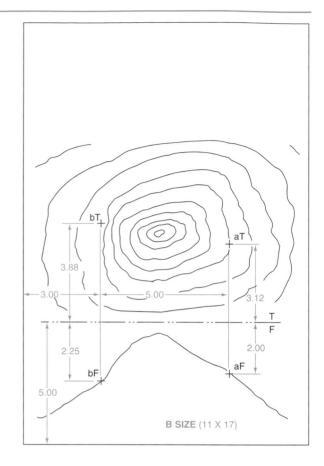

PROBLEM 8-49

A 40′ wide, level road, at elevation 150′, is to be run through a hilly terrain as shown on the contour map. Trace the map and use the accompanying scale to determine cut and fill boundaries for a slope of 1:1 for both the cuts and the fills.

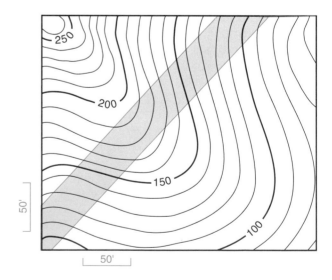

PROBLEM 8-50

Vehicle A is traveling due North at 150 kph (90 mph); vehicle B, due West at 110 kph (66 mph). Vehicle B is 40 kilometers North and 40 kilometers East of vehicle A at the instant shown. What is the shortest distance between the vehicles A and B when they pass each other? An example velocity vector diagram illustrates the consequences of A and B going at the same speed. What should vehicle B's speed be so that the shortest distance between A and B is 15 kilometers when they pass each other?

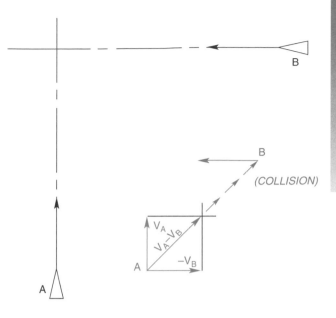

PROBLEM 8-51

An ore outcropping is shown on the topographic contour map. Points A and B are on the upper surface (called the hanging wall) of the vein. A vertical bore hole at point C indicates that the hanging wall is at 877' elevation. The core hole extended at C indicates that the bottom surface of the ore vein (footwall) is at 841' elevation. For the information given for the locations of points A, B, and C and the vein surfaces, determine the following:

1. The strike and dip of the ore vein.
2. The true thickness of the ore vein.

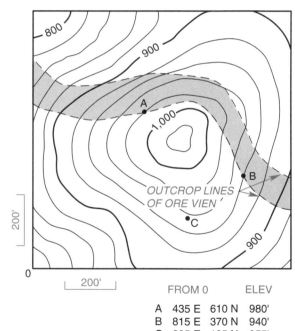

	FROM 0		ELEV
A	435 E	610 N	980'
B	815 E	370 N	940'
C	605 E	195 N	955'

PROBLEM 8-52

The simplified drawing of the four-bar linkage shown (three links and the rigid support) has link A-B rotate 360° around pinned connection A. Link A-B's motion causes link C-D to oscillate about pinned connection D. At the instant shown, a velocity analysis has link C-D rotating CCW at 18.3 radians per second. Note that the velocity of point B of link A-B is V(B) = ω(AB) × (0.03m) = 1.8 m per second. The component of V(B), which is coincident with member B-C, causes link B-C to move to the left with a velocity of V(BC) = 1.2 m per second as found graphically. Next, a component of V(BC) perpendicular to C-D causes C-D to rotate CCW about point D with an angular velocity of ω(CD) = (1.19)/0.065 = 18.3 radians per second. Determine the following:

1. The instantaneous angular velocities of ω(CD) when A-B is at 90°, 180°, and 270°.
2. At what values of 0 will C-D have an angular velocity of zero? Why?

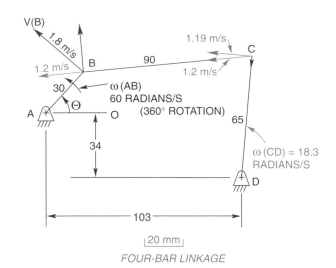

FOUR-BAR LINKAGE

PROBLEM 8-53

The cam and cam follower shown represent a camshaft cam and valve follower in an internal combustion engine. The cam rotates at a constant rate of 2,000 rpm, which is 209 radians per second (2000/60 = 33.33 RPS; and 33.33 × 2 × π = 209 radians per second). The cam has rotated 225° at the instant shown and is transferring a velocity of 120 inches per second to the follower as determined by a graphical analysis. The velocity of the point in contact T at this instant is V(T) = ω(cam) × 0.98 inch = (209 radians per second) × (0.98 inch) = 205 inches per second. Then the resolution of the 205 in./sec gives 120 in./sec to the follower, and the other component indicates a sliding velocity parallel to the bottom surface of the valve. Hence, good lubrication is required.

Lay out the cam from the dimensions given and graphically determine the velocity of the valve throughout one revolution, starting at θ = 0°. Record the results in a graph of V(follower) versus angle of rotation θ. Suggestions: (1) From θ = 180° to 360° determine V(follower) at 15° intervals. (2) Lay out the cam on one sheet of drawing paper and use an overlay sheet of drawing paper for the intervals.

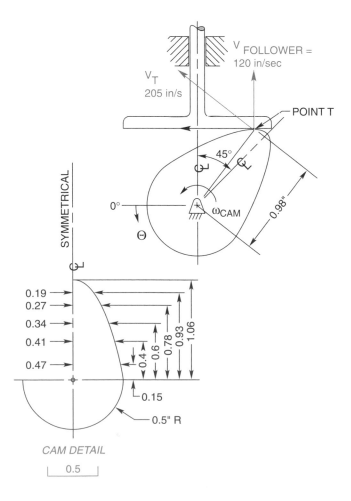

CAM DETAIL

PROBLEM 8-54

The laser beam a-b shines on the metal door shown. As the door swings open, the laser beam traces a path on the door's surface. Determine the path for the door opening 150° and show the path in view F.

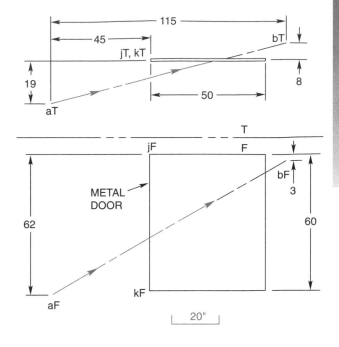

PROBLEM 8-55

A basic toggle mechanism is used to generate a large force from a smaller force for punching holes in metal. For the position shown in the simplified mechanism, what force is generated at punch A, vertically downward? When the 40,000 N force is at B_2, what force at A is generated vertically downward?

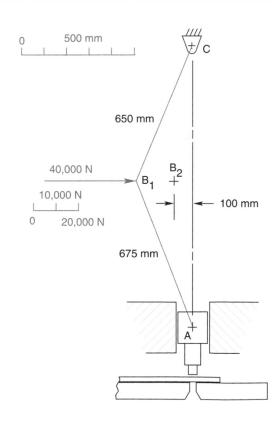

PROBLEM 8-56

For the coplanar-concurrent system shown, determine the forces in members AB and BC using the method of joints.

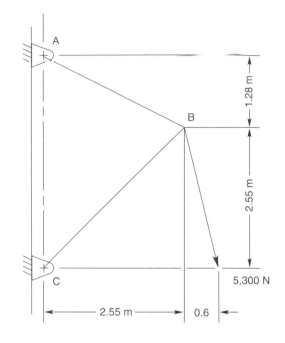

PROBLEM 8-57

Use the method of Maxwell diagram with Bow's notation to determine the forces in each member of the truss shown. Tabulate the magnitudes and whether the members are in tension or compression.

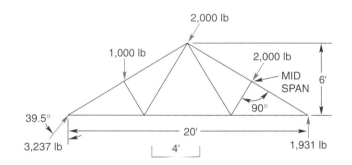

PROBLEM 8-58

Redraw the tripod and the 1,500 lb applied force starting at the lower left-hand area of a drawing sheet. Determine the force in each member by the method shown in this chapter, and use revolution to finally determine the magnitude of each force. *Suggestion:* Use members o-b and o-c to form a plane.

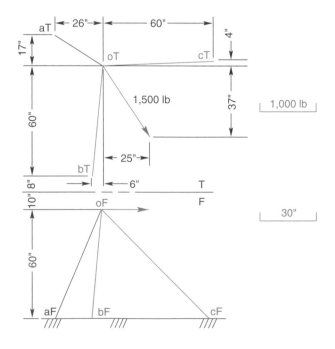

PROBLEM 8-59

From the given information, lay out the four non-coplanar-concurrent force vectors and add them tail to head in any order (the order can differ in each view also) to determine the equivalent resultant of the four forces. Then, find the magnitude of the resultant by revolution. Also, determine the bearing and percent slope of the resultant.

The coordinates of the four forces originating from point o are:

    a. (10,5,5)
    b. (4,–13,–10)
    c. (–8,–8,8)
    d. (–15,5,–10)

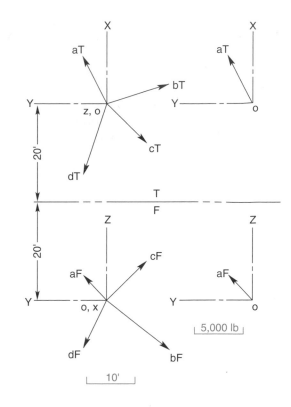

PROBLEMS 8-60

A ham radio operator needs to know the lengths of three guy wires for his antenna installation and where to locate guy wire o-c. At the lower left corner of a sheet of drawing paper, construct a layout of the given roof information and solve the ham's guy wire problem using rotation. Guy wire o-c is to clear corner f of the chimney by one foot. Show all work and tabulate the answers.

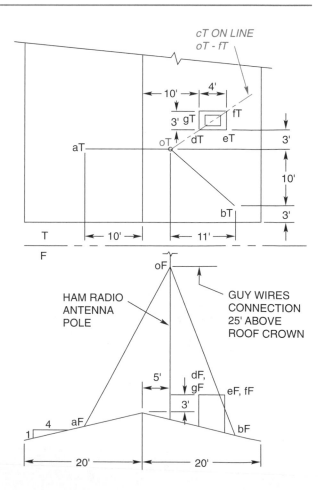

PROBLEM 8-61

A tension spring a-b constrains a 30″ wide door. The spring shown in simplified schematic form has a spring constant of 5 lb/in., i.e., 5 lb will stretch the spring 1 inch. The door shown in the closed position pivots about a hinge located at point c. At the position shown, the spring has an initial tension of 6 lb. What will be the total tension in the spring when the door is opened 90°? Solve the problem using revolution.

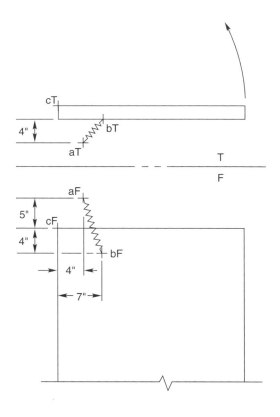

PROBLEM 8-62

The simplified drawing shows a personnel lift, powered by a hydraulic cylinder, DC. At different angles θ the force F(DC) required by the cylinder is calculated by dividing the moment due to the 1,000 lb load (1,000 lb load × the shortest distance to the line of action of the 1,000 lb load from point A), by the shortest distance to the line of action of the cylinder's force F(DC). The shortest distances are called moment arms. Redraw the positions of the lift as specified in the table, locate the cylinder, measure moment arms, and do the calculations to finish the table.

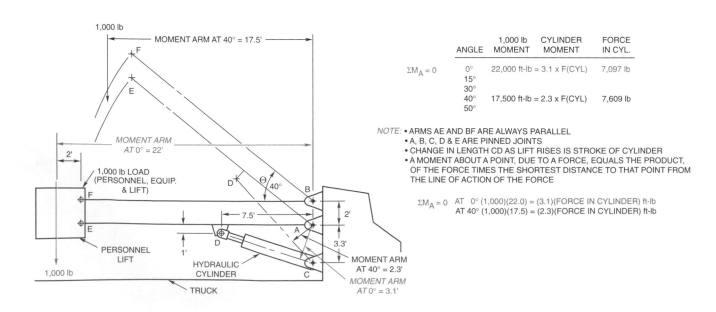

	ANGLE	1,000 lb MOMENT	CYLINDER MOMENT	FORCE IN CYL.
$\Sigma M_A = 0$	0°	22,000 ft-lb = 3.1 x F(CYL)		7,097 lb
	15°			
	30°			
	40°	17,500 ft-lb = 2.3 x F(CYL)		7,609 lb
	50°			

NOTE: • ARMS AE AND BF ARE ALWAYS PARALLEL
• A, B, C, D & E ARE PINNED JOINTS
• CHANGE IN LENGTH CD AS LIFT RISES IS STROKE OF CYLINDER
• A MOMENT ABOUT A POINT, DUE TO A FORCE, EQUALS THE PRODUCT, OF THE FORCE TIMES THE SHORTEST DISTANCE TO THAT POINT FROM THE LINE OF ACTION OF THE FORCE

$\Sigma M_A = 0$ AT 0° (1,000)(22.0) = (3.1)(FORCE IN CYLINDER) ft-lb
AT 40° (1,000)(17.5) = (2.3)(FORCE IN CYLINDER) ft-lb

PROBLEM 8-63

A floating golf-course green, supported by a concrete float, is moved by changing the lengths of four wire ropes powered by winches inside the float. From the position shown, the green is moved 130 yd West and 20 yd North. Use revolution to determine the change in length of each cable, A, B, C, and D. Tabulate the changes. (Ask your golfing friends if they could hit the green at the new position.)

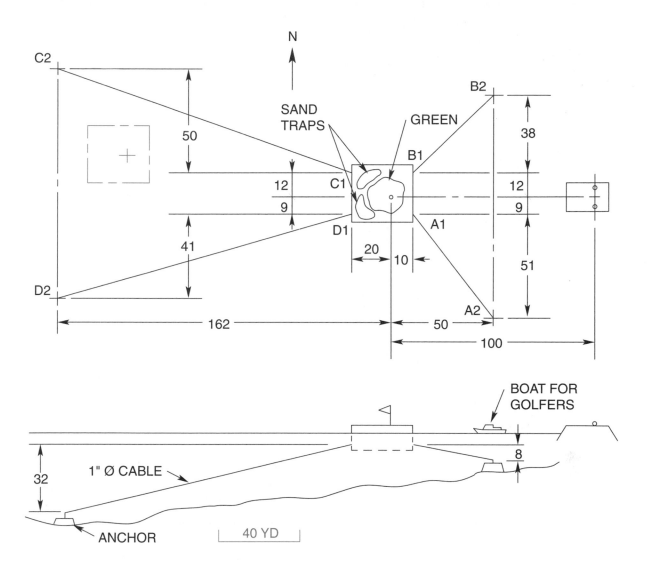

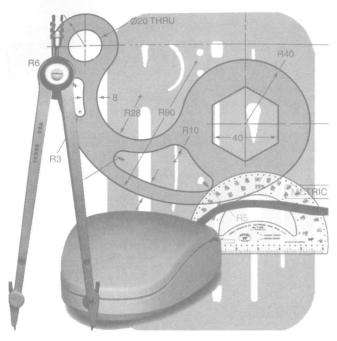

Chapter 9

PATTERNS AND DEVELOPMENTS

CHAPTER OUTLINE

Developments • Parallel line development • Radial line development • Triangulation development • True-length diagrams • Notches • Bends • Summary • Review questions • Chapter 9 problems

CHAPTER OBJECTIVES

Upon completion of this chapter, students should be able to do the following:

■ Define the term *development* and list the major kinds of developments.

■ Demonstrate proficiency in making parallel line developments.

■ Demonstrate proficiency in making radial line developments.

■ Demonstrate proficiency in the application of triangulation developments.

■ Demonstrate proficiency in constructing a true-length diagram.

■ Explain when and how the two major types of notches are used.

■ Demonstrate proficiency in performing bend calculations.

Developments

A *development* is the pattern or template of a shape that is laid out in a single flat plane in preparation for the bending or folding of a material to a required shape. Surface developments are used in many different industries. Some examples of objects requiring developments are cereal boxes, toolboxes, funnels, air-conditioning ducts, and simple mailboxes.

Three major kinds of surface developments are parallel line developments, radial line developments, and triangulation developments. *Parallel line developments* are used for objects having parallel fold lines. *Radial line developments* are those whose fold lines radiate from one point, *Figure 9-1*. *Triangulation development* is the development process of breaking up an object into a series of triangular plane surfaces, *Figure 9-2*. Each kind of development is explained here in full.

SURFACES

A *development surface* is the exterior and/or interior of the sheet material used to form an object. The various kinds of surfaces include plane surface, single-curved surface, and double-curved surface.

If any two points anywhere on a surface are connected to form a straight line, and that line rests upon the surface, it is a *plane surface*. If all points on a surface can be interconnected to form straight lines without excep-

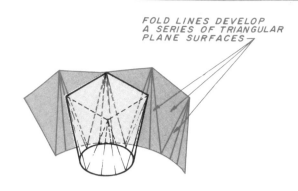

RADIAL LINE DEVELOPMENT

FIGURE 9-2   *Triangulation development*

tion, it is a *flat-plane surface*. The top of a drawing board is an example of a flat-plane surface. A flat-plane surface can have three or more straight edges. Such objects as a cube or a pyramid are bounded by plane surfaces. If a surface can be unrolled to form a plane, it is a *single-curved surface*. A cylinder or a cone is an example of a single-curved surface.

A surface that cannot be developed because it is neither a plane surface nor a curved surface is a *warped surface*. An automobile fender is an example of a warped surface. This kind of surface is usually stamped or pressed into shape. An object fully formed by curved lines with no straight lines is a *double-curved surface*. A sphere is an example of a double-curved surface. This type of surface cannot be developed exactly by using flat patterns; only an approximate development can be made.

LAPS AND SEAMS

Extra material must be provided for laps and *seams*. Many kinds of seams are available, *Figure 9-3*. When making a choice, the drafter must take into account the thickness of the metal so that crowding of the metal at the joints is not a problem. Allowance must be made for the gluing, soldering, riveting, or welding processes for joints. The method of fastening joints together varies with the material and, accordingly, the elimination of rough or sharp edges. Tabs for the drawing problems at the end of this chapter should be designed to support the joining surfaces, *Figure 9-4*.

DESIGN PRACTICES

The drafter should lay out developments according to the dimensions of stock materials for economy and the best use of materials and labor. The stock material area required to cut a pattern should be kept to the smallest convenient size. It is good practice to put the seam at the shortest joint and to attach tops and bottoms along the

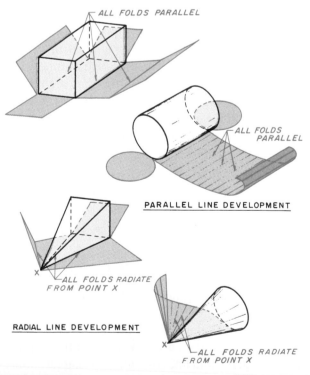

FIGURE 9-1   *Parallel line development and radial line development*

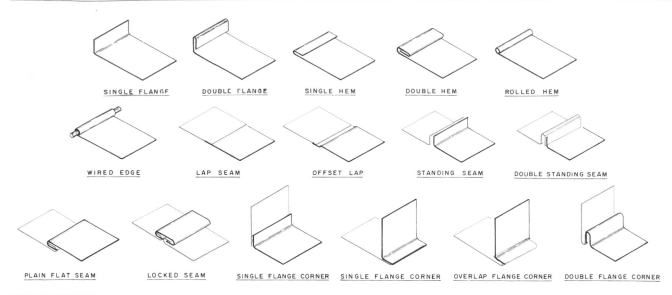

SINGLE FLANGE   DOUBLE FLANGE   SINGLE HEM   DOUBLE HEM   ROLLED HEM

WIRED EDGE   LAP SEAM   OFFSET LAP   STANDING SEAM   DOUBLE STANDING SEAM

PLAIN FLAT SEAM   LOCKED SEAM   SINGLE FLANGE CORNER   SINGLE FLANGE CORNER   OVERLAP FLANGE CORNER   DOUBLE FLANGE CORNER

FIGURE 9-3   *Many types of seams are available*

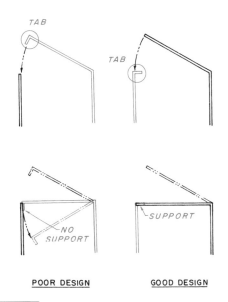

FIGURE 9-4   *Tabs should support the joining surfaces*

longest possible seam or bend to reduce the length requiring soldering, riveting, or welding. It is assumed that the inside surface of the final object is the side that the pattern defines. (The important dimension sizes are usually the inside surfaces.) Fold or bend locations in the material are shown on the pattern with thin, solid lines and are locations that are also assumed to occur on the inside surface of the final object.

THICKNESS OF MATERIAL

The actual thickness of sheet metal is specified by gage numbers. Each gage number indicates a particular thickness of material. *Figure 9-5* lists the gage number and gage sizes for sheet and plate steel. These are nominal

thicknesses, subject to permissible tolerances. Today, there is considerable confusion when using the gage numbering system.

In 1893, Congress established the United States Standard Gage, and it was primarily a *weight gage* rather than a *thickness gage.* It was derived from the weight of wrought iron. At that time, the weight of wrought iron was calculated at 480 pounds per cubic foot; thus, a plate 12 inches square and 1 inch thick weighed 40 pounds. A No. 3 U.S. gage represents a wrought iron plate weighing 10 pounds per square foot. Therefore, if a weight per square foot 1 inch thick is 40 pounds, the plate thickness for a No. 3 gage equals $10 \div 40 = 0.25$ inch, which is the original thickness equivalent for a No. 3 U.S. gage. Since this and all other gage numbers were based on the weight of wrought iron, they are not correct for steel. To add to the confusion, there is considerable variation in the gage thickness for different kinds of material. For example, a gage used for such nonferrous materials as brass and copper is sometimes also used to specify a thickness for steel or vice versa.

Today, to help eliminate the problems, the decimal or metric system of indicating gage size is now replacing the older gage size numbering system.

Parallel Line Development

In parallel line developments all fold lines are parallel. (Refer back to Figure 9-1.)

A three-view drawing of a simple two-inch cube is shown in *Figure 9-6*. A cube is a specific kind of prism. Notice that all the fold lines that form the lateral surfaces of a prism are parallel, *Figure 9-7A*. If the cube were to be unfolded, it would appear as it does in *Figures 9-7B,*

SHEET METAL

GAGE	ALUMINUM		BRASS		STEEL	
	THICKNESS	WT./SQ. FT.	THICKNESS	WT./SQ. FT.	THICKNESS	WT./SQ. FT.
8	.1285	1.812	.1285	5.662	.1644	6.875
9	.1144	1.613	.1144	5.041	.1494	6.250
10	.1019	1.440	.1019	4.490	.1345	5.625
11	.0907	1.300	.0907	3.997	.1196	5.000
12	.0808	1.160	.0808	3.560	.1046	4.375
13	.0720	1.020	.0720	3.173	.0897	3.750
14	.0641	.907	.0641	2.825	.0747	3.125
15	.0571	.805	.0571	2.516	.0673	2.812
16	.0508	.720	.0508	2.238	.0598	2.500
17	.0453	.639	.0453	1.996	.0538	2.250
18	.0403	.580	.0403	1.776	.0478	2.000
19	.0359	.506	.0359	1.582	.0418	1.750
20	.0320	.461	.0320	1.410	.0359	1.500
21	.0285	.402	.0285	1.256	.0329	1.375
22	.0253	.364	.0253	1.119	.0299	1.250
23	.0226	.318	.0226	.996	.0269	1.125
24	.0201	.289	.0201	.886	.0239	1.000
25	.0179	.252	.0179	.789	.0209	.875
26	.0159	.224	.0159	.700	.0179	.750
27	.0142	.200	.0142	.626	.0164	.688
28	.0126	.178	.0126	.555	.0149	.625
29	.0113	.159	.0113	.498	.0135	.562
30	.0100	.141	.0100	.441	.0120	.500
31	.0089	.126	.0089	.392	.0105	.438
32	.0080	.112	.0080	.353	.0097	.406
33	.0071	.100	.0071	.313	.0090	.375
34	.0063	.089	.0063	.278	.0082	.344
35	.0056	.079	.0056	.247	.0075	.312
36	.0050	.071	.0050	.220	.0067	.281
37	.0045	.063	.0045	.198	.0064	.266
38	.0040	.056	.0040	.176	.0060	.250
39	.0035	.050	.0035	.154	–	–
40	.0031	.044	.0031	.137	–	–

FIGURE 9-5 *Thickness of material*

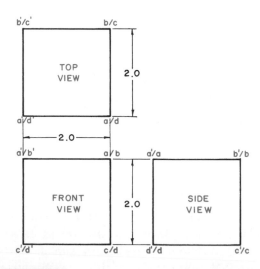

FIGURE 9-6 *Simple multiview drawing of a cube*

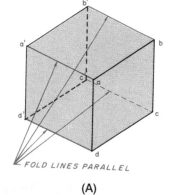

FOLD LINES PARALLEL

(A)

FIGURE 9-7 *(A) All fold lines are parallel*

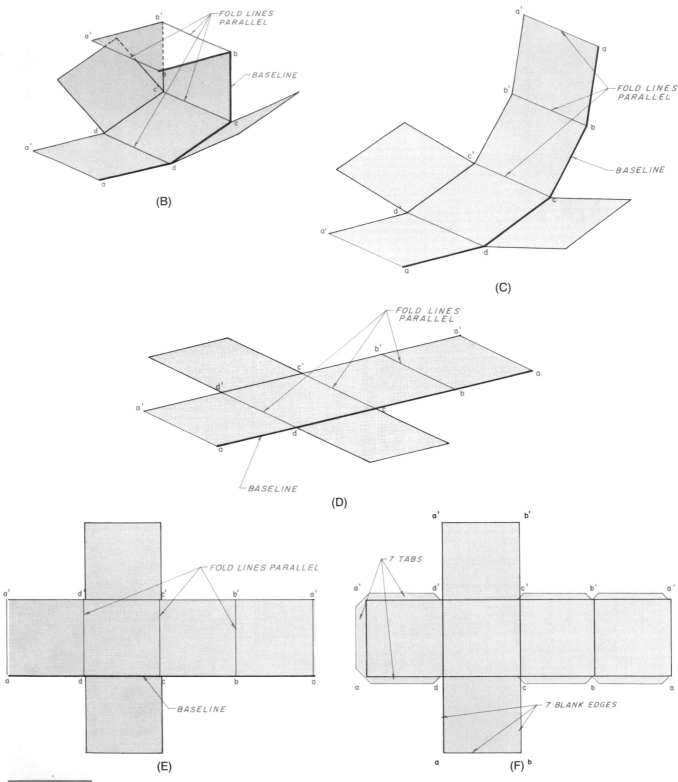

FIGURE 9-7   (Continued) (B) Cube as it unfolds; (C) Cube as it unfolds further; (D) Cube flattened out; (E) Cube in the finished development; (F) Cube flattened out with tabs added

9-7C, and 9-7D. The finished development is shown in *Figure 9-7E.*

Notice that the cube was developed or laid out from a baseline, sometimes referred to as the *stretchout line.* Fold lines are always oriented 90° from the baseline, as illustrat-

ed. If the pattern were to be cut from a flat sheet of material, tabs would be needed to fasten the cube together. Tabs must be positioned so as to make each meet with an untabbed edge, *Figure 9-7F.* Notice that there are seven tabs and seven blank edges. Each tab folds to meet a particular blank edge.

HOW TO DEVELOP A TRUNCATED PRISM USING PARALLEL LINE DEVELOPMENT

GIVEN: A prism with a front view, top view, and auxiliary view, *Figure 9-8A*.

STEP 1 Locate the baseline, which must always be 90° from the parallel fold lines. Label each fold line clockwise in alphabetical or numerical order. Always start with the shortest fold line, *Figure 9-8B*. Either fold line a-a' or b-b' would be eligible, and a-a' is selected. Notice that the starting fold line is also the finishing fold line, and each is labeled at the same location. The finishing fold line is a-a'. All real (or true) lengths of the fold lines are seen in the front view, and the real distances between their locations are seen in the top view.

STEP 2 On a separate sheet of paper, construct a pattern baseline, allowing enough space for the complete development, top and bottom surfaces, and the required tabs. At the left end of the baseline, draw a line perpendicular to the baseline. This is the first fold line and is also a seam edge. Transfer the true length from fold line a-a' (as measured from the three-view drawing in Figure 9-8A). Label this line a-a' also, as illustrated in *Figure 9-8C*.

STEP 3 From the top view of *Figure 9-8A*, obtain the true distance from the first fold line location at a-a' to the second fold line location at b-b'.

On the pattern baseline, transfer this distance, measuring from the first fold line a-a', and draw the next fold line parallel to the first fold line. Label it b-b', *Figure 9-8D*.

STEP 4 Repeat the first three steps, alternating true fold line lengths and true distances between fold lines until all the true lengths and true distances are transferred. Each of the true lengths a-a' through e-e' are transferred from the front view, and the true distances between them are found from the top view of Figure 9-8A. Connect the points as illustrated in *Figure 9-8E*.

STEP 5 The edge selected for attachment of the top surface should be made to keep the rectangle size from which the whole pattern is cut as small as possible. Line a'b' would be the best, except an acute angle cutout c'b'a' is difficult to make. The fold line a'd' is next best, as shown. From Figure 9-8A, transfer the true size and shape of the top surface (auxiliary view) and the bottom surface, *Figure 9-8F*.

STEP 6 Add tabs enough to fasten the edges. The locations should be selected so as not to enlarge the material area needed for the pattern cutout, *Figure 9-8G*. Check to verify that the number of tabs equals the number of blank edges in the same manner as is illustrated in Figure 9-7F. This completes the parallel line development of the object in Figure 9-8A.

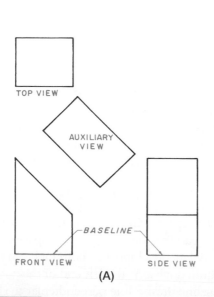

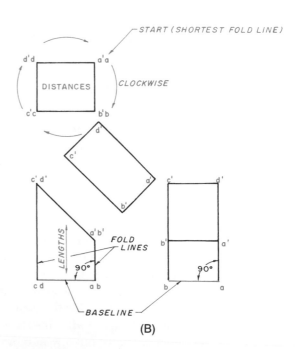

FIGURE 9-8 *(A) Development of a truncated prism using parallel line development; (B) Step 1*

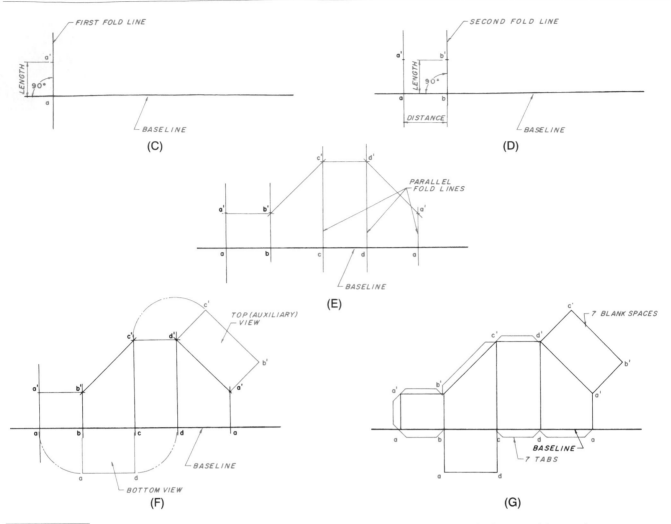

FIGURE 9-8 *(Continued) (C) Step 2; (D) Step 3; (E) Step 4; (F) Step 5; (G) Step 6 (completed development of the prism)*

HOW TO DEVELOP A TRUNCATED CYLINDER USING PARALLEL LINE DEVELOPMENT

GIVEN: A cylinder with a front view, top view, and auxiliary view, *Figure 9-9A*. In developing a cylinder, the same procedure is used as when developing a prism. Because the cylinder has a rounded surface, line segments called station lines must be assumed or chosen on the lateral surface. This is explained in Step 1.

STEP 1 Locate a baseline that must always be 90° from the fold lines, sometimes referred to as *parallel station lines*. The bottom corner of the lateral surface is selected as a convenient location because it already exists. In the view where the cylinder appears as a diameter (the top view in this example), divide the diameter into equal divisions, *Figure 9-9B*. This example uses 12 equal divisions, but any number of appropriate equal divisions would do.

These divisions locate fold line positions, which are used only for construction and layout purposes. Starting with the shortest fold line, consecutively label each division—in this example, a through l. Note that a-a′ marks the beginning and ending of the same line, which meets from opposite directions. The true "distances" between station lines can be approximated in this top view. Project each point into the next (front) view in order to find the true "lengths" of the fold lines. It should be reemphasized that the fold lines must be 90° from the baseline.

STEP 2 On a separate sheet of paper, construct a pattern baseline allowing enough space for the complete development, both ends, and the required tabs. At the left end of the pattern baseline draw a line perpendicular to the baseline. This is the first fold line location

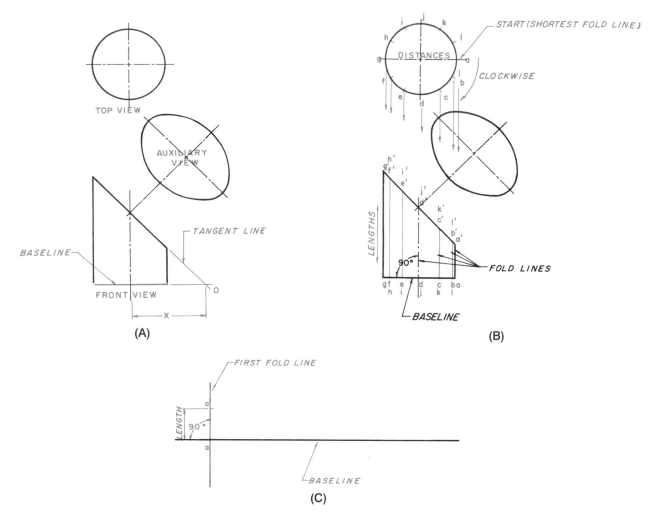

FIGURE 9-9 *(A) Development of a truncated cylinder using parallel line development; (B) Step 1; (C) Step 2*

a-a'. Transfer the true length a-a' from the front view of Figure 9-9B. Label this line as illustrated in *Figure 9-9C*.

STEP 3 Two methods are used for finding the distances between station lines. Method A is an approximate method that assumes the cylinder to be a prism, whereas Method B is mathematically correct.

METHOD A: From the top view of *Figure 9-9B*, obtain the direct chordal distance from the first station line a-a' to the second fold line b-b'. On the pattern baseline, draw the second fold line parallel to the first station line a-a' at the chordal distance found from the top view, and label it b-b', *Figure 9-9D*. Repeat these steps, alternating chordal lengths and true distances until all station lines have been located and drawn. Note that the last fold line length a-a' is a duplicate of the first.

METHOD B: The required length to form the cylinder's circumference can be calculated using the following formula: circumference = cylinder diameter × pi. (Pi = 3.1416, approximately.) The baseline is extended to this length and divided into the same number of divisions as was done at Step 1 in the top view of the cylinder, Figure 9-9B. The division can best be performed as a construction exercise, rather than mathematically. Refer back to Chapter 4, "Geometric Construction." Each of the 13 division locations (from 12 spaces) is the correct location of the required station lines and can be successively labeled.

STEP 4 Add the true size and shape of the bottom surface (bottom view), and the top surface (auxiliary view). The attachment locations of these surfaces are particularly difficult, as the mathematical point of tangency is not easily

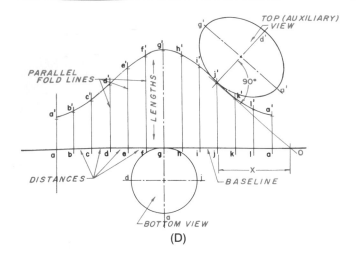

(D)

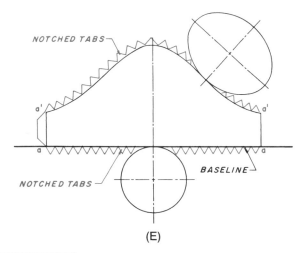

(E)

FIGURE 9-9 *(Continued) (D) Steps 3 and 4; (E) Step 5 (completed development of the cylinder)*

same procedure. *Figure 9-10* shows a prism truncated at both ends, which necessitates creating a new baseline other than a part edge to ensure that all fold lines (instead of station lines) are at 90° to the baseline. *Figure 9-11* is a cylindrical prism, truncated at each end, which requires the selection of station lines instead of using existing fold lines. *Figure 9-12* shows a series of truncated cylinders, similar in construction to the preceding views, whose patterns are used to form an elbow.

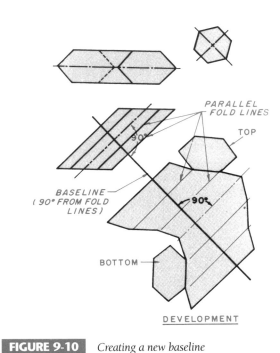

FIGURE 9-10 *Creating a new baseline*

isolated on the pattern. The top surface location is chosen to allow the widest access of cutting to this point. The bottom surface can be added to any of the fold lines. In order to locate the top surface, the distance X from center to point 0 is transferred from Figure 9-9B to the development, and projected 90° as illustrated in Figure 9-9D.

STEP 5 Add the tabs necessary for edge attachments. In cylinders, a notched tab must be used in order to accommodate crowding by surface curvature. A tab must also be added at fold line a-a′ in order to attach edge a-a′ to the other edge a-a′; the top (auxiliary) and bottom fold to meet the notched tabs, *Figure 9-9E*.

Regardless of the shape, whether it be a cylinder or a prism, parallel line developments follow, essentially the

FIGURE 9-11 *Using station lines instead of fold lines*

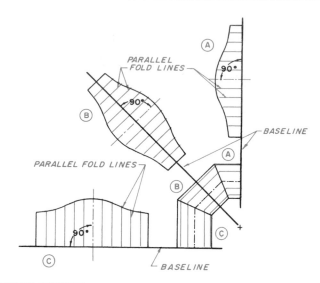

FIGURE 9-12 *Series of patterns used to form an elbow*

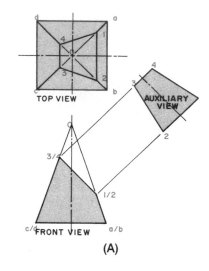

(A)

Radial Line Development

Radial line development is different from parallel line development in that all fold lines or line segments radiate from one point. (Review Figure 9-1.) As in all patterns, true lengths and true distances must be used in laying out the developments.

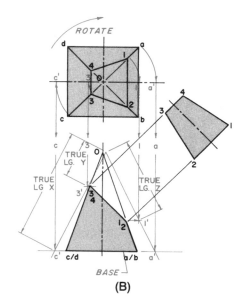

(B)

HOW TO DEVELOP A TRUNCATED PYRAMID USING RADIAL LINE DEVELOPMENT

GIVEN: A two-view drawing and an auxiliary view of a pyramid with the top cut off at an angle (truncated), *Figure 9-13A.* Label points clockwise starting with the shortest fold line. In this example, either a-1 or b-2 could be used. Line a-1 is selected, but the fold line location a-1 is also the meeting corner of edges a-1 and a-1. True distances between the end points of the fold lines are evident in the top view. For example, a-b, b-c, c-d, and d-a are true lengths.

STEP 1 To find the true length of the fold lines, they must be rotated to a position that is parallel to the frontal viewing plane, *Figure 9-13B.* In the top view, point a is rotated as shown and projected into the front view. This revolved line from a' to 0 is its true length. Similarly, point c in the top view is rotated and projected into the front view, giving the true length of c'-0. The same procedure is used to find the true lengths 3'-0 and 1'-0. To construct a development of this object, continue with the following steps.

STEP 2 Using the true length from 0 to a (a'-0), swing the arc as shown in *Figure 9-13C.* On

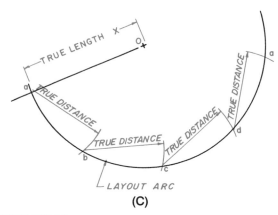

(C)

FIGURE 9-13 *(A) Development of a pyramid using radial line development; (B) Step 1; (C) Step 2*

this arc, mark off as chord lengths the true distances between the fold line end points a to b, b to c, c to d, and d to e, as transferred from the top view of Figure 9-13B. Label all

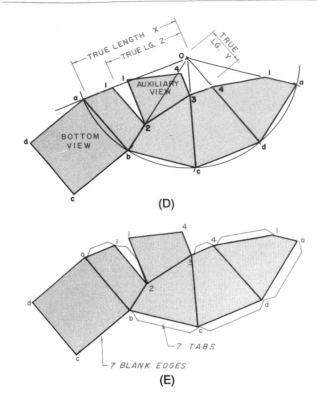

(D)

(E)

FIGURE 9-13 (Continued) (D) Step 3; (E) Step 4 (completed development of a pyramid)

points as shown. Connect the fold lines a to 0, b to 0, c to 0, d to 0, and a to 0.

STEP 3 Using the true lengths found in Figure 9-13B, locate points 1, 2, 3, 4, and 1 on the respective fold lines 0-a, 0-b, 0-c, 0-d, and 0-a. Note that 0-1 is the same length as both 0-2 and 0-1, and 0-3 is the same length as 0-4. Add the auxiliary and bottom views, transferring them directly from the original drawing 9-13A, *Figure 9-13D.*

STEP 4 Add the required tabs as illustrated in *Figure 9-13E.* Check to be sure the number of tabs equals the number of blank edges. In this example, there are seven tabs and seven blank spaces.

HOW TO DEVELOP A TRUNCATED CONE USING RADIAL LINE DEVELOPMENT

As with cylindrical parallel line developments, station lines are line segments that must be positioned on the lateral surface in order to draw a development.

GIVEN: A two-view drawing of a round cone, with the top cut off at an angle (truncated), and having no fold lines, *Figure 9-14A.* It is necessary to first complete the given views and to draw a true view of the inclined flat surface. To pro-

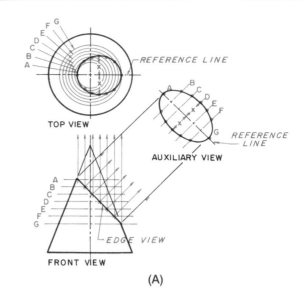

(A)

FIGURE 9-14 (A) Development of a truncated cone using radial line development

ject the flat surface to the top view, the front view is sliced at points A through G. These seven slices are projected into the top view and appear as circles, correspondingly labeled A through G. The points of intersection of the edge view with each slice is projected into the top view. To construct the auxiliary view, project from the seven point intersections perpendicular from the edge view of the flat surface to any convenient distance for the auxiliary view. Draw and use the center line as a reference line, and transfer all point-to-center line distances from the top view to the auxiliary view; refer to reference distances X.

STEP 1 To construct a development of this object, station lines need to be established, and their true lengths and true distances must be determined. In the top view of *Figure 9-14B,* divide the base circle into 12 equal parts, and label each point. In this example, numbers 1 through 12 are arbitrarily used, positioned as illustrated. The station line 0-3 must have two identities because two edges will meet at this point. The true distances between station lines are located between their end points on the cone base, seen in the top view. Project the 12 points down into the base of the front view. These are the end points of the station lines in the front view. Draw a line from each of these end points up to point 0.

STEP 2 There are two methods of drawing the pattern outline. Each begins with swinging an

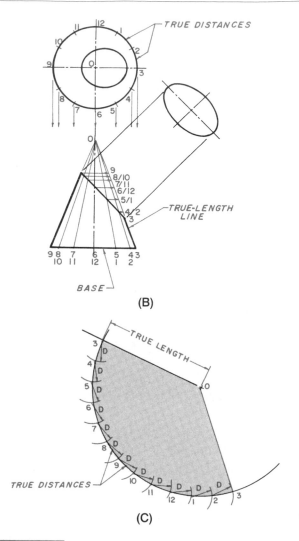

line end points repeatedly to equal the same number of spaces on the arc as the top view of the cone was divided into in Step 1, and ending at point 3, *Figure 9-14C*. Connect the station line end points from point 0 to each of these arc intersections, and proceed to Step 3.

METHOD B: The central angle of the pattern is determined by a pattern arc length needed to equal the circumference of the base of the cone. The ratio of the central angle (A°) to a full-circle 360° is the same as the circumference of the base of the cone is to the full circumference generated by the pattern's radius.

$$\frac{A°}{360°} = \frac{\text{CIRCUMFERENCE OF BASE OF CONE}}{\text{CIRCUMFERENCE OF PATTERN}}$$

$$\frac{A°}{360°} = \frac{\text{PI} \times \text{CONE DIAMETER AT BASE}}{\text{PI} \times \text{PATTERN DIAMETER}}$$

$$\frac{A°}{360°} = \frac{\text{CONE BASE DIAMETER}}{2 \times \text{PATTERN RADIUS}}$$

$$A° = \frac{180 \times \text{CONE BASE DIAMETER}}{\text{PATTERN RADIUS (A' – 0)}}$$

After the central pattern radius A° is found, the angle between successive station lines (a°) is found by dividing the same number of conic divisions done in Step 1 into A°. The chordal lengths (D) of these divisions are found by using the formula:

$$D = 2 \times \text{PATTERN RADIUS} \times [\text{SIN} (\tfrac{1}{2} A°)]$$

This chordal distance can then be struck off on the pattern radius to form the same number of spaces as the conic divisions in Step 1, and labeled accordingly. See Figure 9-14C. Connect these station line end points from point 0 to each of the intersections, and proceed to Step 3.

STEP 3 To draw the edge of the intersection that the cone's lateral surface makes with the upper flat surface, true lengths from the apex to this edge at each station line must be determined. Only 3'-0 and 9'-0 are visible as true lengths in the front view of Figure 9-14B. To find the other true lengths, each segment must be rotated to a position parallel to the frontal viewing plane; in this example, on line 3-0. Project the intersection of each station line at the truncated surface edge to line 3-0 on the cone profile in the front view (on line 3-0).

Starting with the shortest line segments of the cone, swing true lengths arc 0-3' on the corresponding pattern station lines, as

FIGURE 9-14 *(Continued) (B) Step 1; (C) Step 2*

arc whose radius is the true length of the lateral distance from the apex to the base, or length 0-3 or 0-9. The compass radius can be set to this distance on Figure 9-14B. This distance can also be derived mathematically if the diameter of the base of the cone is known and the altitude (perpendicular distance from base to apex) is known. The formula is:

RADIUS = THE SQUARE ROOT OF [(½ CONE BASE DIAMETER)² + (ALTITUDE)²]

After swinging this arc, choose Method A, which approximates the cone as a pyramid, or Method B, which is mathematically correct.

METHOD A: In the top view of Figure 9-14B, the direct chordal distance between any two station line end points, say 5 and 6 (they are all equal), is an approximation of the arc length between them. On the 0-3 radius just drawn, strike off this chordal distance between station

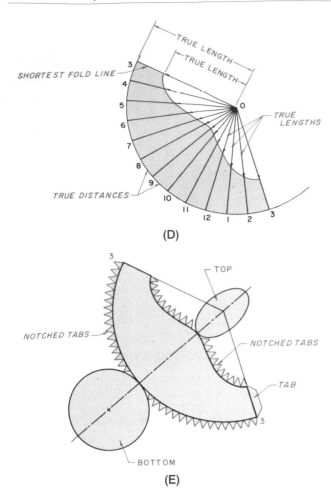

(D)

(E)

FIGURE 9-14 *(Continued) (D) Step 3; (E) Step 4 (completed development of a cone)*

illustrated in *Figure 9-14D*. Swing the apex-to-edge true length found for each station point at the corresponding station line on the pattern. Repeat for all 13 stations, 3 through 12 and back to 3. Label each point as illustrated.

Draw light line segments between each of the station line intersection positions found on the corresponding pattern station lines. This completes the true contour of the top edge, Figure 9-14D.

STEP 4 Add the top, bottom, and split tabs, as illustrated in *Figure 9-14E*.

Triangulation Development

Triangulation is the third major method used to lay out a surface development. *Triangulation* is a method of dividing a surface into a number of triangles and then transferring each triangle's true size and shape to the development. (Review Figure 9-2.) As with parallel line and radial line developments, *true lengths* and *true distances* must be used exclusively in pattern constructions.

True-Length Diagrams

Before any layouts can be started, true lengths and true distances of the object's boundary edges must be determined. A true-length diagram is usually used to develop true lengths and true distances of these edges. A true-length diagram is often a quicker way to obtain the needed projections than descriptive geometry methods.

HOW TO DEVELOP A TRANSITIONAL PIECE USING TRIANGULATION DEVELOPMENT

GIVEN: *Figure 9-15A* describes a transitional piece using a top view, front view, and isometric view. Each of the object's corners are labeled as illustrated. Note that the true distances A to B, B to C, C to D, D to A, 1 to 2, 2 to 3, 3 to 4, and 4 to 1 can be measured directly from the top view. As drawn, the lengths A-1, B-1, B-2, C-2, C-3, D-3, D-4, and A-4 are not shown in their true lengths. A true-length diagram must be used to find these.

STEP 1 A true-length diagram is a combination of two views. In this example, it is a combination of the top view and the front view, *Figure 9-15B*. The height between points A and 1 is shown projected from the front view to the true-length diagram. The distance X between corresponding line end points A and 1 is found in the top view and transferred to the true-length diagram. The illustrated diagonal line, drawn in the true-length diagram, is the true length of A-1. As all the heights and top-view distances between end points are the same for each of the lateral edges, then A-1 is also the same length as B-1, B-2, C-2, C-3, D-3, D-4, and A-4. These true lengths are needed to lay out the development.

STEP 2 Starting from line A-1 and using true lengths and true distances, reconstruct each triangle representing a surface of the object, as illustrated in *Figure 9-15C*. (Review Chapter 3 for aid in transferring a triangle.) Notice in this example that all true distances are transferred from the top view, and all true distances are transferred from the true-length diagram. A tab is added to join A-1 to A'-1'. This completes the transitional piece development as drawn in Figure 9-15A.

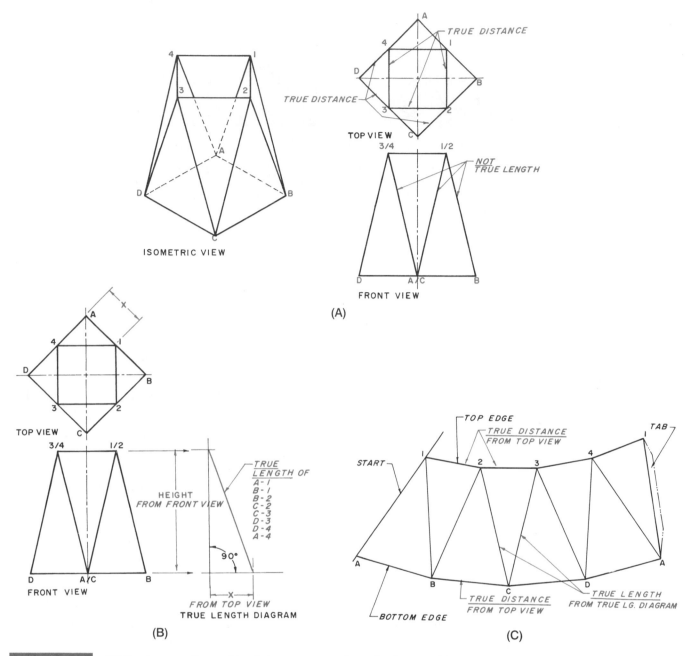

FIGURE 9-15 *(A) Development of a transitional piece using triangulation development; (B) Step 1 (true-length diagram); (C) Step 2*

HOW TO DEVELOP A TRANSITIONAL PIECE WITH A ROUND END USING TRIANGULAR DEVELOPMENT

A transitional piece that is square or rectangular at one end and round at the other is laid out using a procedure very similar to the one outlined previously. The only difference is that station lines must be positioned on the lateral surface of the round-to-corner transitions. *Figure 9-16A* illustrates a two-view transitional piece with a round top end and a rectangular bottom end. In this example, the top end is divided into 12 equal spaces. Each point is numbered clockwise as illustrated. The bottom four corners making up the bottom rectangle are labeled A through D, *Figure 9-16B*. The true distances from A-B, B-C, C-D, and D-A' are found in the top view. Station lines W, X, Y, and Z are not true lengths, but the true lengths must be determined with a true-length diagram. Note that in the true-length diagram, the true length of line W, from A' to 0, is also the same lengths for B-6, C-6, and D-12. Lines X, Y, and Z are also used four times at corresponding locations. Segment the object into triangles and develop the pattern using true lengths and true distances to reconstruct each adjoining segment, *Figure 9-16C*.

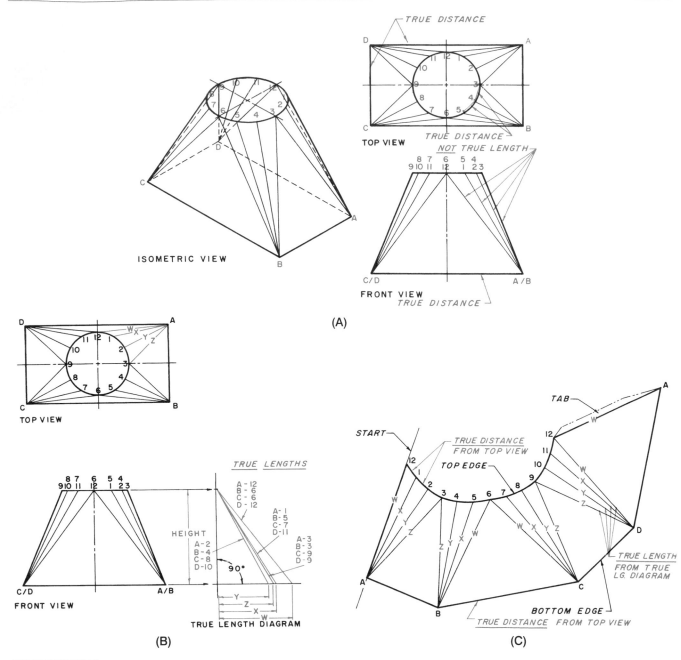

FIGURE 9-16 (A) Development of a transitional piece with a round end using triangulation; (B) True-length diagram; (C) Completed development of a transitional piece with a round end

HOW TO DEVELOP A TRANSITIONAL PIECE WITH NO FOLD LINES USING TRIANGULAR DEVELOPMENT

Figure 9-17A shows a transitional piece with no evident fold lines or line segments. It is developed in the same manner as any other triangulation development.

STEP 1 Divide the top and bottom surfaces into equal spaces as illustrated in *Figure 9-17B*. In this example, the top edge is divided into 12 equal spaces, 0 through 12. The bottom edge is divided into the same number of equal spaces and lettered A through L to A.

STEP 2 Connect points 12 to A, 1 to B, 2 to C, and consecutively to points L to 11 with solid lines, as illustrated in *Figure 9-17C*. Dashed lines are added to segment the object into various triangles.

STEP 3 True distances from 1 to 2, 2 to 3, 3 to 4, and so on through 11 to 12, and the true distances from A to B, B to C through to L to A' are found in the top view, *Figure 9-17C*. The other fold lines connecting the top and bottom are not true lengths and, therefore, true-length diagrams must be made, one for the

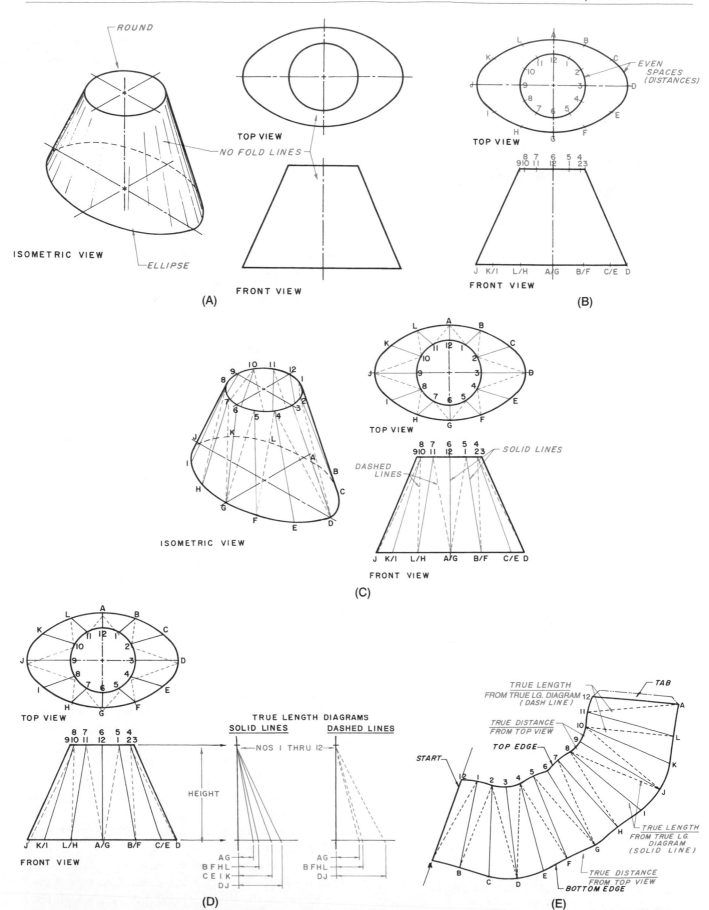

FIGURE 9-17 (A) Development of a transitional piece with no fold lines; (B) Step 1; (C) Step 2; (D) Step 3 (true-length diagram); (E) Completed development of a transitional piece

Industry Application

A WIN-WIN-WIN PARTNERSHIP

Andrews Sheet Metal Fabrication (ASMF) is a successful family business that to date has supported three generations of the Andrews family. ASMF has been in business for more than 60 years. In the old days, all of the key management and technical positions were held by family members who learned their respective trades from grandfathers, fathers, older brothers, and uncles. But in recent years, ASMF's reputation for quality has resulted in rapid expansion. The company must now look outside the family for designers, drafters, and leading technicians.

This is causing a problem. John Andrews, the current CEO of ASMF, explains the situation as follows: "When we started looking outside of the family to fill technical positions—particularly design and drafting positions—we simply could not find anyone qualified. We received plenty of applications from people who had excellent mechanical drafting skills relating to precision metal parts, but most had never heard of sheet metal fabrication. No one knew how to perform bend calculations. Out of frustration, we started calling technical schools and community colleges as far away as 100 miles. Finally, after two weeks of telephone calls we came across a small technical college in a town 75 miles to the south that sounded promising. In fact, it sounded too good to be true."

"Fully expecting to be disappointed, I visited the college and examined the curriculum. To my great surprise, I could have hired any of the school's advanced students. They all knew patterns, developments, and bend calculations. I gave projects that included parallel line development, radial line development, triangulation development, and true-length diagrams. I talked with several students who actually pointed out errors in a set of hand calculations I had brought with me."

"I hired two students on the spot, and worked out an agreement to notify the head of the college's drafting program every time a position became available. We plan to have a long-term relationship with this college. In fact, I'm going to establish the ASMF scholarship for the sheet metal trades to encourage drafting students to study our field. Our new relationship with the college is a Win-Win-Win partnership. ASMF wins, the college wins, and the students win."

solid lines and the other for the dashed lines, *Figure 9-17D*. The development is laid out by constructing connecting triangles using the lengths of each leg of each triangle, *Figure 9-17E*.

Notches

Some developments require notches. Two major types of notches are usually used: a sharp V or a rounded V, *Figure 9-18*. The sharp V is used only if a minimal force is tending to part the material. The sharp point of the V will tear or crack under stress. The rounded V is used where parts would be under greater stress. The radius of the vertex of the rounded V should be at least twice the thickness of the metal used, and larger if possible.

Bends

When sheet metal is bent to form a corner, rib, or design, the outer surface stretches and the inner surface compresses. The length of material needed by each bend must be calculated and added to the straight, unbent portions of the pattern. The material needed by each bend is the length of the neutral axis within the material, where neither compression nor stretching occurs. This is calculated either by formulas or by using various charts available for this purpose.

Bend allowance charts are included in Appendix A and are much faster to use than a formula. Two basic kinds of charts are used to calculate bend allowance, in both the English and metric systems. One type is used for 90° bends; the other for bends from 1° through 180°. The total length of a pattern is called the *developed length*.

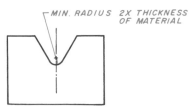

FIGURE 9-18 *Notches*

To determine the developed length, the stretched-out flat pattern must include all straight sides, plus the calculated bend allowances.

HOW TO FIND ALL STRAIGHT SIDES OF A 90° BEND

Figure 9-19A shows a simple 90° bend with an inside radius of .25, a sheet metal thickness of .125, and legs of 2.0 and 3.0 (the English system inch is used in this example). Bend radii are always measured from the surface closest to the bend radius center.

STEP 1 Locate the tangent points at the ends of the straight sides.

STEP 2 Add the thickness of the sheet metal to the bend radius .25 (.125 + .25 = .375).

STEP 3 Subtract the sum of the sheet metal thickness .125 and the radius .25 from the 3.00 overall length of the object (3.00 – .375 = 2.625).

STEP 4 Subtract the sum of the sheet metal thickness .125 and the bend radius .25 from the 2.00 overall height of the object (2.00 – .375 = 1.625).

STEP 5 Add the two straight sides together (2.625 + 1.625 = 4.250). This is the total length of the straight sides of this object, *Figure 9-19B.*

HOW TO FIND THE BEND ALLOWANCE OF A 90° BEND

Review Figure 9-19A.

STEP 1 Note the metal thickness, in this example .125, and the inside radius, in this example .25.

STEP 2 Refer to the bend allowance chart in inches for 90° bends in Appendix A.

STEP 3 The left-hand column gives various metal thicknesses. Go down the left-hand column until the required size or the closest size is found. In this example it is .125.

STEP 4 Along the top of the chart is listed various inside radii; go across the top of the chart to the required size or the closest radius, in this example .25.

STEP 5 From the .125 number in the left-hand column, project across to the right; from the .25 number along the top of the chart, project down to where the two columns intersect. Given is the bend allowance for material .125 thick with a .25 radius. In this example the bend allowance is .480, *Figure 9-19C.*

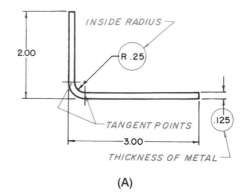

(A)

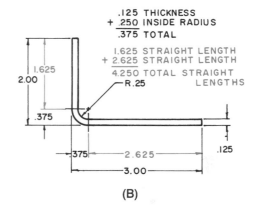

(B)

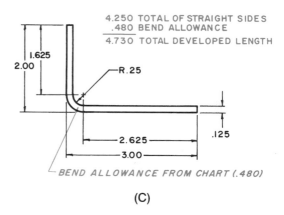

(C)

FIGURE 9-19 *(A) Bend allowance of 90° bend; (B) Total of straight lengths; (C) Total length including straight lengths and bend allowance*

STEP 6 The stretched-out dimension is found by adding the straight sides to the bend allowance; in this example, 4.250 + .480 = 4.730 (see Figure 9-19C).

HOW TO FIND THE TOTAL STRAIGHT-SIDE LENGTH ADJOINING A BEND OTHER THAN 90°

Figure 9-20A shows a simple 30° bend with an inside radius of 6.35 and a sheet metal thickness of 3.175 (the metric system is used in this example).

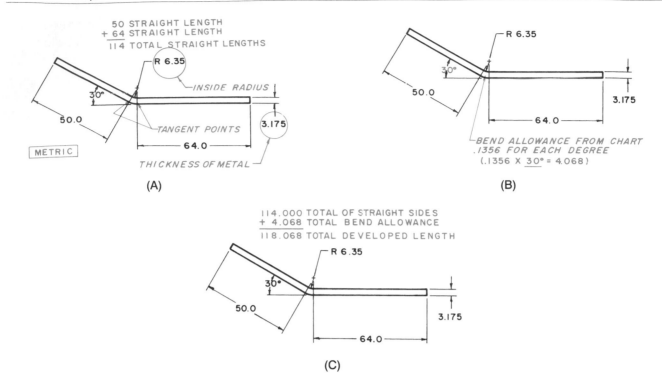

FIGURE 9-20 *(A) Bend allowance of a bend other than 90°—total of straight lengths; (B) Total length of bend allowance; (C) Total length including straight lengths and bend allowance.*

STEP 1 Locate the tangent points at the ends of the straight sides.

STEP 2 Determine the length of the straight sides and add them together (64.0 + 50.0 = 114.0). This is the total length of the straight sides of this object.

HOW TO FIND THE BEND ALLOWANCE OF OTHER THAN A 90° BEND
Review Figure 9-20A.

STEP 1 Note the metal thickness, in this example 3.175, and the inside radius, in this case R6.35.

STEP 2 Refer to the bend allowance chart in millimeters for 1° bends in Appendix A.

STEP 3 The left-hand column gives various metal thicknesses. Go down the left-hand column until the required size or the closest size is found. In this example it is 3.175.

STEP 4 Along the top of the chart is listed various inside radii. Go across the top of the chart to the required size or closest radius, in this example 6.35.

STEP 5 From the 3.175 number in the left-hand column, project across to the right; from the 6.35 number along the top of the chart, project down to where the two columns intersect. Given is the factor used to calculate the bend allowance. In this example .1356.

STEP 6 Multiply this factor times the actual degrees in the bend, from a straight 180° line, *Figure 9-20B.* This is the bend allowance (.1356 × 30° = 4.068).

STEP 7 Add the straight-side lengths to the bend allowance to get the total developed length: 114.0 + 4.068 = 118.068, *Figure 9-20C.*

NOTE: A 30° bend dimension as illustrated in **Figure 9-21A** *is actually 150° (180° – 30° = 150°),* **Figure 9-21B**. *The total bend must be calculated from a straight piece that is actually 180° before bending.*

Summary

• A development is the pattern or template of a shape that is laid out in a single flat plane in preparation for bending or folding into a required shape. The three types of developments are parallel, radial line, and triangulation.

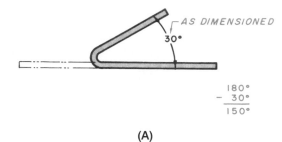

(A)

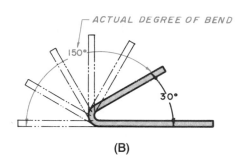

(B)

FIGURE 9-21 *(A) Total bend must be calculated from straight line dimension—given is 30°; (B) As dimensioned, 30° is actually a 150° bend from a straight line*

- In parallel line development, all fold lines are parallel, and the object is laid out from a baseline known as the stretch line.

- In radial line development, all fold lines and line segments radiate from one point. True lengths and true distances must be used in laying out the developments.

- Triangulation is a method of dividing a surface into a number of triangles and then transferring each triangle's true size and shape to the development.

- When a development must be notched, two major types are used: sharp V and rounded V. Parts that will be subjected to greater levels of stress use the rounded V.

- When bending a sheet metal to form a corner, rib, or design, the outer surface stretches and the inner surface compresses. The length of material needed by each bend must be calculated and added to the straight, unbent portions of the pattern.

Review Questions

Answer the following questions either true or false.

1. A transitional piece is usually stamped or pressed into a particular shape.

2. A gage number represents the weight per square foot of a particular material.

3. The importance of developing the pattern or template with the inside surface up is that important dimension sizes and fold or bend locations are assumed to occur on the inside surface of the final object.

4. Tabs are used to fasten items together.

5. Bend allowance refers to the practice of calculating the length of material needed by each bend.

6. A true-length diagram often provides needed projections more quickly than descriptive geometry methods.

7. If a transitional piece does not have fold lines, it is developed in the same manner as any other triangulation development.

8. An automobile fender is an example of a single-curved surface.

Answer the following questions by selecting the best answer.

1. Which kind of development would be used to develop a pattern?
   a. Parallel line development
   b. Truncated line development
   c. Radial line development
   d. Triangulation line development

2. What are the two major kinds of notches used in a development?
   a. Sharp U or rounded U
   b. Sharp V or rounded U
   c. Rounded U or rounded V
   d. Rounded V or sharp V

3. In parallel line development, at which angle to the baseline must the fold lines be projected?
   a. 45°
   b. 60°
   c. 90°
   d. 180°

4. The gage system of calling out the thickness of a material is being phased out because:
   a. It is out of date.
   b. It causes considerable confusion.
   c. It is inaccurate.
   d. None of the above

5. Which of the following elements **must** be located or laid out before a development can actually be started?
   a. True lengths
   b. True distances
   c. Station lines
   d. Both a and b

Chapter 9 Problems

The following problems are intended to give the beginning drafter practice in sketching, laying out auxiliary views if required, and drawing the stretched-out development of various objects. These problems will use one or more of the three standard methods of development: parallel line developments, radial line developments, and triangulation developments. The student will also practice calculating developed lengths using various charts to determine full length before bending.

CONVENTIONAL INSTRUCTIONS

The steps to follow in laying out any object that is to be developed are:

STEP 1 Study the problem carefully.

STEP 2 Determine which method or methods of development will be used (parallel line/radial line/triangulation).

STEP 3 On a scrap sheet of paper, draw to scale the number of views required to lay out **all** true lengths and true distances. Add the end view and auxiliary views, in scale, if necessary. Draw with a **sharp** 4-H lead, and work as accurately as possible.

STEP 4 Label each point, if necessary, in order to keep track of progress.

STEP 5 Check all lengths and auxiliary views if included.

STEP 6 Starting from a baseline, lightly lay out the development. Use true lengths and true distances for all measurements. (Be sure to start the seam at the shortest fold line possible.)

STEP 7 Label each point, if necessary, in order to keep track of progress.

STEP 8 Use phantom lines to represent all fold lines.

STEP 9 Add tabs as required, and check to see that there are enough but without duplications.

STEP 10 Recheck all true lengths and true distances.

STEP 11 Darken in all lines.

CAD Instructions

Student who plan to complete the following problems on a CAD system should begin with these general instructions. Before reading the specific instructions for each activity, go through each step in the following planning checklist. The checklist applies to any CAD system and will help ensure the optimum use of your time and resources.

1. Analyze the problem carefully. Decide exactly what you are being asked to do.

2. Determine what resources and references (if any) you will need in order to complete the problem. Collect those references and resources and have them readily available.

3. Decide if any particular standards apply to the project, and have those standards available.

4. Determine what types of views will be required and how many of each.

5. Determine what the final plotted scale of the drawing will need to be, and select the appropriate paper size for plotting/printing.

6. Plan your drawing sequence. In what order will you develop the drawing (i.e., lines, features, dimension lines, leaders, dimensions)?

7. Review the various CAD commands you will have to use in order to develop the drawing.

8. Examine your CAD system to ensure that everything is in working order, then begin the project.

PROBLEMS 9-1 THROUGH 9-18

Using the parallel line development method, develop each object starting from the given seam. Label each point clockwise; add ends and/or auxiliary views as necessary to develop a complete object. Add tabs .125 (3) × 45° to suit. Use phantom lines for all fold lines.

Extra assignment(s): Cut out the developments and glue or tape them together to prove their accuracy.

Problem 9-1

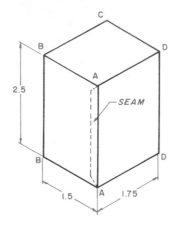

 Problem 9-4

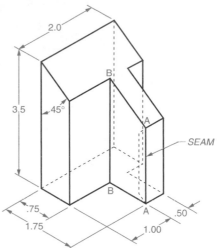

 Problem 9-2

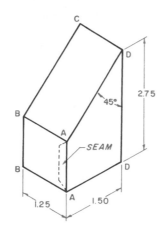

 Problem 9-5

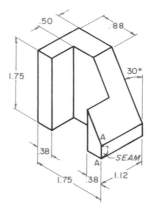

 Problem 9-3

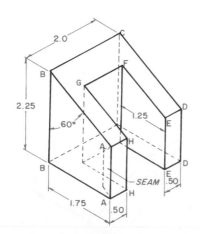

 Problem 9-6

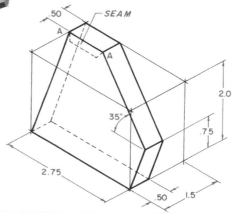

Problem 9-7

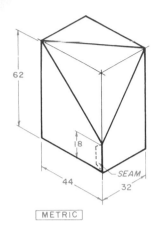

62
18
44
32
SEAM
METRIC

Problem 9-10

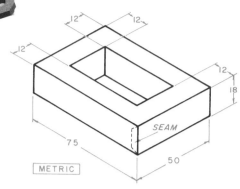

12
12
12
12
18
SEAM
75
50
METRIC

Problem 9-8

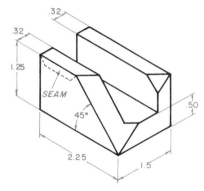

.32
.32
1.25
SEAM
45°
.50
2.25
1.5

Problem 9-11

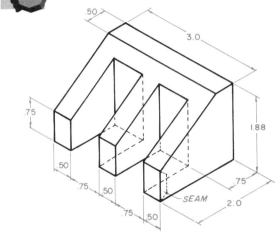

.50
3.0
.75
1.88
.50
.75
.50
.75
.75
.50
SEAM
2.0

Problem 9-9

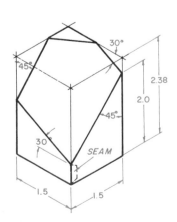

30°
45°
2.38
2.0
45°
30
SEAM
1.5
1.5

Problem 9-12

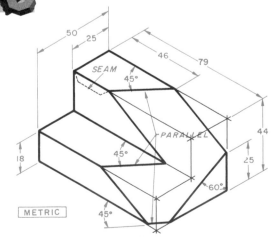

50
25
46
79
SEAM
45°
PARALLEL
44
18
45°
25
60°
METRIC
45°

Problem 9-13

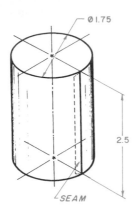

Ø1.75

2.5

SEAM

Problem 9-14

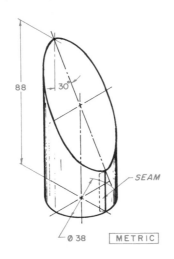

88

30°

SEAM

Ø 38   METRIC

Problem 9-15

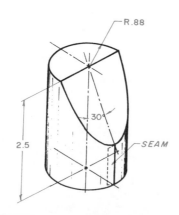

R.88

30°

2.5

SEAM

Problem 9-16

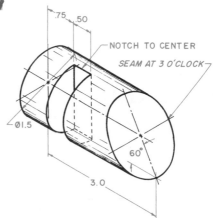

.75   .50

NOTCH TO CENTER

SEAM AT 3 O'CLOCK

Ø1.5

60°

3.0

Problem 9-17

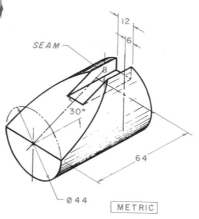

12

6

SEAM

8

30°

64

Ø 44   METRIC

Problem 9-18

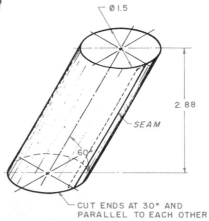

Ø 1.5

2.88

SEAM

60°

CUT ENDS AT 30° AND
PARALLEL TO EACH OTHER

PROBLEMS 9-19 THROUGH 9-23

Using the parallel line development method, develop each object starting from the given seams. Add .125 (3) × 45° tabs as required to hold parts together. Design parts so there are as many identical parts as possible. Use phantom lines for all fold lines.

Extra assignment(s): Cut out the developments and glue or tape them together to prove their accuracy.

Problem 9-19

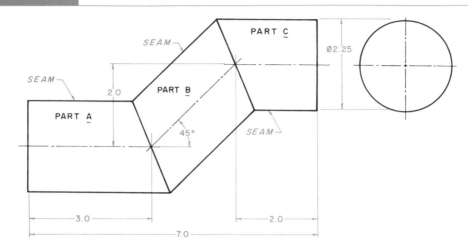

Problem 9-20

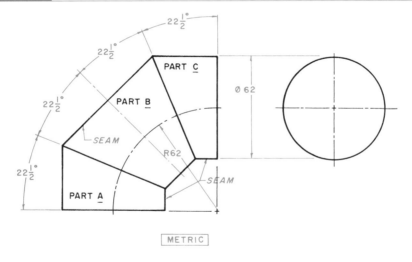

METRIC

Problem 9-21

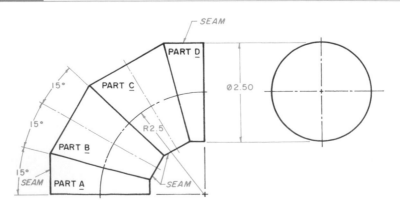

Problem 9-22

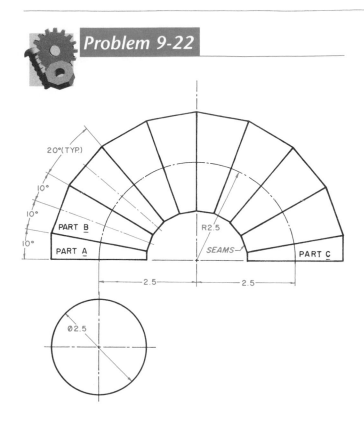

Problem 9-23

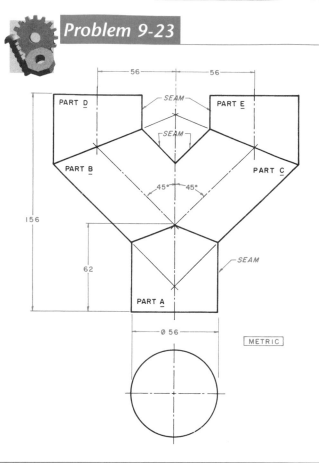

PROBLEMS 9-24 THROUGH 9-26

Develop the objects, using the given dimensions. Using bend allowance charts, design layouts to include material for bends.

Extra assignment(s): Cut out the developments and glue or tape them together to prove their accuracy.

Problem 9-24

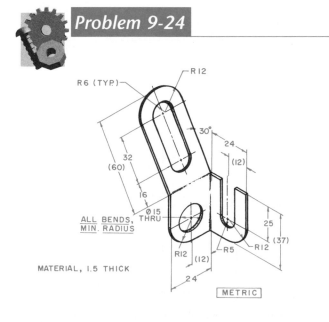

Problem 9-25

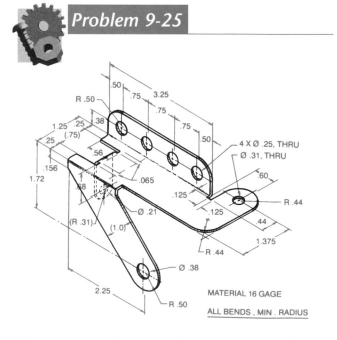

Problem 9-26

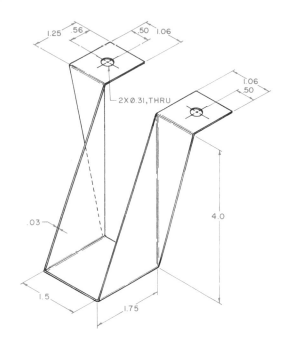

PROBLEMS 9-27 THROUGH 9-30

Carefully lay out a full-size multiview drawing of the object; include all intersection lines. Make a full-size development of the object. Add tabs .125 (3) × 45° as required to hold parts together.

Extra assignment(s): Cut out the developments and glue or tape them together to prove their accuracy.

Problem 9-27

ALL DIMENSIONS ARE TO OUTSIDE LIMITS OF PARTS

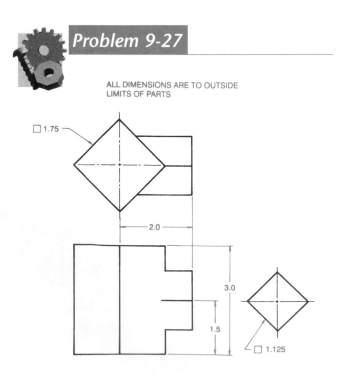

Problem 9-28

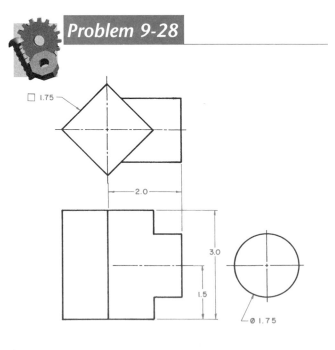

Problem 9-29

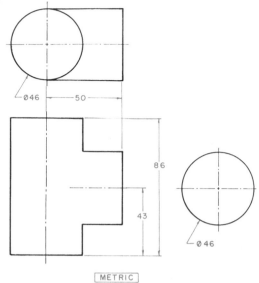

METRIC

Problem 9-30

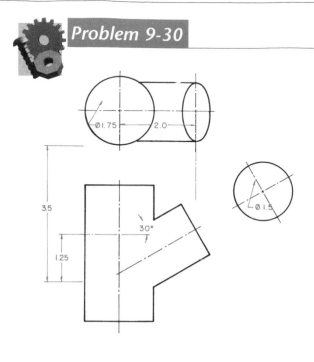

PROBLEMS 9-31 THROUGH 9-42

Using the radial line development method, develop each object starting from the given seam. Label each point clockwise; add ends and/or auxiliary views as necessary to develop a complete object. Add tabs .125 (3) × 45° to suit. Use phantom lines for all fold lines. ***Extra assignment(s):*** Cut out the developments and glue or tape them together to prove their accuracy.

Problem 9-31

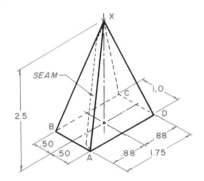

Problem 9-32

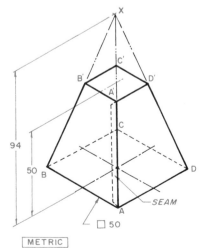

METRIC

Problem 9-33

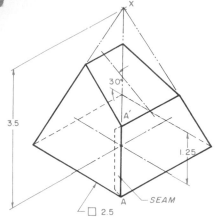

Problem 9-36

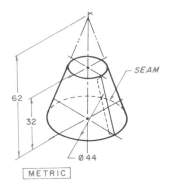

METRIC

Problem 9-34

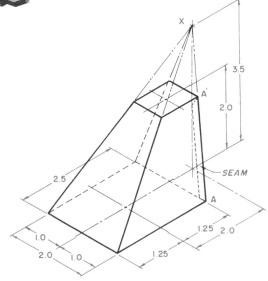

Problem 9-37

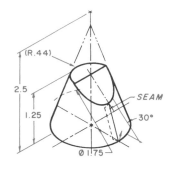

Problem 9-35

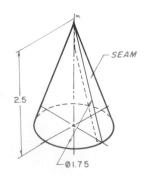

Problem 9-38

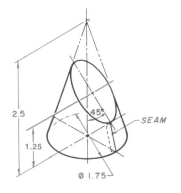

Problem 9-39

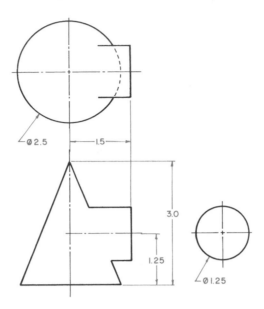

Problem 9-40

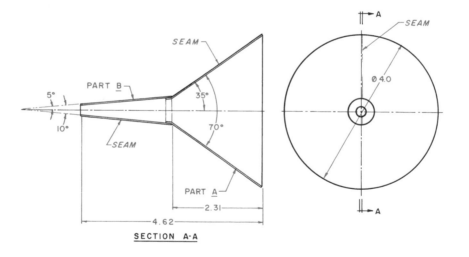

SECTION A-A

Problem 9-41

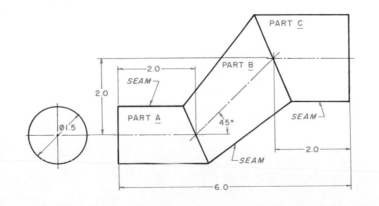

Problem 9-42

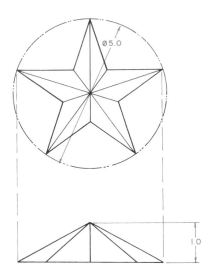

PROBLEMS 9-43 THROUGH 9-46

Using the triangulation development method, develop each object starting from the given seam. Label each point clockwise: add tabs .125 (3) × 45° to suit. Use phantom lines for all fold lines.

Extra assignment(s): Cut out the developments and glue or tape them together to prove their accuracy.

Problem 9-43

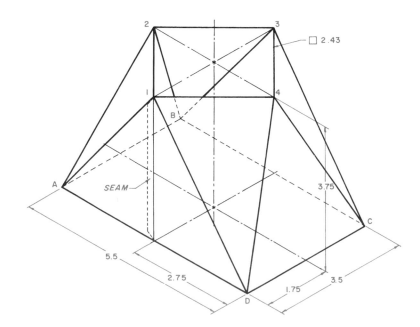

Problem 9-44

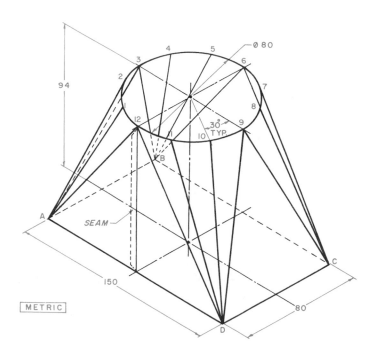

METRIC

Problem 9-45

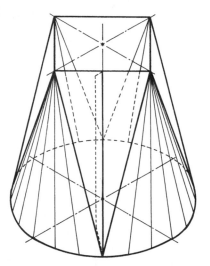

Problem 9-46

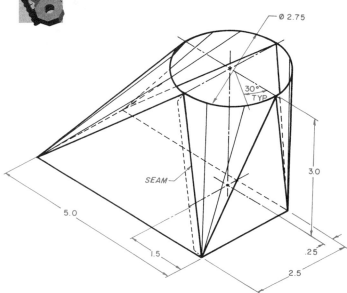

PROBLEMS 9-47 THROUGH 9-49

Using development charts, calculate the true developed length of each object. Round the answer to the nearest three places. Recheck all calculations.

 Problem 9-47

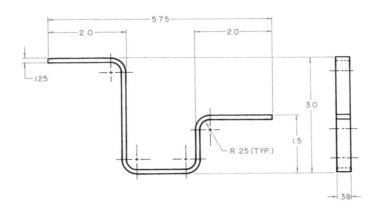

 Problem 9-48

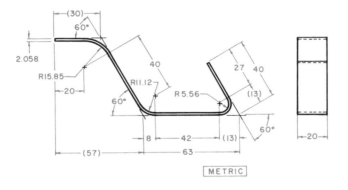

METRIC

 Problem 9-49

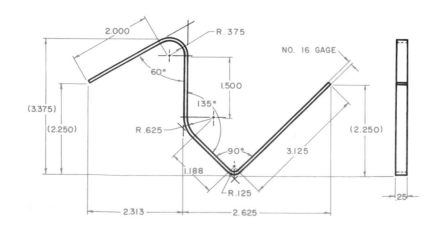

PROBLEMS 9-50 THROUGH 9-54

Using either the parallel line, the radial line, or the triangulation development method, develop each object. Be sure to leave a seam and to add all required tabs to suit as required. Use phantom lines for all fold lines.

Extra assignment(s): Cut out the developments and glue or tape them together to prove their accuracy.

Problem 9-50

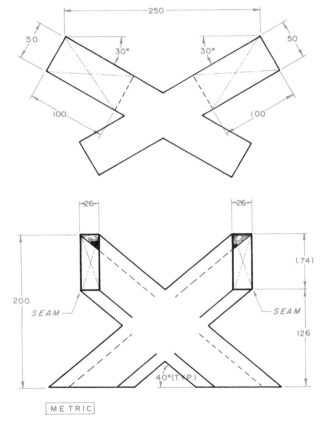

Problem 9-51

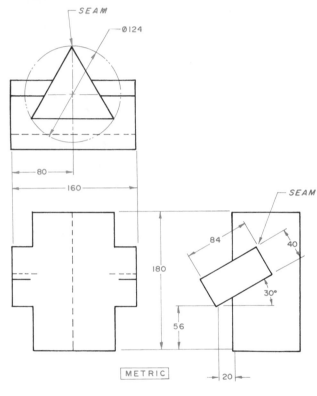

Problem 9-52

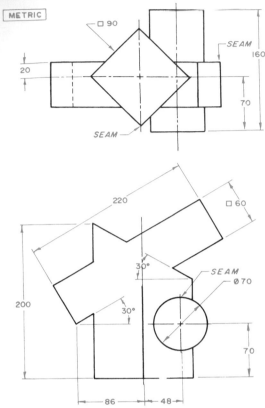

METRIC

□ 90

SEAM

160

20

70

SEAM

220

□ 60

30°

SEAM

Ø 70

200

30°

70

86   48

Problem 9-53

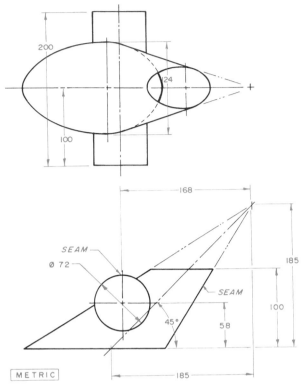

200

24

100

168

SEAM

Ø 72

185

SEAM

100

45°

58

METRIC

185

Problem 9-54

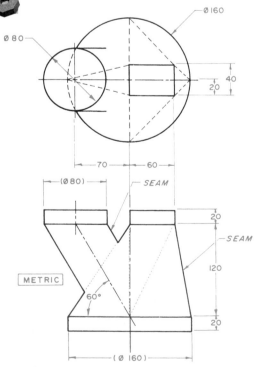

Ø 160

Ø 80

40

20

70   60

(Ø 80)   SEAM

20

SEAM

120

METRIC

60°

20

(Ø 160)

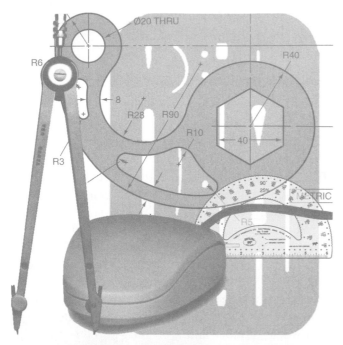

DIMENSIONING AND NOTATION

CHAPTER OUTLINE

Dimensioning systems • Dimension components • Laying out dimensions • Steps in dimensioning • Specific dimensioning techniques • Summary of dimensioning rules • Notation • Rules for applying notes on drawings • Group technology • Summary • Review questions • Chapter 10 problems

CHAPTER OBJECTIVES

Upon completion of this chapter, students should be able to do the following:

■ Explain the three main dimensioning systems used in the United States: metric, decimal-inch, and fractional dimensioning.

■ Demonstrate proficiency in properly using the following dimension components: extrusion lines, dimension lines, leader lines, and arrowheads.

■ Demonstrate the steps used in laying out dimensions for an object.

■ Demonstrate proper application of the dimensioning techniques for the following: dimensioning symbols, dimensioning chords, arcs, angles, radii, curved surfaces, offsets, irregular curves, contours, multiple radii, normal objects by offset, spheres, round holes, making simple hole callouts, applying drill size tolerances, dimensioning hole locations for single and multiple holes, and locating holes about a bolt center.

■ Demonstrate proficiency at applying the following dimensioning techniques: locating holes on center lines and concentric arcs, multiple holes along the same center line, repetitive features, callouts for tapered holes, callouts for countersunk holes, callouts for counterbored holes, dimensioning a cylinder and a square, double dimensioning, reference dimensioning, dimensioning slots and rounded ends, not-to-scale dimensions, nominal dimensions, dimensioning external chambers, internal chambers, necks, undercuts, knurls and keyways, staggered dimensions, and dimensioning flat tapers, round tapers, threads, pads, bosses, sheet metal bends, sectional views, pyramids, and cones.

(continued)

KEY TERMS

Allowance

Arrowhead

Attributes

Chamfer

Clearance fit

Decimal-inch dimensioning

Detail notes

Dimensioning

Dimension line

Extension line

Fractional dimensioning

General notes

Interference fit

Knurl

Leader line

Least material condition (LMC)

Manufacturing part family

Maximum material condition (MMC)

Metric dimensioning

Nominal dimension

Notation

Opitz system

Part family

Tolerance

Transition fit

Verification

Written notes

- Demonstrate proficiency at applying the following dimensioning techniques: dimensioning concentric and nonconcentric shafts, rectangular coordinate dimensioning, dimensioning holes on a bolt center diameter, finish marks, X-Y-Z coordinates, tabular dimensioning, tabular drawing, tolerancing, shaft limits, hole limits, allowance, clearance, interference fit, transition fit, size limits, design size, maximum and minimum sizes, location limits, design location, maximum and minimum location, calculating fits, matching parts, and standard fits.

- Demonstrate proficiency in the proper application of notation.

- Explain the concept of group technology as it relates to design and drafting.

One of the most fundamental drafting tasks is to meet the requirements of the engineering definition of the part while providing for the most economical production process and interchangeability considerations. All of this is accomplished by the use of proper dimensions and notation on drawings. *Dimensioning* is the process whereby size and location data for the subject of a technical drawing are provided. *Notation* is the process whereby needed information not covered by dimensions is placed on a technical drawing.

It is critical that drafters, designers, and engineers be proficient in standard dimensioning practices. The most widely accepted dimensioning standard is American Society of Mechanical Engineers document Y14.5M-1994 (ASME Y14.5M-1994R). Similar standards are produced by the International Standards Organization (ISO). However, unless otherwise specified, ASME Y14.5M-1994 is the standard used for guiding dimensioning practices.

Modern dimensioning practices described in ASME Y14.5M-1994 apply in most instances in which interchangeability of parts is a major consideration. The concept dictates that parts produced from a drawing at one manufacturing site should be interchangeable with those produced at another manufacturing site. Automotive parts are an excellent example of production for interchangeability. Some parts are manufactured in America, some in Europe, and some in Japan, but they must all fit together in one car during assembly. Although interchangeability is not a factor with all parts that are produced, the drafter should still use the basic dimensioning principles of ASME Y14.5M-1994. This is particularly important when the parts will be produced using such ever-increasing automated, semiautomated, or integrated processes as numerical control, computer-aided manufacturing (CAM), or computer-integrated manufacturing (CIM).

Dimensioning Systems

Three dimensioning systems are used on technical drawings in the United States: metric dimensioning, decimal-inch dimensioning, and fractional dimensioning. Certain rules of practice with which drafters should be familiar pertain to each of these dimensioning systems.

METRIC DIMENSIONING

The standard metric unit of measurement for use on technical drawings is the millimeter (.001 meter) or .039 inch. *Figure 10-1* is a chart of various metric units of measurement of less than a meter that shows where the millimeter fits in.

When using *metric dimensioning*, several general rules should be observed. When a dimension is less than 1 millimeter, a zero must be placed to the left of the decimal point, *Figure 10-2A*. When a metric dimension is a whole number, neither the zero nor the decimal is required, *Figure 10-2B*. When a metric dimension consists of a whole number and a decimal portion of another

UNIT	MULTIPLE OF A METER
METER	1.0
DECIMETER	0.1
CENTIMETER	0.01
MILLIMETER	0.001

FIGURE 10-1 *Metric linear measurements*

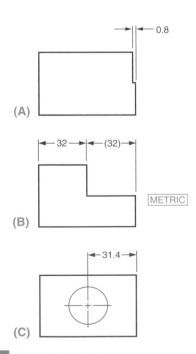

FIGURE 10-2 *Metric dimensioning*

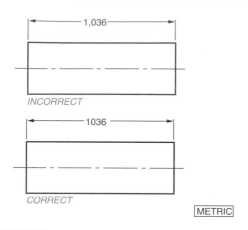

FIGURE 10-3   *No commas in metric dimensions*

millimeter, it is written as follows: whole number first, decimal point second, and finally, the decimal part of the number. The decimal part of the number is not followed by a zero in metric dimensioning, *Figure 10-2C*. Individual digits in metric dimensions are not separated by commas or spaces, *Figure 10-3*. Drawings prepared with metric dimensions are identified with the word "metric" contained in a small rectangle below the part.

DECIMAL-INCH DIMENSIONING

Decimal-inch dimensioning is frequently used in the dimensioning of technical drawings. It is a much less cumbersome system for mechanical drawings than is the fractional system, and it is still used more than the metric system. When using the decimal-inch dimensioning system, several rules should be observed. If a dimension is less than one inch, only a decimal point and the numbers of the decimal fraction are required. A zero is not required preceding the decimal point, *Figure 10-4A*. The number of places beyond the decimal point that a decimal-inch dimension is carried is determined by the specified tolerance for the part in

question, *Figure 10-4B*. In this figure, a tolerance of .002 (±.001), three places to the right of the decimal, is specified. Consequently, the dimension of 1.637 is carried out three places to the right of the decimal.

There are no specified sizes for decimal points, but they should be made dark enough and large enough to be seen and to reproduce through any normal reproduction process (diazo, photocopy, or microfilm).

FRACTIONAL DIMENSIONING

Fractional dimensioning is not frequently used on mechanical technical drawings. Its primary use is on architectural and structural engineering drawings. However, since it is occasionally still used on mechanical drawings, drafters should be familiar with this system. When using fractional dimensions on mechanical drawings, several rules should be observed. The line separating the numerator and denominator of a fraction should be a horizontal line, not an inclined line, *Figure 10-5*. Full-inch dimensions should be a minimum of one-eighth inch in height. The combined height of the numerator, denominator, and horizontal line of a fraction should be one-quarter inch, *Figure 10-6*, for A-, B-, and C-size drawings, and a minimum of five-sixteenths inch for D-, E-, and F-sizes. Also, be sure to leave a visible gap between the numerator, denominator, and horizontal line of the fraction.

ORIENTING DIMENSIONS

Regardless of whether you are dimensioning in the metric, decimal-inch, or fractional dimensioning system, there are two subsystems for orienting dimensions. These orientation subsystems are known as the aligned system and the unidirectional system. The aligned dimensioning system is illustrated in *Figure 10-7*. In this system the dimensions are aligned with the dimension line so that they can be read either from the bottom of the drawing sheet or from the right-hand side. Notice in Figure 10-7 the area that has been marked off to be avoided if possible. Dimensions written in this area are difficult to read from either the bottom or the right-hand side of the drawing sheet and, therefore, should be

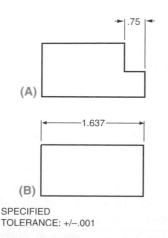

SPECIFIED
TOLERANCE: +/–.001

FIGURE 10-4   *Decimal-inch dimensioning*

$$\frac{1}{8} \qquad \frac{1}{8}$$

CORRECT   *INCORRECT*

FIGURE 10-5   *Horizontal line is the correct method*

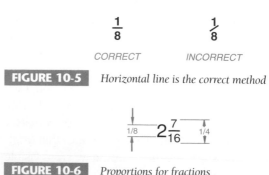

FIGURE 10-6   *Proportions for fractions*

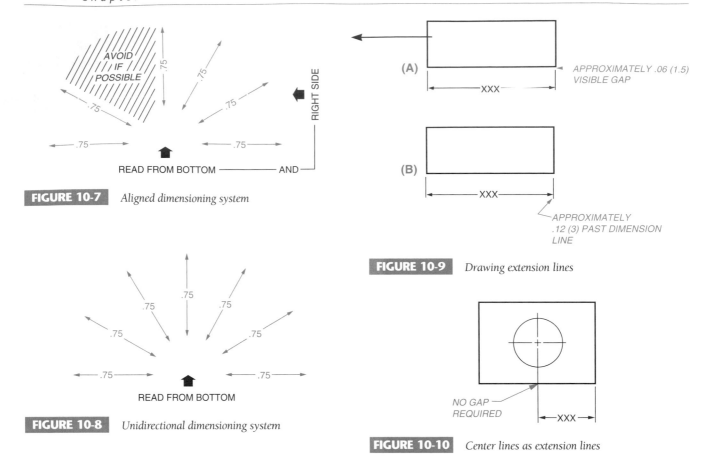

FIGURE 10-7 *Aligned dimensioning system*

FIGURE 10-8 *Unidirectional dimensioning system*

FIGURE 10-9 *Drawing extension lines*

FIGURE 10-10 *Center lines as extension lines*

avoided. The unidirectional dimensioning system is illustrated in *Figure 10-8*. In this system all dimensions are written so that they can be read from the bottom of the drawing sheet. This means that the guidelines used in entering the dimensions should always be parallel with the bottom of the page.

Dimension Components

Several components are common to all dimensioning systems. These include extension lines, dimension lines, arrowheads, leader lines, and the actual numbers or dimensions. Drafters and engineers should be knowledgeable in the proper use of these components.

EXTENSION LINES

An *extension line* is a thin, solid line that extends from the object in question or from some feature of the object. Several rules should be observed when placing extension lines on technical drawings. There should be a small but visible gap (approximately .06″) between the object or object feature and the beginning of an extension line, *Figure 10-9A*. Extension lines should extend uniformly beyond dimension lines for a distance of approximately one-eighth inch, *Figure 10-9B*. Extension lines that originate on the object, such as center lines, may cross visible lines with no gap required, *Figure 10-10. Figure 10-11* is

a fully dimensioned mechanical drawing that summarizes the basic rules drafters and engineers should remember about the spacing of extension lines, dimension lines, and dimensions. Notice in this figure that when center lines become extension lines there is no gap required as they cross object lines. Notice also that leader lines are not broken when they cross object lines. Note the visible gap (approximately .06) between the object itself and the beginning of extension lines. Note the .12 or approximately one-eighth-inch overhang of the extension line beyond dimension lines. Suggested spacing of dimension lines and distances between views are summarized in the chart in Figure 10-11.

DIMENSION LINES

A *dimension line* is a thin, solid line used to indicate graphically the linear distance being dimensioned. Dimension lines are normally broken for placement of the dimension, *Figure 10-12A*. If a horizontal dimension line is not broken, the dimension is placed above the dimension line with guidelines parallel to it, *Figure 10-12B. Figure 10-13* illustrates the proper methods to be used for locating dimensions when space is limited on a drawing. *Notice from this illustration that one dimension can be written and then leaders can be used to point to where the dimension applies. A dimension may be written within the ex-*

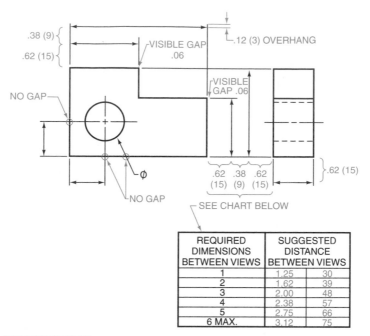

FIGURE 10-11 *Spacing for extension lines, dimension lines, and dimensions*

REQUIRED DIMENSIONS BETWEEN VIEWS	SUGGESTED DISTANCE BETWEEN VIEWS	
1	1.25	30
2	1.62	39
3	2.00	48
4	2.38	57
5	2.75	66
6 MAX.	3.12	75

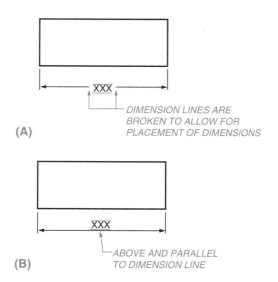

FIGURE 10-12 *Placement of dimensions*

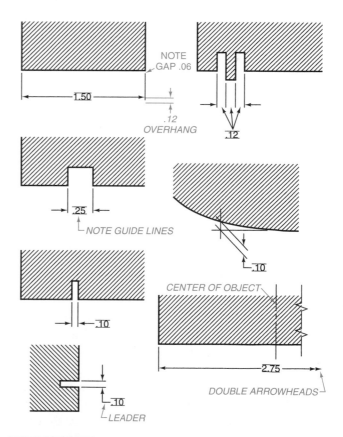

FIGURE 10-13 *Placement of dimensions when space is limited*

tension lines with dimension lines extending inward rather than outward. Dimension lines may be bent at a 90° angle, and extension lines, when necessary, can be drawn at an oblique angle. Double arrowheads can be used on a dimension line when a long or short break line has been used.

When dimensioning multiple features of an object, dimensions should be aligned uniformly rather than staggered or randomly scattered about the object, *Figure 10-14. Figure 10-15* illustrates another rule for the proper placement of dimensions. When dimensioning several views of the same object, always place the dimensions between the views, and always include the overall size, width, height, and depth of the object.

Dimension lines are drawn parallel to the direction of measurement. Sufficient distance between the object and the dimension lines, and between successive dimension lines, is important so that cramped and crowded dimensions do not result. First dimension lines should

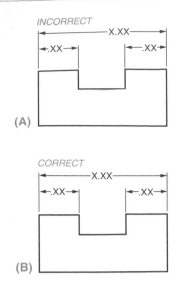

FIGURE 10-14 *Proper placement of dimensions when dimensioning multiple features*

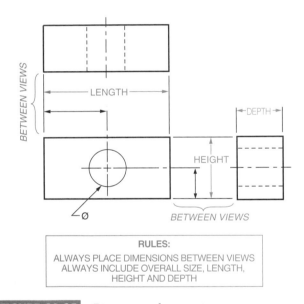

FIGURE 10-15 *Dimensioning between views*

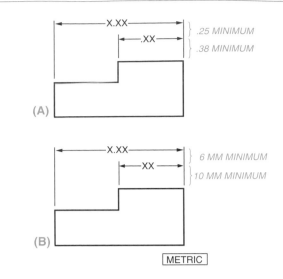

FIGURE 10-16 *Successive dimension lines*

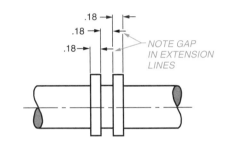

FIGURE 10-17 *Successive dimension lines in close areas*

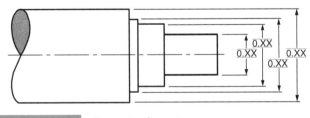

FIGURE 10-18 *Staggering dimensions*

be at least .38 inch from the object, *Figure 10-16A.* Successive dimension lines should be at least .25 inch apart. If using metric dimensions, the first dimension lines should be at least 10 mm away from the object. Successive lines should have at least 6 mm between them, *Figure 10-16B. Figure 10-17* illustrates the methods to be used when successive dimensions are called for in close areas. Notice how the dimension lines and corresponding dimensions are offset, and notice also the gap left in the extension lines to accommodate crossover of dimension lines.

When the shape of an object requires a series of parallel lines, the breaks and dimensions should be slightly staggered to make it easier to read the dimensions, *Figure 10-18. Figure 10-19* illustrates the proper way of

stacking a successive group of dimension lines. Drafters and engineers should always place the shortest dimensions closest to the objects, and always keep dimensions outside and away from the object, as illustrated in Figure 10-19.

LEADER LINES

A *leader line* is a thin line that begins horizontally, breaks at an angle, and terminates in an arrowhead, or, on occasion, a dot, *Figure 10-20.* Leaders that terminate at an edge or at some other specific point on a drawing should have an arrowhead. Leaders that terminate inside an object on a flat surface should end in a dot. The preferred angle of a leader line is from a minimum of 30° to

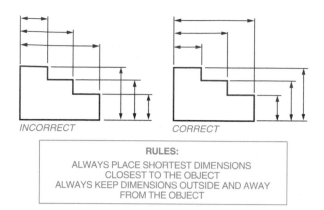

FIGURE 10-19   *Stacking dimension lines*

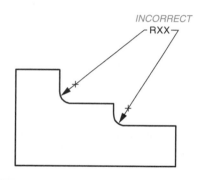

FIGURE 10-20   *Leader lines*

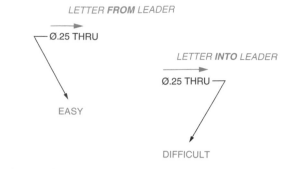

FIGURE 10-22   *Lettering and leaders*

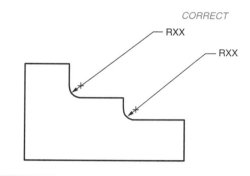

FIGURE 10-23   *One dimension per leader line is preferred*

a maximum of 60°, **Figure 10-21.** The horizontal portion of a leader should be from .125 inch to .25 inch in length, Figure 10-21. To avoid confusion, leader lines should not be drawn parallel to extension lines or dimension lines. When placing a dimension and/or note at the end of a leader line, it is easier to letter from the leader than into the leader, **Figure 10-22.**

When using a leader line to direct a dimension to its appropriate feature on a drawing, one dimension for each leader is the preferred method, **Figure 10-23.** More than one leader line extending from the same dimension can create a confusing situation that is difficult to interpret. Leader lines pointing to the center of a circle should be directed toward but not extended into the circle center, **Figure 10-24. Figure 10-25** illustrates the proper use of leader lines in a variety of dimensioning situations.

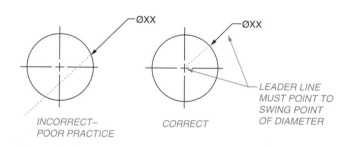

FIGURE 10-24   *Leader lines point at the center of a circle*

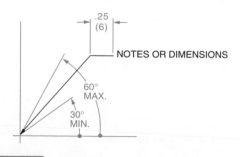

FIGURE 10-21   *Angle of a leader line*

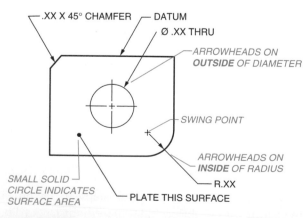

FIGURE 10-25   *Proper use of leader lines*

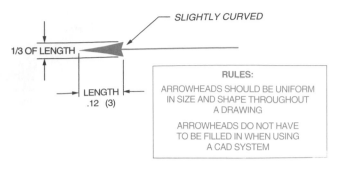

FIGURE 10-26 *Proper size and shape of arrowheads*

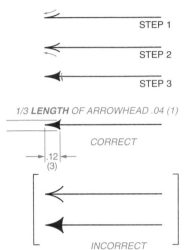

FIGURE 10-27 *Drawing arrowheads*

ARROWHEADS

An *arrowhead* is the most commonly used termination symbol for dimension and leader lines. Arrowheads should be approximately three times as long as they are wide. They should be large enough to be seen but small enough that they do not detract from the appearance of the drawing. A commonly accepted length for arrowheads is .12 inch. Arrowheads may be slightly larger or smaller than this, but, regardless of their size, they should be uniform throughout a drawing, *Figure 10-26*. Although the same standard applies to drawings prepared on a CAD system, arrowheads do not have to be filled in on drawings prepared using automated processes. Some CAD systems fill in arrowheads and some do not. Both open and filled-in arrowheads have been considered acceptable since the advent of CAD.

Figure 10-27 illustrates the three steps used in drawing arrowheads.

Laying Out Dimensions

Figure 10-28 illustrates the four steps in laying out dimensions for an object. First, place light layout lines on the drawing sheet. Second, draw in light guidelines

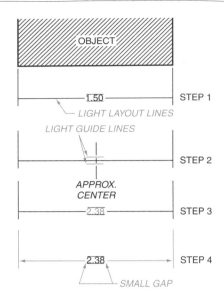

FIGURE 10-28 *Laying out dimensions*

for the dimensions. Third, darken your extension lines, dimension lines, and dimensions. Fourth, add arrowheads.

Steps in Dimensioning

There are two basic steps in dimensioning objects, regardless of the type of object.

STEP 1 Apply the size dimensions. These are dimensions that indicate the overall size of the object and the various features that make up the object.

STEP 2 Apply the locational dimensions. Locational dimensions are dimensions that locate various features of an object from some specified datum or surface. *Figure 10-29* gives examples of size and location dimensions.

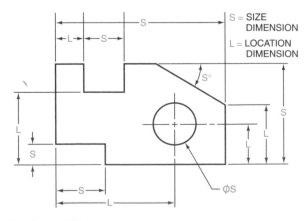

FIGURE 10-29 *Size and location dimensions*

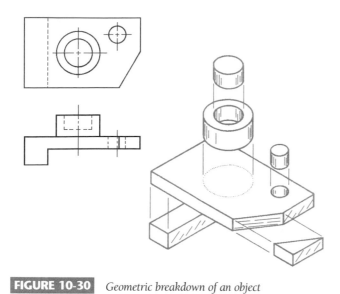

FIGURE 10-30 *Geometric breakdown of an object*

CR	CONTROLLED RADIUS
R	RADIUS
Ø	DIAMETER
2X	NUMBER OF TIMES–PLACES
⟱	DEEP OR DEPTH
∨	COUNTERSINK
⊔	COUNTERBORE OR SPOTFACE
□	SQUARE
( )	REFERENCE DIMENSION
.25	NOT TO SCALE DIMENSION
◿	SLOPE
▷	CONICAL TAPER
⌒	ARC LENGTH
SR	SPHERICAL RADIUS
SØ	SPHERICAL DIAMETER
X	BASIC DIMENSION
⟨ST⟩	STATISTICAL TOLERANCE

FIGURE 10-31 *Dimensioning symbols*

In order to properly dimension an object, drafters and engineers must be able to mentally break the object down into component parts and subelements. *Figure 10-30* illustrates graphically how this is done. On the left side of the figure is a two-dimensional representation of the object containing a top and front view. On the right side of the figure is a three-dimensional geometric breakdown of the object. This three-dimensional breakdown shows the drafter and engineer what geometric shapes must be sized using dimensions and which must be located using dimensions.

Specific Dimensioning Techniques

All of the information on dimensioning so far has been of a general nature. The following sections deal with techniques used for applying dimensions to specific situations that are recurrent in drafting. With a thorough knowledge of the general information presented earlier, and the specific information presented in these sections, drafters and engineers will be able to dimension any situation confronted on technical drawings.

DIMENSIONING SYMBOLS

There are a number of symbols used in different dimensioning situations with which drafters and engineers should be familiar. These symbols are summarized in *Figure 10-31*. Review these symbols carefully and commit them and their uses to memory before proceeding.

DIMENSIONING CHORDS, ARCS, AND ANGLES

Chords, arcs, and angles are dimensioned in a similar manner. When dimensioning a chord, the dimension

line should be perpendicular and the extension lines parallel to the chord. When dimensioning an arc, the dimension line runs concurrent with the arc curve, but the extension lines are either vertical or horizontal. An arc symbol is placed above the dimension. When dimensioning an angle, the extension lines extend from the sides forming the angle, and the dimension line forms an arc. These methods are illustrated in *Figure 10-32*. *Figures 10-33* and *10-34* contain additional information relative to dimensioning angles.

Notice in Figure 10-33 that angles are normally written in degrees, minutes, and seconds. The symbols used to depict degrees, minutes, and seconds are also shown in this figure. Angular measurements may also be stated in decimal form. This is particularly advantageous when they must be entered into an electronic digital calculator. The key to converting angular measurements to decimal form is in knowing that each degree contains 60 minutes, and each minute contains 60 seconds. Therefore, converting a measurement stated in degrees,

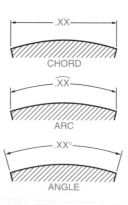

FIGURE 10-32 *Dimensioning chords, arcs, and angles*

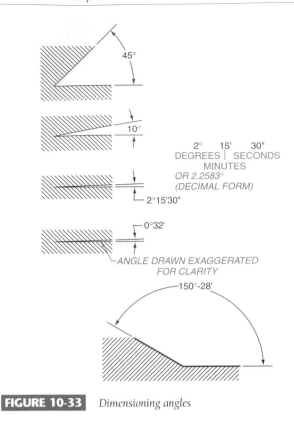

FIGURE 10-33 *Dimensioning angles*

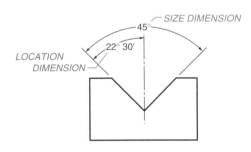

FIGURE 10-34 *Dimensioning angles*

minutes, and seconds into decimal form is a two-step process. Consider the example of the angular measurement 2 degrees, 15 minutes, and 30 seconds. This measurement would be converted to decimal form as follows:

STEP 1 The seconds are converted to decimal form by dividing by 60. Thirty seconds divided by 60 equals .50. The 15 minutes are added to this so that the minutes are expressed in decimal form or 15.50.

STEP 2 The minutes stated in decimal form are converted to decimal degrees by dividing by 60. The number 15.50 divided by 60 equals .25833. The 2 degrees are added to this number to have the measurement stated in decimal form or 2.2583 degrees.

Figure 10-34 illustrates how angles can be dimensioned for size and location. The size dimension gives the

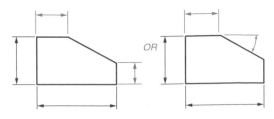

FIGURE 10-35 *Using normal linear dimensions for effecting the proper angle*

overall size of the angular cut. However, it does not locate the angular cut in the object. This is done by using a locational dimension off of the center line of the object. *Figure 10-35* shows how normal linear dimensions can be used for effecting the proper angle. When this method is used in the example on the left, all surfaces are dimensioned except the angular surface. When this method is used the angular surface is defined by default. In the method on the right, three linear dimensions and one angular dimension define the angular surface.

DIMENSIONING A RADIUS

Dimension lines used to specify a radius have one arrowhead, normally at the arc end. An arrowhead should not be used at the center of the arc. Where space permits, the arrowhead should be placed between the arc and the arc center with the arrowhead touching the arc. If space permits, the dimension is placed between the arc center and the arrowhead. If space is not available, the dimension may be placed outside the arc by extending the dimension line into a leader. The arc center of a radius is denoted with a small cross. *Figure 10-36* illustrates the preferred method for dimensioning a radius.

DIMENSIONING A FORESHORTENING RADIUS

On occasion the center of an arc radius will fall outside the drawing or will be so far removed from the drawing as to interfere with other views. When this is the case, the radius dimension line should be foreshortened and the radius center located using coordinate dimensions. This is done by relocating the arc center and placing a zig-zag in the radius dimension line, as shown in *Figure 10-37*. When this method is used it is important that the arc center actually lay upon the real center line of the arc.

DIMENSIONING CURVED SURFACES

Curved surfaces containing two or more arcs are dimensioned by showing the radii of all arcs and locating the arc centers using coordinate dimensions. Other radii may be located using their points of tangency. This method is illustrated in *Figure 10-38*.

RULES:
MOST RADIUS DIMENSION
ARROWHEADS ARE PLACED
INSIDE THE RADIUS EXCEPT
FOR SMALL RADIUS

R.XX
— SMALL RADIUS

OR AS A NOTE PLACED ABOVE TITLE BLOCK
NOTE: ALL UNMARKED RADII = R.XXX

FIGURE 10-36 *Dimensioning a radius*

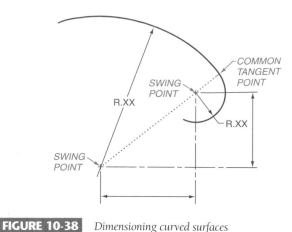

FIGURE 10-38 *Dimensioning curved surfaces*

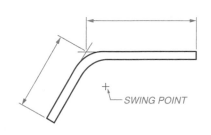

FIGURE 10-39 *Dimensioning offsets*

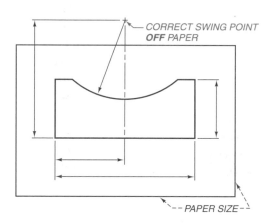

CORRECT SWING POINT
OFF PAPER

- - PAPER SIZE - -

BOTTOM LEG POINTS
TOWARD CORRECT
SWING POINT

INCORRECT SWING POINT
ON PAPER

- - PAPER SIZE - -

FIGURE 10-37 *Dimensioning a foreshortening radius*

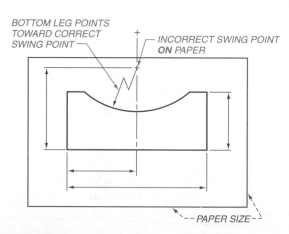

DIMENSIONING OFFSETS

Offsets are dimensioned from the points of intersection of the tangents along one side of the object, *Figure 10-39*. The distance from one end of the offset to the intersection is specified with a coordinate dimension. The distance from the other end to the offset is also specified with a coordinate dimension, as shown in Figure 10-39.

DIMENSIONING IRREGULAR CURVES

You have already seen in Figure 10-38 how to dimension curved surfaces using arc centers, radius dimensions, and tangent points. Irregular curves may also be dimensioned using coordinate dimensions from a specified datum. When this is the case, the coordinate dimensions extend from a common base line to specified points along the curve, *Figure 10-40*.

DIMENSIONING CONTOURS NOT DEFINED AS ARCS

Contours not defined as arcs can be dimensioned by indicating X–Y coordinates at points along the surface of the contour. Each of these points, sometimes referred to as stations, is numbered. The X and Y coordinates for each station are tabulated and placed in table form under the drawing, *Figure 10-41*.

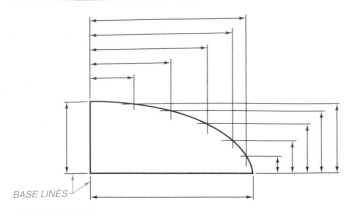

FIGURE 10-40 *Dimensioning irregular curves*

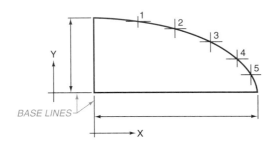

STATION	1	2	3	4	5
X	1.12	2.12	3.00	3.62	4.00
Y	1.75	1.50	1.12	.75	.31

FIGURE 10-41 *Dimensioning contours not defined as arcs*

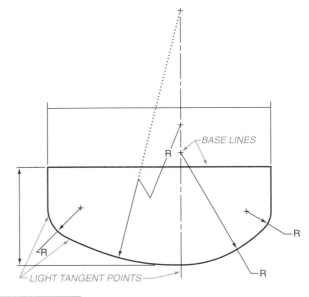

FIGURE 10-42 *Dimensioning multiple radii*

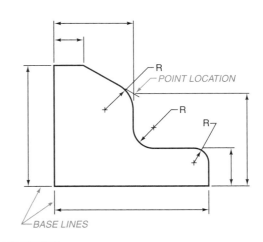

FIGURE 10-43 *Dimensioning multiple radii*

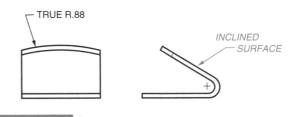

FIGURE 10-44 *Dimensioning multiple radii*

DIMENSIONING MULTIPLE RADII

When dimensioning an object that requires several radii, arcs should be dimensioned showing the radius in a view that gives the true shape of the curve. The dimension lines for a radius should be drawn as a radial line at an angle, rather than horizontally or vertically. Only one arrowhead is used. The R precedes the dimensional value when dimensioning in decimal-inch or the metric system. This method is illustrated in *Figures 10-42, 10-43,* and *10-44.* Notice in Figure 10-44 that where a radius is dimensioned in a view that does not show the true radius, a note should be used to indicate what the true radius is.

DIMENSIONING BY OFFSET (ROUND OBJECTS)

Another way to dimension a round object is the offset method. In this method, dimension lines are used as extension lines. Dimension lines are distributed across the object perpendicular to the center line of the object and spaced using coordinate dimensions, *Figure 10-45.*

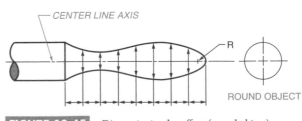

FIGURE 10-45 *Dimensioning by offset (round object)*

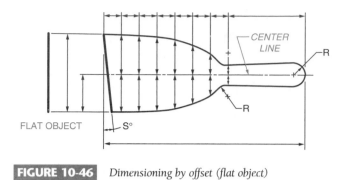

FIGURE 10-46   *Dimensioning by offset (flat object)*

DIMENSIONING BY OFFSET (FLAT OBJECTS)

Flat objects may also be dimensioned using the offset method. Again, dimension lines on the object become extension lines. Dimension lines are spaced across the object perpendicular to the center line of the datum and extended to become extension lines for the coordinate dimensions that space them, *Figure 10-46*.

DIMENSIONING SPHERES

Figure 10-47 illustrates the proper method for dimensioning spheres. When the diameter method is used, a leader points to the center of the sphere, and the diameter note is preceded with a capital S to indicate that it is a spherical diameter. When the radius method is used, a dimension extends from the arc center to the arc and is extended on with a leader, and the note is preceded by a capital SR to indicate a spherical radius.

DIMENSIONING ROUND HOLES

Round holes are dimensioned in the view in which they appear as circles, *Figure 10-48*. Holes may be dimensioned using a leader that points toward the center of the hole in which the note gives the diameter, or extension

RULES:

MOST DIAMETER DIMENSIONS PLACE ARROWHEAD OUTSIDE HOLE – EXCEPT FOR LARGE DIAMETER

ALWAYS CALL OFF DIAMETER SIZE NOT RADIUS SIZE

FIGURE 10-48   *Dimensioning round holes*

lines may be drawn from the circle with a dimension that also indicates the diameter. Larger circles are dimensioned with a dimension line drawn across the circle through its center at an angle with the diameter dimension shown. Except for very large holes, the arrowhead and the dimensional value are placed outside the hole. It is important when dimensioning holes to call off the diameter, not the radius.

SIMPLE HOLE CALLOUTS

Drafters and engineers need to know how to apply simple callouts to both through-holes and blind-holes. A through-hole is one that passes all the way through the object. A blind-hole is one that cuts into but does not pass through the object. Both types of holes, and the callout used for each, are illustrated in *Figure 10-49*. Machine holes are generally drilled to make the rough hole and then reamed to refine the hole. Figure 10-49 shows the difference between a drill and a ream. No hole is only reamed. A hole must be drilled before it can be reamed.

A through-hole callout has a leader line extending toward the center of the hole in the view in which the hole appears as a circle. The note attached to the leader gives the diameter of the hole and the word "thru" to indicate that the hole passes through an object. Blind-hole callouts are similar to through-hole callouts, except that the depth symbol is followed by the actual depth of the hole.

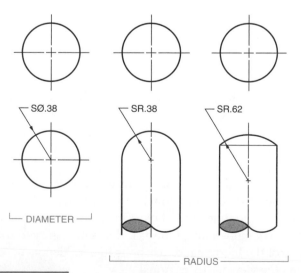

FIGURE 10-47   *Dimensioning spheres*

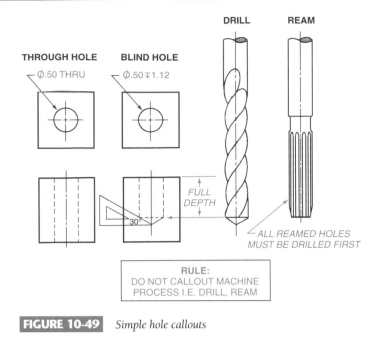

FIGURE 10-49 *Simple hole callouts*

DRILL SIZE TOLERANCING

Holes are not drilled to the exact size specified on a drawing. This is because there are several factors that mitigate against a perfectly sized hole. The accuracy of the drill, the tolerance level of the machine, and the qualifications of the machinist all have an impact on the actual size of a hole once it's drilled. It is accepted in manufacturing that no hole, even with the added accuracy provided by computer-aided manufacturing, is going to be drilled exactly the size specified. Therefore, drafters and engineers need to know how much variation to expect in a hole so that they can decide what limits to give the hole and whether the actual drilled hole will give the fit required in a given situation.

Drill size tolerance charts have been developed to assist drafters and engineers in determining the expected upper and lower limits of a drilled hole. Such charts are contained in Appendix A as Table 16. Turn to this table and you will notice that the standard drill size is given a number/letter, a fractional designation, a decimal designation, and a metric designation. To the right-hand side of the table the tolerance for each drill size is given in decimals. The left-hand column is the plus tolerance, and the right-hand column is the minus. To use this table, apply the following steps:

1. Find the letter or number of the drill in question on the left-hand side of the table.

2. Find the corresponding size for the drill in decimal form.

3. Find the plus tolerance for that size drill and add it to the decimal drill size. This will give you the upper limit for the hole size.

4. Find the minus tolerance for that drill size and subtract it from the decimal drill size. This will give you the lower limit for the hole.

5. Write the upper and lower limits with the smaller number on top separated from the larger number by a horizontal line. For example, an H-size drill has a decimal diameter of 0.2660. To find the upper limits for a hole using this drill size, add the plus tolerance of 0.0064 to this decimal drill size to get 0.2724 as the upper limit for the hole. To find the lower limit, subtract the minus tolerance of .002 from the decimal drill size to get a lower limit of 0.2640.

DIMENSIONING HOLE LOCATIONS

In addition to dimensioning the size of a hole, drafters and engineers must also dimension the locations of holes. *Figure 10-50* illustrates the proper methods for dimensioning the location of a hole. The preferred method is to dimension the location of the hole in the view where the hole appears as a circle. It is common practice to dimension from a reference side of the object to the center line of the hole. One should never dimension to a hidden line, however. If it is necessary to show one of the locational dimensions in a hidden view, convert the view into a sectional view.

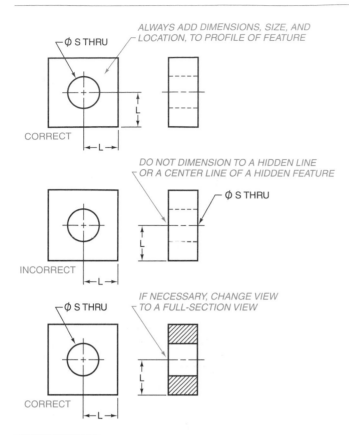

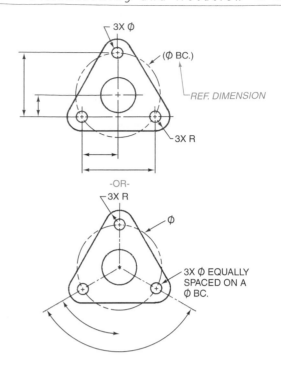

FIGURE 10-52   *Locating holes about a bolt center*

FIGURE 10-50   *Dimensioning hole locations*

DIMENSIONING HOLE LOCATIONS (MORE THAN ONE HOLE)

Figure 10-51 illustrates the proper method of dimensioning the locations of more than one hole within an object. When there is more than one hole in an object, what is important is the relationship of the holes to each other. Because of this, the dimensions applied show the distance horizontally and vertically between the center lines of the holes.

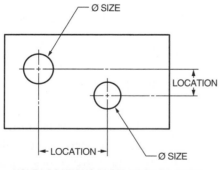

HOLE LOCATIONS IN RELATION TO EACH
OTHER MOST IMPORTANT

FIGURE 10-51   *Dimensioning hole locations (more than one hole)*

LOCATING HOLES ABOUT A BOLT CENTER

A common dimensioning situation is to have holes spaced about a bolt center. When this is the case the holes may be dimensioned using coordinate dimensions or angular dimensions, *Figure 10-52*. For the greatest accuracy, coordinate dimensions can be given from the center line of the bolt's center as well as center-to-center dimensions of the holes. This method is illustrated in the top half of Figure 10-52. Notice that the diameter of the bolt center is specified as a reference dimension. Reference dimensions are given only for informational purposes. They are not intended to be measured or to be used in dictating manufacturing operations. Another method that can be used for locating holes about a bolt center is illustrated in the bottom half of Figure 10-52. With this method the diameter of the bolt circle is shown and a note specifying the number of holes, their diameters, and the indication "equally spaced" is given. If the holes are unequally spaced, the same method can be used except that the note "equally spaced" is dropped and an angular measurement with reference to one of the center lines is indicated.

LOCATING HOLES ON CENTER LINES AND CONCENTRIC ARCS

When locating holes on common center lines, the center lines become base lines from which coordinate measurements can be made, as illustrated in the top half of

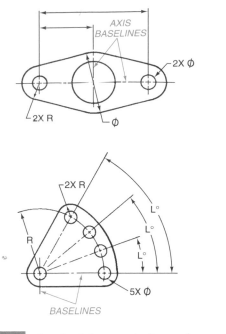

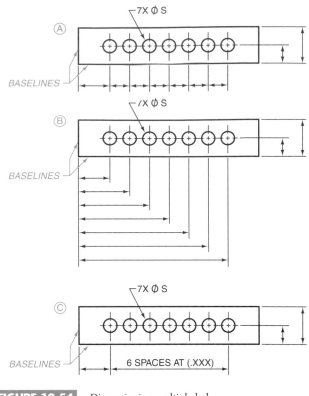

FIGURE 10-53 *Locating holes on center lines and concentric arcs*

FIGURE 10-54 *Dimensioning multiple holes*

Figure 10-53. All three holes in the object in this figure share the same horizontal center line. Consequently, the only dimensions required are dimensions from center to center of the circles. This is accomplished by providing a coordinate dimension from the center of a large circle to the center of one of the smaller circles, and the overall coordinate dimension between the centers of the two small circles, as illustrated. When centers are located on concentric arcs, the polar system of dimensioning can be used, as illustrated in the bottom half of Figure 10-53. When this is the case the radial distances from the point of concurrency and the angular measurements between the holes are used to locate the centers of the holes. The distance from the point of concurrency to the center of each hole is indicated by the radius, as shown.

DIMENSIONING MULTIPLE HOLES ALONG THE SAME CENTER LINE

Figure 10-54 shows the proper method used for dimensioning multiple holes along the same center line. The method shown at the top of Figure 10-54 is known as "chain dimensioning." The center line of the first hole is dimensioned from a reference base line, which is normally one side of the object. Then each successive center line is dimensioned from the immediate preceding center line. A problem with this type of dimensioning is that if one of the features in the chain is improperly located, all successive features will be improperly located. As a result, another method of dimensioning multiple holes or features is often used in place of chain

dimensioning. This dimensioning concept is known as "datum," or "base line" dimensioning. As is shown in Figure 10-54B, each successive center line is dimensioned from a base line or datum. In this way any one of the multiple features can be improperly located by the machinist without affecting the others. When this happens, the error can sometimes be corrected.

Another method that can be used for dimensioning multiple features within an object is illustrated at the bottom of Figure 10-54. With this method the center line of the first feature is dimensioned from a given datum, usually the edge of the object. Remaining features are dimensioned using a note that indicates the number of spaces at a given distance. This method does not have the disadvantage of needing an accurate placement of the first feature in order for each successive feature to be accurately located.

DIMENSIONING REPETITIVE FEATURES

Figure 10-55 illustrates the proper methods for dimensioning an object that contains repetitive features. Repetitive features may be located on a drawing using a note that specifies the number of times a given dimension is repeated, and giving one particular dimension and the total length as a reference. When dealing with repetitive features it is necessary to use both size and locational dimensions. Note the dimensions that have

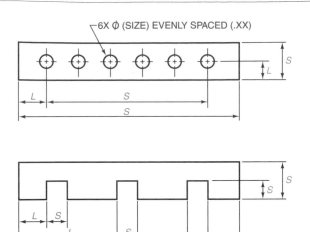

L = LOCATION DIMENSION
S = SIZE DIMENSION

FIGURE 10-55 *Dimensioning repetitive features*

been given in Figure 10-55. Some are size dimensions and some are location dimensions.

CALLOUTS FOR TAPERED HOLES

Occasionally the design of a manufactured part will require a tapered hole. *Figure 10-56* illustrates the proper method for annotating a tapered hole and shows an example of a 4° tapered bit. Callouts for tapered holes are similar to those for regular holes except that the note must also contain the taper symbol and the degree and the length of taper. The callout for the tapered hole in Figure 10-56 means that the subject hole has a .50 diameter at the top. It is drilled through and tapered at 4° for the full length of the hole.

CALLOUTS FOR COUNTERSUNK HOLES

Countersinking is a machining process to form a conical head so that a fastener can fit flush with the top of a part. The callout for a countersunk hole should contain the diameter of the countersink hole, which is the maximum diameter on the surface, and the angle of the countersink. The proper method for calling out a countersunk hole is illustrated in *Figure 10-57*. First a .50 diameter hole is drilled through the part. Then, using a countersink bit, the top of the hole is countersunk at 82° until the head of the countersink has a diameter of 1.12.

CALLOUTS FOR COUNTERBORED HOLES

On occasion a hole in a part is counterbored to allow the head of a bolt or another type fastener to be recessed. The bottom of a counterbore is flat rather than angled as in a countersink. The callout for a counterbored hole shows the diameter of the drilled hole, the counterbore symbol, the diameter of the counterbore itself, and the depth of the counterbore. When a hole is to be counterbored, the hole itself is drilled first and then the counterbore is applied using a special counterbore bit, as shown in *Figure 10-58*.

CALLOUTS FOR SPOTFACES

A spotface is a machining process used to finish the surface around the top of a hole so that a bolt or other type of threaded fastener can feed properly. A spotface resembles a very shallow counterbore. The callout for a spotface contains the diameter of the actual hole, and the diameter of the spotface. The counterbore symbol is also used for a spotface. The depth of a spotface is not normally shown. When a hole is to be spotfaced, the hole

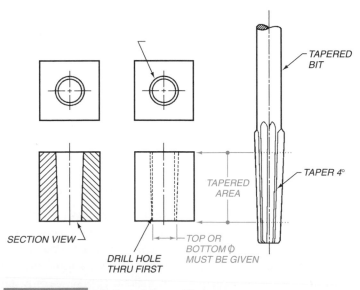

FIGURE 10-56 *Callouts for tapered holes*

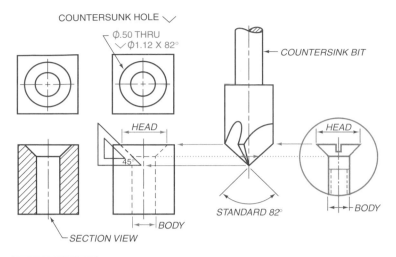

FIGURE 10-57 *Callouts for countersunk holes*

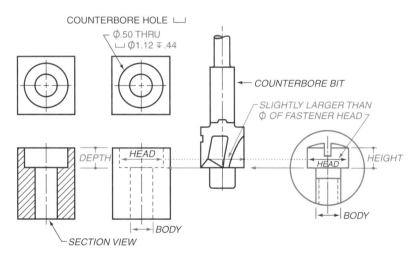

FIGURE 10-58 *Callouts for counterbored holes*

itself is drilled first and then the spotface is applied with a special spotface bit, as illustrated in *Figure 10-59*.

DIMENSIONING A CYLINDER

Figure 10-60 shows the proper way to dimension a cylinder. Notice from this figure that two methods can be used. In the first method the longitudinal view and the end view of the cylinder are given. When this method is used, the length of the cylinder is shown and the diameter dimension is applied between the longitudinal view and the end view. In the second method, the end view is left off. Consequently, the diameter dimension is applied to the longitudinal view. When dimensioning a cylinder, always specify the diameter rather than the radius.

DIMENSIONING A SQUARE

Figure 10-61 illustrates the proper method for dimensioning an object that is square in cross section. No center line on the object indicates that the object is square

rather than round in cross section. This type of object requires a length dimension on the longitudinal view and a height dimension placed between the longitudinal view and the end view, and preceded by the square symbol.

DOUBLE DIMENSIONING

Figure 10-62 illustrates the concept of double dimensioning. This is sometimes referred to as "superfluous" dimensioning. Drafters and engineers should avoid double dimensioning. It is redundant and can lead to confusion.

REFERENCE DIMENSIONING

Figure 10-63 illustrates the concept of reference dimensioning. In the object shown, the .62 dimension in parentheses is a reference dimension. Reference dimensions are not required. They are occasionally given for information purposes only. The old method used the letters REF following the dimension to indicate a reference

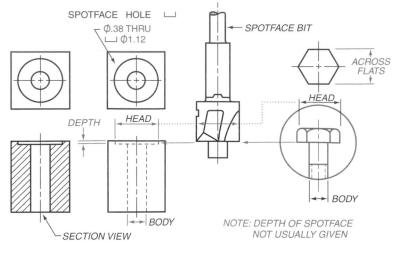

FIGURE 10-59 Callouts for spotfaces

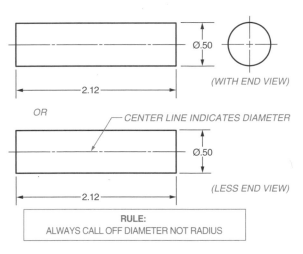

FIGURE 10-60 Dimensioning a cylinder

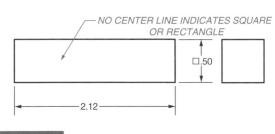

FIGURE 10-61 Dimensioning a square

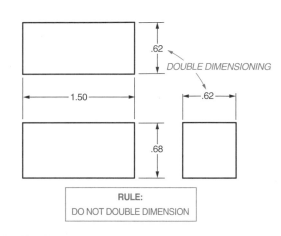

FIGURE 10-62 Double dimensioning

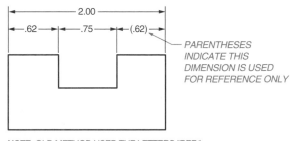

FIGURE 10-63 Reference dimensioning

dimension. The newer and accepted method is to place reference dimensions in parentheses. *Figures 10-64* and *10-65* further illustrate the concept of reference dimensions. In Figure 10-64, the second .25 dimension in the front view of the object is a double dimension; it is superfluous. In Figure 10-65, the problem has been solved by placing the second dimension in parentheses, thereby making it a reference dimension. The drafter or engineer had two options in this case for solving the problem: omit the dimension altogether or make it a reference dimension.

DIMENSIONING INTERNAL SLOTS

Internal slots represent a common manufacturing situation. Drafters and engineers approach internal slots as two partial holes separated by a space. There are a variety of methods for dimensioning internal slots. The preferred

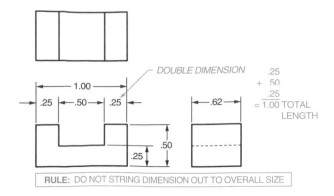

FIGURE 10-64 Reference dimensions

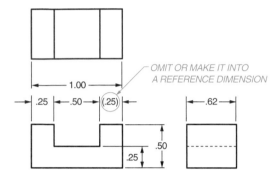

FIGURE 10-65 Reference dimensions

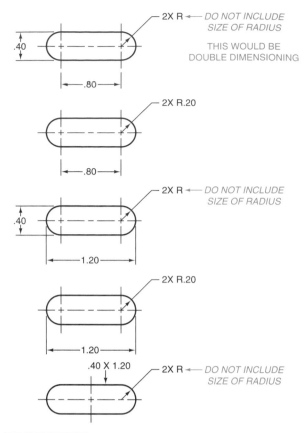

FIGURE 10-66 Dimensioning internal slots

methods are illustrated in *Figure 10-66*. Notice from these examples that when the width of the slot is dimensioned, there is no need to indicate the size of the radius. To do so would be double dimensioning. All of the methods shown in Figure 10-66 are acceptable methods and can be substituted for each other based on the needs of the individual drawing and local preferences.

Figures 10-67 and *10-68* are examples of how to dimension internal slots for two other manufacturing situations. Note in Figure 10-67 that the distance from the concurrent center point to the center line of the slot is indicated with a radius dimension. In addition, the angle formed by the extension of the 2-hole center lines to their intersection at the converging point must also be dimensioned as an angle. Figure 10-68 illustrates the proper method for dimensioning an internal slot when one of the holes is larger than the other. If a width dimension is provided, a radius value is not provided. If a width dimension is not provided, the radius dimension must be.

DIMENSIONING ROUNDED ENDS

Figure 10-69 illustrates the proper method for dimensioning the three most common situations involving external rounded ends. Parts with fully rounded ends should be dimensioned with an overall dimension and the radius should be indicated but not dimensioned, as

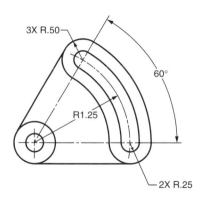

FIGURE 10-67 Dimensioning internal slots

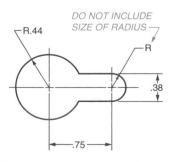

FIGURE 10-68 Dimensioning internal slots

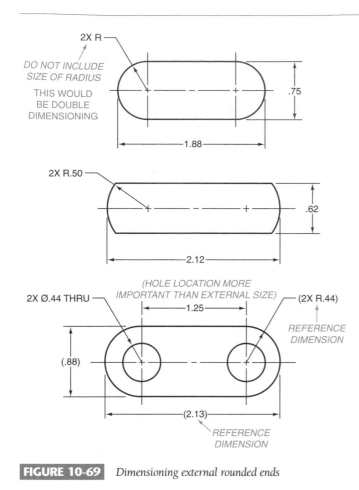

FIGURE 10-69 *Dimensioning external rounded ends*

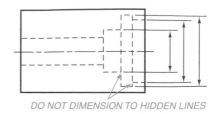

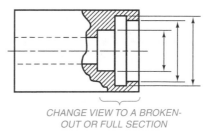

FIGURE 10-70 *Do not dimension to hidden lines*

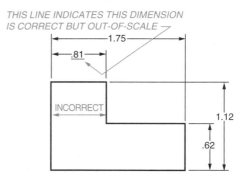

FIGURE 10-71 *Not to scale (NTS) dimensions*

indicated at the top of Figure 10-69. Figures with partially rounded ends, such as the example in the middle of Figure 10-69, require a radius specification. For parts with rounded ends and holes, overall dimensions can be given as reference dimensions, as shown in the bottom example of Figure 10-69.

DIMENSIONING AND HIDDEN LINES

There is only one rule for dimensioning and hidden lines:

DO NOT DIMENSION TO HIDDEN LINES

Figure 10-70 illustrates what drafters and engineers can do when faced with a situation where it is necessary to dimension hidden lines. When this is the case, and it cannot be avoided in any other way, convert the hidden-line view to a sectional view so that you are dimensioning to a solid line.

NOT TO SCALE (NTS) DIMENSIONS

On occasion it is necessary to enter a dimension not to scale. When this is the case, the not-to-scale dimension should be underlined with a solid straight line, as illustrated in *Figure 10-71*. The old method used a wavy

line under the dimension. This is no longer acceptable practice.

NOMINAL DIMENSIONS

A *nominal dimension* is one that represents a general classification size given to a part or product. It may not, and usually does not, express the actual size of the object dimensioned. There is usually a small amount of difference between the actual size of a part and the nominal size, as illustrated in *Figure 10-72*.

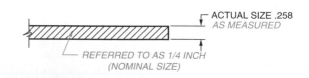

FIGURE 10-72 *Nominal dimensions*

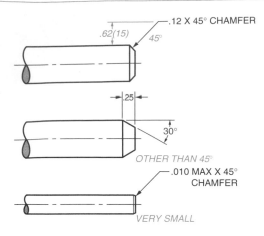

FIGURE 10-73 *Dimensioning external chamfers*

DIMENSIONING EXTERNAL CHAMFERS

A *chamfer* is a beveled edge normally applied to cylindrical parts and a variety of different types of fasteners. Chamfers are used to eliminate rough edges that might break off or impede the assembly of parts. *Figure 10-73* illustrates three accepted procedures for dimensioning external chamfers. The illustration at the top shows how to dimension a chamfer using a leader and note. This is frequently done when the chamfer is a 45° chamfer. The middle illustration shows how to dimension larger chamfers at angles other than 45°. Notice that the length of the chamfer is dimensioned, as is the angle of the chamfer. The bottom illustration shows how to dimension a 45° chamfer that is very small. The dimension is called out and followed by the term "MAX." This tells the machinist that a 45° chamfer is to be applied but its length is to be no more than .010.

DIMENSIONING INTERNAL CHAMFERS

Figure 10-74 illustrates two accepted methods for dimensioning internal chamfers. The example on the left illustrates how to call out a 45° chamfer. The exam-

ple on the right illustrates how to call out a chamfer other than 45°. Notice that when the chamfer is other than 45° the dimension of the top of the hole after the chamfer has been applied must be given, as must the angle of the chamfer.

STAGGERED DIMENSIONS

Figure 10-75 illustrates the proper method for applying staggered dimensions. A succession of parallel dimensions can become confusing and difficult to read. As a result, the accepted practice is to stagger the series of parallel dimensions as shown in the bottom half of Figure 10-75. Staggered dimensions are usually easier to read than aligned parallel dimensions.

DIMENSIONING NECKS AND UNDERCUTS

Necks and undercuts are recessed cuts into cylindrical parts. They are generally used where cylinders or other geometric shapes come together. *Figure 10-76* illustrates the proper method for dimensioning necks and undercuts. Notice from the three examples on the left side of Figure 10-76 that necks and undercuts may be dimensioned using a leader and accompanying note. The three examples on the right side of Figure 10-76 show that necks and undercuts may also be dimensioned using a combination of notes, angular dimensions, and size dimensions.

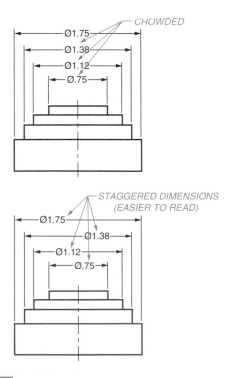

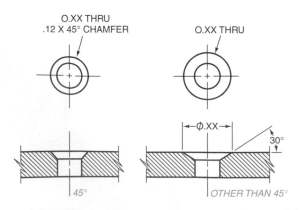

FIGURE 10-74 *Dimensioning external chamfers*

FIGURE 10-75 *Staggered dimensions*

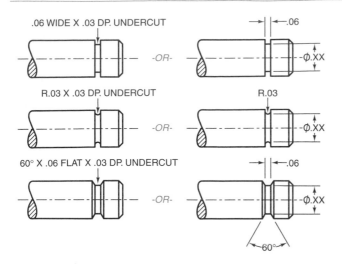

FIGURE 10-76 *Dimensioning necks and undercuts*

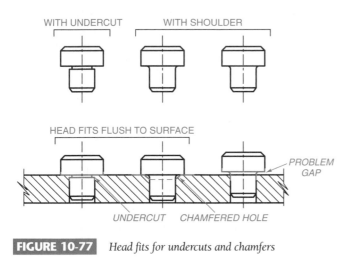

FIGURE 10-77 *Head fits for undercuts and chamfers*

HEAD FITS FOR UNDERCUTS AND CHAMFERS

Figure 10-77 illustrates the proper method for showing head fits for parts with undercuts and shoulders. With an undercut, the head fits flush with the surface. With a shoulder, the head will only fit flush with the surface if the surface has been chamfered.

DIMENSIONING KNURLS

Knurls are diamond-shaped or parallel patterns cut into cylindrical surfaces to improve gripping, to improve the bonding between parts for permanent press fits, and, sometimes, for decoration. The accepted method for dimensioning both parallel and diamond-shaped knurls is illustrated in *Figure 10-78*. Notice in this figure that when knurls are fully dimensioned on a drawing, there is no need to actually show them. The callouts for a knurl should include the type, pitch, and diameter. The initials DP in the knurl callouts stand for diametrical pitch. The most commonly used diametrical pitches for knurling are

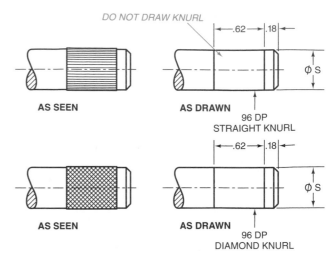

FIGURE 10-78 *Dimensioning knurls*

64 DP, 96 DP, 128 DP, and 160 DP. In addition to the knurling callout, the length of the knurl and the diameter of the cylinder that will receive the knurls must be shown.

DIMENSIONING KEYWAYS

Figure 10-79 illustrates the preferred methods for dimensioning both exterior and interior keyways. Notice that both size and location dimensions are required to completely dimension a keyway. The size dimension gives the overall width of the keyway, and the location dimension locates the keyway on the object in question from a specified datum such as the center line of the object. With an exterior keyway, the depth is shown using a dimension that runs from the bottom of the keyway to the bottom of the part in question and parallel to the center line or axis of the shaft. With an interior keyway, the depth of the keyway is specified using a dimension that runs from the top of the keyway to the bottom of the hole and parallel to the center line of the hole. *Figure 10-80* illustrates the preferred

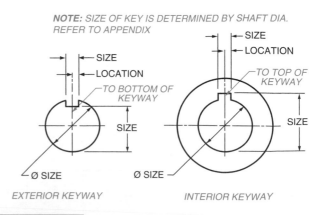

FIGURE 10-79 *Dimensioning keyways*

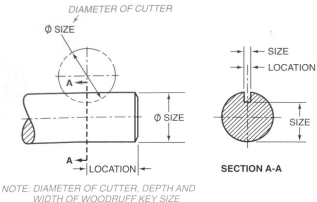

NOTE: DIAMETER OF CUTTER, DEPTH AND
WIDTH OF WOODRUFF KEY SIZE
IS FOUND IN APPENDIX A

FIGURE 10-80 *Dimensioning a woodruff keyway*

method for dimensioning a woodruff keyway. A woodruff keyway requires both size and location dimensions. The size dimensions specify the width and depth of the keyway. The location dimension from a specified datum such as a center line locates the keyway in the part as shown in Figure 10-80. In addition, the location of the center of the keyway from one end of the shaft must also be shown. The diameter of the shaft and the woodruff cutter must also be dimensioned, as shown in Figure 10-80.

DIMENSIONING FLAT TAPERS

Figure 10-81 illustrates the preferred method for dimensioning flat tapers. The method on the left uses 2-size dimensions and a leader with a taper symbol attached. In addition, the taper ratio must be shown. The 5:1 taper ratio called out for the part on the left means that the part tapers one unit in the vertical direction for every five units traveled in the horizontal. The example on the right illustrates another acceptable method for dimensioning flat tapers. In this method, 3-size dimensions provide all of the information needed to manufacture the part with the proper taper.

DIMENSIONING ROUND TAPERS

Figure 10-82 illustrates two acceptable methods for dimensioning round tapers. Round tapers can be dimen-

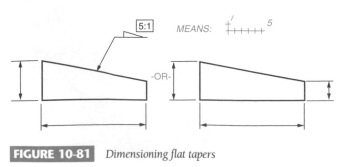

FIGURE 10-81 *Dimensioning flat tapers*

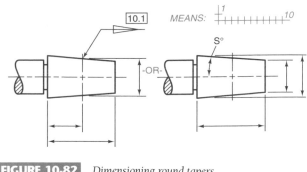

FIGURE 10-82 *Dimensioning round tapers*

sioned by providing the following dimensional data: (a) the diameter or width at both ends of the taper; (b) the length of the tapered feature; and (c) the rate of taper. The example on the right shows the diameter at both ends of the taper. The example on the left shows the length of the tapered feature and is in the taper ratio of 10:1, meaning that the part is tapered 1 unit for every 10 units parallel to the horizontal center line of the shaft.

DIMENSIONING THREADS

Figure 10-83 shows the preferred methods for dimensioning threaded fasteners. Notice that it is not necessary to attempt to draw an actual representation of the thread. Rather, the schematic approach shown in the middle and bottom examples is preferred. The threads themselves are called out in a thread note attached to a leader. In addition, the nominal diameter of the shaft being threaded and the longitudinal distance of the threading must also be dimensioned, as shown in Figure 10-83. A chamfer callout should be shown with the leader line to indicate how the end of the threaded fastener is to be chamfered.

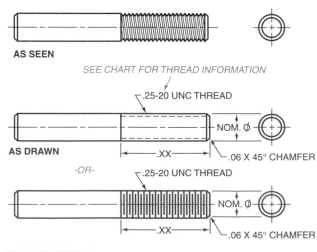

FIGURE 10-83 *Dimensioning threads*

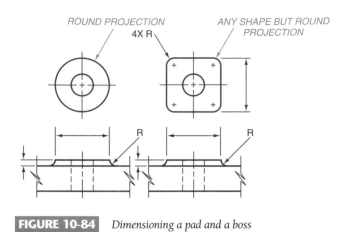

FIGURE 10-84   *Dimensioning a pad and a boss*

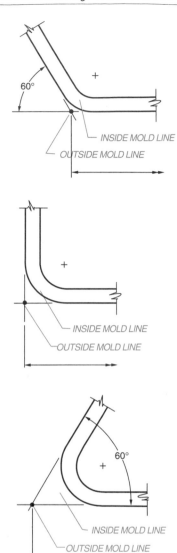

FIGURE 10-85   *Dimensioning sheet metal bends*

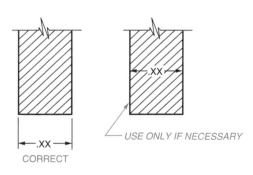

FIGURE 10-86   *Dimensioning sectional views*

DIMENSIONING A PAD AND A BOSS

Figure 10-84 illustrates the preferred method for dimensioning a pad and a boss. A boss is a round projection from a part. A pad is a projection of any shape other than round. Notice from Figure 10-84 that the boss requires a diameter dimension, a dimension showing how far the boss projects out from the part, and a radius indicator. The pad requires 2-size dimensions: one for length and one for width. It also requires a projection dimension and a radius indicator.

DIMENSIONING SHEET METAL BENDS

Figure 10-85 illustrates the different methods for dimensioning sheet metal bends. When dimensioning a sheet metal object, allowances must be made for the bends. The intersection of the surfaces adjacent to a bend is called the "mold line." This is the line that is used for dimensioning purposes. Notice in Figure 10-85 that each bend has an inside mold line and an outside mold line. Dimensions begin at the intersection of the outside mold line, as shown.

DIMENSIONING SECTIONAL VIEWS

Figure 10-86 illustrates the rule for dimensioning sectional views. The dimension should, whenever possible, be placed outside the object. Placing dimensions inside an object causes them to conflict with the section lines. When there is no other way to apply a dimension than to place it inside a sectioned area, remove enough of the section lining to permit the dimension to be easily read and the dimension lines to be easily seen, as shown in the right-hand example in Figure 10-86.

DIMENSIONING PYRAMIDS

Figure 10-87 illustrates the acceptable method for dimensioning a pyramid. A pyramid is dimensioned by showing the height in the front view and the dimensions of the base and the center of the vertex in the top view. If the pyramid base is square, it is only necessary to show the base dimension on one side of the base in the top view and to indicate that the base is square by using the appropriate symbol.

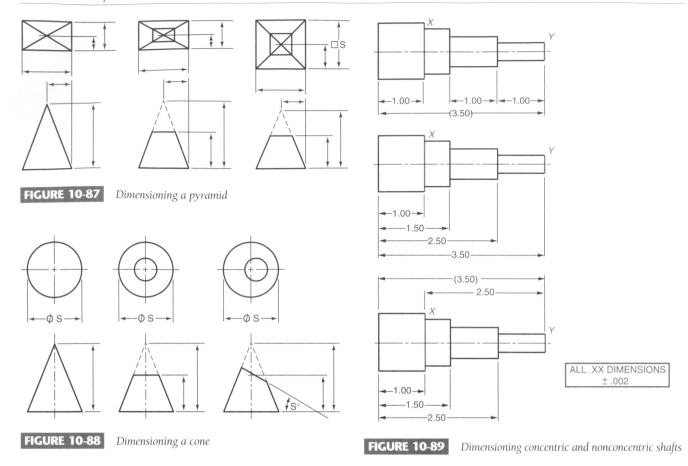

FIGURE 10-87 *Dimensioning a pyramid*

FIGURE 10-88 *Dimensioning a cone*

FIGURE 10-89 *Dimensioning concentric and nonconcentric shafts*

Dimensioning a Cone

Figure 10-88 illustrates the preferred methods for dimensioning a cone. A cone is dimensioned by showing its altitude in the view where the cone appears as a triangle and showing the diameter of the base between the top view and front view. The frustum of a cone is dimensioned by giving the diameter between the top and front view, and both heights in the front view where the cone would appear as a triangle. The example on the right-hand side of Figure 10-88 illustrates how to dimension a cone that is cut off at an angle. The diameter of the cone's base is shown between the top and front views. The height of the cone before it is cut and the height of the cone at the center line of the cone after the cut are shown in the front view where the cone would appear as a triangle. The angle of the cut also is dimensioned in this view.

Dimensioning Concentric and Nonconcentric Shafts

Figure 10-89 illustrates the various methods that can be used for dimensioning concentric and nonconcentric shafts and cylinders. The example at the top uses the chain dimensioning concept. The example in the center uses the datum or baseline dimensioning concept. The

example at the bottom uses the datum dimensioning concept combined with a reference dimension.

Rectangular Coordinate Dimensioning

Figure 10-90 illustrates the proper methods for applying rectangular coordinate dimensioning. Rectangular coordinate dimensioning is a system that locates features of the drawing relative to each other, individually or as a

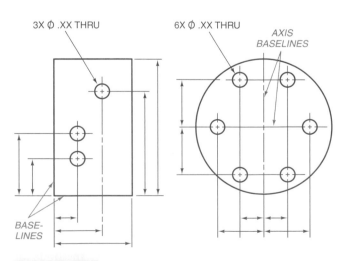

FIGURE 10-90 *Rectangular coordinate dimensioning*

group, from a datum or series of datums. Round holes and other similar features are located by dimensioning distances to their center points. Linear dimensions in the rectangular coordinate dimension system specify distances from two or three mutually perpendicular planes which are used as datums. Coordinate dimensions should clearly indicate what features of the part being dimensioned establish these datums.

DIMENSIONING HOLES ON A BOLT CENTER DIAMETER

Figure 10-91 illustrates the proper methods used in dimensioning holes on a bolt center diameter. In the example on the left, the diameter of the bolt's circle is given and one angle is shown to indicate the number of degrees between the center lines of the six circles. This is sufficient information for the machinist to properly drill the holes. In the example on the right, a note is used to accomplish the same thing. Notice that the note specifies the number and size of the holes. It explains that the holes are drilled through the object and equally spaced on a bolt center. Either of these methods may be used for dimensioning holes around a center. *Figure 10-92* shows two other methods that are used for dimensioning holes around a center. These methods are used when the holes around the bolt center are not equally spaced. The example on the left gives a dimension for the bolt center, a note containing the number of holes and their size, and angles turned from a reference center line for locating the center lines of the other three holes. The example on the right substitutes linear dimensions for angular dimensions.

FINISH MARKS

Figure 10-93 illustrates three different systems for applying finish marks to drawings. The traditional system and

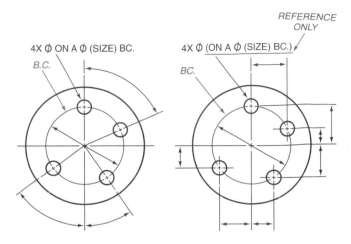

FIGURE 10-92   *Dimensioning holes around a center*

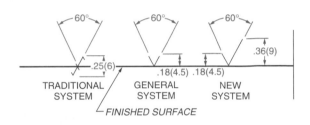

FIGURE 10-93   *Finish marks*

the general system have been replaced by the new system. However, engineers should be familiar with all three systems so that they are prepared to interpret finish marks applied to drawings produced in years past. Note the sizes specified when applying finish marks. A finish mark in the new system, which is the type of finish mark that should be used on all drawings now, resembles a mechanically produced check mark that forms a 60° "v." Finish marks are used for specifying the type of surface finish to be applied to any machined surface. Finish marks should always appear where the surface in question appears as an edge view. Finish marks are not used when dimensioning notes are given for such machining processes as drilling and boring, as they would be redundant. Finish marks are also omitted when the part in question is made of machined stock. When all surfaces on a part are finished, a note stating "Finish all over" or the letters FAO may be substituted.

Surface imperfections of a part are usually so small that they may be barely visible to the human eye. Consequently, they are measured in microinches. *Figure 10-94* illustrates the concept of the microinch.

Figure 10-95 illustrates the proper method for applying microinch callouts to finish marks. Notice that microinch callouts are not used with the traditional finish mark system.

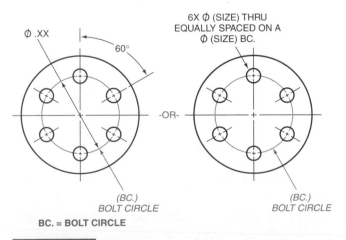

BC. = BOLT CIRCLE

FIGURE 10-91   *Dimensioning holes on a bolt center diameter*

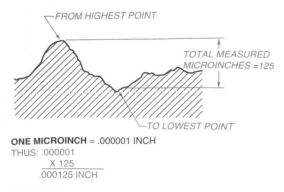

ONE MICROINCH = .000001 INCH
THUS: .000001
        X 125
      .000125 INCH

FIGURE 10-94 *Microinches*

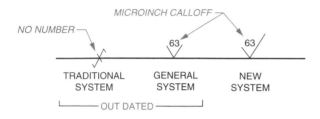

FIGURE 10-95 *Microinch callouts*

Figure 10-96 illustrates the rule for applying finish marks to drawings. Note that finish marks are to be repeated in every view where the surface in question appears as an edge. Table 46 in Appendix A shows the surface roughness textures obtainable by common production methods. Engineers should be familiar with the information in this table. The various manufacturing processes are listed along the left side of the table. The typical applications are listed along the bottom of the table, and the corresponding roughness height rating for each process/application is listed horizontally in a bar-graph format. *Figure 10-97* is a table of typical surface roughness height applications. Microinch ratings are listed in the left-most column of the table. The corresponding micrometer ratings are listed in the next column. Applications are then explained horizontally beside each rating. Engineering students should familiarize themselves with these ratings and their corresponding applications.

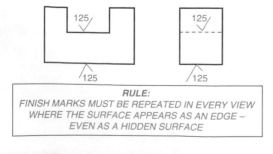

FIGURE 10-96 *Applying finish marks to drawings*

X-Y-Z Coordinates

Figure 10-98 illustrates the concept of X-Y-Z coordinates. In order to be able to fully understand the various dimensioning systems, one must understand the concept of X-Y-Z coordinates. This is particularly true in the case of tabular dimensioning. Note from this illustration that X coordinates move from an origin in a horizontal direction, Y coordinates move from an origin in a vertical direction, and Z coordinates move from an origin in a direction indicating depth.

Tabular Dimensioning

Figure 10-99 illustrates the concept of tabular dimensioning. Tabular dimensioning is a system in which size and location dimensions are given in a table rather than on the drawing itself. Notice that the only dimensions applied to the drawing itself are the overall length, width, and depth dimensions. Each hole is given an alphanumeric designation. Holes with the same letter designation have the same diameter. Holes with the same letter designation but a different numerical suffix have the same diameter but a different depth. The holes are located in the part using X, Y, and Z coordinates from the specified base lines. In certain cases such as the one illustrated in Figure 10-99, tabular dimensioning is a more convenient, less confusing approach.

Tabular Drawing

When parts have similar features but different dimensions, the dimensions can be given in tabular form to decrease the amount of drawing that must be done. With a tabular drawing, one drawing of the part is made with all appropriate dimension lines applied. However, instead of dimension notations, numerical notations are substituted. Then a table of dimensions is prepared showing the part numbers for all of the parts that the drawing applies to and the dimensions for each letter notation, *Figure 10-100*.

Tolerancing

Even the best machinist operating the most accurate machine tool is not able to produce a part as precisely and accurately as specified on a drawing. For this reason it is necessary to apply tolerances when dimensioning parts. A *tolerance* is the total amount of variation permitted from the design size of a part. A basic rule of thumb when applying tolerances is to make them as large as possible and still be able to produce a usable part. Tolerances may be expressed as limits, as shown in *Figures 10-101* and *10-102* or as the design size followed by the tolerance (12.382 ± 0.003). The design size of a part is also known as the nominal size. This is the

MICROMETERS AA RATING	MICROINCHES AA RATING	APPLICATION
25.2	1000	Rough, low-grade surface resulting from sand casting, torch or saw cutting, chipping, or rough forging. Machine operations are not required because appearance is not objectionable. This surface, rarely specified, is suitable for unmachined clearance areas on rough construction items.
12.5	500	Rough, low-grade surface resulting from heavy cuts and coarse feeds in milling, turning, shaping, boring, and rough filing, disc grinding, and snagging. It is suitable for clearance areas on machinery, jigs, and fixtures. Sand casting or rough forging produces this surface.
6.3	250	Coarse production surfaces, for unimportant clearance and cleanup operations, resulting from coarse surface grind, rough file, disc grind, rapid feeds in turning, milling, shaping, drilling, boring, grinding, etc., where tool marks are not objectionable. The natural surfaces of forgings, permanent mold castings, extrusions, and rolled surfaces also produce this roughness. It can be produced economically and is used on parts where stress requirements, appearance, and conditions of operations and design permit.
3.2	125	The roughest surface recommended for parts subject to loads, vibration, and high stress. It is also permitted for bearing surfaces when motion is slow and loads light or infrequent. It is a medium commercial machine finish produced by relatively high speeds and fine feeds taking light cuts with sharp tools. It may be economically produced on lathes, milling machines, shapers, grinders, etc., or on permanent mold castings, die castings, extrusion, and rolled surfaces.
1.6	63	A good machine finish produced under controlled conditions using relatively high speeds and fine feeds to take light cuts with sharp cutters. It may be specified for close fits and used for all stressed parts, except fast rotating shafts, axles, and parts subject to severe vibration or extreme tension. It is satisfactory for bearing surfaces when motion is slow and loads light or infrequent. It may also be obtained on extrusions, rolled surfaces, die castings, and permanent mold castings when rigidly controlled.
0.8	32	A high-grade machine finish requiring close control when produced by lathes, shapers, milling machines, etc., but relatively easy to produce by centerless, cylindrical, or surface grinders. Also, extruding, rolling, or die casting may produce a comparable surface when rigidly controlled. This surface may be specified in parts where stress concentration is present. It is used for bearings when motion is not continuous and loads are light. When finer finishes are specified, production costs rise rapidly; therefore, such finishes must be analyzed carefully.
0.4	16	A high-quality surface produced by fine cylindrical grinding, emery buffing, coarse honing, or lapping, it is specified where smoothness is of primary importance, such as rapidly rotating shaft bearings, heavily loaded bearing, and extreme tension members.
0.2	8	A fine surface produced by honing, lapping, or buffing. It is specified where packings and rings must slide across the direction of the surface grain, maintaining or withstanding pressures, or for interior honed surfaces of hydraulic cylinders. It may also be required in precision gauges and instrument work, or sensitive value surfaces, or on rapidly rotating shafts and on bearings where lubrication is not dependable.
0.1	4	A costly refined surface produced by honing, lapping, and buffing. It is specified only when the design requirements make it mandatory. It is required in instrument work, gauge work, and where packing and rings must slide across the direction of surface grain such as on chrome plated piston rods, etc. where lubrication is not dependable.
0.05	2	Costly refined surfaces produced only by the finest of modern honing, lapping, buffing, and superfinishing equipment. These surfaces may have a satin or highly polished appearance depending on the finishing operation and material. These surfaces are specified only when design requirements make it mandatory. They are specified on fine or sensitive instrument parts or other laboratory items, and certain gauge surfaces, such as precision gauge blocks.
0.025	1	

FIGURE 10-97 *Typical surface roughness height applications*

dimension from which the limits are calculated. To calculate the upper limit of a toleranced dimension, add the allowable tolerance to the nominal size. To calculate the lower limits for such dimensions, subtract the allowable tolerance from the nominal size, Figures 10-101 and 10-102. There are several definitions with which drafters and engineers should be familiar before applying tolerances on drawings. The most important of these are:

Actual Size. Also called the produced size, this is the actual size of the part after it is produced.

Allowance. The amount of room between two main parts, usually considered the tightest possible fit between the two parts, is called the *allowance*.

Bilateral Tolerance. A bilateral tolerance is one in which variation from the nominal dimension is allowed in both the plus and the minus directions (.500 ±.002).

Datum. A theoretically perfect surface, plane, axis, center plane, or point from which dimensions for related features are established.

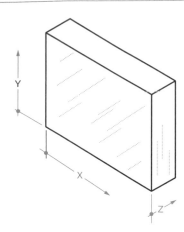

FIGURE 10-98 *X-Y-Z coordinates*

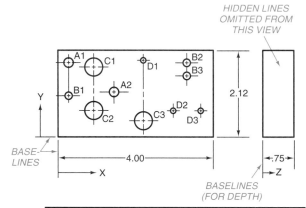

FIGURE 10-99 *Tabular dimensioning*

HOLE	⌀	X	Y	Z	NOTE
A1	.25	.30	1.75	THRU	—
A2		1.38	1.06	.38	—
B1	.18	.30	.94	THRU	—
B2		3.25	1.75	THRU	—
B3		3.25	1.50	THRU	—
C1	.38	.88	1.68	.38	—
C2		.88	.62	.25	—
C3		1.12	.38	.38	—
D1	.08	2.12	1.81	THRU	—
D2		3.00	.62	.38	—
D3		3.62	.62	.38	—

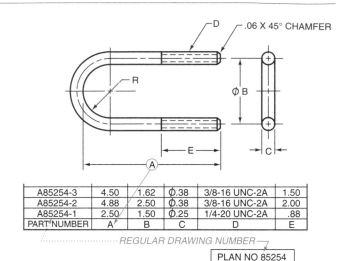

FIGURE 10-100 *Tabular drawing*

A85254-3	4.50	1.62	⌀.38	3/8-16 UNC-2A	1.50
A85254-2	4.88	2.50	⌀.38	3/8-16 UNC-2A	2.00
A85254-1	2.50	1.50	⌀.25	1/4-20 UNC-2A	.88
PART NUMBER	A	B	C	D	E

REGULAR DRAWING NUMBER

PLAN NO 85254

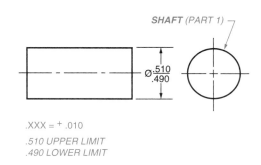

.XXX = + .010
.510 UPPER LIMIT
.490 LOWER LIMIT

FIGURE 10-101 *Shaft limits*

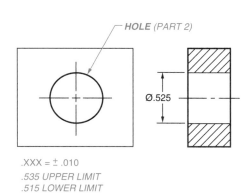

.XXX = ± .010
.535 UPPER LIMIT
.515 LOWER LIMIT

FIGURE 10-102 *Hole limits*

Least Material Condition (LMC). The condition which exists when the least material is available for the features being dimensioned. *Least material condition (LMC)* is the lower limit for an external feature and the upper limit for an internal feature.

Limits. The largest and smallest possible size in which a part may be produced and still be usable. The limits are calculated by applying the specified tolerance to the nominal dimension.

Maximum Material Condition (MMC). The condition of a part in which the most material is present. *Maximum material condition (MMC)* is the upper limit for an external feature and the lower limit for an internal feature.

Nominal Dimension. This is the dimension from which the limits are calculated. By applying the specified tolerance to the nominal dimension one can calculate the upper and lower limits.

Tolerance. The total permissible variation in size for a part. The tolerance is the difference between the upper and lower limits for a part.

Unilateral Tolerance. A tolerance that may vary in only one direction; either in the plus or minus direction but not both:

$$.500 \begin{matrix} +.000 \\ -.002 \end{matrix} \qquad .250 \begin{matrix} +.002 \\ -.000 \end{matrix}$$

In addition to these tolerance-related definitions, drafters and engineers should also be familiar with the types of fits that are used with mating parts. The following definitions relate specifically to fits between mating parts.

Fit. Fit refers to the amount of tightness or looseness between mating parts. There are three types of fits: clearance, interference, and transition.

Clearance Fit. A clearance fit is one in which there is space remaining after mating parts have been assembled.

Interference Fit. The type of fit that occurs when there is a negative allowance, or in other words, when the shaft is larger than the collar with which it must mate.

Transition Fit. The type of fit in which there may be either a clearance or an interference fit after mating parts have been assembled. This occurs when the smallest shaft size within the allowable tolerance will fit within the largest hole size, resulting in a clearance fit, and when the largest shaft size within the allowable tolerance will interfere with the smallest hole within an allowable tolerance, creating an interference fit.

SHAFT LIMITS

Figure 10-101 illustrates the concept of shaft limits. In this figure the shaft has a nominal dimension of .500. The tolerance specified is ±.010. To determine the upper limit for the shaft, the tolerance of .010 is added to the nominal dimension of .500, resulting in an upper limit of .510. To determine the lower limit of the shaft, the tolerance of .010 is subtracted from the nominal dimension of .500, resulting in a lower limit of .490.

HOLE LIMITS

Figure 10-102 illustrates the concept of hole limits. The hole through the part shown has a nominal diameter of .525. The specified tolerance is ±.010. To determine the upper limit for the hole, the tolerance of .010 is added to the nominal dimension of .525, resulting in an upper limit of .535. The lower limit is determined by subtracting the tolerance of .010 from the nominal dimension of .525, resulting in a lower limit of .515.

ALLOWANCE

Figure 10-103 illustrates the concept of allowance or the tightest fit between mating parts. Part 1 is a shaft that, when assembled, will mate with Part 2. To determine the

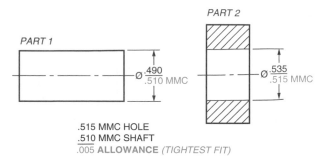

FIGURE 10-103 *Allowance*

allowance between Part 1 and Part 2, the maximum material condition of the shaft (the largest allowable size) is subtracted from the maximum material condition of the hole (the smallest size hole). The maximum material condition of the shaft is .510. The maximum material condition of the hole is .515. Therefore, the allowance between Part 1 and Part 2 is .005.

CLEARANCE FIT

Figure 10-104 illustrates the concept of *clearance fit.* The shaft, Part 1, is to mate with the collar, Part 2, during assembly. To determine the clearance between Part 1 and Part 2, the least material condition of the shaft is subtracted from the least material condition of the hole. The least material condition of the shaft is .490. The least material condition of the hole is .535. Therefore, the clearance between Part 1 and Part 2 is .045.

INTERFERENCE FIT

Figure 10-105 illustrates the concept of *interference fit.* The smallest shaft in this example is larger than the largest corresponding hole. To determine the amount of interference, you can calculate the allowance and the clearance. The allowance, in the case of an interference fit, represents the maximum amount of interference. The clearance, in an interference fit, represents the smallest amount of interference. You can see from the calculations in Figure 10-105 that the maximum

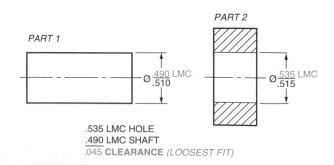

FIGURE 10-104 *Clearance*

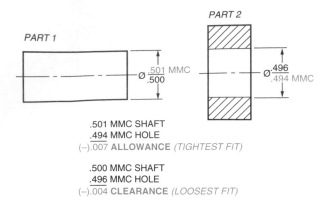

FIGURE 10-105 *Interference fit*

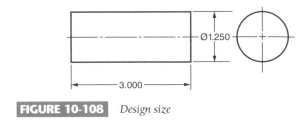

FIGURE 10-108 *Design size*

amount of interference during assembly will be .007, or the allowance. The minimum amount of interference will be .004, or the clearance.

TRANSITION FIT

Figure 10-106 illustrates the concept of *transition fit*. In this example there can be both a clearance fit and an interference fit. The largest shaft in this example would interfere with the smallest hole. However, the smallest shaft will fit through the largest hole. This condition is illustrated in the calculations.

SIZE LIMITS

Figure 10-107 illustrates the concept of size limits. The cylindrical part shown in this figure will be usable if pro-

duced at any size between the upper and lower length limits, and at any diameter between the upper and lower diameter limits, inclusive. Applying size limits such as those illustrated in Figure 10-107 cuts down on waste, requires less time to produce the part, and, therefore, cuts the cost of production and the corresponding costs of the part.

DESIGN SIZE

Figure 10-108 illustrates the concept of design size. Design size is another way of stating nominal size. The design size represents the ideal the drafter or engineer is seeking. Ideally, the produced size will match the design size. However, since this is not practical, tolerances are normally applied.

MAXIMUM AND MINIMUM SIZES

Figure 10-109 illustrates the concept of maximum and minimum sizes. These are the sizes between which the actual length and diameter of the produced part must fall. After all machining processes have been completed, the inspection department will check the part. If the actual produced size falls within the established maximum and minimum sizes (inclusive), the part will be passed as usable. If it is smaller in any dimension than the minimum size, the part will be scrapped as waste. If it is larger in any dimension than the maximum size, it might be salvaged through additional machining.

LOCATION LIMITS

Just as limits are applied to size dimensions, they are also applied to location dimensions, and for the same rea-

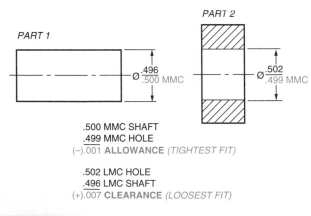

FIGURE 10-106 *Transition fit*

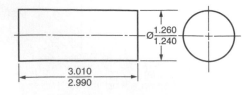

FIGURE 10-107 *Size limits*

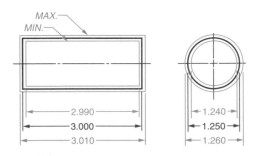

FIGURE 10-109 *Maximum and minimum sizes*

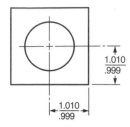

FIGURE 10-110 *Location limits*

sons. Machinists cannot locate features on a part precisely to the design dimension. Consequently, limits must be applied to location dimensions.

Figure 10-110 illustrates how limits are applied to location dimensions. The part shown in Figure 10-110 will be accepted if the central line of the hole falls within the limits shown.

DESIGN LOCATION

Figure 10-111 illustrates the concept of design location. The design location is the location the drafter or engineer actually wants. It is the theoretically ideal location of the feature. The dimensions shown on this figure are the design dimensions, sometimes referred to as nominal dimensions. The specified tolerance is applied to these nominal dimensions to determine the upper and lower limits for locating the feature.

MAXIMUM AND MINIMUM LOCATIONS

Figure 10-112 illustrates the concept of maximum and minimum locations. By applying the specified tolerance to the nominal dimensions, the maximum and minimum locations of the center line of the feature in this figure can be determined. The tolerances applied to the nominal dimension give the machinist latitude in both the X and Y directions. This establishes a square tolerance zone within which the actual center point must fall. So long as the center of the feature falls anywhere within the prescribed tolerance zone, the parts will be accepted as usable.

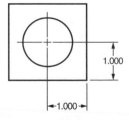

FIGURE 10-111 *Design location*

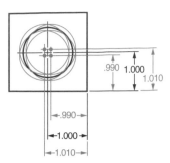

FIGURE 10-112 *Maximum and minimum locations*

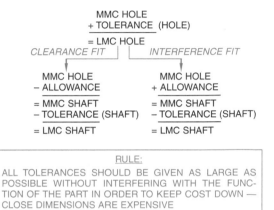

FIGURE 10-113 *Calculating fits*

CALCULATING FITS

Figure 10-113 illustrates the mathematics involved in calculating fits. Take special note of the rule that states "All dimensions should be given as large as possible without interfering with the function of the part in order to keep costs down—close dimensions are expensive." The tighter the fit, the more difficult the part is to manufacture. The more difficult the part is to manufacture, the more it costs to manufacture. Therefore, an engineer and a drafter should never assign a tighter fit than is absolutely necessary for the part to properly function.

MATCHING PARTS

Figure 10-114 illustrates the concept of matching parts. During assembly, Part A must fit over Part B. You can see from this drawing that the hole in Part A must be large enough to accommodate the size and location tolerances of the pins in Part B. The parts must be able to fit if the holes in Part A are bored and located on the low side of the tolerance and the pins in Part B are made and located on the high side of the tolerance. Obviously, determining the dimensions to be used for matching parts takes a good deal of careful planning. It is critical that features on mating parts be located from the same datum as shown in this figure. Some additional factors

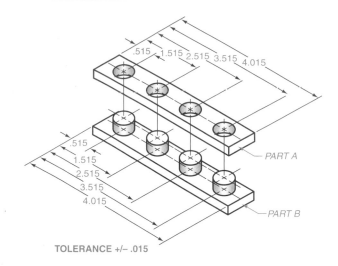

TOLERANCE +/− .015

FIGURE 10-114 *Matching parts*

to consider when planning the dimensions for matching parts are:

1. The length of time the parts are to be engaged
2. The speed at which mating parts will move after assembly
3. The load of the parts after assembly
4. The type of lubrication used if lubrication is necessary
5. The average temperature at which parts will operate
6. The level of humidity during operation
7. The materials of which parts are made
8. The required functional lifespan of the part
9. The capabilities of machine tools and the people that operate them in the manufacturing plant
10. Overall costs of the part

Standard Fits

Figure 10-115 shows the different types of standard fits. The American National Standards Institute document governing fits is ANSI B4.1-1967, R1994. This standard covers the five types of fits listed in Figure 10-115. Each of these fits is explained in this section.

STANDARD FITS	
RC	RUNNING AND SLIDING FITS
LC	CLEARANCE LOCATIONAL FITS
LT	TRANSITION LOCATIONAL FITS
LN	INTERFERENCE LOCATIONAL FITS
FN	FORCE AND SHRINK FITS

FIGURE 10-115 *Standard fits*

Running or Sliding Clearance Fits (RC). Running or sliding fits are used to provide a similar running or sliding performance, with an allowance for suitable lubrication, for all sizes within the specified range. *Figure 10-116* is a table of American National Standard Running and Sliding Fits (ANSI B4.1-1967, R1994). From this table you can see that there are nine classes of running and sliding fits: Classes RC1, RC2, RC3, RC4, RC5, RC6, RC7, RC8, and RC9.

Class RC1 is used with close sliding fits for accurately locating parts that must fit together without any play. Class RC2 is used for sliding fits where accurate location is important, but where greater maximum clearance than is provided by class RC1 fits is necessary. Parts manufactured to this fit do not run freely, but they do move and turn easily. Class RC3 is used for precision running fits. This class ensures the closest possible fit that can be expected to run freely. This fit is used for precision work that will operate at slow speeds under light pressures, but it is not suitable where significant temperature differences are likely to occur.

Class RC4 is used for close running fits on accurate machinery with medium surface speeds and pressure when accurate location and very little play is desired. Classes RC5 and RC6 are used for medium running fits on accurate machinery with higher running speeds and/or heavier pressures. Class RC7 is used for free running fits where accuracy is less important and/or where significant temperature variations are likely. Classes RC8 and RC9 are used for loose running fits when large tolerances are necessary, together with an allowance on the external parts.

Clearance Locational Fits (LC). Clearance locational fits are used with parts that are stationary but must be able to be freely assembled or disassembled. *Figure 10-117* is a table of American National Standard Clearance Locational Fits from ANSI 4.1-1967, R1994.

Transition Locational Fits (LT). Transition locational fits fall on a continuum between clearance and interference fits, and are used in applications where locational accuracy is important but a small amount of clearance or interference is allowable. *Figure 10-118* (page 425) is a table of American National Standard Transition Locational Fits from ANSI B4.1-1967, R1994.

Interference Locational Fits (LN). Interference locational fits are used when locational accuracy is very important and for parts requiring rigidity and alignment but no special requirements for bore pressure. Interference locational fits are not used for parts that are designed to transmit frictional loads to other parts through the tightness of the fits. These types of applications are covered by force fits. *Figure 10-119* (page 426) is a table of American National Standard Interference Locational Fits from ANSI B4.1-1967, R1994.

American National Standard Running and Sliding Fits (ANSI B4.1-1967, R1994)

Tolerance limits given in body of table are added or subtracted to basic size (as indicated by + or – sign) to obtain maximum and minimum sizes of mating parts.

Values shown below are in thousandths of an inch

Nominal Size Range, Inches Over – To	Class RC 1 Clearance*	Class RC 1 Hole H5	Class RC 1 Shaft g4	Class RC 2 Clearance*	Class RC 2 Hole H6	Class RC 2 Shaft g5	Class RC 3 Clearance*	Class RC 3 Hole H7	Class RC 3 Shaft f6	Class RC 4 Clearance*	Class RC 4 Hole H8	Class RC 4 Shaft f7
0– 0.12	0.1 / 0.45	+0.2 / 0	–0.1 / –0.25	0.1 / 0.55	+0.25 / 0	–0.1 / –0.3	0.3 / 0.95	+0.4 / 0	–0.3 / –0.55	0.3 / 1.3	+0.6 / 0	–0.3 / –0.7
0.12– 0.24	0.15 / 0.5	+0.2 / 0	–0.15 / –0.3	0.15 / 0.65	+0.3 / 0	–0.15 / –0.35	0.4 / 1.12	+0.5 / 0	–0.4 / –0.7	0.4 / 1.6	+0.7 / 0	–0.4 / –0.9
0.24– 0.40	0.2 / 0.6	+0.25 / 0	–0.2 / –0.35	0.2 / 0.85	+0.4 / 0	–0.2 / –0.45	0.5 / 1.5	+0.6 / 0	–0.5 / –0.9	0.5 / 2.0	+0.9 / 0	–0.5 / –1.1
0.40– 0.71	0.25 / 0.75	+0.3 / 0	–0.25 / –0.45	0.25 / 0.95	+0.4 / 0	–0.25 / –0.55	0.6 / 1.7	+0.7 / 0	–0.6 / –1.0	0.6 / 2.3	+1.0 / 0	–0.6 / –1.3
0.71– 1.19	0.3 / 0.95	+0.4 / 0	–0.3 / –0.55	0.3 / 1.2	+0.5 / 0	–0.3 / –0.7	0.8 / 2.1	+0.8 / 0	–0.8 / –1.3	0.8 / 2.8	+1.2 / 0	–0.8 / –1.6
1.19– 1.97	0.4 / 1.1	+0.4 / 0	–0.4 / –0.7	0.4 / 1.4	+0.6 / 0	–0.4 / –0.8	1.0 / 2.6	+1.0 / 0	–1.0 / –1.6	1.0 / 3.6	+1.6 / 0	–1.0 / –2.0
1.97 – 3.15	0.4 / 1.2	+0.5 / 0	–0.4 / –0.7	0.4 / 1.6	+0.7 / 0	–0.4 / –0.9	1.2 / 3.1	+1.2 / 0	–1.2 / –1.9	1.2 / 4.2	+1.8 / 0	–1.2 / –2.4
3.15– 4.73	0.5 / 1.5	+0.6 / 0	–0.5 / –0.9	0.5 / 2.0	+0.9 / 0	–0.5 / –1.1	1.4 / 3.7	+1.4 / 0	–1.4 / –2.3	1.4 / 5.0	+2.2 / 0	–1.4 / –2.8
4.73– 7.09	0.6 / 1.8	+0.7 / 0	–0.6 / –1.1	0.6 / 2.3	+1.0 / 0	–0.6 / –1.3	1.6 / 4.2	+1.6 / 0	–1.6 / –2.6	1.6 / 5.7	+2.5 / 0	–1.6 / –3.2
7.09– 9.85	0.6 / 2.0	+0.8 / 0	–0.6 / –1.2	0.6 / 2.6	+1.2 / 0	–0.6 / –1.4	2.0 / 5.0	+1.8 / 0	–2.0 / –3.2	2.0 / 6.6	+2.8 / 0	–2.0 / –3.8
9.85–12.41	0.8 / 2.3	+0.9 / 0	–0.8 / –1.4	0.8 / 2.9	+1.2 / 0	–0.8 / –1.7	2.5 / 5.7	+2.0 / 0	–2.5 / –3.7	2.5 / 7.5	+3.0 / 0	–2.5 / –4.5
12.41–15.75	1.0 / 2.7	+1.0 / 0	–1.0 / –1.7	1.0 / 3.4	+1.4 / 0	–1.0 / –2.0	3.0 / 6.6	+2.2 / 0	–3.0 / –4.4	3.0 / 8.7	+3.5 / 0	–3.0 / –5.2
15.75–19.69	1.2 / 3.0	+1.0 / 0	–1.2 / –2.0	1.2 / 3.8	+1.6 / 0	–1.2 / –2.2	4.0 / 8.1	+2.5 / 0	–4.0 / –5.6	4.0 / 10.5	+4.0 / 0	–4.0 / –6.5

See footnotes at end of table.

Values shown below are in thousandths of an inch

Nominal Size Range, Inches Over – To	Class RC 5 Clearance*	Class RC 5 Hole H8	Class RC 5 Shaft e7	Class RC 6 Clearance*	Class RC 6 Hole H9	Class RC 6 Shaft e8	Class RC 7 Clearance*	Class RC 7 Hole H9	Class RC 7 Shaft d8	Class RC 8 Clearance*	Class RC 8 Hole H10	Class RC 8 Shaft c9	Class RC 9 Clearance*	Class RC 9 Hole H11	Class RC 9 Shaft
0– 0.12	0.6 / 1.6	+0.6 / 0	–0.6 / –1.0	0.6 / 2.2	+1.0 / 0	–0.6 / –1.2	1.0 / 2.6	+1.0 / 0	–1.0 / –1.6	2.5 / 5.1	+1.6 / 0	–2.5 / –3.5	4.0 / 8.1	+2.5 / 0	–4.0 / –5.6
0.12– 0.24	0.8 / 2.0	+0.7 / 0	–0.8 / –1.3	0.8 / 2.7	+1.2 / 0	–0.8 / –1.5	1.2 / 3.1	+1.2 / 0	–1.2 / –1.9	2.8 / 5.8	+1.8 / 0	–2.8 / –4.0	4.5 / 9.0	+3.0 / 0	–4.5 / –6.0
0.24– 0.40	1.0 / 2.5	+0.9 / 0	–1.0 / –1.6	1.0 / 3.3	+1.4 / 0	–1.0 / –1.9	1.6 / 3.9	+1.4 / 0	–1.6 / –2.5	3.0 / 6.6	+2.2 / 0	–3.0 / –4.4	5.0 / 10.7	+3.5 / 0	–5.0 / –7.2
0.40– 0.71	1.2 / 2.9	+1.0 / 0	–1.2 / –1.9	1.2 / 3.8	+1.6 / 0	–1.2 / –2.2	2.0 / 4.6	+1.6 / 0	–2.0 / –3.0	3.5 / 7.9	+2.8 / 0	–3.5 / –5.1	6.0 / 12.8	+4.0 / 0	–6.0 / –8.8
0.71– 1.19	1.6 / 3.6	+1.2 / 0	–1.6 / –2.4	1.6 / 4.8	+2.0 / 0	–1.6 / –2.8	2.5 / 5.7	+2.0 / 0	–2.5 / –3.7	4.5 / 10.0	+3.5 / 0	–4.5 / –6.5	7.0 / 15.5	+5.0 / 0	–7.0 / –10.5
1.19– 1.97	2.0 / 4.6	+1.6 / 0	–2.0 / –3.0	2.0 / 6.1	+2.5 / 0	–2.0 / –3.6	3.0 / 7.1	+2.5 / 0	–3.0 / –4.6	5.0 / 11.5	+4.0 / 0	–5.0 / –7.5	8.0 / 18.0	+6.0 / 0	–8.0 / –12.0
1.97– 3.15	2.5 / 5.5	+1.8 / 0	–2.5 / –3.7	2.5 / 7.3	+3.0 / 0	–2.5 / –4.3	4.0 / 8.8	+3.0 / 0	–4.0 / –5.8	6.0 / 13.5	+4.5 / 0	–6.0 / –9.0	9.0 / 20.5	+7.0 / 0	–9.0 / –13.5
3.15– 4.73	3.0 / 6.6	+2.2 / 0	–3.0 / –4.4	3.0 / 8.7	+3.5 / 0	–3.0 / –5.2	5.0 / 10.7	+3.5 / 0	–5.0 / –7.2	7.0 / 15.5	+5.0 / 0	–7.0 / –10.5	10.0 / 24.0	+9.0 / 0	–10.0 / –15.0
4.73– 7.09	3.5 / 7.6	+2.5 / 0	–3.5 / –5.1	3.5 / 10.0	+4.0 / 0	–3.5 / –6.0	6.0 / 12.5	+4.0 / 0	–6.0 / –8.5	8.0 / 18.0	+6.0 / 0	–8.0 / –12.0	12.0 / 28.0	+10.0 / 0	–12.0 / –18.0
7.09– 9.85	4.0 / 8.6	+2.8 / 0	–4.0 / –5.8	4.0 / 11.3	+4.5 / 0	–4.0 / –6.8	7.0 / 14.3	+4.5 / 0	–7.0 / –9.8	10.0 / 21.5	+7.0 / 0	–10.0 / –14.5	15.0 / 34.0	+12.0 / 0	–15.0 / –22.0
9.85–12.41	5.0 / 10.0	+3.0 / 0	–5.0 / –7.0	5.0 / 13.0	+5.0 / 0	–5.0 / –8.0	8.0 / 16.0	+5.0 / 0	–8.0 / –11.0	12.0 / 25.0	+8.0 / 0	–12.0 / –17.0	18.0 / 38.0	+12.0 / 0	–18.0 / –26.0
12.41–15.75	6.0 / 11.7	+3.5 / 0	–6.0 / –8.2	6.0 / 15.5	+6.0 / 0	–6.0 / –9.5	10.0 / 19.5	+6.0 / 0	–10.0 / –13.5	14.0 / 29.0	+9.0 / 0	–14.0 / –20.0	22.0 / 45.0	+14.0 / 0	–22.0 / –31.0
15.75–19.69	8.0 / 14.5	+4.0 / 0	–8.0 / –10.5	8.0 / 18.0	+6.0 / 0	–8.0 / –12.0	12.0 / 22.0	+6.0 / 0	–12.0 / –16.0	16.0 / 32.0	+10.0 / 0	–16.0 / –22.0	25.0 / 51.0	+16.0 / 0	–25.0 / –35.0

All data above heavy lines are in accord with ABC agreements. Symbols H5, g4, etc. are hole and shaft designations in ABC system. Limits for sizes above 19.69 inches are also given in the ANSI Standard.

* Pairs of values shown represent minimum and maximum amounts of clearance resulting from application of standard tolerance limits.

FIGURE 10-116 *ANSI standard running and sliding fits (Courtesy ASME)*

American National Standard Clearance Locational Fits (ANSI B4.1-1967, R1994)

Tolerance limits given in body of table are added or subtracted to basic size (as indicated by + or − sign) to obtain maximum and minimum sizes of mating parts.

Nominal Size Range, Inches (Over — To)	Class LC 1 Clearance*	LC 1 Std. Tol. Hole H6	LC 1 Std. Tol. Shaft h5	Class LC 2 Clearance*	LC 2 Hole H7	LC 2 Shaft h6	Class LC 3 Clearance*	LC 3 Hole H8	LC 3 Shaft h7	Class LC 4 Clearance*	LC 4 Hole H10	LC 4 Shaft h9	Class LC 5 Clearance*	LC 5 Hole H7	LC 5 Shaft g6
						Values shown below are in thousandths of an inch									
0– 0.12	0 / 0.45	+0.25 / 0	0 / −0.2	0 / 0.65	+0.4 / 0	0 / −0.25	0 / 1	+0.6 / 0	0 / −0.4	0 / 2.6	+1.6 / 0	0 / −1.0	0.1 / 0.75	+0.4 / 0	−0.1 / −0.35
0.12– 0.24	0 / 0.5	+0.3 / 0	0 / −0.2	0 / 0.8	+0.5 / 0	0 / −0.3	0 / 1.2	+0.7 / 0	0 / −0.5	0 / 3.0	+1.8 / 0	0 / −1.2	0.15 / 0.95	+0.5 / 0	−0.15 / −0.45
0.24– 0.40	0 / 0.65	+0.4 / 0	0 / −0.25	0 / 1.0	+0.6 / 0	0 / −0.4	0 / 1.5	+0.9 / 0	0 / −0.6	0 / 3.6	+2.2 / 0	0 / −1.4	0.2 / 1.2	+0.6 / 0	−0.2 / −0.6
0.40– 0.71	0 / 0.7	+0.4 / 0	0 / −0.3	0 / 1.1	+0.7 / 0	0 / −0.4	0 / 1.7	+1.0 / 0	0 / −0.7	0 / 4.4	+2.8 / 0	0 / −1.6	0.25 / 1.35	+0.7 / 0	−0.25 / −0.65
0.71– 1.19	0 / 0.9	+0.5 / 0	0 / −0.4	0 / 1.3	+0.8 / 0	0 / −0.5	0 / 2	+1.2 / 0	0 / −0.8	0 / 5.5	+3.5 / 0	0 / −2.0	0.3 / 1.6	+0.8 / 0	−0.3 / −0.8
1.19– 1.97	0 / 1.0	+0.6 / 0	0 / −0.4	0 / 1.6	+1.0 / 0	0 / −0.6	0 / 2.6	+1.6 / 0	0 / −1	0 / 6.5	+4.0 / 0	0 / −2.5	0.4 / 2.0	+1.0 / 0	−0.4 / −1.0
1.97– 3.15	0 / 1.2	+0.7 / 0	0 / −0.5	0 / 1.9	+1.2 / 0	0 / −0.7	0 / 3	+1.8 / 0	0 / −1.2	0 / 7.5	+4.5 / 0	0 / −3	0.4 / 2.3	+1.2 / 0	−0.4 / −1.1
3.15– 4.73	0 / 1.5	+0.9 / 0	0 / −0.6	0 / 2.3	+1.4 / 0	0 / −0.9	0 / 3.6	+2.2 / 0	0 / −1.4	0 / 8.5	+5.0 / 0	0 / −3.5	0.5 / 2.8	+1.4 / 0	−0.5 / −1.4
4.73– 7.09	0 / 1.7	+1.0 / 0	0 / −0.7	0 / 2.6	+1.6 / 0	0 / −1.0	0 / 4.1	+2.5 / 0	0 / −1.6	0 / 10.0	+6.0 / 0	0 / −4	0.6 / 3.2	+1.6 / 0	−0.6 / −1.6
7.09– 9.85	0 / 2.0	+1.2 / 0	0 / −0.8	0 / 3.0	+1.8 / 0	0 / −1.2	0 / 4.6	+2.8 / 0	0 / −1.8	0 / 11.5	+7.0 / 0	0 / −4.5	0.6 / 3.6	+1.8 / 0	−0.6 / −1.8
9.85–12.41	0 / 2.1	+1.2 / 0	0 / −0.9	0 / 3.2	+2.0 / 0	0 / −1.2	0 / 5	+3.0 / 0	0 / −2.0	0 / 13.0	+8.0 / 0	0 / −5	0.7 / 3.9	+2.0 / 0	−0.7 / −1.9
12.41–15.75	0 / 2.4	+1.4 / 0	0 / −1.0	0 / 3.6	+2.2 / 0	0 / −1.4	0 / 5.7	+3.5 / 0	0 / −2.2	0 / 15.0	+9.0 / 0	0 / −6	0.7 / 4.3	+2.2 / 0	−0.7 / −2.1
15.75–19.69	0 / 2.6	+1.6 / 0	0 / −1.0	0 / 4.1	+2.5 / 0	0 / −1.6	0 / 6.5	+4 / 0	0 / −2.5	0 / 16.0	+10.0 / 0	0 / −6	0.8 / 4.9	+2.5 / 0	−0.8 / −2.4

See footnotes at end of table.

Nominal Size Range, Inches (Over — To)	Class LC 6 Clearance*	LC 6 Hole H9	LC 6 Shaft f8	Class LC 7 Clearance*	LC 7 Hole H10	LC 7 Shaft e9	Class LC 8 Clearance*	LC 8 Hole H10	LC 8 Shaft d9	Class LC 9 Clearance*	LC 9 Hole H11	LC 9 Shaft c10	Class LC 10 Clearance*	LC 10 Hole H12	LC 10 Shaft	Class LC 11 Clearance*	LC 11 Hole H13	LC 11 Shaft
						Values shown below are in thousandths of an inch												
0– 0.12	0.3 / 1.9	+1.0 / 0	−0.3 / −0.9	0.6 / 3.2	+1.6 / 0	−0.6 / −1.6	1.0 / 2.0	+1.6 / 0	−1.0 / −2.0	2.5 / 6.6	+2.5 / 0	−2.5 / −4.1	4 / 12	+4 / 0	−4 / −8	5 / 17	+6 / 0	−5 / −11
0.12– 0.24	0.4 / 2.3	+1.2 / 0	−0.4 / −1.1	0.8 / 3.8	+1.8 / 0	−0.8 / −2.0	1.2 / 4.2	+1.8 / 0	−1.2 / −2.4	2.8 / 7.6	+3.0 / 0	−2.8 / −4.6	4.5 / 14.5	+5 / 0	−4.5 / −9.5	6 / 20	+7 / 0	−6 / −13
0.24– 0.40	0.5 / 2.8	+1.4 / 0	−0.5 / −1.4	1.0 / 4.6	+2.2 / 0	−1.0 / −2.4	1.6 / 5.2	+2.2 / 0	−1.6 / −3.0	3.0 / 8.7	+3.5 / 0	−3.0 / −5.2	5 / 17	+6 / 0	−5 / −11	7 / 25	+9 / 0	−7 / −16
0.40– 0.71	0.6 / 3.2	+1.6 / 0	−0.6 / −1.6	1.2 / 5.6	+2.8 / 0	−1.2 / −2.8	2.0 / 6.4	+2.8 / 0	−2.0 / −3.6	3.5 / 10.3	+4.0 / 0	−3.5 / −6.3	6 / 20	+7 / 0	−6 / −13	8 / 28	+10 / 0	−8 / −18
0.71– 1.19	0.8 / 4.0	+2.0 / 0	−0.8 / −2.0	1.6 / 7.1	+3.5 / 0	−1.6 / −3.6	2.5 / 8.0	+3.5 / 0	−2.5 / −4.5	4.5 / 13.0	+5.0 / 0	−4.5 / −8.0	7 / 23	+8 / 0	−7 / −15	10 / 34	+12 / 0	−10 / −22
1.19– 1.97	1.0 / 5.1	+2.5 / 0	−1.0 / −2.6	2.0 / 8.5	+4.0 / 0	−2.0 / −4.5	3.0 / 9.5	+4.0 / 0	−3.0 / −5.5	5.0 / 15.0	+6 / 0	−5.0 / −9.0	8 / 28	+10 / 0	−8 / −18	12 / 44	+16 / 0	−12 / −28
1.97– 3.15	1.2 / 6.0	+3.0 / 0	−1.0 / −3.0	2.5 / 10.0	+4.5 / 0	−2.5 / −5.5	4.0 / 11.5	+4.5 / 0	−4.0 / −7.0	6.0 / 17.5	+7 / 0	−6.0 / −10.5	10 / 34	+12 / 0	−10 / −22	14 / 50	+18 / 0	−14 / −32
3.15– 4.73	1.4 / 7.1	+3.5 / 0	−1.4 / −3.6	3.0 / 11.5	+5.0 / 0	−3.0 / −6.5	5.0 / 13.5	+5.0 / 0	−5.0 / −8.5	7 / 21	+9 / 0	−7 / −12	11 / 39	+14 / 0	−11 / −25	16 / 60	+22 / 0	−16 / −38
4.73– 7.09	1.6 / 8.1	+4.0 / 0	−1.6 / −4.1	3.5 / 13.5	+6.0 / 0	−3.5 / −7.5	6 / 16	+6 / 0	−6 / −10	8 / 24	+10 / 0	−8 / −14	12 / 44	+16 / 0	−12 / −28	18 / 68	+25 / 0	−18 / −43
7.09– 9.85	2.0 / 9.3	+4.5 / 0	−2.0 / −4.8	4.0 / 15.5	+7.0 / 0	−4.0 / −8.5	7 / 18.5	+7 / 0	−7 / −11.5	10 / 29	+12 / 0	−10 / −17	16 / 52	+18 / 0	−16 / −34	22 / 78	+28 / 0	−22 / −50
9.85–12.41	2.2 / 10.2	+5.0 / 0	−2.2 / −5.2	4.5 / 17.5	+8.0 / 0	−4.5 / −9.5	7 / 20	+8 / 0	−7 / −12	12 / 32	+12 / 0	−12 / −20	20 / 60	+20 / 0	−20 / −40	28 / 88	+30 / 0	−28 / −58
12.41–15.75	2.5 / 12.0	+6.0 / 0	−2.5 / −6.0	5.0 / 20.0	+9.0 / 0	−5 / −11	8 / 23	+9 / 0	−8 / −14	14 / 37	+14 / 0	−14 / −23	22 / 66	+22 / 0	−22 / −44	30 / 100	+35 / 0	−30 / −65
15.75–19.69	2.8 / 12.8	+6.0 / 0	−2.8 / −6.8	5.0 / 21.0	+10.0 / 0	−5 / −11	9 / 25	+10 / 0	−9 / −15	16 / 42	+16 / 0	−16 / −26	25 / 75	+25 / 0	−25 / −50	35 / 115	+40 / 0	−35 / −75

All data above heavy lines are in accordance with American-British-Canadian (ABC) agreements. Symbols H6, H7, s6, etc. are hole and shaft designations in ABC system. Limits for sizes above 19.69 inches are not covered by ABC agreements but are given in the ANSI Standard.
* Pairs of values shown represent minimum and maximum amounts of interference resulting from application of standard tolerance limits.

FIGURE 10-117 *ANSI standard clearance locational fits* (Courtesy ASME)

American National Standard Transition Locational Fits (ANSI B4.1-1967, R1994)

Nominal Size Range, Inches Over / To	Class LT 1			Class LT 2			Class LT 3			Class LT 4			Class LT 5			Class LT 6		
	Fit*	Hole H7	Shaft js6	Fit*	Hole H8	Shaft js7	Fit*	Hole H7	Shaft k6	Fit*	Hole H8	Shaft k7	Fit*	Hole H7	Shaft n6	Fit*	Hole H7	Shaft n7
	Values shown below are in thousandths of an inch																	
0– 0.12	−0.12 +0.52	+0.4 0	+0.12 −0.12	−0.2 +0.8	+0.6 0	+0.2 −0.2							−0.5 +0.15	+0.4 0	+0.5 +0.25	−0.65 +0.15	+0.4 0	+0.65 +0.25
0.12– 0.24	−0.15 +0.65	+0.5 0	+0.15 −0.15	−0.25 +0.95	+0.7 0	+0.25 −0.25							−0.6 +0.2	+0.5 0	+0.6 +0.3	−0.8 +0.2	+0.5 0	+0.8 +0.3
0.24– 0.40	−0.2 +0.8	+0.6 0	+0.2 −0.2	−0.3 +1.2	+0.9 0	+0.3 −0.3	−0.5 +0.5	+0.6 0	+0.5 +0.1	−0.7 +0.8	+0.9 0	+0.7 +0.1	−0.8 +0.2	+0.6 0	+0.8 +0.4	−1.0 +0.2	+0.6 0	+1.0 +0.4
0.40– 0.71	−0.2 +0.9	+0.7 0	+0.2 −0.2	−0.35 +1.35	+1.0 0	+0.35 −0.35	−0.5 +0.6	+0.7 0	+0.5 +0.1	−0.8 +0.9	+1.0 0	+0.8 +0.1	−0.9 +0.2	+0.7 0	+0.9 +0.5	−1.2 +0.2	+0.7 0	+1.2 +0.5
0.71– 1.19	−0.25 +1.05	+0.8 0	+0.25 −0.25	−0.4 +1.6	+1.2 0	+0.4 −0.4	−0.6 +0.7	+0.8 0	+0.6 +0.1	−0.9 +1.1	+1.2 0	+0.9 +0.1	−1.1 +0.2	+0.8 0	+1.1 +0.6	−1.4 +0.2	+0.8 0	+1.4 +0.6
1.19– 1.97	−0.3 +1.3	+1.0 0	+0.3 −0.3	−0.5 +2.1	+1.6 0	+0.5 −0.5	−0.7 +0.9	+1.0 0	+0.7 +0.1	−1.1 +1.5	+1.6 0	+1.1 +0.1	−1.3 +0.3	+1.0 0	+1.3 +0.7	−1.7 +0.3	+1.0 0	+1.7 +0.7
1.97– 3.15	−0.3 +1.5	+1.2 0	+0.3 −0.3	−0.6 +2.4	+1.8 0	+0.6 −0.6	−0.8 +1.1	+1.2 0	+0.8 +0.1	−1.3 +1.7	+1.8 0	+1.3 +0.1	−1.5 +0.4	+1.2 0	+1.5 +0.8	−2.0 +0.4	+1.2 0	+2.0 +0.8
3.15– 4.73	−0.4 +1.8	+1.4 0	+0.4 −0.4	−0.7 +2.9	+2.2 0	+0.7 −0.7	−1.0 +1.3	+1.4 0	+1.0 +0.1	−1.5 +2.1	+2.2 0	+1.5 +0.1	−1.9 +0.4	+1.4 0	+1.9 +1.0	−2.4 +0.4	+1.4 0	+2.4 +1.0
4.73– 7.09	−0.5 +2.1	+1.6 0	+0.5 −0.5	−0.8 +3.3	+2.5 0	+0.8 −0.8	−1.1 +1.5	+1.6 0	+1.1 +0.1	−1.7 +2.4	+2.5 0	+1.7 +0.1	−2.2 +0.4	+1.6 0	+2.2 +1.2	−2.8 +0.4	+1.6 0	+2.8 +1.2
7.09– 9.85	−0.6 +2.4	+1.8 0	+0.6 −0.6	−0.9 +3.7	+2.8 0	+0.9 −0.9	−1.4 +1.6	+1.8 0	+1.4 +0.2	−2.0 +2.6	+2.8 0	+2.0 +0.2	−2.6 +0.4	+1.8 0	+2.6 +1.4	−3.2 +0.4	+1.8 0	+3.2 +1.4
9.85–12.41	−0.6 +2.6	+2.0 0	+0.6 −0.6	−1.0 +4.0	+3.0 0	+1.0 −1.0	−1.4 +1.8	+2.0 0	+1.4 +0.2	−2.2 +2.8	+3.0 0	+2.2 +0.2	−2.6 +0.6	+2.0 0	+2.6 +1.4	−3.4 +0.6	+2.0 0	+3.4 +1.4
12.41–15.75	−0.7 +2.9	+2.2 0	+0.7 −0.7	−1.0 +4.5	+3.5 0	+1.0 −1.0	−1.6 +2.0	+2.2 0	+1.6 +0.2	−2.4 +3.3	+3.5 0	+2.4 +0.2	−3.0 +0.6	+2.2 0	+3.0 +1.6	−3.8 +0.6	+2.2 0	+3.8 +1.6
15.75–19.69	−0.8 +3.3	+2.5 0	+0.8 −0.8	−1.2 +5.2	+4.0 0	+1.2 −1.2	−1.8 +2.3	+2.5 0	+1.8 +0.2	−2.7 +3.8	+4.0 0	+2.7 +0.2	−3.4 +0.7	+2.5 0	+3.4 +1.8	−4.3 +0.7	+2.5 0	+4.3 +1.8

All data above heavy lines are in accord with ABC agreements. Symbols H7, js6, etc. are hole and shaft designations in ABC system.
* Pairs of values shown represent maximum amount of interference (−) and maximum amount of clearance (+) resulting from application of standard tolerance limits.

FIGURE 10-118 *ANSI Standard Transition Locational Fits* (Courtesy ASME)

Force and Shrink Fits (FN). Force or shrink fits are a special type of interference fit used when maintenance of constant bore pressure for all sizes within the specified range is important. The amount of interference varies with the diameter, and the difference between minimum and maximum values is small to ensure maintenance of the resulting pressures within reasonable limits. *Figure 10-120* is a table of American National Standard Force and Shrink Fits from ANSI B4.1-1967, R1994.

Using Fit Tables

Figures 10-116 through 10-120 are fit tables taken from ANSI B4.1-1967, R1994. It is important for engineers and drafters to know how to use tables such as these. Look at Figure 10-116. The nine classes explained earlier are arranged across the top of the table. Under each class column there are three other columns: one giving the clearance, one giving the standard tolerance limits for the hole, and one giving the standard tolerance limits for the shaft. On the extreme left-hand side of the chart the nominal size range in inches is given. To use this table, as well as the others contained in this chapter,

select the appropriate class across the top of the table and locate the nominal size at the extreme left-hand side of the table. Moving across the row that contains the appropriate nominal size range to the class column you have selected, you will find the tolerance limits you need. These tolerance limits are added to or subtracted from the basic size to obtain maximum and minimum sizes of mating parts. Each tolerance is preceded by either a plus or a minus to indicate whether it is added to or subtracted from the basic size. Values must be changed to inches before adding and subtracting.

EXAMPLE:

Using fit tables:

Values are limits in thousandths of an inch.

Basic size: .0156

Nominal size range: .0–0.12

Limits of clearance: 0.1 = 000.1 = .0001

0.45 = 000.45 = .00045

Standard limits: (Move decimal three places to the left to obtain values in inches.)

Tolerance limits given in body of table are added or subtracted to basic size (as indicated by + or − sign) to obtain maximum and minimum sizes of mating parts.

Nominal Size Range, Inches		Class LN 1			Class LN 2			Class LN 3		
		Limits of Inter-ference	Standard Limits		Limits of Inter-ference	Standard Limits		Limits of Inter-ference	Standard Limits	
			Hole H6	Shaft n5		Hole H7	Shaft p6		Hole H7	Shaft r6
Over	To	Values shown below are given in thousandths of an inch								
0–	0.12	0 0.45	+ 0.25 0	+ 0.45 + 0.25	0 0.65	+ 0.4 0	+ 0.65 + 0.4	0.1 0.75	+ 0.4 0	+ 0.75 + 0.5
0.12–	0.24	0 0.5	+ 0.3 0	+ 0.5 + 0.3	0 0.8	+ 0.5 0	+ 0.8 + 0.5	0.1 0.9	+ 0.5 0	+ 0.9 + 0.6
0.24–	0.40	0 0.65	+ 0.4 0	+ 0.65 + 0.4	0 1.0	+ 0.6 0	+ 1.0 + 0.6	0.2 1.2	+ 0.6 0	+ 1.2 + 0.8
0.40–	0.71	0 0.8	+ 0.4 0	+ 0.8 + 0.4	0 1.1	+ 0.7 0	+ 1.1 + 0.7	0.3 1.4	+ 0.7 0	+ 1.4 + 1.0
0.71–	1.19	0 1.0	+ 0.5 0	+ 1.0 + 0.5	0 1.3	+ 0.8 0	+ 1.3 + 0.8	0.4 1.7	+ 0.8 0	+ 1.7 + 1.2
1.19–	1.97	0 1.1	+ 0.6 0	+ 1.1 + 0.6	0 1.6	+ 1.0 0	+ 1.6 + 1.0	0.4 2.0	+ 1.0 0	+ 2.0 + 1.4
1.97–	3.15	0.1 1.3	+ 0.7 0	+ 1.3 + 0.8	0.2 2.1	+ 1.2 0	+ 2.1 + 1.4	0.4 2.3	+ 1.2 0	+ 2.3 + 1.6
3.15–	4.73	0.1 1.6	+ 0.9 0	+ 1.6 + 1.0	0.2 2.5	+ 1.4 0	+ 2.5 + 1.6	0.6 2.9	+ 1.4 0	+ 2.9 + 2.0
4.73–	7.09	0.2 1.9	+ 1.0 0	+ 1.9 + 1.2	0.2 2.8	+ 1.6 0	+ 2.8 + 1.8	0.9 3.5	+ 1.6 0	+ 3.5 + 2.5
7.09–	9.85	0.2 2.2	+ 1.2 0	+ 2.2 + 1.4	0.2 3.2	+ 1.8 0	+ 3.2 + 2.0	1.2 4.2	+ 1.8 0	+ 4.2 + 3.0
9.85–	12.41	0.2 2.3	+ 1.2 0	+ 2.3 + 1.4	0.2 3.4	+ 2.0 0	+ 3.4 + 2.2	1.5 4.7	+ 2.0 0	+ 4.7 + 3.5
12.41–	15.75	0.2 2.6	+ 1.4 0	+ 2.6 + 1.6	0.3 3.9	+ 2.2 0	+ 3.9 + 2.5	2.3 5.9	+ 2.2 0	+ 5.9 + 4.5
15.75–	19.69	0.2 2.8	+ 1.6 0	+ 2.8 + 1.8	0.3 4.4	+ 2.5 0	+ 4.4 + 2.8	2.5 6.6	+ 2.5 0	+ 6.6 + 5.0

All data in this table are in accordance with American-British-Canadian (ABC) agreements. Limits for sizes above 19.69 inches are not covered by ABC agreements but are given in the ANSI Standard.

Symbols H7, p6, etc. are hole and shaft designations in ABC system.

* Pairs of values shown represent minimum and maximum amounts of interference resulting from application of standard tolerance limits.

FIGURE 10-119　*ANSI Standard Interference Locational Fits* (Courtesy ASME)

Hole:　+0.2　= 000.2　= +.0002

　　　　−0　　　= −0

Shaft:　−0.1　= −000.1　= −.0001

　　　　−0.25　= −000.25　= −.00035

Calculating for tolerance dimensions:

Hole:　　.0156　basic hole size
　　　　+.0002　standard limits in thousands
　　　　.0158　upper limit for hole

.0156　basic hole size
−0　　　standard limits in thousands
.0156　lower limit for hole

Shaft:　.0156　basic hole size
　　　−.0001　standard limits for shaft
　　　.0155　upper limit for shaft

.0156　basic hole size
−.00025　standard limit for shaft
.01535　lower limit for shaft

Nominal Size Range, Inches	Class FN 1			Class FN 2			Class FN 3			Class FN 4			Class FN 5		
		Standard Tolerance Limits			Standard Tolerance Limits			Standard Tolerance Limits			Standard Tolerance Limits			Standard Tolerance Limits	
Over To	Interference*	Hole H6	Shaft	Interference*	Hole H7	Shaft s6	Interference*	Hole H7	Shaft t6	Interference*	Hole H7	Shaft u6	Interference*	Hole H8	Shaft x7
	Values shown below are in thousandths of an inch														
0–0.12	0.05 0.5	+0.25 0	+0.5 +0.3	0.2 0.85	+0.4 0	+0.85 +0.6				0.3 0.95	+0.4 0	+0.95 +0.7	0.3 1.3	+0.6 0	+1.3 +0.9
0.12–0.24	0.1 0.6	+0.3 0	+0.6 +0.4	0.2 1.0	+0.5 0	+1.0 +0.7				0.4 1.2	+0.5 0	+1.2 +0.9	0.5 1.7	+0.7 0	+1.7 +1.2
0.24–0.40	0.1 0.75	+0.4 0	+0.75 +0.5	0.4 1.4	+0.6 0	+1.4 +1.0				0.6 1.6	+0.6 0	+1.6 +1.2	0.5 2.0	+0.9 0	+2.0 +1.4
0.40–0.56	0.1 0.8	+0.4 0	+0.8 +0.5	0.5 1.6	+0.7 0	+1.6 +1.2				0.7 1.8	+0.7 0	+1.8 +1.4	0.6 2.3	+1.0 0	+2.3 +1.6
0.56–0.71	0.2 0.9	+0.4 0	+0.9 +0.6	0.5 1.6	+0.7 0	+1.6 +1.2				0.7 1.8	+0.7 0	+1.8 +1.4	0.8 2.5	+1.0 0	+2.5 +1.8
0.71–0.95	0.2 1.1	+0.5 0	+1.1 +0.7	0.6 1.9	+0.8 0	+1.9 +1.4				0.8 2.1	+0.8 0	+2.1 +1.6	1.0 3.0	+1.2 0	+3.0 +2.2
0.95–1.19	0.3 1.2	+0.5 0	+1.2 +0.8	0.6 1.9	+0.8 0	+1.9 +1.4	0.8 2.1	+0.8 0	+2.1 +1.6	1.0 2.3	+0.8 0	+2.3 +1.8	1.3 3.3	+1.2 0	+3.3 +2.5
1.19–1.58	0.3 1.3	+0.6 0	+1.3 +0.9	0.8 2.4	+1.0 0	+2.4 +1.8	1.0 2.6	+1.0 0	+2.6 +2.0	1.5 3.1	+1.0 0	+3.1 +2.5	1.4 4.0	+1.6 0	+4.0 +3.0
1.58–1.97	0.4 1.4	+0.6 0	+1.4 +1.0	0.8 2.4	+1.0 0	+2.4 +1.8	1.2 2.8	+1.0 0	+2.8 +2.2	1.8 3.4	+1.0 0	+3.4 +2.8	2.4 5.0	+1.6 0	+5.0 +4.0
1.97–2.56	0.6 1.8	+0.7 0	+1.8 +1.3	0.8 2.7	+1.2 0	+2.7 +2.0	1.3 3.2	+1.2 0	+3.2 +2.5	2.3 4.2	+1.2 0	+4.2 +3.5	3.2 6.2	+1.8 0	+6.2 +5.0
2.56–3.15	0.7 1.9	+0.7 0	+1.9 +1.4	1.0 2.9	+1.2 0	+2.9 +2.2	1.8 3.7	+1.2 0	+3.7 +3.0	2.8 4.7	+1.2 0	+4.7 +4.0	4.2 7.2	+1.8 0	+7.2 +6.0
3.15–3.94	0.9 2.4	+0.9 0	+2.4 +1.8	1.4 3.7	+1.4 0	+3.7 +2.8	2.1 4.4	+1.4 0	+4.4 +3.5	3.6 5.9	+1.4 0	+5.9 +5.0	4.8 8.4	+2.2 0	+8.4 +7.0
3.94–4.73	1.1 2.6	+0.9 0	+2.6 +2.0	1.6 3.9	+1.4 0	+3.9 +3.0	2.6 4.9	+1.4 0	+4.9 +4.0	4.6 6.9	+1.4 0	+6.9 +6.0	5.8 9.4	+2.2 0	+9.4 +8.0

See footnotes at end of table.

Nominal Size Range, Inches	Class FN 1			Class FN 2			Class FN 3			Class FN 4			Class FN 5		
		Standard Tolerance Limits			Standard Tolerance Limits			Standard Tolerance Limits			Standard Tolerance Limits			Standard Tolerance Limits	
Over To	Interference*	Hole H6	Shaft	Interference*	Hole H7	Shaft s6	Interference*	Hole H7	Shaft t6	Interference*	Hole H7	Shaft u6	Interference*	Hole H8	Shaft x7
	Values shown below are in thousandths of an inch														
4.73– 5.52	1.2 2.9	+1.0 0	+2.9 +2.2	1.9 4.5	+1.6 0	+ 4.5 + 3.5	3.4 6.0	+1.6 0	+ 6.0 + 5.0	5.4 8.0	+1.6 0	+ 8.0 + 7.0	7.5 11.6	+2.5 0	+11.6 +10.0
5.52– 6.30	1.5 3.2	+1.0 0	+3.2 +2.5	2.4 5.0	+1.6 0	+ 5.0 + 4.0	3.4 6.0	+1.6 0	+ 6.0 + 5.0	5.4 8.0	+1.6 0	+ 8.0 + 7.0	9.5 13.6	+2.5 0	+13.6 +12.0
6.30– 7.09	1.8 3.5	+1.0 0	+3.5 +2.8	2.9 5.5	+1.6 0	+ 5.5 + 4.5	4.4 7.0	+1.6 0	+ 7.0 + 6.0	6.4 9.0	+1.6 0	+ 9.0 + 8.0	9.5 13.6	+2.5 0	+13.6 +12.0
7.09– 7.88	1.8 3.8	+1.2 0	+3.8 +3.0	3.2 6.2	+1.8 0	+ 6.2 + 5.0	5.2 8.2	+1.8 0	+ 8.2 + 7.0	7.2 10.2	+1.8 0	+10.2 + 9.0	11.2 15.8	+2.8 0	+15.8 +14.0
7.88– 8.86	2.3 4.3	+1.2 0	+4.3 +3.5	3.2 6.2	+1.8 0	+ 6.2 + 5.0	5.2 8.2	+1.8 0	+ 8.2 + 7.0	8.2 11.2	+1.8 0	+11.2 +10.0	13.2 17.8	+2.8 0	+17.8 +16.0
8.86– 9.85	2.3 4.3	+1.2 0	+4.3 +3.5	4.2 7.2	+1.8 0	+ 7.2 + 6.0	6.2 9.2	+1.8 0	+ 9.2 + 8.0	10.2 13.2	+1.8 0	+13.2 +12.0	13.2 17.8	+2.8 0	+17.8 +16.0
9.85–11.03	2.8 4.9	+1.2 0	+4.9 +4.0	4.0 7.2	+2.0 0	+ 7.2 + 6.0	7.0 10.2	+2.0 0	+10.2 + 9.0	10.0 13.2	+2.0 0	+13.2 +12.0	15.0 20.0	+3.0 0	+20.0 +18.0
11.03–12.41	2.8 4.9	+1.2 0	+4.9 +4.0	5.0 8.2	+2.0 0	+ 8.2 + 7.0	7.0 10.2	+2.0 0	+10.2 + 9.0	12.0 15.2	+2.0 0	+15.2 +14.0	17.0 22.0	+3.0 0	+22.0 +20.0
12.41–13.98	3.1 5.5	+1.4 0	+5.5 +4.5	5.8 9.4	+2.2 0	+ 9.4 + 8.0	7.8 11.4	+2.2 0	+11.4 +10.0	13.8 17.4	+2.2 0	+17.4 +16.0	18.5 24.2	+3.5 0	+24.2 +22.0
13.98–15.75	3.6 6.1	+1.4 0	+6.1 +5.0	5.8 9.4	+2.2 0	+ 9.4 + 8.0	9.8 13.4	+2.2 0	+13.4 +12.0	15.8 19.4	+2.2 0	+19.4 +18.0	21.5 27.2	+3.5 0	+27.2 +25.0
15.75–17.72	4.4 7.0	+1.6 0	+7.0 +6.0	6.5 10.6	+2.5 0	+10.6 + 9.0	9.5 13.6	+2.5 0	+13.6 +12.0	17.5 21.6	+2.5 0	+21.6 +20.0	24.0 30.5	+4.0 0	+30.5 +28.0
17.72–19.69	4.4 7.0	+1.6 0	+7.0 +6.0	7.5 11.6	+2.5 0	+11.6 +10.0	11.5 15.6	+2.5 0	+15.6 +14.0	19.5 23.6	+2.5 0	+23.6 +22.0	26.0 32.5	+4.0 0	+32.5 +30.0

All data above heavy lines are in accordance with American-British-Canadian (ABC) agreements. Symbols H6, H7, s6, etc. are hole and shaft designations in ABC system. Limits for sizes above 19.69 inches are not covered by ABC agreements but are given in the ANSI standard.

* Pairs of values shown represent minimum and maximum amounts of interference resulting from application of standard tolerance limits.

FIGURE 10-120 *ANSI Standard Force and Shrink Fits* (Courtesy ASME)

Tolerance dimensions for basic size of .0156:

Hole: .01580   Shaft: .01550
          .01560   .01535

To check calculations:

Subtract: .01580 upper limit of hole
          −.01535 lower limit of shaft
           .00045 limit of clearance

           .01560 lower limit of hole
          −.01550 upper limit of shaft
           .0001   limit of clearance

Limit of Clearance:

.0001   = .1 from chart

.00045 = .45 from chart

MANUFACTURING PRECISION

Figure 10-121 is a table showing the levels of tolerances engineers and drafters can expect from various common manufacturing processes. It is important for engineers and drafters to know what to expect from the manufacturing processes that will be used in converting their design into an actual working part. It will serve no purpose to specify a tolerance greater than that which can be expected of a given manufacturing process.

Summary of Dimensioning Rules

This section summarizes most of the dimensioning rules with which engineers and drafters should be familiar. This section can be used as a checklist for ensuring that drawings you do throughout this book are properly dimensioned. Use this summary as a checklist until you are so familiar with these rules that the checklist is no longer necessary.

1. Give all dimensions necessary to accurately communicate the design to manufacturing personnel, but only those dimensions necessary.

2. Apply dimensions in such a way that they cannot be misinterpreted.

3. Show dimensions between surfaces or points that have a functional relationship.

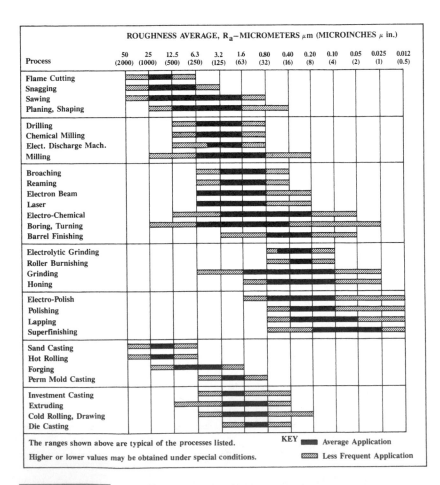

FIGURE 10-121 *Manufacturing precision* (From Machinery's Handbook, 24th edition, © 1992 by Industrial Press Inc. Used with permission)

4. Do not dimension to or from rough surfaces. Always attempt to dimension from a finished surface.

5. Give complete dimensions, so that manufacturing personnel do not have to interrupt their work to calculate or otherwise determine a dimension they need.

6. Apply dimensions in the views where the shape of the part is shown most accurately.

7. Apply dimensions in those views where the part being dimensioned appears true size and true shape.

8. Do not dimension to hidden lines. If it becomes necessary to dimension to hidden lines, change the view to a sectional view.

9. Attempt to place dimensions that apply to two adjacent views between the views.

10. Avoid crossing extension lines with dimension lines.

11. Clearly dimension and annotate all features. Never expect manufacturing personnel to assume anything.

12. Do not apply a complete chain of detail dimensions. Omit the last one or indicate that the last dimension in the chain is a reference dimension.

13. Avoid placing dimensions on the part itself.

14. Always place dimension lines uniformly throughout the drawing.

15. Never use a line that constitutes a portion of the drawing of a part as a dimension line.

16. Avoid crossing dimension lines wherever possible.

17. Extension lines that cross other extension lines or object lines should not be broken.

18. A center line may be used as an extension line.

19. Center lines should not extend from view to view.

20. Leader lines should point to the center of a hole or circular feature.

21. Leader lines should be straight and broken at precise angles. They should never curve.

22. Leader lines should begin approximately at the center of a note height and extend either from the beginning or end of the note.

23. If it becomes necessary to letter within a section-lined area, erase enough of the section lines to leave a clear space so that the dimension can be easily read.

24. Fraction bars should be horizontal rather than angled.

25. The numerator and denominator of a fraction should not touch the horizontal fraction bar.

26. Finish marks should be applied on views in which the surface appears as an edge.

27. Finish marks are not necessary on parts made from rolled stock.

28. If a part is finished all over, finish marks are not used. Substitute the letters FAO or apply a note that says "Finish All Over."

29. Dimension a cylinder by showing the diameter and length on the rectangular view or by showing the diameter as a diagonal dimension in the circular view.

30. Always specify drill sizes in decimals.

31. A metric or decimal-inch diameter dimension should be preceded by the diameter symbol.

32. A metric or decimal-inch radius dimension should be preceded by a capital R.

33. Dimensions that are not to scale should be underlined with a straight line.

34. Decimal dimensioning is preferred for mechanical and manufacturing-related drawings.

Notation

A good drawing may be defined as one that contains all of the information required by the various design and manufacturing people who will use it in producing the subject part. Most of this information can be conveyed graphically, using standard dimensioning practices. However, it is not uncommon to encounter a situation in which all of the needed information cannot be communicated graphically. In these cases, notes are used to communicate or clarify the designer's intent.

Notes are brief, carefully worded statements placed on drawings to convey information not covered or not adequately explained using graphics, *Figure 10-122.* Notes should be clearly worded so as to allow only one correct interpretation.

There are no ANSI standards specifically governing the use of notes on technical drawings. However, several rules of a general nature should be observed. These rules apply to both general notes and the more specific detail notes.

Rules for Applying Notes on Drawings

Notes may be lettered freehand, entered using a keyboard in a computer-aided drafting (CAD) system, or added through any one of several mechanical lettering processes. Sample notes are included on the drawings in *Figures 10-123* and *10-124.* In any case, regardless of

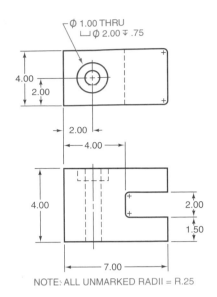

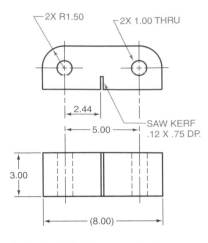

NOTE: FAO TO 63 MICROINCHES

FIGURE 10-122 *Notes on drawings improve communication*

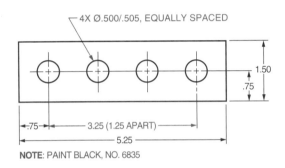

NOTE: PAINT BLACK, NO. 6835

FIGURE 10-123 *Sample note lettered mechanically*

how they are put on the drawing, notes should be oriented horizontally on the drafting sheet, *Figure 10-125*. General notes should be located directly above the title block, *Figure 10-126* (page 433). When using manual processes to apply general notes such as those in Figure 10-126, the first note is placed directly above the title block, the second is placed on top of it, and so on up the line. This allows notes to be added as needed without renumbering. However, when using a CAD system, this is not necessary since one of the advantages of CAD is that notes can be renumbered and rearranged automatically. Detail notes should be located as close as possible to the detail they are describing and connected to it by a leader line, *Figure 10-127*(page 434).

Notes should be applied to drawings after all graphics have been completed. This prevents notes and graphics from overlapping, and minimizes other technique problems such as smudging the worksheet and frequent erasing of notes as changes are required on a drawing.

Two basic types of notes are used on technical drawings: general notes and detail notes. They serve different purposes and, therefore, must be examined separately.

GENERAL NOTES

General notes are broad items of information that have a job- or project-wide application rather than relating to just one single element of a project or a part. They are usually placed immediately above the title block on the drawing sheet and numbered sequentially.

Information placed in general notes includes such characteristics of a product as finish specifications; standard sizes of fillets, rounds, and radii; heat treatment specifications; cleaning instructions; general tolerancing data; hardness testing instructions; and stamping specifications. *Figures 10-128* (page 435) and *10-129* (page 436) are examples of drawings containing general notes.

DETAIL NOTES

Detail notes are specific notes that pertain to one particular element or characteristic of a part. They are placed as near as possible to the characteristic to which they apply and are connected using a leader line, *Figure 10-130* (page 437).

Detail notes should not be placed on views, *Figure 10-131* (page 437). The only exception to this rule is in cases where a great deal of open space exists on a view, but very little around it, *Figure 10-132* (page 437). Notes should never be superimposed over other data such as dimensions, lines, or symbols, *Figure 10-133* (page 437).

Common sense is the best rule to follow in applying detail notes. Since detail notes are used to more completely communicate or to clarify intent, they should be placed as close as possible to the element to which they pertain and in such a way as to be easily read.

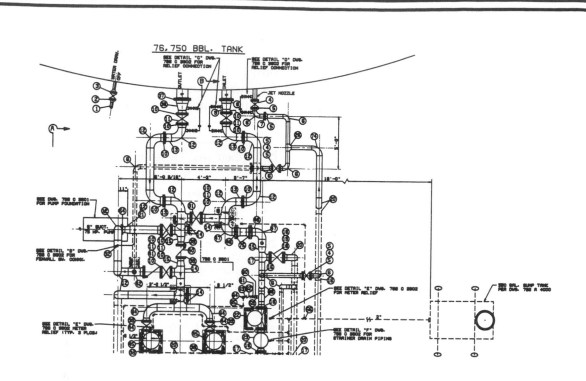

PIPING

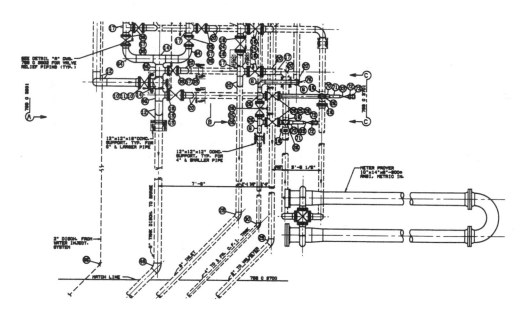

DRAWN BY: **ICS** COMPUTER AIDED DRAFTING SYSTEM
MANUFACTURED BY: INTERACTIVE COMPUTER SYSTEMS, INC.
P. O. BOX 14908 BATON ROUGE, LOUISIANA 70898
(504) 292-7570

FIGURE 10-124 *Notes lettered on a CAD system* (Courtesy Interactive Computer Systems Inc.)

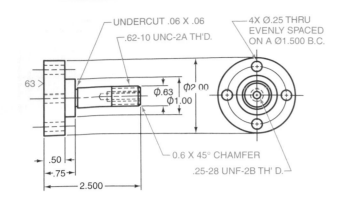

UNDERCUT .06 X .06
.62-10 UNC-2A TH'D.
4X Ø.25 THRU EVENLY SPACED ON A Ø1.500 B.C.

63

Ø.63
Ø2.00
Ø1.00

.50
.75
2.500

0.6 X 45° CHAMFER
.25-28 UNF-2B TH' D.

FIGURE 10-125 *Notes are placed horizontally*

WRITING NOTES

The written word lends itself to interpretations that may vary. Thus, *written notes* used on drawings must convey the exact intent of the designer. Consequently, drafters must be especially careful in the preparation of notes.

The first step is to place in a legend all abbreviations used in the notes, *Figure 10-134* (page 437). This ensures that all readers of the notes interpret the abbreviations consistently and correctly.

Another technique that will limit misinterpretation of notes is punctuation or indentation. Notes containing more than one sentence should be properly punctuated so that readers know where one sentence leaves off and the next one begins, *Figure 10-135* (page 437). Indentation is used when the length of a note requires more than one line. The second line is indented so that readers know it is part of the first line, *Figure 10-136* (page 437).

Another technique is particularly useful for beginning drafters. It is called *verification*. Notes to be placed on drawings are first jotted down, legibly, on a print of the drawing. Another drafter, preferably one with experience, is then asked to read the notes to ensure that they are not open to multiple interpretations.

NOTE SPECIFICATIONS

A number of specifications should be observed when writing notes. The most widely accepted lettering style for notes is uppercase (all capital) block Gothic letters in a vertical format. Some companies accept uppercase block Gothic lettering in an inclined or slanted format. The proper lettering height is one-quarter inch for titles such as GENERAL NOTES, and one-eighth to three-sixteenths inch for actual notes, *Figure 10-137* (page 437).

Spacing between successive lines of the same note should be approximately one-sixteenth inch. Spacing between separate notes should be approximately one-eighth inch, *Figure 10-138* (page 437).

There are no hard and fast rules governing the length of a line of notation, but from four to six words per line is widely accepted, *Figure 10-139* (page 438). When a note string will contain more than six words, it should be divided into more than one line. The number of words in the multiple line note should be divided so as to balance the finished note. One long line followed by a drastically shorter line is bad form, *Figure 10-140* (page 438).

On occasion, the last word in a note string will have to be hyphenated. There are rules governing the dividing of words, according to syllables. If the proper point of division is not obvious from the makeup of the word, consult a dictionary. Do not hyphenate abbreviated words.

SAMPLE NOTES

The following is a list of sample notes of the types frequently used on technical drawings:

- All fillets and rounds R1/8
- 45 degree chamfer all edges
- Surfaces A & B parallel
- Stock .125 thick
- Heat treat
- All internal radii R.0625
- FAO
- Ream for 1/4 dowels
- All bends R5/32
- CBORE from bottom
- Identical bolts—both sides
- 3/32 × 1/16 oil groove
- 87 CSK 3/4 dia—4 holes
- Rounds R.125 unless otherwise specified
- 4 holes equally spaced
- 1/32 × 45 chamfer
- #7 Drill .75 deep
- Machine steel—4 reqd.
- Power brush all ext surfaces
- Sandblast before painting

Notice that these sample notes are very brief, concise, and to the point. Words such as "a," "an," "the," and "are" are used only sparingly. The notes are not always complete sentences in terms of proper grammar, but they are complete thoughts in terms of communication. Notice also the use of abbreviations such as FAO (finish all over), CBORE (counterbore), and CSK (countersink). Abbreviations can be used frequently to cut down on the length of notes and the time required to letter them. However, when used, abbreviations should always be placed in a legend to ensure consistency of interpretation.

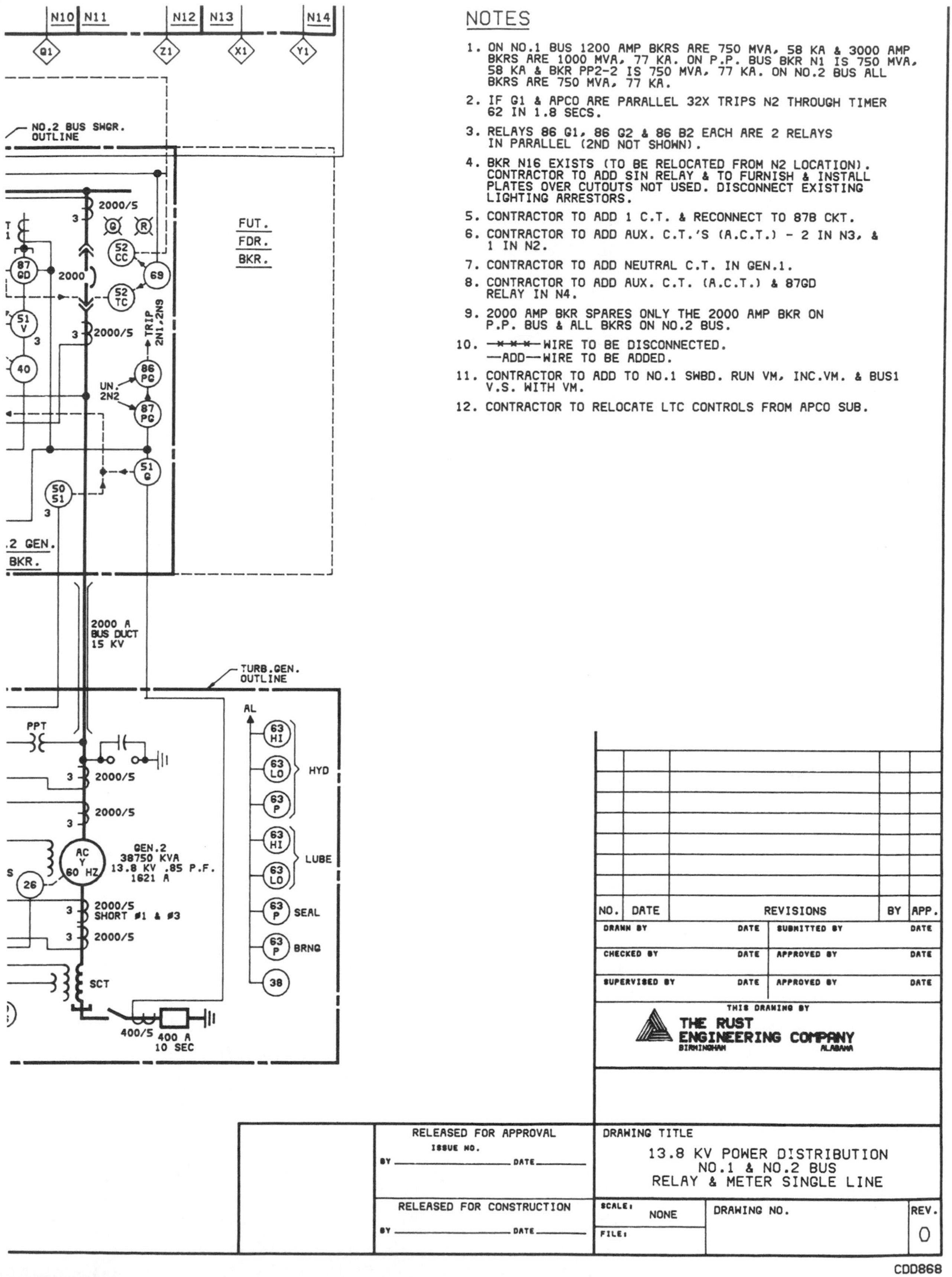

NOTES
1. ON NO.1 BUS 1200 AMP BKRS ARE 750 MVA, 58 KA & 3000 AMP BKRS ARE 1000 MVA, 77 KA. ON P.P. BUS BKR N1 IS 750 MVA, 58 KA & BKR PP2-2 IS 750 MVA, 77 KA. ON NO.2 BUS ALL BKRS ARE 750 MVA, 77 KA.
2. IF G1 & APCO ARE PARALLEL 32X TRIPS N2 THROUGH TIMER 62 IN 1.8 SECS.
3. RELAYS 86 G1, 86 G2 & 86 B2 EACH ARE 2 RELAYS IN PARALLEL (2ND NOT SHOWN).
4. BKR N16 EXISTS (TO BE RELOCATED FROM N2 LOCATION). CONTRACTOR TO ADD SIN RELAY & TO FURNISH & INSTALL PLATES OVER CUTOUTS NOT USED. DISCONNECT EXISTING LIGHTING ARRESTORS.
5. CONTRACTOR TO ADD 1 C.T. & RECONNECT TO 87B CKT.
6. CONTRACTOR TO ADD AUX. C.T.'S (A.C.T.) - 2 IN N3, & 1 IN N2.
7. CONTRACTOR TO ADD NEUTRAL C.T. IN GEN.1.
8. CONTRACTOR TO ADD AUX. C.T. (A.C.T.) & 87GD RELAY IN N4.
9. 2000 AMP BKR SPARES ONLY THE 2000 AMP BKR ON P.P. BUS & ALL BKRS ON NO.2 BUS.
10. —×—×—WIRE TO BE DISCONNECTED.
—ADD—WIRE TO BE ADDED.
11. CONTRACTOR TO ADD TO NO.1 SWBD. RUN VM, INC.VM. & BUS1 V.S. WITH VM.
12. CONTRACTOR TO RELOCATE LTC CONTROLS FROM APCO SUB.
N10 N11 N12 N13 N14
Q1 Z1 X1 Y1
NO.2 BUS SWGR. OUTLINE
FUT. FDR. BKR.
2000/5
52 CC
52 TC
69
87 GD
2000
51 V
40
TRIP 2N1,2N9
86 PG
87 PG
UN. 2N2
51 G
50 51
.2 GEN. BKR.
2000 A BUS DUCT 15 KV
TURB.GEN. OUTLINE
PPT
AL
63 HI
63 LO
63 P
HYD
LUBE
SEAL
BRNG
38
GEN.2 38750 KVA 13.8 KV .85 P.F. 1621 A
AC Y 60 HZ
26
2000/5 SHORT #1 & #3
SCT
400/5
400 A 10 SEC
NO. DATE REVISIONS BY APP.
DRAWN BY DATE SUBMITTED BY DATE
CHECKED BY DATE APPROVED BY DATE
SUPERVISED BY DATE APPROVED BY DATE
THIS DRAWING BY
THE RUST ENGINEERING COMPANY
BIRMINGHAM ALABAMA
RELEASED FOR APPROVAL
ISSUE NO.
BY DATE
RELEASED FOR CONSTRUCTION
BY DATE
DRAWING TITLE
13.8 KV POWER DISTRIBUTION
NO.1 & NO.2 BUS
RELAY & METER SINGLE LINE
SCALE: NONE
FILE:
DRAWING NO.
REV. 0
CDD868

FIGURE 10-126 *General notes are located over the title block* (Courtesy the Rust Engineering Co).

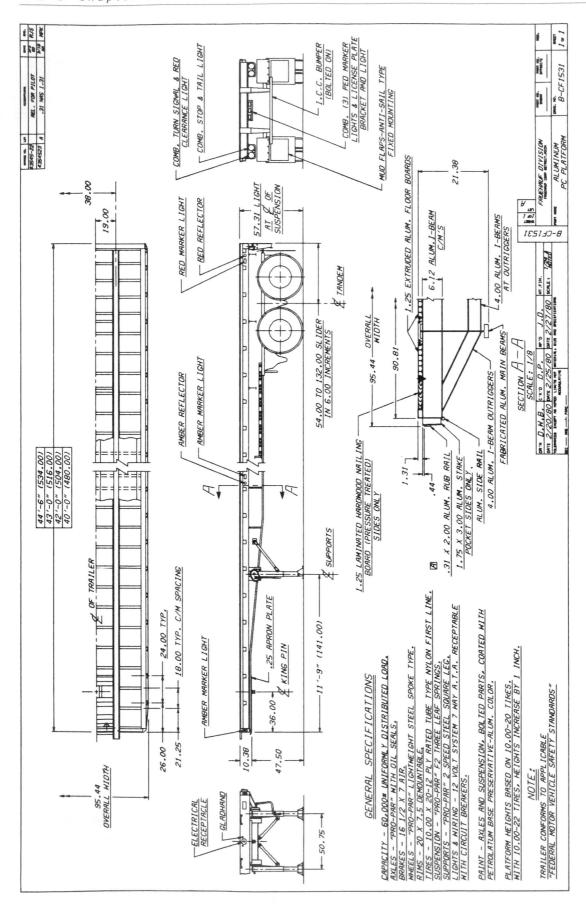

FIGURE 10-127 *Detail notes using leader lines (Courtesy Fruehauf Division, Fruehauf Corp)*

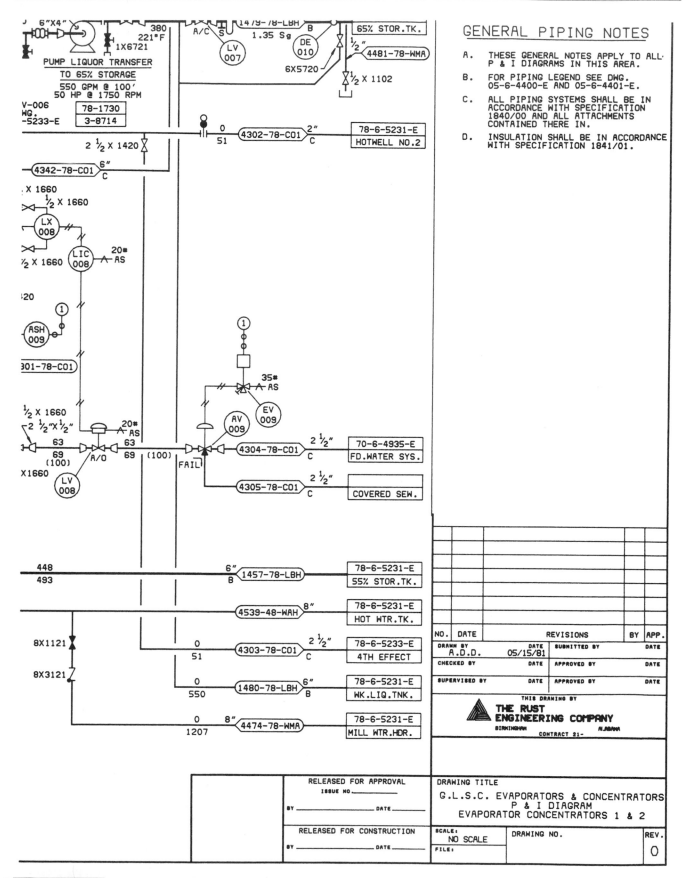

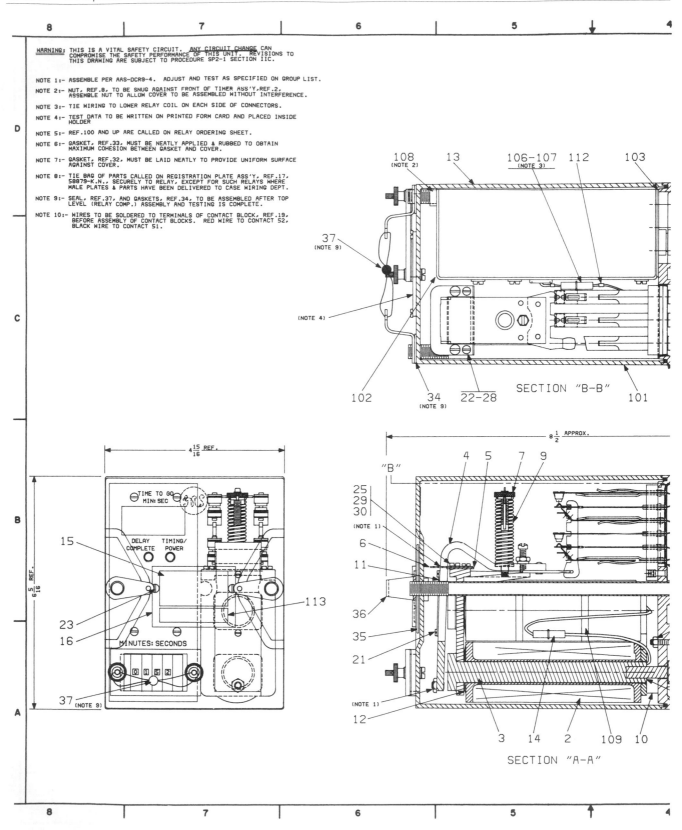

WARNING: THIS IS A VITAL SAFETY CIRCUIT. ANY CIRCUIT CHANGE CAN COMPROMISE THE SAFETY PERFORMANCE OF THIS UNIT. REVISIONS TO THIS DRAWING ARE SUBJECT TO PROCEDURE SP2-1 SECTION IIC.

NOTE 1:- ASSEMBLE PER AAS-DCR9-4. ADJUST AND TEST AS SPECIFIED ON GROUP LIST.

NOTE 2:- NUT, REF.8, TO BE SNUG AGAINST FRONT OF TIMER ASS'Y,REF.2, ASSEMBLE NUT TO ALLOW COVER TO BE ASSEMBLED WITHOUT INTERFERENCE.

NOTE 3:- TIE WIRING TO LOWER RELAY COIL ON EACH SIDE OF CONNECTORS.

NOTE 4:- TEST DATA TO BE WRITTEN ON PRINTED FORM CARD AND PLACED INSIDE HOLDER

NOTE 5:- REF.100 AND UP ARE CALLED ON RELAY ORDERING SHEET.

NOTE 6:- GASKET, REF.33, MUST BE NEATLY APPLIED & RUBBED TO OBTAIN MAXIMUM COHESION BETWEEN GASKET AND COVER.

NOTE 7:- GASKET, REF.32, MUST BE LAID NEATLY TO PROVIDE UNIFORM SURFACE AGAINST COVER.

NOTE 8:- TIE BAG OF PARTS CALLED ON REGISTRATION PLATE ASS'Y, REF.17, 58879-K.N., SECURELY TO RELAY, EXCEPT FOR SUCH RELAYS WHERE MALE PLATES & PARTS HAVE BEEN DELIVERED TO CASE WIRING DEPT.

NOTE 9:- SEAL, REF.37, AND GASKETS, REF.34, TO BE ASSEMBLED AFTER TOP LEVEL (RELAY COMP.) ASSEMBLY AND TESTING IS COMPLETE.

NOTE 10:- WIRES TO BE SOLDERED TO TERMINALS OF CONTACT BLOCK, REF.19, BEFORE ASSEMBLY OF CONTACT BLOCKS. RED WIRE TO CONTACT 52, BLACK WIRE TO CONTACT 51.

SECTION "B-B"

SECTION "A-A"

FIGURE 10-129 *General notes on drawings* (Courtesy General Railway Signal Corp)

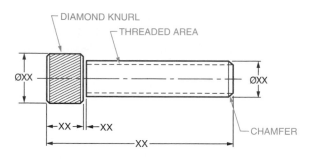

FIGURE 10-130   *Detail note connected with a leader line*

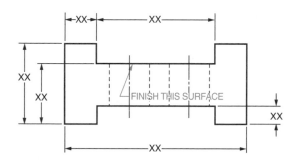

FIGURE 10-131   *Poor placement of a detail note*

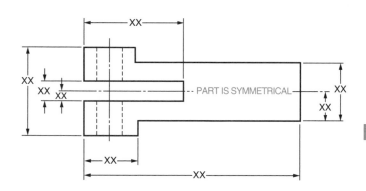

FIGURE 10-132   *Acceptable placement of a detail note when there is space*

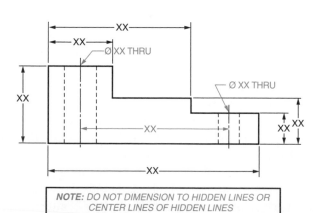

NOTE: DO NOT DIMENSION TO HIDDEN LINES OR CENTER LINES OF HIDDEN LINES

FIGURE 10-133   *Superimposing a note is poor practice*

LEGEND

1.	ASSY	=	ASSEMBLY
2.	FAO	=	FINISH ALL OVER
3.	MS	=	MACHINE STEEL
4.	FAB	=	FABRICATE
5.	CARB	=	CARBURIZE
6.	DVTL	=	DOVETAIL
7.	HT TR	=	HEAT TREAT
8.	KST	=	KEYSEAT
9.	TPR	=	TAPER
10.	SF	=	SPOTFACED

FIGURE 10-134   *Abbreviations are explained in a legend*

NOTE

POWER BRUSH ALL SURFACES. APPLY THREE COATS PAINT. STAMP PART WITH NUMBER 01929612.

FIGURE 10-135   *Proper punctuation of notes is important*

NOTE

ALL DIMENSIONS AND CONDITIONS SHALL BE FIELD CHECKED AND VERIFIED BY PROPER TRADES

FIGURE 10-136   *Indenting for clarity*

GENERAL NOTES ⟵ LETTERS .25 HIGH

ALL FILLETS AND ROUNDS .125 DIA. ⟵ LETTERS .125–
ALL SURFACES MUST BE FREE OF BURRS. ⟵ .1875 HIGH

FIGURE 10-137   *Proper lettering height is important*

NOTES

1. SPACING BETWEEN LINES OF THE SAME NOTE SHOULD BE .0625.

2. SPACING BETWEEN SEPARATE NOTES SHOULD BE .125∅.

FIGURE 10-138   *Spacing of notes*

NOTES

NOTES SHOULD BE LIMITED TO
4 TO 6 WORDS PER LINE

FOUR TO SIX WORDS

FIGURE 10-139 *Number of words per line limitations are important*

NOTES

AVOID LONG LINE IN A NOTE FOLLOWED BY
ONE SHORT LINE

FIGURE 10-140 *Lines in notes should be approximately equal*

GROUP TECHNOLOGY

Group technology is a manufacturing concept in which similar parts are grouped together in parts groups or families. Parts may be alike in two ways:

1. In their design characteristics

2. In the manufacturing processes required to produce them

By grouping similar parts into families, manufacturing personnel can improve efficiency. Such improvements are the result of advantages gained in such areas as setup time, standardization of processes, and scheduling.

Group technology can also improve the productivity of design personnel by decreasing the amount of work and time involved in designing a new part. Chances are a new part will be similar to an existing part in a given family. When this is the case, the new part can be developed by simply modifying the design of the existing part. Design modifications tend to require less time and work than new design. This is especially true in the age of CAD. The advantages of group technology are dealt with at greater length later in this chapter.

HISTORICAL DEVELOPMENT

Ever since the Industrial Revolution, manufacturing and engineering personnel have been searching for ways to optimize manufacturing processes. There have been numerous developments over the years since the Industrial Revolution mechanized production. Mass production and interchangeability of parts in the 1800s were major steps forward in optimizing manufacturing.

However, even with mass production and assembly lines, most manufacturing is done in small batches ranging from one workpiece to two or three thousand. In fact, even today over 70 percent of manufacturing involves batches of less than three thousand workpieces. Historically, less has been done to optimize small-batch production than has been done for assembly line work.

There have been attempts to standardize a variety of design tasks and some working queuing and sequencing in manufacturing, but until recent years design and manufacturing in small-batch settings has been somewhat random.

The underlying problem that has historically prevented significant improvements to small-batch manufacturing is that any solution must apply broadly to general production processes and principles rather than to a specific product. This is a difficult problem because the various workpieces in a small batch can be so random and different.

When manufacturing entered the age of automation and computerization, such developments as scheduling software, sequencing software, and materials requirements planning (MRP) systems became available to improve the production of both small and large batches. Even these developments have not optimized the production of small batch manufacturing lots. The problem has become even more critical because, since the end of World War II, the trend has been toward smaller batches.

In recent years the problems of small-batch production have finally begun to receive the attention necessary to bring about improvements. A major step is the ongoing development of group technology.

PART FAMILIES

It has already been stated that parts may be similar in design (size and shape) and/or in the manufacturing processes used to produce them. A group of such parts is called a *part family.* It is possible for parts in the same family to be very similar in design yet radically different in the area of production requirements. The opposite may also be true.

Figure 10-141 contains examples of two parts from the same family. These parts were placed in the same family based on design characteristics. They have exactly the same shape and size. However, you will notice that they differ in the area of production processes.

Part 1, after it is drilled, will go to a painting station for two coats of primer. Its dimensions must be held to a tolerance of plus or minus .125 inch. Part 2, after it is drilled, will go to a finishing station for sanding and buffing. Its dimensions require more restrictive toler-

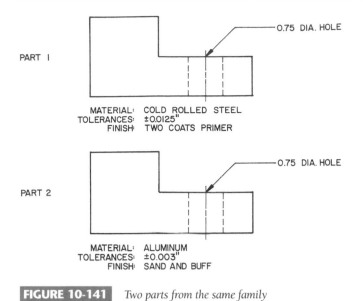

FIGURE 10-141   *Two parts from the same family*

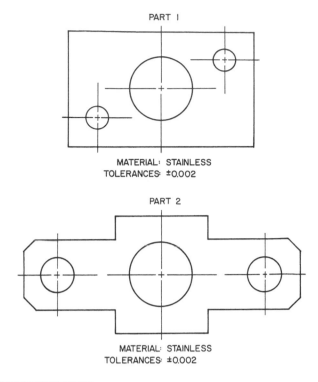

FIGURE 10-142   *Two parts from the same family*

ances of plus or minus .003 inch. The parts differ in material. The material for Part 1 is cold-rolled steel; Part 2 is aluminum.

Figure 10-141 contains examples of two parts from the same family. Although the design characteristics of these two parts are drastically different (i.e., different sizes and shapes), a close examination will reveal that they are similar in the area of production processes. Part 1 is made of stainless steel. Its dimensions must be held to a tolerance of plus or minus .002 inch, and three holes must be drilled through it. Part 2, in spite of the differences in its size and shape, has exactly the same manufacturing characteristics.

The parts shown in Figure 10-141, because of their similar design characteristics, were grouped in the same family. Such a family is referred to as a design part family. Those in *Figure 10-142* were grouped together because of similar manufacturing characteristics. Such a family is referred to as a *manufacturing part family*. The characteristics used in classifying parts are referred to as attributes.

By grouping parts into families, manufacturing personnel can cut down significantly on the amount of materials handled and movement wasted in producing them. This is because manufacturing machines can be correspondingly grouped into specialized work cells instead of the traditional arrangement of machines according to function (i.e., mills together, lathes together, drills together, etc.).

Each work can be specially configured to produce a given family of parts. When this is done, the number of setups, the amount of materials handling, the length of lead time, and the amount of in-process inventory are all reduced.

GROUPING PARTS INTO FAMILIES

Part grouping is not as simple a process as one might think. You already know the criteria used: design similarities and manufacturing similarities. But how does the actual grouping take place?

There are three methods that can be used for grouping parts into families, *Figure 10-143*.

1. Sight inspection
2. Route sheet inspection
3. Parts classification and coding

All three methods require the expertise of experienced manufacturing personnel.

Sight inspection is the simplest, least sophisticated method. It involves looking at parts, photos of parts, or drawings of parts. Through such an examination, experienced personnel can identify similar characteristics and group the parts accordingly. This is the easiest approach, especially for grouping parts by design attributes, but it is also the least accurate of the three methods.

The second method involves inspecting the routing sheets used to route the parts through the various operations to be performed. This can be an effective way to group parts into manufacturing part families, provided the routing sheets are correct. If they are, this method is more accurate than the sign inspection approach. This method is sometimes referred to as the PFA or production flow analysis method.

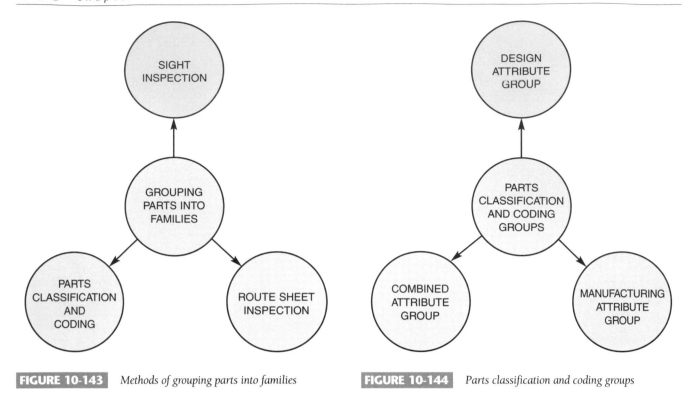

FIGURE 10-143 *Methods of grouping parts into families*

FIGURE 10-144 *Parts classification and coding groups*

The most widely used method for grouping parts is the third method: parts classification and coding. This is also the most sophisticated, most difficult, and most time-consuming method. Parts classification and coding is complex enough to require a more in-depth treatment than the other two methods.

PARTS CLASSIFICATION AND CODING

Parts classification and coding is a method in which the various design and/or manufacturing characteristics of a part are identified, listed, and assigned a code number. Recall that these characteristics are referred to as *attributes*. This is a general approach used in classifying and coding parts. There are many different systems that have been developed for actually carrying out the process, none of which has emerged as the standard.

The many different classification and coding systems that have been developed all fall into one of three groups, *Figure 10-144*.

1. Design attribute group
2. Manufacturing attribute group
3. Combined attribute group

Students of CAD/CAM should be familiar with how systems are grouped.

Design Attribute Group. Classification and coding systems that use design attributes as the qualifying criteria fall into this group. Commonly used design attri-

butes include dimensions, tolerances, shape, finish, and material.

Manufacturing Attribute Group. Classification and coding systems that use manufacturing attributes as the qualifying criteria fall into this group. Commonly used manufacturing attributes include production processes, operational sequence, production time, tools required, fixtures required, and batch size.

Combined Attribute Group. There are advantages in using design attributes and advantages in using manufacturing attributes. Systems that fall into the design attribute group are particularly advantageous if the goal is design retrieval. Those in the manufacturing group are better if the goal is any of a number of production-related functions. However, there is a need for systems that combine the best characteristics of both. Such systems are both design and manufacturing attributes.

Sample Parts Classification and Coding System. Some companies develop their own parts classification and coding systems. But this can be an expensive and time-consuming approach. The more widely used approach is to purchase a commercially prepared system. There are several such systems available. However, the most widely used of these is the Opitz system. It is a good example of a parts classification and coding system and how one works.

OPITZ SYSTEM

Classification and coding systems use alphanumeric symbols to represent the various attributes of a part. One of the many classification and coding systems is the Opitz system. This system uses characters in 13 places to code the attributes of parts, and hence, to classify them. These digit places are represented as follows:

12345 6789 ABCD

The first five digits (12345) code the major design attributes of a part. The next four digits (6789) are for coding manufacturing-related attributes and are called the supplementary code. The letters (ABCD) code the production operation and sequence.

The alphanumeric characters shown represent places. For example, the actual numeral used in each place can be 0–9. The numeral used in the 1 place indicates the length-to-diameter ratio of the part. The numeral used in the 2 place indicates the external shape of the part. The numeral used in the 3 place indicates the internal shape of the part. The numeral used in the 4 place indicates the type of surface machining. The numeral used in the 5 place indicates gear teeth and auxiliary holes. With such a system, a part might be coded as follows:

20801

The 2 means that the part has a certain length-to-diameter ratio. The first 0 means the part has no external shape elements. The 8 means the part has an internal thread. The second 0 means no surface machining is required. The 1 means the part is axial, not on pitch.

Figure 10-145 is a chart that shows the basic overall structure of the Opitz system for parts classification and coding. *Figure 10-146* is a chart used for assigning an Opitz code to rotational parts. *Figure 10-147* is a draw-

ing of an actual part that has been assigned an Opitz code of 15400. This code was arrived at as follows:

STEP 1 The total length of the part is divided by the overall diameter:

$$\frac{1.75(L)}{1.25(D)} = 1.40$$

Since 1.40 is greater than .5 but less than 3, the first digit in the code is 1.

STEP 2 An overall description of the external shape of the part would read ". . . a rotational part that is stepped on both ends with one stepped end threaded." Consequently, the most appropriate second digit in the code is 5.

STEP 3 A description of the internal shape of the part would read ". . . a through hole." Consequently, the third digit in the code is 4.

STEP 4 Upon examination, it can be seen that no surface machining is required. Therefore, the fourth digit in the code is 0.

STEP 5 It can also be seen that no auxiliary holes or gear teeth are required. Therefore, the fifth digit in the code is also 0.

ADVANTAGES AND DISADVANTAGES OF GROUP TECHNOLOGY

There are several advantages that can be realized through the application of group technology. These advantages include:

1. Improved design

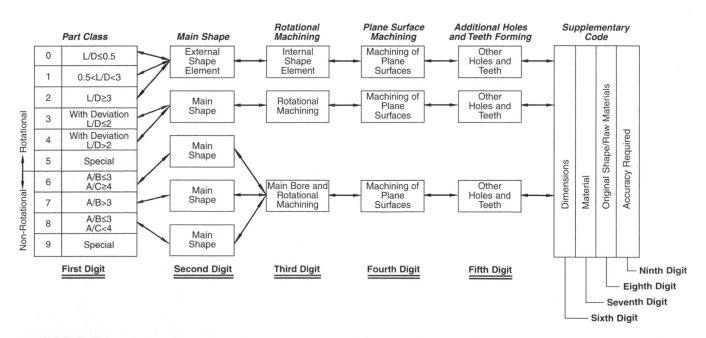

FIGURE 10-145 *Basic structure of the Opitz system*

PART CLASS		MAIN SHAPE EXTERNAL		ROTATIONAL MACHINING INTERNAL		PLANE SURFACE MACHINING	ADDITIONAL HOLES AND TEETH
	L/D<0.5	Smooth, no shape elements		No hole, no breakthrough		No surface machining	No auxiliary hole
1	0.5<L/D<3	Stepped to one end or smooth	No shape elements	Smooth or stepped to one end	No shape elements	Surface plane and/or curved in one direction, external	Axial, not on pitch circle diameter
2	L/D≤3		Thread		Thread	External plane surface related by graduation around a circle	Axial on pitch circle diameter
3	—		Functional groove		Functional groove	External groove and/or slot	Radial, not on pitch circle diameter
4	—	Stepped to both ends	No shape elements	Stepped to both ends	No shape elements	External spline (polygon)	Axial and/or radial and/or other direction
5	—		Thread		Thread	External plane surface and/or slot, external spline	Axial and/or radial on PCD and/or other directions
6	—		Functional groove		Functional groove	Internal plane surface and/or slot	Spur gear teeth
7	—	Functional cone		Functional cone		Internal spline (polygon)	Bevel gear teeth
8	—	Operating thread		Operating thread		Internal and external polygon, groove and/or slot	Other gear teeth
		First Digit		**Second Digit** → Third Digit		**Fourth Digit**	**Fifth Digit**

Rotational ↔ Non-Rotational

No gear teeth / With gear teeth

FIGURE 10-146 *Assigning a code using the Opitz system*

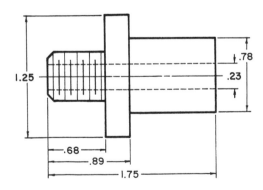

FIGURE 10-147 *Sample part that can be coded using the Opitz system*

2. Enhanced standardization
3. Reduced materials handling
4. Simplified production scheduling
5. Improved quality control

IMPROVED DESIGN

Group technology allows designers to use their time more efficiently and productively by decreasing the amount of new design work required each time a part is to be designed. When a new part is needed, its various attributes can be listed. Then, an existing part with as many of these attributes as possible can be identified and retrieved. The only new design required is that which relates to attributes of the new part not contained in the existing part. Because this characteristic of group technology tends to promote design standardization, additional design benefits accrue.

ENHANCED STANDARDIZATION

Parts are classified into groups according to their similarities. The more similarities, the better. Consequently, enhanced standardization is promoted by group tech-

Industry Application

GROUP TECHNOLOGY AND GLOBAL COMPETITION

"We need to adopt the group technology concept," said Juan Martinez. Martinez is the chief designer for Magna-Tech Manufacturing, a leading producer of metal components used in the manufacture of automobiles. In recent years Magna-Tech's competition has become both global and fierce. In order to survive and prosper, the company needs to optimize the performance of every employee and every process from design and drafting to manufacturing.

As the design and drafting department's representative on the company's management team, Martinez is expected to propose strategies for cutting costs and improving productivity, and he is expected to propose strategies to help Magna-Tech become more competitive on a continual basis. The company's CEO has made it clear to all of his managers that he is looking for strategies that will give Magna-Tech a sustainable competitive advantage in the global marketplace. This is why Juan Martinez scheduled the current departmental meeting. He needs input from his design and drafting personnel.

"I want us to brainstorm this morning," said Martinez. "Before the next manager's meeting, I need a list of benefits our company can gain by adopting group technology." "The first benefit is that group technology will cut down on the amount of time needed to design a part," said Margaret Flagstone. "Right now we practically start over every time we have a new part we have already designed. With group technology we could place the new part in a part family. Once we know what family it fits in, much of the design is already done."

"Excellent point," said Martinez. "I'll tell you what we might do. What if I write four or five benefits on the flip chart and see if you agree with them? Then when we have our top five benefits we can divide ourselves into five small groups and explore each benefit in depth. I'd like to have some concrete examples to offer at the manager's meeting." With that, Martinez began writing benefits on a flip chart.

nology. As you have already seen, design factors account for part of the enhanced standardization. Setups and tooling also become more standardized because similar parts require similar setups and similar tooling.

REDUCED MATERIALS HANDLING

One of the major elements of group technology is the arrangement of machines into specialized work cells, each cell producing a given family of parts. Traditionally, machines are arranged by function (i.e., lathes together in one group, mills together in another, etc.). A functional arrangement causes excessive movement of parts from machine to machine and back. Specialized work cells reduce such movement to a minimum and, in turn, reduce materials handling.

SIMPLIFIED PRODUCTION SCHEDULING

With machines grouped into specialized work cells and arrangement of parts into families, there is less work center scheduling requiring less scheduling overall. Group technology involves less sophisticated scheduling.

IMPROVED QUALITY CONTROL

Quality control is improved through group technology because each work cell is responsible for a specific family of parts. This does two things, both of which tend to improve the quality of parts: (1) it promotes pride in the work among employees assigned to each work cell, and (2) it makes production errors and problems easier to trace to a specific source. In traditional manufacturing shops, each work center performs certain tasks, but no one center is responsible for an entire part. Such an approach does not promote quality control.

In spite of these advantages, there are several disadvantages associated with group technology. Transitioning from machines in a traditional shop into specialized work cells can cause a major disruption. Such disruptions translate quickly into monetary losses. As a consequence, some managers are reluctant to pursue group technology.

Another problem with group technology is that parts classification coding is a time-consuming, difficult, and expensive process, especially in the initial stages. This causes some managers to balk at the expense.

Summary

- The three dimensioning systems used on technical drawings in the United States are the metric, decimal-inch, and fractional. In all of these systems, dimensions are oriented to be read from the bottom or right-hand side of the drawing sheet.

- The components that are common to all dimensioning systems are extension lines, dimension lines, arrowheads, leader lines, and dimensions.

- The four steps in laying out dimensions are as follows: (1) place eight layout lines on the drawing sheet; (2) lightly draw in guidelines for the dimensions; (3) darken the extension lines and dimension lines and add dimensions; and (4) add arrowheads.

- Tolerancing-related concepts with which students should be familiar are actual size allowance, bilateral tolerance, datum, least material condition, limits, maximum material condition, nominal dimension, tolerance, unilateral tolerance, fit, clearance fit, interference fit, and transition fit.

- Not all information needed to convey a design can be given graphically. In such cases, notes are used to provide the additional information. Notes are brief, carefully worded statements placed on drawings to convey information not adequately explained graphically.

- Group technology is a manufacturing concept in which similar parts are grouped together in parts families. Such parts may be alike in two ways: (1) in their design characteristics and (2) in the manufacturing processes required to produce them. Advantages of group technology include improved design, enhanced standardization, reduced materials handling, simplified production scheduling, and improved quality control.

Review Questions

Answer the following questions either true or false.

1. The most frequently used dimensioning standard is fractional dimensioning.

2. The proper way to indicate on a drawing that a dimension is NTS is to put a wavy line under the dimension.

3. An extension line should extend beyond the dimension line approximately 1/8″.

4. The key word to remember when making arrowheads, regardless of the size, is "uniform."

5. Unidirectional dimensioning is aligning all dimensions with the dimension line.

6. Notes are brief, carefully worded statements placed on drawings to convey information not covered in the graphic.

7. Notes should be oriented on a drafting sheet horizontally.

8. The term *verification* as it relates to notes is printing a list of your notes and asking an experienced drafter to ensure that they are not open to multiple interpretations.

9. Group technology is a concept by which similar parts are grouped together in part groups or families.

10. Attributes are characteristics used to group parts.

11. General notes are specific and pertain to a particular element or characteristic of a part.

12. The proper height for lettering note titles is 1/4″.

13. Parts are grouped into families by their design characteristics or the manufacturing process required to produce them.

Answer the following questions by selecting the best answer.

1. Of the dimensioning systems used, which one is most frequently used on mechanical technical drawings?
   a. Metric dimensioning
   b. Decimal-inch dimensioning
   c. Fractional dimensioning
   d. Proportional dimensioning

2. Which is the correct rule when labeling for a metric dimension that is less than one millimeter?
   a. A decimal is not required
   b. Neither a zero nor a decimal is required
   c. Place a zero to the right of the decimal
   d. None of the above

3. Which is the correct way to show the line separating the numerator and denominator of a fraction?
   a. /
   b. —
   c. \
   d. All of the above

4. Which is not a component of all dimensioning systems?
   a. Arrowheads
   b. Extension lines
   c. Leader lines
   d. Orientation lines

5. The recommended length/width ratio of arrowheads is that they are:
   a. Twice as long as they are wide.
   b. Three times as long as they are wide.

c. Twice as wide as they are long.

d. Three times as wide as they are long.

6. Which is the proper location for general notes?
    a. In the lower left-hand corner
    b. In the upper left-hand corner
    c. Above the title block
    d. Below the title block

7. Which of the following is not true regarding notation?
    a. The number of words in a multiple-line note should be divided to balance the finished note.
    b. Four to six words per line is widely accepted.
    c. There are hard and fast rules regarding notation.
    d. Hyphenation, by syllables, is accepted.

8. Which of the following is **not** a method for grouping parts into families?
    a. Size and shape
    b. Sight inspection
    c. Route sheet inspection
    d. Parts classification and coding

9. Which of the following is **not** a commonly applied design attribute?
    a. Dimensions
    b. Weight
    c. Tolerances
    d. Shape

10. Which of the following is **not** a commonly applied manufacturing attribute?
    a. Production processes
    b. Operation sequence
    c. Sight inspection
    d. Tools required

11. Which of the following is **not** an advantage of group technology?
    a. Improved design
    b. Reduced materials handling
    c. Easier parts classification coding
    d. Simplified production scheduling

Chapter 10 Problems

The following problems are intended to give beginning drafters and engineers practice in applying the various dimensioning techniques required on technical drawings in industry. Students should follow the general instructions that apply to all problems, and any additional instructions provided for special problems. All problems should be dimensioned in strict accordance with the rules set forth in this chapter.

General Instructions

1. Study each problem carefully.

2. Make a complete sketch of the problem, applying all dimensions on the sketch. Note on the sketch the proper spacing between views, between the object and first dimension lines, and between successive dimension lines.

3. Compare your fully dimensioned sketch against all applicable sections of this chapter and "The Summary of Dimensioning Rules."

NOTE: These instructions do not apply to Problems 10-47 through 10-50.

CAD INSTRUCTIONS

Students who plan to complete the following problems on a CAD system should begin with these general instructions. Before reading the specific instructions for each activity, go through each step in the following planning checklist. The checklist applies to any CAD system and will help ensure the optimum use of your time and resources.

1. Analyze the problem carefully. Decide exactly what you are being asked to do.

2. Determine what resources and references (if any) you will need in order to complete the problem. Collect those references and resources and have them readily available.

3. Decide if any particular standards apply to the project and have those standards available.

4. Determine what types of views will be required and how many of each.

5. Determine what the final plotted scale of the drawing will need to be, and select the appropriate paper size for plotting/printing.

6. Plan your drawing sequence. In what order will you develop the drawing (i.e., lines, features, dimension lines, leaders, dimensions)?

7. Review the various CAD commands you will have to use in order to develop the drawing.

8. Examine your CAD system to ensure that everything is in working order, then begin the project.

PROBLEMS 10-1 THROUGH 10-24

1. Use the grid to determine all dimensions.
2. Redraw the views shown, and fully dimension your drawing.

PROBLEM 10-1

Ⓐ Ⓑ Ⓒ

Ⓓ Ⓔ Ⓕ

Ⓖ Ⓗ Ⓘ

Ⓙ Ⓚ Ⓛ

PROBLEM 10-2

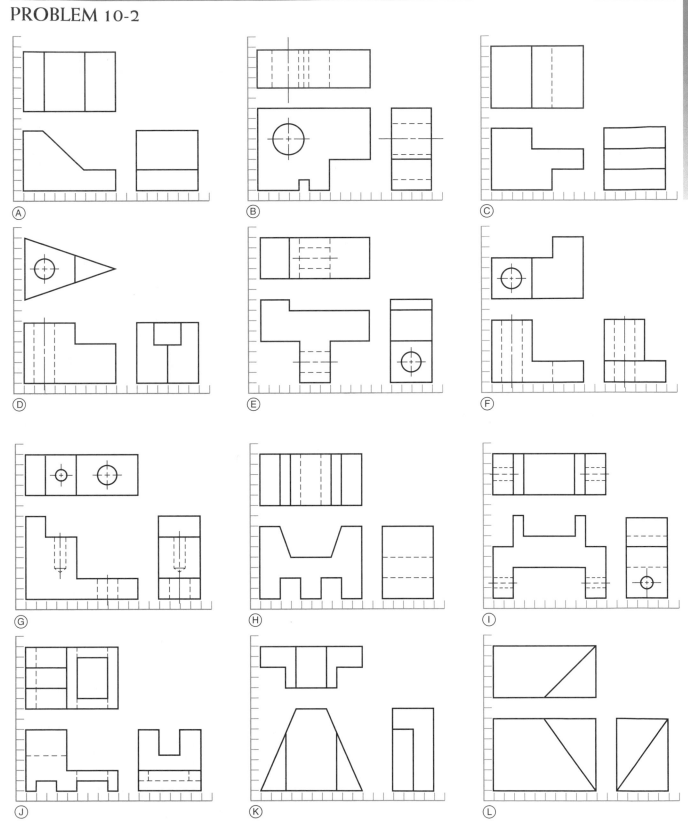

PROBLEM 10-3

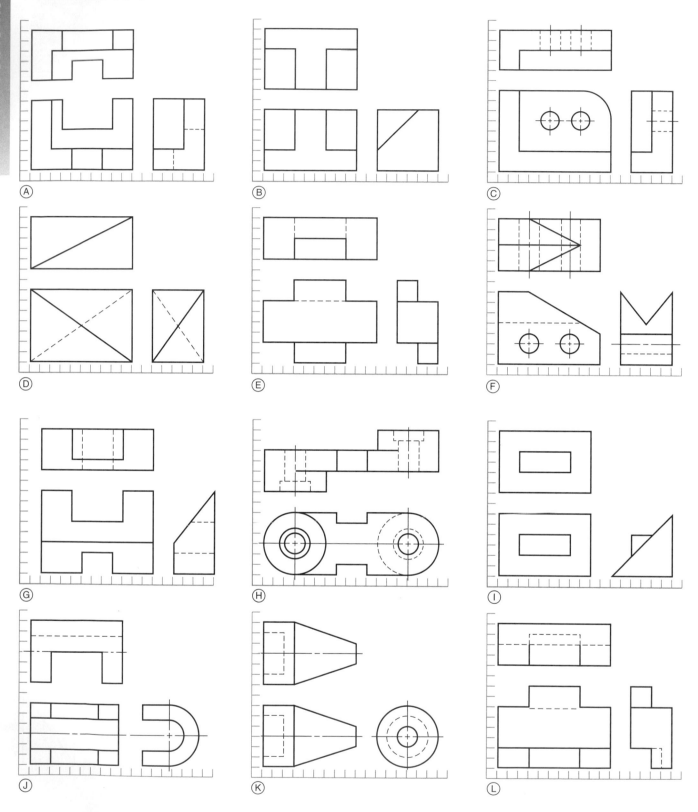

PROBLEM 10-4

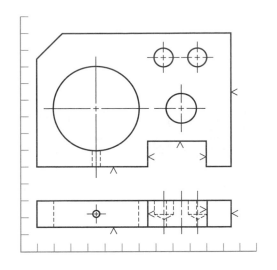

PROBLEM 10-7

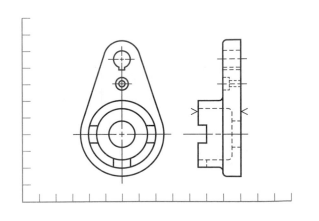

PROBLEM 10-5

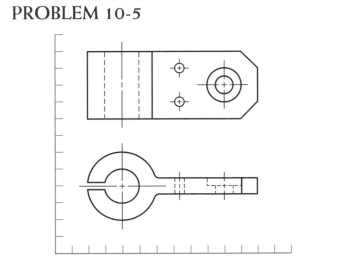

PROBLEM 10-8

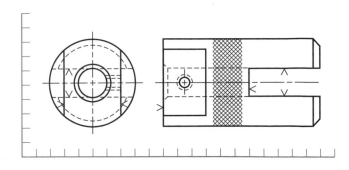

PROBLEM 10-9

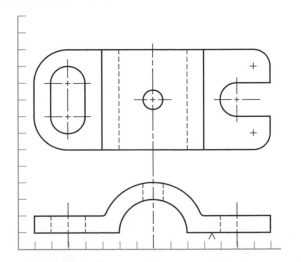

PROBLEM 10-6

PROBLEM 10-10

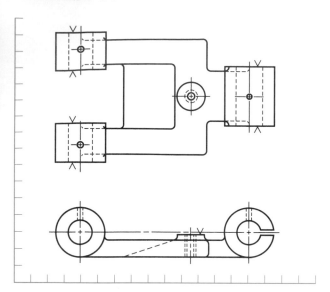

PROBLEM 10-13

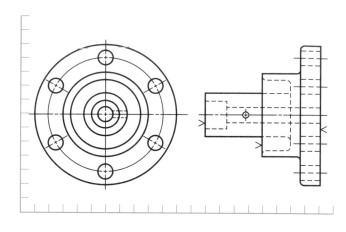

PROBLEM 10-11

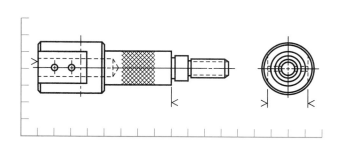

PROBLEM 10-14

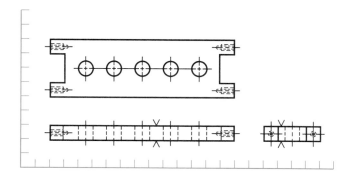

PROBLEM 10-12

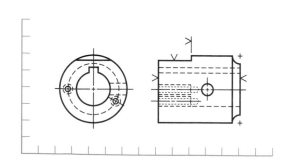

PROBLEM 10-15

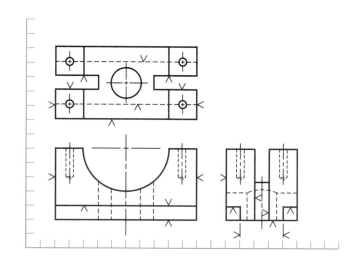

PROBLEM 10-16

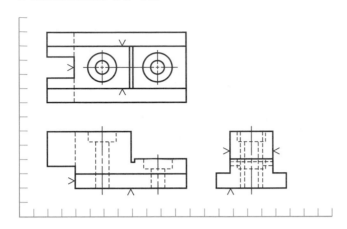

PROBLEM 10-19

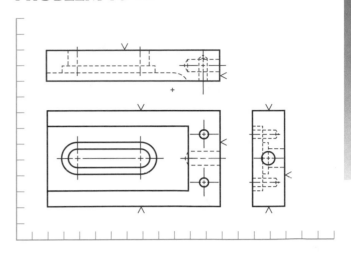

PROBLEM 10-17

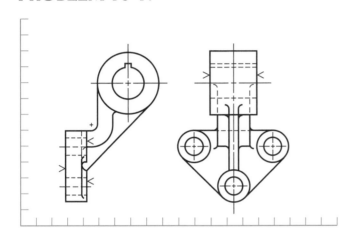

PROBLEM 10-20

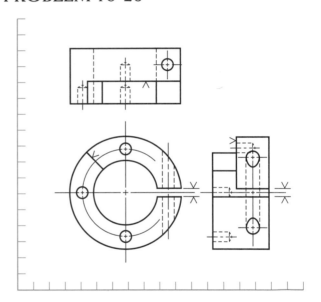

PROBLEM 10-18

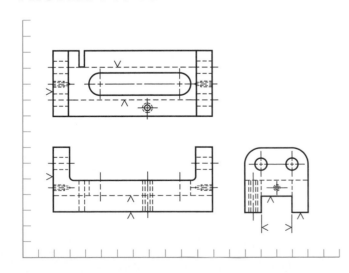

PROBLEM 10-21

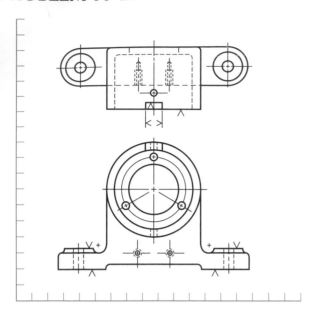

PROBLEM 10-23

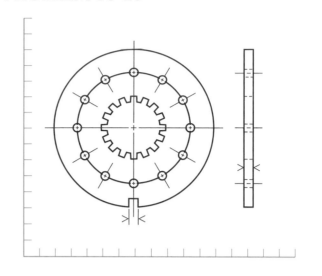

PROBLEM 10-22

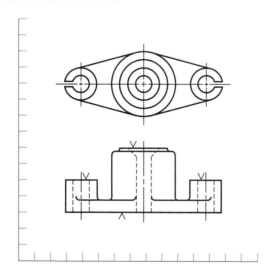

PROBLEM 10-24

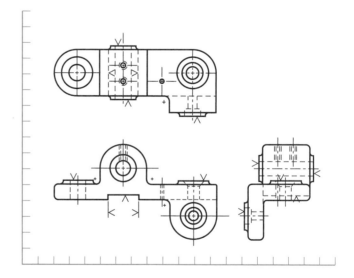

PROBLEMS 10-25 THROUGH 10-46

1. Convert the isometric drawings provided into orthographic drawings, showing as many views and/or sections as necessary to communicate the design.
2. Fully dimension your drawing.

PROBLEM 10-25

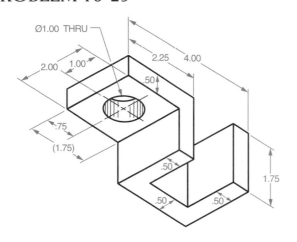

PROBLEM 10-26

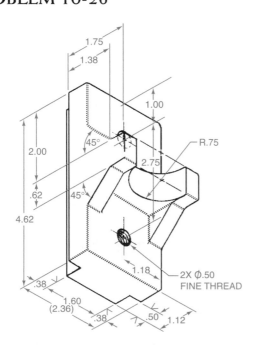

PROBLEM 10-27

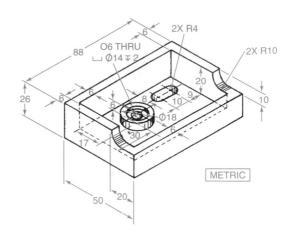

PROBLEM 10-28

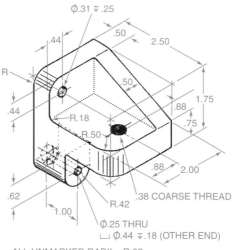

ALL UNMARKED RADII = R.09

PROBLEM 10-29

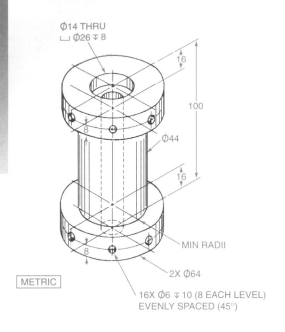

Ø14 THRU
⌴ Ø26 ↧ 8

16

100

8

Ø44

16

MIN RADII

8

2X Ø64

16X Ø6 ↧ 10 (8 EACH LEVEL)
EVENLY SPACED (45°)

METRIC

PROBLEM 10-31

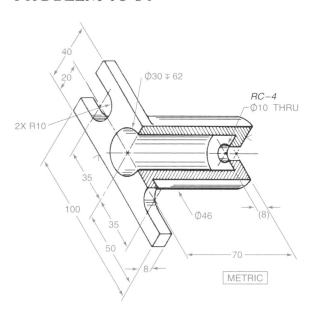

40

20

Ø30 ↧ 62

RC–4
Ø10 THRU

2X R10

35

100

35

Ø46

50

(8)

8

70

METRIC

PROBLEM 10-30

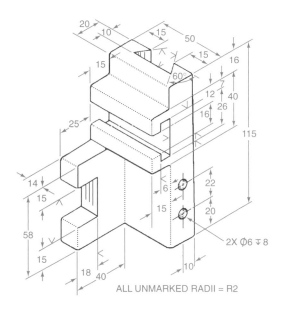

20

10

15

50

15

15

16

15

60°

12

40

16

26

25

115

14

6

22

15

15

20

58

2X Ø6 ↧ 8

15

18

40

10

ALL UNMARKED RADII = R2

PROBLEM 10-32

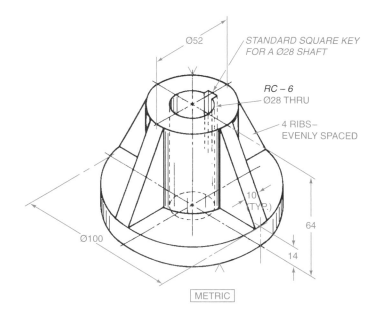

Ø52

STANDARD SQUARE KEY
FOR A Ø28 SHAFT

RC – 6
Ø28 THRU

4 RIBS–
EVENLY SPACED

10
(TYP.)

64

Ø100

14

METRIC

PROBLEM 10-33

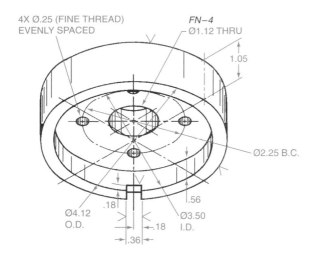

4X Ø.25 (FINE THREAD)
EVENLY SPACED

FN–4
Ø1.12 THRU

1.05

Ø2.25 B.C.

.56

.18

Ø4.12
O.D.

Ø3.50
I.D.

.18

.36

PROBLEM 10-34

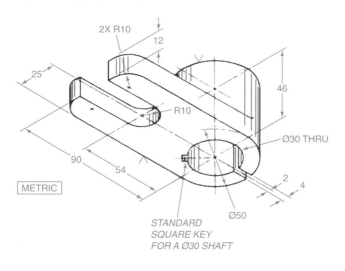

2X R10

12

25

46

R10

Ø30 THRU

90

54

2

4

METRIC

Ø50

*STANDARD
SQUARE KEY
FOR A Ø30 SHAFT*

PROBLEM 10-35

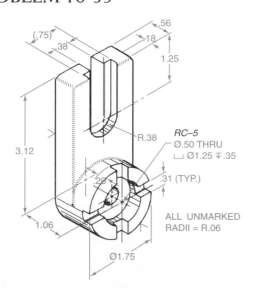

.56

(.75)

.18

.38

1.25

3.12

R.38

RC–5
Ø.50 THRU
⌴ Ø1.25 ⍌ .35

.26

.31 (TYP.)

1.06

*ALL UNMARKED
RADII = R.06*

Ø1.75

PROBLEM 10-36

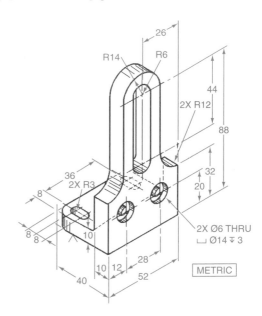

26

R14

R6

44

2X R12

88

36
2X R3

32

8

20

8
8

10

2X Ø6 THRU
⌴ Ø14 ⍌ 3

10 12

28

40

52

METRIC

PROBLEM 10-37

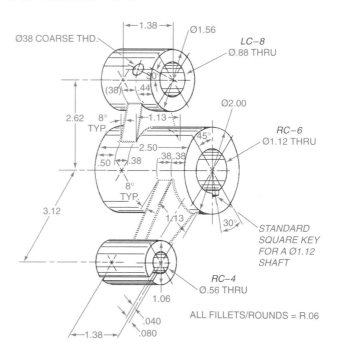

Ø38 COARSE THD.

1.38

Ø1.56

LC–8
Ø.88 THRU

30°

(38)

.44

2.62

8°
TYP.

1.13

Ø2.00

RC–6
Ø1.12 THRU

45°

2.50

.50

.38

38 38

8°
TYP.

1.13

30°

*STANDARD
SQUARE KEY
FOR A Ø1.12
SHAFT*

3.12

RC–4
Ø.56 THRU

1.06

.040

.080

1.38

ALL FILLETS/ROUNDS = R.06

PROBLEM 10-38

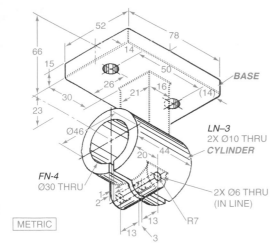

BASE

LN–3
2X Ø10 THRU
CYLINDER

FN-4
Ø30 THRU

2X Ø6 THRU
(IN LINE)

R7

METRIC

NOTE: CYLINDER IS CENTERED ON THE BASE

PROBLEM 10-40

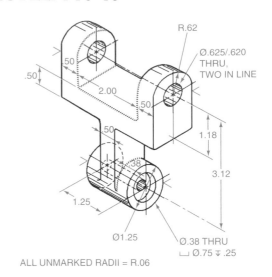

R.62

Ø.625/.620
THRU,
TWO IN LINE

ALL UNMARKED RADII = R.06

Ø.38 THRU
⌴ Ø.75 ⫱ .25

Ø1.25

PROBLEM 10-39

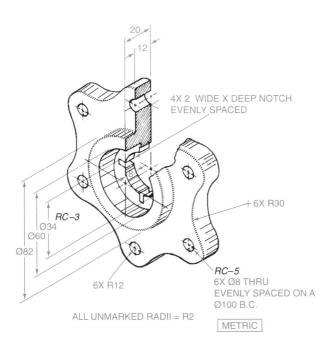

4X 2 WIDE X DEEP NOTCH
EVENLY SPACED

6X R30

RC–3
Ø34
Ø60
Ø82

RC–5
6X Ø8 THRU
EVENLY SPACED ON A
Ø100 B.C.

6X R12

ALL UNMARKED RADII = R2   METRIC

PROBLEM 10-41

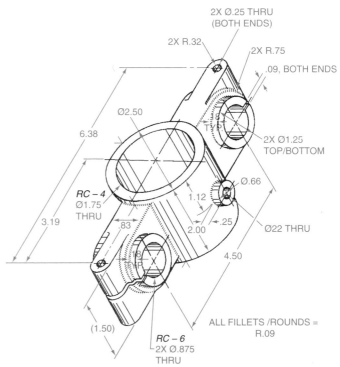

2X Ø.25 THRU
(BOTH ENDS)

2X R.32

2X R.75

.09, BOTH ENDS

Ø2.50

6.38

2X Ø1.25
TOP/BOTTOM

RC – 4
Ø1.75
THRU

Ø.66

3.19

1.12

.83

.25

2.00

Ø22 THRU

4.50

(1.50)

RC – 6
2X Ø.875
THRU

ALL FILLETS /ROUNDS =
R.09

PROBLEM 10-42

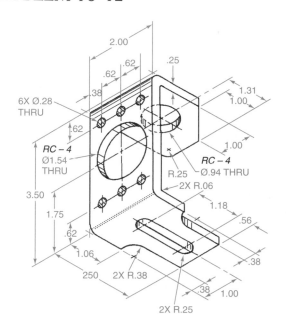

PROBLEM 10-44

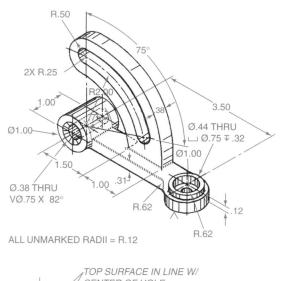

ALL UNMARKED RADII = R.12

TOP SURFACE IN LINE W/
CENTER OF HOLE

PROBLEM 10-43

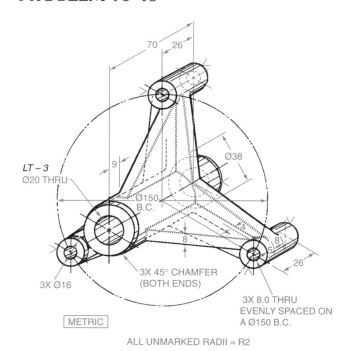

METRIC

ALL UNMARKED RADII = R2

PROBLEM 10-45

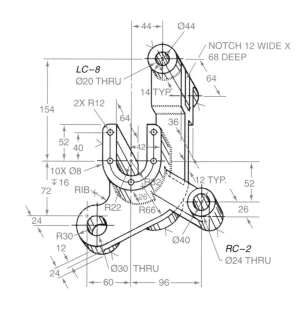

PROBLEM 10-46

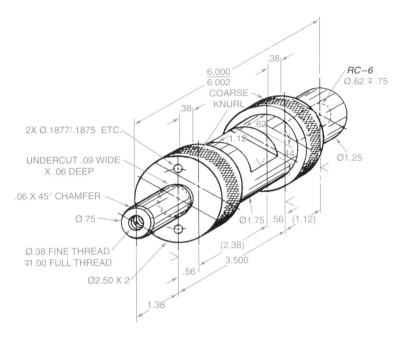

PROBLEM 10-47

Convert Problem 10-47 into a tabular drawing.

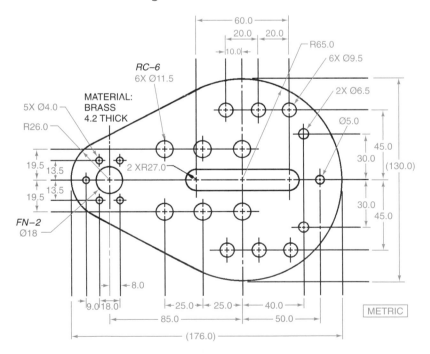

PROBLEM 10-48

Convert Problem 10-48 into a standard coordinate dimensioning system drawing.

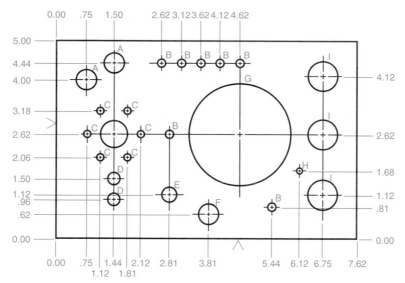

MATERIAL: STEEL .12 THICK

HOLE	SIZE
I	.750/.752
H	.120
G	2.562/2.565
F	.510/.512
E	.375/.380
D	.320
C	.188
B	.218/.220
A	.500

PROBLEM 10-49

Convert Problem 10-49 into a tabular drawing.

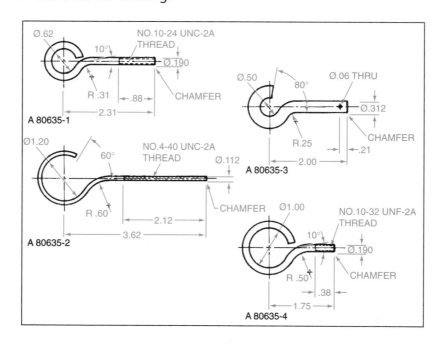

PROBLEM 10-50
Draw parts AO608-5 and AO608-8 full size.

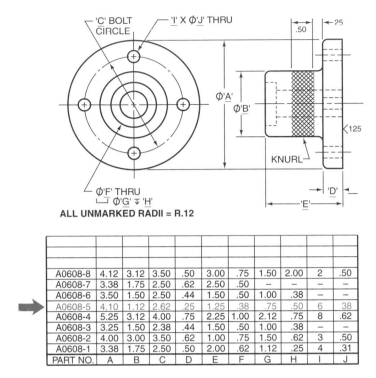

ALL UNMARKED RADII = R.12

PART NO.	A	B	C	D	E	F	G	H	I	J
A0608-8	4.12	3.12	3.50	.50	3.00	.75	1.50	2.00	2	.50
A0608-7	3.38	1.75	2.50	.62	2.50	.50	–	–	–	–
A0608-6	3.50	1.50	2.50	.44	1.50	.50	1.00	.38	–	–
A0608-5	4.10	1.12	2.62	.25	1.25	.38	.75	.50	6	38
A0608-4	5.25	3.12	4.00	.75	2.25	1.00	2.12	.75	8	.62
A0608-3	3.25	1.50	2.38	.44	1.50	.50	1.00	.38	–	–
A0608-2	4.00	3.00	3.50	.62	1.00	.75	1.50	.62	3	.50
A0608-1	3.38	1.75	2.50	.50	2.00	.62	1.12	.25	4	.31

PROBLEMS 10-51 AND 10-52
Carefully fill in all blank lines.

PROBLEM 10-51

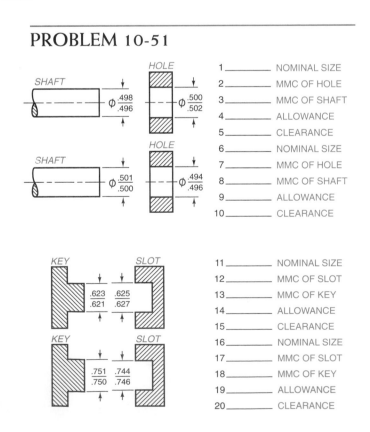

1	_____	NOMINAL SIZE
2	_____	MMC OF HOLE
3	_____	MMC OF SHAFT
4	_____	ALLOWANCE
5	_____	CLEARANCE
6	_____	NOMINAL SIZE
7	_____	MMC OF HOLE
8	_____	MMC OF SHAFT
9	_____	ALLOWANCE
10	_____	CLEARANCE
11	_____	NOMINAL SIZE
12	_____	MMC OF SLOT
13	_____	MMC OF KEY
14	_____	ALLOWANCE
15	_____	CLEARANCE
16	_____	NOMINAL SIZE
17	_____	MMC OF SLOT
18	_____	MMC OF KEY
19	_____	ALLOWANCE
20	_____	CLEARANCE

PROBLEM 10-52

PROB	NOM SIZE	BASIC SIZE	DIMENSION OF HOLE	DIMENSION OF SHAFT	HOLE TOLERANCE	SHAFT TOLERANCE	ALLOWANCE	CLEARANCE
1			.625 / .630	—		.005	+.002	
2	1.25		—	—	.001	.002	+.001	
3			.312 / .314	.316 / .314				
4	.75		—	—	.003	.002	+.001	
5			.812 / .813	.815 / .814				
6			22.2 / 22.3	22.5 / 22.4				
7	20		—	—	0.4	0.3	+0.2	
8			9.6 / 9.4	9.8 / 9.4				
9	30		—	—	0.2	0.4	+0.2	
10			15.6 / 15.4	—		0.6	+0.4	

(INCH: problems 1–5; METRIC: problems 6–10)

PROBLEM 10-53

Refer to the mounting guide drawing, No. 392450, and answer the following questions:

1. What is dimension A?
2. What tolerance is expected for two-place dimensions?
3. What are the dimension limits (maximum and minimum) of dimension B?
4. What size tap hole is specified for the 5/8 threaded hole?
5. What thread series is indicated for the 5/8 threaded hole?
6. When was the drawing last revised?
7. What are the dimension limits (maximum and minimum) of dimension C?
8. What is the maximum dimension of F?
9. For one assembly of the mounting guide, two stud bolts that are supposed to go through the two holes D are found to be spaced too far apart. The two stud bolts have a center-to-center distance of 1.625" and are .373" in diameter. The .422" D holes are 1.563" apart. One remedy for this type of problem is to enlarge the mating holes. Therefore, what is the minimum diameter for the two holes D that would permit the assembly?

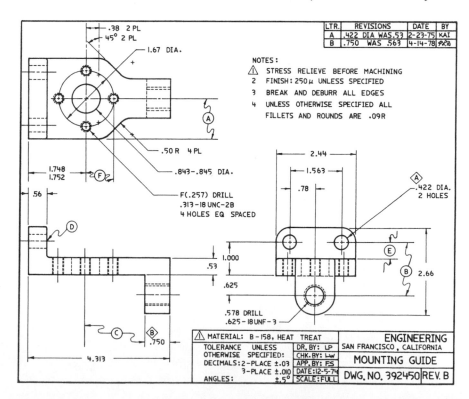

Chapter 11

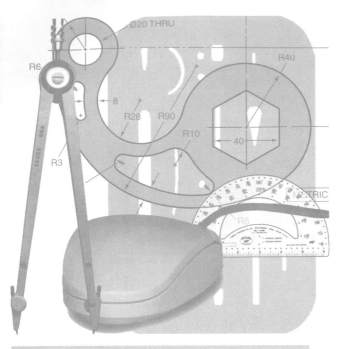

SOLID/3D MODELING: COMPUTATIONAL DESIGN AND ANALYSIS

KEY TERMS

Boolean operators

Computational fluid dynamics (CFD)

Finite-element analysis (FEA)

Finite-element model (FEM)

Gouraud

Holography

Modeling

Parametric solid modeling

Phong

Photorealism

Radiosity

Ray tracing

Rendering

Solid modeling

Tesselation

Texture mapping

Thermal analysis

Virtual reality

Visualization

CHAPTER OUTLINE

Factors in using visualization • Understanding data • Modeling techniques • Types of visualization images • Methods of representing variables • Relationships • Engineering visualization applications • Case studies • Summary • Review questions

CHAPTER OBJECTIVES

Upon completion of this chapter, students should be able to do the following:

■ Explain the advantages and disadvantages of using visualization.

■ Describe the most widely used three-dimensional imaging techniques.

■ Explain the various methods used to represent variables.

■ Explain the various categories of engineering visualization.

Visualization is the pictorial representation of data for the purpose of more clearly and concisely communicating information, *Figure 11-1*. The computer has made it possible to create pictorial representations of data that were virtually impossible with previous graphics techniques. The purpose of this chapter is to define and describe existing and emerging techniques for creating and analyzing data by means of computer-generated visualization. The old saying that a picture is worth a thousand words has never been more true. Today, however, the saying is more appropriately stated as: "A picture is worth 786,432 (1,024 × 768) pixels."

Visualization falls roughly into two categories. In some cases, visualization is used as an interactive design tool. An example of this is in mechanical CAD, where a clear image makes the design process easier. In other cases, visualization is used to represent complex data in a way that makes it easier to understand.

One example of representing complex data is in analyzing satellite telemetry, such as that sent by the Voyager and Explorer space missions. *Figure 11-2* illustrates a composite of true-color and filtered images of Jupiter that have been extracted from an animation of the planet rotating on its axis. These images represent digital data transmitted by the Voyager satellites that were used to render a three-dimensional model that was then animated to rotate on the planet's axis. The ability to view side-by-side animations of the real and filtered images greatly improved research on cloud composition compared with the previous method of viewing individual static images of small portions of the planet.

Increasingly complex engineering problems generate corresponding increases in the amount of information produced. The improvements in products made possible

FIGURE 11-2   *Jupiter: True color and spectral classification image* (Courtesy Planetary Arts and Sciences)

by more accurate data justify the large amounts of data to be utilized. In many cases the amount of data to be analyzed is so great that it is almost impossible to comprehend it in numerical form. The extremely large amount of data is one of the major reasons for utilizing visualization techniques. Stacks of printouts consisting of numerical data are frequently impossible to comprehend without the benefit of visualization.

Factors in Using Visualization

Several factors are leading to the increased use of visualization techniques and output:

- More complex engineering problems are generating increasingly large amounts of information.
- Increased computational speed makes it possible to perform visualization tasks on workstations and per-

FIGURE 11-1   *Harrier jet* (Courtesy NASA)

sonal computers instead of being limited to super-computers.

- Increasing performance/price ratio makes visualization hardware and software affordable for more people and companies.

- Improved user interfaces make the operation of systems and the production of visualization output easier to perform. Rather than being so complex and unique that only a specialist can use a system to create visualization output, it has become increasingly easier for greater numbers of people to productively utilize visualization technology and techniques.

Even with tremendous improvements in performance and ease of use, major barriers to the adoption of visualization techniques still exist. Among the more critical factors are the following:

- In some cases, the complexity of visualization techniques and technology continue to make it difficult to learn and use. This is less true than it has been in the past, but it still takes substantial effort and training to produce high-quality visualization output.

- Visualization generally involves three-dimensional data. This represents a major increase in complexity from two-dimensional data and requires a solid understanding of three-dimensional information.

- An unlimited number of ways to represent any given data set presents complex choices. The decision of how to represent the data is subjective. Frequently, no standard rules exist to govern selection.

Understanding Data

Throughout this chapter, you will see a number of examples of powerful visualization techniques, from mechanical CAD to NASA satellite data, to digitized biomedical data. In all cases, these examples use underlying 3D data. The differences between them is how the 3D data is obtained. In some cases, the 3D data is the product of complex calculations and is generated to represent something that would not be otherwise visible. In other cases, the data is the direct result of a designer creating a 3D model. The same data used for visualization can be used to create the tools to make a product.

Modeling Techniques

Modeling is the term used to describe the process of creating data that represents a real-world engineering problem. The term is used rather pervasively in a number of disciplines, so you should be comfortable with it. In art, the process of creating a sculpture is called modeling. In market research, the process of building data that represents how a demographic group responds is called modeling. And in CAD, the process of creating a 2D or 3D data set that represents a real object or system is likewise called modeling.

In the case of 3D CAD, the process used to create the model limits the visualization techniques which can be used to represent it. The most primitive modeling technique is 3D wireframe. In this method, the geometry of 3D surfaces are represented, but the actual topology of those surfaces is not applied. (You can think of geometry as the edges, and topology as the skins.) *Figures 11-3A and 11-3B* show a cubelike shape, represented both with and without topology. Without topology, there is no way to know which side the hole through the middle goes through.

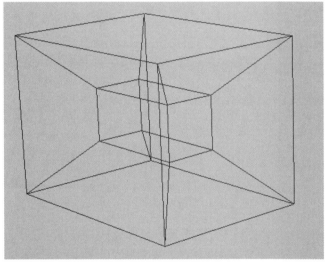

(A)

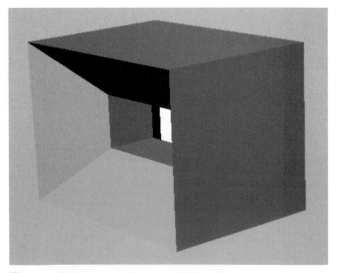

(B)

FIGURE 11-3 *(A) Ambiguous wireframe cube shape; (B) Non-ambiguous shaded cube shape* (Courtesy The Yares Organization)

Several years back, 3D wireframe modeling was more accepted because it required less in the way of software, hardware, and training. Today, CAD software and computers are powerful enough that there is no justification for using 3D wireframe modeling.

The next logical step from 3D wireframe is surface modeling. In this method, skins are applied to the edges of surfaces. This modeling method has been popular for many years in the aerospace and automotive industries, as it can represent complex objects with great precision. It has a further advantage over 3D wireframe in that, because the surface topology is known, the computer can calculate which lines or edges will be obscured in any particular view. This is known as a hidden-line calculation, *Figure 11-4.* It is actually a computationally intense process (particularly with complex or wavy surfaces), but the result is a pleasing image.

Surface modeling, *Figure 11-5,* is powerful, but it creates problems during the editing process. Changes made to one surface propagate to others—it's rather like the problems a tailor faces when fitting a sleeve onto a suit. From a designer's standpoint, building a surface model can be tedious, and there are no guarantees that all the surfaces in a model will fit together correctly when it is done. A surface model can be very accurate, but it takes great care on the part of the designer to assure this.

SOLID MODELING

A solid model, *Figures 11-6* and *11-7,* is simply a surface model where the computer system has kept track of the surface topology, made sure that all the seams between surfaces match, and made sure that the entire model is closed. Sometimes, designers talk about a solid model "holding water." This means there are no gaps at the seams and no openings between surfaces. This makes solid models quite accurate.

There are a number of processes for creating solid models, but the end result is always the same—a topo-

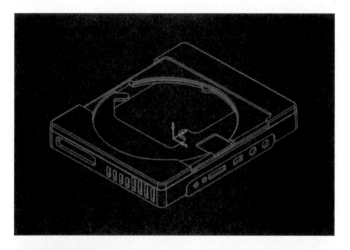

FIGURE 11-4 *Wireframe with hidden lines removed* (Courtesy Matra Datavision Inc)

FIGURE 11-6 *Solid* (Courtesy Matra Datavision Inc)

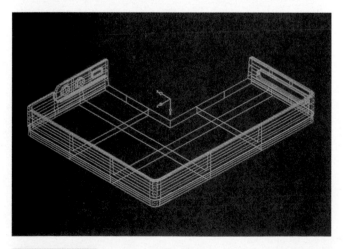

FIGURE 11-5 *Surface model made by EUCLID CAD/CAM* (Courtesy Matra Datavision Inc)

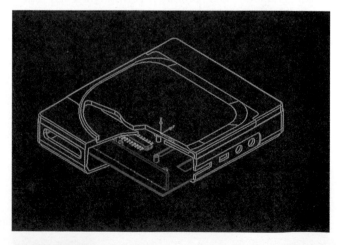

FIGURE 11-7 *Sectioned solid* (Courtesy Matra Datavision Inc)

logically closed solid model. From a visualization stand-point, there is usually little difference between a surface model and a solid model. From a creation standpoint, solid models are simpler to create and require more computer horsepower. Fortunately, this last point isn't the problem it used to be.

Surface and solid models can be represented as 3D wireframe, with or without hidden lines removed. They can also be represented in a shaded form, where each surface is assigned a shading color. Depending on a computer's graphics power, it may be possible to create shaded views as fast or faster than hidden-line views. Some moderately priced graphics subsystems are able to manipulate shaded views in real time.

As computers increase in speed and capability, design methodology will change. The use of 3D technology, particularly solid modeling, is becoming more and more prevalent. A number of powerful yet easy-to-use solid modeling programs are available, and this has made a noticeable change in the acceptance of solid modeling for mainstream design tasks.

Solid models are formed by joining various geometric shapes such as cones, prisms, pyramids, boxes, wedges, and cylinders. It is as if the computer mathematically "welds" the shapes together to form the solid model. Examine the solid models in Figures 11-6 and 11-7 closely and identify as many geometric shapes as you can. The computer uses *Boolean operators* to join the various shapes that make up the solid model. The most commonly used Boolean operators are:

- Union (also called *addition*)
- Difference (also called *subtraction*)
- Intersection

Figure 11-8 illustrates how these Boolean operators work. In Figure 11-8A, solid X has been added to (union) solid Y to form a new solid. This new solid has a volume equal to the combined volume of solids X and

Y before the union. In Figure 11-8B, solid X has been subtracted (difference) from the new solid formed in Figure 11-8A. Notice that the volume where solids X and Y overlapped has been removed from solid Y. That volume appears in Figure 11.8C as the intersection of solids X and Y. The volume of the solid in Figure 11-8C is equal to the overlap between solids X and Y.

A more advanced form of solid modeling is *parametric solid modeling*. With this advanced form of the concept, the model's geometry is controlled design parameters—hence the name parametric solid modeling—and curved paths are controlled mathematically rather than using coordinates. The benefit of parametric solid modeling is that as changes in the design are made, the model and corresponding drawings are updated automatically. This aspect of parametric solid modeling can shorten the design-prototyping-manufacturing process significantly.

Types of Visualization Images

STILL IMAGES

Single-frame, still images remain one of the most prevalent types of images used for visualization. The exploded view of a fully shaded lifting block assembly in *Figure 11-9* presents a clear means of communicating product details. The ability to separate product components demonstrates an additional benefit of 3D modeling. This same 3D database may be used to generate multiple views of the product by simply designating an alternate viewing location.

There are many types of still images. We've already spoken of wireframe and hidden-line images. Shaded images can be created with various algorithms to obtain differing levels of realism. The simplest shading method is flat shading, where any curved surfaces on a model are broken up into facets (a process called *tessellation*), and each surface is assigned a color based on its angle with respect to the viewer. This gives the model a decidedly

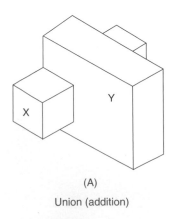

(A)

Union (addition)

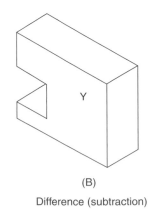

(B)

Difference (subtraction)

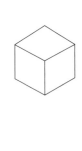

(C)

Intersection

FIGURE 11-8 *Boolean operators in solid modeling*

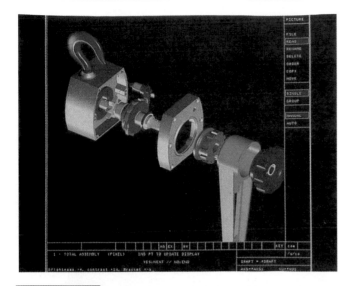

FIGURE 11-9 *Lifting block: Exploded assembly* (Courtesy IBM

FIGURE 11-10 *Lifting block: Transparent* (Courtesy IBM

chunky and faceted look. Flat shading is fairly rare these days, as the computation power required to generate a better image is negligible. The more common method is *Gouraud* shading, which improves on flat shading by interpolating the colors between adjacent facets. This gives rather a nice, smooth look—in fact, Gouraud shading is often called smooth shading. Most quality graphics boards in modern personal computers can Gouraud-shade images in real time. For very complex images, there are special 3D graphics boards that are optimized for displaying shaded images.

The next step up in shaded image quality is *Phong* shading, which adds specular reflection. This is a mirror-like glint that makes objects look shiny. Phong shading is possible on higher-end graphics systems, and it is useful for simulations that must be very lifelike.

Beyond simple shading, there are a number of methods that can be used to enhance an image. It is possible, when assigning shading colors to a model, to specify a transparency. This makes it possible to view internal details of parts. The lifting block in *Figure 11-10* illustrates how transparency can be used to clearly show the internal detail of assembled products.

Photorealism is the term used to describe images that are created in such a fashion that they might fool the viewer into thinking that they are photographs. In general, photorealistic images include multiple light sources, shadow casting, transparency, reflections, and refractions. As a rule, they also include *texture mapping*, where patterns (such as images of wood, stone, or fabric) are applied to object surfaces. *Figure 11-11* shows a typical photorealistic image. Photorealistic images are also often called rendered images, with the process of creating them being called *rendering*.

FIGURE 11-11 *Typical photorealistic image* (Courtesy Autodesk)

As a rule, every improvement in image quality with photorealistic images comes at the cost of substantial preparation work and processing time. Accurate shadow casting and reflection/refraction calculation require a rendering process called *ray tracing*, which literally involves tracing the rays of light from the eye of the viewer through the scene. Diffuse light calculations (involving, for example, the wash of light off a wall, or through a stained glass window) require a rendering process called *radiosity*. *Figure 11-12* shows an image created using both ray tracing and radiosity. Note the realistic and accurate lighting.

ANIMATION

Animation provides a means of representing time-dependent variables and active visualization of objects or events. Products consisting of moving parts and

FIGURE 11-12 *An image created using both ray tracing and radiosity* (Courtesy Autodesk)

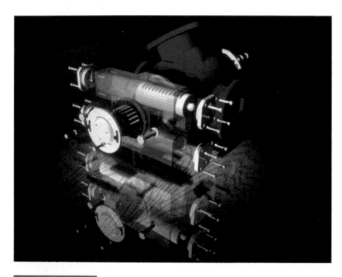

FIGURE 11-13 *A still frame from an animation depicting the motion of parts within an assembly*

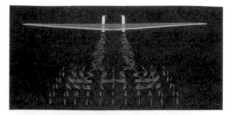

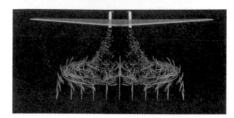

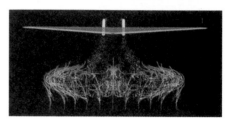

FIGURE 11-14 *Delta wing animation series (three views)* (Courtesy NASA)

active functions often cannot be clearly represented by still images. Animation is becoming increasingly useful for a variety of innovative applications. In addition to playing a programmed animation, it is now possible to move through 3D models interactively. Advancements in visualization technology will continue to increase the reality with which 3D models can be viewed and toured electronically.

Figure 11-13 shows a still frame from an animation depicting the motion of parts within an assembly. This animation incorporates photorealistic rendering and was created from parts modeled in a solid modeling CAD program.

The NASA Ames Research Center utilizes animation for a variety of design and analysis purposes. One project utilizing animation involves the flowfield about a delta wing equipped with thrust-reverser jets. Powered-

lift flight in close vicinity to the ground can have serious implications for aircraft functionality. The three images in *Figure 11-14* have been extracted from an animated sequence to show the progression of the jet and the formation of a ground vortex. Particle traces are colored by time at release. Although the three images give a general understanding of the flowfield, they do not communicate the continuous and real-time progression of a live video.

STEREO IMAGES

Stereo images are a special type of 3D model imaging. Stereo images consist of pairs of images that replicate views according to what the left and right eyes see. Since there is a slightly different angle of view between left and right eyes, it is necessary to create two images with slightly different views of an object. These matched sets of images must be viewed in a manner that limits the left eye to the left-view image and the right eye to the right-view image. The benefit of stereo images is their ability to re-create a view conveying depth similar to what would be seen in our true 3D world.

The images in *Figure 11-15* represent a stereo pair of an F-18 jet aircraft. They may be viewed with a stereo viewer to create a stereo image. In place of a stereo viewer, the images may be viewed by placing a business-size envelope between the images with the long dimension projecting out from the images. Place the end of the

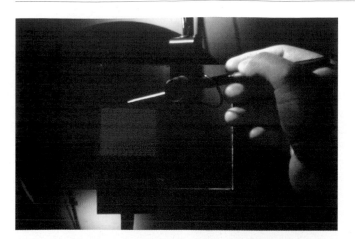

FIGURE 11-15 *Holography: System* (Courtesy MIT: Media Lab) © Hiroshi Nishikawa

envelope against your nose and relax your eyes until the images merge. There are numerous methods of creating stereo images. A very basic way of viewing 3-D stereo images is with the use of red and blue lens glasses similar to those used with 3D movies. More sophisticated methods include computer and glasses combinations that can also create 3D stereo animations.

HOLOGRAPHY

Holography produces images that simulate the effect of seeing the three-dimensionality of an object from multiple viewpoints. Holographic images have become fairly common for a variety of applications. Perhaps the most prevalent application has been the use of holographic images on credit cards. In the past, holographic images have been limited to physical objects that could be photographed in a manner that permits compositing multiple images into a holographic image.

It is now possible to produce holographic images from computer-generated 3D models. In their most basic form, these images are static images on sheet material. One advancement over a static image is the combination of computer and holographic technology to create holographic images in 3D space. *Figure 11-16* illustrates a system that integrates a computer-generated 3D model of an automobile with holography equipment to produce an interactive hologram. The holographic image may be seen from different viewpoints by rotating the 3D model within the computer system. *Figure 11-17* represents a series of wireframe views of the automobile ending on the right with a surfaced model.

The complexity and computational requirements involved in generating a holographic image in this manner are extremely demanding. For this reason, the model is initially viewed as a wireframe and then rendered as a surfaced model when the desired viewpoint is established. As computational speeds increase, it will be possible to generate increasingly realistic images.

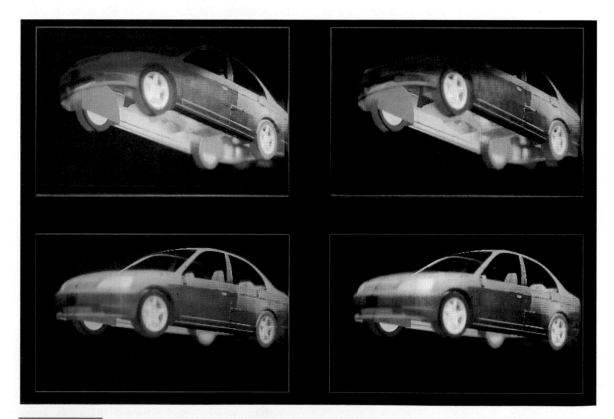

FIGURE 11-16 *Holography: Multiple views of a vehicle* (Courtesy MIT: Media Lab) © Hiroshi Nishikawa

FIGURE 11-17 *Holography: Multiple views of vehicle* (Courtesy MIT: Media Lab) © Quesada/Burke Studio, NY. Courtesy Scientific American

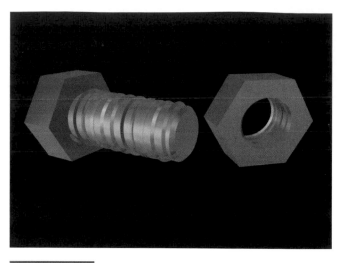

FIGURE 11-19 *With VMRI, a user can interactively navigate around models, but they cannot move individual parts.* (Courtesy Autodesk)

Increased size and the addition of color now exist for computer-generated holograms. Faster computational speeds will also improve the capability to interact with the model in real time.

VIRTUAL REALITY

Virtual reality is a technology designed to simulate either a physical or hypothetical environment. While the emphasis in this chapter is on visualization, it is important to note that virtual reality is intended to involve all the natural senses. Virtual reality systems can incorporate stimulus for all the five basic senses, but most commonly they deal only with sight and sound.

Virtual reality has a wealth of applications. It can create an environment that permits a sense of how a particular engineering design may feel or function. By being able to experience such an environment, the researcher can test and analyze alternative designs before investing in physical prototypes. Where multiple senses are important to a design project, virtual reality provides a more complete and realistic simulation.

Virtual reality also has great potential for hypothetical and hazardous environments. These environments either don't exist, are too dangerous for humans to enter physically, or may be beyond the reach of humans. The virtual reality system illustrated in *Figure 11-18* is designed to provide the user with a simulated experience of walking on the planet Mars. The view the operator sees is replicated on the computer screen in order that observers may see what the operator is seeing. The operator is equipped with sensors that signal locational and relationship data to the simulation. All of these components work together to make it possible for the operator to interact with the computational world in a manner that simulates being in the real environment.

In the last few years, the VRML (Virtual Reality Modeling Language) standard has become popular. Strictly speaking, it is not a virtual reality language. It is a language to model static scenes, which users can navigate through using whatever input and output devices they choose. It doesn't provide for object interaction or motion within a scene. Still, it is an effective tool for interactively viewing complex models. It doesn't provide for motion within a scene, but it does allow a user to navigate around a static 3D model, *Figure 11-19*.

Methods of Representing Variables

Engineering projects involve a large variety of variables. Visualization may be called upon to represent either single or multiple variables in a given visualization image. Numerous means of representing product characteristics and data are available. The methods of representing variables may be used either individually or in combination, depending on the project requirements.

FIGURE 11-18 *Virtual reality* (Courtesy NASA)

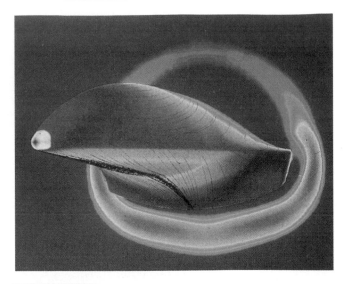

FIGURE 11-20 *Hypersonic vehicle* (Courtesy NASA)

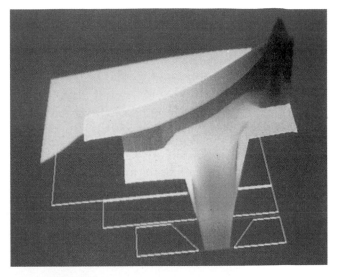

FIGURE 11-21 *Convective thermal field* (Courtesy Xerox Corporation)

COLOR

Color is one of the most common and useful means of representing data with visualization. It may be assigned to numerous types of variables including temperature, function, and pressure. *Figure 11-20* illustrates a hypersonic vehicle and related flight pressures. Pressures are shown for both the vehicle and a cross section of the shock layer. The color hues range from magenta to blue and represent highest to lowest pressures, respectively.

In addition to using ranges of color hue to represent a variable, color can be used more directly to make visual representations of 3D models easier to understand. Look back through this chapter and you'll see many images that, if reproduced in black and white, would be much less clear.

SHAPE

Shape may be used to indicate both visible and invisible variables. This can be anything from pressure levels to molecular bonding forces. An object may in reality be physically deformed or may be represented as deformed to emphasize force. Images may also represent forces that are not visible to the human eye.

The perspective view in *Figure 11-21* represents the convective thermal field computed for a cross section (two-dimensional) of the thermal printhead geometry in an inkjet printing system. The model incorporates composite material structures, convective flow of the ink, and selective heating levels from the thermal device. The problem cross section is outlined in the lower area. The temperature field is mapped into the third (vertical) dimension. This image also uses color and is mapped so that blue-yellow-red corresponds to a range of low- to high-temperature values. In this case the surface plot clearly emphasizes the jump discontinuities in temperature gradient where crossing material interfaces with differing conductivities.

SYMBOLS

Numerous types of symbols have been developed to represent variable characteristics. Some symbols have become standards and have very logical applications. Arrows are one example of a symbol with standard applications. They may be used to clearly indicate vectors and direction.

The types of symbols available for visualization and their possible combinations are infinite. Common types of variable symbols include points, lines, (+) and (–) symbols, and 2D and 3D polygons and shapes. The challenge of visualization is to develop the clearest and most intuitive representations possible to convey the information being described. A number of symbols are illustrated in examples throughout this chapter.

Relationships

There are a variety of relationships between variable symbols that may be visually represented to assist with engineering visualization. Similar to the infinite types of variable symbols, there are numerous ways of representing relationships. The three most common types are:

- **Direction:** The visualization of the aircraft hull illustrated in *Figure 11-22* utilizes arrows to indicate load distributions. The arrows indicate directional load forces on the hull.

**Pressure Loading on the Fuselage
of Finite Element Model**

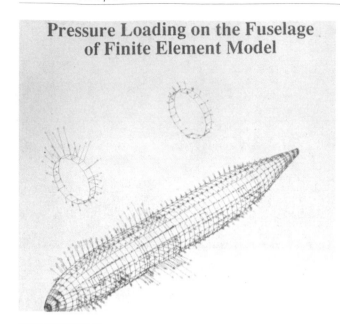

FIGURE 11-22 *Boeing aircraft vectors: Direction and size*
(Courtesy Boeing)

- **Size:** The same arrows used to indicate directional load forces in Figure 11-22 also provide an indication of the amount of load forces on the hull at specific locations. This is achieved by varying the length of the arrows so that longer arrows indicate greater amounts of load force.
- **Density:** Another common method of representing a variable is by the density of symbols. In this case the arrows could all be the same length but more closely packed where greater load forces exist.

The decision of how to best represent variables is very subjective. The person responsible for visually representing the data being viewed must select the method of representation based on experience. The more knowledge a designer has about the alternative visualization techniques available, the better that person will be able to communicate necessary information.

MOTION

Many engineering applications involve movement. Movement may be either moving parts or various types of flowing matter. The use of animation to represent moving entities has become increasingly practical in terms of both cost and production. Motion is vitally important to a vast number of engineering visualization applications. Aero-dynamics is a prime example of an application that must include an understanding of moving matter, specifically air. The ability to portray airflow has dramatically improved the means of evaluating and optimizing the aerodynamics of aircraft. As an example, the image of the YAV-8B Harrier jet in Figure 11-1 may

be viewed as an animation with the airflow progressing through simulated paths.

Engineering Visualization Applications

Engineering visualization involves numerous types of design and research information. Each of these types of information requires unique and specific means of representation in order to most clearly and concisely convey selected information. The following examples represent major types of engineering visualization.

DESIGN

The biggest application of engineering visualization is in design—whether of microchips or mile-long bridges. In the past, visualization was often seen as a marketing tool, primarily because the cost and difficulty involved couldn't be otherwise justified. As computers have become more powerful, visualization has become integrated into the design process. With this integration, it is possible not only to improve a product's time to market, but also to optimize that product's cost and performance.

FINITE-ELEMENT ANALYSIS

Finite-element analysis (FEA) is a vital means of analyzing the structural response of objects. In this method, a model of a part is broken up into a finite number of polygonal elements, called a mesh. If the elements of the mesh are small enough, a structural analysis of this *finite-element model (FEM)* corresponds rather closely with the real-world response of the part. The golf club in *Figure 11-23* has been created as a finite-element model. In

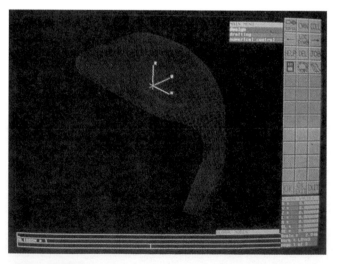

FIGURE 11-23 *Finite-element analysis model* (Courtesy Karsten Manufacturing Corporation/Ping Golf Clubs)

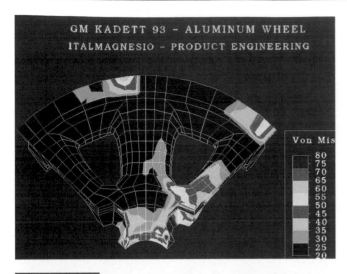

FIGURE 11-24 *Results of a finite-element analysis on a wheel section* (Courtesy The Yares Organization)

this example the model is being analyzed to determine the structural and impact characteristics of the golf club.

Stress is a factor in engineering designs of all types. While mechanical engineers spend quite some time understanding stress, you can simply envision it as the internal forces in an object that result from an external load. It may be concentrated in a small area or distributed over a large span. A localized concentration of stress is called a stress riser and can lead to part failure. In designs where failure due to stress is a concern, FEA helps the designer reach an optimal solution much faster than building prototypes or models would. *Figure 11-24* shows the results of a finite-element analysis, where the colors represent the stress in individual sections.

In the past, FEA was the domain of stress specialists—typically PhDs in mechanical engineering. Recently, FEA programs have become easy enough to use that experienced designers can use them to get a quick read-out on whether a part is likely to meet spec. Naturally, critical parts are still referred to the stress specialists. No sane person would want to fly in a plane in which the critical parts were not analyzed by an expert.

THERMAL ANALYSIS

Thermal analysis is a means of evaluating the thermal characteristics of a part or product. Thermal analysis is vital to numerous types of design applications. A clear understanding of a product's thermal characteristics can improve both efficiency and longevity. It may, in fact, prevent product failure.

Thermal analysis may be used for applications where thermal characteristics are the only design variable. It may also be used for design projects where heat or cold is just one of many variables, and where each of the multiple variables has a direct impact on related variables.

The heat may be flowing through a conduit or around the exterior of an object. Other products may be solid objects involving critical thermal transfer through a specific material or materials.

One very common application for thermal analysis is the design of heat ducts. These same principles apply to a variety of similar types of heating and cooling systems. The cross-sectional shape and internal volume of a duct significantly impact thermal transfer characteristics.

Thermal analysis systems employ visualization techniques that produce color-coded images that clearly illustrate temperature distributions. These simulations permit quick and easy evaluation of alternative conduit designs.

The two heat duct designs illustrated in *Figures 11-25* and *11-26* depict two design configurations. The thermal color coding shows distinctly different thermal characteristics. Depending on the desired thermal characteristics, one design may provide significantly better results.

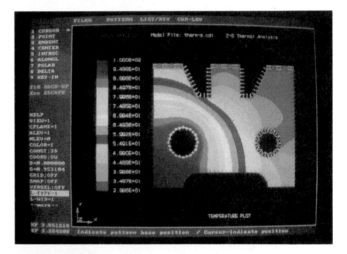

FIGURE 11-25 *Heat duct: Thermal analysis* (Courtesy Colorado State University, Department of Industrial Sciences)

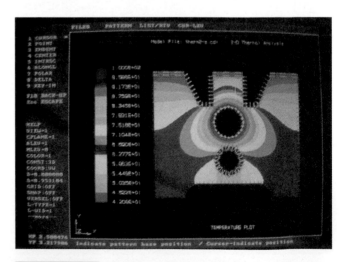

FIGURE 11-26 *Heat duct: Thermal analysis* (Courtesy Colorado State University, Department of Industrial Sciences)

COMPUTATIONAL FLUID DYNAMICS

Computational fluid dynamics (CFD) is another means of using computers to mathematically simulate physical events. This involves the study of fluids and their characteristics under a variety of conditions. Fluids include both gases and liquids. Computer programs and algorithms have evolved to a level that make them highly accurate for predicting fluid flow characteristics for mechanical and electrical products, architectural structures, medicine, and the automotive and aerospace industries.

CFD plays a vital role in the aerospace industry. The design of airplanes depends upon CFD to produce efficient planes that also perform optimally. *Figure 11-27* illustrates the use of CFD to determine airflow over an F-16A aircraft. The particle traces illustrate the simulated flowfield about the F-16A. The colors are used to indicate height in relation to the wing. The red color is closest to the horizontal center plane of the wing, and red is highest above the plane. While it may not eliminate the need for testing a physical prototype in a wind tunnel, it significantly reduces the number of physical prototypes that must be built and tested.

The image of the space shuttle in *Figure 11-28* illustrates how computational and physical prototypes may be compared. The left half of the shuttle image is the result of a CFD simulation representing surface pressures with color. The right half of the image illustrates the corresponding surface pressures obtained with a physical model in a wind tunnel. The simulation approximates the physical results with sufficient accuracy to justify the use CFD for design purposes.

In addition to the aerospace industry, there are growing numbers of other industries discovering innovative applications for CFD visualization. Applications include

FIGURE 11-28   *Shuttle: Simulated/physical* (Courtesy NASA)

sports equipment, architecture, automobiles, and even consumer products, where CFD can help designers improve internal airflow and cooling.

Medical science is developing numerous innovative applications using CFD. One such application is in the design of artificial hearts. This requires an understanding of liquid fluid flows and involves several variables. The heart illustrated in *Figure 11-29* is being studied to determine vorticity. Excessive vorticity may result in high shear regions within the heart. This is critical since high shear may damage blood cells as they travel through the heart. In this example, vorticity magnitude is represented by colors, with blue being low magnitude and white being high magnitude. The green cubes represent blood cell locations that have reached a specified threshold of vorticity.

PRE-ASSEMBLY ANALYSIS

On complex products, assembly is never simply a matter of bolting together a bunch of parts. In aircraft, for example, physically fitting systems and their components into an aircraft is a monumental task. Not only must the systems fit into limited space, they are also constrained by aerodynamic considerations for the shape of the aircraft. Visualization techniques make it possible to locate potential system interferences during the design stage and eliminate costly revisions. Boeing has been at the forefront of applying visualization technology to reduce prototyping costs.

Space allocation in the area beneath the flight deck is a complex task involving several groups within Boeing. Computer technology and visualization provide a

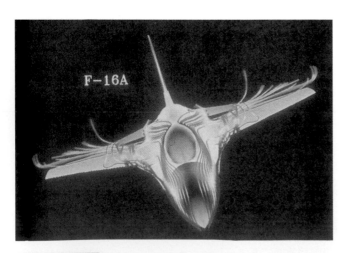

FIGURE 11-27   *F-16A aircraft* (Courtesy NASA)

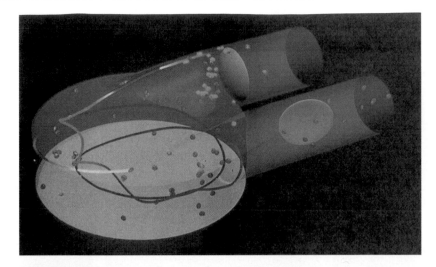

FIGURE 11-29 *Artificial heart* (Courtesy NASA)

common database that can be used interactively by multiple groups to resolve problems in the early stages of design. This concurrent engineering procedure saves a tremendous amount of time and resources. Equally as important, it results in optimum solutions to design criteria.

Figure 11-30 illustrates an interference in the cargo compartment located beneath the passenger compartment. The interference occurs where the left end of the cream-colored channel meets the green mounting bracket. The channel serves as a mounting device for a light fixture used to illuminate the cargo compartment.

Configuration management software developed by Boeing is used to manage and "assemble" individual 3D solid-modeled parts into a digital mockup, which can be used instead of a physical prototype. In this case, the interference was corrected by replacing the original channel with a shorter channel. The correctly fitted channel is illustrated in *Figure 11-31*. By detecting the interference early in the design process, it was possible to eliminate the interference before spending money on tooling.

KINEMATIC SIMULATIONS

Kinematics is an excellent application for visualization and simulation. Linkages and mechanisms are present in numerous products that we use daily. Computers are ideal for solving the mathematical calculations necessary. Once the calculations have been performed, images may be generated that clearly illustrate resulting designs. Technology exists that makes it easy to computationally build kinematic systems. Once a system has been built, it may easily be animated and tested for performance. *Figure 11-32* illustrates three stages of an animated sequence representing a linkage system and relevant system characteristics.

FIGURE 11-30 *Boeing structural: Interference* (Courtesy Boeing)

FIGURE 11-31 *Boeing structural: Clearance* (Courtesy Boeing)

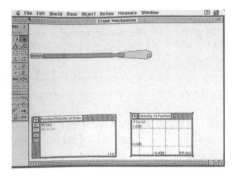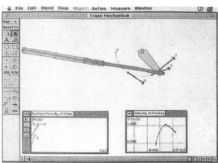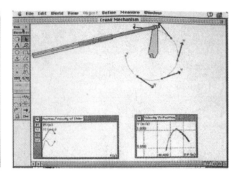

FIGURE 11-32 *Kinematic sequence* (Courtesy Knowledge Revolution)

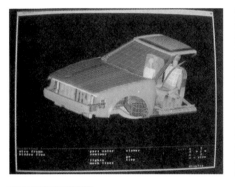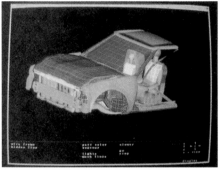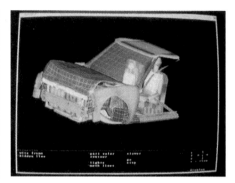

FIGURE 11-33 *Vehicle crash testing: Real and simulated* (Courtesy Altair Engineering with CRAY XMS Supercomputer)

AUTOMOTIVE CRASH SIMULATIONS

The automotive industry is using visualization in computational simulations to better understand structural damage resulting from crashes. This crash information is vital to improving the safety of vehicles. In the past, such information was only available from physically crashing real vehicles. *Figure 11-33* illustrates computer-simulated crash results. These three images have been extracted from an animation sequence that permits the study of progressive stages of crash impact. The ability to interactively study computational crashes permits a better understanding of the crash results. This information may then be used to optimize designs first with simulations. This reduces the need for crash testing real vehicles.

Case Studies

Engineering visualization is one aspect of product design. In reality there are multiple phases that occur between product conception and the final product. The following two examples illustrate how visualization is utilized to improve products, their production, and their cost effectiveness. These two examples also consist of several types of visualization images and considerations that need to be addressed.

BIOMECHANICAL VISUALIZATION EXAMPLE

Biomechanical technology is incorporating ever-increasing use of engineering and visualization techniques to improve the quality and capabilities of work in this field. The following example describes work being done with hip and knee implant replacement techniques. In addition to demonstrating benefits of modern technology, it exemplifies very real benefits on a personal level. Prior to the use of this technology, a typical hip replacement had a life expectancy of four to five years. With the increased information and accuracy of visualization and related technologies, the life expectancy of the replacement has increased to 10 years and continues to improve. In simple terms, it reduces return surgery to one-half the previous frequency.

There are several steps involved in producing implant prosthetics. The first step in the process involves obtaining three-dimensional data, which can be used to create a 3D FEM of the problem area. This information is typically obtained by techniques such as magnetic resonance imaging (MRI). The 3D FEM may be used for a variety of design and analysis purposes.

The hip implant illustrated in *Figure 11-34* is being optimized relative to stresses in the region of the bone near the implant interface. The FEM on the right is the initial prosthetic design and the FEM on the left is the

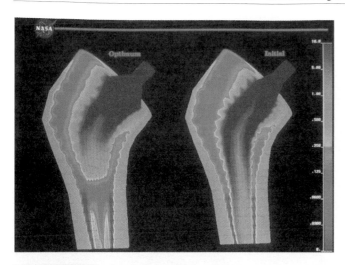

FIGURE 11-34 *Hip implant: Optimal/initial* (Courtesy NASA and Case Western Reserve University)

FIGURE 11-36 *Knee implant: Asymmetrical loads* (Courtesy NASA and Case Western Reserve University)

optimized prosthetic design. The material selected for this particular implant was titanium. Implant material selection is critical to hip replacement functionality. This type of surgery involves several critical factors, including bone material stress and growth characteristics. Understanding stress and growth characteristics has proven to be one of the major advancements related to increased prosthetic implant life expectancy.

Orthopedic knee implants involve techniques similar to hip implants. Since a knee is subjected to both symmetrical and asymmetrical loading, it is necessary to evaluate the effects of both types of loading. *Figure 11-35* illustrates equivalent stress distributions in the cortical bone at three points in the design evolution for symmetric loads. *Figure 11-36* illustrates similar stress distributions for asymmetrical loads.

The knee implant in this case involves multiple materials in order to optimize functionality and longevity. *Figure 11-37* illustrates idealized morphology and material regions for a typical adult male tibia with an implant. Once the 3D FEM has been optimized for fit and func-

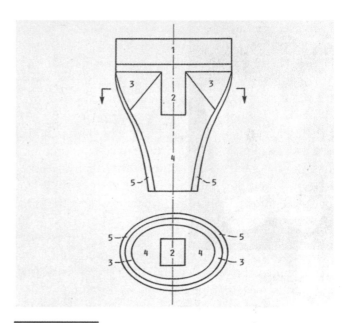

FIGURE 11-37 *Knee implant: Composite materials* (Courtesy NASA and Case Western Reserve University)

tionality, it must next be produced. It is here that the relationship between design and production becomes evident. Typical knee implant components are illustrated in *Figure 11-38*. The production of implant components is now possible with automated machining equipment. The same 3D model that was used for design purposes may be used to generate codes to control machining operations. The automated machining equipment may then produce the prosthetic implant.

PRODUCT DESIGN EXAMPLE

Product design techniques are changing dramatically. Competition is demanding faster design cycles and improved design solutions. The following example describes how Xerox utilizes computer technology for product design. In addition to visualization techniques, methods of conducting real-time interactive design with

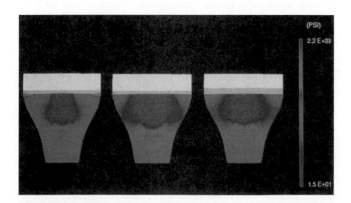

FIGURE 11-35 *Knee implant: Symmetrical loads* (Courtesy NASA and Case Western Reserve University)

FIGURE 11-38 *Knee implant: Components* (Courtesy NASA and Case Western Reserve University)

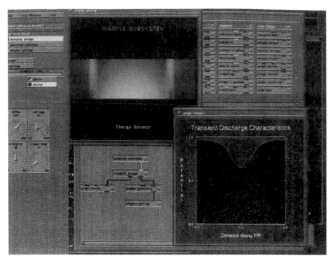

FIGURE 11-39 *Xerox interactive display* (Courtesy Xerox Corporation)

FIGURE 11-40 *Xerox laser exposure of photoreceptor* (Courtesy Xerox Corporation)

variable controls on the computer screen are discussed. The goal is improved product quality and design time.

The use of physical prototypes is both time consuming and expensive. One means of significantly improving design solutions and time requirements is to perform initial design using computational prototypes. This permits ease of testing design alternatives and promotes investigating more designs than traditional physical prototyping methods. Testing more designs provides increased information from which to select the optimum design solution.

Xerographic printing and copying is pervasive throughout the world. The process of xerography is a very complex process demanding exacting standards. The design of xerography systems is accordingly complicated and presents an important case for the need to develop computational prototyping techniques. It is equally important that researchers and designers are provided fully integrated computer systems with highly functional interactive user interfaces.

One critical aspect of the xerography process is understanding the effects of laser exposure on the photoreceptor. *Figure 11-39* is a screen-captured image from the designer's workstation. It represents an interactive session for analyzing the xerography imaging process. The upper window shows a cross section of the photoreceptor with space charge densities color mapped as blue-yellow-red to indicate zero, moderate, and high levels, respectively. The lower-right window shows the evolution of the surface voltage on the photoreceptor. As the image discharges subsequent to the laser illumination, the photo-generated charge from the bottom moves upward to lower the surface voltage. Development of the image is voltage sensitive.

The designer's screen illustrates built-in capabilities permitting interactive changing of variables. On the left are four dials that can be used to change photoreceptor

velocity, dielectric constants, backplate voltage, and wire voltage. These dials let the designer computationally test varying levels and combinations of critical variables.

Further clarification of the effects of the laser exposure on the photoreceptor are illustrated in *Figure 11-40*. This model graphically shows the scattering and attenuation of plane waves representing the laser illumination as it moves through the photoreceptor. In this instance, intensity is color mapped so that blue, yellow, and red indicate zero, moderate, and high levels of intensity, respectively. The resulting model indicates that light is a wavelike phenomenon. The plane waves are incident on the left and become compressed and wrapped around the toner particles on the surface of the photoreceptor. It may be noted that the plane waves are similar to water waves as they flow around poles placed in water.

Industry Application

TO SOLID MODEL OR NOT? A DEBATE

The design team for Thompson Engineering International (TEI) had been debating the issue of visualization for over a week. All members of the team agreed that TEI's current 2D CAD capabilities were inadequate. The new space station project, the company's largest and most complex project ever, would be a particular challenge. The parts TEI will design for the joint U.S./European space station must be perfect; there is no room for error. This is why the topic of visualization is now being debated so intensely.

Several members of the team argue for a minimalist approach, combining 2D, 3D wireframe, and surface modeling where necessary. They argue that it will require less computing power, and a shorter learning curve for design and drafting personnel. Plus, they won't have to throw away the investment they've made in their existing systems. Other members of the team favor moving completely to solid modeling; they counter the arguments of the wireframe advocates, claiming that approach won't yield sufficient information. In their words, "A wireframe image just is not descriptive enough to do us any good."

"Solid modeling," claims its advocates, "provides the most accurate, most comprehensive information possible, and the image is realistic." They add that "Revisions are particularly easy with solid modeling." The supporters of the minimalist approach counter by saying "We can't do solid modeling without an expensive hardware upgrade and it will take our team six months to develop modeling skills. This space-station project doesn't have a six-month learning curve built into it."

This type of debate happens frequently in engineering companies. Join the debate. What is your opinion?

Summary

- Visualization is the pictorial representation of data for the purpose of more clearly and concisely communicating information.

- The increased use of visualization can be attributed to the increasing complexity of engineering problems, the improved capabilities of computers, relative affordability of computer hardware and software, and the relative ease of learning to create visualization output.

- Three-dimensional modeling techniques include wireframe, surface modeling, and solid modeling.

- Types of visualization images include still images, transparency, photorealism, animation, stereo images, holography, and virtual reality.

- Variables in engineering projects may be represented using color, shape, symbols, size, density, and motion.

- Engineering visualization categories include design, finite-element analysis, thermal analysis, computational fluid dynamics, kinematic simulation, and automotive crash simulations.

Review Questions

Answer the following questions either true or false.

1. Visualization is the pictorial representation of data for the purpose of more clearly and concisely communicating information.

2. Visualization is beneficial because it allows extremely large amounts of data to be seen and therefore comprehended.

3. Computer technology and visualization have made it possible to resolve problems that were previously too complex to solve.

4. The wireframe method of modeling is the most time-consuming computationally.

5. A disadvantage of solid modeling is that it is not accurate.

6. Only high-dollar problems, such as those faced by NASA, justify the effort involved in visualization.

7. A disadvantage of visualization is that the hardware and software are not affordable for most people and companies.

8. Normal personal computers are capable of displaying and manipulating shaded images in real time.

Answer the following questions by selecting the best answer.

1. Which of the following is **not** one of the basic methods of engineering modeling?
   a. Wireframe
   b. Surface
   c. Photorealism
   d. Solid modeling

2. Which of the following is **not** a type of visualization image?
   a. Animation
   b. Stereo images
   c. Transparency
   d. Intrinsic

3. Which of the following is the appropriate type of visualization image to represent an item in a physical or hypothetical environment?
   a. Virtual reality
   b. Holography
   c. Photorealism
   d. Animation

4. Which of the following is **not** a method for representing visualization variables?
   a. Color
   b. Motion
   c. Symbols
   d. Thermal analysis

5. Which of the following is **not** one of the five engineering visualization categories?
   a. Holography
   b. Finite-element analysis
   c. Thermal analysis
   d. Computational fluid dynamics

6. An example of an application using computational fluid dynamics is:
   a. The design of artificial turf.
   b. The design of artificial intelligence.
   c. The design of artificial knees.
   d. The design of artificial hearts.

7. Which of the following is **not** a factor in using visualization?
   a. More complex engineering problems are generating increasingly large amounts of information.
   b. Increased computational speed makes it possible to perform visualization tasks on workstations and personal computers instead of being limited to supercomputers.
   c. The decision of how to represent the data is subjective.
   d. Improved user interfaces make the operation of systems and the production of visualization output easier to perform.

8. Which of the following is **not** true regarding stress analysis?
   a. Stress is a critical factor in engineering designs of all types.
   b. Trained designers may be able to do basic stress analyses, but critical analyses should be left for specialists.
   c. Prior to being analyzed for stress, a model of a part must be broken up into a finite number of polygonal elements.
   d. Stress risers are desirable to assure that parts will not fail.

SECTION

3

Design Drafting Applications

Career Profile: Mary Trahan

Mary started her career with drafting classes in high school and went on to a technical college, where she trained further on the subject. Coming out of the program, she joined the Consolidated Waterworks District 1 in Terrebonne Parish, Louisiana as a drafter. At that time, her drawings were done with pen and paper. Things have change a lot in 22 years!

Doing things on paper meant that when they had to replace valves or when a main line broke, they would have to redraft the map to account for the change. Thankfully, in the 1980s, the Waterworks started digitizing the maps, tracing aerial photographs electronically into computer-aided design (CAD) software. It was a big step up, but the program was not great for mapping. The roads, landmarks, and water mains are not merely straight lines, which CAD programs are best at. Often the curves did not match up. And technical specifications (like the depth at which a pipe is buried) had to be kept in a separate database.

By the 1990s, the Waterworks began working with other utilities, such as gas, electric, in order to fund and create a parish geographic information system (GIS). It became a function way to view each other's information on intelligent maps. With the North American Datum coordinates, the maps can more closely approximate the spherical shape of the earth (versus the flatness of CAD). GIS also combined sophisticated graphics with the spatial integrity of tabular data. That means that Trahan can see if a specific pipe has had several leaks in the past few years, making it easier to prevent major problems before they happen. It also helps with safeguards: If there is a chemical leak at the plant, drafters can draw a polygon in the program and get customers' information for that area to notify them.

On a daily basis, Trahan is responsible for maintaining all the water-related changes on the map, whether they are water lines or fire hydrants. For instance, in the case of a new subdivision, she will go out after it is completed to make a final inspection and to ensure that main lines and valves operate as designed and comply with her system. Once she and the board approve the job, the engineer will submit digital copies (usually in AutoCAD), and Trahan will transfer those changes into the system map. Most of the drawings are flat, from a perspective above, but the benefit of the software is that she can also do other views. A profile, looking from the top of the earth into the depths, is especially necessary for new construction, so engineers can ensure that all utilities cross within proper distances. Trahan's insertions will make it easier for these other utilities to know where they can (and cannot) run lines.

Chapter 12

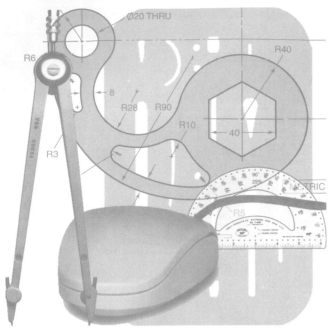

GEOMETRIC DIMENSIONING AND TOLERANCING

KEY TERMS

All around symbol

Angularity

Basic dimension

Between symbol

Bilateral tolerance

Circularity

Cylindricity

Datum

Datum feature

Datum feature simulator

Datum feature symbol

Datum plane

Datum reference frame

Datum surface

Datum target symbol

Feature control symbol

Flatness

Free-state variation

General tolerancing

Geometric dimensioning and tolerancing

Least material condition (LMC)

Limit dimensioning

Maximum material condition (MMC)

Modifiers

Parallelism

Perpendicularity

Positional tolerancing

Profile

Profile of a line

Profile of a surface

Projected tolerance zone

Regardless of feature size (RFS)

Rule #1

Runout

Size tolerance

Statistical tolerancing symbol

Straightness

Tangent plane

Tolerancing

True position

Unilateral tolerance

Virtual condition

CHAPTER OUTLINE

Summary of geometric dimensioning and tolerancing terms • Geometric dimensioning and tolerancing defined • Modifiers • Feature control symbol • True position • Circularity (roundness) • Cylindricity • Angularity • Parallelism • Perpendicularity • Profile • Runout • Concentricity • Summary • Review questions • Chapter 12 problems

CHAPTER OBJECTIVES

Upon completion of this chapter, students should be able to do the following:

■ Describe what is meant by the term *general tolerancing*.

■ Define the concept *geometric dimensioning and tolerancing*.

■ Explain the purpose of a modifier.

■ Distinguish between the concepts *maximum material condition (MMC)* and *regardless of feature size (RFS)*.

■ Explain the concept *least material condition (LMC)*.

■ Describe what is meant by *projected tolerance zone*.

■ Make a sketch that illustrates the concept of datums.

■ Demonstrate how to establish datums.

■ Apply feature control symbols when dimensioning objects.

■ Explain the concept of *true position*.

Summary of Geometric Dimensioning and Tolerancing Terms

Actual Local Size. The value of any individual distance at any cross section of a feature.

Actual Mating Size. The dimensional value of the actual mating envelope.

Actual Size. Actual measured size of a feature.

Allowance. The difference between the larger shaft size limit and the smallest hole size limit.

Angularity. Tolerancing of a feature at a specified angle other than 90° from a referenced datum.

Basic Dimension. A theoretically "perfect" dimension similar to a reference or nominal dimension. It is used to identify the exact location, size, shape, or orientation of a feature. Associated tolerances are applied by notes, feature control frame, or other methods, excluding tolerance within title blocks.

Bilateral Tolerances. Tolerances that are applied to a nominal dimension in the positive and negative directions.

Bonus Tolerance. The permitted allowable increase in tolerance as the feature departs from the material condition identified within the feature control frame.

Circular Runout. A tolerance that identifies an infinite number of single circular elements measured at cross sections on a feature when the feature is rotated 360° for each cross section.

Circularity. A tolerance that controls the circular cross section of round features that is independent of other features. The tolerance zone boundary is formed by two concentric perfect circles.

Clearance Fit. A condition between mating parts in which the internal part is always smaller than the external parts it fits into.

Coaxiality. The condition of two or more features having coincident axes.

Compound Datum Features. Two datum features used to establish a datum or axis plane.

Concentricity. A tolerance in which the axis of a feature must be coaxial to a specified datum regardless of the datum's and the feature's size. The lack of concentricity is eccentricity.

Cylindricity. A tolerance that simultaneously controls a surface of revolution for straightness, parallelism, and circularity of a feature, and is independent of any other features on a part. The tolerance zone boundary is composed of two concentric perfect cylinders.

Datum. Reference points, lines, planes, cylinders, and axes that are assumed to be exact. They are established from datum features.

Datum Axis. The axis of a referenced datum feature such as a hole or shaft.

Datum Feature. A feature that is used to establish a datum.

Datum Feature of Size. A feature that has size, such as a shaft, that is used to establish a datum.

Datum Identification Symbol. A special rectangular box that contains the datum reference letter and a dash on either side of the letter. It is used to identify datum features.

Datum: Feature Simulator. A surface of adequately precise form (such as a surface plate, a gage surface, or a mandrel) contacting the datum feature(s) and used to establish the simulated datum(s).

Datum: Reference. Entering a datum reference letter in a compartment of the feature control frame following the tolerance value.

Datum: Reference Frame. Three mutually perpendicular planes that establish a coordinate system. It is created by datum references in a feature control frame or by a note.

Datum: Simulated. A point, axis, or plane established by processing or inspection equipment, such as the following: simulator, surface plate, a gage surface, or a mandrel.

Datum Simulation. The use of a tool contacting a datum feature used to simulate a true geometric counterpart of the feature.

Datum Simulator. A tool used to contact a datum feature.

Datum Target. Specified points, lines, or areas on a feature used to establish datums.

Datum Target Area. A specified area on a part that is contacted to establish a datum.

Datum Target Line. A line on a surface that is contacted to establish a datum.

Datum Target Point. A specified point on a surface used to establish a datum.

Datum Target Symbol. A circle divided horizontally into halves containing a letter and number to identify datum targets.

Envelope, Actual Mating. The term is defined according to the type of features as follows:

(a) *For an External Feature.* A similar perfect feature counterpart of smallest size that can be circumscribed about the features so that it just contacts the surface at the highest points. For example, a smallest cylinder of perfect form or two parallel planes of perfect form at minimum separation that just contact(s) the highest points of the surface(s).

For features controlled by orientation or positional tolerances, the actual mating envelope is orientated relative to the appropriate datum(s), for example, perpendicular to a primary datum plane.

(b) *For an Internal Feature.* A similar perfect feature counterpart of largest size that can be inscribed within the feature so that it just contacts the surface at the highest points. For example, a largest cylinder of perfect form or two parallel planes of perfect form at maximum separation that just contact(s) the highest points of the surface(s).

For features controlled by orientation or positional tolerance, the actual mating envelope is oriented relative to the appropriate datum(s).

Feature. A component of a part such as a hole, slot, surface, pin, tab, or boss.

Feature of Size. One cylindrical or spherical surface, or a set of two opposed elements or opposed parallel surfaces, associated with a size dimension.

Feature, Axis of. A straight line that coincides with the axis of the true geometric counterpart of the specified feature.

Feature, Center Plane of. A plane that coincides with the center plane of the true geometric counterpart of the specified feature.

Feature, Derived Median Line of. An imperfect line (abstract) that passes through the center points of all cross sections of the feature. These cross sections are normal to the axis of the actual mating envelope. The cross section center points are determined as per ANSI B89.3.1.

Feature, Derived Median Plane of. An imperfect plane (abstract) that passes through the center points of all line segments bounded by the feature. These line segments are normal to the actual mating envelope.

Fit. A term used to describe the range of assembly that results from tolerances on two mating parts.

Flatness. A tolerance that controls the amount of variation from the perfect plane on a feature independent of any other features on the part.

Form Tolerance. A tolerance that specifies the allowable variation of a feature from its perfect form.

Free-State Variation. The condition of a part that permits its dimensional limits to vary after removal from manufacturing or inspection equipment.

Least Material Condition (LMC). A condition of a feature in which it contains the least amount of material relative to the associated tolerances. Examples are maximum hole diameter and minimum shaft diameter.

Limit Dimensions. A tolerancing method showing only the maximum and minimum dimensions that establish the limits of a part size or location.

Limits. The maximum and minimum allowable sizes of a feature.

Location Tolerance. A tolerance that specifies the allowable variation from the perfect location of a feature relative to datums or other features.

Maximum Material Condition (MMC). A condition in which the feature contains the maximum amount of material relative to the associated tolerances. Examples are maximum shaft diameter and minimum hole diameter.

Modifier. The application of MMC or LMC to alter the normally implied interpretation of a tolerance specification.

Parallelism. A tolerance that controls the orientation of interdependent surfaces and axes that must be of equal distance from a datum plane or axis.

Perpendicularity. A tolerance that controls surfaces and axes that must be at right angles with a referenced datum.

Position Tolerance. A tolerance that controls the position of a feature relative to the true position specified for the features, as related to a datum or datums.

Primary Datum. The first datum reference in a feature control frame. It is normally elected because it is most important to the design criteria and function of the part.

Profile of a Line. A tolerance that controls the allowable variation of line element in only one direction on a surface along an elemental tolerance zone with regard to a basic profile.

Profile of a Surface. A tolerance that controls the allowable variation of a surface from a basic profile or configuration.

Profile Tolerance Zone. A tolerance zone that can control the form of an individual feature and provide for a composite control of form, orientation, and location.

Projected Tolerance Zone. A tolerance zone that applies to the location of an axis beyond the surface of the feature being controlled.

Reference Dimension. A non-tolerance zone or location dimension used for information purposes only that does not govern production or inspection operations.

Regardless of Feature Size (RFS). A condition of a tolerance in which the tolerance must be met regardless of the produced size of the feature.

Runout. The composite surface variation from the desired form of a part of revolution during full rotation of the part on a datum axis.

Secondary Datum. The second datum reference in a feature control frame. Established after the primary datum, it has less design influence and functionality.

Size, Virtual Condition. The actual value of the virtual condition boundary.

Straightness. A tolerance that controls the allowable variation of a surface or an axis from a theoretically perfect line.

Symmetry. A condition for which a feature (or features) is equally disposed or shaped about the center plane of a datum feature.

Tangent Plane. A theoretically exact plane derived from the true geometric counterpart of the specified feature surface by contacting the high points on the surface.

Tertiary Datum. The third datum reference in a feature control frame. Established after the secondary datum, it has the least amount of design influence or functionality.

Tolerance. The acceptable dimensional variation or allowance of a part.

Total Runout. A tolerance that provides for a composite control of all surface elements as the part is rotated 360° about a datum axis.

Transition Fit. A condition in which the prescribed limits of mating parts produce either a clearance or an interference when the parts are assembled.

True Geometric Counterpart. The theoretically perfect boundary (virtual condition or actual mating envelope) or best-fit (tangent) plane of a specified datum feature.

True Position. The theoretically exact location of a feature.

Unilateral Tolerance. A tolerance that allows variations in only one direction.

Virtual Condition. A constant boundary produced by the combined effects of the maximum material condition size and geometric tolerance. It represents the worst-case condition of assembly at MMC.

Zero Tolerance at MMC or LMC. A tolerancing method where no tolerance is shown in the feature control frame. The tolerance allowed is totally dependent on the size of the feature departure from MMC or LMC.

GENERAL TOLERANCING

The Industrial Revolution created a need for mass production, assembling interchangeable parts on an assembly line to turn out great quantities of a given finished product. Interchangeability of parts was the key. If a particular product was composed of 100 parts, each part could be produced in quantity, checked for accuracy, stored, and used as necessary.

Since it was humanly and technologically impossible to have every individual part produced exactly alike (it still is), the concept of geometric and positional tolerancing was introduced. *Tolerancing* means setting acceptable limits of deviation. For example, if a mass-produced part is to be 4″ in length under ideal condi-

tions, but is acceptable as long as it is not less than 3.99″ and not longer than 4.01″, there is a tolerance of plus or minus .01″, *Figure 12-1*. This type of tolerance is called a *size tolerance.*

There are three different types of size tolerances: unilateral and bilateral, shown in *Figure 12-2,* and limit dimensioning. When a *unilateral tolerance* is applied to a dimension, the tolerance applies in one direction only (for example, the object may be larger but not smaller, or it may be smaller but not larger). When a *bilateral tolerance* is applied to a dimension, the tolerance applies in both directions, but not necessarily evenly distributed. In *limit dimensioning,* the high limit is placed above the low value. When placed in a single line, the low limit precedes the high limit and the two are separated by a dash.

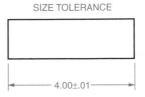

FIGURE 12-1 *Size tolerance*

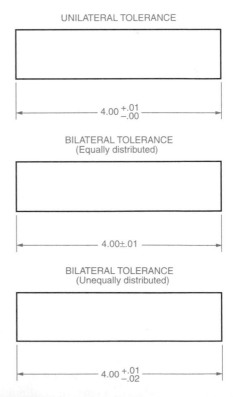

FIGURE 12-2 *Two types of tolerances*

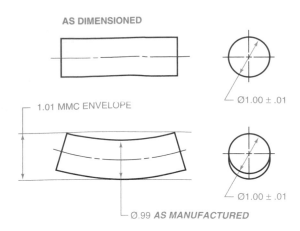

AS DIMENSIONED

1.01 MMC ENVELOPE

∅1.00 ± .01

∅1.00 ± .01

∅.99 *AS MANUFACTURED*

FIGURE 12-3   *Tolerance of form*

FOR INDIVIDUAL FEATURES	FORM
FOR INDIVIDUAL OR RELATED FEATURES	PROFILE
FOR RELATED FEATURES	ORIENTATION LOCATION RUNOUT

FIGURE 12-4   *Types of tolerances*

Tolerancing size dimensions offers a number of advantages. It allows for acceptable error without compromises in design, cuts down on unacceptable parts, decreases manufacturing time, and makes the product less expensive to produce. However, it soon became apparent that in spite of advantages gained from size tolerances, tolerancing only the size of an object was not enough. Other characteristics of objects also needed to be toleranced, such as location of features, orientation, form, runout, and profile.

In order for parts to be acceptable, depending on their use, they need to be straight, round, cylindrical, flat, angular, and so forth. This concept is illustrated in *Figure 12-3*. The object depicted is a shaft that is to be manufactured to within plus or minus .01 of 1.00 inch in diameter. The finished product meets the size specifications but, since it is not straight, the part might be rejected.

The need to tolerance more than just the size of objects led to the development of a more precise system of tolerancing called geometric dimensioning and *positional tolerancing*. This new practice improved on conventional tolerancing significantly by allowing designers to tolerance size, form, orientation, profile, location, and runout, *Figure 12-4*. In turn, these are the characteristics that make it possible to achieve a high degree of interchangeability.

Geometric Dimensioning and Tolerancing Defined

Geometric dimensioning and tolerancing is a dimensioning practice that allows designers to set tolerance limits not just for the size of an object, but for all of the various critical characteristics of a part. In applying geometric dimensioning and tolerancing to a part, the designer must examine it in terms of its function and its relationship to mating parts.

Figure 12-5 is an example of a drawing of an object that has been geometrically dimensioned and toleranced. It is taken from the dimensioning standards as defined by the American National Standards Institute (ANSI), written by the American Society of Mechanical Engineers (ASME) or ASME Y14.5M—1994. This manual is a necessary reference for drafters and designers involved in geometric dimensioning and positional tolerancing.

The key to learning geometric dimensioning and positional tolerancing is to learn the various building blocks that make up the system, as well as how to properly apply them. *Figure 12-6* contains a chart of the building blocks of the geometric dimensioning and tolerancing system. In addition to the standard building blocks shown in the figure, several modifying symbols are used when applying geometric tolerancing, as discussed in detail in upcoming paragraphs.

Another concept that must be understood in order to effectively apply geometric tolerancing is the concept of datums. For skilled, experienced designers, the geometric building blocks, modifiers, and datums blend together as a single concept. However, for the purpose of learning, they are dealt with separately and undertaken step-by-step as individual concepts. They are presented now in the following order: modifiers, datums, and geometric building blocks.

ANSI's dimensioning standards manual (Y14.5 series) changes from time to time as standards are updated. For example, the Y14.5 manual became Y14.5M in 1982 to accommodate metric dimensioning. Revised again in 1988, it became Y14.5M-R1988. In the latest edition, the standard takes on the name of the developing agency, the American Society of Mechanical Engineers (ASME). ASME Y14.5M—1994 is the latest edition in the ongoing revision process of the standard. This chapter helps students learn the basics of geometric dimensioning and

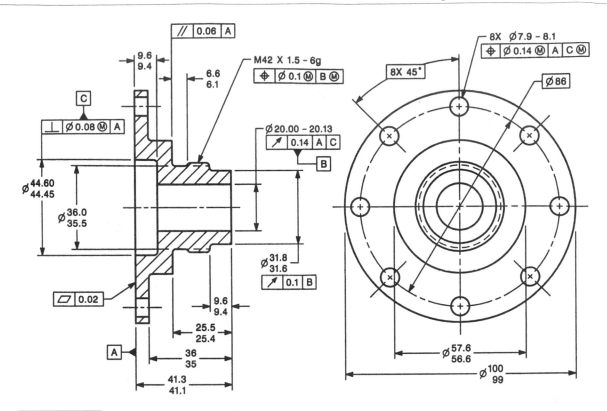

FIGURE 12-5 *Geometrically dimensioned and toleranced drawing (From ASME Y14.5M—1994)*

SYMBOL	CHARACTERISTIC	GEOMETRIC TOLERANCE
—	STRAIGHTNESS	FORM
▱	FLATNESS	
○	CIRCULARITY	
⌭	CYLINDRICITY	
⌒	PROFILE OF A LINE	PROFILE
⌓	PROFILE OF A SURFACE	
∠	ANGULARITY	ORIENTATION
⊥	PERPENDICULARITY	
//	PARALLELISM	
⌖	TRUE POSITION	LOCATION
◎	CONCENTRICITY	
≡	SYMMETRY	
* ↗	CIRCULAR RUNOUT	RUNOUT
* ⌰	TOTAL RUNOUT	

* MAY BE FILLED IN

FIGURE 12-6 *Building blocks*

positional tolerancing so they will be able to apply the latest standards set forth by ASME at any point in time and in accordance with any edition of the manual that is specified. Students should not use this chapter as a reference in place of the ASME standard. Always refer to the latest edition of the standard for specifics that go beyond the basics covered herein.

Modifiers

Modifiers are symbols that can be attached to the standard geometric building blocks to alter their application or interpretation. The proper use of modifiers is fundamental to effective geometric tolerancing. Various modifiers are often used: maximum material condition, least material condition, projected tolerance zone, free-state variation, tangent plane, all around, between symbol, and statistical tolerance, *Figure 12-7A, Figure 12-7B,* and *Figure 12-7C.*

MAXIMUM MATERIAL CONDITION

Maximum material condition (MMC) is the condition of a characteristic when the most material exists. For

Ⓜ	MAXIMUM MATERIAL CONDITION
Ⓛ	LEAST MATERIAL CONDITION
Ⓟ	PROJECTED TOLERANCE ZONE
Ⓕ	FREE STATE VARIATION
Ⓣ	TANGENT PLANE
⌖	ALL AROUND
↔	BETWEEN SYMBOL
Ⓢ⊤	STATISTICAL TOLERANCE

THE RFS SYMBOL Ⓢ CAN STILL BE USED BUT THE PREFERRED PRACTICE IS TO OMIT IT.

FIGURE 12-7A *Modifiers used when applying geometric tolerancing*

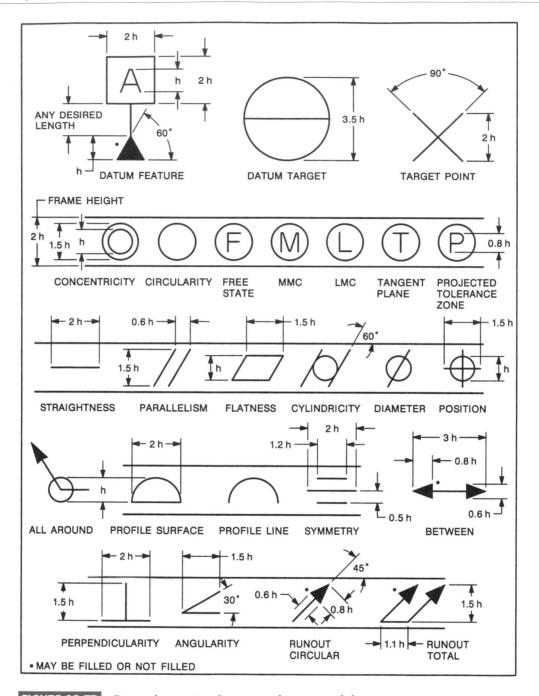

FIGURE 12-7B *Form and proportion of geometric tolerancing symbols*

example, MMC of the external feature in *Figure 12-8* is .77″. This is the MMC because it represents the condition where the most material exists on the part being manufactured. The MMC of the internal feature in the figure is .73″. This is the MMC because the most material exists when the hole is produced at the smallest allowable size.

In using this concept, the designer must remember that the MMC of an internal feature is the smallest allowable size. The MMC of an external feature is the largest allowable size within specified tolerance limits

inclusive. A rule of thumb to remember is that MMC means most material.

REGARDLESS OF FEATURE SIZE

Regardless of feature size (RFS), tells machinists that a tolerance of form or position or any characteristic must be maintained regardless of the actual produced size of the object. Geometric tolerances are understood to apply regardless of feature size where the modifiers M or L are not used. It is permissible to show the RFS modi-

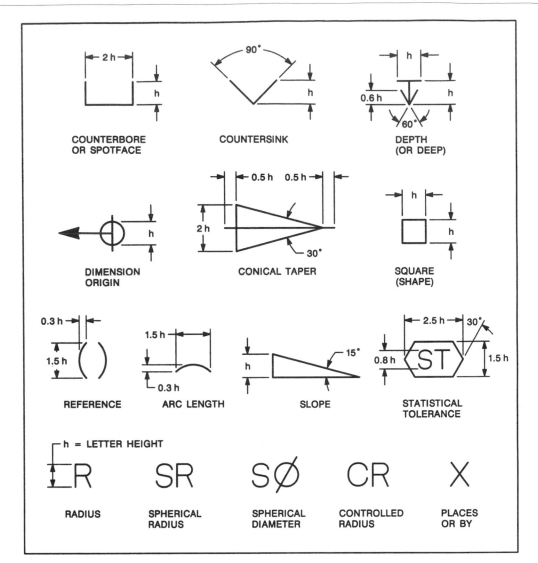

FIGURE 12-7C   *Form and proportion of dimensioning symbols and letters*

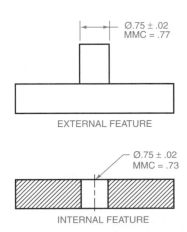

FIGURE 12-8   *MMC of an external and an internal feature*

fier; however, it is redundant and the preferred practice is to omit it. The RFS concept is illustrated in *Figure 12-9*. In the RFS example, the object is acceptable if produced in sizes from 1.002″ to .998″ inclusive. The form control is axis straightness to a tolerance of .002″ regardless of feature size. This means that the .002″ axis straightness tolerance must be adhered to, regardless of the produced size of the part.

Contrast this with the MMC example. In this case, the produced sizes are still 1.002″ to .998″. However, because of the MMC modifier, the .002″ axis straightness tolerance applies only at MMC or 1.002″.

If the produced size is smaller, the straightness tolerance can be increased proportionally. Of course, this makes the MMC modifier more popular with machinists for several reasons: (1) it allows them greater room for error without actually increasing the tolerance, (2) it

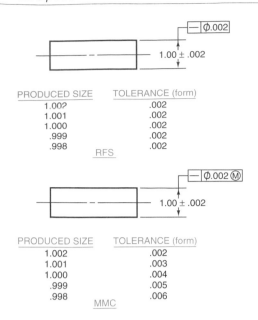

FIGURE 12-9 *Regardless of feature size (RFS)*

decreases the number of parts rejected, (3) it cuts down on unacceptable parts, (4) it decreases the number of inspections required, and (5) it allows the use of functional gaging. All of these advantages translate into substantial financial savings while, at the same time, making it possible to produce interchangeable parts at minimum expense.

LEAST MATERIAL CONDITION

Least material condition (LMC) is the opposite of MMC. It refers to the condition in which the least material exists. This concept is illustrated in *Figure 12-10*.

In the top example, the external feature of the part is acceptable if produced in sizes ranging from .98" to

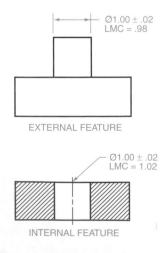

FIGURE 12-10 *Least material condition (LMC)*

1.02" inclusive. The least material exists at .98". Consequently, .98" is the LMC.

In the bottom example, the internal feature (hole) is acceptable if produced in sizes ranging from .98" to 1.02" inclusive. The least material exists at 1.02". Consequently, 1.02 represents the LMC.

PROJECTED TOLERANCE ZONE

Projected tolerance zone is a modifier that allows a tolerance zone established by a locational tolerance to be extended a specified distance beyond a given surface. This concept is discussed further later in this chapter under the heading "True Position."

FREE-STATE VARIATION

Free-state variation is the concept that some parts cannot be expected to be contained within a boundary of perfect form. Some parts may vary in form beyond the MMC size limits after forces applied during manufacture are removed. For example, a thin-walled part shape may vary in its free state due to stresses being released in the part. This variation may require that the part meet its tolerance requirements while in its free state.

Parts that are subject to free-state variation do not have to meet the Rule #1 requirement of perfect form at MMC. These parts are standard stock such as bars, sheets, tubes, extrusions, structural shapes, or other items produced to established industry or government standards. The appropriate standard would govern the limits of form variation allowed after manufacture.

The free-state symbol specifies the maximum allowable free-state variation. It is placed within a feature control frame, following the tolerance and any modifiers, *Figure 12-11*.

TANGENT PLANE

The *tangent plane* concept uses a modifying symbol with an orientation tolerance to modify the intended control of the surface. When an orientation tolerance is applied to a surface, the primary control is equivalent to the symbology used. An example of the primary control of a parallel callout is parallelism. However, when applied, the specified symbol controls not only parallelism but other form variations such as concavity, convexity, waviness, flatness, and other imperfections as well.

FIGURE 12-11 *Feature control frame with free-state symbol*

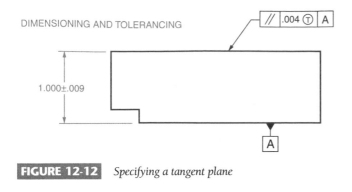

DIMENSIONING AND TOLERANCING

1.000±.009

FIGURE 12-12 *Specifying a tangent plane*

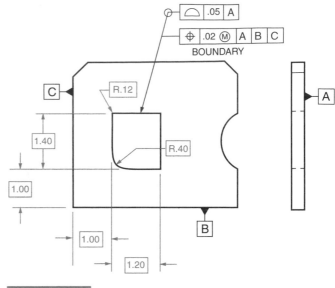

BOUNDARY

R.12

R.40

1.40

1.00

1.00

1.20

FIGURE 12-13 *All around symbol*

If two such controlled surfaces are assembled, the abrupt variation in the surfaces can cause different mating effects and assembly conditions. There are several ways to control the effects of surface conditions when applying orientation tolerances. The obvious method is to refine the surface control with a form tolerance such as flatness. This is permissible because the orientation tolerance controls flatness to the extent of the specified tolerance value.

Another method is to modify the orientation tolerance to apply a tangent plane. When the modifier is applied, the orientation tolerance zone for a tangent plane is identical to any other orientation tolerance zones with one exception. The orientation tolerance no longer controls the form of the surface. The surface of the controlled feature must be within the specified limits of size, but it is not required to fall within the parallelism tolerance zone boundary. Only a plane tangent to the high points on the surface must be within the tolerance zone boundary. The symbol is placed within the feature control frame following the stated tolerance, *Figure 12-12*.

ALL AROUND SYMBOL

The *all around symbol* is the symbolic means of indicating that the specified tolerance applies all around the part. The normal tolerance zone of a geometric callout extends the length of the feature in question. If there is an abrupt change in surface condition, such as an offset, the tolerance zone would conclude at the beginning of the offset. Applying the all around symbol extends the tolerance zone all around the feature to include abrupt surface variations, *Figure 12-13*. This concept will be discussed further later in this chapter under the heading "Profile."

BETWEEN SYMBOL

The *between symbol* is a symbolic means of indicating that the stated tolerance applies to a specified segment of a surface between designated points. The normal toler-

ance zone of a geometric callout extends the length of the feature in question. Application of this symbol can be used to limit the tolerance zone to a specified area. It can also be used to clarify the extent of the profile tolerance when it is not clearly visible due to surface variations. *Figure 12-14* illustrates the use of this symbol.

STATISTICAL TOLERANCING SYMBOL

The *statistical tolerancing symbol* is a symbolic means of indicating that the stated tolerance is based on statistical process control (SPC). The symbol can be applied in one of two ways. When the tolerance is a statistical size tolerance, the symbol is placed next to the size dimension as shown in *Figure 12-15*. When the tolerance is a statistical geometric tolerance, the symbol is placed in the feature control frame as shown in *Figure 12-16*.

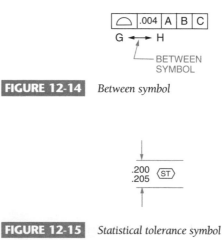

G ← → H

BETWEEN SYMBOL

FIGURE 12-14 *Between symbol*

.200
.205 (ST)

FIGURE 12-15 *Statistical tolerance symbol*

$$\boxed{\oplus \; | \; \varnothing \; .005 \; \text{(M)} \; \langle\text{ST}\rangle \; | \; A \; | \; B \; | \; C}$$

FIGURE 12-16 *Symbol indicating the specified tolerance is a statistical geometric tolerance*

DATUMS

Datums are theoretically perfect points, lines, axes, surfaces, or planes used for referencing features of an object. They are established by the physical datum features that are identified on the drawing. Identification of datum features is done by using a *datum feature symbol*. This symbol consists of a capital letter enclosed in a square frame. A leader line extends from the frame to the selected feature. A triangle is attached to the end of the leader and is applied in the appropriate way to indicate a datum feature. The symbols should only be applied to physical features. They should not be attached to center lines, axes, center planes, or other theoretical entities. *Figure 12-17* shows two ways in which datum feature symbols are placed on drawings. The datum symbol is attached to an extension line of the feature outline, clearly separated from the dimension line when the datum feature is a surface or placed on the visible outline of a feature surface.

In *Figures 12-18A, 12-18B,* and *12-18C,* the datum feature symbol is placed on an extension of the dimension line of a feature of size when the datum is an axis or center plane. In *Figures 12-18D, 12-18E,* and *12-18F,*

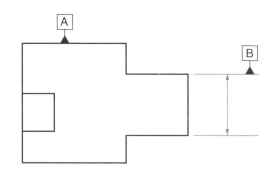

FIGURE 12-17 *Datum feature symbols on a feature surface and an extension line*

the datum is an axis. The symbol can be placed on the outline of a cylindrical surface or an extension line of the feature outline, separated from the size dimension. Figure 12-18F shows one arrow of the dimension line being replaced by the datum feature triangle when space is limited. If no feature control frame is used, the symbol is placed on a dimension leader line to the feature size dimension as seen by the example of Datum B in *Figure 12-19.* In *Figure 12-20* the symbol is attached to the feature control frame below (or above) when the feature(s) controlled is a datum center plane.

ESTABLISHING DATUMS

In establishing datums, designers must consider the function of the part, the manufacturing processes that

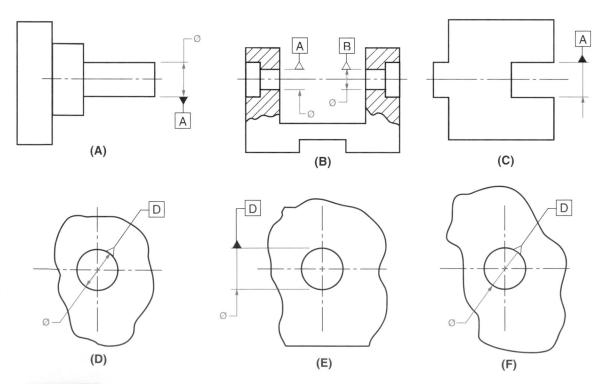

FIGURE 12-18 *Placement of datum feature symbols on features of size*

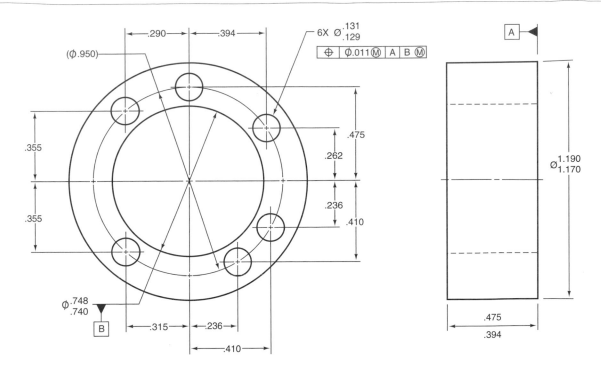

NOTE: UNTOLERANCED DIMENSIONS LOCATING TRUE POSITION ARE BASIC

FIGURE 12-19   *Datum reference on dimension leader line*

DIMENSIONING AND TOLERANCING

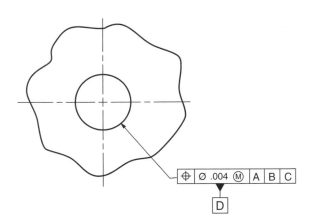

FIGURE 12-20   *Placement of datum feature symbol in conjunction with a feature control frame*

will be used in producing the part, how the part will be inspected, and the part's relationship to other parts after assembly. Designers and drafters must also understand the difference between a datum, datum feature, datum feature simulator, datum surface, datum plane, and a datum feature of size.

A datum is theoretical in nature and is located by the physical datum features identified on the drawing. A datum is considered to be the true geometric counterpart of the feature. It is the origin from which measurements are made, or that provides geometrical references

to which other features are established. A *datum feature* is the physical feature on a part used to establish a datum, *Figure 12-21*. It is identified on a drawing by use of a datum feature symbol, *Figure 12-22*.

A *datum feature simulator* is a surface, the form of which is of such precise accuracy (such as a surface plate, a gage surface, or a mandrel) that it is used to simulate the datum. The datum feature simulator contacts the datum feature(s) and simulates the theoretical datum. Simulation is necessary because measurements cannot be made from the theoretical true geometric counterpart. It is therefore necessary to use high-quality geometric features to simulate datums. Although the features are not perfect, they are of such a quality that they can be used for that purpose. *Figures 12-23* and *12-24* illustrate this concept with respect to a surface and a feature of size.

A *datum surface* (feature) is the inexact surface of the object used to establish a datum plane. A *datum plane* is a theoretically perfect plane from which measurements are made. Since inaccuracies and variations in the surface condition of the datum surface make it impractical to take measurements from, then a theoretically perfect plane must be established from which measurements are made. To establish this datum plane, the high points of the datum surface are brought into contact with, in this case, a surface plate, which simulates the datum plane. This concept is illustrated in Figure 12-21.

THIS ON THE DRAWING

MEANS THIS

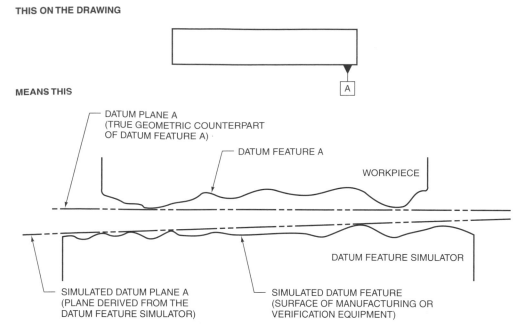

(A) WORKPIECE & DATUM FEATURE SIMULATOR PRIOR TO CONTACT

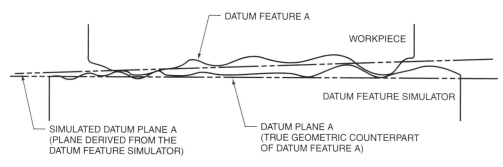

(B) WORKPIECE & DATUM FEATURE SIMULATOR IN CONTACT

FIGURE 12-21 *Datum feature, simulated datum, and theoretical datum plane*

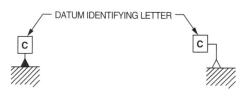

DATUM FEATURE TRIANGLE MAY BE FILLED OR NOT FILLED.
LEADER MAY BE APPROPRIATELY DIRECTED TO A FEATURE.

FIGURE 12-22 *Datum feature symbol*

Notice the irregularities on the datum surface. The high points on the datum surface actually establish the datum plane, which, in this case, is the top of the manufacturing equipment. All measurements referenced to DATUM A are measured from the theoretically perfect datum plane. High point contact is used for establishing datums when the entire surface in question will be a machined surface.

A datum feature of size is established by associating the datum feature symbol with the size dimension of the selected feature size. When identified, the theoretical datum is the axis, center line, or center plane of the true geometric counterpart. It is simulated by the processing equipment (such as a chuck, vise, or centering device). The datum feature simulator establishes the datum axis, center line, or center plane from which measurements can be referenced. This concept is illustrated in Figures 12-23 and 12-24.

DATUM TARGETS

On rougher, more irregular surfaces, such as those associated with castings, specified points, lines, or area contacts are used for establishing datums. Datum targets

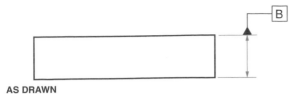

AS DRAWN

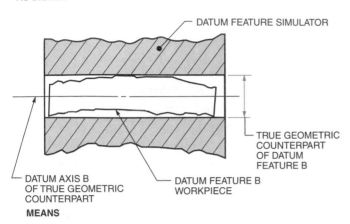

MEANS

FIGURE 12-23 *Primary external datum diameter with datum feature simulator*

AS DRAWN

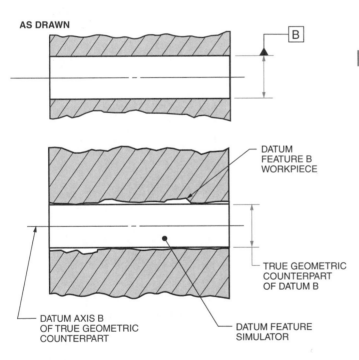

FIGURE 12-24 *Primary internal datum diameter with datum feature simulator*

designate specific points, lines, or areas of contact on a part that are used in establishing a datum. They are used when it is not always practical to identify an entire surface as a datum feature.

A *datum target symbol* is used to identify datum targets. It consists of a circle divided in half with a horizontal line. The lower portion contains the datum identifying letter followed by a datum target number. The

numbers are sequential, starting with one for each datum. The letter and number establish a target label to identify planes or axes as datums. The upper half of the symbol is normally empty except when one uses a diameter symbol followed by a value to identify the shape and size of the target area, *Figure 12-25*. *Figure 12-26* shows a part using the datum's target areas to establish a datum plane.

Dimensions used to locate targets may be *basic dimensions* or toleranced dimensions. A basic dimension is a theoretically perfect dimension, much like a nominal or

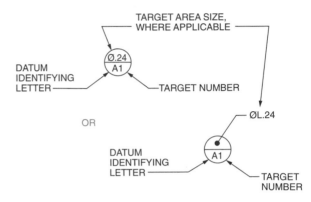

FIGURE 12-25 *Datum target symbol*

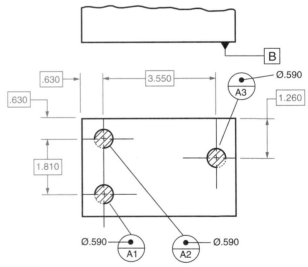

AS DRAWN

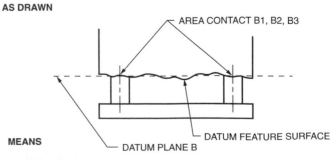

MEANS

FIGURE 12-26 *Primary datum plane established by three datum target areas*

design dimension. The dimension is identified by enclosing the value in a rectangular box as shown in *Figure 12-27*. Tolerances placed in general notes or within the title block do not apply to basic dimensions. In *Figure 12-28*, the datum targets are located using basic dimensions. Points are located relative to one another and dimensioned to show the relationship between targets.

When specific datum target points are used for establishing datums, a minimum of three points, not in a straight line, are required for the primary datum, a minimum of two for the secondary, and a minimum of one for the tertiary, Figure 12-28. In Figure 12-28, primary datum plane A is the top of the object, and it is established by points A1, A2, and A3. Secondary datum plane

B is the front of the object, and tertiary datum plane C is the right side. The datum feature symbol is placed on a drawing in the view where the surface in question appears as an edge.

Notice also that the secondary datum must be perpendicular to the first, and the tertiary datum must be perpendicular to both the primary and secondary datums. These three mutually perpendicular datum planes establish what is called the datum reference frame. The *datum reference frame* is a hypothetical, three-dimensional frame that establishes the three axes of an X, Y, and Z coordinate system into which the object being produced fits and from which measurements can be made. *Figure 12-29* shows an object located within a datum reference frame. For features that have sides (for example, rectangular and square objects), it takes three datums to establish a datum reference frame.

For cylindrical features, a complete reference frame is established with two datum references. *Figure 12-30* shows an object within a reference frame. Datum D is

2.50

FIGURE 12-27 *Basic dimension symbol*

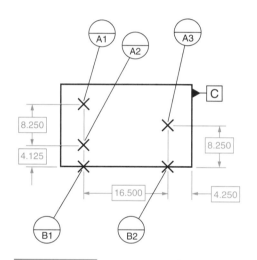
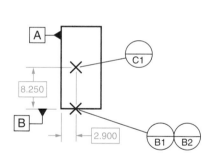

FIGURE 12-28 *Dimensioning datum targets*

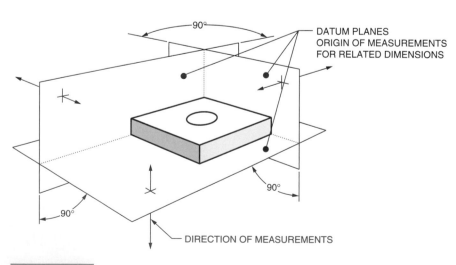

FIGURE 12-29 *Datum reference frame*

DIMENSIONING AND TOLERANCING

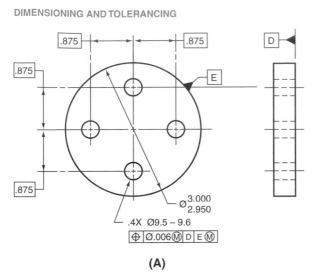

(A)

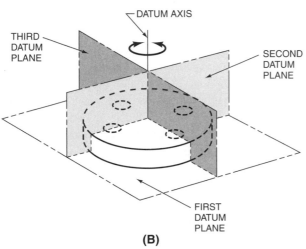

(B)

FIGURE 12-30   *Part with cylindrical datum feature*

the primary datum feature and is used to establish datum plane K. Notice that datum feature E is established by two theoretical planes intersecting at right angles on the datum axis. The datum axis becomes the origin of measurements to locate other features on the object. Datum feature E uses the second and third planes to locate the datum axis. The reference frame is thus established using two datums.

Figure 12-31 is an example of a "basic dimension." A basic dimension is a theoretically perfect dimension, much like a nominal or design dimension, that is used to locate or specify the size of a feature. Basic dimensions are enclosed in rectangular boxes, as shown in Figure 12-31.

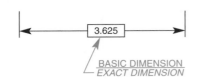

FIGURE 12-31   *Basic dimensions*

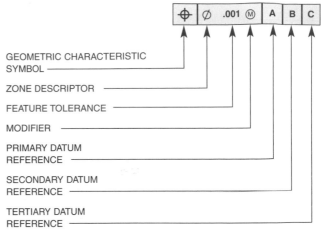

FIGURE 12-32   *Order of elements in a feature control symbol*

Feature Control Symbol

The *feature control symbol* is a rectangular box in which all data referring to the subject feature control are placed, including: the symbol, datum references, the feature control tolerance, and modifiers. These various feature control elements are separated by vertical lines. (Figure 12-5 contains a drawing showing how feature control symbols are actually composed.)

The order of the data contained in a feature control frame is important. The first element is the feature control symbol. Next is the zone descriptor, such as a diameter symbol where applicable. Then, there is the feature control tolerance, modifiers when used, and datum references listed in order from left to right, *Figure 12-32*.

Figures 12-33 through *12-37* illustrate how feature control symbols are developed for a variety of design situations. Figure 12-33 is a feature control symbol that specifies a .005 tolerance for flatness and no datum reference. Figure 12-34 specifies a tolerance of .005 for the true position of a feature relative to Datum A. Figures 12-35 and 12-36 show the proper methods for constructing feature control symbols with two and three datum references, respectively. Figure 12-37 illustrates a feature control symbol with a modifier and a controlled datum added.

True Position

True position is the theoretically exact location of the center line of a product feature such as a hole. The tolerance zone created by a position tolerance is an imaginary cylinder, the diameter of which is equal to the stated position tolerance. The dimensions used to locate a feature that is to have a position tolerance must be basic dimensions.

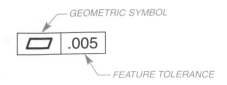

GEOMETRIC SYMBOL

FEATURE TOLERANCE

FIGURE 12-33 *Feature control symbol with no datum reference*

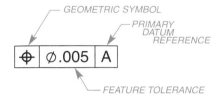

GEOMETRIC SYMBOL

PRIMARY DATUM REFERENCE

FEATURE TOLERANCE

FIGURE 12-34 *Feature control symbol with one datum reference*

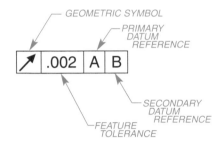

GEOMETRIC SYMBOL

PRIMARY DATUM REFERENCE

SECONDARY DATUM REFERENCE

FEATURE TOLERANCE

FIGURE 12-35 *Feature control symbol with two datum references*

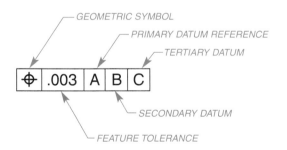

GEOMETRIC SYMBOL

PRIMARY DATUM REFERENCE

TERTIARY DATUM

SECONDARY DATUM

FEATURE TOLERANCE

FIGURE 12-36 *Feature control symbol with three datum references*

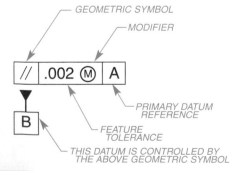

GEOMETRIC SYMBOL

MODIFIER

PRIMARY DATUM REFERENCE

FEATURE TOLERANCE

THIS DATUM IS CONTROLLED BY THE ABOVE GEOMETRIC SYMBOL

FIGURE 12-37 *Feature control symbol with a modifier*

Figure 12-38 contains an example of a part with two holes drilled through it. The holes have a position tolerance relative to three datums: A, B, and C. The holes are located by basic dimensions. The feature control frame

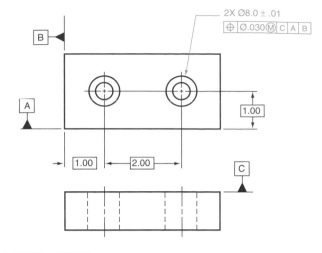

FIGURE 12-38 *True position*

states that the positions of the center lines of the holes must fall within cylindrical tolerance zones having diameters of .030″ at MMC relative to DATUMS A, B, and C. The modifier indicates that the .030″ tolerance applies only at MMC. As the holes are produced larger than MMC, the diameter of the tolerance zones can be increased correspondingly.

Figure 12-39 illustrates the concept of the cylindrical tolerance zone from Figure 12-38. The feature control frame is repeated showing a .030″ diameter tolerance zone. The broken-out section of the object from Figure 12-38 provides the interpretation. The cylindrical tolerance zone is shown in phantom lines. The center line of the hole is acceptable as long as it falls anywhere within the hypothetical cylinder.

USING THE PROJECTED TOLERANCE ZONE MODIFIER

ASME recommends the use of the projected tolerance zone concept when the variation in perpendiculars of

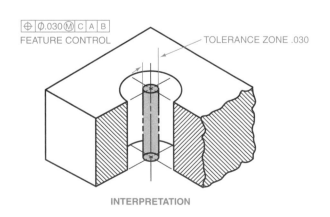

FEATURE CONTROL

TOLERANCE ZONE .030

INTERPRETATION

FIGURE 12-39 *Cylindrical tolerance zone*

threaded or press-fit holes could cause fasteners, such as screws, studs, or pins, to interfere with mating parts.

The attitude of a threaded fastener is controlled by the inclination of the threaded hole into which it will assemble. There are instances where the inclination can be such that the fastener interferes with the mating feature. One method of overcoming this problem is to use a projected tolerance zone. When projected, the tolerance zone's intended outcome is to decrease the inclination of the fastener passing through the mating part. It is often thought that the tolerance zone extends through the feature being controlled to a point beyond the part equal to the projection, but this is not the case. Instead, the controlled feature has no internal tolerance; the zone is totally outside of the feature being controlled. The height of the zone is equal to the value specified within the feature control frame. *Figure 12-40* illustrates this concept.

The projected tolerance zone symbol is a capital P enclosed by a circle. It is placed within the feature control frame following the tolerance value or modifier where applicable. The projection height is placed after the projected tolerance zone symbol, as illustrated in Figure 12-40. When a projected tolerance zone modifier is used, the surface from which the tolerance is projected is identified as a datum and the length of the projected tolerance zone is specified. In cases where it is not clear from which surface the projection extends, such as a through hole, a heavy chain line is used with a dimension applied to it, as illustrated in *Figure 12-41*. The resultant tolerance zone lies totally outside the feature being controlled.

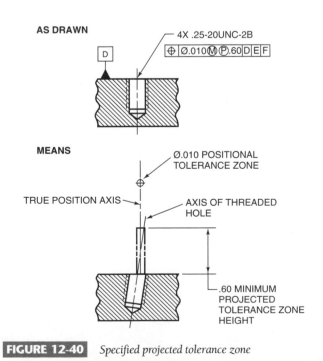

FIGURE 12-40 *Specified projected tolerance zone*

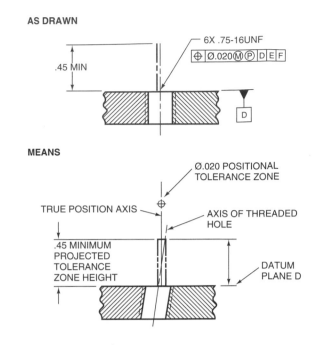

FIGURE 12-41 *Projected tolerance zone using chain line*

FLATNESS

Flatness is a feature control of a surface that requires all elements of the surface to lie within two hypothetical parallel planes. When flatness is the feature control, a datum reference is neither required nor proper.

Flatness is applied by means of a leader pointing to the surface or by an extension line of the surface. It cannot be attached to the size dimension. The modifiers M or L cannot be used with flatness because it is a surface control only. The flatness tolerance is not additive and must be less than the tolerance of size of the part unless the appropriate note is added exempting it from Rule #1 requirements.

Figure 12-42 shows how flatness is called out in a drawing and the effect such a callout has on the produced part. The surface indicated must be flat within a tolerance zone of .010, as shown in Figure 12-42.

Flatness is specified when size tolerances alone are not sufficient to control the form and quality of the surface and when a surface must be flat enough to provide a stable base or a smooth interface with a mating part.

Flatness is inspected for a full indicator movement (FIM) using a dial indicator. FIM is the newer term that has replaced the older "total indicator movement" or TIR. FIM means that the swing of the indicator needle from one extreme to the other cannot exceed the amount of the specified tolerance. The dial indicator is set to run parallel to a surface table that is a theoretically perfect surface. The dial indicator is mounted on a stand or height gage. The machined surface is run under

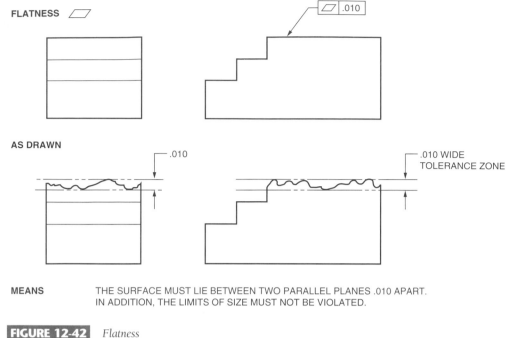

FLATNESS

AS DRAWN

.010

.010 WIDE
TOLERANCE ZONE

MEANS THE SURFACE MUST LIE BETWEEN TWO PARALLEL PLANES .010 APART.
IN ADDITION, THE LIMITS OF SIZE MUST NOT BE VIOLATED.

FIGURE 12-42 *Flatness*

it, allowing the dial indicator to detect irregularities that fall outside of the tolerance zone.

STRAIGHTNESS

A *straightness* tolerance can be used to control surface elements, an axis, or a center plane. When used to control single elements for a flat surface, it is applied in the view where the element to be controlled is a straight line. When applied, it controls line elements in only one direction. It differs from flatness in that flatness covers an entire surface rather than just single elements on a surface. A straightness tolerance yields a tolerance zone of a specified width, within which all points on the line in question must lie. Straightness is generally applied to longitudinal elements.

Another difference between straightness and flatness concerns the application of the feature control frame. The method in which the feature control frame is applied, determines the intended control. If the feature control frame is attached to an extension line of the surface or attached to a leader pointing to the surface, the

intended control is to the surface, *Figure 12-43A*. However, if the feature control frame is attached to a dimension line or adjacent to a dimension, the intended control is an axis or center plane, *Figure 12-43B*. Drastically different results are realized based on the application method.

STRAIGHTNESS OF A FLAT SURFACE

Figure 12-44 shows how a straightness tolerance is applied on a drawing to the elements of a flat surface. The straightness tolerance applies only to the top surface. The bottom surface straightness error is controlled by the limits of size. In this case, the straightness tolerance is used as a refinement for the top surface only. The feature control frame states that any longitudinal element for the referenced surface, in the direction indicated, must lie between two parallel straight lines that are .002" apart.

STRAIGHTNESS OF A CYLINDRICAL SURFACE

Straightness applied to the surface of a cylindrical feature is shown in *Figure 12-45*. It is similar to that of a flat

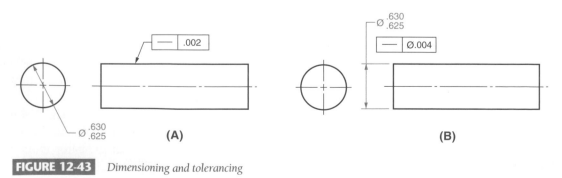

FIGURE 12-43 *Dimensioning and tolerancing*

STRAIGHTNESS

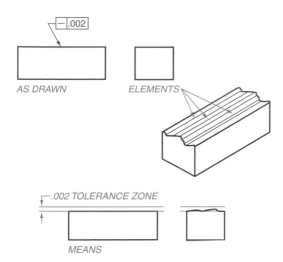

AS DRAWN   ELEMENTS

.002 TOLERANCE ZONE

MEANS

FIGURE 12-44 *Straightness of a flat surface*

AS DRAWN

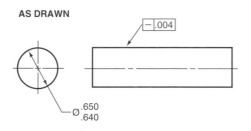

MEANS

(A)

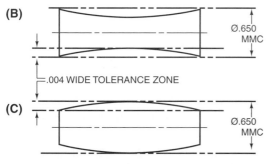

EACH LONGITUDINAL ELEMENT OF THE SURFACE MUST LIE BETWEEN TWO PARALLEL LINES (.004 APART) WHERE THE TWO LINES AND THE NOMINAL AXIS OF THE PART SHARE A COMMON PLANE. THE FEATURE MUST BE WITHIN THE SPECIFIED LIMITS OF SIZE AND THE BOUNDARY OF PERFECT FORM AT MMC (.650).

WAISTING (B) OR BARRELING (C) OF THE SURFACE MUST BE WITHIN THE STRAIGHTNESS TOLERANCE AND THE LIMITS OF SIZE OF THE FEATURE.

FIGURE 12-45 *Straightness of surface elements*

surface, with one exception. Since the surface is round, opposing surface line elements must also be considered when verifying straightness. The full straightness tolerance may not be available for these elements due to conditions such as wasting or barreling of the surface. Additionally, the straightness tolerance is not additive to the size tolerance and must be contained within the limits of size. This means that if the part is made at MMC, no straightness tolerance is available because any variation in surface straightness would cause the part to exceed the MMC boundary of size. *Figure 12-46* illustrates the relationship between a straightness tolerance and a size tolerance of a part. Remember, each element of the surface must stay within the specified straightness tolerance zone and within the size tolerance envelope. Straightness is affected by running the single-line elements of a surface under a dial indicator for a full indicator movement (FIM).

Figures 12-47 through 12-52 further illustrate the concept of straightness. *Figure 12-47* shows a part with a size tolerance, but no feature control tolerance. In this example, the form of the feature is controlled by the size tolerance. The difference between maximum and minimum limits defines the maximum form variation that is allowed. ASME Y14.5M outlines the requirements of form control for individual features controlled only with a size dimension. This requirement is known as *Rule #1*. According to the standard, Rule #1 states: "Where only a tolerance size is specified, the limits of size of an individual feature define the extent to which variations in its geometric form, as well as size, are allowed." This means that the size limits of a part determine the maximum and minimum limits (boundaries) for that part. The MMC limit establishes a boundary limit of perfect form. If a part is at MMC, it must have perfect form. No variation in form is allowed. As the part varies in size toward

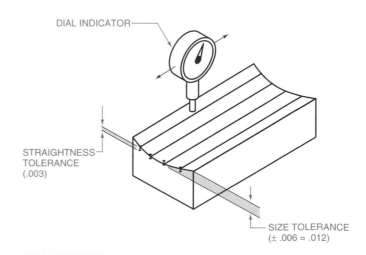

FIGURE 12-46 *Straightness interpreted*

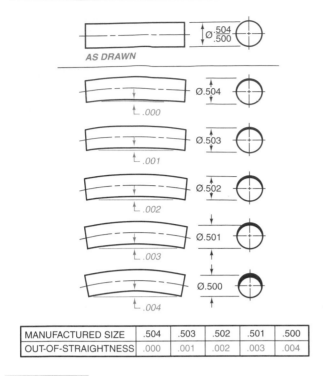

MANUFACTURED SIZE	.504	.503	.502	.501	.500
OUT-OF-STRAIGHTNESS	.000	.001	.002	.003	.004

FIGURE 12-47 *Object with no feature control symbol (Rule #1 applies)*

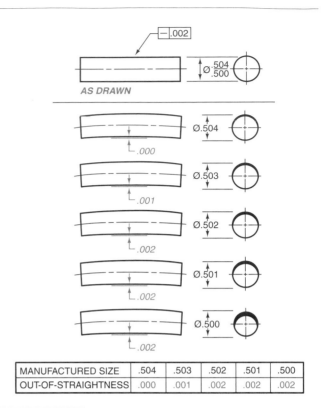

MANUFACTURED SIZE	.504	.503	.502	.501	.500
OUT-OF-STRAIGHTNESS	.000	.001	.002	.002	.002

FIGURE 12-48 *Straightness at RFS*

LMC, the form of the part is allowed to vary equal to the variation in size from MMC. When the part is made at LMC, the form variation is equal to the difference between the MMC and LMC sizes as illustrated in Figure 12-47.

Figure 12-48 is the same part with a straightness tolerance of .002 regardless of feature size tolerance. The implied regardless of feature size tolerance limits the amount the surface can be out of straightness to a maximum of .002 regardless of the produced size of the part. However, because the straightness control is on a cylindrical surface, the .002 tolerance might not be available as the part approaches MMC. The drawing at the top of the figure illustrates how the part would be drawn. The five illustrations below the part as drawn illustrate the actual shape of the object with each corresponding produced size and the available tolerance.

STRAIGHTNESS OF AN AXIS OR CENTER PLANE

To locate the axis of a part, the size of the part must be known. To locate the center plane of two parallel features, the distance between the features must be known. These are two examples of what is known as features of size. Logically, then, to control the axis of a part the feature control frame must be applied to the size dimension of that part, or to control the center plane of a rectangular part it must be applied to the size dimension, Figure 12-43B. When straightness is applied to control the axis of the feature, the tolerance zone is cylindrical and

extends the full length of the controlled feature. Straightness applied to control the center plane of a non-cylindrical feature is shown in *Figure 12-49*. It is similar to that of straightness of a cylindrical feature, except that the tolerance zone is a width and no diameter symbol is used within the feature control frame.

Straightness applied to the axis or center plane of a feature creates a boundary condition known as *virtual condition*. Virtual condition in ASME Y14.5 is defined as follows: "A constant boundary generated by the collective effects of a size feature's specified MMC or LMC and the geometric tolerance for that material condition." This means that you are allowed to add the straightness tolerance to the MMC size for a shaft and subtract the straightness tolerance from the MMC size for a hole. The resultant boundary represents the extreme form variation allowed for the part. Although this boundary is theoretical, it represents the size boundary of mating features. Unlike straightness of a surface control, a straightness

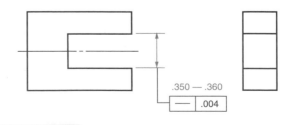

FIGURE 12-49 *Straightness*

control of an axis or center plane allows for the availability of straightness tolerance even when the part is made at MMC. Axis or center plane control of a feature becomes more desirable because of the increased availability of tolerance and better control of mating features.

Figure 12-50 is the same part as in the previous examples with a straightness tolerance of .002 at maximum material condition applied. The use of the MMC modifier is limited to tolerances controlling the axis or center plane of features. It specifies the tolerance allowed when part is produced at MMC. The drawing at the top of the figure illustrates how the part would be drawn. The five illustrations below the part as drawn illustrate the actual shape of the object with each corresponding produced size. A virtual condition boundary of .506 is created. When the part is at .504, the .506 virtual condition boundary allows for straightness of .002 at MMC. Since the .002 straightness tolerance applies at maximum material condition, the amount that the part can be out of straightness increases correspondingly as the produced size decreases. The table at the bottom of Figure 12-50 summarizes the manufactured sizes and the corresponding amounts that the part can be out of straightness for each size.

Figure 12-51 is an example of the same part with a .002 straightness tolerance at least material condition (LMC). It specifies the tolerance allowed when the part is produced at LMC. This results in the opposite effect of

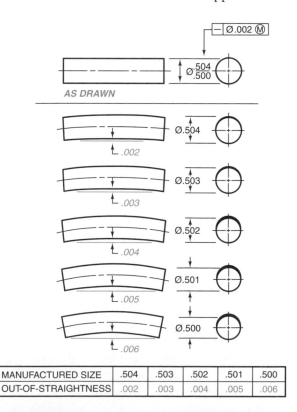

MANUFACTURED SIZE	.504	.503	.502	.501	.500
OUT-OF-STRAIGHTNESS	.006	.005	.004	.003	.002

FIGURE 12-51   *Straightness of an axis at LMC*

what occurred in Figure 12-49. Notice that the .002 straightness tolerance applies at the least material condition. As the actual produced size increases, the amount of out of straightness allowed increases correspondingly.

Figure 12-52 illustrates the same part from a .002 straightness tolerance and a regardless of feature size tolerance. Notice in this example that the .002 straightness tolerance applies regardless of the actual produced size of the part.

Circularity (Roundness)

Circularity, sometimes referred to as roundness, is a feature control for a surface of revolution (cylinder, sphere, cone, and so forth). It specifies that all points of a surface must be equidistant from the center line or axis of the object in question. The tolerance zone for circularity is formed by two concentric and coplanar circles between which all points on the surface of revolution must lie.

Figures 12-53 and *12-54* illustrate how circularity is called out on a drawing and provides an interpretation of what the circularity tolerance actually means. At any selected cross section of the part, all points on the surface must fall within the zone created by the two concentric circles. At any point where circularity is measured, it must fall within the size tolerance. Notice that a circularity tolerance cannot specify a datum reference.

MANUFACTURED SIZE	.504	.503	.502	.501	.500
OUT-OF-STRAIGHTNESS	.002	.003	.004	.005	.006

FIGURE 12-50   *Straightness of an axis at MMC*

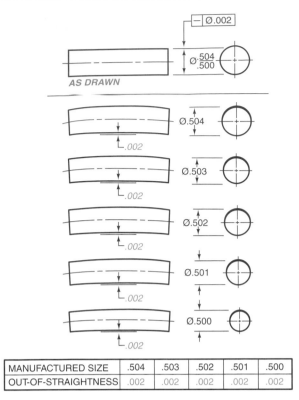

MANUFACTURED SIZE	.504	.503	.502	.501	.500
OUT-OF-STRAIGHTNESS	.002	.002	.002	.002	.002

FIGURE 12-52 *Straightness of an axis at RFS*

CIRCULARITY ○

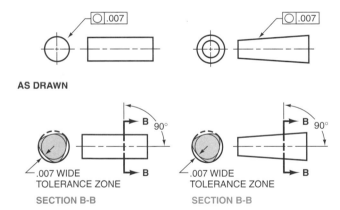

MEANS   EACH CIRCULAR ELEMENT OF THE SURFACE IN A PLANE PERPENDICULAR TO AN AXIS MUST LIE BETWEEN TWO CONCENTRIC CIRCLES, ONE HAVING A RADIUS .007 LARGER THAN THE OTHER. EACH CIRCULAR ELEMENT OF THE SURFACE MUST BE WITHIN THE SPECIFIED LIMITS OF SIZE.

FIGURE 12-53 *Circularity for a cylinder or cone*

Circularity establishes elemental single-line tolerance zones that may be located anywhere along a surface. The tolerance zones are taken at any cross section of the feature. Therefore, the object may be spherical, cylindrical, tapered, or even hourglass shaped so long as the cross section for inspection is taken at 90° to the nominal axis

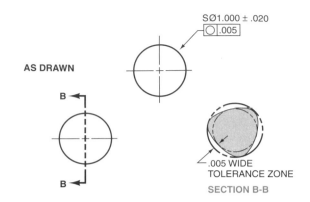

AS DRAWN

EACH CIRCULAR ELEMENT OF THE SURFACE IN A PLANE PASSING THROUGH A COMMON CENTER MUST LIE BETWEEN TWO CONCENTRIC CIRCLES, ONE HAVING A RADIUS 0.25 LARGER THAN THE OTHER. EACH CIRCULAR ELEMENT OF THE SURFACE MUST BE WITHIN THE SPECIFIED LIMITS OF SIZE.

FIGURE 12-54 *Circularity for a sphere*

of the object. A circularity tolerance is inspected using a dial indicator and making readings relative to the axis of the feature. In measuring a circularity tolerance, the full indicator movement (FIM) of the dial indicator should not be any larger than the size tolerance, and there should be several measurements made at different points along the surface of the diameter. All measurements taken must fall within the circularity tolerance.

Cylindricity

Cylindricity is a feature control in which all elements of a surface of revolution form a cylinder. It gives the effect of circularity extended the entire length of the object, rather than just a specified cross section. The tolerance zone is formed by two hypothetical concentric cylinders.

Figure 12-55 illustrates how cylindricity is called out on a drawing. Notice that a cylindricity tolerance does not require a datum reference.

Figure 12-55 also provides an illustration of what the cylindricity tolerance actually means. Two hypothetical concentric cylinders form the tolerance zone. The outside cylinder is established by the outer limits of the object at its produced size within specified size limits. The inner cylinder is smaller (on radius) by a distance equal to the cylindricity tolerance.

Cylindricity requires that all elements on the surface fall within the size tolerance and the tolerance established by the feature control.

A cylindricity tolerance must be less than the size tolerance, and it is not additive to the maximum material condition of the feature. Cylindricity is inspected using a dial indicator and making readings relative to the axis of the feature. In measuring cylindricity tolerance, the

CYLINDRICITY

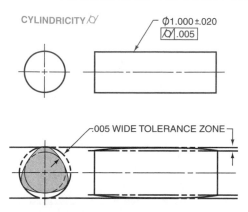

THE CYLINDRICAL SURFACE MUST LIE BETWEEN TWO CONCENTRIC CYLINDERS, ONE HAVING A RADIUS .005 LARGER THAN THE OTHER. THE SURFACE MUST BE WITHIN THE SPECIFIED LIMITS OF SIZE.

FIGURE 12-55 *Specifying cylindricity*

full indicator movement (FIM) of the dial indicator should not be any larger than the size tolerance. The part is rotated as the indicator is moved simultaneously along the surface of the diameter. At no point can the full indicator movement exceed the tolerance specified.

Angularity

Angularity is a feature control in which a given surface, axis, or center plane must form a specified angle other than 90° with a datum. Consequently, an angularity tolerance requires one or more datum references. The tolerance zone formed by an angularity callout consists of two hypothetical parallel planes that form the specified angle with the datum. All points on the angular surface or along the angular axis must lie between these parallel planes.

ANGULARITY OF A SURFACE

Figure 12-56 illustrates how an angularity tolerance on a surface is called out on a drawing. Notice that the specified angle is basic. This is required when applying an angularity tolerance. Figure 12-56 also provides an interpretation of what the angularity tolerance actually means. The surface must lie between two parallel planes of .02 apart that are inclined at a 30° basic angle to datum plane A.

Angularity also controls the flatness of the surface to the same extent that it controls the angular orientation. When it is required that the flatness of the feature be less than the orientation, a flatness callout can be used as a refinement of the orientation callout. When using flatness as a refinement, the tolerance is less than the orientation tolerance. The feature control frame is normal-

AS DRAWN

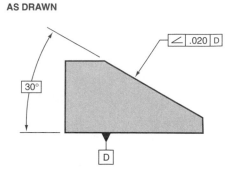

MEANS

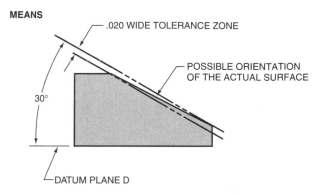

THE SURFACE MUST LIE BETWEEN TWO PARALLEL PLANES .020 APART WHICH ARE INCLINED AT 30° BASIC TO DATUM PLANE D. THE SURFACE MUST BE WITHIN THE SPECIFIED LIMITS OF SIZE.

FIGURE 12-56 *Specifying angularity for a surface*

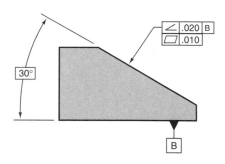

FIGURE 12-57 *Angularity with flatness refinement*

ly placed on an extension line below the orientation control, *Figure 12-57*.

ANGULARITY OF AN AXIS OR CENTER PLANE

An angularity callout can also be used to control the axis or center plane of a feature. This is done by placing the feature control frame with the size dimension in an appropriate manner as seen in *Figure 12-58*. The tolerance zone for an axis control can be cylindrical in shape or two parallel planes. When the diameter symbol is used within the feature control frame, the tolerance zone is cylindrical. When no diameter is used, the tolerance zone shape is two parallel planes, *Figure 12-59*.

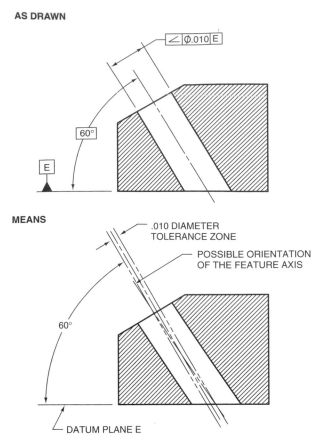

AS DRAWN

MEANS

REGARDLESS OF FEATURE SIZE, THE FEATURE AXIS MUST LIE
WITHIN A .010 DIAMETER CYLINDRICAL ZONE INCLINED 60° TO
DATUM PLANE A. THE FEATURE AXIS MUST BE WITHIN THE
SPECIFIED TOLERANCE OF LOCATION.

FIGURE 12-58 *Angularity for an axis (cylindrical tolerance zone)*

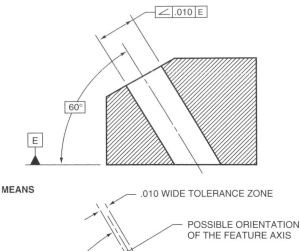

AS DRAWN

MEANS

REGARDLESS OF FEATURE SIZE, THE FEATURE AXIS
MUST LIE BETWEEN TWO PARALLEL PLANES .010
APART THAT ARE INCLINED 60° TO DATUM PLANE A.
THE FEATURE AXIS MUST BE WITHIN THE SPECIFIED
TOLERANCE OF LOCATION.

NOTE: THIS CONTROL APPLIES ONLY TO THE VIEW
ON WHICH IT IS SPECIFIED.

FIGURE 12-59 *Angularity for an axis (two parallel planes)* (From
ASME Y14.5M–1994)

Parallelism

Parallelism is a feature control that specifies that all
points on a given surface, axis, line, or center plane must
be equidistant from a datum. Consequently, a parallelism tolerance requires one or more datum references.
A parallelism tolerance zone is formed by two hypothetical parallel planes that are parallel to a specified datum.
They are spaced apart at a distance equal to the parallelism tolerance.

PARALLELISM OF A SURFACE

Figure 12-60 illustrates how a parallelism is called out on
a drawing and provides an interpretation of what the parallelism tolerance actually means. Notice that all elements
of the toleranced surface must fall within the size limits.

Notice in Figure 12-60 that the .006 parallelism tolerance is called out relative to Datum A. You must specify a datum when calling out a parallelism tolerance.
Parallelism should be specified when features such as
surfaces, axes, and planes are required to lie in a com-

mon orientation. Parallelism is inspected by placing the
part on an inspection table and running a dial indicator
a full indicator movement across the surface of the part.

Parallelism also controls the flatness of the surface to
the same extent that it controls parallel orientation.
When it is required that the flatness of the feature be less
than the orientation, a flatness callout can be used as a
refinement of the orientation callout. When using flatness as a refinement, the tolerance is less than the orientation tolerance. The feature control frame is normally placed on an extension line below the orientation
control, *Figure 12-61*.

PARALLELISM OF AN AXIS OR CENTER PLANE

Parallelism can be used to control the orientation of an
axis to a datum plane, an axis to an axis, or the center
plane of noncylindrical parts. When applied to control
the axis or center plane, the feature control frame must

PARALLELISM //

AS DRAWN

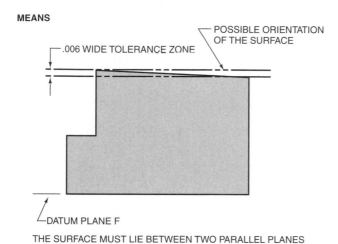

MEANS

.006 WIDE TOLERANCE ZONE

POSSIBLE ORIENTATION OF THE SURFACE

DATUM PLANE F

THE SURFACE MUST LIE BETWEEN TWO PARALLEL PLANES 0.12 APART WHICH ARE PARALLEL TO DATUM PLANE A. THE SURFACE MUST BE WITHIN THE SPECIFIED LIMITS OF SIZE.

FIGURE 12-60 *Parallelism for a surface to datum plane*

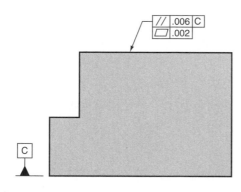

FIGURE 12-61 *Parallelism with flatness refinement*

AS DRAWN

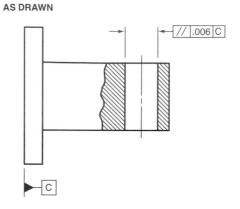

MEANS

.006 WIDE TOLERANCE ZONE

POSSIBLE ORIENTATION OF FEATURE AXIS

DATUM PLANE C

REGARDLESS OF FEATURE SIZE, THE FEATURE AXIS MUST LIE BETWEEN TWO PARALLEL PLANES .006 APART THAT ARE PARALLEL TO DATUM PLANE C. THE FEATURE AXIS MUST BE WITHIN THE SPECIFIED TOLERANCE OF LOCATION.

FIGURE 12-62 *Parallelism for an axis to datum plane*

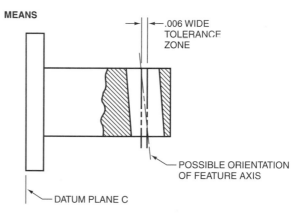

be placed with the size dimension in the appropriate fashion. When used to control an axis to a datum plane, the tolerance zone shape is two parallel planes separated by the amount of the stated tolerance. The tolerance control is only applicable relative to the specified datum surface, *Figure 12-62*. Virtual condition exists for the controlled feature, which allows for the availability of additional tolerance. The M and L modifiers can be used because we are controlling a feature of size.

When used to control an axis to a datum axis, the tolerance zone shape is cylindrical and the diameter is equal to the amount of the stated tolerance. The tolerance control is three-dimensional, allowing the axis to float relative to the orientation of the datum, *Figure 12-63*. Virtual condition exists for the controlled feature, which allows for the availability of additional tolerance. The M and L modifiers can be used because we are controlling a feature of size.

When used to control a center plane to a datum plane or a center plane to a center plane, the tolerance zone is similar to that of an axis to a surface or an axis to a datum axis. However, the tolerance zone shape is never cylindrical. The shape is two parallel planes separated by the amount of the stated tolerance. Virtual condition exists for the controlled feature, which allows for the availability of additional tolerance. The M and L modifiers can be used because we are controlling a feature of size.

Perpendicularity

Perpendicularity is a feature control that specifies that all elements of a surface, axis, center plane, or line form a

AS DRAWN

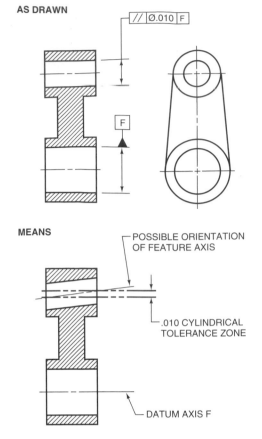

MEANS

REGARDLESS OF FEATURE SIZE, THE FEATURE AXIS MUST LIE WITHIN A .010 DIAMETER CYLINDRICAL ZONE PARALLEL TO DATUM AXIS A. THE FEATURE AXIS MUST BE WITHIN THE SPECIFIED TOLERANCE OF LOCATION.

FIGURE 12-63 *Parallelism for an axis to datum axis*

90° angle with a datum. Consequently, a perpendicularity tolerance requires a datum reference. A perpendicularity tolerance is formed by two hypothetical parallel planes that are at 90° to a specified datum. They are spaced apart at a distance equal to the perpendicularity tolerance.

PERPENDICULARITY OF A SURFACE

Figure 12-64 illustrates how a perpendicularity tolerance is called out on a drawing and provides an interpretation of what the perpendicularity tolerance actually means. The elements of the toleranced surface must fall within the size limits and between two hypothetical parallel planes that are a distance apart equal to the perpendicularity tolerance.

The perpendicularity of a part such as the one shown in Figure 12-64 could be inspected by clamping the part to an inspection angle. The datum surface should rest against the inspection angle. Then a dial indicator should be passed over the entire surface for a full indicator movement to determine if the perpendicularity tolerance has been complied with.

AS DRAWN

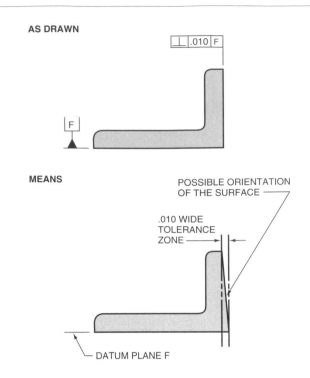

MEANS

THE SURFACE MUST LIE BETWEEN TWO PARALLEL PLANES .010 APART THAT ARE PERPENDICULAR TO DATUM PLANE F. THE SURFACE MUST BE WITHIN THE SPECIFIED LIMITS OF SIZE.

FIGURE 12-64 *Perpendicularity for a surface to a datum plane*

Perpendicularity also controls the flatness of the surface to the same extent that it controls orientation. When it is required that the flatness of the feature be less than the orientation, a flatness callout can be used as a refinement of the orientation callout. When using flatness as a refinement, the tolerance is less than the orientation tolerance. The feature control frame is normally placed on an extension line below the orientation control, *Figure 12-65.*

PERPENDICULARITY OF AN AXIS OR CENTER PLANE

Perpendicularity can be used to control the orientation of an axis to a datum plane, an axis to an axis, or the cen-

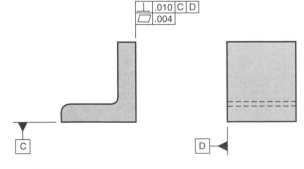

FIGURE 12-65 *Perpendicularity with flatness refinement*

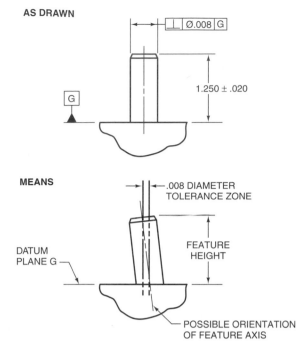

AS DRAWN

MEANS

.008 DIAMETER
TOLERANCE ZONE

DATUM
PLANE G

FEATURE
HEIGHT

POSSIBLE ORIENTATION
OF FEATURE AXIS

REGARDLESS OF FEATURE SIZE, THE FEATURE AXIS MUST
LIE WITHIN A CYLINDRICAL ZONE .008 DIAMETER THAT IS
PERPENDICULAR TO AND PROJECTS FROM DATUM PLANE
G FOR THE FEATURE HEIGHT. THE FEATURE AXIS MUST BE
WITHIN THE SPECIFIED TOLERANCE OF LOCATION.

FIGURE 12-66   *Perpendicularity for an axis to a datum plane*

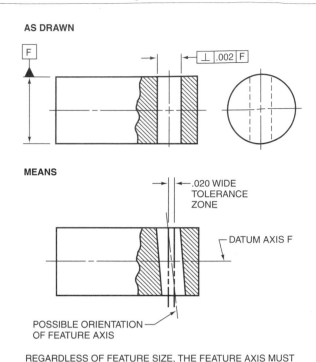

AS DRAWN

MEANS

.020 WIDE
TOLERANCE
ZONE

DATUM AXIS F

POSSIBLE ORIENTATION
OF FEATURE AXIS

REGARDLESS OF FEATURE SIZE, THE FEATURE AXIS MUST
LIE BETWEEN TWO PARALLEL PLANES .020 APART THAT ARE
PERPENDICULAR TO DATUM AXIS F. THE FEATURE AXIS
MUST BE WITHIN THE SPECIFIED TOLERANCE OF LOCATION.

FIGURE 12-67   *Perpendicularity for an axis to a datum axis*

ter plane of noncylindrical parts. When applied to control the axis or center plane, the feature control frame must be placed with the size dimension in the appropriate fashion. When used to control an axis to a datum plane, the tolerance zone shape is cylindrical, and its diameter equals the amount of the stated tolerance. The tolerance control is three-dimensional, allowing the axis to be at any orientation relative to the specified datum surface, *Figure 12-66*. Virtual condition exists for the controlled feature, which allows for the availability of additional tolerance. The M and L modifiers can be used because we are controlling a feature of size.

When used to control an axis to a datum axis, the tolerance zone shape is two parallel planes that are separated by a distance equal to the amount of the stated tolerance. The tolerance control is only applicable relative to the orientation of the datum, *Figure 12-67*. Virtual condition exists for the controlled feature, which allows for the availability of additional tolerance. The M and L modifiers can be used because we are controlling a feature of size.

When used to control a center plane to a datum plane or a center plane to a center plane, the tolerance similarity to that of an axis to a surface or an axis to a datum axis. However, the tolerance zone shape is never cylindrical. The shape is two parallel planes separated by the amount of the stated tolerance. Virtual condition exists

for the controlled feature, which allows for the availability of additional tolerance. The M and L modifiers can be used because we are controlling a feature of size.

Profile

Profile is a feature control that specifies the amount of allowable variance of a surface or line elements on a surface. There are three different variations of the profile tolerance: unilateral (inside), unilateral (outside), and bilateral (unequal distribution), *Figure 12-68*. A profile tolerance is normally used for controlling arcs, curves, and other unusual profiles not covered by the other feature controls. It is a valuable feature control for use on objects that are so irregular that other feature controls do not easily apply.

When applying a profile tolerance, the symbol used indicates whether the designer intends profile of a line or profile of a surface, *Figures 12-69* and *12-70* (pages 513–514). *Profile of a line* establishes a tolerance for a given single element of a surface. *Profile of a surface* applies to the entire surface. The difference between profile of a line and profile of a surface is similar to the difference between circularity and cylindricity.

When using a profile tolerance, drafters and designers should remember to use phantom lines to indicate whether the tolerance is applied unilaterally up or unilaterally down. A bilateral profile tolerance requires no

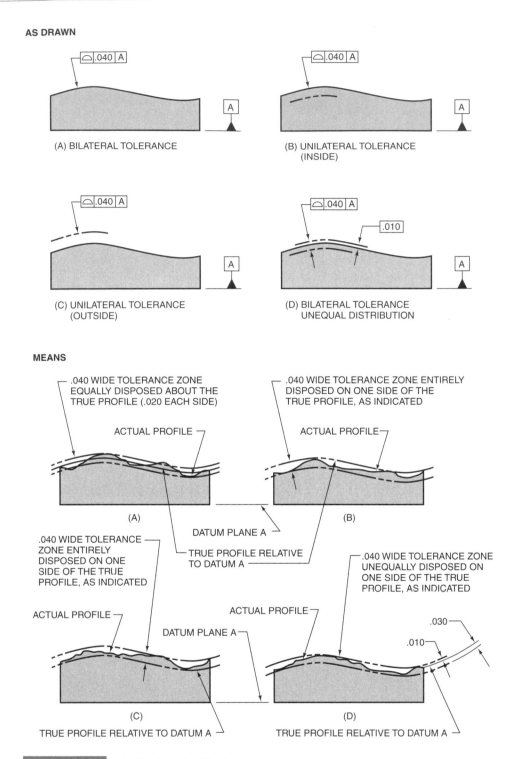

FIGURE 12-68 *Application of profile of a surface*

phantom lines. An ALL AROUND symbol should also be placed on the leader line of the feature control frame to specify whether the tolerance applies ALL AROUND or between specific points on the object, *Figure 12-71*.

Figure 12-72 provides an interpretation of what the BETWEEN A & B profile tolerance in Figure 12-71 actually means. The rounded top surface, and only the top surface, of the object must fall within the specified

tolerance zone. *Figure 12-73* provides an interpretation of what the ALL AROUND profile tolerance in Figure 12-71 actually means. The entire surface of the object, all around the object, must fall within the specified tolerance zone.

Profile tolerances may be inspected using a dial indicator. However, because the tolerance zone must be measured at right angles to the basic true profile and

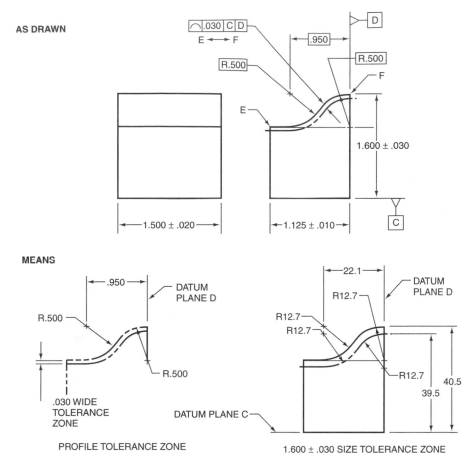

AS DRAWN

MEANS

PROFILE TOLERANCE ZONE

1.600 ± .030 SIZE TOLERANCE ZONE

EACH LINE ELEMENT OF THE SURFACE BETWEEN POINTS E AND F, AT ANY CROSS
SECTION, MUST LIE BETWEEN TWO PROFILE BOUNDARIES .030 APART IN RELATION TO
DATUM PLANES C AND D. THE SURFACE MUST BE WITHIN THE SPECIFIED LIMITS OF SIZE.

FIGURE 12-69 *Profile of a line with size control*

perpendicular to the datum, the dial indicator must be set up to move and read in both directions. Other methods of inspecting profile tolerances are becoming more popular, however. Optical comparators are becoming widely used for inspecting profile tolerances. An optical comparator magnifies the silhouette of the part and projects it onto a screen where it is compared to a calibrated grid or template so that the profile and size tolerances may be inspected visually.

Runout

Runout is a feature control that limits the amount of deviation from perfect form allowed on surfaces or rotation through one full rotation of the object about its axis. Revolution of the object is around a datum axis. Consequently, a runout tolerance does require a datum reference.

Runout is most frequently used on objects consisting of a series of concentric cylinders and other shapes of revolution that have circular cross sections; usually, the

types of objects manufactured on lathes, *Figures 12-74* and *12-75*.

Notice in Figures 12-74 and 12-75 that there are two types of runout: circular runout and total runout. The circular runout tolerance applies at any single-line element through which a section passes. The total runout tolerance applies along an entire surface, as illustrated in Figure 12-75. Runout is most frequently used when the actual produced size of the feature is not as important as the form, and the quality of the feature must be related to some other feature. Circular runout is inspected using a dial indicator along a single fixed position so that errors are read only along a single line. Total runout requires that the dial indicator move in both directions along the entire surface being toleranced.

Concentricity

It is not uncommon in manufacturing to have a part made up of several subparts all sharing the same center line or axis. Such a part is illustrated in *Figure 12-76*. In

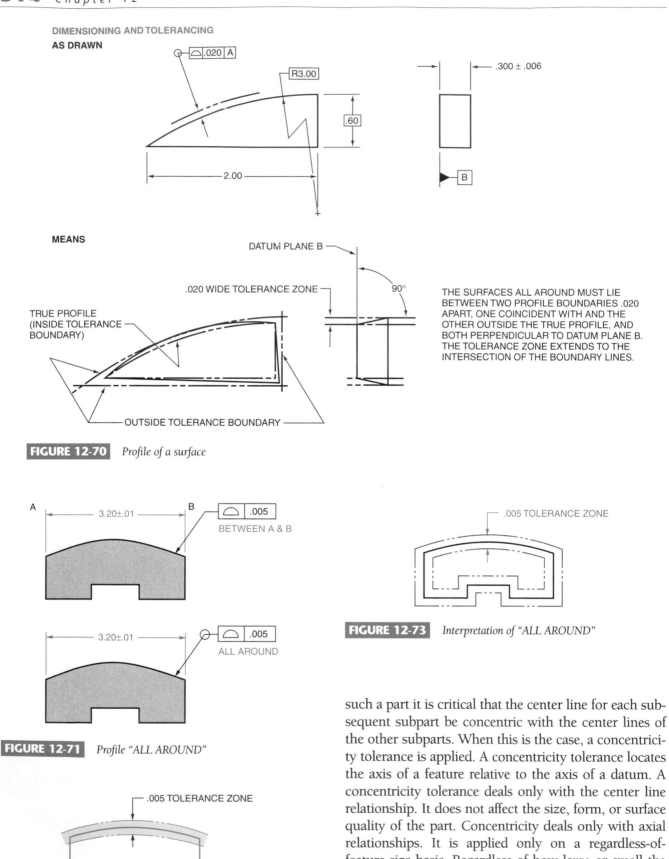

DIMENSIONING AND TOLERANCING

AS DRAWN

⌖ | △ .020 | A

R3.00

.300 ± .006

.60

2.00

▶ | B

MEANS

DATUM PLANE B

.020 WIDE TOLERANCE ZONE

90°

TRUE PROFILE (INSIDE TOLERANCE BOUNDARY)

OUTSIDE TOLERANCE BOUNDARY

THE SURFACES ALL AROUND MUST LIE BETWEEN TWO PROFILE BOUNDARIES .020 APART, ONE COINCIDENT WITH AND THE OTHER OUTSIDE THE TRUE PROFILE, AND BOTH PERPENDICULAR TO DATUM PLANE B. THE TOLERANCE ZONE EXTENDS TO THE INTERSECTION OF THE BOUNDARY LINES.

FIGURE 12-70 *Profile of a surface*

A   3.20±.01   B   △ | .005
BETWEEN A & B

FIGURE 12-71 *Profile "ALL AROUND"*

3.20±.01   ⌀ | △ | .005
ALL AROUND

.005 TOLERANCE ZONE

FIGURE 12-72 *Interpretation of "BETWEEN A & B"*

.005 TOLERANCE ZONE

FIGURE 12-73 *Interpretation of "ALL AROUND"*

such a part it is critical that the center line for each subsequent subpart be concentric with the center lines of the other subparts. When this is the case, a concentricity tolerance is applied. A concentricity tolerance locates the axis of a feature relative to the axis of a datum. A concentricity tolerance deals only with the center line relationship. It does not affect the size, form, or surface quality of the part. Concentricity deals only with axial relationships. It is applied only on a regardless-of-feature-size basis. Regardless of how large or small the various subparts of an overall part are, only their axes are required to be concentric. A concentricity tolerance creates a cylindrical tolerance zone in which all center-

AS DRAWN

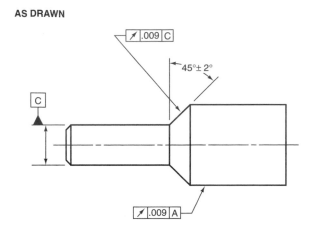

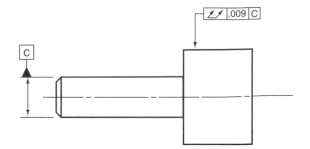

MEANS

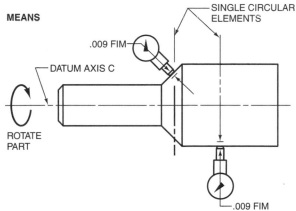

AT ANY MEASURING POSITION, EACH CIRCULAR ELEMENT OF THESE SURFACES MUST BE WITHIN THE SPECIFIED RUNOUT TOLERANCE (.009 FULL INDICATOR MOVEMENT) WHEN THE PART IS ROTATED 360° ABOUT THE DATUM AXIS WITH THE INDICATOR FIXED IN A POSITION NORMAL TO THE TRUE GEOMETRIC SHAPE. THE FEATURE MUST BE WITHIN THE SPECIFIED LIMITS OF SIZE.

FIGURE 12-74 *Specifying circular runout relative to a datum diameter*

MEANS

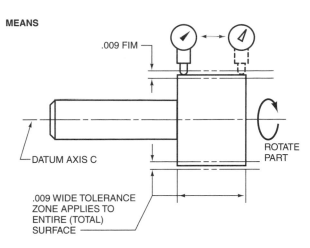

THE ENTIRE SURFACE MUST LIE WITHIN THE SPECIFIED RUNOUT TOLERANCE ZONE (.009 FULL INDICATOR MOVEMENT) WHEN THE PART IS ROTATED 360° ABOUT THE DATUM AXIS WITH THE INDICATOR PLACED AT EVERY LOCATION ALONG THE SURFACE IN A POSITION NORMAL TO THE TRUE GEOMETRIC SHAPE WITHOUT RESET OF THE INDICATOR. THE FEATURE MUST BE WITHIN THE SPECIFIED LIMITS OF SIZE.

FIGURE 12-75 *Specifying total runout relative to a datum diameter*

lines for each successive subpart of an overall part must fall. This concept is illustrated in *Figure 12-77*. A concentricity tolerance is inspected by a full indicator movement of a dial indicator.

SYMMETRY

Parts that are symmetrically disposed about the center plane of a datum feature are common in manufacturing settings. If it is necessary that a feature be located symmetrically with regard to the center plane of a datum feature, a symmetry tolerance may be applied, *Figure 12-78*. The part in Figure 12-78 is symmetrical about a center plane. To ensure that the part is located symmetrically with respect to the center plane, a .030 symmetry tolerance is applied. This creates a .030 tolerance zone within which the center plane in question must fall, as illustrated in the bottom portion of Figure 12-78.

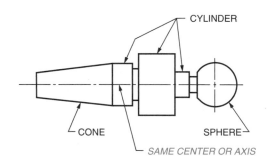

FIGURE 12-76 *Part with concentric subparts*

POSITIONING

Position tolerancing is used to locate features of parts that are to be assembled and mated. Position is symbolized by a circle overlaid by a large plus sign or cross. This symbol is followed by the tolerance, a modifier when appropriate, and a reference datum, *Figure 12-79*. *Figures 12-80* and *12-81* illustrate the difference

AS DRAWN

MEANS

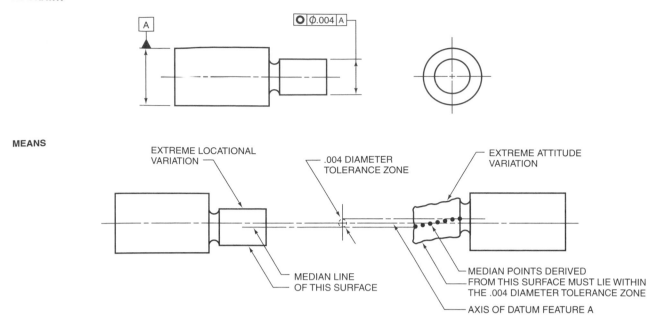

EXTREME LOCATIONAL VARIATION

.004 DIAMETER TOLERANCE ZONE

EXTREME ATTITUDE VARIATION

MEDIAN LINE OF THIS SURFACE

MEDIAN POINTS DERIVED FROM THIS SURFACE MUST LIE WITHIN THE .004 DIAMETER TOLERANCE ZONE

AXIS OF DATUM FEATURE A

WITHIN THE LIMITS OF SIZE AND REGARDLESS OF FEATURE SIZE, ALL MEDIAN POINTS OF DIAMETRICALLY OPPOSED ELEMENTS OF THE FEATURE MUST LIE WITHIN A Ø.004 CYLINDRICAL TOLERANCE ZONE. THE AXIS OF THE TOLERANCE ZONE COINCIDES WITH THE AXIS OF DATUM FEATURE A. THE SPECIFIED TOLERANCE AND THE DATUM REFERENCE APPLY ONLY ON AN RFS BASIS.

FIGURE 12-77   *Concentricity tolerancing*

between conventional and position dimensioning. The tolerance dimensions shown in Figure 12-80 create a square tolerance zone. This means that the zone within which the center line being located by the dimensions must fall takes the shape of a square. As you can see in Figure 12-81, the tolerancing zone is round when position dimensioning is used. The effect of this on manufacturing is that the round tolerancing zone with position dimensioning increases the size of the tolerance zone by 57%, *Figure 12-82*. This means that for the same tolerance the machinist has 57% more room for error without producing an out-of-tolerance part.

When using position dimensioning, the tolerance is assumed to apply regardless of the feature size unless modified otherwise. *Figure 12-83* illustrates the effect of modifying a position tolerance with a maximum material condition modifier. In this example, a hole is to be drilled through a plate. The maximum diameter is .254 and the minimum diameter is .250. Therefore, the maximum material condition of the part occurs when the hole is drilled to a diameter of .250. Notice from this example that as the hole increases, the positional tolerance increases. At maximum material condition (.250 diameter), the tolerance zone has a diameter of .042. At least material condition (.254 diameter), the tolerance zone increases to .046 diameter. The tolerance zone diameter increases correspondingly as the hole size increases.

REVIEW OF DATUMS

Fundamental to an understanding of geometric dimensioning and tolerancing is an understanding of datums. Since many engineering and drafting students find the concept of datums difficult to understand, this section will review the concept in depth. It is important to understand datums because they represent the starting point for referencing dimensions to various features on parts and for making calculations relative to those dimensions. Datums are usually physical components. However, they can also be invisible lines, planes, axes, or points that are located by calculations or as they relate to other features. Features such as diameters, widths, holes, and slots are frequently specified as datum features.

Datums are classified as primary, secondary, or tertiary, *Figure 12-84*. Three points are required to establish a primary datum. Two points are required to establish a secondary datum. One point is required to establish a tertiary datum, *Figure 12-85*. Each point used to establish a datum is called off by a datum target symbol, *Figure 12-86*. The letter designation in the datum target symbol is the datum identifier. For example, the letter A in *Figure 12-87* is the datum designator for Datum A. The number 2 in *Figure 12-88* is the point designator for Point 2. Therefore, the complete designation of A2 means Datum A-Point 2.

AS DRAWN

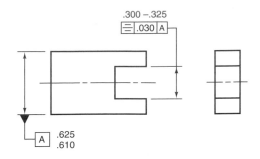

MEANS

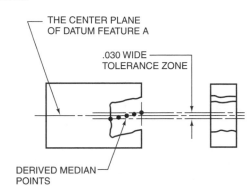

THE CENTER PLANE
OF DATUM FEATURE A

.030 WIDE
TOLERANCE ZONE

DERIVED MEDIAN
POINTS

WITHIN THE LIMITS OF SIZE AND REGARDLESS OF FEATURE
SIZE, ALL MEDIAN POINTS OF OPPOSED ELEMENTS OF THE
SLOT MUST LIE BETWEEN TWO PARALLEL PLANES .030 APART,
THE TWO PLANES BEING EQUALLY DISPOSED ABOUT DATUM
PLANE A. THE SPECIFIED TOLERANCE AND THE DATUM
REFERENCE CAN ONLY APPLY ON AN RFS BASIS.

FIGURE 12-78 *Symmetry tolerancing*

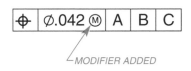

FIGURE 12-79 *Position symbology*

CONVENTIONAL DIMENSIONING

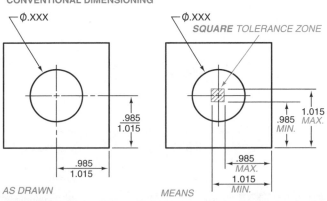

FIGURE 12-80 *Conventional dimensioning*

TRUE POSITIONING

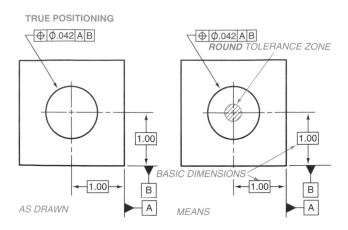

FIGURE 12-81 *Position dimensioning*

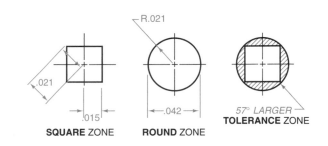

FIGURE 12-82 *Comparison of tolerance zones*

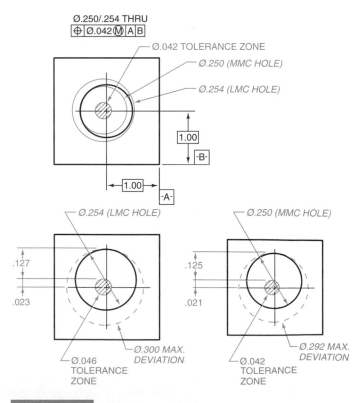

FIGURE 12-83 *Positioning at MMC*

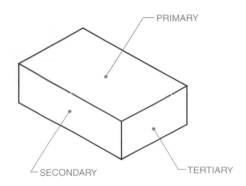

FIGURE 12-84 *Datums*

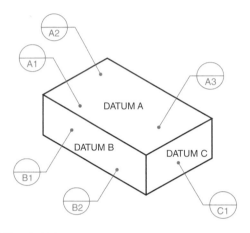

FIGURE 12-85 *Establishing datums*

EACH POINT IS CALLED OFF BY A **DATUM TARGET SYMBOL**

FIGURE 12-86 *Datum target symbol*

THE 'A' INDICATES THE **DATUM**

FIGURE 12-87 *Datum designation*

THE '2' INDICATES THE **DATUM**

FIGURE 12-88 *Point designator*

Figure 12-89 illustrates how the points that establish datums should be dimensioned on a drawing. In this illustration, the three points that establish Datum A are dimensioned in the top view and labeled using the

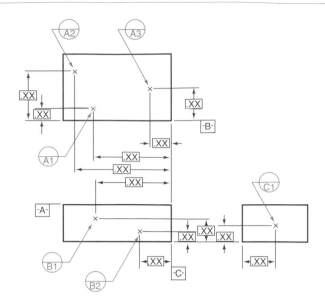

FIGURE 12-89 *Dimensioning datum points*

datum target symbol. The two points that establish Datum B are dimensioned in the front view. The one point that establishes Datum C is dimensioned in the right-side view. *Figure 12-90* illustrates the concept of datum plane and datum surface. The theoretically perfect plane is represented by the top of the machine table. The less perfect actual datum surface is the bottom surface of the part. *Figure 12-91* shows how the differences between the perfect datum plane and the actual datum surface are reconciled. The three points protruding from the machine table correspond with the three points that establish Datum A. Once this difference has been reconciled, inspections of the part can be carried out.

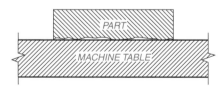

FIGURE 12-90 *Datum plane versus datum surface*

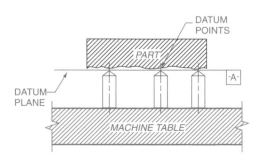

FIGURE 12-91 *Reconciling the datum surface to the datum plane*

Summary

- General tolerancing involves setting acceptable limits of deviation for manufactured parts.

- Geometric dimensioning and tolerancing involves setting tolerance limits for all characteristics of a part.

- Modifiers are symbols that can be attached to the standard geometric building blocks to alter their application or interpretation.

- MMC is when the most material exists in the part. RFS means that a tolerance of form or position or any characteristic must be maintained regardless of the actual produced size of the object.

- Projected tolerance zone is a modifier that allows a tolerance zone established by a locational tolerance to be extended a specified distance beyond a given surface.

- Datums are theoretically perfect points, lines, axes, surfaces, or planes used for referencing features of an object.

- True position is the theoretically exact location of the center line of a product feature such as a hole.

Review Questions

Answer the following questions either true or false.

1. Tolerancing means setting acceptable limits of deviation.

2. The three types of size tolerances are unilateral, location, and runout.

3. The need to tolerance more than just the size of an object led to the development of geometric dimensioning and tolerancing.

4. Geometric dimensioning specifies the allowable variation of a feature from perfect form.

5. The term *regardless of feature size* is a modifier that tells machinists that a tolerance of form or position or any characteristic must be maintained, regardless of the actual produced size of the object.

6. Datums are components of a part such as a hole, slot, surface, or boss.

7. A datum is established on a cast surface by a "flag" or symbol.

Answer the following questions by selecting the best answer.

1. Which of the following is the identification for the ASME standard on dimensioning?
   a. ASME Y14.5 M—1994
   b. ASME Y24.5 M—1992
   c. ASME Y34.5 M—1990
   d. ASME Y44.5 M—1988

2. Which of the following has the incorrect symbol?
   a. Flatness ▱
   b. Circularity ○
   c. Straightness —
   d. True position ◆

3. Which of the following has the incorrect symbol?
   a. Perpendicularity =
   b. Straightness —
   c. Parallelism //
   d. Angularity ∠

4. Which of the following is **not** true regarding *flatness?*
   a. It differs from straightness.
   b. The term *flatness* is interchangeable with the term *straightness.*
   c. When flatness is the feature control, a datum reference is neither required nor proper.
   d. Flatness is specified when size tolerances alone are not sufficient to control the form and quality of the surface.

5. The term *least material condition* means:
   a. The opposite of MMC.
   b. A condition of a feature in which it contains the least amount of material.
   c. The theoretically exact location of a feature.
   d. Both a and b

6. Which of the following is not true regarding feature control symbols?
   a. The order of data in a feature control frame is important.
   b. The first element is the feature control symbol.
   c. Various feature control elements are separated by //.
   d. Datum references are listed in order from left to right.

7. Which of the following feature controls must have a datum reference?
   a. Flatness
   b. Straightness
   c. Cylindricity
   d. Parallelism

Chapter 12 Problems

The following problems are intended to give beginning drafters practice in applying the principles of geometric dimensioning and tolerancing.

CONVENTIONAL INSTRUCTIONS

The steps to follow in completing the problems are:

STEP 1 Study the problem carefully.

STEP 2 Make a checklist of tasks you will need to complete.

STEP 3 Center the required view or views in the work area.

STEP 4 Include all dimensions according to ASME Y14.5M—1994.

STEP 5 Recheck all work. If it's correct, neatly fill out the title block using light guidelines and free-hand lettering.

NOTE: These problems do not follow current drafting standards. You are to use the information shown here to develop properly drawn, dimensioned, and toleranced drawings.

CAD INSTRUCTIONS

Students who plan to complete the following problems on a CAD system should begin with these general instructions. Before reading the specific instructions for each activity, go through each step in the following planning checklist. The checklist applies to any CAD system and will help ensure the optimum use of your time and resources.

1. Analyze the problem carefully. Decide exactly what you are being asked to do.

2. Determine what resources and references (if any) you will need in order to complete the problem. Collect those references and resources and have them readily available.

3. Decide if any particular standards apply to the project, and have those standards available.

4. Determine what types of views will be required and how many of each.

5. Determine what the final plotted scale of the drawing will need to be, and select the appropriate paper size for plotting/printing.

6. Plan your drawing sequence. In what order will you develop the drawing (i.e., lines, features, dimension lines, leaders, dimensions)?

7. Review the various CAD commands you will have to use in order to develop the drawing.

8. Examine your CAD system to ensure that everything is in working order, then begin the project.

Problem 12-1

Apply tolerances so that this part is straight to within .004 at MMC.

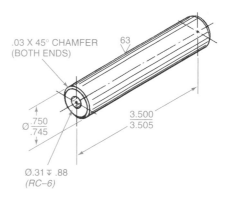

Problem 12-2

Apply tolerances so that the top surface of this part is flat to within .001 and the two sides of the slot are parallel to each other within .002 RFS.

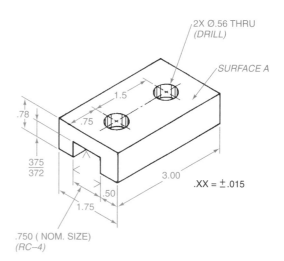

Problem 12-3

Apply tolerances so that the smaller diameter has a cylindricity tolerance of .005 and the smaller diameter is concentric to the larger diameter to within .002. The shoulder must be perpendicular to the axis of the part to within .002.

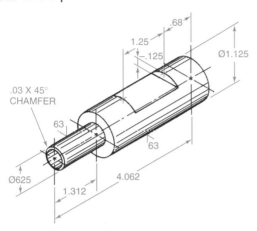

Problem 12-5

Apply angularity, true position, and parallelism tolerances of .001 to this part. Select the appropriate datums. The parallelism tolerances should be applied to the sides of the slot.

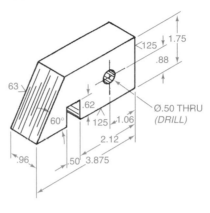

Problem 12-4

Apply tolerances to locate the holes using position and basic dimensions relative to datums A-B-C.

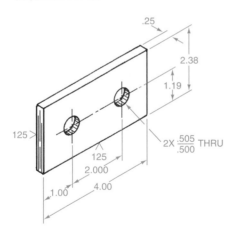

Problem 12-6

Apply tolerances so that the outside diameter of the part is round to within .004 and the ends are parallel to within .001 at maximum material condition.

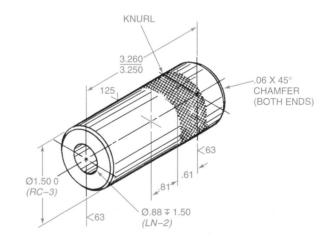

Problem 12-7

Apply a line profile tolerance to the top of the part between points X and Y of .004. Apply position tolerances to the holes of .021, and parallelism tolerances of .001 to the two finished sides.

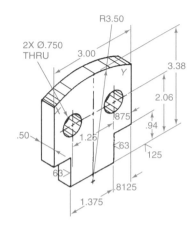

Problem 12-9

Select datums and apply tolerances in such a way as to ensure that the slot is symmetrical to within .002 with the .625 diameter hole, and the bottom surface is parallel to the top surface to within .004.

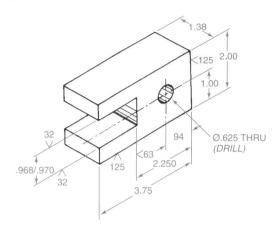

Problem 12-8

Use the bottom of the part as Datum A and the right side of the part as Datum B. Apply surface profile tolerances of .001 to the top of the part between points X and Y.

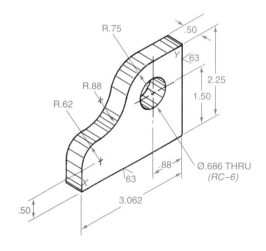

Problem 12-10

Apply tolerances to this part so that the tapered end has a total runout of .002.

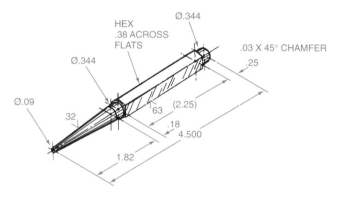

Problem 12-11

Apply tolerances to this part so that diameters X and Z have a total runout of .02 relative to Datum A (the large diameter of the part) and line runout of .004 to the two tapered surfaces.

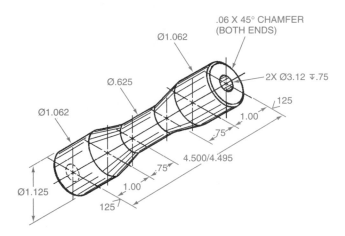

Problem 12-12

Select datums and apply a positional tolerance of .001 at MMC to the holes, and a perpendicularity tolerance of .003 to the vertical leg of the angle.

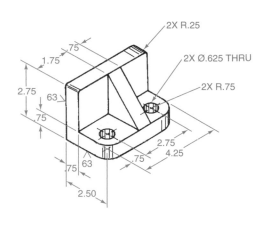

PROBLEMS 12-13 THROUGH 12-30

For each of the remaining geometric dimensioning and tolerancing problems, examine the problem closely with an eye to the purpose that will be served by the part. Then select datums, tolerances, and feature controls as appropriate, and apply them properly to the parts. In this way you will begin to develop the skills required of a mechanical designer. Do not overdesign. Remember, the closer the tolerances, and the more feature control applied, the more expensive the part. Try to use the rule of thumb that says: "Apply only as many feature controls and tolerances as absolutely necessary to ensure that the part will properly serve its purpose after assembly."

Problem 12-13

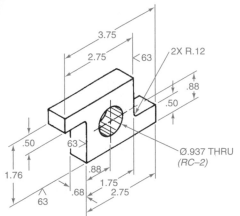

Problem 12-14

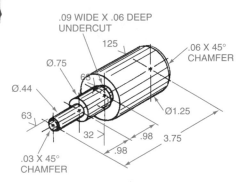

Problem 12-15

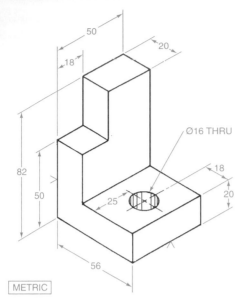

50
18
20
Ø16 THRU
82
50
18
20
25
56
METRIC

Problem 12-18

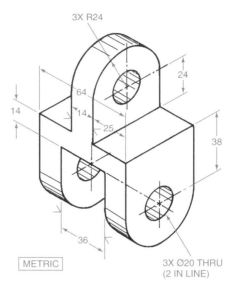

3X R24
24
64
14
14
25
38
36
METRIC
3X Ø20 THRU
(2 IN LINE)

Problem 12-16

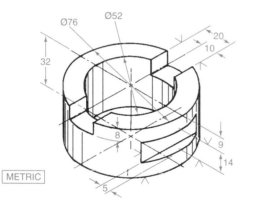

Ø76   Ø52
32
20
10
8
9
14
5
METRIC

Problem 12-19

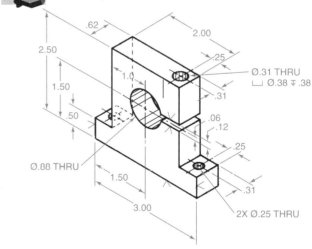

.62
2.00
2.50
.25
1.0
Ø.31 THRU
⌴ Ø.38 ⤓ .38
1.50
.31
.50
.06
.12
.25
Ø.88 THRU
1.50
.31
3.00
2X Ø.25 THRU

Problem 12-17

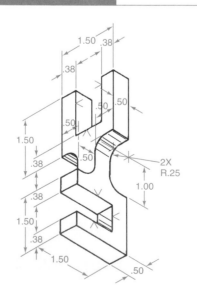

1.50   .38
.38
.50   .50
.50
1.50
.38
.50
2X
R.25
1.00
.38
1.50
.38
1.50
.50

Problem 12-20

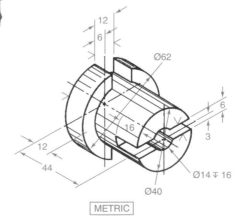

12
6
Ø62
6
16
3
12
44
Ø14 ⤓ 16
Ø40
METRIC

Problem 12-21

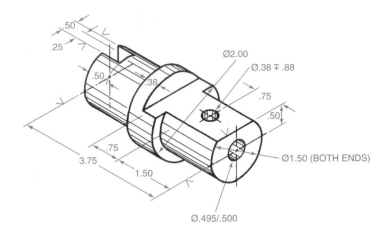

Problem 12-22

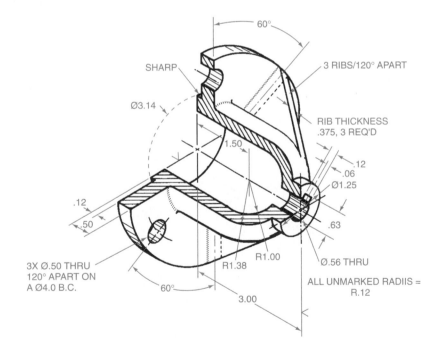

Problem 12-23

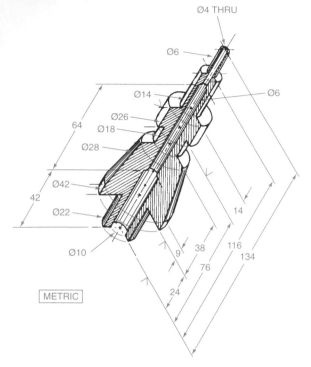

Ø4 THRU
Ø6
Ø14
Ø26
Ø6
Ø18
Ø28
64
Ø42
42
Ø22
14
Ø10
9
38
116
76
134
24

METRIC

Problem 12-24

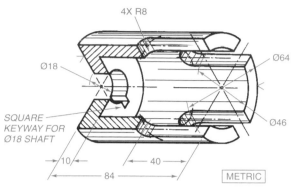

4X R8
Ø18
Ø64
SQUARE KEYWAY FOR Ø18 SHAFT
Ø46
10
40
84

METRIC

Problem 12-25

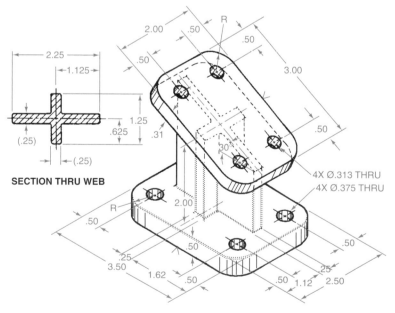

2.00
.50
R
.50
2.25
.50
1.125
3.00
1.25
.625
.31
.50
(.25)
30°
(.25)
SECTION THRU WEB
4X Ø.313 THRU
4X Ø.375 THRU
R
2.00
.50
.50
.25
3.50
.50
.50
1.12
2.50
1.62
.50
.25
.50

ALL UNMARKED RADII = R.06

Problem 12-26

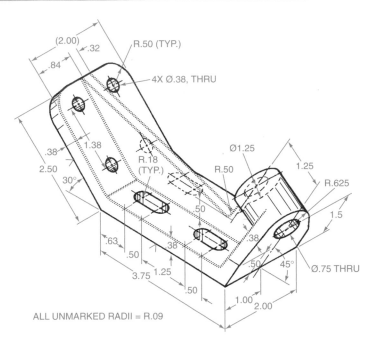

(2.00)
.32
.84
R.50 (TYP.)
4X Ø.38, THRU
1.38
.38
R.18 (TYP.)
2.50
Ø1.25
R.50
1.25
30°
R.625
50
1.5
.63
.38
.50
.50
.38
45°
3.75
1.25
Ø.75 THRU
.50
1.00
2.00

ALL UNMARKED RADII = R.09

Problem 12-27

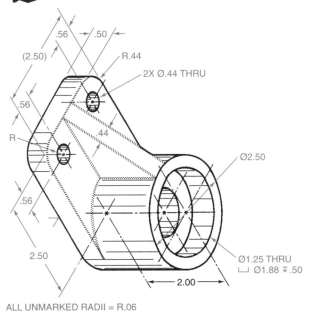

.56
.50
(2.50)
R.44
2X Ø.44 THRU
.56
44
R
.56
Ø2.50
2.50
Ø1.25 THRU
⌴ Ø1.88 ⍗.50
2.00

ALL UNMARKED RADII = R.06

Problem 12-28

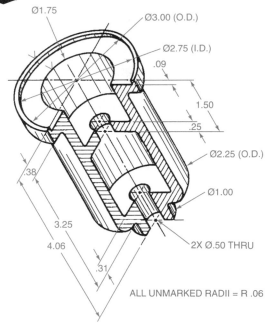

Ø1.75
Ø3.00 (O.D.)
Ø2.75 (I.D.)
.09
1.50
.25
.38
Ø2.25 (O.D.)
Ø1.00
3.25
4.06
2X Ø.50 THRU
.31

ALL UNMARKED RADII = R .06

Problem 12-29

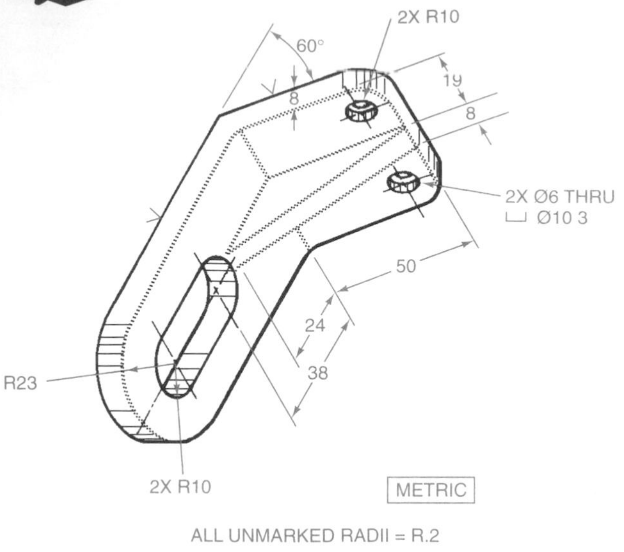
2X R10
60°
8
19
8
2X Ø6 THRU
⌴ Ø10 3
50
24
38
R23
2X R10
METRIC
ALL UNMARKED RADII = R.2

Problem 12-30

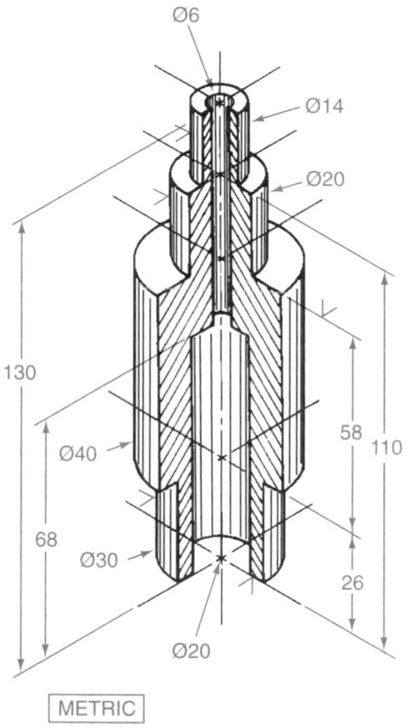
Ø6
Ø14
Ø20
130
Ø40
58
110
68
Ø30
26
Ø20
METRIC

PROBLEMS 12-31

This problem deals with feature control symbols. In items 1–19, explain what each symbol means. In items 20–30, draw the required symbols.

MEANS

1) ______
2) ______
3) ______
4) ______
5) ______
6) ______
7) ______

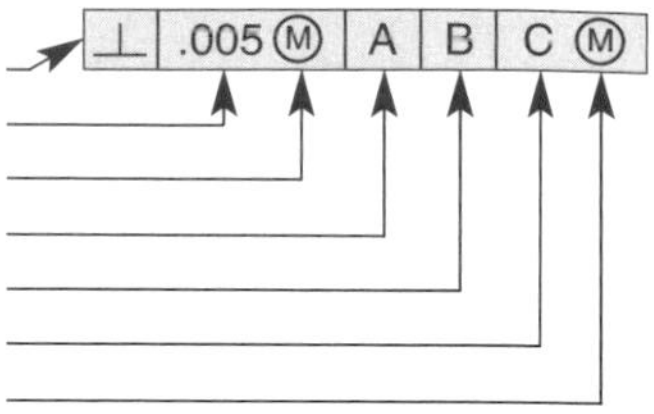
⊥ .005 Ⓜ A B C Ⓜ

SYMBOL	MEANS	SYMBOL	MEANS
8) Ⓜ	______	14) ○	______
9) [-A-]	______	15) ⌭	______
10) ( )	______	16) ⌖	______
11) R	______	17) Ⓢ	______
12) ↧	______	18) ∅	______
13) ∨	______	19) ⌓	______

	SYMBOL
20) ANGULARITY	______
21) POSITION	______
22) FLATNESS	______
23) PROFILE OF A SURFACE	______
24) PERPENDICULARITY	______
25) CIRCULAR RUNOUT	______
26) STRAIGHTNESS	______
27) TOTAL RUNOUT	______
28) PROFILE OF A LINE	______
29) CYLINDRICITY	______
30) CIRCULARITY	______

Chapter 13

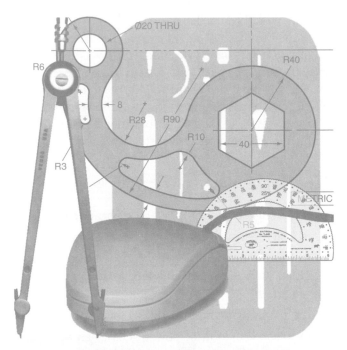

FASTENERS

KEY TERMS

Angle of thread	Lead
Axis	Machine screw
Beveled rings	Major diameter
Blind hole	Minor diameter
Bolt	Multiple thread
Bowed rings	Permanent fasteners
Butt joint	Pitch
Cap screws	Pitch diameter
Crest	Retaining rings
Depth of thread	Rivets
Expanded diameter	Root
External rings	Screw
External thread	Self-locking rings
Form	Series of thread
Full threads	Set screw
Grooved fasteners	Single thread
Grooved studs	Spring pin
Internal rings	Stud
Internal thread	Temporary fasteners
Key	Thread relief
Keyseat	Through hole
Lap joint	

CHAPTER OBJECTIVES

Upon completion of this chapter, students should be able to do the following:

- Explain the classifications of fasteners.
- Explain the four uses of threads.
- List the various screw thread forms.
- Distinguish between the tap and die processes.
- Demonstrate two methods for measuring threads per inch.
- Define the term *pitch*.
- Distinguish between single and multiple threads.
- Demonstrate how to distinguish between right- and left-hand threads.
- Demonstrate the various methods of thread representation.
- Explain the concept of *thread relief*.
- Distinguish among the following: screw, bolt, and stud.
- Demonstrate the proper methods for drawing screws, bolts, and studs.
- Define the term *rivet*.
- Explain the concept of keys and keyseats.
- Explain how grooved fasteners are used.
- Describe the uses of spring pins.
- Explain the concept of fastening systems.
- Describe how retaining rings are used.

529

Classifications of Fasteners

As a new product is developed, determining how to fasten it together is a major consideration. The product must be assembled quickly, using standard, easily available, low-cost fasteners. Some products are designed to be taken apart easily—others are designed to be permanently assembled. Many considerations are required as to what kind, type, and material of fastener is to be used. Sometimes the stress load upon a joint must be considered. There are two major classifications of fasteners: permanent and temporary. *Permanent fasteners* are used when parts will not be disassembled. *Temporary fasteners* are used when the parts will be disassembled at some future time.

Permanent fastening methods include welding, brazing, stapling, nailing, gluing, and riveting. Temporary fasteners include screws, bolts, keys, and pins.

Many temporary fasteners include threads in their design. In early days, there was no such thing as standardization. Nuts and bolts from one company would not fit nuts and bolts from another company. In 1841, Sir Joseph Whitworth worked toward some kind of standardization throughout England. His efforts were finally accepted, and England came up with a standard thread form called Whitworth Threads.

In 1864, the United States tried to develop a standardization of its own but, because it would not interchange with the English Whitworth Threads, it was not adopted at that time. It was not until 1935 that the United States adopted the American Standard Thread. It was actually the same 60° V-thread form proposed back in 1864. Still, there was no standardization between countries. This created many problems, but nothing was done until World War II forced the issue of interchangeability on the Western Allies. As a result of problems experienced by the Allies, in 1948 the United States, Canada, and Great Britain developed the Unified Screw Thread. It was a compromise between the newer American Standard Thread and the old Whitworth Threads.

Today, with the changeover to the metric system, new standards are being developed. The International Organization for Standardization (ISO) was formed to develop a single international system using metric screw threads. This new ISO standard will be united with the American National Standards Institute (ANSI) standards. At the present time, we are in a transitional period, and a combination of both systems is still being used.

Threads

Threads are used for four different applications:

1. to fasten parts together, such as a nut and a bolt.

2. for fine adjustment between parts in relation to each other, such as the fine adjusting screw on a surveyor's transit.

3. for fine measurement, such as a micrometer.

4. to transmit motion or power, such as an automatic screw threading attachment on a lathe or a house jack.

There are many types and sizes of fasteners, each designed for a particular function. Permanently fastening parts together by welding or brazing is discussed in Chapter 19. Although screw threads have other important uses, such as adjusting parts and measuring and transmitting power, only their use as fasteners and only the most-used kinds of fasteners are discussed in this chapter.

THREAD TERMS

Refer to *Figure 13-1* for the following terms.

• *External thread*—Threads located on the outside of a part, such as those on a bolt.

• *Internal thread*—Threads located on the inside of a part, such as those on a nut.

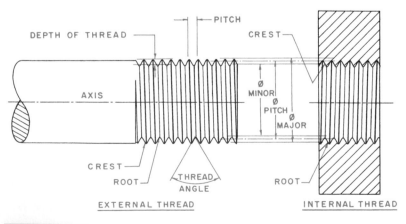

FIGURE 13-1 *Thread terms*

- *Axis*—A longitudinal center line of the thread.
- *Major diameter*—The largest diameter of a screw thread, both external and internal.
- *Minor diameter*—The smallest diameter of a screw thread, both external and internal.
- *Pitch diameter*—The diameter of an imaginary diameter centrally located between the major diameter and the minor diameter.
- *Pitch*—The distance from a point on a screw thread to a corresponding point on the next thread, as measured parallel to the axis.
- *Root*—The bottom point joining the sides of a thread.
- *Crest*—The top point joining the sides of a thread.
- *Depth of thread*—The distance between the crest and the root of the thread, as measured at a right angle to the axis.
- *Angle of thread*—The included angle between the sides of the thread.
- *Series of thread*—A standard number of threads per inch (TPI) for each standard diameter.

Screw Thread Forms

The *form* of a screw thread is actually its profile shape. There are many kinds of screw thread forms. Seven major kinds are discussed next.

UNIFIED NATIONAL THREAD FORM

The Unified National thread form has been the standard thread used in the United States, Canada, and the United Kingdom since 1948, *Figure 13-2A*. This thread form is used mostly for fasteners and adjustments.

ISO METRIC THREAD FORM

The ISO metric thread form is the new standard to be used throughout the world. Its form or profile is very similar to that of the Unified National thread, except that the thread depth is slightly less, *Figure 13-2B*. This thread form is used mostly for fasteners and adjustments.

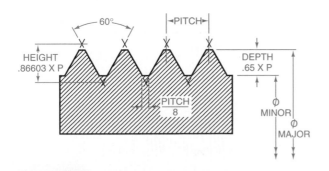

FIGURE 13-2A *Unified National thread form (UN)*

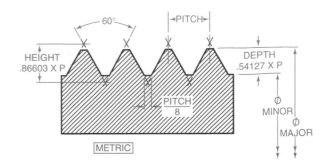

FIGURE 13-2B *ISO metric thread form*

SQUARE THREAD FORM

The square thread's profile is exactly as its name implies; that is, square. The faces of the teeth are at right angles to the axis and, theoretically, this is the best thread to transmit power, *Figure 13-2C*. Because this thread is difficult to manufacture, it is being replaced by the Acme thread.

ACME THREAD FORM

The Acme thread is a slight modification of the square thread. It is easier to manufacture and is actually stronger than the square thread, *Figure 13-2D*. It, too, is used to transmit power.

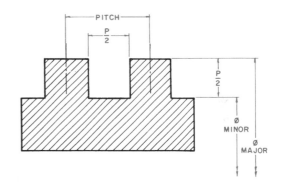

FIGURE 13-2C *Square thread form*

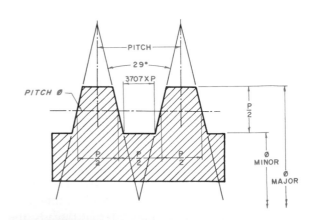

FIGURE 13-2D *Acme thread form*

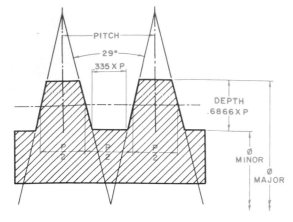

FIGURE 13-2E   *Worm thread form*

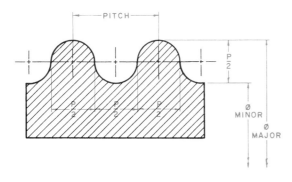

FIGURE 13-2F   *Knuckle thread form*

WORM THREAD FORM

The worm thread is similar to the Acme thread, and it is used primarily to transmit power, *Figure 13-2E*.

KNUCKLE THREAD FORM

The knuckle thread is usually rolled from sheet metal and is used, slightly modified, in electric light bulbs, electric light sockets, and sometimes for bottle tops. The knuckle thread is sometimes cast, *Figure 13-2F.*

BUTTRESS THREAD FORM

The buttress thread has certain advantages in applications involving exceptionally high stress along its axis in one direction only. Examples of applications are the breech assemblies of large guns, airplane propeller hubs, and columns for hydraulic presses, *Figure 13-2G*.

Tap and Die

Various methods are used to produce inside and outside threads. The simplest method uses thread-cutting tools called taps and dies. The tap cuts internal threads; the die cuts external threads, *Figure 13-3*. In making an

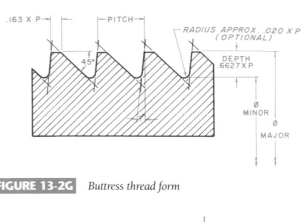

FIGURE 13-2G   *Buttress thread form*

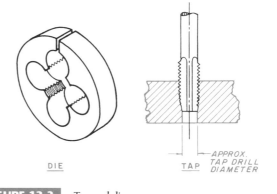

FIGURE 13-3   *Tap and die*

internal threaded hole, a tap-drilled hole must be drilled first. This hole is approximately the same diameter as the minor diameter of the threads.

Notice how the tap is tapered at the end; this taper allows the tap to start into the tap-drilled hole. This tapered area contains only partial threads.

Threads per Inch (TPI)

One method of measuring threads per inch (TPI) is to place a standard scale on the crests of the threads, parallel to the axis, and count the number of full threads within one inch of the scale, *Figure 13-4*. If only part of an inch of stock is threaded, count the number of full threads in one-half inch and multiply by two to determine TPI.

A simpler, more accurate method of determining threads per inch is to use a screw thread gage, *Figure 13-5*. By trial and error, the various fingers or leaves of the gage are placed over the threads until one is found that fits

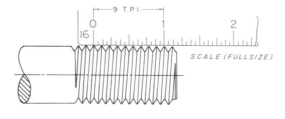

FIGURE 13-4   *Use of a scale to calculate threads per inch (TPI)*

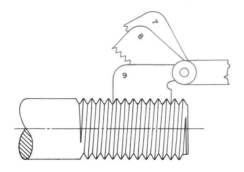

FIGURE 13-5   *Use of a screw pitch gage*

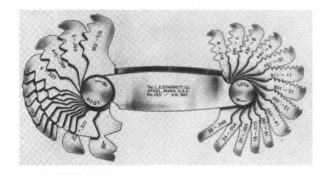

FIGURE 13-6   *Reading screw pitch gage* (Courtesy the L.S. Starrett Co.)

exactly into all the threads. Threads per inch are then read directly on each leaf of the gage, *Figure 13-6*.

Pitch

The pitch of any thread, regardless of its thread form or profile, is the distance from one point on a thread to the corresponding point on the adjacent thread as measured parallel to its axis, *Figure 13-7*. Pitch is found by dividing the TPI into one inch.

In this example, a coarse thread pitch, there are 10 threads in one measured inch; 10 TPI divided into one inch equals a pitch of 10. In a fine thread of the same diameter there are 20 threads in one measured inch; 20 TPI divided into one inch equals a pitch of 20, *Figure 13-8*.

For metric threads, the pitch is specified in millimeters. Pitch for a metric thread is included in its call-off designation. For example: M10 × 1.5. The 1.5 indicates the pitch; therefore, it does not as a rule have to be calculated.

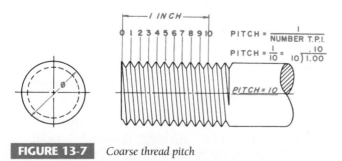

FIGURE 13-7   *Coarse thread pitch*

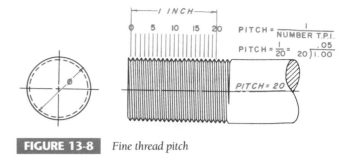

FIGURE 13-8   *Fine thread pitch*

Single and Multiple Threads

A *single thread* is composed of one continuous ridge. The lead of a single thread is equal to the pitch. *Lead* is the distance a screw thread advances axially in one full turn. Most threads are single threads.

Multiple threads are made up of two or more continuous ridges following side-by-side. The lead of a double thread is equal to twice the pitch. The lead of a triple thread is equal to three times the pitch, *Figure 13-9*.

Multiple threads are used when speed or travel distance is an important design factor. A good example of a double or triple thread is found in an inexpensive ballpoint pen. Take a ballpoint pen apart and study the end of the external threads. There will probably be two or three ridges starting at the end of the threads. Notice

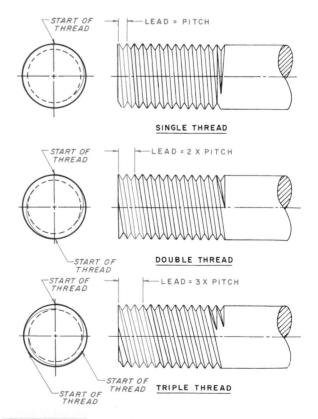

FIGURE 13-9   *Single and multiple threads*

how fast the parts screw together. The speed, not power, is the characteristic of multiple threads.

Right-Hand and Left-Hand Threads

Threads can be either right-handed or left-handed. To distinguish between a right-hand and a left-hand thread, use this simple trick. A right-hand thread winding tends to lean toward the left. If the thread leans toward the left, the right-hand thumb points in the same direction. If the thread leans to the right, *Figure 13-10*, the left-hand thumb leans in that direction indicating that it is a left-hand thread.

THREAD REPRESENTATION

The top illustration of *Figure 13-11* shows a normal view of an external thread. To draw a thread exactly as it will actually look takes too much drafting time. To help speed up the drawing of threads, one of two basic systems is used, and each is described and illustrated. The schematic system of representing threads was developed approximately in 1940, and it is still used somewhat today. The simplified system of representing threads was developed 15 years later, and it is actually quicker and in greater use today.

HOW TO DRAW THREADS USING THE SCHEMATIC SYSTEM

STEP 1 Refer to *Figure 13-12*. Lightly draw the major diameter, and locate the approximate length of full threads.

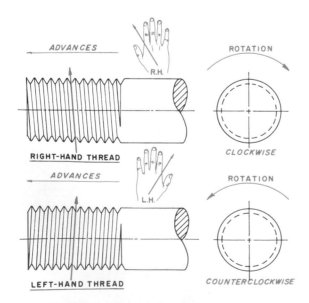

FIGURE 13-10 *Right-hand and left-hand threads*

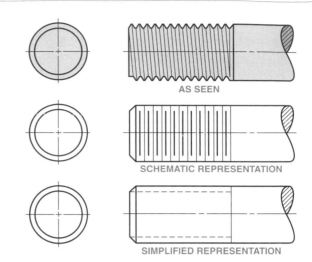

FIGURE 13-11 *Thread representation*

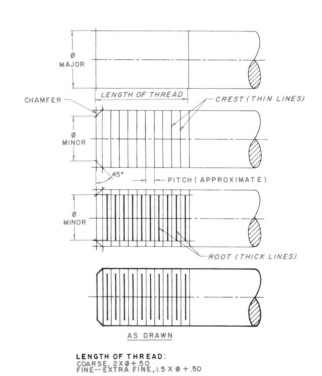

FIGURE 13-12 *How to draw threads using the schematic system*

STEP 2 Lightly locate the minor diameter and draw the 45° chamfered ends as illustrated. Draw lines to represent the crest of the threads spaced approximately equal to the pitch.

STEP 3 Draw slightly thicker lines centered between the crest lines to the minor diameter. These lines represent the root of the threads.

STEP 4 Check all work and darken in. Notice the crest lines are **thin** black lines and the root lines are **thick** black lines.

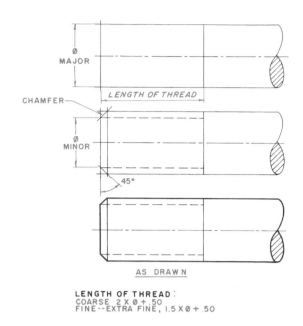

LENGTH OF THREAD:
COARSE 2 X Ø + .50
FINE--EXTRA FINE, I.5 X Ø + .50

FIGURE 13-13 *How to draw threads using the simplified system*

EXTERNAL THREADS

AS SEEN

SCHEMATIC SYSTEM

SCHEMATIC SYSTEM (SECTION)

SIMPLIFIED SYSTEM

SIMPLIFIED SYSTEM (SECTION)

FIGURE 13-14 *Standard external thread representation*

HOW TO DRAW THREADS USING THE SIMPLIFIED SYSTEM

STEP 1 Refer to *Figure 13-13*. Lightly draw the major diameter and locate the approximate length of full threads.

STEP 2 Locate the minor diameter and draw the 45° chamfered ends as illustrated. Draw dash lines along the minor diameter. This represents the root of the threads.

STEP 3 Check all work and darken in. The dash lines are thin black lines.

STANDARD EXTERNAL THREAD REPRESENTATION

The most recent standard to illustrate external threads using either the schematic or simplified system is illustrated in *Figure 13-14*. Note how section views are illustrated using schematic and simplified systems.

STANDARD INTERNAL THREAD REPRESENTATION

There are two major kinds of interior holes: through holes and blind holes. A *through hole,* as its name implies, goes completely through an object. A *blind hole* is a hole that does **not** go completely through an object. In the manufacture of a blind hole, a tap drill must be drilled into the part first, *Figure 13-15*. To illustrate a tap drill, use the 30°–60° triangle. This is **not** the actual angle of a drill point, but it is close

enough for illustration. The tap is now turned into the tap drill hole. Because of the taper on the tap, *full threads* do not extend to the bottom of the hole (refer back to Figure 13-3). The drafter illustrates the tap drill and the full threaded section as shown to the right in Figure 13-15.

Using the schematic system to represent a through hole is illustrated in *Figure 13-16*. A blind hole and a section view are drawn as illustrated in *Figure 13-17*.

Using the simplified system to represent a through hole is illustrated in *Figure 13-18*. A blind hole and a section view are drawn as illustrated in *Figure 13-19*.

Thread Relief (Undercut)

On exterior threads, it is impossible to make perfectly uniform threads up to a shoulder; thus, the threads tend to run out, as illustrated in *Figure 13-20*. Where mating parts must be held tightly against the shoulder, the last one or two threads must be removed or **relieved**. This is usually done no farther than to the depth of the threads so as not to weaken the fastener. The simplified system of thread representation is illustrated at the bottom of Figure 13-20.

Full interior threads cannot be manufactured to the end of a blind hole. One way to eliminate this problem is to call-off a thread relief or undercut, as illustrated in *Figure 13-21*. The bottom illustration is as it would be drawn by the drafter.

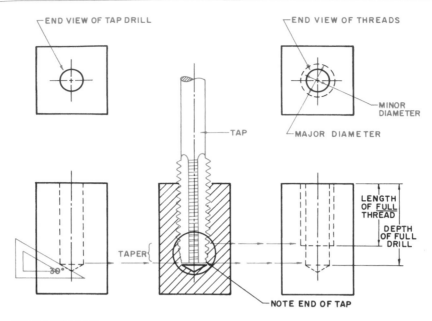

FIGURE 13-15 *Standard internal thread representation*

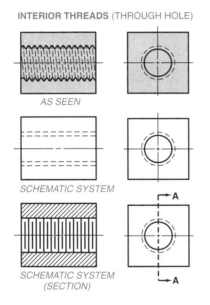

FIGURE 13-16 *Standard internal thread representation for a through hole (schematic system)*

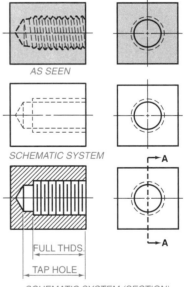

FIGURE 13-17 *Standard internal thread representation for a blind hole (schematic system)*

Screw, Bolt, and Stud

Figure 13-22 illustrates and describes a screw, a bolt, and a stud. A *screw* is a threaded fastener that does not use a nut and is screwed directly into a part.

A *bolt* is a threaded fastener that passes directly through parts to hold them together and uses a nut to tighten or hold the parts together.

A *stud* is a fastener that is a steel rod with threads at both ends. It is screwed into a blind hole and holds other parts together by a nut on its free end. In general practice, a stud has either fine threads at one end and coarse threads at the other, or Class 3-fit threads at one end and Class 2-fit threads at the other end. Class of fit is fully explained later in this chapter under "Classes of Fit."

The minimum full thread length for a screw or a stud is:

In steel: equal to diameter.

In cast iron, brass, bronze: equal to 1.5 times the diameter.

In aluminum, zinc, plastic: equal to 2 times the diameter.

INTERIOR THREADS (THROUGH HOLE)

AS SEEN

SIMPLIFIED SYSTEM

A

SIMPLIFIED SYSTEM (SECTION)

A

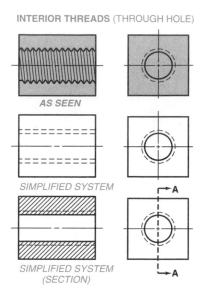

FIGURE 13-18 *Standard internal thread representation for a through hole (simplified system)*

INTERIOR THREADS (BLIND HOLE)

AS SEEN

SIMPLIFIED SYSTEM

A

FULL THD.

TAP HOLE

SIMPLIFIED SYSTEM (SECTION)

A

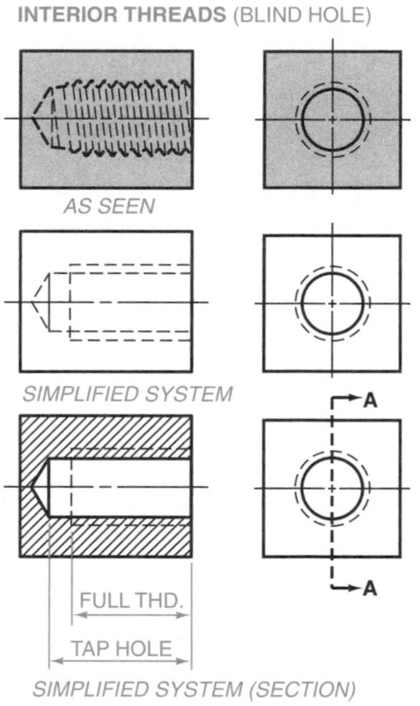

FIGURE 13-19 *Standard internal thread representation for a blind hole (simplified system)*

The clearance hole for holes up to .375 (9) diameter is approximately .03 oversize; for larger holes, .06 oversize.

MACHINE SCREWS

Machine screw sizes run from .021 (.3) to .750 (20) in diameter. There are eight standard head forms. Four major kinds are illustrated in *Figure 13-23*. *Machine screws* are used for screwing into thin materials. Most machine screws are threaded within a thread or two to the head. Although these are screws, machine screws

EXTERIOR THREAD RELIEF (UNDERCUT)

NOTE THREAD RUNOUT

DEPTH OF THREAD

THREAD RELIEF

AS SEEN MINIMUM OF 1 OR 2 THREADS
THREAD RELIEF .06 X THD. DEPTH

AS DRAWN

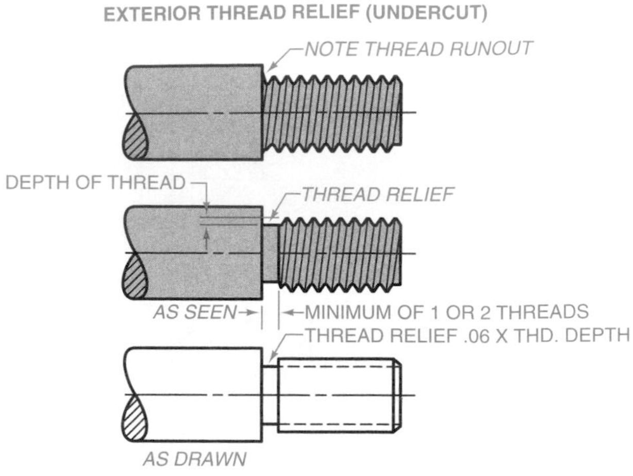

FIGURE 13-20 *External thread relief (undercut)*

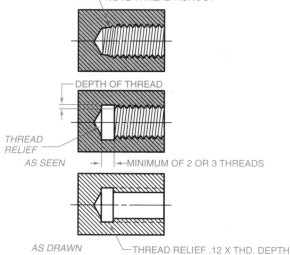

INTERIOR THREAD RELIEF (UNDERCUT)

NOTE THREAD RUNOUT

DEPTH OF THREAD

THREAD
RELIEF

AS SEEN MINIMUM OF 2 OR 3 THREADS

AS DRAWN THREAD RELIEF .12 X THD. DEPTH

FIGURE 13-21 *Internal thread relief (undercut)*

sometimes incorporate a hex-head nut to fasten parts together.

The length of a machine screw is measured from the bottom of the head to the end of the screw (refer again to Figure 13-23).

CAP SCREWS

Cap screw sizes run from .250 (6) and up. There are five standard head forms, *Figure 13-24*. A *cap screw* is usually used as a true screw, and it passes through a clearance hole in one part and screws into another part.

HOW TO DRAW A MACHINE SCREW OR CAP SCREW

The exact dimensions of machine screws and cap screws are given in Appendix A but, in actual practice, they are

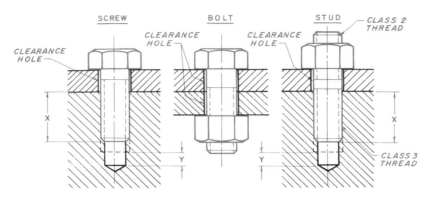

X = MINIMUM THREADS REQUIRED :
      STEEL, X = OUTSIDE DIAMETER
      CAST IRON / BRASS / BRONZE , X = 1.5 X OUTSIDE DIAMETER
      ALUMINUM / ZINC / PLASTIC, X = 2 X OUTSIDE DIAMETER

Y = MINIMUM SPACE = 2 X PITCH LENGTH

CLEARANCE HOLE : 0 TO .375 (9) = .03 (I) LARGER THAN OUTSIDE DIAMETER
                          .375 (9) UP   = .06 (2) LARGER THAN OUTSIDE DIAMETER

FIGURE 13-22   *Screw, bolt, and stud*

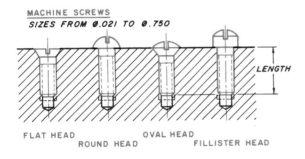

FIGURE 13-23   *Machine screws*

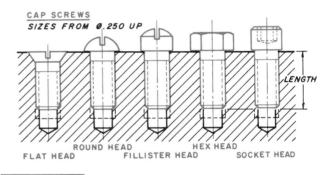

FIGURE 13-24   *Cap screws*

seldom used for drawing purposes. *Figures 13-25* and *13-26* show the various sizes as they are proportioned in regard to the diameter of the fastener. Various fastener templates are now available to further speed up drafting time.

SET SCREWS

A *set screw* is used to prevent motion between mating parts, such as the hub of a pulley on a shaft. The set screw is screwed into and through one part so that it

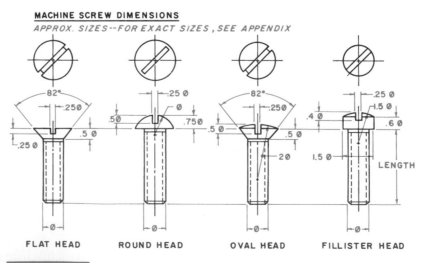

FIGURE 13-25   *Approximate sizes for machine screws or cap screws*

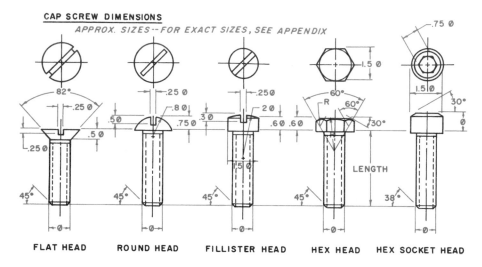

FIGURE 13-26 *Approximate sizes for machine screws or cap screws*

applies pressure against another part, thus preventing motion. Set screws are usually manufactured of steel, and they are hardened to make them stronger than the average fastener.

Set screws have various kinds of heads and many kinds of points. *Figure 13-27* illustrates a few of the more common kinds of set screws. Set screws are manufactured in many standard lengths of very small increments, so almost any required length is probably "standard." Exact sizes and lengths can be found in Appendix A. As with machine screws and cap screws, in actual practice, the drawing of set screws is done using their proportions in relationship to their diameters.

HOW TO DRAW SQUARE- AND HEX-HEAD BOLTS

Exact dimensions for square- and hex-head bolts are given in Appendix A, but in actual practice, they are drawn using the proportions as given in *Figures 13-28* and *13-29*. Notice that the heads are shown in the profile so three surfaces are seen in the front view. In the event a square- or hex-head bolt must be illustrated 90°, the proportions as illustrated in *Figure 13-30* are used.

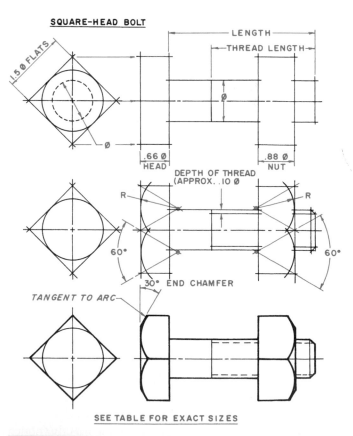

FIGURE 13-27 *Approximate sizes for set screws and set screw points*

FIGURE 13-28 *Approximate sizes for a square-head bolt*

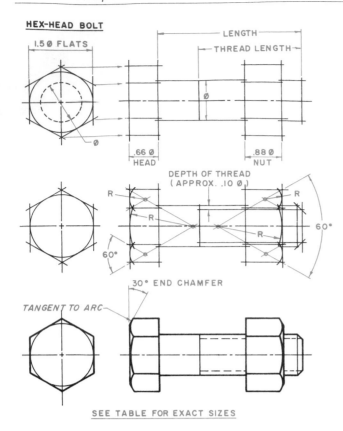

Approximate sizes for a hex-head bolt

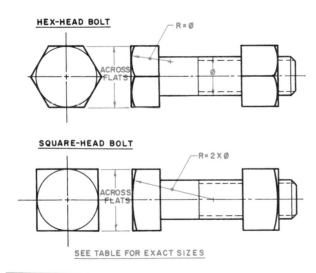

FIGURE 13-30 *Side view of hex- and square-head bolts*

NUTS, BOLTS, AND OTHER FASTENERS IN SECTION

If the cutting plane passes through the axis of any fastener, the fastener is **not** sectioned. It is treated exactly as a shaft and drawn exactly as it is viewed. Refer to *Figure 13-31*. The illustration at the left is drawn correctly. The figure at the right is drawn incorrectly (notice how difficult it is to understand, especially the nut).

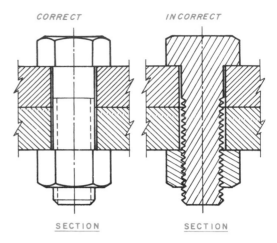

FIGURE 13-31 *Fasteners in section*

THREAD CALL-OFFS

Although not all companies use the same call-offs for various fasteners, it is important that all drafters within one company use the same method. One method used to call off fasteners is illustrated in *Figure 13-32*. Regardless of which system is used, the first line contains the fastener's general identification, type of head,

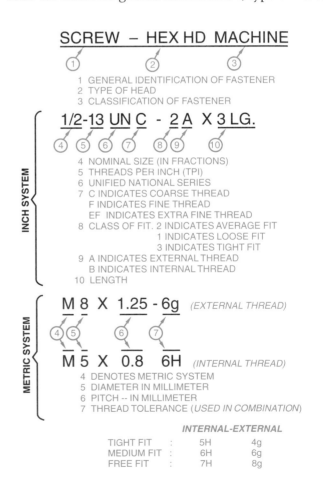

FIGURE 13-32 *Thread call-off*

SLOTTED PHILLIPS^R HEX CAP TORX^R CLUTCH TYPE G SCRULOX^R MULTI-SPLINE

TRIPLE SQUARE TRI-WING^R CLUTCH TYPE A TORQ-SET^R SLAB HEAD POZIDRIV^R REED & PRINCE (FREARSON)

FIGURE 13-33 *Kinds of screw heads*

and classification. The second line contains all the exact detailed information. All threads are assumed to be right hand (R.H.), unless otherwise noted. If a thread is to be left hand (L.H.), it is noted at the end of the second line.

VARIOUS KINDS OF HEADS

Many different kinds of screw heads are used today. *Figure 13-33* illustrates a few of the standard heads.

Rivets

Rivets are permanent fasteners, usually used to hold sheet metal together. Most rivets are made of wrought iron or soft steel and, for aircraft and space missiles, copper, aluminum, alloy, or other exotic metals.

Riveted joints are classified by applications, such as pressure vessels, structural, and machine members. For data concerning joints for pressure vessels refer to such sources as ASME boiler codes. For data concerning larger field structural rivets, such as bridges, buildings, and ships, see ANSI standards or a *Machinery's Handbook*. This chapter covers information for small-size rivets for machine-member riveted joints used for lighter mass-produced applications.

Two kinds of basic rivet joints are the lap joint and the butt joint, *Figure 13-34*. In the *lap joint,* the parts overlap each other and are held together by one or more rows of rivets. In the *butt joint,* the parts are butted and

are held together by a cover plate or butt strap that is riveted to both parts.

Factors to consider are: type of joint, pitch of rivets, type and diameter of rivet, rivet material, and size of clearance holes, *Figure 13-35*. The diameter of a rivet is calculated from the thickness of metal and commonly ranges between

$$d = 1.2 \sqrt{t} \quad \text{AND} \quad d = 1.4 \sqrt{t}$$

where d is the diameter and t is the thickness of the plate.

SIZE AND TYPE OF HOLE

Rivet holes must be punched, punched and then reamed, or drilled. As a general rule, holes are usually made .06 (1.5 mm) larger in diameter than the nominal rivet diameter.

RIVET SYMBOLS

Rivets applied in mass-produced applications are represented in *Figure 13-36A*, illustrating the kind of rivet, to which side it is applied, if it is to be countersunk, and so forth.

KINDS OF RIVETS

There are many different kinds of small-size rivets. The five major kinds are truss head, button head, pan head, countersunk head, and flat head. The countersunk-head rivet is not as strong as the other kinds of rivets; therefore, more rivets must be used to gain strength equal to the other types.

DRAWING OF RIVETS

American standard small solid rivets are shown in their approximate standard proportions in Figure 13-36A. These sizes are close enough to be used for drawing the rivet if necessary. For exact sizes, data must be obtained from other sources.

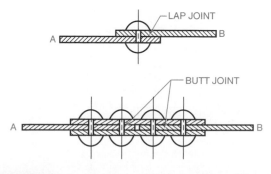

FIGURE 13-34 *Two basic rivet joints (lap joint and butt joint)*

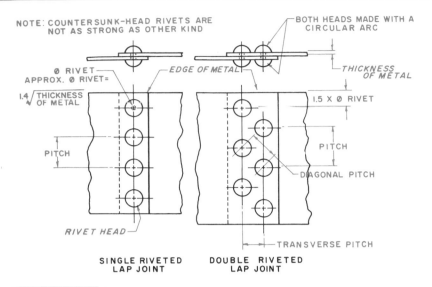

NOTE: COUNTERSUNK–HEAD RIVETS ARE NOT AS STRONG AS OTHER KIND

BOTH HEADS MADE WITH A CIRCULAR ARC

Ø RIVET
APPROX. Ø RIVET=
$1.4 \sqrt{\text{THICKNESS OF METAL}}$

EDGE OF METAL

THICKNESS OF METAL

1.5 X Ø RIVET

PITCH

PITCH

DIAGONAL PITCH

RIVET HEAD

TRANSVERSE PITCH

SINGLE RIVETED LAP JOINT

DOUBLE RIVETED LAP JOINT

FIGURE 13-35   *Factors to consider in rivet joints*

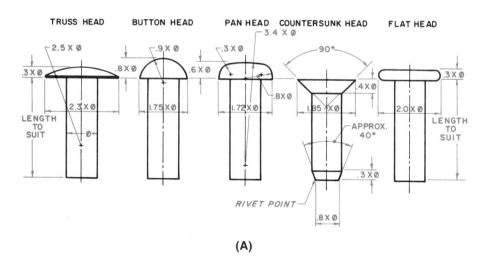

AMERICAN STANDARD SMALL RIVETS (FROM .06 TO .44)

TRUSS HEAD   BUTTON HEAD   PAN HEAD   COUNTERSUNK HEAD   FLAT HEAD

3.4 X Ø

2.5 X Ø   .9 X Ø   .3 X Ø   90°

.3 X Ø   .8 X Ø   .6 X Ø   .4 X Ø   .3 X Ø

.8 X Ø

2.3 X Ø   1.75 X Ø   1.72 X Ø   1.85 X Ø   2.0 X Ø

LENGTH TO SUIT

Ø

APPROX. 40°

LENGTH TO SUIT

RIVET POINT

.3 X Ø

.8 X Ø

(A)

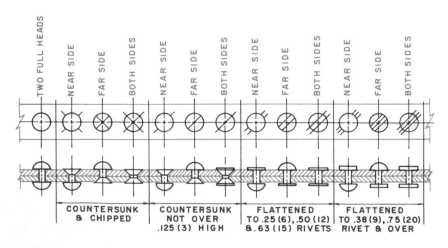

TWO FULL HEADS   NEAR SIDE   FAR SIDE   BOTH SIDES   NEAR SIDE   FAR SIDE   BOTH SIDES   NEAR SIDE   FAR SIDE   BOTH SIDES   NEAR SIDE   FAR SIDE   BOTH SIDES

COUNTERSUNK & CHIPPED

COUNTERSUNK NOT OVER .125 (3) HIGH

FLATTENED TO .25 (6), .50 (12) & .63 (15) RIVETS

FLATTENED TO .38 (9), .75 (20) RIVET & OVER

(B)

FIGURE 13-36   *(A) Approximate sizes for standard small rivets; (B) Illustrated rivet code*

END POINTS

The end of a standard rivet is usually cut off straight If a point is required, a standard size point is illustrated at the lower end of a countersunk head (refer back to Figure 13-36A).

The drafter must indicate what kind of rivet is to be used, and on which side the head is to be positioned. To illustrate this, the drafter uses a code such as that illustrated in *Figure 13-36B*.

Keys and Keyseats

KEY

A *key* is a demountable part that provides a positive means of transferring torque between a shaft and a hub.

Keys are used to prevent slippage and to transmit torque between a shaft and a hub. There are many kinds of keys. The five major kinds used in industry today are illustrated in *Figure 13-37*: square key, flat key, gib head key, Pratt & Whitney key, and Woodruff key. Where a lot of torque is present, a double key and keyseat are often used. In extreme conditions, a spline is machined into the shaft and into the hub. For spline information, refer to a *Machinery's Handbook*.

KEYSEAT

A *keyseat* is an auxiliary-located rectangular groove machined into the shaft and/or hub to receive the key.

CLASSES OF FIT

There are three classifications of fit:

CLASS 1 A side surface clearance fit obtained by using bar stock key and keyseat tolerances. This is a relatively free fit.

CLASS 2 A possible side surface interference or side surface clearance fit obtained by using bar stock key and keyseat tolerances. This is a relatively tight fit.

CLASS 3 A side surface interference fit obtained by interference fit tolerances. This is a very tight fit and has not been generally standardized.

KEY SIZES

For a general rule, the key width is about one-fourth the nominal diameter of the shaft. For exact recommended key sizes, refer to Appendix A or a *Machinery's Handbook*.

DIMENSIONING KEYSEATS

Methods of dimensioning a stock key are shown in *Figure 13-38*.

For dimensioning a Woodruff keyseat, the key number must be included, *Figure 13-39*. Refer to the key size in Appendix A to obtain the exact sizes. To dimension a Pratt & Whitney keyseat, see *Figure 13-40*.

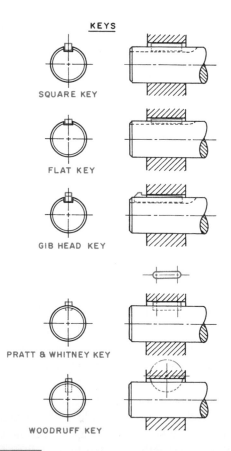

FIGURE 13-37 *Keys and keyseats*

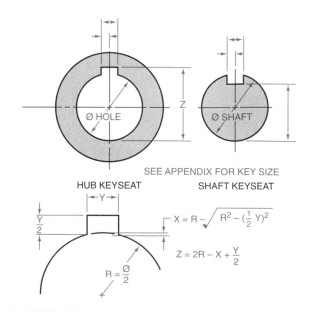

$$X = R - \sqrt{R^2 - (\tfrac{1}{2}Y)^2}$$

$$Z = 2R - X + \frac{Y}{2}$$

$$R = \frac{\varnothing}{2}$$

FIGURE 13-38 *Dimensioning keyseats*

Industry Application

A FASTENER PROBLEM

The problem the design team at Reynolds Tool and Die Company faced at the beginning of the Anderson project had been fasteners. Specifically, the team had to decide how to anchor several springs that are critical components in a family of fixtures to be designed and manufactured for Anderson Aviation Company.

Using a small bolt was discussed and dismissed as being impractical because there would be a blind hole in the fixture. Without a through hole, a bolt could not be used. Various types of screws were discussed. One line of thought was that the head of the screw would anchor the end loop of the tension spring. In addition, a screw would work with a blind hole. The only negative brought up in the discussion of using a screw was the need for threads in the blind hole. The threading process would increase the cost of each fixture by 12 percent, a cost that had not been anticipated when the design team originally estimated the cost of the Anderson project.

The Reynolds design team was almost resigned to the 12 percent loss on the project when one member suggested using a grooved pin. After researching the recommendation, the team decided to use a Type G Number 2 pin 3/8″ in diameter. Such a pin has locking grooves on one end to fit into the blind hole and an angular groove at the other end to hold the end loop of a tension spring.

The decision to use a grooved pin turned out to be serendipitous. Not only did the groove perform as needed, it also decreased the anticipated cost of assembling the fixtures by 9 percent. Consequently, Reynolds Tool and Die Company earned more than anticipated on the Anderson project.

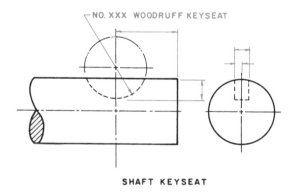

FIGURE 13-39 *Woodruff keyseat*

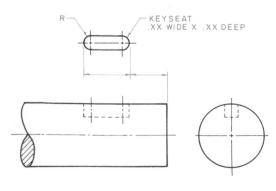

FIGURE 13-40 *Pratt & Whitney keyseat*

Grooved Fasteners

Many types of fasteners are used in industry, each with its own application. Threaded fasteners, such as nuts and bolts, are used to hold parts in tension. *Grooved fasteners* are used to solve metal-to-metal pinning needs with shear application, *Figure 13-41*.

Grooved fasteners have great holding power and are resistant to shock, vibration, and fatigue. They are available in a wide range of types, sizes, and materials. A grooved fastener often has a better appearance than most other methods of fastening. This can be important to the overall design, if the fastener is visible.

Grooved fasteners have three parallel grooves, equally spaced, impressed longitudinally on their exterior surface. To make these grooves, a grooving tool is pressed **below** the surface to displace a carefully determined amount of material. Nothing is removed. The metal is displaced to each side, forming a raised portion or flute extending along each side of the groove, *Figure 13-42*. The crest of the flute constitutes the *expanded diameter* (Dx). The expanded diameter (Dx) is a few thousandths larger than the nominal diameter (D) of the stock.

Grooved fasteners can be custom manufactured in order to meet most any application. For example, a custom-grooved pin can be made with a cross-drilled hole, a groove for a snap ring, or a threaded hole in one end. The groove length can also be varied and placed anywhere along the length of the pin.

FIGURE 13-41 *Grooved fastener*

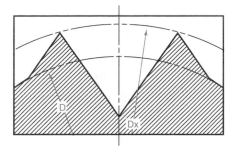

FIGURE 13-42 *Enlarged sectional view of one of the grooves before inserting fastener*

INSTALLATION

The grooved fastener is forced into a drilled hole slightly larger than the nominal or specified diameter of the pin, *Figure 13-43*. The crest or flutes are forced back into the grooves when the fastener is driven into the hole. The resiliency of the metal forced back into the grooves creates powerful radial forces against the hole wall.

In many cases, grooved fasteners are lower in cost than knurled pins, taper pins, pins with cotter pins, rivets, set screws, keys, or other methods used to fasten metal-to-metal parts together. The installation cost of grooved pins is invariably lower because of the required hole tolerances, and no special guides are required at assembly.

MATERIAL

Standard grooved fasteners are made of cold-drawn, low-carbon steel. The physical properties of this material are more than enough for ordinary applications. Alloy steel, hard brass, silicon bronze, stainless steel, and other exotic metals may also be specially ordered. These special materials are usually heat treated for optimum physical qualities.

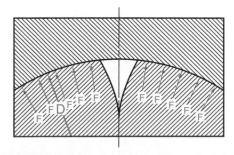

FIGURE 13-43 *The same view after insertion. A powerful radial thrust is obtained.*

FINISH

Standard grooved fasteners have a finish of zinc electroplate, deposited approximately .00015 inch deep. Chromate, brass, cadmium, and black oxide can also be specially ordered.

STANDARD TYPES

Study the various types of grooved fasteners and the related technical data in *Figures 13-44, 13-45, 13-46, 13-47* (page 548), and *13-48* (page 548). Note the various types of fasteners, how each functions, and what each replaces. For example, Type A is used in place of taper pins, rivets, set screws, and keys.

STANDARD SIZES

Refer to the standard size charts in *Figures 13-49* (page 549), *13-50* (page 550), *13-51* (page 551), and *13-52* (page 552). Across the top of each chart are the nominal sizes from 1/16″ diameter to 1/2″ diameter. At the left side of each chart are listed the standard lengths from 1/4″ long to 4 1/2″ long. Various other technical information can be derived from the charts. For drilling procedure, hole tolerances, application data, and high-alloy pin applications, see *Figures 13-53* (page 553), *13-54* (page 553), *13-55* (page 554).

GROOVED STUDS

Grooved studs are widely used for fastening light metal or plastic parts to heavier members or assemblies, *Figures 13-56* (page 554) and *13-57* (page 555). They replace screws, rivets, peened pins, and many other types of fasteners. The grooves function as any grooved fastener.

Spring Pins

Another type of fastener is the *spring pin*, *Figure 13-58* (page 555). Spring pins are manufactured by cold-forming strip metal in a progressive roll-forming operation. After forming, the pins are broken off and deburred to eliminate any sharp edges. They are then heat treated to a R/C 46/53. This develops spring qualities or resiliency in the metal.

In the free state, the pins are larger in diameter than the hole into which they are to be inserted. The pins compress themselves as they are driven into the hole, thus exerting radial forces around the entire circumference of the hole.

TYPE A

The Type A DRIV-LOK pin has three **full-length tapered** grooves. This popular type is used in applications requiring excellent locking effect and ease of assembly. It is widely used in place of taper pins, rivets, set screws and keys. Typical applications include keying sprockets, gears, collars, knobs, handles, levers and wheels to shafts.

Locking Collar to Shaft

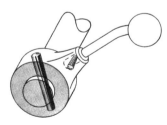

Lever and Shaft Assembly

TYPE A3

Type A3 pins have three **full-length parallel** grooves with a short pilot end to insure easy starting. They are recommended for applications requiring maximum locking effect where severe vibration and/or shock loading are present.

Keying Gear to Shaft

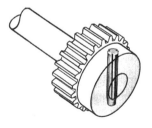

Locking Gear to Shaft

TYPE U

Type U has three **parallel** grooves extending the **full length** of the pin. A short pilot at each end permits hopper or automatic feeding. Identical ends speed manual insertion as operator need not examine pin to determine proper end to start. Full-length parallel grooves provide maximum locking effect. Typical applications include keying gears, collars, knobs, handles, etc. to shafts.

Attaching Knob to Shaft

Pinning "V" Pulley to Shaft

FIGURE 13-44 *Typical applications for grooved pin types A, A3, and U*

TYPE B

Type B pins have three **tapered** grooves extending **one-half** the length of the pin. This type is widely used as a hinge or pivot pin. Driven or pressed into a straight drilled hole, the grooved portion locks in one part, while the ungrooved portion will remain free. Also excellent for dowel and locating applications.

TYPE C

The Type C Pin has three **parallel** grooves extending **one-quarter** its overall length. It is ideally suited for linkage or pivot applications, especially where a relatively short locking section and longer free length are required. Widely used in certain types of hinge applications. The long lead permits easy insertion.

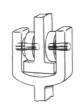

Roller Pins

Control Valve Hinge Assembly

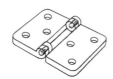

Hinge Pins

Linkage or Hinge Pin

FIGURE 13-45 *Typical applications for grooved pin types B and C*

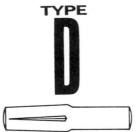

TYPE D

Type D has three **reverse** taper grooves extending **one-half** the length of the pin. Recommended for use in blind holes as a stop pin, roller pivot, dowel, or for certain hinge or linkage applications. Reverse taper grooves permit easy insertion in blind holes.

TYPE E

Type E has three **parallel half-length** grooves located equidistant from each end. Widely used as T handle on valves and tools. Also used as a cross pin, cotter pin, pivot pin, etc. where center locking is required.

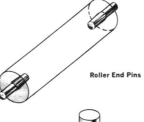

Roller End Pins

Linkage Pin

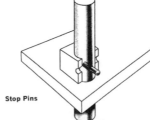

Stop Pins

T Handle for Valve

FIGURE 13-46 *Typical applications for grooved pin types D and E*

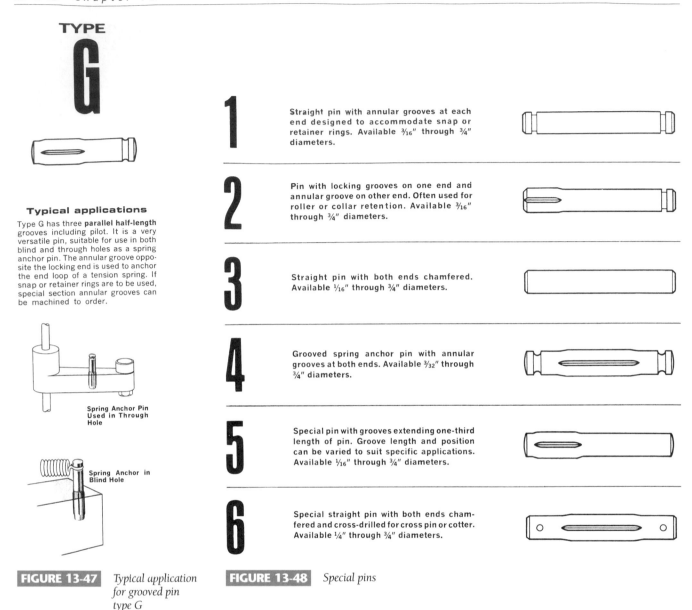

TYPE G

Typical applications

Type G has three **parallel half-length** grooves including pilot. It is a very versatile pin, suitable for use in both blind and through holes as a spring anchor pin. The annular groove opposite the locking end is used to anchor the end loop of a tension spring. If snap or retainer rings are to be used, special section annular grooves can be machined to order.

Spring Anchor Pin Used in Through Hole

Spring Anchor in Blind Hole

FIGURE 13-47 *Typical application for grooved pin type G*

1 Straight pin with annular grooves at each end designed to accommodate snap or retainer rings. Available 3/16" through 3/4" diameters.

2 Pin with locking grooves on one end and annular groove on other end. Often used for roller or collar retention. Available 3/16" through 3/4" diameters.

3 Straight pin with both ends chamfered. Available 1/16" through 3/4" diameters.

4 Grooved spring anchor pin with annular grooves at both ends. Available 3/32" through 3/4" diameters.

5 Special pin with grooves extending one-third length of pin. Groove length and position can be varied to suit specific applications. Available 1/16" through 3/4" diameters.

6 Special straight pin with both ends chamfered and cross-drilled for cross pin or cotter. Available 1/4" through 3/4" diameters.

FIGURE 13-48 *Special pins*

Spring pins are made from a high-carbon steel—1074, stainless steel, either 420 heat treated or 300 series cold worked. Some spring pins are also made from brass or beryllium copper. Spring pins meet the demand of many industrial applications. *Figures 13-59* (page 555) and *13-60* (page 556) list some suggested applications.

Fastening Systems

Fastening systems play a critical role in most product design. They often do more than position and secure components. In many cases, fastening systems have a direct effect upon the product's durability, reliability, size, and weight. They affect the speed with which the product may be assembled and disassembled, both during manufacturing and later in field service. Fastening systems also affect cost, not only for the fasteners, but also for the machining and assembly operations they require.

Unless a drafter has had a great deal of experience in using different fastening devices and techniques, it is often difficult to choose a fastener that combines optimum function with maximum economy. A fastening system that is best for one product may not be desirable for another.

(Text continues on page 557)

TYPE A

TYPE A3

TYPE U

STANDARD SIZES

Nominal diameter and recommended drill sizes	1/16	5/64	3/32	7/64	1/8	5/32	3/16	7/32	1/4	5/16	3/8	7/16	1/2
Dec. Equivalents	.0625	.0781	.0938	.1094	.1250	.1563	.1875	.2188	.2500	.3125	.3750	.4375	.5000
Crown Height, In.	.0065	.0087	.0091	.0110	.0130	.0170	.0180	.0220	.0260	.0340	.0390	.0470	.0520
Radius, In. ±.010	5/64	3/32	1/8	9/64	5/32	3/16	1/4	9/32	5/16	3/8	15/32	17/32	5/8
Pilot Length, In. (Ref.)	1/32	1/32	1/32	1/32	1/32	1/16	1/16	1/16	1/16	3/32	3/32	3/32	3/32
Chamfer Length, In. (Type U Only)	1/64	1/64	1/64	1/64	1/64	1/32	1/32	1/32	1/32	3/64	3/64	3/64	3/64

"Dx" EXPANDED DIAMETER—CAN BE DETERMINED ACCURATELY ONLY WITH RING GAGES

LENGTH OF PIN IN INCHES

Length of pin	1/16	5/64	3/32	7/64	1/8	5/32	3/16	7/32	1/4	5/16	3/8	7/16	1/2
1/4 (.250)	.068	.084	.101	.117	.134								
3/8 (.375)	.068	.084	.101	.117	.134	.166	.198						
1/2 (.500)	.068	.084	.101	.117	.134	.166	.198	.230	.263				
5/8 (.625)	.068	.084	.101	.117	.134	.166	.198	.230	.263	.329			
3/4 (.750)	.068	.084	.101	.116	.134	.166	.198	.230	.263	.329	.394		
7/8 (.875)	.068	.084	.101	.116	.133	.165	.198	.230	.263	.329	.394	.459	
1 (1.000)	.068	.084	.101	.115	.133	.165	.198	.230	.263	.329	.394	.459	.525
1 1/4 (1.250)			.101	.115	.132	.164	.197	.230	.263	.329	.394	.459	.525
1 1/2 (1.500)					.132	.164	.197	.229	.262	.329	.394	.459	.525
1 3/4 (1.750)						.163	.197	.229	.262	.328	.393	.459	.525
2 (2.000)						.163	.196	.229	.262	.328	.393	.458	.525
2 1/4 (2.250)							.196	.229	.262	.328	.393	.458	.525
2 1/2 (2.500)								.229	.262	.328	.393	.458	.524
2 3/4 (2.750)								.228	.261	.327	.393	.458	.524
3 (3.000)								.227	.260	.327	.392	.457	.523
3 1/4 (3.250)									.260	.326	.392	.457	.523
3 1/2 (3.500)										.326	.391	.456	.522
3 3/4 (3.750)											.391	.456	.522
4 (4.000)											.390	.455	.521
4 1/4 (4.250)											.390	.455	.521
4 1/2 (4.500)												.454	.520

Nom. Diam. (D)	Exp. Diam. (Dx) reduced by
1/16	.002
5/64	.002
3/32	.002
7/64	.002
1/8	.002
5/32	.002
3/16	.003
7/32	.003
1/4	.003
5/16	.004
3/8	.005
7/16	.006
1/2	.006

TOLERANCES:

On Nominal Diameter "D"
+.000—.001 up to 7/64" diameter
+.000—.002 7/64" and above

On Expanded Diameter "Dx"
±.001 up to 7/64" diameter
±.002 7/64" and above

On over-all Length "L"
±.010 for all diameters

◄ For stainless steels and other special materials, the expanded diameters shown in above table are reduced by amounts shown at left.

Note: Intermediate pin lengths, pin diameters up to 3/4", groove lengths, and groove positions to order as specials. When ordering, specify type, diameter, length, and special finishes.

FIGURE 13-49 *Standard size chart of grooved pin types A, A3, and U*

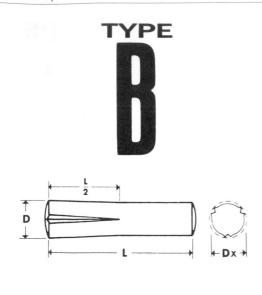

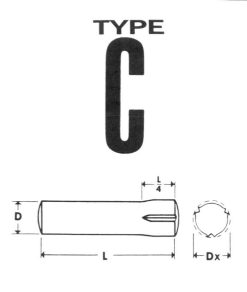

STANDARD SIZES

Nominal diameter and recommended drill sizes	1/16	5/64	3/32	7/64	1/8	5/32	3/16	7/32	1/4	5/16	3/8	7/16	1/2
Dec. Equivalents	.0625	.0781	.0938	.1094	.1250	.1563	.1875	.2188	.2500	.3125	.3750	.4375	.5000
Crown Height, In.	.0065	.0087	.0091	.0110	.0130	.0170	.0180	.0220	.0260	.0340	.0390	.0470	.0520
Radius, In. ±.010	5/64	3/32	1/8	9/64	5/32	3/16	1/4	9/32	5/16	3/8	15/32	17/32	5/8

"Dx" EXPANDED DIAMETER—CAN BE DETERMINED ACCURATELY ONLY WITH RING GAGES

LENGTH OF PIN IN INCHES	1/16	5/64	3/32	7/64	1/8	5/32	3/16	7/32	1/4	5/16	3/8	7/16	1/2
1/4 (.250)	.068	.084	.101	.117	.134								
3/8 (.375)	.068	.084	.101	.117	.134	.166	.198						
1/2 (.500)	.068	.084	.101	.117	.134	.166	.198	.230	.263				
5/8 (.625)	.068	.084	.101	.117	.134	.166	.198	.230	.263	.329			
3/4 (.750)	.068	.084	.101	.117	.134	.166	.198	.230	.263	.329	.394		
7/8 (.875)	.068	.084	.101	.117	.134	.166	.198	.230	.263	.329	.394	.459	
1 (1.000)	.068	.084	.101	.117	.134	.166	.198	.230	.263	.329	.394	.459	.525
1 1/4 (1.250)			.101	.117	.134	.166	.198	.230	.263	.329	.394	.459	.525
1 1/2 (1.500)					.134	.166	.198	.230	.263	.329	.394	.459	.525
1 3/4 (1.750)						.165	.198	.230	.263	.329	.394	.459	.525
2 (2.000)						.165	.198	.230	.263	.329	.394	.459	.525
2 1/4 (2.250)							.197	.230	.263	.329	.394	.459	.525
2 1/2 (2.500)								.230	.263	.329	.394	.459	.525
2 3/4 (2.750)								.229	.262	.329	.394	.459	.525
3 (3.000)								.229	.262	.329	.394	.459	.525
3 1/4 (3.250)									.262	.328	.393	.459	.525
3 1/2 (3.500)										.328	.393	.459	.525
3 3/4 (3.750)											.393	.458	.525
4 (4.000)											.393	.458	.525
4 1/4 (4.250)											.393	.458	.524
4 1/2 (4.500)												.458	.524

Nom. Diam. (D)	Exp. Diam. (Dx) reduced by
1/16	.002
5/64	.002
3/32	.002
7/64	.002
1/8	.002
5/32	.002
3/16	.003
7/32	.003
1/4	.003
5/16	.004
3/8	.005
7/16	.006
1/2	.006

TOLERANCES:

On Nominal Diameter "D"
+.000 −.001 up to 7/64" diameter
+.000 −.002 7/64" and above

On Expanded Diameter "Dx"
±.001 up to 7/64" diameter
±.002 7/64" and above
On over-all Length "L"
±.010 for all diameters

◄ For stainless steels and other special materials, the expanded diameters shown in above table are reduced by amounts shown at left.

Note: Intermediate pin lengths, pin diameters up to 3/4", groove lengths, and groove positions to order as specials. When ordering, specify type, diameter, length, and special finishes.

FIGURE 13-50 *Standard size chart of grooved pin types B and C*

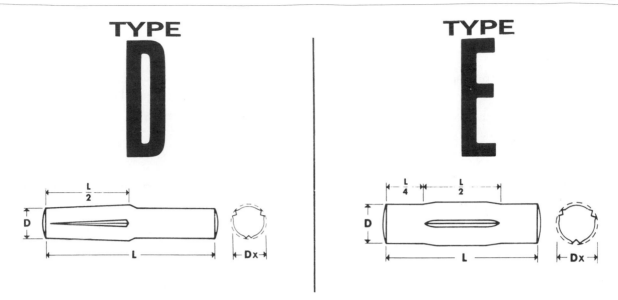

TYPE D   TYPE E

STANDARD SIZES

Nominal diameter and recommended drill sizes	1/16	5/64	3/32	7/64	1/8	5/32	3/16	7/32	1/4	5/16	3/8	7/16	1/2
Dec. Equivalents	.0625	.0781	.0938	.1094	.1250	.1563	.1875	.2188	.2500	.3125	.3750	.4375	.5000
Crown Height, In.	.0065	.0087	.0091	.0110	.0130	.0170	.0180	.0220	.0260	.0340	.0390	.0470	.0520
Radius, In. ±.010	5/64	3/32	1/8	9/64	5/32	3/16	1/4	9/32	5/16	3/8	15/32	17/32	5/8

"Dx" EXPANDED DIAMETER—CAN BE DETERMINED ACCURATELY ONLY WITH RING GAGES

LENGTH OF PIN IN INCHES:

Length	1/16	5/64	3/32	7/64	1/8	5/32	3/16	7/32	1/4	5/16	3/8	7/16	1/2
1/4 (.250)	.068	.084	.101	.117	.134								
3/8 (.375)	.068	.084	.101	.117	.134	.166	.198						
1/2 (.500)	.068	.084	.101	.117	.134	.166	.198	.230	.263				
5/8 (.625)	.068	.084	.101	.117	.134	.166	.198	.230	.263	.329			
3/4 (.750)	.068	.084	.101	.117	.134	.166	.198	.230	.263	.329	.394		
7/8 (.875)	.068	.084	.101	.117	.134	.166	.198	.230	.263	.329	.394	.459	
1 (1.000)	.068	.084	.101	.117	.134	.166	.198	.230	.263	.329	.394	.459	.525
1 1/4 (1.250)			.101	.117	.134	.166	.198	.230	.263	.329	.394	.459	.525
1 1/2 (1.500)					.134	.166	.198	.230	.263	.329	.394	.459	.525
1 3/4 (1.750)						.165	.198	.230	.263	.329	.394	.459	.525
2 (2.000)						.165	.198	.230	.263	.329	.394	.459	.525
2 1/4 (2.250)							.197	.230	.263	.329	.394	.459	.525
2 1/2 (2.500)								.230	.263	.329	.394	.459	.525
2 3/4 (2.750)								.229	.262	.329	.394	.459	.525
3 (3.000)								.229	.262	.329	.394	.459	.525
3 1/4 (3.250)									.262	.329	.394	.459	.525
3 1/2 (3.500)										.328	.393	.459	.525
3 3/4 (3.750)										.328	.393	.459	.525
4 (4.000)											.393	.458	.525
4 1/4 (4.250)											.393	.458	.524
4 1/2 (4.500)												.458	.524

Inset table:

Nom. Diam. (D)	Exp. Diam. (Dx) reduced by
1/16	.002
5/64	.002
3/32	.002
7/64	.002
1/8	.002
5/32	.002
3/16	.003
7/32	.003
1/4	.003
5/16	.004
3/8	.005
7/16	.006
1/2	.006

TOLERANCES:

On Nominal Diameter "D"
+.000—.001 up to 7/64" diameter
+.000—.002 7/64" and above

On Expanded Diameter "Dx"
±.001 up to 7/64" diameter
±.002 7/64" and above

On over-all Length "L"
±.010 for all diameters

◄ For stainless steels and other special materials, the expanded diameters shown in above table are reduced by amounts shown at left.

Note: Intermediate pin lengths, pin diameters up to 3/4", groove lengths, and groove positions to order as specials. When ordering, specify type, diameter, length, and special finishes.

FIGURE 13-51 *Standard size chart of grooved pin types D and E*

TYPE G

(combined former F and G)

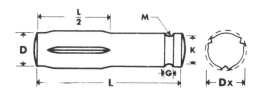

STANDARD SIZES

Nominal diameter and recommended drill sizes	3/32	7/64	1/8	5/32	3/16	7/32	1/4	5/16	3/8	7/16	1/2
Dec. Equivalents	.0938	.1094	.1250	.1563	.1875	.2188	.2500	.3125	.3750	.4375	.5000
Crown Height, In.	.0091	.0110	.0130	.0170	.0180	.0220	.0260	.0340	.0390	.0470	.0520
Radius, In., ±0.010	1/8	9/64	5/32	3/16	1/4	9/32	5/16	3/8	15/32	17/32	5/8
Pilot Length, In. (Ref.)	1/32	1/32	1/32	1/16	1/16	1/16	1/16	3/32	3/32	3/32	3/32
M Neck Radius (Ref.)	1/64	1/64	1/32	1/32	1/32	3/64	3/64	1/16	1/16	3/32	3/32
G Neck Width (Ref.)	1/32	1/32	1/16	1/16	1/16	3/32	3/32	1/8	1/8	3/16	3/16
Shoulder Width +.010—.000	1/32	1/32	1/32	3/64	3/64	1/16	1/16	3/32	1/8	1/8	1/8
K Neck Diameter ±.005	.062	.078	.083	.104	.125	.146	.167	.209	.250	.293	.312

"Dx" EXPANDED DIAMETER—CAN BE DETERMINED ACCURATELY ONLY WITH RING GAGES

LENGTH OF PIN IN INCHES

Length	3/32	7/64	1/8	5/32	3/16	7/32	1/4	5/16	3/8	7/16	1/2
3/8 (.375)	.101	.117	.134	.166	.198						
1/2 (.500)	.101	.117	.134	.166	.198	.230	.263				
5/8 (.625)	.101	.117	.134	.166	.198	.230	.263	.329			
3/4 (.750)	.101	.117	.134	.166	.198	.230	.263	.329	.394		
7/8 (.875)	.101	.117	.134	.166	.198	.230	.263	.329	.394	.459	
1 (1.000)	.101	.117	.134	.166	.198	.230	.263	.329	.394	.459	.525
1¼ (1.250)	.101	.117	.134	.166	.198	.230	.263	.329	.394	.459	.525
1½ (1.500)			.134	.166	.198	.230	.263	.329	.394	.459	.525
1¾ (1.750)				.165	.198	.230	.263	.329	.394	.459	.525
2 (2.000)				.165	.198	.230	.263	.329	.394	.459	.525
2¼ (2.250)					.197	.230	.263	.329	.394	.459	.525
2½ (2.500)						.230	.263	.329	.394	.459	.525
2¾ (2.750)						.229	.262	.329	.394	.459	.525
3 (3.000)						.229	.262	.329	.394	.459	.525
3¼ (3.250)							.262	.328	.393	.459	.525
3½ (3.500)								.328	.393	.459	.525
3¾ (3.750)									.393	.458	.525
4 (4.000)									.393	.458	.525
4¼ (4.250)									.393	.458	.524
4½ (4.500)										.458	.524

Nom. Diam. (D)	Exp. Diam. (Dx) reduced by
3/32	.002
7/64	.002
1/8	.002
5/32	.002
3/16	.003
7/32	.003
1/4	.003
5/16	.004
3/8	.005
7/16	.006
1/2	.006

TOLERANCES:

On Nominal Diameter "D"
+.000—.001 up to 7/64" diameter
+.000—.002 7/64" and above

On Expanded Diameter "Dx"
±.001 up to 7/64" diameter
±.002 7/64" and above

On over-all Length "L"
±.010 for all diameters

◄ For stainless steels and other special materials, the expanded diameters shown in above table are reduced by amounts shown at left.

Note: Intermediate pin lengths, pin diameters up to 3/4", groove lengths, and groove positions to order as specials. When ordering, specify type, diameter, length, and special finishes.

FIGURE 13-52 *Standard size chart of grooved pin type G*

Pin Diameter	Decimal Equivalent	Recommended Drill Size	Hole Tolerances ADD To Nominal Diameter
1/16"	.0625	1/16"	.002"
5/64	.0781	5/64	.002"
3/32	.0938	3/32	.003"
7/64	.1094	7/64	.003"
1/8	.1250	1/8	.003"
5/32	.1563	5/32	.003"
3/16	.1875	3/16	.004"
7/32	.2188	7/32	.004"
1/4	.2500	1/4	.004"
5/16	.3125	5/16	.005"
3/8	.3750	3/8	.005"
7/16	.4375	7/16	.006"
1/2	.5000	1/2	.006"

Tolerances for drilled holes shown in table are based on depth-to-diameter ratio of approximately 5 to 1. Higher ratios may cause these figures to be exceeded, but in no case should they be exceeded by more than 10%. Specifications for holes having a depth-to-diameter ratio of 1 to 1 or less should be held extremely close; 60% of figures shown in table is recommended.

Undersized drills should never be used to produce holes for Driv-Lok Pins. This malpractice results from the false assumption that the pins will "hold better." Instead, they bend, damage the hole wall, crack castings, and peel their expanded flutes, thus reducing their retaining characteristics and preventing their reuse.

Holes made in hardened steel or cast iron are recommended to have a slight chamfer at the entrance. This eliminates shearing of the flutes as the pins are forced in.

Care should be exercised in all drilling for Driv-Lok Pins. Drills used should be new or properly ground with the aid of an approved grinding fixture. The drilling machine spindle must be in good condition and operated at correct speeds and feeds for the metal being drilled, and suitable coolant is always recommended. Drill jigs with accurate bushings always facilitate good drilling practice.

FIGURE 13-53 *Drilling procedure and hole tolerance*

RECOMMENDED PIN DIAMETER FOR VARIOUS SHAFT SIZES AND TORQUE TRANSMITTED BY PIN IN DOUBLE SHEAR

Shaft Size	Pin Diameter	Torque Inch Lbs.	H.P. at 100 R.P.M.	Shaft Size	Pin Diameter	Torque Inch Lbs.	H.P. at 100 R.P.M.
3/16	1/16	4.6	.007	7/8	1/4	347	.555
7/32	5/64	8.4	.013	15/16	5/16	580	.927
1/4	3/32	13.7	.022	1	5/16	618	.990
5/16	7/64	23.6	.038	1 1/16	5/16	657	1.05
3/8	1/8	37.2	.060	1 1/8	3/8	1010	1.61
7/16	5/32	67.6	.108	1 3/16	3/8	1065	1.70
1/2	5/32	77.2	.124	1 1/4	3/8	1120	1.79
9/16	3/16	125.0	.200	1 5/16	7/16	1590	2.55
5/8	3/16	139.0	.222	1 3/8	7/16	1670	2.67
11/16	7/32	207.0	.332	1 7/16	7/16	1740	2.79
3/4	1/4	297.0	.476	1 1/2	1/2	2380	3.81
13/16	1/4	322	.516				

This table is a guide in selecting the proper size Driv-Lok Grooved Pin to use in keying machine members to shafts of given sizes and for specific load requirements. Torque and horsepower ratings are based on pins made of cold finished, low carbon steel and a safety factor of 8 is assumed.

MINIMUM SINGLE SHEAR VALUES (LBS.) OF DRIV-LOK PINS OF VARIOUS MATERIALS

DRIV-LOK PIN DIAM.	MATERIAL				
	Cold Finished 1213 Steel	Shear-Proof® ALLOY STEEL R.C. 40 - 48	Brass	Silicon Bronze	Heat Treated Stainless Steel
1/16	200	363	124	186	308
5/64	312	562	192	288	478
3/32	442	798	272	408	680
7/64	605	1091	372	558	933
1/8	800	1443	492	738	1230
5/32	1240	2236	764	1145	1910
3/16	1790	3220	1100	1650	2750
7/32	2430	4386	1495	2240	3740
1/4	3190	5753	1960	2940	4910
5/16	4970	8974	3060	4580	7650
3/8	5810	12960	4420	6630	11050
7/16	7910	17580	6010	9010	15000
1/2	10300	23020	7850	11800	19640

FIGURE 13-54 *Grooved pin application/engineering data*

MATERIAL HANDLING EQUIPMENT—The Type E pin provides positive locking with a half-length groove in the center of the pin. Extreme shear is exerted in this application, yet the Shear-Proof pin is used with complete safety for both men and materials. Type E is a special pin.

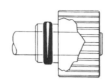

HEAVY-DUTY GEAR AND SHAFT ASSEMBLY—Type A Shear-Proof pin as specified for this application to give maximum locking power over the entire pin and gear hub area. The Type A pin, with grooves the full length of the pin, is the standard stock pin which meets most applications.

AUTOMATIC TRANSMISSION IN AUTOMOBILES—Special Type C was selected as a shaft in this transmission servo to replace a cross-drilled shaft with a cross pin for holding shaft in position. This eliminated a costly drilling operation and the cross pin.

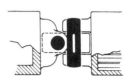

UNIVERSAL JOINTS IN HAND TOOLS—Special Type E pin with center groove eliminates costly staking and grinding operations and improves product appearance. This pin is easily installed, fits flush, and permits plating before assembly.

EYE BOLT HINGE PIN — Type C pin, with quarter-length grooves, provides maximum ease of assembly. There is no interference until three-fourths of the pin is in position. The high safety factor inherent in Shear-Proof pins makes them practical and efficient for such constant-shear applications. Type C is a special pin.

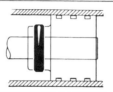

HIGH-PRESSURE PISTON AND ROD ASSEMBLY—Type B pin was used here because the half-length grooves simplified the job of starting the pin into the hole. Ease of assembly was matched with sufficient locking power even when subjected to continuous, strong reciprocating forces. Type B is a special pin.

FIGURE 13-55 *High alloy shear-proof pins*

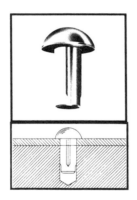

TYPICAL APPLICATIONS

Fastening knobs
handles, etc.

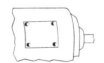

Attaching nameplates,
instruction panels

Widely used for
fastening brackets

Fastening spring assemblies
or control arms

STANDARD SIZES and SPECIFICATIONS

Stud Number	Nominal Shank Diameter	Recommended Drill Size	Head Dia. Max.	Head Dia. Min.	Head Height Max.	Head Height Min.
0	.067	51	.130	.120	.050	.040
2	.086	44	.162	.146	.070	.059
4	.104	37	.211	.193	.086	.075
6	.120	31	.260	.240	.103	.091
7	.136	29	.309	.287	.119	.107
8	.144	27	.309	.287	.119	.107
10	.161	20	.359	.334	.136	.124
12	.196	9	.408	.382	.152	.140
14	.221	2	.457	.429	.169	.156
16	.250	¼"	.472	.443	.174	.161

Maximum Expansion—Standard Lengths

Stud No.	⅛"	³⁄₁₆"	¼"	⁵⁄₁₆"	⅜"	½"
0	.074	.074	.074			
2	.096	.096	.095			
4		.115	.113	.113		
6			.132	.130	.130	
7				.147	.147	.144
8					.155	.153
10					.173	.171
12						.206
14						.234
16						.263

TOLERANCES:
On length - .010
On Exp. Diameter - .002
On Nominal Diameter + .000
        - .002

Note: The expanded diameter can be determined accurately only with ring gages.

FIGURE 13-56 *Standard studs*

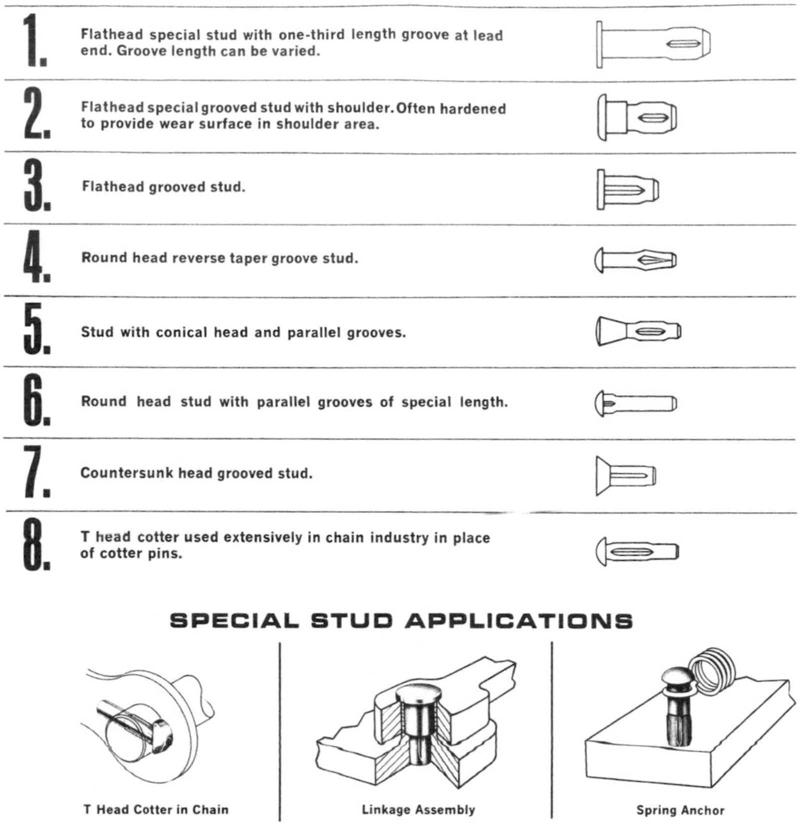

1. Flathead special stud with one-third length groove at lead end. Groove length can be varied.

2. Flathead special grooved stud with shoulder. Often hardened to provide wear surface in shoulder area.

3. Flathead grooved stud.

4. Round head reverse taper groove stud.

5. Stud with conical head and parallel grooves.

6. Round head stud with parallel grooves of special length.

7. Countersunk head grooved stud.

8. T head cotter used extensively in chain industry in place of cotter pins.

SPECIAL STUD APPLICATIONS

T Head Cotter in Chain   Linkage Assembly   Spring Anchor

FIGURE 13-57   *Special studs*

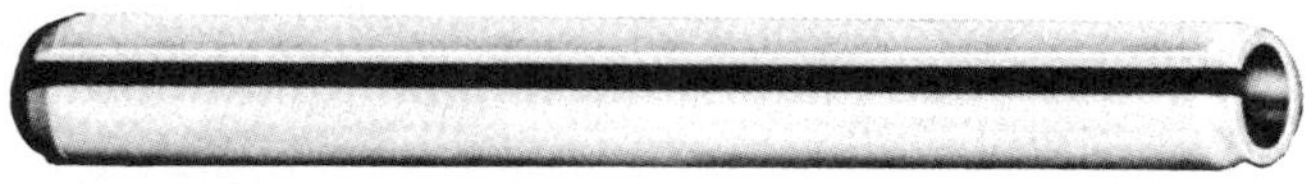

FIGURE 13-58   *Spring pin*

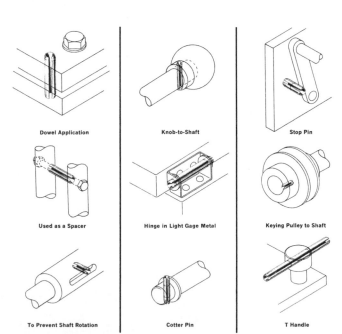

Dowel Application   Knob-to-Shaft   Stop Pin

Used as a Spacer   Hinge in Light Gage Metal   Keying Pulley to Shaft

To Prevent Shaft Rotation   Cotter Pin   T Handle

FIGURE 13-59   *Spring pin application*

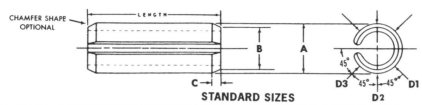

STANDARD SIZES

Nominal	A Minimum ⅓(D₁+D₂+D₃)	A Maximum (Go Ring Gage)	B Max.	C Min.	C Max.	WALL THICK-NESS	RECOMMENDED HOLE SIZE Min.	RECOMMENDED HOLE SIZE Max.	MINIMUM DOUBLE SHEAR STRENGTH POUNDS Carbon Steel and Stainless Steel
.062	.066	.069	.059	.007	.028	.012	.062	.065	425
.078	.083	.086	.075	.008	.032	.018	.078	.081	650
.094	.099	.103	.091	.008	.038	.022	.094	.097	1,000
.125	.131	.135	.122	.008	.044	.028	.125	.129	2,100
.156	.162	.167	.151	.010	.048	.032	.156	.160	3,000
.187	.194	.199	.182	.011	.055	.040	.187	.192	4,400
.219	.226	.232	.214	.011	.065	.048	.219	.224	5,700
.250	.258	.264	.245	.012	.065	.048	.250	.256	7,700
.312	.321	.328	.306	.014	.080	.062	.312	.318	11,500
.375	.385	.392	.368	.016	.095	.077	.375	.382	17,600
.437	.448	.456	.430	.017	.095	.077	.437	.445	20,000
.500	.513	.521	.485	.025	.110	.094	.500	.510	25,800

All dimensions listed on this page are in accordance with National Standards. Wall thicknesses within the Spring Pin industry are standard.

SPRING PIN WEIGHT PER 1000 PIECES
Material—Steel · Nominal Diameter

LENGTH	.062	.078	.094	.125	.156	.187
0.187	.10	0.18				
0.250	.15	0.22	0.33			
0.312	.19	0.28	0.41			
0.375	.23	0.34	0.50	0.89		
0.437	.27	0.40	0.58	1.00	1.50	
0.500	.30	0.46	0.66	1.20	1.70	2.50
0.562	.34	0.51	0.75	1.30	1.90	2.90
0.625	.38	0.57	0.83	1.50	2.10	3.20
0.687	.42	0.63	0.92	1.60	2.40	3.50
0.750	.46	0.69	1.00	1.80	2.60	3.80
0.812	.49	0.74	1.10	1.90	2.80	4.10
0.875	.53	0.80	1.20	2.10	3.00	4.50
0.937	.57	0.86	1.30	2.20	3.20	4.80
1.000	.61	0.92	1.40	2.40	3.40	5.10
1.125		1.00	1.50	2.70	3.80	5.70
1.250		1.20	1.70	3.00	4.30	6.40
1.375		1.30	1.80	3.30	4.70	7.00
1.500		1.40	2.00	3.60	5.10	7.60
1.625				3.90	5.50	8.30
1.750				4.20	6.00	8.90
1.875				4.50	6.40	9.60
2.000				4.70	6.80	10.0
2.250					7.80	12.0
2.500					8.60	13.0

LENGTH	.219	.250	.312	.375	.437	.500
0.562	4.00					
0.625	4.50	5.30				
0.687	4.90	5.90				
0.750	5.30	6.30	9.90	15.0		
0.812	5.80	6.90	11.0			
0.875	6.20	7.40	12.0			
0.937	6.70	8.00	12.0			
1.000	7.00	8.50	13.0	20.0	24.0	
1.125	7.90	9.50	14.0			
1.250	8.80	11.0	16.0	24.0	30.0	41.0
1.375	9.70	12.0	18.0			
1.500	11.0	13.0	19.0	29.0	36.0	49.0
1.625	12.0	14.0	21.0			
1.750	12.0	15.0	22.0	34.0	42.0	57.0
1.875	13.0	16.0	24.0			
2.000	14.0	17.0	25.0	38.0	48.0	65.0
2.250	16.0	19.0	29.0	43.0	54.0	73.0
2.500	18.0	21.0	32.0	48.0	60.0	81.0
2.750	19.0	23.0	35.0	53.0	66.0	89.0
3.000	21.0	25.0	38.0	58.0	72.0	97.0
3.250		28.0	41.0	62.0	77.0	105.0
3.500		31.0	44.0	67.0	83.0	114.0
3.750			48.0	72.0	89.0	122.0
4.000			51.0	77.0	95.0	130.0

FIGURE 13-60 *Spring pin data*

Retaining Rings

Retaining rings are precision-engineered fasteners. They provide removable shoulders for positioning or limiting the movement of parts in an assembly, *Figure 13-61*. Applications range from miniaturized electronic systems to massive earth-moving equipment. Retaining rings are used in automobiles, business machines, and complex components for guided missiles. They are found in such commonplace items as doorknobs and in sophisticated underwater seismic cable connectors.

Typical ring applications are shown in *Figures 13-62* through *13-66*. From these figures, the drafter may examine the design of various types of rings, determine their purpose and function, and decide how they may be used to the best advantage.

Most retaining rings are made of materials that have good spring properties. This permits the rings to be deformed elastically to a substantial degree, yet still spring back to their original shape during assembly and disassembly. This allows most rings to function in one of two ways: (1) they may be sprung into a groove or other recess in a part, or (2) they may be seated on a part in a deformed condition so that they grip the part by frictional means. In either case, the rings form a fixed shoulder against which other components may be abutted and prevented from moving.

Unlike wire-formed rings, which have a uniform section height, stamped rings have a tapered radial width. The width decreases symmetrically from the center section to the free ends. The tapered section permits the rings to remain circular after they have been compressed for insertion into a bore or expanded for assembly over a shaft. Most rings are designed to be seated in grooves. This constant circularity assures maximum contact surface with the bottom of the groove. It is also an important factor in achieving high static and dynamic thrust load capacities.

FIGURE 13-62 *Ring application in a precision differential* (Courtesy Koh-I-Noor Rapidograph)

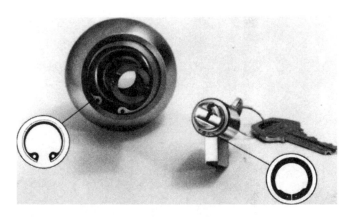

FIGURE 13-63 *Ring application in a cylindrical lockset* (Courtesy Koh-I-Noor Rapidograph)

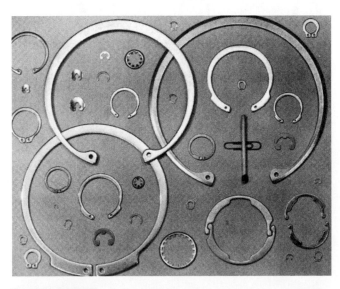

FIGURE 13-61 *Retaining rings* (Courtesy Koh-I-Noor Rapidograph)

FIGURE 13-64 *Ring application in an electromagnetic clutch brake* (Courtesy Koh-I-Noor Rapidograph)

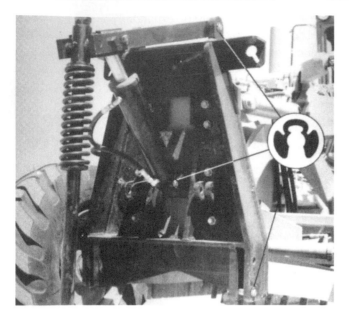

FIGURE 13-65 *Ring application in a road grader* (Courtesy Koh-I-Noor Rapidograph)

TYPES OF RETAINING RINGS

A great number of fastening requirements are involved in product design. This factor has led to the development of many different types of retaining rings. Standard rings are shown in *Figure 13-67*.

Limited space prevents describing in detail all the different ring types available. In general, however, retaining rings can be grouped into two major categories.

Internal rings for axial assembly are compressed for insertion into a bore or housing. They generally have a large gap and holes in the lugs, located at the free ends, for pliers which are used to grasp the rings securely during installation and removal, *Figure 13-68*.

External rings for axial assembly are expanded with pliers, *Figure 13-69*, so they may be slipped over the end of a shaft, stud, or similar part. They have a small gap, and the lug position is reversed from that of the internal ring. Radially assembled external rings do not have holes for pliers. Instead, the rings have a large gap and are pushed into the shaft directly in the plane of the groove with a special application tool, *Figure 13-70*.

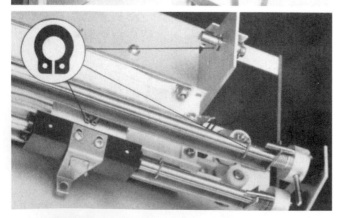

FIGURE 13-66 *Ring application in strip chart recorder* (Courtesy Koh-I-Noor Rapidograph)

Standard Truarc® Retaining Ring Series

BASIC *internal series* **N5000**	CRESCENT® *external series* **5103**	HEAVY DUTY *external series* **5160**
BOWED *internal series* **N5001**	CIRCULAR PUSH-ON *external series* **5105**	TRIANGULAR NUT *external series* **5300**
BEVELED *internal series* **N5002**	INTERLOCKING *external series* **5107**	KLIPRING® *external series* **5304** **T-5304**
CIRCULAR PUSH-ON *internal series* **5005**	INVERTED *external series* **5108**	TRIANGULAR PUSH-ON *external series* **5305**
INVERTED *internal series* **5008**	REINFORCED CIRCULAR PUSH-ON *external series* **5115**	GRIPRING® *external series* **5555** D5555•G5555
BASIC *external series* **5100**	BOWED E-RING *external series* **5131** X5131	MINIATURE HIGH-STRENGTH *external series* **5560**
BOWED *external series* **5101**	E-RING *external series* **5133** X5133•Y5133	PERMANENT SHOULDER *external series* **5590**
BEVELED *external series* **5102**	PRONG-LOCK® *external series* **5139**	PRECISION SUPPORT WASHER **5900**
	REINFORCED E-RING *external series* **5144**	

FIGURE 13-67 *Standard retaining rings series* (Courtesy Koh-I-Noor Rapidograph)

FIGURE 13-68 *Internal ring pliers* (Courtesy Koh-I-Noor Rapidograph)

FIGURE 13-69 *External ring pliers* (Courtesy Koh-I-Noor Rapidograph)

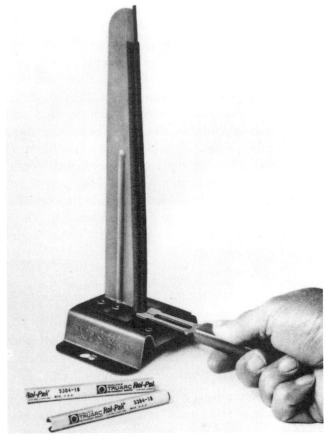

FIGURE 13-70 *Applicator and dispenser for retaining rings* (Courtesy Koh-I-Noor Rapidograph)

In addition to the tools shown here, retaining rings can be installed with automatic equipment. This equipment can be designed for specific, high-speed, automatic assembly lines. Retaining ring grooves serve two purposes: (1) they assure precise seating of the ring in the assembly, and (2) they permit the ring to withstand heavy thrust loads. The grooves must be located accurately and precut in the housing or shaft before the rings are assembled. Shaft grooves often can be made at no additional cost during the cut-off and chamfering operations.

Self-locking rings do not require any grooves because they exert a frictional hold against axial displacement. They are used mainly as positioning and locking devices where the ring will be subjected to only moderate or light loading.

Bowed rings differ from conventional types in that they are bowed around an axis perpendicular to the diameter bisecting the gap. The bowed construction permits the rings to function as springs as well as fasteners. This provides resilient end play take-up in the assembly.

Beveled rings have a 15° bevel on the groove-engaging edge. They are installed in grooves having a comparable bevel on the load-bearing wall. When the ring is seated in the groove, it acts as a wedge against the retained part. Sometimes play develops between the ring and retained part because of accumulated tolerances or wear in the assembly. If play develops, the spring action of the ring causes the fastener to seat more deeply in its groove and move in an axial direction, automatically taking up the end play. (Because self-locking rings can be seated at any point on a shaft or in a bore, they too can be used to compensate for tolerances and eliminate end play.)

MATERIALS AND FINISHES

As indicated previously, retaining rings are made of materials having good spring properties. Some also have high tensile and yield strengths. They must also have adequate ratio or ultimate tensile strength to elasticity. This permits the required deformation without too much permanent set. A ratio of 1:100 is satisfactory for most rings having the tapered-section design.

Standard material for most rings is carbon spring steel (SAE 1060-1090). For special applications, rings are also available in stainless steel (PH 15-7 Mo), beryllium copper (Alloy #25), and aluminum (Alclad 7075-T6). Rings are normally phosphate coated. Cadmium, zinc, and other platings and finishes are used for assemblies where extra corrosion resistance is needed or where the rings must withstand other unusual environmental conditions. Selection of the ring material and finish for a specific product design should be based upon the operating conditions under which the ring must function. These may include temperature, the presence of corrosive elements, thrust loads, and other factors.

SELECTION CONSIDERATIONS

Selecting the best ring for a product load capacity is a critical factor in some product designs. There are other factors the drafter should consider, however, before selecting specific ring types for a given product.

- Will there be adequate clearance to assemble the ring with pliers or other tools?

- Must the ring take up accumulated tolerances, either resiliently or rigidly?

- Is it possible to machine a ring groove on the shaft or inside a bore?

- Should the ring be adjustable to several positions on a shaft?

MAKING THE CHOICE

After considering all these conditions, the drafter may find that more than one ring type is suitable. How, then, can the drafter make the best choice?

The most important design criterion is the ability of the ring to do the fastening job required. Before a final selection of ring type is made, the drafter should consider savings that may be possible in various parts of the assembly. These include:

- The cost of installing the ring.

- Whether or not a groove is required.

- If the ring can be installed **permanently** or if it may have to be removed for field service.

A self-locking ring, for example, eliminates the need for the ring grooves. If the ring will be subjected to only moderate loading, this may be ideal for the assembly. But

most self-locking rings must be destroyed for removal. If field service is anticipated, another style of ring should be adopted.

The ideal ring is the one that will function adequately and provide the most economical means of fastening. Retaining rings are designed primarily as shoulders for positioning and retaining machine components on shafts and in housings and bores. Different rings have been developed and manufactured to meet specific fastening needs and problems.

To ensure correct selection of the proper type for any individual application, rings have been grouped according to their basic function. The selector guides, *Figures 13-71* and *13-72,* provide a visual index to all standard types.

Figures 13-73 and *13-74* (pages 564 through 567) are from the latest *Truarc Technical Manual.* Figure 13-73 is a sample of an internal series (N5000); Figure 13-74 is from an external series (5100).

EXAMPLE (REFER TO FIGURE 13-73):

A shaft of 1.000 (24.4) diameter would use a No. N5000-100.
- Free diameter = $1.111 \pm \begin{smallmatrix} .015 \\ .010 \end{smallmatrix}$

- Thickness = .042 ± .002
- Lug size = .155 ± .005
- Plier hole diameter = .062

The required groove must be:
- Diameter = 1.066 ± .003 (.004 TIR)

- Width = $.046 \pm \begin{smallmatrix} .003 \\ .000 \end{smallmatrix}$

EXAMPLE (REFER TO FIGURE 13-74):

A shaft of 4.000 (101.6) diameter would use a No. 5100-400.

- Free diameter = $3.700 \pm \begin{smallmatrix} .020 \\ .030 \end{smallmatrix}$

- Thickness = .109 ± .003
- Lug size = .352 ± .005

- Plier hole size = $.125 \pm \begin{smallmatrix} .015 \\ .002 \end{smallmatrix}$

The required groove must be:
- Diameter = 3.792 ± .006 (.006 TIR)

- Width = $.120 \pm \begin{smallmatrix} .005 \\ .000 \end{smallmatrix}$

(Text continues on page 568)

DESIGN FEATURES

RING TYPES FOR AXIAL ASSEMBLY

Series N5000, 5100: Tapered section assures constant circularity and groove pressure. Secure against heavy thrust loads and high rotational speeds.

Series 5008, 5108: Lugs inverted to abut groove bottom. Rings form high circular shoulder, concentric with bore or shaft. Good for parts having large corner radii or chamfers.

Series 5160: Heavy-duty ring resists high thrust, impact loads. Eliminates spacer washers in bearing assemblies.

Series 5560: New miniature, high-strength ring. Forms tamper-proof shoulder on small diameter shafts subject to heavy thrust loads.

Series 5590: Permanent-shoulder ring for small diameter shafts. When compressed into groove, notches deform to close gaps, reducing both I.D. and O.D.

RING TYPES FOR RADIAL ASSEMBLY

Series 5103: Forms narrow, uniformly concentric shoulder. Excellent for assemblies where clearance is limited.

Series 5133: Provides large shoulder on small diameter shafts. Installed in deep groove for added thrust capacity.

Series 5144: Reinforced to provide five times greater gripping strength, 50% higher rpm limits than conventional E-rings. Secure against rotation.

Series 5107: Two-part ring balanced to withstand high rpm's, heavy thrust loads, relative rotation between parts.

Series 5304: New high-strength ring for large bearing surface. Can be installed quickly with pliers or mallet, removed with ordinary screw driver.

Series T5304: Thinner model of 5304. Can be seated in same width grooves as E-rings, has more gripping power. Good for cast or molded grooves.

RING TYPES FOR TAKING UP END-PLAY

Series N5001, 5101: Bowed cylindrically to accommodate large tolerances, provide resilient end-play take-up.

Series N5002, 5102: Rings beveled 15° on groove-engaging edge for use in groove with similar bevel. Wedge action provides rigid end-play take-up.

Series 5131: Provides large shoulder on small diameter shafts. Bowed for resilient end-play take-up.

Series 5139: Bowed ring designed for use as shoulder against rotating parts. Prongs lock against shaft, prevent ring from being forced from groove.

SELF-LOCKING-TYPE RINGS (No groove required)

Series 5115: Push-on-type fastener for ungrooved shafts and studs. Has arched rim for extra strength, long prongs for wide shaft tolerances.

Series 5105, 5005: Flat rim, shorter prongs, smaller O.D. than 5115. For flat contact surface, better clearance.

Series 5555: Secure against axial displacement from either direction. No groove needed. Adjustable, reusable.

Series 5305: Dished body, three heavy prongs lock on shaft under spring tension. Withstands heavy thrust loads.

Series 5300: Free-spinning nut. Dished body flattens under torque, eliminating need for separate lock washers.

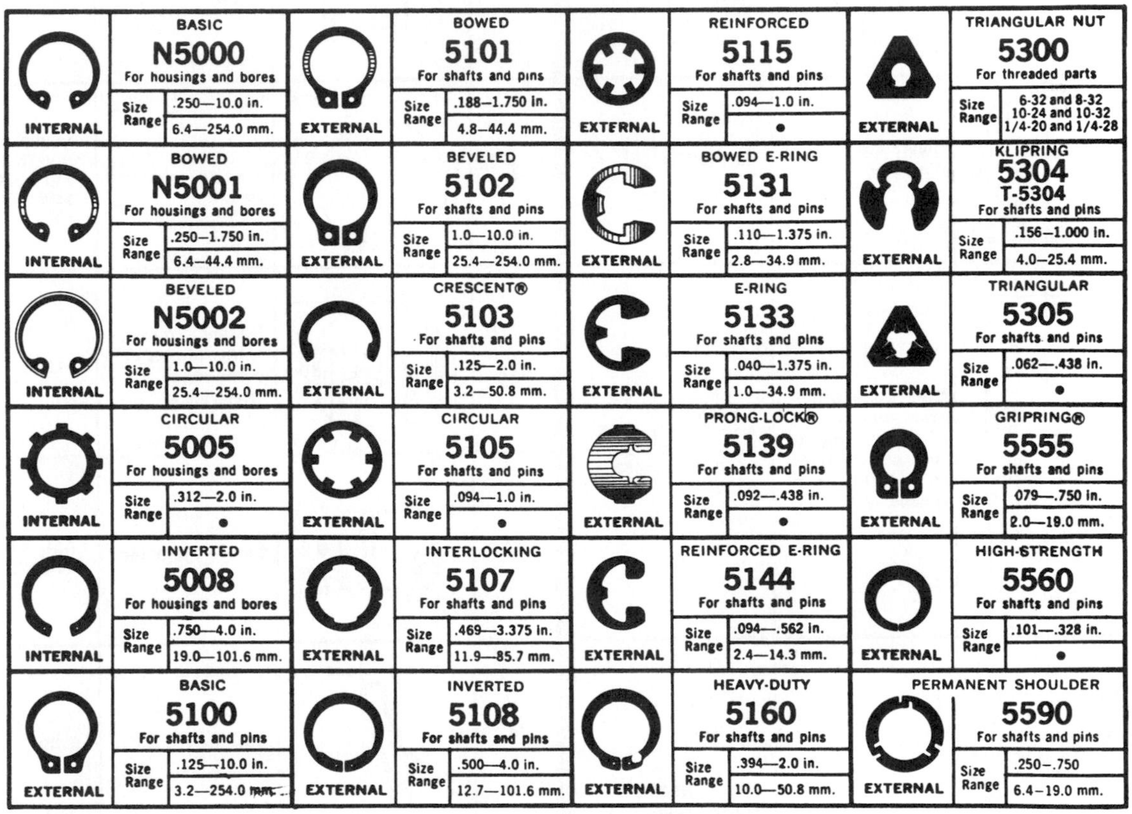

	BASIC **N5000** For housings and bores			BOWED **5101** For shafts and pins			REINFORCED **5115** For shafts and pins			TRIANGULAR NUT **5300** For threaded parts
INTERNAL	Size Range	.250—10.0 in. / 6.4—254.0 mm.	EXTERNAL	Size Range	.188—1.750 in. / 4.8—44.4 mm.	EXTERNAL	Size Range	.094—1.0 in. / ●	EXTERNAL	Size Range	6-32 and 8-32 / 10-24 and 10-32 / 1/4-20 and 1/4-28
	BOWED **N5001** For housings and bores			BEVELED **5102** For shafts and pins			BOWED E-RING **5131** For shafts and pins			KLIPRING **5304** T-5304 For shafts and pins
INTERNAL	Size Range	.250—1.750 in. / 6.4—44.4 mm.	EXTERNAL	Size Range	1.0—10.0 in. / 25.4—254.0 mm.	EXTERNAL	Size Range	.110—1.375 in. / 2.8—34.9 mm.	EXTERNAL	Size Range	.156—1.000 in. / 4.0—25.4 mm.
	BEVELED **N5002** For housings and bores			CRESCENT® **5103** For shafts and pins			E-RING **5133** For shafts and pins			TRIANGULAR **5305** For shafts and pins
INTERNAL	Size Range	1.0—10.0 in. / 25.4—254.0 mm.	EXTERNAL	Size Range	.125—2.0 in. / 3.2—50.8 mm.	EXTERNAL	Size Range	.040—1.375 in. / 1.0—34.9 mm.	EXTERNAL	Size Range	.062—.438 in. / ●
	CIRCULAR **5005** For housings and bores			CIRCULAR **5105** For shafts and pins			PRONG-LOCK® **5139** For shafts and pins			GRIPRING® **5555** For shafts and pins
INTERNAL	Size Range	.312—2.0 in. / ●	EXTERNAL	Size Range	.094—1.0 in. / ●	EXTERNAL	Size Range	.092—.438 in. / ●	EXTERNAL	Size Range	.079—.750 in. / 2.0—19.0 mm.
	INVERTED **5008** For housings and bores			INTERLOCKING **5107** For shafts and pins			REINFORCED E-RING **5144** For shafts and pins			HIGH-STRENGTH **5560** For shafts and pins
INTERNAL	Size Range	.750—4.0 in. / 19.0—101.6 mm.	EXTERNAL	Size Range	.469—3.375 in. / 11.9—85.7 mm.	EXTERNAL	Size Range	.094—.562 in. / 2.4—14.3 mm.	EXTERNAL	Size Range	.101—.328 in. / ●
	BASIC **5100** For shafts and pins			INVERTED **5108** For shafts and pins			HEAVY-DUTY **5160** For shafts and pins			PERMANENT SHOULDER **5590** For shafts and pins
EXTERNAL	Size Range	.125—10.0 in. / 3.2—254.0 mm.	EXTERNAL	Size Range	.500—4.0 in. / 12.7—101.6 mm.	EXTERNAL	Size Range	.394—2.0 in. / 10.0—50.8 mm.	EXTERNAL	Size Range	.250—.750 / 6.4—19.0 mm.

FIGURE 13-71 *Selector guide: Standard ring series* (Courtesy Koh-I-Noor Rapidograph)

The symbols listed below are used in the data charts for various ring types. Ring, groove, and retained part dimensions are in inches; allowable thrust loads are in pounds.

SYMBOL		DEFINITION	RING SERIES WHERE APPLICABLE	SYMBOL		DEFINITION	RING SERIES WHERE APPLICABLE
A		Minimum gap width: internal ring installed in groove	N5000, N5001, N5002, 5008	Ch max.		Maximum allowable chamfer height of retained part	N5000, N5001, N5002, 5008, 5100, 5101, 5102, 5103, 5107, 5108, 5131, 5133, 5144, 5160, 5555, 5560
B		Lug height	N5000, N5001, N5002, 5100, 5101, 5102, 5160, 5555	D		Free diameter	N5000, N5001, N5002, 5008, 5100, 5101, 5102, 5103, 5107, 5108, 5131, 5133, 5144, 5160, 5304, 5555, 5560, 5590
b	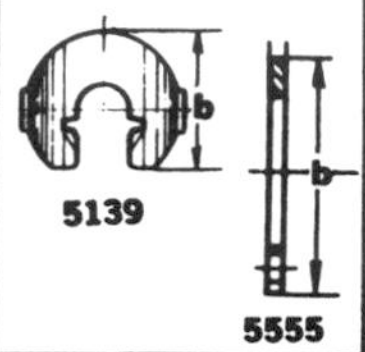	Ring height	5139, 5555, 5305, 5300, 5304	d		Nominal groove depth	ALL RINGS USED IN GROOVES
C	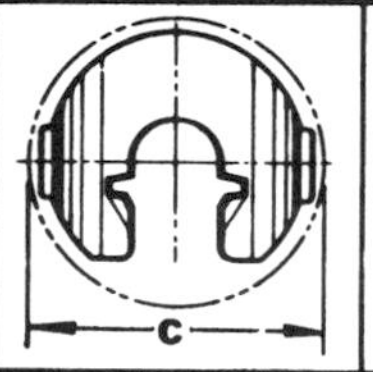	Clearance diameter	5139, 5590	E		Large section height	N5000, N5001, N5002, 5008, 5100, 5101, 5102, 5103, 5107, 5108, 5160, 5560
C₁	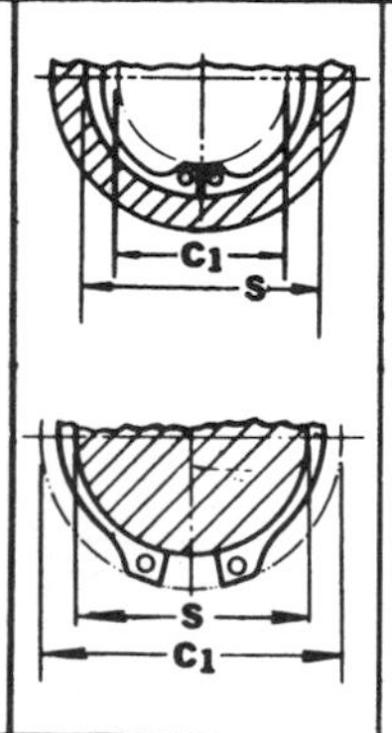	Clearance diameter: ring sprung into housing or over shaft, prior to installation in groove	N5000, N5001, N5002, 5008, 5100, 5101, 5102, 5108, 5160, 5555, 5560, 5590	e REF.		Distance from center of ring to outer edge (Reference)	5139
				G		Groove diameter	ALL RINGS USED IN GROOVES
C₂	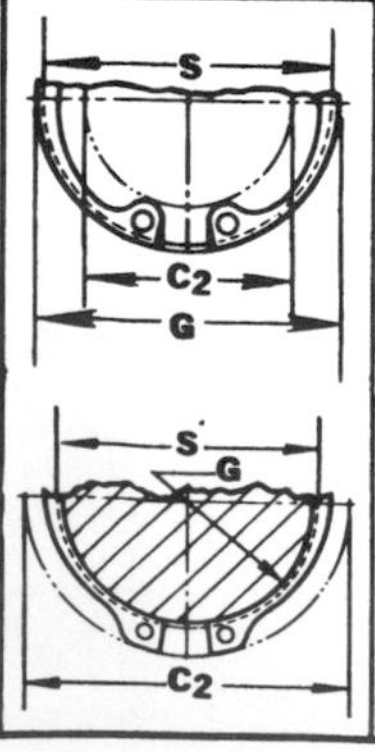	Clearance diameter: ring installed in groove	N5000, N5001, N5002, 5008, 5100, 5101, 5102, 5103, 5107, 5108, 5131, 5133, 5144, 5160, 5560	H	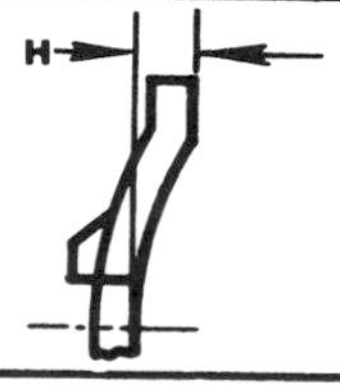	Bow height	5139
				h		Ring height	5115

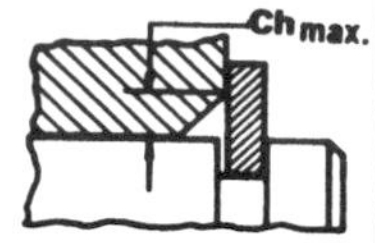

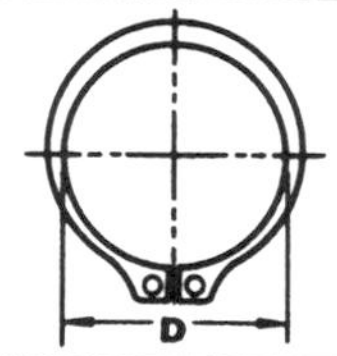

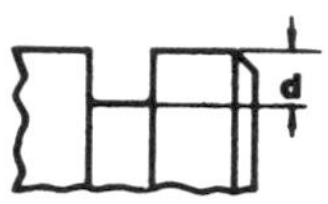

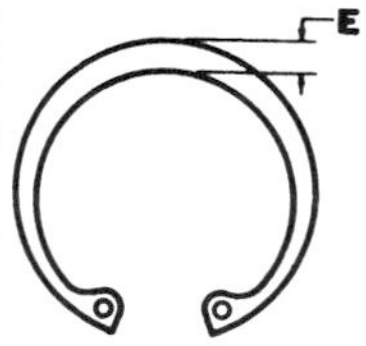

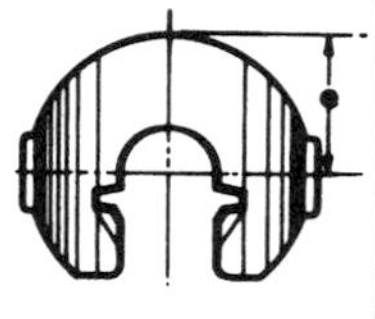

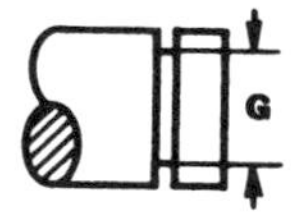

FIGURE 13-72 *Definition symbols* (Courtesy Koh-I-Noor Rapidograph)

SYMBOL		DEFINITION	RING SERIES WHERE APPLICABLE	SYMBOL		DEFINITION	RING SERIES WHERE APPLICABLE
J	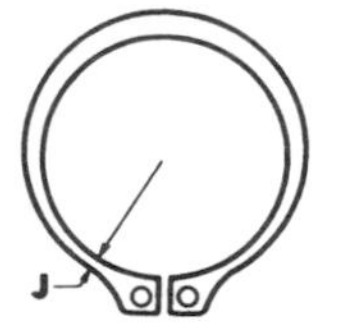	Small section height	N5000, N5001, N5002, 5008, 5100, 5101, 5102, 5108, 5160, 5560	S	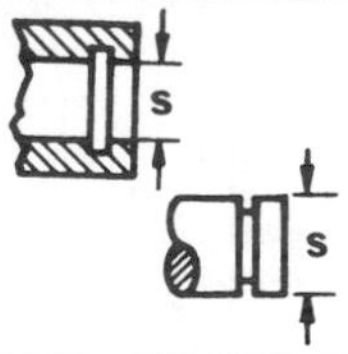	Shaft or housing diameter	ALL RINGS
K	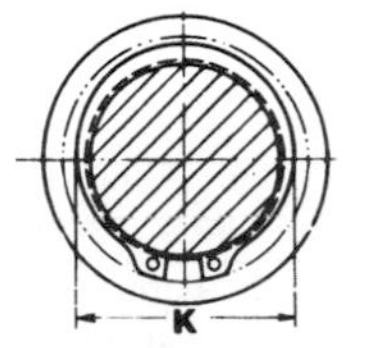	Maximum gaging diameter: ring installed in groove	5100, 5101, 5102, 5108, 5160, 5560	t		Ring thickness	ALL RINGS
k	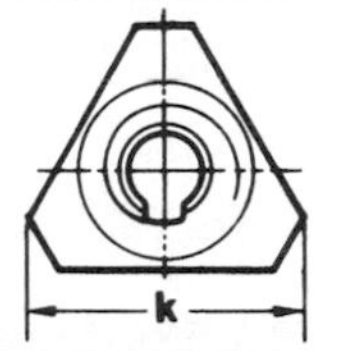	Overall ring width	5305, 5300	U	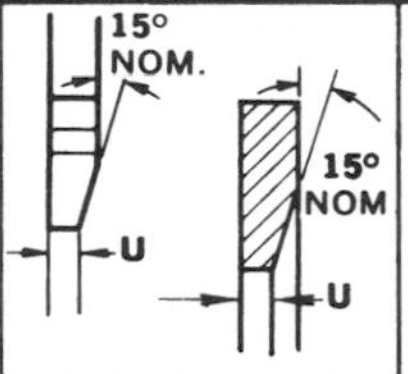	Ring thickness at beveled edge	N5002, 5102
L M	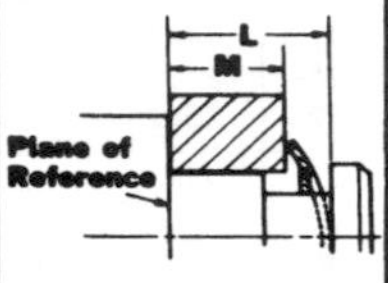	L: Location of outer groove wall from plane of reference M: Width of retained part	N5001, N5002, 5101, 5102, 5131, 5139	V	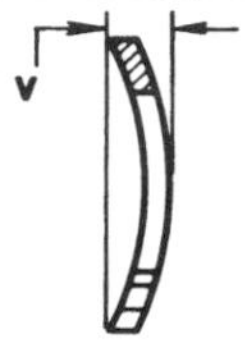	Overall bow height	N5001, 5101, 5131
P	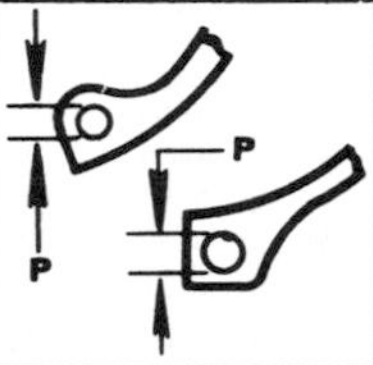	Pliers hole diameter	N5000, N5001, N5002, 5008, 5100, 5101, 5102, 5108, 5160, 5555	W	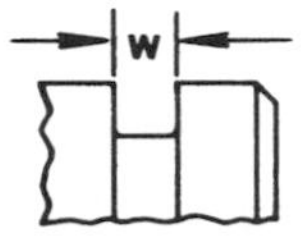	Groove width	ALL RINGS USED IN GROOVES
p	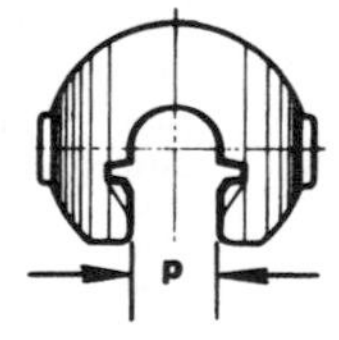	Gap width	5139	X	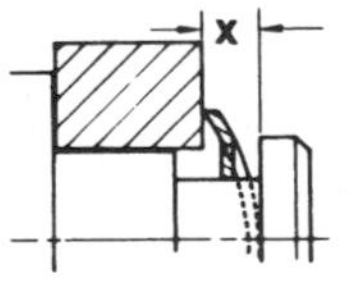	Distance from outer groove wall to face of retained part	N5001, 5101, 5131, 5139
P_r P'_r P_g	Allowable thrust load for ring (lbs.) Allowable assembly load with maximum corner radius or chamfer Allowable thrust load for groove (lbs.)		ALL RINGS USED IN GROOVES	Y	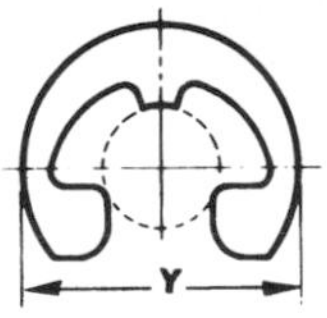	Free outside diameter	5103, 5133, 5144, 5131, 5304
R	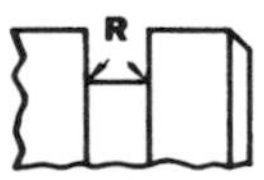	Radius of groove bottom	ALL RINGS USED IN GROOVES	Z	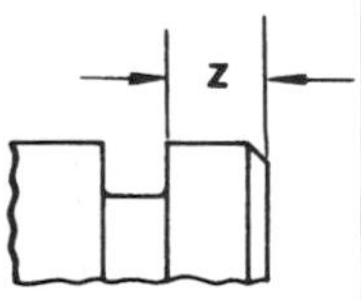	Edge margin	ALL RINGS USED IN GROOVES
$R_{max.}$	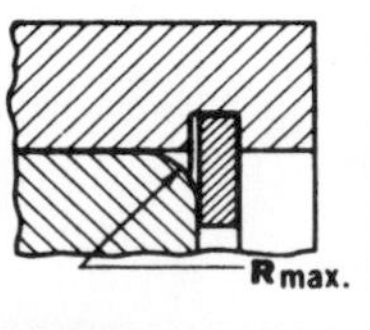	Maximum allowable corner radius of retained part	N5000, N5001, N5002, 5008, 5100, 5101, 5102, 5103, 5107, 5108, 5131, 5133, 5144, 5160, 5555, 5560	Z_1	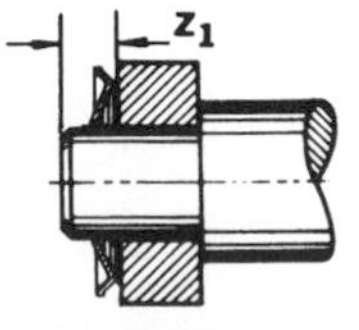	Minimum distance from face of retained part to end of shaft or housing	5115, 5005, 5105, 5305

FIGURE 13-72   *(Continued)*

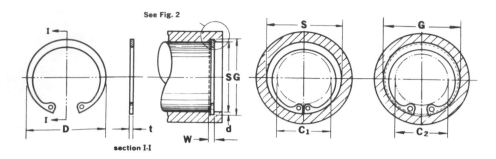

HOUSING DIA.			**INTERNAL SERIES N5000**	**TRUARC RING DIMENSIONS**					**GROOVE DIMENSIONS**					**APPLICATION DATA**			
				FREE DIA.		THICKNESS		Approx. weight per 1000 pieces	DIAMETER		WIDTH		Nominal groove depth	CLEARANCE DIAMETER		ALLOW. THRUST LOAD (lbs.) Sharp corner abutment	
														When sprung into housing S	When sprung into groove G	RINGS (Standard material) Safety factor = 4	GROOVES (Cold rolled steel bores and housings) Safety factor = 2
Dec. equiv. inch S	Approx. fract. equiv. inch S	Approx. mm S	size—no.	D	tol.	t	tol.	lbs.	G	tol.	W	tol.	d	C_1	C_2	P_r	P_g
.250	¼	6.4	N5000-25	.280		.015		.08	.268	±.001 .0015 T.I.R.	.018	+.002 −.000	.009	.115	.133	420	190
.312	5/16	7.9	N5000-31	.346		.015		.11	.330		.018		.009	.173	.191	530	240
.375	⅜	9.5	N5000-37	.415		.025		.25	.397	±.002	.029		.011	.204	.226	1050	350
.438	7/16	11.1	N5000-43	.482		.025		.37	.461	.002	.029		.012	.23	.254	1220	440
.453	29/64	11.5	N5000-45	.498		.025		.43	.477	T.I.R.	.029		.012	.25	.274	1280	460
.500	½	12.7	N5000-50	.548	+.010 −.005	.035		.70	.530		.039		.015	.26	.29	1980	510
.512	—	13.0	N5000-51	.560		.035		.77	.542	±.002 .004 T.I.R.	.039		.015	.27	.30	2030	520
.562	9/16	14.3	N5000-56	.620		.035		.86	.596		.039		.017	.275	.305	2220	710
.625	⅝	15.9	N5000-62	.694		.035		1.0	.665		.039		.020	.34	.38	2470	1050
.688	11/16	17.5	N5000-68	.763		.035		1.2	.732		.039	+.003 −.000	.022	.40	.44	2700	1280
.750	¾	19.0	N5000-75	.831		.035		1.3	.796		.039		.023	.45	.49	3000	1460
.777	—	19.7	N5000-77	.859		.042		1.7	.825		.046		.024	.475	.52	4550	1580
.812	13/16	20.6	N5000-81	.901		.042		1.9	.862		.046		.025	.49	.54	4800	1710
.866	—	22.0	N5000-86	.961		.042		2.0	.920		.046		.027	.54	.59	5100	1980
.875	⅞	22.2	N5000-87	.971	+.015 −.010	.042		2.1	.931	±.003 .004 T.I.R.	.046		.028	.545	.60	5150	2080
.901	—	22.9	N5000-90	1.000		.042	±.002	2.2	.959		.046		.029	.565	.62	5350	2200
.938	15/16	23.8	N5000-93	1.041		.042		2.4	1.000		.046		.031	.61	.67	5600	2450
1.000	1	25.4	N5000-100	1.111		.042		2.7	1.066		.046		.033	.665	.73	5950	2800
1.023	—	26.0	N5000-102	1.136		.042		2.8	1.091		.046		.034	.69	.755	6050	3000
1.062	1 1/16	27.0	N5000-106	1.180		.050		3.7	1.130		.056		.034	.685	.75	7450	3050
1.125	1⅛	28.6	N5000-112	1.249		.050		4.0	1.197		.056		.036	.745	.815	7900	3400
1.181	—	30.0	N5000-118	1.319		.050		4.3	1.255		.056		.037	.79	.86	8400	3700
1.188	1 3/16	30.2	N5000-118	1.319		.050		4.3	1.262		.056		.037	.80	.87	8400	3700
1.250	1¼	31.7	N5000-125	1.388		.050		4.8	1.330		.056		.040	.875	.955	8800	4250
1.259	—	32.0	N5000-125	1.388	+.025 −.020	.050		4.8	1.339		.056		.040	.885	.965	8800	4250
1.312	1 5/16	33.3	N5000-131	1.456		.050		5.0	1.396		.056		.042	.93	1.01	9300	4700
1.375	1⅜	34.9	N5000-137	1.526		.050		5.1	1.461	±.004 .005 T.I.R.	.056		.043	.99	1.07	9700	5050
1.378	—	35.0	N5000-137	1.526		.050		5.1	1.464		.056	+.004 −.000	.043	.99	1.07	9700	5050
1.438	1 7/16	36.5	N5000-143	1.596		.050		5.8	1.528		.056		.045	1.06	1.15	10200	5500
1.456	—	37.0	N5000-145	1.616		.050		6.4	1.548		.056		.046	1.08	1.17	10300	5700
1.500	1½	38.1	N5000-150	1.660		.050		6.5	1.594		.056		.047	1.12	1.21	10550	6000
1.562	1 9/16	39.7	N5000-156	1.734		.062		8.9	1.658		.068		.048	1.14	1.23	13700	6350
1.575	—	40.0	N5000-156	1.734		.062		8.9	1.671	±.005 T.I.R.	.068		.048	1.15	1.24	13700	6350
1.625	1⅝	41.3	N5000-162	1.804	+.035 −.025	.062	±.003	10.0	1.725		.068		.050	1.15	1.25	14200	6900
1.653	—	42.0	N5000-165	1.835		.062		10.4	1.755		.068		.051	1.17	1.27	14500	7200
1.688	1 11/16	42.9	N5000-168	1.874		.062		10.8	1.792		.068		.052	1.21	1.31	14800	7450

Notes (from column headers):

MIL-R-21248 / MS 16625

Thickness t applies only to unplated rings. For plated and stainless steel (Type H) rings, add .002" to the listed maximum thickness. Maximum ring thickness will be at least .0002" less than the listed minimum groove width (W).

T.I.R. (total indicator reading) is the maximum allowable deviation of concentricity between groove and housing.

FIGURE 13-73 *Internal series N5000 (Courtesy Koh-I-Noor Rapidograph)*

FIG. 1:
MAXIMUM ALLOWABLE
CORNER RADIUS (R_{max.})
AND CHAMFER (Ch_{max.})

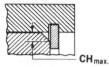

R_{max.}

CH_{max.}

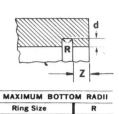

FIG. 2:
ENLARGED DETAIL
OF GROOVE PROFILE
AND EDGE MARGIN (Z)

MAXIMUM BOTTOM RADII	
Ring Size	R
-25 thru -100	.005
-102 thru -1000	.010

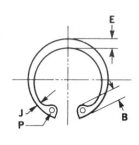

FIG. 3:
SUPPLEMENTARY
RING DIMENSIONS

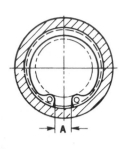

FIG. 4:
MINIMUM GAP WIDTH
(Ring installed in groove)

SUPPLEMENTARY APPLICATION DATA				SUPPLEMENTARY RING DIMENSIONS								
INTERNAL SERIES **N5000**	Maximum allowable corner radii and chamfers of retained parts (Fig. 1)		Allow. assembly load with R max. or Ch max.	Edge margin (Fig. 2)	LUG		LARGE SECTION		SMALL SECTION		HOLE DIAMETER		MIN. GAP WIDTH (Fig. 4) Ring installed in groove
size—no.	R_{max.}	Ch_{max.}	P'_{r(lbs.)}	Z	B	tol.	E	tol.	J	tol.	P	tol.	A
N5000-25	.011	.0085	190	.027	.065		.025	±.002	.015	±.002	.031		.047
N5000-31	.016	.013	190	.027	.066		.033		.018		.031		.055
N5000-37	.023	.018	530	.033	.082		.040		.028		.041		.063
N5000-43	.027	.021	530	.036	.098	±.003	.049	±.003	.029	±.003	.041		.063
N5000-45	.027	.021	530	.036	.098		.050		.030		.047		.071
N5000-50	.027	.021	1100	.045	.114		.053		.035		.047		.090
N5000-51	.027	.021	1100	.045	.114		.053		.035		.047		.092
N5000-56	.027	.021	1100	.051	.132		.053		.035		.047		.095
N5000-62	.027	.021	⊦100	.060	.132		.060	±.004	.035	±.004	.062	+.010 −.002	.104
N5000-68	.027	.021	1100	.066	.132		.063		.036		.062		.118
N5000-75	.032	.025	1100	.069	.142		.070		.040		.062		.143
N5000-77	.035	.028	1650	.072	.146		.074		.044		.062		.145
N5000-81	.035	.028	1650	.075	.155		.077		.044		.062		.153
N5000-86	.035	.028	1650	.081	.155		.081		.045		.062		.172
N5000-87	.035	.028	1650	.084	.155		.084	±.005	.045	±.005	.062		.179
N5000-90	.038	.030	1650	.087	.155		.087		.047		.062		.188
N5000-93	.038	.030	1650	.093	.155		.091		.050		.062		.200
N5000-100	.042	.034	1650	.099	.155		.104		.052		.062		.212
N5000-102	.042	.034	1650	.102	.155		.106		.054		.062		.220
N5000-106	.044	.035	2400	.102	.180		.110		.055		.078		.213
N5000-112	.047	.036	2400	.108	.180	±.005	.116		.057		.078		.232
(S=1.181) N5000-118	.047	.036	2400	.111	.180		.120		.058		.078		.226
(S=1.188) N5000-118	.047	.036	2400	.111	.180		.120		.058		.078		.245
(S=1.250) N5000-125	.048	.038	2400	.120	.180		.124		.062		.078		.265
(S=1.259) N5000-125	.048	.038	2400	.120	.180		.124	±.006	.062	±.006	.078		.290
N5000-131	.048	.038	2400	.126	.180		.130		.062		.078		.284
(S=1.375) N5000-137	.048	.038	2400	.129	.180		.130		.063		.078	+.015 −.002	.297
(S=1.378) N5000-137	.048	.038	2400	.129	.180		.130		.063		.078		.305
N5000-143	.048	.038	2400	.135	.180		.133		.065		.078		.313
N5000-145	.048	.038	2400	.138	.180		.133		.065		.078		.320
N5000-150	.048	.038	2400	.141	.180		.133		.066		.078		.340
(S=1.562) N5000-156	.064	.050	3900	.144	.202		.157		.078		.078		.338
(S=1.575) N5000-156	.064	.050	3900	.144	.202		.157		.078		.078		.374
N5000-162	.064	.050	3900	.150	.227		.164	±.007	.082	±.007	.078		.339
N5000-165	.064	.050	3900	.153	.227		.167		.083		.078		.348
N5000-168	.064	.050	3900	.156	.227		.170		.085		.078		.357

FIGURE 13-73 *(Continued)*

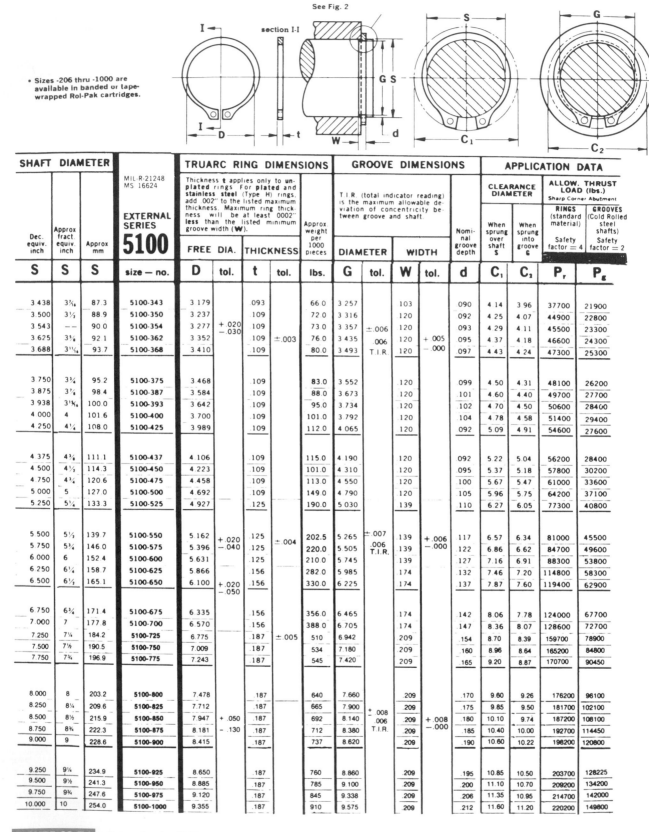

- Sizes -206 thru -1000 are available in banded or tape-wrapped Rol-Pak cartridges.

TRUARC RING DIMENSIONS — Thickness **t** applies only to unplated rings. For **plated and stainless steel** (Type H) rings, add .002" to the listed maximum thickness. Maximum ring thickness will be at least .0002" **less** than the listed minimum groove width (**W**).

GROOVE DIMENSIONS — T.I.R. (total indicator reading) is the maximum allowable deviation of concentricity between groove and shaft.

APPLICATION DATA

SHAFT DIAMETER			MIL-R-21248 MS 16624 EXTERNAL SERIES **5100**	FREE DIA.		THICKNESS		Approx weight per 1000 pieces	DIAMETER		WIDTH		Nominal groove depth	CLEARANCE DIAMETER		RINGS (standard material) Safety factor = 4	GROOVES (Cold Rolled steel shafts) Safety factor = 2
Dec. equiv. inch **S**	Approx fract. equiv. inch **S**	Approx mm **S**	size — no.	**D**	tol.	**t**	tol.	lbs.	**G**	tol.	**W**	tol.	**d**	When sprung over shaft **C₁**	When sprung into groove **C₂**	**Pᵣ**	**P₉**
3.438	3 7/16	87.3	5100-343	3.179	+.020 −.030	.093	±.003	66.0	3.257	±.006 .006 T.I.R.	103	+.005 −.000	.090	4.14	3.96	37700	21900
3.500	3½	88.9	5100-350	3.237		.109		72.0	3.316		120		.092	4.25	4.07	44900	22800
3.543	--	90.0	5100-354	3.277		.109		73.0	3.357		120		.093	4.29	4.11	45500	23300
3.625	3⅝	92.1	5100-362	3.352		.109		76.0	3.435		120		.095	4.37	4.18	46600	24300
3.688	3 11/16	93.7	5100-368	3.410		.109		80.0	3.493		120		.097	4.43	4.24	47300	25300
3.750	3¾	95.2	5100-375	3.468		.109		83.0	3.552		120		.099	4.50	4.31	48100	26200
3.875	3⅞	98.4	5100-387	3.584		.109		88.0	3.673		120		.101	4.60	4.40	49700	27700
3.938	3 15/16	100.0	5100-393	3.642		.109		95.0	3.734		120		.102	4.70	4.50	50600	28400
4.000	4	101.6	5100-400	3.700		.109		101.0	3.792		120		.104	4.78	4.58	51400	29400
4.250	4¼	108.0	5100-425	3.989		.109		112.0	4.065		120		.092	5.09	4.91	54600	27600
4.375	4⅜	111.1	5100-437	4.106		.109		115.0	4.190		120		.092	5.22	5.04	56200	28400
4.500	4½	114.3	5100-450	4.223		.109		101.0	4.310		120		.095	5.37	5.18	57800	30200
4.750	4¾	120.6	5100-475	4.458		.109		113.0	4.550		120		.100	5.67	5.47	61000	33600
5.000	5	127.0	5100-500	4.692		.109		149.0	4.790		120		.105	5.96	5.75	64200	37100
5.250	5¼	133.3	5100-525	4.927		.125		190.0	5.030		139		.110	6.27	6.05	77300	40800
5.500	5½	139.7	5100-550	5.162	+.020 −.040	.125	±.004	202.5	5.265	±.007 .006 T.I.R.	139	+.006 −.000	.117	6.57	6.34	81000	45500
5.750	5¾	146.0	5100-575	5.396		.125		220.0	5.505		139		.122	6.86	6.62	84700	49600
6.000	6	152.4	5100-600	5.631		.125		210.0	5.745		139		.127	7.16	6.91	88300	53800
6.250	6¼	158.7	5100-625	5.866		.156		282.0	5.985		174		.132	7.46	7.20	114800	58300
6.500	6½	165.1	5100-650	6.100	+.020 −.050	.156		330.0	6.225		174		.137	7.87	7.60	119400	62900
6.750	6¾	171.4	5100-675	6.335		.156		356.0	6.465		174		.142	8.06	7.78	124000	67700
7.000	7	177.8	5100-700	6.570		.156		388.0	6.705		174		.147	8.36	8.07	128600	72700
7.250	7¼	184.2	5100-725	6.775		.187	±.005	510	6.942		209		.154	8.70	8.39	159700	78900
7.500	7½	190.5	5100-750	7.009		.187		534	7.180		209		.160	8.96	8.64	165200	84800
7.750	7¾	196.9	5100-775	7.243		.187		545	7.420		209		.165	9.20	8.87	170700	90450
8.000	8	203.2	5100-800	7.478	+.050 −.130	.187		640	7.660	+.008 .006 T.I.R.	209	+.008 −.000	.170	9.60	9.26	176200	96100
8.250	8¼	209.6	5100-825	7.712		.187		665	7.900		209		.175	9.85	9.50	181700	102100
8.500	8½	215.9	5100-850	7.947		.187		692	8.140		209		.180	10.10	9.74	187200	108100
8.750	8¾	222.3	5100-875	8.181		.187		712	8.380		209		.185	10.40	10.00	192700	114450
9.000	9	228.6	5100-900	8.415		.187		737	8.620		209		.190	10.60	10.22	198200	120800
9.250	9¼	234.9	5100-925	8.650		.187		760	8.860		209		.195	10.85	10.50	203700	128225
9.500	9½	241.3	5100-950	8.885		.187		785	9.100		209		.200	11.10	10.70	209200	134200
9.750	9¾	247.6	5100-975	9.120		.187		845	9.338		209		.206	11.35	10.95	214700	142000
10.000	10	254.0	5100-1000	9.355		.187		910	9.575		209		.212	11.60	11.20	220200	149800

FIGURE 13-74 *External series 5100* (Courtesy Koh-I-Noor Rapidograph)

FIG. 1:
MAXIMUM ALLOWABLE CORNER RADIUS (R_max.) AND CHAMFER (Ch_max.)

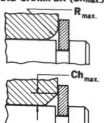

FIG. 2:
ENLARGED DETAIL OF GROOVE PROFILE AND EDGE MARGIN (Z)

MAXIMUM BOTTOM RADII	
Ring size	R
-12 thru -23	Sharp corners
-25 thru -35	.003
-37 thru -100	.005
-102 thru -200	.010

FIG. 3:
SUPPLEMENTARY RING DIMENSIONS

FIG. 4:
MAXIMUM GAGING DIAMETER
(Ring installed in groove)

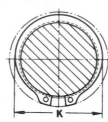

FIG. 5:
LUG DESIGN
Sizes -12 thru -23

	SUPPLEMENTARY APPLICATION DATA					SUPPLEMENTARY RING DIMENSIONS								
EXTERNAL SERIES **5100**	Maximum allowable corner radii and chamfers of retained parts (Fig. 1)		Allow. assembly load with R max. or Ch_max.	Edge margin (Fig. 2)	Calculated RPM limits (Std. ring mat'l.) Apply req'd. safety factor	(Fig. 3)								MAX. GAGING DIA. Ring installed in groove (Fig. 4)
						LUG		LARGE SECTION		SMALL SECTION		HOLE DIAMETER		
size — no.	R_max.	Ch_max.	P'_r (lbs.)	Z		B	tol.	E	tol.	J	tol.	P	tol.	K
5100-343	.129	.077	7350	.270	5900	.308		.292		.148		.125		3.712
5100-350	.122	.073	10500	.276	5900	.328		.285		.148		.125		3.764
5100-354	.123	.074	10500	.279	5800	.328	±.005	.288	±.008	.149	±.008	.125		3.809
5100-362	.127	.076	10500	.285	5700	.328		.296		.153		.125		3.898
5100-368	.1295	.078	10500	.291	5600	.330		.302		.156		.125	+.015 −.002	3.966
5100-375	.133	.080	10500	.297	5500	.332		.310		.160		.125		4.037
5100-387	.137	.082	10500	.303	5100	.330		.318		.163		.125		4.169
5100-393	.137	.082	10500	.306	5200	.342		.318		.163		.125		4.230
5100-400	.135	.081	10500	.312	5000	.352		.318		.163		.125		4.288
5100-425	.146	.088	10500	.276	4800	.395		.318		.176		.125		4.558
5100-437	.146	.088	10500	.276	4700	.395		.318		.181		.125		4.683
5100-450	.102	.061	10500	.285	4500	.404		.285		.128		.125		4.730
5100-475	.115	.069	10500	.300	4200	.429		.303		.136		.125		4.996
5100-500	.165	.099	10500	.315	4000	.450	±.008	.360	±.010	.194	±.010	.156		5.346
5100-525	.169	.101	13500	.330	3900	.472		.372		.211		.156		5.605
5100-550	.175	.105	13500	.351	3700	.497		.390		.209		.156		5.867
5100-575	.184	.110	13500	.366	3500	.518		.408		.220		.156		6.134
5100-600	.143	.086	13500	.381	3400	.540		.381		.171		.156		6.302
5100-625	.148	.089	21000	.396	3100	.561		.396		.176		.156		6.568
5100-650	.191	.114	21000	.411	3000	.586		.438		.236		.156		6.905
5100-675	.200	.120	21000	.426	3000	.608		.456		.246		.187		7.172
5100-700	.208	.125	21000	.441	2900	.629		.474		.256		.187		7.439
5100-725	.214	.128	30000	.460	2800	.660		.490		.267		.187		7.700
5100-750	.220	.132	30000	.480	2700	.676		.507		.277		.187		7.963
5100-775	.227	.136	30000	.495	2600	.660		.523		.285		.187		8.228
5100-800	.235	.141	30000	.510	2500	.735		.540		.294		.187		8.493
5100-825	.242	.146	30000	.525	2400	.735		.556		.304		.187		8.758
5100-850	.250	.150	30000	.540	2300	.735	±.012	.573	±.015	.314	±.015	.187	+.020 −.005	9.023
5100-875	.258	.155	30000	.555	2200	.735		.591		.322		.187		9.280
5100-900	.267	.160	30000	.570	2200	.735		.609		.333		.187		9.557
5100-925	.274	.164	30000	.585	2100	.735		.625		.341		.187		9.830
5100-950	.281	.168	30000	.600	2100	.735		.642		.350		.187		10.086
5100-975	.287	.172	30000	.618	2000	.735		.658		.358		.187		10.340
5100-1000	.294	.176	30000	.636	2000	.735		.675		.367		.187		10.610

FIGURE 13-74 *(Continued)*

Summary

- Threads are used for four applications: (1) to fasten parts together, (2) for fine adjustments between parts, (3) for fine measurements, and (4) to transmit motion or power.

- Screw thread forms include the unified national, ISO, metric, square, acme worm, knuckle, and buttress.

- Inside and outside threads are produced using taps and dies. Taps are used to cut internal threads; dies are used to cut external threads.

- Pitch is the distance from one point on a thread to the corresponding point on the adjacent thread as measured parallel to its axis.

- A single thread is composed of one continuous ridge. Lead is the distance a screw thread advances axially in one full turn. Multiple threads are made up of two or more continuous ridges side-by-side.

- A right-hand thread rotates in a clockwise direction. A left-hand thread rotates in a counterclockwise direction.

- Threads may be represented graphically in the following ways: schematic or simplified.

- A screw is a threaded fastener that does not use a nut and is screwed directly into a part. A bolt is a threaded fastener that passes directly through parts to hold them together and uses a nut. A stud is a fastener that is a steel rod with threads at both ends.

- Rivets are permanent fasteners, usually used to hold sheet metal together.

- A key is a demountable part that provides a positive means of transferring torque between a shaft and a hub.

- Grooved fasteners are used to solve metal-to-metal pinning needs with shear application.

- Spring pins are manufactured by cold-forming strip metal in a progressive roll-forming operation. They are slightly larger in diameter than the hole into which they are inserted.

- Retaining rings are precision-engineered fasteners that provide removable shoulders for positioning or limiting movement in an assembly.

Review Questions

Answer the following questions either true or false.

1. Two other uses for threads besides fastening things together are adjusting parts and measuring and transmitting power.

2. If you have a 1/4 – 20 UNC threaded screw and rotate it 10 full turns, the end will travel 1/2 inch.

3. A 5115 series retaining ring would be recommended to provide resilient end play take-up.

4. The free diameter of an external 5100-775 retaining ring is 7.243.

5. The depth of a thread is figured by measuring the distance between the crest and the root of the thread, as measured at a right angle to the axis.

Answer the following questions by selecting the best answer.

1. What #5100 series retaining ring would be recommended for a shaft diameter of 203.2 mm?
   a. 5100-343
   b. 5100-437
   c. 5100-675
   d. 5100-800

2. What is **not** a factor to be considered before selecting any type of retaining ring, other than load capacity?
   a. Lug size
   b. Adequate clearance
   c. The cost of installing the ring
   d. Whether or not a groove is required

3. What does ⅝ – 11 UNC – 2A × 2½ LG mean?
   a. ⅝″ long, right-hand thread, average fit
   b. ⅝″ nominal size, 11 TPI, Unified National Series-Coarse, average fit, external thread, 2½″ long
   c. ⅝″ diameter, 11″ long, coarse series
   d. 11 fit, coarse × 2½″ long × ⅝″ diameter

4. What is the recommended hole size for a .312″ (8) diameter spring pin?
   a. Min .312, Max .318
   b. Min .321, Max .328
   c. Min .322, Max .325
   d. Min .315, Max .318

5. A screw that is 3″ long, ⅜″ in diameter, 15 threads per inch, unified coarse, right-hand threads, average fit, and rounded-head style cap screw is called out by:
   a. ⅜ × 15 UNC – RDH Cap Screw × 3.0 LG
   b. RDH Cap Screw – 15 UNC ⅜ × 3.0 LG
   c. ⅜ – 15 UNC – 2A × 3.0 LG RDH
   d. 3″ Long Round Head Cap Screw (⅜″ diameter, coarse)

6. The housing diameter for an internal series N5000-25 ring is:
   a. .688
   b. .453
   c. .250
   d. .875

7. Which statement is **not** true of a series 5115 self-locking ring?
   a. Is used for flat contact surfaces
   b. Withstands heavy thrust loads
   c. Should not be used with ungrooved shafts
   d. Has arched rim for extra strength

Chapter 13 Problems

The following problems are intended to give the beginning drafter practice using various size charts of fastener and, by using these charts, practice in laying out and drawing the many fasteners used.

CONVENTIONAL INSTRUCTIONS

The steps to follow in laying out fasteners are:

STEP 1 Study all required specifications.

STEP 2 Using the appropriate size chart for size dimension, lightly lay out fastener in place.

STEP 3 Check each dimension.

STEP 4 Darken in the fastener, starting with diameters and arcs first to correct line thickness. Use simplified thread representation.

STEP 5 Neatly add all required call-off specifications, using the latest drafting standard.

CAD Instructions

Students who plan to complete the following problems on a CAD system should begin with these general instructions. Before reading the specific instructions for each activity, go through each step in the following planning checklist. The checklist applies to any CAD system and will help ensure the optimum use of your time and resources.

1. Analyze the problem carefully. Decide exactly what you are being asked to do.

2. Determine what resources and references (if any) you will need in order to complete the problem. Collect those references and resources and have them readily available.

3. Decide if any particular standards apply to the project and have those standards available.

4. Determine what types of views will be required and how many of each.

5. Determine what the final plotted scale of the drawing will need to be, and select the appropriate paper size for plotting/printing.

6. Plan your drawing sequence. In what order will you develop the drawing (i.e., lines, features, dimension lines, leaders, dimensions)?

7. Review the various CAD commands you will have to use in order to develop the drawing.

8. Examine your CAD system to ensure that everything is in working order, then begin the project.

Problem 13-1

On an A-size, 8½ × 11 sheet of paper, lay out the center lines per the given dimensions.

Calculate standard thread lengths.

Draw the following fasteners on the center lines as illustrated.

A. Flathead cap screw—⅜–16 UNC–2A × 2.0 lg.

B. Roundhead cap screw—1/2–13 UNC–2A × 2.25 lg.

C. Fillister-head cap screw—⅝–11 UNC–2A × 2.5 lg.

D. Oval-head cap screw—⅜–24 UNF–2A × 1.75 lg.

E. Hex-head cap screw with hex nut—1–8 UNC–2A × 2.5 lg.

F. Socket-head cap screw—½–20 UNF–2A × 2.0 lg.

G. Cotter pin—⅜ dia. × 3.5 lg.

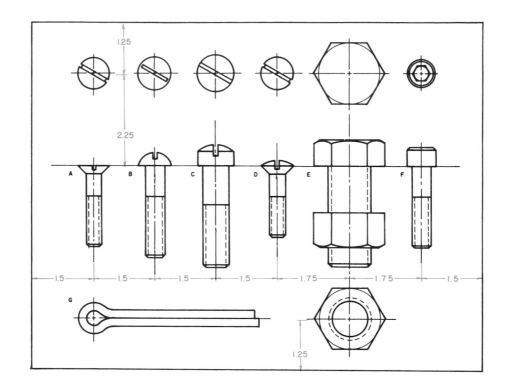

Problem 13-2

On an A-size, 8½ × 11 sheet of paper, lay out the center lines per the given dimensions.

Calculate standard thread lengths.

Draw the following fasteners on the center lines as illustrated.

A. Hex-head bolt—1⅜–UNC × 3.0 lg.
B. Lock washer (for a ⅞ dia. cap screw).
C. Square nut—1–12 UNF 2B.
D. Square-head cap screw—¾–10 UNC × 3.0 lg.

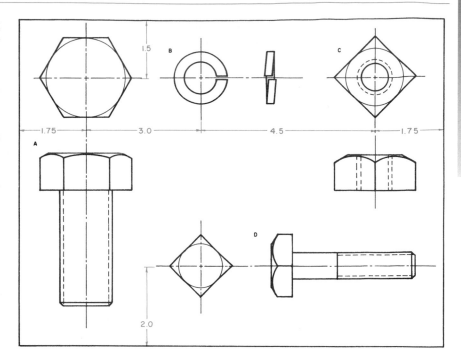

Problem 13-3

On an A-size, 8½ × 11 sheet of paper, lay out the center lines per the given dimensions.

A. Calculate the required size of square key needed for a 2.25 ID collar. Draw collar and keyway, and dimension per drafting standards.

B. Calculate the required size of square key needed for the matching shaft of the collar. Draw the end view of the shaft and dimension per drafting standards.

C. Calculate the **minimum** length hex-head cap screw, ¾–10 UNC –2A required to safely secure part X to part Y (material: aluminum). Illustrate and call off the required pilot drill size and recommended depth into part Y. Illustrate and call off the required size and depth of **full** threads for the hex-head cap screw.

Illustrate and call off the diameter for clearance hole in part X.

D. Draw a square-head set screw with a full dog point, ⅝–18 UNF × 3.0 lg.

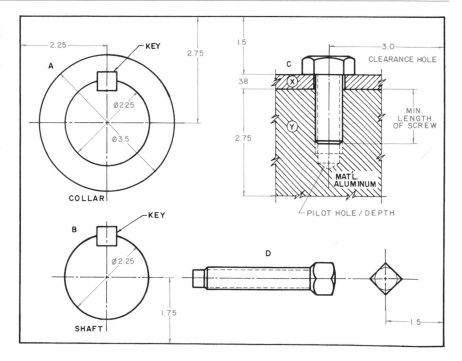

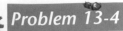

Problem 13-4

On an A-size, 8½ × 11 sheet of paper, lay out the center lines, part X, and part Y per the given dimensions.

Calculate the standard thread lengths.

Draw the following fasteners on the center lines as illustrated.

Give **all** specific hole call-off information for each fastener in the space above each as illustrated. Include clearance hole sizes, counterbore specifications, countersink specifications, spotface specifications, required tap drill size and depth for 75% thread; and tap size and required depth per the latest drafting standards.

A. Socket-head cap screw—⅞–9 UNC × 1.75 lg (calculate counterbore information).
B. Oval-head cap screw—¾–10 UNC × 2.5 lg (calculate countersink information).
C. Hex-head cap screw—1–12 UNF × 2.5 lg (calculate spotface information).
D. Fillister-head cap screw with plain washer and hex-head nut—4–16 UNF × length to suit. (Calculate clearance hole required.)

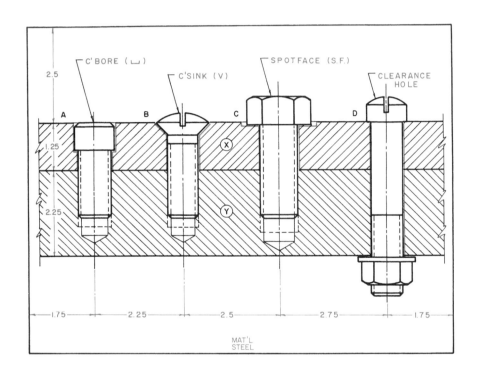

Problem 13-5

Given; an illustration of an old design (Figure 13-5A) that uses a washer and cotter pin to hold the .3543 diameter shaft in place. Using this design, it was necessary to locate and drill a hole for the cotter pin. The new design incorporated the use of one retaining ring to achieve the same function at a much lower cost. Choose the correct type and size retaining ring and fill in the required dimensions at A, B, and C (see Figure 13-6A). Neatly letter all retaining ring data below the dimensioned shaft as illustrated.

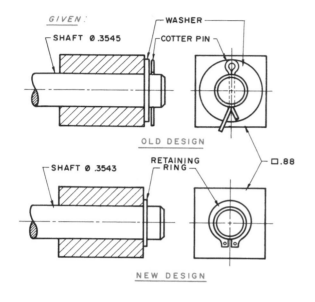

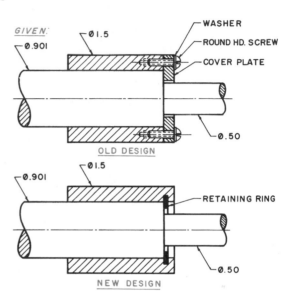

Problem 13-6

Given; an illustration of an old design (Figure 13-5B) that uses a cover plate with 4 holes for 4 washers and 4 round-head machine screws to hold the shaft in place. Using this design required drilling 4 blind tap-drill holes that had to be tapped for the 4 round-head screws. The new design incorporated the use of one retaining ring to achieve the same function at a much lower cost. Choose the correct type and size retaining ring and fill in the required dimensions at A, B, and C (see Figure 13-6B). Neatly print all retaining ring data below the dimensioned collar as illustrated.

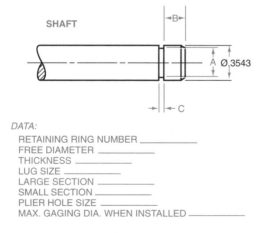

DATA:
RETAINING RING NUMBER _____
FREE DIAMETER _____
THICKNESS _____
LUG SIZE _____
LARGE SECTION _____
SMALL SECTION _____
PLIER HOLE SIZE _____
MAX. GAGING DIA. WHEN INSTALLED _____

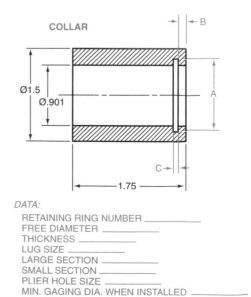

DATA:
RETAINING RING NUMBER _____
FREE DIAMETER _____
THICKNESS _____
LUG SIZE _____
LARGE SECTION _____
SMALL SECTION _____
PLIER HOLE SIZE _____
MIN. GAGING DIA. WHEN INSTALLED _____

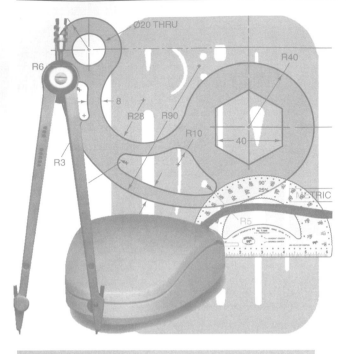

SPRINGS

KEY TERMS

Active coils	Inside diameter
Coil (turn)	Loaded length
Compression spring	Mean diameter
Direction of winding	Outside diameter
Extension spring	Solid length
Flat spring	Spring
Free length	Torsion spring
Helical spring	Wire size

CHAPTER OUTLINE

Spring classification • Helical springs • Flat springs • Terminology of springs • Required spring data • Other spring design layout • Standard drafting practices • Section view of a spring • Isometric views of a spring • Summary • Review questions • Chapter 14 problems

CHAPTER OBJECTIVES

Upon completion of this chapter, students should be able to do the following:

■ Explain the classifications of springs.

■ Describe the three different kinds of helical springs and their functions.

■ Define the function of flat springs.

■ Define the various terms associated with springs and their design.

■ List the spring data that should accompany spring drawings.

■ Demonstrate proficiency in drawing the various types of springs.

■ Demonstrate proficiency in applying standard drafting practices when drawing springs.

■ Demonstrate proficiency in drawing sectional views of springs.

■ Demonstrate proficiency in drawing isometric views of springs.

A *spring* is a mechanical device that is used to store and apply mechanical energy. A spring can be designed and manufactured to apply a pushing action, a pulling action, a torque or twisting action, or a simple power action.

Product manufacturers usually use standard-size springs that are purchased from companies that specialize in making springs. These springs are usually mass produced and are, therefore, relatively inexpensive. Occasionally, however, a special spring must be designed to perform a special function. To be able to design special-function springs, the drafter or designer must know the many terms associated with springs and how to design and construct special springs.

Spring Classification

A spring is generally classified as either a helical spring or a flat spring. *Helical springs* are usually cylindrical or conical; the *flat springs*, as their name implies, are usually flat.

Helical Springs

Three types of helical springs are compression springs, extension springs, and torsion springs, *Figures 14-1A, 14-1B, and 14-1C.*

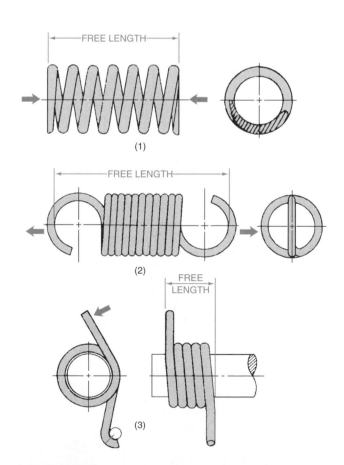

FIGURE 14-1A *Types of helical springs (1) compression spring, (2) extension spring, (3) torsion spring*

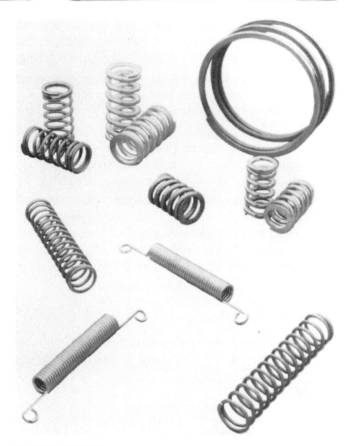

FIGURE 14-1B *Varieties of compression and extension springs* (Courtesy AMETEK Inc., Hunter Spring Division)

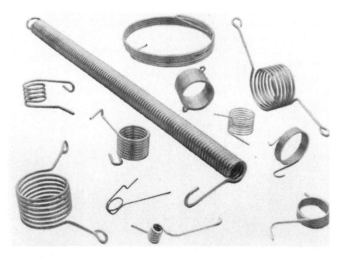

FIGURE 14-1C *Varieties of torsion springs* (Courtesy AMETEK Inc., Hunter Spring Division)

COMPRESSION SPRING

A *compression spring* offers resistance to a compressive force or applies a pushing action. In its free state, the coils of the compression spring do not touch. The types of ends of a compression spring are illustrated and described as follows:

Plain open-end—Figure 14-2A. The plain open-end spring is very unstable, has only point contact, and has ends that are not perpendicular to the axis of the spring.

Plain closed-end—Figure 14-2B. The plain closed-end spring is a little more stable and provides a round, parallel surface contact perpendicular to the axis of the spring.

Ground open-end—Figure 14-2C. The ground open-end spring is much more stable and provides a flat surface contact perpendicular to the axis of the springs.

Ground closed-end—Figure 14-2D. The ground closed-end spring is the most stable, and the best design. This design provides a large, flat surface contact perpendicular to the axis of the spring.

A good example of a compression spring is the kind that is used in the front end of an automobile.

EXTENSION SPRING

An *extension spring* offers resistance to a pulling force. In its free state, the coils of the extension spring are usually

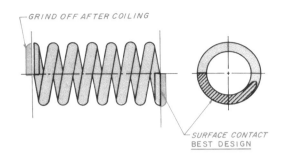

FIGURE 14-2D   *Ground closed-end compression spring*

either touching or very close. The ends of an extension spring are usually a hook or loop, *Figure 14-3.* Illustrated are a few of the many kinds of ends that are used. The loop or hook can be over the center or at the side. Each end is designed for a specific application or assembly. Regardless of the shape of the loop or hook, the overall length of an extension spring is measured from the inside of the loops or hook. A good example of an extension spring is the kind that is used to counterbalance a garage door.

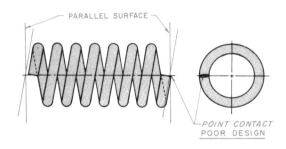

FIGURE 14-2A   *Plain open-end compression spring*

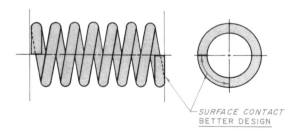

FIGURE 14-2B   *Plain closed-end compression spring*

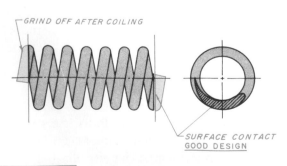

FIGURE 14-2C   *Ground open-end compression spring*

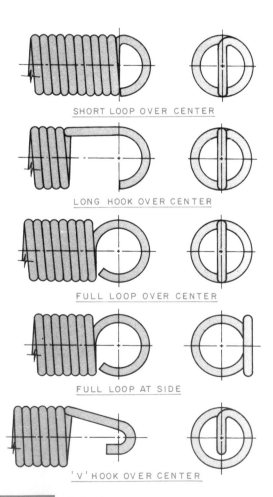

FIGURE 14-3   *Ends of extension spring*

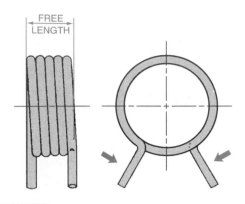

FIGURE 14-4   *Torsion spring*

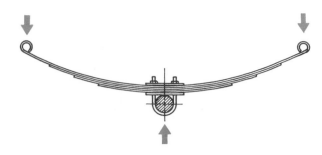

FIGURE 14-6   *Leaf spring*

TORSION SPRING

A *torsion spring* offers resistance to a torque or twisting action. In its free state, the coils of a torsion spring are usually either touching or very close. The ends of a torsion spring are usually specially designed to fit a particular mechanical device, *Figure 14-4*. A good example of a torsion spring is the kind that is used to return a doorknob to its original position.

Flat Springs

The other classification or type of spring is the flat spring. The flat spring is made of spring steel and is designed to perform a special function. Flat springs are not standard and must be designed and manufactured to fit each particular need. The flat spring is considered a power spring, and it resists a pressure. A good example of a flat spring is the kind that is used for a door latch on a cabinet, *Figure 14-5,* or a leaf spring as used in a truck, *Figure 14-6*. A flat spiral spring used to power a wind-up clock or toy is another example, *Figure 14-7*.

Terminology of Springs

It is important to know and fully understand the various terms associated with springs and their design, *Figure 14-8*.

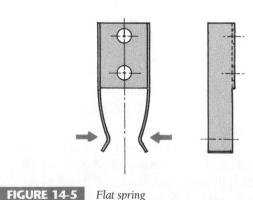

FIGURE 14-5   *Flat spring*

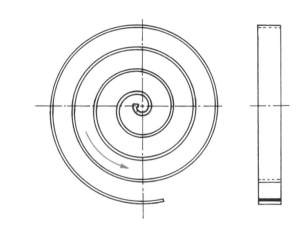

FIGURE 14-7   *Special spring*

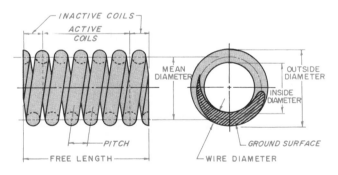

FIGURE 14-8   *Spring terminology*

FREE LENGTH

Free length is the overall length of the spring when it is in its free state of unloaded condition. Free length of a compression spring is measured from the extreme ends of the spring. In an extension spring, the free length is measured inside the loops or hooks.

SOLID LENGTH

Solid length of a compression spring is the overall length of the spring when **all** coils are compressed together so they touch. This can be mathematically calculated. If the wire diameter is .500, the overall solid length is 3.25.

OUTSIDE DIAMETER

Outside diameter (OD) is the outside diameter of the spring (2 × wire size **plus** inside diameter).

WIRE SIZE

Wire size is the diameter of the wire used to make up the spring.

INSIDE DIAMETER

Inside diameter (ID) is the inside diameter of the spring (2 × wire size **minus** outside diameter).

ACTIVE COILS

Active coils in a compression spring are usually the total coils minus the two end coils. In a compression spring, the coil at each end is considered nonfunctional. In an extension spring, active coils include **all** coils.

LOADED LENGTH

Loaded length is the overall length of the spring with a special given or designed load applied to it.

MEAN DIAMETER

Mean diameter is the theoretical diameter of the spring measured to the center of the wire diameter. This theoretical diameter is used by the drafter to lay out and draw a spring. To mathematically calculate the mean diameter, subtract the wire diameter from the outside diameter.

COIL

Coil or *turn* is one full turn or 360° of the wire about the center axis, *Figure 14-9*.

DIRECTION OF WINDING

Direction of winding is the direction in which the spring is wound. It can be wound right hand or left hand, *Figure 14-10*. The thumbs of your hands point toward the direction in which the coil windings are leaning. If the coil windings slant to the right, as the example to the left illustrates, it agrees with the direction of your left-hand thumb; thus, a left-hand winding. If the coil windings slant to the left as the example to the right illus-

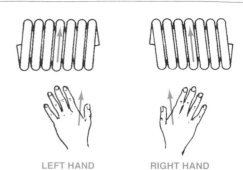

FIGURE 14-10 *Direction of winding*

trates, it agrees with the direction of your right-hand thumb; thus, a right-hand winding. If the coil winding direction is not called off on the drawing, it will be manufactured with right-hand winding.

Required Spring Data

Each drawing must include complete dimensions and specific data. Besides the regular dimensions, that is, outside diameter (OD) and/or inside diameter (ID), wire size, free length, and solid length (compression spring), the following data should be called off at the lower right side of the work area:

- Material
- Number of coils (including active and inactive coils)
- Direction of winding (if not noted, it is assumed to be right hand)
- Torque data (if torsion spring)
- Finish
- Heat treatment specification
- Any other required data

HOW TO DRAW A COMPRESSION SPRING

GIVEN: Plain open ends, 1.50 outside diameter, .25 diameter wire size, 4.00 free length, oil-tempered spring steel wire, 8 total coils (6 active coils), left-hand winding, heat treatment: heat to relieve coiling stresses, finish: black paint. Make a rough sketch of the spring using the given specifications, *Figure 14-11A*. A rough sketch is extremely important in developing a new spring design.

STEP 1 Measure and construct two vertical lines the required overall length (front view). Divide the space between the two vertical lines into 17 equal spaces (front view) 2 × total coils plus 1; in this example, 2 × 8 + 1 = 17, *Figure 14-11B*. Measure and construct the

1 TURN   2 TURNS   3 TURNS

FIGURE 14-9 *Coils or turns*

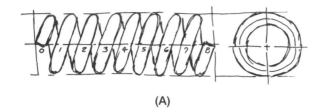

(A)

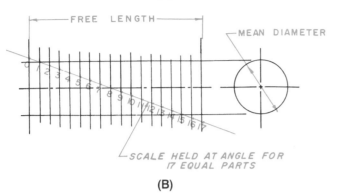

(B)

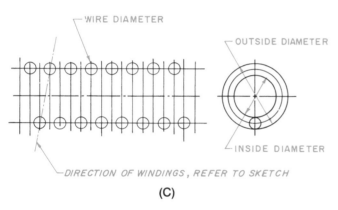

(C)

| **FIGURE 14-11** | (A) How to draw a compression spring with plain open ends; (B) Step 1; (C) Step 2 |

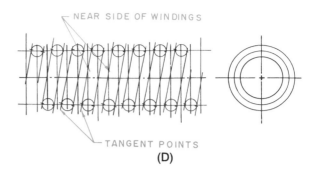

(D)

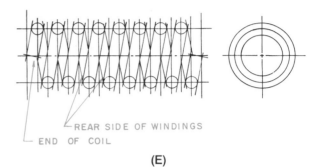

(E)

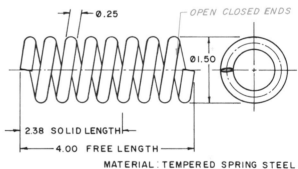

MATERIAL : TEMPERED SPRING STEEL
8 TOTAL COILS (6 ACTIVE)
LEFT-HAND COILS
HEAT TO RELIEVE COILING STRESSES
FINISH : BLACK

(F)

| **FIGURE 14-11** | (Continued) (D) Step 3; (E) Step 4; (F) Step 5 |

mean diameter. To calculate the mean diameter, subtract the wire diameter from the outside diameter (end view).

STEP 2 Determine the direction of windings; refer to the sketch and lightly construct the wire diameter accordingly (front view). The wire diameters are constructed on the mean diameter. Lightly draw the outside diameter and inside diameter (end view), *Figure 14-11C*.

STEP 3 Lightly construct the near side of the spring. Add short, light tangent points, *Figure 14-11D*.

STEP 4 Lightly construct the rear side of the spring. Add ends of spring; in this example, plain open ends. Check all work against given requirements, *Figure 14-11E*.

Recheck all work. Darken in spring using the latest drafting standards. Add end of spring, right end, dimensions, and all required spring data, *Figure 14-11F*.

NOTICE: Outside diameter, wire size, free length, and solid length are given as regular dimensions. The material, number of coils, direction of winding, heat treatment specifications, and finish requirements are listed below and to the right as a note. In this example, the outside diameter is the controlling dimension; that is, the outside dimension is more important than the inside dimension. Inside and outside dimensions should not both be added to the same drawing.

HOW TO DRAW A COMPRESSION SPRING (PLAIN CLOSED ENDS)

GIVEN: Plain closed ends, 20 mm inside diameter, 5 mm diameter wire size, 100 mm free length, material: hard drawn steel spring wire, 5 active coils (total 7 coils), left-hand winding, heat treatment: heat to relieve coiling stresses, finish: zinc plate. Make a rough sketch of the spring using the given specifications, *Figure 14-12A*.

STEP 1 Measure and construct two vertical lines the required overall length (front view). Draw a line **inside** the two vertical lines equal to half the wire diameter from each end. Divide the space between the two vertical lines into 13 equal spaces (front view) 2 × total coils minus 1; in this example, 2 × 7 − 1 = 13, *Figure 14-12B*. Measure and construct the mean diameter. To calculate the mean diameter, add the wire diameter to the inside diameter (end view).

STEP 2 Determine the direction of winding; refer to the sketch and lightly construct the wire diameter accordingly (front view). Lightly draw the outside diameter and inside diameter (end view), *Figure 14-12C*.

STEP 3 Lightly construct the near side of the spring. Add short tangent points, *Figure 14-12D*.

STEP 4 Lightly construct the rear side of the spring. Add ends of springs. In this example, plan closed ends. Check all work against given requirements, *Figure 14-12E*.

STEP 5 Recheck all work. Darken in spring using the latest drafting standards. Add all required dimensions and spring data, *Figure 14-12F*.

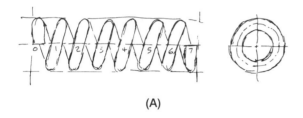

(A)

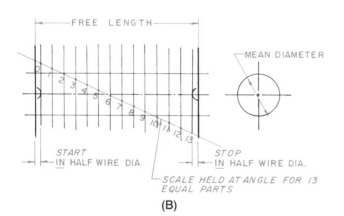

(B)

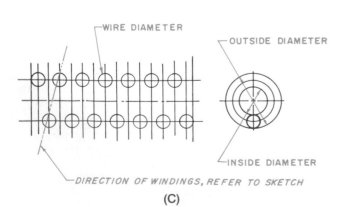

(C)

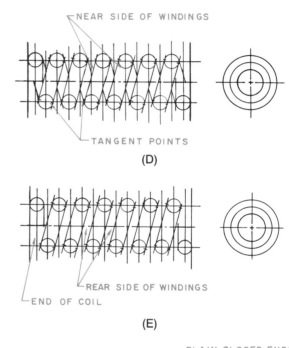

(D)

(E)

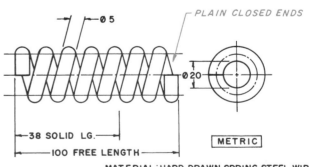

(F)

FIGURE 14-12 *(A) How to draw a compression spring with plain closed ends; (B) Step 1; (C) Step 2; (D) Step 3; (E) Step 4; (F) Step 5*

HOW TO DRAW A COMPRESSION SPRING (GROUND CLOSED ENDS)

(The following illustrations are for a ground closed-end compression spring. The same steps can be used for a ground open-end compression spring.)

GIVEN: Ground closed ends, 1.125 outside diameter, .188 diameter wire size, 3.00 free length, material: hard drawn steel spring wire, 7 total coils (5 active coils), right-hand winding, heat treatment: heat to relieve coiling stresses, finish: zinc plate. Make a rough sketch of the spring using the given specifications, *Figure 14-13A*.

STEP 1 Measure and construct two vertical lines the required free length (front view). Divide the space between the two vertical lines into 13 equal spaces (front view), 2 × total coils minus 1; in this example, 2 × 7 − 1 = 13, *Figure 14-13B*. Measure and construct the mean diameter. To calculate the mean diameter, subtract the wire diameter from the outside diameter (end view).

STEP 2 Determine the direction of winding; refer to the sketch and lightly construct the wire diameter accordingly (front view). Lightly draw the outside diameter and inside diameter (end view), *Figure 14-13C*.

STEP 3 Lightly construct the rear side of the spring. Add short tangent points, *Figure 14-13D*.

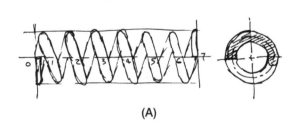

(A)

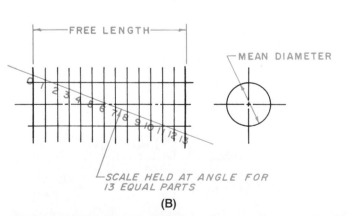

FREE LENGTH

MEAN DIAMETER

SCALE HELD AT ANGLE FOR 13 EQUAL PARTS

(B)

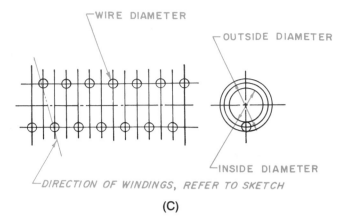

WIRE DIAMETER

OUTSIDE DIAMETER

INSIDE DIAMETER

DIRECTION OF WINDINGS, REFER TO SKETCH

(C)

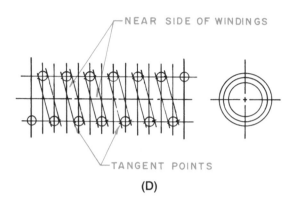

NEAR SIDE OF WINDINGS

TANGENT POINTS

(D)

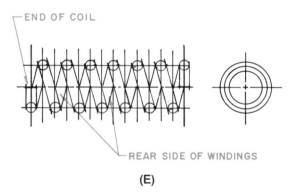

END OF COIL

REAR SIDE OF WINDINGS

(E)

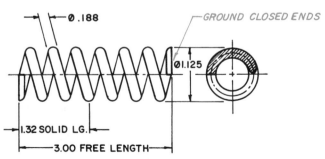

Ø.188

GROUND CLOSED ENDS

Ø1.125

1.32 SOLID LG.

3.00 FREE LENGTH

MATERIAL: HARD DRAWN SPRING STEEL WIRE
7 TOTAL COILS (5 ACTIVE COILS)
HEAT TO RELIEVE COILING STRESSES
FINISH: ZINC PLATE

(F)

FIGURE 14-13 *(A) How to draw a compression spring with ground closed ends; (B) Step 1*

FIGURE 14-13 *(Continued) (C) Step 2; (D) Step 3; (E) Step 4; (F) Step 5*

STEP 4 Lightly construct the rear side of the spring. Add ends of spring; in this example, ground closed end. Check all work against given requirements, *Figure 14-13E*.

STEP 5 Recheck all work. Darken in spring using the latest drafting standards. Add all required dimensions and spring data, *Figure 14-13F*.

HOW TO DRAW AN EXTENSION SPRING

GIVEN: Full loop, over center each end 1.625 outside diameter, .188 diameter wire size, 4.750 approximate free length, material: hard drawn spring steel wire, heat treatment: heat to relieve coiling stresses, finish: black paint, windings as tight as possible. Make a rough sketch of the spring using the given specifications, *Figure 14-14A*.

STEP 1 Draw the outside diameter, inside diameter, and mean diameter (end view). Lightly draw the required free length. In the design of an extension spring, it is almost impossible to arrive at the **exact** free length. In designing an extension spring, it is best to design it slightly **shorter** than actually required. Construct the end loops with the **inside** diameter within the specified free length, *Figure 14-14B*. Note in this example the loops are the exact diameter as the spring itself.

STEP 2 Starting at the intersection of the mean diameter of the right end loop and the mean diameter

of the spring, lower side, locate and draw the first lower wire diameter, *Figure 14-14C*. Project directly up to the mean diameter of the spring, upper side, and draw the first upper wire diameter **over** to the left, half a wire diameter as illustrated. Lay out, side-by-side, wire diameters touching, as illustrated; in this example, 10 1/2 coils. Notice the coils must end on the mean diameter, left end. This may mean adjusting slightly the location of the left end loop. It is best to bring it in slightly, if necessary.

STEP 3 Locate points X and Y, as illustrated in *Figure 14-14D*, and adjust the drafting machine at this angle and lock on this angle. Draw the wire loops.

STEP 4 Lightly draw the loop in the right-side view. To calculate just what the loop will actually look like, number various points along the loop in the front view; in this example, point 1, starting point; point 2, up; point 3, over; and point 4, around. Project these points to the end view, as illustrated in *Figure 14-14E*.

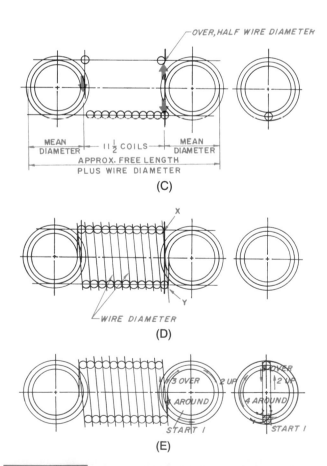

(C)

(D)

(E)

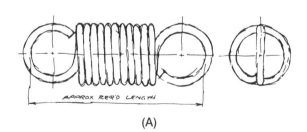

(A)

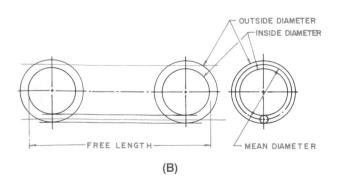

(B)

FIGURE 14-14 *(A) How to draw an extension spring with full loop ends; (B) Step 1*

FIGURE 14-14 *(Continued) (C) Step 2; (D) Step 3; (E) Step 4*

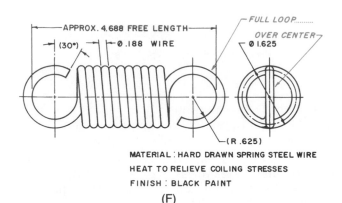

MATERIAL : HARD DRAWN SPRING STEEL WIRE
HEAT TO RELIEVE COILING STRESSES
FINISH : BLACK PAINT

(F)

(Continued) (F) Step 5

STEP 5 Recheck all work. Darken in spring using the latest drafting standards. Add all required dimensions and spring data, *Figure 14-14F.*

Other Spring Design Layout

When drawing any specially designed spring for a particular function, one should make a rough sketch of it, including all required specifications. A torsion spring is developed in a manner very similar to an extension spring with its tight coils. Usually, the required torque pressure is included in the given spring data. The actual designed deflection of the torsion spring is illustrated by phantom lines, and the angle is noted.

Standard Drafting Practices

Some company standards specify that the drafter not draw the complete spring because of the time and cost involved. A shortcut method for drawing a spring is shown in *Figure 14-15.* At top is the conventional method of representing a compression spring; below it is the same spring drawn by the schematic drawing system.

Another method of drawing a spring, especially a long spring, is illustrated in *Figure 14-16.* The uncom-

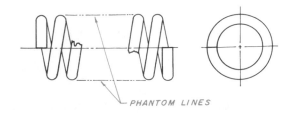

Incompleted coils with phantom lines

pleted coils are illustrated by phantom lines. If the company does not have a standard method of drawing springs it is the drafter's decision as to which method of representation is used. In most cases, it is best to draw the object, in this case the spring, in such a way that there is no question whatsoever as to exactly what is required. Of course, regardless of which system is used, full dimensions and specification data must be included.

Section View of a Spring

If a cutting-plane line passes through the axis of a spring, the spring is drawn in one of two ways. A small spring is drawn with the back coils showing, as illustrated in *Figure 14-17A,* with the section lining of the coils filled in solid. A larger spring is drawn the same way, but uses standard section lining, *Figure 14-17B.* Notice that both illustrations are right-hand springs, but because the front half has been removed, only the rear half is seen, which appears as left hand.

Isometric Views of a Spring

A template, such as the one shown in *Figure 14-18,* makes it quick and simple to draw isometric views of

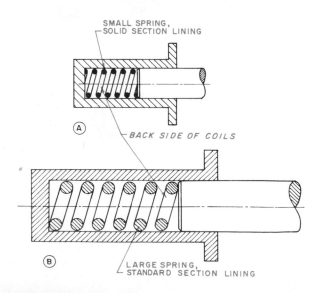

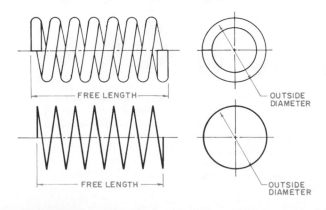

Schematic system of representing a compression spring

Section view of a spring

Industry Application

SAVING DRAFTING TIME

John Dugan had been the outstanding drafting student at Minot Technical Institute. His work as a student had won him a closet full of awards and much recognition. Consequently, nobody was surprised when Dugan secured a job in the drafting department of Minot Machine Tool Company, an excellent employer.

John Dugan's talent was apparent from the outset, and within weeks he was assigned to a team that took on the company's most difficult projects. His first assignment on the "A" team was to draw the plans for several different extension springs. Springs had been Dugan's specialty at Minot Tech. In the words of his professor, "Dugan can draw springs that look so realistic you can lift them off the page and use them."

Consequently, Dugan was taken aback when his team leader stopped him after he had completed only one of the five spring drawings he was assigned. "What in the world are you doing?" asked the team leader, clearly flabbergasted. Dugan, unsure of what was wrong and shocked by the first negative reaction he had ever had to his work, could only stammer, "What do you mean? I'm drawing the springs you assigned me." "I asked you to draw them, not make them," said the team leader. "We don't do pictorial spring drawings here. This is the drafting department, not the art department. We use schematic representation with all the critical data called out in the notes. It takes time to create pictorial representations of springs, and time is money." "Besides," said the team leader, "there is no need to give the machining department anything but a schematic and some notes. That is all they need to make the springs, and it is our job to give them that much information and nothing more. Every minute you spend creating these beautiful pictorials is time that could have been put to better use on another project." John Dugan had just learned a valuable lesson that eventually every drafter must learn.

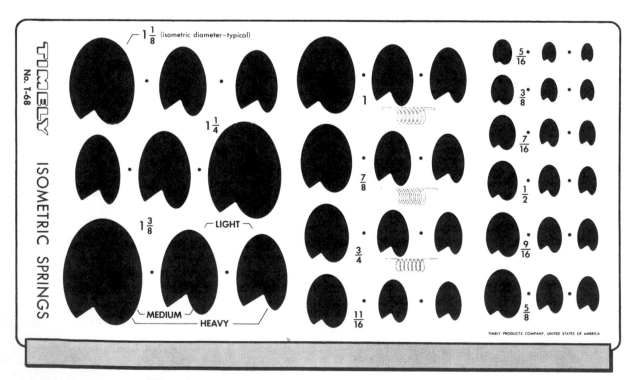

This template makes it quick and simple to draw isometric views of all light, medium, or heavyweight open springs. Lightweight springs are slightly smaller than the dimension printed on the template because they are drawn with the two smallest cutouts in any set. The thick lower edge and pinpoint centers insure precise positioning to repeat each loop for any length spring.

FIGURE 14-18   *Isometric spring template* (Courtesy Timely Products Co.)

light, medium, or heavyweight open springs. Light-weight springs are slightly smaller than the dimension printed on the template because they are drawn with the two smallest cutouts in any set. The thick lower edge and pinpoint centers ensure precise positioning to repeat each loop for any length spring.

Summary

- Springs can be classified as helical or flat. There are three types of helical springs; compression, extension, and torsion.

- Key terms associated with spring design include the following: free length, solid length, outside diameter, wire size, inside diameter, active coils, loaded length, wear diameter, coil, and direction of winding.

- Required spring data include the following: material, number of coils, direction of winding, torque data, finish, and heat treatment specifications.

- Some companies specify the use of schematic representation of springs as a way to save time and money. Another time-saving method is to draw only those coils that are at each end of the spring and connect them with phantom lines.

- If a cutting-plane line passes through the axis of a spring, the spring may be drawn in one of two ways. A small spring is drawn with the section lines filled in. A larger spring uses standard section lining.

Review Questions

Answer the following questions either true or false.

1. It is incorrect to dimension both the inside diameter and the outside diameter of a spring on the same drawing.
2. The two general classifications of springs are helical and flat.
3. The solid length of a compression spring is the length of the spring stretched out straight.
4. Section lining is illustrated in a large spring where the cutting-plane line passes through the center axis of the spring by filling the coils in solid.
5. The free length of a spring means the overall length of a spring when it is in its free state or unloaded condition.

Answer the following questions by selecting the best answer.

1. If the coil winding direction is not called off on the drawing, it will be manufactured:

a. Left-hand winding.
b. Right-hand winding.
c. The same as other springs in the drawing.
d. None of the above

2. Which dimension should be given for the diameter of a spring on a drawing?
a. Inside
b. Outside
c. Radial
d. Both a and b

3. Other than the required dimensions, what spring data must be noted below and to the right?
a. Material
b. Number of coils
c. Direction of winding
d. All of the above

4. Which of the following is **not** a type of helical spring?
a. Compression spring
b. Flat spring
c. Torsion spring
d. Extension spring

5. Which of the following is **not** a common use for a torsion spring?
a. A doorknob
b. A clock
c. A truck
d. A wind-up toy

Chapter 14 Problems

The following problems are intended to give the beginning drafter practice in drawing various kinds of springs with special required specifications.

CONVENTIONAL INSTRUCTIONS

The steps to follow in laying out springs are:

STEP 1 Study all required specifications.

STEP 2 Make a rough sketch of the spring, incorporating **all** required specifications and required dimensions.

STEP 3 Calculate the size of the basic shape of the front and end views.

STEP 4 Center the two views within the work area with correct spacing for dimensions between views.

STEP 5 Starting with the end view, or the view that contains the diameter, lightly draw the spring adhering to **all** specific required specifications. Take care to show correct direction of winding, correct type of ends, and total count of windings.

STEP 6 Check to see that both views are centered within the work area.

STEP 7 Check all dimensions in both views.

STEP 8 Darken in both views, starting first with all diameters and arcs.

STEP 9 Neatly add all dimensioning as required to fully describe the object using the latest drafting standards.

STEP 10 Neatly add all required specifications under the two views. The drawing must include the following:

Free length
Wire size
Total number of coils
Type of ends
Solid length
Outside diameter
Inside diameter
Direction of windings
Material (usually, hard drawn steel, spring wire)
Required heat-treating process (usually, heat to relieve coiling stresses)
Any other special requirements

STEP 11 For torsion-type springs, illustrate working location and travel with phantom lines and dimension.

CAD Instructions

Students who plan to complete the following problems on a CAD system should begin with these general instructions. Before reading the specific instructions for each activity, go through each step in the following planning checklist. The checklist applies to any CAD system and will help ensure the optimum use of your time and resources.

1. Analyze the problem carefully. Decide exactly what you are being asked to do.

2. Determine what resources and references (if any) you will need in order to complete the problem. Collect those references and resources and have them readily available.

3. Decide if any particular standards apply to the project and have those standards available.

4. Determine what types of views will be required and how many of each.

5. Determine what the final plotted scale of the drawing will need to be, and select the appropriate paper size for plotting/printing.

6. Plan your drawing sequence. In what order will you develop the drawing (i.e., lines, features, dimension lines, leaders, dimensions)?

7. Review the various CAD commands you will have to use in order to develop the drawing.

8. Examine your CAD system to ensure that everything is in working order, then begin the project.

PROBLEMS 14-1 THROUGH 14-7

Construct a 2-view drawing of each spring using the listed steps. Center the views on an A-size, 8½ × 11 sheet of paper with a 1″ (25 mm) space between views. Use correct line thicknesses and all drafting standards. Add all required dimensions and specifications per the latest drafting standards.

Problem 14-1

Compression-type spring
Free length 7.00
Wire size .25
12 active coils
  (14 **total** coils)
Plain open ends
OD 2.00/ID 1.50
R.H. winding
Calculate solid length

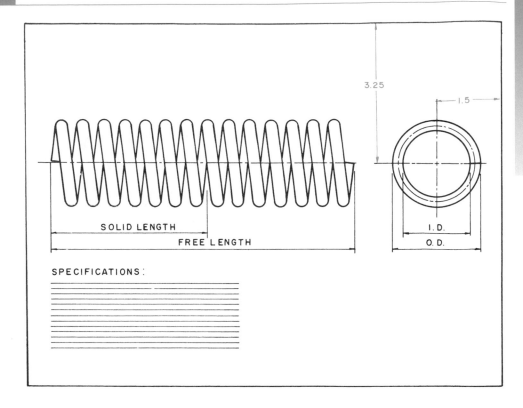

SPECIFICATIONS:

Problem 14-2

Compression-type spring
Free length 5.25
Wire size .375
4 active coils
  (6 **total** coils)
Ground open ends
OD 2.75/ID 2.00
L.H. winding
Calculate solid length

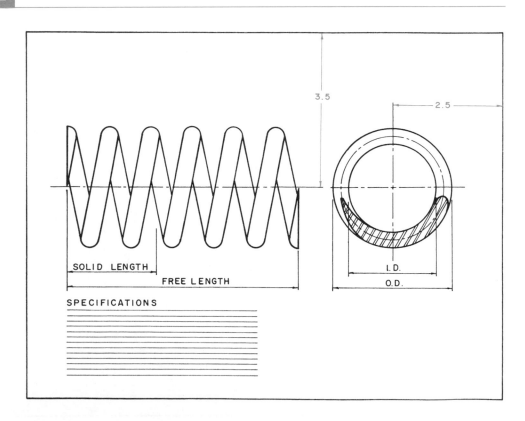

SPECIFICATIONS

Problem 14-3

Compression-type spring
Free length 6.25
Wire size .375
9 total coils
   (7 **active** coils)
Plain closed ends
OD 2.00/ID 1.25
L.H. winding
Calculate solid length

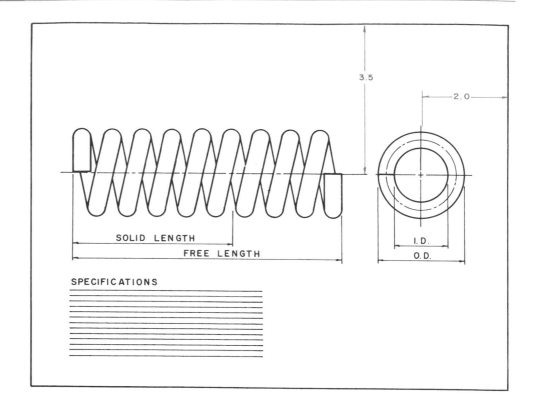

SOLID LENGTH

FREE LENGTH

I.D.

O.D.

3.5

2.0

SPECIFICATIONS

Problem 14-4

(Metric)
Compression-type spring
Free length 90
Wire size 12
4 total coils
   (2 **active** coils)
Ground closed ends
OD 96/ID 72
R.H. winding
Calculate solid length

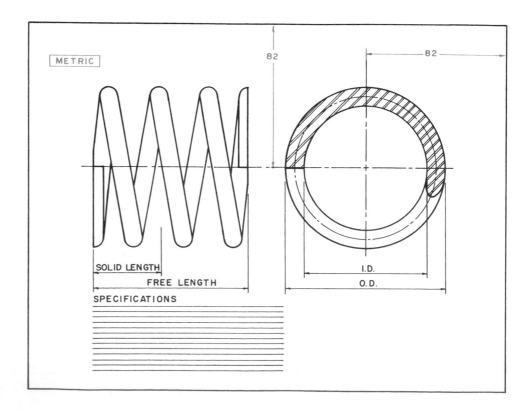

METRIC

82

82

SOLID LENGTH

FREE LENGTH

I.D.

O.D.

SPECIFICATIONS

Problem 14-5

Extension-type spring
Full loop over center, both
 ends
Approx. free length 7.0
Wire size .25
OD 2.0/ID 1.50
R.H. winding (This is
 standard for all
 extension springs)

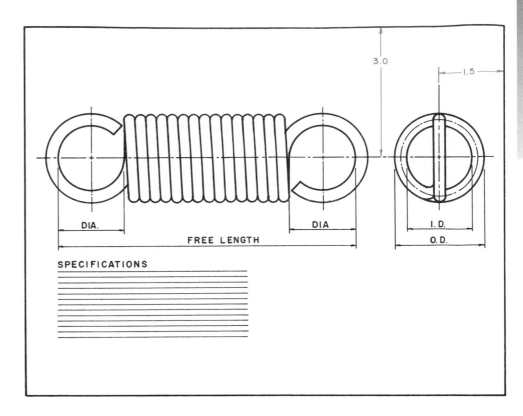

Problem 14-6

(Metric)
Extension-type spring
Full loop over center
 (right end)
Long hook over center
 (left end)
Free length 150
Wire size 5
12 total coils
65 coil length
OD 42/ID 32
Standard R.H. winding

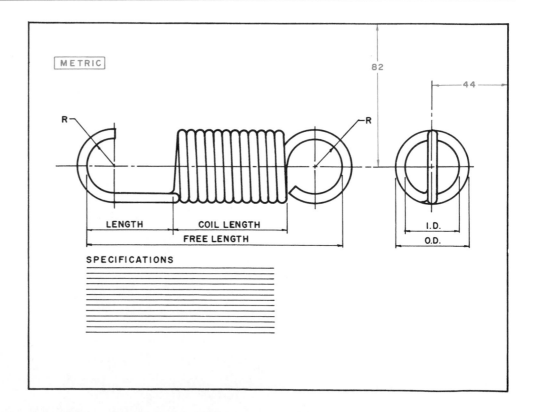

Problem 14-7

Torsion-type spring
Free length 3.625
Wire size .25
13 total coils
OD 2.0/ID 1.50
R.H. standard winding
90° maximum working flex
    (counterclockwise)
Arm lengths 2.0 (as shown)

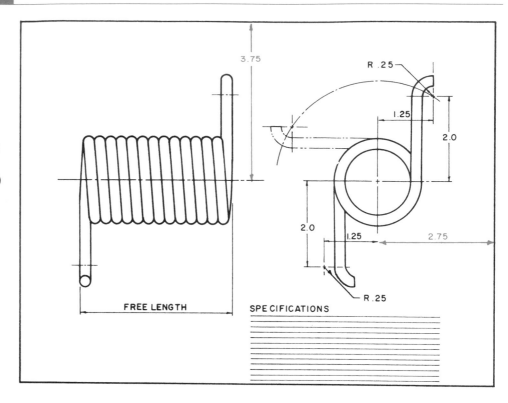

CAMS

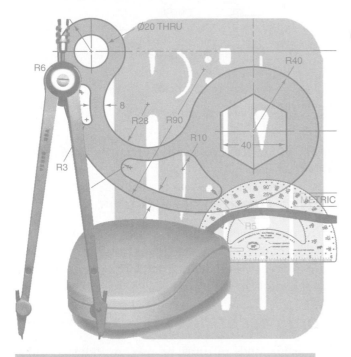

KEY TERMS

Base circle	Parallel
Baseline	Perpendicular
Cam	Radial arm
Cylindrical	Rotation
Displacement diagram	Time interval
Dwell	Timing diagram
Harmonic motion	Uniform acceleration
Lowest position	Uniform velocity
Modified uniform velocity	Working circle

CHAPTER OUTLINE

Cam principle • Basic types of followers •
Cam mechanism • Cam terms • Cam motion
• Laying out the cam from the displacement diagram
• Timing diagram • Dimensioning a cam
• Summary • Review questions •
Chapter 15 problems

CHAPTER OBJECTIVES

Upon completion of this chapter, students should be able to do the following:

■ Explain the cam principle.

■ Describe the four basic types of followers.

■ Distinguish between the two major kinds of cams used in industry.

■ Define the most important cam terms.

■ Compare and contrast the four principal types of cam motion.

■ Demonstrate proficiency at laying out a cam from a displacement diagram.

■ Demonstrate proficiency in drawing various types of cams.

■ Demonstrate proficiency in developing timing diagrams.

■ Demonstrate proficiency in dimensioning cams.

Cam Principle

A *cam* produces a simple means to obtain irregular or specified predictable, designed motion. These motions would be very difficult to obtain in any other way.

Figure 15-1 illustrates the basic principle and terms of a cam. In this example, a rotating shaft has an irregularly shaped disc attached to it. This disc is the cam. The follower, with a small roller attached to it, pushes against the cam. As the shaft is rotated, the roller follows the irregular surface of the cam, rising or falling according to the profile of the cam. The roller is held tightly against the cam either by gravity or a spring.

Figure 15-2 illustrates a simple cam in action. Notice how the rotation of the shaft converts into an up-and-down motion of the follower. A cam using a flat-faced follower is shown in *Figure 15-3.* This type of cam, for example, is used to raise and lower the valve in an automobile engine. A modified cam follower is used to change the rotary motion of shaft A into an up-and-

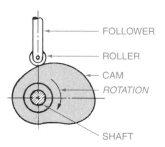

| FIGURE 15-1 | *Basic cam principle and terms* |

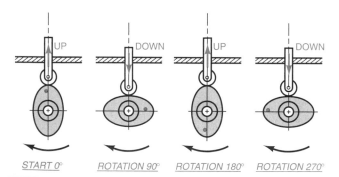

| FIGURE 15-2 | *Simple cam in action* |

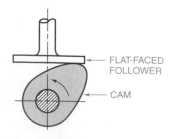

| FIGURE 15-3 | *Flat-faced follower* |

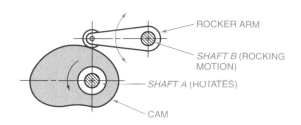

| FIGURE 15-4 | *Rocker arm follower with wheel* |

down motion of the rocker arm, and then into a rocking motion of shaft B, *Figure 15-4.*

Basic Types of Followers

Speed of rotation and the actual load applied upon the lifter determine the type of follower to be used. There are various basic types of followers: roller, pointed, flat-faced, and spherical-faced, *Figure 15-5.*

Cam Mechanism

Two major kinds of cams are used in industry: radial arm design and cylindrical design. The *radial arm* design changes a rotary motion into either an up-and-down motion or a rocking action as discussed previously. In the *cylindrical* design, a shaft rotates exactly as it does in the radial design, but the action or direction of the follower differs greatly. In the radial arm design, the follower operates *perpendicular* to the camshaft. In the cylindrical design, the follower operates *parallel* to the camshaft. See *Figure 15-6* for a simplified example of a cylindrical design cam. Shaft A has an irregular groove cut into it. As the cam rotates, the follower traces along

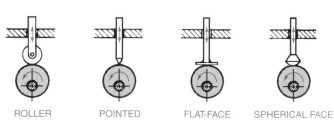

| FIGURE 15-5 | *Basic types of followers* |

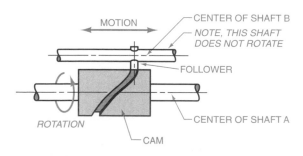

| FIGURE 15-6 | *Cylindrical cam mechanism* |

the groove. The follower is directly attached to a shaft that moves to the left and right. Notice this shaft does **not** rotate; the motion is parallel to the axis of the rotating camshaft.

Cams produce one of three kinds of motion: uniform velocity, harmonic, or uniform acceleration. Each is discussed in full later.

Cam Terms

WORKING CIRCLE

The *working circle* is considered to be a distance equal to the distance from the center of the camshaft to the highest point on the cam, *Figure 15-7*.

DISPLACEMENT DIAGRAM

The *displacement diagram* is a designed layout of the required motion of the cam. It is laid out on a grid, and its length represents one complete revolution of the cam. The length of the displacement diagram is usually drawn equal in length to the circumference of the working circle. This is not absolutely necessary, but it will give an in-scale idea of the cam's profile. The height of the displacement diagram is equal to the radius of a working circle, *Figure 15-8*. The bottom line of the displacement diagram is the *baseline*. All dimensions should be measured upward from the baseline. On the cam itself, think of the center of the cam as the baseline.

The length of the displacement diagram is divided into equal lines or grids, each of which represents

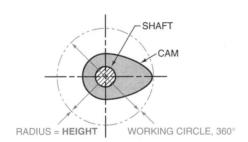

FIGURE 15-7 *Cam terms and working circle*

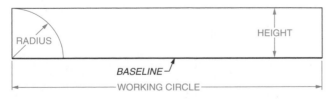

WORKING CIRCLE = CIRCUMFERENCE OF CIRCLE = 360°
BASELINE = CENTER OF WORKING CIRCLE

FIGURE 15-8 *Displacement diagram (basic layout)*

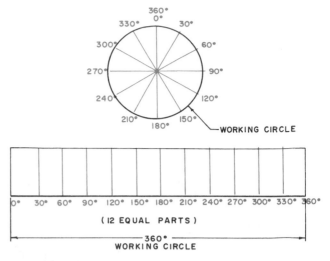

FIGURE 15-9 *Displacement diagram with 30° increments*

degrees around the cam. These divisions can be 30°, 15°, or even 10°. The finer the divisions, the more accurate is the final cam profile, *Figure 15-9*.

DWELL

Dwell is the period of time during which the follower does not move. This is shown on the displacement diagram by a straight horizontal line throughout the dwell angle. On the cam, the dwell is drawn by a radius.

TIME INTERVAL

The *time interval* is the time it takes the cam to move the follower to the designed height.

ROTATION

Rotation of the cam is either clockwise or counterclockwise. The actual cam profile is laid out opposite of the rotation.

BASE CIRCLE

The *base circle* is used to lay out an offset follower that is illustrated in detail later. The base circle is a circle with a radius equal to the distance from the center of the shaft to the center of the follower wheel at its *lowest position*. On an offset follower, the base circle replaces the working circle.

Cam Motion

Four major types of curves are usually employed. Special irregular curves other than the four major types can be designed to meet a specific movement, if necessary.

The four types are: uniform velocity, modified uniform velocity, harmonic motion, and uniform acceleration. For

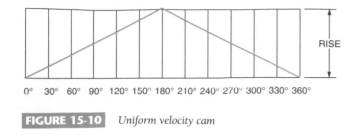

0° 30° 60° 90° 120° 150° 180° 210° 240° 270° 300° 330° 360°

FIGURE 15-10 *Uniform velocity cam*

comparison, the following four examples use the same working circle, same shaft, and same rise.

UNIFORM VELOCITY

In this type of motion, the cam follower moves with a *uniform velocity;* that is, it rises and falls at a constant speed, but the start and stop are very abrupt and rough, *Figure 15-10.*

MODIFIED UNIFORM VELOCITY

Because of the abrupt and rough start and stop of the uniform velocity, it is modified slightly. *Modified uniform velocity* smoothes out the roughness slightly by adding a radius at the ends of the high and low points. This radius is equal to 1/3 the rise or fall, *Figure 15-11.* This radius smoothes out the start and stop somewhat and is good for slow speed.

HARMONIC MOTION

Harmonic motion is very smooth, but the speed is not uniform. Harmonic motion has a smooth start and stop and is good for fast speed.

To lay out a harmonic motion on the displacement diagram, draw a semicircle whose diameter is equal to the designed rise or fall. Divide the semicircle into equal divisions; in this example, 30°. Divide the overall hori-

zontal distance of the rise or fall equally into increments of the same number as the semicircle; in this example, six equal divisions. Projection points on the semicircle are projected horizontally to the corresponding vertical lines. For example, the first point on the semicircle is projected to the first vertical line, the second point on the semicircle is projected to the second vertical line, and so forth. Connect the curve with an irregular curve. This completes the harmonic curve, *Figure 15-12.*

UNIFORM ACCELERATION

Uniform acceleration is the smoothest motion of all cams, and its speed is constant throughout the cam travel. The uniform acceleration curve is actually a parabolic curve; the first half of the curve is exactly the reverse of the second half. This form is best for high speed.

To lay out a uniform acceleration motion on the displacement diagram, divide the designed rise or fall into 18 equal parts. To do this, place the edge of a scale with its "0" on the starting level of the rise or fall, and equally space 18 units of measure on the high or low elevation of the rise or fall, *Figure 15-13.* Mark off points 1, 4, 9, 4 (14), and 1 (17), and draw horizontal lines. Note that 14 is actually 4, and 17 is actually 1. Divide the overall horizontal distance of the rise or fall into 12 equal increments. From points 1, 4, 9, 4 (14), and 1 (17), project points to the corresponding vertical lines. For example, the first point is from point 1 on the division line to the first vertical line, and then from point 4 on the division line to the second vertical, and so forth. Connect the curve with an irregular curve. This completes the uniform acceleration curve.

COMBINATION OF MOTIONS

The illustrations thus far have covered the four major types of cams: uniform velocity, modified uniform veloc-

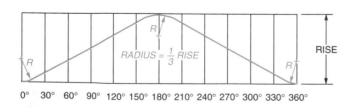

0° 30° 60° 90° 120° 150° 180° 210° 240° 270° 300° 330° 360°

FIGURE 15-11 *Modified uniform velocity cam*

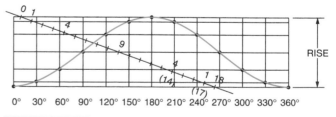

0° 30° 60° 90° 120° 150° 180° 210° 240° 270° 300° 330° 360°

FIGURE 15-13 *Uniform acceleration cam*

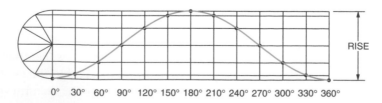

0° 30° 60° 90° 120° 150° 180° 210° 240° 270° 300° 330° 360°

FIGURE 15-12 *Harmonic motion cam*

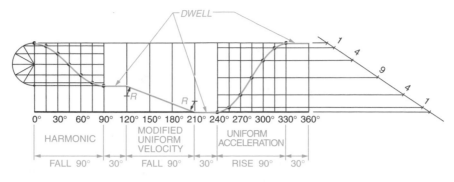

FIGURE 15-14 *Combination of motions*

ity, harmonic motion, and uniform acceleration. Any combinations of these can be designed for a particular function or requirement. *Figure 15-14* illustrates a displacement diagram with a 90° fall using a harmonic motion, followed by a 30° dwell. The cam then falls again 90° using a modified uniform velocity followed by a 30° dwell. The cam then rises using uniform acceleration motion followed by another 30° dwell.

Laying Out the Cam from the Displacement Diagram

Regardless of which type of motion is used, the layout process is exactly the same. The working circle is drawn equal to the height of the displacement diagram. The circle of the cam must be divided into the same number of equal divisions as the displacement diagram, *Figure 15-15A*. Label the increments on both the displacement diagram and the working circle of the cam; in this example, harmonic motion is illustrated.

Notice that the even increment on the cam layout is labeled opposite of the rotation of the cam. In this example, rotation is clockwise; therefore, the labeling of the radial lines should be counterclockwise. Transfer each distance from the displacement diagram to the corresponding radial line, *Figure 15-15B*. Using an irregular curve, complete the cam layout.

OFFSET FOLLOWER

The cam follower is usually in line with the same center line as the cam (see Figure 15-1). Occasionally, however, when position or space is a problem, the follower can be designed to be on another center line, or offset as illustrated in *Figure 15-16*. In order to lay out an offset follower, it is necessary to use slightly different steps than those used for a regular cam.

HOW TO DRAW A CAM WITH AN OFFSET FOLLOWER

GIVEN: Cam data: harmonic motion cam, follower offset .625, base circle .88 radius, rise of 1.50,

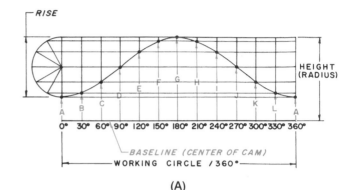

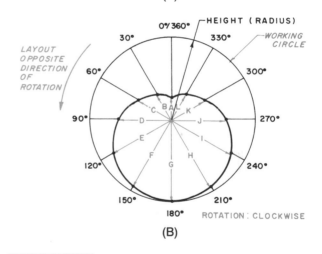

FIGURE 15-15 *(A) Displacement diagram example; (B) Heights transferred from displacement diagram*

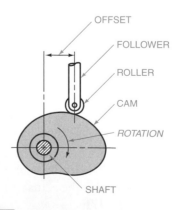

FIGURE 15-16 *Associated cam terms*

shaft diameter .375, direction of rotation counterclockwise.

Follower data: Follower width .250, roller .375 diameter.

STEP 1 Design and lay out a displacement diagram exactly as with any cam, *Figure 15-17A.*

STEP 2 From the center of the shaft of the cam, construct the offset circle. The offset circle has a radius equal to the required follower offset. From the same swing point, lay out the base circle. The radius of the base circle is equal to the distance from the center of the shaft to the center of the follower wheel at its **lowest** position, *Figure 15-17B.*

STEP 3 The point where it crosses a constructed vertical line from the center line of the follower is point A on the displacement diagram, *Figure 15-17C.* This represents the lowest position of the follower wheel. This vertical line is the measuring line and will be used to lay out the other points around the cam.

STEP 4 Divide the offset circle into the same equal divisions as used on the displacement diagram; in this example, 12 equal parts. Lightly and consecutively, letter or number the divisions on the offset circle. Start at the lowest point of the cam; in this example, point A. Letter or number in a direction opposite to that of the cam rotation. Note that point A is the exact location where the center line of the follower is tangent with the offset circle. Construct a layout projection line 90° from the tangent point of the circle, *Figure 15-17D.*

STEP 5 Transfer the distances from the displacement diagram to the measuring line, and lightly letter or number each point. With the compass point on the center of the shaft of the cam, swing these distances to the corresponding radial lines (refer back to Figure 15-17D). Lightly letter or number each point. Do not forget to *lay out the cam opposite to the direction of rotation.* These points represent the path of the center of the follower wheel. Lightly connect these points together, *Figure 15-17E.*

STEP 6 With the compass set at the radius for the roller, lightly draw the roller around the path of the follower. The inside, high points of these arcs represent the actual profile of the cam, *Figure 15-17F.*

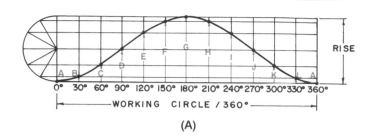

(A)

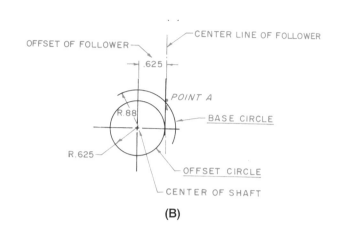

(B)

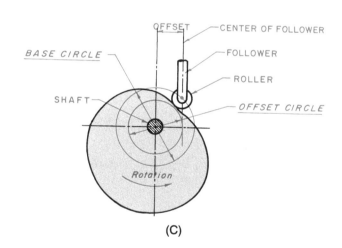

(C)

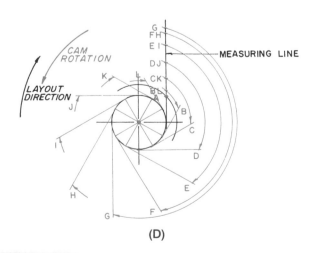

(D)

FIGURE 15-17 *(A) Step 1 (drawing a cam with an offset follower design layout); (B) Step 2; (C) Step 3; (D) Step 4*

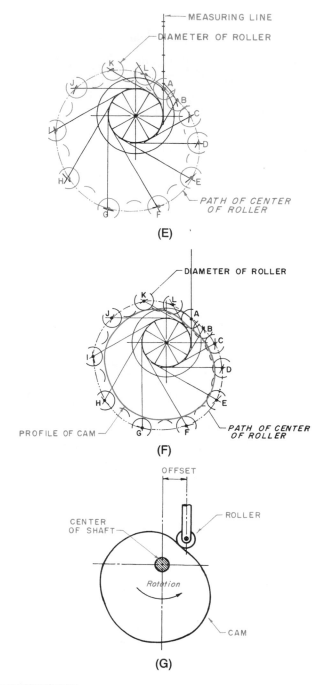

(E)

(F)

(G)

FIGURE 15-17 *(Continued) (E) Step 5; (F) Step 6; (G) Step 7*

STEP 7 Check all construction work. If correct, darken and complete the cam, *Figure 15-17G*.

HOW TO DRAW A CAM WITH A FLAT-FACED FOLLOWER

GIVEN: Cam data: harmonic motion cam, base circle .75 radius, rise of 1.50, shaft diameter .375, direction of rotation clockwise.

Follower data: flat-faced follower, 1.25 diameter distance across the flat surface. Design

and lay out a displacement diagram exactly as with any cam. In this example, the displacement diagram in Figure 15-17A will be used.

STEP 1 Construct the base circle. The base circle is drawn with a radius equal to the distance from the center of the shaft to the face of the follower in its lowest position. Divide the base circle into the same equal divisions as used on the displacement diagram. Lightly and consecutively letter or number the divisions on the base circle, *Figure 15-18A*.

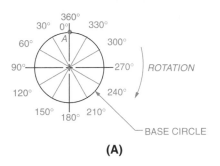

(A)

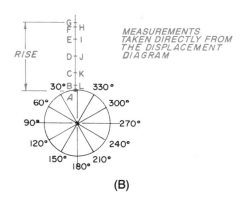

(B)

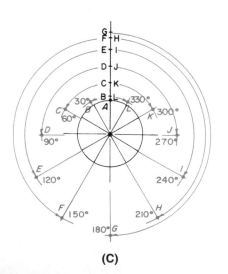

(C)

FIGURE 15-18 *(A) Step 1 (drawing a cam with a flat-faced follower); (B) Step 2; (C) Step 3*

STEP 2 Locate point A, which represents the location of the flat-faced follower at its lowest point. Construct a measuring line starting from the base circle and transfer the distances from the displacement diagram to the measuring line. Letter or number each point, *Figure 15-18B*. Do not forget to lay out the cam *opposite to the direction of rotation*.

STEP 3 Swing these distances to the corresponding radial lines, *Figure 15-18C*.

STEP 4 From these distances, construct lines perpendicular from these radial lines at the point of intersection, *Figure 15-18D*.

STEP 5 Lightly construct the cam within these perpendicular lines, *Figure 15-18E*.

STEP 6 Check all construction work. If correct, darken and complete the cam, *Figure 15-18F*. Notice the position of the flat-faced follower at the 0°, 90°, and 150° positions.

Timing Diagram

Many times, more than one cam is attached to the same shaft. In this case, each cam works independently but must function in relation to the other cams. In order to be able to visualize graphically and study the interaction of the cams on the same shaft, place the various displacement diagrams in a column with the starting points and ending points in line, as illustrated in *Figure 15-19*. This represents one full revolution and is referred to as a *timing diagram*. For example, at 240° it is easy to visualize the exact position of each cam roller.

Dimensioning a Cam

A cam must be fully dimensioned. The cam size dimensions are constructed radiating from the center of the shaft by both radii length and degrees from the starting point, as illustrated in *Figure 15-20*. Other important dimensions pertain to the shaft hole size and keyway size, and are dimensioned according to most recent drafting standards.

Try not to place dimensions on the cam itself; place them around the perimeter, if possible.

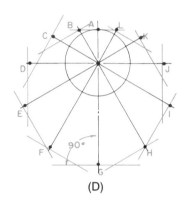

(D)

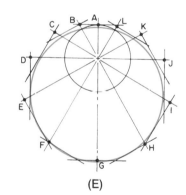

(E)

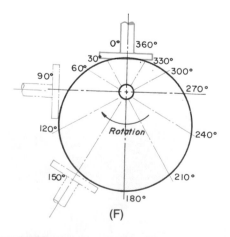

(F)

FIGURE 15-18 *(Continued) (D) Step 4; (E) Step 5; (F) Step 6*

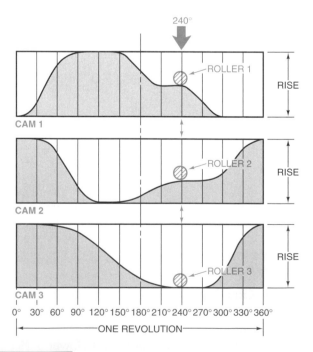

FIGURE 15-19 *Timing diagram*

Industry Application

LEARNING ON THE JOB

"This new job has several cams in it," said Marge McCrory. "Does anybody know how to draw cams?" When she got nothing but blank stares from the members of her drafting team, McCrory said, "All right then, we'll just have to learn." She distributed copies of several different diagrams among team members.

"This is what we start with. These are displacement diagrams. Don't ask me how just yet, but we use these to lay out the cams. There is one displacement diagram for each cam." Then McCrory told her team members to go to the company's Engineering Library and start researching cams. "I want to know what all of these terms on the diagrams mean, how we get a cam layout from a diagram, and how to completely detail a cam including dimensions," said McCrory.

When the team gathered in the Engineering Department conference room the next day, McCrory could tell immediately that their research had been successful. Her team looked confident and anxious to get started. "Who wants to demonstrate how to lay out a cam from one of these displacement diagrams?" When several hands went up simultaneously, McCrory called on Doug Fulton, the team's senior drafter.

Fulton went to the marker board and quickly sketched his displacement diagram. Then, under it, he sketched the layout for the corresponding cam. He began the cam layout by drawing a circle, which he labeled "working circle." The radius of the circle was equal to the height of the displacement diagram. He divided the circle into 30° segments to match the 30° increments on the displacement diagram. He then transferred each displacement distance from the diagram onto the radial lines of the working circle, beginning at the center of the circle and working outward along each respective radial line.

Within just a few minutes Fulton had laid out his cam. McCrory then asked another team member to use Fulton's diagram to illustrate cam terminology. Another team member put a transparency on an overhead projector to illustrate dimensioning techniques. McCrory was pleased with her team's work and told them so.

This case study from the workplace illustrates an important point. Students seldom graduate knowing everything they will need to know on the job. Even if they did, what must be known today will change tomorrow. Like Marge McCrory's team members, the modern drafting technician must be able to learn on the job.

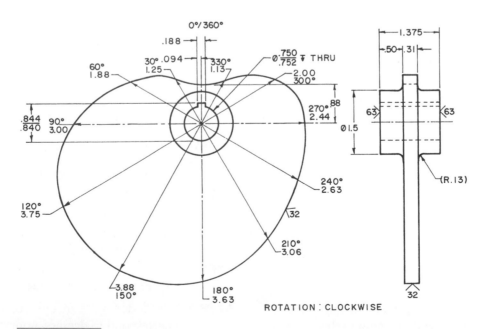

FIGURE 15-20 *Dimensioning a cam*

Summary

- A cam produces a simple means to obtain irregular or specified predictable, designed motion.

- Cam followers are of four types: roller, pointed, flat-face, and spherical-face.

- The two main types of cams are radial arm design and cylindrical design. Important cam-related terms are *working circle, displacement diagram, time interval, rotation,* and *base circle*.

- The four main types of cam motion are uniform velocity, modified uniform velocity, harmonic velocity, and uniform acceleration.

- By placing the displacement diagrams for different cams that are attached to the same shaft in a column, we can produce a timing diagram. A timing diagram allows you to visualize graphically the interaction of the cams.

Review Questions

Answer the following questions either true or false.

1. Speed of rotation and the actual load applied on the lifter determine the type of follower to be used.

2. The two major kinds of cams used in industry are radial arm and parallel.

3. A working circle is considered a distance equal to the distance from the center of the camshaft to the highest point on the cam.

4. A displacement diagram is a designed layout of the required motion of the cam.

5. Dwell is the period of time during which the follower moves.

Answer the following questions by selecting the best answer.

1. Which of the following is **not** a type of motion produced by a cam?
   a. Harmonic
   b. Uniform acceleration
   c. Uniform velocity
   d. Harmonic displacement

2. Which of the following is **not** true regarding the design and layout of the cam displacement diagram?
   a. The bottom line of the displacement diagram is the baseline.
   b. All dimensions should be measured upward from the baseline.
   c. The length of the displacement diagram is usually drawn equal in length to the circumference of the working circle.

   d. A displacement diagram is a designed layout that measures the applied load.

3. Which of the following is **not** true regarding a timing diagram?
   a. It is used when more than one cam is attached to the same shaft.
   b. Each cam works independently.
   c. Each cam works dependently.
   d. Each cam must function in relation to the other cams.

4. Which of the following is **not** true regarding cam motion?
   a. With uniform velocity, there is a very rough stop and start.
   b. Harmonic motion is very smooth, but the speed is not uniform.
   c. Harmonic motion is very smooth, and the speed is uniform.
   d. Uniform acceleration is the smoothest motion.

5. Which of the following is **not** true regarding the length of the displacement diagram?
   a. The length of the displacement diagram represents one complete revolution of the cam.
   b. The length of the displacement diagram is usually drawn equal in length to the circumference of the working circle.
   c. The height of the displacement diagram is equal to the radius of the working circle.
   d. The height of the displacement diagram is equal to the circumference of the working circle.

Chapter 15 Problems

The following problems are intended to give the beginning drafter practice in designing and layout of cam displacement diagrams according to the required specifications to do a particular function. The cam displacement diagram will then be transferred to an actual cam profile layout. The drafter will practice laying out uniform velocity, harmonic motion, and uniform acceleration motions.

CONVENTIONAL INSTRUCTIONS

The steps to follow in laying out cams are:

STEP 1 Study all required specifications.

STEP 2 Make a rough sketch of the cam displacement diagram, incorporating **all** required specifications and required dimensions.

STEP 3 Lightly lay out the displacement diagram. Space out horizontally as far as possible.

STEP 4 Break down the area into equal spaces that correspond to degrees around the cam. Label each line, indicating the degrees.

STEP 5 Following the required specifications, lightly develop the cam's travel on the displacement diagram. (Show all light layout work—do not erase.)

STEP 6 Check all work.

STEP 7 Darken in the displacement diagram.

STEP 8 To lay out the cam profile, locate the center of the cam.

STEP 9 Draw the shaft, hub, working circle, and key, if required, in place.

STEP 10 Divide the working circle into equal degrees as required.

STEP 11 Note direction of travel. Lightly letter the degrees around the working circle opposite of the direction of travel.

STEP 12 Locate the starting elevation and draw the cam follower from the starting point.

STEP 13 Lightly draw in the rest of the cam, following details as required.

STEP 14 Transfer all distances from the cam displacement diagram to the corresponding degree line around the cam layout.

STEP 15 Lightly connect all points using an irregular curve.

STEP 16 Check all work.

STEP 17 Darken in all work.

STEP 18 Neatly add all required specifications and dimensions to the drawing.

The drawing must include the following:

Shaft diameter

Hub size

Key or set screw size

Direction of rotation

Working circle size

All dimensions according to the most recent drafting standards. (For these problems, dimension only the front view.)

CAD Instructions

Students who plan to complete the following problems on a CAD system should begin with these general instructions. Before reading the specific instructions for each activity, go through each step in the following planning checklist. The checklist applies to any CAD system and will help ensure the optimum use of your time and resources.

1. Analyze the problem carefully. Decide exactly what you are being asked to do.

2. Determine what resources and references (if any) you will need in order to complete the problem. Collect those references and resources and have them readily available.

3. Decide if any particular standards apply to the project and have those standards available.

4. Determine what types of views will be required and how many of each.

5. Determine what the final plotted scale of the drawing will need to be, and select the appropriate paper size for plotting/printing.

6. Plan your drawing sequence. In what order will you develop the drawing (i.e., lines, features, dimension lines, leaders, dimensions)?

7. Review the various CAD commands you will have to use in order to develop the drawing.

8. Examine your CAD system to ensure that everything is in working order, then begin the project.

PROBLEMS 15-1A AND 15-1B

Construct a cam displacement diagram on an A-size, 8½ × 11 sheet of paper per the dimensions given. Lay out a cam with the following specifications:

1. Working circle 5.0 dia.
2. Shaft dia. .75
3. Hub dia. 1.25
4. Square key .18
5. Rotation: counterclockwise
6. Follower dia. 1.0
7. Start 1.0 from center

Required motion:

1. Rise 90°, modified uniform velocity 1.5
2. Dwell 60°
3. Fall 60°, modified uniform velocity .75
4. Dwell 30°
5. Fall 60°, modified uniform velocity .75
6. Dwell 60°

Construct the cam and follower on an A-size, 8½ × 11 sheet of paper per the dimensions given, and transfer all dimensions from the displacement diagram to the cam layout.

Check all work and add all required dimensions and specifications according to the most recent drafting standards.

Problem 15-1A

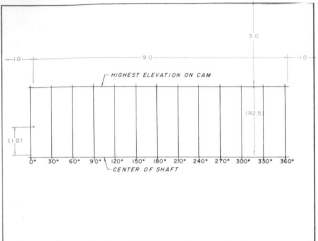

Problem 15-1B

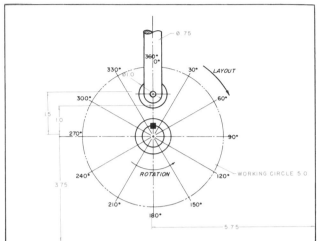

PROBLEMS 15-2A AND 15-2B

Construct a cam displacement diagram on an A-size, 8½ × 11 sheet of paper per the dimensions given. Lay out a cam with the following specifications:

1. Working circle 5.0 dia.
2. Shaft dia. .75
3. Hub dia. 1.25
4. Square key .18

5. Rotation: clockwise
6. Follower dia. 1.0
7. Start 1.0 from center

Required motion:

1. Rise 90°, harmonic motion 1.00
2. Dwell 30°

3. Rise 90°, harmonic motion .50
4. Dwell 30°
5. Fall 90°, harmonic motion 1.50
6. Dwell 30°

Construct the cam and follower on an A-size, 8½ × 11 sheet of paper per the dimensions given, and transfer all dimensions from the displacement diagram to the cam layout.

Check all work and add all required dimensions and specifications according to the most recent drafting standards.

Problem 15-2A

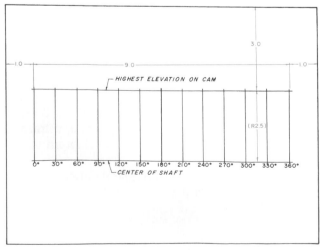

Problem 15-2B

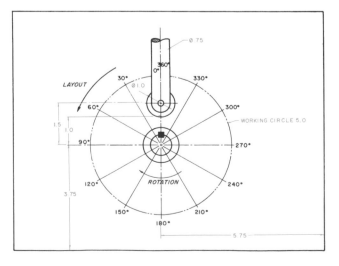

PROBLEMS 15-3A AND 15-3B

Construct a cam displacement diagram on an A-size, 8½ × 11 sheet of paper per the dimensions given. Lay out a cam with the following specifications:

1. Working circle 5.0
2. Shaft dia. .63
3. Hub dia. 1.0
4. Square key to suit
5. Rotation: clockwise
6. Follower dia. .75
7. Start 2.5 from center

Required motion:

1. Fall 180°, uniform acceleration 1.50

2. Dwell 45°
3. Rise 90°, uniform acceleration 1.50
4. Dwell 45°

Construct the cam and follower on an A-size, 8½ × 11 sheet of paper per the dimensions given, and transfer all dimensions from the displacement diagram to the cam layout.

Check all work and add all required dimensions and specifications according to the most recent drafting standards.

Problem 15-3A

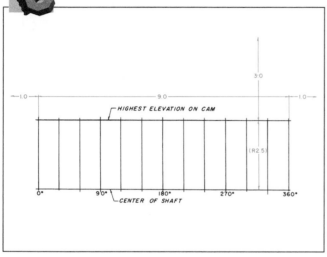

Problem 15-3B

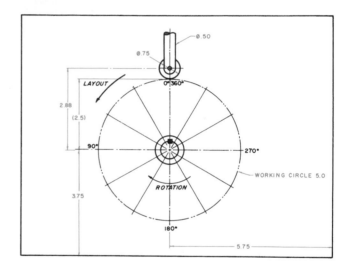

PROBLEMS 15-4A AND 15-4B

Construct a cam displacement diagram on an A-size, 8½ × 11 sheet of paper per the dimensions given. Lay out a cam with the following specifications (metric):

1. Working circle 245
2. Shaft dia. 28
3. Hub dia. 44
4. Square key to suit
5. Rotation: clockwise
6. Follower dia. 22
7. Start 32 from center

Required motion:

1. Rise 60°, harmonic motion 25
2. Dwell 15°
3. Rise 90°, harmonic motion 38
4. Dwell 15°

5. Fall 45°, harmonic motion 44
6. Dwell 15°
7. Rise 30°, harmonic motion 18
8. Dwell 15°
9. Fall to starting level 75°, harmonic motion 38

Construct the cam and follower on an A-size, 8½ × 11 sheet of paper per the dimensions given, and transfer all dimensions from the displacement diagram to the cam layout.

Check all work and add all required dimensions and specifications according to the most recent drafting standards.

Problem 15-4A

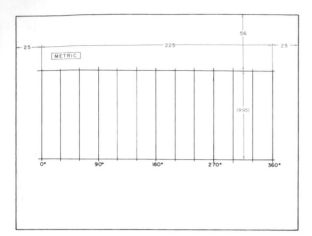

Problem 15-4B

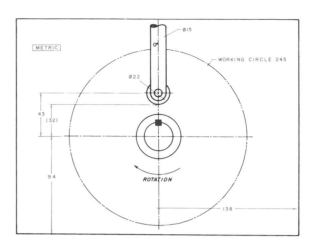

PROBLEMS 15-5A AND 15-5B

Construct a cam displacement diagram on an A-size, 8½ × 11 sheet of paper per the dimensions given. Lay out a cam with the following specifications:

1. Working circle 7.0
2. Shaft dia. .625
3. Hub dia. 1.0
4. Square key to suit

5. Rotation: counterclockwise
6. Follower dia. 1.0
7. Start 1.0 from center

Required motion:

1. Rise 120°, modified uniform velocity 2.5
2. Dwell 60°
3. Fall 30°, modified uniform velocity .75
4. Dwell 30°
5. Fall to starting level 90°, modified uniform velocity 1.75
6. Dwell 30°

Construct the cam and follower on an A-size, 8½ × 11 sheet of paper per the dimensions given, and transfer all dimensions from the displacement diagram to the cam layout.

Check all work and add all required dimensions and specifications according to the most recent drafting standards.

Problem 15-5A

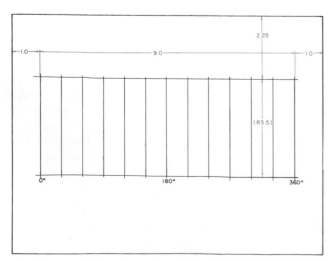

Problem 15-5B

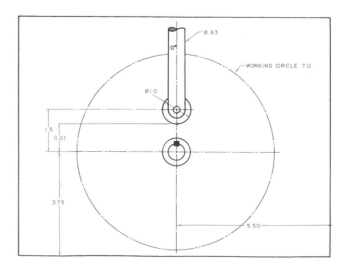

PROBLEMS 15-6A AND 15-6B

Construct a cam displacement diagram on an A-size, 8½ × 11 sheet of paper per the dimensions given.
Lay out a cam with the following specifications:

1. Working circle 7.75
2. Shaft dia. .56
3. Hub dia. .88
4. Square key to suit

5. Rotation: clockwise
6. Follower dia. .375
7. Start 1.625 from center

Required motion:

1. Fall 90°, uniform acceleration 3.0
2. Dwell 15°

3. Rise 60°, uniform acceleration 2.0
4. Dwell 105°
5. Rise 90°, uniform acceleration to starting level (1.0)

Construct the cam and follower on an A-size, 8½ × 11 sheet of paper per the dimensions given, and transfer all dimensions from the displacement diagram to the cam layout.

Check all work and add all required dimensions and specifications according to the most recent drafting standards.

Problem 15-6A

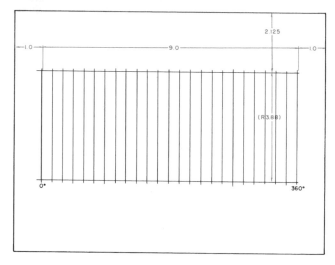

Problem 15-6B

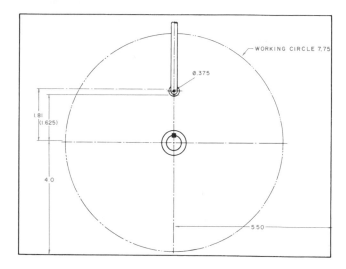

PROBLEMS 15-7A AND 15-7B

Construct a metric cam displacement diagram per the dimensions given.
Lay out a cam with the following specifications:

1. Rise/fall 56
2. Shaft dia. 16
3. Hub dia. 32
4. Square key to suit
5. Rotation: clockwise

6. Follower: flat-faced* (34 dia.)
7. Base circle 26 radius
8. Harmonic motion/one full turn

Construct the cam and flat-faced follower per the dimensions given.

Check all work and add required dimensions and specifications according to the most recent drafting standards and using the metric system.

*Design the flat-faced follower to suit.

Problem 15-7A

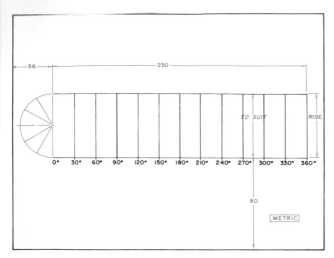

Problem 15-7B

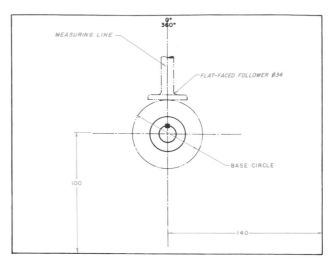

PROBLEMS 15-8A AND 15-8B

Construct a cam displacement diagram per the dimensions given.

Lay out an offset cam with the following specifications:

1. Offset circle dia. 2.00
2. Shaft dia. .75
3. Hub dia. 1.25
4. Square key to suit
5. Rotation: clockwise
6. Follower dia. .75
7. Follower offset 1.00
8. Base circle 1.375 radius

Required motion:

1. Rise 75°, uniform acceleration 1.00
2. Dwell 10°

3. Rise 35°, harmonic motion .75
4. Dwell 5°
5. Fall 70°, modified uniform velocity 1.25
6. Dwell 15°
7. Rise 45°, harmonic motion .75
8. Dwell 15°
9. Fall 60°, uniform acceleration 1.26
10. Dwell 30°

Construct the cam and offset follower per the dimensions given, and transfer all dimensions from the displacement diagram to the cam layout.

Check all work and add all required dimensions and specifications according to the most recent drafting standards.

Problem 15-8A

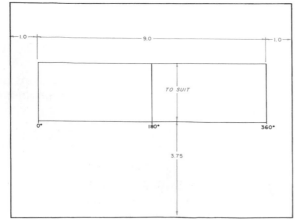

Problem 15-8B

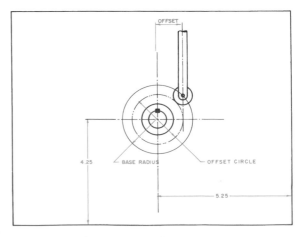

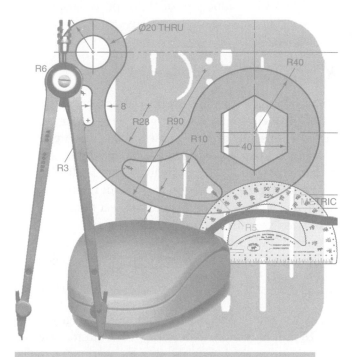

GEARS

KEY TERMS

Addendum

Angle gear

Base circle

Bevel gear

Center-to-center distance

Chain and sprockets

Chordal addendum

Chordal thickness

Circular pitch

Clearance

Dedendum

Diametral pitch

Gear tooth caliper

Gear train

Helical gear

Herringbone gear

Involute curve

Miter gear

Outside diameter

Pinion gear

Pitch diameter

Rack gear

Ring gear

Root diameter

Spiral bevel gear

Spur gear

Train value

Whole depth

Working depth

Worm

Worm gear

CHAPTER OUTLINE

Kinds of gears • Gear ratio • The involute curve • Pitch diameter and basic terminology • Pitch diameter (D) • Diametral pitch (P) • Gear blank • Pressure angle, base circle, and center-to-center distance • Rack • Gear train • Required tooth-cutting data • Measurements required to use a gear tooth caliper • Bevel gear • Worm and worm gear • Materials • Design and layout of gears • Summary • Review questions • Chapter 16 problems

CHAPTER OBJECTIVES

Upon completion of this chapter, students should be able to do the following:

■ Identify different kinds of gears.

■ Comment briefly on flexible drives such as chain and belt.

■ Discuss terms related to the profiles and kinematics of gears.

■ Determine the gear ratio between two gears.

■ Calculate train values of gear trains.

■ Construct an involute curve on a base circle.

■ Calculate the center distance between two gears in contact.

■ Prepare working drawings of gears.

Spur gears are the chief focus of this chapter; however, we plan to introduce you to various kinds of gears, gear nomenclature, technical drawing requirements for gears, the kinematics (study of motion without considering the forces) of gear trains, and related topics such as belt and chain drives. Brief considerations of forces on spur gear teeth are mentioned with respect to the pressure line between gears. Further inclusion of force topics appear in the problems at the end of the chapter.

Gear analysis and design is complex and beyond the scope of this text. For example, gear designers utilize equations that calculate stresses induced in gear teeth due to the forces transmitted and adverse operating conditions. Even more detailed analysis is required whenever wear on gear tooth surfaces is the determining design criterion.

Gears transfer rotary motion from one shaft to another shaft. Gears can change direction of rotation, speed up or slow down rotation, transmit rotational power (torque), and change rotary motion to straight-line motion. There are various kinds of gears, each with their own function, *Figure 16-1*.

Drafters must be able to identify each kind of gear, know the various functions of each, and be able to prepare working drawings of the various gears using correct terminology.

Kinds of Gears

Gears are usually classified by the position or location of the shafts they connect. A spur gear, pinion gear, or helical gear is usually used to connect shafts that are parallel to each other. Intersecting shafts at 90° are usually connected by a beveled gear or angle gear. Shafts that are not parallel to each other and that do not intersect use a worm and worm gear. In order to connect rotary motion into a reciprocating or back and forth motion, a rack gear and pinion would be used. (Note that the rack gear and pinion are also known as a rack and pinion gear).

SPUR GEAR

The *spur gear* is the most commonly used gear, *Figure 16-2*. It is cylindrical in form, with teeth that are cut straight across the face of the gear. All teeth are parallel to the axis of the shaft. The spur gear is usually considered the driven gear.

PINION GEAR

The *pinion gear* is exactly like a spur gear but it is usually smaller and has fewer teeth, Figure 16-2. The pinion gear is normally considered the drive gear.

RACK GEAR

The *rack gear* is a type of spur gear, but its teeth are in a straight line or flat instead of in a cylindrical form, *Figure 16-3*. The rack gear is used to transfer circular motion into straight-line motion.

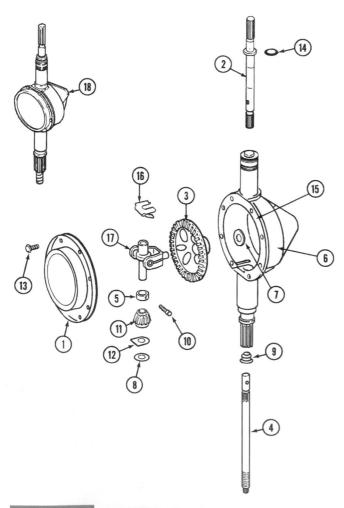

FIGURE 16-1 *Exploded assembly of gears in use* (Courtesy The Maytag Company)

FIGURE 16-2 *Spur gear (left), pinion gear (right)* (Courtesy Boston Gear)

FIGURE 16-3 *Rack gear* (Courtesy Boston Gear)

FIGURE 16-5 *Bevel gear* (Courtesy Boston Gear)

FIGURE 16-4 *Ring gear* (Courtesy Boston Gear)

FIGURE 16-6 *Miter gear* (Courtesy Boston Gear)

RING GEAR

The *ring gear* is similar to the spur, pinion, and rack gears, except that the teeth are internal, *Figure 16-4*.

BEVEL GEAR

A bevel gear is another gear commonly used, *Figure 16-5*. A *bevel gear* is cone shaped in form with straight teeth that are on an angle to the axis of the shaft. Bevel gears are used to transmit power and motion between intersecting shafts that are at 90° to each other.

ANGLE GEAR

The *angle gear* is similar to a bevel gear, except that the angles are at other than 90° to each other.

MITER GEAR

The *miter gear* is exactly the same as a bevel gear, except that both mating gears have the same number of teeth. The shafts are at 90° to each other, *Figure 16-6*.

SPIRAL BEVEL OR MITER GEARS

Any bevel gears with curved teeth are called *spiral bevel gears*, *Figure 16-7*.

WORM GEAR

Worm gears are used to transmit power and motion at a 90° angle between nonintersecting shafts, *Figure 16-8*. They are normally used to reduce speed. The worm gear is round like a wheel. The *worm* is shaped like a screw,

FIGURE 16-7 *Spiral bevel gear* (Courtesy Boston Gear)

FIGURE 16-8 *Worm and worm gear* (Courtesy Boston Gear)

with threads (or teeth) wound around it. Because one full turn of the worm is required to advance the worm gear one tooth, a high-ratio speed reduction is achieved. The worm drives the larger worm gear.

HELICAL GEARS

Helical gears operate on parallel shafts and transmit rotary motion in a manner identical to spur gears. The teeth have an involute shape but are not parallel to their rotational axis, *Figure 16-9*. Helical gear sets can transmit heavy loads at high speeds due to gradual engagement of the teeth and the smooth transfer of load from one tooth to another.

HERRINGBONE GEARS

Since helical gear teeth are cut at an angle to the axis of the gear, a force is generated along the axis. To compensate for this force, left-hand and right-hand teeth are combined, *Figure 16-10*. Large ships, for example, use *herringbone gears* in the drive train between the turbine and the propeller.

CHAIN AND SPROCKETS

A *chain and sprockets* are used to transmit motion and power to shafts that are parallel to each other, *Figure 16-11*.

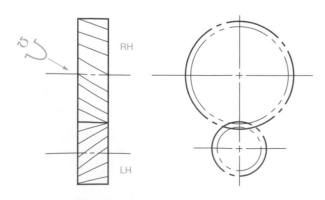

FIGURE 16-9 *Helical gears*

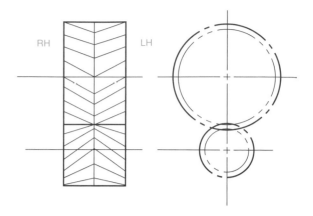

FIGURE 16-10 *Herringbone gears*

FIGURE 16-11 *Chain and sprockets* (Courtesy Boston Gear)

Sprockets are similar to spur and pinion gears, for the teeth are cut straight across the face of the sprocket and are parallel to the shaft. There are many types of chains and sprockets but all use the same terms, formulas, ratios, and so forth.

BELT DRIVES

Belt drives, like the belts used on a typical automobile engine, are similar to chain-and-sprocket drives; however, slippage can occur between belts and pulleys. Belt drives are employed to start large masses such as ventilating fans so some slippage can occur if needed.

Gear Ratio

Friction wheels can be used to illustrate the basic kinetic relationship between a driver gear and a driven gear, *Figure 16-12*. Wheel A is on a shaft parallel to the shaft of wheel B and it drives wheel B through friction at the point of contact. If wheel A rotates clockwise, then wheel B will rotate counterclockwise. Furthermore, if the circumference of wheel B is two times the circumference of wheel A, wheel A must make two complete revolutions to cause wheel B to make one (assuming no slippage).

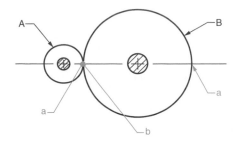

FIGURE 16-12 *Relationship of two wheels of different diameters*

The train value (TV) of this drivetrain is calculated by the equation:

$$TV = \frac{\text{driver circumference}}{\text{driven circumference}} = 0.5$$

For example, if the rotational speed of A is 30 RPM (revolutions per minute) then the rotational speed of B is calculated as follows:

$$B(RPM) = TV \times A(RPM) = 0.5 \times 30 = 15 \text{ RPM}$$

Assume that the drivetrain now is made up of gear A (20 teeth) rotating at 30 RPM and driving gear B (40 teeth). The train value (TV) is calculated by:

$$TV = \frac{\text{Number of Teeth on Driver}}{\text{Number of Teeth on Driven}} = \frac{20}{40} = 0.5$$

Then the rotational speed of gear B = $0.5 \times 30 = 15$ RPM.

The Involute Curve

Spur gears use the involute curve as the profile of their gear teeth because of the involute's unique property of being able to transfer motion and forces in a uniform manner.

HOW TO CONSTRUCT AN INVOLUTE CURVE ON A BASE CIRCLE

An *involute curve* may be considered to be the path traced by the end of a string as it is unwound from a circle called a *base circle*.

EXAMPLE:

GIVEN: Portion of a base circle m-n, which has a convenient radius of 3.2″ to fit on your paper, *Figure 16-13*.

STEP 1 Locate a point 6 to anchor a 3″ long theoretical string. Call the string 6-0 and label it 6, 5, 4, 3, 2, 1, and 0 at 0.5″ intervals, *Figure 16-13A*.

STEP 2 Beginning at point 1, construct a tangent to the 3.2″ curve (a base circle for the involute

curve), lay out the distance 1-0 (0.5″) along the tangent from the point of tangency 1, and label it 1-0(1), *Figure 16-13B*.

STEP 3 At point 2, construct a tangent to the base circle and lay out a distance from the point of tangency 2 equal to two times the distance 1-0. Label it 2-0(2), *Figure 16-13C*.

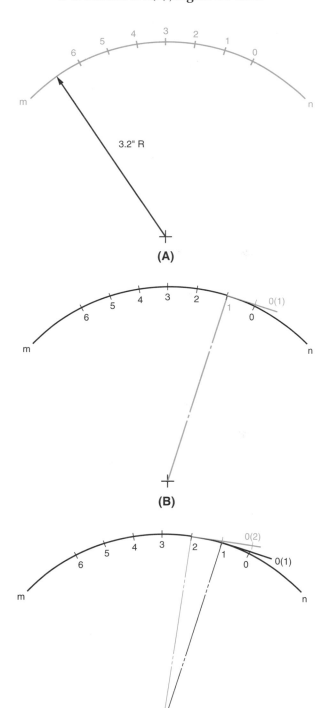

FIGURE 16-13 *How to construct an involute curve on a base circle; (A) Step 1; (B) Step 2; (C) Step 3*

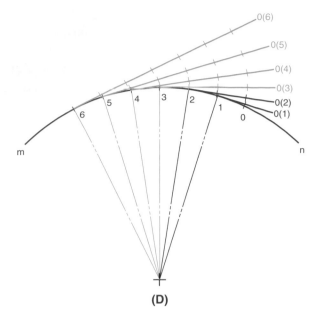

(D)

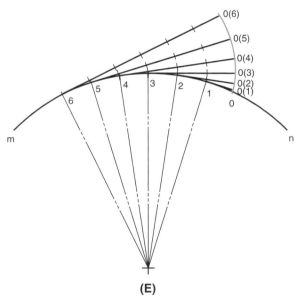

(E)

FIGURE 16-13 *(Continued) (D) Step 4; (E) Step 5*

STEP 4 From points 3, 4, 5, and 6 lay off in a similar manner distances along the tangents and label them 3-0(3), 4-0(4), 5-0(5), and 6-0(6), respectively, *Figure 16-13D.*

STEP 5 Finally connect the points 0(1)-0(6) with a French curve to produce the involute curve, *Figure 16-13E.*

Pitch Diameter and Basic Terminology

The drafter must know and understand all major terms associated with various kinds of gears used today in

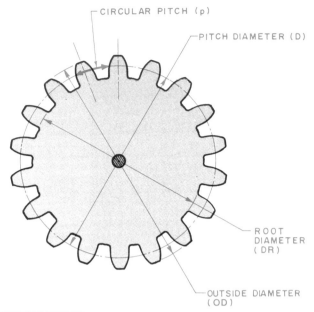

FIGURE 16-14 *Basic gear terminology*

industry. Illustrated in *Figure 16-14* are the basic terms associated with most gears, regardless of type.

Pitch Diameter (D)

The *pitch diameters* (D) of two spur gears exist theoretically only when the two gears are in contact. When the point of contact between a tooth on gear A and a tooth on gear B falls on the line of centers between the two gears, then that point determines the radii of imaginary pitch circles of gear A and gear B, respectively, *Figure 16-15.*

Figure 16-16 illustrates additional terms that will help you to prepare working drawings of spur gears. Circular pitch (p) and the terms are explained in the paragraphs that follow.

CIRCULAR PITCH (P)

Circular pitch (p) is the distance between corresponding points of adjacent teeth measured on the circumference of the pitch diameter, Figure 16-16.

Diametral Pitch (P)

Diametral pitch (P) is a ratio equal to the number of teeth (N) per inch of pitch diameter (D). Follow the derivation that follows to better understand the term.

$$(p) \times (N) = \text{the circumference of the pitch circle}$$

$$(D) \times \pi = \text{the circumference of the pitch circle}$$

$$\text{so } (p) \times (N) = (D) \times \pi$$

$$\text{and } (p) \times \frac{(N)}{(D)} = \pi$$

$$\text{then } (p) \times (P) = \pi$$

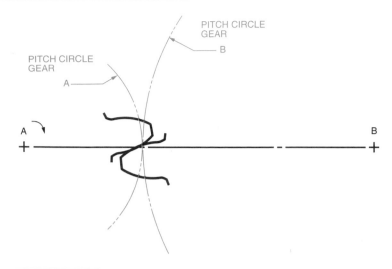

FIGURE 16-15 *Pitch circles determined from point of contact on line of centers*

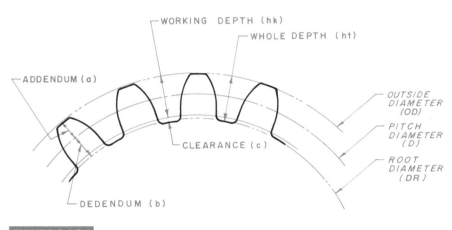

FIGURE 16-16 *Major gear terms*

EXAMPLE:

A gear with 48 teeth on a 3.0″ pitch diameter (D), would have a diametral pitch (P) of 16.

NOTE: All pairs of spur gears must have the same diametral pitch (P)!

WORKING DEPTH (HK)

The *working depth* is the distance that a tooth projects into the mating space. It is equal to gear addendum (a) plus pinion addendum (a).

ADDENDUM (A)

The *addendum* is the radial distance from the pitch diameter to the top of the tooth. It is equal to 1/P.

DEDENDUM (B)

The *dedendum* is the radial distance from the pitch diameter to the bottom of the tooth. It is equal to 1.157/P, 1.250/P, or 1.350/P; 1.250/P is preferred for 20° teeth.

CLEARANCE (C)

Clearance is the space between the working depth and the whole depth. It is equal to dedendum (b) minus addendum (a).

WHOLE DEPTH (HT)

The *whole depth* is the total depth of a tooth space. It is equal to the addendum (a), plus the dedendum (b), plus the clearance (c).

NOTE: All notations in parentheses ( ) correspond to those used in the most recent edition of Machinery's Handbook *so that they agree if a more detailed formula is required by the drafter.*

OUTSIDE DIAMETER (OD)

The *outside diameter* is the measurement over the extreme outer edge of teeth. It is equal to the pitch diameter, plus 2 times the addendum (a).

Root Diameter (DR)

The *root diameter* is the measurement over the extreme inner edges of teeth. It is equal to the pitch diameter minus 2 times the dedendum (b).

HOW TO DRAW A SPUR GEAR TOOTH

Draw a spur gear tooth for the experience and for learning spur gear nomenclature. Later you will discover that working drawings of gears and gear sets seldom require drawing the tooth profile.

GIVEN: Spur gear specifications of D = 6.0″, P = 4, T = 24, and ϕ = pressure angle of 20°.

STEP 1 Construct an arc representing the basic circle, *Figure 16-17A*. The base circle radius = D cos ϕ/2 = 6.0 × cos 20°/2 = 2.819″.

STEP 2 At an arbitrary point B on the base circle, draw an involute curve one of several ways. Your choices are the basic method shown in Figure 16-13, the approximate method (D/8 = approximate involute curve), or the simplest method using a template, *Figure 16-18*. A fourth choice would be to use a CAD program. For this example use the approximate method, *Figure 16-17B*. Then D/8 = 6.0″/8 = 0.75″.

STEP 3 Draw an arc representing the pitch circle, D/2 = 6.0″/2 = 3.0″ and label point T where the pitch circle intersects the approximate involute curve, *Figure 16-17C*.

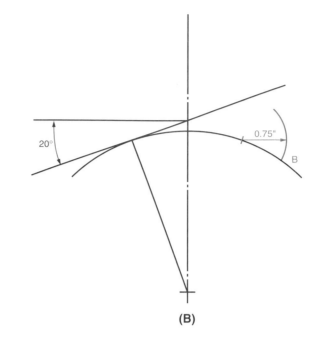

(B)

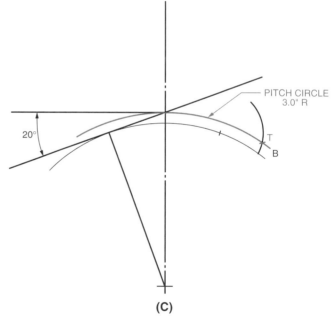

(C)

FIGURE 16-17 *(Continued) (B) Step 2; (C) Step 3*

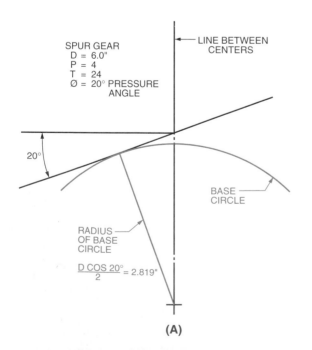

(A)

FIGURE 16-17 *How to draw a spur gear tooth; (A) Step 1*

STEP 4 Locate the tooth width along the pitch circle, *Figure 16-17D,* and label the width T-S. The width T-S equals half the circular pitch p. The circular pitch p = π/P = π/4 = 0.785″. The tooth width then is 0.785″/2 = 0.392″. Construct an approximate involute curve from the base circle and going through point S.

STEP 5 Finish the tooth profile by constructing the addendum and dedendum circles plus two radial lines, *Figure 16-17E*. Add the addendum 1/P = 1/4 = 0.25″ to the pitch circle arc,

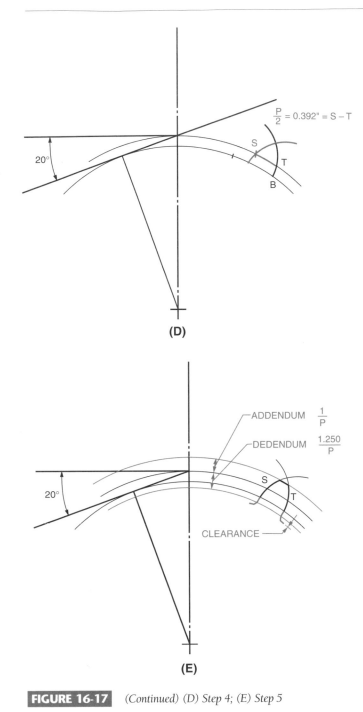

$$\frac{P}{2} = 0.392" = S - T$$

(D)

ADDENDUM $\frac{1}{P}$

DEDENDUM $\frac{1.250}{P}$

CLEARANCE

(E)

FIGURE 16-17 *(Continued) (D) Step 4; (E) Step 5*

FIGURE 16-18 *Gear template—spur/rack/worm* (Courtesy Modern School Supply)

	Inch 20° 24T P = 4	Metric 20° 24T P = 4
	SPUR GEAR TEETH Comparison	
Pitch Diameter	6.0"	152.4 mm
Circular Pitch	π/P = 0.785"	19.95 mm
Base Circle Diameter	D cos 20° = 5.638"	143.2 mm
Module(MDL) = (length of pitch diameter per tooth)		pitch dia/24 = 152.4/24 = 6.35 mm
Addendum	1/P = 0.25"	MDL = 6.35 mm
Dedendum	1.157/P = 0.289" or 1.250/P = 0.312	1.157 x MDL = 7.35 mm 1.25 x MDL = 7.94 mm

FIGURE 16-19 *Comparison of metric and inch spur gears*

drafter must first calculate the diametral pitch (P). The numbers next to each tooth indicate the diametral pitch. The lower part of the template opening is used to draw spur gear teeth. The top portion of the opening is used to draw teeth of a rack or worm.

METRIC SPUR GEARS

Metric spur gears and inch spur gears are essentially identical. For a comparison of an inch gear with a metric one, a gear with a 20° pressure angle, 24 teeth, and a diametral pitch of P = 4 is shown in *Figure 16-19*.

Gear Blank

Gears are usually cut from a gear blank. The gear blank must allow sufficient space for the gears and a method to attach the gear to the shaft. Illustrated in *Figure 16-20* is a common gear blank with an outside diameter and face width for the gears to be cut into, a hub, and a hole for the set screw to secure the gear blank to the shaft. A half-section is used to illustrate the gear blank in this example. Two complete sets of dimensions must be applied to the gear drawing; those relating to the gear blank, as illustrated, and those relating to the teeth. Gear

and subtract the dedendum 1.250/P = 0.312″ from the pitch circle arc. Finally, from where the base circle intersects the tooth profile, draw radial segments from the base circle to the dedendum arc. To provide a small amount of clearance, Figure 16-16, add small fillets.

GEAR TEMPLATE

A quick and efficient method of illustrating teeth is by using a gear template, Figure 16-18 (shown at approximately half size). When using a gear template, the

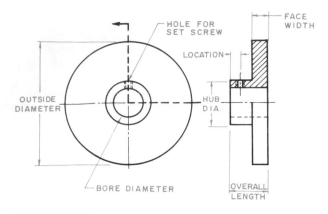

FIGURE 16-20 *Gear blank*

dimensions do not hold tight tolerancing, but all gear tooth dimensions must hold very tight tolerancing.

Pressure Angle, Base Circle, and Center-to-Center Distance

The pressure angle determines the diameters of the base circles, which in turn determine the tooth forms. To illustrate this principle, the steps to prepare a drawing of two spur teeth in contact will be discussed.

HOW TO DRAW TWO SPUR GEAR TEETH IN CONTACT

GIVEN: A 20-tooth pinion gear to drive a 40-tooth gear, P = 5. The pressure angle ϕ is to be 20°.

STEP 1 Start the drawing by laying out the center-to-center distance, *Figure 16-21A*. The *center-to-center distance* equals the sum of the radius of the pinion plus the radius of the gear.

D (pinion) = N/P = 20/5 = 4″

R (pinion) = 4/2 = 2″

D (gear)   = N/P = 40/5 = 8″

R (gear)   = 8/2 = 4″

C-to-C   = 2 + 4 = 6″

Label the juncture of the two radii, T. Construct a perpendicular to the line between centers at point T.

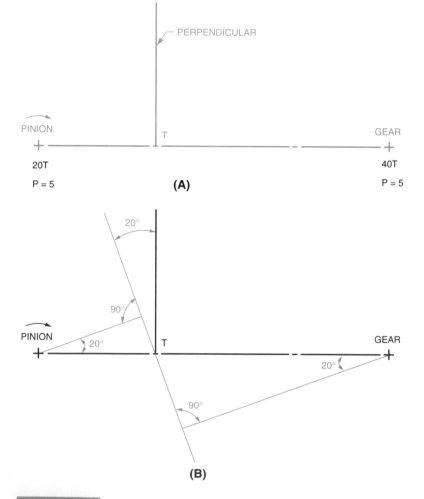

FIGURE 16-21 *How to draw two spur gear teeth in contact; (A) Step 1; (B) Step 2*

STEP 2 Lay out the pressure line at 20° to the perpendicular at T assuming that the pinion is rotating clockwise and driving the gear, *Figure 16-21B*. Next construct perpendiculars from the pinion and gear centers to the pressure line.

STEP 3 Use the perpendiculars that represent the radii of the respective base circles to draw the base circles, *Figure 16-21C*.

NOTE: *The lengths of the radii of the base circles equal the radii of the pitch circles times cos 20°.*

STEP 4 To save time, use the approximate method (D/8, Figure 16-17) to draw the tooth profiles

from each base circle, the profiles to go through point T, *Figure 16-21D*. Finish the two teeth and note that the base of the tooth on the gear is wider than the base on the pinion.

STEP 5 Assume that the pinion rotates approximately one tooth width clockwise (p/2) and construct a new pinion tooth, *Figure 16-21E*. Draw a gear tooth in contact and observe that the two teeth are in contact on the pressure line. The true involute profile ensures that the teeth will always be in contact along the pressure line—hence the smooth transfer of rotation and power.

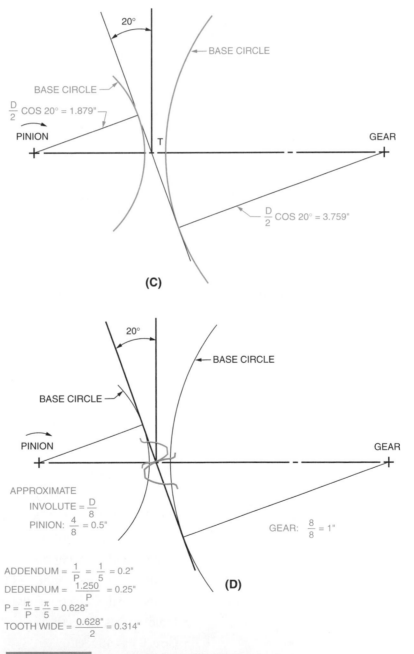

FIGURE 16-21   *(Continued) (C) Step 3; (D) Step 4*

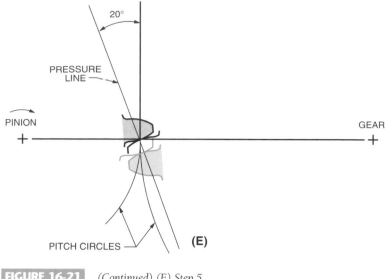

FIGURE 16-21 *(Continued) (E) Step 5*

Rack

A rack is simply a gear with teeth formed on a flat surface, *Figure 16-22*. A rack changes rotary motion into reciprocating motion. All terms and formulas associated with spur and pinion gears apply to the rack. The sides of the teeth are straight, not involute as on a spur gear. The teeth are inclined at the same angle as the pressure angle of the mating pinion gear. Note that the circular pitch of the pinion gear is the same as the linear pitch of the rack. All dimensions for heights of depth, pitch, and addendum lines are calculated from a datum or reference line. The rack is usually manufactured out of rectangular stock but, occasionally, it is manufactured out of round stock to meet a specific design.

The required tooth-cutting data for a rack are:

- Pressure angle (same as mating pinion gear)
- Tooth thickness at pitch line (same as mating pinion gear)
- Whole depth (same as mating pinion gear)
- Maximum allowance pitch variation
- Accumulated pitch error max. (over total length)

Gear Train

A *gear train* is two or more gears used to achieve a designed RPM. The ratio of the RPM of the last gear to the RPM of the first gear is called the *train value* of the gear train. Calculate the train value as follows:

$$\text{Train Value (TV)} = \frac{\text{product of number of teeth on driving gears}}{\text{product of number of teeth on driven gears}}$$
and
$$\text{RPM (last gear)} = \text{TV} \times \text{RPM first gear}$$

EXAMPLE:

Assume that a 150 RPM gear motor with a 15-tooth pinion drives a 40-tooth gear mounted on a conveyor driveshaft. The speed of the driveshaft would be:

$$\text{RPM (driveshaft)} = \frac{\begin{array}{c}15 \text{ teeth}\\ \text{on driver}\end{array}}{\begin{array}{c}40 \text{ teeth}\\ \text{on driven}\end{array}} \times \text{RPM (gear motor)}$$

$$\text{RPM (driveshaft)} = 0.375 \times 150 = 56.25 \text{ RPM}$$

Consider a more complicated gear train, *Figure 16-23A*.

- Shafts 1 and 2 are connected by pinion A and gear B
- Shafts 2 and 3 are connected by pinion C and gear D
- Shafts 3 and 4 are connected by pinion E and gear F
- Pinion A has 12 teeth
- Gear B has 45 teeth
- Pinion C has 20 teeth

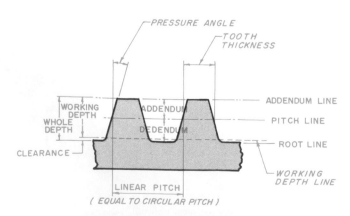

FIGURE 16-22 *Gear terminology for a rack*

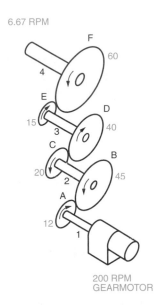

FIGURE 16-23A *A more complicated gear train*

- Gear D has 40 teeth
- Pinion E has 15 teeth
- Gear F has 60 teeth

PROBLEM:

For an input speed of 200 RPM at shaft 1, what is the output speed of shaft 4 in RPM?

$$\text{Train Value (TV)} = \frac{12 \times 20 \times 15}{45 \times 40 \times 60} = 0.033$$

Output speed shaft 4 = 0.033 x 200 = 6.67 RPM

Observe the use of arrows in Figure 16-23A to show the direction of gear rotation. Gear A is rotating clockwise; therefore, gear B is rotating counterclockwise.

Occasionally, gear trains are drawn with their centers on a horizontal line, *Figure 16-23B.* The upper view clarifies the relationships of the gears and the directions of rotation.

The equation for calculating the train value of a spur gear train applies to other gear trains such as bevel, worm, and helical, plus combinations of all of them.

Required Tooth-Cutting Data

Overall gear blank dimensions are shown on the detail drawing (as illustrated in Figure 16-20) and can hold loose tolerancing. The actual gear dimensions must hold tight tolerancing.

Twelve essential items of information must be calculated and listed on the gear detail drawing. This list is usually located at the lower right-hand side of the drawing, and generally consists of the data in *Figure 16-24,* but these items may vary slightly from company to com-

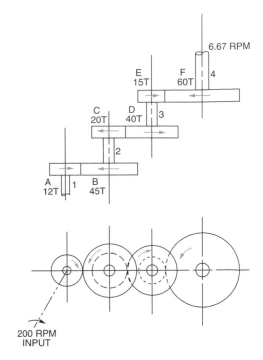

FIGURE 16-23B *Optional methods of drawing gear trains*

pany. The required material, heat-treating process, and other important information must also be included, usually in the title block.

In actual practice, the gear teeth are not drawn because it takes too much drawing time. A simplified method is used to illustrate the actual gear tooth, similar to that used to illustrate threads of a fastener, *Figure 16-25.* The outside diameter and the root diameter are illustrated by a thin phantom line. The pitch diameter is illustrated by a thin center line. Actual teeth are not drawn, except to illustrate some special feature in relationship to a tooth, a spline, a keyway, or a locating point. In this example, all dimensions in color are related to the gear blank only. All gear dimensions are called-off in the cutting data box underneath.

Recent practice of spur gear design favors 20° and 25° pressure angles; however, 14.5° full depth and 20° stub (shorter addendum) are not recommended for new designs according to the *Machinery's Handbook, 25th Edition.*

Refer to the *Machinery's Handbook* for more extensive information on gearing and gearing standards, including ANSI B6.1-1968, R1974, and ANSI B6.7-1977.

Measurements Required to Use a Gear Tooth Caliper

A *gear tooth caliper, Figure 16-26,* is used to check and measure an individual gear tooth. Gear tooth calipers measure two parts of the tooth profile: the chordal

REQUIRED CUTTING DATA

ITEM	TO FIND:	HAVING	FORMULA
1	Number of teeth (N)	D & P	D x P
		DO & P	(DO x P) – 2
2	Diametral pitch (P)	p	$\frac{3.1416}{p}$
		D & N	$\frac{N}{D}$
		DO & N	$\frac{N+2}{DO}$
3	Pressure angle (ø)	–	20° STANDARD 14°-30' OLD STANDARD
4	Pitch diameter (D)	N & P	$\frac{N}{P}$
		N & DO	$\frac{N \times DO}{N+2}$
		DO & P	$DO - \frac{2}{P}$
5	Whole depth (ht)	a & b	a + b
		P	$\frac{2.157}{P}$
6	Outside diameter (OD)	N & P	$\frac{N+2}{P}$
		D & P	$D + \frac{2}{P}$
		D & N	$\frac{(N+2) \times D}{N}$
7	Addendum (a)	P	$\frac{1}{P}$
8	Working depth (hk)	a	2 x a
		P	$\frac{2}{P}$
9	Circular thickness (t)	P	$\frac{3.1416}{2 \times P}$
10	Chordal thickness	N, D & a	$\sin\left(\frac{90°}{N}\right) \times D$
11	Chordal addendum	N, D & a	$\left[1-\cos\left(\frac{90°}{N}\right)\right] \times \frac{D}{2} + a$
12	Dedendum (b)	P	$\frac{1.157}{P}$

FIGURE 16-24 *Required tooth-cutting data for spur and pinion gears*

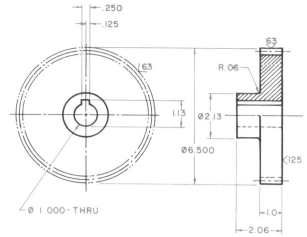

CUTTING DATA	
NUMBER OF TEETH	24
DIAMETRAL PITCH	4
PRESSURE ANGLE	20°
PITCH DIAMETER	6.000
WHOLE DEPTH	0.5393
OUTSIDE DIAMETER	6.500
ADDENDUM	.250
WORKING DEPTH	.500
CIRCULAR THICKNESS	.3925
CHORDAL THICKNESS	.2566
CHORDAL ADDENDUM	.3924
DEDENDUM	.289

FIGURE 16-25 *Example of a spur gear detail drawing*

FIGURE 16-26 *Using a gear tooth caliper* (Courtesy L. S. Starrett Co.)

thickness and the chordal addendum, *Figure 16-27*. The center slide is set to the desired chordal addendum using the vertical vernier and scale. The tips of the two jaws are then used to measure the chordal thickness read on the horizontal vernier and scale.

CHORDAL THICKNESS

The *chordal thickness* is the length of the chord measured straight across at the pitch diameter. It is measured straight across the tooth, not along the pitch diameter as is the circular thickness.

$$\text{Chordal thickness} = \sin\left(\frac{90°}{N}\right) \times \text{Pitch dia.}$$

CHORDAL ADDENDUM

The *chordal addendum* is the distance from the top of the tooth to the point at which the chordal thickness is measured.

$$\text{Chordal addendum} = \left[1 - \cos\left(\frac{90°}{N}\right)\right] \times \frac{\text{Pitch dia.}}{2} + \text{addendum}$$

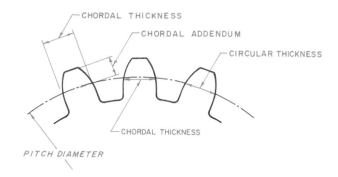

FIGURE 16-27 *Measurements required to use a gear tooth caliper*

Bevel Gear

Bevel gears transmit power and motion between intersecting shafts at right angles to each other. *Figure 16-28* illustrates the various terms associated with bevel gears. The required tooth-cutting data for bevel gears are similar to spur/pinion gears. These must be calculated and listed on the detail drawing. *Figure 16-29* shows the order in which the data and formulas should be listed on the drawing. Bevel gears must be designed and drawn in pairs to ensure a perfect fit.

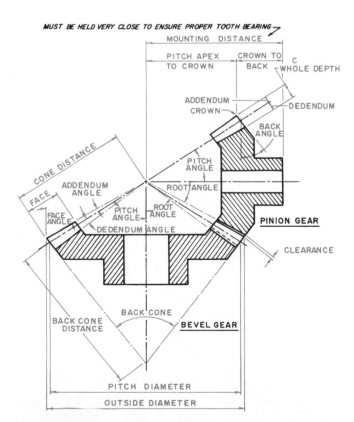

FIGURE 16-28 *Gear terminology for bevel gears*

	REQUIRED CUTTING DATA			
ITEM	**TO FIND:**	**HAVING**	**FORMULA**	
			SPUR	**PINION**
1	Number of teeth (N)	–	AS REQ'D.	
2	Diametral pitch (P)	p	$\dfrac{3.1416}{p}$	
3	Pressure angle (ø)	–	20° STANDARD 14°-30' OLD STANDARD	
4	Cone distance (A)	D & d	$\sin d \sqrt{\dfrac{D}{2}}$	
5	Pitch distance (D)	p	$\dfrac{N}{p}$	
6	Circular thickness (t)	p	$\dfrac{1.5708}{p}$	
7	Pitch angle (d)	N & d (of pinion)	90°–d(pinion)	$\tan d \dfrac{N\ pinion}{N\ gear}$
8	Root angle (γ R)	d & δ	d – δ	
9	Addendum (a)	p	$\dfrac{1}{p}$	
10	Whole depth (ht)	p	$\dfrac{2.188}{p}$ + .002	
11	Chordal thickness (C)	D & d	$\dfrac{1}{2}\left(\dfrac{D}{\cos d}\right)$	$1 - \cos\left(\dfrac{\frac{90°}{N}}{\cos d}\right) + a$
12	Chordal addendum (aC)	d	$\sin\left(\dfrac{\frac{90°}{N}}{\cos d}\right)$	
13	Dedendum (bC)	P	$\dfrac{2.188}{P} - a(pinion)$	$\dfrac{2.188}{P} - a(gear)$
14	Outside diameter (OD)	D, a & d	D + (2 x a) x cos d	
15	Face	A	1/3 A (max.)	
16	Circular pitch (p)	p & N	$\dfrac{3.1416 \times p}{N}$	
17	Ratio	N gear & N pinion	$\dfrac{N\ gear}{N\ pinion}$	
18	Back angle (γ O)	–	SAME AS PITCH ANGLE	
19	Angle of shafts	–	90°	
20	Part number of mating gear	–	AS REQ'D.	
21	Dedendum angle δ	A & b	$\dfrac{b}{A} = \tan \delta$	

FIGURE 16-29 *Required tooth-cutting data for bevel gears*

Worm and Worm Gear

The worm and worm gear are used to transmit power between nonintersecting shafts at right angles to each other. With the worm and worm gear, a large speed ratio

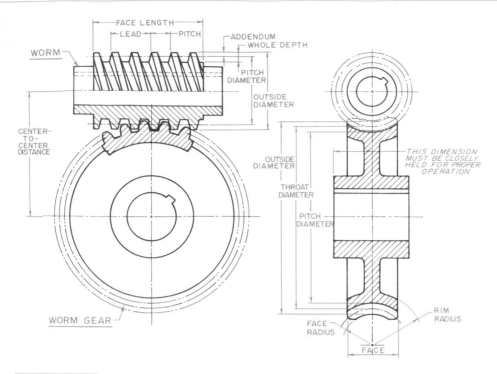

FIGURE 16-30 *Examples of worm and worm gear data*

is possible because one revolution of a single-thread worm turns the worm gear only one tooth and one space.

The worm's thread is similar in form to a rack tooth. The worm gear is similar in form to a spur gear, except that the teeth are twisted slightly and curved to fit the curvature of the worm.

When drawing the worm and worm gear, an approximate representation is used, *Figure 16-30*. Cutting data for the worm and worm gear must be listed on the drawing in the lower-right side in the order illustrated in *Figure 16-31* for the worm gear, and in *Figure 16-32* for the worm.

> NOTE: *It is important that the mounting of a worm and gear set ensures that the central plane of the gear passes essentially through the axis of the worm. This may be accomplished by adjusting the gear axially at assembly by means of shims. When properly mounted and lubricated, worm gear sets will become more efficient after the initial breaking-in period.*

Other information that must be included on a worm/worm gear detail drawing includes:

- Gear blank information
- Tooth-cutting data
- Reference to mating part

Materials

Gears are made of many materials, such as brass, cast iron, steel, and plastic, to mention but a few. Many bevel

REQUIRED CUTTING DATA

ITEM	TO FIND:	HAVING	FORMULA
1	Number of teeth (N)	–	AS REQ'D.
2	Pitch diameter (D)	N & p	$\frac{N \times p}{3.1416}$
3	Addendum (a)	p	p x .3181
		P	$\frac{1}{P}$
4	Whole depth (ht)	p	p x .6866
		P	$\frac{2.157}{P}$
5	Lead (L) Right-Left	p & N	p x N
6	Worm part no.	–	AS REQ'D.
7	Pressure angle ø	–	20° STANDARD 14°-30' OLD STANDARD
8	Outside diameter (OD)	Dt & Pa	Dt + .4775 x Pa
9	*Circular pitch (p)	P	$\frac{3.1416}{P}$
		L & N	$\frac{L}{N}$
10	Diametral pitch (P)	p	$\frac{3.1416}{p}$
11	Throat diameter (Dt)	D & Pa	D + .636 x Pa
12	Ratio of worm/worm gear	N worm & N worm gear	$\frac{N \text{ worm gear}}{N \text{ gear}}$
13	Center-to-center distance between worm & worm gear	D worm & D worm gear	$\frac{D \text{ worm} + D \text{ worm gear}}{2}$

*Circular pitch (p) must be same as worm axial pitch (Pa)

FIGURE 16-31 *Required tooth-cutting data for a worm gear*

ITEM	TO FIND:	HAVING	FORMULA
	REQUIRED CUTTING DATA		
1	Number of teeth (N)	P	$\dfrac{3.1416}{P}$
2	Pitch diameter (D)	Pa	(2.4 x Pa) + 1.1
		DO & a	DO – (2 x a)
3	*Axial pitch (Pa)	–	Distance from a point on one tooth to same point on next tooth
4	Lead (L) Right or Left	p & N	p x N
5	Lead angle (La)	L & D	$\dfrac{L}{3.1416 \times D}$ = tan La
6	Pressure angle (ø)	–	20° STANDARD 14°-30' OLD STANDARD
7	Addendum (a)	p	p x .3183
		P	$\dfrac{1}{P}$
8	Whole depth (ht)	Pa	.686 x Pa
9	Chordal thickness	N, D & a	$\left[1 - \cos\left(\dfrac{90°}{N}\right)\right] \times \dfrac{D}{2} + a$
10	Chordal addendum	N, D & a	$\sin\left(\dfrac{90°}{N}\right) \times D$
11	Outside diameter (OD)	D & a	D + (2 x a)
12	Worm gear part no.	–	AS REQ'D.

*Axial pitch (Pa) must be same as worm gear circular pitch (p)

FIGURE 16-32 *Required tooth-cutting data for worm*

gears are forged, some are stamped from thin material, some are cast as a blank and machined, and others are die-cast to the exact size and shape.

Each application must be carefully analyzed. Metal gears have been used for years, but, if the load is not excessive, plastic gears have many advantages. Plastic runs quieter, has a self-lubricating effect, weighs much less, and costs much less. In addition, complicated multiple gears can be molded into a single piece, which further reduces costs.

Design and Layout of Gears

The initial design of gears starts with the nominal size of the pitch diameters. The required speed ratio, loading, space limitations, and center-to-center distances are also important factors to consider. Mating gears and pinions must have equal diametral pitch in order to correctly mesh; therefore, the diametral pitch should be one of the first considerations. A complete analysis and design of a gear or a complete gear chain are very complex, and far beyond the scope of this text. Most designers try to use standard gears from a company that specializes in gears, gear chains, and gear design. Most gear manufacturing companies can be of assistance in designing gears for

special applications. These companies can usually manufacture gears at a lower price than if they were made in-house.

Further analysis, in-depth study, and design data can be found in the most recent edition of *Machinery's Handbook*.

Summary

- Gears are classified by the position or location of the shafts they connect. The most widely used types of gears are spur, pinion, rack, ring, bevel, angle, miter, worm, helical, and herringbone.

- Diametral pitch is a ratio equal to the number of teeth (N) per inch of pitch diameter (D). Circular pitch is the distance between corresponding points of adjacent teeth measured on the circumference of the pitch diameter.

- A gear blank is the material from which the gear is cut. The gear blank must allow sufficient space for the gears and a method to attach the gear to the shaft.

- A rack is a gear with teeth formed on a flat surface. A rack changes rotary motion into reciprocating motion.

- A gear train is two or more gears used to achieve a designed RPM.

- A tooth gear caliper is used to check and measure an individual gear tooth.

- Bevel gears transmit power and motion between intersecting shafts at right angles to each other.

- The worm and worm gear are used to transmit power between nonintersecting shafts at right angles to each other.

- Gears are made of many materials such as brass, cast iron, steel, and plastic. Gears may be machined, forged, stamped, or cast.

Review Questions

Answer the following questions either true or false.

1. The outside diameter of a spur gear having a pitch diameter of 1.500 and 48 teeth is 1.68.

2. The pitch diameter of a spur gear having 40 teeth and a diametral pitch of 20 is 19.68.

3. The circular pitch of a bevel pinion having a 16 diametral pitch and 20 teeth is 1.43.

4. A ratio of 2:1 between gear A and gear B means that gear B rotates two times every time gear A rotates once.

Industry Application

GEARING UP FOR A JOB

The want ad read in part as follows: *"Opening for entry-level mechanical drafting technician. Must have training in laying out and drawing gears."* It was the sentence about gears that gave Peter Walton a problem. He had been an excellent student of Florida Community College (FCC). His field of study was mechanical drafting, and he had graduated with honors. But gears? Walton didn't remember studying gears. His program at FCC had focused on tool design. Walton knew a great deal about jigs and fixtures, but next to nothing about gears.

But Walton needed a drafting job, and, at least for the time being, could not relocate. His mother was in poor health and his father had died when Walton was a young child. Consequently, he needed to stay at home and look after his mother. There weren't many drafting jobs available in Walton's little town of Long Bend, Florida. Since graduating from FCC he had been working at a convenience store. Walton had earned a degree in drafting, and he wanted to put it to work.

Knowing that he might not get another chance without relocating, Walton made up his mind. "I am going to get this job." Walton called and made an appointment for an interview. Regarding the gear issue, he told the company personnel manager, "I will bring several gear drawings that will verify my ability to lay out various types of gears." To himself he said, "I may not be an expert on gears now, but I will be when I interview for the job!"

Walton had just one week to learn about gear design and to prepare several drawings. He didn't waste any time. Walton drove to FCC and met with his former drafting professor who quickly pulled several books off the shelf and showed him what he would have to learn. Walton began his crash course immediately.

He read about spurs, pinion, rack, ring, bevel, miters, and worm gears. He learned about gear ratio, pressure angle, and diametral pitch. He learned how to calculate the value of the gear train. Walton studied until his head hurt and his vision was blurry; then he studied some more. After three days he was ready to start drawing. After two days of drawing he had a comprehensive portfolio that included a spur gear, two bevel gears, and a worm gear.

Two days later the interview went well and Walton got the job. In fact, the company officials were so impressed with Walton's knowledge of gears that they increased the starting salary for the job!

5. The pitch diameter of a gear can be thought of as the outside diameter of the friction wheel.

6. The root diameter of a spur gear having a pitch diameter of 2.000 and 32 teeth is 1.75.

Answer the following questions by selecting the best answer.

1. Diametral pitch can be defined as:
   a. Diameter of the gear blank divided by two.
   b. A ratio equal to the number of teeth on a gear per inch of pitch diameter.
   c. The chordal thickness divided by the pressure angle.
   d. Number of teeth divided by the dedendum.

2. What is the dedendum of a spur gear having a pitch diameter of 3.000 and 48 teeth?
   a. 0.3857
   b. 0.2692
   c. 1.4617
   d. 0.3889

3. How many teeth are on a spur gear having a pitch diameter of 1.750, diametral pitch of 20, and outside diameter of 1.850?
   a. 3.24
   b. 37
   c. 16
   d. 35

4. What is the diametral pitch of a pinion gear having 75 teeth and 3.208 outside diameter? (Use the formula $P = N/D$)
   a. 19.68
   b. 23.38
   c. 24.83
   d. 23.21

5. When cutting the teeth for bevel gears, how does one compute diametral pitch (P)?
   a. $.5 (2.188) + \sin d$
   b. $1 \div p$
   c. $3.1416 \times \sin d$
   d. $3.1416 \div p$

6. When cutting the teeth for a worm gear, how does one compute the pitch diameter?
   a. D + (2 × a)
   b. .686 × Pa
   c. P × N
   d. OD − (2 × a)

7. A gear train can best be described as:
   a. A series of gears linked together.
   b. Two or more gears used to achieve a designed RPM.
   c. A power transmission device for transferring motion.
   d. A series of gear pinions.

8. How does one compute the center-to-center distance between worm and worm gear?
   a. P × .3181
   b. D + .636 × Pa
   c. Nworm gear ÷ Dworm
   d. Dworm + Dworm gear 2

Chapter 16 Problems

The following problems are intended to give the beginning drafter practice in using the many formulas associated with gears and practice in laying out finished professional detail drawings of major kinds of gears.

CONVENTIONAL INSTRUCTIONS

The steps to follow in laying out gears are:

STEP 1 Study the problem carefully.

STEP 2 Make a sketch if neccessary.

STEP 3 Do all math required for each problem. Keep all math work for rechecking.

STEP 4 Center the required views within the work area.

STEP 5 Include all dimensioning according to the most recent drafting standards.

STEP 6 Add all required gear cutting specifications in the lower right side of the paper.

STEP 7 Recheck all work, and, if correct, neatly fill out the title block using light guidelines and neat lettering.

CAD Instructions

Students who plan to complete the following problems on a CAD system should begin with these general instructions. Before reading the specific instructions for each activity, go through each step in the following planning checklist. The checklist applies to any CAD system and will help ensure the optimum use of your time and resources.

1. Analyze the problem carefully. Decide exactly what you are being asked to do.

2. Determine what resources and references (if any) you will need in order to complete the problem. Collect those references and resources and have them readily available.

3. Decide if any particular standards apply to the project and have those standards available.

4. Determine what types of views will be required and how many of each.

5. Determine what the final plotted scale of the drawing will need to be, and select the appropriate paper size for plotting/printing.

6. Plan your drawing sequence. In what order will you develop the drawing (i.e., lines, features, dimension lines, leaders, dimensions)?

7. Review the various CAD commands you will have to use in order to develop the drawing.

8. Examine your CAD system to ensure that everything is in working order, then begin the project.

PROBLEM 16-1

Use the following specifications to lay out a single-view drawing of a 2:1 ratio spur gear and pinion gear in mesh.

Spur gear:	Pinion gear:
88 teeth	hub diameter 1.625
pressure angle 20°	bore of 1.000 diameter
pitch diameter 5.500	pressure angle 20°
hub diameter 2.625	
bore of 2.000 diameter	

Use all standard drafting methods to illustrate the outside diameter, pitch diameter, and root diameter. Calculate and add the center-to-center distance between the shafts.

PROBLEM 16-2

Draw a spur gear having the following specifications: 56 teeth, pressure angle 20°, pitch diameter 3.500, hub diameter 1.00 gear blank with overall size (width) 1.00, face size .50, bore .5625; use a 1/8″ set screw.

Complete two views using the half-section method of representing gears. Do not show the gear teeth; use the conventional method of illustrating teeth.

Dimension per all standard practices. Calculate and add all standard required cutting data to the lower-right side of the work area. Material S.A.E. #1320 cast steel. Heat treatment: carburize .015 – .020 deep.

PROBLEM 16-3

Complete a 2-view detail drawing of a pinion gear having the following specifications (calculate and add all standard cutting data): 30 teeth, pressure angle 20°, pitch diameter 6.000, hub diameter 2.000, heat treatment carburize .050 deep, gear blank with overall size (width) 3.500, face size 1.250, bore 1.000, material S.A.E. #4620 steel.

PROBLEM 16-4

Make a detail drawing, completely dimensioned, of a spur gear with the following specifications: diametral pitch 16, pressure angle 20°, pitch diameter 5.500, whole depth .1348, outside diameter 5.625, addendum .0625, working depth .125, circular thickness .098, chordal thickness .0633, chordal addendum .0985, whole depth .1348, and dedendum .0723, material cast brass.

Design a simple gear blank with a set screw to fasten the gear to a .75 dia. shaft.

PROBLEM 16-5

Design and draw a detailed drawing, completely dimensioned, of a pair of bevel gears having teeth of a 4.0 diametral pitch, the gear with 25 teeth, the pinion gear with 13 teeth, face width of 1.00, gear shaft 1.25 diameter, and pinion shaft .875 diameter. (Calculate for FN-2 fits with the shaft.)

Design the hub diameter to be approximately twice the shaft diameter. Backing for the gear 1.375 and for the pinion .75. Design and dimension the remaining portion to suit, using standard drafting practices. Add all cutting data to the lower-right side of the work area. Material: S.A.E. #3120 cast steel, heat treatment: carburize .015/.020 deep.

PROBLEM 16-6

Make a design layout of a worm and worm gear with the following specifications: shaft diameter 1.25, single-thread worm, lead .75, worm gear with 28 teeth. Add all dimensions and cutting data as required.

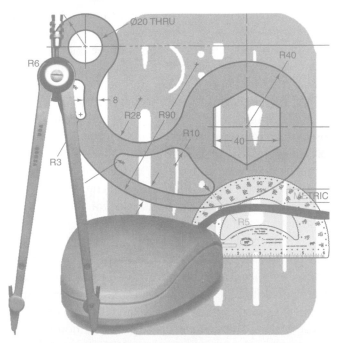

ASSEMBLY AND DETAIL DRAWINGS FOR DESIGN

KEY TERMS

Application engineering	ECR (engineering change request)
Assembly drawing	Invention agreement
Conceptual design	Mass production company
Contract	Parts list
Design development layout drawings	Patent drawings
Design procedure	Proposal
Design section	Purchased part
Detail drawings	Quotation department
DFA (design for assembly) procedure	Scientific design
DFM (design for manufacturing) procedure	Selling price
	Sketches
ECO (engineering change order)	Subassembly drawing
	Title block

CHAPTER OUTLINE

Technical drawings and the engineering department • The engineering department • Systems company • Design section • Working drawings: detail and assembly • Title block • Numbering system • Drawing revisions: ECRs and ECOs • Application engineering • Quotation department • Mass production company • Design for manufacturability (DFM) • DFM guidelines • Design for assembly (DFA) • DFA guidelines • Personal technical file • Invention agreement • Patent drawings • Summary • Review questions • Chapter 17 problems

CHAPTER OBJECTIVES

Upon completion of this chapter, students should be able to do the following:

■ Discuss the use of sketches, layout drawings, detail drawings, subassembly drawings, and final assembly drawings.

■ Summarize the contents of a parts lists or a bill of materials, including purchased parts.

■ Prepare sketches, layout drawings, detail drawings, subassembly drawings, and final assembly drawings.

■ Describe a typical engineering department and the functions of a design section, an application department, and a quotation department.

■ Describe the design procedure used in industry and identify your own potential role in the procedure.

■ Compare the similarities and differences between a proposal and a contract.

■ Process ECRs and ECOs.

■ Describe how a selling price is established for a product.

■ Start your own technical file in a systematic manner.

■ Explain DFM and DFA.

■ Recognize invention agreements and patent drawings.

Technical Drawings and the Engineering Department

A representative engineering department and a flow-chart showing lines of communication, *Figure 17-1,* provide a framework for discussing technical drawings and related engineering practices. Technical drawings include sketches and layout, detail, and assembly drawings. Related engineering practices encompass the following: design procedure, parts lists, title blocks, engineering revisions, invention agreements, patents, personal technical files, designing for manufacturing and assembly, and quotation-proposals. In the problems at the end of the chapter you will have the opportunity to do basic designing using the procedures, drawings, and practices discussed. You will be able to utilize topics from preceding chapters, such as dimensioning, standards, fasteners, tolerances, and machine elements.

The Engineering Department

Engineering departments vary from company to company, but most of them have functions in common. An example engineering department in a medium-sized engineering company, Figure 17-1, has a variety of sections usually directed by a chief engineer. Our example

has three sections: design, application, and quotation. The design section is responsible for new products and minor design improvements for existing products. The application engineering section handles special-order modifications to existing products of the company. For example, if our company manufactured commercial scales for weighing trucks, the application section would modify scales to meet specific customer requirements. The quotation department estimates costs of new or modified existing products and prepares quotations of selling prices to present to potential customers. In larger engineering companies, design and application sections may be divided into fields of engineering such as mechanical, civil, electrical, electronic, structural, materials, agricultural, petroleum, and so forth.

Directly under the chief engineer are the section heads, usually engineers or designers with extensive design-engineering experience plus an ability to lead teams of skilled personnel. The teams could consist of engineers, designers, drafters, detailers, checkers, technical writers, illustrators, and typists. Although most engineering departments have definite departmental structures, many companies move drafters, designers, and engineers back and forth among the various sections as needed. The moves help new personnel to learn the departmental structure and operation of their orga-

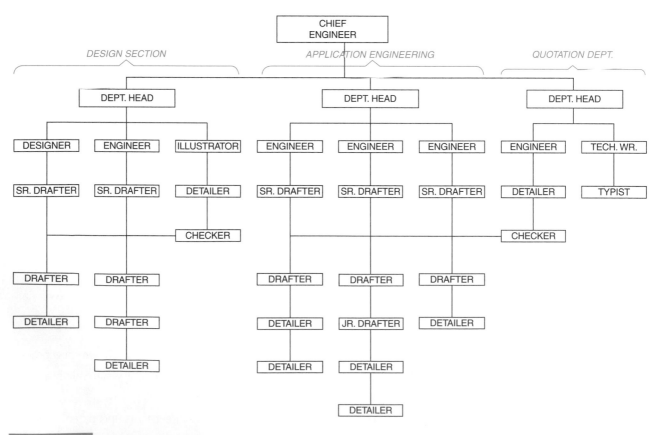

FIGURE 17-1 *Chain of command in an engineering department*

nization. Another exercise to gain an understanding of how an engineering organization functions is to follow paper documents along the chains of command as indicated in the flowchart, Figure 17-1.

Hence, for illustrative purposes in this chapter, two hypothetical engineering organizations are discussed. The first company produces one-of-a-kind systems to meet special needs of customers—in this case, a customer who produces electrical power. The second is a company that mass produces a device used for positioning a small crosshair cursor or a pointer on a computer screen—the ubiquitous mouse.

Systems Company

Assume that a systems company has received a request for proposal (RFP) from a solid-waste handling company that needs to convert burnable waste into electrical power. The proposal is to be submitted to the solid-waste handling company in two months. The president of the systems company sends a memo to his chief engineer with directions to process the RFP. Consequently, the chief calls a meeting of the design, application, and quotation department heads. They decide that the design section will need to design a new furnace and one new conveyor for handling solid waste coming from a shredder. The application engineering group will need to modify several designs of existing conveyors. A steam turbine design and an electrical generator design will only require small changes. The quotation department has the task of preparing a proposal to be submitted to the waste-handling company before the two-month deadline. The proposal will include the selling price of the complete energy-generating system, the delivery dates, and descriptions of the system components. Then, if the waste-handling company accepts the proposal and signs it, the proposal becomes a binding contract on both companies.

Design Section

The *design section* will be discussed first to illustrate the uses of the drawings and the related engineering practices listed earlier, which include preparing assembly, detail, and layout drawings, following design procedures, making comprehensive parts lists, executing engineering revisions, handling invention agreements, keeping personal technical files, and preparing quotations. Application engineering will be discussed later.

DESIGN PROCEDURE

A design process or *design procedure* used by the design section acts as a general guideline for a team of design-

ers, engineers, and drafters, *Figure 17-2*. Their preliminary design tasks are to design a new conveyor and create a new furnace. Detailed information on these two system components needs to be developed so the quotation department can prepare cost estimates on them.

No one design process serves an individual designer, design team, or organization. Each individual, team, and organization develops a design process to suit their respective needs. (Similar-but-different design processes also are discussed in the Introduction and Chapters 1 and 24.)

The actual process of developing a new product from its established need (the RFP in this example) to its final production, consumes many hours and involves a variety of highly trained, skilled personnel. The flowchart in Figure 17-2 shows the overall process to production; however, in this preliminary RFP phase, the design team only goes through to the cost study stage to prepare information for quoting purposes. Later, if the proposal is accepted by the customer, the design section team would develop the conveyor and furnace through to the production stage.

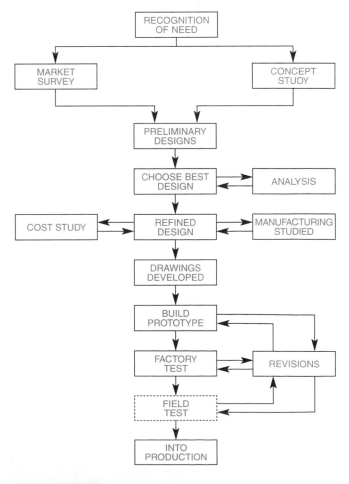

FIGURE 17-2 *Design procedure*

Nevertheless, to continue describing the design procedure followed by the design section team for the new conveyor and new furnace, concept-study *sketches* would be done first. (Note that a market survey would not be required for this project because the RFP stated the need.) Sketches would be made of conveyor ideas for moving the waste from a shredder to the furnace, *Figure 17-3*. Next, where and how to burn the waste in the furnace and where to utilize the hot gases to generate steam would be considered. During these studies, many ideas are presented, usually in sketches. Designers are encouraged to generate and sketch ideas without restraint; that is, without being influenced by existing designs, in order to arrive at a fresh and perhaps unique solution. The more ideas that are proposed in the time available, the greater the chances of developing the best solution to the problem. These brainstorming sessions are conducted in an atmosphere of "no criticism" so members will not feel inhibited. Analysis of the ideas is scheduled after the free-wheeling sessions.

The design section team employs two effective design approaches. In *conceptual design*, much use is made of known technical information from such sources as reference books, technical journals, handbooks, and manufacturers' catalogs. (Refer to Chapters 13, 21, and 24 for examples.) All members of the team should be aware of the latest technical information available. In *scientific design*, use is made of concepts from physics, mechanics, materials science, chemistry, and mathematics.

During the analysis phase the best ideas are identified and developed further in sketches. Suitable engineering materials are investigated. Loads on the structures are analyzed and energy balances (output versus input) are estimated.

LAYOUT DRAWINGS

The department head in the design section follows the team's progress and directs them to prepare *design development layout drawings* for a preliminary design review. The team draws the best sketches to scale either by pencil or a computer. For example, *Figure 17-4* shows a sketch of a preferred drag-type conveyor for delivering burnable shredded waste to the furnace, and *Figure 17-5* shows portions of the preferred conveyor drawn to scale in a layout drawing. Note the informality of the design development layout drawing. Standard dimensioning practices are encouraged, but not required. Clear handwritten notes are acceptable. Drafters, designers, and engineers do layout drawings as needed. The layouts may not be completely dimensioned, but they have overall dimensions and important center lines located.

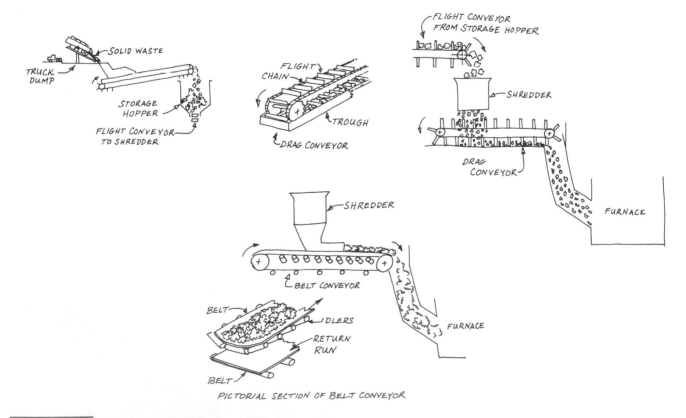

FIGURE 17-3 *Sketch for conveyor ideas for conveying waste to the furnace*

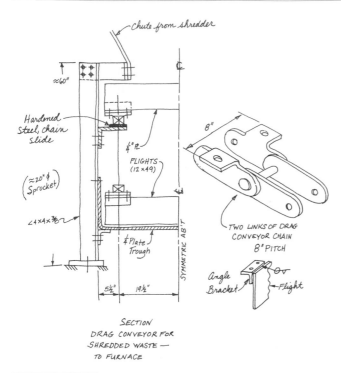

FIGURE 17-4 *Sketch of preferred conveyor for conveying waste to the furnace*

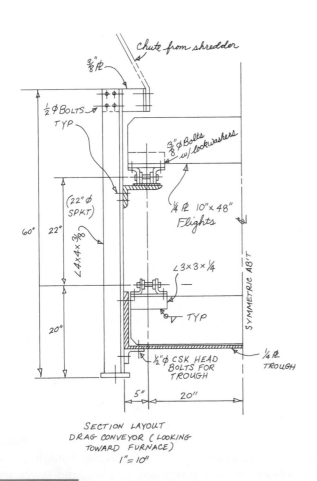

FIGURE 17-5 *Layout drawing of preferred conveyor*

Informal notes specify tolerances, materials, special manufacturing processes, and related pertinent information. Title blocks are optional.

During the design review, the layout drawings help the team to decide on the final designs to consider. The quotation department assists the team in preliminary cost analyses. Finally, the conveyor design and a new furnace concept that best meet the requirements are selected. Manufacturing personnel who also have participated in these discussions recommend appropriate manufacturing processes for the conveyor and furnace.

Assume now that the quotation department has received all the technical data and specifications they need from the design section and application engineering departments to prepare a cost estimate to satisfy the RFP. A proposal is written and presented to the waste-handling company just before the two-month deadline. In addition to the written proposal and several systems site–drawings, three types of illustrative material may be presented for better visualization of the system. An illustrator may prepare a watercolor perspective drawing of the whole system or a computer-generated 3D picture. A third type that might be presented is an architectural-type scale model made of cardboard and balsa wood of the whole system showing dumping sites, chutes, conveyors, the furnace and boiler, and the electricity-generating equipment.

The waste-handling company representatives may ask for further information and explanations, such as brief descriptions of system components and when specific components will be delivered, installed, and operating. Finally, after further comments and questions, the proposal is accepted and signed. It is now a binding, legal contract on both companies. *Figure 17-6* shows an example signature page from a proposal-contract form.

Working Drawings: Detail and Assembly

The systems company president notifies the chief engineer that the proposal is now a contract. The chief directs the designers and engineers to forward the design development layout drawings of the new conveyor and the furnace to drafters and detailers who use these drawings to create working drawings; that is, detail and assembly drawings for manufacturing. There are a number of different kinds of working drawings associated with the development and production of a product, so the drafter must be familiar with each kind, know the company standards, and be able to draw each kind. The major kinds of working drawings are:

• Assembly drawings

• Subassembly drawings

COMPANY

20—APPROVAL: There are no agreements or oral understandings outside of this proposal, which shall become a contract only when accepted by you, _____

_____, and approved in writing by a plant manager or an executive officer of Company.

Prepared and submitted to you for
Company

by _____

Accepted this _____ day of _____

19____ for

_____

by _____

Title _____

Approved for   Company at

_____

this _____ day of _____ 19 _____

by _____

Title _____

FIGURE 17-6 *Proposal-contract signature page*

- Detail drawings
- Modified purchased drawings

ASSEMBLY DRAWING

Any product that has more than one part must have an assembly drawing. The *assembly drawing* illustrates how a product is assembled when completed. The assembly drawing can have one, two, three, or more views as needed. Often, a full section view is used to show how the parts are assembled.

Assembly drawings usually are not dimensioned, except for general overall dimensions or to indicate the capabilities of the assembly. For example, a clamp assembly drawing might have a dimension to illustrate how wide it opens. (Refer to Problem 17-7.) Assembly drawings do not contain hidden lines unless they are absolutely necessary to illustrate some important feature that the reader might otherwise miss.

An assembly drawing must call off each part that is used to make up the assembly. Each part must be called-off by its part number and title, and the total number required to make up one complete assembly. Companies most often use one of two methods to call off the parts. In Method A, each part has a leader line with the part number, title, and number required, *Figure 17-7*. In Method B, each part has a leader line with a balloon call-off number inside, *Figure 17-8*. This method uses a tabulation chart to list the part number, title, and number required.

SUBASSEMBLY DRAWING

A subassembly is composed of two or more parts permanently fastened together. A *subassembly drawing* is very similar to an assembly drawing. All standard practices associated with assembly drawings apply to subassembly drawings. That is, full section views, few overall dimensions, hidden lines only if necessary, and call-offs using either method one or two are employed.

Subassemblies sometimes require machining operations after assembly, *Figure 17-9*. For instance, a hole is to be drilled through the parts after assembly. Drilling each individual part and then aligning the holes at assembly would be much more difficult and costly. Thus, the required dimensions to drill the hole are added to the subassembly drawing.

There can be any number of subassemblies in the final assembly. For example, in an automobile, the engine is a subassembly, the alternator is a subassembly, the transmission is a subassembly, and so forth. These subassemblies, together with various single parts, make up the final assembly of the automobile. Note that subassemblies are often purchased, stocked, and stored for later final assemblies.

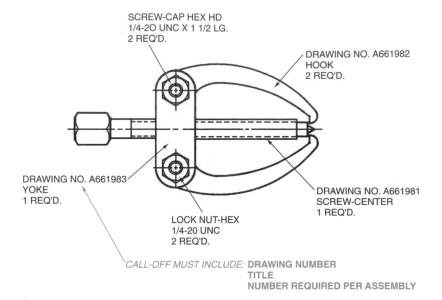

SCREW-CAP HEX HD
1/4-2O UNC X 1 1/2 LG.
2 REQ'D.

DRAWING NO. A661982
HOOK
2 REQ'D.

DRAWING NO. A661983
YOKE
1 REQ'D.

DRAWING NO. A661981
SCREW-CENTER
1 REQ'D.

LOCK NUT-HEX
1/4-20 UNC
2 REQ'D.

CALL-OFF MUST INCLUDE: **DRAWING NUMBER**
TITLE
NUMBER REQUIRED PER ASSEMBLY

FIGURE 17-7 *Assembly call-off (Method A)*

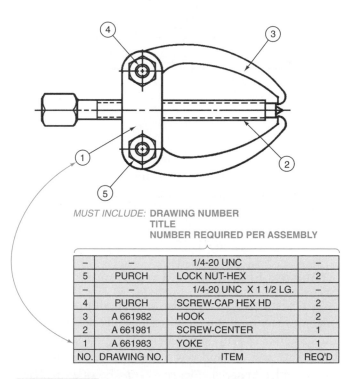

MUST INCLUDE: **DRAWING NUMBER**
TITLE
NUMBER REQUIRED PER ASSEMBLY

–	–	1/4-20 UNC	–
5	PURCH	LOCK NUT-HEX	2
–	–	1/4-20 UNC X 1 1/2 LG.	–
4	PURCH	SCREW-CAP HEX HD	2
3	A 661982	HOOK	2
2	A 661981	SCREW-CENTER	1
1	A 661983	YOKE	1
NO.	DRAWING NO.	ITEM	REQ'D

FIGURE 17-8 *Assembly call-off (Method B)*

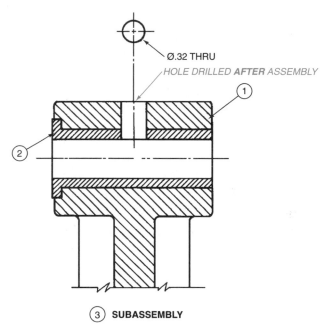

Ø.32 THRU
*HOLE DRILLED **AFTER** ASSEMBLY*

③ **SUBASSEMBLY**

FIGURE 17-9 *Subassembly drawing*

DETAIL DRAWINGS

Every manufactured part must have its own fully dimensioned *detail drawing,* with its own drawing number and title block, *Figure 17-10.* All information needed to manufacture the part is contained in the detail drawing, including as many views as needed to fully illustrate and dimension the parts.

The general practice in technical drawing is to draw only one part to one sheet of paper, regardless of how small the part is, because individual parts are often made on different machines or by different processes, and the bookkeeping plus costing is easier for one part per drawing.

Most companies use detail drawings as just described. However, some companies separate the detail drawings into various manufacturing processes, such as:

• Pattern detail drawings for castings

• Casting detail drawings

• Machining detail drawings

• Forging detail drawings

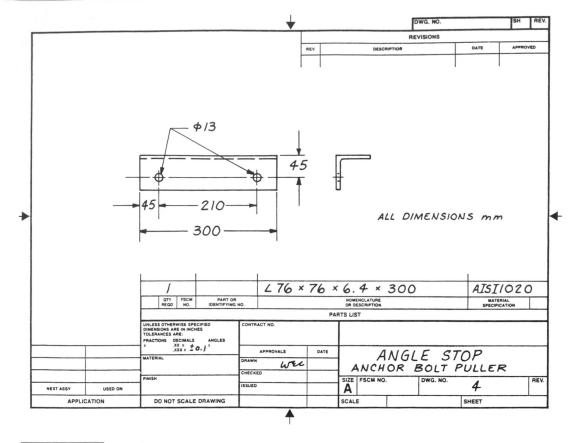

- Welding detail drawings
- Stamping detail drawings

Two observations regarding technical drawings are added here for your information.

1. Fold lines, construction lines, reference planes, and reference numbers and letters are almost never included on technical drawings. They may have been utilized to prepare views and sections, but they are omitted from the drawing.

2. A seldom-discussed drawing technique designed to clarify portions of drawings may be used effectively. Usually a small portion of a large drawing is removed, enlarged, and labeled, e.g., DETAIL A. Refer to figures in the latter portion of Chapter 21 for examples.

PURCHASED PARTS

Most manufacturing companies cannot make such purchased items as electric motors, bearings, shafting, bolts, nuts, screws, and washers as cheaply as companies that specialize in making these items. Therefore, designers should specify these standard items whenever possible. *Purchased parts* are drawn in assembly drawings, usually in outline form, to show their function and relationship

to the product. All specific information for a purchased part should be listed on the drawing or parts list, including size, material, special features, vendor part number, performance, and so forth. For example, specifications for a DC motor might read as follows: "12 volt DC motor, GE no. 776113 or equal."

MODIFIED PURCHASED PARTS

A purchased part that needs to be modified, however slightly, must have the modification fully described and dimensioned as with any detailed part, except that the purchased part size and description are listed in the title block under Material. *Figure 17-11* shows a standard fastener, M20 × 2.5-6g × 70 long hex-head machine screw, which is to have a tip specially turned and a 4 mm diameter hole drilled through the tip. The standard fastener is drawn with the modifications added and dimensioned. Then it would be called out as a standard metric fastener in the title block under Material. This directs the machinist to make the part from a standard fastener.

Title Block

Title blocks vary from company to company, although most companies follow the ANSI standard illustrated in

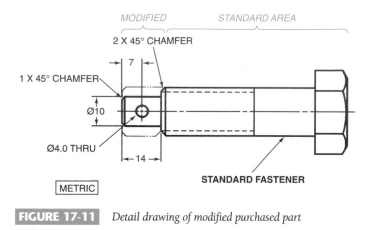

FIGURE 17-11 *Detail drawing of modified purchased part*

Figure 17-10. The primary function of a *title block* is to provide information not normally given on the drawing, such as:

- Name and address of the company
- Drawing title
- Drawing or part number
- Scale of the drawing
- Name of the drafter and date completed
- Name of the checker and date checked
- Name of the person who approved the drawing and date
- Material the part is to be made of
- Tolerances/limits of dimensions
- Heat treatment requirements if not specified on the drawing
- Finish requirements if not specified on the drawing

SIZE OF LETTERING WITHIN THE TITLE BLOCK

- General information, .125 (3) freehand/.120 mechanical
- Drawing title, .250 (6) freehand/.240 mechanical
- Drawing number, .312 (8) freehand/.350 mechanical

CHECKING DRAWINGS

The importance of absolute accuracy in engineering drawings cannot be stressed enough. The slightest error could cause large and unnecessary expenses. In some fields of engineering, such as in aircraft and space vehicles, an error could cause tremendous consequences—lost lives! Therefore, every precaution should be taken to avoid errors. In most engineering departments, checkers are employed to verify all dimensions and to see that standard materials and features (holes, threads, keyways, etc.) are used wherever possible. The checker's responsibility is to check the entire drawing and sign it

before the drawing is finally approved (usually by a chief engineer) and released.

Conscientious drafters check their drawings before submitting them to the checkers to ensure accuracy and to save the checker's time. A partial checklist for drafters is as follows:

- Are all dimensions included?
- Are there unnecessary dimensions?
- Can the part be manufactured in the most economical way?
- Are dimensions and instructions clear and understandable?
- Will the part assemble with mating parts?
- Have all limits, tolerances, and allowances been properly analyzed for all moving parts?
- Have undesirable accumulations of tolerances been analyzed?
- Are all notes added?
- Are finish-texture symbols added?
- Are the material and treatment for each part specified?
- What is the drawing's general appearance? (legibility, neatness, arrangement of views)
- Does it follow all drawing and company standards?
- Is the title block complete? (Title, part number, drafter's name, etc.)

PARTS LIST

A *parts list* may contain the assembly drawing number, all subassembly drawing numbers, all detail drawing numbers, all purchased parts, material of each part, and the quantity of individual parts needed for one complete assembly.

A drafter usually prepares the parts list while following company guidelines. Some companies list the parts in order of size, some in the order of importance in the manufacturing process, and others in order of assembly. The assembly method is preferred due to the time savings later in the assembly area. An assembly method parts list is shown in *Figure 17-12,* for a machine vise. The parts are listed in the order used for assembling the machine vise.

Note that the purchased parts do not have a drawing number; therefore, they are listed under the column titled Drawing No. as PURCH., indicating that these are standard parts to be purchased, not made, by the company. Additional typical purchases for most companies are motors, bearings, switches, gears, chains, belts, and controls. Complete specifications must be listed for purchased parts so that there is no question as to what must be purchased. The purchasing agent for the company may not have a technical background and therefore will rely on the parts list specifications for purchasing "buyouts." In this example, items numbered 7, 8, 17, 23, and 24 are purchased parts.

Some companies use a system of indents to list the parts, Figure 17-12. There are four indents, indicating

the various kinds of drawings. The first indent indicates the main assembly drawing. The second indent indicates all subassembly drawings used to make up the assembly. The third indent indicates all detail drawings and modified purchase parts with drawing numbers to make up the assembly. The fourth indent indicates all purchased parts to make up the assembly.

Not all companies use this system, but using the indent system gives a quick indication of which parts are used to make up the various subassemblies. Note on the example list that item number 3 is indented one place, indicating a subassembly. Item numbers 4, 5, and 6 are detail drawings used to make up the subassembly item number 3. Item numbers 7 and 8 are purchased parts used to complete the subassembly item number 3.

For additional examples of parts lists or bills of materials, refer to Figures I-31, I-36, 21-35, 21-46, 24-16, and 24-17.

Numbering System

The numbering system for identifying and recording technical drawings varies from company to company. Although no standard system is used by all companies, most companies assign a sequential number for each drawing as it is finished. Some also add a prefix letter to indicate the drawing size. A drawing on a C-size sheet with an assigned number of 114937 would be indicated on the drawing by C–114937. The drawing number could be done in freehand, 0.312 (8) high, or mechanically lettered, 350 (9) high.

Drawing Revisions: ECRs and ECOs

After any technical drawing has been signed off (drawn, checked, and approved) and incorporated into the company's system, changes in the drawing are usually accomplished through *ECRs* and *ECOs.* Anyone having a suggestion or need to improve a part or assembly within the product, or an error to correct on the drawing, must bring it up in an *engineering change request (ECR)* procedure meeting, *Figure 17-13,* and request an *engineering change order (ECO).* No change can be made in a drawing after it has been released (signed off) to manufacturing, even by the drafter or designer who drew it. All departments concerned must agree to the change, Figure 17-13. The ECR and ECO explain what change is to be made, what it was before, why the change is to be made, whether it affects other parts, and who suggested the change. When the engineering change request has been approved, the committee issues the engineering change order. The ECR and ECO

INDENT 1 = ASSEMBLY DRAWING
INDENT 2 = SUBASSEMBLY DRAWINGS
INDENT 3 = DETAIL DRAWINGS
INDENT 4 = PURCHASED PARTS

NO	DRAWING NO.	DESCRIPTION	MATERIAL	QUAN
		MASTER PARTS LIST		
1	D77942	VISE ASSEMBLY-MACHINE	AS NOTED	1
2				
3	C77947	BASE-VISE	AS NOTED	1
4	B77952	BASE-LOWER	C.I.	1
5	A77951	BASE-UPPER	C.I.	1
6	A77946	SPACER-BASE	STEEL	1
7	PURCH.	BOLT 1/2-13 UNC-2.0 LG.	STEEL	1
8	PURCH.	NUT 1/2-13 UNC	STEEL	1
9				
10	C77955	JAW-SLIDING	STEEL	1
11				
12	A77954	SCREW-VISE	STEEL	1
13	A77953	ROD-HANDLE	STEEL	1
14	A77956	BALL-HANDLE	STEEL	2
15				
16	A77961	PLATE-JAW	STEEL	2
17	PURCH.	SCREW 1/4-20 UNC-1.0 LG.	STEEL	4
18				
19	A77962	COLLAR	STEEL	1
20				
21	A77841	KEY-SPECIAL VISE	STEEL	2
22				
23	PURCH.	BOLT 1/2-13 UNC-4.0 LG.	STEEL	4
24	PURCH.	NUT 1/2-13 UNC	STEEL	4

JAN ENGINEERING PETERBOROUGH, NH	Model No. 160	Parts Lister NELSON	Date 6 AUG 98	
Title VISE ASSEMBLY-MACHINE		PAGE 1 OF 1 PAGES	Drawing No. A1198891	

FIGURE 17-12 *Parts list*

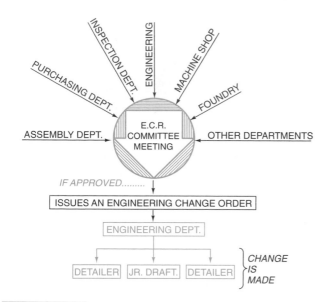

FIGURE 17-13   *Engineering change request order procedure*

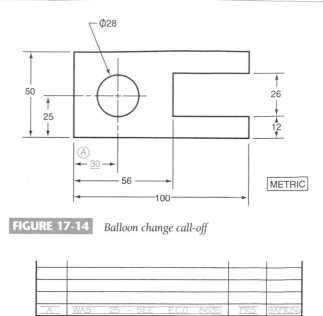

FIGURE 17-14   *Balloon change call-off*

LET	REVISION	BY	DATE
A	WAS 25 — SEE E.C.O. 6635	PK5	9APR86

FIGURE 17-15   *Revision block* (Courtesy Bishop Graphics Accupress)

list the departments in the company that will be directly affected by the change.

Beginning drafters, detailers, designers, or newly hired engineering graduates may "get their feet wet" in the company by being assigned the task of incorporating ECOs in the company's technical drawings. These assignments are made primarily to introduce the employees, new to the company, to technical documents that flow through the company and to how the company functions. Introducing new employees to the flow of technical documents illustrates an admonition of a commencement speaker who said to the graduates, "Get dirt under your fingernails so you'll know what's going on!"

Minor changes on original drawings may be just erasures or additions made directly on the original, whether on paper or in a computer. If a dimension is not noticeably affected by the change, the dimension is simply corrected and underlined with a heavy black line. This indicates that the dimension is correct, but slightly out of scale, *Figure 17-14*.

If the change is extensive, the drawing must be redrawn. The original drawing is not destroyed: it is stamped OBSOLETE and includes the note "SUPERSEDED BY (the number of the corrected drawing)." The new drawing must have a note referring back to the superseded drawing. The change on the drawing must be clearly marked where the change was made, and it must clearly describe what was changed. Each company has its own standards to indicate this information. The most common method is to place a letter or number in a small balloon near a changed dimension or dimensions, or near the change made to the drawing. The letter or number must be listed in the revision block, *Figure 17-15*. The revision block, located either to the

left of the title block or on the top right corner of the drawing sheet, shows the letter or number of the change, what was changed, the date of the change, and who made it. If the change is extensive, only the ECO number is listed. Anyone needing to know why the change was made can refer back to the ECO or the ECR.

Some companies add the last change letter to the drawing number, which, in effect, actually changes the part number. For example, part number C–65498 would become C–65498–A for the first change. Then, if changed again at a later date, it would become C–65498–B, and so on.

In certain cases, appropriate administrative changes may be made without an engineering change order. Such simple changes as misspelled words or incorrect references are nonengineering changes and probably would not affect other departments.

The next section in the system company's structure, application engineering, primarily processes existing designs.

Application Engineering

Preliminary activity in the *application engineering* section proceeds to the cost studies step in the design procedure, Figure 17-2, in a manner similar to the design section. The procedure acts as a guide for an application engineering team of engineers, designers, and drafters who consider designs of conveyors, steam turbines, and electrical generators the company has produced in the past. Also, the team considers concepts that already exist in industry. The team's primary task is to modify and

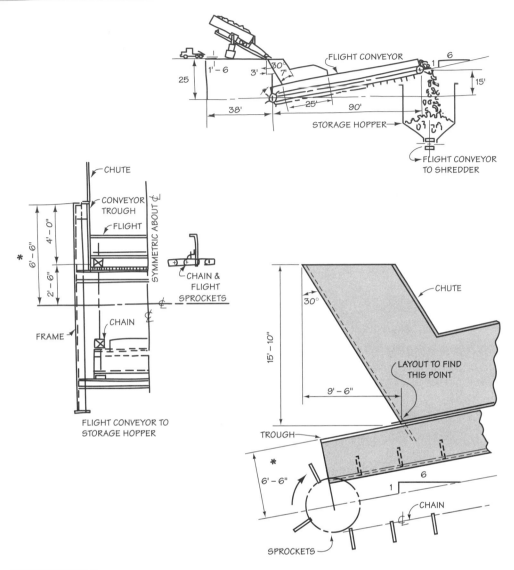

FIGURE 17-16 *Sketches and layout drawing of chute*

incorporate these existing concepts to meet the design requirements for the waste-burning system. For example, a scale is modified to monitor the amount of waste being delivered. A platform over the scale must be modified to accommodate large trucks. After being weighed, the trucks would dump the waste into a chute going to a flight conveyor, a standard product of the company, which would carry the waste to a hopper where the waste would be stored until conveyed to a shredder. From the shredder the waste would be conveyed to the furnace on the new drag conveyor created in the design section, Figures 17-4 and 17-5. Drawings of a steam turbine would be reviewed in an ECR meeting where ECOs would be written to modify the existing specifications and drawings of the turbine. Drawings of an electrical generator system also would be modified from ECOs. Drafters and detailers would process the ECOs.

Several new designs are developed in application engineering, but they are minor and are based on the designer's previous experience. For example, sketches and a layout drawing to determine overall dimensions of the chute previously mentioned are shown in *Figure 17-16*.

Finally, specifications and appropriate layout drawings are forwarded to the quotation department for their use in the cost study.

Quotation Department

During the design and development of the waste-burning furnace and during the processing of the conveyors, steam turbine, and electrical generator, the *quotation department* closely followed the design section's progress and the application engineering's activities. Thus an estimate of the selling price for the entire system was nearly

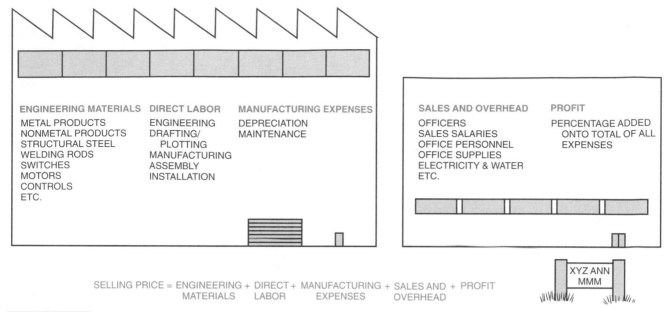

FIGURE 17-17 *Selling price components*

completed for the RFP by the time design and application finished their respective tasks. The selling price is the major dollar figure submitted to the customer, but subtotals for component systems, such as the furnace, the steam turbine, and the generator may be included. The *selling price* is prepared by the quotation department in steps illustrated in *Figure 17-17.* Basic costs include labor and engineering materials. Basic labor (sometimes labeled direct labor) includes estimates of hours for engineering, drafting, manufacturing, assembling, and installing. Engineering materials include steel plates, structural steel shapes, fasteners, welding rods, switches, motors, controls, and so forth. Manufacturing expenses, such as depreciation of machines and maintenance, add to the basic costs to make up manufacturing costs. Sales salaries and expenses plus general overhead costs (electricity, water, office supplies, etc.) are added to manufacturing costs to arrive at the total cost to produce the energy system. Profit added to total cost establishes the selling price.

The selling price, specifications, delivery schedules, and the "boilerplate" make up the *proposal.* A typical boilerplate would be stored in a computer ready for use in any new proposal. It includes introductory comments and sections intended to promote the company to the prospective customer, such as descriptions of past systems designed and installed and summaries of the expertise of the personnel who would handle the job if the proposal were accepted. Another section of the boilerplate protects the company with legal statements referring to late delivery due to strikes, acts of God (earthquakes, tornadoes, etc.), unavailability of materials, and so forth.

A cover letter with a bound copy of the proposal would be mailed to the prospective customer before the due date, or delivered in person if an oral presentation was arranged. Oral presentations give the recipient an opportunity to ask questions, suggest changes, and correct any misunderstandings. When the customer accepts the proposal and signs it, it becomes a binding legal *contract* on both parties. It also becomes a document that immediately generates continuing work for the design and application sections of the systems company.

This continuing work for the design section was discussed earlier. Working drawings were prepared, checked, signed, and sent to manufacturing. In a similar manner, application engineering would produce working drawings, have them checked, signed, and sent on to manufacturing.

Mass Production Company

A *mass production company* that produces large quantities of a product would have an engineering departmental structure similar to the systems company discussed earlier, Figure 17-1; however, the mass production company would have a somewhat different approach to producing a product, *Figure 17-18.* The company's design, application-engineering, quotation, and manufacturing departments would follow this design procedure for mass producing one product while working together more closely than departments in the company that produced a one-of-a-kind system. To illustrate these different procedures, assume that the mass production company primarily produces mouse units, *Figure 17-19,* for use with a computer. Essentially, a mouse provides a quick way for computer operators to position a small crosshair cursor or a small pointer on a video screen by

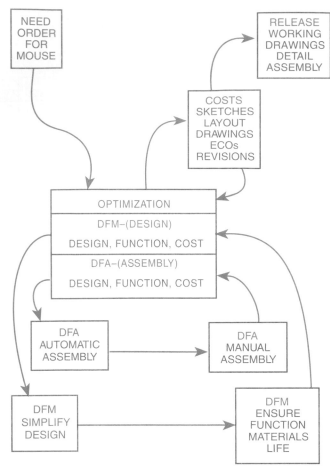

FIGURE 17-18 *DFM-DFA procedures*

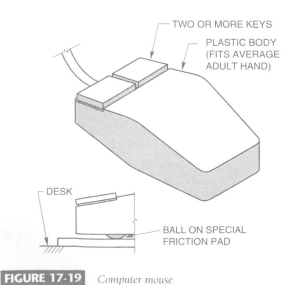

FIGURE 17-19 *Computer mouse*

moving the mouse across a special desk pad. Then once the crosshair or pointer has been positioned, one key atop the mouse is clicked to initiate a basic computer operation.

Assume that the mass production company has received an order for two million mouse units and that each unit requires more functions than their standard mouse. The major challenge to the company is to design and mass produce these units as efficiently as possible and still have a reliable product. The company's design efforts are focused on two important areas of mass production. The first, DFM (design for manufacturability) emphasizes design of the parts that make up a product. The second, DFA (design for assembly) concentrates on assembly of the parts that make up a product.

Design for Manufacturability (DFM)

The *DFM procedure* is an iterative loop and acts as a guide for a design team, Figure 17-18. The primary goals of the DFM team are to develop the new mouse design, for more efficient manufacturing, at the least cost, within a reasonable time, and to ensure its function for a reasonable life. The team made up of engineers, designers, drafters, and manufacturing personnel pursue these goals as illustrated by the iterative path shown. Start with optimizing (design, function, and cost), then go on to simplifying (DFM guidelines), ensuring (function, life, materials), and back to optimizing. One factor not shown in Figure 17-18 is the time constraint for making decisions. This can affect the success of the DFM process. Decisions must be made at each step in the loop, based on information available at the time and the judgment of the team members.

DFM Guidelines

Guidelines for DFM evolved over a period of time, essentially due to an increasing need for collaboration on the design of products between a company's engineering department and manufacturing. To be competitive, companies needed to be more efficient and to produce better products. Some DFM guidelines from DFM literature are:

Minimize the number of parts.

Minimize part variations.

Design parts to have more than one function or use.

Design parts for ease of manufacturing.

Use fasteners that handle easily and can be readily installed.

Arrange the assembly of parts so they assemble in layer fashion.

Avoid flexible parts.

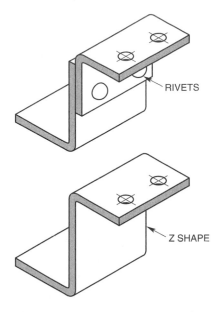

FIGURE 17-20 *Combining two parts into one*

MINIMIZE NUMBER OF PARTS

One approach is to combine two parts into one, thus eliminating fasteners. An obvious example is shown in *Figure 17-20*.

MINIMIZE PART VARIATIONS

If fasteners are required, use all the same size to save storage space and sorting-and-assembly costs. Use standard off-the-shelf parts where possible.

DESIGN PARTS TO BE MULTIFUNCTIONAL

For example, a leaf spring designed to exert a small force could also conduct an electrical current. A structural part could have ridges for guiding a second part into position. A spacer post could be used for a fastener, *Figure 17-21*. One shaft of a specific length and diameter could be used in a variety of products. One-size parts such as bearings, springs, or gears could be used in a variety of products in a company.

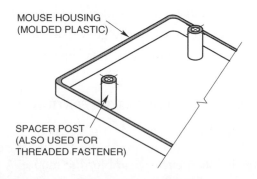

FIGURE 17-21 *Post used for spacer and fastener*

DESIGN PARTS FOR EASE OF MANUFACTURING

Precision cast gears or stamped gears need little or no machining or surface finishing. Basic materials with relatively expensive finishes, such as anodized aluminum, may ultimately save painting, polishing, and machining later in the manufacturing processes, thus keeping labor costs down.

USE FASTENERS EASILY HANDLED AND INSTALLED

Snap-in fasteners save assembly and screwdriver time, *Figure 17-22*.

ARRANGE PARTS ASSEMBLY FOR LAYERING

Assembly of the mouse shown in *Figure 17-23* illustrates layering; that is, essentially, parts are stacked from the bottom up.

AVOID FLEXIBLE PARTS

Use circuit boards instead of flexible, difficult-to-assemble wires for electrical circuits where possible. Of course, the mouse, Figure 17-19, needs one flexible connector for easy movement on a desk pad.

The next step in the loop, Figure 17-18, is to ensure function and life, and to select the best materials. Most mouse units utilize a printed circuit on a board with electronic components, to integrate the digital signals generated by light passing through rotating slotted disks inside the mouse, *Figure 17-24*. The shafts of the two slotted disks are driven by a small ball, mostly inside the body of the mouse, as it rolls on a special desk pad in a direction determined by the computer user, who observes a crosshair or pointer on the video screen.

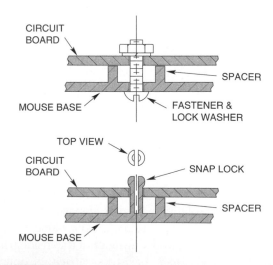

FIGURE 17-22 *Snap-in fastener for circuit board*

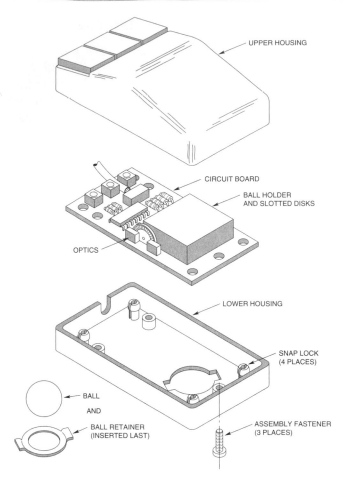

FIGURE 17-23 *Layering for mouse assembly*

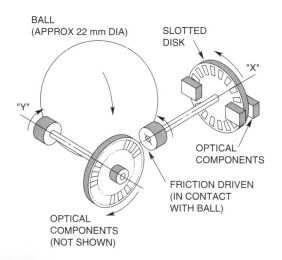

FIGURE 17-24 *Slotted disks inside a mouse*

Since the disks are mounted in "X" and "Y" orientations, the sum of their rotations (digital signals) is used to position the crosshair or pointer on the video screen. The body of the mouse and a number of the inside parts are made of molded plastic. Resistors, capacitors, optical components, and switches are purchased buyouts.

Stainless steel shafts support the disks; bearing supports for the shafts are part of the plastic body, thus no lubrication is required. The keys for initiating basic operations are molded plastic with narrow portions acting as hinges. The only maintenance required for the mouse is to occasionally remove the ball (approximately 22 cm in diameter), wash it in soapy water, rinse, dry, and reinstall. The additional functions ordered by the customer for the new mouse design were incorporated by adding one key and one switch, and expanding the printed circuit. These additions, of course, required engineering change orders (ECOs) to modify the company's existing mouse design.

After the team was satisfied that their designs for manufacturing parts of the mouse were optimum, they concentrated their efforts on design of the parts for assembly (DFA).

Design for Assembly (DFA)

The *DFA procedure* is an iterative loop that occurs simultaneously with the DFM loop, Figure 17-18, but the emphasis is on assembly of parts vis-à-vis the design and manufacturing of parts. The team has similar goals in the DFA loop—develop the best design of parts for assembly of the mouse, at the least cost, within a reasonable time, and ensure its function. They follow the DFA loop shown: first, consider which parts should be assembled automatically, and second, consider which parts need to be assembled manually.

DFA Guidelines

A partial list of DFA guidelines from DFA literature is as follows:

Use pick-and-place robots for simple, routine placement of parts.

Keep the number of parts to a minimum (DFM guideline, also).

Use gravity and vibration for moving parts into assembly areas.

Avoid parts that tangle (hang up on each other).

Use tapers and chamfers to guide parts into position.

Avoid time-consuming manual operations such as screwdriving, soldering, and riveting.

USE ROBOTS FOR SIMPLE OPERATIONS

Robots can be programmed to pick electronic components off a conveyor or from a feeder and place them accurately into a circuit board for the mouse. A mouse may require 25 to 50 electronic components, depending on the basic functions it is to perform.

Industry Application

UNDERSTANDING YOUR CUSTOMER'S NEEDS CAN PAY OFF

Mark and John are about the same age. They graduated from the same technical college at the same time and both went to work for the same company. Both were excellent students, and both have done well on the job. But Mark just beat John in an internal competition for a team leader position. As a result, Mark now earns more than John, has a new office, and is on the fast track for future promotions.

When the time came to select a new team leader, the difference between Mark and John could be stated in just three words: "design for manufacturability." Mark was a proponent of DFM who had proven he was adept at applying its principles. In fact, he always referred to the manufacturing personnel who would use his drawings as his customers. Mark was often seen on the shop floor talking with manufacturing personnel and observing their processes. Over and over Mark could be heard asking shop floor personnel the same questions: "What can I do to make our products easier to manufacture?" As a result, Mark had become a skilled practitioner of the art and science of design for manufacturability or DFM.

John, on the other hand, never visited the shop floor unless it was absolutely necessary. To him the shop floor was a dirty place best avoided. John's attitude can be seen in one of his favorite statements: "I am a designer, not a machinist. If I wanted to know about machining, I would have majored in shop."

John's mistake was to view design and drafting as an island that stands by itself rather than as a part of a larger process that includes design, manufacturing, and all of the other activities necessary to convert an idea into a product that can succeed in the marketplace.

Mark's first decision as a new team leader was to include manufacturing personnel on his team. He wanted to eliminate the "us-against-them" mentality that can grow between design and manufacturing so that his team's designs would be not just functional but also manufacturable.

MINIMIZE THE NUMBER OF PARTS

Combine several components into one molded part, *Figure 17-25*.

MOVE AND ORIENT PARTS USING GRAVITY AND VIBRATION

Figure 17-26 shows one method of orientating spacers and steps to avoid tangling of lock washers and tangling of rivets.

USE TAPERS AND CHAMFERS FOR POSITIONING

An example of tapered fasteners and a receptacle with a tapered hole, both for ease of assembly, are shown in *Figure 17-27*.

AVOID TIME-CONSUMING OPERATIONS

The snap locks in the body shown in Figure 17-22 allow for quick assembly and accurate positioning of the circuit board.

Finally, after the DFA team members agree on the best designs for manufacturability and the best designs for assembly for the new mouse, and the DFM team concludes its optimization efforts, costs for producing the

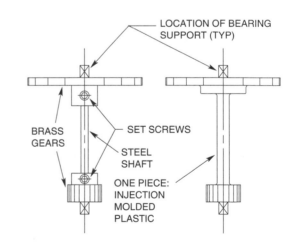

FIGURE 17-25   *Combine several parts into one*

new mouse are checked. If the cost analysis results in further modifications, sketches and layout drawings may be prepared and ECOs may be issued. To continue the design procedure, the new mouse design is formally released. Designers and drafters prepare working drawings. Eventually, all the drawings are released to manufacturing supervisors who have anticipated the drawings because they had worked closely with the design teams during the development of the new mouse design.

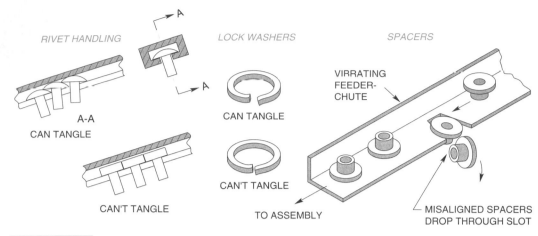

FIGURE 17-26 *Spacers handling and antitangling techniques*

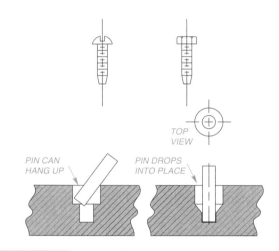

FIGURE 17-27 *Tapered fasteners and a tapered hole for ease of assembly*

As the new mouse goes into production, the personnel of the mass production company realize that even a small savings in cost resulting from following DFM and DFA guidelines is worthwhile when millions of units are involved. Moreover, the company finds that following the guidelines not only results in cost savings but also improves the design and function of products.

Personal Technical File

During the various design and development activities, each engineer, designer, detailer, checker, technical writer, and illustrator should develop and maintain a technical file to enhance his or her involvement in the company. The objective is to minimize the time spent locating technical information pertinent to their respective jobs. Usually a company manufactures products and performs engineering in specific areas of technology, so files can be concentrated in these areas for efficient

retrieval of information. An efficient technical filing system uses an alphabetical-numerical scheme. For example, the first part of a filing scheme for a drafter-designer in a company specializing in conveyors would be an alphabetical listing of technical topics with numerical references. This file should be kept in a three-ringed binder or a computer so that additional information can be added easily. Example entries would be:

Section A			
Aluminum Products	112	Channel Sections	134
ANSI Standards	410	Conveyors	500
Section B		Codes and Standards	410
Bar Stock	132	Controls	210
Beams	133	*Section D*	
Belt Conveyors	510	Drives	600
Belt Materials	122	*Section E*	
Bucket Elevators	530	Electrical Components	200
Section C		Controls	210
Castings	140	Switches	230
Chain Conveyors	520	And so on...	

The numerical part of the file, kept in a file cabinet, would contain tab-numbered dividers, numbered folders, and numbered materials in the folders. Example tabs, folders, and materials would be numbered as follows. (The letters in parentheses are for discussion purposes only: T = Tab number for a divider, F = Folder number, and M = Material labeled with a number.)

100	Engineering Materials	(T)
110	Metals	(F)
111	Ferrous Based	(M)
112	Aluminum	(M)
120	Nonmetals	(F)
121	Ceramics	(M)
122	Rubber (belts)	(M)
130	Structural Shapes	(F)
131	Angles	(M)

132	Bar Stock	(M)
133	Beams	(M)
134	Channels	(M)
200	Electrical Components	(T)
210	Controls	(F)
220	Motors	(F)
221	AC Motors	(M)
222	DC Motors	(M)
230	Switches	(F)
300	Materials Handling	(T)
310	Solid Waste	(F)
400	Codes and Standards	(T)
410	ANCI Standards	(F)
411	ANSI Y14.5M–1982	(M)
500	Conveyors	(T)
510	Belt	(F)
520	Chain	(F)
530	Bucket	(F)
600	Drives	(T)
610	Gear Box	(F)
620	Roller Chain and Sprockets	(F)
630	V-Belt and Pulleys	(F)

And so on...

Start a technical file now, as a student, and expand it later, to suit your technical jobs in industry.

In addition to your personal file, when you are working in a design section, the section usually has an extensive vendor-catalog library and related references. There will most likely be a *Thomas Register* set (lists of vendors, their products, and addresses) and a set of ANSI (American National Standards Institute) standards related to your company's products. The vendor catalog collection would cover components such as the following:

- Bearings
- Belts
- Brakes
- Cams
- Clutches
- Controls
- Couplings
- Fasteners
- Gears
- Hydraulic power (liquid)
- Hydraulic power (pneumatic)
- Jigs and fixtures
- Sprockets
- Structural steel and aluminum
- Keys and pins
- Lubricants
- Motors
- O-rings
- Piping lumber
- Plastics
- Programmable logic circuits
- Pulleys
- Pumps
- Reducers
- Retaining rings
- Roller chain
- Shafts
- Switches
- Tools

Occasionally, a drafter, designer, or engineer may be required by an employer to keep a bound-page notebook for patent development. The pages are bound together, numbered, and each one signed by the individual inventor. Also, each page needs to be signed by at least two witnesses who understand the area of technology of the potential patent. Furthermore, the individual inventor may be required to sign an invention agreement.

Invention Agreement

Anyone who is creative and working for a company, such as a drafter, designer, or engineer, usually must sign an *invention agreement* form giving the company the right to any new invention designed while one is working for the company. The form specifies that the employee will not reveal any of the company's discoveries or projects. The invention agreement is binding on the employee from six months to two years after the employee leaves the company, so that employees will not invent something, quit, and patent the invention themselves.

A company is not obligated to, so it usually does not give extra pay for an invention developed by an employee. This is what the employee is paid to do. However, companies recognize talent and will reward that talent with a promotion, stocks, bonds, or possibly a raise. Signing an employee/employer invention agreement is a normal request in any company. Of course, prospective employees should read the agreement carefully and understand it before signing. An applicant who refuses to sign the agreement may be asked to look elsewhere for employment.

Patent Drawings

All patent applications for a new idea or an invention must include a *patent drawing* to illustrate and explain its function. Patent drawings must be mechanically correct and must adhere to strict patent drawing regulations, *Figure 17-28*. For example, patent drawings are more pictorial and explanatory than regular detail or assembly drawings. Center lines, dimensions, and notes are omitted. All features and parts are identified by numbers that refer to written explanations and descriptions. Line shading is used to improve readability.

All patent drawings must be made so that they reproduce clearly, which essentially means that the drawings must be done in ink. Current requirements specify that India ink or its equivalent be used, and that the drawings be done on strong, white paper, 8.5 × 13 (21.6 × 33.1 cm) or 8.5 × 14 (21.6 × 35.6 cm). A precise "sight" area (20.3 × 29.8 cm) must be used for the drawing itself

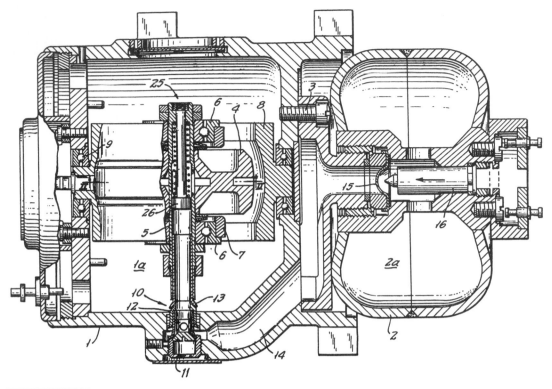

FIGURE 17-28   *Patent drawing* (Courtesy Timex Corp.)

on the strong, white paper. (For further requirements, refer to *37 CFR* Ch 1 7-1-91 edition.) The publication *A Guide for Patent Drawings* may be obtained from the Superintendent of Documents, U.S. Government Printing Office, Washington, DC 20402.

Summary

- A typical engineering department has three sections: design, application, and quotation. The design procedure used in industry and the basis for establishing the selling price of a system or product should be understood by design and drafting students.

- Technical drawing types used in industry include freehand sketches, layout drawings, detail drawings, and assembly drawings. Fold lines are seldom included on technical drawings in industry. Procedures for handling ECRs and ECOs on drawings should be understood by design and drafting students.

- Students should understand mass production and the role of DFM and DFA in the manufacturing process in a mass production company. Design for manufacturability (DFM) involves considering both function and production during the design process. Design for assembly (DFA) requires the consideration of both function and the assembly process.

Review Questions

Answer the following questions either true or false.

1. The application engineering section typically handles special-order modifications of existing products.

2. The only members of the design team should be designers and drafters.

3. Producing computer mouse units is an example of a one-of-a-kind system.

4. RFP stands for request for proposal.

5. In a layout drawing, a freehand sketch is redrawn to scale.

6. A design procedure acts as a general guideline for the design team.

7. Conceptual design is based on materials science, chemistry, and mathematics.

8. Any product that has more than two parts requires an assembly drawing.

9. A company uses purchased parts made by another company because it is cheaper.

10. The numbering system for identifying and recording technical drawings is standardized and used throughout the U.S.

Answer the following questions by selecting the best answer.

1. The two types of illustrative material used by a systems company to help a customer visualize the overall system are:
   a. Assembly and detail.
   b. Working and assembly.
   c. Perspective and scale model.
   d. Assembly and subassembly.

2. Which of the following is **not** part of the information usually included in a title block?
   a. Material the part is to be made of
   b. Tolerances/limits of dimensions
   c. Part or drawing number
   d. Reference numbers of subassembly drawings

3. Which of the following is **not** true regarding checking drawings?
   a. Slight errors can cause large expenses.
   b. A good drafter thoroughly inspects his/her own drawings once completed.
   c. If the drafter does a thorough inspection of his/her own drawing, it is not necessary to send it to the checker.
   d. Every precaution should be taken to avoid errors.

4. Which of the following is **not** true regarding drawing revisions?
   a. A drawing can be modified once an ECR is signed and submitted.
   b. All departments concerned must agree to the change.
   c. A drafter can change his/her own drawing after it has been released and signed off.
   d. Minor changes to drawings could be simple erasures or additions made directly on the original.

5. Which of the following would be part of preparing a selling price?
   a. Engineering materials
   b. Basic labor (direct labor)
   c. Profit
   d. All of the above

6. Which of the following is **not** part of a typical boiler plate?
   a. Introductory comments
   b. Descriptions of past systems designed/installed
   c. An invention agreement for signature
   d. A legal section

7. Which of the following is **not** a guideline from DFM literature?
   a. Minimize number of parts.
   b. Minimize multifunctional parts.
   c. Minimize part variation.
   d. Minimize number of flexible parts.

8. Which of the following is **not** true regarding a mass production company?
   a. Its approach to producing a product would be different than a one-of-a-kind system.
   b. Their internal departments would work together more closely than those in a one-of-a-kind system.
   c. Their internal departments would work together less closely than those in a one-of-a-kind system.
   d. None of the above

Chapter 17 Problems

The major drawings that were illustrated and emphasized in this chapter included freehand sketches, layout, detail, and assembly drawings, the types of drawings used in industry to develop a product. Two completed sets of these drawings are included here for your information and guidance. Then, the problems for Chapter 17 follow.

EXAMPLE NUMBERS 17-1-1 THROUGH 17-1-6 (SUBASSEMBLY NOT REQUIRED)

Objective: Design a hold-down clamp that:
1. bolts through a ¾″ thick worktable and clamps materials 1″ to 1½″ thick.
2. stays in "open" (up) position when not clamping.

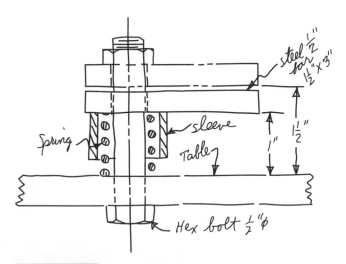

EX 17-1-1   *Freehand sketch of hold-down clamp*

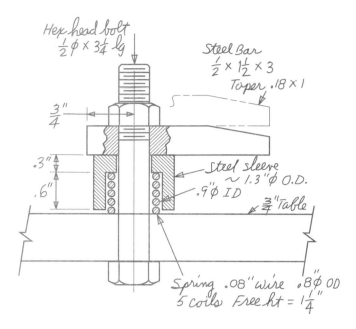

Hex head bolt
½ φ × 3¼ lg

Steel Bar
½ × 1½ × 3

Taper .18 × 1

¾"

.3"

.6"

Steel sleeve
~ 1.3"φ O.D.
.9"φ ID

¾" Table

Spring .08" wire .8" φ OD
5 coils Free ht = 1¼"

EX 17-1-2   *Layout drawing of hold-down clamp*

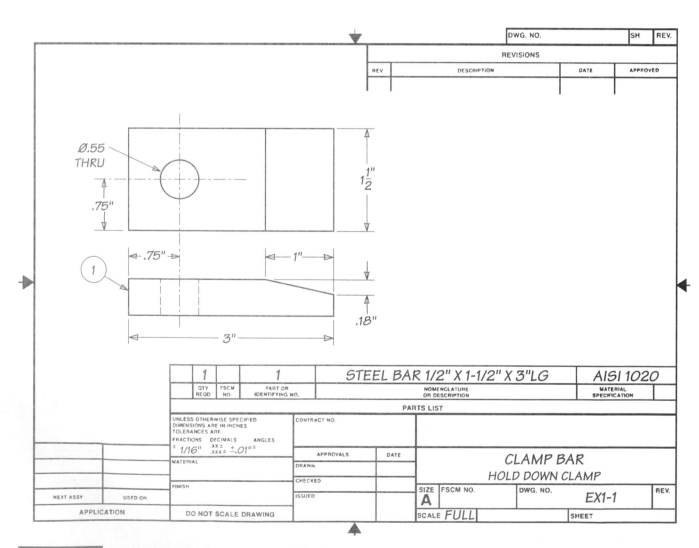

Ø.55
THRU

1½"

.75"

.75"   1"

1

.18"

3"

1		1	STEEL BAR 1/2" X 1-1/2" X 3"LG	AISI 1020	
QTY REQD	FSCM NO.	PART OR IDENTIFYING NO.	NOMENCLATURE OR DESCRIPTION	MATERIAL SPECIFICATION	

PARTS LIST

UNLESS OTHERWISE SPECIFIED DIMENSIONS ARE IN INCHES TOLERANCES ARE:	CONTRACT NO.		
FRACTIONS DECIMALS ANGLES			
± 1/16" .XX± ± .01"± .XXX±	APPROVALS	DATE	CLAMP BAR HOLD DOWN CLAMP
MATERIAL	DRAWN		
	CHECKED		
FINISH	ISSUED	SIZE **A** FSCM NO.	DWG. NO. EX1-1 REV.
NEXT ASSY USED ON		SCALE *FULL*	SHEET
APPLICATION	DO NOT SCALE DRAWING		

EX 17-1-3   *Detail drawing of clamp bar*

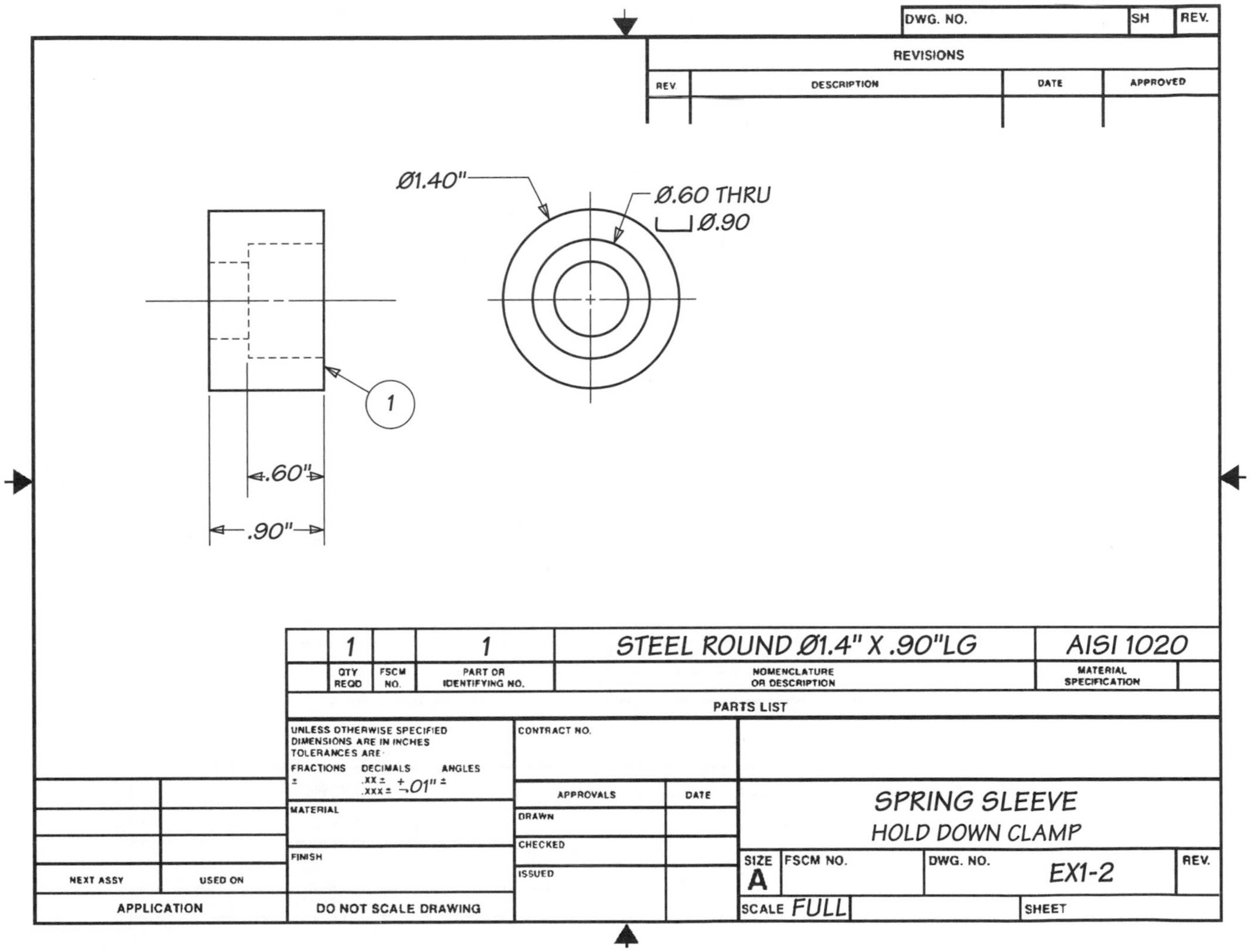

	DWG. NO.		SH	REV.

REVISIONS

REV.	DESCRIPTION	DATE	APPROVED

Ø1.40"

Ø.60 THRU
⌴ Ø.90

1

◄.60"►

◄ .90" ►

1		1	STEEL ROUND Ø1.4" X .90"LG	AISI 1020	
QTY REQD	FSCM NO.	PART OR IDENTIFYING NO.	NOMENCLATURE OR DESCRIPTION	MATERIAL SPECIFICATION	

PARTS LIST

UNLESS OTHERWISE SPECIFIED
DIMENSIONS ARE IN INCHES
TOLERANCES ARE:
FRACTIONS DECIMALS   ANGLES
±   .XX ±   ÷.01" ±
.XXX ±

CONTRACT NO.

APPROVALS | DATE

DRAWN

CHECKED

ISSUED

MATERIAL

FINISH

NEXT ASSY | USED ON

APPLICATION

DO NOT SCALE DRAWING

SPRING SLEEVE
HOLD DOWN CLAMP

SIZE	FSCM NO.	DWG. NO.		REV.
A			EX1-2	

SCALE *FULL* | SHEET

EX 17-1-4   *Detail drawing of spring sleeve*

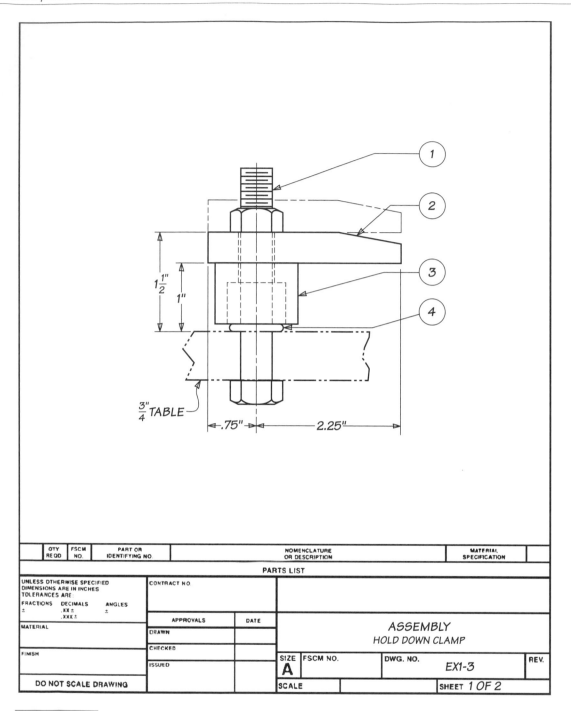

QTY REQD	FSCM NO.	PART OR IDENTIFYING NO.	NOMENCLATURE OR DESCRIPTION	MATERIAL SPECIFICATION	
			PARTS LIST		

UNLESS OTHERWISE SPECIFIED DIMENSIONS ARE IN INCHES TOLERANCES ARE:

FRACTIONS   DECIMALS   ANGLES
±   .XX ±   ±
          .XXX ±

MATERIAL

FINISH

DO NOT SCALE DRAWING

CONTRACT NO.

APPROVALS	DATE
DRAWN	
CHECKED	
ISSUED	

ASSEMBLY
HOLD DOWN CLAMP

SIZE	FSCM NO.	DWG. NO.		REV.
A			*EX1-3*	
SCALE			SHEET *1 OF 2*	

EX 17-1-5   *Assembly drawing of hold-down clamp*

			.08" OD, FREE HT 1¹/₄"		
			.08" Ø5 COILS, PLAIN ENDS		
1		4	SPRING: HARD DRAWN SPRING STEEL WIRE RH		
1		3	SPRING SLEEVE	DWG EX1-2	
1		2	CLAMP BAR	DWG EX1-1	
1		1	HEX HEAD BOLT ¹/₂" DIA X 3¹/₄" LG. HEX NUT		
QTY REQD	FSCM NO.	PART OR IDENTIFYING NO.	NOMENCLATURE OR DESCRIPTION		MATERIAL SPECIFICATION

PARTS LIST

UNLESS OTHERWISE SPECIFIED DIMENSIONS ARE IN INCHES TOLERANCES ARE:	CONTRACT NO.				
FRACTIONS ± / DECIMALS .XX ± / .XXX ± / ANGLES ±				ASSEMBLY HOLD DOWN CLAMP	
	APPROVALS	DATE			
MATERIAL	DRAWN				
	CHECKED				
FINISH	ISSUED		SIZE **A** / FSCM NO.	DWG. NO. EX1-3	REV.
DO NOT SCALE DRAWING			SCALE		SHEET 2 OF 2

EX 17-1-6   *Parts list for hold-down clamp*

EXAMPLE NUMBERS 17-2-1 THROUGH 17-2-8 (SUBASSEMBLY REQUIRED)

Objective: Design an articulated parallelogram, approximately 7" long and 2" wide, made of brass, which:

1. will maintain parallelism between two end members of the device and provide space in the end members for the customer's machining.
2. uses ball bearings at the pivot points and uses retaining rings for the pivot pins.

NOTE: Articulated parallelograms are used in pick-and-place robots and in goniometers (devices to measure angular displacements of knee joints).

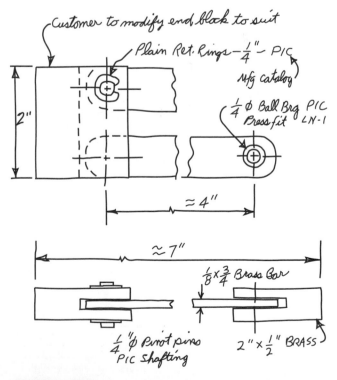

EX 17-2-1   *Freehand sketch of articulated parallelogram*

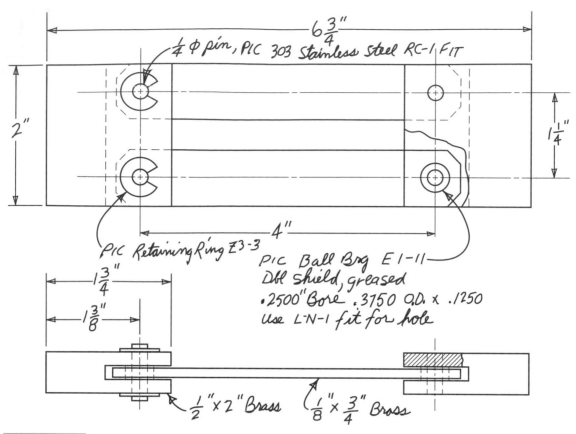

$6\frac{3}{4}$"

$\frac{1}{4}\,\phi$ *pin, PIC 303 Stainless Steel RC-1 FIT*

2"

$1\frac{1}{4}$"

4"

PIC Retaining Ring Z3-3

PIC Ball Brg E1-11
Dbl shield, greased
.2500" Bore .3750 O.D. x .1250
Use L-N-1 fit for hole

$1\frac{3}{4}$"

$1\frac{3}{8}$"

$\frac{1}{2}$" x 2" *Brass*

$\frac{1}{8}$" x $\frac{3}{4}$" *Brass*

EX 17-2-2   *Layout drawing of articulated parallelogram*

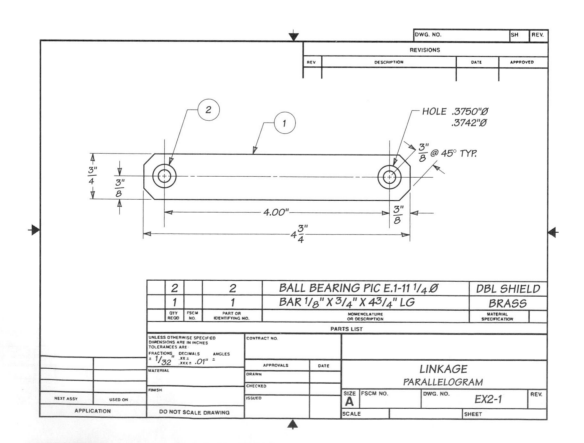

			DWG. NO.		SH	REV.

REVISIONS

REV	DESCRIPTION	DATE	APPROVED

2

1

HOLE .3750"Ø
.3742"Ø

$\frac{3}{8}$" @ 45° TYP.

$\frac{3}{4}$"

$\frac{3}{8}$"

4.00"

$\frac{3}{8}$"

$4\frac{3}{4}$"

2	2	BALL BEARING PIC E.1-11 $^{1}/_{4}$ Ø	DBL SHIELD
1	1	BAR $^{1}/_{8}$" X $^{3}/_{4}$" X $4^{3}/_{4}$" LG	BRASS
QTY REQD	FSCM NO.	PART OR IDENTIFYING NO. NOMENCLATURE OR DESCRIPTION	MATERIAL SPECIFICATION

PARTS LIST

UNLESS OTHERWISE SPECIFIED
DIMENSIONS ARE IN INCHES
TOLERANCES ARE:
FRACTIONS   DECIMALS   ANGLES
± 1/32   .XX ±   ±
.XXX ± .01"

CONTRACT NO.

MATERIAL

APPROVALS   DATE

DRAWN

FINISH

CHECKED

ISSUED

LINKAGE
PARALLELOGRAM

NEXT ASSY   USED ON

APPLICATION

DO NOT SCALE DRAWING

SIZE **A**   FSCM NO.   DWG. NO.   *EX2-1*   REV.

SCALE   SHEET

EX 17-2-3   *Subassembly detail drawing of linkage*

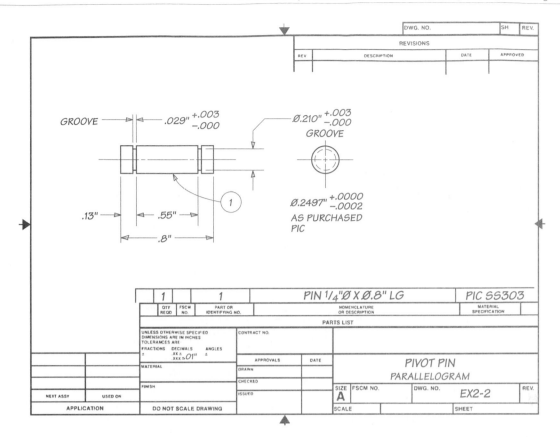

	1		1	PIN 1/4"Ø X Ø.8" LG	PIC SS303	
	QTY REQD	FSCM NO.	PART OR IDENTIFYING NO.	NOMENCLATURE OR DESCRIPTION	MATERIAL SPECIFICATION	
				PARTS LIST		

UNLESS OTHERWISE SPECIFIED
DIMENSIONS ARE IN INCHES
TOLERANCES ARE
FRACTIONS DECIMALS ANGLES
± .XX ± ±
.XXX ≤ .01"

CONTRACT NO.

MATERIAL

FINISH

APPROVALS   DATE
DRAWN
CHECKED
ISSUED

PIVOT PIN
PARALLELOGRAM

SIZE **A** | FSCM NO. | DWG. NO. | *EX2-2* | REV.
SCALE | | SHEET

NEXT ASSY   USED ON
APPLICATION   DO NOT SCALE DRAWING

EX 17-2-4   *Detail drawing of pivot pin*

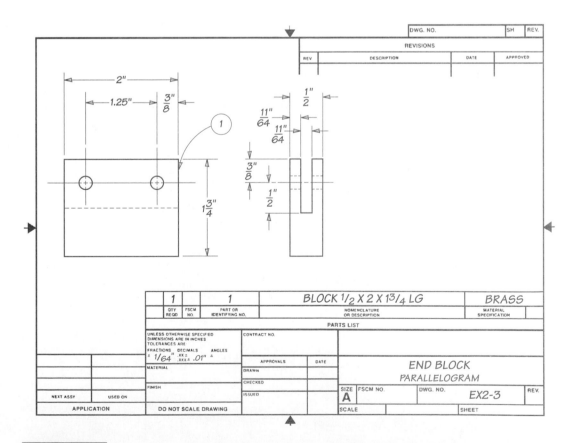

	1		1	BLOCK 1/2 X 2 X 1³/₄ LG	BRASS	
	QTY REQD	FSCM NO.	PART OR IDENTIFYING NO.	NOMENCLATURE OR DESCRIPTION	MATERIAL SPECIFICATION	
				PARTS LIST		

UNLESS OTHERWISE SPECIFIED
DIMENSIONS ARE IN INCHES
TOLERANCES ARE
FRACTIONS DECIMALS ANGLES
± 1/64" .XX ± ±
.XXX ± .01"

CONTRACT NO.

MATERIAL

FINISH

APPROVALS   DATE
DRAWN
CHECKED
ISSUED

END BLOCK
PARALLELOGRAM

SIZE **A** | FSCM NO. | DWG. NO. | *EX2-3* | REV.
SCALE | | SHEET

NEXT ASSY   USED ON
APPLICATION   DO NOT SCALE DRAWING

EX 17-2-5   *Detail drawing of end block*

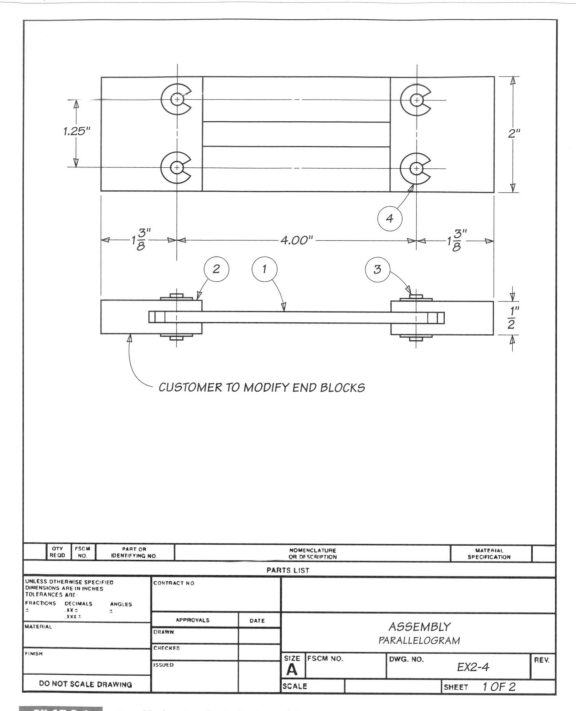

CUSTOMER TO MODIFY END BLOCKS

QTY REQD	FSCM NO.	PART OR IDENTIFYING NO.		NOMENCLATURE OR DESCRIPTION	MATERIAL SPECIFICATION	
				PARTS LIST		

UNLESS OTHERWISE SPECIFIED
DIMENSIONS ARE IN INCHES
TOLERANCES ARE:

FRACTIONS	DECIMALS	ANGLES
±	.XX ±	±
	.XXX ±	

CONTRACT NO.

MATERIAL

FINISH

APPROVALS	DATE
DRAWN	
CHECKED	
ISSUED	

ASSEMBLY
PARALLELOGRAM

SIZE	FSCM NO.	DWG. NO.		REV.
A			EX2-4	

SCALE		SHEET	1 OF 2

DO NOT SCALE DRAWING

EX 17-2-6   *Assembly drawing of articulated parallelogram*

QTY REQD	FSCM NO.	PART OR IDENTIFYING NO.	NOMENCLATURE OR DESCRIPTION		MATERIAL SPECIFICATION	
8		4	RETAINING RING		PIC Z3-3	
4		3	PIVOT PIN		DWG EX2-2	
2		2	END BLOCK		DWG EX2-3	
2		1	LINKAGE		DWG EX2-1	

PARTS LIST

UNLESS OTHERWISE SPECIFIED DIMENSIONS ARE IN INCHES TOLERANCES ARE: FRACTIONS ± DECIMALS .XX ± .XXX ± ANGLES ±	CONTRACT NO.				
MATERIAL	APPROVALS	DATE	ASSEMBLY PARALLELOGRAM		
	DRAWN				
	CHECKED				
FINISH	ISSUED		SIZE A	FSCM NO.	DWG. NO. EX2-4 REV.
DO NOT SCALE DRAWING			SCALE		SHEET 2 OF 2

EX 17-2-7 *Parts list for articulated parallelogram*

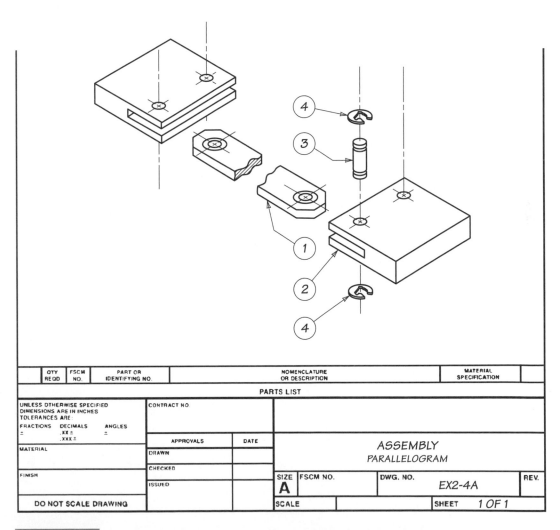

QTY REQD	FSCM NO.	PART OR IDENTIFYING NO.	NOMENCLATURE OR DESCRIPTION		MATERIAL SPECIFICATION	

PARTS LIST

UNLESS OTHERWISE SPECIFIED DIMENSIONS ARE IN INCHES TOLERANCES ARE: FRACTIONS ± DECIMALS .XX ± .XXX ± ANGLES ±	CONTRACT NO.				
MATERIAL	APPROVALS	DATE	ASSEMBLY PARALLELOGRAM		
	DRAWN				
	CHECKED				
FINISH	ISSUED		SIZE A	FSCM NO.	DWG. NO. EX2-4A REV.
DO NOT SCALE DRAWING			SCALE		SHEET 1 OF 1

EX 17-2-8 *Exploded assembly drawing of articulated parallelogram (an alternative type of assembly drawing)*

Problems

Unless otherwise directed, do the problems following the steps indicated using traditional technical drawing instruments or a CAD system as assigned.

CONVENTIONAL INSTRUCTIONS

STEP 1 Prepare a freehand sketch(es) of the device or component to be designed or changed. Refer to Figure 17-4, Example 17-1-1, and Example 17-2-2 for example freehand sketches.

STEP 2 Prepare a layout drawing(s) of the freehand sketch(es) from Step 1. Refer to Figure 17-5, Example 17-2-1, and Example 17-2-2 for example layout drawings.

STEP 3 Prepare the necessary detail drawings of each component of the device, one component per drawing. Refer to Figure 17-10, Figure 17-11, Example 17-1-3, Example 17-1-4, Example 17-2-4, and Example 17-2-5 for examples. Include required fits and tolerances as required.

STEP 4 Prepare a subassembly drawing if needed. Refer to Figure 17-9 and Example 17-2-3 for example subassembly drawings.

STEP 5 Prepare the final assembly drawing with the appropriate parts list. Refer to Figure 17-7 for a Method A assembly drawing, and to Figure 17-8 for a Method B drawing. Also refer to Example 17-1-5, Example 17-1-6, Example 17-2-6, and Example 17-2-7 for final assembly drawings with parts lists on separate sheets. Example 17-2-8 is an example of an exploded assembly drawing.

CAD Instructions

Students who plan to complete the following problems on a CAD system should begin with these general instructions. Before reading the specific instructions for each activity, go through each step in the following planning checklist. The checklist applies to any CAD system and will help ensure the optimum use of your time and resources.

1. Analyze the problem carefully. Decide exactly what you are being asked to do.

2. Determine what resources and references (if any) you will need in order to complete the problem. Collect those references and resources and have them readily available.

3. Decide if any particular standards apply to the project and have those standards available.

4. Determine what types of views will be required and how many of each.

5. Determine what the final plotted scale of the drawing will need to be, and select the appropriate paper size for plotting/printing.

6. Plan your drawing sequence. In what order will you develop the drawing (i.e., lines, features, dimension lines, leaders, dimensions)?

7. Review the various CAD commands you will have to use in order to develop the drawing.

8. Examine your CAD system to ensure that everything is in working order, then begin the project.

Problem 17-1

Prepare a freehand sketch of a step stand, two levels, for general use in homes, institutions (libraries, schools, etc.), and business offices. Use black anodized aluminum sheet and structural angles. Use any combination of 1 × 1 × ⅛, 1½ × 1½ × ⅛, or 2 × 2 × ⅛ angles and assume bolted construction.

Problem 17-2

The winch shown is to be used to pull boats up a ramp, and it needs a support base so it can be mounted on a horizontal wooden platform 1" thick and about 4' high.

a. Prepare a freehand sketch of the support base.

b. Prepare a layout drawing of your support base.

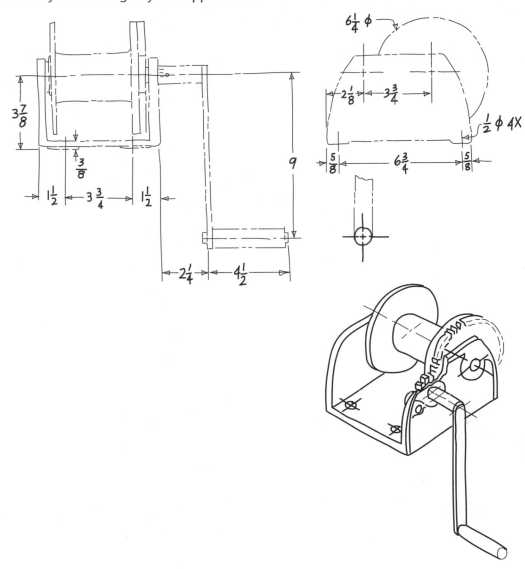

Problem 17-3

Prepare a freehand sketch of a dolly approximately 2′ × 3′ using two fixed and two swivel casters. Fasteners should not project above the top surface. A convenient means to hand-carry the dolly should be provided.

Prepare layout, detail, and assembly drawings as assigned.

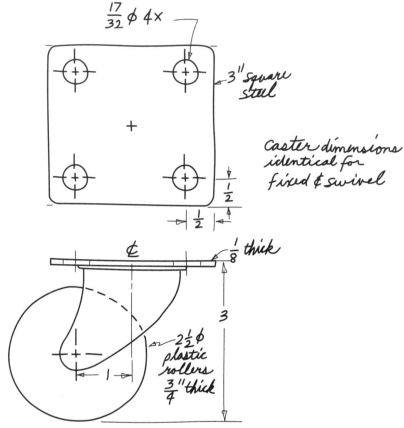

Problem 17-4

The torque wrench shown in the freehand sketch needs to be developed further and sent on to be manufactured in the shop.

   a. Prepare a layout drawing of the torque wrench.
   b. Prepare detail drawings of the torque wrench.
   c. Prepare an assembly drawing and parts list for the torque wrench.

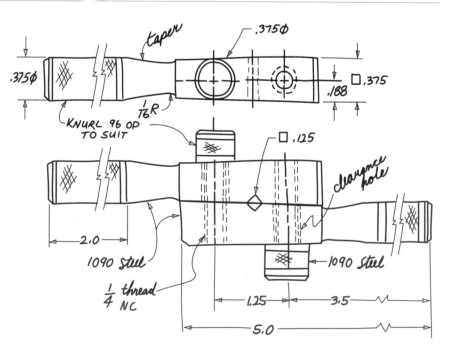

Problem 17-5

The clamp shown in the freehand sketch needs to be developed further and sent on to be manufactured in the shop.
- **a.** Prepare a layout drawing of the clamp.
- **b.** Prepare detail drawings of the clamp.
- **c.** Prepare an assembly drawing and a parts list for the clamp.

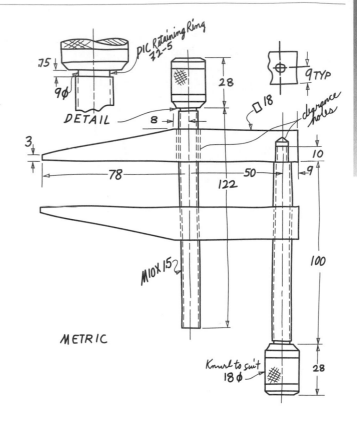

Problem 17-6

GIVEN: A layout drawing of adjustable dividers. Prepare detail and assembly drawings as assigned.

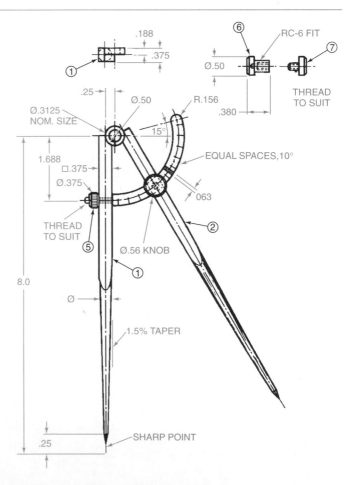

Problem 17-7

GIVEN: A layout drawing of a wood shop clamp. Prepare detail and assembly drawings as assigned.

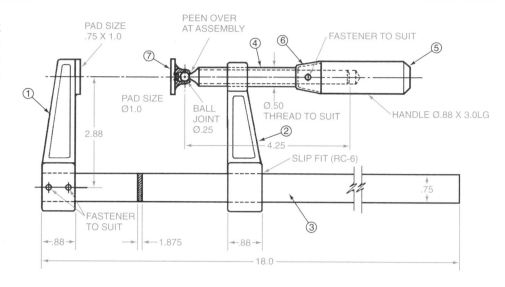

Problem 17-8

Conceptual views of a toggle clamp are shown in the freehand sketches. When pivot B on the toggle clamp arm ABT passes below the line of centers between A and C, the pressure at P causes the elastic head to compress. Then B is "locked" in position and the object under the elastic head at P is clamped securely. A typical compression distance is approximately 1.5 mm. The adjustments at E accommodate different sizes of objects.

  a. From the information given, prepare a layout drawing to establish the remaining dimensions of the components, the pivot point locations, and the bolt hole locations.
  b. Prepare detail drawings of the components.
  c. Prepare an assembly drawing and parts list for the toggle clamp.

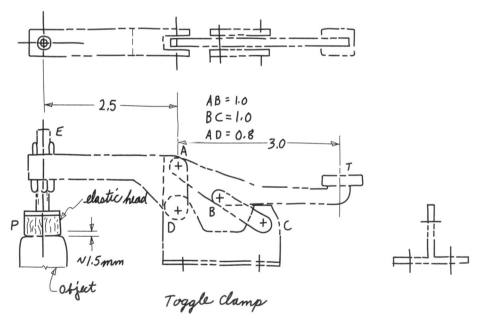

Problem 17-9

GIVEN: A layout drawing of a jack. Prepare detail and assembly drawings as assigned.

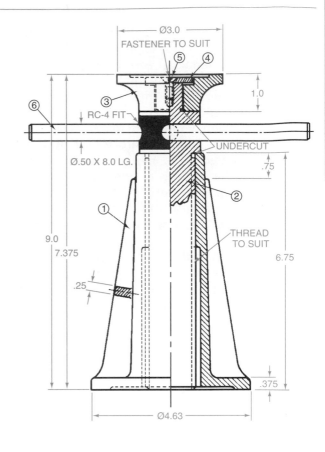

Problem 17-10

GIVEN: A layout drawing of an adjustable compass. Prepare detail and assembly drawings as assigned.

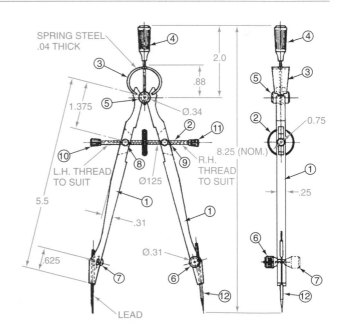

Problem 17-11

Limit switches, which are basically electrical on-off switches, limit the travel of machine parts, act as interlocks, and indicate positions of devices. For example, the machines in the photographs in Chapter 20 use limit switches to limit travel. Limit switches activated by open doors prevent an elevator from moving. Positions of remote devices in hostile environments, such as control rods in nuclear reactors, are indicated safely by limit switches. Rotating antennas can be stopped at preset positions by limit switches to keep wires from tangling.

Use information in the figure to do a layout drawing to determine the dimensions of an actuating shoe that would cause the limit switch to operate and stop the rotation of the antenna. Also, do a layout drawing of the support plate for the actuating shoe. Include a hub and set screw. Hub diameters may average approximately 2 times shaft diameters and are about 2 times shaft diameters long. The hub is to be welded to the support plate. Use low-carbon steel. (Refer to Table 47 in Appendix A.)

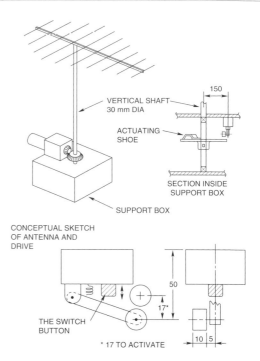

Problem 17-12

An air-actuated cylinder is to be used to bend a specimen in a fatigue test apparatus. The air cylinder and the specimen are shown in the partially completed layout drawing. You are to devise a means to support the cylinder and the specimen and to provide adjustments as indicated.

    a. Complete the layout drawing with a design that provides the support and adjustments.

    b. Prepare detail drawings of your design.

    c. Prepare an assembly drawing with a parts list of your design. When you draw the cylinder in the assembly drawing, use phantom lines. Purchased parts such as motors, reducers, generators, and cylinders often are drawn with phantom lines.

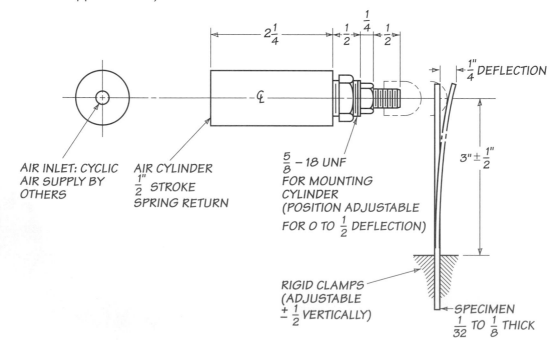

Problem 17-13

GIVEN: A layout drawing of an adjustable lifting clamp. Prepare detail and assembly drawings as assigned.

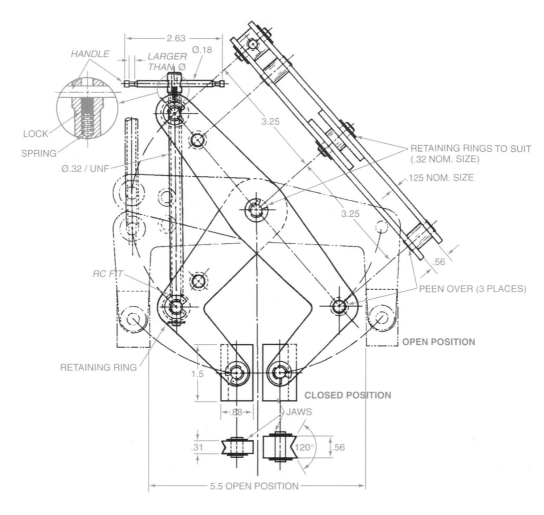

Problem 17-14

GIVEN: A layout drawing of a lever-controlled valve. Prepare detail and assembly drawings as assigned.

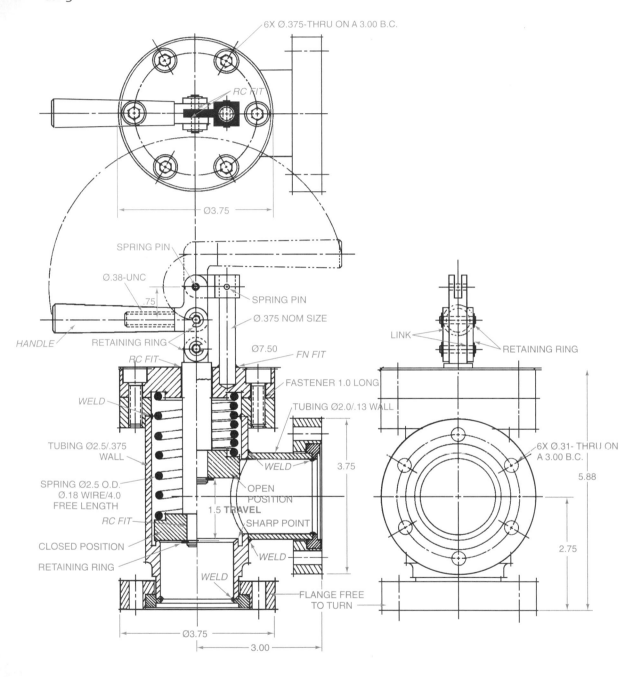

Problem 17-15

A hand-powered winch is shown in the isometric sketch. Using the information in the freehand sketches of the components of the winch, 17-15A through 17-15H:

a. Prepare detail drawings of each component. (Note that layout drawings are omitted in this exercise due to the completeness of the freehand sketches. Discrepancies, if any, will be found while doing the final assembly drawing.)

b. Prepare subassembly drawings of the drum 17-15B and the handle 17-15H.

c. Prepare a final assembly drawing with a parts list for the winch.

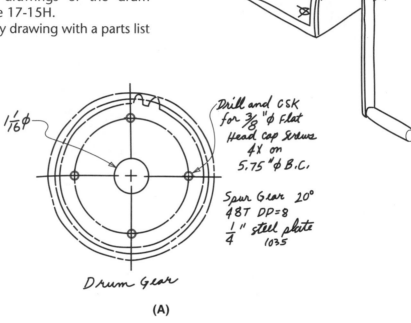

Drill and CSK for ⅜"∅ Flat Head cap Screws 4X on 5.75"∅ B.C.

Spur Gear 20° 48T DP=8 ¼" steel plate 1035

1 1/16 ∅

Drum Gear

(A)

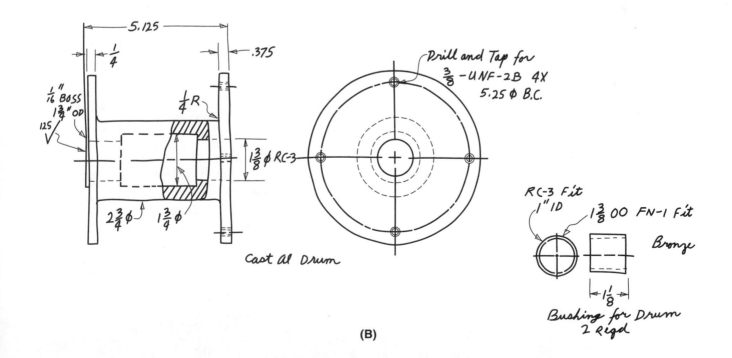

5.125

¼

.375

1/16" BOSS
1⅜" OD
125

¼ R

2¾ ∅   1¾ ∅

1⅜ ∅ RC-3

Cast Al Drum

Drill and Tap for ⅜-UNF-2B 4X 5.25 ∅ B.C.

RC-3 Fit
1" ID   1⅜ OO FN-1 Fit

Bronze

1⅛

Bushing for Drum
2 Reqd

(B)

Problems

Problem 17-15 (continued)

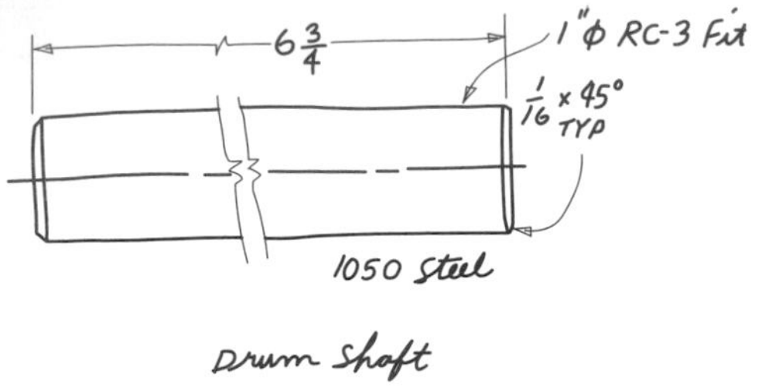
6 3/4
1"ϕ RC-3 Fit
1/16 x 45° TYP
1050 Steel
Drum Shaft

(C)

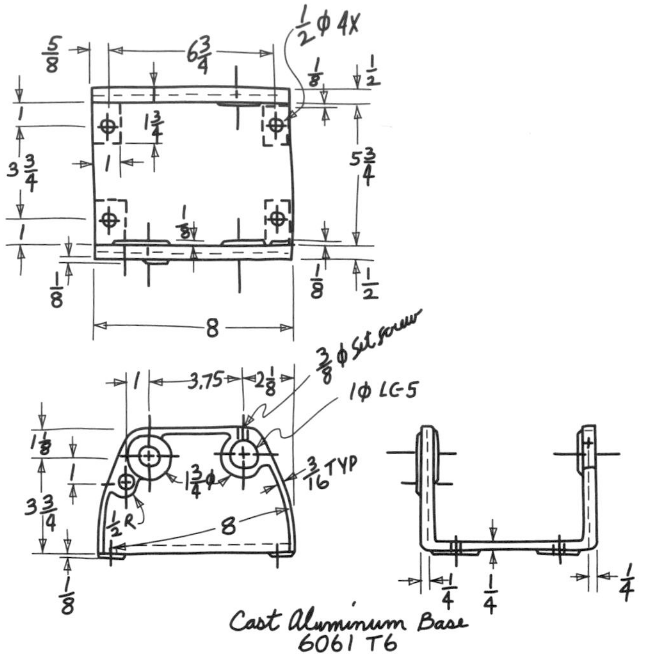
1/2 ϕ 4X
6 3/4
8
Cast Aluminum Base
6061 T6

(G)

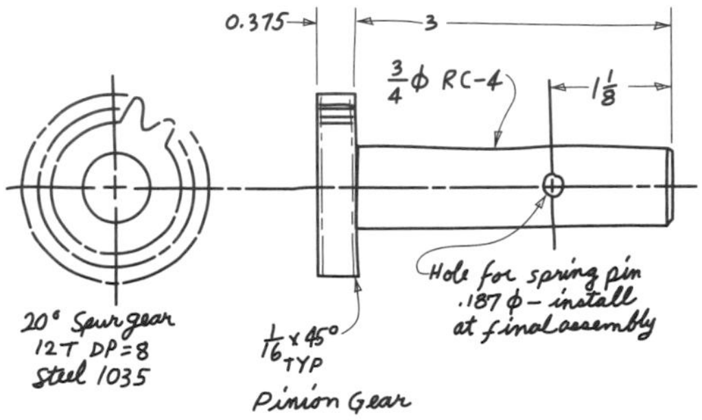
0.375
3
3/4 ϕ RC-4
1 1/8
20° Spur gear
12T DP=8
Steel 1035
1/16 x 45° TYP
Pinion Gear
Hole for spring pin
.187 ϕ – install
at final assembly

(D)

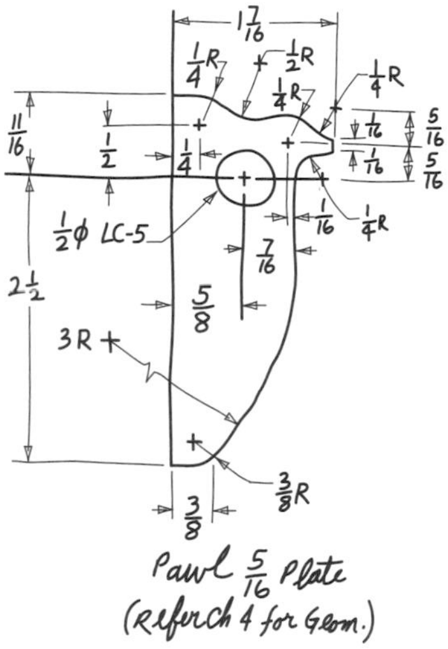
1 7/16
1/2ϕ LC-5
3R
3/8R
Pawl 5/16 Plate
(Refer Ch 4 for Geom.)

(E)

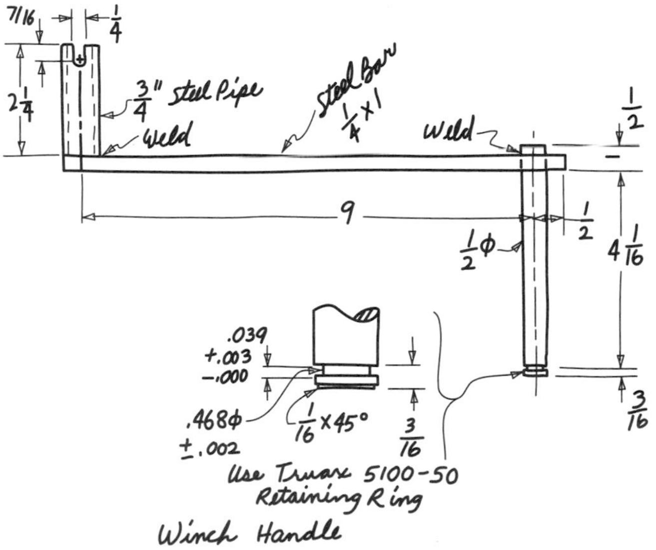
3/4" Steel Pipe
Weld
Steel Bar 1/4 x 1
Weld
9
1/2ϕ
4 1/16
.039 +.003 −.000
.468ϕ ±.002
1/16 x 45°
Use Truarc 5100-50
Retaining Ring
Winch Handle

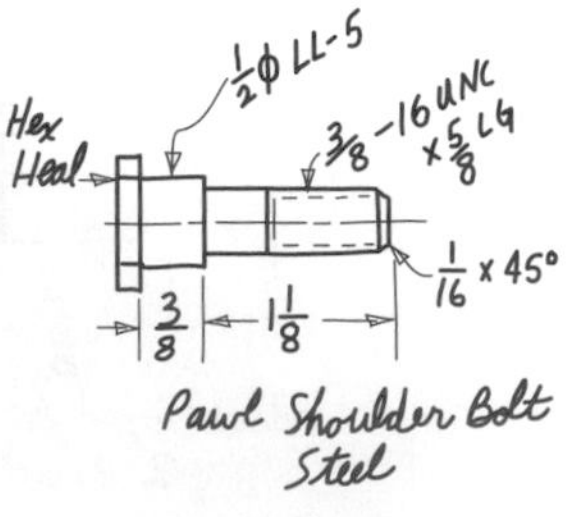
Hex Head
1/2ϕ LL-5
3/8 -16UNC x 5/8 LG
1/16 x 45°
3/8
1 1/8
Pawl Shoulder Bolt
Steel

(F)

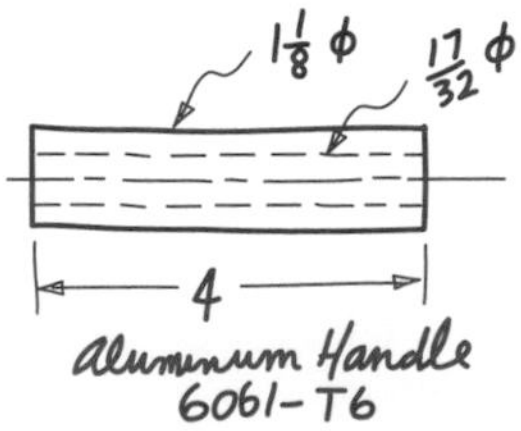
1 1/8 ϕ
17/32 ϕ
4
Aluminum Handle
6061-T6

(H)

Problem 17-16

The corrosion bath shown is rocked back and forth by an arm that is coupled to a variable-speed drive mechanism. Corrosive liquid is sloshed over the test specimens. The equipment was originally built by a skilled mechanic from sketches that have been misplaced or destroyed. Detail and assembly drawings of this testing equipment are needed by another company to build their own bath. From the information given in the sketches, do the following:

a. Prepare a layout drawing of the bath equipment.
b. Prepare detail drawings of the components.
c. Prepare subassembly drawings of the shaft *A* supports and the container for the plastic bath.
d. Prepare an assembly drawing with a parts list for the bath equipment.

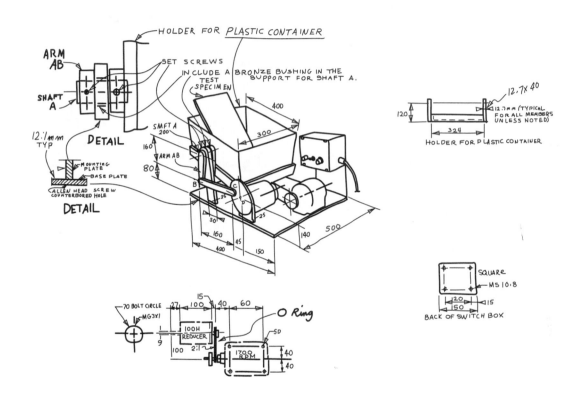

Problem 17-17

Using detail drawings 17-17A through 17-17O, prepare a 3-view assembly drawing and a parts list. Also, the following is a summary of the FITS specified for the various components. (Calculate these and show your work.)

(1) FN-1/1.062 Parts 1, 12

(2) FN-2/.688 Parts 1, 2

(3) LC-4/1.5 Parts 1, 11

(4) RC-5/.562 Parts 3, 4

(5) RC-5/.375 Parts 4, 5

(6) LC-2/.125 Parts 4, 8

(7) FN-2/.250 Parts 9, 14

(8) RC-6/.938 Parts 13, 14, 15

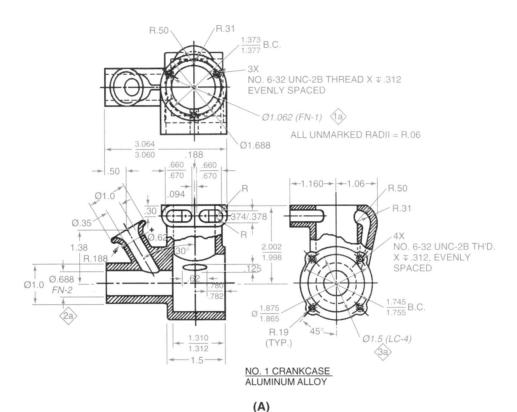

NO. 1 CRANKCASE
ALUMINUM ALLOY

(A)

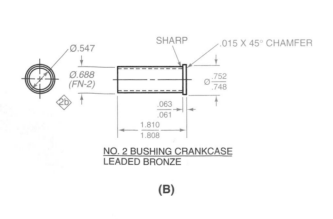

NO. 2 BUSHING CRANKCASE
LEADED BRONZE

(B)

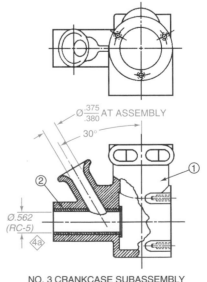

NO. 3 CRANKCASE SUBASSEMBLY
MAT'L AS NOTED

(C)

Problem 17-17 *(continued)*

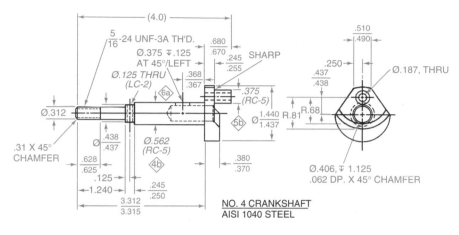

NO. 4 CRANKSHAFT
AISI 1040 STEEL

(D)

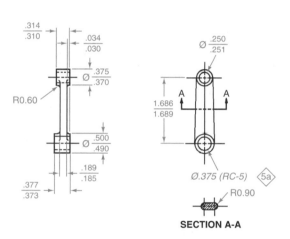

SECTION A-A

NO. 5 ROD-CONNECTING
ALUMINUM ALLOY 7075-T6

(E)

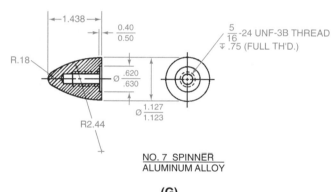

NO. 7 SPINNER
ALUMINUM ALLOY

(G)

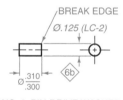

NO. 8 PIN-DRIVE WASHER
1010 STEEL

(H)

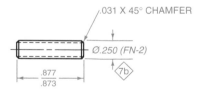

NO. 9 PIN PISTON
1010 STEEL

(I)

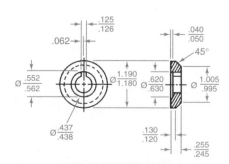

NO. 6 DRIVE WASHER
AISI C1018 STEEL

(F)

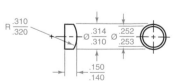

NO. 10 SPACER-PISTON
BRASS

(J)

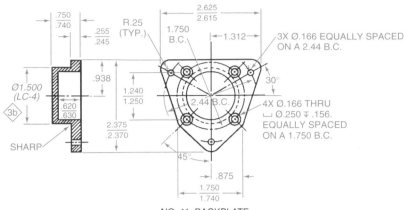

NO. 11 BACKPLATE
CAST ALUMINUM

(K)

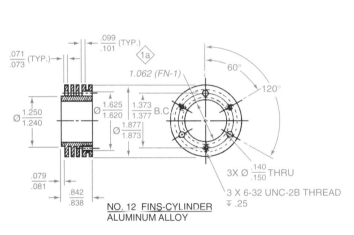

NO. 12 FINS-CYLINDER
ALUMINUM ALLOY

(L)

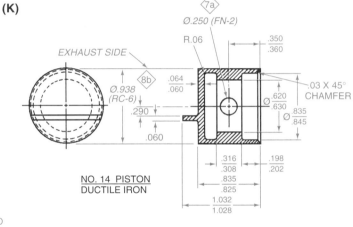

NO. 14 PISTON
DUCTILE IRON

(N)

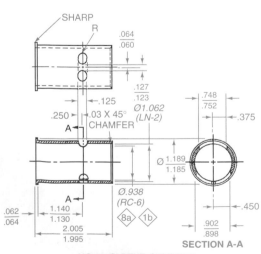

NO.13 SLEEVE-CYLINDER
DUCTILE IRON

(M)

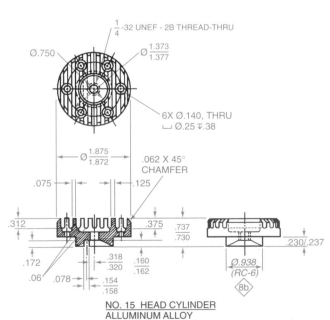

NO. 15 HEAD CYLINDER
ALLUMINUM ALLOY

(O)

Problem 17-18: Reverse Engineering

Reverse engineering basically involves
1. "taking apart" a manufactured device and
2. analyzing each component of the device for design features, size, materials, function, manufacturing processes, and any other factors such as aesthetics, color, and surface finishes.

Valuable insights can be learned from the exercise of "working backwards" as compared to "working forward."
1. Select or be assigned one of the following to be "taken apart" and analyzed as described above:
   a. component(s) from an electronic product such as a hand calculator, a printing calculator, a VCR, a TV, a light switch, or a timer (Note that electronic repair shops are excellent sources of discarded products.)
   b. Automotive devices
   c. Appliances
   d. Exercise and sports equipment
   e. Hardware stores
2. Prepare a freehand sketch of the component selected or assigned, and list the design features, size, material(s), function, and other properties.

Problem 17-19

Assume that you are a designer-drafter employed by a large company that designs and produces yard tools for average homeowners. Your company has received a request for a bid on 100,000 wheelbarrows, of approximately a 5 cubic foot capacity. You are given a freehand sketch of a wheelbarrow to more accurately determine dimensions, sizes, and the buyout requirements (fasteners and the wheel).

Assume that the time constraints require that you must establish a selling price for the wheelbarrow because your company's quotation department must meet a major deadline on another project. An approximate method of quickly setting a selling price is to calculate a "price-per-pound" of existing wheelbar-

rows already on the market. Then multiply the weight of your wheelbarrow design times the price per pound for an approximate selling price.

You may need to visit a hardware store to obtain data or find an appropriate catalog. One of your colleagues said that his wheelbarrow was approximately $2 per pound.

Time is a constraint so:
a. Prepare a layout drawing of the wheelbarrow and estimate its weight.
b. Calculate a price per pound of existing wheelbarrows and establish an approximate selling price.
c. Prepare an assembly drawing with a parts list for the wheelbarrow.

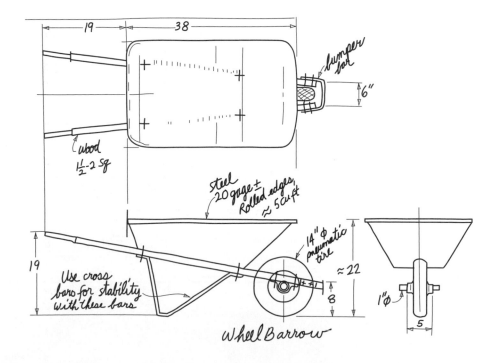

WheelBarrow

Problem 17-20

Attribute listing is similar to reverse engineering, but not as comprehensive. (An explanation of the attribute listing technique is in Chapter 24. A relatively simple example of attribute documenting of a caster is shown in the drawing.) Problem 17-20 combines the attribute listing procedure with technical drawings.

Find a music stand (or an object assigned by your instructor) and do the following:

a. Measure the components of the stand, note the materials used, determine the manufacturing processes, list the function(s) of the stand, and comment on any other facets such as color, weight, and stability. Prepare freehand sketches of the components. Incorporate any design changes you deem appropriate.

b. Prepare layout drawings of your freehand sketches, especially if you recommend changes.

c. Prepare detail drawings of the components.

d. Prepare subassembly and assembly drawings as needed.

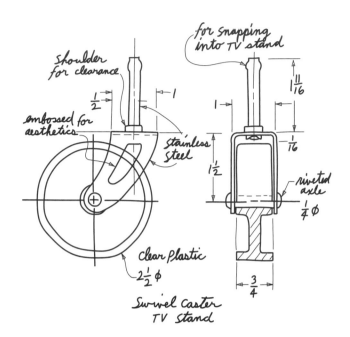

Problem 17-21

The angle stop shown in the drawing is to be changed according to ECO Number C159, originated by GNC on Sept. 11, and justified because of a new model of anchor-bolt puller to be manufactured. The description of change on a typical ECO form indicates that the 210 mm dimension changes to 250 and the holes change from a diameter of 13 to 19. An ECO committee determined that the change was necessary and that the change would not affect performance, reliability, or safety. Each member of the committee signed off the ECO (see Figure 17-13). Make a machine copy of the drawing as it is shown. Make the changes. (Refer to Figure 17-14.) Record the changes in the revisions block (upper right-hand corner).

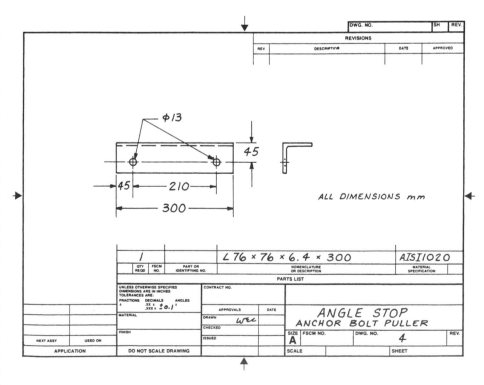

Problem 17-22

An engineering change request (ECR) involving the articulated parallelogram in Example 17-2-1 (page 651) has evolved into an engineering change order (ECO). The device is to be modified as follows:

1. Replace the ball bearings in the linkages with bronze bushings.

2. Use aluminum 6061-T6 instead of brass for the linkages and the end blocks.
   a. Prepare a layout drawing for the changes.
   b. Prepare detail drawings of the new design.
   c. Prepare an assembly drawing with a parts list for the new design.

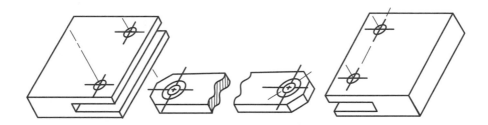

Problem 17-23

Use the five steps listed on page 656 to incorporate an ECO that requires that ball bearings replace the sliding bearings, Part #4.

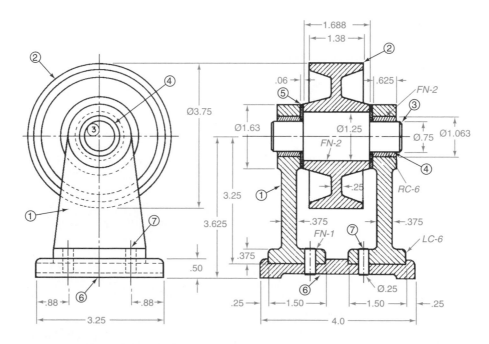

Problem 17-24

Follow the five steps listed on page 656 to develop the puller shown. For this problem, use the information from Figures 17-7 and 17-8 to prepare freehand sketches for each part. Assume an overall length of 9″ for the assembly of the yoke, the screw-center, and the hooks. Make the yoke and hooks of medium carbon steel, AISI 1045. (Refer to Table 47 in Appendix A.) Make the hooks at least ¼″ thick, and use ½″-20UNF threads or larger for the screw-center.

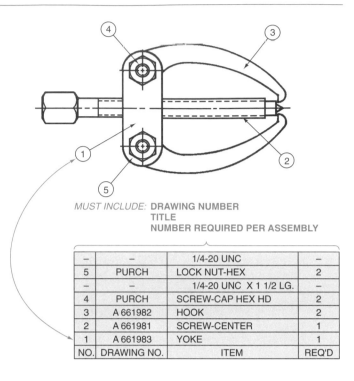

MUST INCLUDE: DRAWING NUMBER
TITLE
NUMBER REQUIRED PER ASSEMBLY

NO.	DRAWING NO.	ITEM	REQ'D
–	–	1/4-20 UNC	–
5	PURCH	LOCK NUT-HEX	2
–	–	1/4-20 UNC X 1 1/2 LG.	–
4	PURCH	SCREW-CAP HEX HD	2
3	A 661982	HOOK	2
2	A 661981	SCREW-CENTER	1
1	A 661983	YOKE	1

Problem 17-25

The pressure transducer shown indicates pressure by the distortion of the piezoelectric crystal that generates a voltage proportional to the distortion. A digital voltmeter is used to read the voltage. Assembly of the transducer could be simplified if the tubular spacer were part of the hex-shaped fitting with the ½″-UNF threads. Prepare a freehand sketch of your idea to simplify the assembly, and follow the five steps listed on page 656 to develop the transducer. The crystal is a purchased item. Use stainless steel. (Refer to Table 47 in Appendix A.)

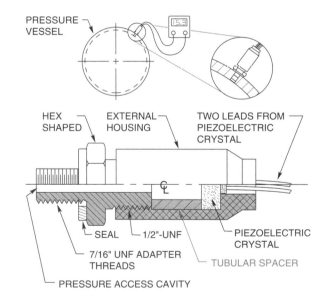

Problem 17-26

Manufacturing and assembly of the adjustable roller support shown could be improved upon. Prepare a freehand sketch of your idea for improvement and then follow the five steps listed on page 656 to develop the adjustable roller support. Note that slotted holes are used frequently for adjustments.

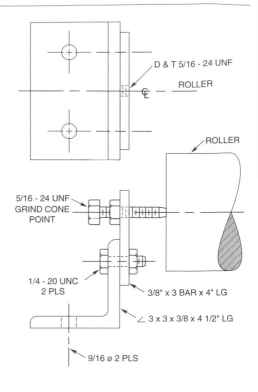

D & T 5/16 - 24 UNF

ROLLER

ROLLER

5/16 - 24 UNF
GRIND CONE
POINT

1/4 - 20 UNC
2 PLS

3/8" x 3 BAR x 4" LG

∠ 3 x 3 x 3/8 x 4 1/2" LG

9/16 ø 2 PLS

PROBLEM 17-27

Start your own technical file by selecting manufacturing catalogs that list products in areas of your interests. Label the catalogs with "100" numbers like the listings in this chapter in the section on Technical Files. Select at least 10 catalogs as a beginning.

As a suggestion, visit one or more of the following locations where you would most likely find a library of catalogs to help start your list: your own drafting classroom, an engineering library, maintenance shops on your campus, or local manufacturing firms.

PROBLEM 17-28

Drawings for patents of mechanical devices must clearly depict the assembly and function of the devices. Peruse a collection of patents or patent gazettes and list the patents with comments on the different types of drawings used to communicate information. List at least 10 patents (patent number, date, and a brief description of the device). Suggestion: librarians can not only help you find the patents but can help you find patents in your areas of technical interest.

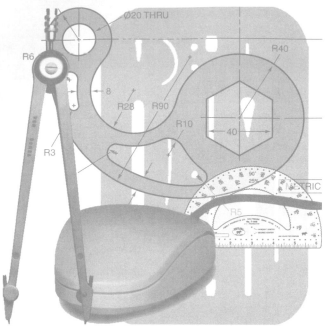

PICTORIAL DRAWINGS

KEY TERMS

Cabinet drawing

Cavalier drawing

Dimetric drawing

Exploded isometric drawing

General oblique drawing

Isometric assembly

Isometric drawing

Isometric projection

Nonisometric line

Oblique drawing

Offset measurement

Perspective drawing

Trimetric drawing

CHAPTER OUTLINE

Purpose of pictorial drawings • Oblique drawings • Isometric drawings • Axonometric: drawing and projection • Other oblique drawings • Perspective drawings • Summary • Review questions • Chapter 18 problems

CHAPTER OBJECTIVES

Upon completion of this chapter, students should be able to do the following:

■ Explain the purpose of pictorial drawings.

■ Demonstrate proficiency in developing oblique drawings, including cavalier and cabinet.

■ Demonstrate proficiency in developing isometric drawings, including diametric and trimetric.

■ Demonstrate proficiency in drawing circles, curves, and irregular shapes in isometric and on flat and inclined surfaces.

■ Demonstrate proficiency in developing one-, two-, and three-point perspective drawings.

Purpose of Pictorial Drawings

Pictorial drawings facilitate the communication of technical information during all phases of the design process—from early sketching of product ideas for future uses, to talking through concepts with colleagues, and finally, to the more formal pictorial drawings (for selling products, maintenance manuals, catalogs, assembly instructions, etc.). This chapter begins with the three most common types of pictorial drawings used by drafters, designers, and engineers—oblique, isometric, and perspective. Sketching these three types was introduced in Chapter 3.

After the discussion of oblique, isometric, and perspective drawings, more complete discussions will follow on additional types of pictorial drawings and projections.

Oblique Drawings

Oblique drawing is the easiest of the three to draw. In oblique drawing, one surface of the object, usually the front view, is drawn exactly as it would appear in a multiview drawing. Circles are true circles, and rectangles parallel to the frontal plane are in true size, *Figure 18-1*. Circles are drawn with a compass or circle template, and rectangles have square corners.

CONSTRUCTING OBLIQUE DRAWINGS

Oblique drawings use three axes; two at right angles (for example, X and Y in Figure 18-1) and one, the Z axis, usually at an angle of the convenient 30° or 45° standard triangle. Measurements are taken from the multiview drawing; X and Y dimensions match those in front views, and Z dimensions, depth, are laid off parallel to the third axis.

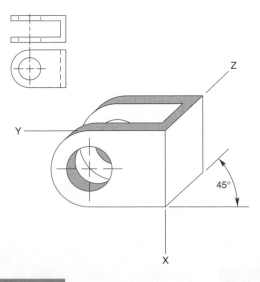

FIGURE 18-1 *Axes for oblique drawings*

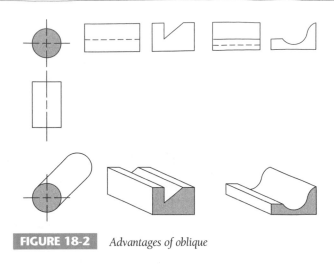

FIGURE 18-2 *Advantages of oblique*

CYLINDERS, ANGLES, AND CURVES IN OBLIQUE DRAWINGS

Figure 18-2 has further illustrations of the primary advantages of oblique drawings, especially starting with the front view. Cylinders are round, angles are in true size, and irregular curves are not distorted.

Isometric Drawings

Isometric drawings are the most used of the three common pictorials. They are easier to draw than perspective drawings and they illustrate objects almost as well as perspective drawings.

Isometric drawings use three axes, sometimes called principal edges, at 120° to each other, *Figure 18-3*.

HOW TO MAKE AN ISOMETRIC DRAWING

GIVEN: A multiview drawing of an object, *Figure 18-4A*.

STEP 1 Locate the starting point and lightly draw the three principal edges at 120° using a 30°–60° triangle, *Figure 18-4B*.

STEP 2 Transfer the depth, height, and width directly from the multiview drawing, *Figure 18-4C*, to the three principal edges. Measure full size directly along each edge. Lightly construct an isometric, basic outline, or "box," of the object, Figure 18-4C.

STEP 3 Transfer true-size lengths, lengths parallel to the principal axes, from the various features of the object, *Figure 18-4D*.

STEP 4 Check all lines, correct if necessary, and darken in the object using correct line thickness, *Figure 18-4E*.

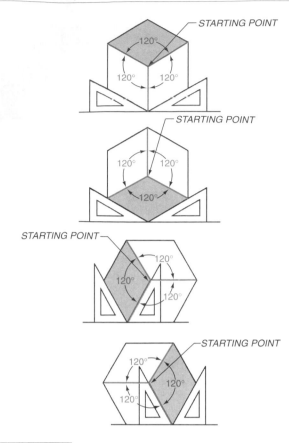

FIGURE 18-3 *Isometric principles*

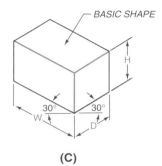

(C)

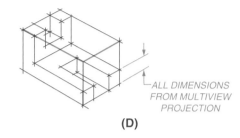

—ALL DIMENSIONS FROM MULTIVIEW PROJECTION

(D)

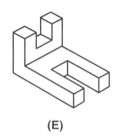

(E)

FIGURE 18-4 *(Continued) (C) Step 2; (D) Step 3; (E) Completed isometric view*

HOW TO CENTER AN ISOMETRIC OBJECT

Save time in locating isometric drawings in or near the center of a drawing space by following these steps.

GIVEN: A multiview of an object, *Figure 18-5A*.

STEP 1 Locate the center point of the work area and do the following: go North 1/2 H, go S60° W a distance of 1/2 D, and go S60° E, 1/2 W to establish a starting point, *Figure 18-5B*.

GIVEN: MULTIVIEW PROJECTION

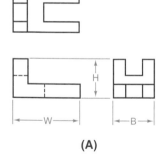

(A)

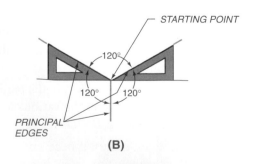

FIGURE 18-4 *(A) How to draw an isometric drawing; (B) Step 1*

GIVEN: MULTIVIEW DRAWING

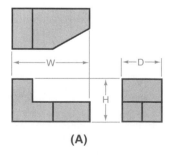

(A)

FIGURE 18-5 *(A) Centering an isometric view within the work area*

STEP 2 From the starting point, draw the basic shape of the object, *Figure 18-5C*.

STEP 3 Construct the object within the basic shape, *Figure 18-5D*.

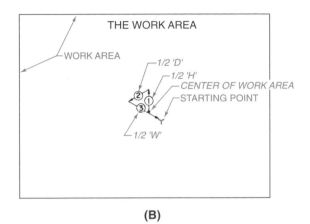

(B)

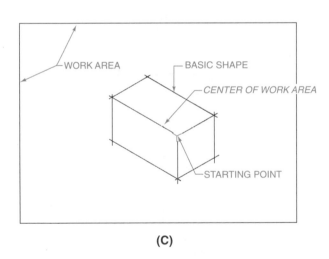

(C)

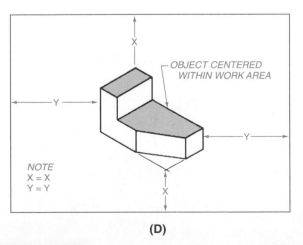

(D)

FIGURE 18-5 (Continued) (B) Step 1; (C) Step 2; (D) Completed isometric view

OFFSET MEASUREMENTS

Offset measurements are used to locate one feature in relationship to another. Measure offsets parallel or perpendicular to one of the three standard planes—frontal, top, or profile—from the multiview of the object being drawn, *Figure 18-6A*.

Develop the isometric drawing as before; first, the basic shape or "box" with dimensions W, D, and H, and then, the offset measurements A, B, and C parallel or perpendicular to the frontal, top, or profile planes, *Figure 18-6B*. Note lines that are truly parallel in the multiview drawing must be parallel in the isometric drawing.

GIVEN: MULTIVIEW PROJECTION

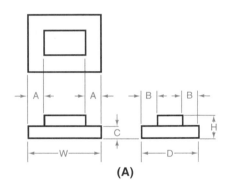

(A)

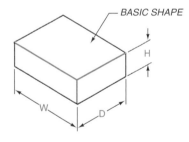

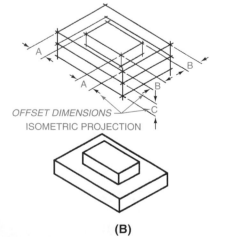

(B)

FIGURE 18-6 (A) Mutiview projection; (B) Offset measurements

Nonisometric Lines

If a line is not parallel or perpendicular to any of the three principal edges in an isometric drawing, it is a *nonisometric line*. *Figure 18-7A* shows a multiview drawing of an object with a 30° inclined surface, labeled as shown with a, b, a', and b'.

Construct the isometric drawing as before; "box" first, then the feature using the X and Y dimensions, and finally, heavy in the lines of the object, *Figure 18-7B*. Line a-b is not in the true length, and the 30° angle is distorted.

Box Construction for Irregular Objects

Some objects, *Figure 18-8A,* have lines that do not line up with the isometric axes, so use a box or basic shape to establish offset dimensions for drawing the object, *Figure 18-8B*.

Next, draw an isometric box, *Figure 18-8C*. Locate key points of the object using offset measurements, *Figure 18-8D*. Finish the object, *Figure 18-8E*.

GIVEN: MULTIVIEW PROJECTION

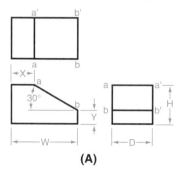

(A)

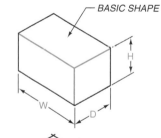

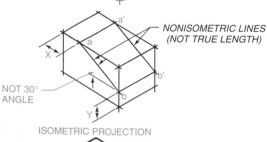

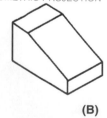

(B)

FIGURE 18-7   *(A) Nonisometric lines; (B) Locating points*

GIVEN: MULTIVIEW

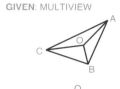

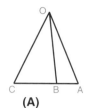

(A)

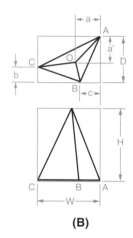

(B)

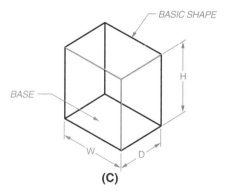

(C)

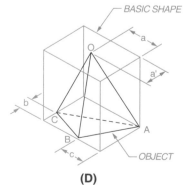

(D)

FIGURE 18-8   *(A) Box construction; (B) Basic shape; (C) Isometric basic shape; (D) Locating points of object*

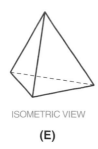

ISOMETRIC VIEW

(E)

FIGURE 18-8   *(Continued) (E) Isometric view completed*

HIDDEN LINES

Hidden lines are not used in isometric drawings unless needed to show an important feature that otherwise would not be seen. Try to orient objects so hidden lines are not needed, not only in isometric drawings, but in all pictorials.

ISOMETRIC CURVES

Construct isometric curves in isometric drawings using offsets or a grid. For a series of offset distances, use a number of evenly spaced lines, all parallel to one of the principal edges, *Figure 18-9A*. Transfer these lines to the isometric layout and transfer each distance, respectively, to locate points on the curve, *Figure 18-9B*. Give depth to the drawing by projecting D distances at 30°, which is parallel to an isometric axis, from each point on the curve. Use a French curve to connect the points.

For transferring a curve using a grid, choose a grid spacing that will give a sufficient number of points along the curve to give as much detail of the curve as needed, *Figure 18-10A*.

Construct the basic shape of the object using the W, H, and D dimensions. Transfer the grid, *Figure 18-10B*. Finally, transfer the irregular curve, square by square. Use a French curve to connect the points on the curve.

IRREGULARLY SHAPED OBJECTS

Transfer and draw irregularly shaped objects by dividing the object into a series of convenient sections. Draw each

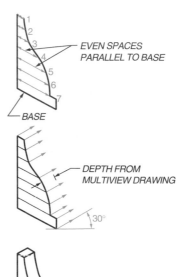

FIGURE 18-9B   *Drawing an isometric curve (offset distances and depth)*

GIVEN: MULTIVIEW

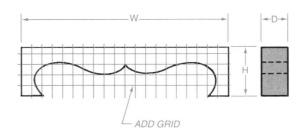

FIGURE 18-10A   *Drawing an isometric curve (grid method)*

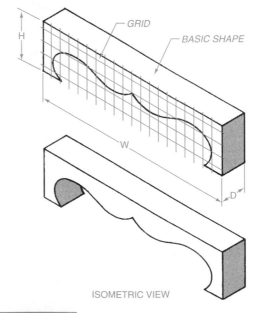

ISOMETRIC VIEW

FIGURE 18-10B   *Isometric grid layout*

GIVEN: MULTIVIEW

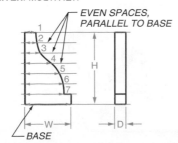

FIGURE 18-9A   *Drawing an isometric curve (offset method)*

section as an isometric, *Figure 18-11A*. The two-view drawing of the boat has been simplified for this example. Divide the object into a series of sections, A through H. Use a logical baseline (in this case a center line). Locate and draw each section, *Figure 18-11B*. Connect all the sections and complete the boat, *Figure 18-11C*.

ISOMETRIC CIRCLES AND ARCS

Drawing isometric circles and arcs, without using templates, takes too much time for isometric drawing. Use an isometric ellipse template, *Figure 18-12*. The template has eight hash marks printed around the perimeter

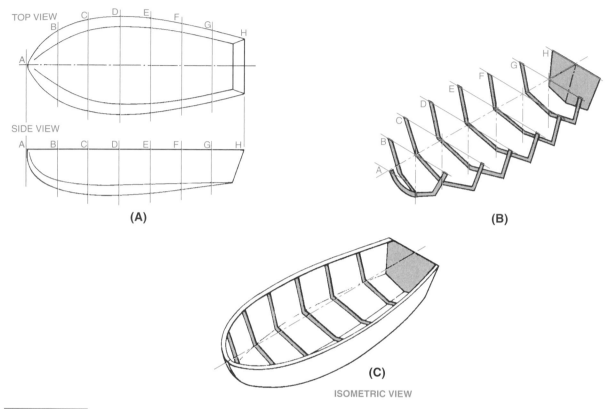

FIGURE 18-11 *(A) Irregularly shaped object; (B) Drawing each section; (C) Completed object*

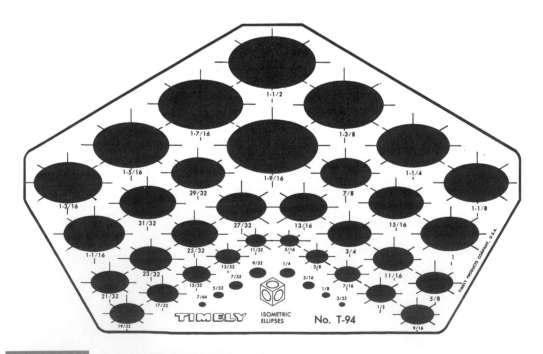

FIGURE 18-12 *Isometric ellipse template* (Courtesy Timely Products Co.)

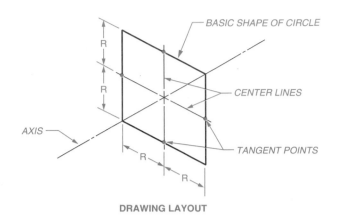

FIGURE 18-13A *Basic shape of the circle*

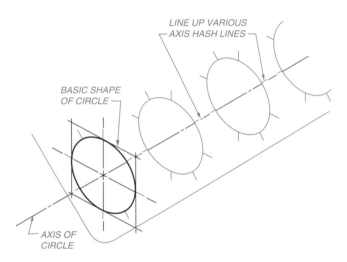

FIGURE 18-13B *Draw the ellipse within the basic shape*

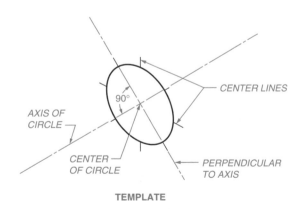

FIGURE 18-13C *Isometric ellipse template aligned on axis of circle*

of each ellipse. Four indicate the isometric center lines of the circle or arc. For example, an ellipse for a one-inch circle would be labeled number –1– and it would measure 1″ across the ellipse along the isometric center line. These same hash marks also indicate the four tangent points of the isometric circle with the isometric basic

shape of the circle, *Figures 18-13A* and *18-13B*. The other four hash marks line up with the axes of the circle; two are coincident with the axis of the circle and are lined up across the narrow part of the ellipse, and two are perpendicular to the axis and are lined up across the longest part of the ellipse, *Figure 18-13C*. The longest distance, across the one-inch ellipse, actually measures more than one inch due to the way the basic shape is constructed for isometric drawing.

HOW TO USE ISOMETRIC ELLIPSE TEMPLATES FOR ISOMETRIC DRAWINGS

STEP 1 For the object given, *Figure 18-14A*, locate the center of the circle or arc to be drawn using offset measurements, and construct the basic shape for the circle so that the sides of the basic shape are parallel to the two appropriate principal edges.

STEP 2 Line up the 1″ ellipse template so that the isometric center line hash marks touch the tangent points 1, 2, 3, and 4 of the basic shape, *Figure 18-14B*. Trace the ellipse. If some doubt exists as to which hash marks to line up, refer to *Figure 18-15* for guidance.

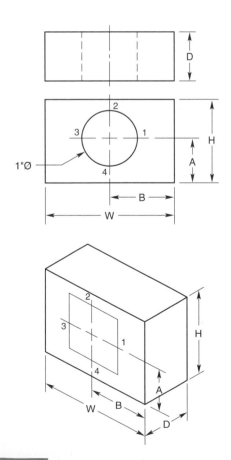

FIGURE 18-14A *How to use an isometric ellipse template*

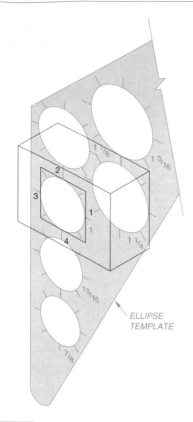

FIGURE 18-14B *Line up hash marks on tangent points*

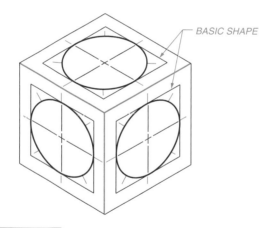

FIGURE 18-15 *Isometric circles in various planes*

HOW TO DRAW TRUE ISOMETRIC CIRCLES

Assume that a relatively accurate isometric drawing of a circle is required, but a template is not available for the size needed.

GIVEN: A regular circle in its basic shape, and the same circle as it would be drawn in an isometric drawing, in its basic shape, *Figure 18-16A.* In both drawings the tangent points are given "clock" numbers.

STEP 1 Use the circle in the basic shape, a square, with the tangent points 3, 6, 9, and 12. Use the

remaining clock numbers and divide the circle into the rectangles shown, *Figure 18-16B.*

STEP 2 Draw the basic shape of the circle and use the offset method to locate points 1, 5, 7, and 11 using the X and Y dimensions, *Figure 18-16C.*

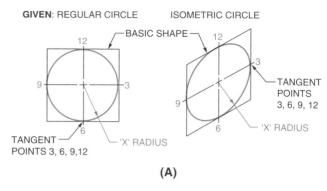

(A)

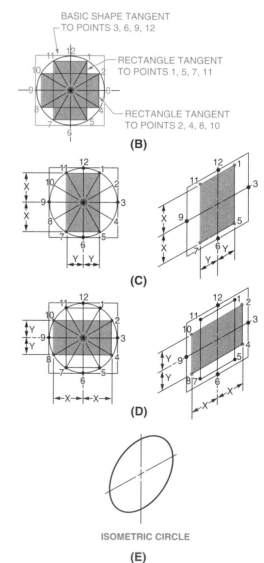

FIGURE 18-16 *(A) Regular circle and isometric circle; (B) Step 1; (C) Step 2; (D) Step 3; (E) Completed isometric circle*

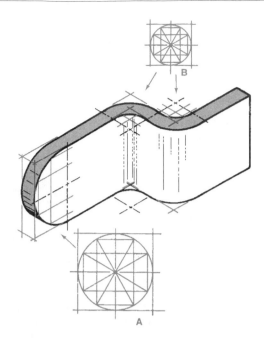

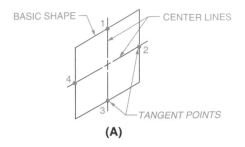

(A)

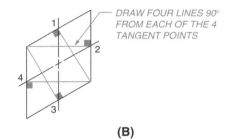

(B)

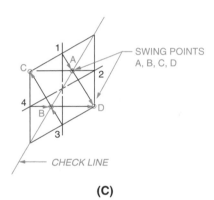

(C)

FIGURE 18-17 *How to draw an isometric arc (offset method)*

STEP 3 Locate points 2, 4, 8, and 10 in a similar manner, *Figure 18-16D*.

STEP 4 Connect the 12 points with a French curve and finish the isometric drawing of the circle, *Figure 18-16E*.

To draw an arc of a circle, use these steps and use the portion of the circle needed, *Figure 18-17*.

HOW TO DRAW APPROXIMATE ISOMETRIC CIRCLES

Assume that an approximate isometric circle is needed in an isometric drawing, but a template is not available with the size required. The four-center method described in the following steps can be used.

GIVEN: A regular circle with a radius X, Figure 18-16A.

STEP 1 Draw the basic shape of the circle in the appropriate location in the drawing. Label the four tangent points 1, 2, 3, and 4, *Figure 18-18A*.

STEP 2 From each of the four tangent points, draw a line at 90° to the opposite sides of the basic shape, *Figure 18-18B*. These lines intersect at four locations called the swing points or centers. Label them A, B, C, and D, *Figure 18-18C*. A line can be used to check the location centers A and B.

STEP 3 Construct a radius at each of the four centers, A, B, C, and D as shown in *Figures 18-18D* and *18-18E*.

STEP 4 Darken in the completed isometric circle and center lines, *Figure 18-18F*.

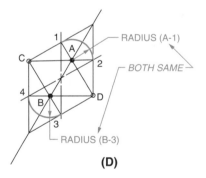

(D)

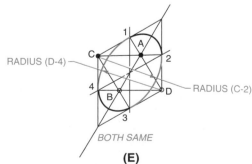

(E)

FIGURE 18-18 *(A) Step 1 (how to draw an approximate isometric circle); (B) and (C) Step 2; (D) and (E) Step 3*

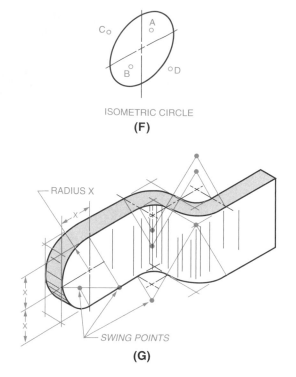

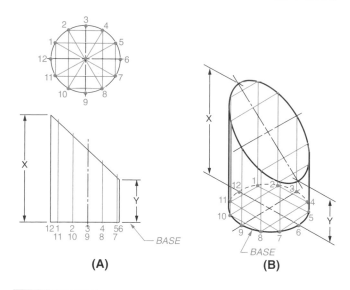

ISOMETRIC CIRCLE
(F)

RADIUS X

SWING POINTS
(G)

FIGURE 18-18 *(Continued) (F) Completed isometric circle; (G) Drawing isometric arcs (four-center method)*

Construct arcs using the four-center method and the radius X, discussed in these four steps, *Figure 18-18G*.

HOW TO DRAW AN ELLIPSE ON AN INCLINED PLANE

GIVEN: A 2-view drawing of a cylinder cut at an angle, *Figure 18-19A*.

STEP 1 Divide the view of the cylinder, showing it as a circle, in 12 equally spaced arcs (1 through 12). Project the points into the view showing the base of the cylinder as an edge. Locate the true-length line segments from each point, Figure 18-19A.

STEP 2 Construct an isometric circle of the base and locate each of the 12 points using the offset technique, *Figure 18-19B*.

STEP 3 Project each true-length line segment up from the base and connect the end points to form the boundary of the elliptically shaped surface. Use a French curve.

ISOMETRIC INTERSECTIONS

An isometric drawing can be constructed from a completed multiview drawing, showing an intersection(s), by following the steps outlined in this chapter. First, construct a basic shape, box, and then transfer a sufficient

(A)   **(B)**

FIGURE 18-19 *(A) How to draw an ellipse on an inclined plane. Given: two views; (B) An isometric ellipse on an inclined plane*

number of points from the multiview drawing, using the offset method, to complete the isometric drawing.

ISOMETRIC DIMENSIONING

Use the same dimensioning standards, rules, and methods discussed in Chapter 10 to dimension isometric drawings. Furthermore, use unidirectional dimensioning for clarity and ease of placing the dimensions, *Figure 18-20*. Note that extension lines and dimension lines are parallel to one of the three axes.

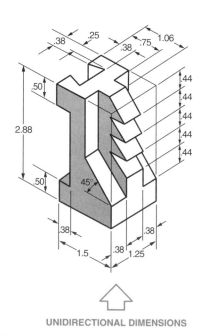

UNIDIRECTIONAL DIMENSIONS

FIGURE 18-20 *Isometric dimensioning*

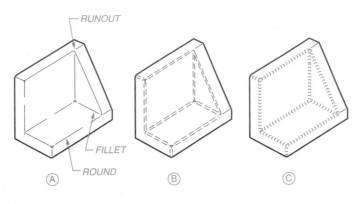

FIGURE 18-21   *Isometric rounds and fillets*

ISOMETRIC ROUNDS AND FILLETS

Rounds and fillets in isometric drawings can be shown three ways, *Figure 18-21A, B,* and *C.* Use whichever way, or combinations of ways, that best shows the fillets or rounds and where they are located.

ISOMETRIC KNURLS, SCREW THREADS, AND SPHERES

Isometric knurls are usually drawn with straight lines, *Figure 18-22A* and *B.* Part A shows the location of a proposed knurl and how to indicate a chamfer. Part B shows the finished drawing.

    Isometric threads are readily represented by a series of ellipses drawn along the center line of the fastener, parallel to each other, and approximating the thread spacing, *Figure 18-23.*

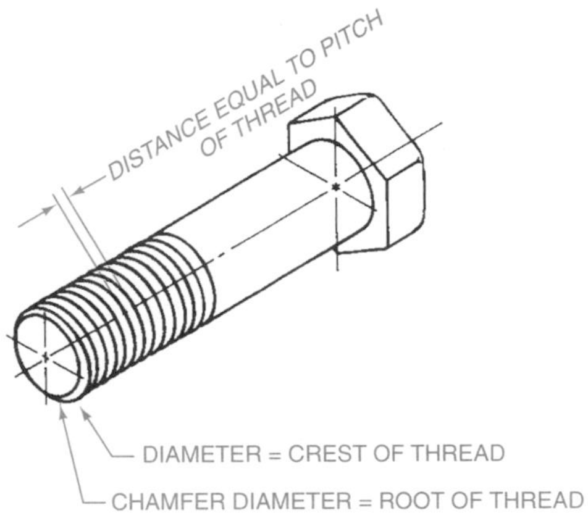

FIGURE 18-23   *Isometric threads*

    Isometric spheres require three ellipses to be drawn, *Figure 18-24*—one along each axis, and one circumscribing the first two. If only a portion of a sphere is wanted, then the first two ellipses can be used to outline a half sphere, *Figure 18-25;* a quarter sphere, *Figure 18-26;* or a three-quarter sphere, *Figure 18-27.*

    Based on the 2-view drawing in *Figure 18-28,* draw a flat surface in a sphere by following the implied steps in *Figure 18-29.*

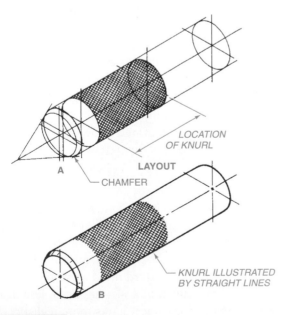

FIGURE 18-22   *Isometric knurl*

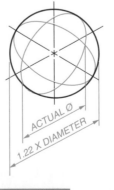

FIGURE 18-24   *Isometric sphere*

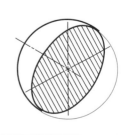

FIGURE 18-25   *Isometric half sphere*

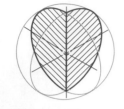

FIGURE 18-26 *Isometric quarter sphere*

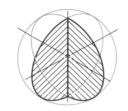

FIGURE 18-27 *Isometric three-quarter sphere*

GIVEN: MULTIVIEW DRAWING

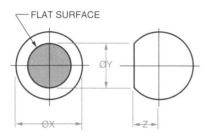

FIGURE 18-28 *Two-view drawing of sphere*

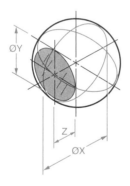

FIGURE 18-29 *Flat surface on a sphere*

Axonometric: Drawing and Projection

Drafters, designers, and engineers use the pictorial drawings—oblique, isometric, and perspective—discussed in the first part of this chapter because approximations of these drawings are easy to sketch, and relatively accurate drawings of these types are easy to construct. Additional pictorial drawings and several projections will be considered next.

ISOMETRIC PROJECTION

Isometric projection causes dimensions along the three principal edges to be foreshortened to approximately

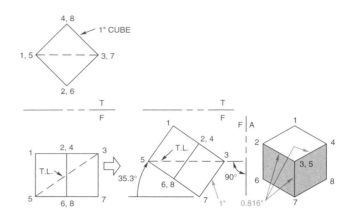

FIGURE 18-30 *Isometric projection*

82% of their true length, *Figure 18-30*. Note that the diagonal of the 1″ cube is seen as a point in the A view, the three edges 3-2, 3-4, and 3-7 equal 0.816″, and the angles between edges 3-2, 3-4, and 3-7 are all equal to 120°.

ISOMETRIC ASSEMBLIES

Isometric assembly drawings appear in two popular formats, exploded and assembled.

Exploded isometric drawings show all the components of an assembly as individual isometric drawings, including fasteners; however, the components are aligned in space in the same way they are to be assembled. See Problem 18-52 for an exploded-assembly sketch. Exploded-assembly instructional drawings usually accompany products that are to be assembled after the buyer purchases the product. Exploded-assembly drawings are used extensively in automotive maintenance manuals.

Assembled isometric assembly drawings show how a single product looks when assembled. For example, refer to the isometric drawing of a robot in Chapter 20. Other isometric assembly drawings may show a large number of individual components, for example, a CIM system (computer-integrated manufacturing) in Chapter 20, and a piping-system drawing in Chapter 21.

DIMETRIC DRAWING

Dimetric drawing differs from isometric drawing in that only two angles are equal. *Figure 18-31* shows a number of orientations for drawing dimetric views; lines of an object that are parallel to the first two axes are drawn full size, lines parallel to the third are drawn to a different scale. Because two different scales are used, less distortion is apparent and an object looks more realistic. One

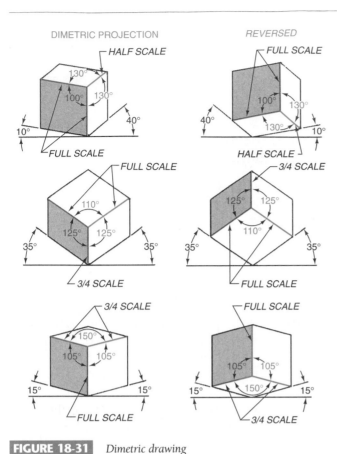

FIGURE 18-31 *Dimetric drawing*

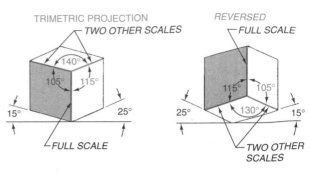

FIGURE 18-32 *Trimetric drawing*

disadvantage, however, is that circles are difficult to draw in dimetric views.

TRIMETRIC DRAWING

Trimetric drawing uses axes that are rotated so that each axis is drawn at a different angle, and each axis uses a different scale, *Figure 18-32.* Trimetric drawings can appear more realistic than any of the other types, but they are the most time consuming to draw, so they aren't used by drafters.

Other Oblique Drawings

CAVALIER DRAWING

Cavalier drawings have a receding axis drawn between 30° and 60° to the horizon, and the receding distances are drawn full size, *Figures 18-33* and *18-34.*

CABINET DRAWING

Cabinet drawings have the same range of receding axes as cavalier drawings, but the receding distances are drawn to half size, *Figure 18-35.*

GENERAL OBLIQUE DRAWING

General oblique drawing is similar to cabinet drawing except that the lines drawn along the direction of the receding distances may be at any scale between half and full, *Figure 18-36.*

OBLIQUE PROJECTIONS

PARALLEL

PARALLEL

PARALLEL

FULL SCALE

HALF SCALE

ANY APPROPRIATE SCALE BETWEEN HALF AND FULL

REGULAR FRONT VIEW

REGULAR FRONT VIEW

REGULAR FRONT VIEW

CAVALIER DRAWING

CABINET DRAWING

GENERAL DRAWING

FIGURE 18-33 *Oblique drawing*

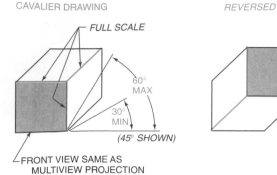

CAVALIER DRAWING

REVERSED

FULL SCALE

60° MAX

30° MIN

(45° SHOWN)

FRONT VIEW SAME AS MULTIVIEW PROJECTION

FIGURE 18-34 *Cavalier drawing*

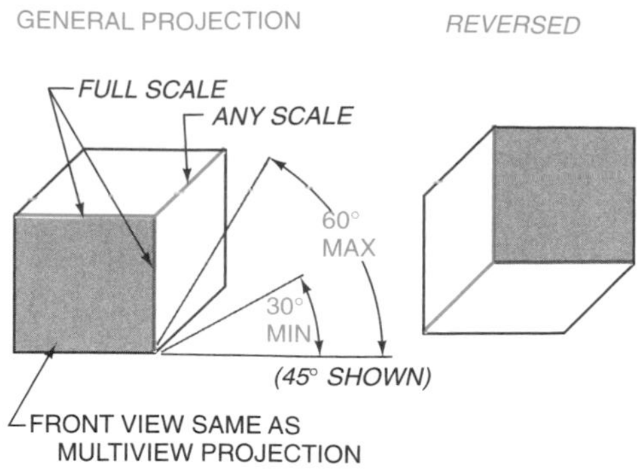

GENERAL PROJECTION

REVERSED

FULL SCALE — ANY SCALE

60° MAX

30° MIN

(45° SHOWN)

FRONT VIEW SAME AS MULTIVIEW PROJECTION

FIGURE 18-36 *General oblique*

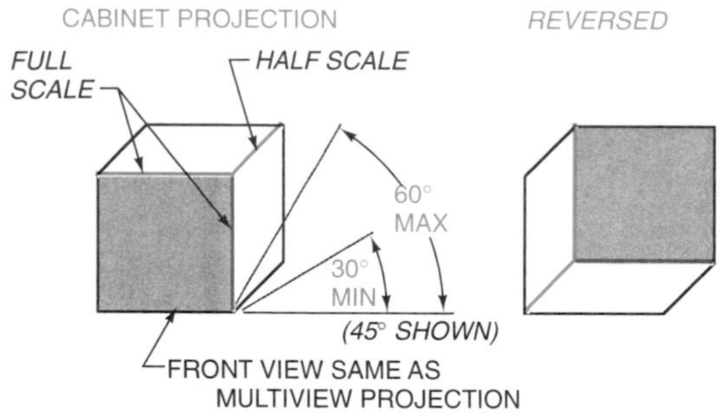

CABINET PROJECTION

REVERSED

FULL SCALE — HALF SCALE

60° MAX

30° MIN

(45° SHOWN)

FRONT VIEW SAME AS MULTIVIEW PROJECTION

FIGURE 18-35 *Cabinet drawing*

Perspective Drawings

A *perspective drawing* is a three-dimensional drawing of an object or scene that shows visible lines almost as a person's eyes would perceive them from one particular viewing location. A perspective drawing is somewhat like a photograph, *Figure 18-37*. The upper drawing shows the lines of a building from the outside; the lower one shows an inside view of a room. Both views are drawn in two-point perspective like several of the sketches in Chapter 3.

OUTSIDE VIEW IN TWO-POINT PERSPECTIVE

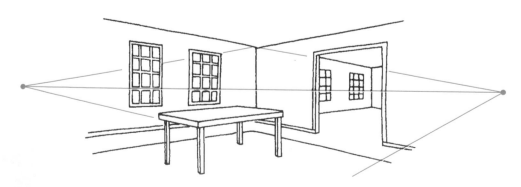

INSIDE VIEW IN TWO-POINT PERSPECTIVE

FIGURE 18-37 *Outside and inside perspective views*

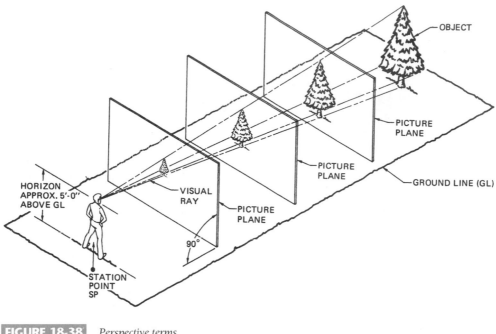

FIGURE 18-38 *Perspective terms*

PERSPECTIVE DRAWING TERMS

Learn the terms used in perspective drawing while studying the figures and procedures that follow.

Station Point (SP). The exact location from which the observer views the object, ***Figure 18-38***.

Ground Line (GL). The edge view of the surface upon which the object to be viewed rests, Figure 18-38.

Horizon (H). A line at the level of the viewer's eyes, Figure 18-38, and the line where the vanishing points are located.

Vanishing Point (VP). A point on the horizon line to which all other lines are projected, ***Figure 18-39***. Six cubes, A through F, are used to show which sides of an object would be seen for various left and right positions of the cubes in perspective drawings. Note that if the dimension X shown is too great, a two-point perspective drawing would be needed.

Picture Plane (PP). Like the projection plane in multiview drawings. An imaginary plane perpendicular to the ground line, Figure 18-38.

Visible Rays (VR). Lines from the eyes of the viewer to each and every point on the object to be viewed. The visible rays intersecting the picture plane create the perspective drawing, Figure 18-38.

Measuring Line (M). Actually used as a vertical line, but it appears horizontal because of the way it is established. The line originates at the projection plane and is 90° to it. Referring ahead, Figures 18-40D, G, and H

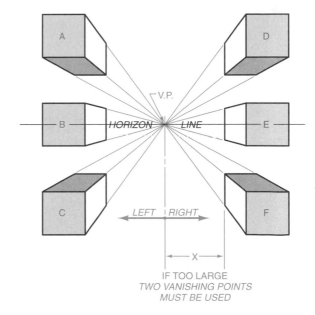

FIGURE 18-39 *Vanishing point and horizon line*

show the measuring line as a vertical line used to measure true heights.

PRESKETCHING FOR PERSPECTIVE DRAWINGS

Make simple, quick sketches of an object to be drawn in two-point perspective to determine the best arrangement of the horizon, station point, ground line, and vanishing points, Figures 18-40A and B. Two of the multiviews are required for the two-point perspective (usually a top view and a front or side view).

GIVEN: MULTIVIEW DRAWING

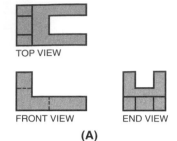

TOP VIEW

FRONT VIEW   END VIEW

(A)

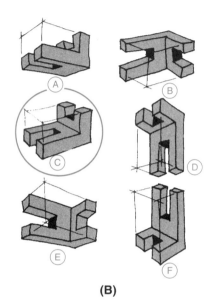

(B)

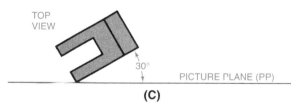

TOP
VIEW

30°

PICTURE PLANE (PP)

(C)

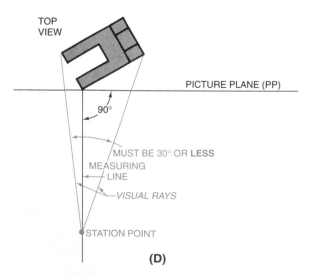

TOP
VIEW

PICTURE PLANE (PP)

90°

MUST BE 30° OR **LESS**
MEASURING
LINE

VISUAL RAYS

STATION POINT

(D)

FIGURE 18-40   *(A) How to draw a two-point perspective drawing; (B) Sketch of object in various positions; (C) Step 1; (D) Step 2*

HOW TO DRAW A TWO-POINT PERSPECTIVE

GIVEN: Multiview drawing *Figure 18-40A* and the sketches of the object in *Figure 18-40B.* Sketch C is selected as the best position to show the features of the object.

STEP 1   On the upper portion of a drawing sheet, draw the top view at any orientation, but attempt to match the best position from the sketches, *Figure 18-40C.* Locate the front edge of the object tangent to the picture plane.

STEP 2   From the front edge of the object, where it is tangent to the picture plane line, project the measuring line at 90° to the picture plane, *Figure 18-40D.* Locate the station point on the measuring line so the inclusive angle of view is 30° or less.

STEP 3   From the station point, construct lines parallel to the two edges of the top view, *Figure 18-40E,* and label their intersections with the picture plane, 1 and 2.

STEP 4   Draw the horizontal line at a convenient location, preferably below the station point to avoid a cluttered drawing, *Figure 18-40F.* From points 1 and 2 on the picture plane, project downward to the horizon line to locate the left and right vanishing points. Locate a ground line and draw a side or front view on the line as shown.

STEP 5   From the view selected for the ground line, in this example the right-side view, project the true height of the object to the measuring line, *Figure 18-40G.* From the true height, project lines to the left and right vanishing points, and then construct a perspective basic shape of the object. Note how the vertical boundaries of the basic shape are established by the top view.

STEP 6   Project the true heights of each feature from the right-side view to the measuring line and back toward the appropriate vanishing point, to a point where the same feature intersects the picture plane line, *Figure 18-40H.* Develop the various features of the object in light lines until satisfied that the drawing is correct.

STEP 7   Finally, darken the lines of the object, *Figure 18-40I.* Hidden lines are not usually drawn except to show an important feature that otherwise would not be seen.

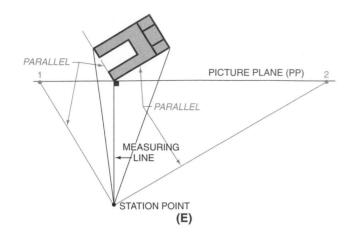

(E)

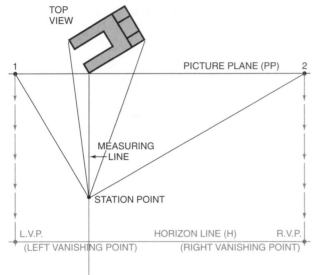

(F)

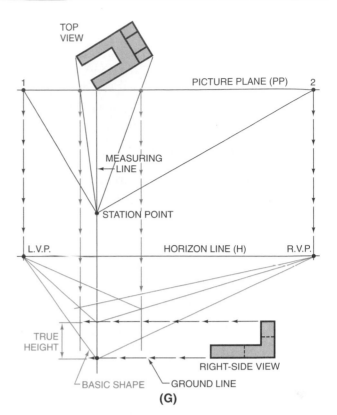

(G)

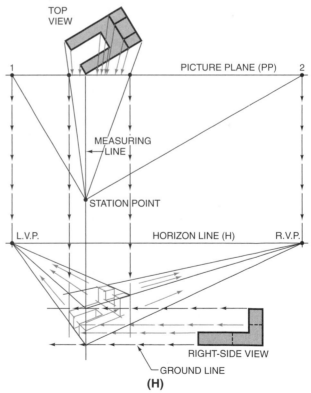

(H)

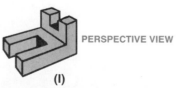

PERSPECTIVE VIEW

(I)

FIGURE 18-40 *(Continued) (E) Step 3; (F) Step 4; (G) Step 5; (H) Step 6; (I) Step 7*

TYPES OF PERSPECTIVE VIEWS

Three types of perspective views are shown in *Figure 18-41*; one-, two-, and three-point, respectively. Only two-point perspective is discussed in this chapter because it is the type most often used in industry.

PERSPECTIVE IRREGULAR CURVES

Use a multiview drawing and a grid as was done for the isometric drawing. Then follow the steps in the previous example to construct a perspective drawing, *Figure 18-42*.

PERSPECTIVE CIRCLES AND ARCS

Divide the circle (or arc) into equal segments, *Figure 18-43*. Follow the steps from the previous example to construct the perspective drawing.

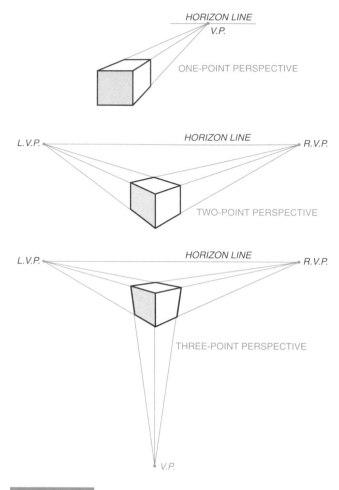

FIGURE 18-41 *Kinds of perspective drawings*

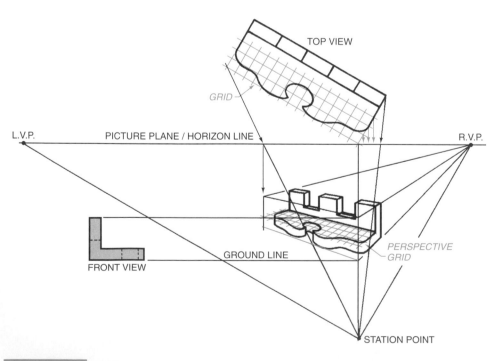

FIGURE 18-42 *Perspective irregular curves*

Industry Application

DRAFTING STUDENTS ARE A VALUABLE RESOURCE

"Who remembers how to make isometric drawings?" asked Mark Malone—Chief Drafter for Pensacola Prestressed Concrete Company (PPCC). Malone had been asked by the foreman of PPCC's construction crew to develop pictorial drawings of several complicated connection joints for a bridge the company was building. The joints in question were so complicated that PPCC's construction crew was having trouble reading the plans. These joints had to be fitted together properly. The structural integrity of the bridge would depend on them.

Malone had agreed to provide pictorials without considering who would draw them. He could not remember the last time pictorial drawings had been called for. Prestressed concrete plans were almost always orthographic. In fact, he had not made a pictorial drawing himself since coming to PPCC 12 years ago. But he did remember being good at it as a drafting student at Pensacola Junior College.

That was when the idea came to him. "I'll ask the drafting class at PJC to take on this project." In the past he had given the class projects when PPCC's drafting team got overloaded and deadlines were fast approaching. The students had always done an excellent job, and they appreciated the real-world experience.

Two days later, Malone had several outstanding isometric joint details, PJC's drafting students had additional real-world experience, the construction crew had drawings they could read, and Malone had found two talented students he could hire in the near future.

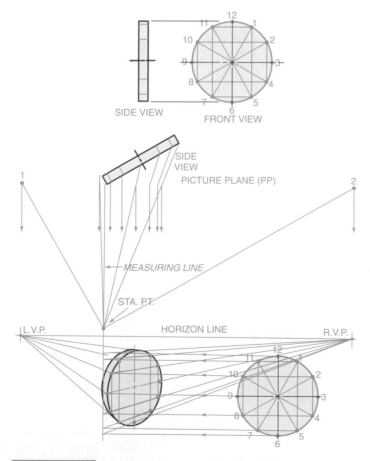

FIGURE 18-43   *Perspective circles or arcs*

Summary

- The most common types of pictorial drawings used by drafters, designers, and engineers are oblique, isometric, and perspective.

- Oblique drawings are the easiest type of pictorial drawing to make. The surface of the object is drawn exactly as it would appear in a multiview drawing. The depth axis is drawn at either 30° or 45°.

- Isometric drawings use three axes, sometimes called principal edges, at 120° to each other. They are the most frequently used pictorial drawings because they depict an object as well as a perspective drawing, but they are easier to make.

- A perspective drawing is a three-dimensional drawing of an object or scene that shows visible lines almost as a person's eye would perceive them from one particular viewing location. Key perspective concepts are the station point, ground line, horizon, vanishing point, picture plane, visible rays, and measuring line. There are one-, two-, and three-point perspective drawings.

- Dimetric drawings are similar to isometric except that only two angles are equal.

- Trimetric drawings are similar to isometric except that all three angles are different.

- A cavalier drawing is a type of oblique drawing in which depth is measured at full scale.

- A cabinet drawing is a type of oblique drawing in which depth is measured at half scale.

- General oblique drawings allow any appropriate scale to be used for the depth.

Review Questions

Answer the following questions either true or false.

1. The true isometric circle method is preferred over the four-center isometric circle method when time is an important factor.

2. An isometric drawing is dimensioned by using unidirectional dimensioning.

3. Trimetric drawings use axes that are rotated so that each axis is drawn at a different angle, and each axis uses a different scale.

4. Isometric, dimetric, and trimetric drawings are classified as perspective drawings.

5. Hidden lines are only used in pictorial drawings to show an important feature that otherwise would not be seen.

6. Isometric drawings use four axes sometimes called principal edges.

7. An isometric projection causes dimensions along the three principal edges to be foreshortened to approximately 87% of their true length.

8. Offset measurements are used to locate the center point of the work area.

Answer the following questions by selecting the best answer.

1. Which of the following is a type of pictorial drawing not frequently used in industry?
   a. Cabinet
   b. Trimetric
   c. Isometric
   d. Two-point perspective

2. Which of the following is the perspective view most used in industry?
   a. One-point
   b. Two-point
   c. Three-point
   d. Four-point

3. Which type of pictorial drawing best illustrates the object as it would be viewed by the eye?
   a. Oblique
   b. Isometric
   c. Perspective
   d. Dimetric

4. Which of the following is the correct definition of a nonisometric line?
   a. A line that is not parallel to any of the three principal edges in an isometric drawing
   b. A line that is not perpendicular to any of the three principal edges in an isometric drawing
   c. A line that is not parallel or perpendicular to any of the three principal edges in an isometric drawing
   d. None of the above

5. Which of the following is **not** true regarding an oblique drawing?
   a. It is easier to draw than isometric or perspective drawings.
   b. Usually the front view is drawn exactly as it would appear in a multiview drawing.
   c. Oblique drawings use three axes.
   d. Oblique drawings are the most used of the three common pictorials.

Chapter 18 Problems

The following problems are intended to give the beginning drafter practice in laying out and developing isometric and perspective renderings of various objects.

CONVENTIONAL INSTRUCTIONS

The steps to follow in laying out the drawings in this chapter are:

STEP 1 Study the problem carefully.

STEP 2 Position the object so it **best** illustrates the most features.

STEP 3 Make a pictorial sketch of the object.

STEP 4 Calculate the basic shape size.

STEP 5 Find the center of the work area.

STEP 6 Lightly center and lay out the basic shape within the work area.

STEP 7 Lightly develop the object within the basic shape.

STEP 8 Check to see that the completed object is centered within the work area.

STEP 9 Check all size dimensions, and check for accuracy of the object.

STEP 10 Darken in the object.

STEP 11 Recheck all work and, if correct, neatly fill out the title block using light guidelines and neat lettering.

CAD Instructions

Students who plan to complete the following problems on a CAD system should begin with these general instructions. Before reading the specific instructions for each activity, go through each step in the following planning checklist. The checklist applies to any CAD system and will help ensure the optimum use of your time and resources.

1. Analyze the problem carefully. Decide exactly what you are being asked to do.

2. Determine what resources and references (if any) you will need in order to complete the problem. Collect those references and resources and have them readily available.

3. Decide if any particular standards apply to the project and have those standards available.

4. Determine what types of views will be required and how many of each.

5. Determine what the final plotted scale of the drawing will need to be, and select the appropriate paper size for plotting/printing.

6. Plan your drawing sequence. In what order will you develop the drawing (i.e., lines, features, dimension lines, leaders, dimensions)?

7. Review the various CAD commands you will have to use in order to develop the drawing.

8. Examine your CAD system to ensure that everything is in working order, then begin the project.

PROBLEMS 18-1 THROUGH 18-27

Construct an isometric drawing of the object in Problems 18-1 through 18-27. (Note that these problems have only straight lines.)

PROBLEM 18-1

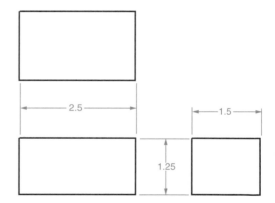

PROBLEM 18-2

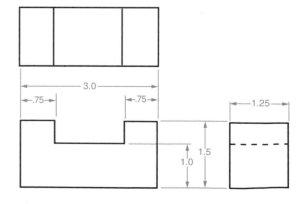

PROBLEM 18-3

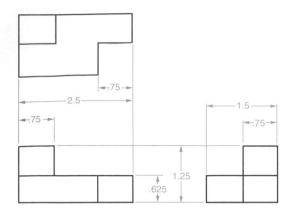

PROBLEM 18-6

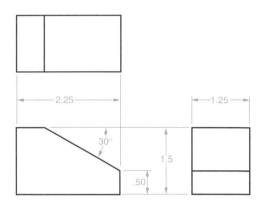

PROBLEM 18-4

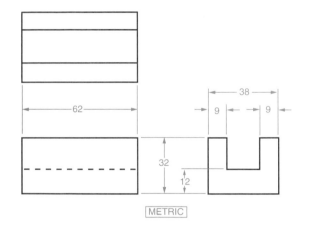

METRIC

PROBLEM 18-7

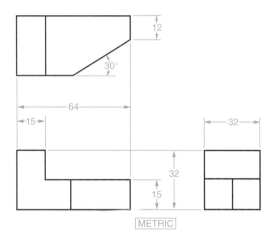

METRIC

PROBLEM 18-5

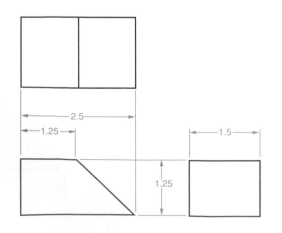

PROBLEM 18-8

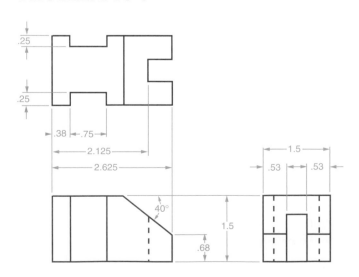

PROBLEM 18-9

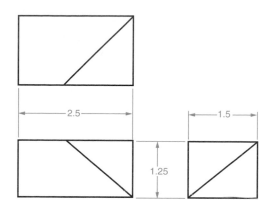

PROBLEM 18-11

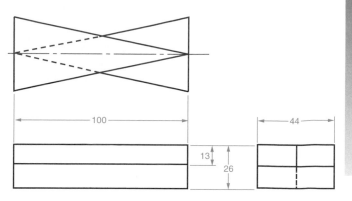

PROBLEM 18-10

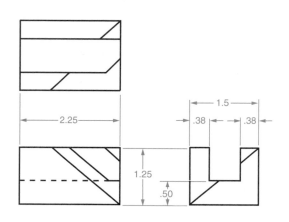

PROBLEM 18-12

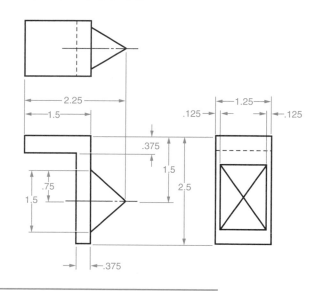

PROBLEM 18-13

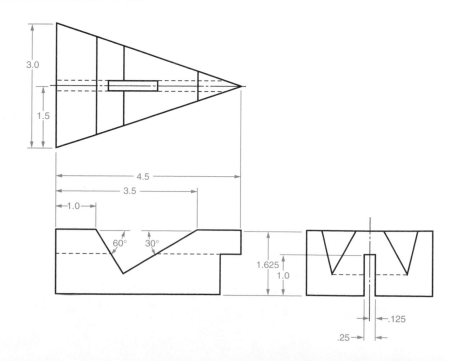

PROBLEM 18-14

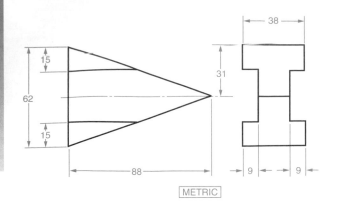

METRIC

PROBLEM 18-15

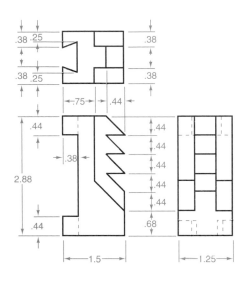

PROBLEM 18-16

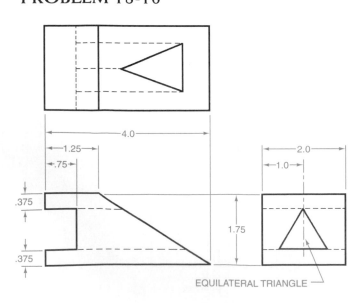

EQUILATERAL TRIANGLE

PROBLEM 18-17

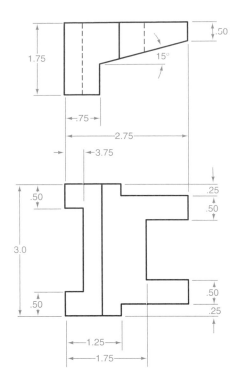

PROBLEM 18-18

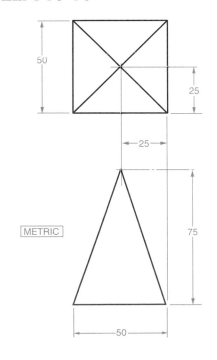

METRIC

PROBLEM 18-19

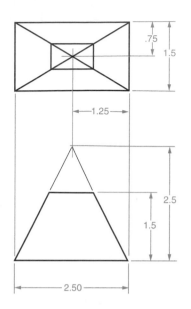

PROBLEM 18-20

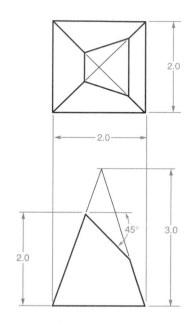

PROBLEM 18-21

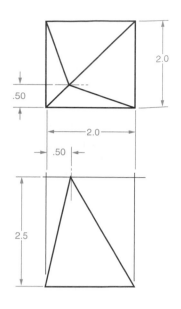

PROBLEM 18-22

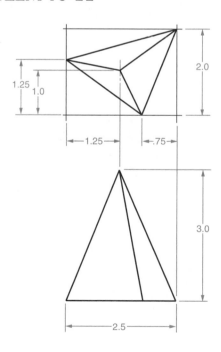

PROBLEM 18-23

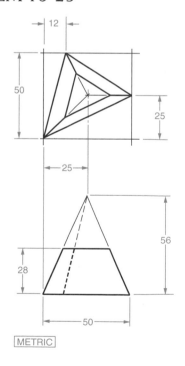

METRIC

PROBLEM 18-24

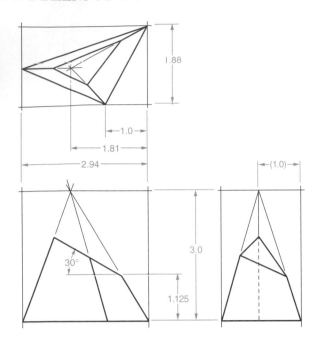

PROBLEM 18-25

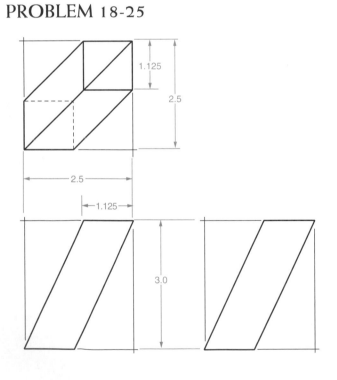

PROBLEM 18-26

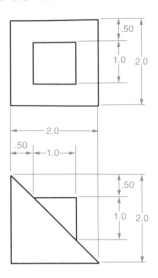

PROBLEM 18-27

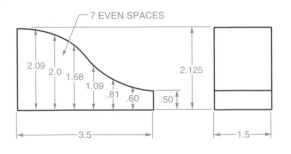

PROBLEMS 18-28 THROUGH 18-51

Construct an isometric drawing of the object in Problems 18-28 through 18-51. (Note that these problems have straight lines, arcs, and circles.)

PROBLEM 18-28

PROBLEM 18-29

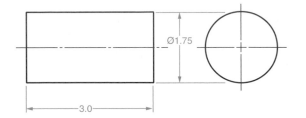

PROBLEM 18-30

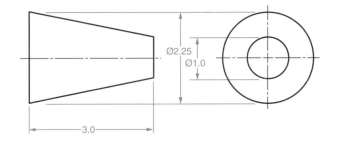

PROBLEM 18-31

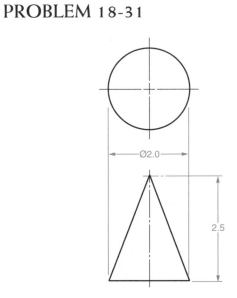

PROBLEM 18-32

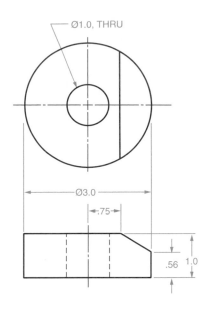

PROBLEM 18-33

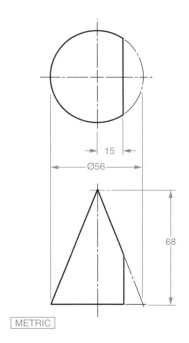

METRIC

Problems

PROBLEM 18-34

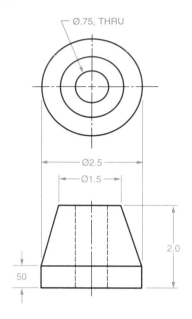

PROBLEM 18-35

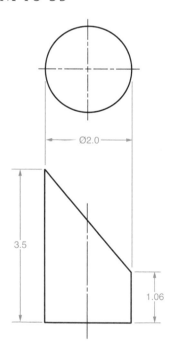

PROBLEM 18-36

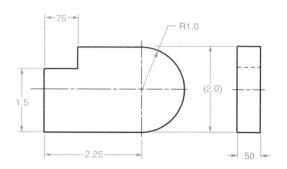

PROBLEM 18-37

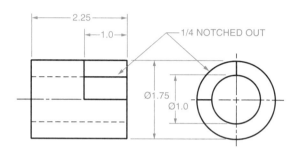

PROBLEM 18-38

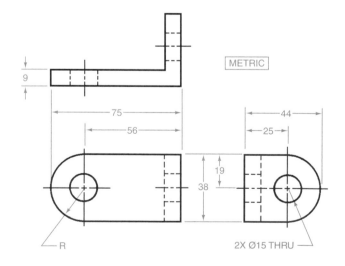

PROBLEM 18-39

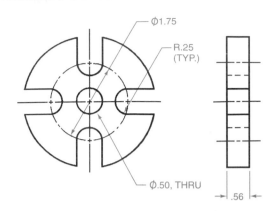

PROBLEM 18-40

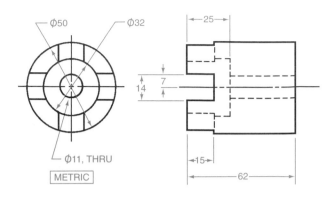

PROBLEM 18-41

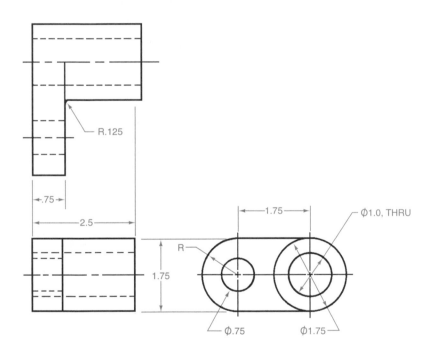

PROBLEM 18-42

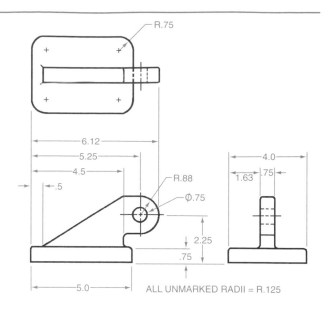

ALL UNMARKED RADII = R.125

PROBLEM 18-43

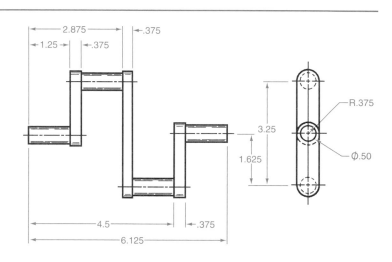

PROBLEM 18-44

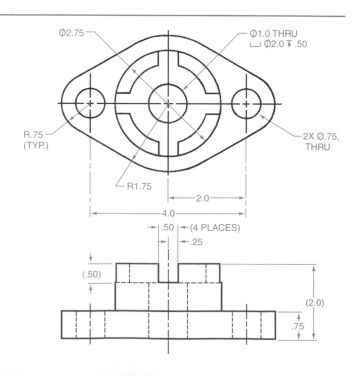

PROBLEM 18-45

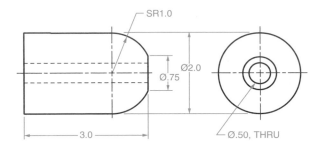

PROBLEM 18-48

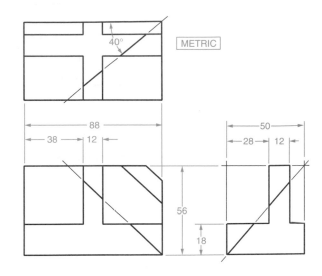

PROBLEM 18-46

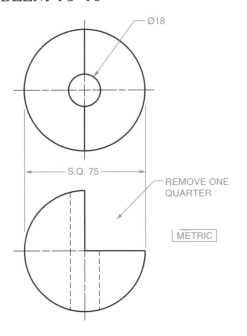

PROBLEM 18-49

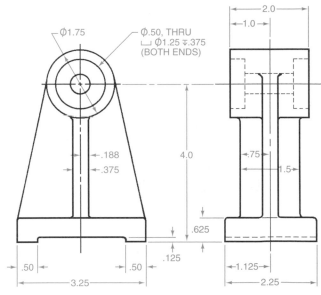

ALL UNMARKED RADII = R.09

PROBLEM 18-47

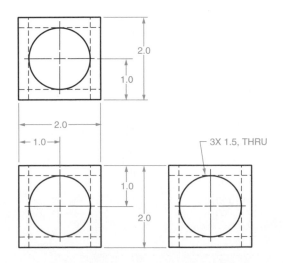

Problem 18-50

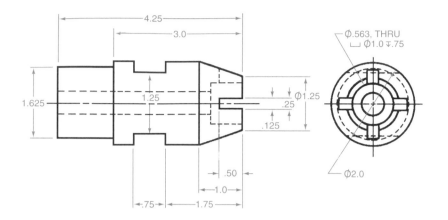

Problem 18-51

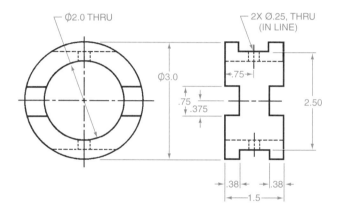

Problem 18-52 *(A to I)*

Construct an exploded isometric drawing of the assembly of a vise, Problem 18-52A. The vise consists of seven different parts and various fasteners, Problems 18-52B through 18-52I. Illustrate the exploded view with the parts in relationship to its assembly. Space parts with an approximate 1″ (25 mm) space between parts. Center the completed view within the work area.

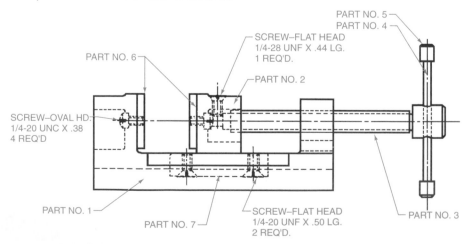

(A)

Problem 18-52 *(continued)*

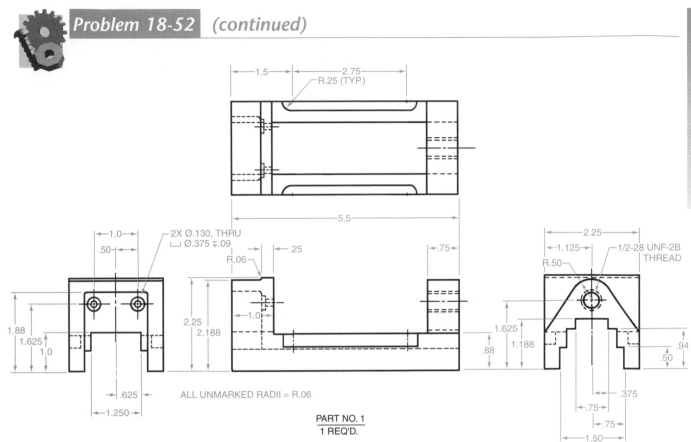

ALL UNMARKED RADII = R.06

PART NO. 1
1 REQ'D.

(B)

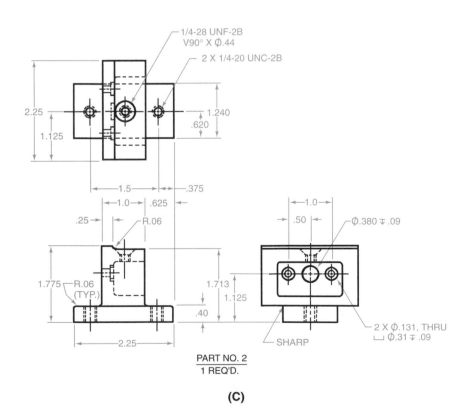

PART NO. 2
1 REQ'D.

(C)

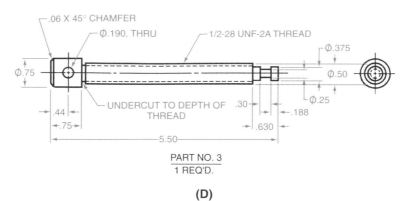

.06 X 45° CHAMFER
Ø.190, THRU
1/2-28 UNF-2A THREAD
Ø.375
Ø.50
Ø.75
Ø.25
UNDERCUT TO DEPTH OF THREAD
.30
.188
.44
.75
.630
5.50

PART NO. 3
1 REQ'D.

(D)

Ø.188
Ø.125
.25
.25
3.25

PART NO. 4
1 REQ'D.

(E)

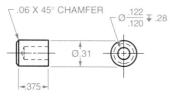

.06 X 45° CHAMFER
Ø $\frac{.122}{.120}$ ∓ .28
Ø.31
.375

PART NO. 5
2 REQ'D.

(F)

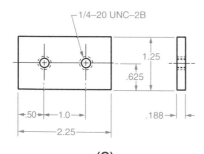

1/4–20 UNC–2B
1.25
.625
.50
1.0
.188
2.25

(G)

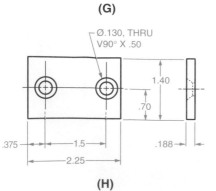

Ø.130, THRU
V90° X .50
1.40
.70
.375
1.5
.188
2.25

(H)

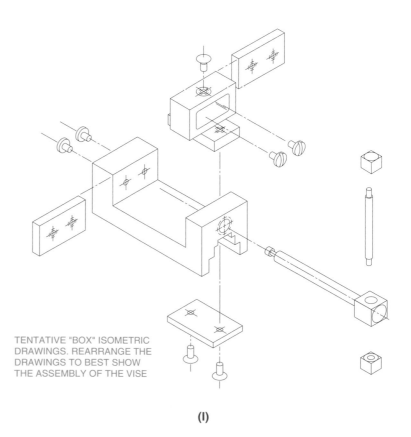

TENTATIVE "BOX" ISOMETRIC
DRAWINGS. REARRANGE THE
DRAWINGS TO BEST SHOW
THE ASSEMBLY OF THE VISE

(I)

Problem 18-53

Using the given multiview drawing, develop a perspective drawing in the positions as illustrated by sketches A, B, C, D, E, and F.

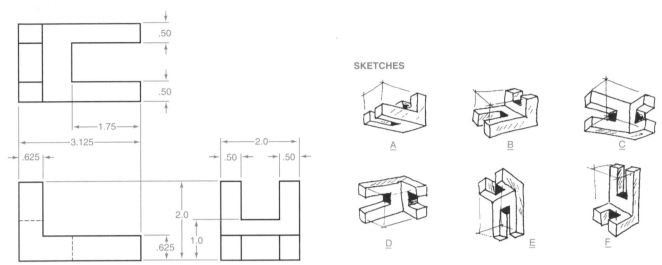

SKETCHES

A   B   C

D   E   F

Problem 18-54

Using the given multiview drawing, develop a perspective drawing in the positions as illustrated by sketches A and B.

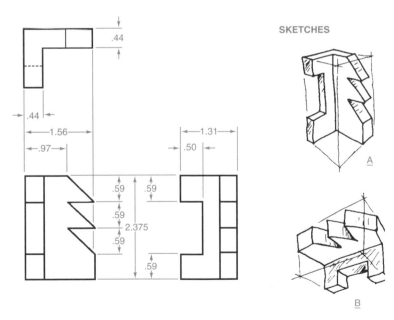

SKETCHES

A

B

PROBLEM 18-55

Choose various objects from Problems 18-1 through 18-51, and develop them into a perspective drawing. Compare the isometric view with the isometric drawing.

PROBLEM 18-56

Prepare an isometric drawing of the object shown in Problem 6-26.

PROBLEM 18-57

Prepare an isometric drawing of the object shown in Problem 9-49.

PROBLEM 18-58

Prepare an oblique drawing of the object shown in Problem 10-13.

PROBLEM 18-59

Prepare an oblique drawing of the object shown in Problem 10-40.

PROBLEM 18-60

Redesign parts 2 and 3 of the vise in Problem 18-52. Propose a more positive means for connecting part 2 to part 3. Show the redesign in any combination of the following formats: multiview drawings, section drawings, isometric drawings, oblique drawings, perspective drawings.

PROBLEM 18-61

Select an assembled product (one that can be disassembled easily, or if necessary, destroyed during disassembly) and draw an exploded, approximate isometric drawing of the components. The objective is to show how the product works. Suggestions: household items, automotive components, office devices (pencil sharpener, file drawer mechanism . . .), sports equipment, weight-training apparatus, laboratory equipment, classroom demonstrations.

PROBLEM 18-62

Design a means, made of simple parts, to raise water one meter from one irrigation canal to another for a third-world country. Employ animal power or a windmill-type apparatus. Prepare an exploded isometric of the components that make up the design.

PROBLEM 18-63

Design a wooden toy (but metal for axles, fasteners, etc.) to be made out of simple shapes; for example, a train engine made of blocks, cylinders, circles, cones, etc. Prepare an exploded isometric showing how the pieces are to go together and how they are to be joined.

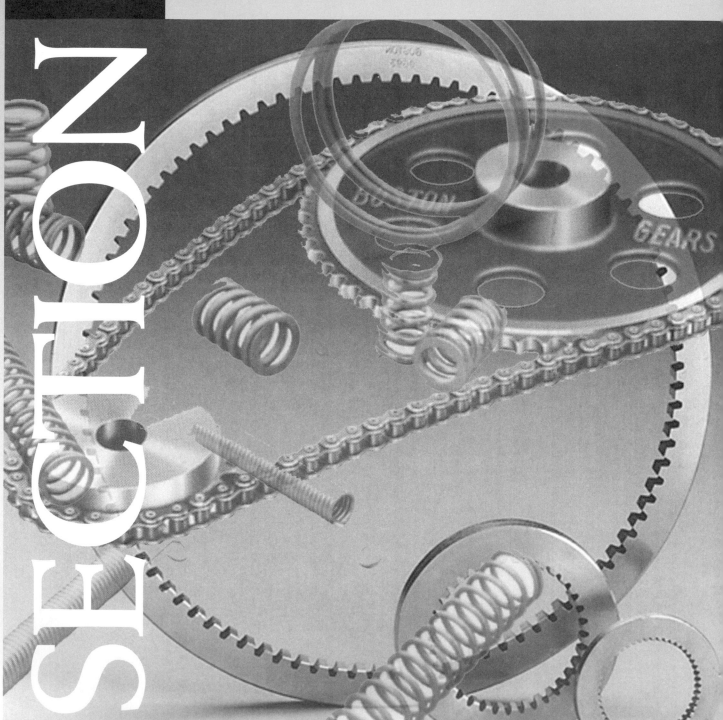

SECTION 4

Related Technologies, Applications, and Process

Career Profile: Terry Schultz

Terry Schultz came upon his career path late in life. At 40 years old, after many years managing restaurants, he was ready for a change. Remembering how much he enjoyed drafting classes in high school, Schultz went back to ITT to get a degree in drafting. After a few years working for others, he decided to go out on his own, with his still-growing small business, Spokane CAD Consultants.

Primarily a structural drafter, Schultz works directly with an engineer on most projects. The engineer will give him sketches, which Schultz looks over to see if there's anything obviously missing. (He recognizes that engineers are not drafters and vice versa, but both need to serve as each other's check and balance.) Sometimes he does his own sketch to see if he understands, before he lays a sketch out in CAD, paying particular attention to detail, scale, and ASME standards.

Often Schultz and the engineer will work with the architect and the architect's drawings. But they also must do their own. Here's the difference: A structural drafter—and the engineer he or she is working with—is most interested in weights, stress loads, and pressures. He or she is thinking, how will I support this building? What beams and supports are necessary? While an archi-

tectural drafter might be looking at the facings of a building, a structural drafter will look at the column supports inside. The drafter also must work with the weldsman—where two pieces of a support will be seamlessly joined, there will be welding. Right now, Schultz and the engineer are creating a basin for a wastewater treatment plant. For this project, Schultz has to make sure the drawing he does outlines a setup that supports 36,000 gallons of water and contains the appropriate piping underneath. In this project, iron is welded to channel; he must know that it is a Phillips weld, what the depth of it is, and what the standard symbol for that is.

The engineer who brought Schultz into this project is someone he has worked with a number of times before. While owning his own business often requires a lot of legwork, he is starting to make contacts who will call him for projects. It is a good sign for the continued expansion of his business; Schultz hopes that within the next few years, he will be managing contract drafters as well. Until then, he can see the day-to-day joy of what he is doing, especially when he gets to see something he drew turn into a sturdy and safe structure.

WELDING

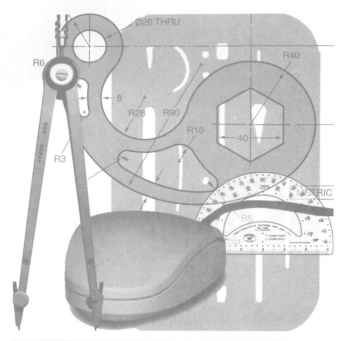

CHAPTER OUTLINE

Welding processes • Basic welding symbol • Size of weld • Length of weld • Placement of weld • Intermittent welds • Process reference • Contour symbol • Field welds • Welding joints • Types of welds • Multiple reference line • Spot weld • Projection weld • Seam weld • Welding template • Design of weldments • Summary • Review questions • Chapter 19 problems

KEY TERMS

Back weld	Intermittent weld
Backing weld	Multiple reference lines
Basic welding symbol	Multiweld
Brazing	Pitch
Chain intermittent weld	Plug weld
Double weld	Projection weld
Field weld	Resistance welding
Fillet weld	Root opening
Flange weld	Seam weld
Forge welding	Section lining
Fusion weld	Slot weld
Fusion welding	Spot welding
Gas welding	Staggered intermittent weld
Groove weld	Surface weld
Increment	Welding

CHAPTER OBJECTIVES

Upon completion of this chapter, students should be able to do the following:

- List the most widely used welding procedure.
- Demonstrate proficiency in the proper use of welding symbols.
- Demonstrate proficiency in the proper dimensioning of welds.
- Demonstrate proficiency in properly indicating weld placement.
- Demonstrate proficiency in properly indicating intermittent welds.
- Explain the proper use of the contour symbols as part of the welding symbol.
- Explain the proper use of the symbology for field welds.
- Describe the five basic types of welded joints.
- Describe the six major types of welds.
- Demonstrate proficiency in the proper use of multiple reference lines.
- Explain the concept of spot welding.
- Explain the differences between spot welds and projection welds.
- Explain the concept of the seam weld.
- Demonstrate proficiency in the use of welding templates.

Welding Processes

Pieces of metal can be fastened together with mechanical fasteners as was discussed in Chapter 13, "Fasteners." They can also be held together by soldering or brazing, and some are fastened together by an adhesive. A permanent way to join pieces of metal together is by *welding*.

One method used to weld parts together is to heat the edges of the pieces to be joined until they melt and join or fuse together. When the pieces cool, they become one homogeneous mass, permanently joined together. A filler rod is sometimes used to mix with the molten metal to make the joint stronger. When a lot of heat is used to melt and fuse the joint, the weld is called a *fusion weld*. Heat for this process can come from a torch, burning gasses, or with a high electric current. Because of the extremely high temperatures involved in welding, the part or parts may be distorted; thus, all machining is usually done after welding.

Another kind of weld used in years past was the pressure weld. In this process, the parts to be welded together are heated to a plastic state, and forced together by pressure or by hammering. This was done by the local blacksmith. This old method was called *forge welding*. Today, faster and better methods are used.

Welding assemblies are usually built-up from stock forms, such as plate steel, square bars, tubing, angle iron and the like. These parts are cut to shape and welded together. The various welding processes are illustrated in *Figure 19-1*. Welded assemblies are much less expensive and more satisfactory than the casting process, especially in a case where only one or only a few identical parts are required.

Welding is used on large structures that would be difficult or impossible to build in a manufacturing plant. Large structures such as building frames, bridges, and ships are welded into one large assembly.

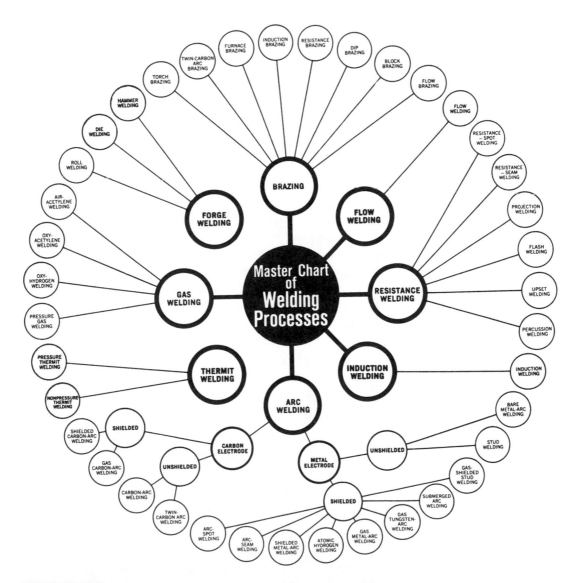

FIGURE 19-1   *Master chart of welding processes* (Courtesy American Welding Society)

TYPE OF WELD	FILLET	GROOVE							BACK OR BACKING	PLUG OR SLOT	SURFAC- ING	FLANGE WELD	
		SQUARE	V	BEVEL	U	J	FLARE V	FLARE BEVEL				EDGE	CORNER
SYMBOL													
WELD ARROW SIDE													
WELD OPPOSITE SIDE													
WELD BOTH SIDES													
NO ARROW/ OPPOSITE SIDE SIGNIFICANCE													
EXAMPLE													

FIGURE 19-2 *Fusion welding*

Major welding methods classified by the American Welding Society include brazing, gas welding, arc welding, resistance welding, and fusion welding. *Brazing* is the process of joining metals together with a nonferrous filler rod. A temperature above 1,000° F, but just below the melting point, is used to join the parts. The most commonly used *gas welding* is oxyacetylene welding. Resistance welding and fusion welding, the two major processes, are explained in this text. Detailed information regarding the other types of welding processes can be obtained by writing to the American Welding Society, P.O. Box 351040, Miami, Florida 33125.

Resistance welding is the process of passing an electric current through the exact location where the parts are to be joined. This is usually done under pressure. The combination of the pressure and the heat generated by the electric current welds the parts together. This process is usually done on thin sheet metal parts.

Fusion welding is usually used for large parts. The fusion welding process uses many standard types of welds, *Figure 19-2.* Each is explained in full.

The next few figures use the fillet welding symbol as an example, but the same method applies, regardless of which welding symbol is used.

Basic Welding Symbol

The *basic welding symbol* consists of a reference line, a leader line and arrow, and, if needed, a tail, *Figure 19-3.* The tail is added for specific information or notes in regard to welding specifications, processes, or reference information. The reference line of the basic welding symbol is usually drawn horizontally. Any welding symbol placed on the upper side of the reference line indi-

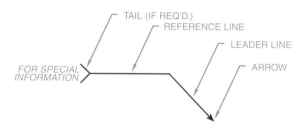

FIGURE 19-3 *Basic welding symbol*

cates WELD OPPOSITE SIDE, *Figure 19-4.* Any welding symbol placed on the lower side of the reference line indicates WELD ARROW SIDE, Figure 19-4. The direction of the leader line and arrow has no significance whatsoever to the reference line.

A fillet weld symbol, Figure 19-2, is added above or below the reference line with the left leg always drawn vertical, *Figure 19-5.* A welding symbol placed above the

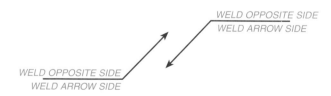

FIGURE 19-4 *Any welding symbol placed on the upper side of the reference line indicates weld OPPOSITE side. Any welding symbol placed on the lower side of the reference line indicates weld ARROW side.*

FIGURE 19-5 *Left leg of welding symbol is always drawn vertical*

reference line means to weld the OPPOSITE side, as illustrated in *Figure 19-6*. A weld symbol placed below the reference line means to weld ARROW side, as illustrated in Figure 19-6. The positions as drawn and as welded are shown in *Figures 19-7* and *19-8*. A weld symbol placed above and below the reference line means to weld both the OPPOSITE side and the ARROW side, as illustrated in *Figure 19-9*.

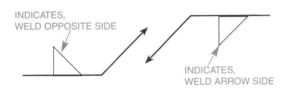

FIGURE 19-6 *Welding symbol placed above the reference line means to weld OPPOSITE side. Welding symbol placed below the reference line means to weld ARROW side.*

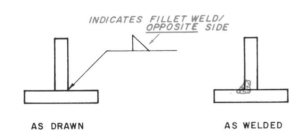

FIGURE 19-7 *Position as drawn and as welded.*

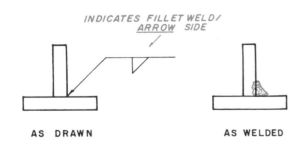

FIGURE 19-8 *Position as drawn and as welded.*

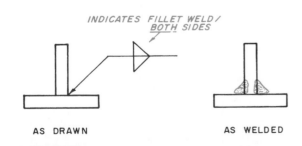

FIGURE 19-9 *Welding symbol placed above and below the reference line means to weld both sides.*

Size of Weld

The weld must be fully dimensioned so that there is no question whatsoever as to its intended and designed size. The size of a weld refers to the length of the leg or side of the weld. The size is placed directly to the left of the welding symbol. The two legs are assumed to be equal in size unless otherwise dimensioned, *Figure 19-10*. If a double weld is needed, the size of each weld must be included as illustrated in *Figure 19-11*.

If the legs or sides of the weld are to be different, the dimensions of the sides must be indicated to the left of the welding symbol in parentheses, as reference-only dimensions, *Figure 19-12*. Because the symbol does not indicate which is the .50 leg and which is the .25 leg, the detail drawing must include the dimensions.

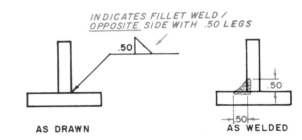

FIGURE 19-10 *Dimension to the left of welding symbol indicates length of leg or side of weld*

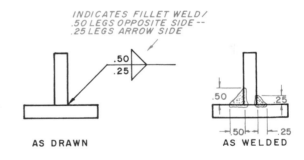

FIGURE 19-11 *Size of each weld is required.*

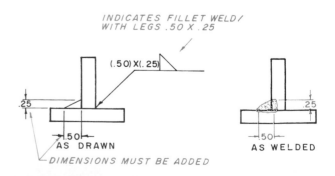

FIGURE 19-12 *Welds with legs or side of different sizes*

Length of Weld

When a weld length is not specified, it is assumed to be continuous the full length, *Figure 19-13*. If a weld must be made to a special length, it must be indicated. This is done by a dimension directly to the right of the weld symbol, *Figure 19-14*.

When a weld is to be made continuously all around an object, it is indicated on the welding symbol by adding a small circle between the reference line and the leader line, *Figure 19-15*.

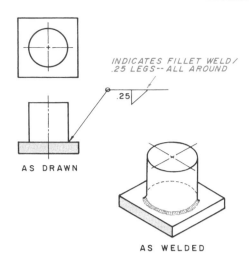

INDICATES FILLET WELD/
.25 LEGS-- ALL AROUND

.25

AS DRAWN

AS WELDED

FIGURE 19-15 *A small circle indicates weld continuously around an object.*

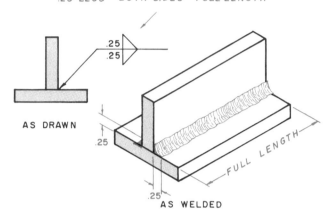

INDICATES FILLET WELD/
.25 LEGS--BOTH SIDES--FULL LENGTH

.25
.25

AS DRAWN

.25

FULL LENGTH

.25

AS WELDED

FIGURE 19-13 *If weld is not specified, it is assumed to be continuous full length.*

Placement of Weld

When a weld is not continuous and is needed in only a few areas, *section lining* is used to indicate where the weld is to be placed, *Figure 19-16*. If a weld is to be placed in a few areas, this is indicated by the use of multiple leader lines and arrows, *Figure 19-17*.

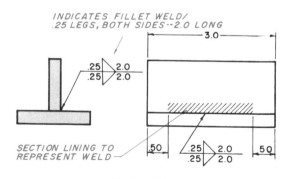

INDICATES FILLET WELD/
.25 LEGS, BOTH SIDES--2.0 LONG

.25 2.0
.25 2.0

3.0

SECTION LINING TO
REPRESENT WELD

.50   .25 2.0   .50
      .25 2.0

AS DRAWN

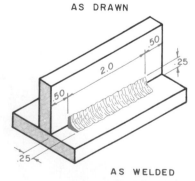

.50
2.0
.50
.25
.25

AS WELDED

FIGURE 19-14 *Length of weld must be noted to the right of the welding symbol.*

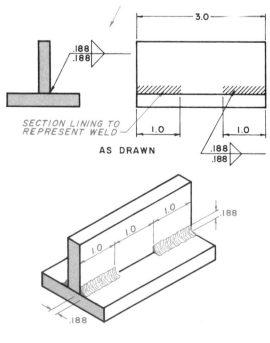

INDICATES FILLET WELD/
.188 LEGS-- BOTH SIDES

.188
.188

3.0

SECTION LINING TO
REPRESENT WELD

1.0   1.0

AS DRAWN

.188
.188

1.0
1.0
1.0
.188

.188

AS WELDED

FIGURE 19-16 *Section lining indicates location of weld.*

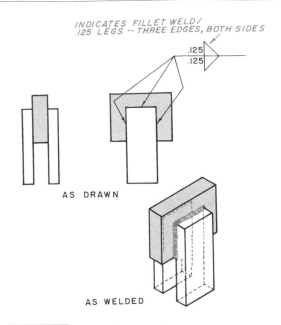

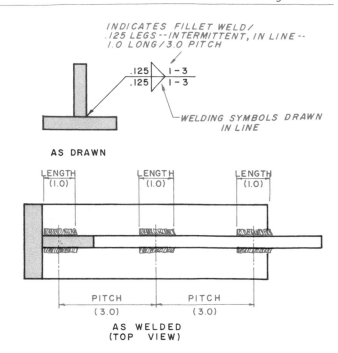

FIGURE 19-17 *Leader line and arrows indicate location of weld.*

FIGURE 19-20 *Chain intermittent welds are applied directly opposite each other.*

Intermittent Welds

Intermittent welds are a series of short welds. When this type of weld is needed, an extra dimension is added to the welding symbol, *Figure 19-18*. The usual size of the leg is noted to the left of the symbol, and the length of the weld is indicated to the right of the welding symbol followed by a dash line and the pitch of the intermittent

welds, *Figure 19-19*. An *increment* is the length of the weld; the *pitch* is the center-to-center distance between increments.

CHAIN INTERMITTENT WELD

A *chain intermittent weld* is a weld in which each weld is applied directly opposite to each other on opposite sides of the joint, *Figure 19-20*. Note that the welding symbol is applied above and below the reference line and that they are in line with each other.

STAGGERED INTERMITTENT WELD

A *staggered intermittent weld* is similar to the chain intermittent weld except the welds are staggered directly opposite each other on opposite sides, *Figure 19-21*. Note the welding symbol is applied above and below the reference line staggered to each other.

Process Reference

There are many welding processes developed by the American Welding Society. Each process has a standard abbreviation designation, *Figure 19-22*. Letter designations are used to specify a method used to obtain a particular contour of a weld, *Figure 19-23*. These abbreviations are taken from the latest American Welding Society's standard #AWS-2.4-79,71.

The standard process abbreviations or letter designations are usually added to the tail of the basic welding

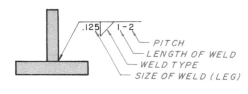

FIGURE 19-18 *Intermittent welds are noted by dimensions to the right of the welding symbol.*

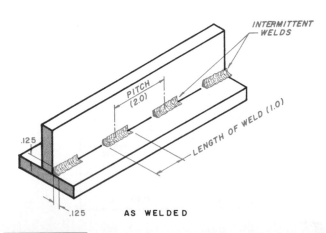

FIGURE 19-19 *Pitch is the center-to-center distance between welds.*

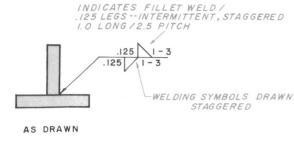

INDICATES FILLET WELD/
.125 LEGS--INTERMITTENT,STAGGERED
1.0 LONG/2.5 PITCH

AS DRAWN

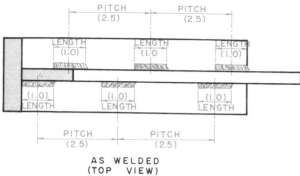

AS WELDED
(TOP VIEW)

FIGURE 19-21 *Staggered intermittent welds are placed staggered opposite each other.*

DESIGNATION OF WELDING AND ALLIED PROCESSES BY LETTERS

AAC	air carbon arc cutting
AAW	air acetylene welding
ABD	adhesive bonding
AB	arc brazing
AC	arc cutting
AHW	atomic hydrogen welding
AOC	oxygen arc cutting
AW	arc welding
B	brazing
BB	block brazing
BMAW	bare metal arc welding
CAC	carbon arc cutting
CAW	carbon arc welding
CAW-G	gas carbon arc welding
CAW-S	shielded carbon arc welding
CAW-T	twin carbon arc welding
CW	cold welding
DB	dip brazing
DFB	diffusion brazing
DFW	diffusion welding
DS	dip soldering
EASP	electric arc spraying
EBC	electron beam cutting
EBW	electron beam welding
ESW	electroslag welding
EXW	explosion welding
FB	furnace brazing
FCAW	flux cored arc welding
FCAW-EG	flux cored arc welding–electrogas
FLB	flow brazing
FLOW	flow welding
FLSP	flame spraying
FOC	chemical flux cutting
FOW	forge welding
FRW	friction welding
FS	furnace soldering
FW	flash welding

FIGURE 19-22 *Designation of welding and allied processes by letters (abbreviations)*

DESIGNATION OF WELDING AND ALLIED PROCESSES BY LETTERS

GMAC	gas metal arc cutting
GMAW	gas metal arc welding
GMAW-EG	gas metal arc welding–electrogas
GMAW-P	gas metal arc welding–pulsed arc
GMAW-S	gas metal arc welding–short circuiting arc
GTAC	gas tungsten arc cutting
GTAW	gas tungsten arc welding
GTAW-P	gas tungsten arc welding–pulsed arc
HFRW	high frequency resistance welding
HPW	hot pressure welding
IB	induction brazing
INS	iron soldering
IRB	infrared brazing
IRS	infrared soldering
IS	induction soldering
IW	induction welding
LBC	laser beam cutting
LBW	laser beam welding
LOC	oxygen lance cutting
MAC	metal arc cutting
OAW	oxyacetylene welding
OC	oxygen cutting
OFC	oxyfuel gas cutting
OFC-A	oxyacetylene cutting
OFC-H	oxyhydrogen cutting
OFC-N	oxynatural gas cutting
OFC-P	oxypropane cutting
OFW	oxyfuel gas welding
OHW	oxyhydrogen welding
PAC	plasma arc cutting
PAW	plasma arc welding
PEW	percussion welding
PGW	pressure gas welding
POC	metal powder cutting
PSP	plasma spraying
RB	resistance brazing
RPW	projection welding
RS	resistance soldering
RSEW	resistance seam welding
RSW	resistance spot welding
ROW	roll welding
RW	resistance welding
S	soldering
SAW	submerged arc welding
SAW-S	series submerged arc welding
SMAC	shielded metal arc cutting
SMAW	shielded metal arc welding
SSW	solid state welding
SW	stud arc welding
TB	torch brazing
TCAB	twin carbon arc brazing
TW	thermit welding
USW	ultrasonic welding
UW	upset welding

FIGURE 19-22 *(Continued) Designation of welding and allied processes by letters (abbreviations)*

LETTER	METHOD
C	CHIPPING
G	GRINDING
H	HAMMERING
R	ROLLING
M	MACHINING

FIGURE 19-23 *Abbreviations of standard method used to obtain a particular contour*

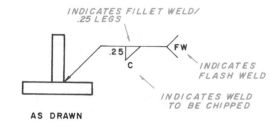

FIGURE 19-24 *Process abbreviations are placed in the tail of the basic welding symbol.*

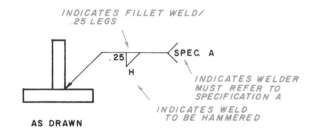

FIGURE 19-25 *Sometimes a note is indicated somewhere on the drawing.*

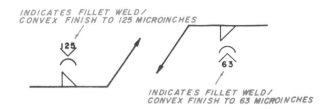

FIGURE 19-28 *Finish number specifies finish required.*

symbol as required, *Figure 19-24.* Sometimes, they are called off in a note somewhere on the drawing, *Figure 19-25.*

Contour Symbol

If a special finish must be made to a weld, a contour symbol must be added to the welding symbol. There are three kinds of contour symbols used: flush, convex, or concave, *Figure 19-26.* The contour symbol is added directly above or below the welding symbol. If the welding symbol is above the reference line, the contour symbol is placed above the welding symbol. If the welding symbol is placed below the reference line, the contour symbol is placed below the welding symbol, *Figure 19-27.* The degree of finish is not usually included, but if a specific finish is desired, a finish number designation must be added, *Figure 19-28.*

FIGURE 19-26 *Contour symbol*

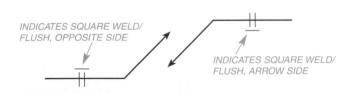

FIGURE 19-27 *Contour symbol is added to the welding symbol.*

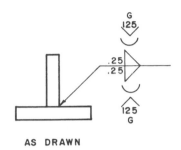

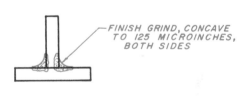

FIGURE 19-29 *Finish weld is to be concave and ground to a 125-microinch surface finish.*

Figure 19-29 represents that the finish weld must be concave and ground to a 125-microinch surface finish. *Figure 19-30* represents that the finish weld must be machined flat on the top surface to a surface finish of 125 microinches. The weld must be convex on the bottom surface with a hammered finish.

Field Welds

Any weld not made in the factory—that is, it is to be made at a later date, perhaps at final assembly on-site—is called a *field weld.* Its symbol is added to the basic weld symbol by a filled-in flag, located between the reference line and the leader line and drawn using a 30°–60° triangle, *Figure 19-31.*

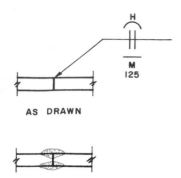

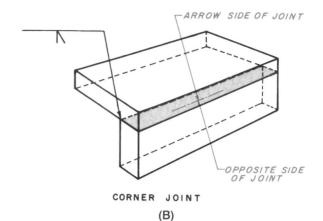

CORNER JOINT

(B)

FIGURE 19-30 *Finish weld is to be machined flat on top surface to 125 microinch and convex on the bottom surface with a hammered finish.*

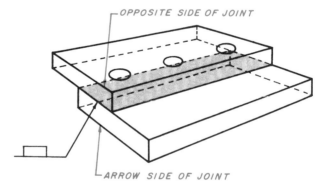

T-JOINT

(C)

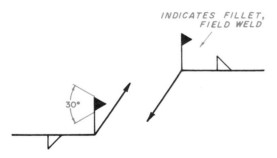

FIGURE 19-31 *Field weld*

Welding Joints

There are five basic kinds of welding joints and they are classified according to the position of the parts that are being joined, *Figures 19-32A* through *19-32E*. There are many types of welds used to weld these joints together. Considerations as to which type of weld to use are based

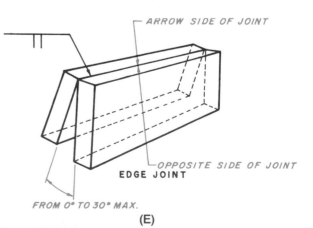

LAP JOINT

(D)

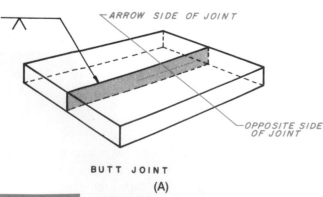

BUTT JOINT

(A)

FIGURE 19-32 *(A) Butt joint*

EDGE JOINT

(E)

FIGURE 19-32 *(Continued)(B) Corner joint; (C) T-joint; (D) Lap joint; (E) Edge joint*

on the particular application, thickness of material, required strength of the joint, and available welding equipment, among others.

Types of Welds

There are six major types of welds (refer back to Figure 19-2): fillet, groove, back or backing, plug or slot, surface, and flange welds.

More than one type of weld may be applied to a single joint—usually for strength or appearance. For example, a groove V weld may be used on one side, and a backing weld on the other side, **Figure 19-33.** This is known as a *multiweld.*

The same type of weld may be used on opposite sides of a single joint. Such a joint is called a *double weld;* for example, double V, double U, or double J weld, **Figure 19-34.** These too, are used for strength and/or appearance.

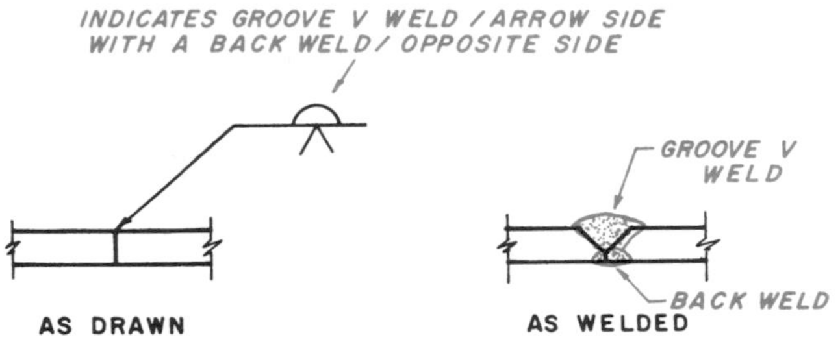

FIGURE 19-33 *Multiwelds (different type of welds)*

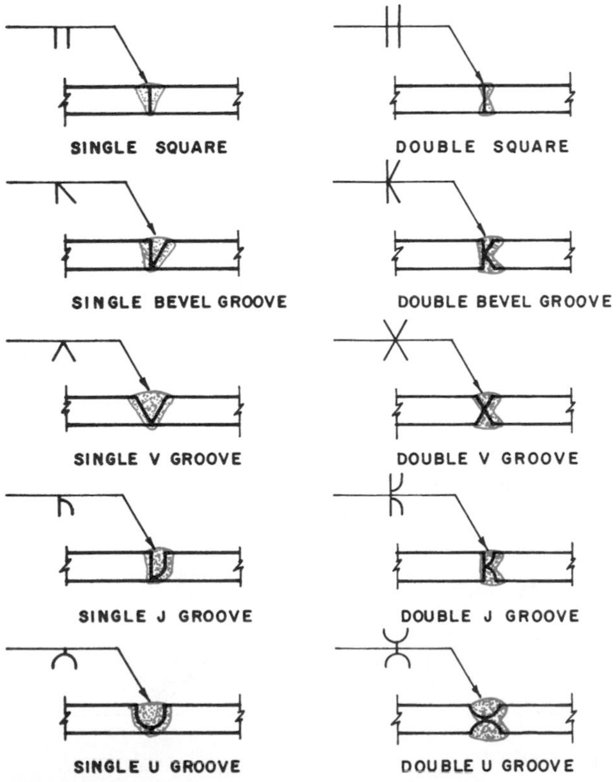

FIGURE 19-34 *Double welds (same type of welds)*

Fillet Welds

Up to this point, the *fillet weld* and its symbol have been used as an example. The other five major types of welding symbols used are very similar.

Groove Weld

Seven types of welds are considered to be *groove welds:* square, V, bevel groove, J, U, flare V, and flare bevel. Each type has its own welding symbol, **Figure 19-35.** In the bevel groove and J welds, only one part is actually chamfered or grooved. To do this, the leader line and arrow point toward the part that has the bevel groove or J groove, **Figure 19-36.** Note that the welding symbol is placed above or below the reference line, as usual, to indicate on which side the chamfer or groove is located.

Size of Groove Weld. In the groove weld, the dimension refers to the actual depth of the chamfer or groove, **not** the size of the weld. With the fillet weld, the size of the chamfer or groove is located directly to the left of the welding symbol and on the same side of the reference line, **Figure 19-37.** If a V groove weld symbol is shown without a size dimension, the size or depth is assumed to be equal to the thickness of the pieces.

Depth or Actual Size of Weld. The depth or actual size of the weld includes the penetration of the complete weld. This is illustrated in **Figure 19-38.** Note in this example that the depth or actual size of the weld is deeper than the depth of the chamfer. Occasionally, the depth or actual size of the weld is less than the size of the chamfer, **Figure 19-39.**

GROOVE WELDS		
TYPE	SYMBOL	AS SEEN
SQUARE	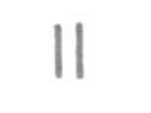	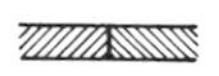
V	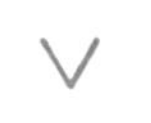	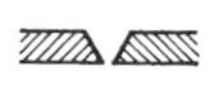
BEVEL GROOVE		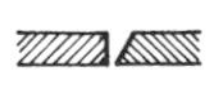
U		
J	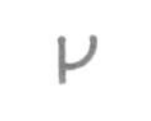	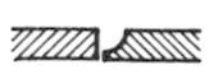
FLARE V	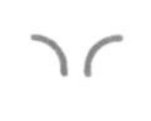	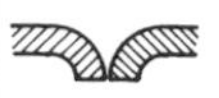
FLARE BEVEL	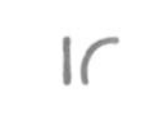	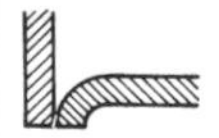

FIGURE 19-35 *Groove welds*

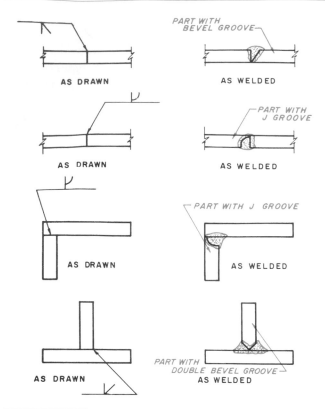

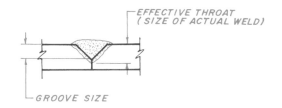

FIGURE 19-38 *Depth or actual size of weld*

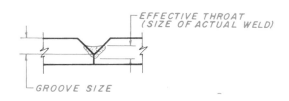

FIGURE 19-39 *Size of weld only*

In order to call off the depth or actual size of the weld and the size of the chamfer or groove, a dual dimension is used and both are added to the left of the welding symbol. In order to differentiate between the two, the size of the chamfer or groove is added to the extreme left of the welding symbol. The size of the depth or actual size of the weld is added directly to the left of the welding symbol, but in parentheses, *Figure 19-40*. Usually the weld is larger than the groove, *Figure 19-41*.

If the weld is less than the size of the chamfer, the dimensions are added to the left of the welding symbol in the same manner, *Figure 19-42*.

Root Opening. The allowed space between parts is called the *root opening*, *Figure 19-43*. The root opening is applied **inside** the welding symbol, *Figure 19-44*.

FIGURE 19-36 *Groove weld illustrations*

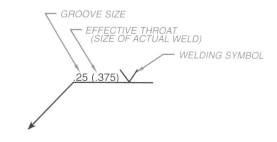

FIGURE 19-40 *Size of weld is added to the left of the welding symbol in parentheses*

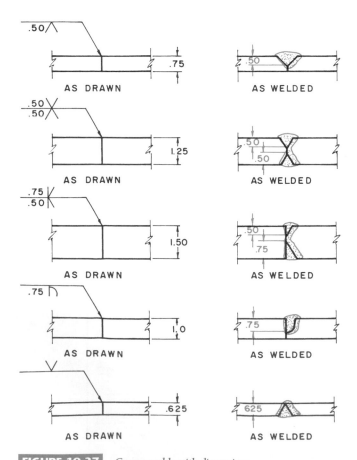

FIGURE 19-37 *Groove welds with dimensions*

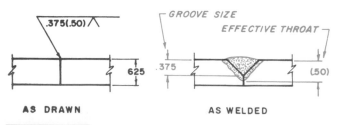

FIGURE 19-41 *Example of effective throat-weld larger than groove*

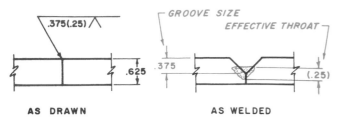

FIGURE 19-42 *Example of effective throat-weld smaller than groove*

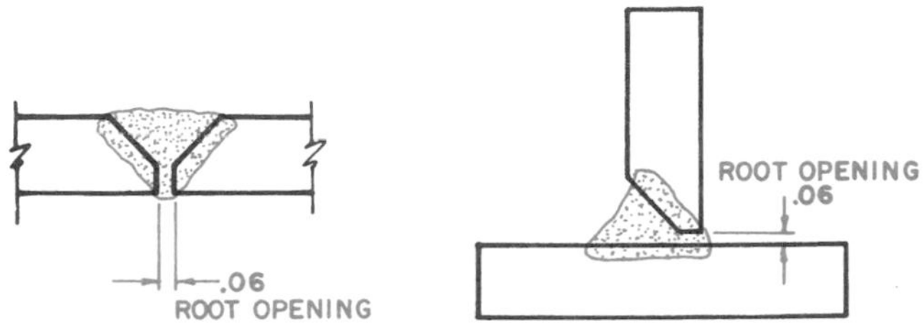

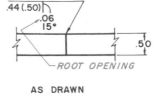

FIGURE 19-46 *Required radius*

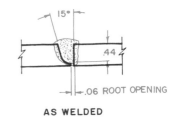

FIGURE 19-43 *Root opening*

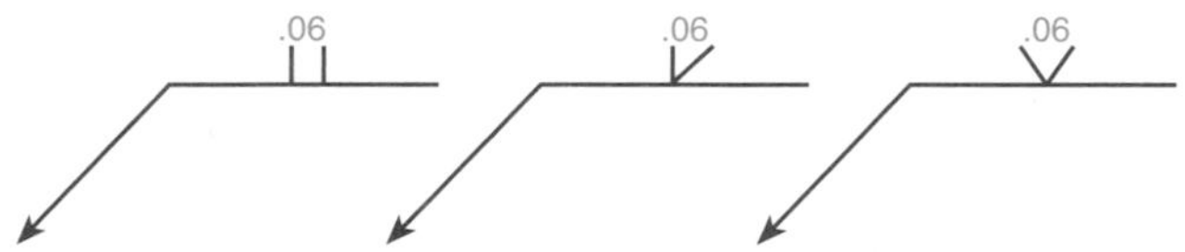

Groove Radius. If a J or U groove weld must be held to tight tolerances, the angle and radius must be included with the size dimensions in order to obtain the exact required geometric size and shape of all manufactured parts. The angle is also added inside the welding symbol and the required radius is called off as a note, *Figure 19-46*. Note that the leader line and arrow point toward the part with the groove. Because the welding symbol is placed below the reference line, the groove and weld are on the side of the arrow.

BACK WELD—BACKING WELD

A *back weld* is a weld applied to the opposite side of the joint **after** the major weld has been applied. A *backing weld* is a weld applied first, followed by the major weld. Both use the same welding symbol and both are used to strengthen a weld (refer back to Figure 19-2). The back weld is also used for appearance.

FIGURE 19-44 *Size of root opening*

The root opening may be called off as a general note. For example, "unless otherwise noted, root opening for all groove welds is .06." If there is **no** opening between parts, a zero is added inside the welding symbol.

Chamfer Angle. In a groove weld, a V or bevel groove must be held to tight tolerances. The angle must also be included with the size dimension in order to obtain the exact required geometric size and shape of all manufactured parts. The angle is also added inside the welding symbol, *Figure 19-45*. Note that the leader line and arrow point toward the part with the chamfer. Because the welding symbol is placed above the reference line, the chamfer and weld are on the side opposite from the arrow.

If a welding melt-through is required, the same symbol is used, except it is filled in solid, *Figure 19-47*. When the back weld or backing weld symbol is used, it is placed on the opposite side of the reference line from the groove weld symbol, *Figures 19-48* and *19-49*.

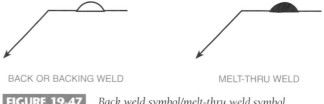

FIGURE 19-47 *Back weld symbol/melt-thru weld symbol*

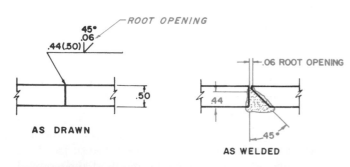

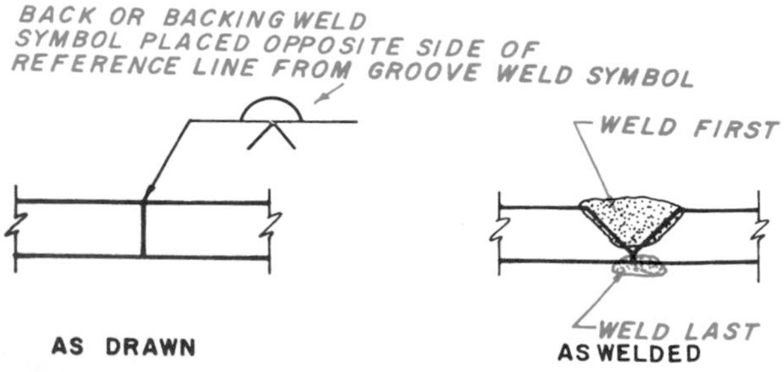

FIGURE 19-45 *Chamfered angle*

FIGURE 19-48 *Illustration of back weld symbol*

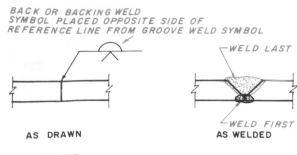

FIGURE 19-49 *Illustration of back weld symbol*

PLUG AND SLOT WELDS

The *plug weld* and *slot weld* both use the same welding symbol. The only difference between the two welds is the shape of the hole through which the weld is applied. A plug weld is made through a round hole. A slot weld is made through an elongated hole (refer back to Figure 19-2).

The plug or slot welding symbol is applied to the basic welding symbol and interpreted in exactly the same way. If the welding symbol is applied above the reference line, the hole or slot is on the opposite side of the arrow. If below the reference line, the hole or slot is on the same side as the arrow, *Figure 19-50*.

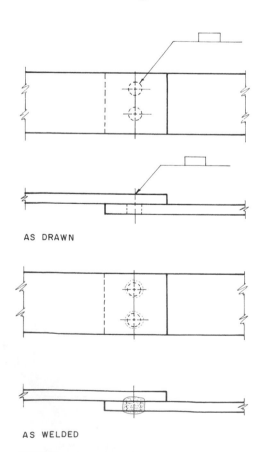

FIGURE 19-50 *Plug and slot welds*

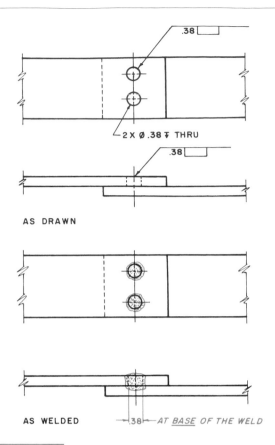

FIGURE 19-51 *Size of plug or slot weld*

Size of Plug or Slot Weld. The given dimension for a plug or slot weld refers to the diameter of the plug or slot at the base of the weld. The required dimension is placed directly to the left of the plug or slot weld symbol, *Figure 19-51*. The size of the slot weld is drawn and dimensioned directly on the detail drawing.

Depth of Weld. Unless noted, the plug or slot weld hole is completely filled by the weld. If filling the hole is not necessary, the full depth of weld must be called off. The required depth of weld is added inside the welding symbol, *Figure 19-52*.

Tapered Plug or Slot. If the plug or slot hole is tapered, the taper inclusive angle is added directly above or below the welding symbol, *Figure 19-53*.

SURFACE WELD

If a surface must have material added to it or must be built up, a *surface weld* is added (refer back to Figure 19-2). This welding symbol is **not** used to indicate the joining of parts together; therefore, the welding symbol is simply added below the reference line, *Figure 19-54*.

Size of Surface Weld. If the surface is to be built up and a special height is **not** required, the welding symbol is drawn as illustrated in Figure 19-54. If the surface is to be built up and a specific height is required, the required height is added directly to the left of the welding sym-

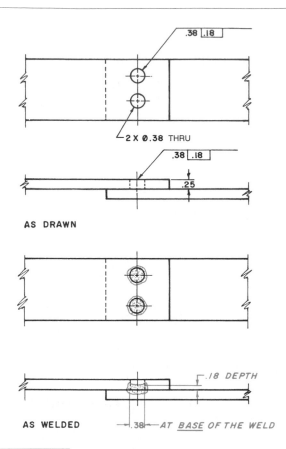

AS DRAWN

AS WELDED

FIGURE 19-52 *Depth of weld*

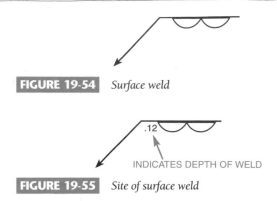

FIGURE 19-54 *Surface weld*

INDICATES DEPTH OF WELD

FIGURE 19-55 *Site of surface weld*

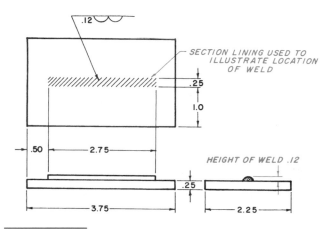

SECTION LINING USED TO ILLUSTRATE LOCATION OF WELD

HEIGHT OF WELD .12

FIGURE 19-56 *Illustration of a surface weld*

bol, *Figure 19-55*. If only a portion of the surface is to be built up, that portion must be illustrated and dimensioned on the detail drawing, *Figure 19-56*. The actual weld is represented by section lining.

FLANGE WELD

A *flange weld* is used to join thin metal parts together. Instead of welding to the surfaces of the parts to be joined, the weld is applied to the edges of thin material. A weld applied to thin metal could burn right through.

There are two distinct flange weld symbols (refer back to Figure 19-2): the first is called an edge-flange weld, *Figure 19-57*; the other is called a corner-flange weld, *Figure 19-58*. The straight line of this welding symbol is always drawn to the left of the partially curved line, regardless of the actual joint being illustrated. The

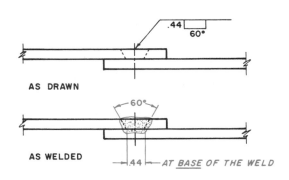

AS DRAWN

AS WELDED

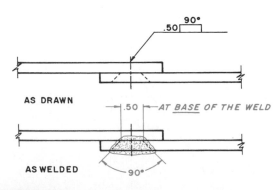

AS DRAWN

AS WELDED

FIGURE 19-53 *Tapered plug or slot*

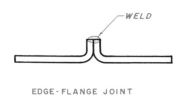

WELD

EDGE-FLANGE JOINT

FIGURE 19-57 *Edge-flange weld*

FIGURE 19-58   *Corner-flange weld*

edge-flange or corner-flange weld symbol is placed above or below the reference line to indicate OPPOSITE side or ARROW side, but it is never on both sides, *Figures 19-59* and *19-60*.

Dimensioning a Flange Weld.   It requires three dimensions to fully dimension a flange weld: the radius of the flange, the height above the point of tangency between parts, and the size of the weld itself, *Figures 19-61* and *19-62*. The three dimensions are added to the welding symbol as indicated previously. The welding symbol should appear at the peak of the joint as illustrated. If

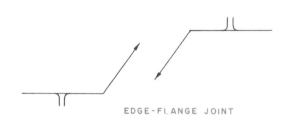

FIGURE 19-59   *Edge-flange weld symbol*

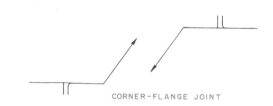

FIGURE 19-60   *Corner-flange weld symbol*

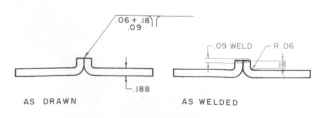

FIGURE 19-61   *Dimensioning a flange weld*

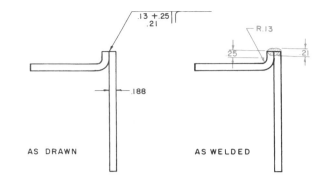

FIGURE 19-62   *Dimensioning a corner-flange weld*

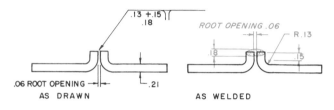

FIGURE 19-63   *Root opening*

there is to be a root opening, that is, a space between parts, the required space must be indicated on the drawing, **not** on the welding symbol, *Figure 19-63*.

Multiple Reference Line

If there is more than one operation on a particular joint, these operations are indicated on *multiple reference lines*. The first reference line closest to the arrow is for the first operation, the second reference line is for the second operation or supplemental data, and the third reference line is for the third operation or test information, *Figure 19-64*.

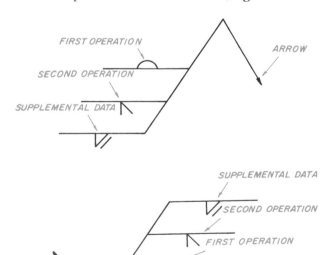

FIGURE 19-64   *Multiple reference line*

Industry Application

ON-THE-JOB TRAINING

Fitness Equipment Corporation (FEC) is one of the largest manufacturers in the United States of heavy-gage fitness machines. Their equipment is used in some of the leading commercial health clubs and by several professional football teams. FEC's products are made primarily of steel tubing. Consequently, welding is one of the company's primary manufacturing processes.

Because welding is such an important process to the company, new hires in the drafting department spend their first six weeks of employment on the shop floor learning about welding. According to Jerry White, FEC's chief drafter "we want our drafting and design personnel to know as much about our welding processes as our welders. Otherwise, how can they properly specify the various welds that might be needed on a given project? They don't have to develop the hand-eye coordination of a welder, but we do want them to know what kind of weld is called for in a given situation, how to properly indicate it using correct symbology, and how to properly dimension welds."

When asked about this practice, FEC's drafters are universally positive in their responses. Carla McDaniels summarized her reaction to six weeks on the shop floor as follows: "I was apprehensive at first. Having never been on the shop floor of a manufacturing plant, I had visions of a dark, dungeon-like atmosphere where men in hoods used torches on steel bars. However, my mentor was a woman, and the welding shop at FEC is cleaner than my house. It turned out to be a great six weeks. I learned a lot about welding, but I also learned what the welders need from me when I prepare a drawing. I also learned why my instructors back at MacArthur Technical Institute had stressed the need to put myself in the shoes of the people who would use my drawings. There are things I would have put in my drawings that are not needed and other things I would have left out that are needed."

Another drafter said, "Let me put it this way; based on how much I learned on that shop floor in six weeks, if I ever decide to design houses for a living I'm going to first spend six weeks working with a carpenter."

Spot Weld

Spot welding is a resistance welding process done by passing an electric current through the exact location where the parts are to be joined. Spot welding is usually done to hold thin sheet metal parts together. The resistance weld process uses three standard types of welds, *Figure 19-65*. Each is explained in full.

The symbol for a spot weld, as illustrated in *Figure 19-66*, is located tangent to or touching the reference line, and it uses the same added supplemental data symbols and dimensions as are illustrated with the fusion welding process, *Figure 19-67*. The tail is *always* included, in order to indicate the process required to make the weld (refer back to Figure 19-22).

DIMENSIONING THE SPOT WELD

The diameter size of the spot weld is added directly to the left of the weld symbol, *Figure 19-68*. Sometimes the diameter size is omitted, and a required shear strength is inserted in its place. In this condition, an explanation of the value is added to the drawing by a note, *Figure 19-69*.

The spacing of the spot weld is called off by a number directly to the right of the weld symbol. This num-

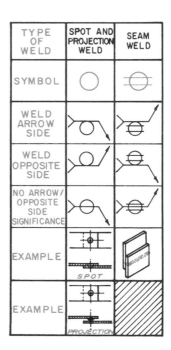

FIGURE 19-65 *Spot welds*

ber is the pitch, and refers to the center-to-center distance between spots, *Figure 19-70*. The number of welds for a particular joint is added above or below the welding symbol in parentheses.

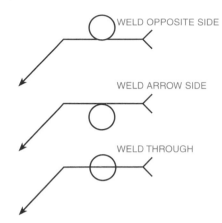

FIGURE 19-66   *Spot weld symbol*

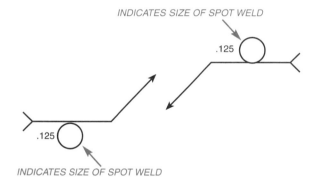

FIGURE 19-67   *Spot weld symbol used with reference line*

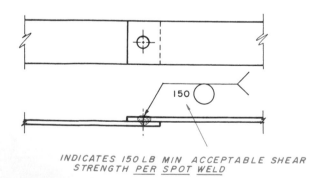

FIGURE 19-68   *Diameter of spot weld*

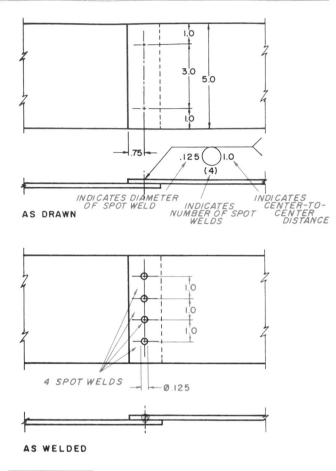

FIGURE 19-70   *Pitch of spot welds*

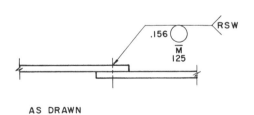

INDICATES 150 LB MIN ACCEPTABLE SHEAR
STRENGTH PER SPOT WELD

FIGURE 19-69   *Shear strength of spot weld*

CONTOUR AND FINISH SYMBOLS

Contour and finish symbols are added to the welding symbol exactly as is done in the fusion welding symbol, *Figure 19-71.*

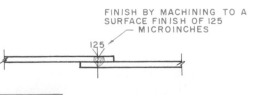

FIGURE 19-71   *Contour and finish symbols*

Projection Weld

A *projection weld* is similar to a spot weld and uses the same welding symbol. One of the two parts has a series of evenly spaced dimples stamped into it. Each stamped dimple is the location of an individual spot weld, *Figure 19-72*. A projection weld allows more penetration and, as a result, is a much better weld. Note that the dimples are **not** illustrated on the detail drawing.

The reference line (OPPOSITE side-ARROW side significance) is changed slightly as used for a projection weld. The OPPOSITE side-ARROW side indicates which of the two parts to be joined has the dimples on it, *Figures 19-73* and *19-74*.

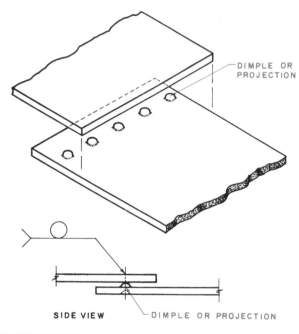

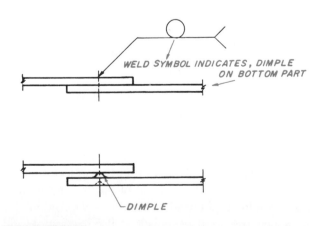

FIGURE 19-72   *Projection weld*

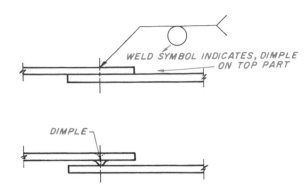

FIGURE 19-74   *Projection weld symbol—dimple on top*

DIMENSIONING, CONTOUR, AND FINISH SYMBOLS

Dimensioning, contour, and finish symbols are applied to projection welds exactly as they are for spot welds.

Seam Weld

A *seam weld* is like a spot weld, except that the weld is actually continuous from start to finish.

The welding symbol is modified slightly to indicate a seam weld, *Figure 19-75*. *Figure 19-76* indicates how the seam weld symbol is applied to the reference line.

Welding Template

In order to simplify and speed up the drawing time for welding symbols, welding templates can be used. The welding template standardizes the size of each welding symbol and provides a quick, ready reference for welding symbols, *Figure 19-77*.

FIGURE 19-75   *Seam weld symbol*

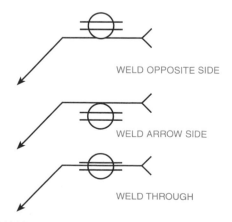

FIGURE 19-76   *Seam weld symbol used with reference line*

FIGURE 19-73   *Projection weld symbol—dimple on bottom*

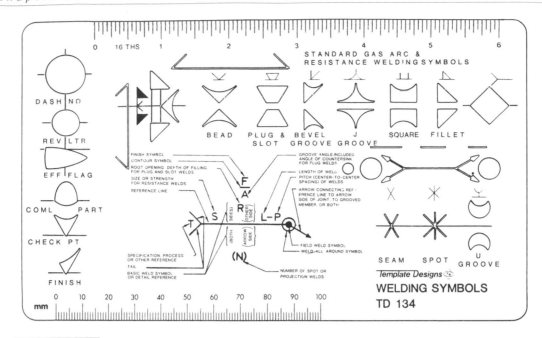

FIGURE 19-77 *Welding template*

Design of Weldments

Welded items are held together by fixtures during the actual welding process. This maintains the proper relationships between the parts. However, once the parts are released from the fixture, the potential for warping and other problems is introduced. This problem can be solved for the most part through proper design practices. Designers should keep the following objectives in mind when designing weldments.

1. Design for the convenience of the welder or welding machine/robot.
2. Design for proper heat control in the area of the weld.
3. Design in appropriate clamping mechanisms to prevent warping.
4. Design in outlets for the welding atmosphere.
5. Design in the appropriate clearances to accommodate for filler material.
6. Consider ease of operation after the weld is completed.

Summary

- Welding is a permanent way to join metal pieces. Welding is less expensive and more satisfactory than the casting process.
- The basic welding symbol consists of a reference line, a leader line and arrow, and, if needed, a tail. The tail is added for specific information or notes in regard to welding specifications, processes, or reference information.
- The size of a weld must be fully dimensioned so that there is no question whatsoever as to its intended and designed size. When a weld length is not specified, it is assumed to be continuous for the full length.
- When a weld is not continuous and is needed in only a few areas, section lining is used to indicate where the weld should be placed.
- Intermittent welds are a series of short welds. An increment is the length of the weld; the pitch is the center-to-center distance between increments.
- If a special finish must be made to a weld, a contour symbol must be added to the welding symbol. Three kinds of contour symbols are used: flush, convex, and concave.
- Any weld not made in the factory is called a field weld.
- There are five basic kinds of welding joints; butt, corner, T-joint, lap, and edge.
- Spot welding is a resistance welding process done by passing an electric current through the exact location where the parts are to be joined.
- A projection weld is similar to a spot weld and uses the same welding symbol. Dimples (projections) are stamped into one of the metal pieces. Each dimple is a location for a spot weld. A projection weld allows for more penetration. A seam weld is also like a spot weld except that the weld is continuous from start to finish.

Review Questions

Answer the following questions either true or false.

1. A plug weld and a slot weld both use the same symbol.

2. A field weld is one of the six major types of welds.

3. An intermittent weld is a series of short welds.

4. A welding symbol placed below the line means to weld the opposite side.

5. A flange weld is used to join thin metal parts together.

6. A process reference is a number that is placed at the bottom of the drawing as a reference.

7. A back weld is a weld applied to the opposite side of the joint after the major weld has been applied.

8. A backing weld is a weld applied to the opposite side of the joint after the major weld has been applied.

Answer the following questions by selecting the best answer.

1. Which of the following is **not** one of the major types of welds?
   a. Fillet
   b. Groove
   c. Joint
   d. Back or backing

2. Which of the following is the correct definition for the term *root opening*?
   a. The allowed space between parts
   b. The actual depth of the chamber or groove
   c. The actual resistance between two parts
   d. None of the above

3. Which of the following correctly identifies the major welding processes?
   a. Resistance and fusion
   b. Field and contour
   c. Fillet and groove
   d. Plug and slot

4. Which of the following is **not** one of the major types of welding joints?
   a. Lap joint
   b. Edge joint
   c. L-joint
   d. T-joint

5. Which of the following is **not** true regarding spot weldings?
   a. It is usually done to hold sheet metal together.
   b. It is stronger than a projection weld.
   c. It is represented by the same symbol as a projection weld.
   d. It is a resistance welding process.

6. Which of the following is **not** one of the types of groove welds?
   a. Bevel groove
   b. Flare V
   c. Square
   d. Flare U

7. What indicates a field weld on the welding symbol?
   a. A filled-in flag
   b. A filled-in circle
   c. A filled-in square
   d. A filled-in triangle

8. What indicates that a weld is to be continuous around an object?
   a. Adding a small square between the reference line and the leader line
   b. Adding a small circle between the reference line and the leader line
   c. Adding a small rectangle between the reference line and the leader line
   d. Adding a small triangle between the reference line and the leader line

Chapter 19 Problems

The following problems are intended to give the beginning drafter practice in studying the many factors involved in dimensioning and calling off welding processes. The student will sharpen skills in centering the views within the work area, line weight, dimensioning, and speed.

CONVENTIONAL INSTRUCTIONS

The steps to follow in laying out the following problems are:

STEP 1 Study the problem carefully.

STEP 2 Choose the view with the most detail as the front view.

STEP 3 Make a sketch of the required views.

STEP 4 Add required dimensions to the sketch per the latest drafting standards.

STEP 5 Add welding symbols as required to the sketch per the latest welding standards.

STEP 6 Lightly lay out all required views and check that they are centered within the work area.

STEP 7 Check that all dimensions are added using the latest drafting standards.

STEP 8 Darken in all views using correct line thickness and add all dimensioning using light guidelines.

STEP 9   Recheck all work and, if correct, neatly fill out the title block using light guidelines and neat lettering.

STEP 10   Recheck that all required welding symbols are added correctly.

CAD Instructions

Students who plan to complete the following problems on a CAD system should begin with these general instructions. Before reading the specific instructions for each activity, go through each step in the following planning checklist. The checklist applies to any CAD system and will help ensure the optimum use of your time and resources.

1. Analyze the problem carefully. Decide exactly what you are being asked to do.

2. Determine what resources and references (if any) you will need in order to complete the problem. Collect those references and resources and have them readily available.

3. Decide if any particular standards apply to the project and have those standards available.

4. Determine what types of views will be required and how many of each.

5. Determine what the final plotted scales of the drawing will need to be, and select the appropriate paper size for plotting/printing.

6. Plan your drawing sequence. In what order will you develop the drawing (i.e., lines, features, dimension lines, leaders, dimensions)?

7. Review the various CAD commands you will have to use in order to develop the drawing.

8. Examine your CAD system to ensure that everything is in working order, then begin the project.

PROBLEMS 19-1 THROUGH 19-11

Follow the listed steps, using the latest drafting standards.

PROBLEM 19-1

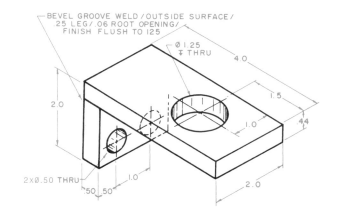

PROBLEM 19-2

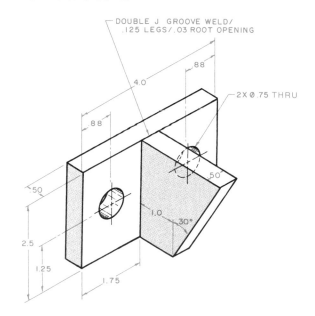

PROBLEM 19-3

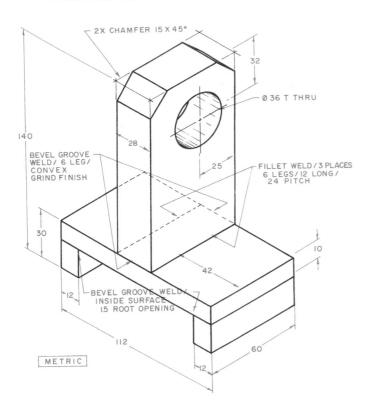

PROBLEM 19-5

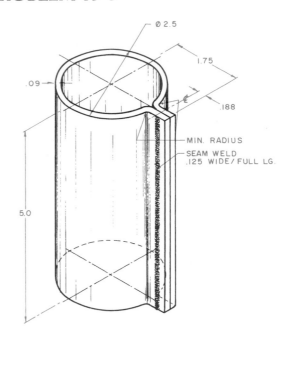

PROBLEM 19-4

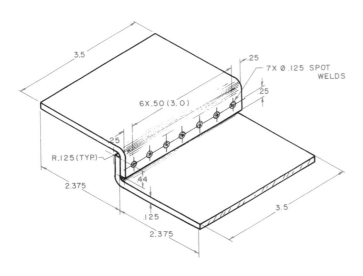

PROBLEM 19-6

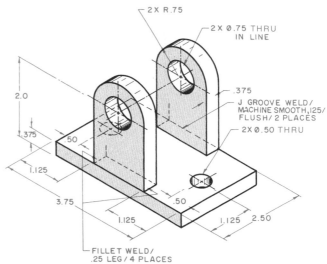

PROBLEM 19-7

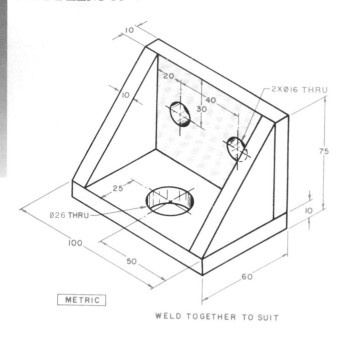

10
20
10
40
30
2X Ø16 THRU
75
25
Ø26 THRU
10
100
50
60

METRIC

WELD TOGETHER TO SUIT

PROBLEM 19-9

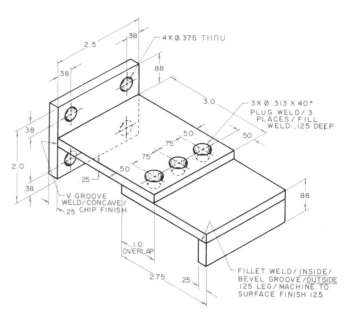

2.5
38
4X Ø.375 THRU
38
88
3.0
3X Ø.313 X 40°
PLUG WELD/ 3
PLACES / FILL
WELD .125 DEEP
38
50
75
50
2.0
75
50
38
25
V GROOVE
WELD/CONCAVE/
.25 CHIP FINISH
1.0
OVERLAP
88
2.75
.25
FILLET WELD/ INSIDE/
BEVEL GROOVE/OUTSIDE
.125 LEG/ MACHINE TO
SURFACE FINISH 125

PROBLEM 19-8

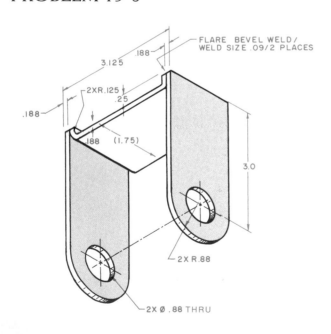

.188
3.125
FLARE BEVEL WELD/
WELD SIZE .09/2 PLACES
2XR.125
.25
.188
.188
(1.75)
3.0
2X R.88
2X Ø .88 THRU

PROBLEM 19-10

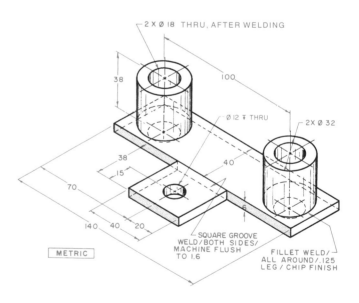

2X Ø 18 THRU, AFTER WELDING
38
100
Ø 12 ▼ THRU
2X Ø 32
38
15
40
70
6
140
40
20
SQUARE GROOVE
WELD/BOTH SIDES/
MACHINE FLUSH
TO 1.6
FILLET WELD/
ALL AROUND/.125
LEG / CHIP FINISH

METRIC

PROBLEM 19-11

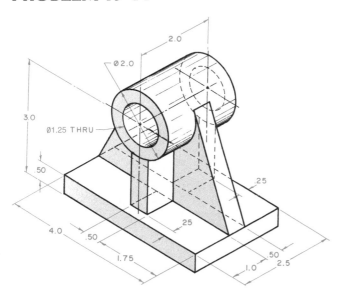

PROBLEMS 19-12 THROUGH 19-15

Convert the following problems from Chapter 5 from casting drawings to weldment drawings:

PROBLEM 19-12

Refer back to Problem 5-27.

PROBLEM 19-13

Refer back to Problem 5-28.

PROBLEM 19-14

Refer back to Problem 5-36.

PROBLEM 19-15

Refer back to Problem 5-37.

Chapter 20

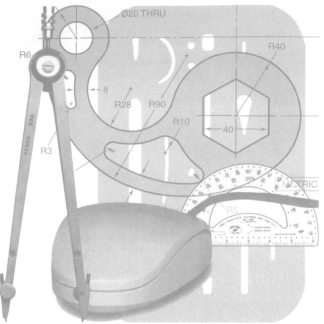

MODERN MANUFACTURING: MATERIALS, PROCESSES, AND AUTOMATION

KEY TERMS

Abrasive water jet machining (AWJM)

Alloy steel

Aluminum

Angle plate fixture

Angle plate jigs

Annealing

Arbor press

Band saw (or contour band machine)

Bending

Blanking

Boring mill

Box jig

Breakdown (or bender)

Broaches

Broaching machine

Built-up tool

Case hardening

Cast irons

Cast steels

Cast tool

Casting

Centrifugal casting

Ceramics

Channel jigs

Cheek

Chemical milling

Chuck

Clapper box

Closed jigs

Composites

Compound

Continuous path

Cope

Copper

Core prints

Cutoff saw

Cutting off

Cylindrical grinder

Die

Die casting

Draft

Drag

Drawing

Drill bushings

Drill press

Drop forging

Duplex fixture

Elastomers

Electrical discharge grinding (EDG)

Electrical discharge machining (EDM)

Electrical discharge wire cutting (EDWC)

Electrochemical discharge grinding (ECDG)

Electrochemical grinding (ECG)

Electrochemical honing (ECH)

Electrochemical machining (ECM)

Electrochemical turning (ECT)

Electromechanical machining (EMM)

Electron beam machining (EBM)

Engineering materials

External broaching

Extruding

Finishing impression

Fixed-renewable bushing

Fixtures

Flask

Foolproofing

Forging

Forming

Fuller

Gate

Green sand

Hardening

Heat treatment

High-performance alloys

Horizontal boring mills

Hurry-up-and-wait

Hydrodynamic machining (HDM)

Indexing fixture

Indexing jig

(Continued)

CHAPTER OUTLINE

Engineering materials • Traditional manufacturing processes • Casting • Forging • Extruding • Stamping • Machining • Special workholding devices • Heat treatment of steels • Nontraditional machining processes • Automation and integration (CAM and CIM/FMS) • Computer-aided manufacturing (CAM) • Industrial robots • CIM/FMS • Just-in-time manufacturing (JIT) • Manufacturing resource planning (MRPII) • Statistical process control (SPC) • Summary • Review questions

CHAPTER OBJECTIVES

Upon completion of this chapter, students should be able to do the following:

- Define the concept of engineering materials.
- Distinguish between the two broad categories of engineering materials.
- List the most widely used types of metals in modern manufacturing.
- List the most widely used nonmetals in modern manufacturing.
- Define the term *composites.*
- Describe the principal types of casting processes.
- Describe the most common types of forging processes.
- Define the term *machining.*
- Describe the most common machining processes.
- Explain the concept of computer-numerically-controlled machines.
- Explain the concept of jigs.
- Explain the concept of fixtures.
- Describe the most widely used nontraditional machining processes.
- Explain the concept of automation.

(Continued)

KEY TERMS

Internal broaching	Open jigs	Progressive die	Statistical process control (SPC)
Investment casting	Parting	Punching	Surface grinder
Jigs	Parting line	Quench	Table jigs
Laser beam machining (LBM)	Pattern	Refractory metals	Tempering
Lathe	Perforating	Renewable bushings	Template jigs
Lead	Photochemical machining	Repeatability	Tin
Lead time	Piercing	Riser	Tolerance
Leaf jig	Planers	Robot	Tool forces
Liner bushings	Plasma	Rolling	Trimming
Lost wax casting	Plasma arc machining (PAM)	Rotary ultrasonic machining (RUM)	Ultrasonic machining (USM)
Machining	Plastics	Sand core	Ultrasonically assisted machining (UAM)
Magnesium	Plate fixtures	Sandwich jigs	Upsetting
Manufacturing	Plate jigs	Shaper	Vertical boring mills
Maraging steel	Point-to-point	Shaving	Vertical turret lathe
Milling machine	Positioned	Shearing	Vise jaw fixture
Modified angle plate fixture	Powder metallurgy	Shrink rule	Waste
Modified angle plate jigs	Powdered metals	Shrinkage	Water jet machining (WJM)
Multistation jig	Power saw	Slip-renewable bushings	Welded tool
Nontraditional machining	Precision grinders	Slitting	Zinc
Normalizing	Press forging	Sprue hole	
Numerically controlled machine tools	Press-fit bushings	Stainless steels	
	Process synchronization	Stamping	

- Explain the concept of computer-aided manufacturing (CAM).
- Define the term *robot*.
- Explain the concept of computer-integrated manufacturing (CIM).
- Explain the concept of the flexible manufacturing system (FMS).

This chapter describes the processes used in a modern manufacturing setting to produce metal and nonmetal products as well as the engineering materials from which these products can be made. It also describes how these processes have been automated and integrated in an effort to improve quality and increase productivity.

The ultimate purpose of any engineering drawing is to provide the information necessary to make or fabricate an object. To achieve this purpose an engineering drawing must completely describe and detail the desired object. The drawing must show the specific size and geometric shape of the object as well as furnishing the related information concerning the material specifications, finish requirements, and any special treatments required.

To properly construct an engineering drawing, the designer must be thoroughly familiar with the materials and methods used to transform the drawn image into the actual object specified in the drawing. By knowing the materials and basic processes used to manufacture an object, the designer can design the object with the manufacturing processes in mind. This will not only make manufacturing easier but will often reduce the cost of a manufactured product. Therefore, the purpose of this chapter is to acquaint the new designer with the basic processes used to fabricate objects, as well as the capabilities and limitations of the machine tools used to machine these objects to their final form.

Engineering Materials

Manufacturing is the enterprise through which raw materials are converted into finished products. This conversion adds value to the raw materials. In a modern engineering and manufacturing setting these materials are referred to as *engineering materials*. Engineering materials can be divided into two broad categories: metals and nonmetals. Metals include steels, alloy steels, and a variety of high-performance alloys. Nonmetals include plastics, composites, elastomers, and ceramics.

METALS

Metals are widely used in modern manufacturing. The most frequently used metals are carbon steels; alloy steels; high-strength, low-alloy steels; stainless steels; maraging steels; cast steels; cast irons; high-performance alloys; tungsten, molybdenum, and titanium; aluminum; copper; magnesium; lead; tin; zinc; and powdered metals. These various metals are described in the following paragraphs.

CARBON STEELS

The main elements in steels are iron and carbon. Carbon steels are divided into three groups based on their carbon content. Low-carbon steels contain from .001 to .30% carbon. Medium-carbon steels contain from .70 to 1.30% carbon. High-carbon steels contain from .70 to 1.30% carbon. Certain grades of carbon steels also contain boron to improve hardenability and aluminum to control grain size and promote deoxidation. The following types of products are made of carbon steels: bar, sheet, strip, plate, wire, tubing, and structural shapes.

ALLOY STEELS

An *alloy steel* is any steel that contains one or more alloying elements added to produce characteristics not available in unalloyed carbon steels. Widely used alloying elements include manganese, silicon, copper, aluminum, chromium, cobalt, columbium, molybdenum, nickel, titanium, tungsten, vanadium, and zirconium. Such alloys are added to improve the mechanical and physical properties of steels.

HIGH-STRENGTH, LOW-ALLOY STEELS

These are steels that are significantly stronger than carbon steels as a result of the addition of small amounts of alloying elements and special processing. These alloying elements include manganese, silicon, phosphorus, copper, aluminum, chromium, niobium, vanadium, titanium, molybdenum, nickel, zirconium, nitrogen, and calcium. These steels are categorized according to composition. The different categories vary as to strength, toughness, formability, weldability, and corrosion resistance.

STAINLESS STEELS

Stainless steels are steels that contain 10.5% or more chromium. The special appearance of stainless steels is the result of a chromium-based oxide film that coats the metal's surface. Stainless steels also contain a number of other elements that are added to improve their machinability and corrosion resistance. These elements include nickel, molybdenum, copper, titanium, silicon, manganese, columbium, aluminum, nitrogen, and sulfur. There are almost 60 standard grades of stainless steel, but there are five main types. These types are austenitic, ferritic, martensitic, prescription-hardening, and duplex.

MARAGING STEELS

Maraging steels are a special class of high-strength steels in which nickel is the primary alloy. The name *maraging* is derived from the term *martensite age hardening*, which is a process of age hardening of a matrix comprised of low-carbon, iron-nickel martensite. The process used to produce maraging steels is a double-vacuum melting process. The first process involves vacuum-induction melting. The second process involves vacuum arc remelting. This double-vacuum process maintains a high level of purity and reduces residual elements. Maraging steels are produced in bar, plate, and sheet form.

CAST STEELS

Cast steels are steels that are produced in ingot form by melting all component elements together and pouring the molten metal into a mold. This is an intermediate step. Cast steel ingots are subsequently hot or cold worked into another shape. Alloying elements include nickel, chromium, vanadium, and copper. The main types of cast steels are carbon, low-alloy, corrosion-resistant, and heat-resistant.

CAST IRONS

Cast irons are metals in a family of cast ferrous metals. Cast irons are distinguished from steels by the relative carbon content (i.e., less than 2% in steels and over 2% in cast irons). In addition, cast irons also contain from 1% to 3% silicon. These differences are the basis of the corresponding differences in how steels and cast irons can be used. The metallurgical properties of cast irons make them easy to melt and cast. These are the following basic types of cast irons: white, malleable, gray, ductile, and compacted-graphite iron.

HIGH-PERFORMANCE ALLOYS

High-performance alloys are a group of metals that have high strength and/or high corrosion resistance over a broad range of temperatures. High-performance alloys are primarily iron-nickel, nickel, cobalt, or iron-based. They typically have a complex composition with at least three and sometimes over 10 component elements. Also known as the super-alloys, their most common elements include chromium, molybdenum, tungsten, columbium, aluminum, cobalt, titanium, carbon, and nitrogen. The most common applications of high-performance alloys are in aircraft and aerospace settings.

TUNGSTEN, MOLYBDENUM, AND TITANIUM

Tungsten and molybdenum are known as *refractory metals* because of their exceptional ability to withstand high temperatures. They also have the following additional characteristics: low coefficient of expansion, high modulus of elasticity, high resistance to thermal shock, high conductivity, and excellent hardness properties at high temperatures. Unlike tungsten and molybdenum, titanium is not a refractory metal, but its properties are similar to such metals. The processes used to produce these

three metals are similar. They are extracted from ore, processed into chemicals, and processed further into powders or sponges. Typical applications of tungsten and molybdenum are the automotive, aerospace, mining, chemical, processing, electrical, nuclear, petroleum processes, ordnance, and metal-processing industries. Titanium is used primarily in the aerospace, chemical, and medical industries.

ALUMINUM

Aluminum is a bright, silvery metal that is very light. A given amount of ferrous metal will weigh three times as much as the same amount of aluminum. Aluminum is an abundant metal, accounting for approximately 8% of the earth's outer surface. Pure aluminum is ductile, soft, and weak compared with other metals. However, with the addition of alloying elements such as copper, magnesium, manganese, and zinc, it can be made much harder and stronger. Aluminum is light, an excellent conductor, and able to resist corrosion. Consequently, it is widely used in both manufacturing and construction.

COPPER

Copper is a heavy metal that is salmon pink in its pure form and abundant. It is nonmagnetic and has excellent conductivity. Only silver is a better conductor. There are over 300 different types of copper and copper alloys. They are produced in a variety of forms including strip, plate, sheet, pipe, tube, rod, forgings, wire, foil, extrusions, and castings. Copper is widely used in both manufacturing and construction.

MAGNESIUM

Magnesium is a soft, silvery-white metal that lacks the strength needed for structural uses. However, it is extremely light and machinable. To increase its strength, magnesium is alloyed with aluminum, manganese, lithium, thorium, zinc, and zirconium. The result is a metal that can be machined at high speeds, has good conductivity, a high level of fatigue, and is dent resistant.

LEAD

Lead is a dark grey metal that has the following characteristics: high density, low melting point, corrosion resistance, chemical stability, malleability, lubricity, and electrical properties. Lead can be easily produced in a variety of forms.

TIN

Tin is a silvery-white metal that is soft and ductile. Its high level of malleability allows tin to be drawn into wire or hammered into extremely thin sheets. Tin is too weak to be used in its pure form, but alloyed with antimony, copper, lead, zinc or silver, tin becomes useful. Its most common use is in coating steel containers used to preserve food. Tin is also used in solder alloys, bronzes, fusible alloys, type metals, pewter, and dental amalgams. Tin chemical compounds are used in the electroplating, ceramics, plastics, pesticide, and antifungal industries.

ZINC

Zinc is a bluish-white metal that is widely used in manufacturing. Only iron, aluminum, and copper are more widely used than zinc. Zinc is produced in castings and is widely used for coating other materials. Zinc alloys offer several advantages over other die-cast metals, including ease of casting, strength, ductility, and the ability to hold close tolerances and to be cast in thin sections.

POWDERED METALS

Powdered metals are exactly what the name says. They are produced by atomization, electrolysis, chemical reduction, or oxide reduction. Various powdered metals are mixed and sintered to produce alloys with an almost endless number of compositions. Sintering produces bonds among the various powdered particles. The properties of products made from powdered metals depend on which metals are mixed, in what amounts, and by what processes. These characteristics can be controlled by varying particle size, particle shape, size distribution, sieve analysis, bulk density of the powder, rate of flow into the die, and the compressibility of the powder in the die.

NONMETALS

Nonmetals are widely used in modern manufacturing. The most widely used nonmetals are plastics, composites, ceramics, and elastomers. These various nonmetals are described in the following paragraphs.

PLASTICS

Plastics are nonmetal materials that are made from either natural or synthetic resins. Most plastics used in modern manufacturing are made from synthetic resins. There are many different kinds of plastics, including polystyrene, high-impact polystyrene, acrylics, poly-carbonate, ABS, acetal, nylon, polypropylene, polyethylene, epoxy, and phenolic. Most plastics used in an engineering or manufacturing setting are made from thermoplastic resins. Such plastics have the following characteristics: (a) thermal, mechanical, chemical, corrosion resistance, and fabricability; (b) ability to sustain high mechanical loads; and (c) predictable performance.

Thermosets. Thermosets are a special subset of engineering plastics that are used in applications where resistance to heat and environmental conditions is important. For example, thermosets have long been used for insulation in electrical and electronics applications. The recipes for thermoset molding compounds can be varied to create just the right characteristics for a given application. Distinctive properties of thermosets include dimensional stability, low-to-zero creep, low water absorption, maximum physical strength, good electrical properties, high heat deflection temperatures, high heat resistance, minimum thermal expansion, and specific gravities in the 1.35 to 2.00 range.

Thermoplastics. Thermoplastics are a special subset of engineering plastics that are very similar to thermosets. However, processing advantages have allowed thermoplastics to replace thermosets in many applications. The benefits of thermoplastics over thermosets are faster molding, lighter weights, ability to have thinner walls that allow for more complex designs, and greater impact resistance. However, thermosets typically experience less creep at higher temperatures.

COMPOSITES

Composites are materials created by combining two or more materials. Perhaps the most readily recognizable composite is plywood. Fiberglass is another example of a composite. Composite technology is divided into four broad areas: organ matrix composites (resins), metal matrix composites, carbon-carbon composites, and ceramic matrix composites. Composites can be divided into two categories in terms of how they are produced; laminates and sandwiches. Laminates consist of two or more layers bonded together. Sandwiches consist of thick, low-density layers (i.e., honeycomb or foam materials) pressed between thin layers of higher strength, higher density material. Applications of composites are extensive and varied. They include aircraft and aerospace applications, athletic equipment, automobiles, prosthetic devices, printed circuit boards, boats, and many others.

CERAMICS

Ceramics are materials made by combining metallic and nonmetallic elements such as oxides, carbides, and nitrites. Glass, bricks, tile, and porcelain are all common ceramics. Refractory ceramics are important in modern manufacturing because of their usefulness in electronics applications. They are used in thermistors, rectifiers, capacitors, transducers, and for high-voltage insulation. Characteristics of ceramics include the fact that they are hard, brittle, have a high melting point, have low conductivity, are chemically and thermally stable, have high compressive strength, and have high creep resistance.

ELASTOMERS

The most recognizable form of elastomer is natural rubber. *Elastomers* are linear polymers that are able to stretch and return to their original shape and size (i.e., a rubber band). Widely used elastomers include polyacrylate, ethylene propylene, neoprene, polysulfide, silicone, and urethane. Common applications of elastomers include oil hoses, o-rings, electric insulation, belts, gaskets, seals, value disks, and foam padding.

Traditional Manufacturing Processes

The term *manufacturing processes* refers to the basic methods used by the shop to make the object described in the engineering drawing. The specific processes used to make the object depend on the object itself. In some cases the part may be cast, while other parts may be forged, extruded, or stamped. In many instances the part, once fabricated into its basic form, must also be machined to maintain a specific degree of accuracy or to produce a feature not possible with other manufacturing processes.

When the manufacturing shop receives the engineering drawing, in the form of a part print, it is first reviewed to ensure all pertinent data and information necessary to make the object are contained in the drawing. During this review, the shop personnel will consider several factors necessary to determine how the part must be made. These factors include: the type and condition of the material used to make the object, the overall size and shape of the part, the types of operations required, the required accuracy of the part, and the number of parts to be made.

The type and condition of the material used to make the part are important considerations. Some parts may be made from solid bar stock while other parts must be extruded, cast, forged, or stamped. In most cases, parts made from bar stock require the least lead time. The *lead time* is the interval from the time the shop receives the drawing until production begins. Most shops maintain a sufficient supply of bar stock to begin production as soon as the drawing is received. However, when a part must be extruded, cast, forged, or stamped, a longer lead time is required to make the necessary molds or dies to fabricate the object.

The size and shape of the object must be considered to determine if the object is within the capabilities of the shop to make. In addition, the size and shape may also determine the size of the machine tools required as well as the datum, or reference, surfaces used to locate the part during manufacture.

The types of operations required determine the types of equipment and machine tools needed to make the part. If, for example, the part requires holes, a drill press, vertical milling machine, or other machine tool may be used. The next consideration, the required accuracy of the part, also determines how the operations are performed. A hole with a required accuracy of .002″ (0.05 mm), for example, would require reaming, while a hole with a required accuracy of .020″ (0.5 mm) could be drilled.

The number of parts to be made will frequently determine if the part should be cast, forged, extruded, or machined from solid stock. Likewise, the number of parts will also determine if any special workholders are to be made. Larger production runs will normally justify more sophisticated tools and processes since the cost can be spread over a larger number of parts. Smaller production runs normally demand the parts be made at the lowest possible cost with little or no investment in special molds, dies, or workholders.

While each of these factors has been separated for the purpose of this discussion, in practice each is considered as part of the other. These factors are so closely related in production that they frequently overlap, and one cannot be considered without the others.

The primary processes used to fabricate manufactured products are casting, forging, extruding, stamping, and machining. To use these processes to their best advantage, the designer must be familiar with the strengths and weaknesses of each process as well as the fundamental aspects of each process.

Casting

Casting is a process of pouring molten metal into a mold that contains the desired shape in the form of a cavity. The principal types of casting used in manufacturing today are sand casting, investment casting, centrifugal casting, and die casting.

SAND CASTING

Sand casting is the most common type of casting method. The major components of the molds used to make sand castings are shown in *Figure 20-1* and include the *flask, pattern,* and *green sand*. The flask is a two-part box, or frame, used to contain the sand. The top half of the flask is called the *cope* and the bottom half is called the *drag*. Occasionally, a third section may be installed between the cope and drag. This section is called a *cheek*, and is used where a deep or complex

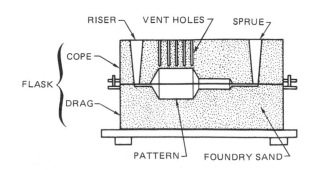

FIGURE 20-1 *Components of a sand casting mold* (From Thode, *Materials Processing*, Delmar Publishers Inc.)

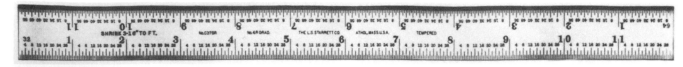

shape must be cast. In sand casting the mold is prepared by ramming the green sand around a model of a part, called the pattern. The model is then removed to form the cavity for the molten metal. The plane of division between the cope and drag is called the *parting line*. In many castings, the parting line occurs at the approximate middle of the part. The parting line can be seen on most castings by a ragged line that is usually ground off.

The molten metal enters the mold through the *sprue hole* and is directed to the cavity by one or more *gates*. The sprue hole is formed by installing a sprue peg in the cope. This peg is then removed after the final ramming of the cope. The *riser* is used to vent the mold and to allow gases to escape. The riser also acts as a small reservoir to keep the cavity full as the metal begins to shrink during the cooling process.

Once the cope and drag are rammed and the pattern is removed, a solid part could be poured in the mold. However, in some cases a hollow part, or one that has large holes, must be poured. To reduce the amount of material needed to fill the cavity and to reduce the time necessary to machine the part, a *sand core* may be installed in the cavity. When sand cores are intended to be installed in a cavity, *core prints* must also be provided to locate and anchor the sand cores during the casting process. Core prints are normally a simple extension of the pattern.

Making Patterns. When patterns are made for casting, two important factors must always be considered. The first is the draft. *Draft* is the slope or taper of the sides of a pattern that permits it to be removed from the cope and drag without disturbing the cavity. The draft also permits the cast part to be removed easily from some molds. The amount of draft necessary will normally depend on the part being cast, but will normally be about 1°.

The second consideration is shrinkage. When metals are cast, a certain amount of *shrinkage* occurs as the metal cools. The specific amount of shrinkage depends on the metal being cast. Steel, for example, shrinks at a rate of approximately 3/16″ per foot, while cast iron shrinks about 1/8″ per foot. To allow for this shrinkage and to make sure the final part is the correct size, the pattern must be made slightly larger than the final size of the desired part.

When making a pattern, the pattern maker will normally use a shrink rule to compensate for the shrinkage.

A *shrink rule*, *Figure 20-2*, is a standard steel rule that has the graduations marked for a specific amount of shrinkage. A shrink rule used for cast iron has an extra 1/8″ added to each foot, while a shrink rule for steel has an additional 3/16″ added.

In most instances, when a pattern must be made, the pattern maker will receive the engineering drawing that contains all the final sizes of the part. The pattern maker will then make all the necessary calculations needed to make the oversized pattern. But, occasionally, the pattern maker will be given a drawing with all the calculations already made to produce the pattern. In either case, the patterns must be made to suit the material being cast.

INVESTMENT CASTING

Investment casting produces parts with great detail and accuracy, while at the same time allowing very thin cross sections to be cast. In this process, the pattern is made by casting wax in the desired form. The pattern is then placed in a sand mold and the mold is fired to melt out the wax pattern. This process is also referred to as *lost wax casting*. The molten metal is then fed into the mold to produce the cast part.

CENTRIFUGAL CASTING

Centrifugal casting, *Figure 20-3*, is a process of pouring a measured amount of molten metal into a rotating mold. This process can be used for a single mold or multiple

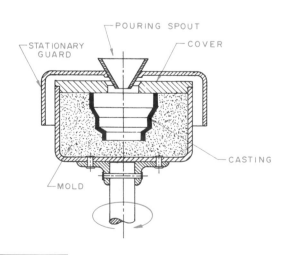

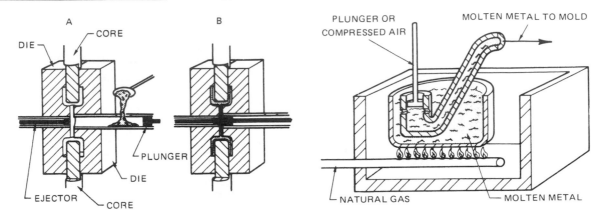

FIGURE 20-4   *Die casting* (From Thode, *Materials Processing,* Delmar Publishers Inc.)

molds. The centrifugal force created by the rotation forces the molten metal to fill the cavity. This same centrifugal force, along with the measured amount of material, also controls the wall thickness of the cast part and results in a less porous cast surface than is possible with sand casting. The molds used for this purpose are usually permanent molds made from metal rather than sand. These molds are used repeatedly and produce highly accurate parts requiring very little machining. However, due to their cost, permanent molds are normally used only for high-volume production.

DIE CASTING

Die casting, **Figure 20-4,** is a process in which molten metal is forced into metal dies under pressure. This process is very well suited for such materials as zinc, aluminum, copper, and magnesium alloys. Parts produced by die casting are superior in appearance and accuracy and require little or no machining to final size.

POWDER METALLURGY

Powder metallurgy, **Figure 20-5,** while not an actual casting process, does have some similarities to casting. In the *powder metallurgy* process, metal particles, or powder, are blended and mixed to achieve the desired composition. The powder is then forced into a die of the desired form under pressure from 15,000 to 100,000 pounds per square inch. The resulting heat fuses the powder into a solid piece that can be machined.

Forging

Forging is the process of forming metals under pressure using a variety of different processes. The most common forms of forging are drop forging and press forging. Other variations of forging include rolling and upsetting.

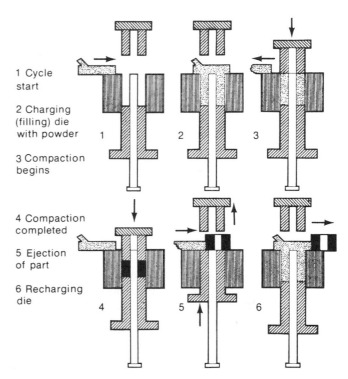

1 Cycle start

2 Charging (filling) die with powder

3 Compaction begins

4 Compaction completed

5 Ejection of part

6 Recharging die

FIGURE 20-5   *Producing parts with powder metallurgy* (Courtesy Metal Powder Industry Foundation)

DROP FORGING

Drop forging is the process of forming a heated metal bar, or billet, in dies. In practice, the heated metal is placed on a lower portion of a forging die and struck repeatedly with the upper die portion. This forces the metal into the shape of the cavity of the dies. The pressure required to form the metal is produced by a drop hammer.

Most drop forge dies contain at least four different stations, or dies, to complete the part, *Figure 20-6.* The first station is called the *fuller.* This station is used to rough-form the part to fit the other cavities. The second station, when used, is called the *breakdown,* or *bender.* This station forms the part into any special contours required by

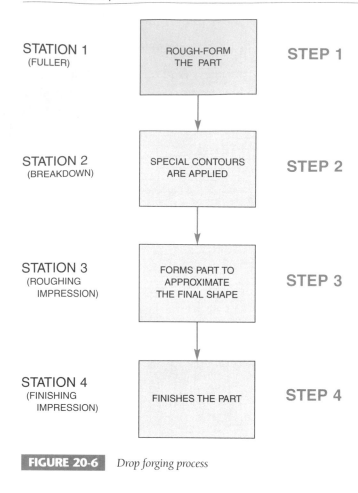

STATION 1 (FULLER) — ROUGH-FORM THE PART — STEP 1

STATION 2 (BREAKDOWN) — SPECIAL CONTOURS ARE APPLIED — STEP 2

STATION 3 (ROUGHING IMPRESSION) — FORMS PART TO APPROXIMATE THE FINAL SHAPE — STEP 3

STATION 4 (FINISHING IMPRESSION) — FINISHES THE PART — STEP 4

FIGURE 20-6 *Drop forging process*

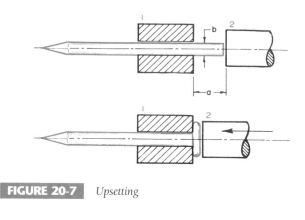

FIGURE 20-7 *Upsetting*

the next station, the roughing impression. The roughing impression rough-forms the part to the desired form. The final station is called the *finishing impression* and is used to finish the part to the desired shape and size. Additional stations, such as a trimmer to remove the fins and a cut-off to sever the forged part from the bar, may also be included on the die if they are necessary.

PRESS FORGING

Press forging is a single-step process in which the heated metal bar or billet is forced into a single die and is com-pleted in a single press stroke. The press normally used for press forging is hydraulic.

ROLLING

Rolling is a process in which the heated metal bar is formed by passing through rollers having the desired form impressed on their surfaces. Rolling is also fre-quently used to flatten and thin out thick sections.

UPSETTING

Upsetting, **Figure 20-7**, is a process used to enlarge selected sections of a metal part. Bolt heads, for example, are normally produced by an upsetting process.

Extruding

Extruding is a process of forcing metal through a die of a desired form and cross section. The bars produced are then cut to the required lengths. Typical examples of extruded parts include parts for aluminum windows and doors. Extrusions provide a nearly final shape, as shown in **Figure 20-8** and, in many cases, only need to be cut off and machined slightly to complete a finished part.

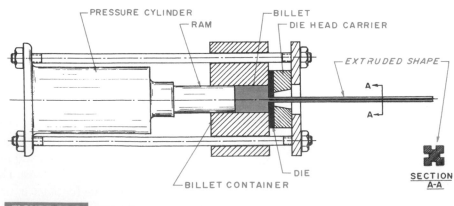

FIGURE 20-8 *Extruding*

Stamping

Stamping is a process of using dies to cut or form metal sheets or strips into a desired form. The main tool used for metal stamping is a die. The term *die* has a double meaning. It can be used to describe the entire assembled tool or the lower cutting part of the tool. The exact meaning of the term can usually be determined from the context of its use. The principal parts of a die are shown in *Figure 20-9*. The upper die shoe is used to mount the punch. The die is mounted on the lower die shoe. The guide pins and guide pin bushings are used to maintain the alignment between the punch and die. The stripper is designed to strip the stock material from the punch after the cutting stroke of the die.

When more than one operation takes place on a part during a single stroke, blanking and punching, for example, the operation is said to be *compound*. However, when several operations take place sequentially in a die, the die is called a *progressive die*.

The principal stamping operations normally performed are shearing, cutting off, parting, blanking, punching, piercing, perforating, trimming, slitting, shaving, bending, forming, drawing, coining, and embossing.

OPERATIONS THAT PRODUCE BLANKS

The operations that produce blanks are shearing, cutting off, parting, and blanking, *Figure 20-10*. *Shearing* is a cutting action performed along a straight line. *Cutting off* is a cutting operation performed on a part to produce an edge other than a straight edge. In many cases, cutting off is used to finish the edges of a part. Like shearing, cutting off does not produce any scrap. *Parting,* on the other hand, is also an operation used to finish the edges of a part, but, unlike the other operations, parting does produce scrap.

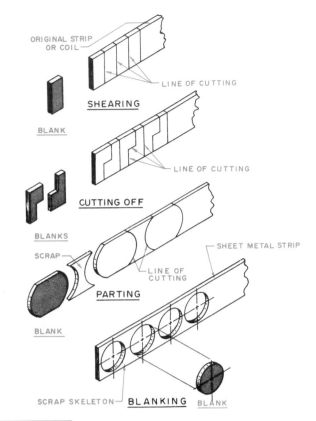

FIGURE 20-10 *Operations that produce blanks*

Blanking is an operation that produces a part cut completely by the punch and die. In blanking operations, the piece that falls through the die is the desired part. The scrap is the skeleton left on the stock strip.

OPERATIONS THAT PRODUCE HOLES

The operations that produce holes are punching, piercing, and perforating. *Punching* is an operation that cuts a hole in a metal sheet or strip, *Figure 20-11A*. In punching, the piece that falls through the die is scrap, and the area on the strip is the desired part. Punching is normally performed on parts that are to be blanked to provide holes for bolts, screws, or similar parts.

Piercing is an operation similar to punching, except that in piercing no scrap is produced, *Figure 20-11B*. A pierced hole is sometimes used to increase the thickness of metal around a hole for tapping threads in a stamped part. *Perforating* is simply a punching operation performed on sheets to produce either a uniform hole pattern or a decorative form on the sheet.

OPERATIONS THAT CONTROL SIZE

The operations used to control size are trimming, slitting, and shaving, *Figures 20-12A* and *20-12B*. *Trimming* is an operation performed on formed or drawn parts to remove the ragged edge of the blank. *Slitting* is

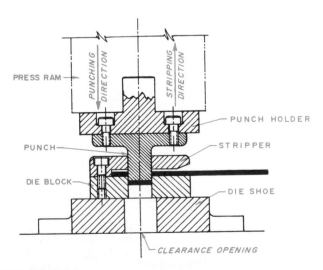

FIGURE 20-9 *Metal stamping die*

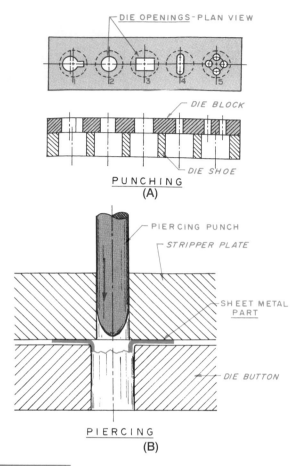

FIGURE 20-11 (A) Punching operation to produce holes;
(B) Piercing operation to produce holes

performed on large sheets to produce thin stock strips. Slitting may be performed by rollers or by straight blades. *Shaving* is an operation performed on a blanked part to produce an exact dimensional size or to square a blanked edge. Shaving produces a part with a close dimensional size, and a cut edge that is almost perpendicular to the top and bottom surfaces of the blanked part.

OPERATIONS THAT BEND OR FORM PARTS

The operations used to bend or form parts are bending, forming, and drawing, *Figure 20-13*. *Bending* is an operation in which a metal part is simply bent to a desired angle. *Forming*, on the other hand, is an operation in which a part is bent or formed into a complex shape. Forming is also a catchall word frequently used to describe any bending operation that does not fall into one of the other categories. *Drawing* is a process of stretching and forming a metal sheet into a shape similar to a cup or top hat.

Machining

Machining is the process of removing metal with machine tools and cutters to achieve a desired form or feature. The principal types of machine tools used to perform these operations are lathes, milling machines, drill presses, saws, and grinders. Other machines that are variations of the basic machines used in the machine shop include shapers and planers, boring mills, broaching machines, and numerically controlled machines.

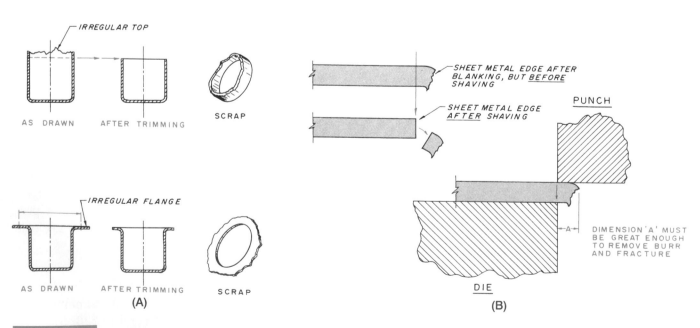

FIGURE 20-12 (A) Operation that controls size-trimming; (B) Operation that controls size-shaving

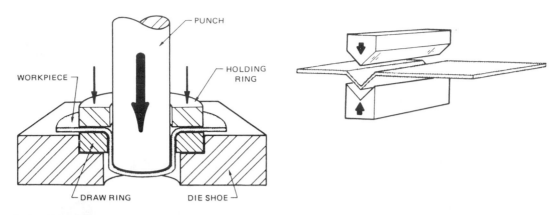

FIGURE 20-13 *Operations that bend or form parts* (From Thode, *Materials Processing,* Delmar Publishers Inc.)

The specific type of machine used to perform a particular machining operation is determined by the type of operation and the degree of accuracy required.

LATHES

A lathe, *Figure 20-14,* is one of the most commonly used and versatile machine tools in the shop. Typically, a *lathe* performs such operations as straight and taper turning, facing, straight and taper boring, drilling, reaming, tapping, threading, and knurling, *Figure 20-15.*

In operation, the workpiece is held in the headstock and supported by the tailstock. The cutting tool is mounted in the tool post and is traversed past the rotating workpiece to machine the desired form, *Figure 20-16.* The most common method used to hold parts in a lathe is with a chuck. A *chuck* is a device much like a vise, having movable jaws that grip and hold the workpiece. The most popular type of chuck is the three-jaw chuck, but there are also two-jaw, four-jaw, and six-jaw chucks

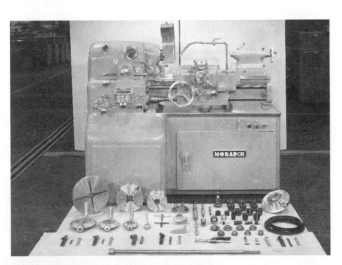

FIGURE 20-14 *Tools for turning* (Courtesy Monarch Co.)

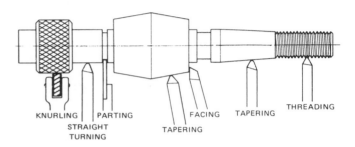

FIGURE 20-15 *Basic lathe operations* (From Thode, *Materials Processing,* Delmar Publishers Inc.)

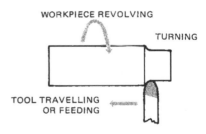

FIGURE 20-16 *Lathe setup for straight turning* (From Krar and Oswald, *Turning Technology,* Delmar Publishers Inc.)

as well. Other types of devices used to hold and drive a workpiece in a lathe include collets, face plates, drive plates, and lathe dogs.

MILLING MACHINES

Milling machines are used to machine workpieces by feeding the part into a rotating cutter. The two basic variations of the milling machine are the vertical milling machine and the horizontal milling machine, *Figure 20-17.*

In operation, the milling cutter is mounted in either the spindle or arbor, and the workpiece is held on the machine table. The most commonly used device to mount the workpiece for milling is the milling machine vise, but the workpiece may also be held directly on the

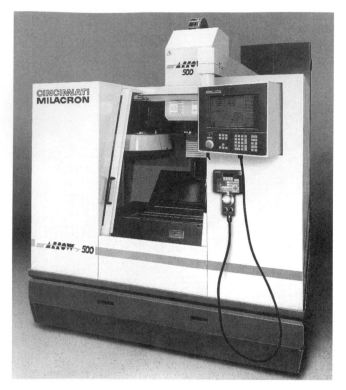

FIGURE 20-17 *CNC milling machine* (Courtesy Cincinnati Milacron)

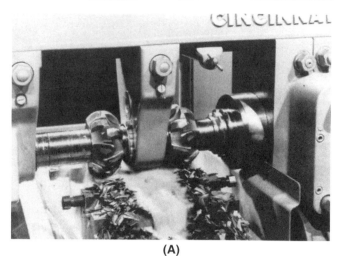

(A)

(B)

FIGURE 20-19 *(A) and (B) Standard milling cutters* (Courtesy Cincinnati Milacron)

CONVENTIONAL MILLING   CLIMB MILLING

FIGURE 20-18 *Basic milling machine operations* (From Thode, *Materials Processing,* Delmar Publishers Inc.)

machine table or in other workholding devices. The operations most frequently performed on a milling machine include plain milling, face milling, end milling, straddle milling, gang milling, form milling, keyseat milling, and gear cutting, *Figure 20-18. Figures 20-19A* and *20-19B* show two types of cutters commonly used for milling operations.

DRILL PRESSES

A *drill press, Figure 20-20,* is a machine that is used mainly for producing holes. The principal types of drill presses used in the shop are the radial drill press and the sensitive drill press.

The operations normally performed on drill presses include drilling, reaming, tapping, chamfering, spot-

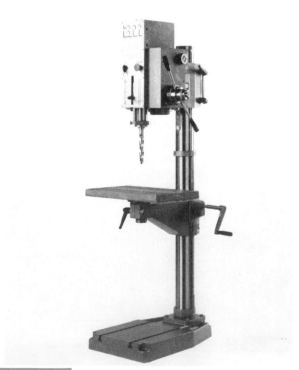

FIGURE 20-20 *Drill press* (Courtesy Wilton Corp.)

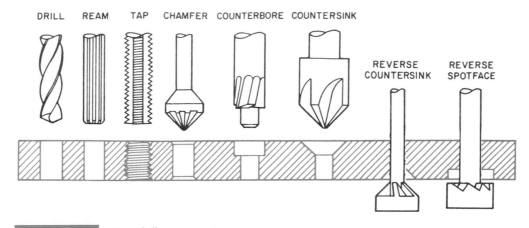

FIGURE 20-21 *Basic drill press operations*

facing, counterboring, countersinking, reverse countersinking, and reverse spotfacing, *Figure 20-21.*

When using a drill press, the workpiece is normally held in a vise or clamped directly to the machine table. The drill or other cutting tool is mounted in the machine spindle and is fed into the workpiece either by hand or with a mechanical feed unit.

POWER SAWS

The *power saws* used most often in the machine shop are the contour band machine or band saw and the cutoff saw, *Figure 20-22.* The contour band machine or *band saw* is used primarily for sawing intricate or detailed shapes; the *cutoff saw* is used to cut rough bar stock to lengths suitable for machining in other machine tools.

PRECISION GRINDERS

Precision grinders are available in several styles and types to suit their many and varied applications. The principal types of precision grinders used in the machine shop are the surface grinder and the cylindrical grinder, among others.

The *surface grinder, Figure 20-23,* is mainly used to produce flat, angular, or special contours on flat workpieces. The most popular variation of this machine consists of a grinding wheel mounted on a horizontal spindle and a reciprocating table that traverses back and forth under the grinding wheel. One other variation of the surface grinder frequently found in the machine shop uses a vertical spindle and a round, rotating table, *Figure 20-24.* The workpiece is generally mounted and held on a magnetic chuck during the grinding operation on both styles of surface grinders.

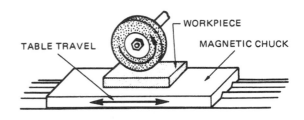

FIGURE 20-23 *Surface grinder* (From Thode, *Materials Processing*, Delmar Publishers Inc.)

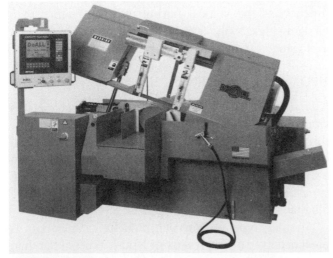

FIGURE 20-22 *CNC machine tool* (Courtesy DoAll Corp.)

FIGURE 20-24 *Vertical spindle-rotating table surface grinder* (From Thode, *Materials Processing*, Delmar Publishers Inc.)

Cylindrical grinders are used to precisely grind cylindrical or conical workpieces. For grinding, the workpiece is mounted either between centers or in a precision chuck, much like a lathe. The grinding wheel is mounted behind the workpiece on a horizontal spindle. The table of the grinder traverses the workpiece past the grinding wheel in a reciprocating motion, while the workpiece rotates at a preset speed.

SHAPERS AND PLANERS

Shapers and *planers* are machine tools that use single-point cutting tools to perform their cutting operations. The shaper uses a reciprocating ram to drive the cutter. The cutting tool on this machine tool is mounted on the end of the ram in a unit called the *clapper box.* The clapper box can be positioned vertically or at an angle, depending on the work to be performed. The depth of cut is adjusted by lowering the vertical slide of the clapper box or by raising the position of the table. The workpiece is mounted on the table or vise and is fed past the reciprocating cutter by the table feed. The length and position of the stroke are regulated by the position of the ram and are easily adjusted to suit the size of the workpiece.

Planers operate in a manner similar to the shaper. The major differences between these two machines are their size and method of cutting. The planer is normally much larger than the shaper and the work, rather than the cutter, moves on the planer. In operation, the workpiece is mounted on the table of the planer, and the table reciprocates under the stationary tool. As the table reciprocates, the tool is moved across the workpiece by the feed unit. The depth of cut is determined by the height of the tool from the table and is adjusted by lowering or raising the cross beam. The length and position of the cut are determined by the position of the table.

These processes are seldom used today.

BORING MILLS

A *boring mill,* **Figures 20-25** and **20-26,** is a machine tool normally used to machine large workpieces. The two common variations of these machines are horizontal and vertical. The distinction between the two is determined by the position of the spindle. *Horizontal boring mills* perform a wide variety of different machining tasks normally associated with a milling machine but on a much larger scale. *Vertical boring mills* are commonly used to turn, bore, and face large parts in much the same way as a lathe. Another variation of the horizontal boring mill sometimes found in the machine shop is the *vertical turret lathe.* This machine serves the same basic

FIGURE 20-25 *Boring mill* (Courtesy Cincinnati Milacron)

FIGURE 20-26 *CNC boring mill* (Courtesy Cincinnati Milacron)

function as the vertical boring mill but, rather than using a single tool, a vertical turret lathe uses a turret arrangement to mount and position the cutting tools.

BROACHING MACHINES

A *broaching machine* is used to modify the shape of a workpiece by pulling tools called *broaches* across or through the part. Both internal and external forms can be broached. *Internal broaching* can produce holes with a wide variety of different forms, **Figure 20-27.** *External broaching* is typically used to produce some gear teeth, plier jaws, and other similar details.

Another type of internal broaching operation is broached in an *arbor press* using a push-type, rather than

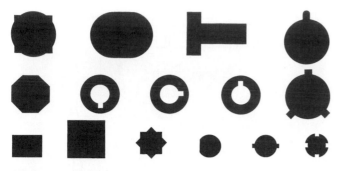

FIGURE 20-27 *Typical internal broached shapes* (Courtesy The DuMont Corp.)

FIGURE 20-28 *CNC machining center* (Courtesy Cincinnati Milacron)

a pull-type broach. This type of broaching is often used in the top to produce keyways or other simple shapes.

NUMERICALLY CONTROLLED MACHINES

Numerically controlled machine tools represent the norm in machine tool design in wide use today. These machine tools are operated by computers, and they can markedly reduce the errors caused by human operators.

The two basic variations of these machines in use today are the point-to-point and the continuous path machines. The principal difference between the two is in the movements of the tool with reference to the workpiece. *Point-to-point* machines operate on a series of pre-programmed coordinates to locate the position of the tool. When the tool finishes at one point, it automatically goes to the next point by the shortest route. This type of control is very useful for drilling machines, but machines such as milling machines require greater control of the tool movement. So, *continuous path* controls are used to control the movement of the tool throughout the cutting cycle. For example, if a circle or radius were to be milled, the operator or programmer would need only to program a few points along the arc. The computer would compute the remaining points and guide the cutting tool throughout the complete cycle.

These controls are frequently used for a wide range of different machine tools, from drill presses and milling machines to lathes and grinders. Due to the advent of these controls, a whole new form of machine tool has evolved: the machining center, *Figure 20-28.* These machines can take a rough, unmachined part and completely machine the whole part without removing it from the machine or changing a setup.

Special Workholding Devices

Special workholding devices are frequently used to produce parts in high-volume production runs. The major

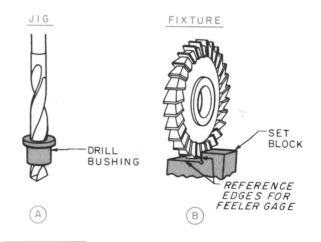

FIGURE 20-29 *Referencing the tool to the workpiece*

categories of special workholding devices commonly used in the shop are jigs and fixtures. The primary function of jigs and fixtures is to transfer the required accuracy and precision from the operator to the tool. This permits duplicate parts to be produced within the specified limits of size without error. Both jigs and fixtures hold, support, and locate the workpiece; the principal difference between these tools is the method used to control the relationship of the tool to the workpiece. *Jigs* guide the cutting tool through a hardened drill bushing during the cutting cycle. *Fixtures,* on the other hand, reference the cutting tool by means of a set block, *Figure 20-29.*

CLASSIFICATION OF JIGS

Jigs are normally classified by the type of operation they perform and their basic construction. Typically, jigs are used to drill, ream, tap, countersink, chamfer, counterbore, and spotface. Jigs are also divided into two

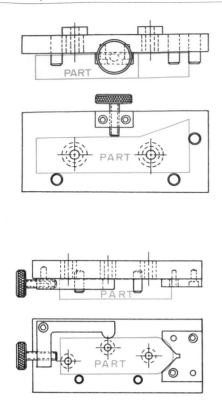

FIGURE 20-30 *Plate jig*

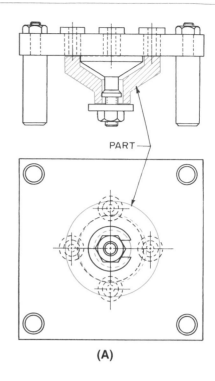

(A)

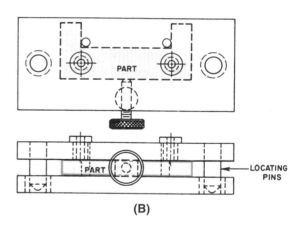

(B)

FIGURE 20-31 *(A) Table jig; (B) Sandwich jig*

general construction categories: open and closed. *Open jigs* are jigs that cover only one side of the part and are used for relatively simple operations. *Closed jigs* are jigs that enclose the part on more than one side and are intended to machine the part on several sides without removing the part from the jig.

The most common types of jigs include plate jigs, angle plate jigs, box jigs, and indexing jigs. While there are several other distinct styles of jigs, these represent the most common forms.

Plate jigs, **Figure 20-30**, are the most common form of jig. These jigs consist of a simple plate that contains the required drill bushings, locators, and clamping elements. Typical variations of the basic plate jig include the *table jig*, **Figure 20-31A**, and the *sandwich jig*, **Figure 20-31B**. Another and even simpler version of the plate jig is the *template jig*, **Figure 20-32**. These jigs are used where accuracy and not speed is the prime consideration. Template jigs may or may not have drill bushings, and do not normally have a clamping device. This form of jig is frequently used for light machining or for layout work.

An *angle plate jig*, **Figure 20-33**, is a modified form of a plate jig in which the surface to be machined is not perpendicular to the locating surface. This type of jig is often used to machine pulleys, gears, hubs, or similar parts. Another variation of this type of jig is the *modified*

angle plate jig. These jigs are used to machine parts at angles other than 90°.

A *box jig*, **Figure 20-34**, is designed to be used for parts that require machining on several sides. With these jigs, the part is mounted in the jig and clamped with a leaf or door. Other variations of the basic box jig include the *channel jig*, **Figure 20-35A**, and the *leaf jig*, **Figure 20-35B**. These jigs are similar in design to the box jig but they machine the part on only two or three sides, rather than all six sides.

An *indexing jig*, **Figure 20-36**, is used primarily to machine parts that have machined details at intervals

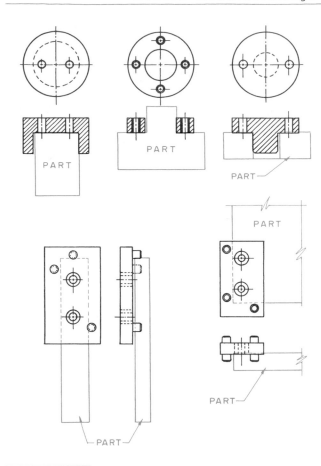

FIGURE 20-32 *Template jigs*

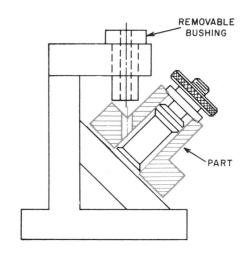

FIGURE 20-33 *Angle plate jig*

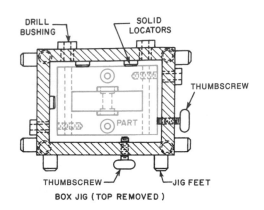

BOX JIG (TOP REMOVED)

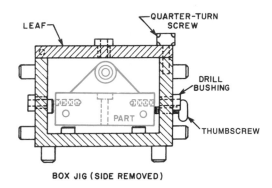

BOX JIG (SIDE REMOVED)

FIGURE 20-34 *Box jig*

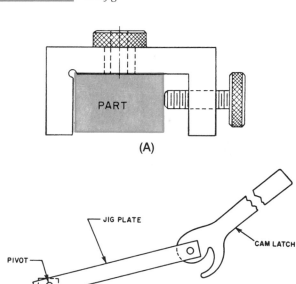

(A)

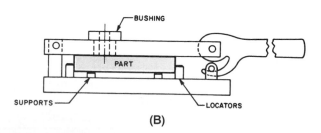

(B)

FIGURE 20-35 *(A) Channel jig; (B) Leaf jig*

around the part. Drilling four holes 90° apart is a typical example of the type of work this jig is best suited to perform. Another type of jig that uses an indexing arrangement to locate the jig, rather than the part, is the *multistation jig*, **Figure 20-37**. These jigs are used to machine several parts at one time.

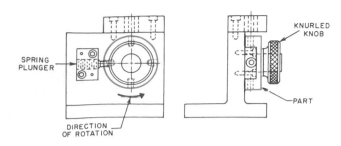

FIGURE 20-36 *Indexing jig*

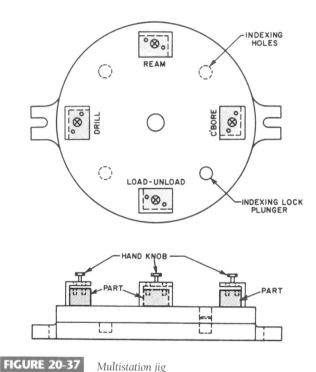

FIGURE 20-37 *Multistation jig*

Classification of Fixtures

Fixtures are classified by the type of machine they are used on, the type of operation performed, and by their basic construction features. The principal types of fixtures normally used in the shop include plate fixtures, angle plate fixtures, vise jaw fixtures, and indexing fixtures. Typically, fixtures are used for milling, turning, sawing, grinding, inspecting, and several other varied operations.

A *plate fixture, Figure 20-38,* is the most common type of fixture. Like plate jigs, plate fixtures are simply a plate containing the locators, set blocks, and clamping devices necessary to locate and hold the workpiece and to reference the cutter. While similar in design to a plate jig, plate fixtures are normally made much heavier than plate jigs to resist the additional cutting forces.

An *angle plate fixture, Figure 20-39A,* and a *modified angle plate fixture, Figure 20-39B,* are simple modifications of the basic plate fixture design. These fixtures are used when the reference surface is at an angle to the surface to be machined.

A *vise jaw fixture, Figure 20-40,* is a useful modification to the standard milling machine vise. With this type of fixture, the standard jaws of a milling machine vise are replaced with specially shaped jaws to suit the part to be machined. The result is an accurate fixture that can be made at a minimal cost. Since the clamping device is contained within the vise, this fixture is very cost effective. Likewise, one vise can be used for a countless number of different fixtures by simply changing the jaws.

Indexing fixtures, Figure 20-41, are mainly used to machine parts that have a repeating part feature. Typical

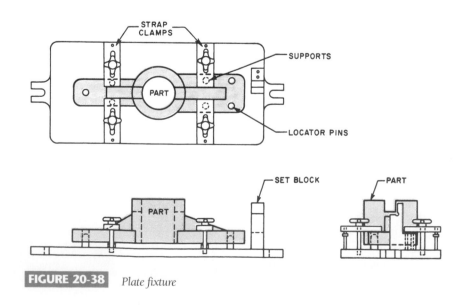

FIGURE 20-38 *Plate fixture*

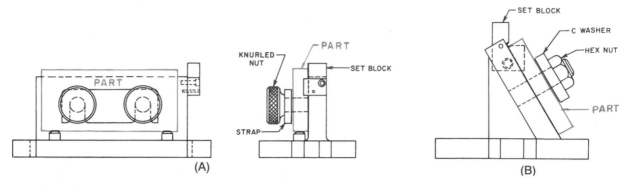

FIGURE 20-39 *(A) Angle plate fixture; (B) Modified angle plate fixture*

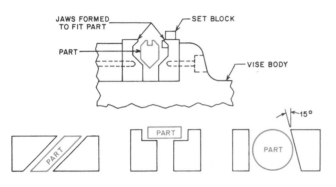

FIGURE 20-40 *Vise jaw fixture*

examples of the types of parts that are machined in an indexing fixture are shown in *Figure 20-42*. Here again, the indexing feature may also be used to position the tool as well as the part. The *duplex fixture* shown in *Figure 20-43* shows a method of indexing the fixture to machine two parts. In use, the first part is machined while the second is loaded. The fixture is then rotated and the second is unloaded and a fresh part loaded. This process is continued throughout the production run.

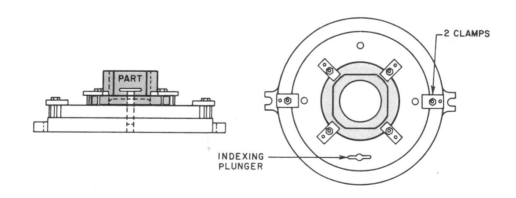

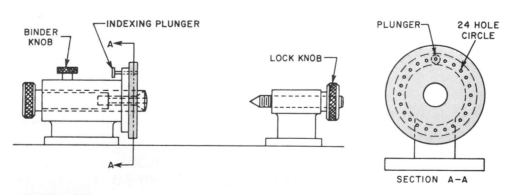

FIGURE 20-41 *Indexing fixture*

FIGURE 20-42 *Typical parts machined in an indexing fixture*

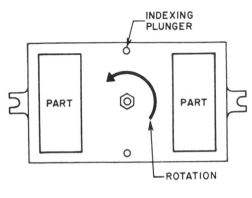

FIGURE 20-43 *Duplex fixture*

LOCATING PRINCIPLES

To properly machine any part in a jig or fixture, the part must first be located correctly. The first rule of locating is repeatability. *Repeatability* is the feature of a jig or fixture that permits parts to be loaded in the tool in the same position part after part. When selecting locators for a part, the prime considerations must be position, foolproofing, tolerance, and elimination of duplicate or redundant locators.

Locators must be *positioned* as far apart as practical, and should contact the workpiece on a reliable surface to ensure repeatability. The locators must also be designed to minimize the effect of chips or dirt. *Figure 20-44* shows a few methods generally used to relieve locators to prevent interference from chips.

Foolproofing is simply a method used to prevent the part from being loaded incorrectly. A simple pin, *Figure 20-45*, is normally enough to ensure proper loading of every part. The *tolerance* of a locator is determined by the specific size of the part to be machined. As a general rule, the tolerance of jigs and fixtures should be approximately 30% to 50% of the part tolerance. For example, if a hole were to be located within .010″ (.25 mm) from an edge, the locators in the jig or fixture should be positioned within .003″ to .005″ (.08 mm to .13 mm) to make sure the part is properly positioned. An overly tight tolerance only adds cost, not quality, to a tool.

Finally, locators should never duplicate any location. As shown in *Figure 20-46*, a part should be located on only one surface. Locating a part on two parallel surfaces only serves to reduce the effectiveness of the location, and could improperly locate the part. First determine the reference surface, and use only that surface for locating.

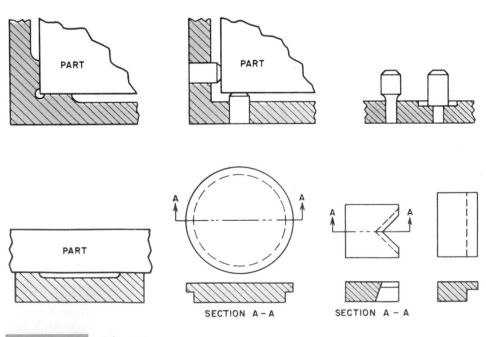

FIGURE 20-44 *Relieving locators*

Industry Application

A MID-CAREER CHANGE

Mark Roberts is the senior tool designer at Reynolds Manufacturing, Inc. (RMI). He is known in design and drafting circles as one of the best tool designers in the business, but it took a career change to help Roberts find his niche. He started work right out of high school as a machining trainee for RMI. In only half the time it normally takes, Roberts had earned his "ticket" as a full-fledged machinist.

Before long he gained a reputation on the shop floor for making special workholding devices that improved the performance of machine tools, particularly on high-volume jobs. Over time, Roberts found himself spending more hours in the day making jigs and fixtures than in doing his machining jobs. After discussing the situation with his supervisor, Roberts decided that he wanted to design jigs and fixtures full time.

Everyone agreed that this would benefit the company. The only problem was that Roberts actually knew very little about design. He knew machines and machining processes. He could visualize what was needed, and he could make excellent jigs and fixtures. But because he knew nothing about design, his workholding devices tended to be cumbersome. In addition, because he could not communicate his ideas on paper, he had to design, make, and modify every device himself. This slowed the process considerably.

Roberts and his supervisor agreed that if he could design workholding devices, and develop understandable drawings, others could actually make them, thereby speeding the process. This realization caused Roberts to enroll in the night design and drafting program at Madison Technical Institute (MTI). He graduated with a certificate in mechanical design and drafting, and was MTI's student of the year. Within six months of his graduation, Roberts was doing so well in the area of jig and fixture design that his company was able to open a tool design division that now designs and manufactures jigs and fixtures for other companies.

Roberts is still the senior tool designer at RMI where the jig and fixture division is a commercial success. He also teaches tool design in the night program at Madison Technical Institute.

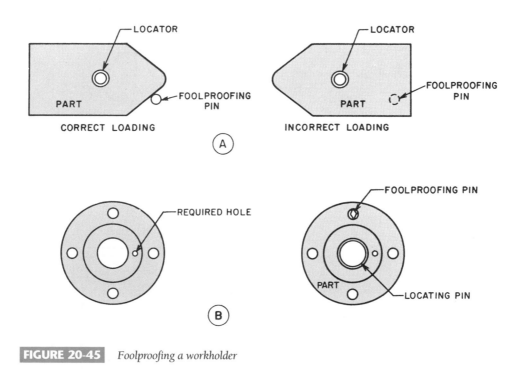

FIGURE 20-45 *Foolproofing a workholder*

Restricting Movement. Every part is free to move in a limitless number of directions if left unrestricted. But, for the purpose of jig and fixture design, the number of directions in which a part can move has been limited to twelve: six axial and six radial, *Figure 20-47*. The methods used to restrict these twelve movements normally depend on the part itself, but the examples in *Figures 20-48A* and *20-48B* show two methods that are frequently used. In the first example, a three-pin base restricts five directions of movement. In the second

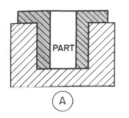
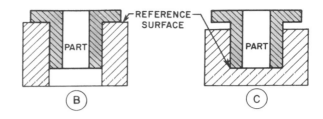

FIGURE 20-46   *Duplicate locating*

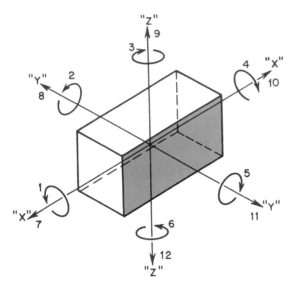

FIGURE 20-47   *Planes of movement*

example, a five-pin base restricts eight directions of movement. In *Figure 20-48C,* a six-pin base restricts nine directions of movement.

Types of Locators.   Locators are commercially available in many styles and types. *Figures 20-49A* through *20-49F* show several of the most common types of locators.

CLAMPING PRINCIPLES

In addition to locators, most jigs and fixtures use some type of clamping device to restrict the directions of movement not contained by the locators. In the design of a clamping arrangement, several factors should be considered. These include the position of the clamps, clamping forces, and tool forces.

Clamps should always be positioned to contact the part at either its most rigid point or a supported point. Clamping a part at any other point could bend or distort the part, as shown in *Figure 20-50.* The clamping forces used to hold a part should always be directed toward the most solid part of the tool. Clamping a part as shown in *Figure 20-51A,* will normally result in an egg-shaped part because the clamping forces are directed only toward each other. However, by clamping the part as

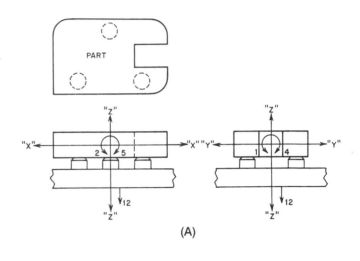

(A)

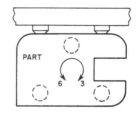
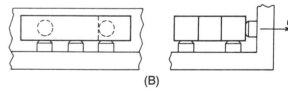

(B)

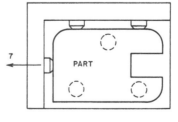
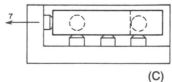
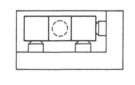

(C)

FIGURE 20-48   *(A), (B), and (C) Restricting part movement*

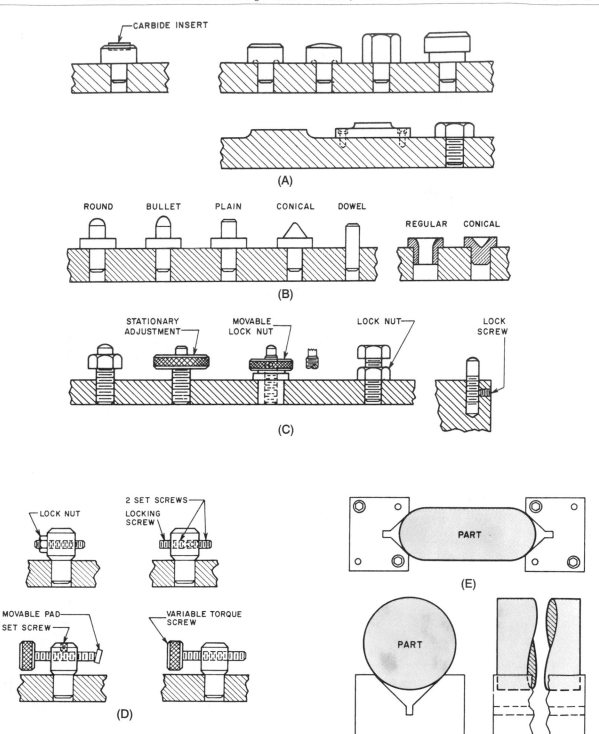

FIGURE 20-49 *(A–F) Types of locators*

shown in *Figure 20-51B,* not only is the part held securely, but the chance of distortion is greatly reduced. As a rule, use only enough clamping force to hold the part against the locators. The locators should always resist the bulk of the tool thrust, not the clamps.

When designing a jig or fixture, remember to direct the tool forces toward the locators, not the clamps. *Tool forces* are those forces generated by the cutting tool during the machining cycle. In most cases, the tool forces can be used to an advantage when holding any part. As shown in *Figure 20-52,* the downward tool forces are actually pushing the part into the tool. The rotational tool forces are contained by the locators. The only force the clamps need to hold are the forces generated by the

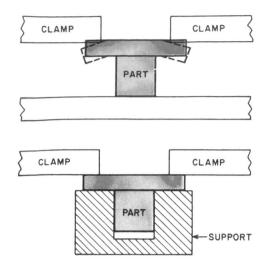

FIGURE 20-50 *Always clamp a part at its most rigid point or a supported point*

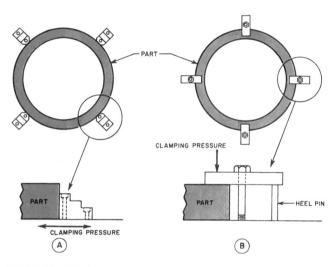

FIGURE 20-51 *Clamping forces*

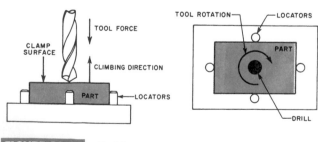

FIGURE 20-52 *Tool forces*

drill as the point breaks out the opposite side of the part, and these forces are only a fraction of the cutting forces. So, when designing a jig or fixture, always direct the tool forces so that they act to hold the part in the tool. Use the clamps only to hold the part against the locators.

Types of Clamps. Clamps, like locators, are commercially available in many styles and types. ***Figures 20-53A***

through *20-53D* (page 765) show some of the more common types of clamps used for jigs and fixtures. The specific type of clamp you should select for any workholding application will normally be determined by the part and the type of holding force desired.

BASIC CONSTRUCTION PRINCIPLES

When designing any jig or fixture, the first consideration is normally the tool body. Tool bodies are generally made in any of three ways: cast, welded or built-up, *Figure 20-54* (page 765). *Cast tool* bodies are generally the most expensive and require the longest lead time to make. They do, however, offer such advantages as good material distribution, good stability, and good vibration-damping qualities. *Welded tool* bodies offer a faster lead time, but they must be machined frequently to remove any distortion caused by the heat of welding. The most popular and common type of tool body in use today is the built-up. *Built-up tool* bodies are made from pieces of preformed stock, such as precision-ground flat stock, ground rod, or plain cold-rolled sections that are pinned and bolted together. The principal advantages of using this type of construction are fast lead time, minimal machining, and easy modification.

Drill Bushings. Drill bushings are used to position and guide the cutting tools used in jigs. The three basic types of drill bushing used for jigs are press-fit bushings, renewable bushings, and liner bushings.

Press-fit bushings, Figure 20-55, as their name implies, are pressed directly into the jig plate. These bushings are useful for short-run jigs or in applications where the bushings are not likely to be replaced frequently. The two principal types of press-fit bushings are the head type and headless type.

Renewable bushings are used in high-volume applications or where the bushings must be changed to suit different cutting tools used in the same hole. The two types of renewable bushings that are commercially available are the slip-renewable type, *Figure 20-56A* (page 766), and the fixed-renewable type, *Figure 20-56B* (page 766).

Slip-renewable bushings, Figure 20-56C (page 766), are used where the bushing must be changed quickly. Typical applications include holes that must be drilled, countersunk, and tapped. The bushings are held in place with a lock screw and need only be turned counterclockwise and lifted out of the hole. The next bushing is inserted in the hole and turned clockwise to lock it in place.

Fixed-renewable bushings are used for applications where the bushings do not need to be changed often, but where they are changed more often than press-fit bushings. High-volume production is a typical example where fixed renewable bushings are often used.

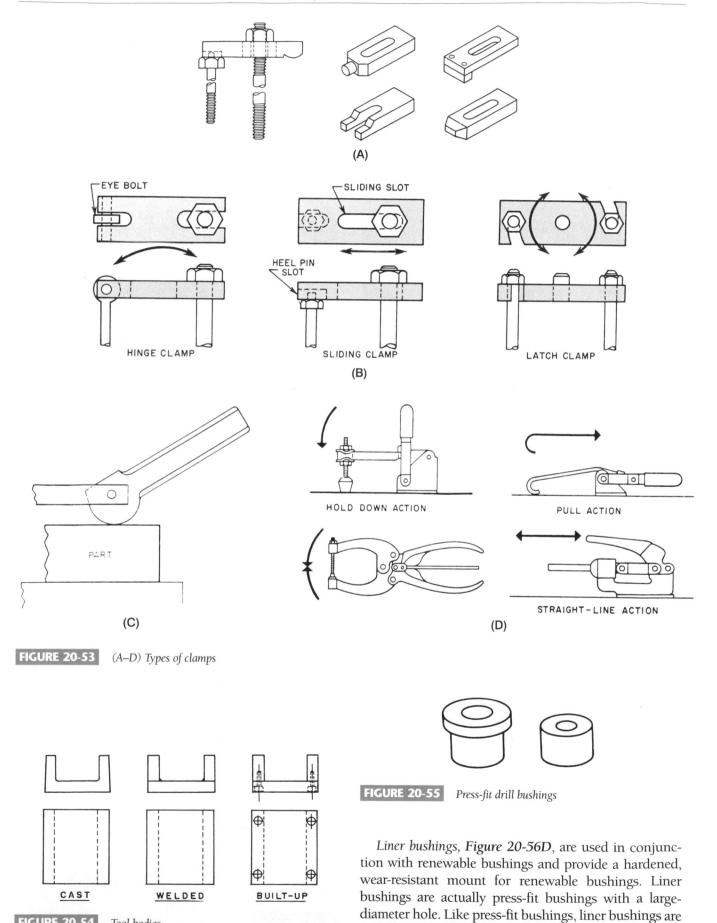

FIGURE 20-53 (A–D) *Types of clamps*

FIGURE 20-54 *Tool bodies*

FIGURE 20-55 *Press-fit drill bushings*

Liner bushings, **Figure 20-56D**, are used in conjunction with renewable bushings and provide a hardened, wear-resistant mount for renewable bushings. Liner bushings are actually press-fit bushings with a large-diameter hole. Like press-fit bushings, liner bushings are available in both a head type and a headless style. The

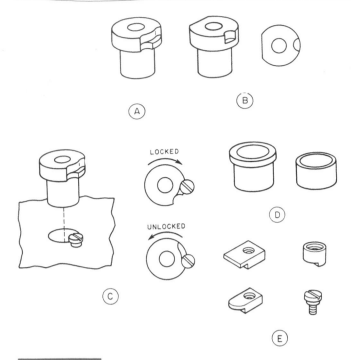

FIGURE 20-56 *(A–E) Renewable drill bushings*

clamps normally used to hold the fixed-renewable bushing in the jig plate are shown in *Figure 20-56E*. The screw is used to mount slip-renewable bushings, whereas the other three styles are used to secure the fixed-renewable bushings.

Several other variations of these basic bushing styles include special-purpose bushings, serrated and knurled bushings, and oil-groove bushings. Oil-groove bushings are shown in *Figure 20-57*.

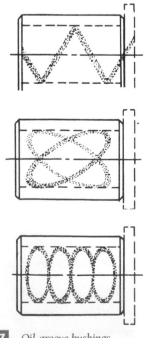

FIGURE 20-57 *Oil-groove bushings*

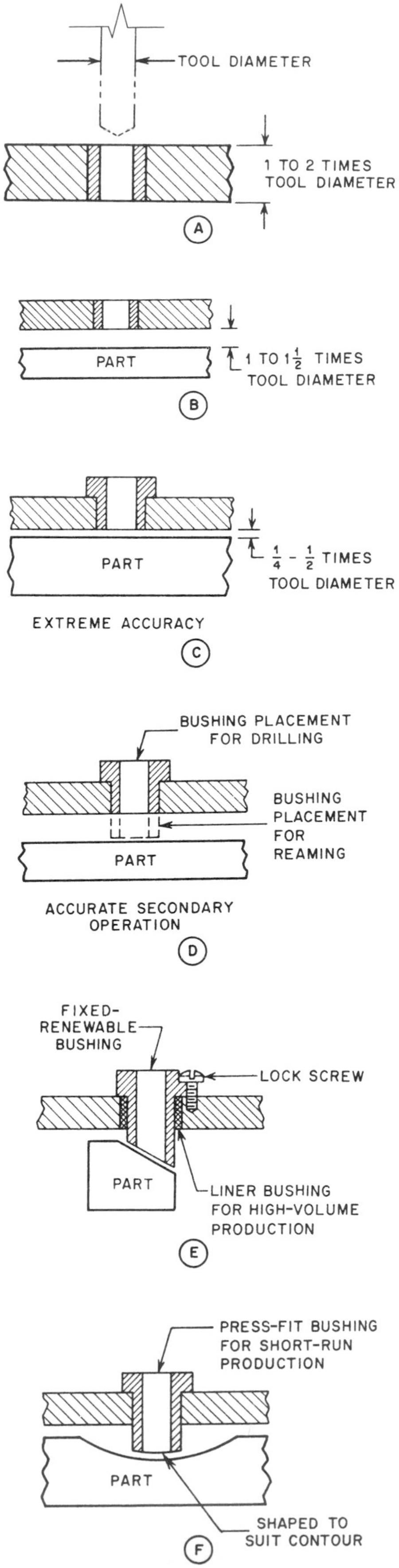

FIGURE 20-58 *(A–F) Mounting drill bushings*

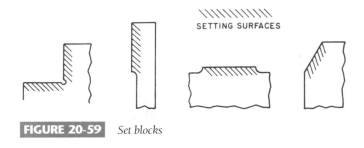

FIGURE 20-59 *Set blocks*

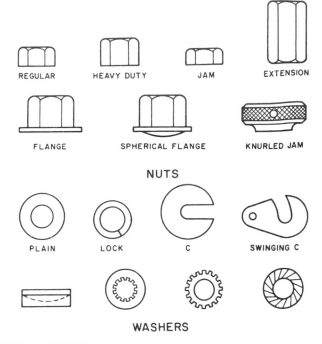

FIGURE 20-60 *Nuts and washers frequently used with jigs and fixtures*

Mounting Drill Bushings. When one is installing a drill bushing in the jig plate, the specific size of the jig plate and spacing from the part are important factors the designer must consider. As shown in *Figure 20-58A*, the jig plate should be one to two times the tool diameter. If a thinner jig plate must be used, a head-type bushing can be used to achieve the added thickness needed to support the cutting tool. The space between the jig plate and the workpiece should be one to one-and-one-half times the tool diameter for drilling, *Figure 20-58B*, and one-quarter to one-half the tool diameter for reaming, *Figure 20-58C*.

In those cases where both drilling and reaming operations are to be performed in the same hole, an arrangement similar to the one shown in *Figure 20-58D* may be used. Here the jig plate is properly positioned for drilling, and the bushing used for reaming is made longer to achieve the desired distance for reaming. When special shapes or contours must be drilled, the ends of the bushings can be modified to suit the contour and maintain the proper support for the cutting tool, *Figures 20-58E* and *20-58F*.

Set Blocks. Set blocks are used along with thickness gages to properly position the cutting tool with a fixture. The specific design of the set block is determined by the part and shape to be machined. The basic set block designs shown in *Figure 20-59* are typical of the styles found on many fixtures.

Fastening Devices. The most common fastening devices used with jigs and fixtures are dowel pins and sockethead cap screws. Other fasteners sometimes used with these tools include the nuts and washers shown in *Figure 20-60*.

Heat Treatment of Steels

Heat treatment is a series of processes used to alter or modify the existing properties in a metal to obtain a specific condition required for a workpiece. The specific properties normally changed by heat treatment include hardness, toughness, brittleness, malleability, ductility, wear resistance, tensile strength, and yield strength. The five standard heat treating operations normally performed on steel parts are hardening, tempering, annealing, normalizing, and case hardening.

Hardening is the process of heating a metal to a predetermined temperature, allowing the part to soak until thoroughly heated, and cooling rapidly in a cooling material called a *quench*. The most common quench media are air, oil, water, and a water and salt mixture called brine. Hardening is mainly used to increase the hardness, wear resistance, toughness, tensile strength, and yield strength of a workpiece.

Tempering is the process of heating a hardened workpiece to a temperature below the hardening temperature, allowing the part to soak for a specific time period, and cooling. Tempering is mainly used to reduce the hardness so the toughness is increased and brittleness is decreased. Almost every hardened part is tempered to control the amount of hardness in the finished part.

Annealing is the process of heating a part, soaking it until it is thoroughly heated, and slowly cooling it by turning off the furnace. Annealing is mainly used to completely remove all hardness in a metal. Frequently, hardened parts are annealed to be remachined or modified and then rehardened.

Normalizing is the process used to remove the effects of machining or cold-working metals. In this process, the metal is heated, allowed to soak, and cooled in still air. This process produces a uniform grain size and eliminates almost all stresses in the metal.

Case hardening is the process of hardening the outside surface of a part to a preselected depth. In most case-hardening operations, carbon is added to the surface of the part by packing the part in a carbonous material, and then heating it so that the carbon is transferred into the surface of the part. Once the carbon is added to the surface of the part, the part is then hardened to produce a hardened shell around a normalized core.

Nontraditional Machining Processes

The term *nontraditional machining* encompasses a number of advanced machining processes that have been developed in response to the demands of modern manufacturing for more effective, more efficient ways to machine the many new materials that have been developed and are still being developed. Although they are still referred to as "nontraditional" processes, they are now common and widely used. The most widely used nontraditional machining processes are explained in this section.

HYDRODYNAMIC MACHINING (HDM)

Hydrodynamic machining (HDM), **Figure 20-61**, removes material using a high-velocity, high-pressure stream of liquid solution. It is used primarily for slitting and contour-cutting many nonmetallic materials, such as wood and paper, asbestos, plastics, gypsum, leather, felt, rubber, nylon, fiberglass, and fiberglass-reinforced plastics. Some very thin workpieces of soft metal can be cut effectively by this process; steel sheet (0.005″, 0.13 mm thick) and aluminum sheet (0.020″, 0.51 mm thick) are processed, but water pressure in excess of 100 ksi (690 MPa) is usually required. Equipment that operates at these higher pressures is not commercially available. The cutting of hard metals using HDM remains mostly experimental.

Practical experience and experimental research has shown that, in some cases, brittle materials such as glass, acrylic, ceramics, and crystal do not appear suitable for cutting by the HDM process because these materials tend to develop severe cracks and may break under processing conditions.

Typically, soft materials are cut easily using HDM, and friable materials can be cut with good edge quality.

At its present stage of development, the industrial use of HDM is limited to the processing of relatively thin materials. When thicker materials are cut, stream lines increase significantly, causing poor edge quality.

ULTRASONIC MACHINING (USM)

Ultrasonic machining (USM) (**Figure 20-62**) removes material using an abrasive slurry driven by a special tool vibrating at a high frequency along a longitudinal axis.

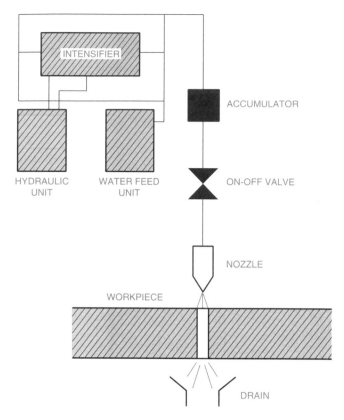

FIGURE 20-61   *Hydrodynamic machining (HDM)*

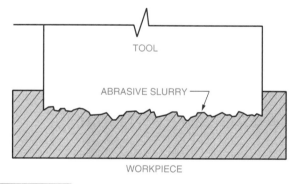

FIGURE 20-62   *Ultrasonic machining (USM)*

It is used primarily for producing blind holes, through holes, slots, and irregular shapes. It is limited in complexity only by the configuration of the tooling. However, in some applications, tool wear and/or taper in the cut may discount the process's effectiveness. The depth-to-width ratio of the cut is usually less than about 3:1. Current practice is limited to 3.5″ (89 mm) diameter tools machining cavities up to about 2.5″ (64 mm) deep.

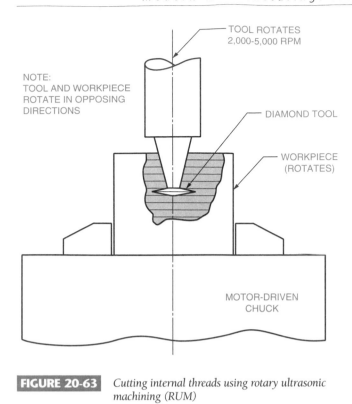

NOTE:
TOOL AND WORKPIECE
ROTATE IN OPPOSING
DIRECTIONS

TOOL ROTATES
2,000-5,000 RPM

DIAMOND TOOL

WORKPIECE
(ROTATES)

MOTOR-DRIVEN
CHUCK

FIGURE 20-63 *Cutting internal threads using rotary ultrasonic machining (RUM)*

ROTARY ULTRASONIC MACHINING (RUM)

Rotary ultrasonic machining (RUM) (*Figure 20-63*) is similar to USM except that: (1) the grinding motion is rotary instead of longitudinal; (2) RUM uses diamond tools instead of steel or Monel; and (3) RUM uses the diamond of the tool to remove material while USM uses an abrasive slurry. RUM is used primarily in the development of prototypes. It is also used for machining ceramics, quartz, glass, ferrite computer parts, and composites.

The applications of RUM are currently limited by tool size. The horn/tool assembly must have a resonant (natural) frequency of about 20 kHz, so tool size is limited. Any variance in tool weight changes the natural frequency of the horn/tool assembly; too heavy a tool increases horn/tool assembly frequency beyond the resonant frequency of the transducer and power supply, causing the tool not to vibrate. Today, the largest vibrating horn feasible is about 1.5″ (38 mm) in diameter. Practical limit to tool weight is about 1.4 oz (40 g).

RUM is used widely in prototype work as well as production applications. The process is effective in producing prototypes because it can make precise parts that can be used to make molds for large-volume production runs.

The process is particularly effective in the machining of sintered materials such as ceramics and ferrites.

Conventionally, these materials are machined and drilled in the "green" state—prior to firing. When fired, the materials experience as much as 16% shrinkage, which destroys the accuracy created during machining. RUM is used to machine materials such as these after firing, and close tolerance relationships are maintained. Some applications include the machining of precision ceramic components, drilling small holes in ceramic printed circuit boards, and drilling small-diameter, deep, intersecting holes in quartz for laser development.

ULTRASONICALLY ASSISTED MACHINING (UAM)

Ultrasonically assisted machining (UAM) involves assisting such traditional machining processes as drilling and turning by coupling the tool with an external source of vibrating energy. It is used in most drilling and turning applications and is effective in reducing the amount of subsurface tearing and plastic flow. It also evens out tool wear.

Under certain conditions, ultrasonic lathe turning has been shown to increase cutting rates by factors of four in aluminum, and five in cutting steel. Nonmetallic materials have also shown marked increases in cutting rates when UAM is used. Alumina can be machined two times faster, and magnesium silicate can be machined up to four times faster in some cases, according to manufacturers of UAM equipment.

Some materials that are too brittle to machine traditionally can be machined effectively with UAM. For example, in one test, low-porosity mullite was machined with a good cut by applying ultrasonic vibrations to the tool post and carbide-tipped tool, but when ultrasonic power was turned off, the workpiece immediately shattered.

Some tests have shown that UAM reduces turning forces in some materials as much as 30–50% as compared to traditional turning methods. In the same tests, it was shown that surfaces produced by UAM exhibited a matte finish, evidence of more complete shearing of the chips from the workpiece. This phenomenon is in sharp contrast to traditional machining, which often produces a glossy surface as a result of tearing, materials enfoldment, and burnishing. Subsurface tearing and plastic flow are reported to be all but eliminated by UAM.

ELECTROMECHANICAL MACHINING (EMM)

Electromechanical machining (EMM) is a process that improves the performance of traditional machining processes. Metal removal is accomplished in the traditional manner (i.e., drilling, milling, turning, etc.) with the exception that the workpiece is electrochemically polarized by applying a controlled voltage across the connection point between the workpiece and an electrolyte. This changes the surface of the workpiece, making it easier to machine.

The limited feasibility of electromechanical turning (EMT) and electromechanical drilling (EMD) has been demonstrated in both laboratory and plant settings. The limited testing conducted to date indicates advantages of surface finish, tool wear, hole tolerance, and chip configuration. However, maximum improvement is tied closely to optimization of electrolytes, and it has been shown that the electrolytes should be modified through the use of inhibitors to minimize corrosion of machine tool components.

ELECTROCHEMICAL DISCHARGE GRINDING (ECDG)

Electrochemical discharge grinding (ECDG) is a combination of two processes: electrochemical grinding (ECG) and electrical discharge grinding (EDG). In ECDG, alternating current or pulsating direct current moves from a bonded graphite wheel (conductor) to a positively charged workpiece. No mechanical contact is made between the workpiece and the wheel. Instead, an electrolyte is pumped into the gap and it serves as the connecting agent. Material removal occurs as follows: (1) the ECG process converts the surface of the metal into an oxide film, and (2) the EDG process removes the oxide film.

The production uses of ECDG are somewhat limited, although certain applications exist that are routinely performed, mostly in the grinding and sharpening of carbide tooling. Nearly any electrically conductive material can be processed, but careful comparison of the relative advantages and disadvantages of ECDG versus processes such as ECG and EDG should be made before specifying the process. For example, ECDG can remove material five times faster than EDG, but it uses up to 15 times the current used in EDG. Typical production tolerances of ±0.001″ (0.03 mm) are achieved with ECDG.

Current successful applications of the process include grinding and sharpening of carbide inserts, generation of delicate profiles using form grinding, grinding of honeycomb materials, and grinding of carbide thread chasers.

ELECTROCHEMICAL GRINDING (ECG)

Electrochemical grinding (ECG) (*Figure 20-64*) combines an electrochemical reaction that occurs on the surface of the metal with abrasion. Subsequently, it is applied to the metal's surface to remove material. The process is typically used on tough, hard metals. The abrasion is

applied by a rotary grinding wheel that grinds off oxidized surface metal.

In operations in which ECG can be applied, it produces results far beyond those that conventional grinding methods can provide. In many cases it can reduce abrasive costs up to 90%. This reduction is most easily observed in connection with diamond wheels and carbide grinding. However, it is also significant with respect to steel and alloy steel grinding with nondiamond wheels.

Also, because it is a cool process, ECG can be used to grind any electrically conductive material without damage to it from heat. Therefore, ECG can simplify fracture-inspection procedures or entirely eliminate scrap due to grinding-heat fractures. In addition, this process can grind steel or alloy steel parts without generating any burr. Thus, the costly operation of subsequent deburring is automatically eliminated.

ELECTROCHEMICAL HONING (ECH)

Electrochemical honing (ECH) is a process similar to ECG. It combines electrochemical oxidation of surface metal and abrasion. The primary difference is that honing is a process of removing material inside a drilled hole. Consequently, the abrasion is applied using nonconductive honing stones attached to a tool that rotates and reciprocates.

To be processed by ECH, workpieces must be conductive. The process is most effective when used to hone hard, tough metals and is well suited for the processing of parts that are susceptible to heat distortion. Electro-

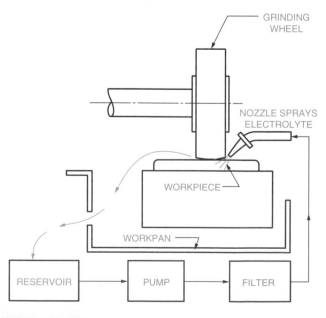

FIGURE 20-64 *Electrochemical grinding (ECG)*

chemical honing causes little heat buildup and no significant stresses, and automatically deburrs the workpiece. The process is particularly effective for parts that require fast stock removal with good surface finish control.

ELECTROCHEMICAL MACHINING (ECM)

Electrochemical machining (ECM) (Figure 20-65) is a broad concept that applies to a number of process variations that all use an electrolytic action in removing metal. With ECM, electricity is combined with a chemical to form a reaction that is the opposite of metal plating. ECM can be used with any conductive workpiece but is used most often with hard metals. Its best application is external shaping of materials.

ECM is used in a wide variety of industries to machine many different metals. Experience has shown, however, that not all materials can be machined successfully (to acceptable metal removal rates and surface finish) using the ECM process. For example, some high-silicon aluminum alloys, such as cast alloys with a substantial silicon content, sometimes cannot be machined with acceptable surface finish. Some troublesome experiences have been documented with SAE 332 aluminum pistons with 10% silicon, for example.

ELECTROCHEMICAL TURNING (ECT)

Electrochemical turning (ECT) is a variation of ECM. The same basic principles apply except that the workpiece rotates. The primary application of ECT is the machining of large, disc-shaped forgings. In some cases, full-face electrodes are plunged into a rotating disc. Bearing races have been finished, with close tolerances and with surface roughness held to less than 5 in. (0.13 μm) R_a. Another application, AISI 316 stainless steel workpieces (2.5″, 6.35 mm diameter), are electrochemically turned, using an electrolyte of NaCl and NaNO$_3$ (3:3), to a surface finish of less than 10 in. (0.25 μm) R_a with out-of-roundness of less than 0.0002″ (0.005 mm) TIR.

ELECTRON BEAM MACHINING (EBM)

Electron beam machining (EBM) is a thermal energy process. This means it uses thermal energy to remove metal. The thermal energy comes in the form of a pulsating stream of high-speed electrons that are focused on a very small area of the workpiece by electrostatic and electromagnetic fields. Material removal occurs as the result of melting or evaporation. EBM is used for drilling small wire-drawing dies, metering holes, and holes for spinnerets used in textile manufacturing.

Any known material, metal or nonmetal, that will exist in high vacuum can be cut, although experience has shown that diamonds do not cut well. Holes with

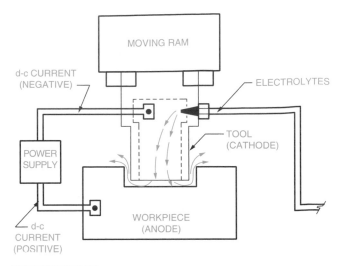

FIGURE 20-65 *Electrochemical machining (ECM)*

depth-to-diameter ratios up to 100:1 can be cut. Limitations include high equipment costs and the need for a vacuum, which usually necessitates batch processing and restricts workpiece size. The process is generally economical only for small cuts in thin parts.

ELECTRICAL DISCHARGE MACHINING (EDM)

Electrical discharge machining (EDM) (Figure 20-66) is a process that removes metal using a series of rapidly recurring electrical discharges between an electrically charged cutting tool and the workpiece submerged in a dielectric fluid.

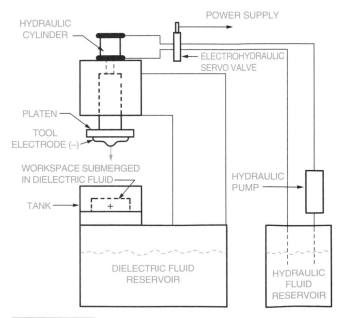

FIGURE 20-66 *Electrical discharge machining (EDM)*

The application of EDM is limited to the machining of electrically conductive workpiece materials, but the process has the capability of cutting these materials regardless of their hardness or toughness. Nonconductors such as glass, ceramics, or plastics cannot be machined using EDM techniques. Machining of hardened steel using EDM eliminates the need for subsequent heat treatment with possible distortion. Complex shapes can be cut in hardened steel or carbide without costly sectional construction being necessary.

The production of stamping dies is a major application afforded by using EDM to match one portion of the die made conventionally. Extruding, heading, drawing, forging, and die casting dies, as well as molds for plastics, are also made using EDM techniques.

ELECTRICAL DISCHARGE WIRE CUTTING (EDWC)

Electrical discharge wire cutting (EDWC) (*Figure 20-67*) is a process that is similar to the band-sawing process. Material removal occurs as the result of an electrical discharge as a wire electrode is drawn through the workpiece. EDWC can be used only with conductive workpieces. Primary applications of EDWC include the production of stamping dies, prototypes, molds, lathe tools, special form inserts, pot and wobble broaches, extrusion dies, and templates.

As is the case with any electrical discharge process, EDWC requires that the workpiece be electrically conductive. The cut produced by the process is free of flar-

ing and is controllable to produce small radii. Workpieces up to about 6″ (152 mm) in thickness can be processed using standard equipment; workpiece stacking effects greater productivity.

Normally produced from hardened metals such as tool steel, stamping dies are routinely cut using EDWC. The process facilitates cutting after heat treatment, thereby eliminating distortion. The use of EDWC in the manufacturing of dies affords significant savings. By conventional methods, dies are sometimes split into two or more sections to facilitate grinding with an optical projection form grinder. Usually, after the contour has been ground in the split sections, the sections are fitted in a holder or adapter. This time-consuming and costly process is eliminated with EDWC employed to produce the dies. Some experts claim that tool components such as dies can be produced in less than one-third the time required by conventional methods.

ELECTRICAL DISCHARGE GRINDING (EDG)

Electrical discharge grinding (EDG) is similar to EDM in that it employs rapidly recurring electrical discharges between an electrode and the workpiece in a dielectric fluid. The difference is that with EDG the electrode is a rotating wheel made of graphite or brass. Primary applications include the grinding of steel and carbide at the same time, thin sections, and brittle and fragile parts.

EDG is generally used for operations such as the following: grinding steel and carbide at the same time without wheel loading, grinding thin sections on which abrasive-wheel pressures might cause distortion, grinding brittle materials or fragile parts on which abrasive materials might cause fracturing, grinding through forms for which diamond-wheel costs would be excessive, and grinding circular forms in direct competition with abrasive-wheel methods.

Electrical discharge grinding is used to grind hard materials such as carbide form tools, hardened steel gear racks, or tungsten carbide inserts. The process is also used to grind hardened lamination dies. Cast iron workpieces are usually not processed using EDG because sand inclusions can damage the graphite grinding wheel.

LASER BEAM MACHINING (LBM)

Laser beam machining (LBM) (*Figure 20-68*) is a process that uses lasers for performing operations that are traditionally performed with cutting tools. These operations include cutting, drilling, slotting, and scribing. The basic principle of LBM is the conversion of electrical energy into coherent light. This coherent light or laser beam is focused on the workpiece where it is converted into

FIGURE 20-67 *Electrical discharge wire cutting (EDWC)*

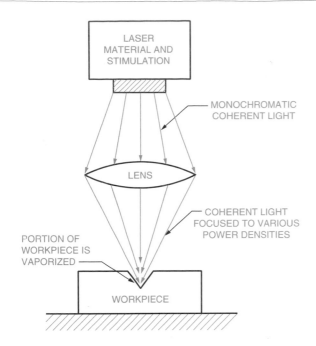

FIGURE 20-68 *Solid state laser beam machining (LBM)*

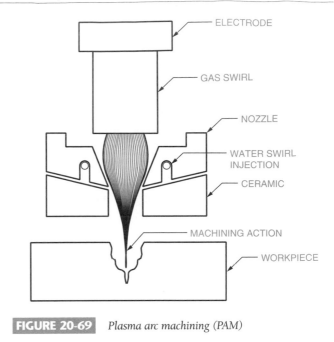

FIGURE 20-69 *Plasma arc machining (PAM)*

thermal energy. This thermal energy melts or vaporizes the material. LBM has broad applications, but it is particularly effective at cutting and drilling.

Any industrial application of the laser should be based on one of three criteria: (1) it can perform a superior job in terms of quality and cost over existing methods, (2) it is the only tool capable or available for the specific job, or (3) it allows restructuring of the manufacturing process, resulting in lower total cost. The cost of using laser systems falls into two categories: (1) the capital investment for the laser system, and (2) the operational cost. At present both the equipment and the direct operational costs of a laser are higher than those of comparable conventional equipment and methods.

Because of the laser's ability to melt or vaporize any known metal and operate in any desired atmospheric environment, it is sometimes preferred over EBM (electron beam machining), which requires a vacuum chamber for certain applications. Other advantages include: (1) the ability to machine areas not readily accessible and extremely small holes, (2) the fact that no direct contact exists between the tool (laser) and the workpiece, (3) small heat-affected zones, and (4) easy control of beam configuration and size of exposed area.

PLASMA ARC MACHINING (PAM)

Plasma arc machining (PAM) (*Figure 20-69*) is a process that makes use of a plasma arc to remove material. *Plasma* is a gas that has been heated to the point that it ionizes and becomes electrically conductive. With PAM,

a stream of ionized particles directed through a nozzle against the workpiece removes material. The primary applications of PAM are hole piercing, stack cutting, gouging, grooving, and bevel cutting.

The plasma arc can be used for "machining" or removing the metal from the surface of a rotating cylinder to simulate a conventional lathe or turning operation. As the workpiece is turned, the torch is moved parallel to the axis of the work. The torch is positioned so the arc will impinge tangentially on the workpiece and remove the outer layer of metal. Cutting can be accomplished with the workpiece rotating in either direction relative to the torch, but best results are obtained when the direction of rotation permits use of the shortest arc length for cutting. The flow of molten metal being removed must be in such a direction that it does not tend to adhere to the hot surface that has just been machined.

CHEMICAL MILLING

Chemical milling is a process that etches or shapes workpieces to close tolerances by chemically induced material removal. Primary applications include metal removal from irregularly shaped workpieces, reduction of web thicknesses, removal of the decarburized layer from low-alloy steel forgings, improved surface finish, and removal of defects.

In general, chemical milling is used to remove metal from a portion or the entire surface of formed or irregularly shaped parts such as forgings, castings, extrusions, or formed wrought stock; reduce web thicknesses below practical machining, forging, casting, or forming limits; taper sheets and preformed shapes; produce stepped webs, resulting in consolidation of several details into one integral piece; remove the decarburized layer from low-alloy steel forgings; remove up to 0.125″ (3.2 mm) per surface of metal to remove decarb and also create finished dimensions of die forgings; improve surface finish; remove surface cracks, laps, and other defects of forgings; remove alpha case from titanium forgings; and improve surface finish and control dimensions of aluminum forgings.

PHOTOCHEMICAL MACHINING

Photochemical machining, also known as chemical blanking, involves producing parts by chemical action. It is accomplished by placing an exact image of the part to be produced on a sheet of metal (the prototype must be chemical-resistant) and immersing them both in a chemical. The chemical action dissolves all of the metal except the desired part. Most photochemically machined parts are thin and flat.

Photochemical machining has a number of applications wherein it provides unique advantages. Some of these include: working on extremely thin materials when handling difficulties and die accuracies preclude the use of normal mechanical methods; working on hardened or brittle materials when mechanical action would cause breakage or stress-concentration points—chemical blanking works well on spring materials and hardened materials that are relatively difficult to punch; production of parts that must be absolutely burr-free; production of extremely complex parts for which die costs would be prohibitive; and producing short-run parts for which the relatively low setup costs and short time from print to production offer advantages. This is especially important in research and development projects and in model shops.

WATER JET MACHINING (WJM)

Unlike the nontraditional processes presented so far, *water jet machining (WJM)* is used with nonmetal workpieces. It involves cutting plastics, nonmetal composites, ceramics, glass, and wood by directing a thin, pressurized jet of water against the workpiece.

ABRASIVE WATER JET MACHINING (AWJM)

Abrasive water jet machining (AWJM) (*Figure 20-70*) is similar to WJM except that an abrasive is added to the water jet after it exits the nozzle. This allows AWJM to be used with brittle and/or heat-sensitive materials such as glass, composites, and plastics.

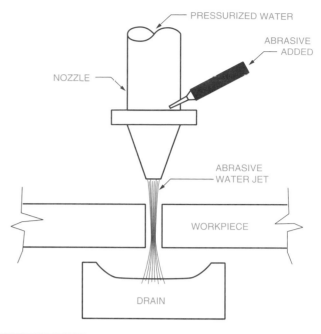

FIGURE 20-70 *Abrasive water jet machining (AWJM)*

Automation and Integration (CAM and CIM/FMS)

There have been four phases in the development of shop and manufacturing processes:

1. Manual phase
2. Mechanization phase
3. Automation phase (CAM)
4. Integration phase (CIM/FMS)

Of course, there has been and continues to be a good deal of overlap among these phases. Manual manufacturing did not suddenly stop when the age of mechanization began, nor did mechanized manufacturing stop when the automation phase began. Correspondingly, the beginning of the integration phase did not bring automated manufacturing to a sudden stop.

It takes many years for a given stage in the development of shop and manufacturing processes to phase out completely. For example, by the year 1900, mechanized manufacturing had replaced manual manufacturing as the dominant approach. However, manual manufacturing was still widely practiced at that time. Today, automated manufacturing has become the dominant approach; nevertheless, mechanized manufacturing is still widely practiced, and even manual manufacturing is still practiced in very isolated cases.

At present, we are seeing the integration phase emerge. It will eventually dominate but not completely replace—at least not for a long time—the other three approaches.

The manual phase preceded the Industrial Revolution. During these early times, manufactured products were produced by hand, using a variety of manual tools. The Industrial Revolution ushered in the mechanization phase. The fundamental component of this phase was the machine. Early machines were steam powered, while later machines, of course, were dependent on electricity. In this phase, machines such as mills, drills, saws, and lathes did the work, but people provided the control.

The automation phase has changed the way machines are controlled. It has also lessened the amount of human involvement in manufacturing. With the automation approach, manufacturing machines are controlled by computer. This has come to be known as CAM or computer-aided manufacturing. CAM machines are operated automatically according to computer programs. People still write the programs, set up the machines, monitor operations, and unload machines.

With CAM, individual operations such as milling, drilling, turning, and cutting are automated, but they still operate independently of each other. The most advanced manufacturing approach to date—computer-integrated manufacturing or CIM—solves this problem by linking automated manufacturing processes with other processes such as materials handling, warehousing, contracts and bids, scheduling, and design. CIM represents a major step toward the wholly automated factory. These two concepts—CAM and its eventual successor CIM—deserve special attention.

Computer-Aided Manufacturing (CAM)

Through CAM, such manufacturing processes as machining and materials handling can be automated. CAM encompasses any manufacturing process that is automated through the use of the computer, but the best examples of CAM are computer numerical control (CNC) of machine tools and robotics.

COMPUTER NUMERICAL CONTROL (CNC)

Numerical control is a method of controlling machine tools by using coded programs. These programs consist of numbers, letters, and special characters that define the path of the machine tool in accomplishing specific tasks. When the job changes, the program must be rewritten to accommodate the change. The programmability feature is the key to the growth of NC. Machine tools that are programmable are more flexible than their nonprogrammable counterparts.

NC is a broad term that encompasses the traditional approach to NC as well as the more modern outgrowths of computer numerical control (CNC) and direct numerical control (DNC).

The first NC machines used punched cards as the programming medium. Later, punched tape was substituted, which was an improvement over manual control but did have some disadvantages. Paper tape was easy to damage and difficult to correct or edit.

Mylar tape, as a less fragile medium, began to replace paper tape. This solved the problem of frequent damage but did not make punched tape any easier to edit. Such problems, coupled with advances in microelectronic technology, led to the use of computers as the means for programming machine tools. Computers solved the damage and editing problems, and this gave birth to both CNC and DNC. CNC involves using a computer in writing, storing, and editing NC programs. DNC involves using a computer as the controller for one or more NC machines, *Figure 20-71.* A more advanced form of DNC is called "distributive numerical control," *Figure 20-72.*

In this concept there is a host coupler and several intermediate computers that are tied to NC machines, robots, and other NC manufacturing devices. The main host computer is the central repository for all programs for all jobs. Specific programs for specific jobs are "dumped" from the host computer to the intermediate computers. This cuts down on the amount of time required to get the programmed instructions from the computer to the machine tool. CNC machines are available for most machining processes, including cleaning and finishing, inspection and quality control, pressing and forming, and material removal.

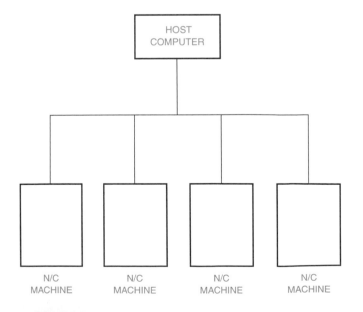

FIGURE 20-71 *Direct numerical control* (Adapted from Seames, *Computer Numerical Control,* Delmar Publishers Inc.)

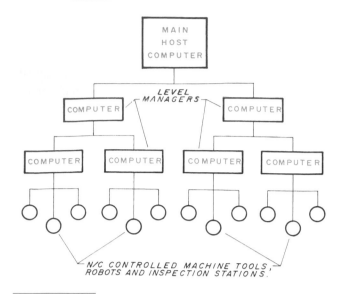

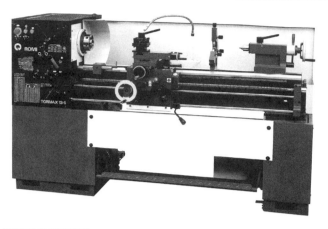

FIGURE 20-74 *Turning process* (Courtesy Bridgeport Machines Inc.)

FIGURE 20-72 *Distributive numerical control* (Adapted from Seames, *Computer Numerical Control*, Delmar Publishers Inc.)

FIGURE 20-75 *CNC machining center* (Courtesy Cincinnati Milacron)

FIGURE 20-73 *CNC machining center* (Courtesy Bridgeport Machines Inc.)

Figure 20-73 is a CNC bed-type vertical machining center.

Figure 20-74 is a CNC milling machine.

Figure 20-75 is a Cincinnati Milacron Maxim 500 Horizontal CNC Machining Center.

All of the CNC machines shown in Figures 20-73 through 20-75 have controllers. There are as many different controllers as there are CNC machines. Regardless of the type of controller, there must be some means of storing and inputting programs that instruct the machines. These means are called input media.

The most frequently used types of input media are punched tape and magnetic tape. Punched tape may be either the paper or mylar (plastic) variety. Mylar is more widely used because it is less prone to damage.

With earlier CNC machines a tape punch was used to put holes in a tape that was fed through it. The holes represented the program code. The punched tape was run through a tape reader that sensed the holes and fed the code into the controller. A problem with this type of input media is that it is easy to make an error when punching the tape and difficult to correct it.

An improvement to this early, cumbersome method of input is to interface a tape punch directly with a microcomputer. The code is typed via the microcomputer's keyboard. As the code is entered, it is stored so that it can be checked for accuracy and edited. Once correct, it is fed to the tape punch, and the tape is prepared. This solves the error correction problem.

Magnetic tape is now more popular than punched tape. With this type of medium the program code is affixed to the tape as magnetic spots rather than as holes punched through it. Because this type of medium is becoming so popular, standards for format and coding have been developed by the Electronics Industries Association.

CNC is the most modern and most technologically advanced method of controlling manufacturing machines. However, it is not the best control method in every case. Like any technological development, CNC has its advantages and disadvantages. There are times when the traditional manual approach is better. CNC is indicated when one or both of the following factors is important:

- Increased productivity
- Decreased labor and production costs

When CNC is indicated, there are advantages and disadvantages with which students of CAD/CAM should be familiar. The advantages include:

1. Better production and quality control
2. Increased productivity, flexibility, accuracy, and uniformity
3. Reduced labor, production, tool, and fixture costs
4. Less parts handling and tool storage

There are other advantages of CNC; however, these are the most important. There are also disadvantages associated with CNC. Those most frequently stated are:

1. High up-front costs
2. Higher operating costs
3. Retraining costs
4. Potential personnel problems

The advantages of CNC outweigh the disadvantages. But the list of disadvantages does make the point that it is not always the appropriate choice.

CNC Applications

CNC applications can be viewed at two levels. First, there is the industry level. At this level, all of the various manufacturing industries use CNC machines. These include:

- Aerospace manufacturers
- Electronics manufacturers
- Automobile, truck, and bus manufacturers
- Appliance manufacturers
- Tooling manufacturers
- Locomotive/train manufacturers

The second application level to consider is the machine level. At this level, CNC applications include:

- Cleaning and finishing applications
- Material removal applications
- Presswork and forming applications
- Inspection/quality control applications

Cleaning and finishing machines perform such tasks as washing, degreasing, blasting, lapping, grinding, and deburring. There are currently several CNC cleaning and finishing machines on the market.

Material removal machines include mills, lathes, grinders, drills, and saws. All such machines may be CNC controlled. In fact, this is the machine level application area most strongly associated with CNC.

Presswork and forming machines perform such tasks as pressing, stamping, blanking, punching, bending, and swagging. There are currently several CNC presswork and forming machines on the market.

Inspection and quality control machines perform such tasks as measuring, gaging, testing, and weighing. There are several CNC inspection and quality control machines on the market.

Industrial Robots

As is sometimes the case with new and emerging technological developments, there are a variety of definitions used for the term "robot." Depending on the definition used, the number of robot installations in this country and others will vary widely. There are a variety of single-purpose machines used in manufacturing plants that, to the lay person, would appear to be robots. These machines are hardwired to perform one function. They cannot be reprogrammed to perform a different function. These single-purpose machines do not fit the definition of industrial robots that is coming to be widely accepted. This definition is the one developed by the Robot Institute of America:

> A robot is a reprogrammable multifunctional manipulator designed to move material, parts, tools, or specialized devices through variable programmed motions for the performance of a variety of tasks.

Notice that the RIA's definition contains the words "reprogrammable" and "multifunctional." It is these two characteristics that separate the true industrial robot from the various single-purpose machines used in modern manufacturing firms. The term "reprogrammable" implies, first, that the robot operates according to a written program, and, second, that this program can be rewritten to accommodate a variety of manufacturing tasks. The term "multifunctional" means that the robot is able, through reprogramming and the use of a variety of end effectors, to perform a number of different manufacturing tasks. Definitions written around these two critical characteristics are becoming the accepted definitions among manufacturing professionals. *Figures 20-76* and *20-77* are examples of modern industrial robots that fit this definition.

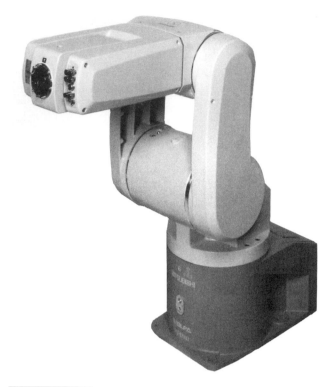

FIGURE 20-76 *Industrial robot* (Courtesy Rixan Associates Inc.)

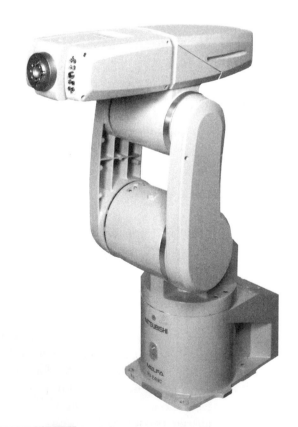

FIGURE 20-77 *Industrial robot* (Courtesy Rixan Associates Inc.)

The microprocessor was the enabling device with regard to the wide-scale development and use of industrial robots. But major technological developments do not take place simply because of a new capability. Something must provide the impetus for taking advantage of the new capability. In the case of industrial robots, the impetus was economics.

In the 1970s it became imperative for the U.S. to produce better products at lower costs in order to be competitive with foreign manufacturers. Other factors, such as the need to find better ways of performing dangerous manufacturing tasks, contributed to the development of industrial robots. However, the principal rationale has always been, and is now, improved productivity.

Industrial robots offer a number of benefits that account for the rapid current and projected growth of industrial robot installations.

1. Increased productivity
2. Improved product quality
3. More consistent product quality
4. Reduced scrap and waste
5. Reduced reworking costs
6. Reduced raw goods inventory
7. Direct labor cost savings
8. Savings in related costs, such as lighting, heating, and cooling
9. Savings in safety-related costs
10. Savings from correctly forecasting production schedules

THE ROBOT SYSTEM

Work in a manufacturing setting is not accomplished by a robot. Rather, it is accomplished by a robot system. A robot system has four major components: the controller, the robot arm or manipulator, the end-of-arm tools, and the power sources, *Figure 20-78*. These components, coupled with the various other pieces of equipment and tools needed to perform the job for which a robot is programmed, are called the robot's work cell.

Figure 20-79 contains a schematic drawing of the work cell for the Cincinnati Milicron Corporation's T3-726 robot, which is used for TIG welding. Notice that the work cell contains not only the robot system but an index table, an operator's safety shield, and special welding equipment.

The contents of a robot's work cell will vary according to the application of the robot. However, the one constant in a robot's work cell is the robot system.

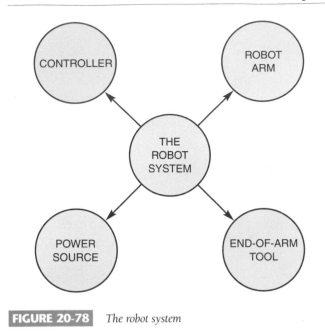

FIGURE 20-78 *The robot system*

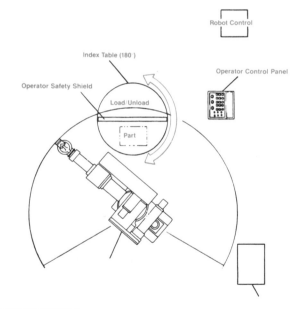

FIGURE 20-79 *Robot work cell* (Courtesy Cincinnati Milacron)

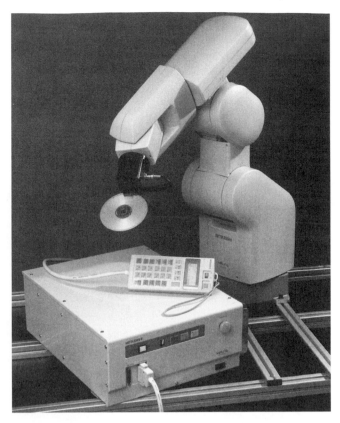

FIGURE 20-80 *Robot with programming device* (Courtesy UNIMATION Inc.)

The Controller. A *robot* is a special-purpose device similar to a computer. As such, it has all the normal components of a computer, including the central processing unit—made up of a control section and an arithmetic/logic section—and a variety of input and output devices.

The controller for a robot system does not look like the microcomputer one is used to seeing on a desk. It must be packaged differently so as to be able to withstand the rigors of a manufacturing environment. Typical input/output devices used in conjunction with a robot controller include teach stations, teach pendants, a display terminal, a controller front panel, and a permanent storage device.

Teach terminals, teach pendants, or front panels are used for interacting with the robot system. These devices allow humans to turn the robot on, write programs, and key in commands to the robot system. Display terminals give operators a soft copy output source. The permanent storage device is a special device on which reusable programs can be stored. *Figure 20-80* is a photograph of a robot system. Notice the teaching box and controller.

Mechanical Arm. The mechanical arm, or manipulator, is the part of a robot system with which most people are familiar. It is the part that actually performs the principal movements in doing manufacturing-oriented work. Mechanical arms are classified according to the types of motions of which they are capable. The basic categories of motion for mechanical arms are rectangular, cylindrical, and spherical. *Figure 20-81* is a drawing of the mechanical arm for the Unimate Puma Series 700 robot.

End-of-Arm Tooling. The human arm by itself can perform no work. It must have a hand, which in turn, grips a tool. The mechanical arm of a robot, like the human arm, can perform no work. It, too, must have a hand, a

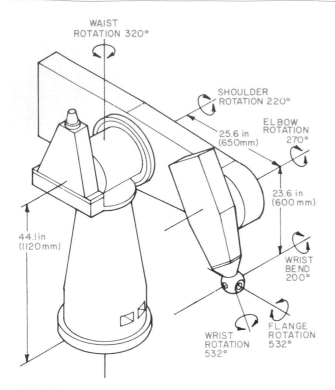

FIGURE 20-81 *Robot mechanical arm* (Courtesy UNIMATION Inc.)

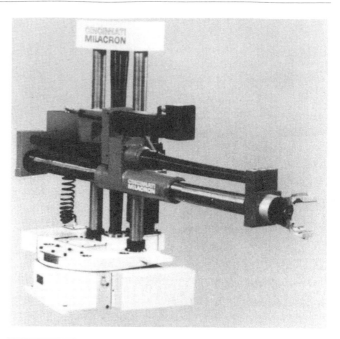

FIGURE 20-82 *End-of-arm-gripper* (Courtesy Cincinnati Milacron) *(Note: Safety equipment may have been removed or opened to clearly illustrate products and must be in place prior to operation.)*

tool, or a device that combines both functions. On industrial robots, the function of the wrist is performed by the tool plate. The tool plate is a special device to which end-of-arm tools are attached. The tools themselves vary according to the types of tasks the robot will perform. Any special device or tool attached to the tool plate to allow the robot to perform some specialized task is classified as an end-of-arm tool or an end effector.

The Power Source. A robot system can be powered by three different types of power. The power sources used in a robot system are electrical, pneumatic, or hydraulic. The controller, of course, is powered by electricity, as is any computer. The mechanical arm and end-of-arm tools may be powered by either pneumatic or hydraulic power. Some robot systems will use all three types of power. For example, a given robot might use electricity to power the controller, hydraulic power to manipulate the arm, and pneumatic power to manipulate the end-of-arm tool. Hydraulic power is fluid based. Pneumatic power comes from compressed gas.

Figure 20-82 shows a widely used industrial robot with a gripper device as an end-of-arm tool. The tool plate is the rectangular plate immediately behind the gripper.

CIM/FMS

Computer-integrated manufacturing is the most modern, most automated form of production. It involves tying different phases of production together into one wholly integrated system. The term "flexible manufacturing system" is sometimes used synonymously with computer-integrated manufacturing system. Actually, however, a flexible manufacturing system is one type of CIM system designed for medium-range production volumes and moderate flexibility. Other types of CIM systems are special systems and manufacturing cells. There are other terms that have been used to describe the same concept, but CIM and FMS are the two that are most widely used today.

There are a number of different types of computer-integrated manufacturing systems. This is because each system is designed to meet the specific needs of the individual manufacturing setting where it will be used. However, in general, a CIM system is any computerized manufacturing system in which numerically controlled machines are joined together and connected by some form of automated material handling system.

Figure 20-83 shows CNC machining centers linked together in an automated cell. In this system, two CNC machining centers are linked together by a cart-type material handling system. The cart system is further linked to four part load/unload stations to permit untended operation.

In a CIM system such as this, the computer is used in several ways: (1) it is used to control CNC machines; (2) it is used to control the materials handling system; and (3) it is used for monitoring all production tasks accomplished by the system.

FIGURE 20-83 *Example of a CIM system* (Courtesy Cincinnati Milacron)

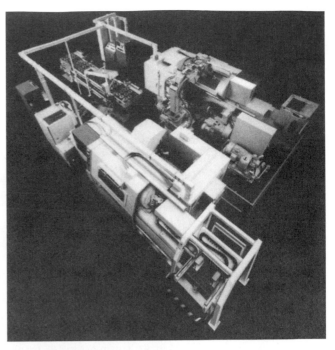

FIGURE 20-84 *FMS* (Courtesy Cincinnati Milacron)

Figure 20-84 is a photograph of a flexible manufacturing machine/robot cell. In this CIM system, rough castings are machined and inspected. The system contains two CNC machines, gaging devices, a conveyor, and two robots. The rough castings are loaded onto the conveyor by a robot, and finished castings are unloaded by a robot.

Figure 20-85 is a photograph of a composite tape laying (CTL) system. The CTL system automatically dispenses tape at speeds of 100 feet per minute. Tape can be automatically laid in 3″, 6″, or 12″ widths, debulked, cut, and overlaid ply-on-ply. Such systems are used in the manufacture of aircraft and space products.

FIGURE 20-85 *CTL system* (Courtesy Cincinnati Milacron)

In spite of advances in manufacturing automation, there is human involvement with a CIM system. Typically, this involvement falls into six broad categories: (1) loading raw stock and materials onto the system for processing; (2) unloading processed workpieces from the system; (3) changing tools on machines within the system; (4) setting tools on machines within the system; (5) continuous maintenance of the system; and (6) occasional repair of the system when there is a breakdown or malfunction.

Stand-alone CNC machines are used in low-volume manufacturing applications that require a high degree of flexibility in order to produce a wide variety of parts. They represent one extreme of the manufacturing spectrum. At the other extreme are transfer lines. Transfer lines are used in high-volume manufacturing applications where all parts produced are identical. Transfer lines are not flexible; they cannot accommodate variety.

Transfer lines are a major component of what is sometimes referred to as "Detroit Automation." A transfer line consists of several workstations linked together by materials-handling devices that transfer workpieces from station to station.

The first station holds the raw material. The last station is a bin for collecting the finished parts. Each station in-between performs some type of operation on the parts as they pass through. The transfer of parts from station to station, and the work performed on them, are automatic. Transfer lines are appropriate in situations that involve the high-volume production of identical parts.

There has always been a gap between these two extremes, and there has always been a need to fill this gap. The medium-volume, moderate-flexibility manufacturing situation has long been a problem in need of a solution. CIM systems are such a solution. CIM fills the void between stand-alone CNC machines and transfer lines, *Figure 20-86*.

CIM offers a number of advantages that, when taken together, form the rationale for this modern approach to medium-volume, flexible production. The most important of these are:

1. Produces families of parts
2. Accommodates the random introduction of parts

3. Requires less lead time
4. Allows a closer relationship between parts to be produced and workpieces loaded onto the system
5. Allows better machine utilization
6. Requires less labor

CIM systems reduce the amount of direct and indirect labor costs associated with finished workpieces. This is because CIM systems require less human involvement in producing them. Ten to 12 traditional CNC machines require 10 to 12 operators. A CIM system with 10 to 12 CNC machines might require as few as four people. This means less direct labor. Indirect labor costs resulting from such tasks as materials handling are also reduced with CIM. This is because most material handling in CIM systems is automated.

TYPES OF CIM SYSTEMS

CIM systems fall between transfer lines and stand-alone CNC machines on a graph of production volume versus part variety flexibility. By applying these same criteria—production volume and part variety—CIM systems can be divided into three categories, *Figure 20-87*.

1. Special systems
2. Flexible manufacturing systems
3. Manufacturing cells

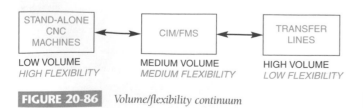

FIGURE 20-86 *Volume/flexibility continuum*

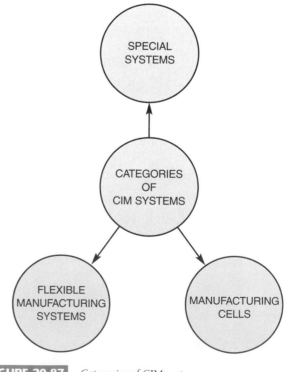

FIGURE 20-87 *Categories of CIM systems*

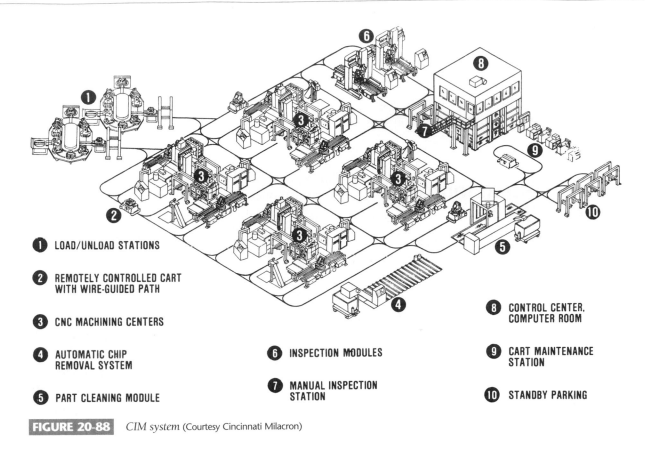

FIGURE 20-88   *CIM system* (Courtesy Cincinnati Milacron)

① LOAD/UNLOAD STATIONS

② REMOTELY CONTROLLED CART WITH WIRE-GUIDED PATH

③ CNC MACHINING CENTERS

④ AUTOMATIC CHIP REMOVAL SYSTEM

⑤ PART CLEANING MODULE

⑥ INSPECTION MODULES

⑦ MANUAL INSPECTION STATION

⑧ CONTROL CENTER, COMPUTER ROOM

⑨ CART MAINTENANCE STATION

⑩ STANDBY PARKING

The preceding systems are listed in order from the least flexible (special systems) to the most flexible (manufacturing cells). Of all CIM systems, those classified as special systems are capable of the most volume production and the least flexibility. Special systems might produce as many as 15,000 parts per year, but they are limited to less than 10 different part types.

Flexible manufacturing systems represent the middle group in CIM systems. An FMS might have a production volume as high as 2,000 parts per year and a capability of handling as many as 200 different parts. *Figure 20-88* is a diagram of an FMS produced by Cincinnati Milacron.

The most flexible of CIM systems are manufacturing cells, which, in turn, are capable of the lowest production volumes. A manufacturing cell might have a production volume as low as 500 parts per year but a capability of handling as many as 500 different part types.

Just-in-Time Manufacturing (JIT)

Just-in-time manufacturing (JIT) is an approach to manufacturing that seeks to completely eliminate waste. The term *waste* is used here in the broadest sense and encompasses rejected parts, wasted personnel time, wasted machine time, excessive inventory, and any other situation that wastes time, material, or any other manufacturing resource. JIT has the following basic characteristics.

PROCESS SYNCHRONIZATION

This characteristic is what gives JIT its name. *Process synchronization* means that all processes involved in producing a product are in balance and synchronized so that production flows smoothly and continuously. Traditionally, production processes have been operated on a *hurry-up-and-wait* basis. While one process is hurrying to complete its task, the next one in succession is idle. Process synchronization smooths this out so that each successive process receives its material just-in-time to be processed.

SIMPLICITY

This characteristic is fundamental to JIT, the view being that the simpler things are, the easier JIT will be to accomplish. With JIT a continuous effort is made to simplify processes so that they can be performed with a continually decreasing amount of resources.

Wholistic View

This characteristic means that JIT cannot be accomplished by any individual unit operating in isolation. The entire company—all units—must be included in the effort. All processes, from purchasing to accounting and production, must be synchronized in order for JIT to work properly.

Manufacturing Resource Planning (MRPII)

Manufacturing resource planning is often referred to as MRPII to distinguish it from its predecessor, which was material requirements planning (MRP). MRP was a way to reduce inventory and waste through systematic planning for material requirements. MRPII is a broader concept that allows companies to systematically plan for all resources needed in manufacturing (i.e., material, capacity of production systems, personnel, time, etc.).

An MRPII system is computer-based and typically consists of at least the following four components: (1) control database; (2) planning database; (3) functional modules; and (4) integrated modules. The control database contains inventory data and a master file on workcenters. The planning database contains a bill of materials file and a process routing file. Functional modules include production scheduling, capacity planning, material planning, and job cost reporting. Integrated modules or modules that might be integrated include such areas as sales, accounting, customer order processing, payroll, etc. Using MRPII properly simplifies the implementation of JIT because it helps synchronize all processes that are either directly or indirectly involved in producing the products in question.

Statistical Process Control (SPC)

Statistical process control (SPC) is a process that uses statistical control charts to monitor production processes in an attempt to reduce the amount of scrap and rework. Tolerances are transformed into upper and lower control limits that are placed on a chart. Periodically, points that correspond to the processing performance of the system are plotted on this chart. The plots are monitored visually or electronically. As long as all points plotted fall within established limits on the chart, the process is under control. If a point falls outside of the limits, cor-

rections should be made immediately to prevent further production of unacceptable parts.

Summary

- Engineering materials are the raw materials that are converted into finished products in manufacturing. They can be divided into two broad categories: metals and nonmetals. Metals include steels, alloy steels, and a variety of high-performance alloys. Nonmetals include plastics, composites, elastomers, and ceramics.

- Casting is a process in which molten material is poured into a mold that contains the desired shape in the form of a cavity. The principal types of casting processes are sand casting, investment casting, centrifugal casting, and die casting.

- Forging is the process of forming metals under pressure using a variety of different processes. The most common forms of forging are drop forging and press forging. Other variations include rolling and upsetting.

- Extruding is a process of forcing metal through a die of a desired form and cross section.

- Stamping is a process of using dies to cut or form metal sheets or strips into a desired form.

- Machining is the process of removing metal with machine tools and cutters to achieve a desired form or feature. The principal types of machine tools are lathes, milling machines, drill presses, saws, and grinders.

- Special work holding devices are used to produce parts in high-volume production runs. These devices are called jigs and fixtures. Both jigs and fixtures hold, support, and locate the workpiece. Jigs guide the cutting tool through a hardened bushing during the cutting cycle. Fixtures reference the cutting tool by means of a set block.

- Heat treatment is a series of processes used to alter or modify the existing properties in a metal to obtain a specific condition required for a workpiece.

- Nontraditional machining encompasses a number of advanced machining processes that have been developed in response to the demands of modern manufacturing for more effective, more efficient ways to machine the many new materials that are now available.

- Computer-aided manufacturing (CAM) encompasses any manufacturing process that is automated through the use of computers. Examples of CAM are computer numerical control of machine tools and robotics.

- A robot is a reprogrammable, multifunctional manipulator designed to move materials, parts, tools, or specialized devices through variable programmed motions for the performance of a variety of tasks.

- Computer-integrated manufacturing (CIM) is the most modern form of production. It involves tying different phases of production together.

- A flexible manufacturing system is one type of CIM system designed for medium-range production volumes and moderate flexibility.

- Just-in-time manufacturing (JIT) is an approach to manufacturing that seeks to eliminate waste. Waste encompasses rejected parts, personnel time, machine time, excessive inventory, or any other resource. The key to JIT is process synchronization.

- Manufacturing resource planning or MRPII is a computer-based approach that allows companies to systematically plan for all resources needed in manufacturing (i.e., material, personnel, time, capacity of production systems, etc.).

Review Questions

Answer the following questions either true or false.

1. The term *lead time* means the interval of time from when the shop receives the drawing until production begins.

2. The cope and the cheek are terms used to describe the top and bottom sections of the flask used for casting.

3. The amount of shrinkage and the amount of expansion must be considered when making a pattern for casting.

4. The approximate percentage of shrinkage for steel is 3/16″ per foot.

5. Investment casting is also referred to as lost wax casting.

6. Centrifugal casting is the casting process in which molten metal is forced into metal dies.

7. There are usually six stations in a drop forge die.

8. Operations such as slitting, trimming, and shaving control size.

9. A drop hammer is the principal tool used in metal stamping.

Answer the following questions by selecting the best answer.

1. Which of the following is **not** one of the major factors to consider before any part is made?
   a. The material the part will be made of
   b. The overall size and shape of the finished object
   c. Types of operations required
   d. The cost of manufacturing the part

2. Which of the following is **not** one of the primary methods used to fabricate manufactured products?
   a. Welding
   b. Extruding
   c. Stamping
   d. Forging

3. Which of the following is **not** one of the common forging processes?
   a. Rough forging
   b. Drop forging
   c. Press forging
   d. Rolling

4. Which of the following is **not** one of the stamping operations that produces blanks?
   a. Cutting off
   b. Shearing
   c. Trimming
   d. Blanking

5. Which of the following stamping operations does not produce holes?
   a. Punching
   b. Piercing
   c. Perforating
   d. All of the above

6. Which of the following is the most common type of fixture?
   a. Plate fixture
   b. Angle plate fixture
   c. Vise jaw fixture
   d. Indexing fixture

7. Which of the following is **not** one of the four operations used to bend or form parts?
   a. Machining
   b. Drawing
   c. Forming
   d. Bending

8. Which of the following is **not** one of the four machine tools used for machining parts?
   a. Drill press
   b. Fixtures
   c. Milling machine
   d. Lathe

9. Which of the following is **not** true regarding the plate jig?
   a. It is the most common form of jig.
   b. It is the least common form of jig.
   c. A typical variation of the plate jig is the table jig.
   d. A typical variation of the sandwich jig is the table jig.

10. Which of the following is **not** one of the operations normally performed on a drill press?
    a. Cutting
    b. Tapping
    c. Reaming
    d. Drilling

11. Which of the following is **not** an element of JIT?
    a. To eliminate waste
    b. Process synchronization
    c. To build up inventory
    d. Each process receives its material just in time

12. Which of the following is **not** an element of MRPII?
    a. Control database
    b. Planning database
    c. Functional database
    d. Integrated database

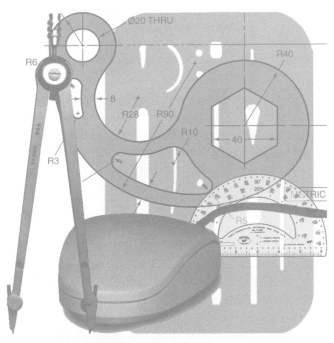

Chapter 21

DRAFTING APPLICATIONS: PIPE, STRUCTURAL, ARCHITECTURAL, CIVIL ENGINEERING, AND GIS

KEY TERMS

Anchor bolt plan

Attribute data

Buffering

Check valves

Contour lines

Eave strut connection details

Electrical plan

Elevations

Extracting

Floor plan

Foundation plan

Gate valves

Generalizing

Georelational data model

Geographic information systems

Girt connection details

Globe valves

Heating-ventilation/air-conditioning plan (HVAC)

Interval

Legal descriptions

Lot and block

Manual of Steel Construction

Mark number

Metes and bounds

Network analysis

Placing drawings

Plot plan

Point-in-polygon responses

Polygon overlay

Pre-engineering

Purlin connection details

Rectangular system

Relational database management systems

Roof plan

Spatial analysis

Spatial data

Spatial queries

Structure

Topology

What-if scenarios

CHAPTER OUTLINE

Types of pipe • Types of joints and fittings • Types of valves • Pipe drawings • Dimensioning pipe drawings • Structural drafting • Pre-engineered metal buildings • Architectural drafting • Civil engineering drafting • Geographic information system • Summary • Review questions • Chapter 21 problems

CHAPTER OBJECTIVES

Upon completion of this chapter, students should be able to do the following:

■ Explain the appropriate applications of the most widely used types of pipe.

■ Describe the three broad classifications of pipe joints and fittings.

■ Explain the most widely used types of pipe valves.

■ Demonstrate proficiency in developing pipe drawings.

■ Describe the most widely used components in structural construction.

■ Demonstrate proficiency in developing structural drawings.

■ Demonstrate proficiency in developing plans for pre-engineered metal buildings.

■ Explain each of the components in a set of residential plans.

■ Compare/contrast residential and commercial architectural plans.

■ Demonstrate proficiency in developing architectural plans.

(continued)

■ Demonstrate proficiency in fundamental civil engineer drawing.

■ Explain the applications of geographic information systems (GIS).

This is an application chapter that allows students to use the design and drafting skills they have learned in developing real-world projects in the areas of pipe, structural, and architectural drafting. Projects developed in this chapter relate directly to specific design and drafting career fields.

Types of Pipe

The wide variety of fluids used by modern society requires a number of different types of pipe. The most widely used of these are:

- Steel pipe
- Cast iron pipe
- Brass and copper pipe
- Copper tubing
- Plastic pipe

STEEL PIPE

Steel pipe is well suited for high-pressure and high-temperature applications. It is frequently used in piping systems that transport water, oil, petroleum, and steam. Steel pipe comes in several cross-sectional configurations. The most widely used of these are the standard, extra strong, and double extra strong configurations, *Figure 21-1*. There are actually 10 different cross-sectional configurations for steel pipe.

Steel pipe is specified by a nominal diameter callout. The actual diameter will vary slightly from the nominal. The American National Standard Institute (ANSI) specifies pipe in 10 different schedules. Each schedule corresponds to a wall thickness. For example, standard and extra strong pipe are Schedules 40 and 80, respectively. The nominal diameter for pipe up to 12″ refers to the inside diameter. Callouts are given in inches or millimeters. The nominal diameter for pipe over 12″ refers to the outside diameter.

CAST IRON PIPE

Cast iron pipe is used most frequently for underground applications—to transport water, gas, and sewage. It is well suited for low-pressure steam connections also.

BRASS AND COPPER PIPE

Brass and copper pipe are used in applications where corrosion will be a problem. Because brass and copper are able to withstand corrosion, the expected lifetime of a piping system in a high-corrosive setting will be longer if brass or copper pipe is used.

COPPER TUBING

Copper tubing is used extensively in applications where vibration and misalignment are important factors. It is used in hydraulic and pneumatic applications, such as industrial settings involving robots, and in automotive settings.

PLASTIC PIPE

Plastic pipe is used extensively in modern piping settings. It is highly resistant to corrosion and chemical degradation. Plastic pipe is flexible and can be easily installed. However, it is not used where heat or pressure are factors. Plastic pipe does not have the chemical makeup to withstand heat or the wall strength to withstand high pressures.

Types of Joints and Fittings

Each kind of pipe explained in the previous section comes in straight length. However, piping systems require turns and branches and changes of size. In every instance where a pipe must change directions or size there is a joint. Joints are accomplished using fittings. There are three broad classifications of fittings:

1. Screwed
2. Flanged
3. Welded

Figure 21-2 illustrates these three types of fittings.

Engineers and drafters need to be able to specify pipe fittings. A given pipe fitting is specified by stating the nominal pipe size, the name of the fitting, and the material out of which the fitting is made. Any fitting that is used to connect different sizes of pipes is referred to as a "reducing fitting." When specifying reducing fittings, you must state the nominal pipe sizes for both the large and small end of the fitting. The large size is stated first. There are enough differences among screwed, flanged, and welded fittings that each must be studied separately.

SCREWED FITTINGS

Screwed fittings are generally used in applications requiring small diameter pipe of 2.5″ or less. The threaded end of the pipe and the internal threads on the fitting are usu-

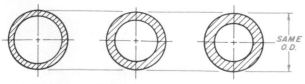

STANDARD   EXTRA STRONG   DOUBLE EXTRA STRONG

FIGURE 21-1   *Schedule of pipe*

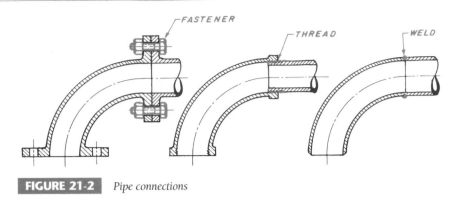

FIGURE 21-2 *Pipe connections*

ally coated with a special lubricant to seal the joint and to ease the connection process. *Figures 21-3, 21-4, 21-5* (page 792), and *21-6* (page 793) contain information on the most commonly used threaded type fittings. Figure 21-3 contains information on the 90° elbow, 90° street elbow, 45° elbow, 45° street elbow, and 90° reducing elbow. The table accompanying each illustration gives the type sizes with which that particular fitting can be used, the various dimensions in which it is available, and the weight of the fitting itself. Figure 21-4 contains information for blind flange, screwed flange, and flat band cap fittings, as well as bushings. Figure 21-5 contains information on union, coupling, cross, tee, and reducing tee fittings. Figure 21-6 contains information on plug, return bend, and lock nut fittings, as well as reducers and nipples.

There are two types of American Standard pipe threads: tapered and straight. Tapered threads are more common. Straight threads are usually used only for special applications. The ANSI Standard for Pipe Threads is ANSI/ASME B1.20.1-1983 (R1992). *Figure 21-7* (page 794) illustrates the conventions used for drawing pipe threads. *Figure 21-8* (page 794) illustrates the American National Standard taper pipe thread notation methods. Both types of threads have the same number of threads per inch.

FLANGED FITTINGS AND JOINTS

Some piping applications require that the piping systems occasionally be disassembled. When this is the case, flanged fittings are appropriate. Flanged fittings and adjoining pipes are bolted together. On occasion, flanged fittings and pipe may be welded together. However, they are normally bolted or glued together. *Figure 21-9* (page 794) contains size, weight, and dimensional data for the most common type of flanged fittings: 90° elbows, 45° elbows, tees, and reducers.

WELDED FITTINGS AND JOINTS

Some piping applications, such as high-pressure and high-temperature systems, require permanent fittings and joints. When this is the case, welded fittings are used. To accommodate the welding process, the connection ends of welded fittings as well as the ends of adjoining pipe are usually beveled. Welded fittings usually weigh less than flanged or screwed fittings, and they are easier to insulate.

Types of Valves

Fluids and gases do not just flow freely through piping systems. They must be regulated, and at certain points stopped. There are a number of different types of valves, including gate, globe, jet, swiveled joint, ball, check, and butterfly. The most frequently used of these are gate, globe, and check valves. These commonly used types of valves are illustrated in *Figures 21-10* (page 795), *21-11* (page 795), and *21-12* (page 795).

GATE VALVES

Gate valves are used to turn the flow of liquids through a pipe on or off without restricting the flow of the liquids through the valve or with as little restriction as possible. Gate valves are not meant to be used to regulate the degree of flow. *Figures 21-13* (page 796) and *21-14* (page 797) contain dimensional data for two commonly used gate valve configurations.

GLOBE VALVES

Globe valves are used to turn the flow of liquids through a pipe on and off, and they are used also to regulate the flow of fluids through the valve to the desired level. *Figures 21-15* (page 798) and *21-16* (page 799) contain dimensional data for two commonly used configurations of globe valves.

CHECK VALVES

Check valves are used to restrict the flow of liquids through a pipe in only one direction. A backward flow is checked by the valve, which is activated by any change in the direction of flow. *Figures 21-17* (page 800) and *21-18* (page 801) contain dimensional data for two commonly used configurations of check valves.

AVAILABLE STYLES AND SIZES

90° ELBOW

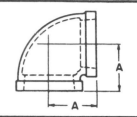

REFERENCE	PIPE SIZE, INCHES														
	1/8	1/4	3/8	1/2	3/4	1	1 1/4	1 1/2	2	2 1/2	3	3 1/2	4	5	6
A	11/16	13/16	15/16	1 1/8	1 5/16	1 7/16	1 3/4	1 15/16	2 1/4	2 11/16	3 1/8	3 7/16	3 3/4	4 1/2	5 1/8
Weight	.055	.060	.095	.145	.210	.355	.705	.790	1.180	1.670	2.590	3.250	4.065	6.900	9.800

90° STREET ELBOW

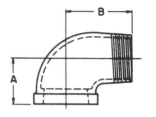

REFERENCE	PIPE SIZE, INCHES										
	1/8	1/4	3/8	1/2	3/4	1	1 1/4	1 1/2	2	2 1/2	3
A	11/16	13/16	15/16	1 1/8	1 5/16	1 7/16	1 3/4	1 15/16	2 1/4	2 11/16	3 1/8
B	1 1/8	1 5/16	1 7/16	1 5/8	1 7/8	2 1/8	2 1/2	2 11/16	3 3/16	3 13/16	4 1/2
Weight	.025	.045	.085	.140	.180	.205	.495	.750	1.250	1.850	2.900

45° ELBOW

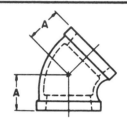

REFERENCE	PIPE SIZE, INCHES														
	1/8	1/4	3/8	1/2	3/4	1	1 1/4	1 1/2	2	2 1/2	3	3 1/2	4	5	6
A	11/16	3/4	13/16	7/8	1	1 1/8	1 5/16	1 7/16	1 11/16	1 15/16	2 3/16	2 3/8	2 5/8	3 1/16	3 15/32
Weight	.040	.040	.075	.115	.200	.260	.455	.605	.970	1.420	1.925	2.530	3.335	5.650	8.700

45° STREET ELBOW

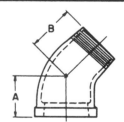

REFERENCE	PIPE SIZE, INCHES									
	1/8	1/4	3/8	1/2	3/4	1	1 1/4	1 1/2	2	2 1/2
A	—	—	—	1 1/16	1 3/16	1 3/8	1 17/32	1 5/16	1 5/8	2 1/16
B	—	—	—	1 3/16	1 3/8	1 17/32	1 23/32	2 1/4	2 3/8	2 1/2
Weight	—	—	—	.140	.180	.205	.495	.700	1.000	1.625

90° REDUCING ELBOW

Size	Weight		Size	Weight		Size	Weight
1/2" x 1/4"	.115		1 1/2" x 1"	.560		4" x 3"	3.065
1/2" x 3/8"	.120		1 1/2" x 1 1/4"	.720			
3/4" x 1/2"	.190		2" x 1 1/4"	1.000			
1" x 1/2"	.260		2" x 1 1/2"	1.030			
1" x 3/4"	.300		2 1/2" x 2"	1.745			
1 1/4" x 3/4"	.405		3" x 2"	2.205			
1 1/4" x 1"	.470		3" x 2 1/2"	2.315			

LATERALS — 1. 45° Y

    a. Size Range - 1/2'' through 3''
    b. Dimensional Standard - F-52618-C (Revision)

NOTE: Any reducing fittings not specifically listed can be produced on a Special Order basis and will be subject to a Special Order charge of 25% of the listed retail price.

(For additional fittings see Page 8)

FIGURE 21-3 *Pipe fittings* (Courtesy Latrobe Foundry Machine and Supply Co.)

AVAILABLE STYLES AND SIZES

FLANGE - Blind

REFERENCE	PIPE SIZE, INCHES										
	½	¾	1	1¼	1½	2	2½	3	3½	4	6
A	3½	3⅞	4¼	4⅝	5	6	7	7½	8½	9	11
B	7/16	7/16	7/16	½	9/16	⅝	11/16	¾	13/16	15/16	1
C	2⅜	2¾	3⅛	3½	3⅞	4¾	5½	6	7	7½	9½
D	⅝	⅝	⅝	⅝	⅝	¾	¾	¾	¾	¾	⅞
E	4	4	4	4	4	4	4	4	8	8	8
Weight	.275	.385	.560	.765	1.015	1.640	2.440	3.075	4.730	5.285	9.250

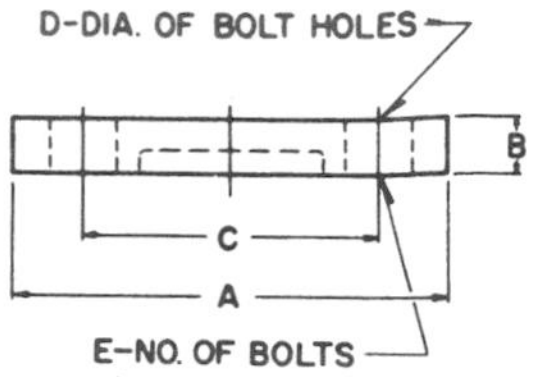

FLANGE - Screwed

FOR REDUCING, SLIP-ON, FLOOR AND WELDING NECK FLANGES SEE PAGE 8.

REFERENCE	PIPE SIZE, INCHES											
	½	¾	1	1¼	1½	2	2½	3	3½	4	5	6
A	3½	3⅞	4¼	4⅝	5	6	7	7½	8½	9	10	11
B	⅝	⅝	11/16	13/16	⅞	1	13/16	1¼	1¼	15/16	17/16	19/16
C	2⅜	2¾	3⅛	3½	3⅞	4¾	5½	6	7	7½	8½	9½
D	⅝	⅝	⅝	⅝	⅝	¾	¾	¾	¾	¾	⅞	⅞
E	4	4	4	4	4	4	4	4	8	8	8	8
Weight	.255	.375	.550	.740	.970	1.530	2.385	2.835	4.250	4.565	5.650	6.850

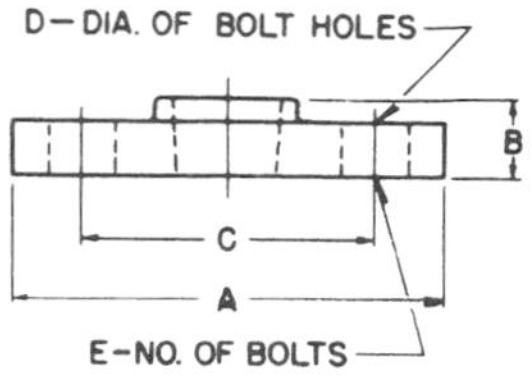

CAP - Flat Band

REFERENCE	PIPE SIZE, INCHES														
	⅛	¼	⅜	½	¾	1	1¼	1½	2	2½	3	3½	4	5	6
A	19/32	25/32	27/32	1 1/16	1 5/32	1¼	1 5/16	1 15/32	1 9/16	2 1/32	2 1/16	2 3/32	2 3/16	2⅜	2⅝
Weight	.015	.025	.050	.075	.100	.150	.290	.350	.450	.760	1.435	1.825	3.080	4.950	6.500

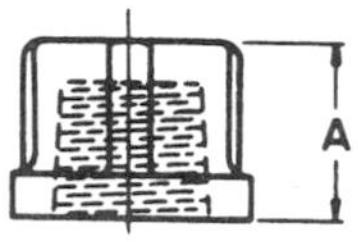

BUSHINGS

Size	Weight	Size	Weight	Size	Weight	Size	Weight
¼" x ⅛"	.010	1" x ¾"	.070	2" x 1¼"	.325	3½" x 3"	.750
⅜" x ⅛"	.020	1¼" x ⅜"	.170	2" x 1½"	.285	4" x 1½"	1.860
⅜" x ¼"	.015	1¼" x ½"	.165	2½" x 1"	.785	4" x 2"	1.720
½" x ⅛"	.045	1¼" x ¾"	.145	2½" x 1¼"	.755	4" x 2½"	1.535
½" x ¼"	.030	1¼" x 1"	.130	2½" x 1½"	.715	4" x 3"	1.255
½" x ⅜"	.030	1½" x ½"	.215	2½" x 2"	.605	5" x 2"	2.600
¾" x ¼"	.055	1½" x ¾"	.210	3" x 1¼"	1.095	5" x 3"	2.350
¾" x ⅜"	.050	1½" x 1"	.205	3" x 1½"	1.015	5" x 4"	2.000
¾" x ½"	.045	1½" x 1¼"	.125	3" x 2"	.895	6" x 3"	3.950
1" x ¼"	.110	2" x ½"	.370	3" x 2½"	.705	6" x 4"	3.600
1" x ⅜"	.105	2" x ¾"	.370	3½" x 2"	1.120	6" x 5"	3.250
1" x ½"	.090	2" x 1"	.365	3½" x 2½"	.920		

FIGURE 21-4 *Pipe fittings* (Courtesy Latrobe Foundry Machine and Supply Co.)

AVAILABLE STYLES AND SIZES

UNION

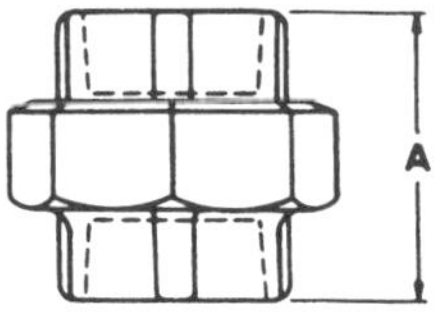

REFERENCE	PIPE SIZE, INCHES												
	1/8	1/4	3/8	1/2	3/4	1	1 1/4	1 1/2	2	2 1/2	3	3 1/2	4
A	1 9/16	1 13/16	1 15/16	2	2 1/8	2 3/4	3	3	3 1/4	3 5/8	4 1/8	4 7/8	5
Weight	.145	.130	.170	.255	.330	.505	.790	.815	1.250	1.745	2.865	5.000	5.800

COUPLING

FOR HALF COUPLINGS AND HEAVY WALL COUPLINGS SEE PAGE 8

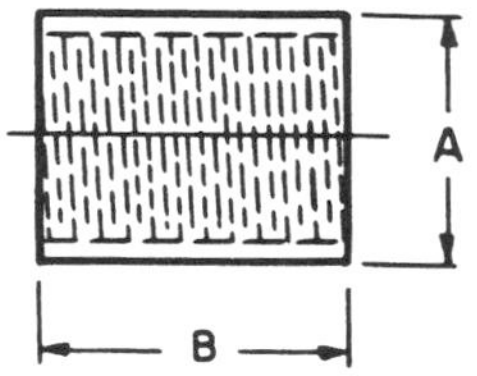

REFERENCE	PIPE SIZE, INCHES														
	1/8	1/4	3/8	1/2	3/4	1	1 1/4	1 1/2	2	2 1/2	3	3 1/2	4	5	6
A	19/32	3/4	29/32	1 1/16	1 11/32	1 5/8	1 31/32	2 15/64	2 23/32	3 5/16	3 15/16	4 7/16	4 15/16	6 1/16	7 3/16
B	15/16	1 1/32	1 5/32	1 5/16	1 9/16	1 13/16	2 1/16	2 5/16	2 9/16	2 7/8	3 1/16	3 7/16	3 7/16	4 1/8	4 1/8
Weight	.020	.025	.035	.060	.100	.135	.230	.310	.500	.730	1.015	1.740	2.040	3.300	4.500

CROSS

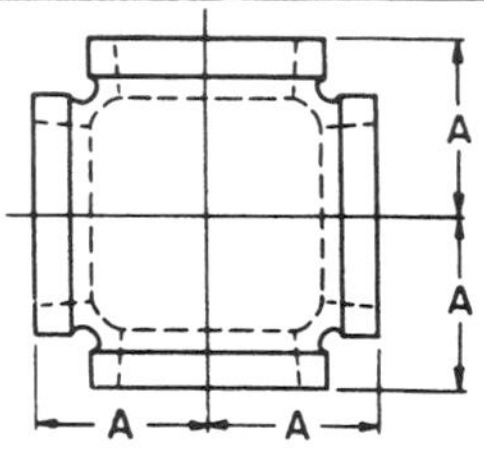

REFERENCE	PIPE SIZE, INCHES												
	1/8	1/4	3/8	1/2	3/4	1	1 1/4	1 1/2	2	2 1/2	3	3 1/2	4
A	11/16	13/16	15/16	1 1/8	1 5/16	1 7/16	1 3/4	1 15/16	2 1/4	2 11/16	3 1/8	3 7/16	3 3/4
Weight	.050	.085	.155	.215	.340	.565	.825	1.115	2.115	2.730	4.000	5.200	6.210

TEE

FOR LATERALS SEE PAGE 2

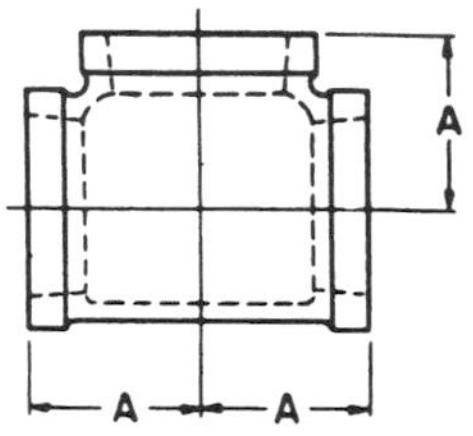

REFERENCE	PIPE SIZE, INCHES														
	1/8	1/4	3/8	1/2	3/4	1	1 1/4	1 1/2	2	2 1/2	3	3 1/2	4	5	6
A	11/16	13/16	15/16	1 1/8	1 5/16	1 7/16	1 3/4	1 15/16	2 1/4	2 11/16	3 1/8	3 7/16	3 3/4	4 1/2	5 1/8
Weight	.060	.070	.130	.180	.285	.435	.745	1.060	1.760	2.250	3.035	4.135	5.580	9.510	13.000

REDUCING TEE

NOTE: In the listing of reducing tees, the last dimension shown is the size of the branch.

Size	Wt.
1/2" x 1/2" x 1/4"	.195
1/2" x 1/2" x 3/8"	.190
1/2" x 1/2" x 3/4"	.320
3/4" x 1/2" x 1/2"	.320
3/4" x 1/2" x 3/4"	.305
3/4" x 3/4" x 3/8"	.315
3/4" x 3/4" x 1/2"	.305
3/4" x 3/4" x 1"	.465
1" x 1/2" x 1/2"	.190
1" x 3/4" x 1/2"	.515
1" x 3/4" x 3/4"	.495
1" x 3/4" x 1"	.465
1" x 1" x 3/8"	.495

Size	Wt.
1" x 1" x 1/2"	.485
1" x 1" x 3/4"	.465
1" x 1" x 1 1/4"	.850
1" x 1" x 1 1/2"	1.300
1 1/4" x 3/4" x 1 1/4"	.830
1 1/4" x 1" x 3/4"	.885
1 1/4" x 1" x 1"	.855
1 1/4" x 1 1/4" x 1/2"	.850
1 1/4" x 1 1/4" x 3/4"	.850
1 1/4" x 1 1/4" x 1"	.800
1 1/4" x 1 1/4" x 1 1/2"	1.170
1 1/2" x 3/4" x 1 1/2"	1.220
1 1/2" x 1" x 1"	1.300

Size	Wt.
1 1/4" x 1" x 1 1/2"	1.180
1 1/2" x 1 1/4" x 1"	1.230
1 1/2" x 1 1/4" x 1 1/4"	1.170
1 1/2" x 1 1/2" x 1/2"	1.245
1 1/2" x 1 1/2" x 3/4"	1.220
1 1/2" x 1 1/2" x 1"	1.180
1 1/2" x 1 1/2" x 1 1/4"	1.115
1 1/2" x 1 1/2" x 2"	1.280
2" x 1 1/2" x 1 1/2"	1.315
2" x 1 1/2" x 2"	1.755
2" x 2" x 1/2"	0.995
2" x 2" x 3/4"	1.290
2" x 2" x 1"	1.300

Size	Wt.
2" x 2" x 1 1/4"	1.325
2" x 2" x 1 1/2"	1.470
2" x 2" x 2 1/2"	1.910
2 1/2" x 2" x 2"	1.925
2 1/2" x 2 1/2" x 2"	2.095
3" x 3" x 2"	2.690
4" x 4" x 2"	4.015
4" x 4" x 3"	5.095
6" x 6" x 4"	10.850

FIGURE 21-5 *Pipe fittings* (Courtesy Latrobe Foundry Machine and Supply Co.)

AVAILABLE STYLES AND SIZES

PLUG — COUNTERSUNK PLUGS

REFERENCE	PIPE SIZE, INCHES														
	1/8	1/4	3/8	1/2	3/4	1	1 1/4	1 1/2	2	2 1/2	3	3 1/2	4	5	6
A	5/8	11/16	13/16	7/8	1	1 3/16	1 7/16	1 5/8	1 9/16	1 11/16	1 13/16	2	2 3/16	2 3/8	2 9/16
Weight	.010	.015	.020	.030	.055	.090	.150	.215	.315	.560	.755	.910	1.500	2.150	3.500

COUNTERSUNK PLUGS - Square Head   Size Range 3/8'' through 4''
Dimension Standard - ANSI   B16.14-1949

RETURN BEND - Close

REFERENCE	PIPE SIZE, INCHES								
	1/8	1/4	3/8	1/2	3/4	1	1 1/4	1 1/2	2
A	–	–	–	1 1/4	1 1/2	1 3/4	2 1/4	2 1/2	3 1/4
B	–	–	–	2 7/8	3 5/32	3 29/32	4 21/32	5 7/32	6 19/32
Weight	–	–	–	.205	.275	.440	.780	1.000	1.840

LOCK NUT

REFERENCE	PIPE SIZE, INCHES										
	1/8	1/4	3/8	1/2	3/4	1	1 1/4	1 1/2	2	2 1/2	3
A	–	5/32	3/16	1/4	5/16	3/8	7/16	1/2	5/8	3/4	15/16
Weight	–	.015	.020	.025	.045	.080	.105	.175	.345	.425	.740

REDUCERS

Size	Weight	Size	Weight	Size	Weight
1/4" x 1/8"	.040	1" x 3/4"	.240	2" x 1 1/2"	.825
3/8" x 1/8"	.060	1 1/4" x 3/4"	.330	2 1/2" x 2"	1.180
3/8" x 1/4"	.060	1 1/4" x 1"	.340	2 1/2" x 1 1/2"	1.200
1/2" x 1/4"	.100	1 1/2" x 3/4"	.480	3" x 2"	1.790
1/2" x 3/8"	.095	1 1/2" x 1"	.490	3" x 2 1/2"	1.750
3/4" x 1/4"	.150	1 1/2" x 1 1/4"	.560	4" x 3"	3.375
3/4" x 3/8"	.130	2" x 1"	.710	5" x 4"	4.75
3/4" x 1/2"	.160	2" x 1 1/4"	.745	6" x 4"	5.00
1" x 1/2"	.215			6" x 5"	5.00

NIPPLES - Close   ADDITIONAL NIPPLES *

Pipe Size	Length	Weight	Pipe Size	Length	Weight	Pipe Size	Length	Weight	Pipe Size	Length	Weight
1/8"	3/4"	.005	3/4"	1 3/8"	.045	1 1/2"	1 3/4"	.135	3"	2 5/8"	.570
1/4"	7/8"	.010	1"	1 1/2"	.075	2"	2"	.210	3 1/2"	2 3/4"	.720
3/8"	1"	.015	1 1/4"	1 5/8"	.105	2 1/2"	2 1/2"	.415	4"	2 7/8"	.895
1/2"	1 1/8"	.030									

* NIPPLES 1. Long Nipples a. Size Range — 1/8" through 4"
b. Length — through 12"

FIGURE 21-6   *Pipe fittings* (Courtesy Latrobe Foundry Machine and Supply Co.)

Pipe Drawings

Pipe drawings are used like any other design drawing, to convey the intentions of the designer and engineer to the tradespeople who will actually construct the piping system. There are two types of pipe drawings: single-line and double-line. *Figure 21-19* (page 802) is an example of a double-line drawing. *Figure 21-20* (page 802) is an example of the single-line version of the same drawing. Single-line drawings, since they are schematic, can be drawn much faster. However, drafters and engineers should be familiar both with double-line and single-line versions of pipe drawings, since both have applications on the job.

SINGLE-LINE DRAWINGS

Single-line pipe drawings consist of schematic, symbolic representations of valve fittings connected by a single line that represents the center line of the pipe. *Figure 21-21* (pages 802 and 803) contains all of the various schematic symbols used for representing pipe fittings and valves on

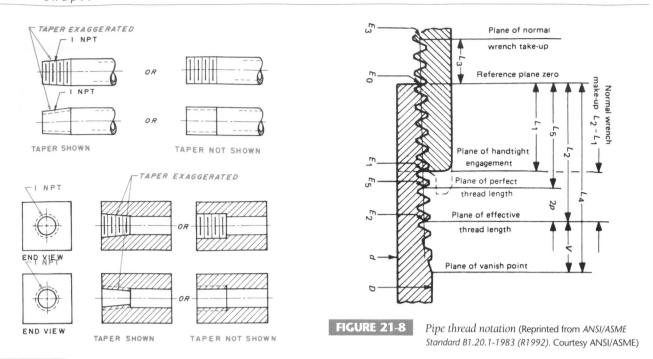

FIGURE 21-7 *Pipe thread conventions*

FIGURE 21-8 *Pipe thread notation* (Reprinted from *ANSI/ASME Standard B1.20.1-1983 (R1992).* Courtesy ANSI/ASME)

AVAILABLE STYLES AND SIZES

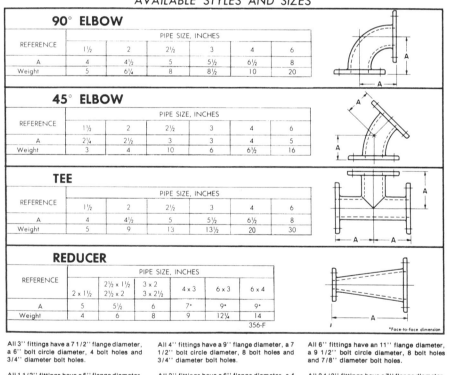

90° ELBOW

REFERENCE	PIPE SIZE, INCHES					
	1½	2	2½	3	4	6
A	4	4½	5	5½	6½	8
Weight	5	6¼	8	8½	10	20

45° ELBOW

REFERENCE	PIPE SIZE, INCHES					
	1½	2	2½	3	4	6
A	2¼	2½	3	3	4	5
Weight	3	4	10	6	6½	16

TEE

REFERENCE	PIPE SIZE, INCHES					
	1½	2	2½	3	4	6
A	4	4½	5	5½	6½	8
Weight	5	9	13	13½	20	30

REDUCER

REFERENCE	PIPE SIZE, INCHES					
	2 x 1½	2½ x 1½ 2½ x 2	3 x 2 3 x 2½	4 x 3	6 x 3	6 x 4
A	5	5½	6	7*	9*	9*
Weight	4	6	8	9	12¼	14

356-F

*Face-to-face dimension

All 3'' fittings have a 7 1/2'' flange diameter, a 6'' bolt circle diameter, 4 bolt holes and 3/4'' diameter bolt holes.

All 4'' fittings have a 9'' flange diameter, a 7 1/2'' bolt circle diameter, 8 bolt holes and 3/4'' diameter bolt holes.

All 6'' fittings have an 11'' flange diameter, a 9 1/2'' bolt circle diameter, 8 bolt holes and 7/8'' diameter bolt holes.

All 1 1/2'' fittings have a 5'' flange diameter, a 3 7/8'' bolt circle diameter, 4 bolt holes and 5/8'' diameter bolt holes.

All 2'' fittings have a 6'' flange diameter, a 4 3/4'' bolt circle diameter, 4 bolt holes and 3/4'' diameter bolt holes.

All 2 1/2'' fittings have a 7'' flange diameter, a 5 1/2'' bolt circle diameter, 4 bolt holes and 3/4'' diameter bolt holes.

FIGURE 21-9 *Pipe fittings* (Courtesy Latrobe Foundry Machine and Supply Co.)

single-line pipe drawings. *Figure 21-22* (page 804) contains the graphic symbols used for representing piping on single-line pipe drawings.

Like mechanical drawings, pipe drawings may be drawn using orthographic or isometric projection. Single-line orthographic projection is the recommended form of representation in most cases. *Figure 21-23* (page 804) is an example of single-line orthographic projection of a pipe drawing. Single-line isometric projection, as illustrated in *Figure 21-24* (page 804), is often used

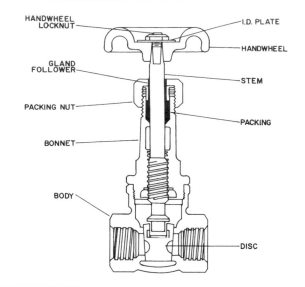

FIGURE 21-10 *Gate valve*

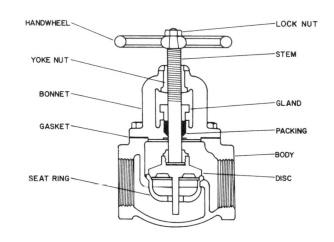

FIGURE 21-11 *Globe valve*

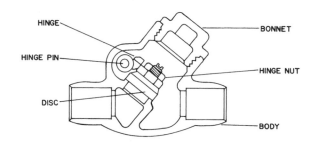

FIGURE 21-12 *Check valve*

for assembly and layout work because the drawing is easier for tradespeople to understand. Regardless of whether you are using orthographic or isometric projection, there are certain drawing conventions with which you should be familiar. The most important of these are crossings, connections, fittings, and machine/devices.

Crossings. When lines representing pipe on a drawing cross but do not intersect or make connections, they are usually drawn without breaks. However, on occasion it

will be necessary to show breaks so that tradespeople using the plans will be absolutely sure as to which pipe is nearest to them and which is farthest away. When such a need arises, breaks can be used as illustrated in *Figure 21-25* (page 805). When one uses break, it should be wide enough to be easily seen but not so wide as to create an inordinate gap. A widely used rule of thumb is to make the break from 5 to 10 times the thickness of the lines representing the pipe.

Connections. All of the various means of making connections at joints and fittings fall into two categories: permanent and detachable. On single-line pipe drawings, permanent connections are indicated by using a heavy dot. Detachable connections are indicated by a single thick line. Both of these methods are illustrated in *Figure 21-26* (page 805).

Fittings. When drawing fittings on single-line drawings it is sometimes necessary to be able to indicate whether a pipe is coming toward the viewer or going away from the viewer. It is also necessary at times to indicate whether the viewer is looking at a front view of the fitting or a rear view. All of the varied types of single-line drawing situations that you may confront when drawing fittings are covered in Figure 21-21. Refer to Figure 21-21 when drawing any type of single-line fitting symbol.

Machines and Devices. Most piping systems will tie into machines, devices, and other types of apparatus, *Figure 21-27* (page 805). When this is the case, the machines, devices, and other types of apparatus can be drawn using thin phantom lines.

Dimensioning Pipe Drawings

There are several dimensioning rules that apply specifically to pipe drawings with which you should be familiar. The most important of these are summarized in the following list.

- Dimensions for pipes and pipe fittings should be shown from center-to-center of pipe and to the outside face of the pipe end or flange.
- The length of pipe is not shown on the drawing except in rare cases.
- Pipe sizes are shown on the drawing using leader lines or as a general note.
- Fitting sizes are shown on the drawing using leader lines or as a general note.
- Pipes with bends should be dimensioned from vertex to vertex.
- The radii of bent pipe should be indicated using a leader line. Supplementary angles of bent pipes should be dimensioned in the normal manner.

(Text continues on page 802)

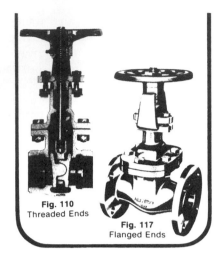

Fig. 110
Threaded Ends

Fig. 117
Flanged Ends

STAINLESS STEEL GATE
CLASS 150
FIGURES: 110 (½″ to 2″), **114** (½″ to 2″), **116** (1½″ to 12″), **117** (½″ to 12″)

DESIGN DESCRIPTION:

Outside Screw and Yoke	Socket Weld Ends (Fig. 114)
Bolted Bonnet	Buttwelding Ends (Fig. 116)
Rising Stem	Flanged Ends (Fig. 117)
Non-Rising Handwheel	Male-Female Bonnet Joint Sizes ½″ to 1″ Incl.
Integral Seats	Double Disc Ball-and-Socket Type Wedge
Threaded Ends (Fig 110)	Flat Faced Flanged Valves, Specify Fig. 117-FF

Fig. 116
1½″ to 12″

b
open

PARTS AND MATERIAL LIST:

DESCRIPTION	ALOYCO 18-8 SMO	DESCRIPTION	ALOYCO 18-8 SMO
1 body*	A351 GR CF8M	15 yoke bushing†	wrought B16 or cast B584-C83600
2 bonnet	A351 GR CF8M		
3 male disc	wrought-type 316 or	16 handwheel	malleable iron
4 female disc	cast A351 GR CF8M	17 yoke nut set screw	type 303
5 disc arm	A351 GR CF8M	18 disc arm pin	type 316
6 gasket†	comp. asbestos or teflon	19 handwheel key	steel
7 stem	type 316	20 yoke bushing nut†	wrought B16 or cast B584-C83600
8 bonnet bolt	A193 GR B8		
9 bonnet bolt nut	A194 GR 8F	21 grease fitting	steel
10 gland plate	A351 GR CF8M	22 yoke	CF8
11 gland follower	wrought-type 316 or cast A351 GR CF8M	23 yoke bolt	A193 GR B8
		24 yoke bolt nut	A194 GR 8F
12 packing†	braided asbestos or teflon	25 gland stud	A193 GR B8
13 gland bolt	A193 GR B8	26 gland stud nut	A194 GR 8F
14 gland bolt nut	A194 GR 8F		

Remarks: *Body material on threaded and welding end valves is ELC grade. Other Alloys available. (Refer to Introduction)

†Unless otherwise specified, these valves may be supplied at the manufacturer's option with: PTFE Gasket and Packing; Type 303 yoke bushing and nut. Such valves are limited to a temperature of 550 F and are so tagged.

Bonnet and Yoke
Detail 8″ to 12″ Incl.

DIMENSIONS IN MM AND INCHES:

SIZE	MM	15	20	25	32	40	50	65	80	100	150	200	250	300
	INCH	½	¾	1	1¼	1½	2	2½	3	4	6	8	10	12
a (110, 114)	mm	70	80	89	108	114	121	...	...	...	...	...	...	...
	inch	2¾	3	3½	4¼	4½	4¾							
a (116)	mm	—	—	—	—	165	216	241	283	305	403	419	457	502
	inch					6½	8½	9½	11⅛	12	15⅞	16½	18	19¾
a (117)	mm	108	118	127	—	165	178	191	203	229	267	292	330	356
	inch	4¼	4⅝	5		6½	7	7½	8	9	10½	11½	13	14
b (110, 114)	mm	210	210	238	277	286	337							
	inch	8¼	8¼	9⅜	10⅞	11¼	13¼							
b (116, 117)	mm	210	210	235	—	291	337	394	432	527	781	1016	1219	1524
	inch	8¼	8¼	9¼		11⁷/₁₆	13¼	15½	17	20¾	30¾	40	48	60
c	mm	89	100	100	127	127	150	178	191	250	305	356	406	457
	inch	3½	4	4	5	5	6	7	7½	10	12	14	16	18
d (114)	mm	13	13	13	13	13	16	—	—	—	—	—	—	—
	inch	½	½	½	½	½	⅝							
d (117)	mm	11	11	11	—	14	16	18	19	24	25	29	30	32
	inch	⁷/₁₆	⁷/₁₆	⁷/₁₆		⁹/₁₆	⅝	¹¹/₁₆	¾	¹⁵/₁₆	1	1⅛	1³/₁₆	1¼

WEIGHTS IN KG AND POUNDS:

		15/½	20/¾	25/1	32/1¼	40/1½	50/2	65/2½	80/3	100/4	150/6	200/8	250/10	300/12
110, 114	kg	2.8	2.9	4.0	5.9	7.0	9.0							
	lb	6.3	6.5	9	13	15.5	20							
116	kg	—	—	—	—	7.0	10	14	16	26	56	91	151	223
	lb					15.5	21	30	35	63	124	201	332	492
117	kg	3.6	4.3	5.4	—	9.9	14	19	24	40	72	124	206	305
	lb	8	9.5	12		22	30	42	51	88	158	275	455	675

Fig. 117

Fig. 114

Depth of Socket

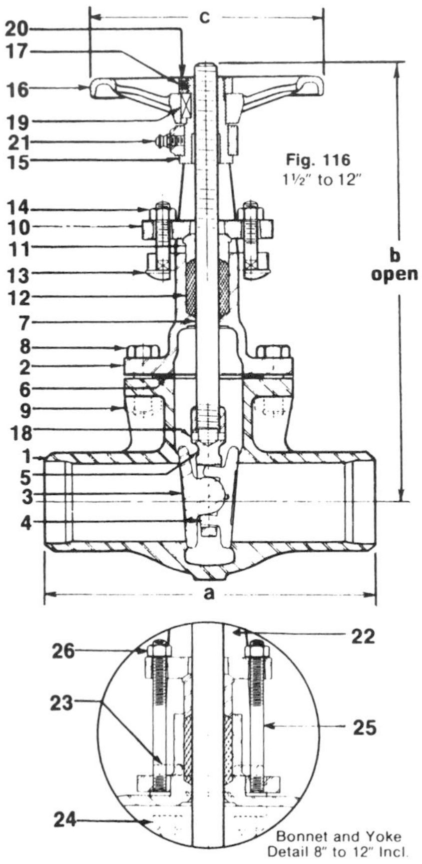

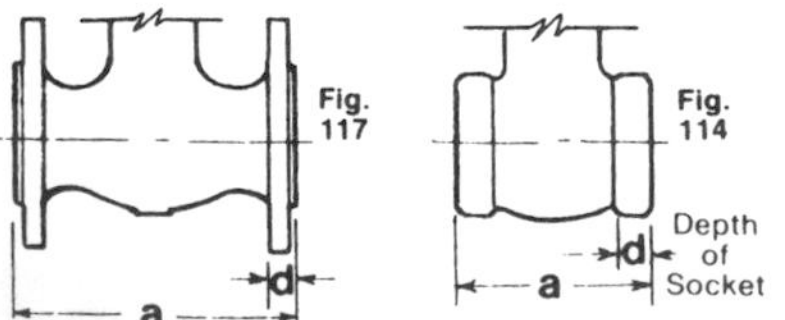

Note: For Valve Services at Temperatures above 700°F, the Aloyco Engineering Department should be consulted for materials and features.

For Engineering Specifications and Data, see Engineering Section of Catalog.

APPLICABLE CLASS 150 STANDARDS:

End Flanges, ANSI B16.5	SW Ends (Bore and Depth), ANSI B16.11
Pipe Threads, ANSI B2.1	BW Ends (Schedule 40), ANSI B16.25 and B16.10
Design, API 603	Face-to-Face, ANSI B16.10
Wall Section, ANSI B16.34	Pressure-Temperature Ratings, ANSI B16.34-1977

FIGURE 21-13 *Gate valve data (Courtesy Crane-Aloyco Inc.)*

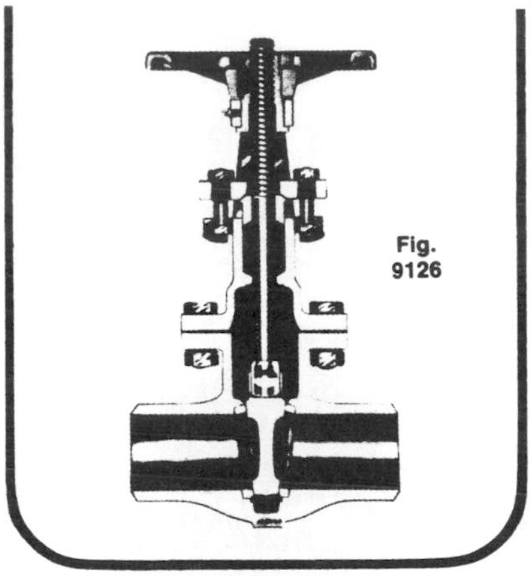

STAINLESS STEEL GATE
CLASS 150
FIGURES: 9126, 9127 (14″ to 24″)

DESIGN DESCRIPTION:

Outside Screw and Yoke
Bolted Bonnet
Rising Stem
Non-rising Handwheel
Solid Wedge

For Flat Faced Flanged Valves, Specify Fig. 9127-FF
Roller Bearing Yoke
Renewable Seat Rings
Buttwelding Ends (Fig. 9126)
Flanged Ends (Fig. 9127)

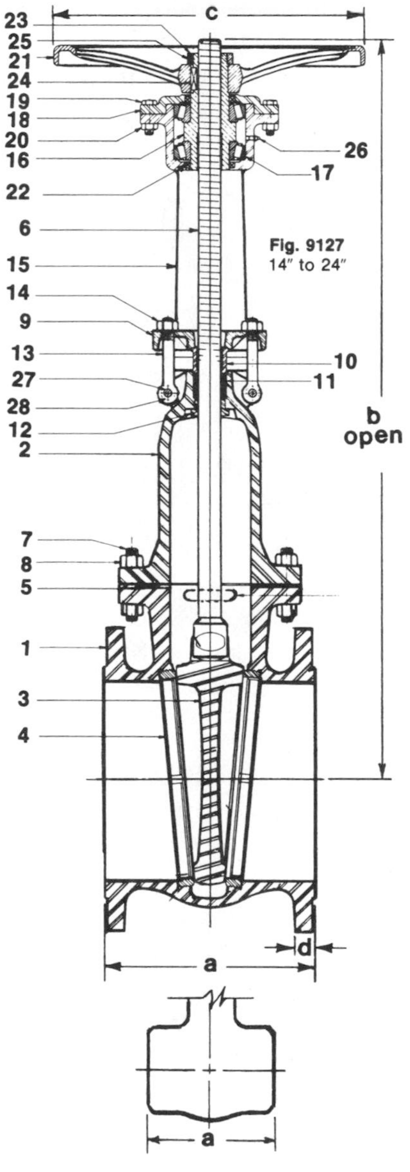

Fig. 9126

PARTS AND MATERIAL LIST:

DESCRIPTION	ALOYCO 18-8 SMO	DESCRIPTION	ALOYCO 18-8 SMO
1 body*	A351 GR CF8M	16 yoke bushing	B584 alloy C86200
2 bonnet	A351 GR CF8M	17 roller bearing	steel
3 wedge	A351 GR CF8M	18 yoke cap	CF8
4 seat ring	A351 GR CF8M	19 yoke cap bolt	A193 GR B8
5 gasket	comp. asbestos	20 yoke cap bolt nut	A194 GR 8F
6 stem	type 316	21 handwheel	malleable iron
7 bonnet studbolt	A193 GR B7	22 oil seal	flax
8 bonnet studbolt nut	A194 2H	23 yoke bushing nut	B584 alloy C86200
9 gland plate	A351 GR CF8M	24 handwheel key	steel
10 gland follower	A351 GR CF8M	25 yoke nut set screw	type 303
11 packing	braided asbestos	26 grease fitting	steel
12 stem hole bushing	A351 GR CF8M	27 hinge bolt	A193 GR B8
13 gland eyebolt	A193 GR B8	28 hinge bolt nut	A294 GR 8F
14 gland eyebolt nut	A194 GR 8F	29 yoke studbolt**	A193 GR B8
15 yoke	CF8	30 yoke studbolt nut**	A194 GR 8F

Remarks: *Body material on welding end valves to be ELC grade.
**Not shown.
Other Alloys Available. Refer to Introduction.

Note: For Valve Services at Temperatures above 700°F, the Aloyco Engineering Department should be consulted for materials and features.

DIMENSIONS IN MM AND INCHES:

SIZE	MM / INCH	350 / 14	400 / 16	450 / 18	500 / 20	600 / 24
a(9126)	mm	571	610	660	711	813
	inch	22½	24	26	28	32
a(9127)	mm	381	406	432	457	508
	inch	15	16	17	18	20
b	mm	1765	1870	2140	2369	2753
	inch	69½	73⅝	84¼	93¼	108⅜
c	mm	508	559	610	660	762
	inch	20	22	24	26	30
d(9127) thickness of flange	mm	35	37	40	43	48
	inch	1⅜	1 7/16	1 9/16	1 11/16	1⅞

WEIGHTS IN KG AND POUNDS:

9126	kg	342	474	535	708	1043
	lb	754	1046	1180	1560	2300
9127	kg	419	559	644	848	1234
	lb	924	1232	1420	1870	2720

APPLICABLE CLASS 150 STANDARDS

End Flanges, ANSI B16.5
Design, ANSI B16.34
Wall Section, API 600

Buttwelding Ends (Schedule 40), ANSI B16.25 and B16.10
Face-to-Face, ANSI B16.10
Pressure-Temperature Ratings, ANSI B16.34-1977

FIGURE 21-14 *Gate valve data* (Courtesy Crane-Aloyco Inc.)

STAINLESS STEEL GLOBE
CLASS 150
FIGURES: 422 (½" to 2"), 427 (½" to 6")

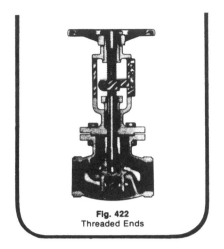

Fig. 422
Threaded Ends

DESIGN DESCRIPTION:

"V" Port Throttling
Bolted Bonnet
Outside Screw and Yoke
Position Indicator
Rising Stem

Non-rising Handwheel
Threaded Ends (Fig. 422)
Flanged Ends (Fig. 427)
When Flat Face End Flanges are
   Required Order as 427-FF

PARTS AND MATERIAL LIST:

DESCRIPTION	ALOYCO 18-8 SMO	DESCRIPTION	ALOYCO 18-8 SMO
1 body	A351 GR CF8M	13 yoke bushing	cast B584 alloy 836 or wrought B16 half hard
2 bonnet	A351 GR CF8M	14 handwheel	malleable iron
3 disc	A351 GR CF8M	15 indicator	CF-16F
4 stem	type 316	16 indicator bolt	A193 GR B8
5 gasket	PTFE	17 escutcheon pins	type 304
6 bonnet bolts	A193 GR B8	18 indicating plate	type 304
7 bonnet bolt nuts	A194 GR 8F	19 yoke bushing nut	cast B584 alloy 836 or wrought B16 half hard
8 gland plate	CF8M	20 handwheel key	steel
9 gland follower	type 316	21 disc pin	type 316
10 packing	PTFE	22 yoke nut set screw	type 303
11 gland bolts	A193 GR B8		
12 gland bolt nuts	A194 GR 8F		

Other Alloys available. (Refer to Introduction)

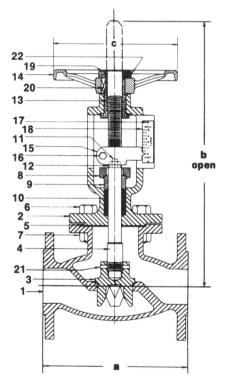

Fig. 427
Sizes ½" to 6"

DIMENSIONS IN MM AND INCHES:

SIZE	MM INCH	15 ½	20 ¾	25 1	40 1½	50 2	65 2½	80 3	100 4	150 6
a	mm	86	95	108	140	165	—	—	—	—
	inch	3⅜	3¾	4¼	5½	6½	—	—	—	—
a (F)	mm	108	117	127	165	203	216	241	292	406
	inch	4¼	4⅝	5	6½	8	8½	9½	11½	16
b	mm	180	235	259	283	308	330	368	419	584
	inch	7¹¹/₁₆	9¼	10³/₁₆	11⅛	12⅛	13	14½	16½	23
c	mm	102	127	127	152	178	191	254	305	356
	inch	4	5	5	6	7	7½	10	12	14

WEIGHTS IN KG AND POUNDS:

422	kg	1.8	2.7	4.5	7.2	9.5	—	—	—	—
	lb	4	6	10	16	21	—	—	—	—
427	kg	2.7	4	5.9	9.5	13	20	26	43	78
	lb	6	9	13	21	29	43	58	95	173

APPLICABLE CLASS 150 STANDARDS:

End Flanges, ANSI B16.5
Wall Section, ANSI B16.34
Face-to-Face, ANSI B16.10
Pressure-Temperature Ratings, ANSI B16.34-1977
Pipe Threads, ANSI B2.1

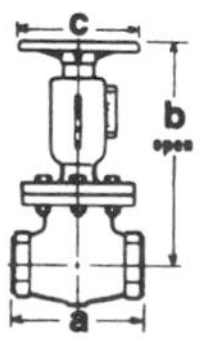

Fig. 422 (½" to 2")

FIGURE 21-15 *Globe valve data* (Courtesy Crane-Aloyco Inc.)

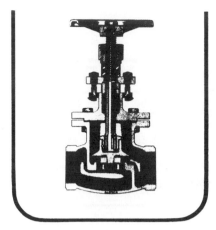

STAINLESS STEEL GLOBE
CLASS 150
FIGURES: 502 (½″ to 2″), 504 (½″ to 2″), 507 (½″ to 10″)

DESIGN DESCRIPTION:

Outside Screw and Yoke
Bolted Bonnet
Renewable PTFE Disc
Retained Gasket
Rising Stem and Handwheel
For Flat Faced Flanged Valve
  Specify Fig. 507 FF

Integral Seat
Threaded Ends (Fig. 502)
Socket Weld Ends (Fig. 504)
Flanged Ends (Fig. 507)
Valves in Sizes 8″ and 10″ Have
  Disc Guide Below Seat

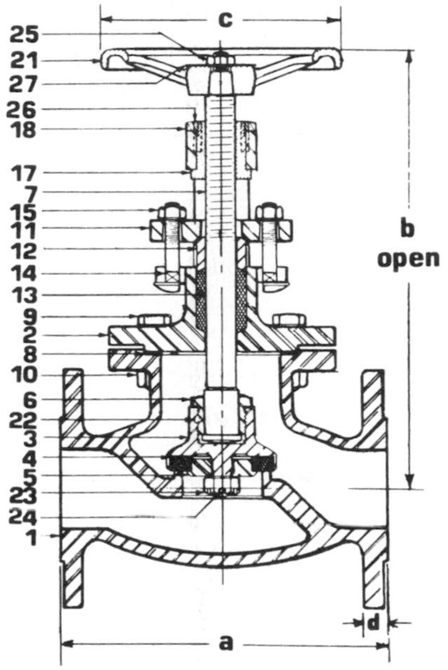

PARTS AND MATERIAL LIST:

DESCRIPTION	ALOYCO 18-8 SMO	DESCRIPTION	ALOYCO 18-8 SMO
1 body*	A351 GR CF8M	14 gland bolt	A193 GR B8
2 bonnet	A351 GR CF8M	15 gland bolt nut	A194 GR 8F
3 disc holder	cast-CF8M, wrought-type 316	16 yoke ***	CF8
4 disc	PTFE	17 yoke bushing	B16 half hard
5 disc holder plate	cast-CF8M, wrought-type 316	18 yoke bushing nut	B16 half hard
6 swivel nut		19 yoke bolt ***	A193 GR B8
7 stem	type 316	20 yoke bolt nut ***	A194 GR 8F
8 gasket	PTFE	21 handwheel	malleable
9 bonnet bolt	A193 GR B8	22 swivel nut pin	type 316
10 bonnet bolt nut	A194 GR 8F	23 disc holder plate nut**	type 316
11 gland plate	CF8M	24 disc holder nut pin	type 316
12 gland follower	cast-CF8M, wrought-type 316	25 handwheel nut	type 303
13 packing	PTFE	26 yoke nut pin	type 303
		27 I.D. plate	type 304

Remarks: *Body material on threaded and welding end valves is ELC grade.
      **Sizes ½″ to 2″ incl.
      ***Sizes 8″ and 10″ only.
      Other Alloys Available. Refer to Introduction.
      **Note: For Valve Services at Temperatures above
      700°F, the Aloyco Engineering Department should be
      consulted for materials and features.**

DIMENSIONS IN MM AND INCHES:

SIZE	MM	15	20	25	40	50	65	80	100	150	200	250
	INCH	½	¾	1	1½	2	2½	3	4	6	8	10
a(502,504)	mm	86	95	108	140	165	—	—	—	—	—	—
	inch	3⅜	3¾	4¼	5½	6½						
a(507)	mm	108	117	127	165	203	216	241	292	406	495	622
	inch	4¼	4⅝	5	6½	8	8½	9½	11½	16	19½	24½
b(502,504)	mm	165	210	248	283	305	—	—	—	—	—	—
	inch	6½	8¼	9¾	11⅛	12						
b(507)	mm	165	210	251	283	305	327	359	425	533	622	822
	inch	6½	8¼	9⅞	11⅛	12	12⅞	14⅛	16¾	21	24½	32⅜
c	mm	67	76	102	127	152	178	203	254	305	406	457
	inch	2⅝	3	4	5	6	7	8	10	12	16	18
d(504) Depth of Socket	mm	10	13	13	13	16	—	—	—	—	—	—
	inch	⅜	½	½	½	⅝						
d(507) Flange Thickness	mm	11	11	11	14	16	17	19	24	25	29	30
	inch	⁷/₁₆	⁷/₁₆	⁷/₁₆	⁹/₁₆	⅝	¹¹/₁₆	¾	¹⁵/₁₆	1	1⅛	1³/₁₆

WEIGHTS IN KG AND POUNDS:

502,504	kg	2.7	5	5	7.3	11	—	—	—	—	—	—
	lb	6	10	10	16	24						
507	kg	5	6.4	6.4	10	15	20.8	26	43	78	128.8	188.7
	lb	10	14	14	22	34	46	58	95	173	284	416

APPLICABLE CLASS 150 STANDARDS:

End Flanges, ANSI B16.5
Wall Section, ANSI B16.34
Face-to-Face, ANSI B16.10
Pipe Threads, ANSI B2.1

Socket Weld Ends (Bore and Depth), ANSI B16.11
Pressure-Temp. Ratings, ANSI B16.34-1977
  Maximum Service Temperature, 450°F

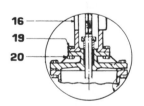

Separable Yoke Detail
Sizes 8″ and 10″ only

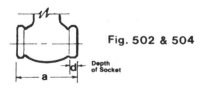

Fig. 502 & 504

FIGURE 21-16   *Globe valve data (Courtesy Crane-Aloyco Inc.)*

Fig. 557
Flanged Ends

STAINLESS STEEL
LIFT CHECK VALVES
CLASS 150

FIGURES: 550 (¼″ to 2″), 554 (¼″ to 2″),
556 (½″ to 6″), 557 (½″ to 6″)*

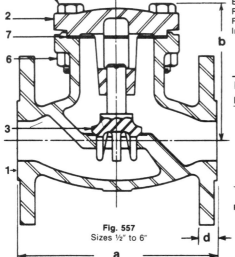

Fig. 557
Sizes ½″ to 6″

DESIGN DESCRIPTION:

Horizontal Type	Threaded Ends (Fig. 550)
Bolted Cover	Socket Weld Ends (Fig. 554)
Retained Gasket	Buttwelding Ends (Fig. 556)
Regrinding Disc	Flanged Ends (Fig. 557)
Integral Seat	

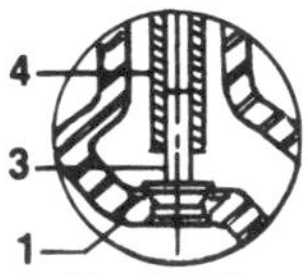

Disc Detail
Sizes ¼″ to ¾″ incl.

PARTS AND MATERIAL LIST:

DESCRIPTION	ALOYCO 18-8 SMO
1 body*	A351 GR CF8M
2 cover	A351 GR CF8M
3 disc	cast CF8M or wrought type 316
4 disc guide**	type 316
5 cover bolt	A193 GR B8
6 cover bolt nut	A194 GR 8F
7 gasket	comp. asbestos

Remarks: *Body material on threaded and welding end valves is ELC grade. Other Alloys available. (Refer to Introduction)
**Sizes ¼″ to ¾″ incl.
†Unless otherwise specified, these valves may be supplied at the manufacturer's option with: PTFE Gasket.

Note: For Valve Services at Temperatures above 700°F, the Aloyco Engineering Department should be consulted for materials and features.

DIMENSIONS IN MM AND INCHES:

SIZE		6	10	15	20	25	40	50	65	80	100	150
	MM	¼	⅜	½	¾	1	1½	2	2½	3	4	6
	INCH											
a (550,554)	mm	79	79	86	95	108	140	165	—	—	—	—
	inch	3⅛	3⅛	3⅜	3¾	4¼	5½	6½				
a (556,557)	mm	—	—	108	117	127	165	203	216	241	292	406
	inch			4¼	4⅝	5	6½	8	8½	9½	11½	16
b	mm	67	67	67	76	86	102	116	133	146	168	200
	inch	2⅝	2⅝	2⅝	3	3⅜	4	4⁹⁄₁₆	5¼	5¾	6⅝	7⅞
d (554)	mm	10	10	10	13	13	13	16	—	—	—	—
	inch	⅜	⅜	⅜	½	½	½	⅝				
d (557)	mm	—	—	11	11	11	14	16	17	19	24	25
	inch			⁷⁄₁₆	⁷⁄₁₆	⁷⁄₁₆	⁹⁄₁₆	⅝	¹¹⁄₁₆	¾	¹⁵⁄₁₆	1

WEIGHTS IN KG AND POUNDS:

550,554	kg	1.4	1.4	1.8	2.3	2.9	5.9	7.3	—	—	—	—
	lb	3	3	4	5	6.3	13	16				
556	kg	—	—	2	2.7	3.4	6.8	10	12	18.6	28	51
	lb			4.5	6	7.5	15	22	26	41	61	112
557	kg	—	—	2.3	3	4.1	8	12.7	14	22	34	68
	lb			5	7	9	17.5	28	31	48	75	150

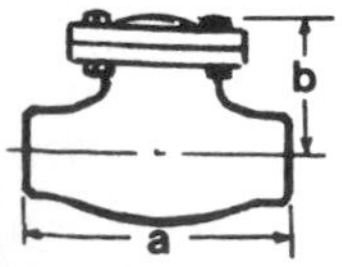

APPLICABLE CLASS 150 STANDARDS:

End Flanges, ANSI B16.5	SW Ends (Bore and Depth), ANSI B16.11
Wall Section, ANSI B16.34	BW Ends (Sch. 40), ANSI B16.25 and B16.10
Face-to-Face, ANSI B16.10	Pressure-Temp. Ratings, ANSI B16.34-1977
Pipe Threads, ANSI B2.1	

Fig. 556
Buttwelding Ends
½″ to 6″

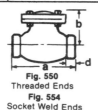

Fig. 550
Threaded Ends
Fig. 554
Socket Weld Ends

FIGURE 21-17 *Check valve data (Courtesy Crane-Aloyco Inc.)*

STAINLESS STEEL
LIFT CHECK VALVES
CLASS 600
FIGURES: 4550-A (½″ to 2″)
4554-A (½″ to 2″)
4557-A (½″ to 2″)

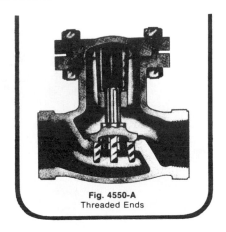

Fig. 4550-A
Threaded Ends

DESIGN DESCRIPTION:
Bolted Cover
Integral Seat
Horizontal Type
Threaded Ends (Fig. 4550-A)
Socket Weld Ends (Fig. 4554-A)
Buttwelding Ends (Fig. 4557-A)

PARTS AND MATERIAL LIST:

DESCRIPTION	ALOYCO 18-8 SMO
1 body	A351 GR CF8M
2 cover	A351 GR CF8M
3 disc	type 316
4 disc guide (½″ and ¾″)	type 316
5 cover studbolt	A193 GR B8
6 cover studbolt nut	A194 GR 8F
7 gasket	comp. asbestos

Other Alloys Available. Refer to Introduction.

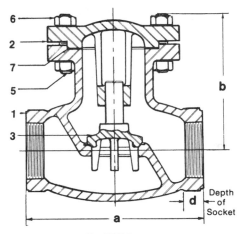

Fig. 4550-A
Sizes ½″ to 2″

DIMENSIONS IN MM AND INCHES:

SIZE		15	20	25	40	50
	MM INCH	½	¾	1	1½	2
a(4550-A, 4554-A)	mm	95	108	127	165	191
	inch	3¾	4¼	5	6½	7½
a(4557-A)	mm	165	191	216	241	387
	inch	6½	7½	8½	9½	15¼
b	mm	94	94	97	148	165
	inch	3¹¹/₁₆	3¹¹/₁₆	3¹³/₁₆	5¹³/₁₆	6½
d(4554-A)	mm	10	13	13	13	16
	inch	⅜	½	½	½	⅝
d(4557-A)	mm	14	16	17	22	25
	inch	⁹/₁₆	⅝	¹¹/₁₆	⅞	1

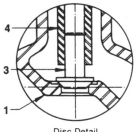

Disc Detail
Sizes ½″ and ¾″

WEIGHTS IN KG AND POUNDS:

4550-A, 4554-A	kg	2.3	3	5	10	14
	lb	5	7	11	22	30
4557-A	kg	5	6.3	10	15	22
	lb	11	14	22	32	48

APPLICABLE CLASS 600 STANDARDS:
End Flanges, ANSI B16.5
Wall Section, ANSI B16.34
Face-to-Face, ANSI B16.10
Buttwelding Ends (Schedule 40),
 ANSI B16.25 and B16.10

Pipe Threads, ANSI B2.1
Pressure-Temperature Rating, ANSI B16.34-1977
Socket Weld Ends (Bore and Depth), ANSI B16.11

Note: For Valve Services at Temperatures above 700°F, the Aloyco Engineering Department should be consulted for materials and features.

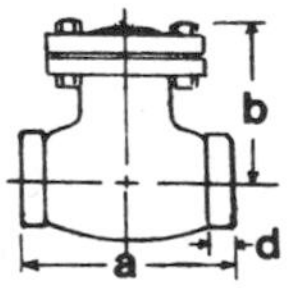

Fig. 4554-A
Socket Weld Ends
½″ to 2″

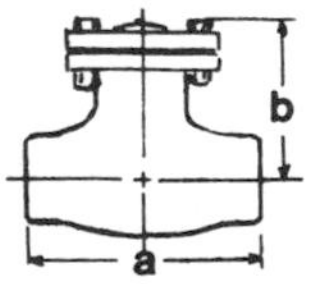

Fig. 4557-A
Buttwelding Ends
½″ to 2″

FIGURE 21-18 *Check valve data (Courtesy Crane-Aloyco Inc.)*

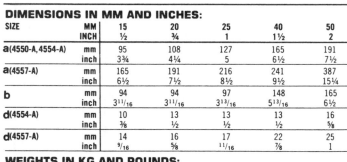

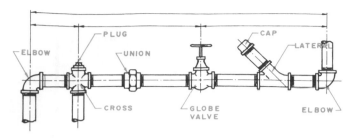

FIGURE 21-19 *Double-line pipe drawing*

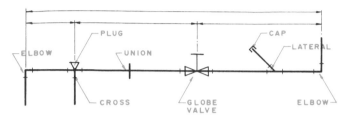

FIGURE 21-20 *Single-line pipe drawing*

- Outside diameters and wall thicknesses of pipe may be indicated on the drawing using a leader line or may be made part of a general note.
- A bill of material should be developed and placed directly on the pipe drawing or attached to it.

SINGLE-LINE ISOMETRIC PIPING DRAWINGS

Isometric piping drawings are prepared in a manner similar to mechanical piping drawings. The lines representing pipes are drawn parallel to either the X, Y, or Z axis, *Figure 21-28* (page 805). Flanges are represented using short strokes of equal thickness, *Figure 21-29* (page 805). Notice that flanges for horizontal pipe are drawn vertically and flanges for vertical pipe are drawn at the appropriate 30° angle to the horizontal.

Figure 21-30 (page 805) illustrates the various methods used for drawing valves. The unidirectional dimensioning illustrated in *Figure 21-31* (page 805) should be used in isometric piping drawings.

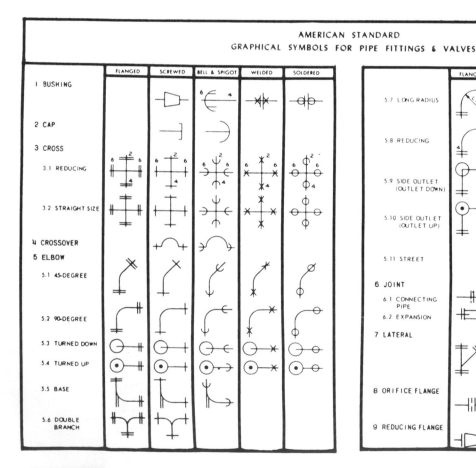

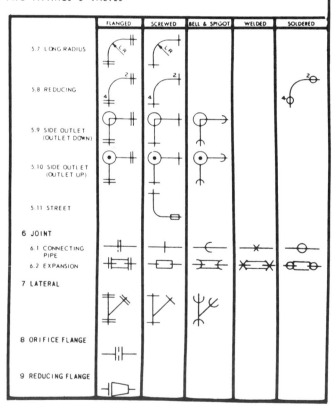

FIGURE 21-21 *Symbols for pipe fittings and valves* (Reprinted from *AZA Z32.2.3-1949 (R1994)*. Courtesy ASME)

AMERICAN STANDARD
GRAPHICAL SYMBOLS FOR PIPE FITTINGS & VALVES

	FLANGED	SCREWED	BELL & SPIGOT	WELDED	SOLDERED
10 PLUGS					
10.1 BULL PLUG					
10.2 PIPE PLUG					
11 REDUCER					
11.1 CONCENTRIC					
11.2 ECCENTRIC					
12 SLEEVE					
13 TEE					
13.1 (STRAIGHT SIZE)					
13.2 (OUTLET UP)					
13.3 (OUTLET DOWN)					
13.4 DOUBLE SWEEP					
13.5 REDUCING					
13.6 SINGLE SWEEP					
13.7 SIDE OUTLET (OUTLET DOWN)					
13.8 SIDE OUTLET (OUTLET UP)					
14 UNION					
15 ANGLE VALVE					
15.1 CHECK					
15.2 GATE (ELEVATION)					
15.3 GATE (PLAN)					
15.4 GLOBE (ELEVATION)					
15.5 GLOBE (PLAN)					
15.6 HOSE ANGLE	SAME AS	SYMBOL	23 1		
16 AUTOMATIC VALVE					
16.1 BY-PASS					

	FLANGED	SCREWED	BELL & SPIGOT	WELDED	SOLDERED
16.2 GOVERNOR-OPERATED					
16.3 REDUCING					
17 CHECK VALVE					
17.1 ANGLE CHECK	SAME AS	SYMBOL	15 1		
17.2 (STRAIGHT WAY)					
18 COCK					
19 DIAPHRAGM VALVE					
20 FLOAT VALVE					
21 GATE VALVE					
*21.1					
21.2 ANGLE GATE	SAME AS	SYMBOLS	15.2 & 15.3		
21.3 HOSE GATE	SAME AS	SYMBOL	23.2		
21.4 MOTOR-OPERATED					
22 GLOBE VALVE					
22.1					
22.2 ANGLE GLOBE	SAME AS	SYMBOLS	15.4 & 15.5		
22.3 HOSE GLOBE	SAME AS	SYMBOL	23.3		
22.4 MOTOR-OPERATED					
23 HOSE VALVE					
23.1 ANGLE					
23.2 GATE					
23.3 GLOBE					
24 LOCKSHIELD VALVE					
25 QUICK OPENING VALVE					
26 SAFETY VALVE					
27 STOP VALVE	SAME AS	SYMBOL	21.1		

*ALSO USED FOR GENERAL **STOP VALVE** SYMBOL WHEN AMPLIFIED BY SPECIFICATION

FIGURE 21-21 *(Continued) Symbols for pipe fittings and valves (Reprinted from AZA Z32.2.3-1949 (R1994). Courtesy ASME)*

AMERICAN STANDARD
GRAPHICAL SYMBOLS FOR PIPING

AIR CONDITIONING

28 BRINE RETURN — — —BR— — —

29 BRINE SUPPLY ————B———

30 CIRCULATING CHILLED OR
   HOT-WATER FLOW ————CH————

31 CIRCULATING CHILLED OR
   HOT-WATER RETURN — — —CHR— — —

32 CONDENSER WATER FLOW ————C————

33 CONDENSER WATER RETURN — — —CR— —

34 DRAIN ————D————

35 HUMIDIFICATION LINE ——— - —H— - ——

36 MAKE-UP WATER ——— - —— - ——

37 REFRIGERANT DISCHARGE ————RD————

38 REFRIGERANT LIQUID ————RL————

39 REFRIGERANT SUCTION — — —RS— — —

HEATING

40 AIR-RELIEF LINE —— —— —— ——

41 BOILER BLOW OFF ——— ——— ———

42 COMPRESSED AIR ————A————

43 CONDENSATE OR VACUUM
   PUMP DICHARGE ——O— —O— —O——

44 FEEDWATER PUMP DISCHARGE ——OO— —OO— —OO——

45 FUEL-OIL FLOW ————FOF————

46 FUEL-OIL RETURN — — —FOR— — —

47 FUEL-OIL TANK VENT — — —FOV— — —

48 HIGH-PRESSURE RETURN ——//— —//— —//——

49 HIGH-PRESSURE STEAM ——//— —//— —//——

50 HOT-WATER HEATING RETURN —— —— —— ——

51 HOT-WATER HEATING SUPPLY ————————

52 LOW-PRESSURE RETURN —— —— —— ——

53 LOW-PRESSURE STEAM ————————

54 MAKE-UP WATER ——— - ——— - ———

55 MEDIUM PRESSURE RETURN ——/— —/— —/—

56 MEDIUM PRESSURE STEAM ———/———/———

PLUMBING

57 ACID WASTE ————ACID————

58 COLD WATER ——— - — - — - —

59 COMPRESSED AIR ————A————

60 DRINKING-WATER FLOW ——— - — - — - —

61 DRINKING-WATER RETURN ——— - —— - —— - ——

62 FIRE LINE ———F———F———

63 GAS ———G———G———

64 HOT WATER ————————

65 HOT-WATER RETURN ——— - ——— - ———

66 SOIL, WASTE OR LEADER
   (ABOVE GRADE) ————————

67 SOIL, WASTE OR LEADER
   (BELOW GRADE) —— —— —— ——

68 VACUUM CLEANING ———V———V———

69 VENT ----------------

PNEUMATIC TUBES

70 TUBE RUNS ════════════

SPRINKLERS

71 BRANCH AND HEAD ———O———O———

72 DRAIN ———S— — —S———

73 MAIN SUPPLIES ————S————

FIGURE 21-22 *Symbols for piping* (Reprinted from *ASA Z332.2.3-1949 (R1994)*. Courtesy ASME)

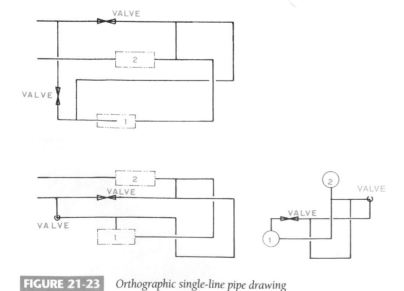

FIGURE 21-23 *Orthographic single-line pipe drawing*

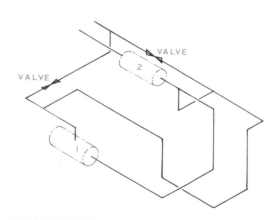

FIGURE 21-24 *Isometric single-line pipe drawing*

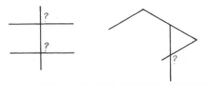

CROSSING OF PIPES WITHOUT BREAKS HARD TO READ

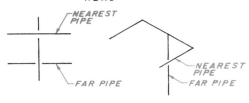

CROSSING OF PIPES WITH BREAK INDICATES NEAREST PIPE

FIGURE 21-25 *Drawing crossing pipe*

PERMANENT   DETACHABLE

FIGURE 21-26 *Drawing pipe connections*

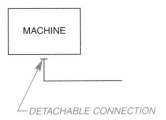

FIGURE 21-27 *Drawing adjoining machinery*

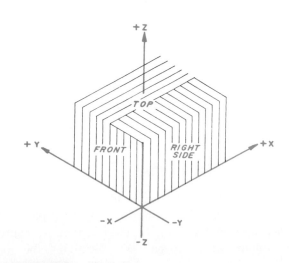

FIGURE 21-28 *Isometric axes*

FLANGES FOR VERTICAL PIPE

FIGURE 21-29 *Drawing flanges*

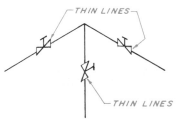

VALVES WITH THREADED CONNECTIONS

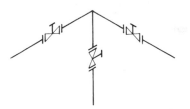

VALVES WITH FLANGE CONNECTIONS

FIGURE 21-30 *Drawing valves*

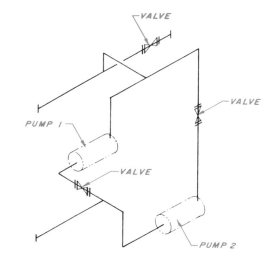

FIGURE 21-31 *Isometric pipe drawing*

Structural Drafting

A *structure* is anything constructed of parts. In heavy construction this encompasses commercial buildings, industrial buildings, bridges, towers, parking decks, stadiums, and numerous other types of structures. *Figures 21-32* and *21-33* are examples of modern

FIGURE 21-32 *Cleveland state office building* (Courtesy Bethlehem Steel Corp.)

FIGURE 21-33 *ADT executive offices* (Courtesy Precast/Prestressed Concrete Institute)

structures. Structures such as those shown in these figures are constructed of many parts. The overall structure must be designed and the design must be documented by engineering drawings. In addition, each individual part must be designed and the design must be documented by shop drawings. Structural drafting consists of preparing the documentation for the design of a structure so that the individual parts can be manufactured and the overall structure erected.

Documentation prepared by structural drafters consists primarily of engineering drawings and shop draw-

ings. Engineering drawings convey structural design information (i.e., what types of products are to be used and in what sizes, how the various parts fit together, how the individual parts are connected, etc.). They consist of various types of framing plans (i.e., anchor bolt, column, beam, floor, and roof framing plans); sections; and connection details. *Figure 21-34* is an example of an engineering drawing. It is a framing plan showing the placement of steel beams and joists for a commercial building. The original material used for preparing structural engineering drawings is typically a set of architectural plans and/or a set of engineering sketches and calculations.

Shop drawings convey all of the information needed to manufacture the individual parts of a structure and are sometimes referred to as detail drawings. They consist of the necessary views of individual members with sectional views added when necessary and a bill of material. *Figure 21-35* is an example of a structural steel shop drawing showing all the information necessary to fabricate several steel beams. It also contains a bill of material showing all of the various materials needed to fabricate the beams shown on this shop drawing.

DRAWING, CHECKING, CORRECTING, AND REVISING

Structural drafters prepare original drawings—engineering and/or shop drawing—using either CAD or manual drafting techniques. The drawings are then checked by a more experienced drafter, known as a checker, and an engineer. Checkers and engineers make sure the drawings conform to the specifications of the original material used to prepare them and that the drafter has correctly interpreted engineering calculations/sketches/instructions. Corrections that need to be made are marked on a checkprint, and the drafter corrects the original.

Drawings that have been produced, checked, and corrected must also be changed on occasion. Such changes are known as revisions. A revision is typically required as the result of an engineering or architectural change, whereas a correction is usually made as the result of a drafting error. Revisions are typically specified on a document called an Engineering Change Order (ECO). *Figure 21-36* (page 809) is an example of an ECO. Normally the cost of making revisions is charged against the company or department that ordered them.

STRUCTURAL PRODUCTS

The products that become the individual components of a structure are standardized for the most part. They are steel and concrete and, to a lesser extent, wood products. *Figure 21-37* (page 809) illustrates the most widely used structural steel shapes. They are W shapes,

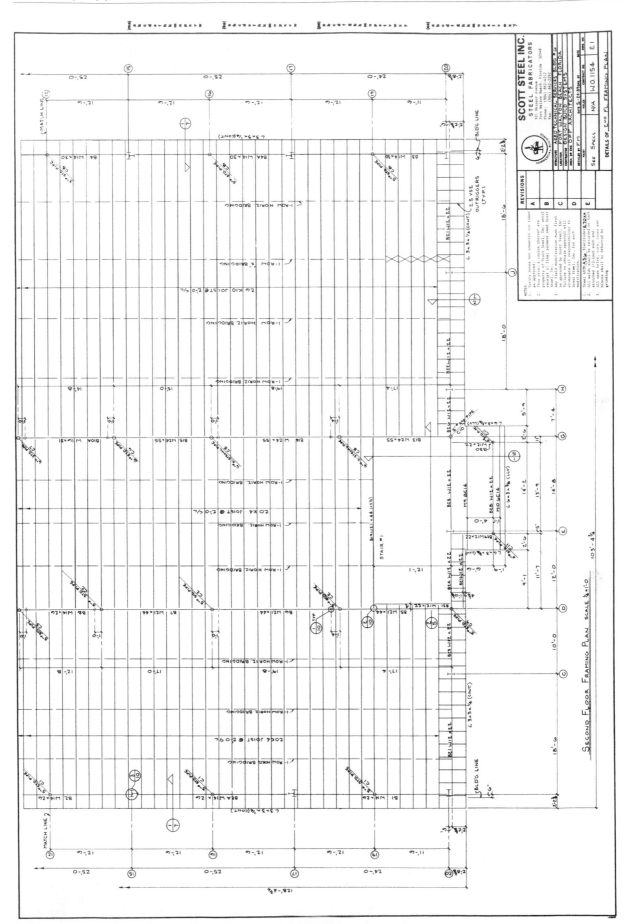

FIGURE 21-34 *Engineering drawing* (Courtesy Scott Steel Inc.)

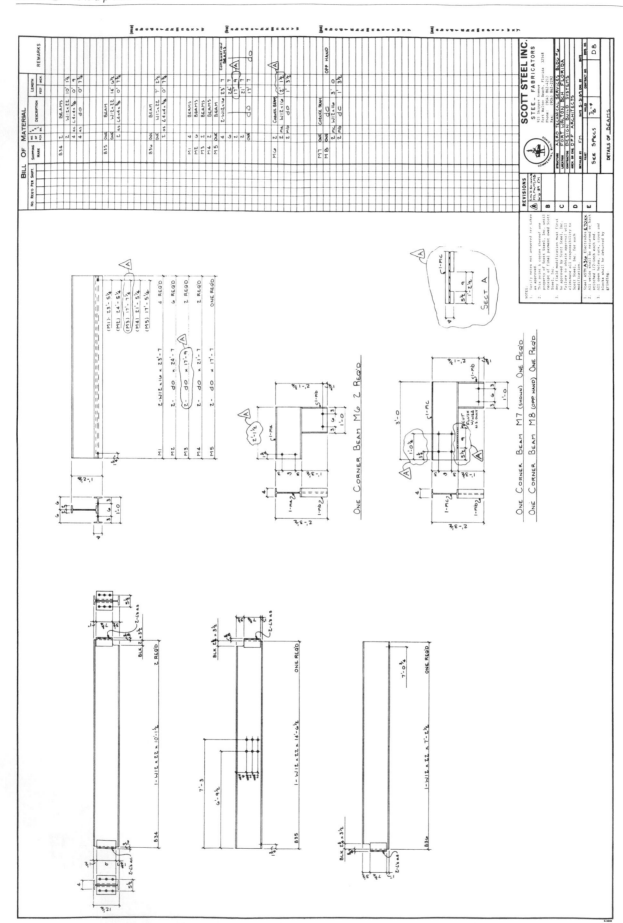

FIGURE 21-35 *Structural steel shop drawing (Courtesy Scott Steel Inc.)*

Scotford Steel International
832 Skipper Avenue
Fort Walton Beach, Florida   32548
904/863-5323

• Engineering Change Order •

Job Number ___*1759-A*___ Date ___*3/11/93*___

Contractor ___*Sharper Construction Co.*___

Architect ___*Rand & Kendrick, AIA*___

Description of Change:

Shorten beams M7 and M8 by 3 inches

___*Wanda Clifford*___
Signature (Requesting)

___*Jefferson Kendrick*___
(Signature (Approving)

FIGURE 21-36   *Engineering drawing* (Courtesy Scott Steel Inc.)

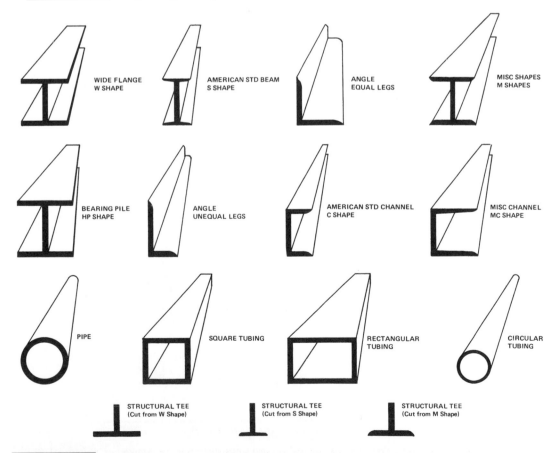

- WIDE FLANGE W SHAPE
- AMERICAN STD BEAM S SHAPE
- ANGLE EQUAL LEGS
- MISC SHAPES M SHAPES
- BEARING PILE HP SHAPE
- ANGLE UNEQUAL LEGS
- AMERICAN STD CHANNEL C SHAPE
- MISC CHANNEL MC SHAPE
- PIPE
- SQUARE TUBING
- RECTANGULAR TUBING
- CIRCULAR TUBING
- STRUCTURAL TEE (Cut from W Shape)
- STRUCTURAL TEE (Cut from S Shape)
- STRUCTURAL TEE (Cut from M Shape)

FIGURE 21-37   *Structural steel shapes* (From Goetsch, *Structural Drafting*, Delmar Publishers Inc.)

S shapes, M shapes, HP shapes, C shapes, MC shapes, equal-legged angles, unequal-legged angles, square tubing, rectangular tubing, circular tubing, pipe, and structural tees.

The various steel shapes shown in Figure 21-37 may be used to form the vertical and horizontal members in the structural skeleton of a building, stadium, tower, or any other type of structure. Vertical members are called columns. Horizontal members may be beams, joists, or girders. The *Manual of Steel Construction* published by the American Institute for Steel Construction (AISC)

contains the dimensions and other design and drafting-related information for structural steel shapes. *Figure 21-38* is a page excerpted from the *Manual of Steel Construction*. Structural steel drafters must have a copy of this manual or ready access to it in order to prepare engineering and/or shop drawings.

Concrete products are either prestressed, precast, post-tensioned, or poured-on-site. Prestressed and post-tensioned products have steel cables running through them that are stretched to create tension that is equal to or greater than the opposite forces they will be sub-

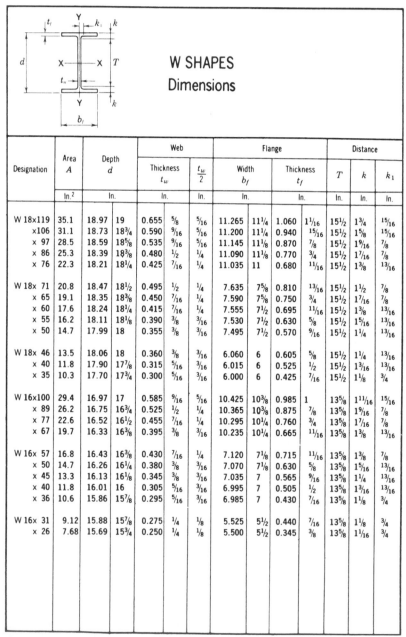

AMERICAN INSTITUTE OF STEEL CONSTRUCTION

FIGURE 21-38 *Page from the* Manual of Steel Construction *(Courtesy American Institute of Steel Construction)*

jected to when incorporated as part of a structure. These cables are in addition to the reinforcing bars, plates, and welded wire mesh normally cast into all structural concrete members. Precast products contain all of these items except the stressed steel cables. Poured-on-site concrete members do not contain the tensioned cables. They are strengthened with reinforcing bars, welded wire mesh, and steel plates only. Precast and poured-on-site members are the same except for where the pouring of concrete takes place; in a plant for precast and on-site for poured-on-site products.

With prestressed concrete products, the steel cables are tensioned before the concrete is poured into the form or before it is extruded in the case of cored flat slabs. With post-tensioned products the cables are placed in the form encased in a special conduit and tensioned after the concrete has cured.

Figures 21-39 and *21-40* illustrate some of the more widely used prestressed and post-tensioned concrete products. They include double-tee members, single-tee members, solid flat slabs, cored flat slabs, L-shaped beams, inverted T-shaped beams, rectangular beams, tee joists, and keystone joists.

CONNECTING STRUCTURAL MEMBERS

The two most widely used methods for connecting individual structural members are bolting and welding. Structural steel construction uses both widely. Concrete construction relies more on welding but does use some bolted connections. The information contained in Chapter 13 applies to the bolts that are used to fasten structural members. *Figure 21-41* shows a structural steel pipe column that will be bolted at the bottom (base plate) and at the top (top plate).

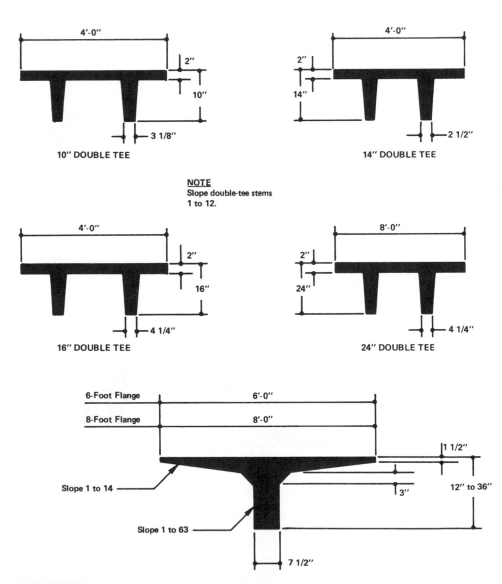

FIGURE 21-39 *Standard precast concrete products* (From Goetsch, *Structural Drafting*, Delmar Publishers Inc.)

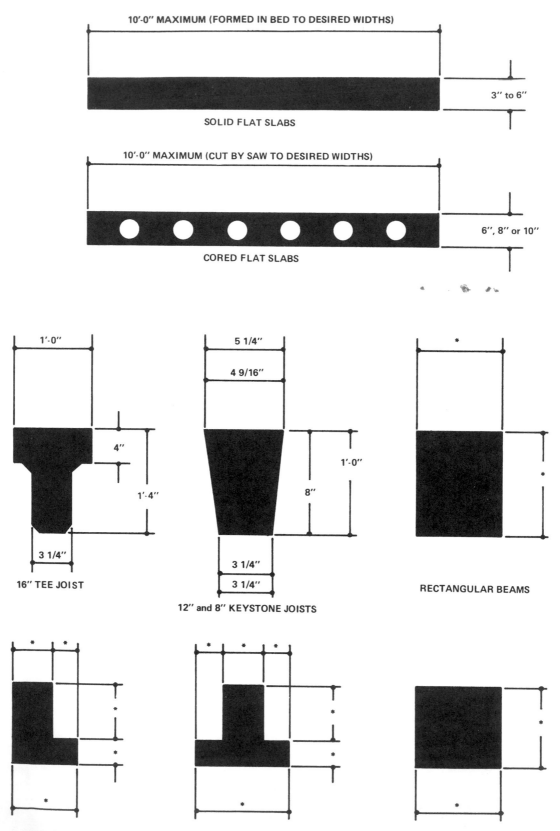

10'-0" MAXIMUM (FORMED IN BED TO DESIRED WIDTHS)

3" to 6"

SOLID FLAT SLABS

10'-0" MAXIMUM (CUT BY SAW TO DESIRED WIDTHS)

6", 8" or 10"

CORED FLAT SLABS

1'-0"

4"

1'-4"

3 1/4"

16" TEE JOIST

5 1/4"

4 9/16"

1'-0"

8"

3 1/4"

3 1/4"

12" and 8" KEYSTONE JOISTS

*

*

RECTANGULAR BEAMS

* DIMENSIONS MAY BE VARIED TO MEET THE REQUIREMENTS OF THE JOB.

FIGURE 21-40 *Standard precast concrete products* (From Goetsch, *Structural Drafting,* Delmar Publishers Inc.)

FIGURE 21-41 *Structural steel pipe fabricated as a column* (Courtesy Scott Steel Inc.)

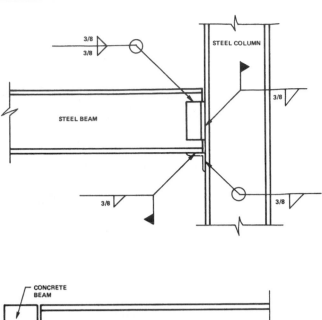

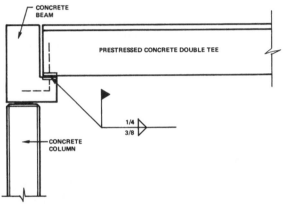

FIGURE 21-42 *Welded connections* (From Goetsch, *Structural Drafting*, Delmar Publishers Inc.)

The information structural drafters need to know about welding and the proper use of welding symbols is covered in Chapter 19. *Figure 21-42* shows both steel and prestressed concrete members connected by welding. *Figure 21-43* shows the welding types and their corresponding symbols as approved by the American Institute of Steel Construction (AISC).

STRUCTURAL STEEL DRAWINGS

The point was made earlier that structural steel drawings consist of both engineering drawings and shop or detail drawings. Engineering drawings are used to communicate what products are to be used; in what sizes and shapes; and how they are to be spaced, connected, and otherwise put together to form a structure. Engineering drawings for structural steel construction consist of framing plans, sections (as needed), and connection details. *Figure 21-44* is an example of a second-floor framing plan for a commercial building. *Figure 21-45* (page 816) is an example of connection details for such a building.

Shop or detail drawings are used to fabricate the individual structural members that are laid out on the engineering drawings. They are very detailed and typically are accompanied by a bill of material. *Figure 21-46* (page 817) is a shop drawing showing the fabrication details for several structural steel beams. The shop drawing contains a bill of material listing all of the material needed to fabricate each beam.

STRUCTURAL CONCRETE DRAWINGS

Structural concrete products may be poured-on-site, precast, prestressed, or post-tensioned. Regardless of the approach, engineering and shop drawings are needed. In the case of concrete members that are poured on-site,

WELDED JOINTS
Standard symbols

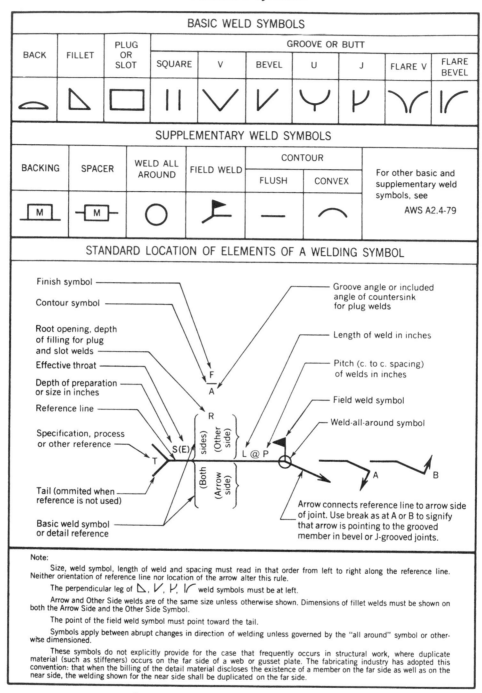

AMERICAN INSTITUTE OF STEEL CONSTRUCTION

FIGURE 21-43 *Welding types of symbols* (Courtesy American Institute of Steel Construction)

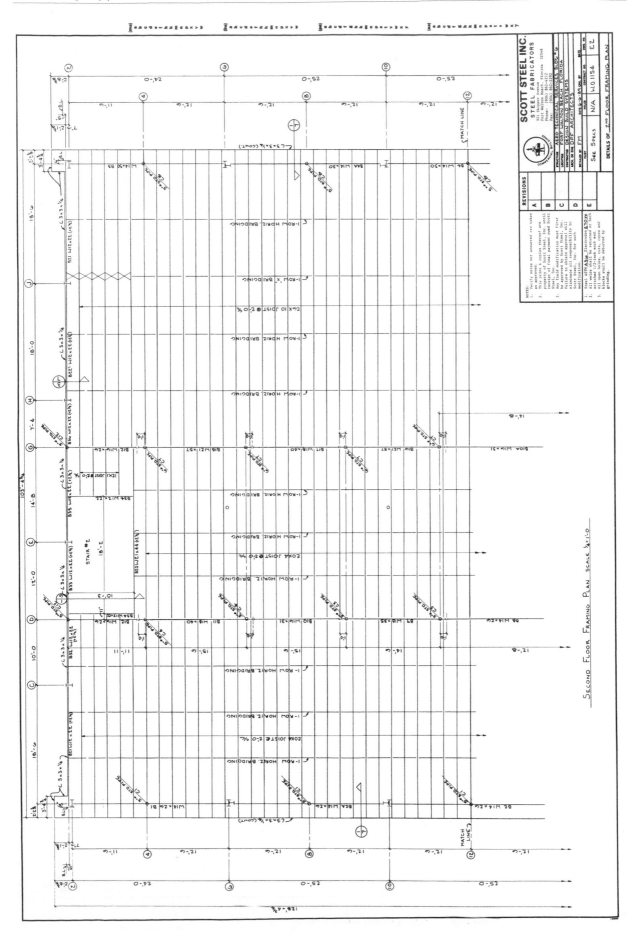

FIGURE 21-44 *Second-floor framing plan* (Courtesy Scott Steel Inc.)

Industry Application

VERSATILITY IS GOOD FOR THE CAREER

As a drafting and design student at Jonesborough Technical Institute, Juanita Vargas had disagreed with what was known as the school's *versatility rule.* The rule reads as follows: "In order to graduate, a student must demonstrate entry-level job skills in at least two fields of drafting." The purpose of the rule is to make JTI's students versatile job seekers once they graduate.

Vargas liked mechanical drafting. She liked the precision of machined parts held to close tolerances. She wanted to make her living as a mechanical design and drafting technician; she had no interest in any other drafting field. However, a rule is a rule, so Vargas also learned structural drafting. She selected this field because structural steel drawings have, in her opinion, some of the characteristics of machine drawings.

Upon graduation Vargas secured a good job with a tool and die manufacturer in a city about 500 miles from her hometown. She liked her job, but was constantly homesick. After she had worked there for three years, her company diversified its holdings by purchasing a structural engineering firm with offices in four different cities, including Vargas's hometown. Soon the word went out that drafting technicians were needed at all four of the newly acquired engineering firms. Vargas and her fellow drafters were told they could transfer if they knew anything about structural drafting, particularly structural steel.

Vargas, who by now had become unbearably homesick, jumped at the chance. Saying a silent "thank you" to Jonesborough Technical Institute for its versatility rule, Vargas packed her bags and moved home. Not only did she get a job in her hometown, Vargas also got a promotion. Versatility is good for the career.

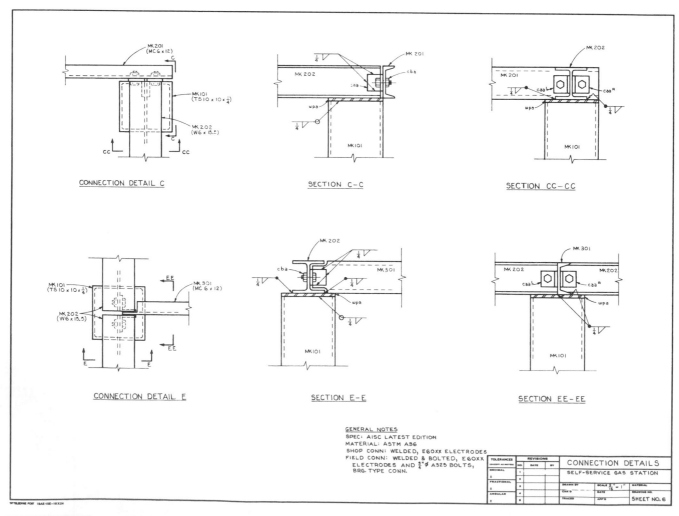

FIGURE 21-45 *Connection details* (From Goetsch, *Structuring Drafting,* Delmar Publishers Inc.)

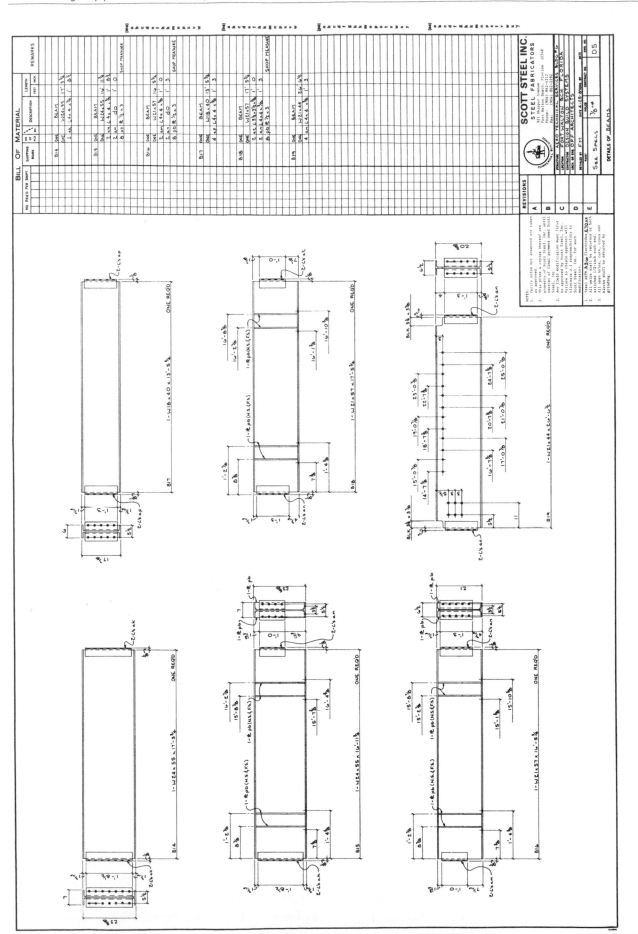

FIGURE 21-46 *Structural steel shop drawing (Courtesy Scott Steel Inc.)*

shop drawings are known as *placing drawings*. Placing drawings show the placement of reinforcing bars in the on-site form. Shop drawings and placing drawings serve the same purpose. The difference is in where the work occurs, in a shop or on-site.

Engineering drawings are used to communicate what products are to be used, in what sizes and shapes, and how they are to be spaced, connected, and otherwise put together to form a structure. Engineering drawings for structural concrete consist of framing plans, sections, (as needed), and connection details. *Figure 21-47* is an example of an engineering drawing for a small industrial building. The drawing contains a roof framing plan and sections. The sections also double as connection details. The roof is to be constructed of 4'-0" wide cored flat slabs.

Shop and/or placement drawings are used to fabricate the individual structural members that are laid out on the engineering drawings. They are very detailed and must communicate every individual piece of material that is to be embedded in the concrete. Such drawings are typically accompanied by a bill of material.

Figure 21-48 (page 820) is a shop drawing for a solid flat slab member with one side beveled. It contains 17 prestressed steel strands that run lengthwise and are wrapped by reinforcing bars spaced lengthwise at 1'-0" intervals (MK443). There are lifting hooks made of bent reinforcing bars (lha) for lifting the member out of the form, placing it on a truck for transportation to the construction site, and erecting it as part of the structure. Plates for welded connections (swpl) are placed along the edges at 7'-0" intervals. There is a bill of material on the right side of the drawing.

Pre-Engineered Metal Buildings

A special subset of the field of structural steel design, drafting, and construction is the pre-engineered metal building. Pre-engineered metal buildings can be used for a variety of applications including warehouses, manufacturing facilities, churches, office facilities, schools, gymnasiums, airplane hangars, retail facilities, ministorage facilities, and numerous other uses.

The principal differences between the construction of a regular structural steel building and a pre-engineered metal building are as follows:

1. With a regular structural steel building, engineers begin with an architect's plan and develop a structural steel framework that will accommodate the building in question. With pre-engineered metal buildings, the engineering calculations for the various building products have already been done. Design personnel need only select the appropriate materials in the proper sizes for the situation in question.

2. With regular structural steel buildings, the steel components are typically (although not always) larger and of a heavier grade. Pre-engineered buildings typically use smaller, lighter steel components for the building frame.

3. With regular steel buildings, there is often much cutting, fitting, and welding to do on-site during the course of constructing the building. With pre-engineered various metal components are not just pre-engineered; they are presized, prefabricated, and numbered.

4. Regular structural steel buildings typically require more equipment, more tools, and more expertise to erect than pre-engineered metal buildings. Regular structural steel buildings also usually take more time to erect than pre-engineered metal buildings.

PRODUCTS USED IN PRE-ENGINEERED METAL BUILDINGS

Pre-engineered metal buildings are constructed from a variety of standard metal products. The following components are commonly used in pre-engineered metal buildings:

Eave strut—A cold-formed C-section fabricated for the proper roof pitch. Eave struts ensure that the metal building is weathertight at the eaves of the roof, *Figure 21-49*.

Purlin and girt—Cold-rolled Z-sections are the principal structural members attached to the building's frame (Figure 21-49). Girts are used in the walls of the building. Purlins are used in the roof. Roof and wall sheeting is attached to the girts and purlins.

Base angle and sheeting angle—Cold-rolled angles. Base angles are attached to the concrete floor of the building and provide a continuous base to which the bottom of the sheeting panels can be attached (Figure 21-49). Sheeting angles are similar but are used for attaching the sheeting panels at the rake of the building.

Roof and wall sheeting panels—There are a number of different types of roof and wall sheeting panels available for use in constructing pre-engineered metal buildings. Some are used primarily for walls, some just for roofs, and some for both; others are architectural or decorative in nature. Some of the more commonly used shapes of sheeting panels are shown in *Figures 21-50* through *Figure 21-53*.

Columns and beams—The rigid-frame system for a pre-engineered metal building consists of columns

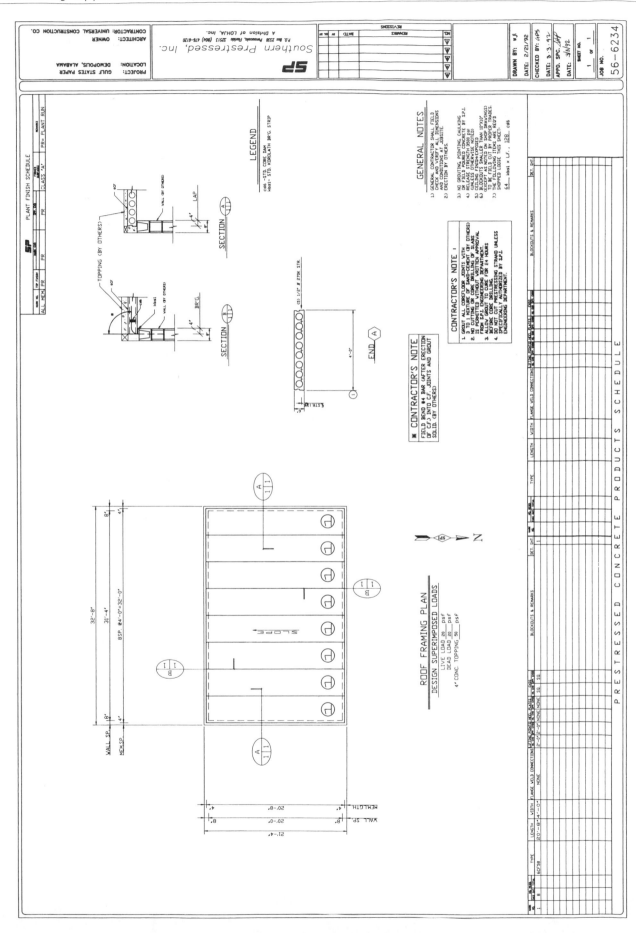

FIGURE 21-47 *Engineering drawing (Courtesy Southern Prestressed Inc.)*

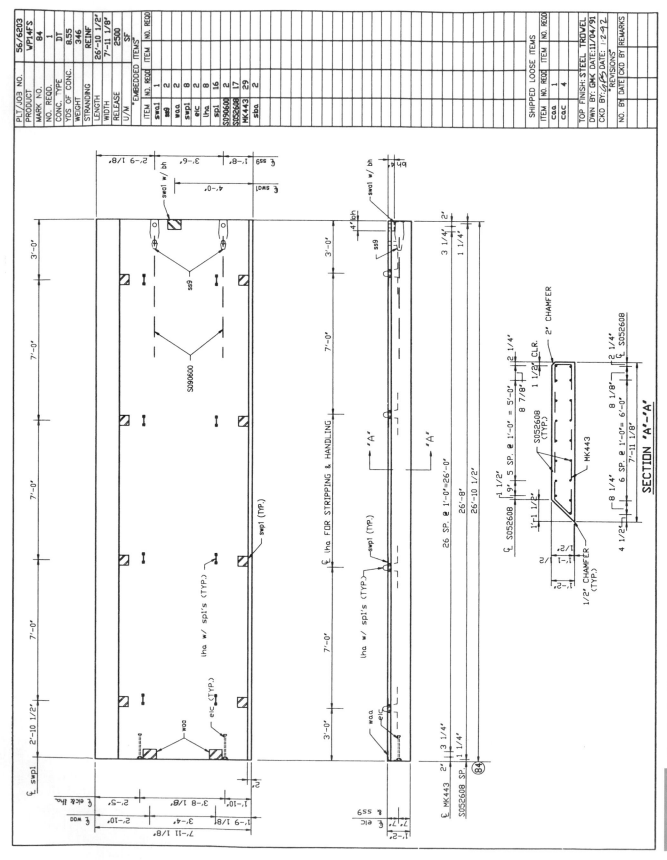

FIGURE 21-48 *Shop drawing* (Courtesy Southern Prestressed Inc.)

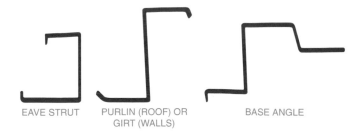

EAVE STRUT | PURLIN (ROOF) OR GIRT (WALLS) | BASE ANGLE

FIGURE 21-49 *Standard components*

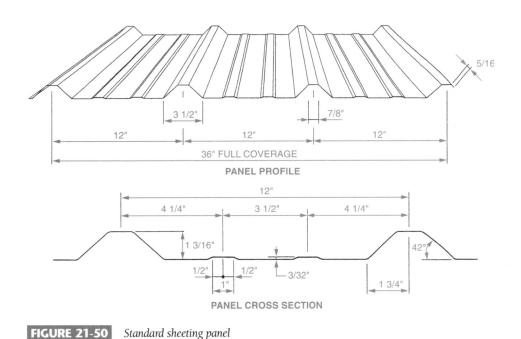

FIGURE 21-50 *Standard sheeting panel*

and beams, as it does with any structural steel building. The types of steel shapes used for columns and beams are the same as those shown earlier in this chapter. *Figure 21-54* is a cross section showing a widely used rigid-frame system for pre-engineered metal buildings.

ANCHOR BOLT PLAN

An *anchor bolt plan* is a symbolic representation of the layout of the anchor bolts in the building's foundation/floor system. It is a plan-view drawing that shows where the anchor bolts are to be located in the floor system. The bases of the various columns used in

the rigid-frame system for the building will be attached to the anchor bolts. Consequently, getting the locations of the anchor bolts correct is critical. *Figure 21-55* is an example of an anchor bolt plan for a pre-engineered metal building. *Figure 21-56* is a sectional view cut through the foundation/floor system to show the vertical placement of anchor bolts and to give various notes relating to the foundation and anchor bolts.

RIGID-FRAME CROSS SECTIONS

The rigid frame for a pre-engineered metal building consists of the columns that are attached to the floor system by anchor bolts and the beams that are attached to the

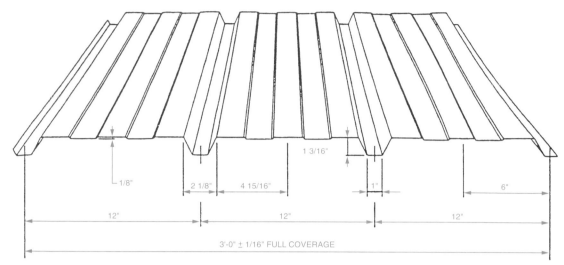

PANEL PROFILE

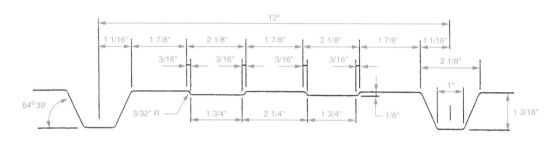

PARTIAL CROSS SECTION

FIGURE 21-51 *Standard sheeting panel*

PANEL PROFILE

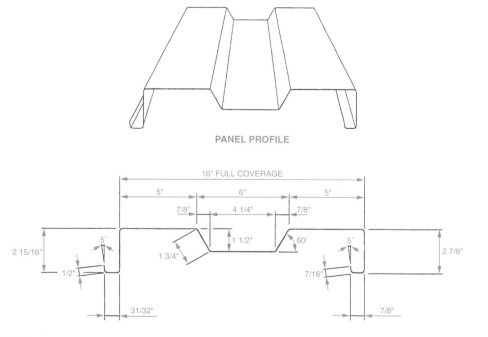

CROSS SECTION

FIGURE 21-52 *Standard sheeting panel*

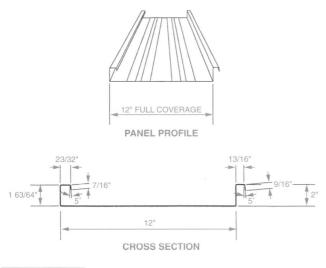

FIGURE 21-53 *Standard sheeting panel*

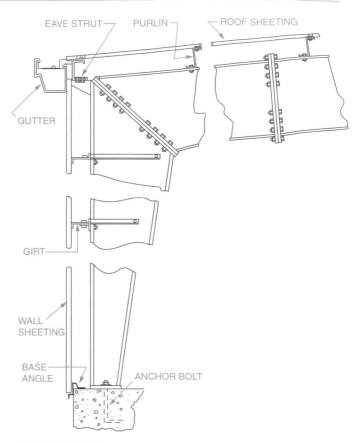

FIGURE 21-54 *Typical rigid-frame section*

columns. A cross section through the building shows the configuration used for the building in question. *Figure 21-57* shows a typical rigid-frame cross section without dimensions. The dimensions are selected by drafting personnel from a chart such as the one shown in *Figure 21-58* based on the desired width and height of the building. These dimensions for various widths and heights of buildings using 20, 30, or 40 psf steel are an example of *pre-engineering.* Rigid-frame cross sections come in a variety of designs.

ROOF FRAMING AND SHEETING PLANS

The roof framing plan shows the arrangement and spacing of the purlins that are attached to the roof beams. The sheeting plan shows the spacing and arrangement of the sheeting panels for the roof. *Figure 21-59* is an example of a roof framing plan for a pre-engineered metal building. *Figure 21-60* is an example of a roof sheeting plan for a metal building.

SIDEWALL AND ENDWALL FRAMING AND SHEETING PLANS

Sidewall framing plans show the location of the girts, eave struts, and base angles to which the side sheeting panels will be attached. Side sheeting plans show the general arrangement of the side sheeting panels for the side of the building in question. Endwall framing and sheeting plans serve the same purpose and show the same information but for the ends of the building.

DRAWING THE SIDEWALL AND ENDWALL FRAMING AND SHEETING PLANS

Figure 21-61 is an example of a sidewall framing plan. *Figure 21-62* is the corresponding sidewall sheeting plan. Very little detail is required in either of these plans. The sidewall framing plan uses single lines to show the spacing of the columns that make up the rigid frame for the building as they sit on the concrete slab. It also shows the location (vertically) of girts and eave struts. It is understood that base angles are located at the top edge of the concrete slab.

The sidewall sheeting plan requires even less detail. It simply shows the number of panels and their arrangement. Length and width dimensions of the individual panels are precalculated and do not need to be shown on the sheeting plan. Notice in Figure 21-62 that each panel has a *mark number.* Panels that are the same width and height have the same mark number. In the case of Figure 21-62 all of the panels with mark number 2 are of one width and length. Those with mark number 1 have a different width than those with mark number 2,

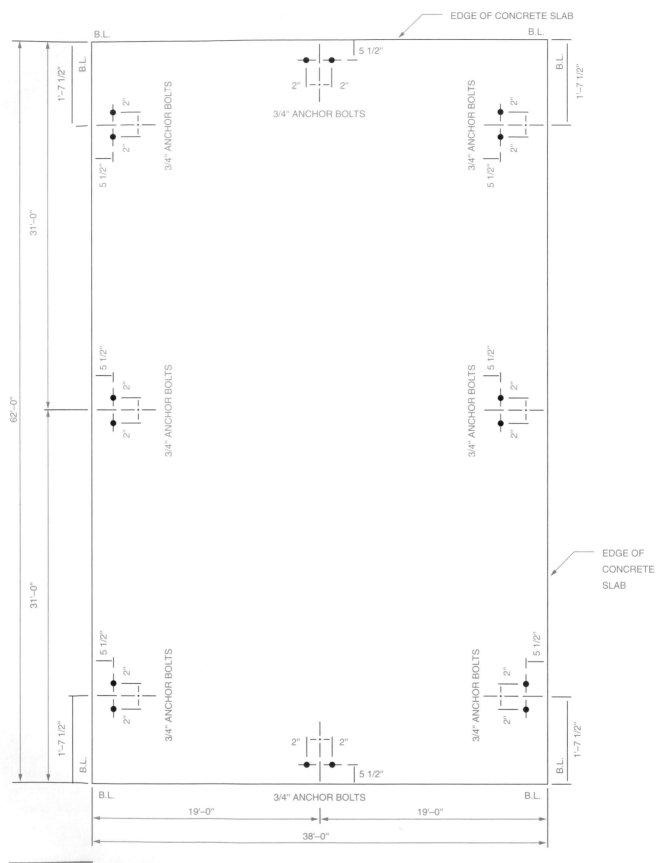

EDGE OF CONCRETE SLAB

B.L.

B.L.

5 1/2"

2" 2"

3/4" ANCHOR BOLTS

B.L.

1'-7 1/2"

3/4" ANCHOR BOLTS

2"

2"

5 1/2"

3/4" ANCHOR BOLTS

2"

2"

5 1/2"

B.L.

1'-7 1/2"

31'-0"

62'-0"

5 1/2"

2"

3/4" ANCHOR BOLTS

2"

5 1/2"

2"

3/4" ANCHOR BOLTS

2"

EDGE OF CONCRETE SLAB

31'-0"

5 1/2"

2"

3/4" ANCHOR BOLTS

2"

5 1/2"

2"

3/4" ANCHOR BOLTS

2"

1'-7 1/2"

B.L.

2" 2"

5 1/2"

1'-7 1/2"

B.L.

B.L.

3/4" ANCHOR BOLTS

B.L.

19'-0"

19'-0"

38'-0"

FIGURE 21-55 *Anchor bolt plan*

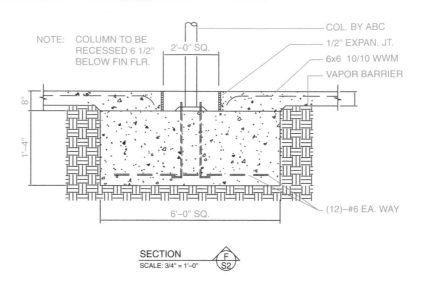

NOTE: COLUMN TO BE RECESSED 6 1/2" BELOW FIN FLR.

2'-0" SQ.

COL. BY ABC

1/2" EXPAN. JT.

6x6 10/10 WWM

VAPOR BARRIER

8"

1'-4"

6'-0" SQ.

(12)–#6 EA. WAY

SECTION
SCALE: 3/4" = 1'-0"

F
S2

FOUNDATION NOTES:

1. CONCRETE SHALL BE 3000 PSI IN 28 DAYS
2. CONCRETE SLAB, APRONS AND DOOR STOOPS SHALL BE REINFORCED WITH 6x6 10/10 WWM.
3. 6 MIL POLY VAPOR BARRIER SHALL BE PLACED BETWEEN FLOOR SLAB AND COMPACTED SUB–GRADE.
4. ALL REINFORCEMENT BARS SHALL BE GRADE 60.
5. VERIFY ANCHOR BOLT SIZE AND LOCATIONS WITH BUILDING MANUFACTURER'S DRAWINGS PRIOR TO CONCRETE PLACEMENT.
6. ANCHOR BOLTS SHALL HAVE A 3" PROJECTION AND EXTEND TO LOWER 1/3 OF FOOTING DEPTH.
7. ANCHOR BOLTS STRENGTH SHALL BE A36 GRADE.

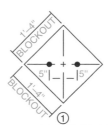

1'-4" BLOCKOUT

1'-4" BLOCKOUT

5" | — — | 5"

①

3/4" ANCHOR BOLTS

NOTE: TOP OF FOOTING AT ANCHOR BOLT ① SHALL BE 8" BELOW FINISH FLOOR.

FIGURE 21-56 *Foundation section accompanying the anchor bolt plan*

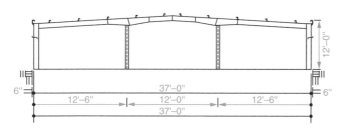

12'-0"

6"

37'-0"

6"

12'-6"   12'-0"   12'-6"

37'-0"

FIGURE 21-57 *Typical rigid-frame cross section*

but both panels numbered with a 1 are the same width and length. The framing plan tells erectors to begin on either end with a mark number 1 panel, attach 23 mark number 2 panels, and finish off at the other end with a mark number 1 panel.

The same rules and practices apply when developing the endwall framing and sheeting plans (*Figure 21-63* and *Figure 21-64*). Notice in Figure 21-64 that, begin-

ning at either the left or right side of the plan, each sheeting panel gets successively longer in the height direction until reaching the two center panels (mark number 12). Each panel is clearly marked to guide the erectors in attaching it to the building frame.

CONNECTION DETAILS

The connection details required in a set of plans for a pre-engineered metal building can vary depending on the nature and characteristics of the building. However, regardless of the size of the building, the configuration of the rigid frame, and the types of sheeting panels selected, at least three connection details are always needed:

- Girt connection details
- Purlin connection details
- Eave strut connection details

TABLE OF ENGINEERING CALCULATIONS								
20 PSE LL					**30 PSE LL**			
WD	HT	B	C	E	WD	HT	B	C	E
20	10	8'–0"	8"	14"	20	10	8'–0"	8"	17"
20	12	10'–0"	8"	14"	20	12	10'–0"	8"	17"
20	14	12'–0"	8"	14"	20	14	12'–0"	8"	17"
20	16	14'–0"	8"	14"	20	16	14'–0"	8"	17"
20	20	18'–0"	8"	14"	20	20	18'–0"	8"	17"
30	10	8'–0"	8"	14"	30	10	8'–0"	8"	17"
30	12	10'–0"	8"	14"	30	12	10'–0"	8"	17"
30	14	12'–0"	8"	14"	30	14	12'–0"	8"	17"
30	16	14'–0"	8"	14"	30	16	14'–0"	8"	17"
30	20	18'–0"	8"	14"	30	20	18'–0"	8"	17"
40	12	10'–0"	9"	15"	40	12	10'–0"	9"	20"
40	14	12'–0"	9"	15"	40	14	12'–0"	9"	20"
40	16	14'–0"	9"	15"	40	16	14'–0"	9"	20"
40	20	18'–0"	9"	15"	40	20	18'–0"	9"	20"

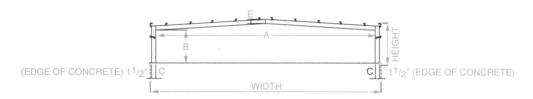

FIGURE 21-58 *Dimensions are pre-engineered for different heights and widths*

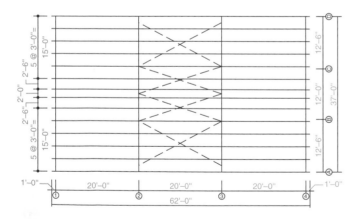

FIGURE 21-59 *Roof framing plan*

GIRT CONNECTION DETAILS

Girt connection details show how the girts are connected to the columns in the rigid frame. A typical girt-to-column connection involves bolting girt clips to the web of the column and bolting the girts to the girts clips (see *Figure 21-65*). Girt connection details should show the connection, give bolt spacing dimensions, and specify the size of bolts to be used. Engineering and design per-

sonnel determine the size and spacing of bolts for the connections and provide this information to drafting personnel either through sketches or pre-engineered tables (although experienced drafting personnel will usually know since connections of pre-engineered metal buildings are typically standardized within companies).

PURLIN CONNECTION DETAILS

Purlin connection details show how purlins are connected to the rafters in the rigid frame. A typical purlin-to-rafter connection involves bolting the purlin to the rafter, overlapping purlins, and bolting the overlapped purlins together (see *Figure 21-66*). Purlin connection details should show the purlin connected to the rafter, the amount of overlap between purlins, and the bolt spacing. Engineering and design personnel determine the size and spacing of bolts and the amount of overlap required and provide this information to drafting personnel through either sketches or pre-engineered tables.

EAVE STRUT CONNECTION DETAILS

Eave strut connection details show how eave struts are connected to the rigid frame. This can be accomplished in various ways depending on the type of rigid frame

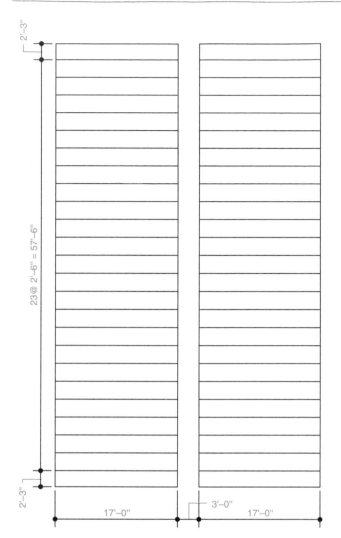

FIGURE 21-60 *Roof sheeting plan*

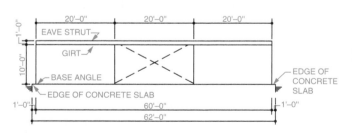

FIGURE 21-61 *Sidewall framing plan*

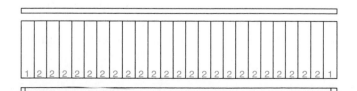

FIGURE 21-62 *Sidewall sheeting plan*

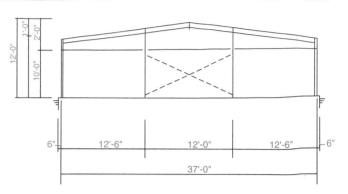

FIGURE 21-63 *Endwall framing plan*

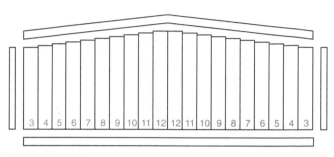

FIGURE 21-64 *Endwall sheeting plan*

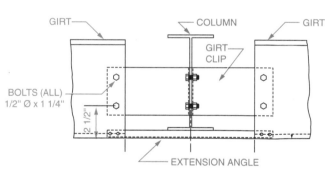

FIGURE 21-65 *Typical grit connection detail (to column)*

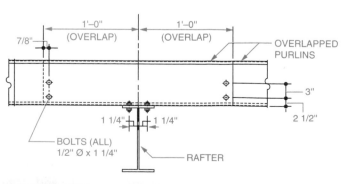

FIGURE 21-66 *Typical purlin connection detail (to rafter) showing overlap*

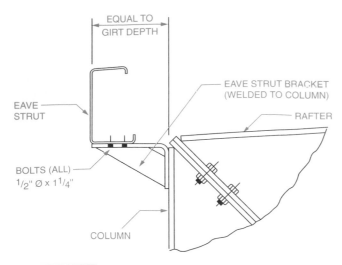

EQUAL TO
GIRT DEPTH

EAVE
STRUT

EAVE STRUT BRACKET
(WELDED TO COLUMN)

RAFTER

BOLTS (ALL)
$1/2"$ Ø x $1^{1}/_{4}"$

COLUMN

FIGURE 21-67   *Typical eave strut connection detail*

used. An eave strut connection detail should show the eave strut connected to the column in the rigid frame or to an eave strut bracket that is connected to the column by a shop weld when the rigid frame is fabricated (see *Figure 21-67*). Engineering and design personnel determine how the actual connection will be made and convey this information to drafting personnel either through sketches or by using standardized details.

Architectural Drafting

Architectural drafting is the process of documenting the designs developed by architects. Architects are called on to plan and design a variety of different types of structures. However, most of an architect's work is typically focused on residential and commercial buildings including single-family houses, town houses, apartments, model communities, and a variety of different types of commercial buildings including office buildings, industrial plants, stores, banks, malls, hospitals, schools, libraries, and so on.

Architectural drafters prepare plans that convey and document an architect's design for any of these types of structures as well as other types. The best way to understand this field is to divide it into two broad categories as follows: residential architectural drafting and commercial architectural drafting.

RESIDENTIAL PLANS

Residential design involves the development of the design for a residential structure (i.e., single-family dwelling, cottage, town house, apartments, etc.). The design process involves such considerations as how the structure will fit into its environment, space and room

planning, aesthetics and function, material selection, code compliance, and many others.

A complete set of plans documenting an architect's design and showing how all of these considerations have been accommodated includes the following: plot plan; foundation plan; floor plan; elevations; heating, ventilation, and air conditioning (HVAC) plan; electrical plan; and framing plans (for roofs, floors, or both). The floor plan also doubles as the plumbing plan in modern construction. The contents of each component is described in the following paragraphs.

PLOT PLAN

A *plot plan* is a plan view of the building site showing where the structure is located on the lot. Surveyors provide the original material for plot plans. It should contain the following information: property lines showing the compass bearing and length of each; shape, location, and size of the building or buildings on the site; elevation of each corner of the site; north arrow; streets, sidewalks, driveways, patios; utilities and easements; wells, septic tanks, and drain fields (as applicable); scale of the drawing and a property description (i.e., lot and block number, street address, or legal description from a survey). Additional information that might be shown on the drawing but is not required includes the following: fences and retaining walls; trees, bushes, shrubs, rivers/streams, and other natural components; and contour lines. *Figure 21-68* is an example of a plot plan that contains the necessary information.

FOUNDATION PLAN

A *foundation plan* is a plan view in section that shows the understructure of the building. It should contain the following information: footings for piers, columns, and foundation walls; foundation walls; columns; piers; retention walls; partition walls, doors, and plumbing fixtures in houses with a basement; windows, vents, doors, and other openings in the foundation walls; beams; pilasters; floor joists (direction, spacing, and size); drains, sump, air-conditioning line; dimension; scale of the drawing; applicable notes; and footing/foundation sections and details. *Figure 21-69* is an example of a foundation plan that contains the necessary information.

FLOOR PLAN

The floor plan is the most important component in a set of residential plans. It is the heart of the plans. A *floor plan* is a plan view of the structure in section with the ceiling and roof removed. It should contain the following information: all necessary dimensions; exterior and interior walls; doors, windows, and other openings; built-in

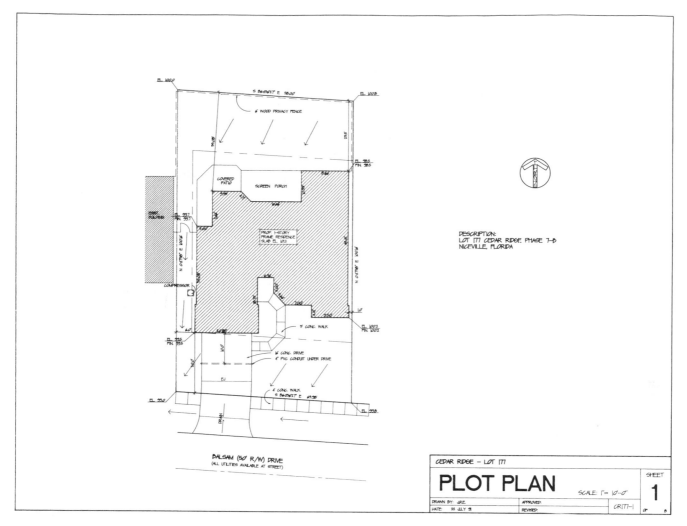

FIGURE 21-68 *Plot plan* (Courtesy Barton Homes Inc.)

cabinets; appliances; plumbing fixtures; stairs; fireplaces; freestanding structures (i.e., garage); decks, patios, and porches; room labels; notes; door schedule; window schedule; area information (i.e., living area, porch area, garage, etc.); and scale of the drawing. *Figure 21-70* is an example of a floor plan that contains the necessary information.

ELEVATIONS

Elevations are orthographic views of the front, back, and sides of the structure. Most elevations are exterior views. However, interior elevations of the kitchen, bathrooms, and utility areas are frequently included in residential plans. Exterior elevations should contain the following information: grade lines, finished floor line, ceiling line, locations of corners of exterior walls, windows, doors, all roof features (i.e., gables, dormers, chimneys, etc.), vertical dimensions, porches, decks, patios, sun-rooms, applicable material symbols, and special details as

required. *Figure 21-71* (page 832) is an example of elevations that contain the necessary information. Notice that the kitchen, bath, and utility area interior elevations are not contained on this sheet. As is often the case, they were drawn on another sheet that had more room (see Sheet 5, **Figure 21-72**, page 833).

HVAC PLAN

The *heating-ventilation/air-conditioning plan* (HVAC) shows the duct work for the house's heating and cooling system superimposed on the floor plan without dimensions. The location of interior and exterior heating/cooling units, size of duct work, direction of air flow, notes, details as needed, hot water heater, thermostat, registers, inlets, and vents should also be shown. Figure 21-72 is an example of an HVAC plan containing the necessary information. Notice that this sheet also contains interior elevations. This is a common practice to cut down on the number of sheets in the overall set of plans.

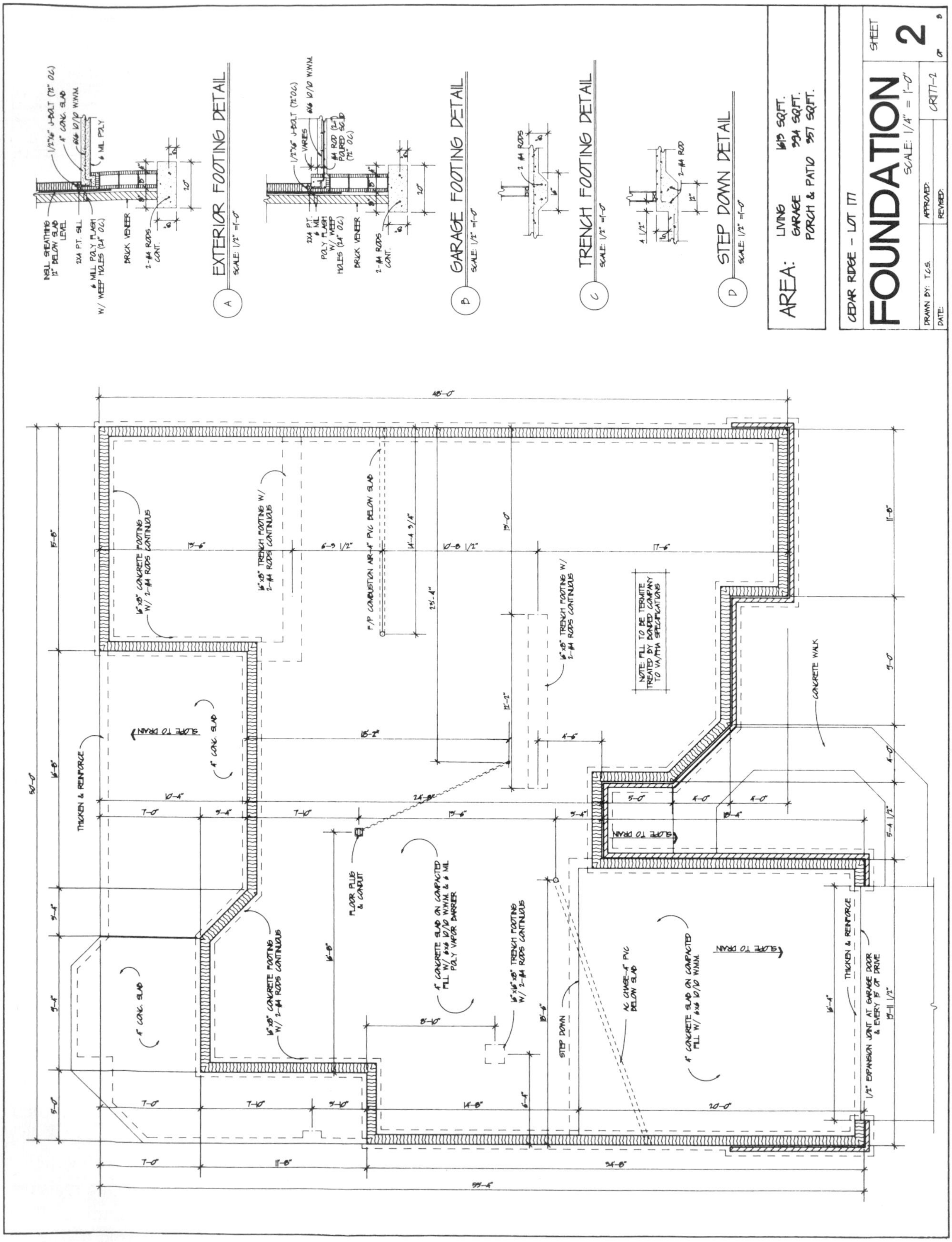

FIGURE 21-69 *Foundation plan (Courtesy Barton Homes Inc.)*

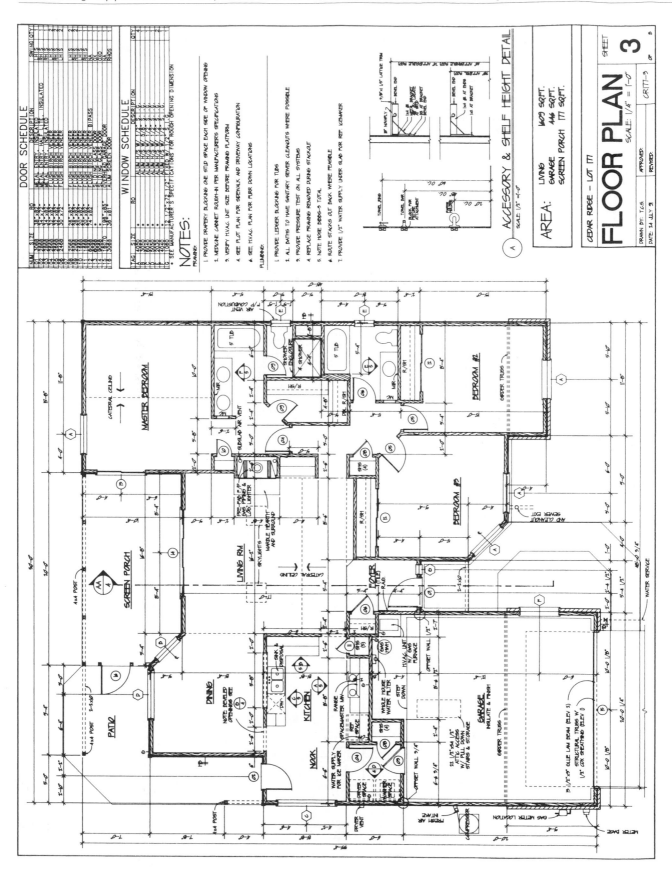

FIGURE 21-70 *Floor plan (Courtesy Barton Homes Inc.)*

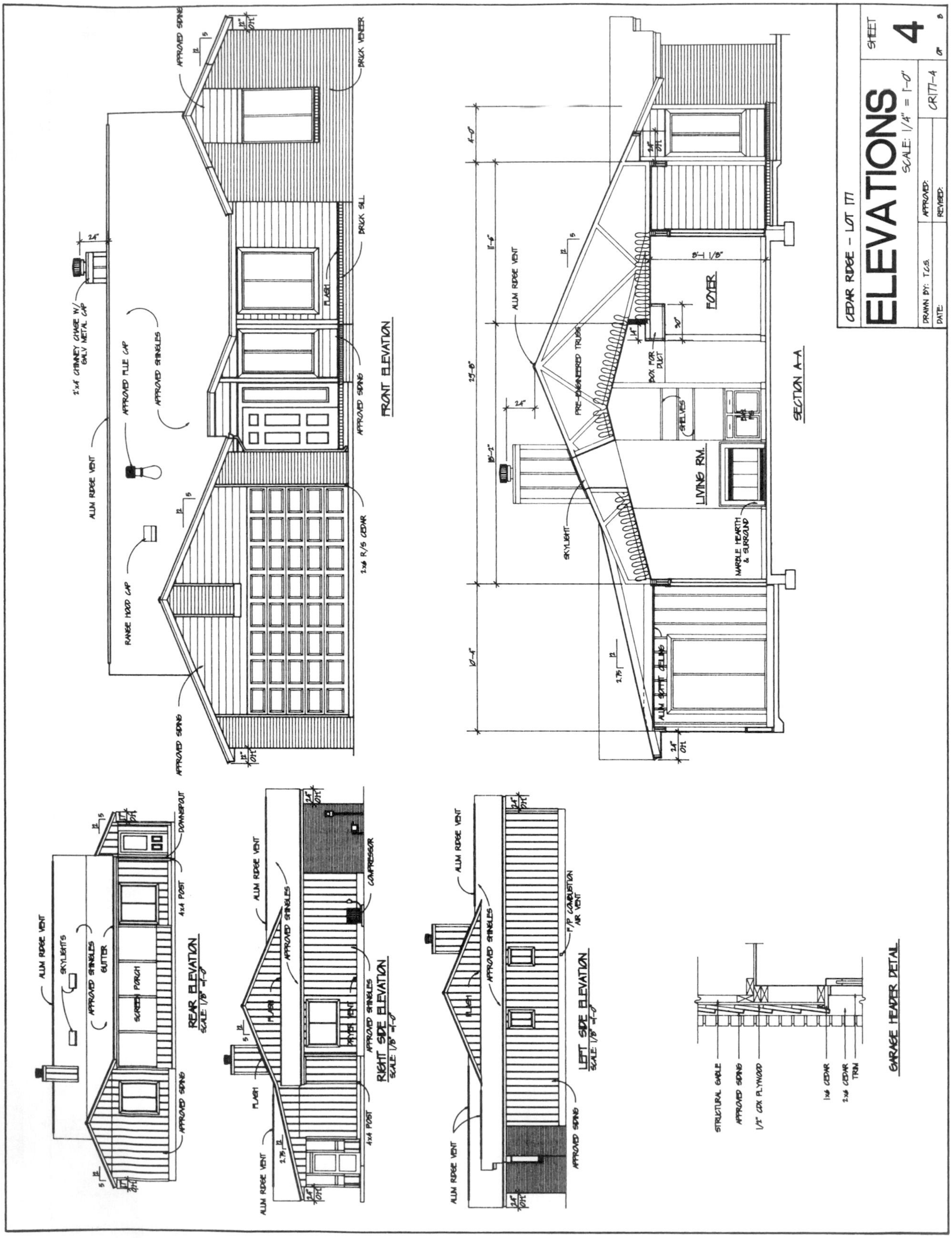

FIGURE 21-71 *Elevations* (Courtesy Barton Homes Inc.)

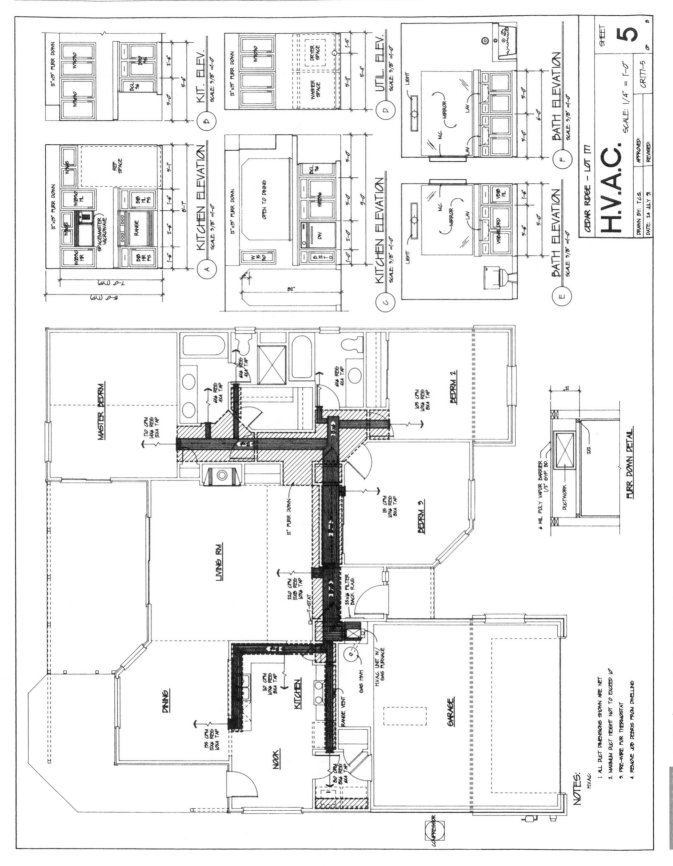

FIGURE 21-72 *HVAC plan* (Courtesy Barton Homes Inc.)

ELECTRICAL PLANS

The *electrical plan* shows the house's electrical system superimposed on a floor plan without dimensions. It should contain the following information: service entrance capacity, meter and distribution panel, location and types of switches, location and types of lighting fixtures, number and types of electrical circuits, details as needed, notes, electrical symbol legend, and an indication of which fixtures are connected to which switches. *Figure 21-73* is an example of an electrical plan with the necessary information.

ROOF PLAN

The *roof plan* is a plan view showing the location and spacing of rafters and/or trusses. It should also contain a truss schedule, details as needed, and truss diagrams as needed. It is superimposed on the floor plan without dimensions. *Figure 21-74* is an example of a roof plan with the necessary information.

COMMERCIAL PLANS

Commercial architectural plans differ somewhat from residential plans. The major difference is in who produces the various components of the plans. With residential plans all components can be produced by an architect, residential designer, and/or architectural drafter including the plot, electrical, and HVAC plans. This is not the case with commercial plans.

A complete set of commercial plans will usually contain architectural, civil engineering, structural (as needed), electrical, and mechanical (HVAC and plumbing) drawings. The architectural component of the plans (i.e., floor plans, elevations, sections, and details) is prepared by an architect or architectural drafter. The component relating to the site is prepared by a civil engineer, the structural component by a structural engineer, the electrical by an electrical engineer, and the mechanical by a mechanical engineer.

Architectural drafters and designers in a commercial setting prepare floor plans, elevations, sections, and details. The basic contents of these plans are similar to those in residential plans. *Figure 21-75* (page 837) is an example of a commercial floor plan. Notice that in addition to the plan view in sections of the building it contains several sectional details.

Each detail has a designator that, when translated, gives its numerical designation, the sheet on which it is cut, and the sheet on which it is drawn. For example, sectional detail "1-A2-A2" can be interpreted as follows:

1 = Detail number (1)
A2 = Sheet on which the detail is cut (A2)
A2 = Sheet on which the detail is drawn (A2)

This floor plan also contains a symbols legend, material legend, equipment legend, keynotes, a scale indicator, and title block information.

Figure 21-76 (page 838) is an example of commercial elevations. In addition to the elevations themselves, it contains two full sections cut through the building, and a roof plan. *Figure 21-77* (page 839) contains numerous other miscellaneous sections and details commonly found in a set of commercial plans plus their corresponding notes.

Civil Engineering Drafting

Civil engineering drafting is the field that deals with mapping, plot plans, earthwork, highways, roads, and bridges, and profiles of the earth. Civil drafting technicians work in civil engineering companies, for surveyors, and in city, county, state, and federal government road and engineering departments. The fundamentals of civil engineering drafting include the following:

• Legal descriptions

• Plot plans

• Contour lines

• Rectangular systems

Other more advanced topics such as profiles, highway layouts, earthwork, and geographic information systems (GIS) will not be dealt with in this section.

LEGAL DESCRIPTIONS

The two most widely used systems for describing property boundaries are the metes and bounds system and the lot and block system. The *metes and bounds* system uses selected known points of origin and then describes property boundaries in relation to this point-of-beginning or POB. The property in question is actually described using compass bearings and distances. *Lot and block* descriptions are used to describe parcels of land that are surveyed and platted on subdivision maps that are officially filed in the office of the clerk of the court (county level).

PLOTTING LEGAL DESCRIPTIONS

All metes and bounds descriptions start at a permanent monument or some other fixed and observable point of beginning. Metes are property lines that are measured in established units (i.e., feet, inches, meters, yards, rods,

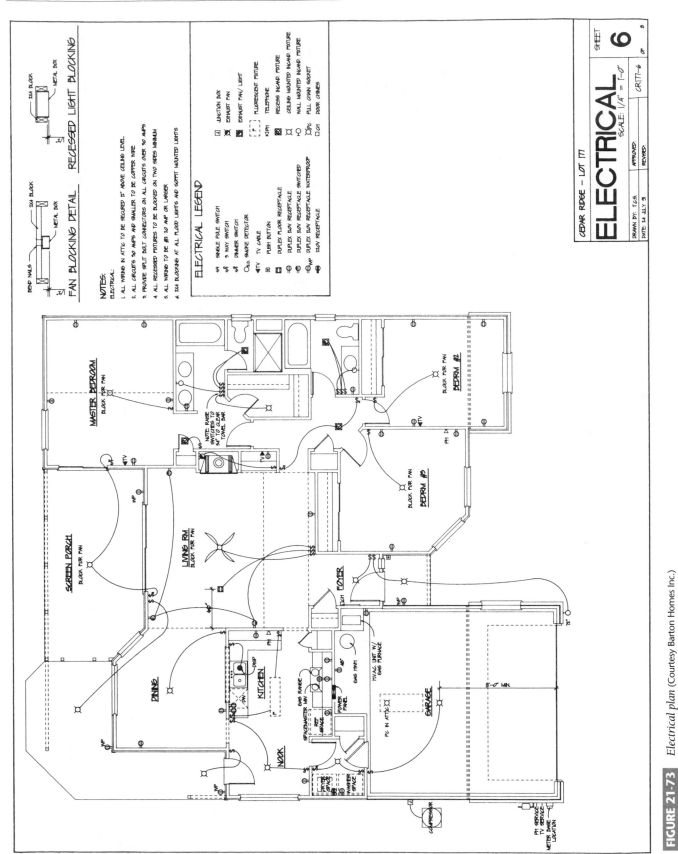

FIGURE 21-73 *Electrical plan* (Courtesy Barton Homes Inc.)

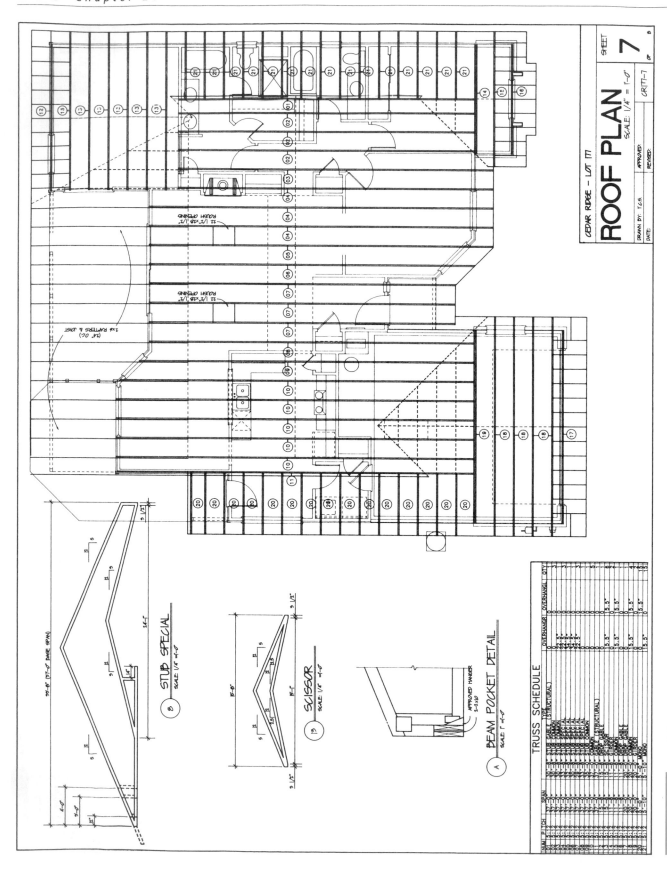

FIGURE 21-74 *Roof plan (Courtesy Barton Homes Inc.)*

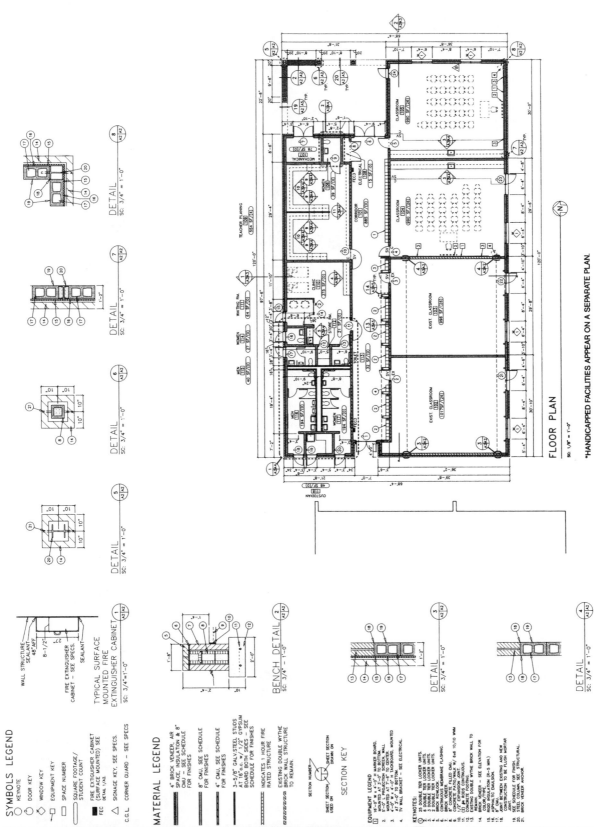

FIGURE 21-75 *Commercial floor plan (Courtesy Kendrick, David, and Dowling Architects Inc.)*

JIM KENDRICK III, AIA
AR0005050
DON W. DAVID, JR. AIA
AR0002932
JAMES R. DOWLING AIA
AR0009795
J. RUSSELL KENDRICK,AIA
AR0011826

ELEVATIONS,
ROOF PLAN &
BUILDING SECTIONS

CLASSROOM BUILDING
FOR
PAXTON SCHOOL
PAXTON, FLORIDA.

PROJECT NO. 9115
DATE DEC. 23, 1991
DRAWN BY WD
CHECKED BY RK

A3

SHT 10 OF 18

BUILDING SECTION
SC 1/8" = 1'-0"

KEYNOTES:

1. STANDING SEAM ROOF SYSTEM - SEE SPECIFICATIONS
2. 2-1/2" RIGID INSULATION (R-18 MIN.) WITH 1/2" NAIL BASE
3. STEEL DECKING - SEE STRUCTURAL
4. STEEL PURLINS - SEE STRUCTURAL
5. STEEL BEAM - SEE STRUCTURAL
6. SUSPENDED ACOUSTIC CEILING SYSTEM
7. CEMENT PLASTER CEILING SYSTEM
8. EXISTING STEEL ROOF STRUCTURE TO REMAIN.
9. INDICATES WALL SYSTEMS THAT EXTEND FULL HEIGHT AND SEAL TO ROOF DECK/STRUCTURE
10. 8" HIGH CAST ALUMINUM LETTERS - SEE SPECIFICATIONS.
11. 1 SOLDIER COURSE OF BRICK - TYPE 2
12. BRICK VENEER - TYPE 1
13. NOT USED
14. METAL GABLE PANEL SYSTEM WITH INVERTED SEAM

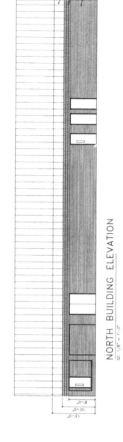

SOUTH BUILDING ELEVATION
SC 1/8" = 1'-0"

NORTH BUILDING ELEVATION
SC 1/8" = 1'-0"

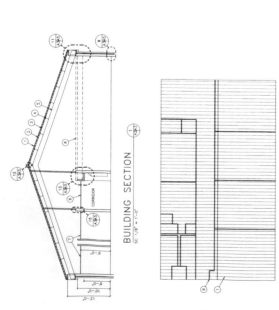

BUILDING SECTION
SC 1/8" = 1'-0"

ROOF PLAN
SC 1/8" = 1'-0"

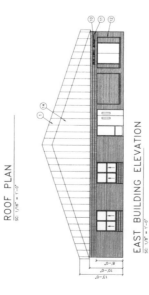

EAST BUILDING ELEVATION
SC 1/8" = 1'-0"

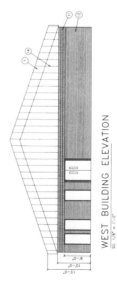

WEST BUILDING ELEVATION
SC 1/8" = 1'-0"

FIGURE 21-76 *Commercial elevations (Courtesy Kendrick, David, and Dowling Architects Inc.)*

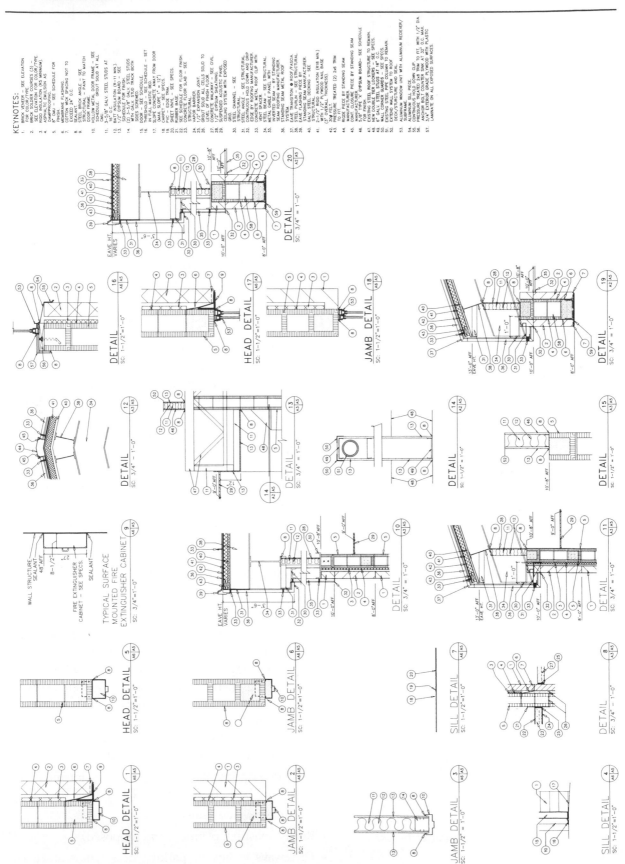

FIGURE 21-77 *Commercial details and sections (Courtesy Kendrick, David, and Dowling Architects Inc.)*

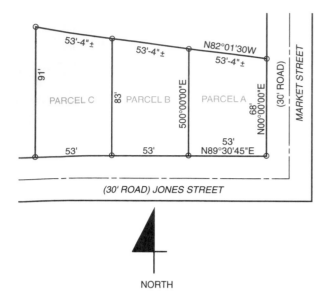

FIGURE 21-78 *Lots with metes and bounds descriptions*

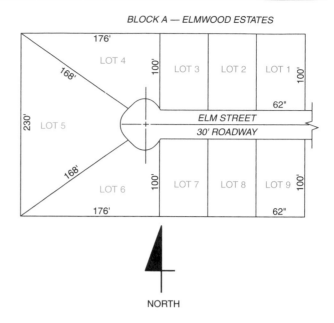

FIGURE 21-79 *Lot and block descriptions*

links, etc.). Bounds are property boundaries that are formed by the lines of adjacent properties, rivers, streams, roads, lakes, bays, etc. *Figure 21-78* contains three parcels of land with metes and bounds descriptions. The description for Parcel A reads as follows:

A parcel of land in Okaloosa County, Florida, beginning at a permanent survey marker located at the intersection of the north right-of-way of Jones Street and the west right-of-way of Market Street. From this point, go N00°00″E 68′ to a metal stake. Then, turn to a heading of N82°01′30″W and go 53′-4″ to a metal stake. Then turn to a heading of S00°00′00″E and go 76′ to a metal stake located on the north right-of-way of Jones Street. Then turn to a heading of N89°30′45″E and go 53′ back to the point of beginning.

With the compass bearings and the distances known, a parcel of property can be drawn. Compass bearings are given in degrees, minutes, and seconds. The compass bearing N82°01′30″W is read as follows:

From north, turn 82 degrees, one minute and 30 seconds toward the west.

Compass bearings used in metes and bounds descriptions will always begin with either north or south and turn toward either the west or the east. For example, the bearing N00°00′00″E is interpreted as follows:

From north, turn zero degrees, zero minutes, and zero seconds toward the east. In other words, go due north. The compass bearing N00°00′00″W is also due north. When going due north or due south, the east and west components can be used interchangeably.

LOT AND BLOCK DESCRIPTIONS

Property that is subdivided, platted, and filed in the official records of the office of the county clerk may be described using lot and block numbers. Subdivisions are typically divided into blocks and the blocks are, in turn, divided into lots, *Figure 21-79*. This figure shows Block A of a subdivision known as Elmwood Estates. The actual metes and bounds description for the entire subdivision as well as a subdivision plat (property map) are on file in the official records of the county in question. Consequently, lots in the subdivision can be described using lot and block designations (i.e., Lot 3, Block A, Elmwood Estates Subdivision). Anyone who needs a more definitive description of a parcel in the subdivision can refer to the subdivision plat in the county's official records.

PLOT PLANS

A set of plans for a commercial building or a residential dwelling must contain a plot plan, *Figure 21-80*. The plot plan locates the building and other key features on the piece of property in question. The requirements for what information must be contained on a plot plan vary from state to state and even from community to community. However, the following list contains the types of information that are typically required:

- Legal description of the property (using lot and block, metes and bounds, or rectangular system)
- North arrow and drawing scale
- Roads (existing, planned, and proposed)
- Driveways, walkways, parking areas, patios, and decks

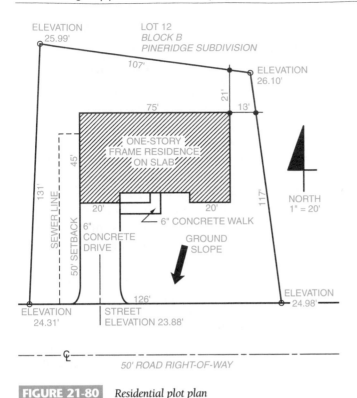

FIGURE 21-80 *Residential plot plan*

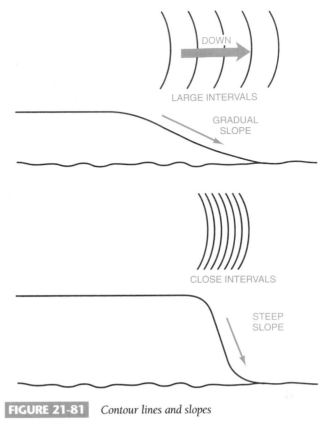

FIGURE 21-81 *Contour lines and slopes*

- Current, planned, and proposed structures on the property
- Locations (as applicable) of wells, water lines, gas lines, power lines, sewer lines, septic tanks, drain fields, leach lines, soil test holes, rain drains, footing drains, and drive/walkway drains
- Elevations of property corners and at the street (measured at the center line of the driveway)
- Ground slope arrow
- Setback dimensions
- Easements

CONTOUR LINES

Contour lines are used on topographical maps to connect points of elevation. To a person who can interpret their meaning, contour lines tell whether the land in question is flat or hilly, gently sloped or radically sloped, concave or convex.

In reading contour lines, a key characteristic to consider is *interval*. The greater the interval between contour lines, the more gradual the slope. A shorter interval indicates a steeper slope, *Figure 21-81*. Streams and ridges are special topographical features that can also be represented using contour lines. *Figure 21-82* shows how contour lines appear at a stream. Notice that they form a V that points upstream. *Figure 21-83* shows how a ridge can be indicated by contour lines.

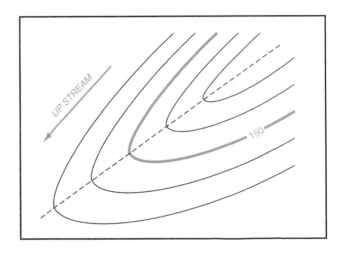

FIGURE 21-82 *Contour lines and streams*

There are three different types of contour lines as shown in *Figure 21-84*. Index contour lines are thicker than the others and are spaced at intervals of five so that every fifth line is an index line. The index line is broken periodically and labeled with the elevation.

Intermediate contour lines are thin, unbroken lines between the interval contour lines. Supplementary contour lines are dashed and typically represent half of the distance between contour lines. Regardless of the

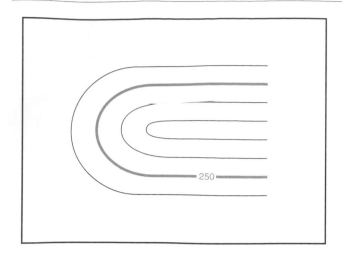

FIGURE 21-83 *Contour lines and ridges*

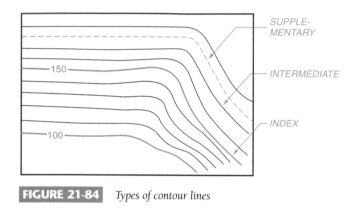

FIGURE 21-84 *Types of contour lines*

relative slope of the land, every fifth contour line is an index contour. Consequently, the contour interval value (i.e., 2, 5, 10 . . . feet) can vary.

RECTANGULAR SYSTEM

The U.S. Bureau of Land Management developed a method of describing land called the *rectangular system*. The system is used in the following states:

- Alabama
- Alaska
- Arizona
- Arkansas
- California
- Colorado
- Florida
- Idaho
- Illinois
- Indiana
- Iowa
- Kansas
- Louisiana
- Michigan
- Minnesota
- Mississippi
- Missouri
- Montana
- Nebraska
- Nevada
- New Mexico
- North Dakota
- Ohio
- Oklahoma
- Oregon
- South Dakota
- Utah
- Washington
- Wisconsin
- Wyoming

Land in these states is divided into a grid of blocks called townships, *Figure 21-85*. Each major segment of land has both a township and a range number. For example, a designation might be T24NR36E. This designation reads as follows: Township 24 North, Range 36 East. Township designations come from the vertical columns that run north and south in this figure. The designations east and west relate to the line marked as the principal meridian. The designations north and south relate to the line marked as the baseline.

SUBDIVIDING TOWNSHIPS

A full-sized township is a tract of land six miles square. Exceptions to this rule are created by geographical features such as shorelines, natural borders, rivers, and large bodies of water. For example, many of the townships in Florida are irregularly shaped and less than full size. Regardless of their actual shape and size, townships are divided into sections as shown in Figure 21-85.

SUBDIVIDING SECTIONS

Sections can be subdivided into an almost unlimited number of smaller units. For example, the section shown in *Figure 21-86* contains several smaller sub-

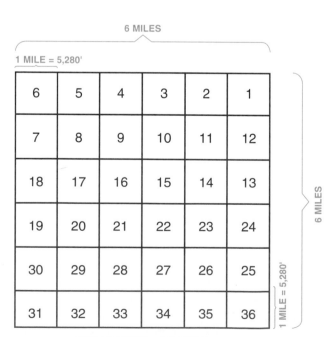

EACH SECTION = 640 ACRES
ONE ACRE = 43,560 SQ. FT.

FIGURE 21-85 *Rectangular system*

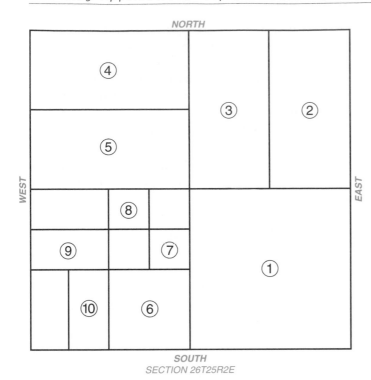

FIGURE 21-86 *Rectangular legal descriptions*

units. Ten of these parcels have been numbered. Using the rectangular system, the *legal descriptions* of these numbered parcels are as follows:

1. The SE ¼ of Section 26T25R2E (Reads as follows: the southeast one-quarter of Section 26 in Township 25, Range 2 East).

2. The eastern ½ of the NE ¼ of Section 26T25R2E.

3. The western ½ of the NE ¼ of Section 26T25R2E.

4. The northern ½ of the NW ¼ of Section 26T25R2E.

5. The southern ½ of the NW ¼ of Section 26T25R2E.

6. The southeast ¼ of the SW ¼ of Section 26T25R2E.

7. The southeast ¼ of the NE ¼ of the SW ¼ of Section 26T25R2E.

8. The northwest ¼ of the NE ¼ of the SW ¼ of Section 26T25R2E.

9. The southern ½ of the NW ¼ of the SW ¼ of Section 26T25R2E.

10. The eastern ½ of the SW ¼ of the SW ¼ of Section 26T25R2E.

The key to reading this type of legal description is to begin at the end and read backwards. This will simplify the process by beginning with the large and working toward the small. For example, look at number 7 in this list of legal descriptions. Reading backwards the component parts are as follows:

• Section 26T25R2E

• SW ¼ of the section

• NE ¼ of that parcel

• SE ¼ of that parcel

Geographic Information System

A *geographic information system* (GIS) is a computer system that collects, stores, retrieves, analyzes, and displays spatial and attribute data relating to features on, under, and above the ground. Spatial data in this sense are data that describe the location of features on and under the ground. These features are represented by points, lines, or polygons that have their own unique location in the world. For example, consider a parcel of land in a typical community and all of the various types of interrelated data that may be collected concerning the parcel, (*Figure 21-87*). The types of data shown in this figure are just a few examples of the types that can be collected and used with a GIS.

Attribute data can be any type of data about a feature other than its location in the world. For example, if the feature is a building, attribute data may include the following:

• Square footage

• Number of floors

• Principal building material (steel, concrete, wood frame)

• Any other data that describe or define the building beyond its location

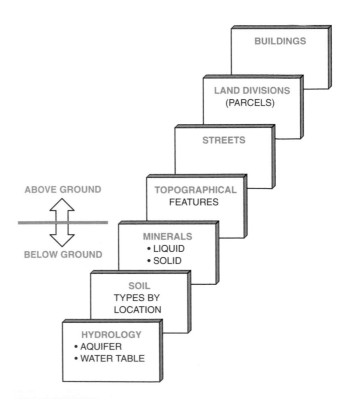

FIGURE 21-87 *Types of data that might be collected and stored in a GIS database*

What makes GIS software special is its ability to examine spatial relations and answer what-if questions. For example, the following types of questions may concern city or county engineers and planners in a given community:

- What if we rezone an area to allow for additional construction?
  - Will there be any negative effects on the water table?
  - Can the area accommodate the new infrastructure that would be needed (e.g., roads, utilities, and drainage)?
  - How will the additional solid and liquid waste treatment requirements affect the area?
- What if drilling for oil is permitted in a given area?
  - How will the local river be affected?
  - What impact will drilling have on the nearby children's park?
  - Will underground utilities have to be rerouted?

A GIS map file can represent the entire world or any part of it to desired level of detail. The part of the world put in the system can have many different types of attributes (e.g., political boundaries, population density, zip codes, census designations, streets, and bridges). A GIS map file can be created for any type of data that have geographic linkages.

Everything that is under, on, or above the land is interrelated, *Figure 21-88*. Any action that affects one feature in a spatial system can affect other features. A major benefit of a GIS is that it can help prevent decisions that create spatial conflicts that can have adverse and unintended consequences.

One of the difficulties with making land-use decisions is that they cannot be made in a vacuum. One seemingly isolated and well-intended activity such as constructing a culvert system to prevent flooding may unwittingly cause an even worse problem than flooding. For example, it may cut through underground utility, water, and/or sewer lines that engineers forgot were there or it may transfer toxic runoff into the community's principal water source. A GIS allows planners to consider the potential domino effects one action can have on all features within a spatial system (Figure 21-88).

KEY GIS CONCEPTS

The first concept to understand about GIS is the *georelational data model*. This model serves as the tie between *spatial data* and *attribute data*. Any attribute data that can be linked to geographic locations is linked by the georelational data model. For example, a GIS map may show the geographic locations of the largest electricity or water users in the community.

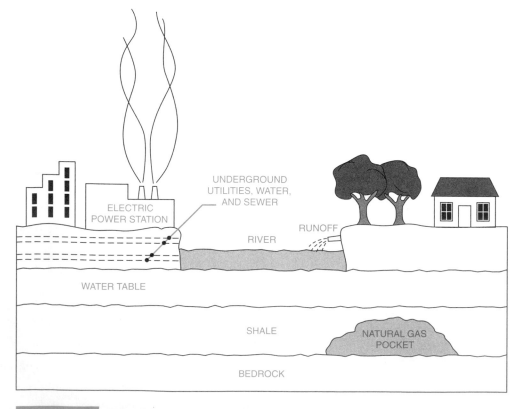

FIGURE 21-88 *Related features in a selected location*

Spatial data are stored as points, lines, and polygons that occupy a specific location in space. In this way, planners are able to relate various features to each other as well as attributes relating to those features. For example, the county's records for solid waste disposal can be linked by the GIS with the locations of residential, commercial, and/or government buildings in the county.

Attribute data can be almost any kind of information relating to what is under, on, and/or above the ground in any specified location or anything that is inside a given component. For example, say the feature in question is a tree. What kind of tree is it? How tall is it? When was it last harvested? How old is it? The answers to these questions are attribute data. A GIS uses a type of software called *relational database management systems* (RDBMS) to relate data stored in the database. The relationship between and among files is based on a common element contained in the files. The computer will key on this common element when retrieving data in response to *what-if scenarios* posed by planners, engineers, and technicians.

A final but fundamental GIS concept is *topology*. This concept is a mathematic system that defines relationships among geometric objects/shapes (such as polygons that represent parcels of land, buildings, or other features). These relationships include adjacency, connectivity, contiguity, and proximity. Are several parcels (building, features, etc.) adjacent? Are they connected? Are they contiguous? Humans can see these spatial relationships, but computers are blind in the human sense. Consequently, they must determine spatial relationships mathematically. Topology gives computers this capability. *Figure 21-89* is an example of a GIS database that was developed for specific applications. The database in Figure 21-89 was developed to assist in the launch of missiles into space. The desired capabilities on the left are supported by the data on the right.

HOW A GIS IS USED

The best way to understand a GIS is to consider an actual example of how one is used. *Figure 21-90* contains a map of a college campus. A GIS could be used by the college's Facilities Management Office for maintaining the master plan for maintenance, renovation, remodeling, and numerous other functions.

The map shows the location on the campus of all facilities, including buildings, roads, parking lots, and athletic facilities. The legend shows the letter designation for each building, which, in turn, has a descriptive name based on its function (e.g., Administration, Gymnasium/Wellness Center, and Arts Center). All pertinent locational data relating to buildings and facilities are stored in the GIS database.

In addition to the locational data, there are also attribute data for all pertinent features below and on the ground. For example, consider Building E, the Learning Resources Center, from the campus map in Figure 21-90. Attribute data for this building may include the following:

- *Floor plans for each level*
  - Maintenance data for all rooms on all floors
  - Past, present, and planned renovation data for all rooms
  - Contents of each room (chairs, tables, computers, audiovisual equipment, etc.)
- *Air conditioning and heating plan for the building*
  - Maintenance schedules
  - Equipment specifications
- *Electrical plan for the building*
  - Maintenance schedules
  - Specifications for electrical equipment, lines, and fixtures

---

SPACE LAUNCH FACILITIES APPLICATION

Desired Capability	Relevant Data Stored
■ Telemetry Data Display	■ Tracking station locations, 3-D spatial relationships
■ Range Tracking	■ Moving location, dynamic zones of impact probability
■ Command Destruct Criteria	■ Graded overall constraints zones
■ Down Range Depiction	■ Visualization of safety constraints, real-time tracking
■ Safety Buffer Zone Analysis (in Real Time)	■ Dynamic quantity/distance explosive zones
■ Large Vehicle Routing	■ Alternative route selections, automated evacuation
■ Security Planning	■ Security zones, SCIP locations, alarmed areas

FIGURE 21-89 *GIS database for a space launch facilities application*

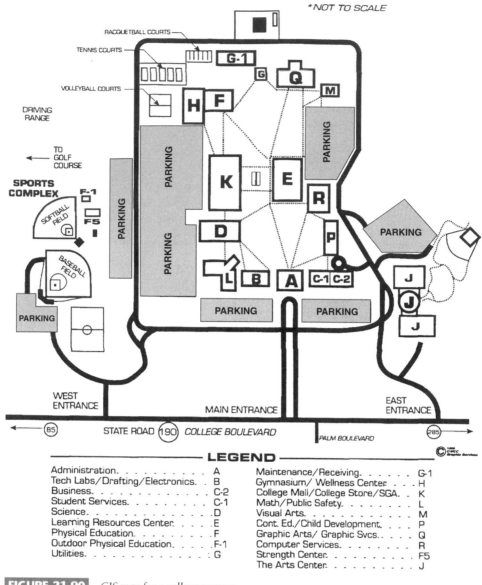

LEGEND

Administration.	A	Maintenance/Receiving.	G-1
Tech Labs/Drafting/Electronics.	B	Gymnasium/ Wellness Center	H
Business.	C-2	College Mall/College Store/SGA.	K
Student Services.	C-1	Math/Public Safety.	L
Science.	D	Visual Arts.	M
Learning Resources Center	E	Cont. Ed./Child Development.	P
Physical Education.	F	Graphic Arts/ Graphic Svcs.	Q
Outdoor Physical Education.	F-1	Computer Services.	R
Utilities.	G	Strength Center.	F5
		The Arts Center.	J

FIGURE 21-90 _GIS map for a college campus_

- _Data and voice lines plan for the building_
  - Maintenance schedules
  - Specifications for the system and all components
- _Fire alarm plan for the building_
  - Maintenance schedules
  - Specifications for the system and all components
- _Sprinkler (fire suppression)_
  - Maintenance schedule
  - Specifications for the system

The same information could be stored about all of the buildings on campus. Similar information could be stored about all features on the campus. Having this information readily available—along with all of the locational data—gives the college's facilities management personnel powerful capabilities.

GIS CAPABILITIES: SPATIAL ANALYSIS

Every GIS has certain standard capabilities. This section explains traits of those standard capabilities that fall under the heading of _spatial analysis._ _Spatial queries_ allow users to pose what-if scenarios concerning features on the map. The GIS bases its responses to spatial queries on points, lines, and polygons. _Point-in-polygon responses_ locate specific points that lie within specific polygons. For example, say a company needed to show the local fire department the location of all fire extinguishers in a given building. The polygon would be the building, and each fire extinguisher would show up as a point superimposed on the building's floor plan as displayed on a computer terminal screen or printed in hard form.

Polygon overlay involves the generation of a new polygon by overlaying two or more polygons. For example, say company personnel want to make sure that the installation of fiber optic cables will not interfere with the existing electrical system. The GIS could overlay the existing electrical system (one polygon) over the floor plan of the building in question (a second polygon). It could then overlay the proposed fiber optic system plan over the first two. The newly created polygon when displayed would point out visually any potential wiring conflicts.

Buffering generates new polygons around a set of points, lines, or other polygons. Nearest neighbor is a GIS capability that identifies the closest features to another feature. This capability is especially valuable in keeping track of features as they relate to other features. For example, say the college in Figure 21-90 wants to add a new business and industry training center in the area between its west entrance and its main entrance. What features are the nearest neighbors on the ground and underground? If one of the nearest neighbors turns out to be a large tract of environmentally sensitive wetlands in this area, the college may decide to put the new training center elsewhere.

Network analysis is a GIS capability that allows users to perform flow analysis, routing determinations, and/or orders stops. The college in Figure 21-90 may use this capability to analyze traffic flow patterns, student routing patterns between classes, or storm drainage flow analysis.

GIS Capabilities: Attribute Queries

Attribute queries pull information about the attributes of features that are shared in the database. How they are organized and displayed depends on the nature of the specific inquiry in question.

Extracting is a GIS capability that selects a subset of data from a broader set. For example, say the college in Figure 21-90 wants to know how many classrooms it has that have wall-mounted videotape players. This would be the subset. The broad set of data includes all classrooms.

Generalizing is a capability that combines polygon features based on common or similar attributes. For example, the college in Figure 21-90 may want to see a list of all rooms on campus that have been recarpeted

GIS APPLICATIONS CHECKLIST FOR A COMPANY
■ Comprehensive/Master Planning
■ Facilities Management
■ Facilities Siting
■ Utilities Planning
■ Water Resource Projects
■ Natural Resources Management
■ Wetlands Inventory
■ Environmental Restoration/Compliance
■ Hazardous Material/Waste Management
■ Environmental Health Risk Assessment
■ Historical/Archaeological Site Protection
■ Fire and Emergency Vehicle Dispatch
■ Maintenance Team Dispatch
■ Security Response and Planning
■ Space Utilization
■ Pavement Management
■ Floodplain Management
■ Terrain Analysis
■ Communication Network Planning
■ Logistics Planning

FIGURE 21-91 *Uses of a GIS at a company*

within the last six months. A floor plan could be displayed with the subject rooms highlighted.

GIS and CAD

The wedding of GIS and CAD makes for a perfect marriage. GIS is, at its heart, a database system. As such, a GIS is best at storing, analyzing, and displaying data. CAD, on the other hand, is graphics based. It displays pictures. Put these two concepts together, and users get the best of two worlds—conveniently organized data that are augmented visually by drawings. For example, the campus map in Figure 21-90 is a product of the marriage of CAD and GIS. The ability to display data in a graphic format simplifies matters when it comes to the user interface. Even nontechnical users can understand a picture. *Figure 21-91* shows typical uses of GIS for a given company.

Summary

- The most widely used types of pipe are steel, cast iron, brass, copper, copper tubing, and plastic.

- The three broad classifications of fittings are screwed, flanged, and welded.

- Fluids and gases flowing through pipe are regulated or stopped at certain points by valves. The most widely used types of valves are gate, globe, jet, swiveled joint, ball, check, and butterfly.

- Pipe drawings serve the same purpose as other types of technical drawings. They convey the intentions of the designer and engineer to the tradespeople who will construct the pipe system. There are two types of pipe drawings: single-line and double-line.

- A structure is anything constructed of parts such as commercial buildings, bridges, towers, parking decks, and stadiums.

- Documentation prepared by structural drafters consists primarily of engineering drawings and shop drawings. Engineering drawings convey structural design information. Shop drawings convey all of the information needed to manufacture the individual parts of the structure and are sometimes called detail drawings. Pre-engineered metal buildings require plans that include anchor bolt plans, rigid-frame cross-sections, and framing and sheeting plans.

- Architectural drafting is the process of documenting the design developed by an architect. Residential plans include: a plot plan; foundation plan; floor plan; elevations; heating, ventilation, and air conditioning (HVAC) plan; electrical plan; and framing plans.

- Plans for commercial buildings contain architectural, civil engineering, electrical, and mechanical components. The architectural component includes floor plans, elevations, sections, and details.

- The basics of civil engineering drafting include legal descriptions (metes and bounds, lot and block), plot plans, contour lines, and the rectangular system.

- GIS systems collect, store, retrieve, analyze, and display spatial and attribute data relating to features on, under, and above the ground.

Review Questions

Answer the following questions either true or false.

1. It is important for engineering and drafting students to be proficient in pipe drawing.

2. A correction is required to do an engineering or architectural change.

3. A revision is usually done when a mistake or drafting error must be fixed.

4. Bolting and welding are the two most widely used methods for connecting structural members.

5. Engineering drawings convey structural design information.

6. Shop drawings convey all information needed to manufacture the individual parts of a structure and are sometimes referred to as detail drawings.

7. All metes and bounds descriptions start at a permanent monument or some other fixed point of beginning.

8. Prefabricated metal buildings cannot be used for anything but warehouses.

9. GIS is just another application of CAD.

Answer the following questions by selecting the best answer.

1. Which of the following is **not** one of the most widely used types of pipe?
   a. Steel
   b. Aluminum
   c. Copper
   d. Brass

2. Which of the following is **not** one of the three broad classifications of pipe?
   a. Standard
   b. Extra strong
   c. Double extra strong
   d. Triple extra strong

3. Which of the following is **not** one of the three most widely used categories of valves?
   a. Swivel
   b. Gate
   c. Check
   d. Globe

4. Which of the following is **not** among the typical components in a set of residential architectural drawings?
   a. Plot plan
   b. Foundation plan
   c. Elevations
   d. Civil engineering plans

5. Which of the following is **not** among the typical components in a set of commercial architectural drawings?
   a. Civil engineering plans
   b. Mechanical drawings

c. Electrical drawings

d. Baseline drawings

6. Property that is subdivided, plotted, and filed in the official records may be described using the following method:

a. Metes and bounds

b. Rectangular

c. Lot and block

d. Contour

7. Which of the following is part of a typical set of plane for a pre-engineered metal building?

a. Meter and bounds plan

b. Orthographic plan

c. Rigid-frame cross section

d. Lot and block plan

8. Select all of the following that are common application of GIS.

a. Fire alarm system in a building

b. Electrical plan for a building

c. Identifying spatial conflicts

d. Extracting subset data

CAD Instructions

Students who plan to complete the following problems on a CAD system should begin with these general instructions. Before reading the specific instructions for each activity, go through each step in the following planning checklist. The checklist applies to any CAD system and will help ensure the optimum use of your time and resources.

1. Analyze the problem carefully. Decide exactly what you are being asked to do.

2. Determine what resources and references (if any) you will need in order to complete the problem. Collect those references and resources and have them readily available.

3. Decide if any particular standards apply to the project and have those standards available.

4. Determine what types of views will be required and how many of each.

5. Determine what the final plotted scale of the drawing will need to be, and select the appropriate paper size for plotting/printing.

6. Plan your drawing sequence. In what order will you develop the drawing (i.e., lines, features, dimension lines, leaders, dimensions)?

7. Review the various CAD commands you will have to use in order to develop the drawing.

8. Examine your CAD system to ensure that everything is in working order, then begin the project.

Chapter 21 Problems

Problem 21-1

Problem 21-1 is a single-line isometric pipe drawing. Convert this drawing into a single-line orthographic drawing showing the following views of the system: top, front, right side, and left side.

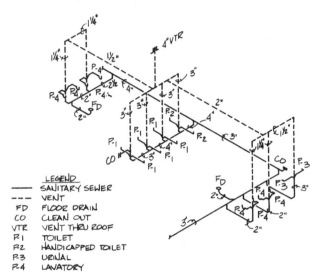

SANITARY SEWER RISER DIAGRAM
NTS

Problem 21-2

Problem 21-2 is a double-line orthographic pipe drawing. Convert it into a single-line orthographic drawing.

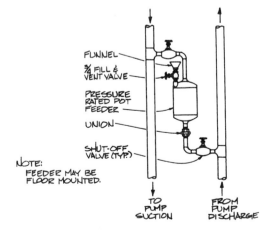

CHEMICAL POT FEEDER DETAIL
N.T.S.

Problem 21-3

Problem 21-3 is a double-line orthographic pipe drawing. Convert it into a single-line orthographic drawing.

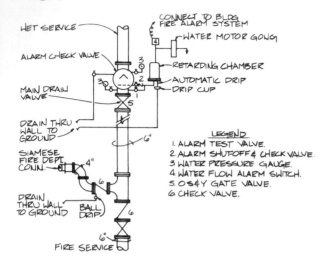

WET SERVICE

CONNECT TO BLDG FIRE ALARM SYSTEM

WATER MOTOR GONG

ALARM CHECK VALVE

RETARDING CHAMBER

MAIN DRAIN VALVE

AUTOMATIC DRIP

DRIP CUP

DRAIN THRU WALL TO GROUND

SIAMESE FIRE DEPT. CONN.

DRAIN THRU WALL TO GROUND

BALL DRIP.

FIRE SERVICE

LEGEND
1. ALARM TEST VALVE.
2. ALARM SHUTOFF & CHECK VALVE.
3. WATER PRESSURE GAUGE.
4. WATER FLOW ALARM SWITCH.
5. OS&Y GATE VALVE.
6. CHECK VALVE.

WET PIPE SPRINKLER SERVICE
NTS

Problem 21-4

Problem 21-4 is a double-line orthographic pipe drawing. Convert it into a single-line orthographic drawing.

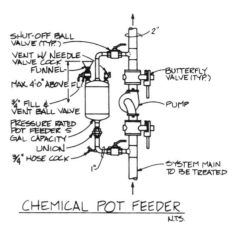

SHUT-OFF BALL VALVE (TYP.)

VENT W/ NEEDLE VALVE COCK FUNNEL

MAX 4'-0" ABOVE FL.

3/4" FILL & VENT BALL VALVE

PRESSURE RATED POT FEEDER 5 GAL. CAPACITY

UNION

3/4" HOSE COCK

2"

BUTTERFLY VALVE (TYP.)

PUMP

SYSTEM MAIN TO BE TREATED

CHEMICAL POT FEEDER
N.T.S.

Problem 21-5

Problem 21-5 is a double-line representation of a piping system. Reconstruct the drawing as shown in single-line representation.

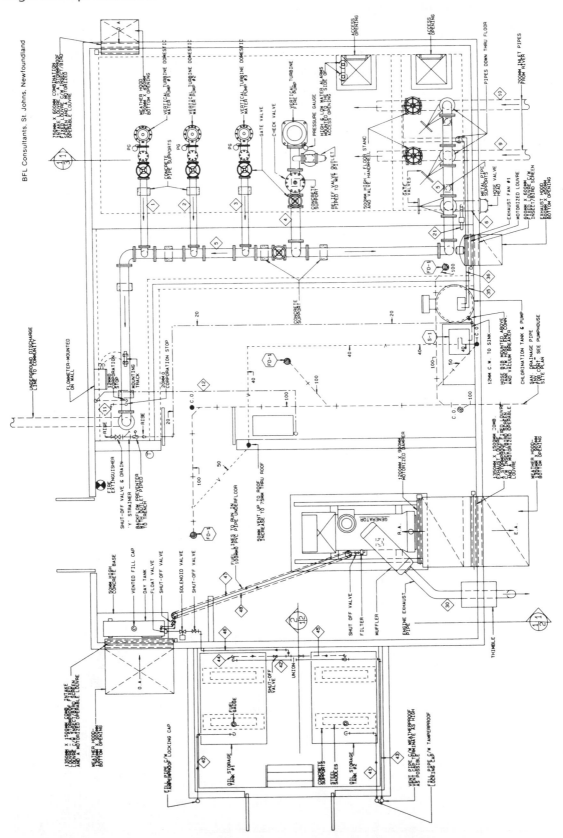

Problems

Problem 21-6

Problem 21-6 is an engineering drawing for a set of bleachers. Reconstruct the entire drawing making the following changes: (a) add 21'–6" to the *Out-to-Out* dimension and adjust all others accordingly; and (b) add one more row of bleachers at the bottom.

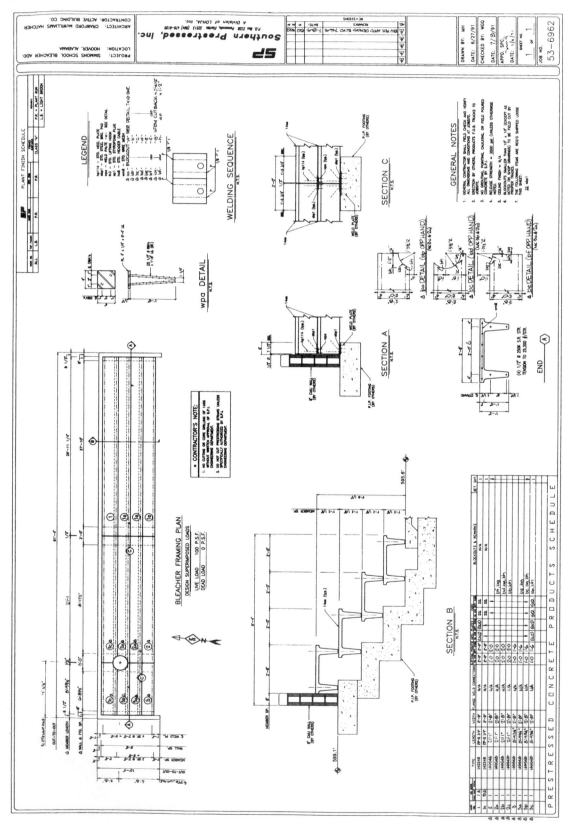

Problem 21-7

Problem 21-7 is an engineering drawing for a pedestrian bridge. Reconstruct the entire drawing making the following change: add one more 33′–2″ bay to the bridge and adjust all dimensions accordingly.

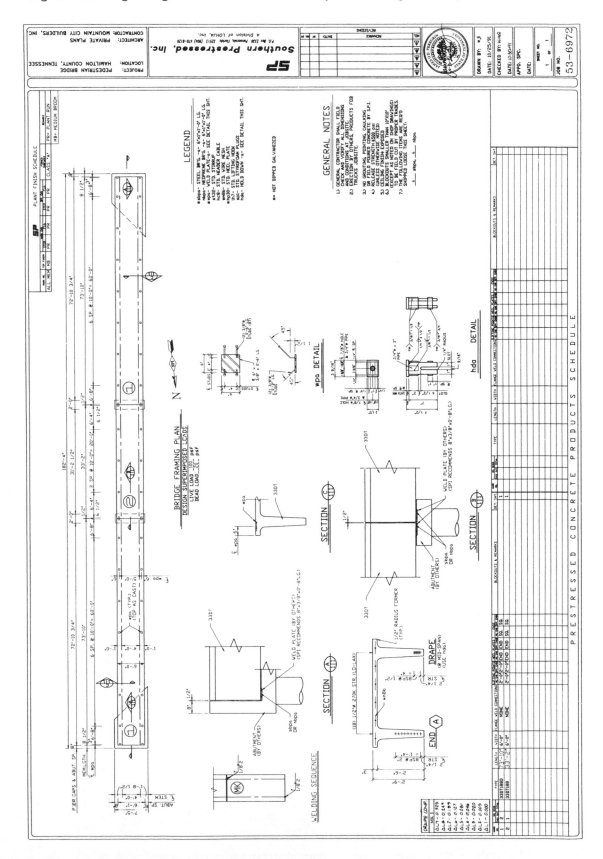

Problem 21-8

Problem 21-8 is a shop drawing for a prestressed concrete beam. Reconstruct the entire drawing making the following change: change the depth of the beam to 1'–6" and adjust accordingly.

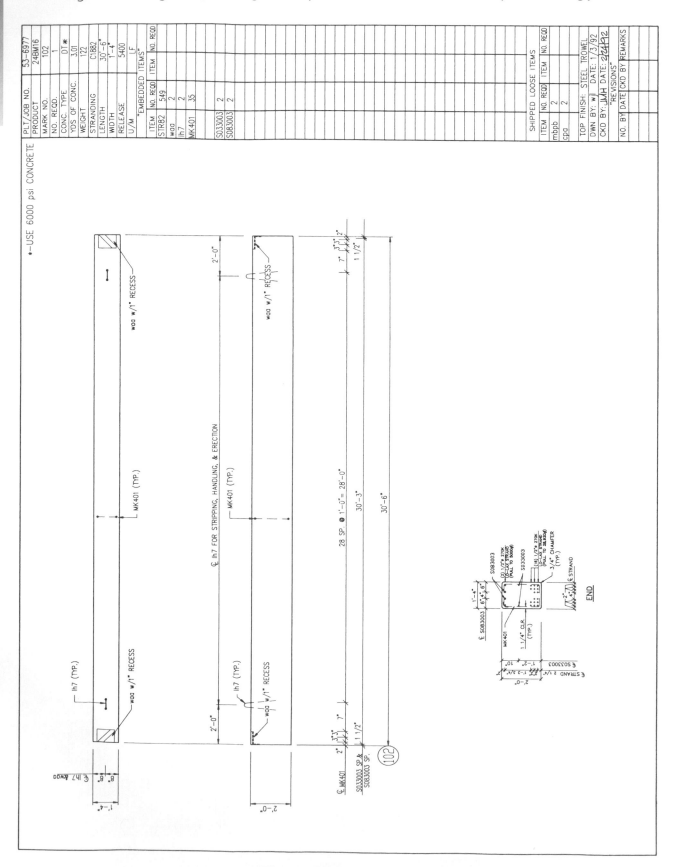

Problem 21-9

Problem 21-9 is a shop drawing for a precast concrete flat slab. Reconstruct the entire drawing making the following change: decrease the width by 1'–1" and the length by 9" and adjust accordingly.

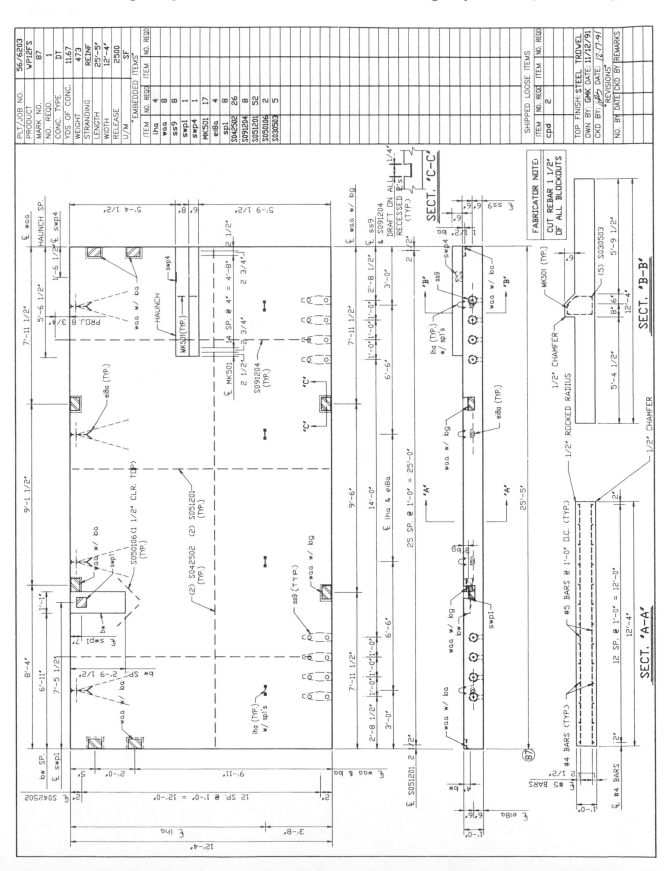

PLT/JOB NO.	56/6203
PRODUCT	WP12FS
MARK NO.	87
NO. REQD.	1
CONC. TYPE	DT
YDS OF CONC.	11.67
WEIGHT	473
STRANDING	REINF
LENGTH	25'-5'
WIDTH	12'-4'
RELEASE	2500
U/M	SF

"EMBEDDED ITEMS"

ITEM	NO. REQD	ITEM	NO. REQD
lha	4		
waa	8		
ss9	8		
swp1	1		
swp4	1		
MK501	17		
ei8a	4		
sp1	8		
S042502	26		
S091204	8		
S051201	52		
S050106	2		
S030503	5		

SHIPPED LOOSE ITEMS

ITEM	NO. REQD	ITEM	NO. REQD
cpd	2		

TOP FINISH: STEEL TROWEL
DWN BY: GMK DATE: 11/12/91
CKD BY: ___ DATE: 12/7-9/

"REVISIONS"

NO.	BY	DATE	CKD BY	REMARKS

SECT. 'C-C'

& S091204
DRAFT ON ALL
RECESSED P's
(TYP.)

FABRICATOR NOTE:
CUT REBAR 1 1/2"
OF ALL BLOCKOUTS

SECT. 'B-B'

1/2" CHAMFER
1/2" ROCKED RADIUS
1/2" CHAMFER

#5 BARS @ 1'-0" O.C. (TYP.)

SECT. 'A-A'

Problem 21-10

Figure 21-71 is a floor plan for a residential dwelling. Subtract 11″ from the overall length and width of this figure. Now make all necessary adjustments to the floor plan.

Problem 21-11

Using the floor plan developed in the previous problem as the nucleus, develop a complete set of residential plans.

Problem 21-12

Figures 21-75, 21-76, and 21-77 contain the basic architectural plans for a classroom building. Add one additional classroom (106) and revise the entire set of plans accordingly.

Problem 21-13

Draw the following parcel at a scale of 1″ = 30′. Beginning at a permanent concrete marker that is the southeast corner of the parcel described herein, turn N7°00′00″W and travel 117.01′. Then turn N82°30′30″W and travel 107.02′; then turn S52°00′15″N and travel 131.03 feet. Then turn due east to the point of beginning (a distance of 126′).

Problem 21-14

Figure 21-78 shows three parcels described using the metes and bounds method. Write a legal description for Parcel B.

Problem 21-15

Redraw the three parcels and roads as shown in Figure 21-78, adding a fourth parcel (Parcel D) to the west of Parcel C. Select an appropriate scale. The parcel should be 53′ wide along Jones Street.

Problem 21-16

• At a scale of 1″ = 300′, draw a complete section as shown in Figure 21-86. Divide it into quarter sections. Refer to Figure 21-85 for dimensions. Plot the following parcels on your section map.
• The north half of the SE ¼ of the section.
• The SE ¼ of the NW ¼ of the section.
• The SW ¼ of the NE ¼ of the SW ¼ of the section.
• The southern ½ of the SE ¼ of the NE ¼ of the SW ¼ of the section.

Problem 21-17

Using Figures 21-68 through 21-86 as a guide, develop a set of plans for a pre-engineered metal building that will be built on a concrete slab that is 40′ wide and 60′ long.

Chapter 22

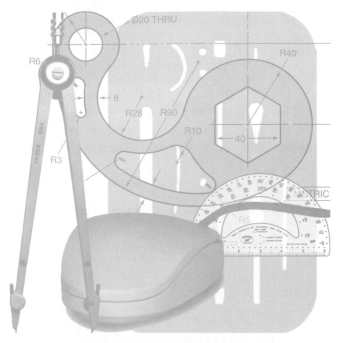

DRAFTING APPLICATIONS: ELECTRONICS AND PRINTED CIRCUIT BOARDS

KEY TERMS

Artwork

Baseline diagrams

Block diagrams

Connection diagrams

Double-sided boards

Drill plan

Gate

Highway diagrams

Lineless diagrams

Logic diagrams

Multilayered boards

Point-to-point diagrams

Printed circuit board

Schematic diagrams

Single-sided board

CHAPTER OUTLINE

Electronics symbols • Schematic diagrams • Connection diagrams • Block diagrams • Logic diagrams • Printed circuit board drawings • Summary • Review questions • Chapter 22 problems

CHAPTER OBJECTIVES

Upon completion of this chapter, students should be able to do the following:

■ Identify the most widely used electronics symbols.

■ Demonstrate proficiency in drawing schematic diagrams.

■ Demonstrate proficiency in drawing connection diagrams.

■ Demonstrate proficiency in drawing block diagrams.

■ Demonstrate proficiency in drawing logic diagrams.

■ Demonstrate proficiency in developing artwork for printed circuit boards.

Electronics Symbols

Electronic products are made of numerous different types of electronic components. Each of these components can be represented symbolically and pictorially. The most frequently used components are antennas, batteries, capacitors, diodes, grounds, inductors, meters, resistors, switches, transformers, and transistors. Other less frequently used components include amplifiers, bells, buzzers, circuit breakers, connectors, crystals, delays, envelopes, fuses, headsets, lamps, motors, magnets, phase shifters, pickup heads, rectifiers, relays, relay coils, repeaters, safety interlocks, speakers, terminal boards, thermal elements, and thermocouples. *Figure 22-1* illustrates the symbols for these electronic components.

Schematic Diagrams

Schematic diagrams are used to show what components are contained on a circuit board and which components connect with which, *Figure 22-2*. It should be noted that the actual location of a component on a circuit board cannot be determined by looking at a schematic diagram. Schematics show components and connections, but they do not show their actual locations.

RULES FOR DRAWING SCHEMATICS

Drafters prepare finished schematic diagrams from sketches provided by engineers. *Figure 22-3* is an example of an engineer's sketch and the corresponding finished schematic diagram. The general rules that should be applied in drawing schematic diagrams are:

1. Inputs go left.
2. Outputs go right.
3. Break connectors and jumble pins.
4. Grounds point downward.
5. Curved side of capacitors point to ground.
6. Arrange relay symbols.
7. Remove doglegs.
8. Avoid crossovers.
9. Don't cramp and crowd components.

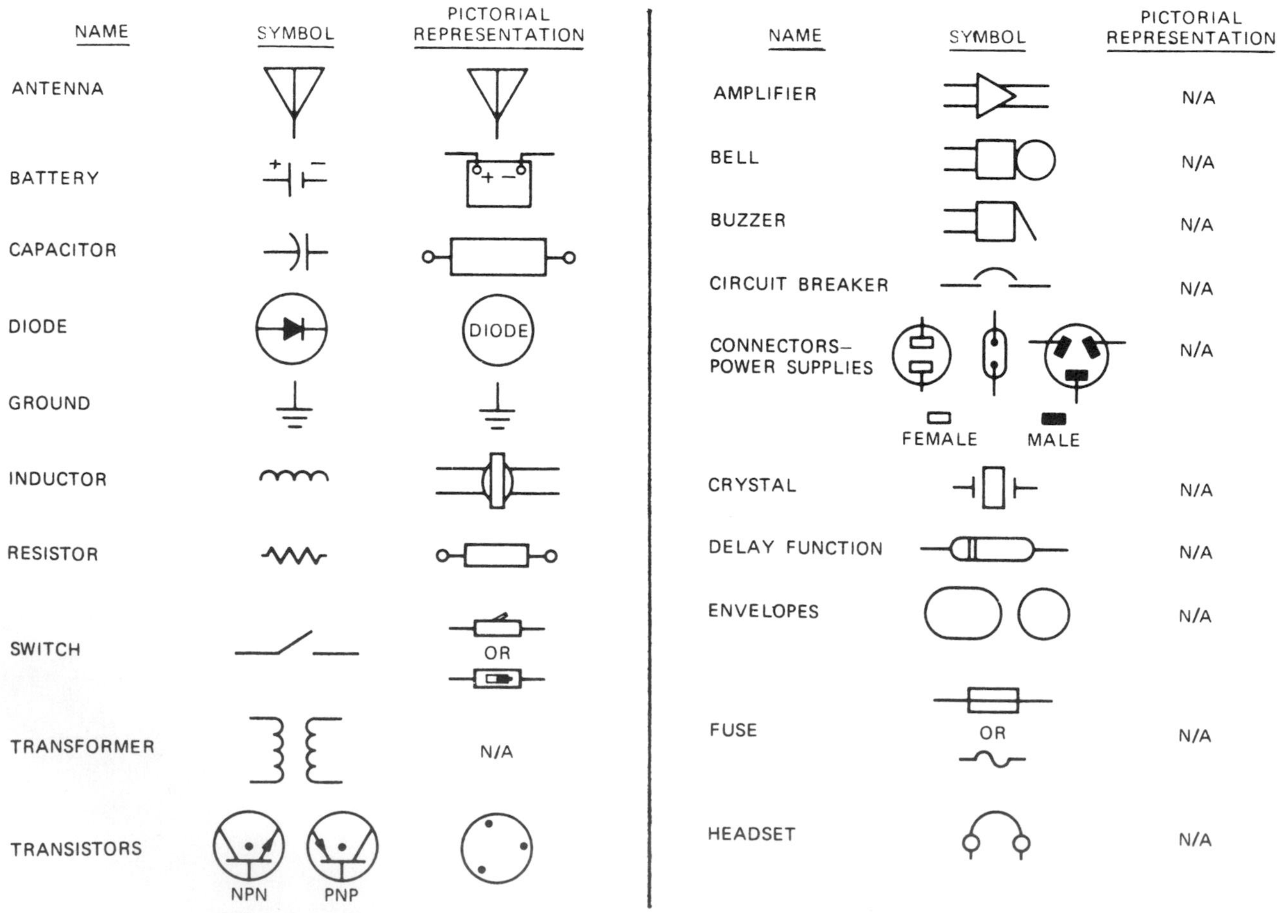

FIGURE 22-1 *Electronics drafting symbols*

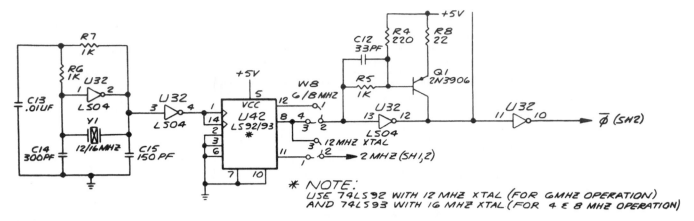

FIGURE 22-2 *Schematic diagram* (From Kirkpatrick, *Electronic Drafting and Printed Circuit Board Design,* 2nd Edition, copyright Delmar Publishers Inc.)

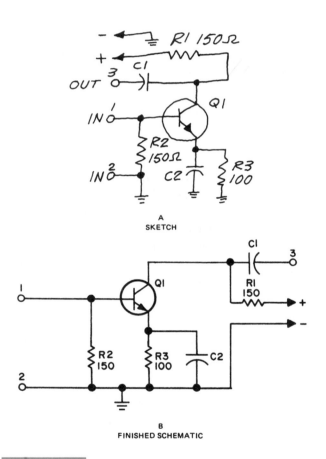

FIGURE 22-3 *Sketch with finished schematic* (From Kirkpatrick, *Electronic Drafting and Printed Circuit Board Design,* 2nd Edition, copyright Delmar Publishers Inc.)

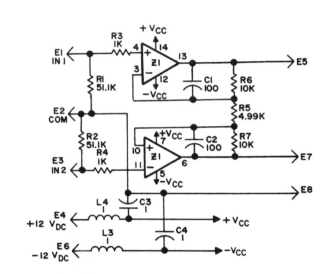

FIGURE 22-4 *Schematic diagram* (From Kirkpatrick, *Electronic Drafting and Printed Circuit Board Design,* 2nd Edition, copyright Delmar Publishers Inc.)

10. Group leads.
11. Remove four-way ties.
12. Number symbols from left to right and top to bottom.

Figure 22-4 is an example of a schematic diagram drawn according to the rules stated previously.

Connection Diagrams

Connection diagrams are used to show how the components in an electronic system are connected. They are used as a guide in assembling electronic systems and maintaining electronic equipment. You might have used a connection diagram in connecting a new stereo, VCR, or compact disc system. There are four types of connection diagrams:

1. Point-to-point diagrams
2. Baseline diagrams
3. Highway diagrams
4. Lineless diagrams

POINT-TO-POINT DIAGRAMS

This type of connection diagram shows the various terminal connection locations and the routing paths of

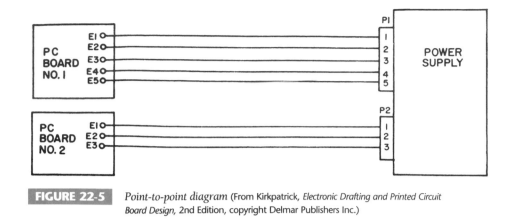

FIGURE 22-5 *Point-to-point diagram* (From Kirkpatrick, *Electronic Drafting and Printed Circuit Board Design*, 2nd Edition, copyright Delmar Publishers Inc.)

every wire in an electronic system. *Point-to-point diagrams* are used primarily with simple circuits. They can be difficult to read when circuits are complex and contain many wires. *Figure 22-5* is an example of a point-to-point diagram. The steps in drawing such a diagram are:

1. Draw the various electronic components in the actual locations they hold in the circuit.

2. Draw the wire paths.

3. Label the wire colors.

BASELINE DIAGRAMS

This type of connection diagram shows all wire paths running into a single baseline. *Baseline diagrams* are easier to read and follow than point-to-point diagrams, but they can be misleading because they do not show the electronic components in their proper positions. *Figure 22-6* is an example of a baseline diagram. The following steps are used in drawing such a diagram:

1. Draw a thick single line that will serve as the baseline.

2. Draw the electronic components, placing half above the baseline and half below it.

3. Draw lines from each connection point to the baseline. The lines must be perpendicular to the baseline.

4. Label each wire path with a destination code.

HIGHWAY DIAGRAMS

This type of connection diagram collects wires that run along similar paths and combines them into groups of wires called "highways." As with point-to-point diagrams, in *highway diagrams* the components are located in the positions they will hold in the actual circuit. *Figure 22-7* is an example of a highway diagram. The following steps are used in drawing such a diagram:

1. Draw the electronic components in their proper positions.

2. Lightly sketch in the wire paths to determine where potential highways exist.

3. Darken the highways, the lines from the connection points to the highways, and the components.

4. Label the wire paths with a destination code, component number, and wire color.

LINELESS DIAGRAMS

This type of connection diagram shows the electronic components with the connection points labeled, but it does not show wires. Instead of wires, the components are accompanied by a table that contains a designation, color, and destination code for each wire. *Figure 22-8* is an example of a *lineless diagram*. The following steps are used in drawing such a diagram:

1. Draw the electronic components in the approximate locations they will hold in the actual circuit.

2. Construct the wiring table.

FIGURE 22-6 *Baseline diagram*

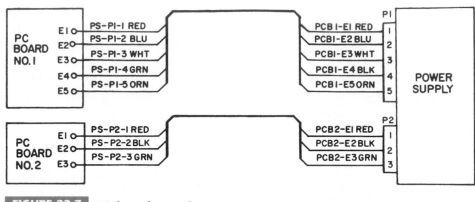

FIGURE 22-7 *Highway diagram* (From Kirkpatrick, *Electronic Drafting and Printed Circuit Board Design,* 2nd Edition, copyright Delmar Publishers Inc.)

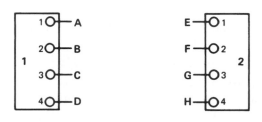

WIRE I.D.	WIRE COLOR	FROM	TO
A	W	1–1	2–4
B	V	1–2	2–3
C	Y	1–3	2–2
D	BL	1–4	2–1

FIGURE 22-8 *Lineless diagram*

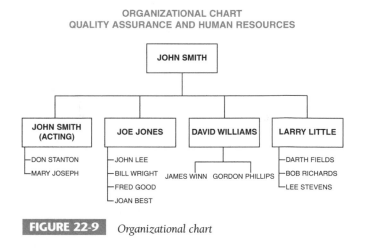

FIGURE 22-9 *Organizational chart*

6. Draw arrowheads.

7. Letter text moving from left to right.

Block Diagrams

Block diagrams are the most widely used types of diagrams in electronics engineering settings. They are also used a great deal in other applications. There are three types of block diagrams:

1. Organizational charts

2. Process flow charts

3. Functional block diagrams

   Figures 22-9, 22-10, and *22-11* illustrate the three types of block diagrams. Block diagrams are drawn according to the following steps:

1. Lightly lay out the boxes at an approximate size.

2. Lightly lay out connecting lines.

3. Darken all horizontal lines.

4. Darken all vertical lines.

5. Draw circles (when required).

RULES FOR DRAWING BLOCK DIAGRAMS

There are several rules of thumb that should be observed when one is drawing block diagrams. These rules lend a degree of standardization to block diagrams and make them easier to read:

1. Whenever possible, draw all boxes the same size.

2. Whenever possible, show the flow from left to right, *Figure 22-12.*

3. When a left-to-right flow is not possible, top-to-bottom flow is acceptable.

4. Crossovers and doglegs should be avoided, *Figure 22-13.*

Logic Diagrams

Logic diagrams are used in conjunction with the design of integrated circuits (ICs). They are most widely used with electronic circuits that are based on the binary

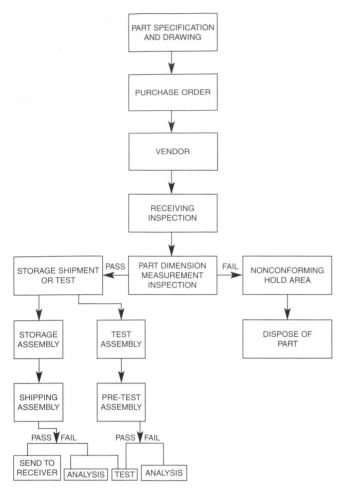

numbering concept, such as those found in computers. *Figure 22-14* is an example of a logic diagram.

The principal components in logic diagrams are called "gates." A *gate* is a miniature circuit that performs a specific function within a larger circuit. The most widely used gates are illustrated in *Figure 22-15*. The gates are:

1. AND gate

2. OR gate

3. NAND gate (negative AND gate)

4. NOR gate (negative OR gate)

Other symbols that are frequently used with these gates are those for AMPLIFIER, FLIP FLOP, DELAY, SINGLE SHOT, GENERAL and SCHMITT TRIGGER. These symbols are also illustrated in Figure 22-15.

The number and letter tables that accompany each gate symbol in Figure 22-15 are called "truth tables." Truth tables are used for analyzing the various outputs that a particular gate will produce as a result of different combinations of inputs. For example, an AND gate such as the one shown in Figure 22-15 can be analyzed using its accompanying truth table. An input of 1 at A and 1 at B will produce an output of 1 at F. Logic diagrams are drawn using special templates or symbols menus in the case of CAD.

FIGURE 22-10 *Process flow chart*

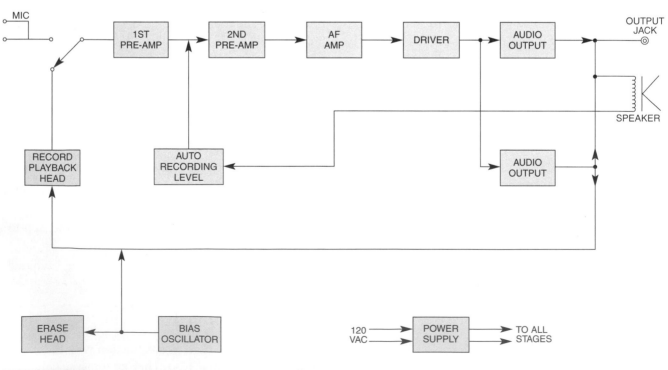

FIGURE 22-11 *Functional block diagram*

Industry Application

LEARNING ELECTRONICS DRAFTING

John Hanscomb, drafting and design instructor at Ohio State Technical Institute (OSTI), was having problems with the printed circuit board components of the curriculum. Students just weren't getting it. They did well in the other components of their studies, which included mechanical, architectural, and structural drafting. When the students got to their lessons on electronics drafting, however, progress came to a sudden halt. After considering the situation for a while and discussing it with colleagues, Hanscomb finally decided that the problem had nothing to do with drafting.

The problem, Hanscomb decided, was electronics. His drafting and design students had little or no understanding of electronics. Consequently, they could not relate to the development of printed circuit board drawings. With mechanical drafting the students could see, touch, and feel the various components of mechanical products. Many were accustomed to working on the mechanical components of their cars, or taking mechanical things apart and putting them back together.

With architectural and structural drafting, students saw buildings, houses, bridges, towers, stadiums, and other structures all the time. Many had had the experience of watching houses, buildings, and structures being constructed. Consequently, students seemed to have an almost intuitive grasp of mechanical, architectural, and structural drafting. Not so with electronics drafting. Hanscomb realized that his students could not see electricity, nor had they ever seen a printed circuit board manufactured. This, he decided, was the problem.

To solve this problem, Hanscomb met with his colleagues in the Electronics Technology Program at OSTI and asked for their help. After discussing the situation at length, OSTI developed a three-week instructional unit entitled, "Electronics for Drafting and Design." This new unit was taught by the school's electronics instructors for several years. Over time, the drafting and electronics faculty converted the electronics unit into a computer-based unit. Now all of OSTI's drafting students complete the unit in a self-paced format before learning how to develop printed circuit board drawings. According to Hanscomb, the electronics unit has made a significant difference in the performance of his students. They can now relate to electronics drafting in the same way that they relate to the mechanical, architectural, and structural components of the curriculum.

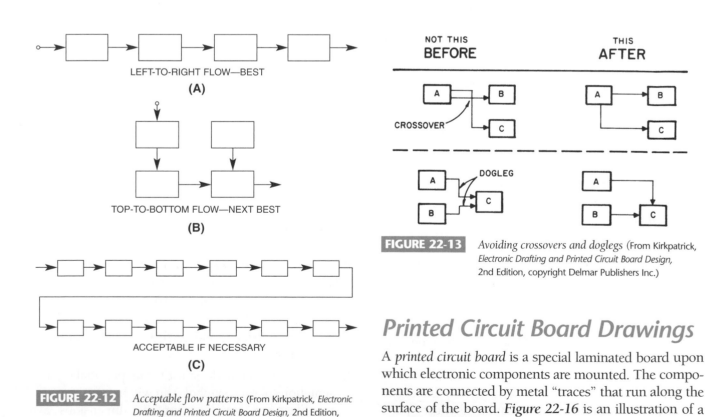

LEFT-TO-RIGHT FLOW—BEST

(A)

TOP-TO-BOTTOM FLOW—NEXT BEST

(B)

ACCEPTABLE IF NECESSARY

(C)

FIGURE 22-12   *Acceptable flow patterns* (From Kirkpatrick, *Electronic Drafting and Printed Circuit Board Design,* 2nd Edition, copyright Delmar Publishers Inc.)

NOT THIS
BEFORE

THIS
AFTER

CROSSOVER

DOGLEG

FIGURE 22-13   *Avoiding crossovers and doglegs* (From Kirkpatrick, *Electronic Drafting and Printed Circuit Board Design,* 2nd Edition, copyright Delmar Publishers Inc.)

Printed Circuit Board Drawings

A *printed circuit board* is a special laminated board upon which electronic components are mounted. The components are connected by metal "traces" that run along the surface of the board. *Figure 22-16* is an illustration of a printed circuit board. This is a single-sided printed circuit

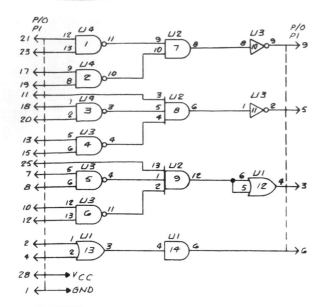

FIGURE 22-14 *Logic diagram (From Kirkpatrick,* Electronic Drafting and Printed Circuit Board Design, *2nd Edition, copyright Delmar Publishers Inc.)*

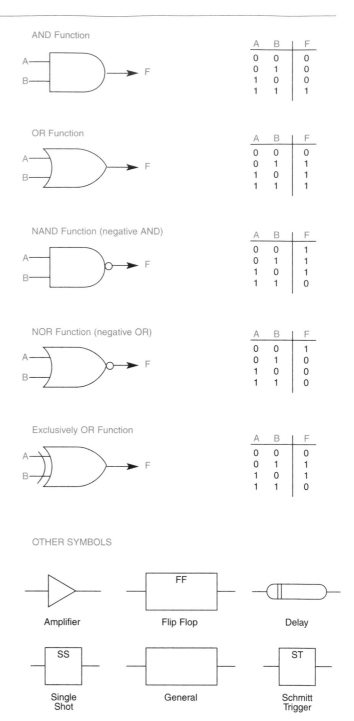

FIGURE 22-15 *Logic symbols*

board. In the top view, the electronics components can be seen mounted on the board. In the bottom view, the traces that connect the components can be seen. There are three main types of printed circuit boards:

1. Single-sided boards
2. Double-sided boards
3. Multilayered boards

Single-sided boards have the electronic components mounted on one side and the connecting circuit etched on the other side. *Double-sided boards* have components mounted on one side also, but the circuits are etched on both sides. *Multilayered boards* are made up of two or more single- and/or double-sided boards.

The most frequently made types of printed circuit board drawings are the drill plan and artwork. The *drill plan* is a typical mechanical drawing that shows where holes are to be drilled through the board. The drill plan includes a hole schedule that contains at least three components:

1. Hole designation (usually a letter)
2. Hole size or description
3. Number of holes of each designation

Figure 22-17 is an example of a drill plan for a printed circuit board.

Artwork for a printed circuit board may be produced on a photoplotter, on a CAD system, or with tape. The artwork shows the circuit that will be etched on the printed circuit board, *Figure 22-18*. Artwork is usually laid out at least twice the size of the actual board and reduced photographically. Figure 22-18 also shows the component

side of the same board. *Figure 22-19* shows how components are actually mounted on a printed circuit board.

The process of preparing printed circuit board drawings involves four steps:

1. Study the schematic and make a list of all electronic components it contains.
2. Redraw the schematic, substituting pictorial representations of each symbol. You will have to consult manufacturers' catalogs for the dimensions of components.

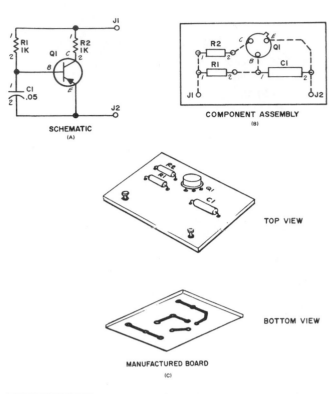

FIGURE 22-16   *Printed circuit boards* (From Kirkpatrick, *Electronic Drafting and Printed Circuit Board Design,* 2nd Edition, copyright Delmar Publishers Inc.)

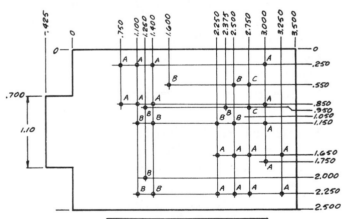

FIGURE 22-17   *Drill plan* (From Kirkpatrick, *Electronic Drafting and Printed Circuit Board Design,* 2nd Edition, copyright Delmar Publishers Inc.)

3. Rearrange the components in such a way as to produce the best conductor paths for the circuit. Avoid arrangements that will cause sharp turns, acute angles, and doglegs in conductor paths. Keep the paths as short as possible

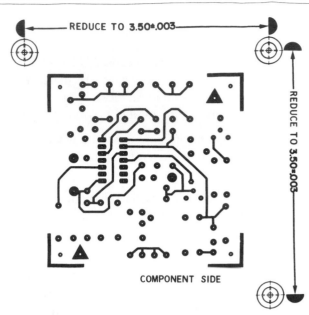

FIGURE 22-18   *Artwork*

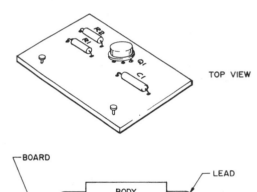

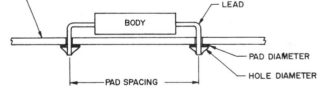

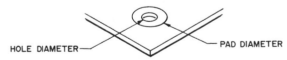

FIGURE 22-19   *Mounting components* (From Kirkpatrick, *Electronic Drafting and Printed Circuit Board Design,* 2nd Edition, copyright Delmar Publishers Inc.)

4. Prepare the printed circuit board drawings at a scale of 2, 3, 4, or 5 to 1.

Summary

- The most frequently used electronic components are antennas, batteries, capacitors, diodes, rounds, inductors, meters, resistors, switches, transformers, and transistors.

- Less frequently used electronic components are amplifiers, bells, buzzers, circuit breakers, connectors, crystals, delays, envelopes, fuses, headsets, lamps, motors, magnets, phase shifters, pickup heads, rectifiers, relays, relay coils, repeaters, safety interlocks, speakers, terminal boards, thermal elements, and thermocouplers.

- Schematic diagrams are used to show what components are contained on a circuit board and which components connect with which.

- Connection diagrams are used to show how the components in an electronic system are connected. There are four types of connection diagrams: point-to-point, baseline, highway, and lineless.

- Block diagrams are the most widely used types of diagrams in electronics engineering settings. There are three types of block diagrams: organizational charts, process flow charts, and functional block.

- A printed circuit board is a special laminated board upon which electronic components are mounted. The components are connected by metal traces that run along the surface of the board. There are three main types of printed circuit boards: single-sided, double-sided, and multilayered.

Review Questions

Answer the following questions either true or false.

1. Schematic diagrams are used to show what components are contained on a circuit board and which components connect with which.

2. The preferred direction of flow on a block diagram is right to left.

3. The second preferred direction of flow on a block diagram is top to bottom.

4. The components of a drill plan for a printed circuit board are hole designation, hole size or description, and number of holes of each designation.

Answer the following questions by selecting the best answer.

1. Which of the following is **not** one of the four types of connection diagrams?
   a. Point-to-point
   b. Baseline
   c. Highway
   d. Line

2. Which of the following is **not** one of the three types of block diagrams?
   a. Organization charts
   b. Logic diagrams

   c. Functional block diagrams
   d. Process flow charts

3. Which of the following is **not** one of the four most widely used gates in logic diagrams?
   a. AND gate
   b. OR gate
   c. NAND gate
   d. BAND gate

4. Which of the following is **not** one of the three main types of printed circuit boards?
   a. Multilayered boards
   b. Triple layered boards
   c. Double-sided boards
   d. Single-sided boards

5. Which of the following is the correct symbol for a capacitor?
   a. ⊨
   b. .=.
   c. ⇁⊢
   d. ✕

CAD Instructions

Students who plan to complete the following problems on a CAD system should begin with these general instructions. Before reading the specific instructions for each activity, go through each step in the following planning checklist. The checklist applies to any CAD system and will help ensure the optimum use of your time and resources.

1. Analyze the problem carefully. Decide exactly what you are being asked to do.

2. Determine what resources and references (if any) you will need in order to complete the problem. Collect those references and resources and have them readily available.

3. Decide if any particular standards apply to the project and have those standards available.

4. Determine what types of views will be required and how many of each.

5. Determine what the final plotted scale of the drawing will need to be, and select the appropriate paper size for plotting/printing.

6. Plan your drawing sequence. In what order will you develop the drawing (i.e., lines, features, dimension lines, leaders, dimensions)?

7. Review the various CAD commands you will have to use in order to develop the drawing.

8. Examine your CAD system to ensure that everything is in working order, then begin the project.

Chapter 22 Problems

PROBLEMS 22-1 AND 22-2

To follow are engineer's sketches containing several logic gates and other electronic components. Convert the sketches into finished schematics.

a. Redraw the schematics using pictorial representations.
b. Use the pictorial schematics as guides in developing drill plans and printed circuit board artwork.

PROBLEM 22-1

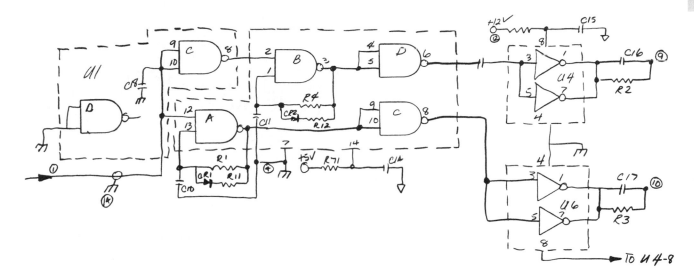

PROBLEM 22-2

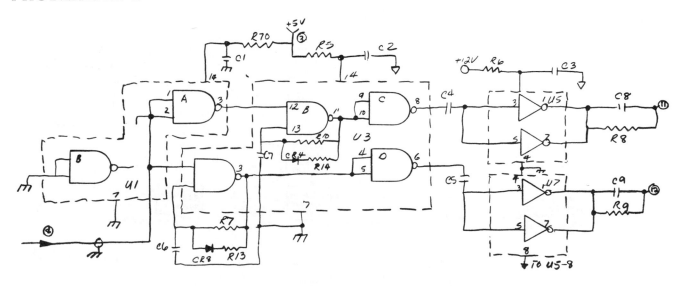

PROBLEM 22-3

Problem 22-3 is a sketch of a pictorial schematic. Using it as a guide, complete the following tasks:

a. Develop the symbolic schematic plan from which such a pictorial schematic might have been drawn.

b. Make a drill plan and printed circuit board artwork using the pictorial schematic as a guide.

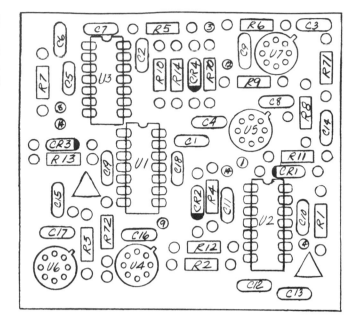

PROBLEM 22-4

Following the rules and steps set forth in this chapter, lay out and draw finished diagrams from the sketches provided.

a. Block diagram **c.** Logic diagram
b. Schematic diagram

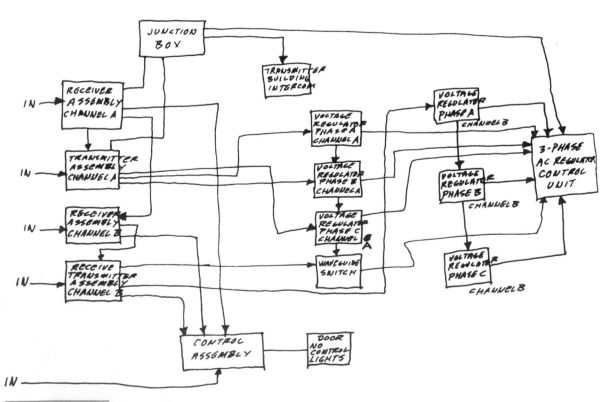

PROBLEM 22-4A (From *Electronic Drafting and Printed Circuit Board Design*, 2nd Edition, by James M. Kirkpatrick, copyright Thomson Delmar Learning)

PROBLEM 22-4 (CONTINUED)

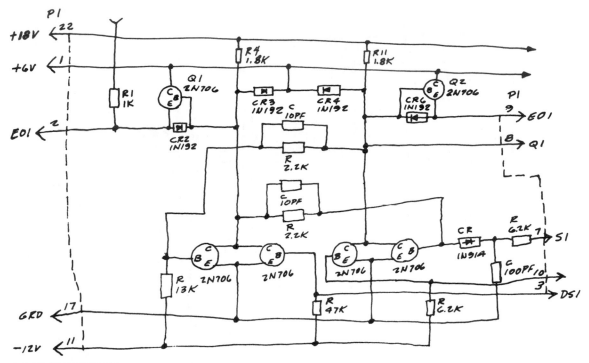

PROBLEM 22-4B (From *Electronic Drafting and Printed Circuit Board Design,* 2nd Edition, by James M. Kirkpatrick, copyright Thomson Delmar Learning)

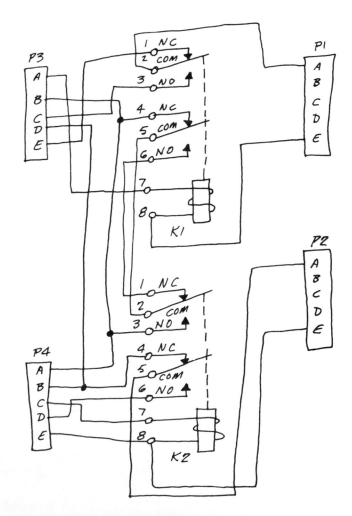

PROBLEM 22-4C (From *Electronic Drafting and Printed Circuit Board Design,* 2nd Edition, by James M. Kirkpatrick, copyright Thomson Delmar Learning)

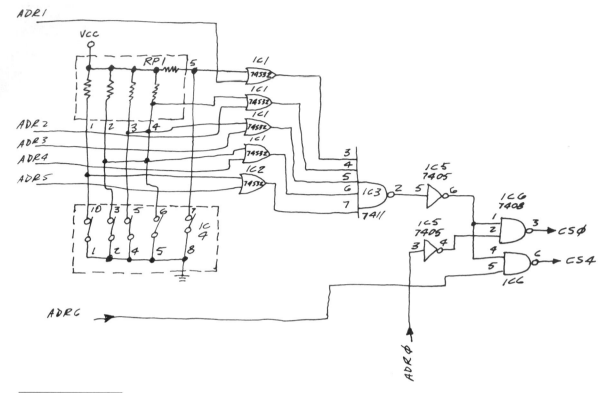

PROBLEM 22-4D (From *Electronic Drafting and Printed Circuit Board Design,* 2nd Edition, by James M. Kirkpatrick, copyright Thomson Delmar Learning)

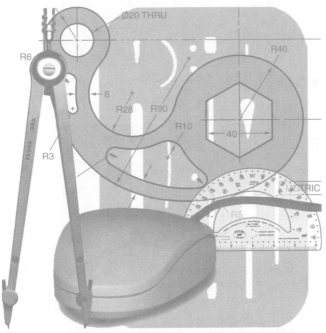

CHARTS
AND GRAPHS

CHAPTER OUTLINE

Five basic components • Functional classes:
an overview • Specific charts and graphs
• Conclusion • Summary • Review
questions • Chapter 23 Problems

KEY TERMS

Adjacent alignment chart

Bar chart

Beam diagram

Concurrency chart

Curve problem

Data problem

Empirical equations

Equilibrant

Equilibrium

Flow chart

Force

Force polygon

Graphical differentiation

Graphical integration

Isopleth

Linear coordinate graphs

Linear graphs

Log-linear (or semi-logarithmic) graphs

Log-log (or logarithmic) graphs

Logarithm

Magnitude

Method of averages

Method of least squares

Method of selected points

Method of slope-intercept

Nomogram

Pictorial chart

Pie chart

Polar graphs

Pole and ray method

Profile chart

Resultant

Schematic chart

Spline

Straight-line equation

Timing graphs

Trilinear graph

Vector

CHAPTER OBJECTIVES

Upon completion of this chapter students should be able to do the following:

■ Incorporate the five basic parameters—grid, scales, labels, plot, and title—for preparing effective charts and graphs.

■ Differentiate between charts and graphs for communicating and generating information.

■ Appreciate the application of computers for making charts and graphs.

■ Understand the concept of optimization in design.

■ Construct vector polygrams to generate information.

■ Read and prepare adjacent-parallel alignment charts, nonadjacent alignment charts, and concurrency charts.

■ Employ equation-matching techniques to empirical data, and check the resulting equation for accuracy.

■ Do graphical differentiation and integration, and derive the associated scales.

■ Relate graphical differentiation and integration to beam diagrams.

The primary objectives of technical charts and graphs—effective communication and interpretation—are identical to the primary objectives of technical drawing. To ensure those objectives when preparing a chart or graph, you should include the five basic components described in the following discussion.

Five Basic Components

First, consider the five basic components of almost all charts and graphs.

1. The grid
2. The scales
3. The scales' labels
4. The plot
5. The title

These are identified in *Figure 23-1*. Selecting and developing all five requires a number of steps that act as a checklist to ensure professional charts and graphs.

THE STEPS

Assuming that the data to be plotted has been collected, and that the function of the chart or graph has been decided, do the following:

1. Examine the data to determine the ranges to be plotted.

2. Use scales to present the data for optimum communication or generation of information, *Figure 23-2*. To help users interpolate between scale numbers, use scales that are multiples of 1, 2, 5, or 10 (unless the intervals are topical, such as days, weeks, months, years, or geographical locations), *Figure 23-3*. Use horizontal scales for independent variables and vertical scales for dependent variables.

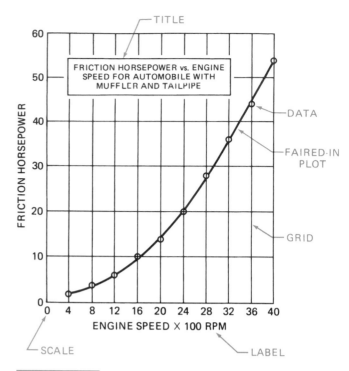

FIGURE 23-1   *Basic components*

3. Select commercial grid sheets to save time. Grids familiar to readers save them time. Moreover, commercial grids were developed for specific, frequently needed functions. Your experiences and observations will help you to select the best formats for presenting data.

4. Plots usually start at the lower left unless the points are negative or the function of the chart or graph requires a different format. Figure 23-1 shows a typical plot. *Figure 23-4* shows a different format.

5. Use symbols for specific points when plotting data from laboratory experiments. Circles, squares, and triangles are common. Curves do not go through the symbols. See Figure 23-1.

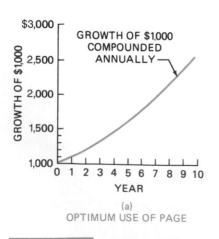

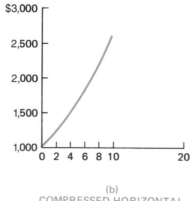

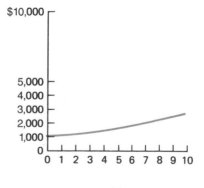

FIGURE 23-2   *Optimum use of page area*

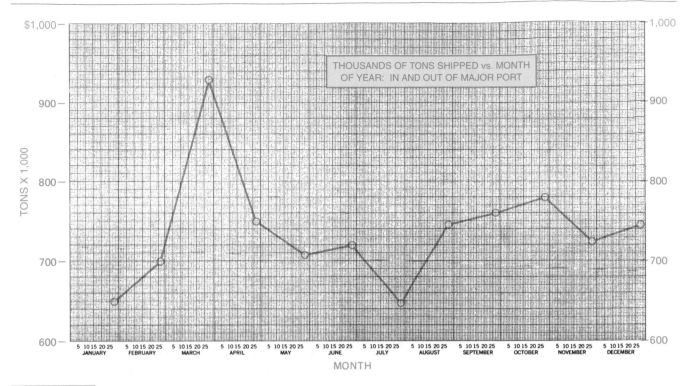

FIGURE 23-3 *Topical scales*

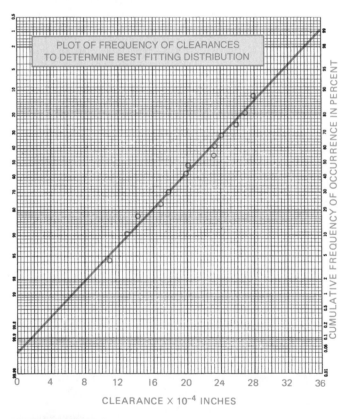

FIGURE 23-4 *Different format*

for experimental data may miss some of the points. See Figure 23-1. Employ line segments for connecting discontinuous data, ***Figure 23-5***. *Note:* Curves representing changing physical phenomena (mechanical, electrical, and chemical) usually appear as smooth plots.

7. Clearly identify multiple plots on the same grid by labels and by a variety of different lines, ***Figure 23-6***.

8. Label each axis by describing the variable and the unit of measurement. See Figure 23-1.

9. As a minimum, include in the title the dependent variable as a function of the independent variable.

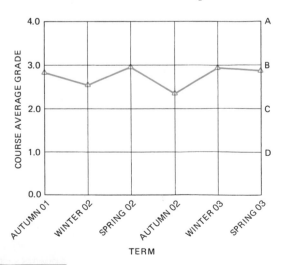

6. Draw all curves lightly before finally darkening them in. Use smooth curves, without symbols, for equations. Best-fit curves through points (symbols)

FIGURE 23-5 *Discontinuous data*

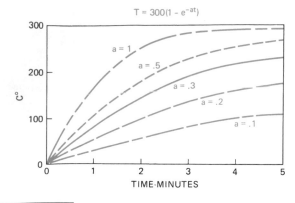

$$T = 300(1 - e^{-at})$$

FIGURE 23-6 *Variety of lines*

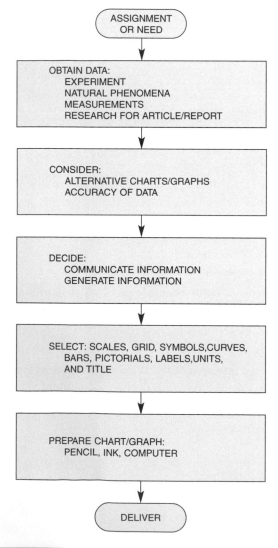

FIGURE 23-7 *Design procedure: Charts and graphs*

Also include the source of the data, the date, and your name. Arrange all labels, titles, and notes to be read either parallel to the bottom of the page or from the right side. Use a cleared space for the title if it appears in the grid area. Titles may be placed outside the grid area.

10. Step back and ask if the final form of the chart or graph communicates effectively or assists in generating the intended information. Successful charts and graphs are three orders of magnitude ($10**3 = 1{,}000$) better at conveying information than words!

The preceding discussion on preparing graphs and charts is summarized in *Figure 23-7*.

Further examples of uses and instructions for preparing specific charts and graphs constitute the remainder of the chapter. The examples illustrate student activities, classroom assignments in technical courses, and industrial applications.

Functional Classes: An Overview

The two major classes of charts and graphs to be discussed are those used to communicate information and those used to generate information.

CHARTS AND GRAPHS USED TO COMMUNICATE INFORMATION

Pie Charts. *Pie charts* look like circular pies and usually represent 100% of a budget, *Figure 23-8*. Hence, the cliche "That's how they slice the (money) pie." School newspapers show these annually. A pie chart appears later in this chapter in conjunction with spreadsheets.

Bar Charts. *Bar charts*, either horizontal or vertical, most often show relative amounts, *Figure 23-9A*; however, they are frequently used to show the relative timing of activities in the form of a planning chart, *Figure 23-9B*.

Pictorial Charts. *Pictorial charts* employing varying sizes of figures seem to be favored by newspapers as a means of illustrating relative amounts to the general public, *Figures 23-10A* and *23-10B*.

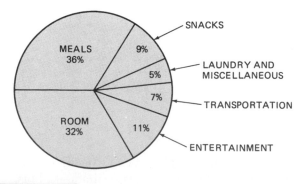

FIGURE 23-8 *Pie chart*

TYPICAL DISTRIBUTION OF TENSILE
TESTS FOR YIELD STRENGTH OF A STEEL

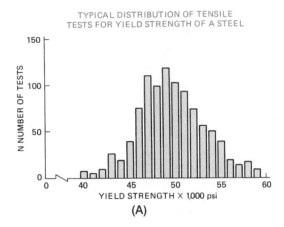

(A)

CAREER PLANNING CHART

(B)

FIGURE 23-9 *(A) Bar chart; (B) Planning chart*

FAMILY OF BOEING COMMERCIAL AIRCRAFT

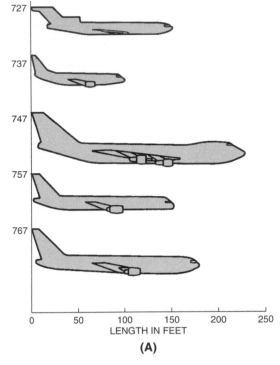

(A)

GROWTH OVER 3 YEARS

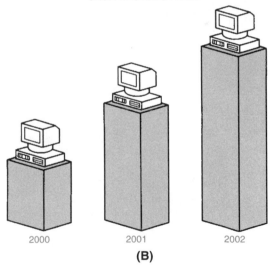

(B)

FIGURE 23-10 *(A) Pictorial bar charts; (B) Pictorial bar charts*

Profile Charts. Material science texts use *profile charts* to illustrate the roughness of the surface of a metal by slicing the metal at a shallow angle and showing the cut surface in a magnified plot, *Figure 23-11A.* A cross section of a portion of a continent, with condensed scales, is used to show terrain characteristics in a profile chart. Civil engineers prepare profile charts to determine how much earth to remove or replace (cuts and fills) when constructing a highway through a hilly terrain, *Figure 23-11B.*

Schematic Charts. Electrical and electronic *schematic charts* of circuits are the most common, *Figure 23-12A;* however, there are others, such as piping, hydraulic power, *Figure 23-12B,* logic, fault tree, and failure tree schematics.

Flow Charts. Students encounter *flow charts* in computer programming courses, *Figure 23-13A.* Other uses include flow of information in an organization, flow of materials in a manufacturing plant, and flow of energy uses in an agricultural facility, *Figure 23-13B.*

Linear Coordinate Graphs. Basic science (mathematics, physics, biology, and chemistry) and engineering science (statics, dynamics, kinematics, strength of materials, materials, and thermodynamics) textbooks save

thousands of words by using *linear coordinate graphs,* in both two-dimensional and three-dimensional formats, *Figures 23-14A* and *23-14B.* Newspapers use them to save space and to communicate effectively. One important use in technology is to show ranges of values or error bars at data points, *Figures 23-14C* (page 878) and *23-14D* (page 878). The range of the Dow Jones Average during any one day is an excellent illustration. Error bars on data points representing an experiment show the ranges of accuracy that can be expected for certain conditions.

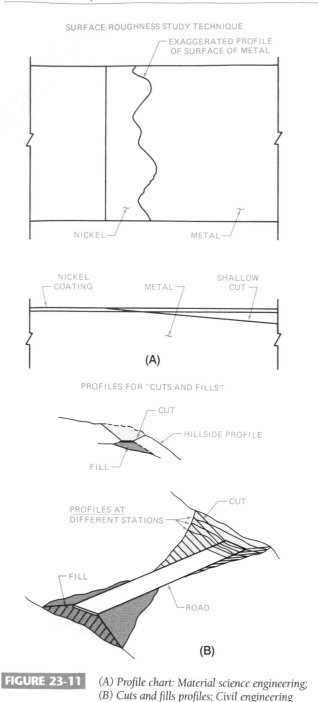

FIGURE 23-11   (A) Profile chart: Material science engineering; (B) Cuts and fills profiles; Civil engineering

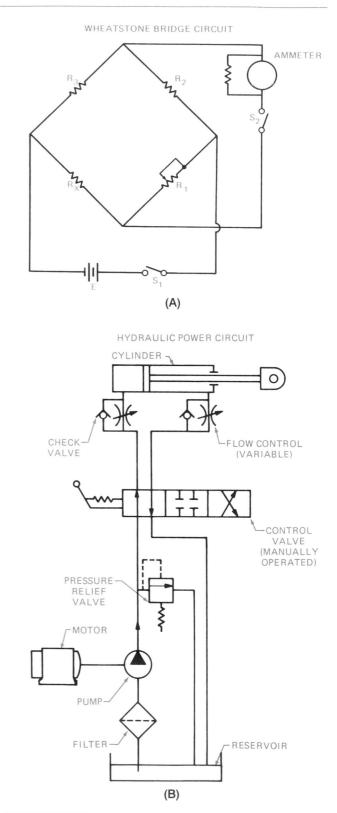

FIGURE 23-12   (A) Electrical schematic; (B) Hydraulic power schematic

Polar Graphs.   Polar graphs are 360° maps usually used to show the effect of a point source of energy (heat, light, electromagnetic sound) at varying distances from the source. Radio and TV stations use these to determine their local coverage. The use of polar graph paper saves hours in preparing these plots, *Figure 23-15* (page 879).

CHARTS AND GRAPHS USED TO GENERATE INFORMATION

Trilinear Graphs.   Metallurgical and chemical personnel use these equilaterally shaped graphs (*trilinear graphs*) to illustrate how differing percentages of three elements of a mixture behave or should behave. For example, the sum of the three perpendiculars, one to each side, from a point equal to the altitude of the trian-

COMPUTER PROGRAM PLAN

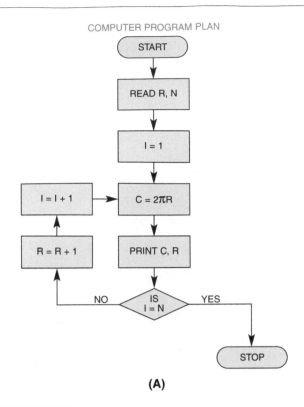

(A)

FIGURE 23-13 *(A) Flowchart: Computer program*

PROPOSED PROTEIN IN PRODUCING PLAN

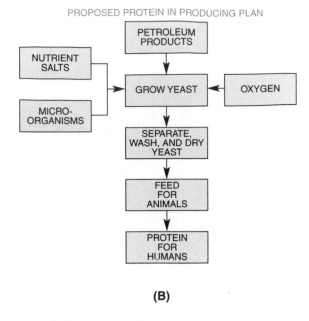

(B)

FIGURE 23-13 *(Continued) (B) Flowchart: Agricultural process*

gle would be 100% in all cases. Commercial grids are available, *Figure 23-16* (page 880).

Linear Graphs. Grids for *linear graphs* are ruled uniformly along the horizontal and the vertical axes. This

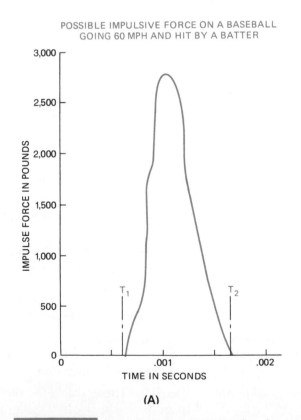

(A)

FREEBODY DIAGRAM OF STEERING WHEEL AND SHAFT

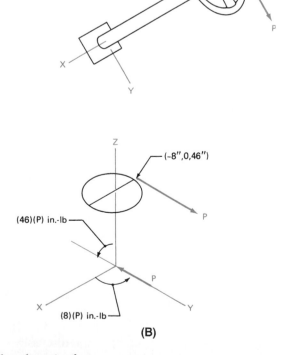

(B)

FIGURE 23-14 *Linear coordinates: (A) Impulsive force vs. time; (B) Three-dimensional*

DAILY RANGES OF DOW JONES TYPE AVERAGES

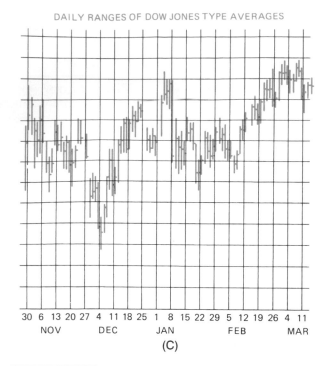

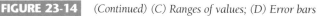

30 6 13 20 27 4 11 18 25 1 8 15 22 29 5 12 19 26 4 11
NOV   DEC   JAN   FEB   MAR

(C)

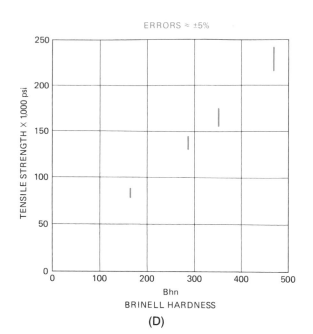

(D)

FIGURE 23-14   *(Continued) (C) Ranges of values; (D) Error bars*

type is used here to introduce the general equation of a straight line, $y = mx + b$, where m is the slope, x the independent variable, y the dependent variable, and b the intercept along the vertical axis. Furthermore, $y = mx + b$ appears here because it is used in conjunction with graphs to generate information, *Figures 23-17A, 23-17B,* and *23-17C.* A more complete discussion appears later in this chapter in a section on empirical equations. Log-linear and log-log applications are also included.

Log-Linear Graphs. *Log-linear* or *semilogarithmic graphs* are linear, or uniformly ruled, along the horizontal axis and logarithmic, or ruled proportionately to the logarithms of numbers, along the vertical axis, *Figure 23-18* (page 881). Curves with equations $y = b(e^{**}mx)$ plot as straight lines on log-linear graphs. Therefore, if data from an experiment is plotted on a log-linear grid, and the data forms a straight line, then an equation for the data may be generated in the form $y = b(e^{**}mx)$. The variables are defined in the linear graph discussion. The number $e = 2.718282$.

Log-Log Graphs. *Log-log* or *logarithmic graphs* are ruled proportionately to the logarithms of numbers along both the horizontal and vertical axes. Curves with equations $y = b(x^{**}m)$, plot as straight lines on log-log grid sheets, *Figures 23-19A* (page 881) and *23-19B* (page 882). Therefore, if data from an experiment plots as a straight line on log-log paper, an equation for the data may be generated in the form $y = b(x^{**}m)$. The variables are defined in the linear graph discussion.

Nomographs. Nomographs or *nomograms* are graphical representations of formulas along straight or curved, graduated lines that are used to find an unknown value of one variable, given the others, *Figure 23-20A* (page 882), *23-20B* (page 883), and *23-20C* (page 883). Not only do nomographs save time, but they allow the user to visualize ranges of possible values. For example, a designer can generate a number of alternatives quickly and work toward an optimum solution. Several nomograms are discussed later in this chapter.

Graphical Calculus: Integration. *Graphical integration* provides designers with a visual, reasonably accurate, approach to integration. Analytical, mechanical, and computer techniques for integrating are also available; however, the graphical approach helps the designer to "see" the results. The application, shown in *Figure 23-21* (page 884), generates the energy within a pressure-volume plot. The technique is discussed later in this chapter.

Graphical Calculus: Differentiation. The technique for doing *graphical differentiation* and the advantages to a designer are similar to graphical integration; however, obtaining reasonably accurate results requires more skill than integrating graphically, *Figure 23-22* (page 885).

Timing Graphs. *Timing graphs* aid designers in coordinating electrical and electronics components in a system driven by clock pulses, *Figure 23-23* (page 886).

Topographic Maps (Charts). These speak for themselves, *Figure 23-24* (page 886). The pilots of space, air,

LUMENS/FT² FOR A LIGHT FIXTURE @ 3 METERS

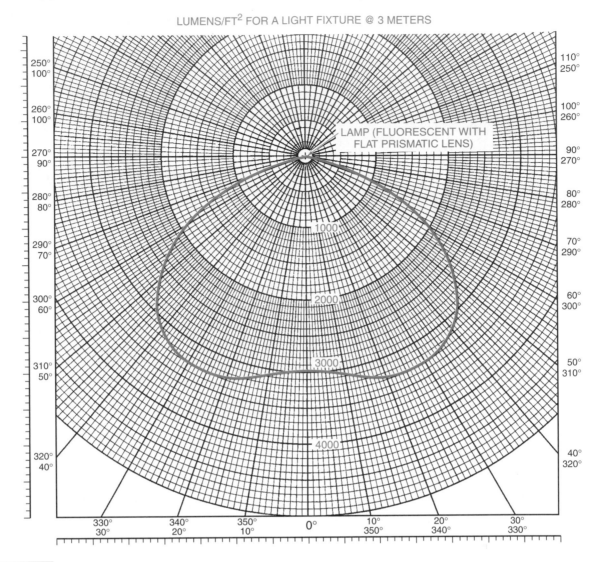

FIGURE 23-15 *Polar coordinates graph*

land, sea, and undersea vehicles use maps and sophisticated devices to locate their positions. Civil, agricultural, mining, and petroleum engineers generate information for their use and explorations from maps.

Force Maps. Force polygons drawn on force maps (same scale in all directions) are included here because of their utility in science and technology courses, *Figure 23-25* (page 886). Force polygons are discussed later in reference to an elementary structures design.

Beam Diagrams. Beams of varying shapes, materials, and sizes occur in so many designs that a brief introduction to *beam diagrams* in this chapter seems appropriate, *Figure 23-26* (page 886). These diagrams are discussed later.

Knowing the major functions of these graphs and charts, and having the skill and knowledge to prepare them, will enhance your ability to read and use them in courses and in industry.

Specific Charts and Graphs

From the preceding overview of the functions and formats of charts and graphs, six types are designated for further discussions.

- Pie Charts and Bar Charts: generated from a spreadsheet in a personal computer.

- Force Polygons: used in an optimization investigation.

- Nomograms: several alignment charts and concurrency charts to graphically represent equations.

- *Empirical Equations:* $y = mx + b$, $y = b(e^{**}mx)$, and $y = b(x^{**}m)$ for matching equations to data.

- Graphical Integration and Differentiation: Pole and ray method.

- Beam Diagrams: free-body, shear, moment, slope, and deflection diagrams.

TRILINEAR CHART ILLUSTRATION

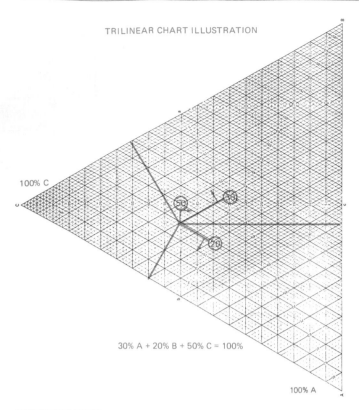

100% C

30% A + 20% B + 50% C = 100%

100% A

FIGURE 23-16   *Trilinear graph*

PIE CHARTS AND BAR CHARTS
FROM A SPREADSHEET

Assume that a student in an orientation-to-higher-education class has prepared a weekly activity schedule

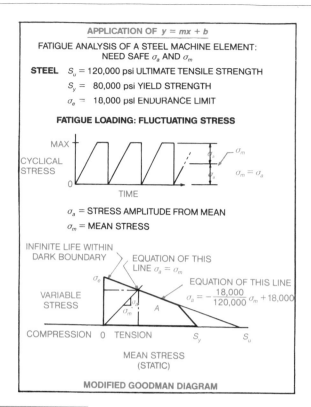

APPLICATION OF $y = mx + b$

FATIGUE ANALYSIS OF A STEEL MACHINE ELEMENT:
NEED SAFE σ_a AND σ_m

STEEL S_u = 120,000 psi ULTIMATE TENSILE STRENGTH

S_y = 80,000 psi YIELD STRENGTH

σ_e = 18,000 psi ENDURANCE LIMIT

FATIGUE LOADING: FLUCTUATING STRESS

σ_a = STRESS AMPLITUDE FROM MEAN

σ_m = MEAN STRESS

INFINITE LIFE WITHIN DARK BOUNDARY

EQUATION OF THIS LINE $\sigma_a = \sigma_m$

EQUATION OF THIS LINE $\sigma_a = -\dfrac{18,000}{120,000}\,\sigma_m + 18,000$

VARIABLE STRESS

COMPRESSION   0   TENSION S_y S_u

MEAN STRESS (STATIC)

MODIFIED GOODMAN DIAGRAM

FIGURE 23-17B   *Straight-line equations: Fatigue analysis*

for the first term as shown in *Figure 23-27* (page 887). The student's anticipated schedule of eating, sleeping, playing, attending class, studying, and everything else has been typed into a spreadsheet format in a personal computer. The format resembles pigeon holes for stuff-

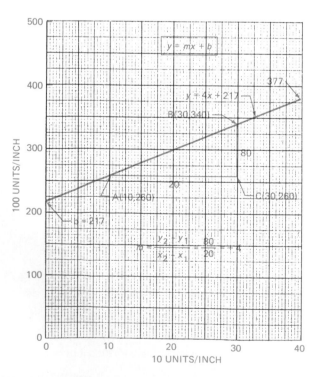

FIGURE 23-17A   *Straight-line equation: $y = mx + b$*

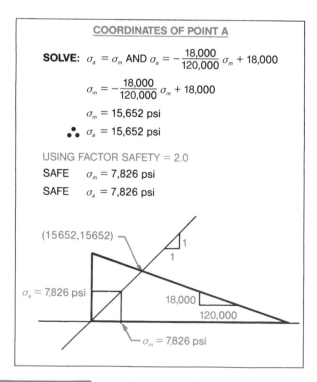

COORDINATES OF POINT A

SOLVE: $\sigma_a = \sigma_m$ AND $\sigma_a = -\dfrac{18,000}{120,000}\,\sigma_m + 18,000$

$\sigma_m = -\dfrac{18,000}{120,000}\,\sigma_m + 18,000$

$\sigma_m = 15,652$ psi

$\therefore$ $\sigma_a = 15,652$ psi

USING FACTOR SAFETY = 2.0

SAFE $\sigma_m = 7,826$ psi

SAFE $\sigma_a = 7,826$ psi

(15652,15652)

$\sigma_a = 7,826$ psi

18,000

120,000

$\sigma_m = 7,826$ psi

FIGURE 23-17C   *Straight-line equations: Fatigue analysis*

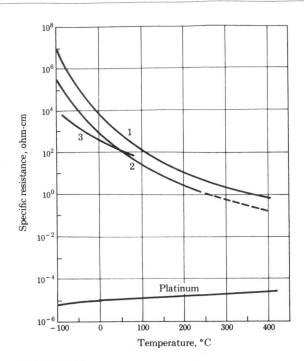

FIGURE 23-18 *Log-linear graph (semilogarithmic)* (From Holman, *Experimental Methods for Engineers,* © 1966, McGraw-Hill Book Co. Used with permission.)

sleep time, (PL) playtime, (EA) eating time, and (MI) miscellaneous—transportation, laundry, club meetings, and church.

A program generated by the student, or one already contained in software, sorts through the seven days and totals the number of hours for each category. Then, instructions are typed or selected from a menu to generate and plot the pie chart and bar chart shown in *Figure 23-28* (page 887). Later, or for the next term, the student can, with a push of a button, easily adjust the schedule and have new plots made.

FORCE POLYGONS IN AN OPTIMIZATION INVESTIGATION

Force polygons, vector diagrams, and vector polygons all refer to the same graphical technique. *Force polygons* are used in this discussion. However, before proceeding to the optimization investigation, several assumptions and definitions need to be stated. This discussion assumes that the reader has had a brief introduction to forces and vectors in Chapter 8 but will appreciate a condensed review of some basic concepts. Only solid bodies at rest with coplanar, concurrent (all in one plane going through one point) vectors are considered. Furthermore, the weights of the bodies, or members, are considered to be negligible (zero) compared to the forces applied.

ing information into, arranged in columns and rows. One column lists each of the 24 hours; seven columns list the activities for each day of the week, coded in categories as follows: (ST) study time, (CL) class time, (SL)

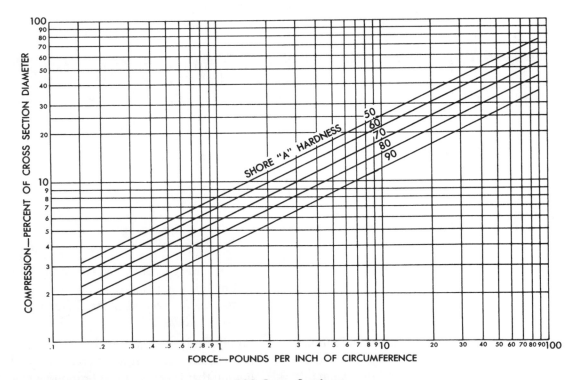

.139 Cross Section

FIGURE 23-19A *Logarithmic graph: O-ring compression* (Courtesy Parker Seal)

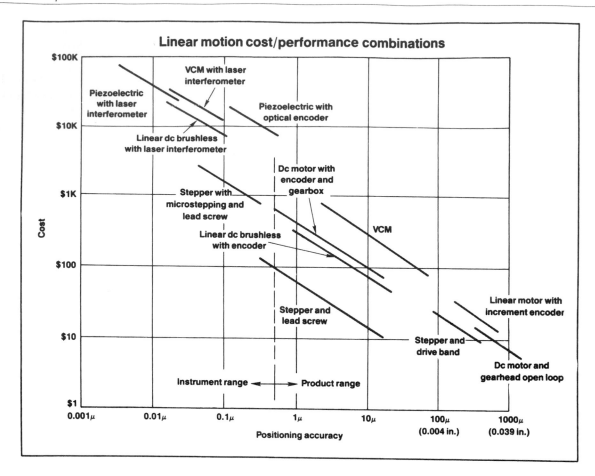

FIGURE 23-19B   *Logarithmic graph: Cost vs. accuracy* (Reprinted from *Machine Design,* Oct 27, 1987, © 1987 Penton Publishing Inc.)

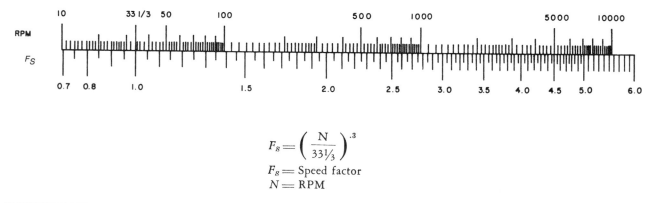

$$F_s = \left(\frac{N}{33\frac{1}{3}} \right)^{.3}$$

$F_s =$ Speed factor
$N =$ RPM

FIGURE 23-20A   *Adjacent alignment chart* (Courtesy McGill Manufacturing Co. Inc.)

1. **Vector:** A *vector* is a graphical symbol that represents direction by means of an arrowhead at one end of a straight line whose scaled length represents a *magnitude*.

2. **Physical Simplifications:** A physical simplification describes the essentials of a real phenomenon. For example, a vector is a graphical model that can represent a weight, a force, a velocity, or an acceleration of a body.

3. **Force:** A force is a directed action of one body on a second one that tends to change the state of motion, or rest, of the second body; therefore, a force is a vector quantity.

4. **Equilibrium:** The sum of forces acting on a solid body, in any direction, add to zero on a body in *equilibrium* that is either at rest or moving at a constant velocity. The following equation of mechanics states this principle succinctly: $\Sigma F = 0$.

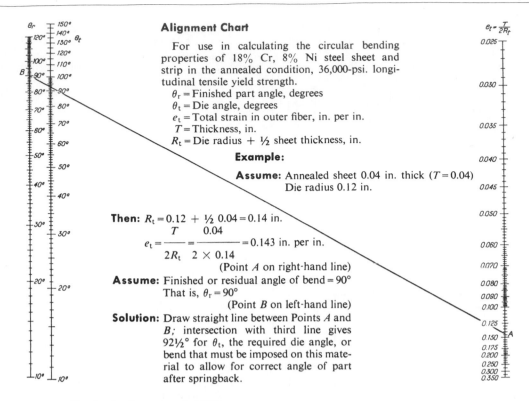

Alignment Chart

For use in calculating the circular bending properties of 18% Cr, 8% Ni steel sheet and strip in the annealed condition, 36,000-psi. longitudinal tensile yield strength.

θ_r = Finished part angle, degrees
θ_t = Die angle, degrees
e_t = Total strain in outer fiber, in. per in.
T = Thickness, in.
R_t = Die radius + ½ sheet thickness, in.

Example:

Assume: Annealed sheet 0.04 in. thick ($T = 0.04$)
Die radius 0.12 in.

Then: $R_t = 0.12 + \frac{1}{2}\, 0.04 = 0.14$ in.

$$e_t = \frac{T}{2R_t} = \frac{0.04}{2 \times 0.14} = 0.143 \text{ in. per in.}$$

(Point *A* on right-hand line)

Assume: Finished or residual angle of bend = 90°
That is, $\theta_r = 90°$

(Point *B* on left-hand line)

Solution: Draw straight line between Points *A* and *B*; intersection with third line gives 92½° for θ_t, the required die angle, or bend that must be imposed on this material to allow for correct angle of part after springback.

Circular bends on annealed 18-8.

FIGURE 23-20B   *Parallel alignment chart: Three variables* (Courtesy Allegheny Ludlum Corp.)

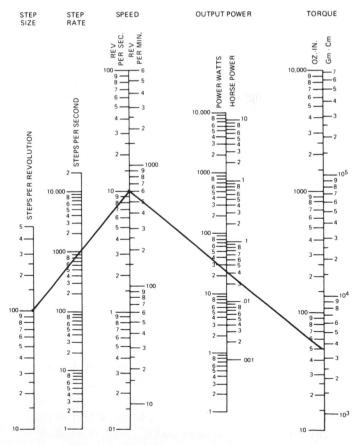

FIGURE 23-20C   *Parallel alignment chart: Four variables*
(Courtesy Pacific Scientific)

Angle – Speed – Power

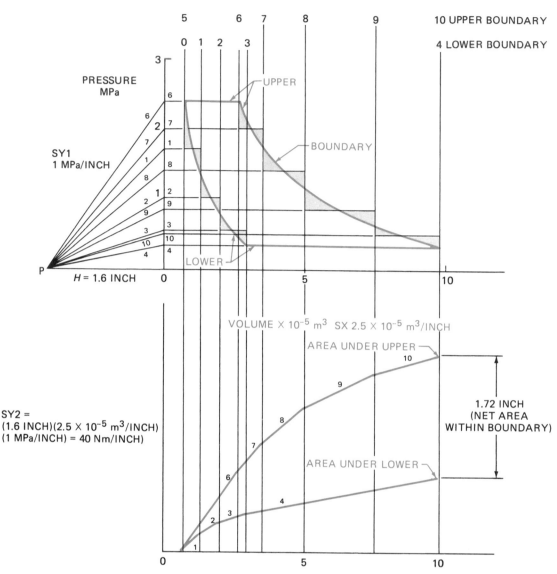

FIGURE 23-21 *Graphical calculus: Integration*

5. **Two-Force Members:** The majority of two-force members are found in structures, frames, trusses, and flexible members in tension. The two forces are equal and opposite in direction, colinear, and applied at the ends of the member. *See Figure 23-29* (page 887) for two examples, one in tension and one in compression.

6. **Vector Addition and Subtraction:** Two vectors that are not parallel and not colinear but are acting on the same body and going through the same point may be added by the parallelogram-law technique. Refer to *Figure 23-30* (page 888) for an example showing the addition of two vectors A and B to obtain a single vector R, which is equivalent to the effect of the two vectors and is called a *resultant*. Figure 23-30 also shows the subtraction of vector B from vector A. A vector labeled *equilibrant* is the

reaction to the resultant vector R and is equal and opposite to R, and its line of action goes through the same point. Of course, the sum of the resultant R and the equilibrant equals zero. (*Note:* the term *equilibrant* is used mostly in definitions, but not much in actual problems in mechanics.)

More than two vectors—such as A, B, and C—acting on a body and going through a common point add, "tail to head," in any order to create a resultant and an equilibrant, *Figure 23-31* (page 888). Figure 23-31 also illustrates that the sum of vectors A, B, and C plus the equilibrant equals zero. Moreover, when all the force vectors (three or more) acting through a common point on a body that is in equilibrium are added "tail to head," the sum must equal zero.

7. **Free-Body Diagrams (FBD):** A body in equilibrium, free from its supports, or a portion of the body

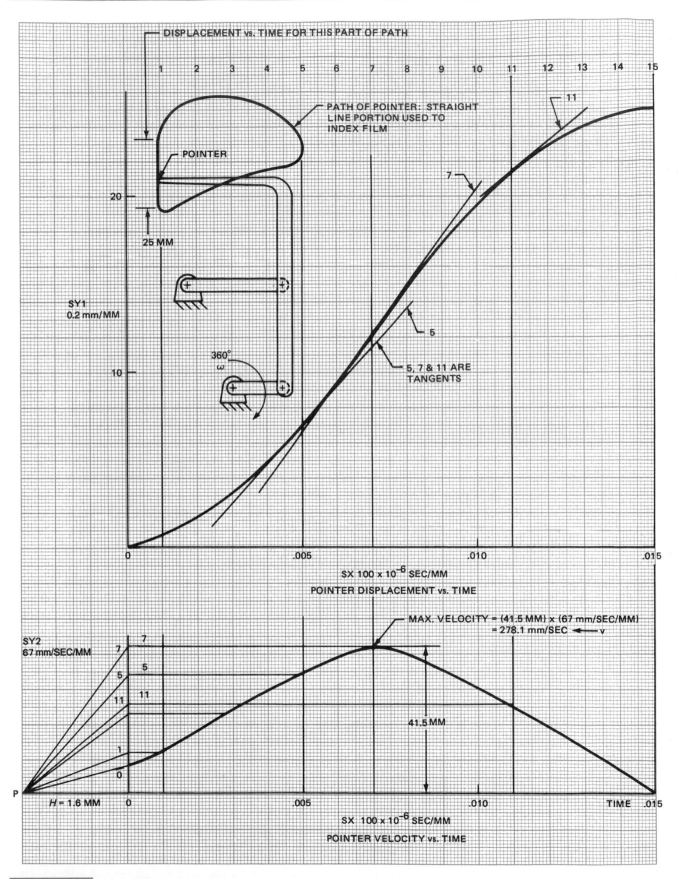

FIGURE 23-22 *Graphical calculus: Differentiation*

INPUT TO LOGIC CIRCUIT FOR DRIVING A STEPPING MOTOR

OUTPUT TO WINDINGS OF STEPPING MOTOR

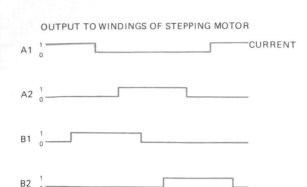

FIGURE 23-23 *Timing graphs*

ABBREVIATED TOPOGRAPHIC MAP

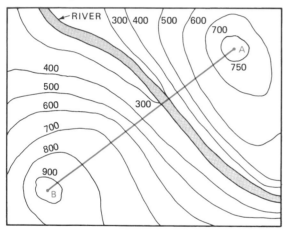

EXAMPLE PROFILE: STATION A TO STATION B

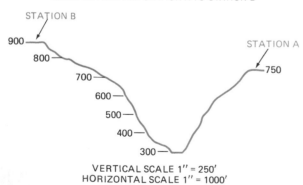

VERTICAL SCALE 1″ = 250′
HORIZONTAL SCALE 1″ = 1000′

FIGURE 23-24 *Topographic map*

shown separately, is a free-body diagram when all the forces acting on the body are shown. *Figure 23-32* (page 888) shows a complete body in equilibrium, and it is drawn as an FBD. The isolated pinned joint at B is also drawn as an FBD. Isolating joints saves space and time and allows for both external forces and internal forces to be shown.

The forces on the isolated joint B are assumed to go through the "pin" at the joint. Joints in trusses

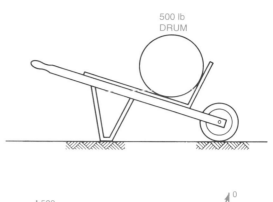

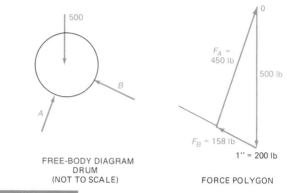

FIGURE 23-25 *Force maps*

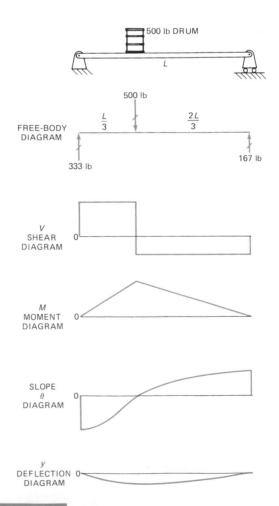

FIGURE 23-26 *Beam diagrams*

HOUR	SUN	MON	TUE	WED	THU	FRI	SAT
0000	SL	SL	SL	SL	SL	SL	SL
0100	SL	SL	SL	SL	SL	SL	SL
0200	SL	SL	SL	SL	SL	SL	SL
0300	SL	SL	SL	SL	SL	SL	SL
0400	SL	SL	SL	SL	SL	SL	SL
0500	SL	SL	SL	SL	SL	SL	SL
0600	SL	SL	SL	SL	SL	SL	SL
0700	SL	MI	MI	MI	MI	MI	SL
0800	SL	EA	EA	EA	EA	EA	SL
0900	MI	CL	CL	CL	CL	CL	MI
1000	EA	CL	MI	CL	MI	CL	EA
1100	MI	CL	CL	CL	CL	CL	ST
1200	MI	EA	EA	EA	EA	EA	ST
1300	EA	ST	CL	ST	CL	MI	EA
1400	ST	ST	CL	ST	CL	MI	PL
1500	ST	ST	CL	ST	CL	PL	PL
1600	ST	ST	MI	ST	MI	PL	PL
1700	ST	MI	MI	MI	MI	PL	PL
1800	EA	EA	EA	EA	EA	EA	PL
1900	EA	MI	ST	MI	ST	PL	EA
2000	PL	ST	ST	ST	ST	PL	PL
2100	ST	ST	ST	ST	ST	PL	PL
2200	ST	ST	ST	ST	ST	PL	PL
2300	ST	SL	SL	SL	SL	PL	PL

FIGURE 23-27 *Week activities: spreadsheet*

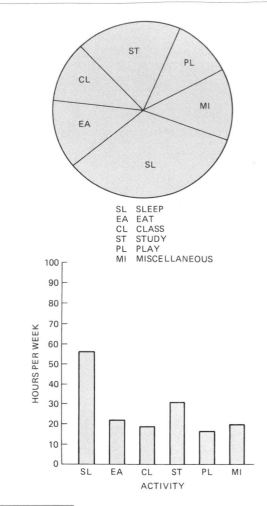

SL SLEEP
EA EAT
CL CLASS
ST STUDY
PL PLAY
MI MISCELLANEOUS

FIGURE 23-28 *Pie and bar charts from spreadsheet*

and frames are usually considered to be pinned even though the joint may be welded, riveted, or bolted. This practice is a simplification in mechanics to facilitate joint analyses with FBDs.

Having the skill to draw accurate free-body diagrams will significantly increase your success in solving force problems.

8. **Force Polygon for Analysis:** The force polygon in *Figure 23-33* was used to determine the direction and magnitude of the forces F(CB) and F(AB) for the isolated joint B in Figure 23-32. Refer to the boom AB and the strut CB, which have been drawn to scale, and follow the steps in the procedure that follows:

a. Start at reference point O and construct to scale the force vector W pointing vertically downward.

b. Since the two remaining forces F(CB) and F(AB) must add on to W and finally end at the starting point O, one light line through the head of W is drawn parallel to the strut CB. A second line is drawn through point O parallel to boom AB. The intersection point of the two lines determines the

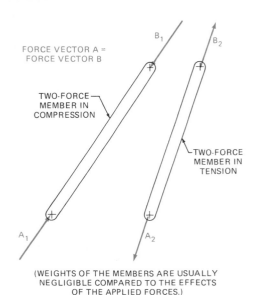

FORCE VECTOR A =
FORCE VECTOR B

TWO-FORCE MEMBER IN COMPRESSION

TWO-FORCE MEMBER IN TENSION

(WEIGHTS OF THE MEMBERS ARE USUALLY NEGLIGIBLE COMPARED TO THE EFFECTS OF THE APPLIED FORCES.)

FIGURE 23-29 *Two-force members*

magnitudes of F(CB) and F(AB); also, the "tail to head" pattern determines their directions. (Only two unknown forces may be determined using this technique in coplanar problems).

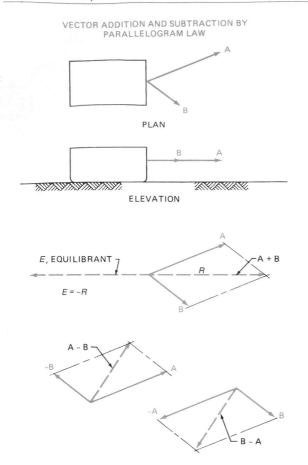

VECTOR ADDITION AND SUBTRACTION BY PARALLELOGRAM LAW

PLAN

ELEVATION

E, EQUILIBRANT

$E = -R$

R

$A + B$

$A - B$

$B - A$

FIGURE 23-30 *Parallelogram law: Vector addition and subtraction*

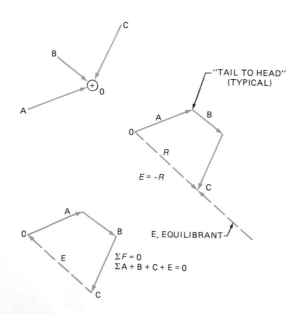

"TAIL TO HEAD" (TYPICAL)

$E = -R$

R

E, EQUILIBRANT

$\Sigma F = 0$
$\Sigma A + B + C + E = 0$

FIGURE 23-31 *Addition: Three or more vectors*

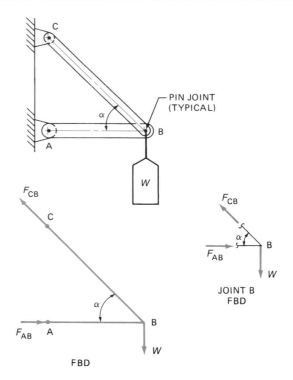

PIN JOINT (TYPICAL)

W

F_{CB}

F_{AB}

W

FBD

F_{CB}

F_{AB}

W

JOINT B
FBD

FIGURE 23-32 *FBD (free-body diagrams)*

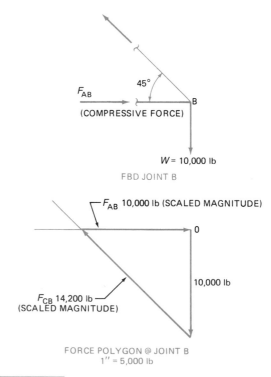

F_{AB}
(COMPRESSIVE FORCE)

45°

$W = 10,000$ lb

FBD JOINT B

F_{AB} 10,000 lb (SCALED MAGNITUDE)

10,000 lb

F_{CB} 14,200 lb
(SCALED MAGNITUDE)

FORCE POLYGON @ JOINT B
$1'' = 5,000$ lb

FIGURE 23-33 *Force polygon application*

c. The direction of $F(CB)$ is "away" from the joint B; therefore, $F(CB)$ is a tensile force. The direction of $F(AB)$ "into" the joint indicates a compressive force. W is an externally applied force, so it is neither tensile nor compressive. $F(AB)$

could have been considered first in Step b and the results would be identical. Now that definitions and assumptions have been stated, the optimization investigation utilizing force polygons comes next.

9. **Optimization Investigation Using Force Polygons:** Optimizing a design usually entails finding maximum performance or minimum cost or both. The objective in this elementary example is to determine the angle θ between the boom AB and the strut CB that will give the minimum cost for the member CB, with CB still supporting the 10,000-pound load safely, *Figure 23-34A*.

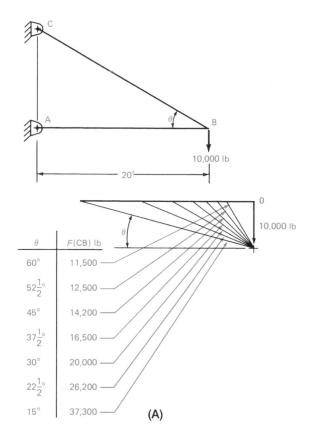

θ	F(CB) lb
60°	11,500
52½°	12,500
45°	14,200
37½°	16,500
30°	20,000
22½°	26,200
15°	37,300

(A)

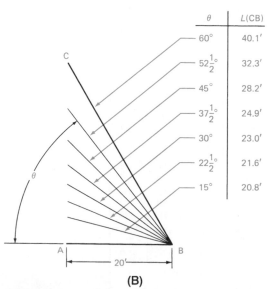

θ	L(CB)
60°	40.1'
52½°	32.3'
45°	28.2'
37½°	24.9'
30°	23.0'
22½°	21.6'
15°	20.8'

(B)

FIGURE 23-34 *(A) Force polygon for F(BC); (B) Space diagrams for L(BC)*

a. Member CB is to be an equal leg, structural angle made of mild steel. Boom AB is 20 feet long and is assumed to be satisfactory for any angle θ between 15° and 60°. Weights of the members are negligible.

b. Force polygons shown in Figure 23-34A determine the tensile forces F(CB) in member CB for different angles θ. Space diagrams (same linear scale in all directions) in *Figure 23-34B* give the required lengths L(CB) for member CB for the different angles θ.

c. Next, the forces are entered in a stress equation from the theory of mechanics of materials to select the equal leg structural angles. The equation $S = (F)/(A)$ states that:

stress in pounds per square inch

(psi) =

$$\frac{\text{pounds force}}{\text{cross-sectional area in square inches}}$$

The *force* is assumed to be normal to the cross-sectional area, and each fiber in the entire member is stressed the same. One type of steel used for structural members has a yield strength of 36,000 psi. This means that the member could withstand tensile forces, in a linear manner, *Figure 23-35*, without excessive yielding (stretching), up to 36,000 psi. Therefore, a designer would divide the yield stress by a factor of safety to obtain an allowable or working stress and specify that the stress in the member must

STRESS-STRAIN CURVE FOR MILD STEEL

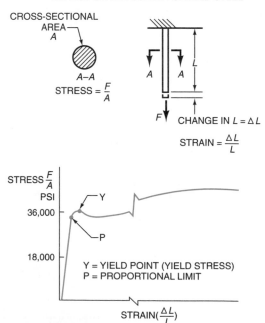

FIGURE 23-35 *Stress vs. strain*

not exceed it. A typical factor of safety for this application would be 2.0, so the allowable stress would be 18,000 psi; that is, halfway down the linear plot in Figure 23-35. This allowable stress is used in the stress equation.

The stress equation may now be rearranged to solve for area.

Cross-sectional area = pounds force/18,000 psi

d. *Figure 23-36* contains an excerpt from an AISC Handbook that lists equal leg structural angles, cross-sectional areas in square inches,

and weight per foot of each angle. The following example-calculation for an angle θ of 30° summarizes several of the steps already described and adds several more steps to arrive at cost.

(i) *L*(CB) is found in a space diagram to be 23 feet long for θ = 30°.

(ii) *F*(CB) is found in a force polygon to be 20,000 pounds for θ = 30°.

(iii) *A*(CB), the required cross-sectional area to limit the stress to 18,000 psi, is found from the equation:

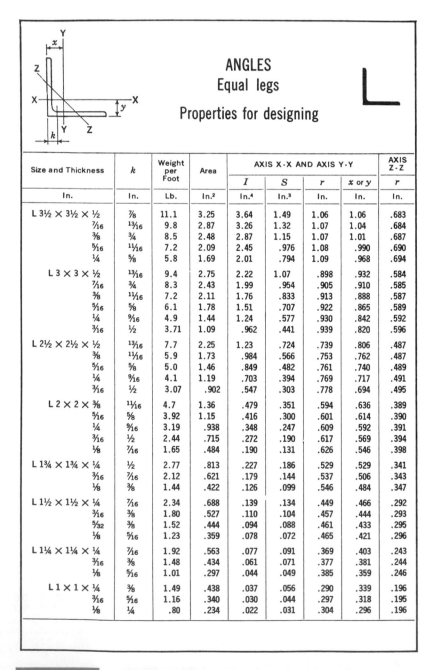

ANGLES
Equal legs
Properties for designing

Size and Thickness		k	Weight per Foot	Area	AXIS X-X AND AXIS Y-Y				AXIS Z-Z
					I	S	r	x or y	r
In.		In.	Lb.	In.²	In.⁴	In.³	In.	In.	In.
L 3½ × 3½ × ½		⅞	11.1	3.25	3.64	1.49	1.06	1.06	.683
	⁷⁄₁₆	¹³⁄₁₆	9.8	2.87	3.26	1.32	1.07	1.04	.684
	⅜	¾	8.5	2.48	2.87	1.15	1.07	1.01	.687
	⁵⁄₁₆	¹¹⁄₁₆	7.2	2.09	2.45	.976	1.08	.990	.690
	¼	⅝	5.8	1.69	2.01	.794	1.09	.968	.694
L 3 × 3 × ½		¹³⁄₁₆	9.4	2.75	2.22	1.07	.898	.932	.584
	⁷⁄₁₆	¾	8.3	2.43	1.99	.954	.905	.910	.585
	⅜	¹¹⁄₁₆	7.2	2.11	1.76	.833	.913	.888	.587
	⁵⁄₁₆	⅝	6.1	1.78	1.51	.707	.922	.865	.589
	¼	⁹⁄₁₆	4.9	1.44	1.24	.577	.930	.842	.592
	³⁄₁₆	½	3.71	1.09	.962	.441	.939	.820	.596
L 2½ × 2½ × ½		¹³⁄₁₆	7.7	2.25	1.23	.724	.739	.806	.487
	⅜	¹¹⁄₁₆	5.9	1.73	.984	.566	.753	.762	.487
	⁵⁄₁₆	⅝	5.0	1.46	.849	.482	.761	.740	.489
	¼	⁹⁄₁₆	4.1	1.19	.703	.394	.769	.717	.491
	³⁄₁₆	½	3.07	.902	.547	.303	.778	.694	.495
L 2 × 2 × ⅜		¹¹⁄₁₆	4.7	1.36	.479	.351	.594	.636	.389
	⁵⁄₁₆	⅝	3.92	1.15	.416	.300	.601	.614	.390
	¼	⁹⁄₁₆	3.19	.938	.348	.247	.609	.592	.391
	³⁄₁₆	½	2.44	.715	.272	.190	.617	.569	.394
	⅛	⁷⁄₁₆	1.65	.484	.190	.131	.626	.546	.398
L 1¾ × 1¾ × ¼		½	2.77	.813	.227	.186	.529	.529	.341
	³⁄₁₆	⁷⁄₁₆	2.12	.621	.179	.144	.537	.506	.343
	⅛	⅜	1.44	.422	.126	.099	.546	.484	.347
L 1½ × 1½ × ¼		⁷⁄₁₆	2.34	.688	.139	.134	.449	.466	.292
	³⁄₁₆	⅜	1.80	.527	.110	.104	.457	.444	.293
	⁵⁄₃₂	⅜	1.52	.444	.094	.088	.461	.433	.295
	⅛	⁵⁄₁₆	1.23	.359	.078	.072	.465	.421	.296
L 1¼ × 1¼ × ¼		⁷⁄₁₆	1.92	.563	.077	.091	.369	.403	.243
	³⁄₁₆	⅜	1.48	.434	.061	.071	.377	.381	.244
	⅛	⁵⁄₁₆	1.01	.297	.044	.049	.385	.359	.246
L 1 × 1 × ¼		⅜	1.49	.438	.037	.056	.290	.339	.196
	³⁄₁₆	⁵⁄₁₆	1.16	.340	.030	.044	.297	.318	.195
	⅛	¼	.80	.234	.022	.031	.304	.296	.196

FIGURE 23-36   *Equal leg structural angles* (Courtesy American Institute of Steel Construction)

$$A(CB) = \frac{F(CB)}{18,000} = \frac{20,000}{18,000}$$

$$= 1.111 \text{ square inches.}$$

(iv) From the AISC Handbook data, the nearest angle with a cross-sectional area of 1.111 or more is Angle 2-½ × 2-½ × ¼ with an area of 1.19 and a weight of 4.1 pounds/foot. Select the lightest member of a group in the handbook, because it is usually the one readily available in warehouses.

(v) Finally, assuming that the cost is $2.00 per pound, this angle would cost: (23 feet)

(4.1 pounds/foot) (2.00/pound) = to the nearest dollar, $189.00.

(vi) The rest of the structural angles were selected in a similar manner and tabulated in *Figure 23-37A*.

(vii) *Figure 23-37B* shows three curves, each a function of angle θ. The optimizing curve, $(CB) versus θ, shows the lowest (optimum) cost at approximately 48°; therefore, the best angle to specify is angle 2 × 2 × 3/16 × 30 feet long.

θ	L(CB) ft	F(CB) lb	A(CB) Calculated	Angle ∠	A(CB) Handbook	lb/ft	Wt(CB) lb	$(CB)
60°	40.1	11,500	0.639	2 x 2 x ³⁄₁₆	0.715	2.44	97.8	196
52½°	32.3	12,500	0.694	2 x 2 x ³⁄₁₆	0.715	2.44	78.8	158
45°	28.2	14,200	0.789	2½ x 2½ x ³⁄₁₆	0.902	3.07	86.6	173
37½°	24.9	16,500	0.917	3 x 3 x ³⁄₁₆	1.09	3.71	92.4	184
30°	23.0	20,000	1.111	2½ x 2½ x ¼	1.19	4.1	94.3	189
22½°	21.6	26,200	1.456	2½ x 2½ x ⁵⁄₁₆	1.46	5.0	108.0	216
15°	20.8	37,200	2.067	3½ x 3½ x ⁵⁄₁₆	2.09	7.2	149.8	300

FIGURE 23-37A   *Table: Optimization investigation*

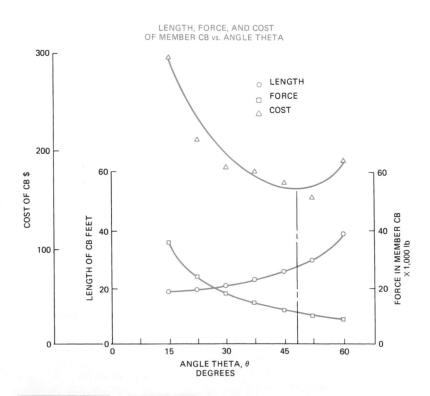

FIGURE 23-37B   *Graphs: Optimization investigation*

NOMOGRAMS: ALIGNMENT CHARTS AND CONCURRENCY CHARTS

Nomograms represent equations graphically as shown in Figure 23-20. Alignment charts, one class of nomograms, appear in periodicals, vendor catalogs, and textbooks. These charts are designed to help you visualize the information to be generated, and to obtain the information quickly and accurate enough for most design calculations. As a further aid, the charts contain an example or two showing how to use them. Thus, designers can use the charts easily without having to know how the chart was constructed.

However, if you know the theory and how to construct alignment charts, you will use them more effectively and confidently. Moreover, you will be able to decide whether the effort and time needed to prepare alignment charts will ultimately save time. You may need to construct alignment charts for reports and manuals.

First, consider a few reminders regarding mathematics, followed by a few definitions; then, consider a number of charts:

1. **Common Logarithm.** A common *logarithm* is an exponent of the base number 10, so that when 10 is raised to that exponent a specific number is generated. For example, the common logarithm of the number 25, written log 25, is equal to 1.397940; that is, $10^{**}1.397940 = 25$. The table in *Figure 23-38* shows an interesting comparison of "the old way" of obtaining logs and the current source, hand calculators.

2. **Function of a Variable:** $y = f(x)$. In the equation $y = x^{**}2 + 3x + 10$, y is a function of the variable x, written in short form as $y = f(x)$. Occasionally, in alignment chart derivations a single variable, such as u, that may have a range of numerical values is written as $f(u) = u$.

3. **Straight-Line Equation** $y = mx + b$. The *straight-line equation*, $y = mx + b$, specifies that a dependent variable y is a function of an independent variable x, m is the slope of the straight line, and b is the intercept along the vertical axis. Assume that the coordinate axes follow a traditional pattern: y is plotted in the vertical direction (ordinate axis) and x is plotted in the horizontal direction (abscissa), as shown in *Figure 23-39*.

4. **Similar Triangles.** Similar triangles form the basis of almost all alignment chart designs. Some properties of the triangles are summarized in *Figure 23-40*.

5. **Alignment Chart.** An alignment chart is a graphical representation of the relationship of two or more variables, such as $y = \log X$, $u + v = w$, $uv = w$, and $u = v/w$.

N Number	Log N Characteristic from Five-Place Table and Mantissa by Observation		Log N Complete Log from Hand Calculator	
	mantissa+characteristic			
1	0.	.00000	0.000000	00
7560	3.	.87852	3.878522	00
756	2.	.87852	2.878522	00
75.6	1.	.87852	1.878522	00
7.56	0.	.87852	8.78522	–01
.756	7.56/10 $0.87852 - 1.00000 = -.12148$		–1.214782	–01
.0756	7.56/100 $0.87852 - 2.00000 = -1.12148$		–1.121478	00

Note:

(a) Logarithms obey exponent laws:

$$\frac{10^m}{10^n} = 10^{m-n} \text{ AND } 10^m \cdot 10^n = 10^{m+n}$$

(b) Mantissa: one less than number of digits before decimal

FIGURE 23-38 *Logarithms*

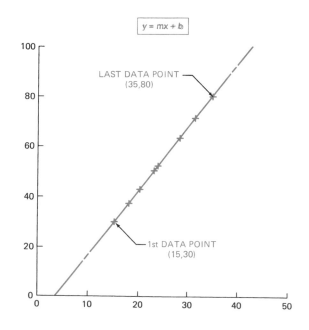

$y = mx + b$

SLOPE $m = \dfrac{y_2 - y_1}{x_2 - x_1} = \dfrac{80 - 30}{35 - 15} = \dfrac{50}{20} = 2.5$

$\therefore y = 2.5x + b$

USE ONE DATA POINT TO FIND b.
$30 = 2.5(15) + b$
$b = -7.5$

FINALLY $y = 2.5x - 7.5$

(HOWEVER, THE EQUATION IS GOOD ONLY OVER THE RANGE OF THE DATA.)

FIGURE 23-39 $y = mx + b$; *Straight-line equation*

SIMILAR TRIANGLES' CHARACTERISTICS

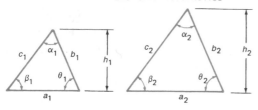

a_1 IS PARALLEL TO a_2

b_1 IS PARALLEL TO b_2

c_1 IS PARALLEL TO c_2

$\alpha_1 = \alpha_2$

$\beta_1 = \beta_2$

$\theta_1 = \theta_2$

$$\frac{a_1}{a_2} = \frac{b_1}{b_2} = \frac{c_1}{c_2} = \frac{h_1}{h_2}$$

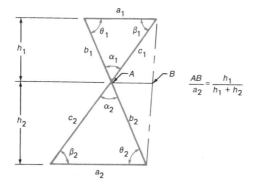

$$\frac{AB}{a_2} = \frac{h_1}{h_1 + h_2}$$

FIGURE 23-40   *Similar triangles*

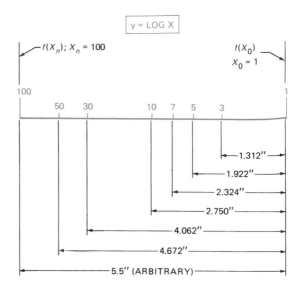

$y = \text{LOG } X$

$L = m[f(X_n) - f(X_0)]$
$L = 5.5''$
$m = 2.75$ INCHES PER DIFFERENCE IN $f(X)$
(SCALE MODULUS)

FIGURE 23-41   *Functional scales:* $y = f(x)$

$$5.5 = m[\log 100 - \log 1]$$

$$m = \frac{5.5}{(2 - 0)}$$

$$= 2.75'' \text{ per difference in } f(X).$$

b. Next, distances to each label were calculated; for example, when $X = 30$:

$$L = (2.75)(\log 30 - \log 1) = 4.062''$$

This distance and several others are shown in Figure 23-41.

8. **Adjacent Alignment Chart for** $f(u) = f(v)$
   a. Two related variables plotted along the same line, but on opposite sides, make up an *adjacent alignment chart*. A temperature conversion chart for relating degrees Celsius to degrees Fahrenheit is an excellent example. Assume that 8″ is arbitrarily designated as the scale length and do the following to create the conversion scale shown in *Figure 23-42*:

   Note: For the remainder of the discussions on alignment charts, note that when you encounter the scale modulus m written, for example, as $m(C)$ or as $m(R)$, the (C) or (R) refers only to the parameter C or R; m is not a function of C or R, etc.

   (i) $L(\text{Celsius}) = L(C) = m(C)(100 - 0)$

   $$8 = m(C)(100)$$

   $m(C) = 0.08$ inches per difference in degrees Celsius

   thus, $L(C) = 0.08C$ is the scale equation.

6. **Scale Equation** $L = mf(x)$. The scale equation fits functions to the page area. In the equation: $L = m[f(X_n) - f(X_0)]$, $L =$ the distance between the final value X_n and the initial value X_0. Designers select distance L arbitrarily: $m =$ scale modulus, which makes the scale fit the page. The quantity $f(X_n) =$ the value of the function of X (when X_n is the largest number of the given range of numbers) and $f(X_0) =$ the lowest value of the function of X.

7. **Functional Scale for** $y = f(X)$. A functional scale is a line graduated and labeled according to a functional relationship. Distances to the labels are calculated using the scale equation, but the distances are not shown on the scale.

The functional scale in *Figure 23-41* is labeled for the function $y = \log X$. The distances to the labels were calculated using the following steps:

   a. Since X ranges from $X = 1$ to $X = 100$, and 5.5″ was arbitrarily selected to fit the scale on the page, the scale modulus was calculated.

   $$L = m[f(X_n) - f(X_0)]$$

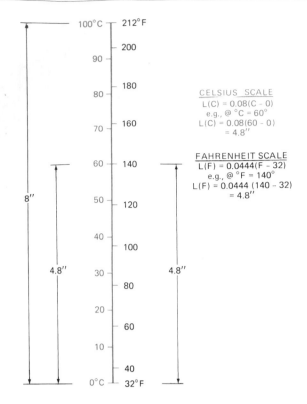

FIGURE 23-42 *Adjacent alignment chart: f(u) = f(v)*

b. Another adjacent alignment chart is the speed-ometer in most automobiles which has MPH and kmPH graduated along the same curved line.

c. A third example solves a designer's needs. The stress equation used in the force polygons discussion can be plotted along one line, as shown in *Figure 23-43*. The steps to produce the adjacent plots for the equation "stress = Force/ Area" are as follows:

(i) Assume an arbitrary scale length of 6″ and the equation restated to read

cross sectional area = pounds force/18,000 psi

(ii) $L(area) = m(A) [(F/18,000) - (0)]$

Assume that the largest value of F will be 45,000 pounds force. (Refer to Figure 24-34A.)

Then $m(A) = 6/[(45,000/18,000) - (0)]$

$m(A) = 2.4$

$L(A) = 2.4[(F/18,000) - (0)]$

$L(A) = (F/7,500 - 0)$

(iii) Since F was found from the force polygons and used in the equation, rewrite the stress equation for F to be $F = 18,000A$

Then $L(F) = m(F) (18,000A)$

The maximum value of A is
$A = 45,000/18,000 = 2.5$

Thus, $m(F) = \dfrac{6}{(18,000) \ (2.5)} = \dfrac{1}{7,500}$

$L(F) = \dfrac{1}{7,500} (18,000A)$

Finally, $L(F) = 2.4A$

(ii) $L(\text{Fahrenheit}) = L(F) = m(F) \dfrac{5}{9} (F - 32)$

$8 = m(F) \dfrac{5}{9} (F - 32)$

when $F = 212$ degrees

$m(F) = \dfrac{8}{\dfrac{5}{9} (212 - 32)} = 0.08$

and $L(F) = (0.08) \dfrac{5}{9} (F - 32)$,

the scale equation.

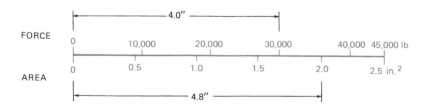

L(F) = 2.4A @ 30,000 lb

$\dfrac{30,000 \text{ lb}}{18,000 \text{ psi}} = 1.67 \text{ in.}^2$ AND $L(F) = 2.4(1.67) = 4.0''$

$L(A) = \dfrac{F}{7,500}$ @ $A = 2.0$ in.^{2}

$F = 18,000A$; $A = 2.0$
$F = 36,000$

$L(A) = \dfrac{36,000}{7,500} = 4.8''$

FIGURE 23-43 *Adjacent alignment chart: stress = (force)/(area)*

(iv) Calculate and plot distances $L(A)$ for a number of values of force between 0 and 45,000 pounds.

(v) Calculate and plot distances $L(F)$ for a number of values of area between 0 and 2.5. Compare several values of area opposite forces, from the chart with the tabulated values in Figure 23-37A.

9. **Nonadjacent Alignment Chart for $f(u) = f(v)$.** Refer to the adjacent alignment chart discussion and visualize that the two scales are separated by a convenient distance to fit a page. Rotate one scale 180° and connect the lower value end of each with a straight line. All that remains to be done to create a nonadjacent alignment chart is to locate a pivot point K to "hinge" straight lines called "isopleths." *Isopleths* are the lines used to locate values in a nomogram.

a. The steps for deriving the technique for preparing nonadjacent alignment charts are as follows:

(i) Locate two parallel lines at an arbitrary distance apart to fit the page. Select $L(u)$ and $L(v)$ as convenient lengths also to fit the page. Construct the similar triangles as shown in *Figure 23-44* and label intersections of the lines. Assume that the initial values of u and v are equal to zero—to facilitate the derivation.

(ii) From similar triangles
$$\frac{L(u)}{L(v)} = \frac{BK}{KC} = \frac{a}{b}$$
From the scale equation
$$\frac{L(u)}{L(v)} = \frac{m(u)\,f(u)}{m(v)\,f(v)} = \frac{a}{b}$$

And since $f(u) = f(v)$
$$\frac{m(u)}{m(v)} = \frac{a}{b}$$

(iii) Determine $m(u)$ and $m(v)$ so the ratio a/b can be found, which is used as a guide to divide the arbitrary distance between the scales in proportion to a and b. Point K locates the division and also acts as the pivot point for isopleths.

b. Use these steps to prepare a nonadjacent alignment chart for the equation $C = 2\pi R$ where circumference is a function of radius R, or $f(C) = f(R)$. R varies from 0 to 20 inches, and $L(R)$ is arbitrarily set at $4''$, *Figure 23-45*.

(i) Determine $m(R)$ from
$$L(R) = m(R)\,[f(R_m) - f(R_o)]$$
$$L(R) = m(R)\,[20 - o]$$
$$m(R) = \frac{4}{20} = 0.2$$
Thus, $L(R) = 0.2R$

(ii) Assume that $L(C)$ is arbitrarily $6''$ long and that $L(C)$ and $L(R)$ are separated by $4''$, ($L(R)$ on the left side), to fit the page. Now determine:
$m(c)$ using $R = C/2\pi$ because the values of R are specified.

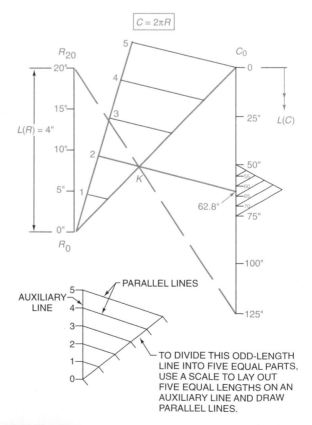

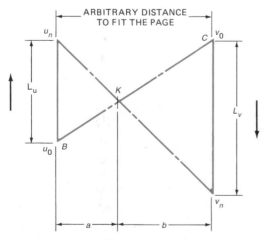

NOTE: L_u IN THIS FIGURE IS THE $L(u)$ IN THE TEXT, ETC.

FIGURE 23-44 *Nonadjacent scales*

FIGURE 23-45 $C = 2(3.14)R$

$$L(C) = m(C) [f(C(20) - f(C(0))]$$

$$= m(C) \left(\frac{2\pi 20}{2\pi} - 0 \right)$$

$$L(C) = m(C) (20)$$

$$m(C) = \frac{6}{20} = 0.3$$

Thus, $L(C) = 0.3 \frac{C}{2\pi}$

or $L(C) = 0.0478C$

(iii) Locate point K from

$$\frac{m(R)}{m(C)} = \frac{0.2}{0.3} = \frac{2}{3}$$

Divide line $R(0) - C(0)$ into five parts and locate K a distance of two parts from $R(0)$.

(iv) Graduate the two scales, $L(R)$ and $L(C)$, using scale equations. A helpful graphical technique for dividing lines is shown in the Figure 23-45.

(v) Use several isopleths to verify the chart, e.g., when $R = 10$, C should equal 62.8 inches.

10. **Parallel Alignment Chart for $f(r) + f(t) = f(s)$**

a. The procedure for preparing a set of three parallel scales for a three-variable equation, $r + t = s$, or more generally $f(r) + f(t) = f(s)$, is derived using similar triangles and scale equations. The derivation aims at obtaining relationships from two of the scales so that the third can be located and graduated.

b. The following steps in the derivation are illustrated in *Figure 23-46*.

(i) Preliminary construction requirements and assumptions are summarized in Figure 23-46.

(ii) To facilitate the derivation assume that:
—Initial values of the variables are equal to zero: $r(0) = t(0) = s(0) = 0$
—The scales $L(r)$ and $L(t)$ have been graduated and that $m(r)$ and $m(t)$ are known.
$L(r) = m(r) f(r)$
$L(t) = m(t) f(t)$
To be determined:
$L(s) = m(s) f(s)$

(iii) From similar triangles:

$$\frac{L(r) - L(s)}{a} = \frac{L(s) - L(t)}{b}$$

Substitute scale equations:

$$\frac{m(r) f(r) - m(s) f(s)}{a} = \frac{m(s) f(s) - m(t) f(t)}{b}$$

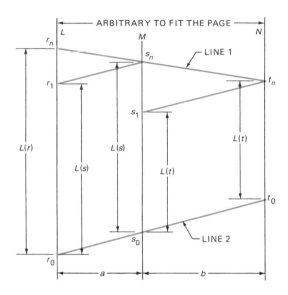

CONSTRUCTION DETAILS:
1. LINES L, M, AND N ARE PARALLEL.
2. $L(r)$, $L(t)$, AND $L(s)$ SATISFY $r + t = s$ OR $f(r) + f(t) = f(s)$.
3. LINES $s_n - r_1$ AND $t_n - s_1$ ARE PARALLEL TO LINE 2.
4. LINES 1 AND 2 ARE NOT PARALLEL.

FIGURE 23-46   *Parallel alignment chart: $f(r) + f(t) = f(s)$*

Cross multiply, collect terms and get $m(r)f(r) + a/b\, m(t)f(t) = (a + b)/b\, m(s)f(s)$. Then since $f(r) + f(t) = f(s)$, the coefficients must all be equal; that is, $m(r) = a/b\, m(t) = (a + b)/b\, m(s)$. From step (iii): $m(r)/m(t) = a/b$ which is used to locate the third scale, $L(s)$.

Also from Step (iii):

$$m(r) = \frac{a + b}{b} m(s) = (1 + \frac{a}{b}) m(s)$$

Since $\frac{a}{b} = \frac{m(r)}{m(t)}$

$$m(r) = (1 + \frac{m(r)}{m(t)}) m(s)$$

Finally, $m(s) = \dfrac{m(r)\, m(t)}{m(r) + m(t)}$

which is the scale modulus for $L(s)$.

11. **Example Parallel Alignment Chart for $r + t = s$.** *Figure 23-47* shows the chart for the equation $r + t = s$ where $r = 0$ to 20 and $t = 0$ to 50. Do the following to construct the chart:

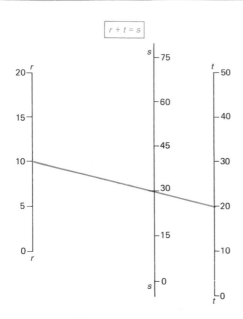

$$r + t = s$$

FIGURE 23-47 *Parallel alignment chart: r + t = 5*

a. Draw the parallel scales for r and t to fit the page and leave room between them for scale s. Arbitrarily select $L(r) = 4''$, $L(t) = 5''$. Locate their zero values at the bottom. $L(r)$ goes on the left.

b. Calculate scale moduli for $m(r)$ and $m(t)$.
$L(r) = m(r) (20 - 0)$

$$m(r) = \frac{4}{20} = 0.2$$

$L(t) = m(t) (50 - 0)$

$$m(t) = \frac{5}{50} = 0.1$$

c. Locate a vertical line for the center scale, $L(s)$, from

$$\frac{a}{b} = \frac{m(r)}{m(t)} = \frac{0.2}{0.1} = \frac{2}{1}$$

which places $L(s)$ two-thirds of the distance between $L(r)$ and $L(t)$, from $L(r)$.

d. Locate one value on $L(s)$ by constructing an isopleth between any two values of r and t, e.g., $r = 10$ and $t = 20$, which of course gives $s = 30$. The remainder of $L(s)$ can be completed using the scale equation $L(s) = m(s)f(s)$, where s varies from 0 to 70.

e. Determine $m(s)$ from

$$m(s) = \frac{m(r)\, m(t)}{m(r) + m(t)} = \frac{(.2)\,(.1)}{(.2 + .1)}$$

$m(s) = 0.067$
Thus, $L(s) = 0.067\, f(s)$ for graduating scale s.

12. **Parallel Alignment Chart for** $f(r) \cdot f(t) = f(s)$. This second example is typical of an equation a designer would use. Assume that a nomogram is desired for $E = IR$. The equations can be rewritten as $\log E = \log I + \log R$, which is in the form of $f(r) + f(t) = f(s)$, which was derived earlier.

To construct the nomogram for this basic equation, (volts) = (amps) (ohms), follow the steps below and refer to *Figure 23-48*.

a. Assume the ranges for the variables I and R; $I = 0.1$ to 10 amps and $R = 1$ to 1,000 ohms. Select $L(I) = 5''$ and $L(R) = 5''$: place them 4" inches apart, vertical, parallel, and with the smaller values at the bottom. Locate $L(I)$ on the left side. Graduate them using the scale equations:

$$L(I) = m(I) \cdot f(I) \text{ and } L(R) = m(R) \cdot f(R).$$

The $y = \log X$ scale in Figure 23-41 is a useful graphical aid in graduating logarithmic scales as illustrated in Figure 23-48.

b. Calculate scale moduli $m(I)$ and $m(R)$ to use in determining a/b and $m(E)$.
$L(I) = m(I) (\log 10 - \log 0.1)$

$$m(I) = \frac{5}{(1.0 - (-1.0))} = 2.5$$

Thus, $L(I) = 2.5\, f(I)$
$L(R) = m(R) (\log 900 - \log 1)$

$$m(R) = \frac{5}{2.954243 - 0} = 1.6925$$

Thus, $L(R) = 1.6925\, f(R)$

c. $\dfrac{a}{b} = \dfrac{m(I)}{m(R)} = \dfrac{2.5}{1.6925} = 1.477$
Thus, $a = 1.477\, b$ and $a + b = 4.00$
and $1.477\, b + b = 4.00$
so, $b = 1.615''$ from $L(R)$;
$a = 2.385''$ from $L(I)$.

d. $m(E) = \dfrac{m(I)\, m(R)}{m(I) + m(R)} = \dfrac{(2.5)\,(1.6925)}{(2.5 + 1.6925)}$
$m(E) = 1.009$
Thus, $L(E) = 1.009\, f(E)$
so, $L(E) = 1.009\, (\log 9000 - \log 0.1)$
$= 1.009$ times $(3.954243 - (-1))$
$L(E) = 4.9988''$

e. Construct an isopleth for any value of E from IR, e.g., (0.1 amps) $\times$ (10 ohms) = 1 volt, which locates the scale $L(E)$.

f. Graduate $L(E)$ using the scale equation and the graphical aid, $y = \log x$.

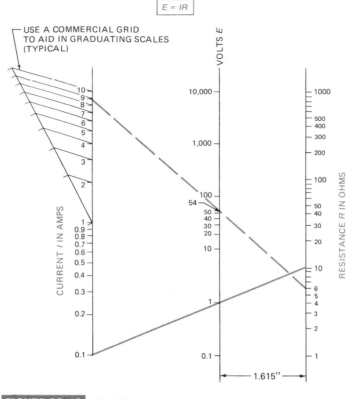

FIGURE 23-48 $E = IR$

13. **Short Method for a Parallel Alignment Chart for $E = IR$.** After constructing a number of alignment charts and learning the theories plus graphical techniques, users will appreciate more practical, faster methods. One method is shown in *Figure 23-49* for the following steps:

   a. On a sheet of commercial three-cycle, log-linear (semi-logarithmic) grid paper:
      Locate $I = 0.1$, 1.0, and 10 amps on the left side of the sheet. Locate $R = 1$, 10, 100, and 1,000 ohms in a similar manner at a convenient distance to the right, say 5″ or 6″.

   b. Do two different calculations for E, but make $E_1 = E_2$, e.g.,
      $E_1 = (10)\,(10) = 100$ volts
      $E_2 = (1)\,(100) = 100$ volts

   c. Connect the I and R values, respectively, with isopleths to locate the E scale where the two isopleths intersect. Label the intersection 100.

   d. Finish the E scale as before.

14. **Concurrency Charts for $f(r) + f(t) = f(s)$.** *Concurrency charts* provide designers with an additional graphical format to visualize equations with three or more variables, and to generate information. The basic format

is the familiar horizontal and vertical cartesian coordinate system. This first concurrency chart example has two objectives: one is to illustrate the format, and the other is to point out that designers may have a number of choices of formats for any one equation.

Note that the equation form $r + t = s$, or $f(r) + f(t) = f(s)$, has been discussed earlier for parallel alignment charts.

Assume that the variables have values: $r = 0$ to 12, and $t = 0$ to 10. *Figure 23-50* shows the concurrency chart for finding the third variable, s. Follow the steps used to prepare the chart:

   a. Observe that the equation $s = r + t$ is a straight-line equation, where s is y, r is x, and t is b in the general straight-line equation $y = mx + b$. The slope $m = 1$.

   b. Divide the horizontal axis in equal segments for the variable r, 0 to 12.

   c. Graduate the vertical axis to account for the minimum and maximum values of s, 0 to 22.

   d. Write equations to account for the intercept t in $s = (1)r + t$. For example,
      When $t = 0$, $s = r$

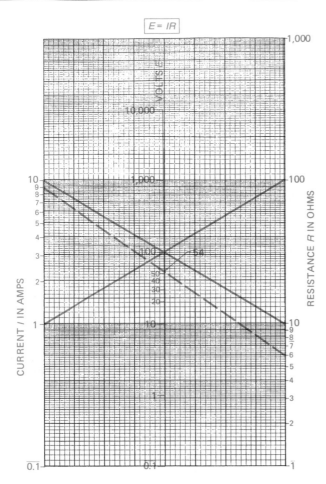

FIGURE 23-49   *Shortcut. Parallel alignment chart*

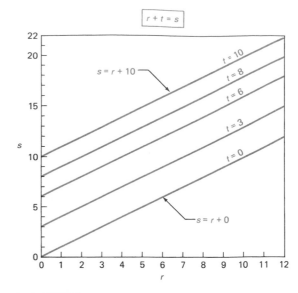

FIGURE 23-50   *Concurrency chart:* $f(r) + f(t) = f(s)$.

a. Straight-line equations can be written to account for the intercept log R, as follows:

When $R = 1$

$t = 1, s = r + 1$
$t = 2, s = r + 2$
.
.
.
$t = 10, s = r + 10$

e. Plot the equations that all have the same slope of 1, and label them with their appropriate *t* value.

15. **Concurrency Charts for** $f(r) \bullet f(t) = f(s)$. A second example of a concurrency chart, shown in *Figure 23-51*, is for the equation $E = IR$, which was used earlier. The equation can be rewritten in the form $\log E = \log I + \log R$, as before. This equation also is patterned after the familiar straight-line equation, because log E is y, log I is x, and log R is b. The slope $m = 1$. The steps to prepare the chart are similar to those for the preceding chart, except that the horizontal and vertical scales are graduated in the logarithmic format. The variables *I* and *R* have the following ranges for this example:

$I = 0$ to 50 amps and $R = 1$ to 10 ohms.

E then will be 0 to 500 volts.

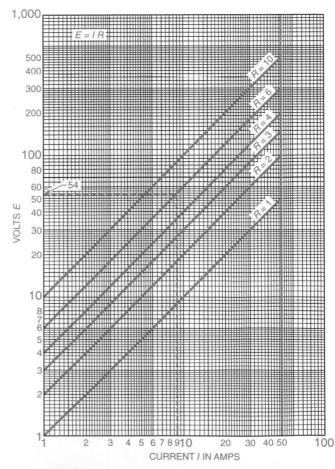

FIGURE 23-51   *Concurrency chart:* $f(r) \bullet f(t) = f(s)$

$\log E = \log I + \log 1$

and $E = I$

When $R = 10$

$\log E = \log I + \log 10$

and $E = 10\ I$

b. Finally, the curves are drawn with a slope of 1, as shown in Figure 23-51.

16. **Concurrency Charts for $f(r) \bullet f(t) = f(s)$.** This chart is an alternative format to the preceding discussion. The same equation, $E = IR$, is used but the steps are different and the chart is different, as seen in *Figure 23-52.* Do the following for this alternative:

a. Observe that $E = IR$ can be compared to the straight-line equations in a slightly different way. In the equations $y = mx + b$ and $E = RI$:

E is y, I is x, and R is the slope of the line.

b. Let the two variables have the same ranges as before, $I = 0$ to 50 amps and $R = 1$ to 10 ohms. Next, the following equations can be written:

When $R = 1$

$E = I$

When $R = 2$

$E = 2I$

.

.

.

When $R = 10$

$E = 10I$

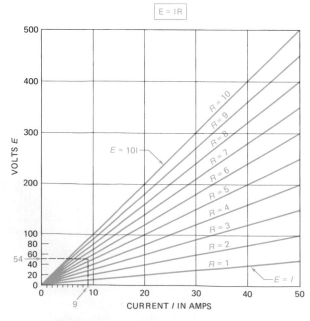

FIGURE 23-52 *Concurrency chart:* $f(r) \bullet f(t) = f(s)$

c. Graduate the horizontal and vertical scales in a linear manner for *I* and *E,* respectively.

The foregoing definitions and discussions of alignment charts and concurrency charts, plus the steps to prepare them, provide sufficient theory and construction guidelines to help designers use these charts effectively to obtain information or to prepare their own. However, further discussions and instructions are available for those who need them. An excellent reference was written by A. S. Levens, *Nomography* (John Wiley and Sons, 1959).

EMPIRICAL EQUATIONS: $y = mx + b$, $y = b(e^{**}mx)$, and $y = b(x^{**}m)$, AND OTHER CURVE-MATCHING TECHNIQUES

Consider two problems involving the need to analytically describe curves, either plotted from data in an experiment or developed during the design of a surface.

1. **Data Problem.** Data involving two changing variables, one independent and the other dependent, has been recorded for an experiment in a lab course. An equation is required to describe the relationship of the two variables so additional data points can be calculated within the range of the data already recorded. What steps should be taken next? How can an equation be generated?

2. **Curve Problem.** This problem differs slightly from the preceding one because the data points are to be part of a design development to generate curved lines for an automobile body surface. Equations of the lines involve the use of a personal computer; however, the problem is still graphical. What steps should be taken? How are the curves generated? Equations?

The *data problem* will be discussed first, primarily from a graphics approach. Then the *curve problem* will be considered from a more analytical basis and involving personal computers. Both problems use physical and graphical illustrations and explanations.

Data Problem. **Empirical Equations for Experimental Data.** The flow chart in *Figure 23-53* depicts the overall approach for deriving and verifying empirical equations. The steps contained in the chart are followed while considering three sets of data.

FIRST SET OF DATA

1. Assume that the data tabulated in *Figure 23-54* has been collected during a lab assignment involving surface factors for various steels, and that the next task is to plot the data on linear coordinate graph paper.

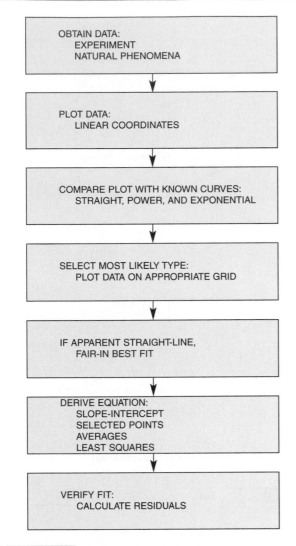

FIGURE 23-53   *Empirical equations: Procedure*

ULTIMATE TENSILE STRENGTH *S* x 1,000 psi	SURFACE FACTOR *k(a)*
100	0.53
120	0.50
140	0.46
160	0.43
180	0.38
200	0.36

FIGURE 23-54   *Data*

2. Before plotting the data, consider the three types of curves that are generally produced from experiments or from observing natural phenomena: (a) straight-line, (b) power, and (c) exponential.

a. Data plotting in a straight line on linear coordinate paper would obviously indicate a linear relationship and an equation would be easily derived;

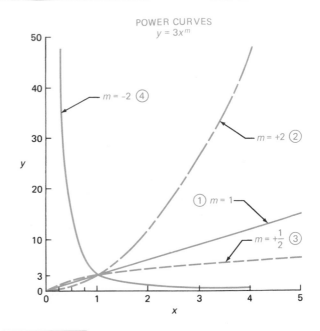

FIGURE 23-55   *Power curves on linear coordinates*

however, be certain to indicate that the line applies only within the range of the data taken.

b. Plots of data tending toward a power curve $y = B(x^{**}m)$ would show patterns similar to those in *Figure 23-55*. When $m = 1$, a unique straight line results that goes through zero and has a slope equal to B. See curve 1 in Figure 23-55. When m equals a positive number greater than 1, curve 2 would be typical. Positive m's less than 1 look like curve 3, which continues in an upward direction. Negative values of m give the hyperbolic shape of curve 4. Curves 1, 2, and 3 go through zero when $x = 0$.

c. Exponential curves, in general, look like those in *Figure 23-56*. When $m = 0$ in the equation $y = B(e^{**}mx)$, the plot is the horizontal line 1 parallel to the X axis. Positive values of m give curve 2, which intercepts the Y axis at the value of B and continues upward. Curve 3 starts downward from B on the Y axis and asymptotically approaches the X axis. Curve 3 is repeated in *Figure 23-57* because 3 and its inverse are the most common types. These two represent phenomena where the rate of change in a variable is proportional to the variable itself. For example, curve 3 illustrates the decay of a radioactive substance N. The equation N(at any time t) = N(originally) $(e^{**} - at)$ can be derived from data. The constant, a, represents 0.693/half life. The inverse curve 4 illustrates the change in temperature of an object being heated. The temperature approaches a maximum value

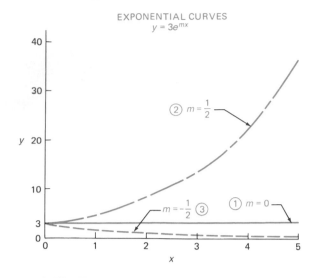

EXPONENTIAL CURVES
$y = 3e^{mx}$

(2) $m = \frac{1}{2}$

(3) $m = -\frac{1}{2}$ (1) $m = 0$

FIGURE 23-56 Exponential curves on linear coordinates

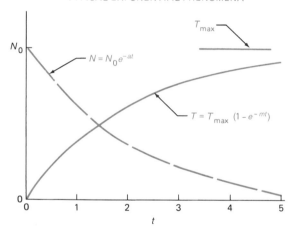

TYPICAL EXPONENTIAL PHENOMENA

T_{max}

$N = N_0 e^{-at}$

$T = T_{max} (1 - e^{-mt})$

FIGURE 23-57 Exponential curves on linear coordinates

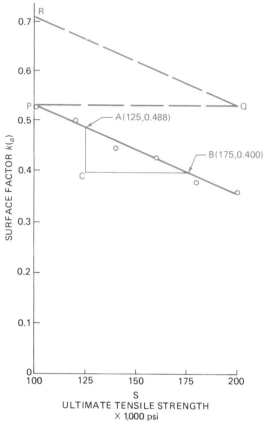

SURFACE FACTOR k_a
vs.
ULTIMATE STRENGTH OF HOT-ROLLED STEEL

A(125,0.488)

B(175,0.400)

SURFACE FACTOR k_a

ULTIMATE TENSILE STRENGTH
× 1,000 psi

S

FIGURE 23-58 Data on linear coordinates

asymptotically as described by T(actual at time t) $= T$(max) $(1 - e^{**} - mt)$.

3. This first set of data plots almost as a straight line of the form $y = mx + b$ on linear coordinate axes, as shown in *Figure 23-58*. A best-fit straight line is drawn by eye through most of the points; that is, a straight line has been "faired-in."

4. Next, derive an equation of the form $y = mx + b$. Of the many approaches available, the four listed here are the ones most often found in graphics texts. In increasing order of difficulty and time involvement, they are:

   a. Method of slope-intercept

   b. Method of selected points

   c. Method of averages

   d. Method of least squares

All four methods are used for this first set of data, and the curve from each method is verified (checked) by the method of residuals.

5. The methods.

   a. *Slope-intercept method*

      (i) Select two points, e.g., A and B in Figure 23-58.

      (ii) Graphically, determine m, negative because of the downward slope, using the distance AC divided by CB. The slope must be compatible with the equation; therefore $AC = 0.088$ surface factor units, $BC = 50$ kpsi, and $AC/BC = m = -.088/50 = -.00176$ surface factor units/kpsi.

      (iii) Thus, the equation is $k(a) = -.00176\, S + b$. Next, graphically determine the intercept b, but note that the S values only go down to 100 kpsi. The intercept must be found when $S = 0$. The graphical method for finding b when the independent variable range does not include zero is to simply project backward to zero, as shown with line PQ and QR in Figure 23-58. The intercept is found to be 0.71.

(iv) Finally the equation is $k(a) = -.0017 6S + 0.71$. The 0.71 value is the intercept at $S = 0$; however, remember to state that the equation is good only between S = 100 kpsi and S = 200 kpsi.

(v) Calculate and sum the residuals, which are equal to observed values of $k(a)$ – calculated values of $k(a)$, using the derived equation, at the six values of S.

S	k(a) obs.	k(a) calc.	Residuals R
100	0.53	0.528	+.002
120	0.50	0.493	+.007
140	0.46	0.458	+.002
160	0.43	0.422	+.008
180	0.38	0.387	−.007
200	0.36	0.352	+.008
		SUM R =	+.020

A positive sum indicates that the curve could be moved upward slightly for a better fit.

b. *Selected points method*

(i) Select two pairs of coordinate points from the best-fit line and substitute them into the basic form of the straight-line equation. Then solve simultaneously for m and b.

The pairs of points are:
$k(a) = 0.5, S = 120$
$k(a) = 0.36, S = 200$

Substitute:
$0.50 = m(120) + b$ (q)
$0.36 = m(200) + b$ (p)

(ii) Solve simultaneously for m and b.
$(p) - (q) = -.14 + m(80)$
$m = -.00175$

and $0.36 = -.00175(200) + b$
$b = 0.71$

(iii) Finally, $k(a) = -.00175 S + 0.71$
between S = 100 kpsi and 200 kpsi

(iv) Calculate residuals R

S	k(a) obs.	k(a) calc.	Residuals R
100	0.53	0.530	+.005
120	0.50	0.496	0
140	0.46	0.462	−.005
160	0.43	0.428	0
180	0.38	0.394	−.015
200	0.36	0.360	0
		SUM R =	−.015

c. *Averages method*

This method doesn't require the direct use of the graphical plot; however, the data must be plot-

ted to decide which type of curve the data points more closely match.

The expression $SUM(y - mx - b) = 0$ says that for the best-fit line, the differences between the observed values of y and the calculated values of $mx + b$ add to zero.

(i) Divide the data into two equal groups, substitute the data into equation $y - mx - b = 0$, and add them.

$$.53 - 100m - b$$
$$.50 - 120m - b$$
$$\underline{.46 - 140m - b}$$
$$1.49 - 360m - 3b = 0 \qquad (r)$$
$$.43 - 160m - b$$
$$.38 - 180m - b$$
$$\underline{.36 - 200m - b}$$
$$1.17 - 540m - 3b = 0 \qquad (s)$$

Solve (r) and (s) simultaneously and find $m = -.00178$ and $b = 0.71$

(ii) The equation of the line becomes $k(a) = -.00178 S + 0.71$ for S from 100 kpsi to 200 kpsi.

(iii) Calculate the residuals in the same way as before and find that SUM R = +.003

d. *Least-squares method*

This method consumes the most time, but it gives the best accuracy of the four methods. The expression $SUM(y - mx - b)^2 = a\ minimum$ says that for the best fit, the differences between the observed data for y and the calculated data $mx + b$, squared, should be a minimum.

(i) Set the derivatives of $SUM(y - mx - b)^2$ with respect to m and to b, equal to zero:

With respect to m:
$SUM [2(y - mx - b) (-x)] = 0$

With respect to b:
$SUM [2(y - mx - b) (-1)] = 0$

(ii) Multiply and rewrite to obtain:
$SUM (x) (y) = b\ SUM(x) + m\ SUM(x**2)$
$SUM (y) = b\ SUM(\text{number of observations}) + m$
$SUM(x)$

(x)(y)	(b)(x)	M(x**2)
100 (.53)	=b100	+m(100)**2
120 (.50)	=b120	+m(120)**2
140 (.46)	=b140	+m(140)**2
160 (.43)	=b160	+m(160)**2
180 (.38)	=b180	+m(180)**2
200 (.36)	=b200	+m(200)**2
386.6	=b900	+m142,000 equa(t)

Y	b	mx
.53	=b	+m100
.50	=b	+m120
.46	=b	+m140
.43	=b	+m160
.38	=b	+m180
.36	=b	+m200
2.66	=6b	+m900 equa(w)

(iv) Solve (t) and (w) simultaneously and find
$m = -.00177$
$b = 0.709$
so, $k(a) = -.00177S + 0.709$

(v) Calculate residuals R and get SUM $R = -.001$, which is the best fit of the four methods summarized here.

    a. Slope-intercept   SUM = +0.014
    b. Selected points   SUM = −0.015
    c. Averages   SUM = +.003
    d. Least squares   SUM = −.001

Second Set of Data. To calibrate a Weir, the following data was taken:

H (feet)	2.	3.	4.	5.	6.
Q (cu. ft/min)	14.2	41.5	80.0	135.2	225.3

H (feet)	7.	8.	9.	10.
Q (cu. ft/min)	330.5	445.5	620.5	790.

A Weir is a rigid plate used to measure the flow of water in open channels (e.g., irrigation channels). The rigid plate goes crosswise in the channel, as shown in *Figure 23-59*, so the water must spill over it. The height of the water, H, above the bottom of the notch indicates the rate of flow, Q.

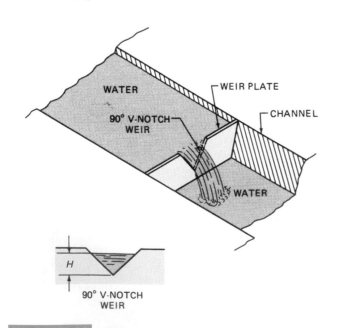

FIGURE 23-59   *Weir*

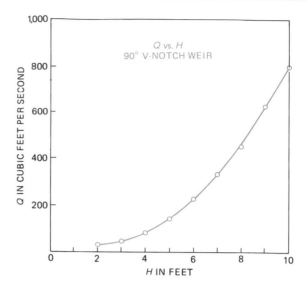

FIGURE 23-60   *Weir data on linear coordinates*

1. Follow the general empirical-equations approach depicted in Figure 23-53 and plot the Weir data on linear coordinate paper to ascertain the type of curve—straight, power, or exponential. The plot of the Weir data in *Figure 23-60* appears similar to the one in Figure 23-55, which seems to be increasing upward as H gets larger. Also, the plot of Q would go through zero for H = 0. Thus, a power-type curve is indicated.

2. Try a logarithmic 2 × 2 cycle commercial grid for obtaining a straight line. *Figure 23-61* indeed shows that the data forms a linear relationship and will fit the logarithmic form of the general power equation, $\log y = \log B + m \log x$.

3. Fair-in a best-fit straight line and prepare to derive a power equation from $\log Q = \log B + m \log H$.

4. The slope-intercept and selected points methods are used for this set of data.

    a. Slope-intercept method

      (i) Select two points on the best-fit line, A and D in Figure 23-61.

      (ii) Determine slope m from

        $m = (\log A - \log C)/(\log C - \log D)$;
        $(\log 450 - \log 80) = 0.7501$
        and $(\log 8 - \log 4) = .301$

        so $m = +(.7501)/(.301)$
        = 2.492 cubic feet per second per foot.

      (iii) The logarithmic equation changes to log Q = log B + 2.492 log H. Next, graphically determine the slope-intercept. Use an offset

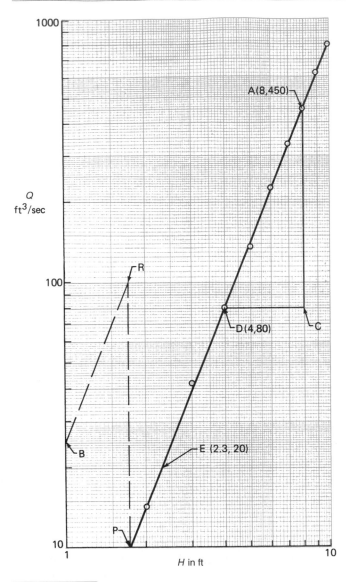

FIGURE 23-61   *Weir data on logarithmic grid*

	6	225.3	217.32	+7.98
	7	330.5	319.10	+11.40
	8	445.5	445.08	+0.42
	9	620.5	596.92	+23.50
	10	790.6	776.14	+14.46
			SUM $R =$	+58.88

The positive sum indicates that the faired-in straight line should be moved slightly upward for a better fit.

b. Selected points method

(i) Select two pairs of coordinates from the best-fit line and substitute them into the equation $\log Q = \log B + m \log H$. From the line, obtain $A(450, 8)$ and $E(20, 2.3)$ to write $\log 450 = \log B + m \log 8$ and $\log 20 = \log B + m \log 2.3$.

(ii) Solve simultaneously:

$\log 450 - \log 20 = m(\log 8 - \log 2, 3)$
$m = 2.498$

and then, $\log 450 = \log B + 2.498 \log 8$, so $B = 2.496$.

(iii) The equation finally is:

$\log Q = \log 2.496 + 2.498 \log H$
and $Q = 2.496 (H^{**}2.498)$

(iv) Calculate residuals to verify the equation.

Third Set of Data. Picture a researcher operating a remote device used to place small canisters containing material into a nuclear reactor used for research. The material is lowered into a flux of neutrons to be irradiated. The material absorbs the neutron energy, but later gives up the energy by emitting gamma rays hundreds of times per second. A graph showing the material's pattern of energy change, E versus time t, is in ***Figure 23-62***. The detention period starting at time $T1$ helps the researcher obtain gamma ray data to be used to determine the

technique to find an intercept for $H = 0$, as shown in Figure 23-61. Upward from P to R and then parallel to the original line from R to B, in effect adds an additional cycle so the intercept can be found: it is $B = 2.5$. Finally, the equation can be written as $Q = 2.5(H^{**}2.492)$ for values of H between 2 feet and 10 feet only. Further measurements should be taken below 2 feet.

H	Q (Observed)	Q (Calculated)	Residual R
2	14.2	14.06	+0.14
3	41.5	38.63	+2.87
4	80.0	79.12	+0.88
5	135.2	137.97	−2.77

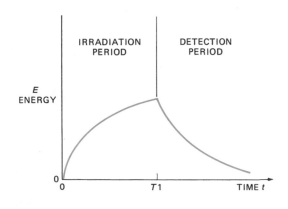

FIGURE 23-62   *Irradiation of sample: Energy level*

half-life of the irradiated material. The data recorded from a gamma ray detector was:

N (counts per second)	2000	210	40	7
t (time in seconds)	0	900	1640	2450

For this set of data, the researcher knew that the decay rate (gamma rays emitted) would be directly proportional to the energy level of the irradiated material. Physical phenomena of this type behave exponentially as $y = B(e^{**}mx)$, so a log-linear grid was selected without bothering to check the data on linear coordinate paper.

1. Follow the general approach in Figure 23-53, starting at the straight-line plotting on a log-linear grid. Use a commercial grid, semilogarithmic, four cycles. *Figure 23-63* shows the plot plus a faired-in best fit straight line. The form of the equation of this line is $\log y = \log B + m \log e\, x$, where the coefficient of x is $m \log e$.

2. Use the slope-intercept and selected points methods for matching an equation to the plot of the data. The straight-line equation for the variables N and t is $\log N = \log B + m \log e\, t$.

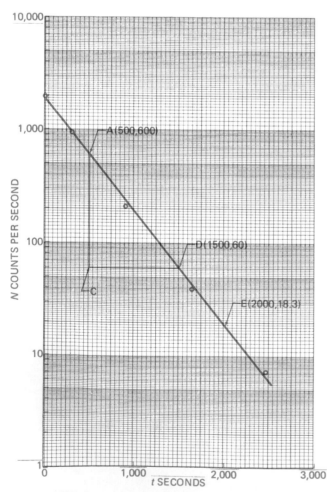

FIGURE 23-63 *Irradiaton data on semilogarithmic grid*

a. Slope-intercept method

  (i) Select points A(500, 600) and D(1500, 60) on the best-fit line.

  (ii) Determine the slope m from

$m \log e =$
$(\log 600 - \log 60)/(500 - 1500)$
$m(.4343) = (2.77815 - 1.77815)/(-1,000)$
$m = -(1)/((1,000)\,(.4343))$
$= -.002303$ counts per second.

  (iii) Then the equation is $\log N = \log B - .002303\,(\log e)t$. The B intercept from the graph is 1,900, so the final equation is

$N = 1,900(e^{**} - .002303\, t)$.

  (iv) The time elapsed to arrive at the half-life $t(HL)$ value of the counts can now be calculated by noting that one-half of the initial count equals one-half of 1,900; therefore,

$1,900/(2) = 1,900\, e^{**} - .002303t(HL)$

and $\tfrac{1}{2} = e^{**} - .002303\, t(HL)$.

Thus, $\log 1 - \log 2 = -.002303(\log e)t\,(HL)$

and $t(HL) = 300.9$ seconds.

The straight-line plot confirms this value of 950 counts per second at approximately 300 seconds.

  (v) Calculate and sum the residuals.

t seconds	N observed	N calculated	Residuals R
0	2,000	1,900	+100
900	210	239.1	−29.1
1640	40	53.5	−3.7
2450	7	6.7	+.3
		SUM R =	67.5

b. Selected points method

  (i) Two pairs of coordinate points from the best-fit line are A(500, 600) and E(2,000, 18.3).

  (ii) Substitute them in $\log N = 1,900 - .002303 (.4343)t$ and solve simultaneously:

$\log 600 = \log B + m\,(.4343)\,(500)$
$\log 18.3 = \log B + m\,(.4343)\,(2,000)$

so $m = -.00233$ counts per second.

Then, to solve for B:

$\log 600 = \log B - .00233\,(\log e)\,500,$
$B = 1,923.6$ counts per second.

  (iii) Finally

$\log N = \log 1,923.6 - .00233\,(\log e)\,t$
so $N = 1,923.6(e^{**} - .00233t)$

(iv) Calculate residuals next to verify the derived equation.

Note how the straight-line equation formed the basis for handling all three sets of data. Additional plotting techniques and analytical approaches are discussed next.

Curve Problem. The goal in the data problem was to find an equation to fit all of the data of an experiment, over the range of the data. The equations that were derived showed the relationships of dependent to independent variables such as $k(a) = -.0017S + .71$, $Q = 2.5(H^{**}2.492)$, and $N = 1,900(e^{**} - .002303t)$—linear, power, and exponential equations, respectively. Finding the proper coordinate grids so that the data plotted in approximately a straight line was the key to deriving the equations.

A number of additional approaches may be used to match equations to curves. One is similar to the selected points technique already discussed, but the basic equation could be third order, such as $y = ax^{**}3 + bx^{**}2 + cx + d$, which requires four sets of selected points. Sometimes the entire curve may be described by a number of equations, piecewise, in series. Still another method describes a curve that may touch the first and last points but only goes near the others. These types of curves are introduced in this section for three main reasons.

1. They require graphical-physical explanations to visualize the concepts.

2. Personal computers make it possible for students and designers to do this type of curve matching and curve generating in a relatively short time.

3. The trend in one type of manufacturing uses these approaches to generate curved surfaces and the related equations, which are then used to program numerically controlled milling machines to cut the surfaces.

Two physical models help to visualize the following curve types:

1. Splines

   a. *Figure 23-64* shows the technique traditionally used by ship builders (before computers) to draw smooth contour lines through a series of points. The "ducks" shown in the figure weigh about five pounds, are cast iron with felt glued to their bottoms, and are used to apply pressure at specified points along a thin metal strip called a *spline*. These pressure points are placed over the points that define a contour line of a ship. Because of the stiffness of the spline, the curve going into a point is exactly in-line with the curve leaving the point. Distances between points are called spans.

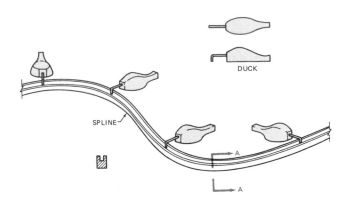

FIGURE 23-64 *Spline*

   b. Cubic splines can be generated which in effect take the place of the metal spline. These equations, one for each span, not only describe the curve over the span but ensure that the span segments are in-line. Software is available to generate cubic splines on personal computers.

2. Bezier (pronounced bay-zee-ay) and B-Splines

   a. Bezier curves: Assume that the thin metal strip shown in *Figure 23-65* is magnetic and without the ducks. The strip is placed on a flat surface between a number of point magnets that pull on the strip but never touch it. The strip is anchored to the end points of a series of points: all points except the ends are the magnets. At the ends, the strip is tangent to a line from the end point to the nearest point. Also, if one magnet along the strip moves, the whole strip may move slightly.

   Designers use Bezier curves in conjunction with a computer screen to generate pleasing curves for the design of automobile bodies. Once the curve satisfies the designer, an equation is automatically generated to match the curve. Ultimately surfaces are generated using curves in horizontal and vertical planes.

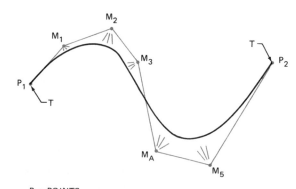

P = POINTS
M = MAGNETS
T = TANGENT POINTS

FIGURE 23-65 *"B-spline" type curve*

One disadvantage of the Bezier curve-generating procedure is that when one point moves, the entire curve can be affected. Usually all but a portion of a curve is OK, and just changing a single portion is desirable.

b. B-Splines: The B-spline curve-making technique allows a single portion of a Bezier-type curve to be modified without affecting the entire curve. Therefore, automobile body designers find that the B-spline technique saves considerable time in designing contour lines and surfaces. Software is also available for generating B-splines on personal computers.

One application of these curve-generating techniques illustrates a trend in some machining operations. Shoe lasts (full-scale models of the left and right feet for a given shoe size) are completely described by spline-type equations. The equations are automatically coded into language that instructs numerically controlled machines to cut the lasts. Thus, no drawings are needed to communicate the information from the designer to the machine.

Conclusion. This curve problem discussion was included in empirical equations to emphasize that the computer is not only affecting the process of making drawings in CAD systems, but it is also being used to help designers to generate graphical curves, surfaces, and models.

The next section contains discussions on graphical calculus, integrating, and differentiating that are also useful to designers.

GRAPHICAL INTEGRATION AND DIFFERENTIATION: POLE AND RAY METHOD

The advantages of knowing how to do graphical calculus, integrating, and differentiating are twofold: (a) they enhance insight and understanding of analytical integrating and differentiating, and (b) they provide a means of handling data and some equations, which would be too difficult and expensive to do analytically. Furthermore, graphical calculus, as do other graphical techniques, provides you with relatively rapid visual techniques for designing, analyzing, and checking. Therefore, this section focuses on the graphical techniques for integrating and differentiating using the *pole and ray method.*

1. First, consider two fundamental ideas, one for integrating, the other for differentiating.

    a. *Figure 23-66* illustrates a graphical explanation of the process of integrating and the resulting integral curve.

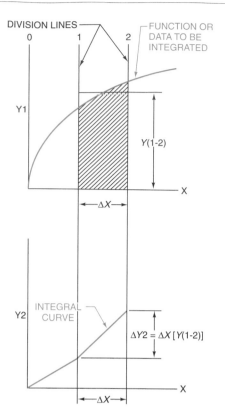

FIGURE 23-66   *Graphical integration: Fundamentals*

  (i) The area between the curve and the horizontal axis in the Y1 versus X graph is divided into segments by vertical division lines 1 and 2.

  (ii) The cross-hatched area between the division lines 1 and 2 represents the change of the integral curve in the vertical direction between the same two division lines. The area shown is above the X axis, so it is positive; therefore, the change in the integral curve is upward between the same two division lines.

  (iii) Items (1) and (2) above restated in the symbols shown would be: the average height of the area, $Y(1-2)$, times the distance between division lines 1 and 2, Δx, equals $\Delta Y2$ of the integral curve between the same two division lines in the Y2 and X graph, i.e., $\Delta Y2 = \Delta X[Y(1-2)]$.

  (iv) Vertical scales Y1 and Y2 are different, but in-line; both horizontal scales are identical.

b. *Figure 23-67* shows a graphical explanation of differentiation and the resulting derived curve.

  (i) Tangents to the curve in the Y1 versus X graph are constructed by eye at the three tangent locator points, and the slopes of

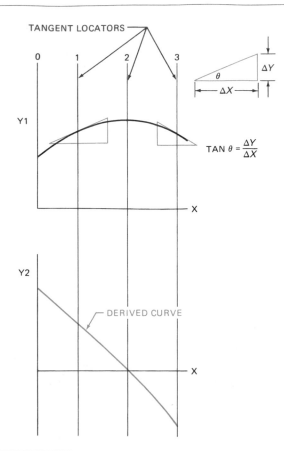

FIGURE 23-67 *Graphical differentiation: Fundamentals*

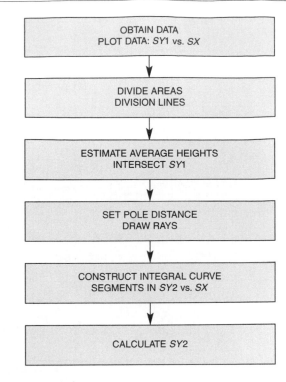

FIGURE 23-68 *Graphical integration: Procedures*

the tangents are determined by a $\Delta Y/\Delta X$ calculation.

(ii) The positive, zero, and negative slopes from points 1, 2, and 3, respectively, are plotted in the Y2 versus X graph.

(iii) Items (1) and (2) restated in the symbols shown would be: the slopes $\Delta Y/\Delta X$ at specific points in the Y1 curve are represented by proportionate vertical distances in the Y2 versus X graph.

(iv) Vertical scales Y1 and Y2 are different but in line: the horizontal scales are identical.

2. Graphical Integration: The overall procedure to do graphical integration follows the flow chart in *Figure 23-68*. Most of the procedure entails graphical construction steps: the last step, deriving a scale, is both analytical and graphical.

   a. Refer to *Figure 23-69* and follow the steps to do the graphical construction. (*Note:* (i) is the circled number 1 in the figure, etc.)

      (i) Plan space for two sets of axes, the upper space for SY1 versus SX and the lower for SY2 versus SX. For illustration purposes, data from an aircraft launch is used. The

following data for acceleration in g's versus time t during the launching cycle for an airplane from an aircraft carrier was recorded in a control center in the carrier.

g's Meters, per sec./per sec.	t seconds
0.00	0.0
0.50	0.2
1.00	0.3
2.55	0.5
4.60	0.6
5.20	0.7
5.40	0.8
5.26	1.0
4.95	1.3
4.60	1.6
4.00	2.0

(ii) Plot the data on a commercial grid 10 × 10 to the inch, and fair-in a best fit curve. Use vertical scale SY1 equal to 2 g/in., and the horizontal scale SX equal to 0.5 s/INCH. (The capital letters, INCH, represent the basic unit of distance on the grid.)

(iii) Use division lines to divide the area between the curve and the horizontal axis. Use more lines during curving portions of the curve for better accuracy. Label the lines 1, 2, 3, 4, 5, and 6, for this plot.

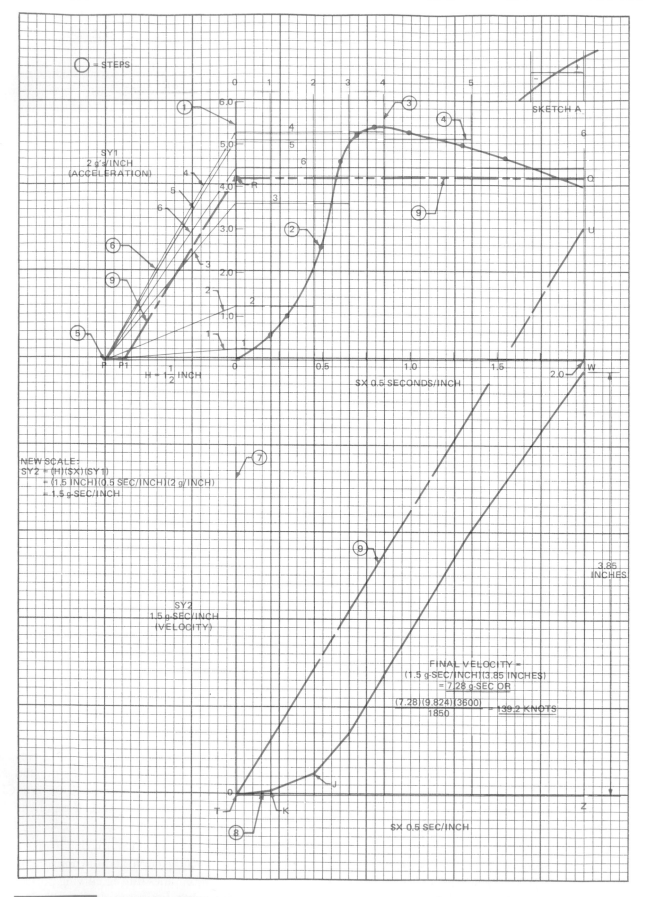

FIGURE 23-69 *Graphical integration: Steps*

(iv) Estimate by eye the average height of each division of area. Sketch A in Figure 23-69 shows that a line for average height is approximated when the negative area is about equal to the positive area. Extend a line to the SY1 axis for each average height, and label each extension.

(v) Set the pole distance H equal to 1.5″ as an extension of the horizontal axis SX, to the left, to point P. The distance H affects the slope of the derived curve: short pole distances cause the curve to be steep; long distances, flat.

(vi) Construct rays from the extension line intersections on axis SY1 to the pole point P.

(vii) Construct and label the second set of axes, SY2 versus SX, so SY2 is aligned with SY1. The calculation for SY2 is discussed after the steps for the graphical construction.

(viii) Start at point T on the SY2 versus SX axes and construct a line TK parallel to the ray associated with the area between the SY1 axis and division line 1. The height of point K above the horizontal axis represents the area between the SY1 axis and division line 1. From K draw a line KJ parallel to the ray associated with the area between division lines 1 and 2. Continue similarly for the remaining rays to point W. Finally, the distance WZ times the scale SY2 represents the total area between the SY1 axis and division line 6.

(ix) To establish a workable pole distance, the following approach works. Construct a trial H distance, say to P1, shown in Figure 23-69. Estimate the average height of the entire area between the SY1 axis and division line 6, which would be line RQ. Draw ray R-P1 and construct a line TU parallel to R-P1 and construct a line TU parallel to R-P1 in the SY2 versus SX graph. Point U indicates the approximate location of where the integral curve will probably terminate; that is, where point W probably will be.

(x) Calculate the scale SY2 for the integral curve graph using the equation $SY2 = (H)(SX)(SY1)$, which is derived as follows:

- *Figure 23-70* shows SY1 versus SX with SY2 versus SX directly below. The first curve is plotted and the area is divided by division lines 1 and 2. *FD* represents the average height of the area between

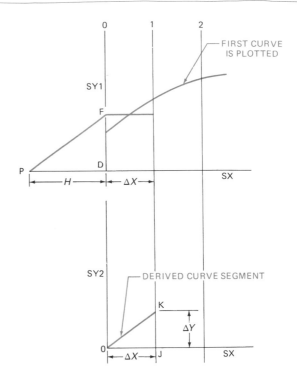

FIGURE 23-70   *Graphical integration: Scale derivation*

the SY1 axis and division line 1. FP is the ray associated with the area. H is the pole distance.

- Line OK in the SY2 versus SX graph is constructed parallel to ray FP, and $KJ = \Delta Y$.

- The desired result of this graphical integration construction is that $(\Delta Y)(SY2) = (FD)(SY1)(\Delta X)(SX)$, which says generally that the area between any two division lines represents the change in height of the integral curve between the same two division lines. Note how the scales must be used in the equation to obtain the correct values.

- Geometrically from similar triangles:

$$\frac{FD}{PD} = \frac{KJ}{OJ} = \frac{\Delta Y}{\Delta X} \text{ and } PD = H$$

$$\text{so,} \quad \frac{FD}{H} = \frac{\Delta Y}{\Delta X}; \text{ then } \Delta Y = \frac{(\Delta X)(FD)}{H}$$

Substitute ΔY in the equation from the preceding step to get

$$\frac{(\Delta X)(FD)}{H} \bullet (SY2) = (FD)(SY1)(\Delta X)(SX)$$

Cancel like quantities and obtain

$$SY2 = (H)(SX)(SY1)$$

b. An additional capability of graphical integration is to divide odd-shaped areas and volumes into equal portions. The following example demonstrates this capability.

   (i) A right-circular-cone-shaped water tank is used for this example to convince the reader that the method works, since the graphical results can be checked analytically. After following this example it should be easy to imagine an odd-shaped fuel tank nestled in a wing section of an airplane and how the tank could be graduated into equal volumes.

  (ii) *Figure 23-71* shows the water tank that needs to be graduated and marked at 3/4, 1/2, and 1/4 FULL.

 (iii) Cross-sectional areas of the 15-foot-tall, 10-foot-diameter tank were calculated at 3-foot intervals and recorded here:

Station S in Feet	0.	3.	6.	9.	12.	15.
Area A (square feet)	0.	3.14	12.6	28.3	50.2	78.5

  (iv) *Figure 23-72* shows how the steps to do graphical integration were followed to plot the data to scale, divide the areas with division lines, estimate the average heights, and transfer those heights to the vertical axis, locate the pole point, draw the rays, and generate the integral curve.

   (v) Distance WZ times $SY2$ scale of 125 cubic feet/INCH equals 390 cubic feet.

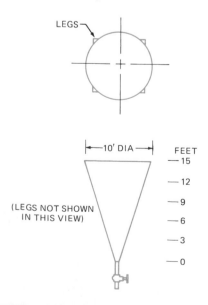

LEGS

10' DIA

(LEGS NOT SHOWN IN THIS VIEW)

FEET
— 15
— 12
— 9
— 6
— 3
— 0

FIGURE 23-71 *Water tank*

  (vi) To obtain the graduations, WZ was divided into four equal lengths: the divisions were then transferred horizontally to the integral curve and, finally, these intersections were projected vertically down to the horizontal axis.

The intercepts on the horizontal axis represent the location along the tank height to mark the ¼, ½, and ¾ FULL, 9.5', 12.2', and 13.5', respectively. Now the following comparison can be made with the analytical solution.

	¼	½	¾	FULL
GRAPHICAL	9.5'	12.2'	13.5'	15'
ANALYTICAL	9.45'	11.91'	13.63'	15'

The close correlation demonstrates the efficacy of this technique to graduate areas or volumes using graphical integration.

3. Graphical Differentiation: Graphical differentiation uses slopes at selected points in a curve instead of adding the areas under the curve as was done in graphical integration. Use the steps in the flowchart shown in *Figure 23-73* as a guide for following the graphical differentiation example discussed next.

   a. The cam shown in *Figure 23-74* (page 915) provides an excellent example of the application of graphical differentiation to visualize the operation of the cam and follower, and to check the cam's operation. Cams convert the rotary motion of a camshaft to reciprocating motion for the follower. If the cam profile changes too rapidly, the follower may separate from the cam surface because of excessive acceleration—an undesirable condition. Therefore, designers need to check their cam profile designs for acceleration problems. Refer to Figure 23-74 for the steps to do a graphical differentiation of the proposed displacement of the follower versus angular positions of the cam. The first differentiation will generate a curve showing the velocity of the follower at any time during the first half of one rotation: the second half is to be a mirror image of the first half. Differentiation of the velocity curve produces an acceleration curve for the follower.

      (i) Prepare space on a commercial metric grid for three sets of axes, the first set in the upper space, for follower displacement versus time. Follower displacement was calculated at every 15° rotation of the cam, which rotated at a constant speed of one revolution every four seconds. The time

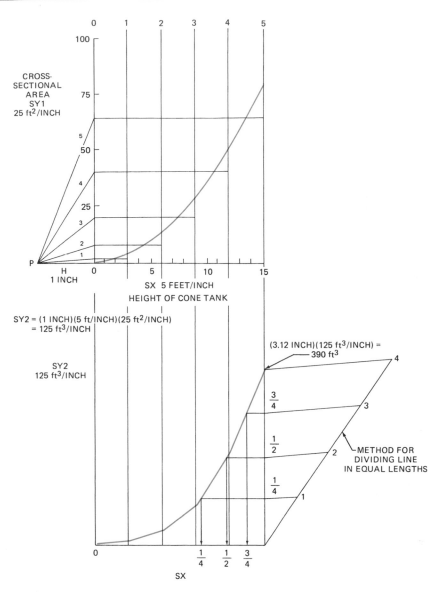

FIGURE 23-72 *Graduating volumes*

elapsed at each 15° was also calculated. The results were as follows:

Vertical Displacement

S in mm	0	.35	1.39	3.13	5.56	8.68

Time in fractions of a second

	0	1/6	1/3	1/2	2/3	5/6

12.5	16.32	19.44	21.87	23.60	24.65	25
1	1-1/6	1-1/3	1-1/2	1-2/3	1-5/6	2

(ii) Plot the results and fair-in a best-fit curve. Use a vertical scale SY1 equal to 0.5 mm/MM and a horizontal scale SX equal to 1/60 second/MM. (The capital letters MM represent the basic unit of distance on the metric grid.)

(iii) Use tangent locators, vertical lines, to intersect the plotted curve at selected points; place more locators where the curve changes rapidly, for better accuracy. Label them 1, 2, 3, 4, 5, and 6 for this example.

(iv) Estimate by eye the slope of the curve, a tangent, at each locator intersection. Label the tangents also.

(v) Set a pole distance $H = 30$ MM as an extension of the horizontal axis SX of the second set of axes. The distance H affects the height of the derived curve: longer pole distances cause higher curves; shorter distances, lower curves.

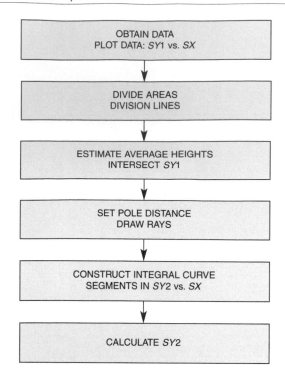

FIGURE 23-73 *Graphical differentiation: Procedure*

(vi) Construct rays parallel to each tangent, from point P to the vertical axis SY2. Label the rays.

(vii) Extend the tangent locators into the area of SY2 versus SX. Then extend horizontal lines from the ray intercepts on the SY2 axis to intersect the related tangent locator.

(viii) Fair-in a best-fit curve, or line, through the intersections on the tangent locators.

(ix) To help establish a workable pole distance, try the following: Determine the steepest slope of the curve being differentiated, and use the ray parallel to it to set a pole distance. Tangent 3 is the steepest in this example. A trial pole location of $H = 60$ MM causes the ray PR to go too high. A second trial $H = 30$ MM was workable.

(x) Calculate the SY2 scale using the equation derived as follows:

$$SY2 = \frac{SY1}{(H)(SX)}$$

• *Figure 23-75* shows SY1 versus SX, with SY2 versus SX directly below. The curve is plotted on axes SY1 versus SX, and then vertical, tangent locators are drawn.

• The slope, tangent to the curve, at location 1 is calculated by $\Delta Y/\Delta X$ from the triangle shown.

• Ray PF is drawn parallel to the slope at locator 1, and a horizontal line is drawn from F to locator 1, point G.

• The desired result of this graphical-differentiation construction is that

$$(GK)(SY2) = \frac{(\Delta K)(SY1)}{(\Delta X)(SX)}$$

From similar triangles:

$$\frac{\Delta Y}{\Delta X} = \frac{FD}{H} = \frac{GK}{H}$$

Substitute and factor.

$$(GK)(SY2) = \frac{(GK)}{(H)} \cdot \frac{(SY1)}{(SX)}$$

$$SY2 = \frac{(SY1)}{(H)(SX)}$$

which is the derivation of the new scale. SY2 for this example is

$$SY2 = \frac{(.5 \text{ mm/MM})}{(30 \text{ MM})(1/60 \text{ sec/MM})}$$

$$SY2 = 1 \text{ mm/second/MM}$$

which is a velocity scale as expected.

(xi) To finish the example, that is, to derive the acceleration curve, repeat the procedures: set axes SY3 versus SX, set a pole distance, determine the slopes, draw rays parallel to the slopes, fair-in the curve, and calculate SY3.

$$SY3 = \frac{1 \text{ mm/second/MM}}{(40 \text{ MM})(1/60 \text{ sec/MM})}$$

$$SY3 = 1 \text{ mm/second/MM}$$

an acceleration as expected.

(xii) The cam profile does cause constant acceleration and deceleration, as shown in Figure 23-75. Their magnitudes are calculated from measurements on the graph to be

acceleration =
$(16 \text{ MM})(1.5 \text{ mm/sec}^2/\text{MM}) = 24 \text{ mm/sec}^2$

deceleration =
$(16.2 \text{ MM})(1.5 \text{ mm/sec}^2/\text{MM}) = 24.3 \text{ mm/sec}^2$

b. Graphical differentiation is more difficult to do accurately than graphical integration because of the possibility of errors in determining the slopes of curves by eye. However, doing and understanding graphical differentiation helps one to visualize that the slope of the higher-order curve at a specific locator is represented by the height from the horizontal axis to the derived lower-order curve at the same locator.

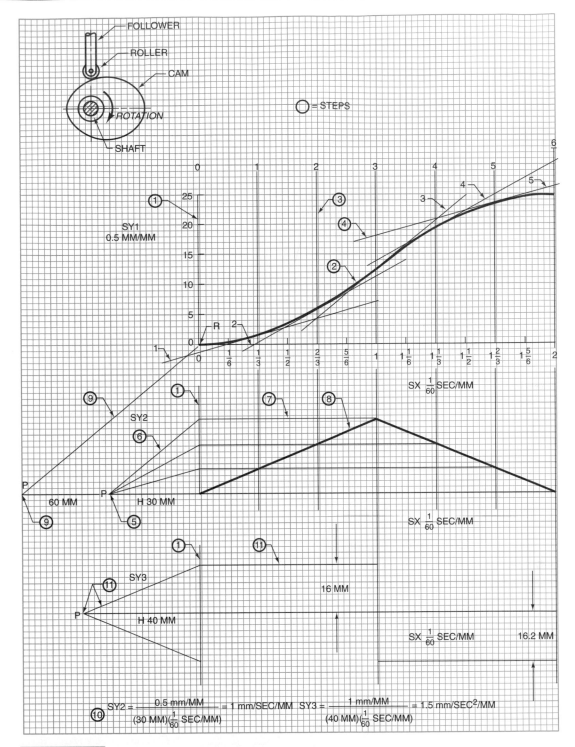

FIGURE 23-74 *Graphical differentiation: Steps*

4. The following are concluding comments on the usefulness of graphical integration.

   a. Some functions such as

   $$\int_0^\pi \sqrt{1 + Cos^2X}\ dx$$

   that have no analytical integral can be integrated using graphical integration. Numerical methods can also be used.

   b. Profiles of cuts through hilly terrains may be impractical to express analytically, so volumes of earth can be calculated. Graphical integration is used. Planimeters, mechanical integrators, are also used.

5. The concepts of graphical differentiation and integration provide a unique insight into analyzing beam diagrams, which are considered in the next section.

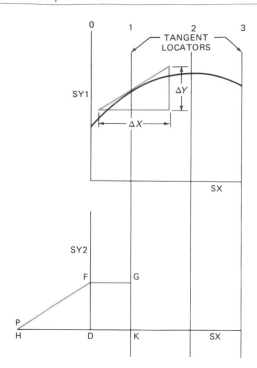

FIGURE 23-75 *Graphical differentiation: Scale derivation*

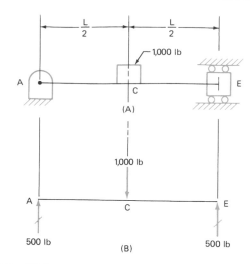

FIGURE 23-76 *(A–B) Centrally loaded beam*

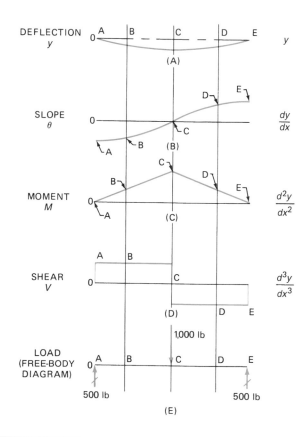

FIGURE 23-77 *(A–E) Centrally loaded beam: Beam diagrams*

BEAM DIAGRAMS

Most students using *Technical Drawing* will have had only a brief introduction to the mechanics of materials, so they may not have been introduced to beam diagrams. However, a brief introduction in this section to beam diagrams related to graphical differentiation and integration should help them understand beam theories in future courses. Review the graphical calculus concepts in the preceding section before perusing this section.

1. Several preliminary concepts:

   a. The centrally loaded beam in *Figure 23-76A* has supports at ends A and E. At A, a symbol shows that the beam is pinned to a support that is fixed to a rigid frame or structure. "Whiskers" represent the rigid frame or structure.

   b. The FBD (free-body diagram) in *Figure 23-76B* is drawn in the traditional simplified-model form used in beam analyses. In this example, the weight of the beam is considered negligible compared to the effect of the applied load, 1,000 pounds. Vectors representing the reactions at A and E, and the applied load at C, are all parallel: Their algebraic sum must equal zero for the beam to be in equilibrium.

2. The 1,000-pound load causes the beam to deflect: the maximum deflection occurring at C is shown in *Figure 23-77A*. This deflection curve is usually the last beam diagram to be derived in beam analyses. However, the deflection diagram is considered first

here to illustrate how graphical differentiation principles apply in deriving beam diagrams.

   a. Differentiate the deflection curve to obtain the slope curve shown in *Figure 23-77B*. From left to right, the slope curve indicates a large negative slope at A, less negative at B, zero at C, positive at D, and larger positive at E. The deflection anywhere along the beam is represented by the variable *y*; slopes are represented by *dy/dx*.

b. The next derived curve is called a moment dia-gram, *Figure 23-77C*. The moment diagram shows that the slope curve has changing positive slopes throughout, and a maximum slope at C. The maximum moment in this beam also occurs at C. A moment about a point is defined as *the force times the shortest distance from the point to the line of action of the force.* Therefore, the vary-ing moment, zero at A and E, and increasing to the maximum at C, varies as a function of the distance from the reactions at A and E. Moment diagrams are second-order derivatives.

c. Finally, in *Figure 23-77D* the shear diagram appears as the derivative of the moment dia-gram. Since the moment diagram has two linear portions, one positive sloped and one negative, the shear diagram appears as two horizontal lines, one above the axis and one below. Shear diagrams are third-order derivatives.

3. Follow the progression of diagrams in *Figure 23-78* for the same beam; however, this time the shear dia-gram is considered first, which is the usual progres-sion in beam analyses.

a. Before constructing the shear diagram, look at the FBD. Upward forces are positive; downward, negative. Start at A and consider the beam from left to right. Figure 23-78B shows the shearing tendency caused by the reaction force at A. This tendency, for this beam, remains constant along the beam until a change in loading occurs at C. From C onward, the effect of the reaction at A plus the larger downward load at C causes the tendency for shear to be downward, and con-stant through to E.

b. The moment diagram in Figure 23-78C, going from left to right, shows the moment increasing as X increases. The conceptual sketch at right isolates a portion of the beam and shows it in

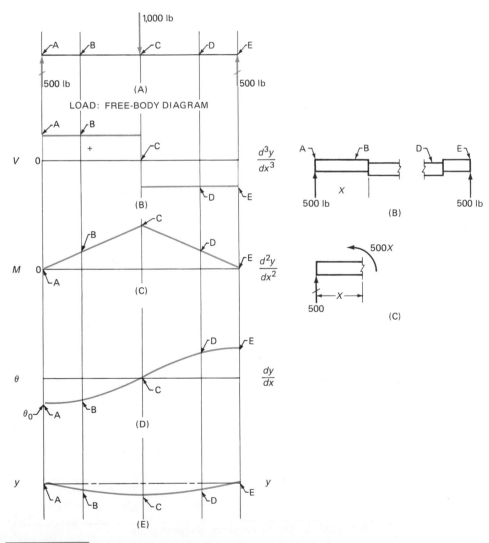

FIGURE 23-78 *(A–E) Centrally loaded beam: Beam diagrams*

equilibrium, because the 500X is balanced by the reaction force of 500 pounds times the moment arm X.

c. Consider the order of the graphical curves compared to the analytical derivatives.

(i) The moment diagram is the integral of the shear diagram. Since the shear diagram has a zero-order slope, the moment diagram is a first-order curve.

d. Figure 23-78D shows the slope diagram, integral of the moment diagram. The slope diagram starts

BEAM DIAGRAMS AND FORMULAS
For various static loading conditions

Equivalent Tabular Load is the uniformly distributed load given in beam tables, pages 2 - 28 to 2 - 81.
For meaning of symbols, see page 2 - 196.

7. SIMPLE BEAM—CONCENTRATED LOAD AT CENTER

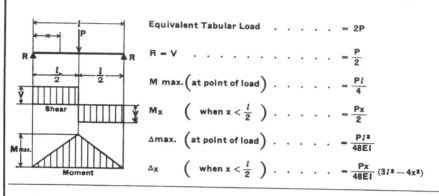

Equivalent Tabular Load . . . . . . $= 2P$

$R = V$ . . . . $= \dfrac{P}{2}$

M max. $\left(\text{at point of load}\right)$ . . . $= \dfrac{Pl}{4}$

M_x $\left(\text{when } x < \dfrac{l}{2}\right)$ . . . . . $= \dfrac{Px}{2}$

Δmax. $\left(\text{at point of load}\right)$ . . . $= \dfrac{Pl^3}{48EI}$

Δ_x $\left(\text{when } x < \dfrac{l}{2}\right)$ . . . . . $= \dfrac{Px}{48EI}(3l^2 - 4x^2)$

8. SIMPLE BEAM—CONCENTRATED LOAD AT ANY POINT

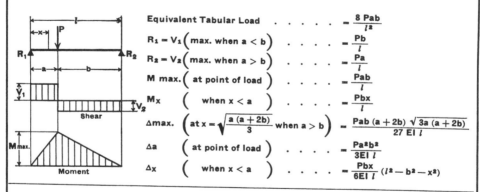

Equivalent Tabular Load . . . . . $= \dfrac{8Pab}{l^2}$

$R_1 = V_1 \left(\text{max. when } a < b\right)$ . . . . $= \dfrac{Pb}{l}$

$R_2 = V_2 \left(\text{max. when } a > b\right)$ . . . . $= \dfrac{Pa}{l}$

M max. $\left(\text{at point of load}\right)$ . . . . $= \dfrac{Pab}{l}$

M_x $\left(\text{when } x < a\right)$ . . . . $= \dfrac{Pbx}{l}$

Δmax. $\left(\text{at } x = \sqrt{\dfrac{a(a+2b)}{3}} \text{ when } a > b\right) = \dfrac{Pab(a+2b)\sqrt{3a(a+2b)}}{27\,EI\,l}$

Δa $\left(\text{at point of load}\right)$ . . . . $= \dfrac{Pa^2b^2}{3EI\,l}$

Δ_x $\left(\text{when } x < a\right)$ . . . . $= \dfrac{Pbx}{6EI\,l}(l^2 - b^2 - x^2)$

9. SIMPLE BEAM—TWO EQUAL CONCENTRATED LOADS SYMMETRICALLY PLACED

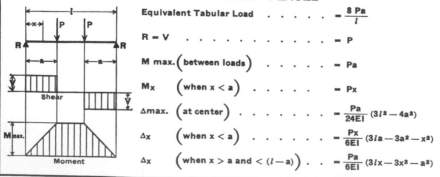

Equivalent Tabular Load . . . . . $= \dfrac{8Pa}{l}$

$R = V$ . . . . $= P$

M max. $\left(\text{between loads}\right)$ . . . . . $= Pa$

M_x $\left(\text{when } x < a\right)$ . . . . . $= Px$

Δmax. $\left(\text{at center}\right)$ . . . . . . $= \dfrac{Pa}{24EI}(3l^2 - 4a^2)$

Δ_x $\left(\text{when } x < a\right)$ . . . . . $= \dfrac{Px}{6EI}(3la - 3a^2 - x^2)$

Δ_x $\left(\text{when } x > a \text{ and } < (l-a)\right)$ . . $= \dfrac{Pa}{6EI}(3lx - 3x^2 - a^2)$

AMERICAN INSTITUTE OF STEEL CONSTRUCTION

FIGURE 23-79 *Beam diagrams: Handbook* (Courtesy American Institute of Steel Construction)

with an initial value of slope at A. This value must be found either analytically or graphically.

  e. Finally, Figure 23-78E shows the deflection curve, the integral of the slope curve. (Moment curves and deflection curves are the ones designers use most: the slope curve is just the intermediate step.) The deflection curve looks as the real beam would with a concentrated load at the center.

4. Compare the beam diagrams shown in *Figure 23-79* and compare them with the preceding discussion. Shear and moment diagrams are shown, but deflections are given as equations only. These are typical of the diagrams that appear in handbooks and texts.

Conclusion

After studying this chapter most readers should agree that communicating information is the popular use of charts and graphs in a variety of media, including textbooks, periodicals, newspapers, and television. However, generating information is the most useful function of charts and graphs for designers.

Summary

- The five basic parameters for preparing effective graphs and charts are grid, scales, labels, plot, and title. The two main types of graphs and charts are those for communicating information and those for generating information.
- Optimization involves achieving the best possible design within the constraints of resources, budget, and other limitations.
- Charts and graphs for generating information include vector polygons, alignment charts, concurrency charts, matching equations to empirical data, graphical differentiation and integration, and beam diagrams.

Review Questions

Answer the following questions either true or false.

1. Similar triangles form the basis for almost all alignment chart designs.

2. An alignment chart is a form of nomogram.

3. A vector is a graphical symbol that represents direction by means of an arrowhead at one end of a straight line whose scaled length represents a magnitude.

4. The equilibrium principle as used in the study of mechanics is: $\Sigma F = 0$.

5. Two-force members are usually found in structures, frames, trusses, and flexible members in tension.

6. FBD stands for free-body diagram.

7. Optimization is a concise statement of maximum requirements.

8. Yield strength means a tolerance that controls the position of a feature.

9. A functional scale is a line graduated and labeled according to a functional relationship.

10. A spline is a thin metal strip.

11. Two related variables plotted on opposite sides of similar triangles make up an adjacent alignment chart.

Answer the following questions by selecting the best answer.

1. Which of the following is **not** one of the five basic components of charts and graphs?
   a. The grid
   b. The scale
   c. The plot
   d. The symbols

2. Which of the following is **not** true regarding a nomograph?
   a. It saves time
   b. It allows the user to visualize ranges of possible values
   c. Creating one is very time consuming
   d. t is a graphical representation of formulas along straight or curved, graduated lines

3. Which of the following is the **correct** definition of an alignment chart?
   a. The graphical representation for a number of values in a particular area
   b. The graphical representation of two or more variables
   c. A line labeled and graduated according to its function
   d. A chart designed to enhance the reader's ability to interpret information

4. The correct definition of an isopleth is:
   a. Lines used to locate values in a nomogram.
   b. Lines used to locate values in similar triangles.
   c. Lines used to locate values in functional scales.
   d. Lines used to locate values in force polygons.

5. Which of the following is the first step in the short method of designing a parallel alignment chart for $E = IR$?
   a. Do two different calculations for E.
   b. Locate $R = 1, 10, 100,$ and $1,000$ ohms on the left side of the sheet.

c. Connect *I* and *R* values.

d. Locate 1 = 0.1, 1.0, and 10 amps on the left side of the sheet.

6. Which of the following is not one of the charts used to communicate information?
   a. Pie chart
   b. Log-linear chart
   c. Profile chart
   d. Bar chart

7. Which of the following is **not** one of the graphs used to generate information?
   a. Trilinear
   b. Log-linear
   c. Linear-coordinate
   d. Log-log

Chapter 23 Problems

Problems for this chapter are divided into two main categories: To communicate information and to generate information. A third group contains a miscellaneous collection of related problems. Refer to appropriate sections of the chapter for guidelines and suggested steps to prepare the charts and graphs.

To Communicate Information

BASIC GRAPHS (REFER TO THE FIVE BASIC COMPONENTS IN THE CHAPTER)

CAD Instructions

Students who plan to complete the following problems on a CAD system should begin with these general instructions. Before reading the specific instructions for each activity, go through each step in the following planning checklist. The checklist applies to any CAD system and will help ensure the optimum use of your time and resources.

1. Analyze the problem carefully. Decide exactly what you are being asked to do.

2. Determine what resources and references (if any) you will need in order to complete the problem. Collect those references and resources and have them readily available.

3. Decide if any particular standards apply to the project and have those standards available.

4. Determine what types of views will be required and how many of each.

5. Determine what the final plotted scale of the drawing will need to be, and select the appropriate paper size for plotting/printing.

6. Plan your drawing sequence. In what order will you develop the drawing (i.e., lines, features, dimension lines, leaders, dimensions)?

7. Review the various CAD commands you will have to use in order to develop the drawing.

8. Examine your CAD system to ensure that everything is in working order, then begin the project.

PROBLEM 23-1

Prepare plots of BHP (brake horsepower) and T (torque), both as a function of RPM (revolutions per minute), for the following data:

RPM	BHP	T
500	15	190
1,000	40	210
1,500	65	230
2,000	90	232
2,200		240
2,500	115	234
3,000	130	230
3,500	145	220
4,000	150	200
4,500	145	170

Problem 23-2

Electronic parts fail over a period of time in a frequency pattern shaped like a bathtub. Plot the given failure-rate data with failure-rate ratios as a function of time.

Period	Failure-Rate Ratio	Time (hours)
Break-in Period		
Initial Failure Rate	2.00	0
	1.56	100
Constant Failure Rate	1.40	200
	1.28	300
	1.19	400
	1.12	500
	1.05	700
	1.00	1,000
Typical Operating Period		
Constant Failure Rate	1.00	1,000 to 5,000

Wear-Out Period

Wear-Out Failure Rate	1.00	50,000
	1.05	51,000
Constant Failure Rate	1.18	52,000
	1.40	53,000
	1.72	54,000
	2.50	55,000

Problem 23-3

Plot Average Cost in dollars versus Weight of Steel Forgings in pounds for the following:

Cost $	4.20	7.50	15.00	30.00	54.00	115.00
Weight lb	2.0	4.0	10.0	20.0	40.0	100.0

PROBLEM 23-4

Prepare a graph of % power as a function of air-fuel ratio for the data representing a typical internal combustion engine.

% Power	90	100	95	87	75	60	43
Air-Fuel Ratio	10:1	12:1	14:1	16:1	18:1	20:1	22:1

Problem 23-5

During a tension test of a steel bar, the following data was calculated and recorded; prepare a graph of Stress in psi versus Strain (stretching) in inches per inch.

Stress lbs/square inch (psi)	Strain inches/inch
00000	0.00000
3,000	0.00010
5,000	0.00020
10,000	0.00030
15,000	0.00050
20,000	0.00070
25,000	0.00080
30,000	0.00100
36,000	0.00120

Problem 23-6

A convenient expression in strength of materials calculations is the relationship of G (shear modulus) as a function of E (modulus of elasticity). The equation is:

$$G = \frac{E}{2(1+0.3)} = \frac{E}{2.6}$$

where the 0.3 is Poisson's ratio.

Prepare a plot of the following data for a variety of engineering materials:

Material	G psi	E psi
Plastic	1,000,000	400,000
Magnesium	2,300,000	6,100,000
Aluminum	3,800,000	10,000,000
Zinc	5,000,000	12,000,000
Copper	6,000,000	16,000,000
Palladium	6,300,000	17,000,000
Platinum	9,300,000	24,000,000
Steel	11,600,000	30,000,000

PIE CHARTS

PROBLEM 23-7

Prepare a weekly schedule of your activities for 24 hours per day. Show the different activities in a pie chart.

Problem 23-8

Use a pie chart to show the distribution of percentages of the sources of electrical power generation as follows: Coal, 56; Nuclear, 17; Hydro-electric, 12; Gas, 10; and Oil, 5.

PROBLEM 23-9

Show in a pie chart the relative importance of the qualifications of a successful, professional, technical person: Character, 40%; Judgment, 18%; Efficiency, 15%; Understanding People, 14%; and Technical Knowledge, 13%.

Problem 23-10

Select a pie or bar chart from your local newspaper or national business publication and revise it based on the ideas presented in a book by E.R. Tufte titled *The Visual Display of Quantitative Information*. Mr. Tufte contends that charts can, and should, be made much simpler and still convey the required information. Some of his suggestions and observations are:

1. Since desktop computers now have the ability to fill in bound spaces with a variety of patterns, pie charts and bar charts are filled with these patterns—because they are there! Mr. Tufte believes that the patterns only clutter up the charts.

2. Scales should only go as far as the data, as shown in the figure below. The location of zero values is implied.

3. Pleasing-to-the-eye dimensions for the overall chart space are approximately equal to the proportions of the golden rectangle: the height-to-width ratio of 1:1.618. (Most ancient Greek and Roman structures seemed to fit these proportions.)

4. Try to be creative in presenting the data; for example, Mr. Tufte displayed a photograph of people lined up in columns of various lengths to represent a statistical histogram to approximate a normal distribution.

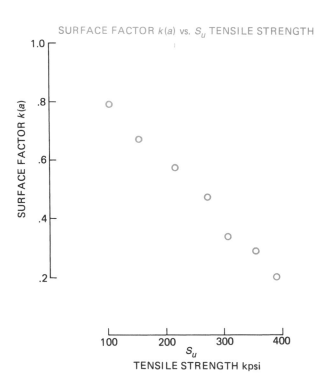

SURFACE FACTOR $k(a)$ vs. S_u TENSILE STRENGTH

BAR CHARTS

Problem 23-11

A local newspaper published data relating to three alternate choices for handling city sewage. The data appeared in tabular form as shown below. Prepare a bar chart to better represent the data.

Monthly Household Sewer Rates (Dollars)

	1995		2005		2030		45 Yr. Avg.	
	Utility Est.	City Est.	Utility Est.	City Est.	Utility Est.	City Est.	Utility Est.	City Est.
Alternative 1	14.95	14.75	9.75	11.45	4.60	6.00	8.85	9.45
Alternative 2	19.70	18.30	11.10	11.45	4.35	6.10	10.45	10.35
Alternative 3	17.65	17.00	10.40	11.20	4.55	6.30	9.75	10.00

PROBLEM 23-12

Prepare a bar chart to represent the population of the United States in Millions versus Year as listed:

Population in Millions	Year
5.3	1800
7.2	1810
9.6	1820
12.9	1830
17.1	1840
23.2	1850
31.4	1860
39.8	1870
50.1	1880
63.0	1890
76.0	1900
92.0	1910
105.7	1920
122.8	1930
131.7	1940
150.7	1950
179.3	1960
203.3	1970
226.5	1980
248.7	1990
360.0 (est.)	2000

PROBLEM 23-13

A recent newspaper article compared a number of baseball pitchers' fastball speeds to speeds of a sneeze, horses, automobiles (freeway speed), jaguars, and horses in MPH. Prepare a creative bar chart which shows the following:

Pitcher/Other	Speed in MPH
A	100
B	98
C	98
D	97
E	96
F	96
G	95
H	95

Pitcher/Other	Speed in MPH
I	94
J	94
K	92
L	88
M	88
N	87
O	86
Sneeze	103
Automobile	65
Jaguar	63
Horse	43

PLANNING CHARTS

PROBLEM 23-14

Construct a planning chart to show the activities required to do an experiment in an eight-hour day. A report also is required for this experiment. The activities necessary to do the experiment and the estimated time for each activity are in the following list. Some activities will overlap.

	Hours
1. Study experiment	.1 hour
2. Set up apparatus	.1-½
3. Take data	.3
4. Plot data	.4-½
5. Write first draft	.1
6. Edit and rewrite on word processor	.1-½
7. Collect parts for report and assemble	.1
8. Put the report in a folder and submit it	.1-½

PROBLEM 23-15

Prepare a planning chart for the activities required to make a float for a parade in your school or community.

 Problem 23-16

Prepare a planning chart for any one of the following:

1. A major repair to an automobile or other major appliance.

2. A field trip for 50 students to a local industry.

3. Manufacturing a product that has several components.

FLOWCHARTS

PROBLEM 23-17

Redraw the rough sketch shown, which shows an example Fault Tree Analysis flowchart. Flow is upward on the page. Lines are either horizontal or vertical, are drawn with a straightedge, and change direction with 90° corners.

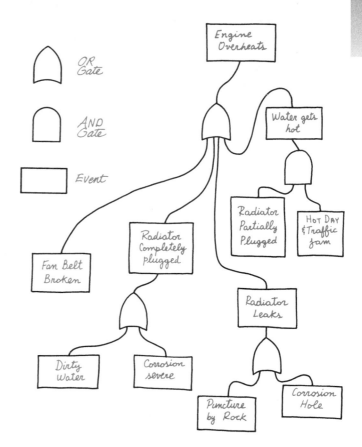

 Problem 23-18

Construct a flowchart of a computer program used in your classes.

SCHEMATICS

Problem 23-19

Using a straightedge and a circle template, redraw the following rough hydraulic power schematic and label each component. Show the directional valve in the position for pushing the piston to the right. Lines in schematics are either horizontal or vertical, and they change direction in 90° corners.

STANDARD GRAPHICAL SYMBOL	EXPLANATION
	Reservoir
	Filter
	Pump
	Pressure Gage
	Spring-loaded pressure relief valve
	Directional valve: hand controlled
	Cylinder: double acting

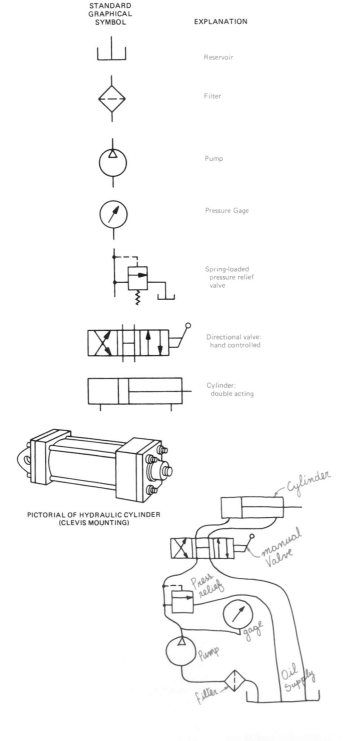

PICTORIAL OF HYDRAULIC CYLINDER
(CLEVIS MOUNTING)

Problem 23-20

Using a straightedge, redraw the rough schematic of an overtone oscillator shown. Generally, signals (voltages, current, and information) travel from left to right; positive voltages are in the upper areas and negative voltages are in the lower.

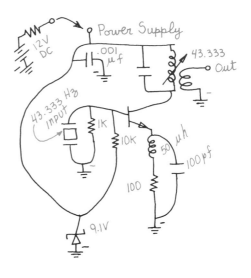

POLAR CHARTS

Problem 23-21

Use a commercial polar grid to plot noise-level values surrounding a jet engine: decibels versus angle.

Angle (Zero Degrees Is North)		Decibels
0	360	20
20	340	76
25	335	85
30	330	93
35	325	97
40	320	99
45	315	96
60	300	92
70	290	91
80	280	94
90	270	95
100	260	96
110	250	95
125	235	100
140	220	90
150	210	95
160	200	90
170	190	82
180	180	65

To Generate Information

NOMOGRAMS: ADJACENT AND NONADJACENT

PROBLEM 23-22

1. Design an adjacent alignment chart for the equation $E = 33,000$ HP, where E is in foot-pounds per minute and HP is horsepower that varies from 2 to 20.
2. Design a nonadjacent chart for the same equation.

Problem 23-23

1. Design an adjacent alignment chart for the equation kW = .746 HP, where HP is horsepower and kW is kilowatts. Let HP vary from 1 to 100.
2. Design a nonadjacent chart for the same equation.

Problem 23-24

1. Design an adjacent alignment chart for the equation 57.3 $R = N$, where R is in radians and N is number of degrees. Let R vary from 0 to 6.28.
2. Design a nonadjacent chart for the same equation.

Problem 23-25

Design adjacent alignment charts for any or all of the following useful conversions:

1. kilograms to pounds
2. pounds to Newtons
3. Pascals to psi
4. Newton-meters to foot-pounds or pound-inches

Problem 23-26

Design an adjacent alignment chart for the conversion of stress in pounds per square inch to MPa, mega-Pascals (Newton per meter squared = Pascal). Let stress vary from 10,000 to 300,000 psi.
Note: 6,895 MPa = 1,000 psi.

NOMOGRAMS: PARALLEL, THREE VARIABLES

Problem 23-27

Design a parallel alignment chart for the equation

$$KE = \frac{(W)\,(V^{**}2)}{(2)\,(g)}$$

where: KE = kinetic energy in foot-pounds
W = weight in pounds
V = velocity in feet/second

Let W vary from 1 to 500 pounds and V from 1 to 100 feet/sec. $g = 32.2$ feet/sec**2.

Problem 23-28

Design a parallel alignment chart for the equation

$$S = \frac{PD}{4T}$$

where: S = circumferential stress in a spherical tank
P = internal gas pressure, psi
D = diameter in inches = 20″
T = thickness of shell

Let P vary from 0 to 150 psi and T from .1″ to .5″.

Problem 23-29

Design a parallel alignment chart for the following helical spring equation:

$$y = \frac{8F\,(D^{**}3)\,N}{(d^{**}4)\,G}$$

where: y = deflection in inches
F = force in lb applied
D = mean diameter of coil in inches = .6″
d = wire diameter in inches = .1″
G = shear modulus 11,500,000 psi
N = number of coils

Let F vary from 0 to 50 pounds and N from 3 to 15.

Problem 23-30

Design a parallel alignment chart for the following shaft design equation:

$$S = \frac{16T}{(3.14)\,d^{**}3}$$

where: S = shear stress, psi
T = torque, inch-pounds (2,000 to 10,000 in-lb)
d = diameter, inches (1 inch to 5 inches)

Note: For a factor of safety of 2, S should not exceed ½ the yield strength of the metal being considered.

Problem 23-31

Design a parallel alignment chart for the following moment of inertia equation for a rectangular cross-sectional member:

$$I = \frac{b\ h^{**}3}{12}$$

where: I = moment of inertia (or actually second moment of area)– – – – –in**4

b = width, inches (0.5 to 3 inches)

h = height, inches (1.5 to 9 inches)

(*I* is a measure of a member's ability to resist bending about an axis parallel to *b* and perpendicular to *h*.)

NOMOGRAMS: CONCURRENCY CHARTS

Problem 23-32

The learning curve premise says that as workers gain experience on producing a new product, they can produce more products in a given length of time. An equation that expresses this improvement in production as learning increases is the following:

$$T = E\ C^{**}m.$$

where: T = hours of production time to produce the cumulated total of batches of products (10 to 1,000)

C = cumulated total of batches of products (1 to 100,000)

E = initial effort in hours to produce the first batch of products (1,000)

m = an exponent that depends on the improvement in learning

For example, assume that the cumulative total of batches of products has doubled, and that the time used for the second half of the double is only 75% of the time used for the first half. Then the exponent *m* is calculated as follows:

$$\frac{.75T}{T} = \frac{E(2C)^{**}m}{E(C)^{**}m} \text{ so } .75 = (2)^{**}m$$

and from log .75 = *m* log 2, *m* = −.415. Then the equation for the learning curve becomes

$T = E\ C^{**}(-.415)$ and log T = log E − .415 log C

which indicates that a straight-line plot could be done on a log-log (logarithmic) grid with a slope of −.415.

Prepare a concurrency chart of percentage learning curves on a 3 x 5 cycles logarithmic grid for 65%, 70%, 75%, 80%, and 85%. Check one of the equations by plotting the curve on linear coordinates until the cumulative total doubles several times.

Problem 23-33

Prepare a concurrency chart of the equation for determining the deflection at the center of a beam that has a concentrated load at the center and is simply supported at the ends. The equation is:

$$y = \frac{W\ L^{**}3}{48\ EI}$$

where: y = deflection in inches

W = the applied concentrated load (1,000 to 5,000 pounds)

L = length between supports (30″ to 100″)

EI = (30,000,000 psi for steel) x (.766 in.**4 for a 2 x 2 x ¼ inch tube) = (2.3) (10**7) lb-in.**2

Problem 23-34

Prepare a parallel alignment chart of the equation in Problem 23-33.

PROBLEM 23-35

Prepare a concurrency chart for an equation from one of your courses.

EMPIRICAL EQUATIONS: SEMILOGARITHMIC

PROBLEM 23-36

Derive an equation for the following data taken in a grassy field downwind from a source of radioactive iodine, over a period of one month.

Activity (counts/second)	Day
102	1
48	5
29	10
13	15
7	20
3	25
2	31

Problem 23-37

Derive an equation for the following data for the change in electrical resistance in a ceramic because of a change in temperature.

Resistance R (ohms)	Temperature (degrees Fahrenheit)
240.0	750
120.0	840
54.0	930
32.0	1,020
15.0	1,100
7.0	1,200
3.7	1,300
1.9	1,370
1.0	1,450

Problem 23-38

Derive an equation that describes the voltage across a resistor in a circuit where a capacitor is being charged, using the following data:

Voltage	Time (tenths of a second)
10.0	0
5.4	2
3.0	4
1.5	6
1.0	8
.6	10

Problem 23-39

Derive an equation that describes the relative response of an electric recording device as a function of frequency.

Relative Response (decibels)	Frequency (Hertz)
−38	10
−20	100
−12	200
−7	300
−5	400
−3	500
−2	600
−1	700

EMPIRICAL EQUATIONS: LOGARITHMIC

Problem 23-40

Verify the Weir equation that was derived in the discussion of empirical equations in the chapter. The data is repeated here for convenience.

Flow Q (cubic feet per minute)	Head H (feet)
14.2	2
41.5	3
80.0	4
135.2	5
225.3	6
330.5	7
445.5	8
620.5	9
790.6	10

Problem 23-41

Derive an equation that describes data taken from film showing height versus time of an object dropped by students from an airplane flying at 10,000 feet elevation.

Airplane Height Minus Object Height (feet)	Time (seconds)
4	.5
17	1.0
120	3.0
400	5.0
1,500	10.0
3,700	15.0
6,500	20.0

Problem 23-42

During a laboratory exercise students measured the deflection of a cantilevered beam due to a concentrated load applied at the end versus the length of the cantilevered beam. Derive an equation from their data to show them how deflection varies as the length is increased for the same cross section of beam.

Deflection (inches)	Beam Length (inches)
.01	10
.09	20
.25	30
.60	40
1.50	50
3.40	70
6.00	80
10.00	100

MISCELLANEOUS

The data for the next seven problems should be plotted first on linear-coordinate paper to determine the type of equation—linear, power, or exponential—the data matches. Then, appropriate grids can be used to derive an equation for the data.

Problem 23-43

In the Saturated Steam: Pressure Table, Specific Volume $V(g)$ is tabulated as a function of Absolute Pressure P in psi. Derive an equation with Specific Volume versus Absolute Pressure for the following steam table data:

P	V
1.000	333.60
5.000	73.53
10.000	38.42
14.696	26.80
15.000	26.29
20.000	20.09
30.000	13.74
40.000	10.50

Problem 23-44

Iron pipe without insulation loses heat to the atmosphere at various rates, depending on the diameter of the pipe. Determine an equation for the following data relating standard pipe sizes to measured heat loss:

Pipe Size P (see std tables for dimensions)	Heat Loss (Btu/foot/hour)
1	79
1-½	114
2	143
2-½	173
3	211
3-½	240
4	271
4-½	300
5	334
6	398
8	518
10	647

Problem 23-45

A typical centrifugal fan creates a pressure differential across the fan that causes air to flow. Derive an equation to relate air flow to revolutions per minute of this squirrel-cage-type fan, from the following data:

Air Flow (cubic feet per min.)	500	800	1,050	1,400	1,900	2,100	2,550	3,000
RPM R	110	290	510	700	1,010	1,190	1,520	1,800

PROBLEM 23-46

A sample exposed to neutron radiation became radioactive. The decay in energy of the sample was recorded hourly. Derive an equation for counts per minute as a function of time. (*Note:* If this data is exponential, a tangent to the beginning of the curve, assuming that the curve is plotted on linear-coordinate paper, will intersect the horizontal axis at a time when the counts equal approximately 36% of the initial value of counts per minute. Try this check when it is believed that the data is exponential.)

Counts Per Minute CPM	Time (minutes)
450	60
215	125
138	175
110	250
95	300
90	355
80	550
75	650

Problem 23-47

The following measurements of diameter versus station (distance in the horizontal direction) were found in a design notebook. Plot the measurements and derive an equation relating diameter to station distance.

Diameter D (mm)	Distance to Station (mm)
3.93	10
3.18	20
2.42	30
1.78	40
1.12	50
0.43	60

PROBLEM 23-48

During a recent consumer product research project an observer noted that the level of popcorn in a box descended rapidly at first, and then the rate of descent decreased slowly. The level of the popcorn was noted by the observer at one-minute intervals. Use the data to derive an equation for popcorn level versus time.

Depth of Popcorn H (mm)	Time (minutes)
250	0
130	1
100	2
70	3
40	4
26	5
17	6
11	7
8	8
5	9
3	10

GRAPHICAL DIFFERENTIATION

Problem 23-49

During a broadcast of a space launch, an armchair observer made notes on the speed of the launch versus time. Plot the observer's data, fair-in a best-fit curve, and differentiate it to determine a curve of acceleration versus time.

Time (minutes)	Speed (kilometers/hour)
0	0
1	1,700
2	5,800
3	11,500
4	15,000

PROBLEM 23-50

Plot the following data related to a field test of an automobile and fair-in a best-fit curve through the data. Since the data is velocity versus time, integrating the curve will give distance versus time, and differentiating the curve will give acceleration versus time. Do both for the data:

Velocity (mph)	Time (sec)
0	0
20	3
29	4
34	5
48	10
58	15
63	20
66	25
67	30
68	35

PROBLEM 23-51

Occasionally, known quantities, known equations, and known results are used to build confidence in techniques new to an individual, such as graphical differentiation and integration. The following data represent the x and y values for the equation $y = (2) (x^{**}2) - (3) (x) + 10$. Plot the points, draw a curve through the points, and then differentiate the curve. Check your results analytically. Where does the curve have zero slope?

x	y
0	10
1	9
2	12
3	19
4	30

One technique for assisting by-eye drawing of tangents to curves while doing graphical differentiation is to use a small mirror so that the plane of the mirror is perpendicular to the curve at the point of tangency. When the reflected curve appears smooth and in-line, use the mirror to construct the perpendicular to the curve. Then draw the tangent at 90°.

GRAPHICAL INTEGRATION

PROBLEM 23-52

Some equations cannot be integrated analytically, so other methods are used, such as numerical integration and graphical integration. Use graphical integration to integrate the expression $y = (5 \sin X)/(X^{**}2)$.

Problem 23-53

The objective in this problem is to use graphical integration to graduate the volume in a water tank. The tank looks like a light bulb upside down. Use the diameters at the given elevations to calculate cross-sectional areas, then plot area versus elevations. Integrate and divide the tank into 1/4, 1/3, 1/2, 2/3, and 3/4 full.

Diameter (feet)	Elevation (feet)
10.0	0
10.0	10
10.0	20
10.2	30
11.8	32
13.0	34
15.0	36
18.4	38
26.4	40
32.8	42
37.0	44
39.0	46
40.2	48
40.0	50
39.0	52
36.6	54
31.0	56
26.0	58
0.0	60

Problem 23-54

Do the aircraft launch example discussed in the chapter and add the speed of the aircraft carrier (24 knots) to the speed of the aircraft attained by graphical integration. Will the aircraft stay airborne?

Problem 23-55

Use the beam shown, but assume that the 1,000-pound load is applied 30" from the left end instead of 45". The reaction at the left end will be 667 pounds, and at the right, 333 pounds. Use graphical integration to do the diagrams, and determine the deflection of the beam at the point where the 1,000-pound force is applied. Check your answer using the beam deflection calculation equation in Figure 23-79:

$$\text{delta } (a) = y = \frac{P (a^{**}2) (b^{**}2)}{3 \, E \, I \, L}$$

where: $P = 1{,}000$ lbs
$\quad a = 30''$
$\quad b = 60''$
$\quad E = 30{,}000{,}000$ psi
$\quad I = .766$ in.**4
$\quad L = 90''$

a. The steps summarized in Figure 23-68 to do graphical integration were followed to prepare the diagrams. Scale calculations are shown at the right. *SY1* was arbitrarily selected; the other values were derived using the equation $SY2 = (H)(SX)(SY1)$. The deflection at C equals the derived scale *SY4* times the actual measurement at C, divided by the product of *E* times *I*.

b. The modulus of elasticity *E* was needed in the deflection calculation because *E* is a measure of the material's ability to resist deforming under loads. *E* for steel equals 30,000,000 pounds per square inch.

(SI units: 210 giga Pascals) The second parameter *I*, moment of inertia in inches to the fourth power, is a geometrical property and is a measure of the beam's tendency to resist bending.

c. Refer to the slope curve for an illustration of the graphical technique used to determine the initial value of the slope. Note that the derived slope curve (integral of the moment curve) begins at zero and then, because of the positive area under the moment curve, moves upward (positively) across the beam. Next, a new horizontal axis is established by eye so that the area under the new horizontal axis approximately equals the area above. If these negative and positive areas are indeed equal, then the deflection curve will start and end at the level of the horizontal reference line for the derived deflection curve. Returning to the same level confirms the new position of the horizontal axis. Furthermore, the initial value of the slope curve is also confirmed.

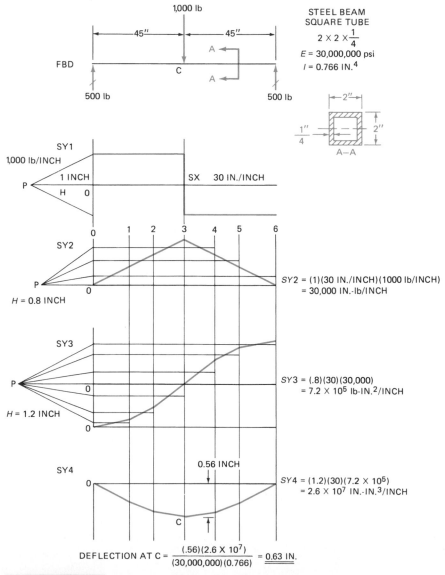

PROBLEM 23-55   *Centrally loaded beam: Graphical integration*

Problem 23-56

Use the figure for Problem 23-55 as a guide for graphically integrating to obtain the deflection at the center of a beam that is 90" long, that is loaded at the center with a 2,500-pound concentrated force, and that is an S beam ("eye" beam) S 3 x 5.7 with a value of I = 2.52 in.**4. Use the equation in Figure 23-79 to check your results:

$$\text{delta (max)} = y = \frac{P(L^{**}3)}{48\,E\,I}$$

where: P = 2,500 lbs
L = 90"
E = 30,000,000 psi
I = 2.52 in.**4

Force Polygons

Problem 23-57

Refer to the figure shown and do the following:

1. Draw the figure to scale using only single lines for members AB and BC.
2. Show a free-body diagram of joint B.
3. Use a force polygon, drawn to scale, to graphically determine the internal forces F (AB) and F (BC).
4. Label the values of F (AB) and F (BC) on the force polygon and indicate also whether these internal forces are going into joint B (compression) or away (tension).

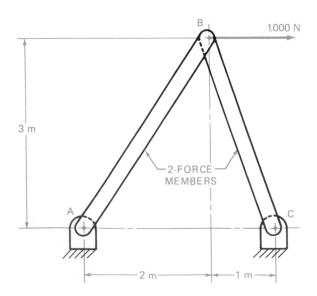

Problem 23-58

The figure shows a bell crank with forces $F(A)$ and $F(B)$ equal to 60 lbs and 100 lbs, respectively. The two applied forces and the reaction at C must go through a common point P for equilibrium conditions. Do the following:

1. Draw the bell crank to scale using single lines for AC and BC.
2. Extend the lines of action of the forces until they intersect; label the intersection point P.
3. From P, construct a line through the pivot point C.
4. Construct a force polygon to scale to determine the magnitude of the reaction at C.

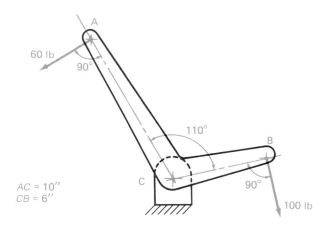

Problem 23-59

The figure shows an excited rider in a carnival-ride chair. The rider plus the chair weigh 750 N, and the angle beta of the cable supporting the chair is 35°. Use a force polygon drawn to scale to determine the tensile force in the cable and the centrifugal force.

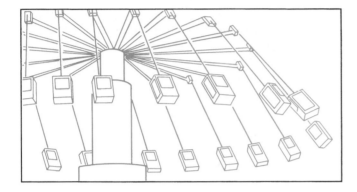

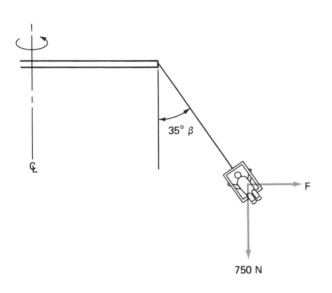

MISCELLANEOUS

Problem 23-60

Optimization problem: The two equations below give costs for insulating steam pipes over a period of a year.

$ (Fixed Cost) = 50t + 50
$ (Heat Loss) = 150/t, where t = insulation thickness

Plot both equations plus their sum as a function of insulation thickness *t*, and determine the optimum thickness to give the least cost.

PROBLEM 23-61

Curve-matching technique: There are a number of curve-matching techniques other than graphical ones; however, to visualize whether the results of a particular technique have indeed matched the curve, a plot of the results is usually done. One technique for matching a cyclic curve—that is, one that repeats periodically—is called a Fourier series. The following series was derived to match a cyclic, square wave function that varied from 1 to 2 every 3.14 radians.

$$y = \frac{3}{2} - \frac{2}{3.14}\left(\sin X + \frac{\sin 3X}{3} + \frac{\sin 5X}{5} + ...\right)$$

Calculate and plot values of *y* for *X* varying from zero to 10 radians. One radian = 57.6°.

PROBLEM 23-62

Curve-matching technique: A quadratic equation is used sometimes to match an equation to a curve. Try three sets of data from the Weir problem discussed in the chapter (*H, Q*: 4, 80; 6, 225.3; 8, 445.5) in the following equation: $Q = A(H^{**}2) + B(H) + C$. Substitute the 3 sets of data to obtain 3 equations and solve them simultaneously for *A, B,* and *C*. Then check the derived equation with additional data, e.g.,

$$H = 5 \text{ and } Q = 135.2.$$

PROBLEM 23-63

Investigate ANSI (American National Standards Institute) Standard ANSI Y15.1M-1979 entitled "Illustrations for Publication and Preparation" and prepare a report for classmates, telling why the standard is useful to designers and showing several illustrations.

Chapter 24

THE DESIGN PROCESS

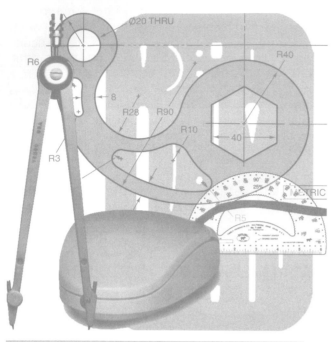

Abstract

Alternatives

Analysis

Attribute listing

Constraints

Costs

Creative process

Decision table

Design file

Design hints

Design process

External clients

Human factors

Internal clients

Iterate

Layout development drawings

Letter of transmittal

Log

Non-routine designs

Oral presentation

Parametric design

Phases (in the design process)

Planning chart

Progress report

Prototypes

Rapid prototyping

Reverse engineering

Routine designs

Safe designs

Solution description

Specifications

Steps (in the design process)

Trade-off

Vendor catalogs

Virtual prototyping

CHAPTER OUTLINE

Time • Learning the design process • The design process: phases and steps • Design projects: routine and non-routine • Modern design practices and standards • Summary • Review questions • Chapter 24 problems

CHAPTER OBJECTIVES

Upon completion of this chapter, students should be able to do the following:

■ Follow the phases and steps in the design process to do a routine design project as an individual effort.

■ Follow the phases and steps in the design process to do a non-routine design project as a member of a team.

■ Use various techniques such as the creative process, attribute listing, a decision table, and model building (actual and computer-generated) to enhance the steps in the design process.

■ Use various information sources such as vendor catalogs, technical texts, technical periodicals, and technical handbooks to augment the steps in the design process.

■ Identify the various types of technical drawings you would use in the design process such as freehand sketches, design development layout drawings, kinematic layout drawings, and detail and assembly drawings.

■ Discuss the objectives of DFM, DFA, DFR, and DFD.

■ Recognize current trends in manufacturing such as TQM, ISO 9000, parametric design, and rapid prototyping.

■ Identify the roles of drafters, designers, and engineers in the steps of the design process.

The *design process* is a plan of action for reaching a goal. The plan, sometimes labeled problem-solving strategy, is used by engineers, designers, drafters, scientists, technologists, and a multitude of professionals. This strategy, the design process, occurs in five *phases* during a product design, and each phase has a number of *steps*. The following phases occur sequentially with some overlapping, as shown in *Figure 24-1*.

Phase I	Recognize and Define the Problem
Phase II	Generate Alternative Solutions: SYNTHESIS
Phase III	Evaluate Alternative Solution Ideas: ANALYSIS
Phase IV	Document, Specify, and Communicate the Solution
Phase V	Develop, Manufacture, and Deliver

The steps, not shown in Figure 24-1, may occur sequentially; however, each phase has a different number of steps depending on the complexity of the problem to be solved and the time available.

Time

The available time, whether hours, days, months, or years, controls the starting and stopping of the phases during the design process. Therefore, successful problem solving requires careful planning. And since the phases must occur within the time span, they must be scheduled as needed to reach the required goal.

Learning the Design Process

Learn the design process by doing it! That is, follow the phases and steps, as a guide, to do at least two design projects. First, individually, do a routine design project. Next, in a group, do a non-routine design project. Use these design experiences to augment and expand your design skills, and to acquire an understanding that the design process can be modified to suit the level and complexity of the problem to be solved. This chapter provides the following to get you started:

1. An outline and discussion of the phases and steps in the design process, *Figure 24-2*

2. A discussion of routine design projects (the type usually assigned to a neophyte drafter, designer, or engineer as their first task in a company)

3. A discussion of non-routine design projects (the type assigned to more experienced drafters, designers, and engineers as they assume more responsibility in their respective jobs)

4. A collection of design hints to enhance your problem-solving skills and design solutions

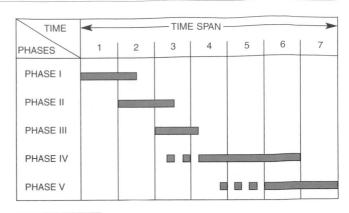

FIGURE 24-1   *The phases in the design process*

5. Examples from a non-routine design project (done by a student group)

Follow the phases and steps in Figure 24-2, generally in the sequence shown. Use the steps as a checklist. Samples from the student group's non-routine design project (#5 above) are included in the following discussion when appropriate.

The Design Process: Phases and Steps

PHASE I RECOGNIZE AND DEFINE THE PROBLEM

STEP 1 Establish Plans for the Phases, Steps, and Records

STEP 2 Gather Information
    a. Primary Sources
    b. Secondary Sources

STEP 3 Combine Information into a Preliminary Definition
    a. Constraints
    b. Specifications
    c. Preliminary Solution Ideas
    d. Iterate

STEP 4 Prepare a Progress Report
    a. Internal Clients
    b. External Clients

Step 1 Establish Plans for the Phases, Steps, and Records. Make a *planning chart*, which is an excellent visual display of when phases and steps are scheduled to begin and end, *Figure 24-3*. Note that over each bar is the name of the individual responsible for that activity, whether the individual does all of the work or not. Refer to Figure 24-21 for the completed chart.

DESIGN PROCESS: PHASES AND STEPS

PHASE I
RECOGNIZE AND DEFINE THE PROBLEM

STEP 1   Establish Plans for the Phases, Steps, and Records

STEP 2   Gather Information
a. Primary Sources
b. Secondary Sources

STEP 3   Combine Information into a Preliminary Definition
a. Constraints
b. Specifications
c. Preliminary Solution Ideas
d. Iterate

STEP 4   Prepare a Progress Report
a. Internal Clients
b. External Clients

PHASE II
GENERATE ALTERNATIVE SOLUTIONS: SYNTHESIS

STEP 5   Generate Solution Ideas
a. Use Systematic Techniques
b. Use Creative Techniques and Attitudes

PHASE III
EVALUATE ALTERNATIVE SOLUTION IDEAS: ANALYSIS

STEP 6   Sift, Combine, and Prepare Solution Ideas for Analysis
a. Preliminary Preparation
b. Specific Preparation

STEP 7   Compare Solution Alternatives to Constraints and Specifications
a. Iterate
b. Persevere
c. Qualitative Evaluation
d. Quantitative Evaluation

STEP 8   Conduct a Design Review
a. In-House Design Reviews
b. Outside Design Reviews

STEP 9   Refine the Final Solution

PHASE IV
DOCUMENT, SPECIFY,
AND COMMUNICATE THE SOLUTION

STEP 10   Prepare Drawings and a Written Report

STEP 11   Give an Oral Presentation

PHASE V
DEVELOP, MANUFACTURE, AND DELIVER

STEP 12   Prepare Technical Detail Drawings of Each Component to Be Manufactured

STEP 13   Schedule Assembly, Final Testing, Packaging, and Delivery

STEP 14   Keep Records

FIGURE 24-2   *Phases and steps in the design process*

Start and keep a *log* of activities and the time spent. Each individual in the group should keep a log. Newspaper-style entries, who—what—where—when—why, are useful because they tell specifically what was done. Supervisors use the information for checking progress and for accounting purposes. But more importantly, the phone numbers, names, dates, and facts may be useful on future projects. Instructors use the log information for checking progress and for grading purposes. See *Figure 24-4* for a sample log from a student's notebook. Refer to Figure 24-20 for a more complete log.

Some companies may require selected individuals to keep a more detailed log-design notebook for legal purposes or for patent development. For patent development, each page is bound, numbered, and then arranged to be signed by a knowledgeable witness.

Start a design file with every page dated to agree with the log, and file the pages chronologically for easy reference. Each individual keeps a *design file,* which includes freehand sketches, machine copies of technical information, copies of related drawings, summaries of telephone conversations and interviews, and vendor brochures. Samples from a student's design file are in *Figures 24-5A* and *24-5B.* A group log and file may be required by a supervisor in industry.

Step 2   Gather Information.

a. *Primary Sources.* Contact the source of the problem, who will provide information to establish most of the project's constraints and specifications, verbally or in writing. Later, in Step 4, confirm any verbal decisions made by the source, in a progress report.

Next, review similar products and experiences for information because most designs are actually new combinations of existing ideas.

b. *Secondary Sources.* Investigate secondary sources, which include a variety of people and references. The following checklist can be expanded to fit the needs of specific problems.
People: shop personnel (machine, structural, electronic, electrical, packaging, shipping, welding . . .)

Practicing professionals: (engineers, lawyers, doctors, technologists, scientists . . .)

City services: (fire, police, waste, utility, library)

Most helpful for students: (Instructors, Librarians, and Shop Personnel)

Patents: Note that most of the information on a patented device only appears in Patent documents.

References:
Books (textbooks on specific topics):
Best listing of references found to date: Kolb, John and Ross, Steven, *Product Safety and Reliability,* 1980,

PLANNING CHART									
TEMPOROMANDIBULAR JOINT FLEXOR									
K. Anderson	M. Bui		A. Hamamoto			W. Peterson			
ACTIVITY	WEEK								
	1	2	3	4	5	6	7	8	9
PHASE I		MB							
Get Organized	KA								
Client/Meeting		AH							
Information Search		WP							
Project Define		WP				⊙ MILESTONE			
Progress Report		MB ◉							

FIGURE 24-3 *Planning chart for student group*

TIME LOG	W. PETERSON		
TEMPOROMANDIBULAR JOINT FLEXOR			
DATE	ACTIVITY	TIME HR	CUM. TIME HR
10-2	Met with Anderson, Bui, Hamamoto- - Exchanged phone numbers - Copied Dr. P's proposal	½	½
10-8	MET - KA MB & AH - discussed proposal - made up questions for Dr. P Discussed planning chart	2 ¾	3 ¼

FIGURE 24-4 *Log*

```
PRELIMINARY DESIGN CONSIDERATIONS

  1. SIZE                    Date 10-10
  2. POWER
     –BELT POWER PACK
     –CHARGER
  3. CONTROLS
     –FREQUENCY
     –OPENING
     –PATIENT OR DR.?
  4. EXPECTED PRODUCTION
     –REUSABLE?
  5. COST
  6. MATERIALS
     –STAINLESS
     –PLASTIC
     –FDA RESTRICTIONS
  7. NOISE LEVEL
  8. FORCES REQUIRED
     –OPENING
     –CLOSING
  9. SAFETY OVERDRIVE
 10. JAW DIMENSIONS
 11. ACCURACY
```

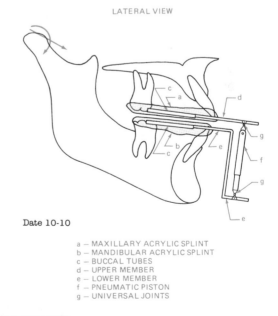

LATERAL VIEW

Date 10-10

a – MAXILLARY ACRYLIC SPLINT
b – MANDIBULAR ACRYLIC SPLINT
c – BUCCAL TUBES
d – UPPER MEMBER
e – LOWER MEMBER
f – PNEUMATIC PISTON
g – UNIVERSAL JOINTS

FIGURE 24-5A *File: Preliminary design considerations*

FIGURE 24-5B *File Information from clients*

McGraw-Hill, Inc. (It also contains more than 45 pages of checklists for designers and installers.)
Handbooks: Each engineering discipline has a handbook, e.g., aeronautical, chemical, civil, electrical, mechanical, materials . . .

Standards and codes (a small sampling):

ANSI American National Standards Institute

ASTM American Society for Testing Metals

Mil Military Standard

UL Underwriters Laboratory

ISO 9001

Catalogs:

Vendor (products, hardware, electrical, electronic, structural shapes . . .)

Thomas Register (listing of companies, what they make, their address, sample catalogs . . .)

McMaster-Carr Supply Co. (Most complete hardware catalog found to date, P.O. Box 54960, Los Angeles, CA 90054, 213/945-2811)

Computer sources (CD's)

Vendor Catalogs. Vendors provide customers with convenient guides to select products and materials from their catalogs, *Figures 24-6A* through *E* (pages 938–942).

a. *Machine Elements.* V-belt drives, ball and roller bearings, roller chain drives, clutches, brakes, couplings, gearmotors (motor and gear reduction box as one unit), shafting, motors (AC and DC), and controls, all have similar guidelines for selecting them. These guidelines include formulas, tables, and charts to convert most customer applications and environmental conditions to numbers or letters in their catalogs.

For example, assume that the belt conveyor shown in *Figure 24-6F* (page 943) needs a roller chain drive to transmit power from the gearmotor's 200 RPM output shaft to the headshaft of the conveyor. Calculations indicate 14 horsepower to operate the conveyor at 500 FPM (feet per minute).

ROLLER CHAIN DRIVE SELECTION

The following considerations are very important in the selection and application of roller chain drives:

HORSEPOWER RATINGS — This catalog lists Horsepower Ratings for ANSI Series, single pitch, single strand chains No. 25 through No. 160 (and lightweight machinery series No. 41 and No. 43). Ratings are listed for various numbers of teeth and speeds of the smaller sprocket. Ratings for intermediate numbers of teeth or RPM may be determined by interpolation. The ratings reflect a service factor of 1, a chain length of approximately 100 pitches, the use of recommended lubrication methods and a drive arrangement where two aligned sprockets are mounted on parallel horizontal shafts. For maximum service life, sprockets with small numbers of teeth, operating at moderate to high speeds or near the rated horsepower should have hardened teeth. Approximately 15,000 hours of service life at full load operation may be expected under these conditions.

NO. OF TEETH — It is good practice to select a pinion sprocket with no less than 17 Teeth, to assure 120° of chain wrap and minimize overhung load. However, certain conditions, i.e., space limitations, light loads, intermittent duty, etc. will permit the use of smaller pinions.

RATIO — Sprocket ratios should not exceed about 6 to 1 for normal chain life.

HARDENED TEETH — Boston Gear steel sprockets can be hardened. Consult the factory for recommended procedure.

CENTER DISTANCE — The correct center distance is very important. In designing chain drives, it is important that the Center Distance should be long enough to provide at least 120° of chain wrap on the smaller sprocket.

RELATIVE SHAFT LOCATIONS — It is desirable that the line between the two shaft centers be as nearly horizontal as possible. If this line is more than 60° from the horizontal, special precautions should be taken.

A roller chain consists essentially of numerous small bearings operating under high pressures and requires adequate lubrication. There are four basic types of lubrication suggested for chain drives, depending on the chain speed and the power transmitted. The Horsepower Rating Tables indicate the type of lubrication recommended.

TYPE I — MANUAL LUBRICATION

Manual lubrication is accomplished by applying oil with a brush or spout can to the inside of the chain at the edges of the side plates. Volume and frequency should be determined by periodic inspection.

TYPE II — DRIP LUBRICATION

Oil is directed between link plate edges from a drip lubricator. Only enough oil to keep the chain moist is necessary and a light metal splash guard will keep the floor and surroundings clean.

TYPE III — BATH OR DISC LUBRICATION

With bath lubrication, the lower strand of the chain runs through a sump of oil. The oil level should reach the pitch line of the chain at its lowest point while operating. With disc lubrication, the chain operates above the oil level. The disc picks up oil from the sump and deposits it on the chain, usually by means of a trough. The disc diameter should be such as to produce rim speeds from 600 minimum to 8000 maximum FPM. This type of lubrication requires that the drive be enclosed in an oil-tight chain case.

TYPE IV — OIL STREAM LUBRICATION

The lubricant is usually supplied by a circulating pump capable of supplying the chain drive with a continuous stream of oil. The oil should be applied inside the chain loop evenly across the chain width, and directed at the lower strand. This type of lubrication requires that the drive be enclosed in an oil-tight chain case.

Recommended lubricant viscosities for various ambient temperatures are listed in the following table:

Temp. Degrees F.	Lubricant	Temp. Degrees F.	Lubricant
20-40	SAE20	100-120	SAE40
40-100	SAE30	120-140	SAE50

SURROUNDING CONDITIONS — Abrasive, corrosive, or high temperature conditions can shorten chain life. If adverse conditions exist, special precautions should be taken. It may be advisable to use a drive with higher capacity than normal, stainless steel chain, etc.

Roller chain drives may be selected with the following procedure:

a. From Table #1 of the Application Classification Chart on Page A94, determine the Service Factor.

b. Multiply the Application HP by the Service Factor to obtain a Design HP.*

c. The Selection Table below may be used to select an appropriate chain size using a sprocket of 17 teeth or larger.

d. From the appropriate horsepower rating table (pages B6-B8) determine the minimum size sprocket needed to provide, at the required speed, a rating equal to (or greater than) the Design horsepower.

e. The tables on pages B9-B11 may then be used to select number of sprocket teeth, shaft center distance and chain length of a drive suitable for the application.

*For Stainless Steel Chains, operating under wet or dry conditions, the Design Horsepower must be multiplied by a Factor (see Table below) for selection purposes.

NOTE: Standard Steel Chains are not recommended for wet or dry applications.

Application Conditions	Factor
Wet (Moisture)	2.0
Dry (Unlubricated)	5.0

Horsepower ratings of Multiple Strand chain may be obtained by multiplying the Single Strand rating by the proper Factor from the following table:

MULTIPLE STRAND RATING FACTORS

Number of Strands	Double	Triple	Quadruple
Rating Factor	1.7	2.5	3.3

These Horsepower Ratings are based on certain operating conditions, see Page B4.

FIGURE 24-6A *Roller chain drive selection considerations* (Courtesy Boston Incom International Inc.)

SELECTION TABLE

RPM of Smaller Sprocket*	DESIGN HORSEPOWER												
	1/2	1	1-1/2	2	3	4	5	7-1/2	10	15	20	25	30
	CHAIN NUMBER												
1800	25	25	35	35	35	40	40	40	50	80	60 — 2	80 — 2	—
1500	25	25	35	35	35	40	40	40	60	60	80	60 — 2	80 — 2
1200	25	35	35	35	40	40	40	50	60	60	60	80	100
1000	25	35	35	35	40	40	40	50	60	60	80	80	80
800	25	35	35	40	40	40	50	50	60	60	80	80	80
700	25	35	35	40	40	50	50	50	60	80	80	80	80
600	35	35	35	40	40	50	50	60	60	80	80	80	100
500	35	35	40	40	50	50	50	60	80	80	80	100	100
400	35	35	40	40	50	50	60	60	80	80	100	100	100
350	35	40	40	40	50	50	60	80	80	80	100	100	100
300	35	40	40	50	50	60	60	80	80	100	100	100	120
250	35	40	40	50	50	60	60	80	80	100	100	120	120
200	35	40	50	50	60	60	80	80	80	100	120 ◄	120	120
175	40	40	50	50	60	60	80	80	80	100	120	120	140
150	40	50	50	60	60	80	80	80	100	120	120	120	140
125	40	50	50	60	80	80	80	100	100	120	120	140	140
100	40	50	60	60	80	80	80	100	100	120	140	140	160
80	40	50	60	80	80	80	100	100	120	140	140	160	160
70	50	60	60	80	80	80	100	120	120	140	160	160	
60	50	60	80	80	80	100	100	120	120	140	160		
50	50	60	80	80	80	100	100	120	140	160	160		
40	50	60	80	80	100	100	120	120	140	160	160		
30	60	80	80	100	100	120	120	140	160				
25	60	80	80	100	120	120	140	140	160				
20	60	80	100	100	120	120	140	160					
15	80	100	100	120	120	140	160						
10	80	100	120	120	140	140							

*Based on 17-Tooth Sprocket.

FIGURE 24-6B *Roller chain drive selection procedure* (Courtesy Boston Gear Incom International Inc.)

The roller chain drive, sprockets, and chain can be selected from a vendor's catalog using the tables in Figures 24-6A-E. Figure 24-6A contains general guidelines for selecting and applying roller chain drives. To select a roller chain drive, follow the five steps in Figure 24-26A. The first step determines the service factor to be 1.25, as indicated in Figure 24-6C, assuming heavy duty, not uniformly fed and operating less than 10 hours per day. Step 2 says that the design horsepower is the calculated horsepower times the service factor, (14) (1.25) = 17.5. The third step is to use the design horsepower in the selection table in Figure 24-6B to determine chain size. Chain size No. 120 with a 17-tooth smaller sprocket will work. The next step (using Figure 24-6D) indicates that an 11-tooth smaller sprocket will theoretically work; however, good practice dictates the use of a 17-tooth sprocket according to comments in Figure 24-6A. Finally, for the fifth step use Figure 24-6E to find the larger sprocket to give a drive ratio of 2.50. A 42-tooth sprocket combined with a 17-tooth gives a ratio of 2.47, which is close enough. The center distance between the two sprockets using a 64-inch-long chain will be 16.777 inches.

b. *Engineering materials and shapes* can be selected from *vendor catalogs.*

(i) The catalogs published by vendors of materials list materials and their typical uses to aid in their selection. To illustrate, one catalog describes 6061 aluminum as being heat treatable, having good formability (will not crack during bending), having good weldability, and being corrosion resistant. Also, 6061 aluminum is used for structural application, boats, and transportation equipment.

(ii) Some materials are selected primarily on the basis of their shape; for example, steel and aluminum are rolled into plates, bars, channels, beams, and angles and are extruded into pipe, square tubes, and rectangular tubes. Square and rectangular shapes are easier to join in fabricating than pipe sections. Pipe sections are equally strong in all directions.

(iii) Materials may be selected on the basis of their weight, rigidity, and coefficient of expansion due to temperature changes. INVAR is used in instrumentation because of its low coefficient of expansion of 1.6×10^{-6} inches per inch per degree Fahrenheit.

(iv) Additional bases for selecting materials are corrosion resistance, machinability, strength, and electrical properties.

APPLICATION CLASSIFICATION FOR VARIOUS LOADS

Type of Machine To Be Driven	Chart I For all Drives		
	Service Factor		
	Loading		
	Not More Than 15 Mins. In 2 Hrs.	Not More Than 10 Hrs. per Day	More Than 10 Hrs. per Day
AGITATORS			
Pure Liquid	0.80	1.00	1.25
Semi-Liquids, Variable Density	1.00	1.25	1.50
BLOWERS			
Centrifugal and Vane	0.80	1.00	1.25
Lobe	1.00	1.25	1.50
BREWING AND DISTILLING			
Bottling Machinery	0.80	1.00	1.25
Brew Kettles — Continuous Duty	—	—	1.25
Cookers — Continuous Duty	—	—	1.25
Mash Tubs — Continuous Duty	—	—	1.25
Scale Hopper — Frequent Starts	—	1.25	1.50
CAN FILLING MACHINES	—	1.00	—
CANE KNIVES	—	1.50	—
CAR DUMPERS	—	1.75	—
CAR PULLERS	—	1.25	—
CLARIFIERS	—	1.00	1.25
CLASSIFIERS	—	1.25	1.50
CLAY WORKING MACHINERY			
Brick Press & Briquette Machine	—	1.75	2.00
Extruders and Mixers	1.00	1.25	1.50
COMPRESSORS			
Centrifugal	—	1.00	1.25
Lobe — Reciprocating, Multi-Cycle	—	1.25	1.50
Reciprocating — Single Cycle	—	1.75	2.00
CONVEYORS — UNIFORMLY LOADED & FED			
Apron	—	1.00	1.25
Assembly-Belt — Bucket or Pan	—	1.00	1.25
Chain — Flight	—	1.00	1.25
Oven — Live Roll — Screw	—	1.00	1.25
CONVEYORS — HEAVY DUTY NOT UNIFORMLY FED			
Apron		1.25	1.50
Assembly-Belt — Bucket or Pan	—	1.25	1.50
Chain — Flight	—	1.25	1.50
Live Roll	—	—	—
Oven — Screw	—	1.25	1.50
Reciprocating — Shaker	—	1.75	2.00
CRANES AND HOISTS			
Main Hoists			
Bridge and Trolley Drive	*	1.00	1.25
CRUSHER			
Ore, Stone	—	1.75	2.00
Sugar	—	1.50	1.50
ELEVATORS			
Bucket — Uniform Load	—	1.00	1.25
Bucket — Heavy Load	—	1.25	1.50
Centrifugal Discharge	—	1.25	1.50
Freight	—	1.25	1.50
Gravity Discharge	—	1.00	1.25
ELEVATORS			
Bucket — Uniform Load	—	1.00	1.25
Bucket — Heavy Load	—	1.25	1.50
Centrifugal Discharge	—	1.25	1.50
Freight	—	1.25	1.50
Gravity Discharge	—	1.00	1.25
FANS			
Centrifugal — Light (Small Diam.)	—	1.00	1.25
Large Industrial	—	1.25	1.50

Type of Machine To Be Driven	Chart I For all Drives		
	Service Factor		
	Loading		
	Not More Than 15 Mins. In 2 Hrs.	Not More Than 10 Hrs. per Day	More Than 10 Hrs. per Day
FEEDERS			
Apron — Belt — Screw	—	1.25	1.50
Disc	—	1.00	1.25
Reciprocating	—	1.75	2.00
FOOD INDUSTRY			
Beet Slicer	—	1.25	1.50
Cereal Cooker	—	1.00	1.25
Dough Mixer —Meat Grinder	—	1.25	1.50
GENERATORS (NOT WELDING)	—	1.00	1.25
HAMMER MILLS	—	1.75	2.00
HOISTS			
Heavy Duty	—	1.75	2.00
Medium Duty and Skip Type	—	1.25	1.50
LAUNDRY TUMBLERS	—	1.25	1.50
LINE SHAFTS			
Uniform Load	—	1.00	1.25
Heavy Load	—	1.25	1.50
MACHINE TOOLS			
Auxiliary Drive	—	1.00	1.25
Main Drive — Uniform Load	—	1.25	1.50
Main Drive — Heavy Load	—	1.75	2.00
METAL MILLS			
Draw Bench Carriers & Main Drive	—	1.25	1.50
SLITTERS	—	1.25	1.50
TABLE CONVEYORS — NON REVERSING			
Group Drives	—	1.25	1.50
Individual Drives	—	1.75	2.00
Wire Drawing, Flattening or Winding	—	1.25	1.50
MILLS ROTARY TYPE BALL & ROD			
Spur Ring Gear and Direct Connected	—	—	2.00
Cement Kilns, Pebble	—	—	1.50
Dryers and Coolers	—	—	1.50
Plain and Wedge Bar	—	—	1.50
Tumbling Barrels	—	—	2.00
MIXERS			
Concrete — Continuous	—	1.25	1.50
Concrete — Intermittent	—	1.25	1.50
Constant Density	—	1.00	1.25
Semi-Liquid	—	1.25	1.50
OIL INDUSTRY			
Oil Well Pumping	—	—	*
Chillers, Paraffin Filter Press	—	1.25	1.50
Rotary Kilns	—	1.25	1.50
PAPER MILLS			
Agitator (Mixer)	—	1.25	1.50
Agitator — Pure Liquids	—	1.00	1.25
Barking Drums — Mechanical Barkers	—	1.75	2.00
Bleacher	—	1.00	1.25
Beater	—	1.25	1.50
Calender Heavy Duty	—	—	2.00
Calender Anti-Friction Brgs.	—	1.00	1.25
Cylinders	—	1.25	1.50
Chipper	—	—	2.00
Chip Feeder	—	1.25	1.50
Coating Rolls — Couch Rolls	—	1.00	1.25
Conveyors — Chips — Bark — Chemical	—	1.00	1.25
Conveyors — Log and Slab	—	—	2.00
Cutter	—	—	2.00
Cylinder Molds, Dryers (Anti-Friction Brg.)	—	—	1.25
Felt Stretcher	—	1.25	1.50
Screens — Chip and Rotary	—	1.25	1.50
Thickener (AC)	—	1.25	1.50
Washer (AC)	—	1.25	1.50
Winder — Surface Type	—	—	1.25

FIGURE 24-6C *Service factors for various applications (Courtesy Boston Gear Incom International Inc.)*

Small Sprocket		HP RATINGS — STANDARD SINGLE * STRAND ROLLER CHAIN — NO. 120 — 1-1/2" PITCH																		
RPM→		10	20	30	50	75	100	125	150	200	250	300	400	500	600	700	800	900	1000	1200
Teeth	P.D.																			
11	5.324"	1.37	2.56	3.68	5.82	8.40	10.9	13.3	15.6	20.3	24.8	29.2	37.8	46.3	54.5	46.3	37.9	31.8	27.1	20.6
13	6.268	1.64	3.06	4.41	6.97	10.1	13.0	15.9	18.7	24.3	29.7	35.0	45.3	55.4	65.3	59.5	48.7	40.8	34.9	26.5
15	7.215	1.91	3.57	5.14	8.13	11.7	15.2	18.6	21.9	28.3	34.7	40.8	52.9	64.6	76.1	73.8	60.4	50.6	43.2	32.9
17	8.164	2.19	4.09	5.88	9.31	13.4	17.4	21.3	25.0	32.4	39.7	46.7	60.5	74.0	87.2	89.0	72.8	61.0	52.1	39.6
19	9.114	2.47	4.61	6.64	10.5	15.2	19.6	24.0	28.2	36.5	44.8	52.7	68.2	83.4	98.3	105	86.1	72.1	61.6	46.8
21	10.064	2.75	5.13	7.39	11.7	16.9	21.8	26.7	31.4	40.7	49.8	58.7	76.0	93.0	110	122	100	83.8	71.6	54.4
Lubrication #		Type I		Type II						Type III				Type IV						

FIGURE 24-6D *Horsepower ratings for No. 120 roller chain* (Courtesy Boston Gear Incom International Inc.)

(v) Combinations of all of the above, plus aesthetics and prior experiences, may be used for selecting materials.

Call or visit a local vendor of materials or machine elements, and request one of their catalogs. Usually, they are eager to have their catalogs in the hands of future buyers, so they are happy to donate one or more.

Furthermore, since most vendors handle a variety of products, a call or a visit may result in a small library!

Step 3 Combine Information into a Preliminary Definition. Prepare a preliminary definition of the problem so that all parties concerned can see the direction and course of the project. The definition mainly consists of two types of statements, *constraints* and *specifications*. These are written as specifically as possible even at this early stage in the project. For example, a vague specification might say that the product is to be safe to use. A more specific specification would say that the product will be safe to use by a 12-year-old child who can read and understand simple directions, and that the directions will be affixed to the product.

a. *Constraints.* List the more restrictive constraints, such as cost, delivery date, availability of materials, codes, and standards. Then, list the less restrictive ones, such as shop capability, available personnel, product life, and aesthetics.

b. *Specifications.* Divide specifications into two main categories, performance and design. Performance specifications describe what the design is to do, what it is to accomplish, what it is to produce, who is to use it, what environment it will operate in, how it will be maintained, how safe it is, and how easy it is to use. Design specifications indicate what materials are to be used and how they are to be fabricated and finished. Purchased products also are included.

Well-written specifications are stated so that they can be tested or measured. For example, the chemistry and strength of a material can be measured and tested, respectively. Resistance, frequency, volume, temperature, and weight can all be measured.

c. *Preliminary Solution Ideas.* Collect preliminary solution ideas that occur while the definition, constraints, and specifications are being developed and written. File these for use in Phase II.

d. *Iterate.* Return to Step 2 to gather further information and to revise the definition. Returning and revising is an iterative process that occurs continually throughout the design process.

Step 4 Prepare a Progress Report. Prepare a *progress report* to the client, who is usually the primary contact during the project. By now, approximately one-fourth of the project time will have been used, so a report to the client should be made preferably in writing. Include the current definition—constraints and specifications—so that if there are any misunderstandings, they can be resolved early in the project. Decisions made verbally should be summarized. Indicate immediate plans and ask for confirmation as soon as possible. The client, a person or an organization, wants to know how the project is progressing and if any problems have arisen.

For the purposes of this discussion, clients are divided into two groups.

a. *Internal Clients* are instructors in classrooms and supervisory persons in industry. And even with this close proximity to the client, important communications should be in writing.

b. *External Clients* comprise a large variety of people and organizations, both in classrooms and in industry. In classroom situations, an external client could be another instructor, a student, a department, a design contest, a handicapped person, a rehabilitation department, a local government, or a national organization. In industry, the external client could be another company; an individual; a department in a city, state, or national government; the military; or even another nation.

SPEED RATIOS — CENTER DISTANCES — CHAIN LENGTHS

RATIO
CENTER DISTANCE — IN PITCHES
CHAIN LENGTH — IN PITCHES

To obtain corresponding values in INCHES, multiply by the appropriate Chain Pitch.

Each cell lists: Ratio / Center Distance (in pitches) / Chain Length (in pitches).

Teeth on DriveN Sprocket	Teeth on DriveR Sprocket														
	11	12	13	14	15	16	17	18	19	20	21	22	23	24	25
15	1.36 / 6.469 / 26	1.25 / 7.235 / 28	1.15 / 6.993 / 28	1.07 / 7.748 / 20	1.00 / 7.500 / 30										
16	1.45 / 7.207 / 28	1.33 / 6.971 / 28	1.23 / 7.736 / 30	1.14 / 7.494 / 30	1.07 / 8.249 / 32	1.00 / 8.000 / 32									
17	1.55 / 7.943 / 30	1.42 / 7.710 / 30	1.31 / 7.473 / 30	1.21 / 8.237 / 32	1.13 / 7.994 / 32	1.06 / 8.749 / 34	1.00 / 8.500 / 34								
18	1.64 / 7.669 / 30	1.50 / 8.446 / 32	1.38 / 8.212 / 32	1.29 / 7.975 / 32	1.20 / 8.737 / 34	1.13 / 8.495 / 34	1.06 / 9.249 / 36	1.00 / 9.000 / 36							
19	1.73 / 8.404 / 32	1.58 / 8.174 / 32	1.46 / 8.949 / 34	1.36 / 8.714 / 34	1.27 / 8.477 / 34	1.19 / 9.238 / 36	1.12 / 8.995 / 36	1.06 / 9.749 / 38	1.00 / 9.500 / 38						
20	1.82 / 8.124 / 32	1.67 / 8.909 / 34	1.54 / 8.679 / 34	1.43 / 9.452 / 36	1.33 / 9.216 / 36	1.25 / 8.978 / 36	1.18 / 9.739 / 38	1.11 / 9.495 / 38	1.05 / 10.249 / 40	1.00 / 10.000 / 40					
21	1.91 / 8.857 / 34	1.75 / 8.632 / 34	1.61 / 9.414 / 36	1.50 / 9.183 / 36	1.40 / 9.955 / 38	1.31 / 9.718 / 38	1.24 / 9.479 / 38	1.17 / 10.239 / 40	1.11 / 9.995 / 40	1.05 / 10.749 / 42	1.00 / 10.500 / 42				
22	2.00 / 9.590 / 36	1.83 / 9.365 / 36	1.69 / 9.139 / 36	1.57 / 9.918 / 38	1.47 / 9.686 / 38	1.37 / 10.457 / 40	1.29 / 10.220 / 40	1.22 / 9.980 / 40	1.16 / 10.740 / 42	1.10 / 10.496 / 42	1.05 / 11.249 / 44	1.00 / 11.000 / 44			
23	2.09 / 9.304 / 36	1.92 / 10.098 / 38	1.77 / 9.872 / 38	1.64 / 9.645 / 38	1.53 / 10.422 / 40	1.44 / 10.189 / 40	1.35 / 10.959 / 42	1.28 / 10.721 / 42	1.21 / 10.481 / 42	1.15 / 11.240 / 44	1.10 / 10.996 / 44	1.05 / 11.749 / 46	1.00 / 11.500 / 46		
24	2.18 / 10.037 / 38	2.00 / 9.815 / 38	1.85 / 10.605 / 40	1.72 / 10.378 / 40	1.69 / 10.150 / 40	1.50 / 10.926 / 42	1.41 / 10.692 / 42	1.33 / 11.461 / 44	1.26 / 11.222 / 44	1.20 / 10.982 / 44	1.14 / 11.741 / 46	1.09 / 11.496 / 46	1.04 / 12.249 / 48	1.00 / 12.000 / 48	
25	2.27 / 9.744 / 38	2.08 / 10.547 / 40	1.92 / 10.324 / 40	1.79 / 11.112 / 42	1.67 / 10.884 / 42	1.56 / 10.654 / 42	1.47 / 11.429 / 44	1.39 / 11.195 / 44	1.31 / 11.963 / 46	1.25 / 11.723 / 46	1.19 / 11.483 / 46	1.14 / 12.241 / 48	1.09 / 11.996 / 48	1.04 / 12.750 / 50	1.00 / 12.500 / 50
30	2.72 / 11.345 / 44	2.50 / 12.161 / 46	2.31 / 11.943 / 46	2.14 / 12.746 / 48	2.00 / 12.522 / 48	1.88 / 12.299 / 48	1.76 / 13.087 / 50	1.67 / 12.858 / 50	1.58 / 13.638 / 52	1.50 / 13.406 / 52	1.43 / 13.172 / 52	1.36 / 13.942 / 54	1.30 / 13.705 / 54	1.25 / 14.469 / 56	1.20 / 14.228 / 56
32	2.91 / 12.812 / 48	2.66 / 12.597 / 48	2.46 / 12.379 / 48	2.28 / 13.188 / 50	2.14 / 12.967 / 50	2.00 / 13.765 / 52	1.88 / 13.539 / 52	1.78 / 13.314 / 52	1.68 / 14.099 / 54	1.60 / 13.869 / 54	1.52 / 14.646 / 56	1.45 / 14.413 / 56	1.39 / 14.178 / 56	1.33 / 14.946 / 58	1.28 / 14.708 / 58
35	3.18 / 13.976 / 52	2.92 / 13.761 / 52	2.69 / 13.546 / 52	2.50 / 14.361 / 54	2.33 / 14.141 / 54	2.19 / 13.921 / 54	2.06 / 14.721 / 56	1.94 / 14.497 / 56	1.84 / 15.288 / 58	1.75 / 15.061 / 58	1.67 / 14.833 / 58	1.59 / 15.613 / 60	1.52 / 15.382 / 60	1.46 / 16.155 / 62	1.40 / 15.921 / 62
36	3.27 / 13.668 / 52	3.00 / 14.495 / 54	2.77 / 14.279 / 54	2.57 / 14.063 / 54	2.40 / 14.874 / 56	2.25 / 14.653 / 56	2.12 / 14.433 / 56	2.00 / 15.230 / 58	1.89 / 15.006 / 58	1.80 / 15.795 / 60	1.71 / 15.567 / 60	1.64 / 15.338 / 60	1.56 / 16.117 / 62	1.50 / 15.886 / 62	1.44 / 16.658 / 64
40	3.64 / 15.561 / 58	3.34 / 15.349 / 58	3.08 / 15.136 / 58	2.86 / 15.961 / 60	2.67 / 15.764 / 60	2.50 / 15.528 / 60	2.35 / 16.339 / 62	2.22 / 16.119 / 62	2.10 / 16.920 / 64	2.00 / 16.697 / 64	1.90 / 16.473 / 64	1.82 / 17.262 / 66	1.74 / 17.035 / 66	1.67 / 17.818 / 68	1.60 / 17.588 / 68
42	3.82 / 15.983 / 60	3.50 / 15.773 / 60	3.23 / 16.605 / 62	3.00 / 16.391 / 62	2.80 / 16.177 / 62	2.62 / 16.994 / 64	2.47 / 16.777 / 64	2.34 / 16.557 / 64	2.21 / 17.364 / 66	2.10 / 17.142 / 66	2.00 / 17.939 / 68	1.91 / 17.714 / 68	1.83 / 17.489 / 68	1.75 / 18.275 / 70	1.68 / 18.047 / 70
45	4.09 / 17.139 / 64	3.75 / 16.930 / 64	3.46 / 16.719 / 64	3.22 / 17.553 / 66	3.00 / 17.340 / 66	2.81 / 18.161 / 68	2.65 / 17.945 / 68	2.50 / 17.728 / 68	2.37 / 18.536 / 70	2.25 / 18.317 / 70	2.14 / 18.096 / 70	2.04 / 18.895 / 72	1.96 / 18.671 / 72	1.88 / 19.463 / 74	1.80 / 19.237 / 74

FIGURE 24-6E *Speed ratios—center distance—chain lengths* (Courtesy Boston Gear Incom International Inc.)

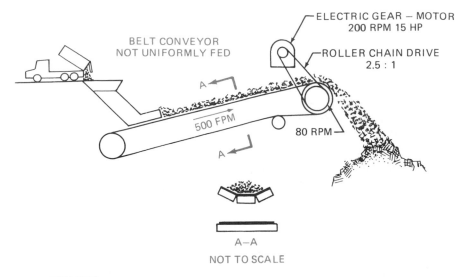

FIGURE 24-6F *Belt conveyor system schematic*

PHASE II GENERATE ALTERNATIVE SOLUTIONS: SYNTHESIS

STEP 5 Generate Solution Ideas
    a. Use Systematic Techniques
    b. Use Creative Techniques and Attitudes

Step 5 Generate Solution Ideas. Review the solution ideas already collected and realize that additional effort and discipline need to be exerted to generate more. Frequently, an idea for a solution occurs early in the process, and there is a tendency—especially among students, because of their time constraints—to select the first idea as the final solution. Successful problem solvers know that additional ideas can be generated beyond the initial obvious ones by continuing to focus on the design problem and employing two types of idea-generating techniques: systematic techniques and creative techniques. Having a creative attitude also helps.

    a. *Use Systematic Techniques.* Verb lists provide a systematic checklist for generating new ideas, primarily building on existing ones. Ask the following questions regarding the idea being considered:

- Can it be modified?

- Can it be made larger? Smaller?

- Can it be used for something else?

- Can other materials be substituted?

- Can components be rearranged?

- Can it be combined with another idea?

Divide an idea into smaller parts so that each part can be systematically examined and developed.

Human Factors. There are published tabulations of body dimensions of average males, females, and children. These are called the *human factors.* This statistical data may be found in textbooks and handbooks: however, for classroom designs use a classmate or yourself for measurements, since most of the adult population is close to the average size. Use children if needed.

For a design file, observe or measure objects designed for average humans. Typical designs include chairs, tables, desks, computer stations, work counters, doorknobs, ladder rungs, switches, handles, and numerous objects encountered daily.

Also, consider and observe the various techniques that are used to interface machines with human operators. A variety of sounds such as voices speaking messages, a number of visual devices using a multitude of colors, numerous tactile devices such as clicks during dialing, and different shapes of knobs are all used.

Try a 180-degree approach. For instance, instead of breaking a walnut shell with a hammer, poke a hollow needle inside and explode the shell with air pressure.

List the obvious attributes of an object to uncover ideas.

- Describe the size using dimensioned sketches.

- List the intended functions of the object.

- Outline how it was probably manufactured.

- List the materials used to make it.

- Describe any other features, such as aesthetics, weight, and color.

Use attribute listing, not only to generate ideas, but also as a checklist for the design of parts on technical drawings. That is, check for complete dimensions to describe the geometry of each part and dimensions to locate each feature, such as bolt holes, slots, special surfaces, and datums. Review appropriate dimensions to ensure the function of the part, for example, clearances and fits. Check the material specified for the part, and whether or not the manufacturability of the part was considered during the design phase. Finally, check all remaining attributes, such as how easily the part can be assembled, surface finishes, heat treatments, and aesthetics. Also list the not-so-obvious attributes, such as other uses that were not intended functions. Developing the habit of proposing these "not-so-obvious" uses for objects will help to generate new ideas for future designs.

Figure 24-7 shows the obvious attributes of a paper clip, and the not-so-obvious ones are listed in *Figure 24-8*.

Browse through technical texts, periodicals, and handbooks that contain numerous ideas for components, mechanisms, and applications of existing products. The list that follows includes the three types.

EXAMPLE OF OBVIOUS ATTRIBUTES

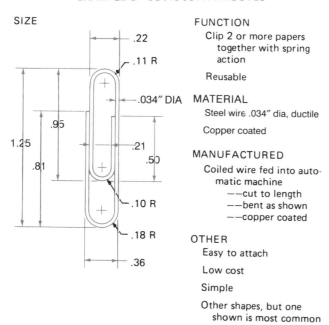

FIGURE 24-7 *Attributes of a paper clip*

EXAMPLES OF NOT-SO-OBVIOUS ATTRIBUTES	
AS IS	**CHANGE**
–dig out ear wax	–fish hook
–make a chain	–cake tester
–balance for paper airplane	–electrical conductor
–clip other items	–chess pieces
–sinker for fishing	–lock pick
–checkers	–artistic shape
–poker chips	–therapy–bend for relaxing
–hanger for small picture	–melt many to make casting
–standard of weight	–fatigue failure demo

FIGURE 24-8 *Not-so-obvious attributes*

- Texts:

Herkimer, Herbert, *The Engineer's Illustrated Thesaurus*, 1952, Chemical Publishing Co.

Schwartz, Otto B./Grafstein, Paul, *Pictorial Handbook of Technical Devices*, 1971, Chemical Publishing Co.

McCormick, Ernest J., *Human Factors Engineering*, 1970, McGraw-Hill, Inc.

Chapanis, A., *Man-Machine-Engineering*, 1965, Wadsworth Publishing Co.

- Periodicals (found in most libraries):

Machine Design

Product Engineering

Test and Measurement World

Control Engineering

Medical Product Manufacturing News

Hydraulics and Pneumatics

Research and Development

Engineer's Digest

- Handbooks:

Shigley, Joseph E., *Standard Handbook of Machine Design*, 1986, McGraw-Hill, Inc.

Laughner, V. H./Hargan, A. D., *Handbook of Fastening and Joining of Metal Parts*, 1956, McGraw-Hill, Inc.

Horger, A. J., *ASME Handbook Metals Engineering Design*, 1965, ASME.

b. *Use Creative Techniques and Attitudes.* A clear understanding of the *creative process* will help to generate ideas. The process—as described by technical people, composers, writers, builders, doctors, hobbyists, and many others—proceeds as follows:

(i) Accumulate all the information about the problem being attacked; that is, thoroughly and conscientiously get involved in the problem, even emotionally if necessary. This involvement, according to creative individuals, causes the subconscious portion of the brain to be alerted that a problem needs solving. Next, relax and let the information collected . . .

(ii) Incubate with other ideas that the subconscious brain retrieves from memory locations. Take a walk, jog. Plop both feet upon a desk and lean back while contemplating the ceiling. (Supervisors may need to be informed that the "process" is working.) Soon, one hopes an idea will surface in the conscious portion of the brain; that is, the idea . . .

(iii) Illuminates its presence and causes exclamations like, "Eureka, I've got it!" Finally, the idea needs to be . . .

(iv) Verified. Does it actually solve the problem? Meet the constraints? Satisfy the specifications?

Develop a creative attitude and behavior by adopting characteristics of creative people. Several are listed here.

1. *Keep an open mind. Be willing to accept change and new ideas.*

2. *Look for similarities and unusual relationships.*

3. *Think positive. Avoid or ignore negative comments such as:* "It's been tried before." "It's not logical." "What a dumb idea." "It'll never work."

4. *Be persistent.*

5. ALWAYS CARRY A NOTEBOOK!!!!!!! Ideas come when least expected and may be lost if not recorded.

Participate in exercises designed to revive creative talents, buried during years of being told what to do and to conform. For example, the late Professor John Arnold, M.I.T. and Stanford University, proposed a fictitious planet, Arcturus IV, populated with creatures who were unusually tall and slow moving, had reaction times one-tenth as fast as humans, had three fingers on each hand, had three eyes, one of which had x-ray vision, and breathed methane. Professor Arnold asked his students to design an automobile-type vehicle for these inhabitants. Tackle this problem before reading on, and compare your design ideas with those of other members of your class. For example, what speed limit should be required due to the slow reaction time of the inhabitants?

Build a simple model related to the design problem being developed. Building a model, or a prototype, almost always produces new ideas because of the simultaneous involvement of the hands, eyes, and brain, plus the three-dimensional insight. Models can be inexpensive—cardboard, balsa wood, tape, etc. Computer graphics can be used.

Do exercises to develop imagination, for the same reason that muscles are exercised: for growth and to avoid atrophying. One convenient source of creative exercises is the daily newspaper, which contains word puzzles, crossword puzzles, and picture puzzles.

Try the following exercises to help develop a creative attitude:

1. Use analogies. For example, pretend to be one of the components in a machine and imagine how each part would function if designed a certain way. What would happen if the design were changed?

2. Systematically propose that a product be changed in any one of the following ways: made bigger, smaller, longer, shorter, colorful, drab, nontraditional, expensive, or cheap.

Read books devoted to developing creativity and try some of their recommendations.

A person's creative potential can be enhanced by taking courses in areas outside of engineering and technology. Take courses to obtain different points of view in other disciplines, such as biology, geology, agriculture, environment, political science, art, philosophy, or oceanography.

Brainstorm according to the ideas of Alex Osborne in his *Applied Imagination*. The procedure is simple to follow. Meet with a group who are willing to abide by the following guidelines:

1. Have one person *record ideas as they are generated,* on a surface visible to all participants. New ideas often stem from the visible ones.

2. Agree that *no one is to criticize* any ideas while the brainstorming is in progress, no matter how ridiculous an idea may seem. Frequently, viable ideas stem from less obvious ones.

3. Meet in as *pleasant surroundings* as practical, and include refreshments. Several one-hour sessions produce better results than one long session.

4. Finally, *meet to critique the ideas* for possible solutions.

Brainstorming can also be done individually. Develop observation skills through freehand sketching by looking at a component while imagining how a sketch of the component would be made. Of course, do the sketch if possible! Looking at an object while imagining how a sketch would be made causes the viewer to see more about the object than he or she would if only a casual glance were given. Try sketching assemblies of components, also.

Assume now that a number of design ideas have been generated. The group usually wants to continue in this creative mode because creating is a satisfying activity. However, since time controls the schedule, the group needs to move on. Students involved in the design process for the first time reluctantly leave this synthesis phase believing that if they had more time they could do a better job. Creative professionals know that they must do the best they can in each phase, in the time available, and then move on to complete their projects within the schedule.

So now is the time to focus on evaluating instead of generating.

PHASE III EVALUATE ALTERNATIVE SOLUTION IDEAS: ANALYSIS

STEP 6 Sift, Combine, and Prepare Solution Ideas for Analysis
   a. Preliminary Preparation
   b. Specific Preparation

STEP 7 Compare Solution Alternatives to Constraints and Specifications
  a. Iterate
  b. Persevere
  c. Qualitative Evaluation
  d. Quantitative Evaluation

STEP 8 Conduct a Design Review
  a. In-House Design Reviews
  b. Outside Design Reviews

STEP 9 Refine the Final Solutuion

Step 6 Sift, Combine, and Prepare Solution Ideas for Analysis. Sift through all the collected ideas and keep those that seem to have potential as a part solution or as a complete solution. File the remaining ideas for possible use in the future.

  a. *Preliminary Preparation.* Expand each potential idea in more detail. Clarify, if possible, vague statements and generalities. Ask each member of the group to develop preliminary statements and sketches of those of his or her ideas that survived the initial sifting and culling.
    Do further research, if needed, to expand the ideas.

  b. *Specific Preparation.* Prepare *layout development drawings* of the preliminary sketches. Layout drawings are essentially a designer's freehand sketch redrawn to scale with instruments or a computer. Designers use two types, design development and kinematic development (the latter is used primarily for mechanical products). Schematic layouts, using lines and symbols, but not to scale, are used for electronic circuits, electrical circuits, piping diagrams, routing paths, processing steps, chemical plants, and logic circuits.

A design development layout drawing is made from a freehand sketch of a design idea for an anchor bolt puller shown in *Figure 24-9A*. The design idea is drawn to scale in the design development layout drawing, *Figure 24-9B*. Notes and comments may be in writing or in casual lettering. Compare this informal approach to the formal technical detail and assembly drawings for shop use, which have standard lettering, lines, and symbols made on standard-size drawing sheets, complete with title blocks.

Use the design development layouts to establish the locations and sizes of parts to be fabricated so as to

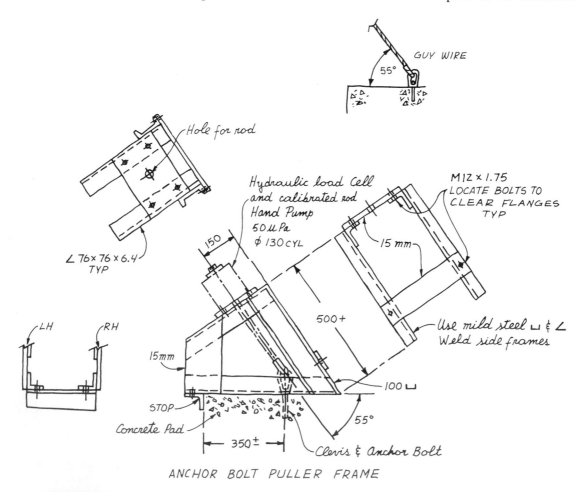

ANCHOR BOLT PULLER FRAME

FIGURE 24-9A *Freehand sketch of a design idea*

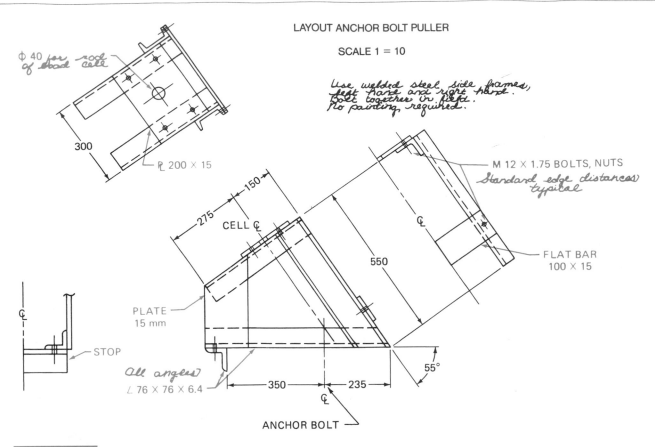

LAYOUT ANCHOR BOLT PULLER

SCALE 1 = 10

Use welded steel side frames, left hand and right hand. Bolt together in field. No painting required.

Φ 40 *for root of stud cell*

300

℄ 200 × 15

150

275

CELL ℄

550

M 12 × 1.75 BOLTS, NUTS
Standard edge distances typical

FLAT BAR
100 × 15

PLATE
15 mm

STOP

all angles
∠ 76 × 76 × 6.4

350   235

55°

℄

ANCHOR BOLT

FIGURE 24-9B *Design development layout*

ensure the functions of the final product. Also use this layout to verify that purchased products will fit where required.

In industry, designers forward their design development layout drawings to drafters who prepare the formal technical detail and assembly drawings. Students in classrooms do both.

A kinematic development layout drawing is shown in *Figure 24-10*. This type of layout shows the relative motions of assembled components, verifies the function of the assembly, checks for interferences, and does velocity plus acceleration analyses. The kinematic analyses are usually done in a two-dimensional format even though the assemblies are, of course, three dimensional.

A useful complement to the kinematic development layout drawing is a simple cardboard-thumbtack model, as illustrated in *Figure 24-11*. CAD systems with programs for kinematic analyses are also useful and available.

Use cardboard-and-glue models (noted earlier) to further complement the layout drawings. Students, designers, drafters, and practicing engineers attest that two major phenomena happen during the process of making a model of a portion or all of a design from a design development layout. One, they discover that the components in the three-dimensional model may not work as planned, so adjustments and refinements are

required. And two, new solution ideas surface, ideas no one had considered. They believe that the combination of using their hands and eyes while shaping and assembling materials triggers an untapped area in their subconscious minds. Furthermore, since ideas evolve as

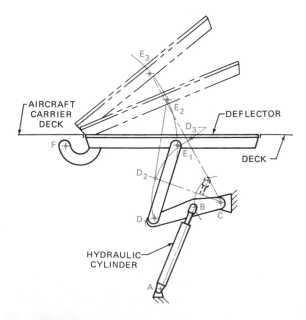

AIRCRAFT CARRIER DECK

DEFLECTOR

DECK

HYDRAULIC CYLINDER

FIGURE 24-10 *Kinematic development layout*

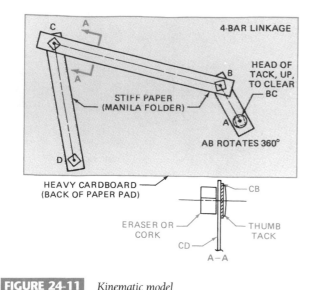

FIGURE 24-11 *Kinematic model*

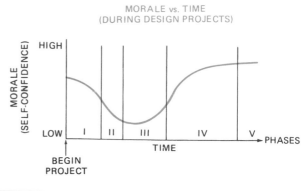

FIGURE 24-12 *Morale curve*

pictures in the mind, sketching, drawing, and model making seem to enhance the forming of these pictures.

Two additional advantages of cardboard models are the relative low cost (compared to prototypes) and the fact that the model may be used later in a design review.

The solution ideas that have been expanded and prepared for further analyses can now be considered as solution alternatives.

Step 7   Compare Solution Alternatives to Constraints and Specifications.

a. *Iterate.* Review Phases I and II for possible additions or deletions to the constraints and specifications. New information may have been discovered. The client may want changes made.

b. *Persevere.* Be cognizant of a morale problem that may occur at about this time in the design process. Imagine that a group, or an individual, has started a project with high hopes and confidence that a solution can be found. However, even though the first five or six steps have been conscientiously done, no real viable solution ideas have been generated. The individual or all of the participants find themselves at a low point in the morale curve shown in *Figure 24-12.* They discover that there is more to the problem than they thought, and useful information is difficult to find. They become discouraged. However, experienced problem solvers know that perseverance will get them through the slump, and that satisfactory results eventually will follow. They know that toward the end of Phase IV their morale and self-confidence for attacking problems will be higher than when they started.

c. *Qualitative Evaluation.* Compare solution alternatives with what has been done before. If a solution alternative is similar to existing designs, either

within the company or in industry, the feasibility of it is relatively easy to determine.

Use a range technique for evaluating preliminary alternatives. Intuition, experience, and judgment help to determine if the size or thickness of a member is within a reasonable range. For example, most students could evaluate design alternatives for home-use ladders based on their own experiences. Similar range approaches have names like back-of-the-envelope calculations and order-of-magnitude analyses. Interestingly, all of these approaches are useful in designing as well as in evaluating.

d. *Quantitative Evaluation.* Build a *prototype* of part or all of a solution alternative. For example, a student group had the problem of compressing a $2' \times 2' \times 4'$ bale of hay into a $2' \times 2' \times 2'$ cube so that their farmer client could ship more hay per shipload from Seattle, Washington, to a port in northern Alaska. The hay compressor was to be part of a complete hay-baler machine. One key factor that they needed for their design was the force required to compress a bale to half size. So they made a crude, but effective, set of rectangular welded-steel boxes, one to telescope within the other, to hold a bale in a testing machine that could generate up to 60,000 pounds of compressive force. See *Figure 24-13* for a conceptual sketch showing the two containers and the bale of hay. The students learned that to compress the bale to half size required 14,000 pounds of force. A hydraulic cylinder was selected to provide the force.

Another type of prototype building is a process called stereolithography. Actual solid models are created layer by layer by an electronic-laser process in a special liquid solution. A focused beam causes the liquid to solidify over exact areas at accurate locations. It adheres to the previous layer, until a complete component is built up.

Two alternatives to actually building a prototype of a proposed design can be done on a computer: wireframe drawing and solid modeling.

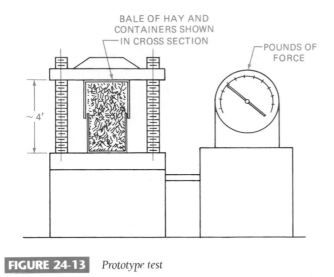

BALE OF HAY AND CONTAINERS SHOWN IN CROSS SECTION

POUNDS OF FORCE

~ 4'

FIGURE 24-13 *Prototype test*

Perspective, wireframe drawings (line drawings in perspective of objects, no hidden lines) on expensive computer systems that can rotate views and zoom into or out of buildings, or marine vessels, are used by architects and marine engineers to visualize what the internal structure would look like.

Solid modeling on a computer screen can be helpful in evaluating a design for shape, structure, and interferences. Also, the solid model can be used as the basis for a finite-element analysis for checking the strength of an object.

Consider the feasibility of each alternative by asking questions such as: How much will it cost? Can the shop make it? Are engineering materials available? Are finished components available? Are sufficient labor hours available? Can the delivery date be met?

Manufacturing Process Costs. The following sample list is headed by the least costly and ends with the most costly manufacturing process. *Costs* are also related to the roughness of the surfaces of the materials. The roughness is a result of the manufacturing or production processes.

Sawing	Rough Surfaces
Cutting (torch)	(lower cost)
Drilling	
Turning	
Punching	
Milling	Average Surface Roughness
Reaming	(Average Cost)
Broaching	
Grinding	
Polishing	
Honing	
Lapping	Smooth Surface (Higher Cost)

Relative Costs of Materials. Use relative cost numbers as an aid to evaluating the feasibility of a design alterna-

tive, and for establishing approximate costs. Mild steel is given a factor of 1.00; all other engineering materials are given relative numbers. Find the current cost of mild steel, and use the relative numbers to establish costs of the other materials. Revise the numbers as needed.

Material	Relative Cost (By Weight)
Mild steel	1.00
Spring steel	2.00
Tool steel	11.00
Cast iron	3.00
Aluminum	6.00
Stainless steel	7.00
Brass	7.00
Copper	7.50
Titanium	130.00
Teflon	67.00
Nylon	19.00
Rubber	4.00
Wood	1.00

Safe Designs. Use the following checklist to provide *safe designs,* safe installations, and protection against negligence suits.

Document the design to show evidence of the following:

a. Accurate calculations

b. Proper use of codes and standards

c. Use of state-of-the-art technology

d. Proper use of materials and hardware

e. Having run tests, if appropriate

f. Use of proper manufacturing processes, proper fasteners, and appropriate heat treatments

g. A quality control program

h. Preparation of instructions for safe use

i. Addition of warning signs to the product

j. Having told future owners of the proper safety devices and practices needed, such as guards around any exposed moving parts of machines, and painted stripes to show danger areas

k. Thorough exploration of all possible uses of the product other than those for which it was designed

Access. Locate fasteners and components for installation and servicing so personnel can have access to them with their hands and tools. Also, provide access to weld areas for welders.

Vibration. Use locking devices such as lock washers and lock nuts on threaded fasteners subjected to vibration. Some high-speed engines have wires passing through holes in the nuts to prevent the nuts from backing off because of vibrating conditions.

Fatigue. Avoid sharp changes of cross-section and small fillet radii on parts subjected to varying loads. See *Figure 24-14* for examples of good and poor practices. The sharp corners and radii (and scored surfaces) cause stress concentrations and therefore, possible fatigue failure.

Cleaning/Corrosion. Provide sloping surfaces wherever standing fluids could cause corrosion. For example, tank bottoms should slope toward a drain hole so that chemicals and cleaning fluids can be completely drained. Upper surfaces should slope to prevent standing rainwater or melting snow.

Temperature. Avoid stringent tolerances on dimensions of parts subjected to extreme temperature changes. Furthermore, check the strength and ductility of materials at extreme temperatures. Also, the temperature of a part may need to be specified for verifying some critical dimensions at that temperature if tolerance on the dimension is less than ±0.0002″.

Prepare a *decision table* to systematically compare each solution alternative with every other alternative as to how well each one satisfies the constraints and specifications from the definition of the problem in Step 3.

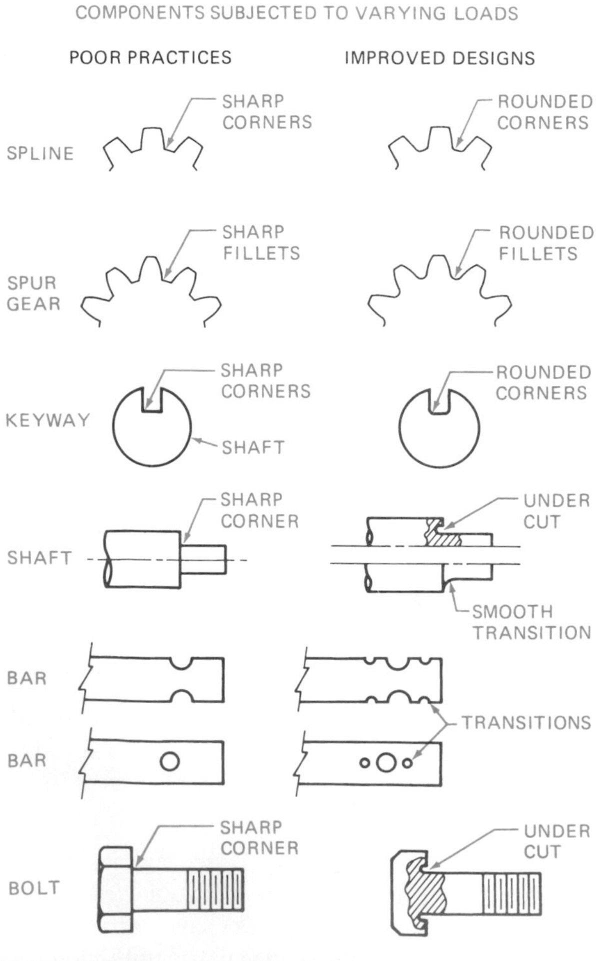

An example decision table is shown in *Figure 24-15*. Although the example is simple, and was made up for illustration purposes, the procedure is the same for more complex analyses. The example is based on the question to a class, "Which alternative for transportation to and from school for a distance of five miles (10 miles round trip) is the best one?" The class was asked to create a decision table for a quantitative evaluation of four transportation alternatives, walk, bike, bus, and car. First, a list of criteria was generated and tabulated in the left-hand column. Next, each criterion was compared to every other criterion and was given a weighting factor for its relative importance according to the students.

The four alternatives then were systematically compared against each other for each criterion. First, the expense criterion was considered. Walking was the least expensive, so it rated a score of 10 as compared to a car, which was given a 4. Biking and bussing scored 7 and 6, respectively.

After the four alternatives were compared in relation to each criterion, products of the weighting factors and the scores were totaled for each alternative. Walking scored the highest; therefore, walking would be the alternative to investigate and develop further. The bus and car are close contenders and should also be looked over once more.

To evaluate solution alternatives in a design project, list the constraints and specifications in the left-hand column and give them weighting factors. Then list the alternatives across the top. Compare and rate every alternative with the others for each constraint and specification. Finally, compute the products of the weighting factors and the alterative scores, total the products for each alternative, and select the alternatives with the higher totals for further study.

Finally, select the best one, realizing that it is the best for the time available, the information collected, and the judgment of the group. Even though it is difficult to do,

COMPONENTS SUBJECTED TO VARYING LOADS

FIGURE 24-14 *Fatigue considerations*

DECISION TABLE									
Criteria	Weighting Factor	Walk		Bike		Bus		Car	
$	10	10	100	7	70	6	60	4	40
Protection from weather	7	3	21	2	14	7	49	8	56
Time spent	6	2	12	5	30	4	24	6	36
Safety	3	7	21	6	18	9	27	8	24
Reliability	8	9	72	8	64	6	48	7	56
			226		196		208		212

FIGURE 24-15 *Decision table*

a decision must be made so that the project stays on schedule.

Schedule a design review next.

Step 8   *Conduct a Design Review.*

a. *In-House Design Reviews.* Prepare drawings, transparencies, slides, models, and appropriate documents to effectively and efficiently communicate the design solution to the attendees.

In a classroom, present sufficient information so that the students in other groups will ask thoughtful questions and make helpful comments. The goal of the review is to benefit from the collective expertise of classmates and to be sure that all facets of the design have been considered.

The instructor may be the only nonstudent present, as clients usually do not attend. Also, the instructor may schedule more than one design review throughout the project or may have conferences with each group.

During the design review, have someone record the questions and comments for modifying or reassessing the solution soon afterwards.

An in-house design review in industry is more comprehensive because, in addition to technical personnel, representatives from accounting, finance, sales, management, manufacturing, shipping, and installation may be present. Furthermore, the design solution discussed may involve a variety of products or an entire complex system.

b. *Outside Design Reviews.* An outside design review provides information to the client at a location selected by the client. Often, the main objective of an outside design review is to "sell" ideas to a client, whether the client is a private citizen, a private company, or a governmental organization. Companies and governmental organizations publish requests for bids or proposals on a variety of projects, such as aircraft, spacecraft, dams, highways, buildings, chemical processes, electronic devices, electrical systems, power plants, vehicles, military devices, ships, submarines, and numerous products. The bidding individual or proposing organization prepares an outside design review, and if the bid or proposal is accepted and a contract is signed, the bidder starts a new design process in more detail than the one leading to the outside review.

Students may give outside design reviews to a client as progress reports; however, most classroom projects require only in-house design reviews that lead to the final solution.

Step 9   *Refine the Final Solution.* Review the questions, comments, and suggestions recorded during the in-house review and decide which ideas to incorporate in design changes or additions. Revise previous steps if necessary; however, time constraints dictate that the next phase should be started soon.

PHASE IV DOCUMENT, SPECIFY, AND COMMUNICATE THE SOLUTION

STEP 10   Prepare Drawings and a Written Report

STEP 11   Give an Oral Presentation

In classroom projects, Phase IV requires about one-half of the time available. Moreover, most classroom projects end with Phase IV because of limited time, limited shop facilities, and limited funds.

In industry, Phase IV and Phase V overlap, and together they consume approximately one-half of the time available.

Step 10   *Prepare Drawings and a Written Report.* Use the informal development layout drawings for preparing the formal technical detail and assembly drawings in standard views, lines, dimensions, and symbols. Only those parts that require shop processes are detailed in technical drawings. Parts—such as fasteners, keys, and pins—that do not require shop processes only appear in the parts list of the assembly drawings. Purchased components such as motors, switches, and bearings appear in assembly drawings in the parts list and are usually drawn in outline form.

Four examples of technical detail drawings of components to be made in a shop are shown in *Figure 24-16* and *Figures 24-17 A, B,* and *C.* Figure 24-16 is a subassembly of structural steel parts, welded together for use in the anchor bolt puller. Figures 24-17A, B, and C show several individual components. *Figure 24-18A* (page 954) is an assembly drawing with a parts list (*Figure 24-18B,* page 954) for the complete anchor bolt puller. Note that the subassembly from Figure 24-16 is given only one part number in the assembly drawing parts list.

Logical Dimensioning. Design components so they can be measured readily and logically—for two main reasons. One, shop personnel need to locate points, planes, and features such as holes from reference planes or surfaces. Two, the function of the component needs to be ensured.

Shop personnel use a large number of measuring devices, some very sophisticated, some relatively simple, but all measure from a reference plane or surface that is usually machined. Some of the more common measuring devices and their measuring capabilities are:

a. Flat, stainless steel scale: to 0.010″

b. Micrometer: inside and outside: to 0.0002″

c. Vernier calipers: inside and outside: to 0.001″

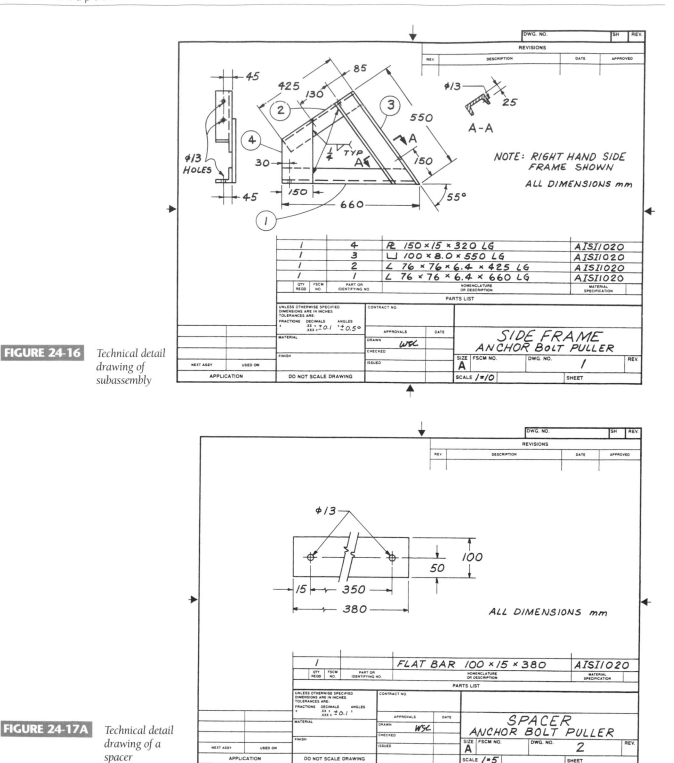

FIGURE 24-16 *Technical detail drawing of subassembly*

FIGURE 24-17A *Technical detail drawing of a spacer*

d. Lathes, milling machines, shapers, boring machines, and gear cutters with calibrated dials to indicate the movement of the cutting tool or the part being machined: to 0.0005″

e. Some newer milling machines and numerically controlled machines with electronic readouts: to 0.0001″

Engineers, designers, and drafters can facilitate these shop processes by dimensioning components from obvious reference planes or surfaces. The plane or surface may need to be created in the manufacturing process.

Ensure the function of a component or an assembly of components by dimensioning to common reference

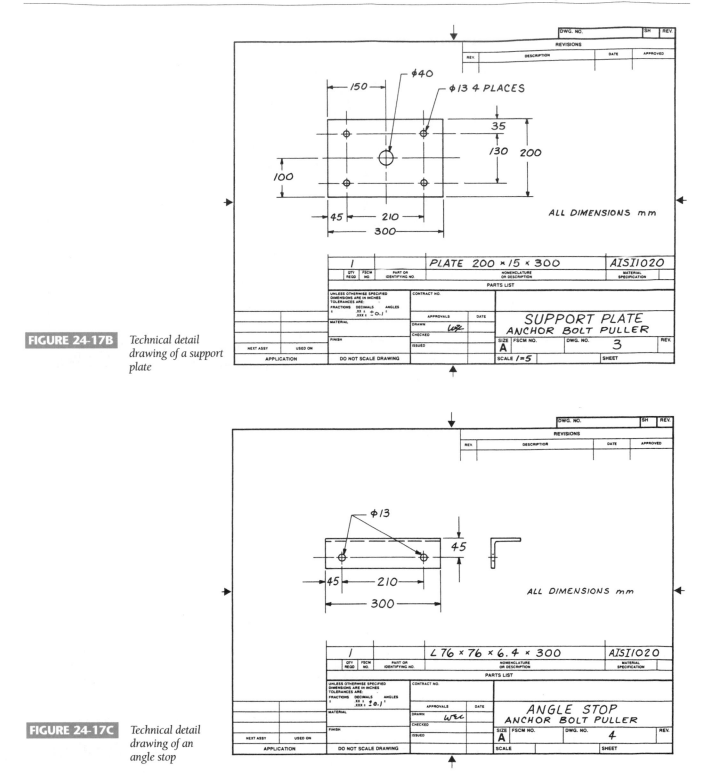

FIGURE 24-17B *Technical detail drawing of a support plate*

FIGURE 24-17C *Technical detail drawing of an angle stop*

planes or surfaces. An example to illustrate reference planes is shown in *Figure 24-19*. The two identical L-shaped brackets are cast iron. Surface no. 1 is milled to establish a plane of reference to locate vertical surface no. 2 to be perpendicular to surface no. 1. Then, the hole at no. 3 is machined perpendicular to surface no. 2 (and hence parallel to surface no. 1) so the shaft for the roller will be horizontal. The two holes in the bracket would

probably be located with respect to vertical surface no. 2 and centered about the center line of the machined hole, surface no. 3.

A more exact dimensioning technique called positional tolerancing and form tolerancing (or geometrical tolerancing) establishes datum surfaces for referencing dimensions and features. The technique requires more skill and time to apply; however, time and money are

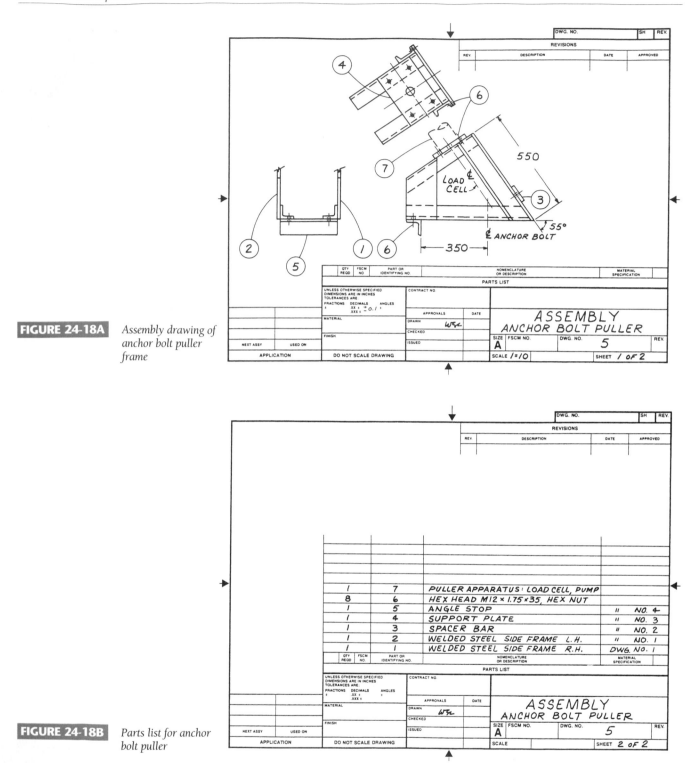

FIGURE 24-18A *Assembly drawing of anchor bolt puller frame*

FIGURE 24-18B *Parts list for anchor bolt puller*

saved because fewer parts are scrapped. They fit better and their function is assured.

This tolerancing technique has grown as machines and systems have become more complex and expensive. Furthermore, when one company manufactures components that will be assembled together with components from another company, the dimensioning and features need to be accurately located so the components will go

together. Refer to Chapter 12, Geometric Dimensioning and Tolerancing, for detailed information.

Technical detail and assembly drawings in industry are similar to the anchor bolt puller drawings, but with several differences. Complex drawing-numbering systems tell the user which ones are details, where individual components are used, which ones are assemblies, and which job or contract used the drawings. These sys-

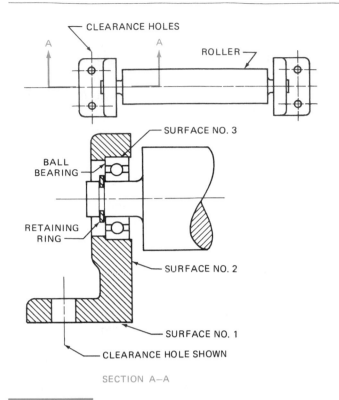

CLEARANCE HOLES

ROLLER

SURFACE NO. 3

BALL BEARING

RETAINING RING

SURFACE NO. 2

SURFACE NO. 1

CLEARANCE HOLE SHOWN

SECTION A—A

FIGURE 24-19 *Reference plane for dimensioning*

tems are as varied as the number and size of industries and organizations. (Refer to Chapter 17.) Usually the drawings are on larger sheets. Also, each individual part needing to be processed in a shop is detailed separately to accommodate accounting procedures and to merge into the flow of information within the company.

DESIGN HINTS

Use *design hints* during all phases of the design process as extensions to the design hints already included. These design hints help in initial designing and become checklists as the project progresses. Keep a file during design problem experiences to expand these lists, and remember: CARRY A NOTEBOOK AT ALL TIMES!!!

Attribute Listing for Checking Drawings. Use the technique of *attribute listing* as a systematic guide for checking the completeness of technical detail drawings. For example, ask the following questions:

a. What is the function? The function dictates the dimensioning plan.

b. What size is it? Are all geometrical shapes dimensioned? Are all the features located?

c. What materials are used?

d. Can it be made? How will it be manufactured? What processes will be used?

e. What other facets need to be checked? Surface finish? Weight? Vibration? Safety? Aesthetics?

"Rules of Thumb." Use "rules of thumb," practical guides based on experience, to establish or check "ball-park" (within + or −10%) quantities. Use them as qualitative checks.

a. *Tolerance.* As a beginning problem solver in technical fields, keep in mind that the tighter the tolerance on the dimensions of an object, the more the object costs. Tolerances should only be as tight (small) as needed to ensure the function of the part. Use as a conceptual reference the thickness of the paper this sentence is written on: It is .003″ to .004″ or 0.08 mm to 0.10 mm thick. Most machinists and machines can easily work to .001″, but .0001″ takes more time and costs more.

b. *Rigidity.* "When in doubt make it stout." This statement is just a reminder that not all designs need to be based on an extensive stress analysis and strength of a material. Rigidity and accurate locations may be more important than strength. For example, a surveyor's transit needs to be rigid to keep the lenses aligned. Instrument panels need to be rigid to support the dials and meters and to prevent vibration. Another practical application of the statement refers to sizing members when budgets are low. That is, a brief stress analysis producing a conservative size (stout) is less expensive than an extensive analysis and a more exact size.

c. *Murphy's Law.* "Anything that can go wrong will go wrong." Use this guide to evaluate how a designed part or assembly might behave. A more important application is to ask, "How could someone misuse the part or assembly?"

d. *Weld Size.* Fillet and butt welds constitute approximately 85% of all welds used in industry. And if the thinnest cross-section dimension of the weld is at least as thick as the thinnest member joined, the weld will be as strong as the members, assuming that the weld is done by a skilled welder.

e. *Threads.* The depth of threads for steel threaded fasteners should be equal to the diameter of the fastener. Use 1.5 to 2.0 times the diameter for the depth of threads for non-ferrous threaded fasteners, or steel fasteners in non-ferrous metals.

f. *Intuitive Ranges.* Early in a design, establish sizes by asking what would be a reasonable range for this size. Establish performance, weights, and function in a similar manner. During an analysis, ask questions such as: Is this a reasonable size? Weight? Function? Performance?

Prepare a written report for the client. Although each client's needs in a report differ, the parts of reports are similar and all the parts should be included. The order is arranged to suit the information to be communicated.

Use the following discussion as a guide for preparing professional, typed project reports to clients. Refer to samples from a student group's report later in the chapter.

a. *Letter of Transmittal.* A letter to the client, separate from but enclosed with the report, saying that the project has been completed, the report is enclosed, and that working with the client on future projects is anticipated. Fee statements may be a part of the letter. Include phone number(s) where the client can call for questions, and an address for mailing fees.

b. *Cover.* Enclose the report in a folder, unless otherwise directed.

c. *Title Page.* Include title, by whom written, to whom submitted, date, and organization (college or company).

d. *Abstract.* Include one or two sentences about the content of the report to capture a potential reader's interest. For product designs, describe the product, specify its performance, and give the cost or price.

e. *Table of Contents.* List the main headings and the subheadings, with appropriate page numbers. List the illustrations, drawings, and tables, plus an appendix if used.

f. *Introduction.* Describe briefly—who, what, when, where, and why—the origin of the project/problem, because even though the client knows this information, other members of the client's organization—who may not—might read the report.

g. *Solution Description.* A written portion should use a liberal number of headings and subheadings and should describe the final solution, including the performance specifications and the design specifications. List the constraints, costs, materials, and special features. Also include the delivery date. Refer to illustrations, figures, charts, graphs, and drawings for efficient communication.

h. *Analyses.* Summarize the major assumptions and calculations. Place detailed *analyses* in an appendix.

i. *Costs.* For a classroom project, which usually ends with Phase IV, list the basic direct costs that the client would have to outlay to have the design made. These costs could include materials, shop labor, and a number of purchased components—fasteners, pins, shafts, switches, controls, electronic circuits, motors, and similar hardware. Even if the classroom project continues through Phase V and a prototype is built, the client will most likely be required to reimburse the students **only** for the actual expenses incurred. The excellent learning experience is payment for the students' labor!

In industry, only the selling price would be given in the report because the basis for arriving at the selling price (labor rates, material costs, overhead, sales, and profit margin) would be proprietary information.

j. *Drawings.* Include the technical assembly drawings in the main report, and the technical detail drawings in an appendix. Pictorial drawings (isometric, oblique, or perspective) may also be appropriate, depending upon the client's technical experience and background.

k. *Illustrations, Figures, Charts, and Graphs.* Check the illustrations, figures, charts, and graphs for complete information. Each one should have a legend. In addition, charts and graphs should have clearly marked and labeled axes.

l. *Alternatives.* Convince the client that a thorough investigation was done by the design group by including a brief summary of the *alternatives* that were considered but discarded. Use partially dimensioned sketches, drawings, and pictorials with appropriate comments as to why the alternatives were not selected.

m. *Recommendations.* Discuss briefly any recommended further developments or investigations that relate to the project. For example, one of the discarded alternatives may have some potential and should be investigated further.

n. *References.* List references and the specific pages that were used. Number the references, and use these numbers in the text of the report.

o. *Appendices.* Place information in the appendices that does not logically fit in the main report. Such information includes extensive calculations, computer printouts, vendor brochures, catalogs, machine copies of referenced material, and technical detail drawings.

In industry, when a company or organization wants to "sell" its expertise to clients it includes a roster of individuals, indicating each individual's technical skills, education, and experience. A summary of projects completed by the company may be included.

Most organizations have an established format for their reports. A number of companies and organizations use computers to store boiler plate information that appears in all of their reports. Standard paragraphs and phrases appear at the beginning and at the end of their reports—hence, the wrap-around image of a boiler. New material fills the blanks.

Rewrite the report text at least two times, whether a classroom report is being prepared or one in industry!! Check for completeness. Check grammar, punctuation, and style. Avoid clichés. Eliminate unnecessary words.

Step 11   Give an Oral Presentation.   Prepare an *oral presentation* of the design solution to the client. The following guidelines for planning a presentation apply to student presentations as well as those done in industry.

a. For a simple, but effective presentation, first tell the audience **what is** going to be presented. Then, present the material. Concluding, tell the audience **what was** presented.

b. Use transparencies on an overhead projector in relatively small classrooms (30 persons) and boardrooms. Cardboard borders for mounting transparencies are useful for notes, key words, and phrases.

c. Use slides in larger rooms, lecture halls, and auditoriums.

d. Make projected words and numbers large enough so everyone can read them. A conservative guide is to make the letters on the screen as high as 1/250 of the distance to the person the farthest away from the screen. Thus a 25-meter distance requires letters on the screen to be 0.1 meter high. Typed material reproduced on transparencies is usually unsatisfactory.

e. Restrict the number of different ideas on a visual to five or less. If more are on the transparency, distribute a copy to the audience beforehand.

f. Use a pointer.

g. Use models large enough to be seen easily.

h. PRACTICE at least once. One or more persons can do the presentation.

i. Allow time for questions.

In industry, a final oral presentation made to a client at the end of Phase IV could be done by a consulting-type firm whose final "product" consists of engineering drawings and a written report.

Oral Presentations.   Public speaking is an important skill for engineers, designers, and drafting technicians.

During the design process discussed in this chapter, oral presentations are scheduled at least two times, the design review and the final presentation. As a guide for giving these two presentations, and others, use the following three practical steps noted earlier.

1. Tell them what you're going to tell them.

2. Tell them.

3. Tell them what you've told them.

Further guidelines are in the exercises at the end of the chapter.

PHASE V DEVELOP, MANUFACTURE, AND DELIVER

STEP 12   Prepare Technical Detail Drawings of Each Component to Be Manufactured

STEP 13   Schedule Assembly, Final Testing, Packaging, and Delivery

STEP 14   Keep Records

Arrange for the necessary paperwork, drawings, materials manufacturing processes, personnel, and funds to develop the device or system to meet the client's requirements.

These requirements may range from modifying an existing device to producing a state-of-the-art system. Thousands of different combinations of requirements occur in industry, so describing them all here is impractical. Consequently, a more appropriate task for this chapter is to present the steps that are common to most industries.

Step 12   Prepare Technical Detail Drawings of Each Component to Be Manufactured.   These drawings "get the wheels turning" within a company. A typical drawing enclosed in a clear-plastic folder may follow the component through a manufacturing area from casting or forging, to machining, to heat treating, to finishing, to assembly and delivery. Additional areas could include checking, quality control, subassembly, and temporary storage.

Step 13   Schedule Assembly, Final Testing, Packaging, and Delivery.   Plan the final processes prior to sending the product to the client. Compare appropriate subassemblies and complete assemblies with the specifications, particularly performance specifications. For example, an electronic device may be tested in a variety of environments before being released for shipment.

Design packaging and shipping containers so that the product will survive shipping without damage. Some devices and assemblies must endure more severe loading enroute than after being installed—for example, large frames, large cylindrical-shaped housings, and pressure vessels.

Write maintenance and service manuals to help prolong the life of the product. Provide instructions for proper and safe use of the product.

Step 14   Keep Records.   Document and file all aspects of the product's development and use. Record the sources and quality of materials used. File the detail and assembly drawings. Summarize the manufacturing processes, and prototype testing results. Add feedback comments from the client. Request cost, accounting, purchasing, and related records from appropriate personnel.

These records are useful for future use if the client wants replacement parts or another complete product. Also, other customers may purchase the product.

Drafters in Industry. Technical drawings were discussed in detail for Step 10 of Phase IV, primarily because student projects usually end with detail and assembly drawings. Drafters in industry would most likely do the detail and assembly drawings during Step 12 of Phase V. The difficulty of their drawing assignments would vary, depending on their experience.

Newly hired drafters soon find that in industry there are a number of sources of assistance. Colleagues in the drafting room can be very helpful by showing new members where to find information on previous jobs done by the company—things like where catalogs are filed, which CDs are available with vendor information, who to contact in the shop for fabricating and assembly practices, and where to have their drawings printed, plotted, and filed.

Experienced drafters not only do the technical drawings, but they become involved in designing. They develop ideas using layout drawings, solid modeling on computers, and frequently use cardboard, glue, and balsa wood to verify their ideas.

Two types of projects for students who are learning the design process are discussed next.

Design Projects: Routine and Non-Routine

ROUTINE DESIGN PROBLEMS

Routine design projects and problems are used to introduce new members of an organization to the various people in the company and to the routes design information needs to flow through. Beginning engineers, designers, technologists, and drafters need to know how jobs are initiated, documented, processed, and completed. An effective way to start these new employees is to assign routine projects and problems to them.

These first assignments usually are assigned to an individual and are characterized as follows:

a. The problem may be a modification to an existing product line, a change in the company's plant facilities, or just a drawing change.

b. The problem difficulty matches the individual's abilities and skills.

c. Information is communicated in a memorandum in a typical industrial format, such as an ECO (engineering change order) or an EDO (engineering design order).

d. Requirements are clearly stated and the information is nearly complete; however, some new data may be needed.

e. An analysis may be required.

f. For guidance, there are similar designs, knowledgeable colleagues, the company library, and vendor catalogs.

g. Technical drawings are required; they become part of the documentation of the modification or change.

h. Labor and materials costs are always present.

i. A condensed design process is followed; however, all the essential steps are followed, including planning, keeping a log, keeping a file, understanding the problem, synthesis, analysis, comparing alternatives, reviewing designs with supervisors, documenting the project in drawings and text, manufacturing, assembly, and installation.

An example first assignment that a newly graduated engineering student received in an EDO was to design a support for a tank containing a liquid for a paper-making process. But before resolving the technical parts of the support, the new engineer, escorted by the chief engineer, was introduced to personnel in drafting, purchasing, the shop, and the paper-making area where the tank was to be installed. Further introductions included individuals in general supplies, copy service, and even those who made coffee.

The technical challenges involved load calculation, sizing beams and columns, and finally, making a design development layout drawing. The chief engineer was consulted before the layout was handed to a drafter. The drafter questioned a few items and then produced technical detail drawings plus an assembly drawing.

The new engineer checked, approved, and signed the drawings. Then he followed them through the shop and ultimately used the drawings at the installation site.

The last step was to document the project. The drawings and vendor information were filed. A short written report summarized the need for the tank support, and a copy of the EDO was included. Cost information from the accounting department completed the file.

NON-ROUTINE DESIGN PROBLEMS

Non-routine design problems are assigned to experienced engineers, designers, technologists, and drafters—their complexity and level of difficulty in direct proportion to the experience, skills, and ability of the recipients. Usually a team of technical personnel attack non-routine design problems.

Non-routine design problems are characterized as follows:

a. The problem originates outside the company, and it stems from the needs of a client who may be an individual, another company, or a governmental organization.

b. A team of individuals with diverse backgrounds or a group with similar backgrounds is formed, depending on the experience, skills, and abilities required.

c. The problem, not clearly defined, must be researched thoroughly so that constraints and specifications can be written.

d. A complete design process, with phases and steps, is required, from the initial planning to the final delivery and installation. More alternatives are generated than for routine problems, and, therefore, more decisions are made.

e. The projects require a longer time span than routine projects.

f. Detailed planning is required for personnel assignments and hours, materials procuring, budget allocations, production processes, and delivery/installation.

g. More engineering trade-offs occur. A *trade-off* is a compromise, usually between costs and materials. For example, an expensive material (titanium) may cost more but weigh less than a less costly material (aluminum) that weighs more.

Since non-routine design problems vary as widely as the number and size of problem sources—individuals to organizations—there are no typical non-routine problems. However, the manner in which they are attacked—the design process—**is** typical. Therefore, follow the phases and steps presented in this chapter to obtain an insight into the efficacy of this process. Then, formulate your own design process as needed.

Examples from a student team member's log, the team's planning chart, and the team's project report are included here for your information. The challenge given to the student team from a doctor was to design a device that would exercise the muscles of a patient's jaw continuously, day and night, to accelerate healing of the jaw muscles after an operation.

The following lists some of the more important aspects of the step process just described. Use it to review the art that follows.

- Log (Step 1) **Figure 24-20**
- Planning Chart (Step 1) **Figure 24-21**

TIME LOG	W. PETERSON		
TEMPOROMANDIBULAR JOINT FLEXOR			
DATE	ACTIVITY	TIME HR	CUM. TIME HR
10-2	Met with Anderson, Bui, Hamamoto- - Exchanged phone numbers - Copied Dr. P's proposal	½	½
10-8	MET - KA MB & AH - discussed proposal - made up questions for Dr. P Discussed planning chart	2 ¾	3 ¼
10-10	Questions for Dr. P MET at Dr. P's Office in MED SCHOOL	2 ½	5 ¾
10-15	Discussed electrical circuits w/ MB	¼	6
10-17	MET w/ group, checked drawings/ went to hobby shop for solenoids Worked on progress report w/MB Typed progress report	2 ¼	8 ¼
10-20	Talked to Mike Wieggand re: valves Group meeting	2	10 ¼
10-30	Contacted vendors for bellows information Prepared transparencies for DESIGN REVIEW - worked with Hamamoto	3	13 ¼
11-1	Trip to surplus yard — found plastics	2 ½	15 ¾

FIGURE 24-20 *Log*

PLANNING CHART

TEMPOROMANDIBULAR JOINT FLEXOR

K. Anerson	M. Bui	A. Hamamoto	W. Peterson

ACTIVITY	WEEK								
	1	2	3	4	5	6	7	8	9
PHASE I		MB							
Get Organized	KA								
Client/Meeting		AH							
Information Search		WP							
Project Define		WP							
Progress Report			MB ⊙		⊙ MILESTONE				
PHASE II				KA					
Generate Alternatives					KA				
PHASE III					AH				
Analysis/Evaluation					AH				
Layout Drawings					KA				
Decision					MB				
Design Review					AH ⊙				
Revise Final Solution						WP			
PHASE IV						WP			
Final Analysis						MB			
Materials						MB			
Rough Draft of Report						WP			
Drawings – Detail Assembly							AH		
Model							KA		
Oral Presentation								KA ⊙	
Final Written								WP ⊙	

FIGURE 24-21 *Planning chart for student group*

PRELIMINARY DESIGN CONSIDERATIONS

1. SIZE   Date 10-10
2. POWER
   –BELT POWER PACK
   –CHARGER
3. CONTROLS
   –FREQUENCY
   –OPENING
   –PATIENT OR DR.?
4. EXPECTED PRODUCTION
   –REUSABLE?
5. COST
6. MATERIALS
   –STAINLESS
   –PLASTIC
   –FDA RESTRICTIONS
7. NOISE LEVEL
8. FORCES REQUIRED
   –OPENING
   –CLOSING
9. SAFETY OVERDRIVE
10. JAW DIMENSIONS
11. ACCURACY

FIGURE 24-22A *File: Preliminary design considerations*

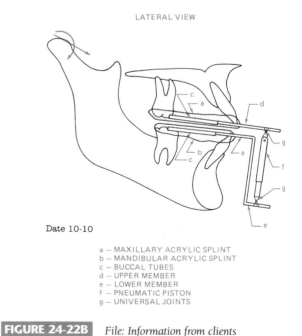

LATERAL VIEW

Date 10-10

a — MAXILLARY ACRYLIC SPLINT
b — MANDIBULAR ACRYLIC SPLINT
c — BUCCAL TUBES
d — UPPER MEMBER
e — LOWER MEMBER
f — PNEUMATIC PISTON
g — UNIVERSAL JOINTS

FIGURE 24-22B *File: Information from clients*

- Information from the Client (Step 1)   **Figures 24-22A, B, and C**
- Progress Report (Step 4)   **Figure 24-23**
- Letter of Transmittal (Step 10)   **Figure 24-24**
- Title Page and Abstract (Step 10)   **Figure 24-25**
- Table of Contents (Step 10)   **Figure 24-26** (page 963)

FRONTAL VIEW

Date 10-10

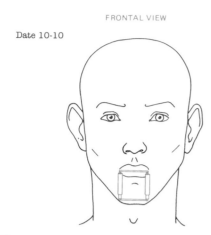

FIGURE 24-22C *File: Information from clients*

- Introduction (Step 10) **Figure 24-27** (page 963)
- Power Pack Assembly Sketch (Step 10) **Figure 24-28** (page 964)
- Alternatives Discarded (Step 10) **Figure 24-29** (page 965)
- Final Recommendations (Step 10) **Figure 24-30** (page 966)
- Appendix: Materials, Electrical Schematics, and Costs (Step 10) **Figures 24-31A, B,** and **C** (pages 966–967)

Modern Design Practices and Standards

Drafters, designers, and engineers entering technical industry today will find that the adjective "changing"

Date 10-10

Department of Mechanical Engineering
(School)
(City, State, Zip Code)

Dr.
University Hospital
(City, State, Zip Code)

Dear Dr.,

This letter is to inform you of our design team's progress and decisions regarding your CPM (continuous passive motion) device. We have discussed the design criteria set forth in your proposal and also examined the additional criteria we discussed with you at our last meeting. Presently, we are considering the feasibility of a pneumatic device with an external power pack. We are currently evaluating this concept and its ability to meet the following criteria:

1) Weight: less than 4 oz
2) Accurate and adjustable movement
3) Frequency in the range of 2–3 cycles per minute with a maximum range of 1–4 cycles per minute
4) Device must be safe for use by a 10-year-old and older for 24 hours at a time
5) Operation must be simple (few moving parts and adjustable by anyone who can read and understand simple directions)
6) Low cost (less than $500, including labor and materials)
7) Parts must be readily available (most from local sources)

Based on our pending evaluation of the above, we will pick what we feel is the best design, and prepare a report outlining our choice. Included in the report will be a cost analysis, detail and assembly drawings, and an overall evaluation. This report will be presented to you at a date and time decided by our instructor.

If you have any questions, please contact us through our instructor.

Sincerely,

W. Peterson

K. Anderson

M. Bui

A. Hamamoto

FIGURE 24-23 *Progress report*

Date 12-6

Department of Mechanical Engineering
(School)
(City, State, Zip Code)

Dr.
University Hospital
(City, State, Zip Code)

Dear Dr.,
This letter is to inform you that we have completed our design for a CPM device to be used in your research project. The enclosed report details the design, giving complete drawings, a cost analysis, and future design considerations.
The design is now ready for the prototype stage.

We are pleased to have had the opportunity to work on this project for you, and we hope we will be able to work with you more in the future.

Sincerely,

W. Peterson

K. Anderson

M. Bui

A. Hamamoto

FIGURE 24-24 *Letter of transmittal*

DESIGN PROJECT: TEMPOROMANDIBULAR JOINT FLEXOR

SUBMITTED TO: DOCTOR

DEPARTMENT OF ORTHODONTICS
UNIVERSITY MEDICAL SCHOOL
(CITY, STATE, ZIP CODE)
Date 12-6

BY: W. PETERSON
     K. ANDERSON
     M. BUI
     A. HAMAMOTO

ABSTRACT

This report contains a description of a continuous passive motion (CPM) device to exercise the temporomandibular joint (TMJ). Briefly, the design incorporates the use of two compressible bellows that are connected by an air hose. While one of the bellows is compressed at a controlled rate, the other bellows expands at a controlled rate. The system is based on user feedback and provides for excellent emergency override, which is essential for safe use. The device costs less than $200.

FIGURE 24-25 *Title page and abstract*

describes current technology better than any other word. A number of these ongoing changes are described briefly in the following paragraphs.

One trend in industries that manufacture devices is to develop prototypes earlier in the design process than in previous years. Two terms describing this trend are *rapid prototyping* and *virtual prototyping*. Stereolithography (SL), mentioned earlier in this chapter, is a rapid prototype builder. SL involves building a solid model layer by layer by an electronic-laser process that uses data from

TABLE OF CONTENTS   PAGES

FIGURE 24-26   *Table of contents*

INTRODUCTION

  This project stems from a graduate research project in the department of orthodontics. Studies have shown that the continuous passive motion (CPM) on joints such as knees, elbows, and fingers have proven to be quite successful in alleviating post-operative pain, swelling, and adhesions. It is the Doctor's hopes as well as ours that a similar application of this research on the temporo-mandibular joint will prove to be equally successful. Our task was to develop a device that would work the jaw up and down to a controlled distance in a slow cyclic manner and at the same time provide a maximum in safety. It was also necessary to use feedback switches to control the inflating and deflating of a bellows that opened and closed the jaw of the patient so that any irregularities or overwhelming resistance to CPM by the individual (sneezing, coughing, yawning, etc.) will immediately be sensed by the system and the necessary override would be provided automatically.

FIGURE 24-27   *Introduction*

3-view drawings. A more recent rapid process uses data from MRI (magnetic resonance imaging) examinations to generate plastic prototypes of human joints that need to be replaced. Doctors use the prototype models for guidance during operations; the data is used also for developing the metallic versions of the joint. This over-all process is so accurate that the replacement joint needs little or no modification before installation.

  An example of virtual prototyping requires a special head-mounted viewer and a hand-mounted positioning monitor for prototype development of a backhoe cab. The person wearing the viewer (probably the customer) can visualize the cab of the machine and the positions of the controls before the backhoe is built. Prior to virtual prototyping, expensive wood models of the cabs were constructed.

  Additional rapid prototyping methods include:

- *Fusion Deposition Modeling* involves controlled placing of low-melting-point thermostatic filament to build up a model.

- *Laminated Object Manufacturing* builds models by fusing layers of plastic-coated papers cut out by a computer-controlled laser.

- *Selective Laser Sintering* prepares "slices" of an object using fused granules.

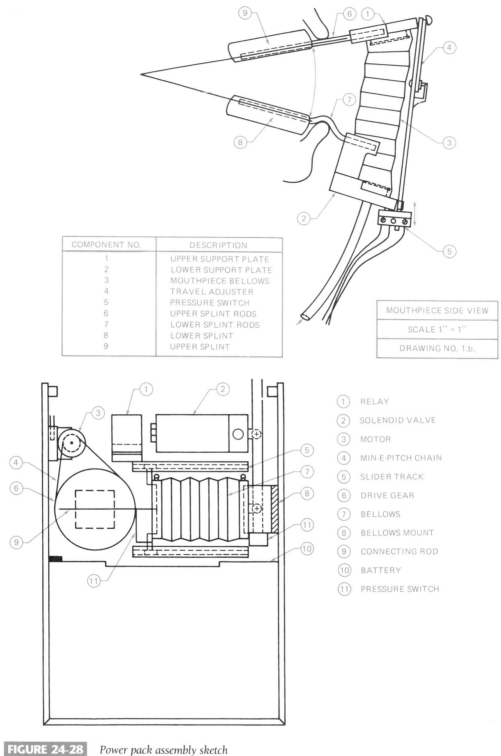

COMPONENT NO.	DESCRIPTION
1	UPPER SUPPORT PLATE
2	LOWER SUPPORT PLATE
3	MOUTHPIECE BELLOWS
4	TRAVEL ADJUSTER
5	PRESSURE SWITCH
6	UPPER SPLINT RODS
7	LOWER SPLINT RODS
8	LOWER SPLINT
9	UPPER SPLINT

MOUTHPIECE SIDE VIEW

SCALE 1″ = 1″

DRAWING NO. 1.b.

1. RELAY
2. SOLENOID VALVE
3. MOTOR
4. MIN-E-PITCH CHAIN
5. SLIDER TRACK
6. DRIVE GEAR
7. BELLOWS
8. BELLOWS MOUNT
9. CONNECTING ROD
10. BATTERY
11. PRESSURE SWITCH

FIGURE 24-28 *Power pack assembly sketch*

• *Solid Ground Curing* uses slices or layers of polymers to build models.

Another trend known as *parametric design* allows drafters and designers to change dimensions in three-dimensional computer models and have all affected dimensions change accordingly. Also, design parameters such as beam sizes, shaft diameters, the number of coils in a spring, and so on, would be modified to suit the dimension changes.

An international trend may affect your future job assignments. The desire of companies—both domestic and international—to sell their products in international markets created a need for international standards to ensure consistent quality and performance. To parti-

ALTERNATIVES CONSIDERED AND DISCARDED

1. Pneumatic Cylinders as Actuators
   Advantages:
   ..small in size
   Disadvantages:
   ..limited control of expansion rate
   ..higher working pressures, noise
   ..difficult to find stroke and size in standard products
   ..lubrication may be a problem
   ..allows minimal user resistance

2. Source of Compressed Air: Bottle and Intricate Valving
   Advantages:
   ..power pack simple and light-weight
   ..no batteries
   ..volume of air easily varied
   Disadvantages:
   ..complicated valving
   ..supplying air bottles cumbersome
   ..compressor needed to charge bottles
   ..noise
   ..difficult to get emergency override

3. Motor-Driven Cam and Linkage
   Advantages:
   ..simple to construct
   Disadvantages:
   ..difficult to interrupt
   ..excessive weight
   ..bulky, physically restrictive
   ..little adjustment
   ..little freedom of jaw movement (patients need rotation, translation, and lateral movements)

FIGURE 24-29   *Alternatives discarded*

cipate and be internationally competitive, your company must obtain certification based on ISO 9000 of the International Standards Organization. Your involvement will most likely be in the technical drawing documentation for certification. Each company must document clearly what it plans to do and then keep copious records to show that it has carried out its plan. An excellent treatise on this subject is *Understanding and Implementing ISO 9000 and ISO Standards,* Goetsch and Davis, 1998, Prentice Hall. The authors digested a somewhat confusing set of ISO standards and substandards and have derived a comprehensive checklist to guide companies seeking registration to and through the proper standards.

Coupled with promoting the requirements of the ISO standards is another trend toward encouraging companies to develop a philosophy of TQM (Total Quality Management). Preparing a company for ISO 9000 certification and developing a TQM philosophy is somewhat similar to the phases in the design process. The planning, preparation, and execution phases in the Goetsch and Davis text involve defining the problems, generating solutions, evaluating them, selecting the best ones, and then implementing them.

An interesting trend in teaching technology has evolved from an industrial practice of thoroughly investigating competitive products by dissecting them and studying the components. Some schools call this procedure *reverse engineering.*

Several trends maturing in past decades involve companies that mass-produce products. Two of the trends, discussed in detail in Chapter 17, are outlined briefly in the next section. Then, two more recent trends are noted.

DESIGN FOR MANUFACTURABILITY (DFM) AND DESIGN FOR ASSEMBLY (DFA)

Design hints for DFM include goals such as the following:

- Minimize the number of parts.
- Minimize part variations.
- Design parts for more than one function or use.
- Design parts for ease of manufacturing.
- Use fasteners that handle easily and can be readily installed.
- Arrange assembly of parts so they assemble in layer fashion.
- Avoid flexible parts.

FINAL RECOMMENDATIONS

ASSUMPTIONS
It is important to note that any state-of-the-art design is subject to change since so many variables, having no cut-and-dried answers, must be dealt with, and values must sometimes be assumed using good engineering judgement. Some of the variables in this design project include average mouth dimensions, forces to open and close the jaw, allowable weights for the mouthpiece and powerpack, influence of an oversize solenoid valve on battery life, and so on.

PROTOTYPE
The presence of these types of variables strongly implies a need for proto-type testing to see if the actual design concept is valid. A second type of testing might be called human factor testing, which involves sizing and comfort factors. Are the dimensions on the mouthpiece satisfactory to accommodate an average-size person? Is the size of the powerpack cumbersome? Are weights unbearable? Are jaw movements restricted?

SOLENOID
It was previously mentioned that the specific solenoid valve was oversized. At this point in time the valve is the smallest one made. Because of its power consumption, which is excessive (7 watts max), the battery must be extraordinarily large. It is hoped that eventually a smaller solenoid valve could be purchased or even designed to be much more economical. This could drastically increase battery life and decrease battery size, ultimately decreasing the size of the powerpack.

CNC LATHE
From a manufacturing viewpoint it is recommended that the prototype have all its plastic parts machined on the CNC lathe, since molds are expensive as used in injection molding. However, if the device is ever mass produced, it is recommended that injection molding be the way to go, since production would be much faster, decreasing labor costs per unit.

FIGURE 24-30 *Final recommendations*

APPENDIX NO. 1

DELRIN

The material chosen for the mouthpiece is Delrin. Delrin is a trade name for an acetal homopolymer. It was chosen because it has many desirable characteristics. Some of the advantages of Delrin are:

1) Good chemical resistance, especially to organic compounds

2) Desired regulatory status:
   a) FDA approved for intermittent contact with aqueous or alcohol-based foods
   b) USDA approved for meat and blood

3) Can be machined like soft brass

4) Can be welded or adhesively bonded

5) Forms good mechanical joints

6) Has good structural integrity

7) Tensile strength (at break) 9700 psi

8) Yield strength 9500 – 1200 psi

9) High impact resistance

10) Rockwell hardness 92 – 94

FIGURE 24-31A *Appendix: Material*

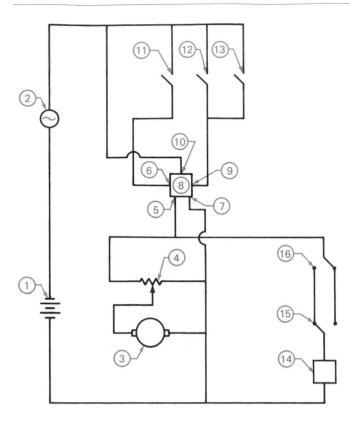

FIGURE 24-31B *Appendix: Electrical schematic*

These goals are discussed early in the design process in meetings where not only designers, drafters, and engineers attend, but manufacturing personnel participate also. Design hints for DFA include goals such as the following:

- Use "pick-and-place" robots for simple, routine placement of parts.

- Keep the number of parts to a minimum (DFM guideline also).

- Use gravity and vibration for moving parts into assembly areas.

- Avoid parts that may tangle (hang up on each other).

- Use tapers and chamfers to guide parts into position.

- Avoid time-consuming manual operations such as screw-driving, soldering, and riveting.

These goals are also discussed early in the design process in the same engineering and manufacturing meetings previously noted.

DESIGNING FOR RECYCLABILITY (DFR) AND DESIGNING FOR DISASSEMBLY (DFD)

Recently, a large manufacturing company that produces automobiles, several companies that produce large appliances, and one large company that produces

QUANTITY	PART NO.	DESCRIPTION	COST
1	NP8-6	Yuasa Battery	$29.55
1	BC-3B	Yuasa Charger	$5.39
1	B11DK 040	Skinner Solenoid Valve	$33.00
1	1615 900-1	Mini Motor (9 rpm) 6v DC	$50.00
1	Rel CY6-5DC-SCO	Relay (Lafayette)	$4.88
2	80511-E	SPST Momentary switch (Lafayette)	$2.94
1	ZC-10B	Fuse Holder (Lafayette)	$1.20
1	3MP10A-10	10-Teeth pinion	$6.34
1	3MP10A-40	40-Teeth gear	$8.59
1	3CCF-50-E	Min-E-Pitch Chain 55 link	$3.24
1	E-62-10K	Unimax Switch	$3.08
1	4PTN-50K	Potentiometer 50 Kohm	$1.98
1	CH3-R10	Charging Jack	$2.99
		Miscellaneous Parts, Tubing, Fittings, Screws	$30.00
		TOTAL	$183.18

FIGURE 24-31C *Appendix: Costs*

computers have initiated programs for DFR and DFD. Their goals are to recover as much material as feasible, mainly metals and plastic, to use again. Not only are their programs economically sound, they acknowledge the fact that landfills are becoming less available, and some basic materials are becoming scarce. Both DFR and DFD programs are considered early in the design process, as are DFM and DFA. Design hints for DFR and DFD include the following:

- Mark plastic molded parts for easy identification.
- Reduce number of different plastics used.
- Make assembly easy with snap-together parts (also easy to disassemble).
- Avoid permanent attachment (adhesives) of dissimilar materials.
- Use materials that can be recycled or used in other products.

The automobile manufacturer found that recycling engine blocks, alternators, starter motors, water pumps, and differentials cut expenses significantly. A manufacturer of large appliances marks all plastic parts (over 50 grams) with an SAE (Society of Automotive Engineers) identification, which gives the type and recyclability of the parts. The computer manufacturer recycles glass and phosphorus from CRTs, plastics from housings (ground up and made into shingles), and metals from parts (steel, aluminum, and copper).

Summary

- The phases and steps in the design process are as follows: recognize and define the problem, generate alternative solutions, evaluate, document, and develop.
- Various techniques for complementing the design process are the creative process, attribute listing, decision tables, and model building.
- The technical drawings used in the design process are the same as those emphasized in Chapter 17; freehand sketches, layout drawings, and detail and assembly drawings.
- Design students should ALWAYS CARRY A NOTEBOOK to record design ideas.
- A well-written specification is written so that it can be tested or measured.
- Sources of design information include vendor catalogs, periodicals, patents, textbooks, and handbooks.

- Modern design practices and standards include DFM, DFA, DFR, DFD, rapid prototyping, parameter design, visual prototyping, ISO 9000 standards, and reverse engineering.

Review Questions

Answer the following questions either true or false.

1. A planning chart provides a visual display of when phases and steps are scheduled to begin and end.
2. One reason why a log is kept is because it could be useful on future projects.
3. Some sources of technical material are reference books, standards and codes, and catalogs.
4. Preliminary and design are the two major categories of specifications.
5. Iteration is the process of returning and revising, which occurs continually throughout the design process.
6. It is standard practice to give the first progress report when the project is 50% complete.
7. An attribute listing is a listing of characteristics.
8. Designers should carry a notebook at all times.
9. A design layout drawing is made from a freehand sketch of a design idea drawn to scale.
10. A decision table is used to determine the date the project will be completed and delivered to the client.

Answer the following questions by selecting the best answer.

1. Which of the following is **not** one of the five phases of the design process?
   a. Recognize and define the problem.
   b. Generate alternative solutions—synthesis.
   c. Evaluate solution ideas—analyze.
   d. Evaluate available time.
2. Which of the following is **not** typically a characteristic of a creative person?
   a. An open mind
   b. Thinks positive
   c. Pessimistic
   d. Accepts change
3. Which of the following is **not** an advantage of making a cardboard-glue model?
   a. There is no use for the model later.
   b. Low cost
   c. Promotes new ideas
   d. Discover if the item will work as planned

4. Which of the following is **not** one of the practical steps for giving an oral presentation?
   a. Tell them what you are going to tell them.
   b. Tell them.
   c. Tell them again.
   d. Tell them what you've told them.

5. Which of the following is **not** a component of a brainstorming session?
   a. Record ideas.
   b. Criticize others' ideas.
   c. Have several sessions if necessary.
   d. Meet to critique ideas.

6. Which of the following is **not** true regarding non-routine design projects?
   a. They are usually given to experienced design project personnel.
   b. They usually require more time than routine design projects.
   c. Detailed planning is required.
   d. Fewer engineering trade-offs occur.

7. Which of the following acronyms has the **incorrect** definition?
   a. DFM—Design for Manufacturability
   b. DFA—Design for Assembly
   c. DFR—Design for Readability
   d. DFD—Design for Disassembly

8. Which of the following is **not** a reason for keeping a log of activities?
   a. For future projects
   b. For accounting purposes
   c. To track progress of the project
   d. A place to keep freehand sketches

Chapter 24 Problems

The following problems complement the design process and provide practice in understanding and learning the steps before attacking routine and non-routine design projects.

CAD Instructions

Students who plan to complete the following problems on a CAD system should begin with these general instructions. Before reading the specific instructions for each activity, go through each step in the following planning checklist. The checklist applies to any CAD system and will help ensure the optimum use of your time and resources.

1. Analyze the problem carefully. Decide exactly what you are being asked to do.

2. Determine what resources and references (if any) you will need in order to complete the problem. Collect those references and resources and have them readily available.

3. Decide if any particular standards apply to the project and have those standards available.

4. Determine what types of views will be required and how many of each.

5. Determine what the final plotted scale of the drawing will need to be, and select the appropriate paper size for plotting/printing.

6. Plan your drawing sequence. In what order will you develop the drawing (i.e., lines, features, dimension lines, leaders, dimensions)?

7. Review the various CAD commands you will have to use in order to develop the drawing.

8. Examine your CAD system to ensure that everything is in working order, then begin the project.

Problem 24-1 *(Step 2)*

Prepare a 5-minute oral presentation on one of the following sources of information. Use transparencies for an overhead projector. Suggestions for transparencies (**Tx**):

Tx 1. Name of source, its location, and name of speaker.

Tx 2. Outline of the talk:
   a. description of the source and its specific location
   b. why source is useful to designers
   c. example of use of source
   d. conclusion

Tx 3. Description of source and its location.

Tx 4. Why source is useful to designers.

Tx 5. Example of use of source.

Tx 6. Conclusion.

Suggested Sources:
1. ASTI (Applied Science and Technology Index)
2. ASTM (American Society for Testing Materials)
3. ANSI (American National Standards Institute)
4. Patents
5. UL (Underwriter's Laboratory)
6. *Machinery's Handbook*
7. Handbook (instructor's or your choice)
8. Vendor's catalog

Problem 24-2 *(Step 3)*

Prepare performance and design specifications for one of the items in the following list. Well-written specifications can be verified by testing. For example, instead of specifying that a bumper jack will be easy to use, lightweight, and made of metal, well-written "specs" would be the following: easy to use by a licensed driver who is not handicapped, a maximum of 20 pounds of force to operate, weighs less than 10 pounds, and is made of SAE 1035 steel. Include a sketch with notes and dimensions.

1. Screwdriver for use in average household
2. Light switch in average household
3. Can opener for kitchen use
4. Knife for kitchen use
5. Pliers for electrician or average household
6. Hammer for carpenter or average household
7. Wood saw
8. Metal saw (hacksaw)

PROBLEM 24-3: (STEP 7)

Prepare a decision table using criteria and weighting factors for the situations listed. Generate at least three alternatives for each situation.

1. Vehicle for town driving
2. Vehicle for country driving
3. Vehicle for backwoods driving
4. Vacation
5. College to attend
6. Job choice in one city
7. Job choice for different locations (West Coast, East Coast, South, North, etc.)
8. Your choice

PROBLEM 24-4: (STEP 6)

Prepare a freehand sketch of a relatively uncomplicated design and then prepare a design development layout drawing of the sketched design. Draw the object to scale, but add notes and dimensions in an informal manner. No title block is needed. Complete information is required. Design suggestions:

1. Wall bracket to support object of your choice. Instructor's choice. Use welded assemblies.
2. Minor modification of a component from:
   a. This text (instructor's choice)
   b. Your choice
   c. Instructor's choice

Problem 24-5 *(Step 6)*

Prepare a kinematic development layout drawing of one of the assemby drawings to follow or of an assembly assigned by your instructor (Problems 24-5A through 24-5E).

Suggestions for Kinematic Analyses:

1. Redraw the figures as single straight lines to represent the linkages.
2. Construct radii about fixed points of rotation.
3. Trace, on a separate sheet of paper, the lines representing the linkages whose motions are to be analyzed.
4. Use the lines as underlays to locate the new positions.

Problem 24-5A *(Step 6)*

The figure to the right shows the linkages and hydraulic cylinder to operate a jet-blast deflector EF. The hydraulic cylinder AB causes linkage CD to rotate CCW (counterclockwise) about point C, which causes linkage DE to move the deflector EF in a CW (clockwise) direction. When EF is in the maximum CW position it acts as a shield for personnel standing behind jet aircraft when the jets are warming up for launching from an aircraft carrier. Determine the maximum CW angle that EF rotates and the change in length (stroke) of cylinder AB to obtain the maximum angle. Draw lines EF, DE, CD, and AB when they are in the position for the maximum CW rotation of EF.

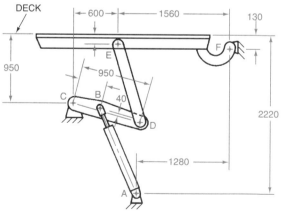

ALL DIMENSIONS IN MM

Problem 24-5B (Step 6)

One application of a four-bar linkage is to control the movement of the hood of an automobile: a pair of four-bar linkages supports the hood. Linkage GH is attached to the hood as shown in the figure. When the hood closes, GH rotates and translates to a horizontal position. Determine the total CCW (counterclockwise) rotation of the hood from the open to the closed position. Also determine the change in length of the spring SP from the open to closed positions. Draw the lines representing the linkages and the spring in the closed position.

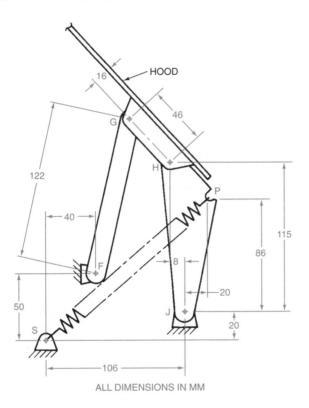

ALL DIMENSIONS IN MM

Problem 24-5C (Step 6)

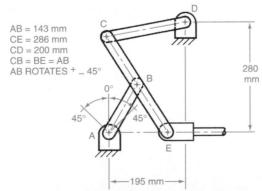

AB = 143 mm
CE = 286 mm
CD = 200 mm
CB = BE = AB
AB ROTATES + − 45°

Problem 24-5D (Step 6)

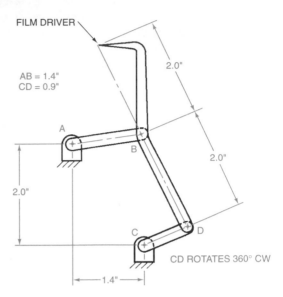

FILM DRIVER

AB = 1.4"
CD = 0.9"

CD ROTATES 360° CW

Problem 24-5E (Step 6)

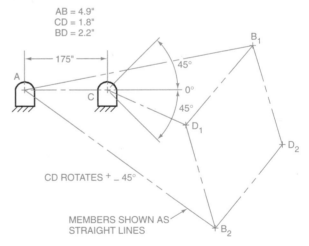

AB = 4.9"
CD = 1.8"
BD = 2.2"

CD ROTATES + − 45°

MEMBERS SHOWN AS STRAIGHT LINES

Problem 24-6 *(Steps 1–11)*

The 19-step Kraft Mill Engineering Roadmap summarizes a design process from a company for improving or changing the company's plant facilities. A recent graduate from an engineering program sent the summary to one of her professors to show what she was doing during the first few months on the job.

Kraft Mill Engineering Roadmap

1. PROJECT SCOPE: Define the problem to be resolved, the changes that are required, and the objective to be achieved.

2. SCOPE REVIEW MEETING: Verify that both Maintenance and Operations approve of the project scope.

3. PRELIMINARY DESIGN: Select several options for consideration. Prepare rough cost estimates, sketch general layouts/schematics, and estimate R01/payback for each option.

4. PRELIMINARY DESIGN REVIEW MEETING: Select one or two options to pursue. Obtain Management's approval to proceed. Determine if project is to be capitalized or expensed.

5. SUBMIT AR OR WO: Write a cost estimate. Use class ±10%, ±20%, or ±40%, as required. Write and submit an AR (Authorization Request) or a WO (work order).

6. RECEIVE APPROVED AR OR WO: No money may be spent until approval is received.

7. DETAILED DESIGN: Make detailed layout drawings and flow schematics. Do energy balance. Develop schedule; make time-bar diagram for larger projects. Determine equipment requirements.

8. PROJECT DESIGN REVIEW MEETING: Obtain "buy off" from Operations and Maintenance of the design and layout. Verify schedule, shut-down periods, etc. Decide whether contractors or the mill will provide labor.

9. BIDS: Prepare bid package for material and for labor. Draw up a bidders list. Send to Purchasing Dept. and go out for bids.

10. REVIEW BIDS: Select the best one or two vendors for each portion of the project. Run background check on contractors. Check references of equipment manufacturers.

11. VENDOR APPROVAL MEETING: Obtain approval to purchase equipment and/or hire contractors.

12. PURCHASE ORDERS: Award bids. Write purchase orders (POs).

13. EQUIPMENT TRACKING: Track equipment as it comes into mill. Store in safe place until installation.

14. CONSTRUCTION: Install equipment.

15. PRE–START-UP: Obtain and file vendor drawings, guarantees, and instruction files. Set up spare parts in the storeroom. Write an operating manual. Develop a start-up plan.

16. PRE–START-UP: Review start-up plan. Set a start-up date.

17. START-UP: Put system into operation.

18. PROJECT REVIEW: Either hold meeting or circulate memo. Restate original project scope. Review the project. Determine that the desired objectives were (or were not) met. Discuss any problems. State final project cost.

19. PROJECT CLOSEOUT: Do necessary paperwork to close out project. Sign the WO as complete, or fill out an AR Capital Job Closure Form.

Note: Some of these events overlap. Depending on the size and type of job, some events may be deleted or some more may need to be added.

Study the 19 steps and do all or any one of the following as assigned by your instructor:

1. Prepare a planning chart assuming that a six-month period is required for a particular plant modification, and that you are doing the job.

2. Relate the 19 steps to the design process discussion in this chapter.

3. List and comment on the different people most likely contacted during the 19 steps.

4. Comment on the apparent meaning of some of the terms and acronyms: for example, scope, ROI, capitalized, expensed, AR, WO, ±10% class, ±20% class, ±40% class, time-bar diagram, flow schematic, buy-off, bid package, background check, PO, tracking, and AR Capital Job Closure Form.

Problem 24-7 (Step 6)

1. Contact local vendors of material to obtain current prices of common ones, such as steel shapes (angles, plates, bars, channels, and beams). Note that prices are usually given in dollars per 100 pounds and that price reductions are given for purchasing large quantities (e.g., a full railroad car).

2. Contact vendors of finished components—such as motors, switches, controls, bearings, chain, and similar products—to obtain costs to several categories of buyers.

   a. Private individuals usually pay the full retail price.

   b. Companies that include the finished components in a kit may get a 15% to 30% discount from the vendors because the companies sell the kits to customers.

   c. Companies that include the finished components as an integral part of an assembled machine are called original equipment manufacturers (OEM). They may receive even larger discounts.

PROBLEM 24-8: (STEP 6)

Frequently, professionals are asked to make quick estimates during conversations over lunch or while phoning a client. This type of quick thinking is also an excellent approach to use for checking final results, whether it is homework in an engineering science course or a system in industry. Ask questions such as, does this look right, does this look reasonable, or does the output seem reasonable for the input?

For practicing "back-of-the-envelope" thinking and calculations, do the following in a time period assigned by your instructor and record the assumptions made:

1. Estimate the quantities of food and the cost of lunch for all of the people in your classroom.

2. Estimate the number of gallons of water used by your school in one day. (Perhaps maintenance personnel can verify the accuracy of the estimate.)

3. Estimate the cost of transporting a college football team 1000 miles, and expenses for a three-day road trip.

4. Estimate the selling price of a new automobile, truck, van, sports car, or motorcycle based on the price per pound of existing vehicles.

5. Estimate selling prices of other products using the approach in #4.

Problem 24-9 (Step 12)

List the manufacturing processes required to make the problem parts shown in the following figures.

Problem 24-9A

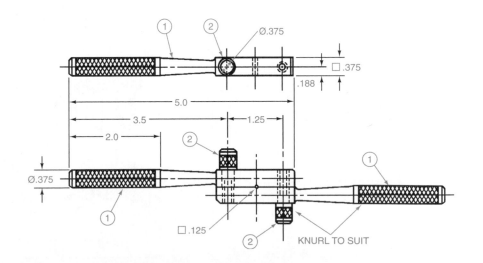

Problem 24-9B

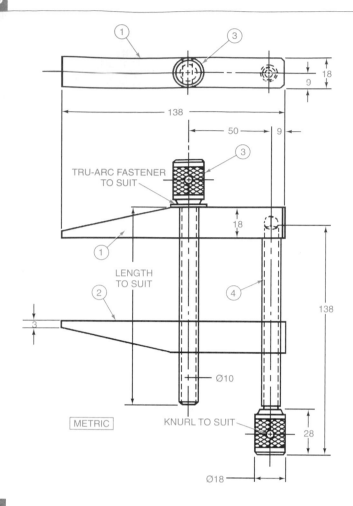

TRU-ARC FASTENER
TO SUIT

LENGTH
TO SUIT

METRIC

KNURL TO SUIT

Ø10

Ø18

Problem 24-9C

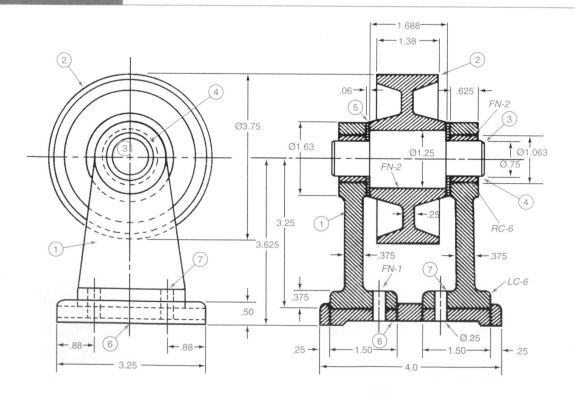

Problem 24-9D

List the manufacturing processes required to make the part or parts assigned by your instructor.

PROBLEM 24-10: (STEP 5)

From memory, make a freehand sketch of an object shown to you for a limited time by your instructor. Then, do a second freehand sketch of the same object while you have a much longer time to view and examine it.

This practice of viewing an object knowing you have to sketch it from memory will train you to see and remember more than you would at a casual glance.

PROBLEM 24-11: (STEP 5)

List the obvious attributes of one of the following items. Include a freehand dimensioned sketch of the object. Then list all the not-so-obvious uses of the object. Be creative.

1. Kitchen table
2. Clothespin
3. Bookshelves
    a. Wood
    b. Metal
4. Cardboard milk carton
5. Plastic container of 35 mm film
6. Screwdriver
7. Nail

ROUTINE DESIGN PROBLEMS (RDP)

The routine design problems listed here are intended to give you experience in the design process while doing problems similar to those assigned to new employees in technical industries. Use the steps, as needed, in the design process to produce the following short report (printed or typed). The steps most needed are 2, 3, 5, 6, and 10.

1. A statement of the problem in your own words
2. A list of information sources used
3. Freehand sketch(es) or preliminary ideas of what needs to be done
4. Design development layout drawings of the ideas in the freehand sketches
5. Kinematic development layout drawings, if required
6. Complete detail and assembly drawings with parts lists
7. Cost analysis

8. Conclusions and recommendations
9. Appendices
    a. Machine copies of 3, 4, 5, and 6 (These copies will show how well pencil drawings reproduce as plotted or printed copies if CAD is used.)
    b. Vendor catalog information
    c. Any additional materials: work order, photographs, etc.
10. A title page and a folder for the report

Problem 24-12 (RDP)

Refer to the torque wrench assembly in the accompanying figure and do the following:

1. Make the wrench just two times as large as shown for all dimensions.
2. Select steel for the handles and the fasteners.
3. Determine the bending stress in the necked-down portion of the handle. Specify as large a radius as practical at the necked-down location to avoid stress concentration. The bending stress will depend on the assumed force F and the length L.

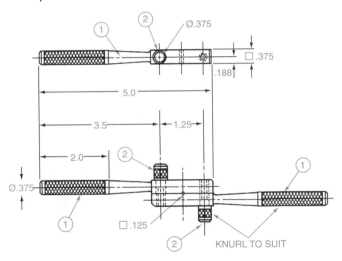

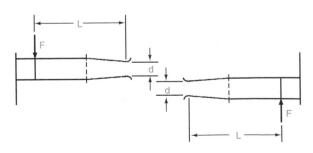

F = FORCE BY OPERATOR IN POUNDS
    (NEED TO ASSUME)
L = ASSUME DISTANCE FROM F TO d IN INCHES
d = DIAMETER OF NECKED–DOWN PORTION IN INCHES

STRESS DUE TO MOMENT F X L AT d IS

$$\text{STRESS} = \frac{(F \times L)\,(d/2)}{(d^4)/64} \text{ IN POUNDS PER SQUARE INCH}$$

Problem 24-13 (RDP)

Refer to the clamp shown in the accompanying figure and assume that parts no. 1 and no. 2 are to be aluminum, and parts no. 3 and no. 4 are steel. When threads in aluminum may be too weak because of the softer metal, a threaded-steel insert is pressed into the aluminum. Use an insert with appropriate threads inside and a nominal outside diameter of 15 mm, but dimensioned for an FN-2 medium-drive fit. Increase the 18 mm dimension for parts no. 1 and no. 2 to 25 mm.

Problem 24-14 (RDP)

The clamp shown in the accompanying figure needs to be modified to clamp rectangular assemblies 50 × 300 mm between the two threaded members instead of between the tapered extensions. The thread size should be increased 100% in diameter. Make any other changes that seem appropriate. Select materials. (*Note:* Part no. 4 must function like part no. 3, and the tapered portions of parts no. 1 and no. 2 are no longer needed.)

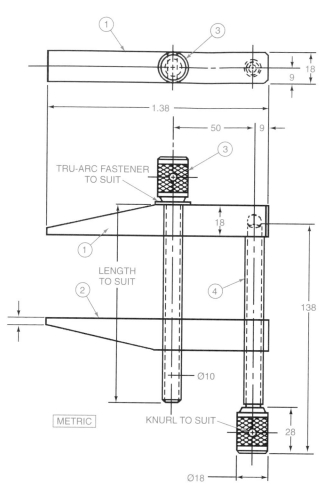

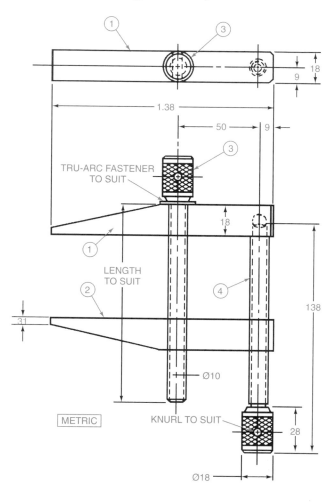

Problem 24-15 (RDP)

Modify the screw jack shown in the accompanying figure to have an eight-inch vertical movement instead of the apparent four-inch vertical one.

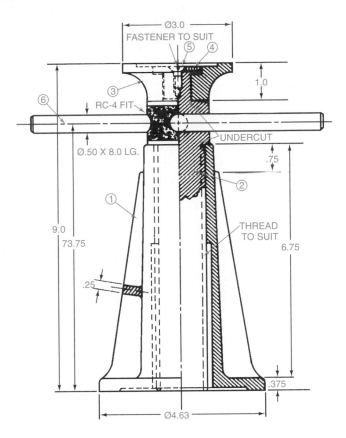

Problem 24-16 (RDP)

Modify the roller assembly shown as follows:

1. Make the shaft, part no. 3, 1" diameter × 3" long.
2. Change the roller, part no. 2, to have a bronze bushing 1" inside diameter, 1.25" outside diameter × 1.6888" long; FN-2 Fit for the 1.25" diameter and RC-3 for the 1" inside diameter.
3. Change the supports, part no. 1, to have set screws to keep the shaft, no. 3, from turning.

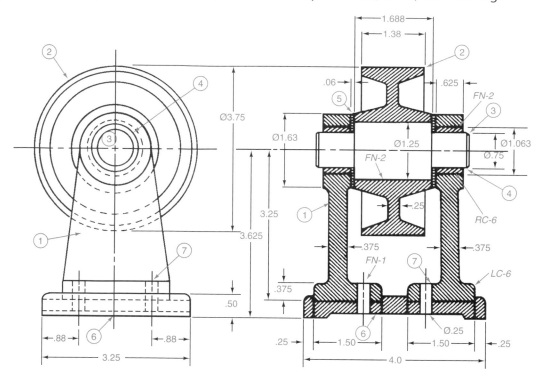

Problem 24-17 (RDP)

Change the vise shown in the accompanying figure to be 50% longer and 75% wider. In a kinematic layout drawing, show the extreme open and closed positions.

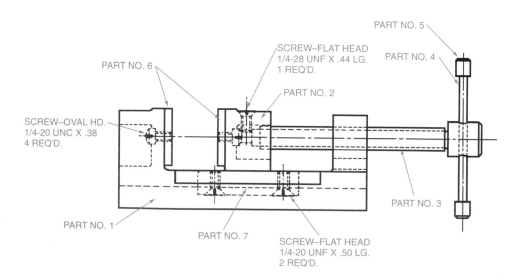

Problem 24-18 (RDP)

Develop the concept of the beverage can crusher shown in the accompanying figure. Use standard materials and parts as indicated. Include a kinematic layout drawing showing extreme positions of the crusher head.

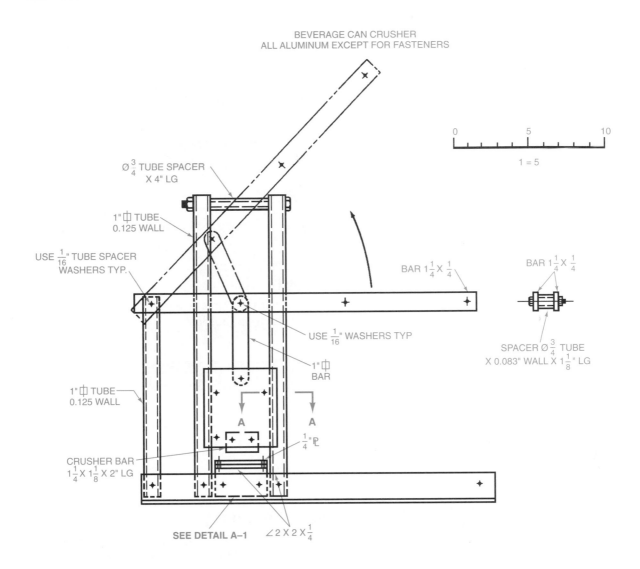

BEVERAGE CAN CRUSHER
ALL ALUMINUM EXCEPT FOR FASTENERS

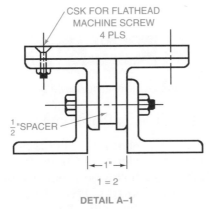

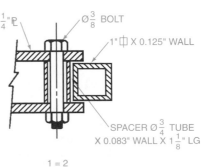

Problem 24-19 (RDP)

Use press-fit drill bushings with heads to make a leaf-type jig for the hole pattern shown in the part to be drilled.

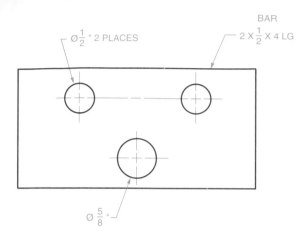

NON-ROUTINE DESIGN PROBLEMS

These non-routine problems are designed to give you experience in the overall design process, but mostly in Phases I–IV, Steps 1–11. When it is feasible, you are encouraged to build prototypes and discuss Steps 12 and 13 in Phase V.

Your instructor will most likely form student design teams to resolve the non-routine design problems, Problems 24-20 through 24-40, and the "real" problems from the suggested sources. Before joining a team, read Chapter 26 (Teamwork on the Job) to obtain insights into what to expect as a team member, whether in school or in industry. Chapter 26 contains informative discussions on the characteristics of successful team members and the characteristics of successful teams. Effective communication between members seems to be the major contributing factor to a team's success. What do you think?

The following groups of non-routine problems fall into three time-related categories: short, medium, and long projects. Suggestions for hours to allot for the three categories are included. Submit all reports in a cover with an appropriate title.

SHORT NON-ROUTINE DESIGN PROBLEMS

Suggestions: Allot 10–15 hours for one person to complete them; allot 5–7 hours for a team of two, etc.

Develop the problem only through Phase I of the design process. That is, prepare a plan of action for one or more students (Step 1 and Step 2), define the problem (Step 3), and prepare a progress report to a real or an assumed client (Step 4). Start a log and design file as assigned.

Problem 24-20

Develop a toy for a ( )-year-old child confined to a bed or a wheelchair. *Note:* Select or be assigned the child's age.

PROBLEM 24-21

Design a small-capacity trailer that can be pulled by an automobile and that can be dismantled and stored in a relatively small space.

Problem 24-22

Design a means of mounting posts on automobiles or pick-up trucks for games such as badminton or volleyball at remote picnic sites.

PROBLEM 24-23

Design a novel or practical use for discarded tires, cars, cartons, cans, or bottles.

Problem 24-24

Design a portable can crusher for aluminum or steel for use in areas where numerous cans are deposited or tossed: for example, on roadsides.

PROBLEM 24-25

Design a teaching aid for one of the courses in your school, in a local high school, or in a local elementary school.

Problem 24-26

Design a portable solar cooker.

MEDIUM NON-ROUTINE DESIGN PROBLEMS

Suggestions: Allot 20–30 hours for one person to complete them; allot 10–15 hours for a team of two, etc. Develop the following problems through Steps 1–8 of the design process. Present your progress report to your classmates, with all members taking part in the presentation. Start a log, start a design file, and prepare a planning chart.

PROBLEM 24-27

Do any one of the Short Non-Routine Design Problems listed, but pursue the problem through Step 8.

Problem 24-28

Design a device to assist an individual who has no legs, but full use of two arms, to get into a wheelchair from the floor, and back to the floor.

Problem 24-29

Design a device to lift a 150-pound handicapped person out of a swimming pool.

Problem 24-30

Design a study or work area for an individual with little income who is confined to a wheelchair.

Problem 24-31

Design a means to use "pedal power" from a typical one-speed bicycle for performing a useful function such as sawing, sewing, mixing, pumping, etc.

LONG NON-ROUTINE DESIGN PROBLEMS

Suggestions: Allot 40–60 hours for one person to complete them; allot 20–30 hours for a team of two, etc.

Develop the following problems through Phase IV of the design process. Note that Phase IV usually requires at least half the total time allotted to a project whenever complete detail and assembly drawings are needed.

PROBLEM 24-32

Do any one of the short or medium non-routine design problems previously listed, but develop the problem through Phase IV of the design process.

Problem 24-33

Design a means to rid aluminum beverage cans of moisture. The cans arrive in truckloads at a recycling center and the driver wants to be paid a realistic price, based on the actual metal turned in, within half an hour.

Problem 24-34

Design an inexpensive fruit dryer.

PROBLEM 24-35

Design a powered vehicle to move one or more employees throughout a huge manufacturing plant or warehouse. The vehicle should be cost effective; that is, enough time (money) should be saved to justify its use.

Problem 24-36

Modify a tricycle to accommodate a child with minimal leg use.

PROBLEM 24-37

Design a safe vehicle for coasting on a track, and the track, for children weighing up to 100 pounds.

Problem 24-38

Design a pool cue for a one-armed person.

Problem 24-39

Design a manually powered device for lifting 25 kilograms and heavier weights, vertically, at least one meter.

Problem 24-40

Design a device that utilizes a typical automobile jack to perform another function(s) such as pressing, clamping, splitting wood, operating a simple "jaws of life" mechanism, etc.

OTHER SOURCES

The following lists describe a number of general sources for finding real, non-routine design problems. These problems are usually arranged for by instructors. After bringing together the source of the real problem (the client) with the design team (or an individual student) the instructor can step back and become an advisor-coach. Real, non-routine design problems are usually longer, often due to the extra time spent visiting and corresponding with clients.

1. School

   Classroom demonstrators of basic principles of chemistry, physics, statics, dynamics, mechanics of materials, engineering sciences, and manufacturing processes.

2. Handicapped

   a. Designs to help individuals confined to a wheelchair to:

   (i) Reach objects and prepare meals in a kitchen

   (ii) Obtain exercise

   (iii) Obtain access to heights higher than curbs

   (iv) Have access to classroom work surfaces

   (v) Have work surfaces on the wheelchair

   b. Designs to continually flex injured elbows and knee joints to accelerate healing.

   c. Designs to aid communications for individuals who have speech difficulties.

3. Hospitals

   a. Games or devices for children who are confined to a bed and have limited use of their arms.

   b. Devices to help injured individuals to perform simple functions such as dress, read, play games, and eat.

4. Civic Organizations

   a. Designs for YWCA or YMCA

   b. Designs to assist scouting programs such as neighborhood improvement projects

5. Local Industry

   a. Projects for the industry

   b. Projects for the industry's client

6. Private individuals (designs for)

   a. Home improvements

   b. Hobbies

   c. Inventions

   d. Invalid care

7. City (designs for)

   a. Better parking meters

   b. Playgrounds

   c. Park facilities

   d. Signs

   e. Junior Achievement

8. State (designs for)

   a. Signs

   b. Parks

   c. Recreational facilities

9. National (designs for)

   a. Design contests (Alcoa, Lincoln Arc Welding, Engineering Societies…)

   b. Environment

10. International (designs for)

   a. Third World countries

   b. To help Peace Corps volunteers

SECTION 5

Employability Skills

Career Profile: Rob Kovarik

While in college in Helena, Montana, Rob Kovarik was in charge of lighting and sound for the band he played in. When he took an acting class to fulfill a course requirement, the director of theater discovered Kovarik's skill and enlisted his help for the university's theatrical productions. He did not end up being a rock star, but his career was just about to take off in a completely different direction.

When he graduated with a dual degree in biology and theater, his portfolio helped him immediately land a job as a technical director for a small theater in Denver. Without a day of formal technical drawing training, he ended up being in way over his head. Fortunately, he had helped to build houses before, so he knew how to *read* plans, and he fudged his way through drafting them by hand.

His second job, however, did not prove so lucky. At age 26, he became technical director of the multi-million-dollar Wollman Street Theatre in Philadelphia. Doing all his drawings by hand on paper, and having 50 people report to him, he was quickly overwhelmed. After two shows he quit, and he promptly applied to a master's program in technical design, eager to attain the skills he lacked.

Now, as technical director at The Theatre School at DePaul University, he coordinates the shows the university puts on (and he also teaches what he has learned to college students). His role in the process is to serve as liaison between the scenic director, who comes up with the artistic idea for the show, and his shop. His drawings are the physical link between the concept and the realization of a set.

One of the lessons he tries to emphasize to his students is that communication is key to successful technical directing. At the beginning of a new project, he will usually read the play, then have a face-to-face meeting with the scene designer who will describe her vision for the set: what elements it will include, when they will need to appear, and what level of automation they will have. The designer may draw an elevation—sometimes as simple as a sketch on a napkin! Kovarik's job is to interpret that and to filter the artistic process through a computer. He says his goal is to make the scene designer happy; if he were to get involved in the artistry, that would merely circumvent the process.

He inserts the dimensions and technical information into his AutoCAD program to create an elevation, and he breaks each element into parts so that the tasks can be divided among crew members to get the job done faster. (This is especially important because his crew is college students who often have only an hour at a time in the shop.) In the 3D program, he will do a plan view, a side view, and a front view of each element. He can take the next step into making it a ladder-rung orthographic image or an isometric image. He will think about materials and safety as he is doing it. For instance, if a platform needs to support 70 people, it will need to be constructed of a sturdy material. He also has to think about the weight and size of the object: Will it fit through the door? Can it be carried on a truck?

Once he gives the technical specifications to the foreman of his shop, they get to work on pricing it out. When the bid is approved, the shop will start the project, which Kovarik supervises. Then the technical director or his assistant usually attends all shows to make sure the equipment operates as it was designed to do. It makes for long days (sometimes 7:00 A.M. to midnight), but for Kovarik, it's not so bad, now that he knows what he is doing!

Chapter 25

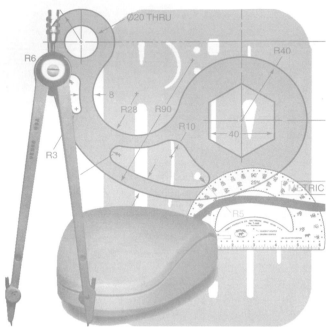

FINDING, GETTING, AND KEEPING A JOB

KEY TERMS

Chamber of Commerce	Portfolio of drawings
Constantly improve	Private employment agencies
Follow-up	Resumé
Government personnel offices	State employment agencies
Interview	Want ads
Letter of introduction	Yellow pages

CHAPTER OUTLINE

Maximizing your employability • Identifying potential job openings • Job-seeking process • Keeping a job and advancing • Summary • Review questions • Chapter 25 problems

CHAPTER OBJECTIVES

Upon completion of this chapter, students should be able to do the following:

- ■ Explain how to maximize one's employability.
- ■ Describe how to identify potential job openings in the field.
- ■ Write a proper letter of introduction.
- ■ Develop an effective resumé.
- ■ Prepare an effective portfolio of work.
- ■ Demonstrate effective interviewing skills.
- ■ Demonstrate how to properly follow up on an interview.
- ■ Explain how to keep a job once hired.

Inexperienced graduates often face a difficult and sometimes frustrating job search in a world of work that caters to experienced applicants. The purpose of this chapter is to help students break through the "experienced only" barrier and begin a career in drafting and design.

The choice of drafting and design as a career field is a wise choice. Present and future employment prospects in this field are excellent. Employment of drafters is expected to increase faster than the average for all occupations. This growth, along with the need to replace those who retire, die, or move into other fields of work, will provide favorable job opportunities for decades to come.

Employment of drafters will continue to increase as a result of the increasingly complex design problems associated with modern products and processes. In addition, more drafters will be needed as support personnel for engineering and scientific occupations.

Obviously, jobs are available. The question is, how to go about getting one? This chapter describes how to identify jobs that are available, and how to get one of these jobs. It also explains how to keep the job once you have gotten it.

Before beginning preparations for the job search, do an appraisal of personal strengths and weaknesses relative to the job market. It will have an effect on your approach. If you are a typical student, there will be a number of attributes to enter in the "plus column." These include one or two years of specialized, in-depth training in drafting and design, an eagerness and ability to learn, and a flexibility in your skills that many experienced employees may have lost. In the "minus column" you may have only one entry; no experience. The idea, then, is to make those entries in the plus column work for you in ways that help overcome that one entry in the minus column. The first step is to make yourself as employable as possible.

Maximizing Your Employability

What follows in this section are five strategies that drafting and design students can use to improve their chances of getting a good job. Students who apply these strategies effectively will improve their prospects in the job market.

1. *Don't be afraid to relocate.* Drafting is an industry-related occupation, so it stands to reason that the biggest demand for drafters will be in those areas with a lot of industry. Most drafters in the private sector work for architects, building contractors, engineering firms, surveyors, and manufacturing firms. Consequently, if you want a job, you may need to go where the jobs are. For graduates with strong ties to a small town or rural community, this advice can be difficult to take. However, loosening the home ties can increase your employment potential markedly.

2. *Learn to market your skills.* Study the strategies in this chapter closely and internalize them. It is almost as important to be a skilled job seeker as it is to be a skilled drafter and designer.

3. *Be willing to start at the bottom and work your way up.* Many places will want to start you off doing such things as making corrections, doing revisions, and even running errands. Don't be offended. Just buckle down, keep your eyes and ears open, ask questions, and learn how to do everything you can as soon as possible. The best combination in drafting and design is your education plus two or three years of experience. Once you have attained this combination, many well-paying, responsible positions will open up to you. If you must start at the bottom, look on your time spent there as an investment in your future.

4. *Be positive and assertive in your job search.* When you interview, avoid appearing as a timid, unsure person. Be positive, well-informed about the position, and assertive enough to say "I would like a job with your company and this is why . . . and I can do this job if you will give me the opportunity."

5. *Expect frustration and overcome it.* It might take 100 interviews to get one job, but don't despair. You only want one job! Do not allow the frustration of a job search to dampen your positive outlook or coerce you into a negative attitude. Go into each and every interview fully prepared and with a positive outlook as if *it* is the one that will pay off.

Identifying Potential Job Openings

The next step in the job search involves locating potential places of employment and job openings. To get the ball rolling, ask yourself this question: What are my first five choices for employment locations? In answering this question you will want to consider several important criteria concerning the cities you choose. How many drafting-related businesses exist in each city? What is the wage scale for drafting positions in each city? What are the living conditions in each city? In addition to these, there are also personal considerations such as location, climate, housing, and the availability of public services, hospitals, and leisure time activities. Most of this information is usually available from the local Chamber of Commerce in each of the cities in question.

More than likely the first city on your list will be your home town. This is certainly understandable, but try to be objective. Ask yourself the same questions about your home town that you would ask about any town to

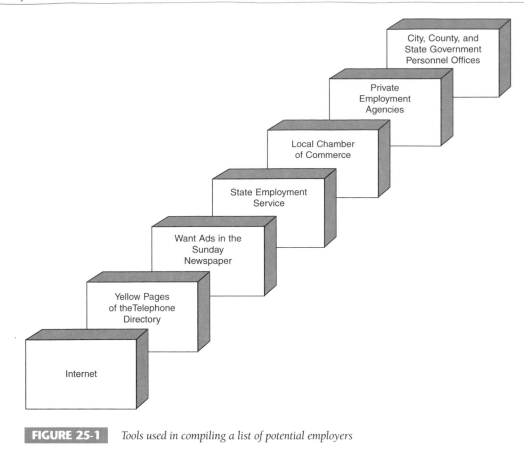

FIGURE 25-1 *Tools used in compiling a list of potential employers*

which you might relocate. Remember, career decisions are some of the most important decisions you will ever make.

Once you have listed the cities in which you plan to seek employment, begin compiling a list of potential employers in each city. Several things can be useful in completing this task. Among the most helpful are: the Internet, the yellow pages in a telephone directory from the city in question; the want ads in the Sunday edition of the local newspaper from the city in question; the local office of the state employment service; private employment agencies; the Chamber of Commerce in the city in question; and city, county, and state *government personnel offices, Figure 25-1.*

USING THE INTERNET AS A TOOL

One of the most effective ways to locate potential jobs is to use the Internet as a tool. The advent of the Internet has simplified the task of finding a job more than any other development to date. All of the other sources of help explained in this chapter are still valid, but the wise job seeker in today's electronic environment will begin with the Internet.

There are several different ways the Internet can help. The most obvious is in making job-search sites available

to you. Some of the many job-search sites that you might find helpful are:

Monster.com *http://www.monster.com*

HotJobs.com: *http://www.hotjobs.com*

CareerBuilder.com: *http://www.careerbuilder.com*

America's Job Bank: *http://www.AJB.dni.us*

In addition to these job-search sites, many newspapers now provide their want ads online, as do state, county, city, and private employment services.

Another way to use the Internet to find potential jobs is to search out the websites of specific companies. Many companies now post their openings on the Human Resources component of their websites. Using this approach to finding a job offers the added benefit of allowing you to learn important facts about the company before deciding whether or not to apply for the job in question. Then, if you decide to apply, you will be well prepared with information about the company.

USING THE YELLOW PAGES AS A TOOL

The *yellow pages* can be helpful to the student seeking employment in drafting. They list every architect, engineering firm, contractor, and surveyor in the city in

Employment Agencies

Progressive Employment Inc.

- Entry Level to Top Management Positions
- Free Service to Employers

Full Service Placement Agency

161 Mary Andrews Parkway 863-6565

FIGURE 25-2 *Sample yellow pages ad*

question. They also list both public and private employment agencies, and sometimes manufacturers. For the time being you need only look through the yellow pages and list the addresses and telephone numbers of employment agencies as well as potential employers that you might wish to contact later, ***Figure 25-2***. Remember, it is not necessary to wait until a job opening is announced in order to apply.

Telephone directories for numerous cities throughout the United States are often collected in public libraries, as well as the libraries of colleges and universities. If you try these places and don't find what you need, go to your local telephone company office—personnel there will be able to help locate the book you need.

Using the Sunday Newspaper as a Tool

Want ads in the Sunday newspaper can be helpful because Sunday is when most employers run their ads. This, in turn, is because employers know that the Sunday paper has the most readers. Want ads are typically printed in one of two ways, depending on the size of the newspaper: in alphabetized career categories such as sales, medical, professional-technical, and trades, or randomly with no specific categories. In either case, you would do well to become a student of the want ads. They can be deceptive to the unseasoned job seeker, but very helpful for those who use them properly.

When looking through randomly listed ads, the only thing to do is check every column. This can be time-consuming, but it is necessary. It's when looking through categorized ads that you must be attentive to tricks of the trade. The categories most likely to turn up jobs are: *Architecture,* for architectural drafters; *Engineering,* for engineering drafters or engineering technicians; *Mechanical,* for mechanical drafters; *Professional-Technical,* for all types of drafters; and, of course, *Miscellaneous.* As was stated earlier, it is not necessary to wait for a job to be advertised before applying. If a company is advertising for engineers, designers, or architects, it stands to reason they will also need drafters—if not now, then later. Put these companies on your list if they look interesting. ***Figure 25-3*** is an example of what one might see when reading the want ads.

Wanted!! Drafting and Design Technicians

Local company has openings for drafting and design technicians with formal education and training. Must have CAD skills and strong mechanical drafting background.

Contact:

Personnel Office
Jones Engineering Company
816 Industrial Park
Ontario, Tennessee 46921
618/281-1971

FIGURE 25-3 *Sample want ad*

State Employment Service

The purpose of *state employment agencies* is to help unemployed people find jobs that are compatible with their skills. They have a responsibility to the unemployed in their respective states and, generally speaking, do an effective job of placing people in positions of employment. However, they also have a responsibility to employers to ensure that only qualified applicants are referred for interviews. Unfortunately, this is an obligation that can cause problems for inexperienced students. More often than not, an employer advertising a drafting position will ask for work experience. This, of course, rules out most students, but don't give up.

Let's take a look at how a typical employment office works. It will help you to understand the overall system. Currently available jobs are usually listed on a computer, and in some cases on microfilm, microfiche, or a printout. Once you have located a job that looks interesting, write down all of the pertinent information such as qualifications required, wages, and the employment office code number. Give this number to the desk clerk and he or she will check to make sure the job has not been filled. If the job is still open, you will be asked to fill out an application or qualifications form. This form will then be passed on to a job counselor who will invite you into his or her office for an interview. One of the requirements on the application form is that you list your three most important jobs relating to drafting. This is where problems can begin. Seeing that you have the training but no experience, the counselor might attempt to turn you away. If this happens, tell the counselor you have reviewed the qualifications for the job and that even though you have limited experience, you have had specialized training. Then ask him or her to call the employer, explain the situation, and see if an interview can be arranged. Often the employer will be receptive and grant you an interview, especially if he or she is having difficulty filling the position. Remember, if you are going to be turned down, it should be by an employer, not the employment office. Many times employers will advertise higher qualifications than they actually expect to get, so don't be easily dissuaded by an employment office counselor.

THE LOCAL CHAMBER OF COMMERCE

This agency can be invaluable to the job seeker. Manufacturing firms are the largest single employers of drafters. The problem is in locating them. They don't always appear conveniently categorized in the yellow pages, and they don't always advertise in the want ads. This is where the *Chamber of Commerce* or Economic Development Council can help. Almost every chamber or EDC office maintains a list of the manufacturing firms located in its city. Not only does the list contain company names, addresses, and telephone numbers, it often gives the product the company manufactures, *Figure 25-4.*

Not all manufacturing firms hire drafters, but most do. If a company looks interesting, add it to your list. To determine if it has a drafting department, call the company and ask for the name of the head of the drafting and design department. If they have one, record his or her name for future reference. If they don't, simply cross the company off your list and move on to the next one.

PRIVATE EMPLOYMENT AGENCIES

These agencies can be very effective at placing graduates in jobs, but students often shy away from them because some charge a fee. This is understandable, but before writing off *private employment agencies,* take a closer look. Remember, the wise job seeker investigates every possibility. Private employment agencies are listed in the yellow pages under "Employment." Each individual agency has its own method of charging for its services. Some charge as little as one week's salary, some one month's salary. Often these agencies charge the hiring company instead of the employee. Usually the employment agency will allow its fee to be paid in monthly installments.

It is not uncommon for a drafting job listed with an employment agency to be classified as "fee paid" or "partial fee paid." This means that the employer will pay the agency's fee, or part of it, in exchange for the agency's locating a qualified applicant and doing the paperwork. In any case, it is worth your while to file an application with one or more private employment agencies, especially ones that specialize in filling technical jobs. Remember, they don't earn a penny until they find you a job, so they will be inclined to try harder to get you one.

CITY, COUNTY, AND STATE GOVERNMENT PERSONNEL OFFICES

Local and state governments are larger employers of drafters than one might expect. Most cities employ drafters in their civil engineering departments, most counties employ drafters in their engineering and property appraiser's offices, and most states employ drafters in their road departments. Moderate-to-large cities will maintain an intergovernmental job office that can be located under "Employment" in the yellow pages, or under its name in the white pages of the telephone book. These offices maintain current information on city, county, state, and federal government jobs. If you happen to live in a town that does not have an intergovernmental job service, go directly to the personnel office of the agency you are interested in and ask how you can apply.

Job-Seeking Process

Getting a job in drafting and design can be a less difficult process if one develops several critical job-seeking skills. These skills include developing a well-written letter of introduction, a comprehensive resumé, and a portfolio of drawing samples. It also means learning how to complete job applications to your best advantage, how to make a positive impression during an interview, and how to follow up on an interview.

THE LETTER OF INTRODUCTION

The *letter of introduction* is used to inquire about out-of-town positions. It serves as a cover letter for your resumé. For example, say you live in Fort Walton Beach, Florida,

---

Manufacturing
Manufacturing Technologies Company
1212 Bayview Boulevard
Fort Walton Beach, Florida 32547
904/672-1247
Contact: Reed F. Worthington
Employees: 316
Product: Electronic Equipment

FIGURE 25-4 *Manufacturing listing*

Industry Application

GETTING YOUR FOOT IN THE DOOR

Mike Jones had been an excellent student. During school he had compiled a strong portfolio of his work, especially in the area of mechanical design and drafting. His CAD skills are strong, and he has a solid understanding of manufacturing processes. Unfortunately, he has been having trouble getting his foot in the door for an interview. Jones has tried the want ads, employment offices, and cold calls to personnel offices. In every case, the message has been the same; *no experience, no interview.*

After two months of frustration, Mike Jones decided to be more creative. He sat down and made a list of the top ten places he would like to work. Then, in order to get past the Personnel Office, he drafted a letter to send directly to the head of the design and drafting department in each company. In his letter, Mike Jones offered the directors 30 days of *free* drafting work as a way to prove his worth. Within a week of sending out his letters, Jones received two offers. He accepted one of them, went to work, and after his thirty-day trial period was hired full time. Mike Jones is now a senior designer for this same company.

but want to inquire about a job in Houston, Texas. A letter of introduction with a resumé attached can take the place of a long drive. The letter should be brief and positively stated. It should say, "I am a drafter. I would like to work for your company and this is why. . . . I have enclosed a resumé of my qualifications. Can we get together for an interview?" *Figure 25-5* is a sample letter of introduction for a drafting and design student.

THE RESUMÉ

A well-written *resumé* is one of the job seeker's best tools. It is simply a categorized summary of all information that is pertinent to employers about the job seeker. A well-written resumé will amplify those qualifications you possess that are most relevant in terms of the job for which you are applying. Most people simply write one resumé and use it for every job they seek. A better approach is to tailor your resumé to the specific job for which you are applying. In other words, if you are applying to an architect, amplify the architectural aspects of your training. If you are applying to a surveyor, amplify the civil aspects of your training. Word processing makes tailoring your resumé a simple enough process. Your resumé should begin with personal information that includes your name, complete mailing address, and telephone number. Next it should list your occupational objective. This component is very important. Your objective should be broad enough to encompass a wide range of employment possibilities, yet specific enough to show that you have a definite direction in terms of your career. An example of a well-stated occupational objective is as follows:

> "*To begin work at a productive level in a drafting position that will allow me to apply my knowledge, skills, and training, and advance at a rate commensurate with my performance.*"

```
Gary Graduate
729 Green Street, Lot No. 1
Fort Walton Beach, Florida 32548

(Today's Date)

Mr. John Q. Employer
Jones Manufacturing Company
Houston, Texas 21609

Dear Mr. Employer:

     I am a drafter with specialized training in mechani-
cal drafting. I plan to move to Houston and would like to
work for your company. My specialized training and
career goals mesh well with the employment needs and
products of your company.

     I have enclosed a resume outlining my qualifications
and will furnish references plus samples of my work
upon request. I look forward to hearing from you.

Sincerely,



Gary Graduate

Encl: As Stated
```

FIGURE 25-5   *Letter of introduction*

The next component should be a summary of your education. It should begin with your highest level of achievement and work backwards. A brief explanation of your specialized training relating to the position in question should be included here. Next, list your work experience from most recent backwards. If you have any type of drafting or drafting-related experience (part-time job, on-the-job training), break this rule and list it first.

Resume
Gary Graduate
729 Green Street, Lot No. 1
Fort Walton Beach, Florida 32548
904/242-9144

Occupational Objective

To begin work at a productive level in a Drafting position that will allow me to apply my knowledge, skills, and training and to advance at a rate commensurate with my performance.

Education

Associate Degree in Drafting and Design
Okaloosa-Walton Community College; Niceville, Florida. .June 1998
My training at Okaloosa-Walton Community College covered all fields of drafting, with specialized, in-depth study in structural and mechanical drafting. I am qualified in the application of geometric tolerancing and dimensioning.

High School Diploma, Fort Walton Beach High School .June 1991

Work Experience

Self-employed installing factory cabinets for Fort Walton Glass,
Fort Walton Beach, Florida 1996–1998

Worked full time following graduation from High School building
custom-made cabinets for Classic Cabinets, Fort Walton Beach, Florida 1991–1993

Worked part time while attending High School building custom-made
cabinets for Classic Cabinets, Fort Walton Beach, Florida 1989–1991

Military Experience

United States Army 1993–1996
Three years' active duty includes 13 months spent in Bosnia as part of NATO peacekeeping forces. Honorably discharged October 1996.

Awards

◆ Graduated on Dean's List with 3.75 grade point average.
◆ Won first place in the Annual Okaloosa-Walton Community College Drafting Contest

Hobbies

Long distance running, tennis, sailing, and fishing.

FIGURE 25-6 *Resumé*

If you are a veteran, list your military information next, especially when applying for civil service jobs. The final two categories are awards and hobbies. Under awards, list anything you think might help you. *Graduated with honors, winner in a drafting contest,* or *made the Dean's List,* are all appropriate entries in the awards category. Under hobbies, list yours. This can be a surprisingly effective way of getting your foot in the door. Remember, employers are people. Your hobbies might interest someone enough to at least give you an interview. *Figure 25-6* is a sample resumé for an inexperienced drafting student.

THE PORTFOLIO OF DRAWINGS

Most employers will want to see samples of your work. For this reason, a comprehensive, attractive *portfolio of drawings* is important. All drawings should be samples of your best efforts in each drafting field studied. Architectural samples should be grouped together, mechanical samples should be grouped together, structural samples should be grouped together, and so on. A good idea is to make a clear copy of each drawing and staple the drawings together along the left-hand side to form an easy-to-examine set. Don't roll your portfolio up; leave it flat. If it must be rolled, do so drawing-side out and it will lie flat when unrolled. Nothing is more aggravating to an employer than to try to examine a portfolio that has become permanently curled from being rolled drawing-side in.

PREPARING FOR THE INTERVIEW

The job seeker who goes into an *interview* well-prepared is more likely to exit with a job. Keeping this in mind, let's take a look at how to prepare for an interview. When a potential employer invites you to an interview, make note of the time and date specified. WRITE IT DOWN! Never be late for an interview! Show up 10 to 15 minutes early; but only 10 to 15 minutes. Before the interview, find out everything you can about the company in question. Determine such things as: where the company is located, what products it manufactures, the name of the person who will conduct the interview, the approximate wage scale for beginning drafters, and any special skills on which the company places a high priority.

The local Chamber of Commerce can give you the general information you need. For specifics, try calling the company's receptionist, or you might even try to contact a person in the company who works in drafting. Chances are that if you stop by the company and speak with its public relations person, you might be able to obtain some brochures or other literature that will help acquaint you with your potential employer. The next step is to formulate answers to questions that are likely to be asked and rehearse them until your answers come naturally. *Figure 25-7* is a list of questions that are typically asked during interviews. Give this list to a friend and conduct several mock interviews. The more you practice, the more natural your answers will sound during the real interview.

THE INTERVIEW

Pull together your resumé and portfolio of drawings, put on a positive attitude, and go to the interview. Through your investigations you should have determined how drafters in this company dress for work. Dress accordingly. Dress nicely, but don't overdo it. Shake hands firmly, but not too hard, and look the potential employer right in the eyes. Answer all questions openly, truthfully, and as best you can. Be positive and confident.

After the interviewer has questioned you, ask questions yourself such as: "What is the salary range?" "Is there room for advancement?" "What type of benefits does the company offer?" When the interview has concluded, if the interviewer tells you he or she will confer with a superior and get back to you, make sure to leave on a positive note. Shake hands again and say, "Thanks

Questions Often Asked in Interviews for Drafting Jobs

1. Why do you want to work for our company? (Reveals if you have been sharp enough to learn anything about the company.)

2. How do you spend your spare time? What are your hobbies? (Shows if your interests are wide.)

3. What type of position are you most interested in? (May indicate if your main interest is making money. Many firms want more than this. Did you give a specific job title?)

4. Are you eager to learn? (Drafters must be lifelong learners.)

5. What do people criticize you for? (This and the next one are good in bringing out personality traits.)

6. What would you say are your best qualities?

7. Why do you think you would be good at this job? (Gives you a chance to tell more about your qualities, but be careful. You don't want to sound like a braggart.)

8. What is the most difficult thing you have ever tackled? The most satisfying? (Can show how high you will reach for achievement.)

9. What subjects did you like in school? (Can indicate personality traits.)

10. What college or school activities did you participate in? (Indicates whether you mixed well and/or how much energy you have.)

11. What sort of progress in our company would seem normal to you? (Indicates basic knowledge of company, realism, goals, and results orientation.)

12. How does this company compare with others you have applied to? (Could show how much shopping around you have done.)

FIGURE 25-7   *Questions typically asked during interviews*

(use the interviewer's name) for your time and consideration. I hope you will give me the opportunity to prove what I can do."

FOLLOW-UP

It is common in an interview to be told, "We will let you know in a week or so." Employers often interview several applicants, and then choose the one they feel is best qualified. If you are put on hold, don't be tentative. Call back to see if a choice has been made. This is called *follow-up*. Make it clear that of all the applicants, you are the one who wants the job most. Don't become a pest, but let them know you want the job.

Keeping a Job and Advancing

Having worked so hard to prepare for your career and to get a job, you will now want to keep the job and advance up the career ladder. This section contains a list of strategies that can help you keep your job and advance in it. Some of the strategies most likely to make you a valued employee are as follows:

- *Be dependable.* Come to work on time, and work while you are there. An employer needs to know that you can be counted on to be available when needed, and that if you are given a task, it will be completed properly and on time.

- *Be a learner.* Don't be afraid to tackle new and unfamiliar assignments. The broader your knowledge and skills become, the more valuable you will be to your employer. Read the literature in your field, attend seminars, take night courses. Never stop learning.

- *Be a worker.* Avoid joining the water cooler clique, coffee pot gang, or being a clock watcher. Everyone needs a break, and you will, too. But don't become the person supervisors bump into every time they pass the coffee pot. Also, be quick to volunteer for overtime and special projects. If you become the supervisor's "go-to" employee, you will become an invaluable asset.

- *Be self-sufficient.* Nothing is wrong with asking questions; in fact, this is a good way to learn. But before asking a question, first try to find the answer for yourself. Look through reference materials, examine similar drawings that were done before, research the files, and read the manual. With this approach, you will soon become the person who others ask when they have questions.

- *Constantly improve.* Make note of every new thing that you learn and internalize it. In today's competitive workplace, good enough today may not be good enough tomorrow. Try to improve constantly, forever.

- *Be a team player.* Getting along with fellow employees is important. Drafting and design projects are completed by a team made up of drafters, checkers, designers, and engineers. Consequently, the ability to work well with people is a must. Teamwork is covered in greater depth in Chapter 26.

Summary

- Five strategies that drafting and design students can use to improve their chances of getting a job include:

do not be afraid to relocate, learn to market your skills, be willing to start at the bottom and work your way up, be positive and assertive in your job search, expect frustration and overcome it.

- The job search involves locating potential places of employment and job openings. This includes using the Internet, the yellow pages, the Sunday newspaper, the state employment service, the local Chamber of Commerce, private employment agencies, and city, county, and state government personnel offices.

- Getting a job in drafting and design can be a less difficult process if one develops several critical job-seeking skills. These skills include developing a well-written letter of introduction, a comprehensive resumé, a comprehensive portfolio of drawing samples, and proper preparation for the interview.

- Strategies that can help you keep your job and advance in it include being dependable, being a learner, being a worker, being self-sufficient, constantly improving, and being a team player.

Review Questions

Answer the following questions either true or false.

1. Students who apply the strategies for maximizing employability effectively will improve their prospects in the job market.

2. Learning how to complete job applications is a critical job-seeking skill.

3. The Sunday want ads are especially effective as a job-seeking tool.

4. A well-written resumé is the job seeker's only tool.

5. The message a letter of introduction should convey is your life history.

Answer the following questions by selecting the best answer.

1. The contents of a well-written resumé are
   a. Personal information.
   b. Occupational objective.
   c. Summary of education.
   d. All of the above.

2. The portfolio of drawings and its purpose is
   a. To show the best samples of your drawings.
   b. To show a summary of your education.
   c. To show a past history of employment.
   d. None of the above.

3. The types of information you should learn about a company before showing up for an interview are
   a. Where the company is located.
   b. What products it manufactures.
   c. The appropriate wage scale for beginners.
   d. All of the above

4. Which of the following strategies will help you to keep a job?
   a. Don't be afraid to relocate.
   b. Be a learner.
   c. Be self-sufficient.
   d. Both b and c

Chapter 25 Problems

PROBLEM 25-1

Select a city you would like to relocate to for a job. Secure the following tools from the city of your choice:
- Telephone directory with yellow pages
- One Sunday newspaper
- Information about the city from its Chamber of Commerce

PROBLEM 25-2

Using the tools collected in Problem 25-1, compile a list of potential employers and employment agencies (private and governmental). Each entry on your list should contain at least the following information:

- Company name, address, and telephone number
- Name of a contact person
- Type of drafters employed

PROBLEM 25-3

Develop a well-written letter of introduction that can be mailed to potential employers upon completion of your education.

PROBLEM 25-4

Develop a comprehensive resumé to share with potential employers upon completion of your education.

PROBLEM 25-5

Compile a comprehensive portfolio of drawings that contains the best samples of your work.

PROBLEM 25-6

Work with a friend to practice your answers to the interview questions listed in Figure 25-7.

Chapter 26

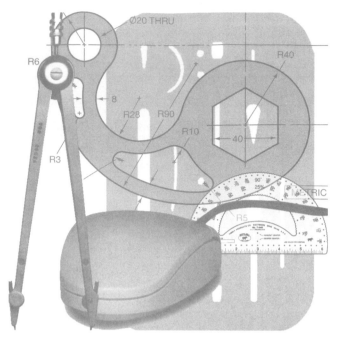

QUALITY, COMPETITION, AND TEAMWORK ON THE JOB

CHAPTER OBJECTIVES

Upon completion of this chapter, students should be able to do the following:

■ Describe the characteristics of successful team players.

■ Describe the characteristics of successful teams.

■ Resolve conflict in teams.

■ Describe how to build an effective team.

■ Explain the importance of continually improving quality and competiveness in a global environment.

The modern business environment is global and intensely competitive. In order to thrive in this environment, drafting technicians must work as effective members of teams, help their employers to continually improve quality, and contribute to making their employers more competitive. Consequently, drafting and design technicians must be skilled team players. An example of a project team of the type drafters and designers can expect to serve on in today's workplace is shown in *Figure 26-1*. Clement Engineering is a small company that has been awarded a contract to design a flexible bicycle carrier. Each member of the project team has specific responsibilities relating to the work the team is expected to accomplish.

The engineer is responsible for performing all of the stress, strain, loading, and bending calculations for the flexible bicycle carrier that Clement Engineering has contracted to design. The product designer is responsible for using the engineer's calculations to design the actual configuration of the carrier. The tool designer will design the jigs and fixtures the manufacturer will use in producing the carrier. The drafters will prepare all necessary drawings, notes, and parts lists to document the design of the carrier and the tool designer's jigs and fixtures. The checker is an experienced drafter who will check and verify the work of the two drafters. Clearly, every member of this team depends in some way on every other member. Consequently, it is important for each individual to be a good team player.

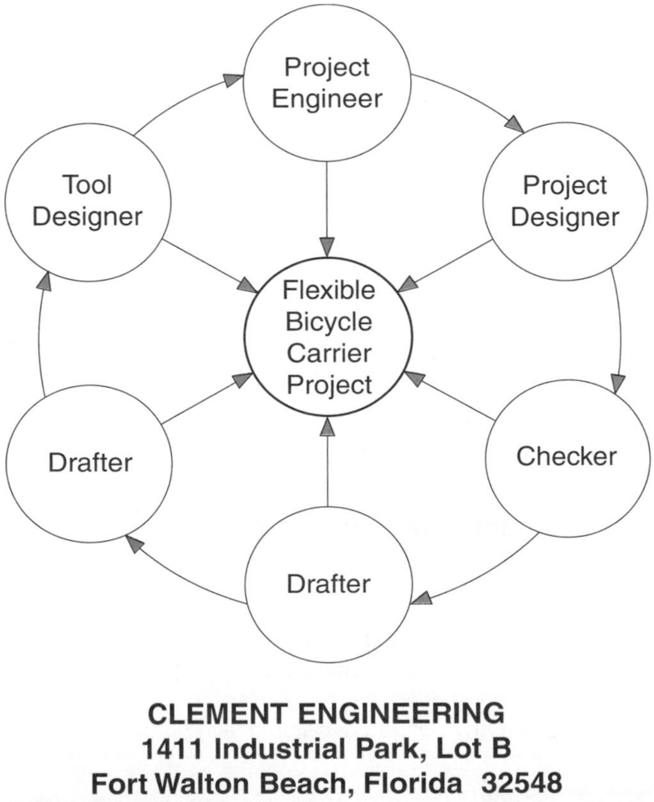

Characteristics of Successful Team Players

Why is it that some work teams succeed while others fail? A study by the Institute for Corporate Competitiveness suggests that one of the key factors that determines the difference between success and failure may be the character traits of individual team members.[1] According to this study, successful teams are composed of members who have the character traits listed in *Figure 26-2*.

- *Honesty.* Team members depend on each other every day. The work of one cannot be properly done without the help of the others. Consequently, it is important for team members to trust each other. Trust is built on a foundation of honesty. Dishonest team members erode the trust of all team members.

- *Selflessness.* Successful team players tend to share the accolades when things go well and take their share of the blame when things go wrong. Selfless team members look out for the team and their fellow team members rather than focusing exclusively on themselves. Selfish team members look out only for themselves.

- *Dependability/Reliability.* Successful team players show up on time, do their work right, and complete all tasks on time. They can be counted on by their teammates. Undependable team members cannot be counted on.

- *Enthusiasm.* Successful team players are enthusiastic. Enthusiasm is catching. This is important because people perform better when they are enthusiastic about their work. Correspondingly, team members who have negative attitudes tend to dampen the spirit of the entire team.

CLEMENT ENGINEERING
1411 Industrial Park, Lot B
Fort Walton Beach, Florida 32548

FIGURE 26-1 *Example of a project team*

Characteristics of Successful Team Players	
✓ Honesty	✓ Initiative
✓ Selflessness	✓ Patience
✓ Dependability/Reliability	✓ Resourcefulness
✓ Enthusiasm	✓ Punctuality
✓ Responsibility	✓ Tolerance/Sensitivity
✓ Cooperativeness	✓ Perseverance

FIGURE 26-2 *Characteristics of successful team players*

[1] Institute for Corporate Competitiveness, *Team Success Study*, 1995, pp. 2–4.

- *Responsibility.* Members of a team typically have well-defined responsibilities. For example, think of a baseball team. Each position has specific responsibilities, and everyone on the team depends on the individuals who play these positions to know what is expected of them, and to satisfy their expectations consistently as well. Successful team players take responsibility for their actions, behavior, and performance. Irresponsible team members tend to avoid responsibility and blame others when things go wrong.

- *Cooperativeness.* Successful team players cooperate with each other in performing their work tasks. Think again of the example of a baseball team. In order to make a double play, the shortstop, second baseman, and first baseman must cooperate with each other. If just one of these players fails to cooperate, the entire process breaks down. Work projects are just like this.

- *Initiative.* Successful team members take the initiative. They never say, "That's not my job," or "I don't feel like it." Taking the initiative means anticipating what needs to be done, and doing it, rather than waiting for someone else to step forward.

- *Patience.* Rugged individualism and self-sufficiency are cultural traits in the United States and Canada that can be traced back to how these countries were settled and by whom. Consequently, working in a team can be difficult. There will be rough spots as employees who are individualists by nature make the difficult transition to being team players. Successful team members are patient with the transitional challenges they face.

- *Resourcefulness.* Successful team players are resourceful. This means they are good at finding ways to get their work done in spite of a lack of resources (e.g., time, funds, equipment, etc.). For example, when deadlines mean there is too much work and too little time, resourceful people are good at finding ways to do more work in less time. Every team needs resourceful members.

- *Punctuality.* Successful team players are punctual. This means they arrive for work on time and complete their work on schedule. This trait is important because the work of each team member depends on that of the other team members. Just one team member who is tardy or gets behind in his or her work can cause the entire team to miss deadlines.

- *Tolerance/Sensitivity.* Successful team players are tolerant and sensitive. This trait is important because the modern workplace is increasingly diverse. This means people in teams are likely to be different in terms of race, gender, religion, cultural heritage, and/or political views. Tolerant/sensitive team players understand that diversity can strengthen a team by providing different viewpoints and perspectives. However, in order to enjoy the benefits of diversity, team members must appreciate and be sensitive to the differences of their fellow team members, whether these differences are based on culture, race, gender, politics, or any of the many other ways people can be different.

- *Perseverance.* Successful team players have perseverance. This means they persist in their work no matter how frustrating the obstacles and roadblocks they encounter. People who persevere provide—by their example—encouragement to team members who might become so frustrated that they want to quit. This is a natural human reaction to difficulty. Consequently, perseverance is an invaluable trait for team members.

Characteristics of Successful Teams

Successful teams are typically composed of members who are good team players—members who have the characteristics explained in the previous section. In addition to the characteristics of its individual members, a team's performance is also affected by several team-wide characteristics, *Figure 26-3.*

DIRECTION

A team's *direction* should come from higher management in the form of a team charter. A team charter consists of the following components:

- Mission
- Broad goals
- Specific activities
- Schedules and time frames

Figure 26-4 contains the charter for the Flexible Bicycle Carrier team from Figure 26-1. This charter makes clear the team's mission. The mission is further classified by four broad goals that communicate more

---

Characteristics of Successful Teams

✓ Direction/Understanding
✓ Accountability
✓ Clearly defined roles
✓ Effective communication
✓ Clearly defined ground rules
✓ Clearly defined decision-making process

FIGURE 26-3 *Characteristics of successful teams*

Team Charter
FLEXIBLE BICYCLE CARRIER TEAM

Team Mission
The mission of the Flexible Bicycle Carrier Team is to design a lightweight, flexible bicycle carrier for the recreational biking market.

Broad Goals
1. To design a lightweight bicycle carrier that is flexible enough to be carried on automobiles, trucks, or vans.
2. To design a bicycle carrier that can be manufactured at a cost of less than $90.
3. To design a bicycle carrier that is rugged enough to last at least five years under normal conditions.
4. Complete the entire project within six months of beginning.

Specific Activities and Time Frames
- Project begins: January 2nd
- Complete initial conceptual design and preliminary design by January 30th.
- Build and field test a prototype of the product by March 30th.
- Make design adjustments by April 15th.
- Update the drawing package by April 30th.
- Complete tool design by May 30th.
- Complete jig and fixture drawing package by June 15th.

FIGURE 26-4 *Example of a team charter*

MASON ENGINEERING, INC.
Waco, Texas

Team Accountability

Team Success
Success will be determined by how effectively and how well the team fulfills its charter.

Evaluation of Team Performance/Progress
Team performance/progress will be monitored continually against the time frame in the team's charter. The team leader will report progress on a weekly basis to MEI's vice-president for engineering.

Team Decisions
Ultimate responsibility for team decisions belongs to the team leader who may delegate as appropriate. However, broad-based input from all team members should be sought before decisions are made. All individual team members affected by a decision will be given a voice in the decision-making process.

Responsibilities of Team Members
Responsibilities of individual team members are specified by the team leader.

Evaluation of Individual Team Members
The team leader will formally evaluate each team member at the conclusion of the project.

FIGURE 26-5 *Team accountability checklist*

specifically what is expected of the team. These broad goals are followed by a list of specific activities, each with a corresponding deadline. With this charter in hand, every team member has clear direction.

ACCOUNTABILITY

While a team charter such as the one in Figure 26-4 gives direction, it does not explain how team members will be held accountable. *Accountability* is an important factor in promoting teamwork. Team members need to know the following:

- How will team success be defined?
- How will team progress/performance be evaluated?
- How will team decisions be made?
- What are the responsibilities of all team members?
- How will individual team members be evaluated?

Figure 26-5 is a team accountability checklist of the kind a team leader might use during an orientation meeting for team members. At such a meeting the team leader distributes copies of the team charter and reviews it. Once all team members understand what is expected of them, both individually and as a team, the team leader uses the checklist in Figure 26-5 as a guide in explaining all applicable accountability measures. The team leader for the Flexible Bicycle Carrier project (Figure 26-1) would be the project engineer.

EFFECTIVE COMMUNICATION

Human communication is an imperfect process at best. The process itself, *Figure 26-6,* is easy enough to understand. It involves the person sending the message (sender), the message itself, the medium used to send the message, and the person who is supposed to receive the message (receiver). However, easy to understand does not mean easy to do. *Figure 26-7* is a list of factors that can inhibit *effective communication*.

"What you heard is not what I meant." "I don't know what you mean." "Do you really mean what you just said?" Statements such as these are common when

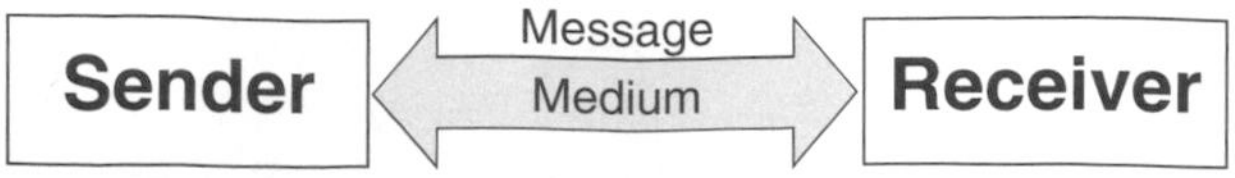

- **Verbal**
  (Telephone, in-person, audio, tape, videotape, compact disc)
- **Nonverbal**
  (Body language)
- **Written**
  (Letters, memorandums, reports, technical papers, specifications, newsletters)
- **Graphic**
  (Sketches, drawings, symbols)

FIGURE 26-6 *Communication process*

Inhibitors of EFFECTIVE COMMUNICATION

- Sender has poor communication skills (verbal, nonverbal, written, or graphic).
- Language differences between sender and receiver.
- Receiver has poor listening skills.
- Problems with the medium (noise, technical malfunctions).

FIGURE 26-7 *List of factors that can inhibit effective communication*

people try to communicate. Sending a message is like throwing a football—the receiver may or may not "get" it. In such cases, the problem is often that the sender has poor communication skills. If the message is verbal, the sender might have an insufficient command of the language, might mumble, might speak too softly, or might have trouble maintaining eye contact with receivers. All of these are common problems that can inhibit effective communication.

Language differences between the sender and receiver can also inhibit effective communication. Obviously there will be problems if the sender and receiver speak different languages. But even people who supposedly speak the same language can experience problems that are the result of age, culture, and gender-related nuances in the language.

For example, parents often have trouble communicating with their teenage children (and vice versa). The difficulties are often the result of the two generations giving different meanings to the same words, or the younger generation introducing new words and phrases with which their parents are unfamiliar. Interestingly, the same parents who don't understand the language of their teenagers now were probably misunderstood themselves when teenagers.

Inner-city employees may have difficulty understanding their rural counterparts (and vice versa). Women and men may interpret the same words or phrases differently. An employee from the Deep South may have difficulty understanding the speech of an employee from the North (and vice versa).

Poor listening skills on the part of the receiver can also inhibit effective communication. This inhibitor is common because people are typically not good listeners. Throughout the school years, students learn to read, write, and perform mathematical calculations—the so-called three Rs. However, little attention is paid to the development of listening skills. Consequently, team members with good listening skills are rare.

Effective communication can also be inhibited by problems with the medium used to carry the message. For example, background noise can override part of a spoken message. If the receivers cannot hear properly, they are likely to misunderstand the message or miss it altogether. Have you ever tried holding a conversation over a cellular phone that is getting beyond its range? You hear static, the words fade in and out, and the message is garbled. Such difficulties with the medium make effective communication unlikely, if not impossible.

Differences in *perceptions* caused by differences in gender, culture, race, and/or experience can make effective communication particularly difficult. This is especially the case when we attach meaning—as we do both consciously and unconsciously—to the nonverbal aspects of communication. A good-natured pat on the back given to one male by another as an expression of encouragement might be interpreted differently by a female team member. Certain words, phrases, or expressions may be interpreted differently by people of different races or cultures.

All of these factors make effective communication a more difficult undertaking than one might expect. In order to communicate effectively with each other, team members must get to know each other well. Doing so is one of the most effective strategies in team building.

Clearly Defined Ground Rules

Successful teams have *clearly defined ground rules* that are thoroughly understood and subscribed to by all members. A team's ground rules should, themselves, be the product of teamwork. Ground rules clarify what

MASON ENGINEERING, INC.
Waco, Texas

Team Ground Rules

Attendance
Attendance at *all* team meetings is mandatory unless excused in advance by the team leader.

Meeting Time and Place
Team meetings take place in Engineering Conference Room A at 8:00 a.m. every Monday and Friday morning. Changes in time, location, and/or day must have prior approval of the team leader.

Promptness
Team meetings begin at 8:00 a.m. All team members should be present and ready to commence business promptly at 8:00 a.m.

Participation
All team members have their individual areas of responsibility (e.g., engineering, design, checking, drafting). However, in team meetings, these distinctions **do not** apply. During meetings, cross-functional participation is welcomed. Suggestions for improvements in any area of team performance are encouraged.

FIGURE 26-8 *Example of the types of team ground rules*

behaviors are expected of team members, including what will be tolerated and what will not. *Figure 26-8* is an example of the types of ground rules commonly used in design and drafting teams.

The ground rules in this figure have four components: attendance, meeting places and time, promptness, and participation. The criteria set forth in each of these components remind team members of the rules they have agreed to follow, and the expectations they have of each other.

Ground rules developed by the team members who will apply them have an advantage over those that are imposed by management. That advantage is known as buy-in. It means that team members accept and subscribe to the ground rules because the rules are **their** rules. Buy-in promotes compliance through voluntary self-policing and peer pressure—the two most effective and positive approaches to enforcement.

CLEARLY DEFINED DECISION-MAKING PROCESS

How are team decisions made? Does the team leader make decisions unilaterally? By consensus? With input from all team members? These are questions that should be answered as soon as a team is formed. The ideal

approach for a design and drafting team is for the team leader to make decisions, but only after soliciting input from all team members. Team members can provide different points of view, different perspectives, and different ideas. By doing so, they can help the team leader make a more informed decision. Correspondingly, by soliciting the input of team members, the team leader can ensure that they buy into the ultimate decision.

This concept of buy-in is critical in promoting teamwork. It is most likely to occur when team members are given a voice and allowed to participate in making decisions they will have to carry out. It does not mean that a team member's ideas, opinions, or suggestions must always be accepted. Rather, it means that they are always welcomed, always solicited, and always given serious consideration. An easy way to understand this phenomenon is to remember the following rule of thumb:

Team members with ideas do not need to win every time, but they do need to be allowed to play the game.

Conflict Resolution in Teams

Even in the most effective teams there will be conflict among the members. Good teams are **not** teams that have no conflict. Rather, they are teams that **know how to handle conflict** effectively. *Figure 26-9* contains several rules of thumb about conflict that people who work in teams should know.

CONFLICT IS COMMON TO ALL TEAMS

Conflict is common to all teams because people have different personalities, ways of doing things, priorities, and ambitions. A quiet, sensitive person might find it difficult to work with a boisterous, outgoing individual. A self-starter who is prone to take the initiative in a

Rules of Thumb
CONFLICT RESOLUTION IN TEAMS

- Conflict is common to all teams.

- Conflict can be a process that develops better communication.

- Conflict can be used as a process that leads to better results.

- Conflict can be used to promote understanding.

- The result of conflict should be better understanding, rather than victory.

FIGURE 26-9 *Rules of thumb concerning conflict*

given situation might have trouble working with a person who needs to be prodded along. Employees with different levels of ambition might not get along easily; a more ambitious employee might see a more easily satisfied colleague as inferior, while an employee with lower ambitions might see a more ambitious colleague as less of a team player, interested only in his or her future.

CONFLICT CAN DEVELOP BETTER COMMUNICATION

On the negative side, conflict can manifest itself in ways such as arguing, bickering, and pettiness. On the positive side, it can manifest itself as differences of opinion about issues by people who can disagree without being disagreeable. In either case, conflict can help team members develop better communication. This does not mean that conflict **will** lead to better communication. It means that it **can.** The key is in how conflict is handled. If team members learn to listen to each other, support ideas rather than individuals, and focus their emotions on solutions rather than each other, they will improve communication as the result of conflict. If team members refuse to listen, support individuals rather than ideas, and focus on who **wins** rather than what idea is **best,** communication will not improve.

CONFLICT CAN LEAD TO BETTER RESULTS

When conflict is a positive, well-intended battle of ideas rather than a battle for power, influence, or advantage, it can lead to better results. When right-minded team members must defend their ideas and opinions in debates with other right-minded team members, their ideas are sharpened. Weaknesses that might not have been thought of are pointed out, viewpoints that have not been heard come out, and ideas that have not been considered get consideration. This type of give and take among team members leads to a better understanding of both problems and solutions. The result, invariably, is better results.

STRATEGIES FOR RESOLVING CONFLICT

In order for conflict to promote better communication and better results, team members must be skilled at applying conflict resolution strategies. *Figure 26-10* is a checklist of strategies that can be used to resolve conflict in teams.

- **Be Partners, Not Enemies.** Good team players can share the same goal (what is best for the organization), but disagree on how best to achieve it. When disagreements occur, good team members remember that teammates are partners, not enemies.

Checklist
CONFLICT RESOLUTION STRATEGIES

- ✓ Before beginning a discussion, say to yourself, "My teammates and I are partners, not enemies."
- ✓ Listen attentively to show that a teammates' concerns are a priority for you.
- ✓ Listen with an attitude of openness to your teammates' concerns.
- ✓ Ask questions to ensure that you understand.
- ✓ Paraphrase to show that you understand.
- ✓ Focus on the message, not on how it is delivered.
- ✓ Focus on the meaning of the message rather than the words used.
- ✓ Ask questions rather than making judgments.
- ✓ Keep personalities out of confrontations; focus on ideas, not individuals.
- ✓ Focus on just **one** issue at a time.
- ✓ Apply only one criterion in choosing from among proposed alternatives: Which one is best for the organization?
- ✓ Deal with specifics, not generalities.
- ✓ Focus on facts, rather than trying to determine motives.
- ✓ Promote mutual understanding, rather than winning and losing.
- ✓ Apologize when wrong and move on.
- ✓ When giving feedback, make it constructive, rather than judgmental or accusatory.
- ✓ When receiving feedback, be open rather than defensive.

FIGURE 26-10 *Checklist of strategies for resolving conflict*

- **Listen Attentively.** When discussing issues of importance to the organization, good team players listen attentively to each other. They don't listen selectively (listening for only what they want to hear). They don't pretend to be interested, and they don't listen protectively (listening only for potentially threatening aspects of the message). Rather, team players show by their attention that the concerns of their teammates are important to them.

- **Listen Openly.** Good teammates are open to good ideas regardless of who originates them. When discussing issues with teammates, good team players remain open-minded and flexible in their thinking.

- **Ask Questions.** Listening attentively and openly does not mean automatically accepting the views of teammates. Misunderstanding undermines teamwork and intensifies conflict. When discussing issues with teammates, ask questions to make sure that you understand what is being proposed. Never **assume** that you understand. Ask.

- **Paraphrase to Show Understanding.** Just as it is important to understand the point a teammate is trying to make, it is also important to **show** the teammate that you understand. An effective way to do this is to paraphrase the message and repeat it back in your own words.

- **Focus on Content.** It is easy to become so focused on **how** teammates convey their messages that you overlook what they are trying to say. In order to discern **what** is being said, you may have to overlook how it is said. If a teammate is angry, disorganized, or inarticulate, overlook these things. How a message is conveyed is just the wrapping paper on the package. Some wrapping paper is attractive and some isn't. Remember, the quality of what is inside the package has nothing to do with the wrapping paper on the outside.

- **Focus on the Meaning.** Some team members are good with words, and some are not. When discussing issues with teammates who have trouble putting their ideas into words, focus on the meaning of the message. Don't get caught up judging the quality of a teammate's verbal skills.

- **Don't Judge.** When the ideas of teammates are poorly conceived, help them come to this realization by asking questions. It is better to help teammates recognize the weaknesses in their proposals than to point them out yourself in a judgmental way. Even if a teammate persists in supporting a bad idea, remember that it is the idea that is bad, not the person promoting it.

- **Focus on Ideas, Not Individuals.** It is impossible to like an idea without liking the person who originates it. There is a natural human tendency to reject the ideas of people we don't like, and to support the ideas of people we do like. Both approaches are mistakes. In a team setting, ideas should be accepted or rejected on their merits, not on the personality or popularity of the person proposing them.

- **Focus on One Issue at a Time.** Mixing multiple issues complicates discussions. Every idea has its own individual strengths and weaknesses. Consequently, it will be less frustrating to discuss one issue at a time. Remember, frustration is a major cause of conflict-inducing behavior.

- **Consider What Is Best for the Organization.** Few things will induce conflict faster than an individual who brings a win-lose attitude to discussions. Such individuals will persist in supporting a bad idea just because it is their own. They want to win, which they mistakenly view as having their idea prevail. This is the antithesis of teamwork. Good team players consider what is best for the organization when supporting or rejecting an idea.

- **Deal with Specifics.** Poorly defined generalities promote conflict. In order to focus on ideas rather than personalities, it is necessary to be specific. When discussing ideas, ask teammates to be specific.

- **Focus on Facts.** "Why is he making this proposal?" "What's in it for him?" These types of questions promote conflict, even when asked only mentally. Ideas should be accepted or rejected on their merits, not on the assumed or supposed motives of the originator.

- **Promote Mutual Understanding.** Misunderstanding causes conflict. Understanding prevents or alleviates conflict. When discussing an issue with a teammate, use the techniques of listening and paraphrasing to ensure that you understand. If teammates have not yet learned these techniques, gently and tactfully prompt them.

- **Apologize and Move On.** People have bad days. People make mistakes. When this happens to you, don't dwell on it. Simply apologize—sincerely—and move on. A sincere apology delivered with a smile and a handshake can be effective in preventing conflict.

- **Forgive, Forget, Move On.** The corollary to apologizing is to forgive, forget, and move on when a teammate apologizes. Teamwork can fall apart quickly when teammates harbor grudges against each other. This advice is easier to follow when decision making in the team is depersonalized.

- **Give Constructive Feedback.** Constructive feedback has two distinguishing characteristics. First, it is sincerely aimed at improving the team's performance, or that of an individual team member. Second, it is given in a positive, non-accusatory, nonjudgmental manner. Good team players give each other positive feedback.

- **Receive Feedback Openly.** The corollary to giving constructive feedback is receiving it. Good team players want to improve their ideas and their performance. Consequently, they keep an open mind and seek out feedback that will help them improve. This can take some practice because the natural human response to criticism, even when it is constructively given, is to become defensive. Avoid this tendency. Otherwise, while defending your ideas, you might miss out on ways to improve them.

Effective Team Building

Good teams don't just happen, they are built. This section describes the four-step approach to team building; an approach that can be used for turning a group of individuals into an effective, productive team. *Figure 26-11* is a team-building checklist based on the four-step approach.

Industry Application

A TEAM IN CONFLICT

McKandle Manufacturing, Inc. (MMI) is a world-class producer of training simulators for the U.S. Department of Defense. As part of a companywide plan to improve performance and, in turn, competitiveness, MMI adopted the teamwork approach in its engineering and manufacturing departments. The concept is working effectively now, but this was not always the case. In fact, when the company first formed its workforce into teams, overall corporate performance actually decreased.

As the company's productivity continued to decline, the vice presidents for engineering and manufacturing both tried to convince MMI's CEO to drop the teamwork concept and go back to the old way of doing things. However, before giving up on teamwork—an approach he had seen work well at other companies—the CEO decided to try another strategy. He brought in a teamwork expert to assess the situation and make recommendations.

In less than a week, the consultant reported her results. MMI didn't have teams. It had groups of individuals who were trying to work together, but didn't know how. They lacked direction, accountability, and communication. In addition, MMI's reward and recognition systems were still based on individual rather than team performance. The consultant set to work immediately to rectify the situation.

She helped each team develop a charter. She helped MMI's key managers design a new team-oriented reward/recognition system. And she conducted numerous seminars and focus group sessions to improve communication companywide. Within three months MMI's declining performance had turned around and was showing significant improvements. Within six months, MMI had become a world-class competitor in the simulator market and was expanding into Canada, Europe, and Asia.

TEAM-BUILDING CHECKLIST
(Four-Step Approach)

♦ **Assess Current Status**
  • Team direction and understanding
  • Characteristics of team members

♦ **Plan Improvement Strategies**
  • Develop or revise the team charter
  • Plan strategies for developing team-friendly characteristics
  • Revise job descriptions if necessary
  • Revise the employee performance evaluation process if necessary

♦ **Execute the Improvement Strategies**

♦ **Evaluate Results and Adjust as Necessary**

FIGURE 26-11 *Team-building checklist*

ASSESS CURRENT STATUS

What is the team's mission? What are the team's goals? What are the team's projected time frames and deadlines? These are questions about direction that every member of a team should be able to answer. A team's mission, goals, and schedules give its members direction, and a team with direction will be a more effective team.

Certain personal characteristics promote teamwork. Do all team members have team-friendly characteristics? Effective team players typically have personal characteristics such as honesty, patience, and reliability.

Accountability is important to people. Students want to know *what will be on the test* and *how their work will be graded*. When people are to be held accountable, they want to know how. This rule of thumb also applies in teams. Team members need to know how they will be evaluated, how often, and by whom.

Teams, whether newly formed or of long standing, should assess their current status concerning team direction, the characteristics of team members, and accountability. *Figure 26-12* is a team assessment checklist that can be used for completing this task. Team members should complete such a checklist in confidence.

PLAN IMPROVEMENT STRATEGIES

Once all team members have completed an assessment checklist, a third party outside of the team should compile the results. Individual checklists are then destroyed, and only the summary is returned to the team. *Figure 26-13* is a summary showing the results of an assessment conducted by an engineering, design, and drafting team. It contains items from the checklist for which the majority of team members checked **No**.

TEAM ASSESSMENT CHECKLIST

Team Direction and Understanding

Yes   No

_____ _____ Does the team have a mission statement?
_____ _____ Do all team members understand the team's mission?
_____ _____ Does the team have a set of goals?
_____ _____ Do all team members understand the goals?
Does the team have a schedule for its work with definite deadlines and
_____ _____ assigned responsibilities?
_____ _____ Do all team members know the team's schedule?

Personal Characteristics of Team Members

_____ _____ Are all team members honest with each other?
Are all team members selfless (do they put the team's needs ahead of
_____ _____ their own)?
_____ _____ Are all team members dependable/reliable (can they be counted on)?
_____ _____ Are all team members enthusiastic about the team's work?
_____ _____ Do all team members take responsibility for their individual performance?
_____ _____ Do all team members share the responsibility for team performance?
_____ _____ Do all team members cooperate with each other in getting the work done?
_____ _____ Do all team members take the initiative in solving problems?
Do all team members take the initiative in getting work done properly and
_____ _____ on time?
_____ _____ Are all team members patient with each other?
Are all team members resourceful in finding ways to overcome barriers,
_____ _____ problems, and unexpected inhibitors?
_____ _____ Are all team members punctual?
Are all team members sensitive to and tolerant of individual differences
_____ _____ among team members?
Do all team members persevere in getting the job done, even when
_____ _____ difficulties are confronted?

Accountability

_____ _____ Do all team members understand how they will be evaluated individually?
Do all team members understand how overall team performance will be
_____ _____ evaluated?
Do all team members understand who will evaluate their performance—
_____ _____ individually and as a team?
_____ _____ Is *teamwork* a criterion when evaluating individual team members?
_____ _____ Are evaluations tied to pay, recognition, and/or promotion?

FIGURE 26-12 *Team assessment checklist*

TEAM ASSESSMENT SUMMARY
(Flexible Bicycle Carrier Team)

Items in this summary received a majority of <u>No</u>
indications during the team assessment process.

No Do all team members understand the team's
mission?

No Do all team members understand the goals?

No Are all team members honest with each other?

No Do all team members cooperate with each other
in getting the job done?

No Is teamwork a criterion when evaluating individual team members?

No Are evaluations tied to pay, recognition, and/or
promotion?

FIGURE 26-13 *Team assessment summary*

The team leader then uses the summary to plan specific improvement activities. *Figure 26-14* is the plan that was developed in response to the *Team Assessment Summary* in Figure 26-13. It contains three activities. The first is a team meeting to discuss the team's charter, mission, and goals. The team in question has a charter, but a majority of its members either don't know the mission and goals or don't understand them. Solving this problem should require nothing more than a meeting.

The second activity will be more difficult. A majority of team members indicated that honesty and cooperation are a problem. Dealing with these issues will require the help of a third party—someone with expertise in group dynamics and human relations. Since there are costs associated with such an activity, the team leader will have to secure the approval of higher management.

TEAM-BUILDING/IMPROVEMENT PLAN
(Flexible Bicycle Carrier Team)

Activity 1: Understanding Team Direction
The team leader will conduct a team meeting to discuss the team's mission and goals. The purpose of the meeting will be to ensure that all team members thoroughly understand the team's mission and goals.

Activity 2: Honesty and Cooperation
The team leader will work with higher management to arrange an off-site seminar/team-building activity for the team. The activity will focus on the issues of honesty and cooperation.

Activity 3: Accountability Issues
The team leader will work with the Human Resources Department to revise the company's standard evaluation instrument to include **teamwork** criteria. The team leader will also ask higher management to convene an **ad hoc** task force to establish real ties between performance evaluations and rewards/recognition.

FIGURE 26-14 *Team-building/improvement plan*

SAMPLE EVALUATION CRITERIA

◆ This employee is an effective team player:
___Always ___Usually ___Sometimes ___Never

◆ This employee enhances overall team performance:
___Always ___Usually ___Sometimes ___Never

FIGURE 26-15 *Sample evaluation criteria*

The third activity will require the team leader to work with higher management and the human resources department. The company's performance evaluation needs to be revised to include at least one criterion that pertains to teamwork. *Figure 26-15* contains examples of such criteria.

Once the performance appraisal instrument has been updated, the team leader will have to work with higher management and human resources personnel to find ways to tie team performance to pay. There are a variety of ways to accomplish this, all of which are outside the scope of this book. It is important for students of this book to understand only that individual pay/recognition should be tied to both individual performance and team performance.

Working in a Globally Competitive Environment

As a result of technological advances in the area of telecommunications and transportation, businesses that once limited their markets to a community, region, or country can now reach customers anywhere on the globe. Engineering, manufacturing, and other technology firms that employ drafting technicians that once competed only against other local firms now find themselves competing against similar companies from Japan, China, India, Indonesia, Korea, and the various countries in the European Union. In order to compete successfully in a global environment, companies must: (1) achieve peak performance from their personnel every day and (2) continually improve their products, services, processes, and people.

Different companies adopt different models for instilling a culture of continual improvement. Two of the widest used models in the United States are the *Baldrige Model* and the *ISO 9000 Model*. Drafting technicians need to know how to function in a work environment in which these types of models are the norm.

BALDRIGE MODEL FOR QUALITY AND CONTINUAL IMPROVEMENT

The National Institute of Standards and Technology (NIST) has as part of its charter the promotion of quality and continual improvement to enhance the economy of the United States. One of the programs instituted by NIST for this purpose is the Malcolm Baldrige National Quality Award. This highly competitive award recognizes organizations for excellence in the areas of quality management and continual improvement. Some companies use the application process as a vehicle for transforming themselves into globally competitive organizations. Others forego the award, but still use the award criteria to improve their performance. The criteria in the Baldrige Model are as follows:

1.0 Leadership
    1.1 Organizational leadership
    1.2 Public responsibility and citizenship
2.0 Strategic Planning
    2.1 Strategy development
    2.2 Strategy deployment
3.0 Customer and Market Focus
    3.1 Customer and market knowledge
    3.2 Customer relations and satisfaction
4.0 Information Analysis
    4.1 Measurement and analysis of organizational performance
    4.2 Information management
5.0 Human Resource Focus
    5.1 Work systems
    5.2 Employee education, training, and development
    5.3 Employee well-being and satisfaction

6.0   Process Management
    6.1   Product and service processes
    6.2   Business processes
    6.3   Support processes
7.0   Business Results
    7.1   Customer-focused results
    7.2   Financial and market results
    7.3   Human resources results
    7.4   Organizational effectiveness results

The work of engineering, CAD, and drafting departments figures prominently into a number of these criteria. For example, under heading 4.1—*Measurement and analysis of organizational performance*—the work of drafting technicians and how well it is done is measured using a variety of different techniques. How long does it take to complete drawing packages? How much time is given to making corrections? What do customers think about the quality of the drafting work? There is constant pressure to do better and better work in less and less time. Drafting technicians must be able to work effectively in an environment in which good enough today is not good enough tomorrow. All work and work processes must get better and better all the time.

ISO 9000 Model

The International Standards Organization created the ISO 9000 series of quality standards as a way to promote uniform quality on an international basis. The idea was to ensure quality products and services that are standardized in terms of what quality actually means on a global basis. Specifically, companies adopt the appropriate ISO 9000 standard for one or more of the following purposes:

1. As a model for continually improving operations.

2. As a model for creating and continually improving the management and quality assurance systems.

3. As a model for continually improving the quality of products and services.

4. As a way to conform to the requirements of major customers.

The most comprehensive of the ISO 9000 standards is ISO 9001. This standard speaks to the full range of functions of engineering and manufacturing companies: design, development, production, installation, servicing, final inspection, and testing. Examples of ISO 9001 criteria that affect drafting technicians fall under the broad heading of design. What follows is an excerpt from the *Design Output* criteria that are found in section 4.4.5 of ISO 9001:

4.4.5   Design Output
    • Is design output documented? (Drawings? Specifications? Notes? etc.)

• Is the documented design output expressed in terms that can be verified against design-input requirements and validated?

• Does the design output do the following: Meet the design-input requirements? Contain or reference acceptance criteria? Identify the characteristics of the design that are crucial to the safe and proper functioning of the product?

• Are design-output documents reviewed before release?

Much of the work of drafting technicians involves documenting the designs developed by engineering personnel. Consequently, drafting and CAD figure prominently in the application of the ISO 9001 model and specific criteria. Therefore, drafting technicians must be prepared to have their work carefully scrutinized and to continually improve the quality of their work.

Summary

• Successful teams are composed of members who have the following character traits: honesty, selflessness, dependability/reliability, enthusiasm, responsibility, cooperativeness, initiative, patience, resourcefulness, punctuality, tolerance/sensitivity, and perseverance.

• In addition to the characteristics of its individual members, a team's performance is also affected by several team-wide characteristics such as direction, accountability, effective communication, clearly defined ground rules, and clearly defined decision-making processes.

• There are several rules of thumb about conflict that people who work in teams should know. They include the facts that conflict is common to all teams, conflict can develop better communication, conflict can lead to better results, and conflict can be used to promote understanding.

• The four-step approach to team building can be used for turning a group of individuals into an effective, productive team. This approach includes assessing current status, planning improvement strategies, executing the improvement strategies, and evaluating results/adjusting as necessary.

Review Questions

Answer the following questions either true or false.

1. Five characteristics of successful team players are perseverance, enthusiasm, honesty, patience, and punctuality.

2. The components of a team charter are the team mission, broad goals, and specific activities/time frames.

3. The term *direction* as it relates to teamwork is the team's mission, goals, and schedules.

4. The term *accountability* relates to the characteristics of each team member.

5. The communication process involves the person sending the message, the message itself, the medium used to send the message, and the receiver of the message.

6. Good teams never have conflicts.

Answer the following questions by selecting the best answer.

1. Two inhibitors of effective communication are:
   a. Poor communication skills and poor listening skills.
   b. Clearly defined ground rules and language differences.
   c. Technical malfunctions and cooperativeness.
   d. None of the above

2. Which of the following are ground rules for a team?
   a. Participation and direction
   b. Attendance and promptness
   c. Resourcefulness and meeting place
   d. Promptness and patience

3. The concept of buy-in as it relates to decision making in teams means that:
   a. A team member's ideas, opinions, or suggestions must always be accepted.
   b. Team members are given a voice and allowed to participate in making decisions they will have to carry out.
   c. By soliciting the input of team members, the team leader can **not** ensure that they buy into the ultimate decision.
   d. None of the above

4. Teammates can use conflict to develop better communication by:
   a. Refusing to listen.
   b. Supporting individuals rather than ideas.
   c. Learning to listen to each other and supporting ideas rather than individuals.
   d. Focusing on who wins rather than what idea is best.

5. Which of the following are strategies for resolving conflict?
   a. Be partners, not enemies.
   b. Listen attentively.
   c. Listen openly.
   d. All of the above

Chapter 26 Problems

PROBLEM 26-1
Select three classmates to join you in a discussion group. Conduct a discussion of the characteristics in Figure 26-2. What kinds of problems have you and your classmates experienced when dealing with people who do not have these characteristics? What problems might such people cause as members of a team? Record the main points made during the discussion.

PROBLEM 26-2
Establish a fictitious drafting and design team. Develop a charter for your team that includes a mission, goals, and specific activities with time frames.

PROBLEM 26-3
Develop a team accountability checklist for your fictitious team from Problem 26-2.

PROBLEM 26-4
Make a list of factors that inhibit communication between yourself and someone you know.

PROBLEM 26-5
Develop a set of ground rules for your fictitious team from Problem 26-2.

PROBLEM 26-6

Think of any team of which you are, or have been, a member. Apply the Team Assessment List in Figure 26-12 to that team.

PROBLEM 26-7

Using the results of the assessment from Problem 26-6, develop a team-building plan for your team.

PROBLEM 26-8

Conduct an Internet search on the topic "Malcolm Baldrige Award." Locate the award application criteria and review them to identify all of the ways drafting and CAD might fit into them.

PROBLEM 26-9

Conduct an Internet search on the topic "ISO9001" and locate the complete set of criteria. Examine all design-related criteria and determine how they affect drafting and CAD.

A p p e n d i x A

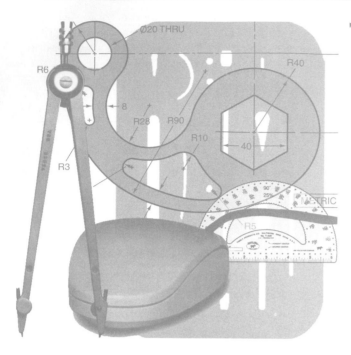

TABLES

Contents

INCHES TO MILLIMETERS

in.	mm	in.	mm	in.	mm	in.	mm
1	25.4	26	660.4	51	1295.4	76	1930.4
2	50.8	27	685.8	52	1320.8	77	1955.8
3	76.2	28	711.2	53	1346.2	78	1981.2
4	101.6	29	736.6	54	1371.6	79	2006.6
5	127.0	30	762.0	55	1397.0	80	2032.0
6	152.4	31	787.4	56	1422.4	81	2057.4
7	177.8	32	812.8	57	1447.8	82	2082.8
8	203.2	33	838.2	58	1473.2	83	2108.2
9	228.6	34	863.6	59	1498.6	84	2133.6
10	254.0	35	889.0	60	1524.0	85	2159.0
11	279.4	36	914.4	61	1549.4	86	2184.4
12	304.8	37	939.8	62	1574.8	87	2209.8
13	330.2	38	965.2	63	1600.2	88	2235.2
14	355.6	39	990.6	64	1625.6	89	2260.6
15	381.0	40	1016.0	65	1651.0	90	2286.0
16	406.4	41	1041.4	66	1676.4	91	2311.4
17	431.8	42	1066.8	67	1701.8	92	2336.8
18	457.2	43	1092.2	68	1727.2	93	2362.2
19	482.6	44	1117.6	69	1752.6	94	2387.6
20	508.0	45	1143.0	70	1778.0	95	2413.0
21	533.4	46	1168.4	71	1803.4	96	2438.4
22	558.8	47	1193.8	72	1828.8	97	2463.8
23	584.2	48	1219.2	73	1854.2	98	2489.2
24	609.6	49	1244.6	74	1879.6	99	2514.6
25	635.0	50	1270.0	75	1905.0	100	2540.0

The above table is exact on the basis: 1 in. = 25.4 mm

MILLIMETERS TO INCHES

mm	in.	mm	in.	mm	in.	mm	in.
1	0.039370	26	1.023622	51	2.007874	76	2.992126
2	0.078740	27	1.062992	52	2.047244	77	3.031496
3	0.118110	28	1.102362	53	2.086614	78	3.070866
4	0.157480	29	1.141732	54	2.125984	79	3.110236
5	0.196850	30	1.181102	55	2.165354	80	3.149606
6	0.236220	31	1.220472	56	2.204724	81	3.188976
7	0.275591	32	1.259843	57	2.244094	82	3.228346
8	0.314961	33	1.299213	58	2.283465	83	3.267717
9	0.354331	34	1.338583	59	2.322835	84	3.307087
10	0.393701	35	1.377953	60	2.362205	85	3.364457
11	0.433071	36	1.417323	61	2.401575	86	3.385827
12	0.472441	37	1.456693	62	2.440945	87	3.425197
13	0.511811	38	1.496063	63	2.480315	88	3.464567
14	0.551181	39	1.535433	64	2.519685	89	3.503937
15	0.590551	40	1.574803	65	2.559055	90	3.543307
16	0.629921	41	1.614173	66	2.598425	91	3.582677
17	0.669291	42	1.653543	67	2.637795	92	3.622047
18	0.708661	43	1.692913	68	2.677165	93	3.661417
19	0.748031	44	1.732283	69	2.716535	94	3.700787
20	0.787402	45	1.771654	70	2.755906	95	3.740157
21	0.826772	46	1.811024	71	2.795276	96	3.779528
22	0.866142	47	1.850394	72	2.834646	97	3.818898
23	0.905512	48	1.889764	73	2.874016	98	3.858268
24	0.944882	49	1.929134	74	2.913386	99	3.897638
25	0.984252	50	1.968504	75	2.952756	100	3.937008

The above table is approximate on the basis: 1 in. = 25.4 mm, 1/25.4 = 0.039370078740+

TABLE 1

INCH/METRIC—EQUIVALENTS

Fraction	Decimal Equivalent Customary (in.)	Decimal Equivalent Metric (mm)	Fraction	Decimal Equivalent Customary (in.)	Decimal Equivalent Metric (mm)
1/64	.015625	0.3969	33/64	.515625	13.0969
1/32	.03125	0.7938	17/32	.53125	13.4938
3/64	.046875	1.1906	35/64	.546875	13.8906
1/16	.0625	1.5875	9/16	.5625	14.2875
5/64	.078125	1.9844	37/64	.578125	14.6844
3/32	.09375	2.3813	19/32	.59375	15.0813
7/64	.109375	2.7781	39/64	.609375	15.4781
1/8	.1250	3.1750	5/8	.6250	15.8750
9/64	.140625	3.5719	41/64	.640625	16.2719
5/32	.15625	3.9688	21/32	.65625	16.6688
11/64	.171875	4.3656	43/64	.671875	17.0656
3/16	.1875	4.7625	11/16	.6875	17.4625
13/64	.203125	5.1594	45/64	.703125	17.8594
7/32	.21875	5.5563	23/32	.71875	18.2563
15/64	.234375	5.9531	47/64	.734375	18.6531
1/4	.250	6.3500	3/4	.750	19.0500
17/64	.265625	6.7469	49/64	.765625	19.4469
9/32	.28125	7.1438	25/32	.78125	19.8438
19/64	.296875	7.5406	51/64	.796875	20.2406
5/16	.3125	7.9375	13/16	.8125	20.6375
21/64	.328125	8.3384	53/64	.828125	21.0344
11/32	.34375	8.7313	27/32	.84375	21.4313
23/64	.359375	9.1281	55/64	.859375	21.8281
3/8	.3750	9.5250	7/8	.8750	22.2250
25/64	.390625	9.9219	57/64	.890625	22.6219
13/32	.40625	10.3188	29/32	.90625	23.0188
27/64	.421875	10.7156	59/64	.921875	23.4156
7/16	.4375	11.1125	15/16	.9375	23.8125
29/64	.453125	11.5094	61/64	.953125	24.2094
15/32	.46875	11.9063	31/32	.96875	24.6063
31/64	.484375	12.3031	63/64	.984375	25.0031
1/2	.500	12.7000	1	1.000	25.4000

TABLE 2 From *Drafting for Trades and Industry—Basic Skills,* Nelson. Delmar Publishers Inc.

METRIC EQUIVALENTS

LENGTH

U.S. to Metric

Metric to U.S.

1 inch = 2.540 centimeters
1 foot = .305 meter
1 yard = .914 meter
1 mile = 1.609 kilometers

1 millimeter = .039 inch
1 centimeter = .394 inch
1 meter = 3.281 feet or 1.094 yards
1 kilometer = .621 mile

AREA

1 inch^2 = 6.451 centimeter^2
1 foot^2 = .093 meter^2
1 yard^2 = .836 meter^2
1 acre^2 = 4,046.873 meter^2

1 millimeter^2 = .00155 inch^2
1 centimeter^2 = .155 inch^2
1 meter^2 = 10.764 foot^2 or 1.196 yard^2
1 kilometer^2 = .386 mile^2 or 247.04 acre^2

VOLUME

1 inch^3 = 16.387 centimeter^3
1 foot^3 = .028 meter^3
1 yard^3 = .764 meter^3
1 quart = .946 liter
1 gallon = .003785 meter^3

1 centimeter^3 = 0.61 inch^3
1 meter^3 = 35.314 foot^3 or 1.308 yard^3
1 liter = .2642 gallons
1 liter = 1.057 quarts
1 meter^3 = 264.02 gallons

WEIGHT

1 ounce = 28.349 grams
1 pound = .454 kilogram
1 ton = .907 metric ton

1 gram = .035 ounce
1 kilogram = 2.205 pounds
1 metric ton = 1.102 tons

VELOCITY

1 foot/second = .305 meter/second
1 mile/hour = .447 meter/second

1 meter/second = 3.281 feet/second
1 kilometer/hour = .621 mile/second

ACCELERATION

1 inch/second^2 = .0254 meter/second^2
1 foot/second^2 = .305 meter/second^2

1 meter/second^2 = 3.278 feet/second^2

FORCE

N (newton) = basic unit of force, kg-m/s^2. A mass of one kilogram (1 kg) exerts a gravitational force of 9.8 N (theoretically 9.80665 N) at mean sea level.

TABLE 3

MULTIPLIERS FOR DRAFTERS

Multiply	By	To Obtain	Multiply	By	To Obtain
Acres	43,560	Square feet	Degrees/sec.	0.002778	Revolutions/sec.
Acres	4047	Square meters	Fathoms	6	Feet
Acres	1.562×10^{-3}	Square miles	Feet	30.48	Centimeters
Acres	4840	Square yards	Feet	12	Inches
Acre-feet	43,560	Cubic feet	Feet	0.3048	Meters
Atmospheres	76.0	Cms. of mercury	Foot-pounds	1.286×10^{-3}	British Thermal Units
Atmospheres	29.92	Inches of mercury	Foot-pounds	5.050×10^{-7}	Horsepower-hrs.
Atmospheres	33.90	Feet of water	Foot-pounds	3.241×10^{-4}	Kilogram-calories
Atmospheres	10,333	Kgs./sq. meter	Foot-pounds	0.1383	Kilogram-meters
Atmospheres	14.70	Lbs./sq. inch	Foot-pounds	3.766×10^{-7}	Kilowatt-hrs.
Atmospheres	1.058	Tons/sq. ft.	Foot-pounds/min.	1.286×10^{-3}	BTU/min.
Board feet	144 sq. in. × 1 in.	Cubic inches	Foot-pounds/min.	0.01667	Foot-pounds/sec.
British Thermal Units	0.2520	Kilogram-calories	Foot-pounds/min.	3.030×10^{-5}	Horsepower
British Thermal Units	777.5	Foot-lbs.	Foot-pounds/min.	3.241×10^{-4}	Kg-calories/min.
British Thermal Units	3.927×10^{-4}	Horsepower-hrs.	Foot-pounds/min.	2.260×10^{-5}	Kilowatts
British Thermal Units	107.5	Kilogram-meters	Foot-pounds/sec.	7.717×10^{-2}	BTU/min.
British Thermal Units	2.928×10^{-4}	Kilowatt-hrs.	Foot-pounds/sec.	1.818×10^{-3}	Horsepower
BTU/min.	12.96	Foot-lbs./sec.	Foot-pounds/sec.	1.945×10^{-2}	Kg-calories/min.
BTU/min.	0.02356	Horsepower	Foot-pounds/sec.	1.356×10^{-3}	Kilowatts
BTU/min.	0.01757	Kilowatts	Gallons	3785	Cubic centimeters
BTU/min.	17.57	Watts	Gallons	0.1337	Cubic feet
Cubic centimeters	3.531×10^{-5}	Cubic feet	Gallons	231	Cubic inches
Cubic centimeters	6.102×10^{-2}	Cubic inches	Gallons	3.785×10^{-3}	Cubic meters
Cubic centimeters	10^{-6}	Cubic meters	Gallons	4.951×10^{-3}	Cubic yards
Cubic centimeters	1.308×10^{-6}	Cubic yards	Gallons	3.785	Liters
Cubic centimeters	2.642×10^{-4}	Gallons	Gallons	8	Pints (liq.)
Cubic centimeters	10^{-3}	Liters	Gallons	4	Quarts (liq.)
Cubic centimeters	2.113×10^{-3}	Pints (liq.)	Gallons-Imperial	1.20095	U.S. gallons
Cubic centimeters	1.057×10^{-3}	Quarts (liq.)	Gallons-U.S.	0.83267	Imperial gallons
Cubic feet	2.832×10^{4}	Cubic cm	Gallons water	8.3453	Pounds of water
Cubic feet	1728	Cubic inches	Horsepower	42.44	BTU/min.
Cubic feet	0.02832	Cubic meters	Horsepower	33,000	Foot-lbs./min.
Cubic feet	0.03704	Cubic yards	Horsepower	550	Foot-lbs./sec.
Cubic feet	7.48052	Gallons	Horsepower	1.014	Horsepower (metric)
Cubic feet	28.32	Liters	Horsepower	10.70	Kg-calories/min.
Cubic feet	59.84	Pints (liq.)	Horsepower	0.7457	Kilowatts
Cubic feet	29.92	Quarts (liq.)	Horsepower	745.7	Watts
Cubic feet/min.	472.0	Cubic cm/sec.	Horsepower-hours	2547	BTU
Cubic feet/min.	0.1247	Gallons/sec.	Horsepower-hours	1.98×10^{6}	Foot-lbs.
Cubic feet/min.	0.4720	Liters/sec.	Horsepower-hours	641.7	Kilogram-calories
Cubic feet/min.	62.43	Pounds of water/min.	Horsepower-hours	2.737×10^{5}	Kilogram-meters
Cubic feet/sec.	0.646317	Millions gals./day	Horsepower-hours	0.7457	Kilowatt-hours
Cubic feet/sec.	448.831	Gallons/min.	Kilometers	10^{5}	Centimeters
Cubic inches	16.39	Cubic centimeters	Kilometers	3281	Feet
Cubic inches	5.787×10^{-4}	Cubic feet	Kilometers	10^{3}	Meters
Cubic inches	1.639×10^{-5}	Cubic meters	Kilometers	0.6214	Miles
Cubic inches	2.143×10^{-5}	Cubic yards	Kilometers	1094	Yards
Cubic inches	4.329×10^{-3}	Gallons	Kilowatts	56.92	BTU/min.
Cubic inches	1.639×10^{-2}	Liters	Kilowatts	4.425×10^{4}	Foot-lbs./min.
Cubic inches	0.03463	Pints (liq.)	Kilowatts	737.6	Foot-lbs./sec.
Cubic inches	0.01732	Quarts (liq.)	Kilowatts	1.341	Horsepower
Cubic meters	10^{6}	Cubic centimeters	Kilowatts	14.34	Kg.-calories/min.
Cubic meters	35.31	Cubic feet	Kilowatts	10^{3}	Watts
Cubic meters	61.023	Cubic inches	Kilowatt-hours	3415	BTU
Cubic meters	1.308	Cubic yards	Kilowatt-hours	2.655×10^{6}	Foot-lbs.
Cubic meters	264.2	Gallons	Kilowatt-hours	1.341	Horsepower-hrs.
Cubic meters	10^{3}	Liters	Kilowatt-hours	860.5	Kilogram-calories
Cubic meters	2113	Pints (liq.)	Kilowatt-hours	3.671×10^{5}	Kilogram-meters
Cubic meters	1057	Quarts (liq.)	Lumber Width (in.) ×		
Degrees (angle)	60	Minutes	$\dfrac{\text{Thickness (in.)}}{12}$	Length (ft.)	Board feet
Degrees (angle)	0.01745	Radians			
Degrees (angle)	3600	Seconds	Meters	100	Centimeters
Degrees/sec.	0.01745	Radians/sec.	Meters	3.281	Feet
Degrees/sec.	0.1667	Revolutions/min.	Meters	39.37	Inches

TABLE 4

MULTIPLIERS FOR DRAFTERS (Concluded)

Multiply	By	To Obtain	Multiply	By	To Obtain
Meters	10^{-3}	Kilometers	Pounds (troy)	373.24177	Grams
Meters	10^{3}	Millimeters	Pounds (troy)	0.822857	Pounds (avoir.)
Meters	1.094	Yards	Pounds (troy)	13.1657	Ounces (avoir.)
Meters/min.	1.667	Centimeters/sec.	Pounds (troy)	3.6735×10^{-4}	Tons (long)
Meters/min.	3.281	Feet/min.	Pounds (troy)	4.1143×10^{-4}	Tons (short)
Meters/min.	0.05468	Feet/sec.	Pounds (troy)	3.7324×10^{-4}	Tons (metric)
Meters/min.	0.06	Kilometers/hr.	Quadrants (angle)	90	Degrees
Meters/min.	0.03728	Miles/hr.	Quadrants (angle)	5400	Minutes
Meters/sec.	196.8	Feet/min.	Quadrants (angle)	1.571	Radians
Meters/sec.	3.281	Feet/sec.	Radians	57.30	Degrees
Meters/sec.	3.6	Kilometers/hr.	Radians	3438	Minutes
Meters/sec.	0.06	Kilometers/min.	Radians	0.637	Quadrants
Meters/sec.	2.237	Miles/hr.	Radians/sec.	57.30	Degrees/sec.
Meters/sec.	0.03728	Miles/min.	Radians/sec.	0.1592	Revolutions/sec.
Microns	10^{-6}	Meters	Radians/sec.	9.549	Revolutions/min.
Miles	5280	Feet	Radians/sec./sec.	573.0	Revs./min./min.
Miles	1.609	Kilometers	Radians/sec./sec.	0.1592	Revs./sec./sec.
Miles	1760	Yards	Reams	500	Sheets
Miles/hr.	1.609	Kilometers/hr.	Revolutions	360	Degrees
Miles/hr.	0.8684	Knots	Revolutions	4	Quadrants
Minutes (angle)	2.909×10^{-4}	Radians	Revolutions	6.283	Radians
Ounces	16	Drams	Revolutions/min.	6	Degrees/sec.
Ounces	437.5	Grains	Square yards	2.066×10^{-4}	Acres
Ounces	0.0625	Pounds	Square yards	9	Square feet
Ounces	28.349527	Grams	Square yards	0.8361	Square meters
Ounces	0.9115	Ounces (troy)	Square yards	3.228×10^{-7}	Square miles
Ounces	2.790×10^{-5}	Tons (long)	Temp. (°C.) + 273	1	Abs. temp. (°C.)
Ounces	2.835×10^{-5}	Tons (metric)	Temp. (°C.) + 17.78	1.8	Temp. (°F.)
Ounces (troy)	480	Grains	Temp. (°F.) + 460	1	Abs. temp. (°F.)
Ounces (troy)	20	Pennyweights (troy)	Temp. (°F.) − 32	5/9	Temp. (°C.)
Ounces (troy)	0.08333	Pounds (troy)	Watts	0.05692	BTU/min.
Ounces (troy)	31.103481	Grams	Watts	44.26	Foot-pounds/min.
Ounces (troy)	1.09714	Ounces (avoir.)	Watts	0.7376	Foot-pounds/sec.
Ounces (fluid)	1.805	Cubic Inches	Watts	1.341×10^{-3}	Horsepower
Ounces (fluid)	0.02957	Liters	Watts	0.01434	Kg.-calories/min.
Ounces/sq. inch	0.0625	Lbs./sq. inch	Watts	10^{-3}	Kilowatts
Pounds	16	Ounces	Watts-hours	3.415	BTU
Pounds	256	Drams	Watts-hours	2655	Foot-pounds
Pounds	7000	Grains	Watts-hours	1.341×10^{-3}	Horsepower-hrs.
Pounds	0.0005	Tons (short)	Watts-hours	0.8605	Kilogram-calories
Pounds	453.5924	Grams	Watts-hours	367.1	Kilogram-meters
Pounds	1.21528	Pounds (troy)	Watts-hours	10^{-3}	Kilowatt-hours
Pounds	14.5833	Ounces (troy)	Yards	91.44	Centimeters
Pounds (troy)	5760	Grains	Yards	3	Feet
Pounds (troy)	240	Pennyweights (troy)	Yards	36	Inches
Pounds (troy)	12	Ounces (troy)	Yards	0.9144	Meters

TABLE 4 *(Concluded)*

CIRCUMFERENCES AND AREAS OF CIRCLES
From 1/64 to 50, Diameter

Dia.	Circum.	Area	Dia.	Circum.	Area	Dia.	Circum.	Area	Dia.	Circum.	Area
1/64	.04909	.00019	8	25.1327	50.2655	17	53.4071	226.980	26	81.6814	530.929
1/32	.09818	.00077	8 1/8	25.5254	51.8485	17 1/8	53.7998	230.330	26 1/8	82.0741	536.047
1/16	.19635	.00307	8 1/4	25.9181	53.4562	17 1/4	54.1925	233.705	26 1/4	82.4668	541.188
1/8	.39270	.01227	8 3/8	26.3108	55.0883	17 3/8	54.5852	237.104	26 3/8	82.8595	546.355
3/16	.58905	.02761	8 1/2	26.7035	56.7450	17 1/2	54.9779	240.528	26 1/2	83.2522	551.546
1/4	.78540	.04909	8 5/8	27.0962	58.4262	17 5/8	55.3706	243.977	26 5/8	83.6449	556.761
5/16	.98175	.07670	8 3/4	27.4889	60.1321	17 3/4	55.7633	247.450	26 3/4	84.0376	562.002
3/8	1.1781	.11045	8 7/8	27.8816	61.8624	17 7/8	56.1560	250.947	26 7/8	84.4303	567.266
7/16	1.3744	.15033	9	28.2743	63.6173	18	56.5487	254.469	27	84.8230	572.555
1/2	1.5708	.19635	9 1/8	28.6670	65.3967	18 1/8	56.9414	258.016	27 1/8	85.2157	577.869
9/16	1.7671	.24850	9 1/4	29.0597	67.2007	18 1/4	57.3341	261.587	27 1/4	85.6084	583.207
5/8	1.9635	.30680	9 3/8	29.4524	69.0292	18 3/8	57.7268	265.182	27 3/8	86.0011	588.570
11/16	2.1598	.37122	9 1/2	29.8451	70.8822	18 1/2	58.1195	268.803	27 1/2	86.3938	593.957
3/4	2.3562	.44179	9 5/8	30.2378	72.7597	18 5/8	58.5122	272.447	27 5/8	86.7865	599.369
13/16	2.5525	.51849	9 3/4	30.6305	74.6619	18 3/4	58.9049	276.117	27 3/4	87.1792	604.806
7/8	2.7489	.60132	9 7/8	31.0232	76.5886	18 7/8	59.2976	279.810	27 7/8	87.5719	610.267
15/16	2.9452	.69029				19	59.6903	283.529	28	87.9646	615.752
1	3.1416	.78540	10	31.4159	78.5398	19 1/8	60.0830	287.272	28 1/8	88.3573	621.262
1 1/8	3.5343	.99402	10 1/8	31.8086	80.5156	19 1/4	60.4757	291.039	28 1/4	88.7500	626.797
1 1/4	3.9270	1.2272	10 1/4	32.2013	82.5159	19 3/8	60.8684	294.831	28 3/8	89.1427	632.356
1 3/8	4.3197	1.4849	10 3/8	32.5940	84.5408	19 1/2	61.2611	298.648	28 1/2	89.5354	637.940
1 1/2	4.7124	1.7671	10 1/2	32.9867	86.5902	19 5/8	61.6538	302.489	28 5/8	89.9281	643.548
1 5/8	5.1051	2.0739	10 5/8	33.3794	88.6641	19 3/4	62.0465	306.354	28 3/4	90.3208	649.181
1 3/4	5.4978	2.4053	10 3/4	33.7721	90.7626	19 7/8	62.4392	310.245	28 7/8	90.7135	654.838
1 7/8	5.8905	2.7612	10 7/8	34.1648	92.8856	20	62.8319	314.159	29	91.1062	660.520
2	6.2832	3.1416	11	34.5575	95.0332	20 1/8	63.2246	318.099	29 1/8	91.4989	666.226
2 1/8	6.6759	3.5466	11 1/8	34.9502	97.2053	20 1/4	63.6173	322.062	29 1/4	91.8916	671.957
2 1/4	7.0686	3.9761	11 1/4	35.3429	99.4020	20 3/8	64.0100	326.051	29 3/8	92.2843	677.713
2 3/8	7.4613	4.4301	11 3/8	35.7356	101.623	20 1/2	64.4027	330.064	29 1/2	92.6770	683.493
2 1/2	7.8540	4.9087	11 1/2	36.1283	103.869	20 5/8	64.7954	334.101	29 5/8	93.0697	689.297
2 5/8	8.2467	5.4119	11 5/8	36.5210	106.139	20 3/4	65.1881	338.163	29 3/4	93.4624	695.127
2 3/4	8.6394	5.9396	11 3/4	36.9137	108.434	20 7/8	65.5808	342.250	29 7/8	93.8551	700.980
2 7/8	9.0321	6.4918	11 7/8	37.3064	110.753	21	65.9735	346.361	30	94.2478	706.858
3	9.4248	7.0686	12	37.6991	113.097	21 1/8	66.3662	350.496	30 1/8	94.6405	712.761
3 1/8	9.8175	7.6699	12 1/8	38.0918	115.466	21 1/4	66.7589	354.656	30 1/4	95.0332	718.689
3 1/4	10.2102	8.2958	12 1/4	38.4845	117.859	21 3/8	67.1516	358.841	30 3/8	95.4259	724.640
3 3/8	10.6029	8.9462	12 3/8	38.8772	120.276	21 1/2	67.5442	363.050	30 1/2	95.8186	730.617
3 1/2	10.9956	9.6211	12 1/2	39.2699	122.718	21 5/8	67.9369	367.284	30 5/8	96.2113	736.618
3 5/8	11.3883	10.3206	12 5/8	39.6626	125.185	21 3/4	68.3296	371.542	30 3/4	96.6040	742.643
3 3/4	11.7810	11.0447	12 3/4	40.0553	127.676	21 7/8	68.7223	375.825	30 7/8	96.9967	748.693
3 7/8	12.1737	11.7932	12 7/8	40.4480	130.191	22	69.1150	380.133	31	97.3894	754.768
4	12.5664	12.5664	13	40.8407	132.732	22 1/8	69.5077	384.465	31 1/8	97.7821	760.867
4 1/8	12.9591	13.3640	13 1/8	41.2334	135.297	22 1/4	69.9004	388.821	31 1/4	98.1748	766.990
4 1/4	13.3518	14.1863	13 1/4	41.6261	137.886	22 3/8	70.2931	393.203	31 3/8	98.5675	773.139
4 3/8	13.7445	15.0330	13 3/8	42.0188	140.500	22 1/2	70.6858	397.608	31 1/2	98.9602	779.311
4 1/2	14.1372	15.9043	13 1/2	42.4115	143.139	22 5/8	71.0785	402.038	31 5/8	99.3529	785.509
4 5/8	14.5299	16.8002	13 5/8	42.8042	145.802	22 3/4	71.4712	406.493	31 3/4	99.7456	791.731
4 3/4	14.9226	17.7206	13 3/4	43.1969	148.489	22 7/8	71.8639	410.972	31 7/8	100.1383	797.977
4 7/8	15.3153	18.6655	13 7/8	43.5896	151.201	23	72.2566	415.476	32	100.5310	804.248
5	15.7080	19.6350	14	43.9823	153.938	23 1/8	72.6493	420.004	32 1/8	100.9237	810.543
5 1/8	16.1007	20.6290	14 1/8	44.3750	156.699	23 1/4	73.0420	424.557	32 1/4	101.3164	816.863
5 1/4	16.4934	21.6476	14 1/4	44.7677	159.485	23 3/8	73.4347	429.134	32 3/8	101.7091	823.208
5 3/8	16.8861	22.6906	14 3/8	45.1604	162.295	23 1/2	73.8274	433.736	32 1/2	102.1018	829.577
5 1/2	17.2788	23.7583	14 1/2	45.5531	165.130	23 5/8	74.2201	438.363	32 5/8	102.4945	835.971
5 5/8	17.6715	24.8505	14 5/8	45.9458	167.989	23 3/4	74.6128	443.014	32 3/4	102.8872	842.389
5 3/4	18.0642	25.9672	14 3/4	46.3385	170.873	23 7/8	75.0055	447.689	32 7/8	103.2799	848.831
5 7/8	18.4569	27.1085	14 7/8	46.7312	173.782	24	75.3982	452.389	33	103.6726	855.299
6	18.8496	28.2743	15	47.1239	176.715	24 1/8	75.7909	457.114	33 1/8	104.0653	861.791
6 1/8	19.2423	29.4647	15 1/8	47.5166	179.672	24 1/4	76.1836	461.863	33 1/4	104.4580	868.307
6 1/4	19.6350	30.6796	15 1/4	47.9094	182.654	24 3/8	76.5763	466.637	33 3/8	104.8507	874.848
6 3/8	20.0277	31.9191	15 3/8	48.3020	185.661	24 1/2	76.9690	471.435	33 1/2	105.2434	881.413
6 1/2	20.4204	33.1831	15 1/2	48.6947	188.692	24 5/8	77.3617	476.258	33 5/8	105.6361	888.003
6 5/8	20.8131	34.4716	15 5/8	49.0874	191.748	24 3/4	77.7544	481.106	33 3/4	106.0288	894.618
6 3/4	21.2058	35.7847	15 3/4	49.4801	194.828	24 7/8	78.1471	485.977	33 7/8	106.4215	901.257
6 7/8	21.5985	37.1223	15 7/8	49.8728	197.933	25	78.5398	490.874	34	106.8142	907.920
7	21.9912	38.4845	16	50.2655	201.062	25 1/8	78.9325	495.795	34 1/8	107.2069	914.609
7 1/8	22.3839	39.8712	16 1/8	50.6582	204.216	25 1/4	79.3252	500.740	34 1/4	107.5996	921.321
7 1/4	22.7765	41.2825	16 1/4	51.0509	207.394	25 3/8	79.7179	505.711	34 3/8	107.9923	928.058
7 3/8	23.1692	42.7183	16 3/8	51.4436	210.597	25 1/2	80.1106	510.705	34 1/2	108.3850	934.820
7 1/2	23.5619	44.1787	16 1/2	51.8363	213.825	25 5/8	80.5033	515.724	34 5/8	108.7777	941.607
7 5/8	23.9546	45.6636	16 5/8	52.2290	217.077	25 3/4	80.8960	520.768	34 3/4	109.1704	948.417
7 3/4	24.3473	47.1730	16 3/4	52.6217	220.353	25 7/8	81.2887	525.836	34 7/8	109.5631	955.253
7 7/8	24.7400	48.7069	16 7/8	53.0144	223.654						

TABLE 5

TRIGONOMETRIC FORMULAS

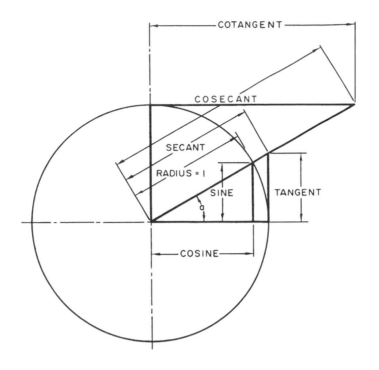

FORMULAS FOR FINDING FUNCTIONS OF ANGLES
$\dfrac{\text{Side opposite}}{\text{Hypotenuse}} = $ **SINE**
$\dfrac{\text{Side adjacent}}{\text{Hypotenuse}} = $ **COSINE**
$\dfrac{\text{Side opposite}}{\text{Side adjacent}} = $ **TANGENT**
$\dfrac{\text{Side adjacent}}{\text{Side opposite}} = $ **COTANGENT**
$\dfrac{\text{Hypotenuse}}{\text{Side adjacent}} = $ **SECANT**
$\dfrac{\text{Hypotenuse}}{\text{Side opposite}} = $ **COSECANT**

FORMULAS FOR FINDING THE LENGTH OF SIDES FOR RIGHT-ANGLE TRIANGLES WHEN AN ANGLE AND SIDE ARE KNOWN	
Length of side opposite	Hypotenuse × Sine Hypotenuse ÷ Cosecant Side adjacent × Tangent Side adjacent ÷ Cotangent
Length of side adjacent	Hypotenuse × Cosine Hypotenuse ÷ Secant Side opposite × Cotangent Side opposite ÷ Tangent
Length of hypotenuse	Side opposite × Cosecant Side opposite ÷ Sine Side adjacent × Secant Side adjacent ÷ Cosine

TABLE 6

RIGHT-TRIANGLE FORMULAS

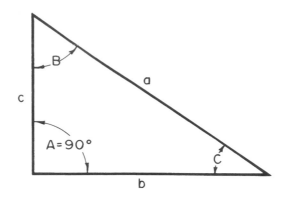

TO FIND ANGLES	FORMULAS	
C	$\frac{c}{a}$ = Sine C	90° — B
C	$\frac{b}{a}$ = Cosine C	90° — B
C	$\frac{c}{b}$ = Tan C	90° — B
C	$\frac{b}{c}$ = Cot C	90° — B
C	$\frac{a}{b}$ = Secant C	90° — B
C	$\frac{a}{c}$ = Csc C	90° — B
B	$\frac{b}{a}$ = Sine B	90° — C
B	$\frac{c}{a}$ = Cosine B	90° — C
B	$\frac{b}{c}$ = Tan B	90° — C
B	$\frac{c}{b}$ = Cot B	90° — C
B	$\frac{a}{c}$ = Secant B	90° — C
B	$\frac{a}{b}$ = Csc B	90° — C

TO FIND SIDES	FORMULAS	
a	$\sqrt{b^2 + c^2}$	
a	c × Csc C	$\frac{c}{\text{sine C}}$
a	c × Secant B	$\frac{c}{\text{Cosine B}}$
a	b × Csc B	$\frac{b}{\text{Sine B}}$
a	b × Secant C	$\frac{b}{\text{Cosine C}}$
b	$\sqrt{a^2 - c^2}$	
b	a × Sine B	$\frac{a}{\text{Cosecant B}}$
b	a × Cos C	$\frac{a}{\text{Secant C}}$
b	c × Tan B	$\frac{c}{\text{Cotangent B}}$
b	c × Cot C	$\frac{c}{\text{Tangent C}}$
c	$\sqrt{a^2 - b^2}$	
c	a × Cos B	$\frac{a}{\text{Secant B}}$
c	a × Sine C	$\frac{a}{\text{Cosecant C}}$
c	b × Cot B	$\frac{b}{\text{Tangent B}}$
c	b × Tan C	$\frac{b}{\text{Cotangent C}}$

TABLE 7

OBLIQUE-ANGLED TRIANGLE FORMULAS

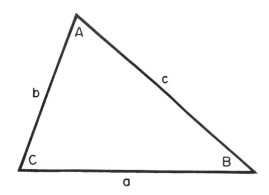

TO FIND	KNOWN	SOLUTION
C	A-B	$180° - (A + B)$
b	a-B-A	$\dfrac{a \times \text{Sin B}}{\text{Sin A}}$
c	a-A-C	$\dfrac{a \times \text{Sin C}}{\text{Sin A}}$
Tan A	a-C-b	$\dfrac{a \times \text{Sin C}}{b - (a \times \text{Cos C})}$
B	A-C	$180° - (A + C)$
Sin B	b-A-a	$\dfrac{b \times \text{Sin A}}{a}$
A	B-C	$180° - (B + C)$
Cos A	a-b-c	$\dfrac{b^2 + C^2 - a^2}{2bc}$
Sin C	c-A-a	$\dfrac{c \times \text{Sin A}}{a}$
Cot B	a-C-b	$\dfrac{a \times \text{Csc C}}{b} - \text{Cot C}$
c	b-C-B	$b \times \text{Sin C} \times \text{Csc B}$

TABLE 8

RUNNING AND SLIDING FITS

VALUES IN THOUSANDTHS OF AN INCH

Nominal Size Range Inches		Class RC1 Precision Sliding			Class RC2 Sliding Fit			Class RC3 Precision Running			Class RC4 Close Running			Class RC5 Medium Running		
		Hole Tol. GR5	Minimum Clearance	Shaft Tol. GR4	Hole Tol. GR6	Minimum Clearance	Shaft Tol. GR5	Hole Tol. GR7	Minimum Clearance	Shaft Tol. GR6	Hole Tol. GR8	Minimum Clearance	Shaft Tol. GR7	Hole Tol. GR8	Minimum Clearance	Shaft Tol. GR7
Over	To	-0		+0	-0		+0	-0		+0	-0		+0	-0		+0
0	.12	+0.15	0.10	-0.12	+0.25	0.10	-0.15	+0.40	0.30	-0.25	+0.60	0.30	-0.40	+0.60	0.60	-0.40
.12	.24	+0.20	0.15	-0.15	+0.30	0.15	-0.20	+0.50	0.40	-0.30	+0.70	0.40	-0.50	+0.70	0.80	-0.50
.24	.40	+0.25	0.20	-0.15	+0.40	0.20	-0.25	+0.60	0.50	-0.40	+0.90	0.50	-0.60	+0.90	1.00	-0.60
.40	.71	+0.30	0.25	-0.20	+0.40	0.25	-0.30	+0.70	0.60	-0.40	+1.00	0.60	-0.70	+1.00	1.20	-0.70
.71	1.19	+0.40	0.30	-0.25	+0.50	0.30	-0.40	+0.80	0.80	-0.50	+1.20	0.80	-0.80	+1.20	1.60	-0.80
1.19	1.97	+0.40	0.40	-0.30	+0.60	0.40	-0.40	+1.00	1.00	-0.60	+1.60	1.00	-1.00	+1.60	2.00	-1.00
1.97	3.15	+0.50	0.40	-0.30	+0.70	0.40	-0.50	+1.20	1.20	-0.70	+1.80	1.20	-1.20	+1.80	2.50	-1.20
3.15	4.73	+0.60	0.50	-0.40	+0.90	0.50	-0.60	+1.40	1.40	-0.90	+2.20	1.40	-1.40	+2.20	3.00	-1.40
4.73	7.09	+0.70	0.60	-0.50	+1.00	0.60	-0.70	+1.60	1.60	-1.00	+2.50	1.60	-1.60	+2.50	3.50	-1.60
7.09	9.85	+0.80	0.60	-0.60	+1.20	0.60	-0.80	+1.80	2.00	-1.20	+2.80	2.00	-1.80	+2.80	4.50	-1.80
9.85	12.41	+0.90	0.80	-0.60	+1.20	0.80	-0.90	+2.00	2.50	-1.20	+3.00	2.50	-2.00	+3.00	5.00	-2.00
12.41	15.75	+1.00	1.00	-0.70	+1.40	1.00	-1.00	+2.20	3.00	-1.40	+3.50	3.00	-2.20	+3.50	6.00	-2.20

Nominal Size Range Inches		Class RC6 Medium Running			Class RC7 Free Running			Class RC8 Loose Running			Class RC9 Loose Running		
		Hole Tol. GR9	Minimum Clearance	Shaft Tol. GR8	Hole Tol. GR9	Minimum Clearance	Shaft Tol. GR8	Hole Tol. GR10	Minimum Clearance	Shaft Tol. GR9	Hole Tol. GR11	Minimum Clearance	Shaft Tol. GR10
Over	To	-0		+0	-0		+0	-0		+0	-0		+0
0	.12	+1.00	0.60	-0.60	+1.00	1.00	-0.60	+1.60	2.50	-1.00	+2.50	4.00	-1.60
.12	.24	+1.20	0.80	-0.70	+1.20	1.20	-0.70	+1.80	2.80	-1.20	+3.00	4.50	-1.80
.24	.40	+1.40	1.00	-0.90	+1.40	1.60	-0.90	+2.20	3.00	-1.40	+3.50	6.00	-2.20
.40	.71	+1.60	1.20	-1.00	+1.60	2.00	-1.00	+2.80	3.50	-1.60	+4.00	6.00	-2.80
.71	1.19	+2.00	1.60	-1.20	+2.00	2.50	-1.20	+3.50	4.50	-2.00	+5.00	7.00	-3.50
1.19	1.97	+2.50	2.00	-1.60	+2.50	3.00	-1.60	+4.00	5.00	-2.50	+6.00	8.00	-4.00
1.97	3.15	+3.00	2.50	-1.80	+3.00	4.00	-1.80	+4.50	6.00	-3.00	+7.00	9.00	-4.50
3.15	4.73	+3.50	3.00	-2.20	+3.50	5.00	-2.20	+5.00	7.00	-3.50	+9.00	10.00	-5.00
4.73	7.09	+4.00	3.50	-2.50	+4.00	6.00	-2.50	+6.00	8.00	-4.00	+10.00	12.00	-6.00
7.09	9.85	+4.50	4.00	-2.80	+4.50	7.00	-2.80	+7.00	10.00	-4.50	+12.00	15.00	-7.00
9.85	12.41	+5.00	5.00	-3.00	+5.00	8.00	-3.00	+8.00	12.00	-5.00	+12.00	18.00	-8.00
12.41	15.75	+6.00	6.00	-3.50	+6.00	10.00	-3.50	+9.00	14.00	-6.00	+14.00	22.00	-9.00

VALUES IN MILLIMETERS

Nominal Size Range Millimeters		Class RC1 Precision Sliding			Class RC2 Sliding Fit			Class RC3 Precision Running			Class RC4 Close Running			Class RC5 Medium Running		
		Hole Tol. H5	Minimum Clearance	Shaft Tol. g4	Hole Tol. H6	Minimum Clearance	Shaft Tol. g5	Hole Tol. H7	Minimum Clearance	Shaft Tol. f6	Hole Tol. H8	Minimum Clearance	Shaft Tol. f7	Hole Tol. H8	Minimum Clearance	Shaft Tol. e7
Over	To	-0		+0	-0		+0	-0		+0	-0		+0	-0		+0
0	3	+0.004	0.003	-0.003	+0.006	0.003	-0.004	+0.010	0.008	-0.006	+0.015	0.008	-0.010	+0.015	0.015	-0.010
3	6	+0.005	0.004	-0.004	+0.008	0.004	-0.005	+0.013	0.010	-0.008	+0.018	0.010	-0.013	+0.018	0.020	-0.013
6	10	+0.006	0.005	-0.004	+0.010	0.005	-0.006	+0.015	0.013	-0.010	+0.023	0.013	-0.015	+0.023	0.025	-0.015
10	18	+0.008	0.006	-0.005	+0.010	0.006	-0.008	+0.018	0.015	-0.010	+0.025	0.015	-0.018	+0.025	0.030	-0.018
18	30	+0.010	0.008	-0.006	+0.013	0.008	-0.010	+0.020	0.020	-0.013	+0.030	0.020	-0.020	+0.030	0.040	-0.020
30	50	+0.010	0.010	-0.008	+0.015	0.010	-0.010	+0.030	0.030	-0.015	+0.040	0.030	-0.030	+0.040	0.050	-0.030
50	80	+0.013	0.010	-0.008	+0.018	0.010	-0.013	+0.030	0.030	-0.020	+0.050	0.030	-0.030	+0.050	0.060	-0.030
80	120	+0.015	0.013	-0.010	+0.023	0.013	-0.015	+0.040	0.040	-0.020	+0.060	0.040	-0.040	+0.060	0.080	-0.040
120	180	+0.018	0.015	-0.013	+0.025	0.015	-0.018	+0.040	0.040	-0.030	+0.060	0.040	-0.040	+0.060	0.090	-0.040
180	250	+0.020	0.015	-0.015	+0.030	0.015	-0.020	+0.050	0.050	-0.030	+0.070	0.050	-0.050	+0.070	0.110	-0.050
250	315	+0.023	0.020	-0.015	+0.030	0.020	-0.023	+0.050	0.060	-0.030	+0.080	0.060	-0.050	+0.080	0.130	-0.050
315	400	+0.025	0.025	-0.018	+0.036	0.025	-0.025	+0.060	0.080	-0.040	+0.090	0.080	-0.060	+0.090	0.150	-0.060

Nominal Size Range Millimeters		Class RC6 Medium Running			Class RC7 Free Running			Class RC8 Loose Running			Class RC9 Loose Running		
		Hole Tol. H9	Minimum Clearance	Shaft Tol. e8	Hole Tol. H9	Minimum Clearance	Shaft Tol. d8	Hole Tol. H10	Minimum Clearance	Shaft Tol. e9	Hole Tol. GR11	Minimum Clearance	Shaft Tol. gr10
Over	To	-0		+0	-0		+0	-0		+0	-0		+0
0	3	+0.025	0.015	-0.015	+0.025	0.025	-0.015	+0.041	0.064	-0.025	+0.060	0.100	-0.040
3	6	+0.030	0.015	-0.018	+0.030	0.030	-0.018	+0.046	0.071	-0.030	+0.080	0.110	-0.050
6	10	+0.036	0.025	-0.023	+0.036	0.040	-0.023	+0.056	0.076	-0.036	+0.070	0.130	-0.060
10	18	+0.040	0.030	-0.025	+0.040	0.050	-0.025	+0.070	0.090	-0.040	+0.100	0.150	-0.070
18	30	+0.050	0.040	-0.030	+0.050	0.060	-0.030	+0.090	0.110	-0.050	+0.130	0.180	-0.090
30	50	+0.060	0.050	-0.040	+0.060	0.080	-0.040	+0.100	0.130	-0.060	+0.150	0.200	-0.100
50	80	+0.080	0.060	-0.050	+0.080	0.100	-0.050	+0.110	0.150	-0.080	+0.180	0.230	-0.120
80	120	+0.090	0.080	-0.060	+0.090	0.130	-0.060	+0.130	0.180	-0.090	+0.230	0.250	-0.130
120	180	+0.100	0.090	-0.060	+0.100	0.150	-0.060	+0.150	0.200	-0.100	+0.250	0.300	-0.150
180	250	+0.110	0.100	-0.070	+0.110	0.180	-0.070	+0.180	0.250	-0.110	+0.300	0.380	-0.180
250	315	+0.130	0.130	-0.080	+0.130	0.200	-0.080	+0.200	0.300	-0.130	+0.300	0.460	-0.200
315	400	+0.150	0.150	-0.090	+0.150	0.250	-0.090	+0.230	0.360	-0.150	+0.360	0.560	-0.230

TABLE 9 From *Drafting for Trades and Industry—Mechanical and Electronic*, Nelson. Delmar Publishers Inc.

LOCATIONAL CLEARANCE FITS

VALUES IN THOUSANDTHS OF AN INCH

Nominal Size Range Inches		Class LC1			Class LC2			Class LC3			Class LC4			Class LC5			Class LC6		
		Hole Tol. GR6	Minimum Clearance	Shaft Tol. GR5	Hole Tol. GR8	Minimum Clearance	Shaft Tol. GR7	Hole Tol. GR10	Minimum Clearance	Shaft Tol. GR9	Hole Tol. GR7	Minimum Clearance	Shaft Tol. GR6	Hole Tol. GR9	Minimum Clearance	Shaft Tol. GR8	Hole Tol. GR9	Minimum Clearance	Shaft Tol. GR8
Over	To	-0		+0	-0		+0	-0		+0	-0		+0	-0		+0	-0		+0
0	.12	+0.25	0	-0.15	+0.4	0	-0.25	+0.6	0	-0.4	+1.6	0	-1.0	+0.4	0.10	-0.25	+1.0	0.3	-0.6
.12	.24	+0.30	0	-0.20	+0.5	0	-0.30	+0.7	0	-0.5	+1.8	0	-1.2	+0.5	0.15	-0.30	+1.2	0.4	-0.7
.24	.40	+0.40	0	-0.25	+0.6	0	-0.40	+0.9	0	-0.6	+2.2	0	-1.4	+0.6	0.20	-0.40	+1.4	0.5	-0.9
.40	.71	+0.40	0	-0.30	+0.7	0	-0.40	+1.0	0	-0.7	+2.8	0	-1.6	+0.7	0.25	-0.40	+1.6	0.6	-1.0
.71	1.19	+0.50	0	-0.40	+0.8	0	-0.50	+1.2	0	-0.8	+3.5	0	-2.0	+0.8	0.30	-0.50	+2.0	0.8	-1.2
1.19	1.97	+0.60	0	-0.40	+1.0	0	-0.60	+1.6	0	-1.0	+4.0	0	-2.5	+1.0	0.40	-0.60	+2.5	1.0	-1.6
1.97	3.15	+0.70	0	-0.50	+1.2	0	-0.70	+1.8	0	-1.2	+4.5	0	-3.0	+1.2	0.40	-0.70	+3.0	1.2	-1.8
3.15	4.73	+0.90	0	-0.60	+1.4	0	-0.90	+2.7	0	-1.4	+5.0	0	-3.5	+1.4	0.50	-0.90	+3.5	1.4	-2.2
4.73	7.09	+1.00	0	-0.70	+1.6	0	-1.00	+2.5	0	-1.6	+6.0	0	-4.0	+1.6	0.60	-1.00	+4.0	1.6	-2.5
7.09	9.85	+1.20	0	-0.80	+1.8	0	-1.20	+2.8	0	-1.8	+7.0	0	-4.5	+1.8	0.60	-1.20	+4.5	2.0	-2.8
9.85	12.41	+1.20	0	-0.90	+2.0	0	-1.20	+3.0	0	-2.0	+8.0	0	-5.0	+2.0	0.70	-1.20	+5.0	2.2	-3.0
12.41	15.75	+1.40	0	-1.00	+2.2	0	-1.40	+3.5	0	-2.2	+9.0	0	-6.0	+2.2	0.70	-1.40	+6.0	2.5	-3.5

Nominal Size Range Inches		Class LC7			Class LC8			Class LC9			Class LC10			Class LC11		
		Hole Tol. GR10	Minimum Clearance	Shaft Tol. GR9	Hole Tol. GR10	Minimum Clearance	Shaft Tol. GR9	Hole Tol. GR11	Minimum Clearance	Shaft Tol. GR10	Hole Tol. GR12	Minimum Clearance	Shaft Tol. GR11	Hole Tol. GR13	Minimum Clearance	Shaft Tol. GR12
Over	To	-0		+0	-0		+0	-0		+0	-0		+0	-0		+0
0	.12	+1.6	0.6	-1.0	+1.6	1.0	-1.0	+2.5	2.5	-1.6	+1.0	4.0	-2.5	+6.0	5.0	-4.0
.12	.24	+1.8	0.8	-1.2	+1.8	1.2	-1.2	+3.0	2.8	-1.8	+5.0	4.5	-3.0	+7.0	6.0	-5.0
.24	.40	+2.2	1.0	-1.4	+2.2	1.6	-1.4	+3.5	3.0	-2.2	+6.0	5.0	-3.5	+9.0	7.0	-6.0
.40	.71	+2.8	1.2	-1.6	+2.8	2.0	-1.6	+4.0	3.5	-2.8	+7.0	6.0	-4.0	+10.0	8.0	-7.0
.71	1.19	+3.5	1.6	-2.0	+3.5	2.5	-2.0	+5.0	4.5	-3.5	+8.0	7.0	-5.0	+12.0	10.0	-8.0
1.19	1.97	+4.0	2.0	-2.5	+4.0	3.6	-2.5	+6.0	5.0	-4.0	+10.0	8.0	-6.0	+16.0	12.0	-10.0
1.97	3.15	+4.5	2.5	-3.0	+4.5	4.0	-3.0	+7.0	6.0	-4.5	+12.0	10.0	-7.0	+18.0	14.0	-12.0
3.15	4.73	+5.0	3.0	-3.5	+5.0	5.0	-3.5	+9.0	7.0	-5.0	+14.0	11.0	-9.0	+22.0	16.0	-14.0
4.73	7.09	+6.0	3.5	-4.0	+6.0	6.0	-4.0	+10.0	8.0	-6.0	+16.0	12.0	-10.0	+25.0	18.0	-16.0
7.09	9.85	+7.0	4.0	-4.5	+7.0	7.0	-4.5	+12.0	10.0	-7.0	+18.0	16.0	-12.0	+28.0	22.0	-18.0
9.85	12.41	+8.0	4.5	-5.0	+8.0	7.0	-5.0	+12.0	12.0	-8.0	+20.0	20.0	-12.0	+30.0	28.0	-20.0
12.41	15.75	+9.0	5.0	-6.0	+9.0	8.0	-6.0	+14.0	14.0	-9.0	+22.0	22.0	-14.0	+35.0	30.0	-22.0

VALUES IN MILLIMETERS

Nominal Size Range Millimeters		Class LC1			Class LC2			Class LC3			Class LC4			Class LC5			Class LC6		
		Hole Tol. H6	Minimum Clearance	Shaft Tol. h5	Hole Tol. H7	Minimum Clearance	Shaft Tol. h6	Hole Tol. H8	Minimum Clearance	Shaft Tol. h7	Hole Tol. H10	Minimum Clearance	Shaft Tol. h9	Hole Tol. H7	Minimum Clearance	Shaft Tol. g6	Hole Tol. H9	Minimum Clearance	Shaft Tol. f8
Over	To	-0		+0	-0		+0	-0		+0	-0		+0	-0		+0	-0		+0
0	3	+0.006	0	-0.004	+0.010	0	-0.006	+0.015	0	-0.010	+0.041	0	-0.025	+0.010	0.002	-0.006	+0.025	0.008	-0.015
3	6	+0.008	0	-0.005	+0.013	0	-0.008	+0.018	0	-0.013	+0.046	0	-0.030	+0.013	0.004	-0.008	+0.030	0.010	-0.018
6	10	+0.010	0	-0.006	+0.015	0	-0.010	+0.023	0	-0.015	+0.056	0	-0.036	+0.015	0.005	-0.010	+0.036	0.013	-0.023
10	18	+0.010	0	-0.008	+0.018	0	-0.010	+0.025	0	-0.018	+0.070	0	-0.040	+0.018	0.006	-0.010	+0.041	0.015	-0.025
18	30	+0.013	0	-0.010	+0.020	0	-0.013	+0.030	0	-0.020	+0.090	0	-0.050	+0.020	0.008	-0.013	+0.050	0.020	-0.030
30	50	+0.015	0	-0.010	+0.025	0	-0.015	+0.041	0	-0.025	+0.100	0	-0.060	+0.025	0.010	-0.015	+0.060	0.030	-0.040
50	80	+0.018	0	-0.013	+0.030	0	-0.018	+0.046	0	-0.030	+0.110	0	-0.080	+0.030	0.010	-0.018	+0.080	0.030	-0.050
80	120	+0.023	0	-0.015	+0.036	0	-0.023	+0.056	0	-0.036	+0.130	0	-0.080	+0.036	0.013	-0.023	+0.090	0.040	-0.060
120	180	+0.025	0	-0.018	+0.041	0	-0.025	+0.064	0	-0.041	+0.150	0	-0.100	+0.041	0.015	-0.025	+0.100	0.040	-0.060
180	250	+0.030	0	-0.020	+0.046	0	-0.030	+0.071	0	-0.046	+0.180	0	-0.110	+0.046	0.015	-0.030	+0.110	0.050	-0.070
250	315	+0.020	0	-0.023	+0.051	0	-0.030	+0.076	0	-0.051	+0.200	0	-0.130	+0.051	0.018	-0.030	+0.130	0.060	-0.080
315	400	+0.036	0	-0.025	+0.056	0	-0.036	+0.089	0	-0.056	+0.230	0	-0.150	+0.056	0.018	-0.036	+0.150	0.060	-0.090

Nominal Size Range Millimeters		Class LC7			Class LC8			Class LC9			Class LC10			Class LC11		
		Hole Tol. H10	Minimum Clearance	Shaft Tol. e9	Hole Tol. H10	Minimum Clearance	Shaft Tol. d9	Hole Tol. H11	Minimum Clearance	Shaft Tol. c10	Hole Tol. GR12	Minimum Clearance	Shaft Tol. gr11	Hole Tol. GR13	Minimum Clearance	Shaft Tol. gr12
Over	To	-0		+0	-0		+0	-0		+0	-0		+0	-0		+0
0	3	+0.041	0.015	-0.025	+0.041	0.025	-0.025	+0.064	0.06	-0.041	+0.10	0.10	-0.06	+0.15	0.13	-0.10
3	6	+0.046	0.020	-0.030	+0.046	0.030	-0.030	+0.076	0.07	-0.46	+0.13	0.11	-0.08	+0.18	0.15	-0.13
6	10	+0.056	0.025	-0.036	+0.056	0.041	-0.036	+0.089	0.08	-0.56	+0.15	0.13	-0.09	+0.23	0.18	-0.15
10	18	+0.070	0.030	-0.040	+0.070	0.050	-0.040	+0.100	0.09	-0.70	+0.18	0.15	-0.10	+0.25	0.20	-0.18
18	30	+0.090	0.040	-0.050	+0.090	0.060	-0.050	+0.130	0.11	-0.90	+0.20	0.18	-0.13	+0.31	0.25	-0.20
30	50	+0.100	0.050	-0.060	+0.100	0.090	-0.060	+0.150	0.13	-0.100	+0.25	0.20	-0.15	+0.41	0.31	-0.25
50	80	+0.110	0.060	-0.080	+0.110	0.100	-0.080	+0.180	0.15	-0.110	+0.31	0.25	-0.18	+0.46	0.36	-0.31
80	120	+0.130	0.080	-0.090	+0.130	0.130	-0.090	+0.230	0.18	-0.130	+0.36	0.28	-0.23	+0.56	0.41	-0.36
120	180	+0.150	0.090	-0.100	+0.150	0.150	-0.100	+0.250	0.20	-0.150	+0.41	0.31	-0.25	+0.64	0.46	-0.41
180	250	+0.180	0.100	-0.110	+0.180	0.180	-0.110	+0.310	0.25	-0.180	+0.46	0.41	-0.31	+0.71	0.56	-0.46
250	315	+0.200	0.110	-0.130	+0.200	0.180	-0.130	+0.310	0.31	-0.200	+0.51	0.51	-0.31	+0.76	0.71	-0.51
315	400	+0.230	0.130	-0.150	+0.230	0.200	-0.150	+0.360	0.36	-0.230	+0.56	0.56	-0.36	+0.89	0.76	-0.56

TABLE 10 From *Drafting for Trades and Industry—Mechanical and Electronic*, Nelson. Delmar Publishers Inc.

LOCATIONAL TRANSITION FITS

VALUES IN THOUSANDTHS OF AN INCH

Nominal Size Range Inches		Class LT1			Class LT2			Class LT3			Class LT4			Class LT5			Class LT6		
		Hole Tol. GR7	Maximum Interference	Shaft Tol. GR6	Hole Tol. GR8	Maximum Interference	Shaft Tol. GR7	Hole Tol. GR7	Maximum Interference	Shaft Tol. GR6	Hole Tol. GR8	Maximum Interference	Shaft Tol. GR7	Hole Tol. GR7	Maximum Interference	Shaft Tol. GR6	Hole Tol. GR8	Maximum Interference	Shaft Tol. GR7
Over	To	-0		+0	-0		+0	-0		+0	-0		+0	-0		+0	-0		+0
0	.12	+0.4	0.10	-0.25	+0.6	0.20	-0.4	+0.4	0.25	-0.25	+0.6	0.4	-0.4	+0.4	0.5	-0.25	+0.6	0.65	-0.4
.12	.24	+0.5	0.15	-0.30	+0.7	0.25	-0.5	+0.5	0.40	-0.30	+0.7	0.6	-0.5	+0.5	0.6	-0.30	+0.7	0.80	-0.5
.24	.40	+0.6	0.20	-0.40	+0.9	0.30	-0.6	+0.6	0.50	-0.40	+0.9	0.7	-0.6	+0.6	0.8	-0.40	+0.9	1.00	-0.6
.40	.71	+0.7	0.20	-0.40	+1.0	0.30	-0.7	+0.7	0.50	-0.40	+1.0	0.8	-0.7	+0.7	0.9	-0.40	+1.0	1.20	-0.7
.71	1.19	+0.8	0.25	-0.50	+1.2	0.40	-0.8	+0.8	0.60	-0.50	+1.2	0.9	-0.8	+0.8	1.1	-0.50	+1.2	1.40	-0.8
1.19	1.97	+1.0	0.30	-0.60	+1.6	0.50	-1.0	+1.0	0.70	-0.60	+1.6	1.1	-1.0	+1.0	1.3	-0.60	+1.6	1.70	-1.0
1.97	3.15	+1.2	0.30	-0.70	+1.8	0.60	-1.2	+1.2	0.80	-0.70	+1.8	1.3	-1.2	+1.2	1.5	-0.70	+1.8	2.00	-1.2
3.15	4.73	+1.4	0.40	-0.90	+2.2	0.70	-1.4	+1.4	1.00	-0.90	+2.2	1.5	-1.4	+1.4	1.9	-0.90	+2.2	2.40	-1.4
4.73	7.09	+1.6	0.50	-1.00	+2.5	0.80	-1.6	+1.6	1.10	-1.00	+2.5	1.7	-1.6	+1.6	2.2	-1.00	+2.5	2.80	-1.6
7.09	9.85	+1.8	0.60	-1.20	+2.8	0.90	-1.8	+1.8	1.40	-1.20	+2.8	2.0	-1.8	+1.8	2.6	-1.20	+2.8	3.20	-1.8
9.85	12.41	+2.0	0.60	-1.20	+3.0	1.00	-2.0	+2.0	1.40	-1.20	+3.0	2.2	-2.0	+2.0	2.6	-1.20	+3.0	3.40	-2.0
12.41	15.75	+2.2	0.70	-1.40	+3.5	1.00	-2.2	+2.2	1.60	-1.40	+3.5	2.4	-2.2	+2.2	3.0	-1.40	+3.5	3.80	-2.2

VALUES IN MILLIMETERS

Nominal Size Range Millimeters		Class LT1			Class LT2			Class LT3			Class LT4			Class LT5			Class LT6		
		Hole Tol. H7	Maximum Clearance	Shaft Tol. js6	Hole Tol. H8	Maximum Clearance	Shaft Tol. js7	Hole Tol. H7	Maximum Clearance	Shaft Tol. k6	Hole Tol. H8	Maximum Clearance	Shaft Tol. k7	Hole Tol. H7	Maximum Clearance	Shaft Tol. n6	Hole Tol. H8	Maximum Clearance	Shaft Tol. n7
Over	To	-0		+0	-0		+0	-0		+0	-0		+0	-0		+0	-0		+0
0	3	+0.010	0.002	-0.006	+0.015	0.005	-0.010	+0.010	0.006	-0.006	+0.015	0.010	-0.010	+0.010	0.013	-0.006	+0.015	0.016	-0.010
3	6	+0.013	0.004	-0.008	+0.018	0.006	-0.013	+0.013	0.010	-0.008	+0.018	0.015	-0.013	+0.013	0.015	-0.008	+0.018	0.020	-0.013
6	10	+0.015	0.005	-0.010	+0.023	0.008	-0.015	+0.015	0.013	-0.010	+0.023	0.018	-0.015	+0.015	0.020	-0.010	+0.023	0.025	-0.015
10	18	+0.018	0.005	-0.010	+0.025	0.008	-0.018	+0.018	0.010	-0.010	+0.025	0.020	-0.018	+0.018	0.023	-0.010	+0.025	0.030	-0.018
18	30	+0.020	0.006	-0.013	+0.030	0.010	-0.020	+0.020	0.015	-0.013	+0.030	0.023	-0.020	+0.020	0.028	-0.013	+0.030	0.036	-0.020
30	50	+0.025	0.008	-0.015	+0.041	0.013	-0.025	+0.025	0.018	-0.015	+0.041	0.028	-0.025	+0.025	0.033	-0.015	+0.041	0.044	-0.025
50	80	+0.030	0.008	-0.018	+0.046	0.015	-0.030	+0.030	0.020	-0.018	+0.046	0.033	-0.030	+0.030	0.038	-0.018	+0.046	0.051	-0.030
80	120	+0.036	0.010	-0.023	+0.056	0.018	-0.036	+0.036	0.025	-0.023	+0.056	0.038	-0.036	+0.036	0.048	-0.023	+0.056	0.062	-0.036
120	180	+0.041	0.013	-0.025	+0.064	0.020	-0.041	+0.041	0.028	-0.025	+0.064	0.044	-0.041	+0.041	0.056	-0.025	+0.064	0.071	-0.041
180	250	+0.046	0.015	-0.030	+0.071	0.023	-0.046	+0.046	0.036	-0.030	+0.071	0.051	-0.046	+0.046	0.066	-0.030	+0.071	0.081	-0.046
250	315	+0.051	0.015	-0.030	+0.076	0.025	-0.051	+0.051	0.036	-0.030	+0.076	0.056	-0.051	+0.051	0.066	-0.030	+0.076	0.086	-0.051
315	400	+0.056	0.018	-0.036	+0.089	0.025	-0.056	+0.056	0.041	-0.036	+0.089	0.062	-0.056	+0.056	0.076	-0.036	+0.089	0.096	-0.056

TABLE 11   From *Drafting for Trades and Industry—Mechanical and Electronic,* Nelson. Delmar Publishers Inc.

LOCATIONAL INTERFERENCE FITS

VALUES IN THOUSANDTHS OF AN INCH

Nominal Size Range Inches		Class LN1 Light Press Fit			Class LN2 Medium Press Fit			Class LN3 Heavy Press Fit			Class LN4			Class LN5			Class LN6		
		Hole Tol. GR6	Maximum Interference	Shaft Tol. GR5	Hole Tol. GR7	Maximum Interference	Shaft Tol. GR6	Hole Tol. GR7	Maximum Interference	Shaft Tol. GR6	Hole Tol. GR8	Maximum Interference	Shaft Tol. GR7	Hole Tol. GR9	Maximum Interference	Shaft Tol. GR8	Hole Tol. GR10	Maximum Interference	Shaft Tol. GR9
Over	To	-0		+0	-0		+0	-0		+0	-0		+0	-0		+0	-0		+0
0	.12	+0.25	0.40	-0.15	+0.4	0.65	-0.25	+0.4	0.75	-0.25	+0.6	1.2	-0.4	+1.0	1.8	-0.6	+1.6	3.0	-1.0
.12	.24	+0.30	0.50	-0.20	+0.5	0.80	-0.30	+0.5	0.90	-0.30	+0.7	1.5	-0.5	+1.2	2.3	-0.7	+1.8	3.6	-1.2
.24	.40	+0.40	0.65	-0.25	+0.6	1.00	-0.40	+0.6	1.20	-0.40	+0.9	1.8	-0.6	+2.2	2.8	-0.9	+2.2	4.4	-1.4
.40	.71	+0.40	0.70	-0.30	+0.7	1.10	-0.40	+0.7	1.40	-0.40	+1.0	2.2	-0.7	+1.6	3.4	-1.0	+2.8	5.6	-1.6
.71	1.19	+0.50	0.90	-0.40	+0.8	1.30	-0.50	+0.8	1.70	-0.50	+1.2	2.6	-0.8	+2.0	4.2	-1.2	+3.5	7.0	-2.0
1.19	1.97	+0.60	1.00	-0.40	+1.0	1.60	-0.60	+1.0	2.00	-0.60	+1.6	3.4	-1.0	+2.5	5.3	-1.6	+4.0	8.5	-2.5
1.97	3.15	+0.70	1.30	-0.50	+1.2	2.10	-0.70	+1.2	2.30	-0.70	+1.8	4.0	-1.2	+3.0	6.3	-1.8	+4.5	10.0	-3.0
3.15	4.73	+0.90	1.60	-0.60	+1.4	2.50	-0.90	+1.4	2.90	-0.90	+2.2	4.8	-1.4	+4.0	7.7	-2.2	+5.0	11.5	3.5
4.73	7.09	+1.00	1.90	-0.70	+1.6	2.80	-1.00	+1.6	3.50	-1.00	+2.5	5.6	-1.6	+4.5	8.7	-2.5	+6.0	13.5	-4.0
7.09	9.85	+1.20	2.20	-0.80	+1.8	3.20	-1.20	+1.8	4.20	-1.20	+2.8	6.6	-1.8	+5.0	10.3	-2.8	+7.0	16.5	-4.5
9.85	12.41	+1.20	2.30	-0.90	+2.0	3.40	-1.20	+2.0	4.70	-1.20	+3.0	7.5	-2.0	+6.0	12.0	-3.0	+8.0	19.0	-5.0
12.41	15.75	+1.40	2.60	-1.00	+2.2	3.90	-1.40	+2.2	5.90	-1.40	+3.5	8.7	-2.2	+6.0	14.5	-3.5	+9.0	23.0	-6.0

VALUES IN MILLIMETERS

Nominal Size Range Millimeters		Class LN1 Light Press Fit			Class LN2 Medium Press Fit			Class LN3 Heavy Press Fit			Class LN4			Class LN5			Class LN6		
		Hole Tol. GR6	Maximum Interference	Shaft Tol. gr5	Hole Tol. H7	Maximum Interference	Shaft Tol. p6	Hole Tol. H7	Maximum Interference	Shaft Tol. t6	Hole Tol. GR8	Maximum Interference	Shaft Tol. gr7	Hole Tol. GR9	Maximum Interference	Shaft Tol. gr8	Hole Tol. GR10	Maximum Interference	Shaft Tol. gr9
Over	To	-0		+0	-0		+0	-0		+0	-0		+0	-0		+0	-0		+0
0	3	+0.006		-0.004	+0.010	0.016	-0.006	+0.010	0.019	-0.006	+0.015	0.030	-0.010	+0.025	0.046	-0.015	+0.041	0.076	-0.025
3	6	+0.008		-0.005	+0.013	0.020	-0.008	+0.013	0.023	-0.008	+0.018	0.038	-0.013	+0.030	0.059	-0.018	+0.046	0.091	-0.030
6	10	+0.010		-0.006	+0.015	0.025	-0.010	+0.015	0.030	-0.010	+0.023	0.046	-0.015	+0.036	0.071	-0.023	+0.056	0.112	-0.036
10	18	+0.010		-0.008	+0.018	0.028	-0.010	+0.018	0.036	-0.010	+0.025	0.056	-0.018	+0.041	0.086	-0.025	+0.071	0.142	-0.041
18	30	+0.013		-0.010	+0.020	0.033	-0.013	+0.020	0.044	-0.013	+0.030	0.066	-0.020	+0.051	0.107	-0.030	+0.089	0.178	-0.051
30	50	+0.015		-0.010	+0.025	0.041	-0.015	+0.025	0.051	-0.015	+0.041	0.086	-0.025	+0.064	0.135	-0.041	+0.102	0.216	-0.064
50	80	+0.018		-0.013	+0.030	0.054	-0.018	+0.030	0.059	-0.018	+0.046	0.102	-0.030	+0.076	0.160	-0.046	+0.114	0.254	-0.076
80	120	+0.023		-0.015	+0.036	0.064	-0.023	+0.036	0.074	-0.023	+0.056	0.122	-0.036	+0.102	0.196	-0.056	+0.127	0.292	-0.102
120	180	+0.025		-0.018	+0.041	0.071	-0.025	+0.041	0.089	-0.025	+0.064	0.142	-0.041	+0.114	0.221	-0.064	+0.152	0.343	-0.114
180	250	+0.030		-0.020	+0.046	0.081	-0.030	+0.046	0.107	-0.030	+0.071	0.168	-0.046	+0.127	0.262	-0.071	+0.178	0.419	-0.127
250	315	+0.030		-0.023	+0.051	0.086	-0.030	+0.051	0.119	-0.030	+0.076	0.191	-0.051	+0.152	0.305	-0.076	+0.203	0.483	-0.152
315	400	+0.036		-0.025	+0.056	0.099	-0.036	+0.056	0.150	-0.036	+0.089	0.221	-0.056	+0.152	0.368	-0.089	+0.229	0.584	-0.152

TABLE 12   From *Drafting for Trades and Industry—Mechanical and Electronic,* Nelson. Delmar Publishers Inc.

FORCE AND SHRINK FITS

VALUES IN THOUSANDTHS OF AN INCH

Nominal Size Range Inches		Class FN1 Light Drive Fit			Class FN2 Medium Drive Fit			Class FN3 Heavy Drive Fit			Class FN4 Shrink Fit			Class FN5 Heavy Shrink Fit		
		Hole Tol. GR6	Maximum Interference	Shaft Tol. GR5	Hole Tol. GR7	Maximum Interference	Shaft Tol. GR6	Hole Tol. GR7	Maximum Interference	Shaft Tol. GR6	Hole Tol. GR7	Maximum Interference	Shaft Tol. GR6	Hole Tol. GR8	Maximum Interference	Shaft Tol. GR7
Over	To	-0		+0	-0		+0	-0		+0	-0		+0	-0		+0
0	.12	+0.25	0.50	-0.15	+0.40	0.85	-0.25				+0.40	0.95	-0.25	+0.60	1.30	-0.40
.12	.24	+0.30	0.60	-0.20	+0.50	1.00	-0.30				+0.50	1.20	-0.30	+0.70	1.70	-0.50
.24	.40	+0.40	0.75	-0.25	+0.60	1.40	-0.40				+0.60	1.60	-0.40	+0.90	2.00	-0.60
.40	.56	+0.40	0.80	-0.30	+0.70	1.60	-0.40				+0.70	1.80	-0.40	+1.00	2.30	-0.70
.56	.71	+0.40	0.90	-0.30	+0.70	1.60	-0.40				+0.70	1.80	-0.40	+1.00	2.50	-0.70
.71	.95	+0.50	1.10	-0.40	+0.80	1.90	-0.50				+0.80	2.10	-0.50	+1.20	3.00	-0.80
.95	1.19	+0.50	1.20	-0.40	+0.80	1.90	-0.50	+0.80	2.10	-0.50	+0.80	2.30	-0.50	+1.20	3.30	-0.80
1.19	1.58	+0.60	1.30	-0.40	+1.00	2.40	-0.60	+1.00	2.60	-0.60	+1.00	3.10	-0.60	+1.60	4.00	-1.00
1.58	1.97	+0.60	1.40	-0.40	+1.00	2.40	-0.60	+1.00	2.80	-0.60	+1.00	3.40	-0.60	+1.60	5.00	-1.00
1.97	2.56	+0.70	1.80	-0.50	+1.20	2.70	-0.70	+1.20	3.20	-0.70	+1.20	4.20	-0.70	+1.80	6.20	-1.20
2.56	3.15	+0.70	1.90	-0.50	+1.20	2.90	-0.70	+1.20	3.70	-0.70	+1.20	4.70	-0.70	+1.80	7.20	-1.20
3.15	3.94	+0.90	2.40	-0.60	+1.40	3.70	-0.90	+1.40	4.40	-0.70	+1.40	5.90	-0.90	+2.20	8.40	-1.40

VALUES IN MILLIMETERS

Nominal Size Range Millimeters		Class FN1 Light Drive Fit			Class FN2 Medium Drive Fit			Class FN3 Heavy Drive Fit			Class FN4 Shrink Fit			Class FN5 Heavy Shrink Fit		
		Hole Tol. GR6	Maximum Interference	Shaft Tol. gr5	Hole Tol. H7	Maximum Interference	Shaft Tol. s6	Hole Tol. H7	Maximum Interference	Shaft Tol. t6	Hole Tol. GR8	Maximum Interference	Shaft Tol. gr7	Hole Tol. H8	Maximum Interference	Shaft Tol. t7
Over	To	-0		+0	-0		+0	-0		+0	-0		+0	-0		+0
0	3	+0.006	0.013	-0.004	+0.010	0.216	-0.006				+0.010	0.024	-0.006	+0.015	0.033	-0.010
3	6	+0.007	0.015	-0.005	+0.013	0.025	-0.007				+0.013	0.030	-0.007	+0.018	0.043	-0.013
6	10	+0.010	0.019	-0.006	+0.015	0.036	-0.010				+0.015	0.041	-0.010	+0.023	0.051	-0.015
10	14	+0.010	0.020	-0.008	+0.018	0.041	-0.010				+0.018	0.046	-0.010	+0.025	0.058	-0.018
14	18	+0.010	0.023	-0.008	+0.018	0.041	-0.010				+0.018	0.046	-0.010	+0.025	0.064	-0.018
18	24	+0.013	0.028	-0.010	+0.020	0.048	-0.013				+0.020	0.053	-0.013	+0.030	0.076	-0.020
24	30	+0.013	0.030	-0.010	+0.020	0.048	-0.013	+0.020	0.053	-0.013	+0.020	0.058	-0.013	+0.030	0.084	-0.020
30	40	+0.015	0.033	-0.010	+0.025	0.061	-0.015	+0.025	0.066	-0.015	+0.025	0.079	-0.015	+0.041	0.102	-0.025
40	50	+0.015	0.036	-0.010	+0.025	0.061	-0.015	+0.025	0.071	-0.015	+0.025	0.086	-0.015	+0.041	0.127	-0.025
50	65	+0.018	0.046	-0.013	+0.030	0.069	-0.018	+0.030	0.082	-0.018	+0.030	0.107	-0.018	+0.046	0.157	-0.030
65	80	+0.018	0.048	-0.013	+0.030	0.074	-0.018	+0.030	0.094	-0.018	+0.030	0.119	-0.018	+0.046	0.183	-0.030
80	100	+0.023	0.061	-0.015	+0.035	0.094	-0.023	+0.035	0.112	-0.023	+0.036	0.150	-0.023	+0.056	0.213	-0.036

TABLE 13   From *Drafting for Trades and Industry—Mechanical and Electronic,* Nelson. Delmar Publishers Inc.

UNIFIED STANDARD SCREW THREAD SERIES

Sizes		Basic Major Diameter	Series with graded pitches			Series with constant pitches								Sizes
Primary	Secondary		Coarse UNC	Fine UNF	Extra fine UNEF	4UN	6UN	8UN	12UN	16UN	20UN	28UN	32UN	
0		0.0600	–	80	–	–	–	–	–	–	–	–	–	0
	1	0.0730	64	72	–	–	–	–	–	–	–	–	–	1
2		0.0860	56	64	–	–	–	–	–	–	–	–	–	2
	3	0.0990	48	56	–	–	–	–	–	–	–	–	–	3
4		0.1120	40	48	–	–	–	–	–	–	–	–	–	4
5		0.1250	40	44	–	–	–	–	–	–	–	–	–	5
6		0.1380	32	40	–	–	–	–	–	–	–	–	UNC	6
8		0.1640	32	36	–	–	–	–	–	–	–	–	UNC	8
10		0.1900	24	32	–	–	–	–	–	–	–	–	UNF	10
	12	0.2160	24	28	32	–	–	–	–	–	–	UNF	UNEF	12
¼		0.2500	20	28	32	–	–	–	–	–	UNC	UNF	UNEF	¼
⁵⁄₁₆		0.3125	18	24	32	–	–	–	–	–	20	28	UNEF	⁵⁄₁₆
³⁄₈		0.3750	16	24	32	–	–	–	–	UNC	20	28	UNEF	³⁄₈
⁷⁄₁₆		0.4375	14	20	28	–	–	–	–	16	UNF	UNEF	32	⁷⁄₁₆
½		0.5000	13	20	28	–	–	–	–	16	UNF	UNEF	32	½
⁹⁄₁₆		0.5625	12	18	24	–	–	–	UNC	16	20	28	32	⁹⁄₁₆
⁵⁄₈		0.6250	11	18	24	–	–	–	12	16	20	28	32	⁵⁄₈
	¹¹⁄₁₆	0.6875	–	–	24	–	–	–	12	16	20	28	32	¹¹⁄₁₆
¾		0.7500	10	16	20	–	–	–	12	UNF	UNEF	28	32	¾
	¹³⁄₁₆	0.8125	–	–	20	–	–	–	12	16	UNEF	28	32	¹³⁄₁₆
⅞		0.8750	9	14	20	–	–	–	12	16	UNEF	28	32	⅞
	¹⁵⁄₁₆	0.9375	–	–	20	–	–	–	12	16	UNEF	28	32	¹⁵⁄₁₆
1		1.0000	8	12	20	–	–	UNC	UNF	16	UNEF	28	32	1
	1¹⁄₁₆	1.0625	–	–	18	–	–	8	12	16	20	28	–	1¹⁄₁₆
1⅛		1.1250	7	12	18	–	–	8	UNF	16	20	28	–	1⅛
	1³⁄₁₆	1.1875	–	–	18	–	–	8	12	16	20	28	–	1³⁄₁₆
1¼		1.2500	7	12	18	–	–	8	UNF	16	20	28	–	1¼
	1⁵⁄₁₆	1.3125	–	–	18	–	–	8	12	16	20	28	–	1⁵⁄₁₆
1⅜		1.3750	6	12	18	–	UNC	8	UNF	16	20	28	–	1⅜
	1⁷⁄₁₆	1.4375	–	–	18	–	6	8	12	16	20	28	–	1⁷⁄₁₆
1½		1.5000	6	12	18	–	UNC	8	UNF	16	20	28	–	1½
	1⁹⁄₁₆	1.5625	–	–	18	–	6	8	12	16	20	–	–	1⁹⁄₁₆
1⅝		1.6250	–	–	18	–	6	8	12	16	20	–	–	1⅝
	1¹¹⁄₁₆	1.6875	–	–	18	–	6	8	12	16	20	–	–	1¹¹⁄₁₆
1¾		1.7500	5	–	–	–	6	8	12	16	20	–	–	1¾
	1¹³⁄₁₆	1.8125	–	–	–	–	6	8	12	16	20	–	–	1¹³⁄₁₆
1⅞		1.8750	–	–	–	–	6	8	12	16	20	–	–	1⅞
	1¹⁵⁄₁₆	1.9375	–	–	–	–	6	8	12	16	20	–	–	1¹⁵⁄₁₆
2		2.0000	4½	–	–	–	6	8	12	16	20	–	–	2
	2⅛	2.1250	–	–	–	–	6	8	12	16	20	–	–	2⅛
2¼		2.2500	4½	–	–	–	6	8	12	16	20	–	–	2¼
	2⅜	2.3750	–	–	–	–	6	8	12	16	20	–	–	2⅜
2½		2.5000	4	–	–	UNC	6	8	12	16	20	–	–	2½
	2⅝	2.6250	–	–	–	4	6	8	12	16	20	–	–	2⅝
2¾		2.7500	4	–	–	UNC	6	8	12	16	20	–	–	2¾
	2⅞	2.8750	–	–	–	4	6	8	12	16	20	–	–	2⅞
3		3.0000	4	–	–	UNC	6	8	12	16	20	–	–	3
	3⅛	3.1250	–	–	–	4	6	8	12	16	–	–	–	3⅛
3¼		3.2500	4	–	–	UNC	6	8	12	16	–	–	–	3¼
	3⅜	3.3750	–	–	–	4	6	8	12	16	–	–	–	3⅜
3½		3.5000	4	–	–	UNC	6	8	12	16	–	–	–	3½
	3⅝	3.6250	–	–	–	4	6	8	12	16	–	–	–	3⅝
3¾		3.7500	4	–	–	UNC	6	8	12	16	–	–	–	3¾
	3⅞	3.8750	–	–	–	4	6	8	12	16	–	–	–	3⅞
4		4.0000	4	–	–	UNC	6	8	12	16	–	–	–	4
	4⅛	4.1250	–	–	–	4	6	8	12	16	–	–	–	4⅛
4¼		4.2500	–	–	–	4	6	8	12	16	–	–	–	4¼
	4⅜	4.3750	–	–	–	4	6	8	12	16	–	–	–	4⅜
4½		4.5000	–	–	–	4	6	8	12	16	–	–	–	4½
	4⅝	4.6250	–	–	–	4	6	8	12	16	–	–	–	4⅝
4¾		4.7500	–	–	–	4	6	8	12	16	–	–	–	4¾
	4⅞	4.8750	–	–	–	4	6	8	12	16	–	–	–	4⅞
5		5.0000	–	–	–	4	6	8	12	16	–	–	–	5
	5⅛	5.1250	–	–	–	4	6	8	12	16	–	–	–	5⅛
5¼		5.2500	–	–	–	4	6	8	12	16	–	–	–	5¼
	5⅜	5.3750	–	–	–	4	6	8	12	16	–	–	–	5⅜
5½		5.5000	–	–	–	4	6	8	12	16	–	–	–	5½
	5⅝	5.6250	–	–	–	4	6	8	12	16	–	–	–	5⅝
5¾		5.7500	–	–	–	4	6	8	12	16	–	–	–	5¾
	5⅞	5.8750	–	–	–	4	6	8	12	16	–	–	–	5⅞
6		6.0000	–	–	–	4	6	8	12	16	–	–	–	6

TABLE 14

DRILL AND TAP SIZES

DECIMAL EQUIVALENTS AND TAP DRILL SIZES
OF WIRE GAGE LETTER AND FRACTIONAL SIZE DRILLS
(TAP DRILL SIZES BASED ON 75% MAXIMUM THREAD)

FRACTIONAL SIZE DRILLS INCHES	WIRE GAGE DRILLS	DECIMAL EQUIVALENT INCHES	SIZE OF THREAD	THREADS PER INCH
	80	.0135		
	79	.0145		
1/64	.....	.0156		
	78	.0160		
	77	.0180		
	76	.0200		
	75	.0210		
	74	.0225		
	73	.0240		
	72	.0250		
	71	.0260		
	70	.0280		
	69	.0292		
	68	.0310		
1/32	.....	.0312		
	67	.0320		
	66	.0330		
	65	.0350		
	64	.0360		
	63	.0370		
	62	.0380		
	61	.0390		
	60	.0400		
	59	.0410		
	58	.0420		
	57	.0430		
	56	.0465		
3/64	.....	.0469	0	80
	55	.0520		
	54	.0550		
	53	.0595	1	64
1/16	.....	.0625		72
	52	.0635		
	51	.0670		
	50	.0700	2	56
	49	.0730		64
	48	.0760		
5/64	.....	.0781		
	47	.0785	3	48
	46	.0810		
	45	.0820	3	56
	44	.0860		
	43	.0890	4	40
	42	.0935	4	48
3/32	.....	.0937		
	41	.0960		
	40	.0980		
	39	.0995		
	38	.1015	5	40
	37	.1040	5	44
	36	.1065	6	32
7/64	.....	.1094		
	35	.1100		
	34	.1110		
	33	.1130	6	40
	32	.1160		
	31	.1200		
1/8	.....	.1250		
	30	.1285		
	29	.1360	8	32
	28	.1405		36

FRACTIONAL SIZE DRILLS INCHES	WIRE GAGE DRILLS	DECIMAL EQUIVALENT INCHES	SIZE OF THREAD	THREADS PER INCH
9/64	.....	.1406		
	27	.1440		
	26	.1470		
	25	.1495	10	24
	24	.1520		
	23	.1540		
5/32	.....	.1562		
	22	.1570		
	21	.1590	10	32
	20	.1610		
	19	.1660		
	18	.1695		
11/64	.....	.1719		
	17	.1730		
	16	.1770	12	24
	15	.1800		
	14	.1820	12	28
	13	.1850		
3/16	.....	.1875		
	12	.1890		
	11	.1910		
	10	.1935		
	9	.1960		
	8	.1990		
	7	.2010	1/4	20
13/64	.....	.2031		
	6	.2040		
	5	.2055		
	4	.2090		
	3	.2130	1/4	28
7/32	.....	.2187		
	2	.2210		
	1	.2280		
	A	.2340		
15/64	.....	.2344		
	B	.2380		
	C	.2420		
	D	.2460		
	E	.2500		
1/4	F	.2570	5/16	18
	G	.2610		
17/64	.....	.2656		
	H	.2660		
	I	.2720	5/16	24
	J	.2770		
	K	.2810		
9/32	.....	.2812		
	L	.2900		
	M	.2950		
19/64	.....	.2969		
	N	.3020		
5/16	.....	.3125	3/8	16
	O	.3160		
	P	.3230		
21/64	.....	.3281		
	Q	.3320	3/8	24
	R	.3390		
11/32	.....	.3437		
	S	.3480		
	T	.3580		

FRACTIONAL SIZE DRILLS INCHES	WIRE GAGE DRILLS	DECIMAL EQUIVALENT INCHES	SIZE OF THREAD	THREADS PER INCH
23/64	.....	.3594		
	U	.3680	7/16	14
3/8	.....	.3750		
	V	.3770		
	W	.3860		
25/64	.....	.3906	7/16	20
	X	.3970		
	Y	.4040		
13/32	.....	.4062		
	Z	.4130		
27/64	.....	.4219	1/2	13
7/16	.....	.4375		
29/64	.....	.4531	1/2	20
15/32	.....	.4687		
31/64	.....	.4844	9/16	12
1/2	.....	.5000		
33/64	.....	.5156	9/16	18
17/32	.....	.5312	5/8	11
35/64	.....	.5469		
9/16	.....	.5625		
37/64	.....	.5781	5/8	18
19/32	.....	.5937		
39/64	.....	.6094		
5/8	.....	.6250		
41/64	.....	.6406		
21/32	.....	.6562	3/4	10
43/64	.....	.6719		
11/16	.....	.6875	3/4	16
45/64	.....	.7031		
23/32	.....	.7187		
47/64	.....	.7344		
3/4	.....	.7500		
49/64	.....	.7656	7/8	9
25/32	.....	.7812		
51/64	.....	.7969		
13/16	.....	.8125	7/8	14
53/64	.....	.8281		
27/32	.....	.8437		
55/64	.....	.8594		
7/8	.....	.8750	1	8
57/64	.....	.8906		
29/32	.....	.9062		
59/64	.....	.9219		
15/16	.....	.9375	1	12
61/64	.....	.9531		
31/32	.....	.9687		
63/64	.....	.9844	1 1/8	7
1	.....	1.0000		

TABLE 15

DRILLED HOLE TOLERANCE (UNDER NORMAL SHOP CONDITIONS)

STANDARD DRILL SIZE				TOLERANCE IN DECIMALS	
DRILL SIZE				PLUS	MINUS
Number	Fraction	Decimal	Metric (mm)		
80		0.0135	0.3412	0.0023	
79		0.0145	0.3788	0.0024	
—	1/64	0.0156	0.3969	0.0025	
78		0.0160	0.4064	0.0025	
77		0.0180	0.4572	0.0026	
76		0.0200	0.5080	0.0027	
75		0.0210	0.5334	0.0027	
74		0.0225	0.5631	0.0028	
73		0.0240	0.6096	0.0028	
72		0.0250	0.6350	0.0029	
71		0.0260	0.6604	0.0029	
70		0.0280	0.7112	0.0030	.0005
69		0.0292	0.7483	0.0030	
68		0.0310	0.7874	0.0031	
—	1/32	0.0312	0.7937	0.0031	
67		0.0320	0.8128	0.0031	
66		0.0330	0.8382	0.0032	
65		0.0350	0.8890	0.0032	
64		0.0360	0.9144	0.0033	
63		0.0370	0.9398	0.0033	
62		0.0380	0.9652	0.0033	
61		0.0390	0.9906	0.0033	
60		0.0400	1.0160	0.0034	
59		0.0410	1.0414	0.0034	
58		0.0420	1.0668	0.0034	
57		0.0430	1.0922	0.0035	
56		0.0465	1.1684	0.0035	
—	3/64	0.0469	1.1906	0.0036	
55		0.0520	1.3208	0.0037	
54		0.0550	1.3970	0.0038	
53		0.0595	1.5122	0.0039	
—	1/16	0.0625	1.5875	0.0039	
52		0.0635	1.6002	0.0039	
51		0.0670	1.7018	0.0040	
50		0.0700	1.7780	0.0041	
49		0.0730	1.8542	0.0041	.001
48		0.0760	1.9304	0.0042	
—	5/64	0.0781	1.9844	0.0042	
47		0.0785	2.0001	0.0042	
46		0.0810	2.0574	0.0043	
45		0.0820	2.0828	0.0043	
44		0.0860	2.1844	0.0044	
43		0.0890	2.2606	0.0044	
42		0.0935	2.3622	0.0045	
—	3/32	0.0937	2.3812	0.0045	
41		0.0960	2.4384	0.0045	
40		0.0980	2.4892	0.0046	
39		0.0995	2.5377	0.0046	
38		0.1015	2.5908	0.0046	
37		0.1040	2.6416	0.0047	
36		0.1065	2.6924	0.0047	
—	7/64	0.1094	2.7781	0.0047	

TABLE 16

DRILLED HOLE TOLERANCE (UNDER NORMAL SHOP CONDITIONS)

STANDARD DRILL SIZE				TOLERANCE IN DECIMALS	
DRILL SIZE					
No./Letter	Fraction	Decimal	Metric (mm)	PLUS	MINUS
35		0.1100	2.7490	0.0047	
34		0.1110	2.8194	0.0048	
33		0.1130	2.8702	0.0048	
32		0.1160	2.9464	0.0048	
31		0.1200	3.0480	0.0049	
—	1/8	0.1250	3.1750	0.0050	
30		0.1285	3.2766	0.0050	
29		0.1360	3.4544	0.0051	
28		0.1405	3.5560	0.0052	
—	9/64	0.1406	3.5719	0.0052	
27		0.1440	3.6576	0.0052	
26		0.1470	3.7338	0.0052	
25		0.1495	3.7886	0.0053	
24		0.1520	3.8608	0.0053	
23		0.1540	3.9116	0.0053	
—	5/32	0.1562	3.9687	0.0053	
22		0.1570	3.9878	0.0053	
21		0.1590	4.0386	0.0054	
20		0.1610	4.0894	0.0054	
19		0.1660	4.2164	0.0055	
18		0.1695	4.3180	0.0055	
—	11/64	0.1719	4.3656	0.0055	
17		0.1730	4.3942	0.0055	
16		0.1770	4.4958	0.0056	.001
15		0.1800	4.5720	0.0056	
14		0.1820	4.6228	0.0057	
13		0.1850	4.6990	0.0057	
—	3/16	0.1875	4.7625	0.0057	
12		0.1890	4.8006	0.0057	
11		0.1910	4.8514	0.0057	
10		0.1935	4.9276	0.0058	
9		0.1960	4.9784	0.0058	
8		0.1990	5.0800	0.0058	
7		0.2010	5.1054	0.0058	
—	13/64	0.2031	5.1594	0.0058	
6		0.2040	5.1816	0.0058	
5		0.2055	5.2070	0.0059	
4		0.2090	5.3086	0.0059	
3		0.2130	5.4102	0.0059	
—	7/32	0.2187	5.5562	0.0060	
2		0.2210	5.6134	0.0060	
1		0.2280	5.7912	0.0061	
A		0.2340	5.9436	0.0061	
—	15/64	0.2344	5.9531	0.0061	
B		0.2380	6.0452	0.0061	
C		0.2420	6.1468	0.0062	
D		0.2460	6.2484	0.0062	
E	1/4	0.2500	6.3500	0.0063	
F		0.2570	6.5278	0.0063	
G		0.2610	6.6294	0.0063	
—	17/64	0.2656	6.7469	0.0064	
H		0.2660	6.7564	0.0064	
I		0.2720	6.9088	0.0064	.002
J		0.2770	7.0358	0.0065	
K		0.2810	7.1374	0.0065	
—	9/32	0.2812	7.1437	0.0065	
L		0.2900	7.3660	0.0066	
M		0.2950	7.4930	0.0066	
—	19/64	0.2969	7.5406	0.0066	

TABLE 16 *(Continued)*

DRILLED HOLE TOLERANCE (UNDER NORMAL SHOP CONDITIONS)

STANDARD DRILL SIZE				TOLERANCE IN DECIMALS	
DRILL SIZE					
Letter	Fraction	Decimal	Metric (mm)	PLUS	MINUS
N		0.3020	7.6708	0.0067	
—	5/16	0.3125	7.9375	0.0067	
O		0.3160	8.0264	0.0068	
P		0.3230	8.2042	0.0068	
—	21/64	0.3281	8.3344	0.0068	
Q		0.3320	8.4328	0.0069	
R		0.3390	8.6106	0.0069	
—	11/32	0.3437	8.7312	0.0070	
S		0.3480	8.8392	0.0070	
T		0.3580	9.0932	0.0071	
—	23/64	0.3594	9.1281	0.0071	
U		0.3680	9.3472	0.0071	
—	3/8	0.3750	9.5250	0.0072	
V		0.3770	9.5758	0.0072	
W		0.3860	9.8044	0.0072	
—	25/64	0.3906	9.9219	0.0073	
X		0.3970	10.0838	0.0073	
Y		0.4040	10.2616	0.0073	
—	13/32	0.4062	10.3187	0.0074	
Z		0.4130	10.4902	0.0074	.002
	27/64	0.4219	10.7156	0.0075	
	7/16	0.4375	10.1125	0.0075	
	29/64	0.4531	11.5094	0.0076	
	15/32	0.4687	11.9062	0.0077	
	31/64	0.4844	12.3031	0.0078	
	1/2	0.5000	12.7000	0.0079	
	33/64	0.5156	13.0968	0.0080	
	17/32	0.5312	13.4937	0.0081	
	35/64	0.5469	13.8906	0.0081	
	9/16	0.5625	14.2875	0.0082	
	37/64	0.5781	14.6844	0.0083	
	19/32	0.5927	15.0812	0.0084	
	39/64	0.6094	15.4781	0.0084	
	5/8	0.6250	15.8750	0.0085	
	41/64	0.6406	16.2719	0.0086	
	21/32	0.6562	16.6687	0.0086	
	43/64	0.6719	17.0656	0.0087	
	11/16	0.6875	17.4625	0.0088	
	45/64	0.7031	17.8594	0.0088	
	23/32	0.7187	18.2562	0.0089	
	47/64	0.7344	18.6532	0.0090	
	3/4	0.7500	19.0500	0.0090	
	49/64	0.7656	19.4469	0.0091	
	25/32	0.7812	19.8433	0.0092	
	51/64	0.7969	20.2402	0.0092	
	13/16	0.8125	20.6375	0.0093	
	53/64	0.8281	21.0344	0.0093	
	27/32	0.8437	21.4312	0.0094	
	55/64	0.8594	21.8281	0.0095	
	7/8	0.8750	22.2250	0.0095	.003
	57/64	0.8906	22.6219	0.0096	
	29/32	0.9062	23.0187	0.0096	
	59/64	0.9219	23.4156	0.0097	
	15/16	0.9375	23.8125	0.0097	
	61/64	0.9531	24.2094	0.0098	
	31/32	0.9687	24.6062	0.0098	
	63/64	0.9844	25.0031	0.0099	
	1	1.0000	25.4000	0.0100	

TABLE 16 *(Concluded)* From *Drafting for Trades and Industry—Mechanical and Electronic,* Nelson. Delmar Publishers, Inc.

INCH-METRIC THREAD COMPARISON

INCH SERIES				METRIC			
Size	Dia.(In.)	TPI		Size	Dia. (In.)	Pitch (mm)	TPI (Approx)
				M1.4	.055	.3 .2	85 127
#0	.060	80					
				M1.6	.063	.35 .2	74 127
#1	.073	64 72					
				M2	.079	.4 .25	64 101
#2	.086	56 64					
				M2.5	.098	.45 .35	56 74
#3	.099	48 56					
#4	.112	40 48					
				M3	.118	.5 .35	51 74
#5	.125	40 44					
#6	.138	32 40					
				M4	.157	.7 .5	36 51
#8	.164	32 36					
#10	.190	24 32					
				M5	.196	.8 .5	32 51
				M6	.236	1.0 .75	25 34
¼	.250	20 28					
5/16	.312	18 24					
				M8	.315	1.25 1.0	20 25
3/8	.375	16 24					
				M10	.393	1.5 1.25	17 20
7/16	.437	14 20					
				M12	.472	1.75 1.25	14.5 20
½	.500	13 20					
				M14	.551	2 1.5	12.5 17
5/8	.625	11 18					
				M16	.630	2 1.5	12.5 17
				M18	.709	2.5 1.5	10 17
¾	.750	10 16					
				M20	.787	2.5 1.5	10 17
				M22	.866	2.5 1.5	10 17
7/8	.875	9 14					
				M24	.945	3 2	8.5 12.5
1"	1.000	8 12					
				M27	1.063	3 2	8.5 12.5

TABLE 17

ISO BASIC METRIC THREAD INFORMATION

Basic Major DIA & Pitch	INTERNAL THREADS			EXTERNAL THREADS		
	Tap Drill DIA	Minor DIA MAX	Minor DIA MIN	Major DIA MAX	Major DIA MIN	Clearance Hole
M1.6 × 0.35	1.25	1.321	1.221	1.576	1.491	1.9
M2 × 0.4	1.60	1.679	1.567	1.976	1.881	2.4
M2.5 × 0.45	2.05	2.138	2.013	2.476	2.013	2.9
M3 × 0.5	2.50	2.599	2.459	2.976	2.870	3.4
M3.5 × 0.6	2.90	3.010	2.850	3.476	3.351	4.0
M4 × 0.7	3.30	3.422	3.242	3.976	3.836	4.5
M5 × 0.8	4.20	4.334	4.134	4.976	4.826	5.5
M6 × 1	5.00	5.153	4.917	5.974	5.794	6.6
M8 × 1.25	6.80	6.912	6.647	7.972	7.760	9.0
M10 × 1.5	8.50	8.676	8.376	9.968	9.732	11.0
M12 × 1.75	10.20	10.441	10.106	11.966	11.701	13.5
M14 × 2	12.00	12.210	11.835	13.962	13.682	15.5
M16 × 2	14.00	14.210	13.835	15.962	15.682	17.5
M20 × 2.5	17.50	17.744	17.294	19.958	19.623	22.0
M24 × 3	21.00	21.252	20.752	23.952	23.577	26.0
M30 × 3.5	26.50	26.771	26.211	29.947	29.522	33.0
M36 × 4	32.00	32.270	31.670	35.940	35.465	39.0
M42 × 4.5	37.50	37.799	37.129	41.937	41.437	45.0
M48 × 5	43.00	43.297	42.587	47.929	47.399	52.0
M56 × 5.5	50.50	50.796	50.046	55.925	55.365	62.0
M64 × 6	58.00	58.305	57.505	63.920	63.320	70.0
M72 × 6	66.00	66.305	65.505	71.920	71.320	78.0
M80 × 6	74.00	74.305	73.505	79.920	79.320	86.0
M90 × 6	84.00	84.305	83.505	89.920	89.320	96.0
M100 × 6	94.00	94.305	93.505	99.920	99.320	107.0

TABLE 18

AMERICAN NATIONAL STANDARD
SQUARE AND HEX BOLTS AND SCREWS – INCH SERIES

ASME/ANSI B18.2.1–1981 (R1992)

DIMENSIONS OF HEX BOLTS

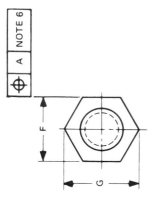

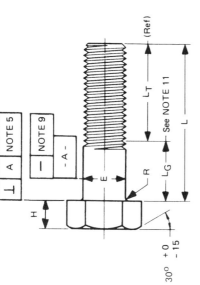

$30° \begin{smallmatrix}+0\\-15\end{smallmatrix}$

Nominal Size or Basic Product Dia. (17)		E Body Dia.(7) Max	F Width Across Flats (4) Basic	F Width Across Flats (4) Max	F Width Across Flats (4) Min	G Width Across Corners Max	G Width Across Corners Min	H Height Basic	H Height Max	H Height Min	R Radius of Fillet Max	R Radius of Fillet Min	L_T Thread Length For Bolt Lengths (11) 6 in. and shorter Basic	L_T Thread Length For Bolt Lengths (11) over 6 in. Basic
1/4	0.2500	0.260	7/16	0.438	0.425	0.505	0.484	11/64	0.188	0.150	0.03	0.01	0.750	1.000
5/16	0.3125	0.324	1/2	0.500	0.484	0.577	0.552	7/32	0.235	0.195	0.03	0.01	0.875	1.125
3/8	0.3750	0.388	9/16	0.562	0.544	0.650	0.620	1/4	0.268	0.226	0.03	0.01	1.000	1.250
7/16	0.4375	0.452	5/8	0.625	0.603	0.722	0.687	19/64	0.316	0.272	0.03	0.01	1.125	1.375
1/2	0.5000	0.515	3/4	0.750	0.725	0.866	0.826	11/32	0.364	0.302	0.03	0.01	1.250	1.500
5/8	0.6250	0.642	15/16	0.928	0.906	1.083	1.033	27/64	0.444	0.378	0.06	0.02	1.500	1.750
3/4	0.7500	0.768	1 1/8	1.125	1.088	1.299	1.240	1/2	0.524	0.455	0.06	0.02	1.750	2.000
7/8	0.8750	0.895	1 5/16	1.312	1.269	1.516	1.447	37/64	0.604	0.531	0.06	0.02	2.000	2.250
1	1.0000	1.022	1 1/2	1.500	1.450	1.732	1.653	43/64	0.700	0.591	0.09	0.03	2.250	2.500
1 1/8	1.1250	1.149	1 11/16	1.688	1.631	1.949	1.859	3/4	0.780	0.658	0.09	0.03	2.500	2.750
1 1/4	1.2500	1.277	1 7/8	1.875	1.812	2.165	2.066	27/32	0.876	0.749	0.09	0.03	2.750	3.000
1 3/8	1.3750	1.404	2 1/16	2.062	1.994	2.382	2.273	29/32	0.940	0.810	0.09	0.03	3.000	3.250
1 1/2	1.5000	1.531	2 1/4	2.250	2.175	2.598	2.480	1	1.036	0.902	0.09	0.03	3.250	3.500
1 3/4	1.7500	1.785	2 5/8	2.625	2.538	3.031	2.893	1 5/32	1.196	1.054	0.12	0.04	3.750	4.000
2	2.0000	2.039	3	3.000	2.900	3.464	3.306	1 11/32	1.388	1.175	0.12	0.04	4.250	4.500
2 1/4	2.2500	2.305	3 3/8	3.375	3.262	3.897	3.719	1 1/2	1.548	1.327	0.19	0.06	4.750	5.000
2 1/2	2.5000	2.559	3 3/4	3.750	3.625	4.330	4.133	1 21/32	1.708	1.479	0.19	0.06	5.250	5.500
2 3/4	2.7500	2.827	4 1/8	4.125	3.988	4.763	4.546	1 13/16	1.869	1.632	0.19	0.06	5.750	6.000
3	3.0000	3.081	4 1/2	4.500	4.350	5.196	4.959	2	2.060	1.815	0.19	0.06	6.250	6.500
3 1/4	3.2500	3.335	4 7/8	4.875	4.712	5.629	5.372	2 3/16	2.251	1.936	0.19	0.06	6.750	7.000
3 1/2	3.5000	3.589	5 1/4	5.250	5.075	6.062	5.786	2 5/16	2.380	2.057	0.19	0.06	7.250	7.500
3 3/4	3.7500	3.858	5 5/8	5.625	5.437	6.495	6.198	2 1/2	2.572	2.241	0.19	0.06	7.750	8.000
4	4.0000	4.111	6	6.000	5.800	6.928	6.612	2 11/16	2.764	2.424	0.19	0.06	8.250	8.500

TABLE 19 From The American Society of Mechanical Engineers—ASME/ANSI B18.2.1–1981 (R1992)

AMERICAN NATIONAL STANDARD
SQUARE AND HEX BOLTS AND SCREWS — INCH SERIES

ASME/ANSI B18.2.1–1981 (R1992)

DIMENSIONS OF HEX CAP SCREWS (FINISHED HEX BOLTS)

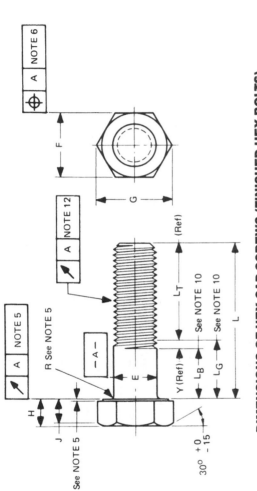

Nominal Size or Basic Product Dia.(18)	E Body Dia.(8) Max	E Body Dia.(8) Min	F Width Across Flats Basic	F Width Across Flats Max	F Width Across Flats Min	G Width Across Corners (4) Max	G Width Across Corners (4) Min	H Height Basic	H Height Max	H Height Min	J Wrenching Height (4) Min	L_T Thread Length For Screw Lengths (10) 6 in. and Shorter Basic	L_T Thread Length For Screw Lengths (10) Over 6 in. Basic	Y Transition Thread Length (10) Max	Runout of Bearing Surface FIM (5) Max
1/4 0.2500	0.2500	0.2450	7/16	0.438	0.428	0.505	0.488	5/32	0.163	0.150	0.106	0.750	1.000	0.250	0.010
5/16 0.3125	0.3125	0.3065	1/2	0.500	0.489	0.577	0.557	13/64	0.211	0.195	0.140	0.875	1.125	0.278	0.011
3/8 0.3750	0.3750	0.3690	9/16	0.562	0.551	0.650	0.628	15/64	0.243	0.226	0.160	1.000	1.250	0.312	0.012
7/16 0.4375	0.4375	0.4305	5/8	0.625	0.612	0.722	0.698	9/32	0.291	0.272	0.195	1.125	1.375	0.357	0.013
1/2 0.5000	0.5000	0.4930	3/4	0.750	0.736	0.866	0.840	5/16	0.323	0.302	0.215	1.250	1.500	0.385	0.014
9/16 0.5625	0.5625	0.5545	13/16	0.812	0.798	0.938	0.910	23/64	0.371	0.348	0.250	1.375	1.625	0.417	0.015
5/8 0.6250	0.6250	0.6170	15/16	0.938	0.922	1.083	1.051	25/64	0.403	0.378	0.269	1.500	1.750	0.455	0.017
3/4 0.7500	0.7500	0.7410	1 1/8	1.125	1.100	1.299	1.254	15/32	0.483	0.455	0.324	1.750	2.000	0.500	0.020
7/8 0.8750	0.8750	0.8660	1 5/16	1.312	1.285	1.516	1.465	35/64	0.563	0.531	0.378	2.000	2.250	0.556	0.023
1 1.0000	1.0000	0.9900	1 1/2	1.500	1.469	1.732	1.675	39/64	0.627	0.591	0.416	2.250	2.500	0.625	0.026
1 1/8 1.1250	1.1250	1.1140	1 11/16	1.688	1.631	1.949	1.859	11/16	0.718	0.658	0.461	2.500	2.750	0.714	0.029
1 1/4 1.2500	1.2500	1.2390	1 7/8	1.875	1.812	2.165	2.066	25/32	0.813	0.749	0.530	2.750	3.000	0.714	0.033
1 3/8 1.3750	1.3750	1.3630	2 1/16	2.062	1.994	2.382	2.273	27/32	0.878	0.810	0.569	3.000	3.250	0.833	0.036
1 1/2 1.5000	1.5000	1.4880	2 1/4	2.230	2.175	2.598	2.480	15/16	0.974	0.902	0.640	3.250	3.500	0.833	0.039
1 3/4 1.7500	1.7500	1.7380	2 5/8	2.625	2.538	3.031	2.893	1 3/32	1.134	1.054	0.748	3.750	4.000	1.000	0.046
2 2.0000	2.0000	1.9880	3	3.000	2.900	3.464	3.306	1 7/32	1.263	1.175	0.825	4.250	4.500	1.111	0.052
2 1/4 2.2500	2.2500	2.2380	3 3/8	3.375	3.262	3.897	3.719	1 3/8	1.423	1.327	0.933	4.750	5.000	1.111	0.059
2 1/2 2.5000	2.5000	2.4880	3 3/4	3.750	3.625	4.330	4.133	1 17/32	1.583	1.479	1.042	5.250	5.500	1.250	0.065
2 3/4 2.7500	2.7500	2.7380	4 1/8	4.125	3.988	4.763	4.546	1 11/16	1.744	1.632	1.151	5.750	6.000	1.250	0.072
3 3.0000	3.0000	2.9880	4 1/2	4.500	4.350	5.196	4.959	1 7/8	1.935	1.815	1.290	6.250	6.500	1.250	0.079

TABLE 20   From The American Society of Mechanical Engineers—ASME/ANSI B18.2.1–1981 (R1992)

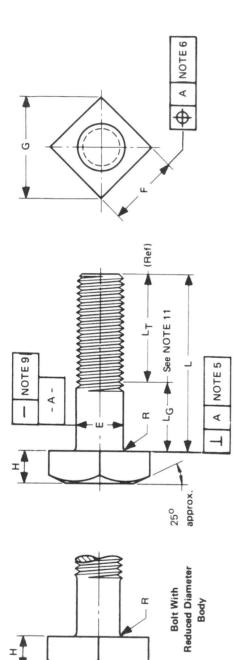

**Bolt With
Reduced Diameter
Body**

DIMENSIONS OF SQUARE BOLTS

Nominal Size or Basic Product Dia. (17)		E Body Dia. (7), (14) Max	F Width Across Flats (4) Basic	F Max	F Min	G Width Across Corners Max	G Min	H Height Basic	H Max	H Min	R Radius of Fillet Max	R Min	L_T Thread Length For Bolt Lengths (11) 6 in. and shorter Basic	L_T over 6 in. Basic
1/4	0.2500	0.260	3/8	0.375	0.362	0.530	0.498	11/64	0.188	0.156	0.03	0.01	0.750	1.000
5/16	0.3125	0.324	1/2	0.500	0.484	0.707	0.665	13/64	0.220	0.186	0.03	0.01	0.875	1.125
3/8	0.3750	0.388	9/16	0.562	0.544	0.795	0.747	1/4	0.268	0.232	0.03	0.01	1.000	1.250
7/16	0.4375	0.452	5/8	0.625	0.603	0.884	0.828	19/64	0.316	0.278	0.03	0.01	1.125	1.375
1/2	0.5000	0.515	3/4	0.750	0.725	1.061	0.995	21/64	0.348	0.308	0.03	0.01	1.250	1.500
5/8	0.6250	0.642	15/16	0.938	0.906	1.326	1.244	27/64	0.444	0.400	0.06	0.02	1.500	1.750
3/4	0.7500	0.768	1 1/8	1.125	1.088	1.591	1.494	1/2	0.524	0.476	0.06	0.02	1.750	2.000
7/8	0.8750	0.895	1 5/16	1.312	1.269	1.856	1.742	19/32	0.620	0.568	0.06	0.02	2.000	2.250
1	1.0000	1.022	1 1/2	1.500	1.450	2.121	1.991	21/32	0.684	0.628	0.09	0.03	2.250	2.500
1 1/8	1.1250	1.149	1 11/16	1.688	1.631	2.386	2.239	3/4	0.780	0.720	0.09	0.03	2.500	2.750
1 1/4	1.2500	1.277	1 7/8	1.875	1.812	2.652	2.489	27/32	0.876	0.812	0.09	0.03	2.750	3.000
1 3/8	1.3750	1.404	2 1/16	2.062	1.994	2.917	2.738	29/32	0.940	0.872	0.09	0.03	3.000	3.250
1 1/2	1.5000	1.531	2 1/4	2.250	2.175	3.182	2.986	1	1.036	0.964	0.09	0.03	3.250	3.500

TABLE 21 From The American Society of Mechanical Engineers—ASME/ANSI B18.2.1–1981 (R1992)

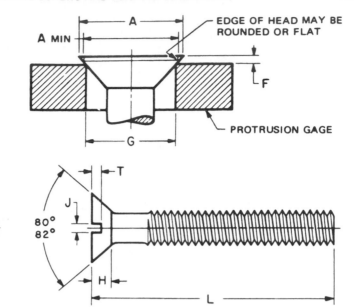

	SLOTTED
	FLAT

Type of Head

AMERICAN NATIONAL STANDARD
MACHINE SCREWS AND MACHINE SCREW NUTS

ANSI B18.6.3–1972–R1991

DIMENSIONS OF SLOTTED FLAT COUNTERSUNK HEAD MACHINE SCREWS

Nominal Size[1] or Basic Screw Diameter		L^2 These Lengths or Shorter are Undercut	A Head Diameter		H^3 Head Height	J Slot Width		T Slot Depth		F^4 Protrusion Above Gaging Diameter		G^4 Gaging Diameter
			Max, Edge Sharp	Min, Edge Rounded or Flat	Ref	Max	Min	Max	Min	Max	Min	
0000	0.0210	—	0.043	0.037	0.011	0.008	0.004	0.007	0.003	*	*	*
000	0.0340	—	0.064	0.058	0.016	0.011	0.007	0.009	0.005	*	*	*
00	0.0470	—	0.093	0.085	0.028	0.017	0.010	0.014	0.009	*	*	*
0	0.0600	1/8	0.119	0.099	0.035	0.023	0.016	0.015	0.010	0.026	0.016	0.078
1	0.0730	1/8	0.146	0.123	0.043	0.026	0.019	0.019	0.012	0.028	0.016	0.101
2	0.0860	1/8	0.172	0.147	0.051	0.031	0.023	0.023	0.015	0.029	0.017	0.124
3	0.0990	1/8	0.199	0.171	0.059	0.035	0.027	0.027	0.017	0.031	0.018	0.148
4	0.1120	3/16	0.225	0.195	0.067	0.039	0.031	0.030	0.020	0.032	0.019	0.172
5	0.1250	3/16	0.252	0.220	0.075	0.043	0.035	0.034	0.022	0.034	0.020	0.196
6	0.1380	3/16	0.279	0.244	0.083	0.048	0.039	0.038	0.024	0.036	0.021	0.220
8	0.1640	1/4	0.332	0.292	0.100	0.054	0.045	0.045	0.029	0.039	0.023	0.267
10	0.1900	5/16	0.385	0.340	0.116	0.060	0.050	0.053	0.034	0.042	0.025	0.313
12	0.2160	3/8	0.438	0.389	0.132	0.067	0.056	0.060	0.039	0.045	0.027	0.362
1/4	0.2500	7/16	0.507	0.452	0.153	0.075	0.064	0.070	0.046	0.050	0.029	0.424
5/16	0.3125	1/2	0.635	0.568	0.191	0.084	0.072	0.088	0.058	0.057	0.034	0.539
3/8	0.3750	9/16	0.762	0.685	0.230	0.094	0.081	0.106	0.070	0.065	0.039	0.653
7/16	0.4375	5/8	0.812	0.723	0.223	0.094	0.081	0.103	0.066	0.073	0.044	0.690
1/2	0.5000	3/4	0.875	0.775	0.223	0.106	0.091	0.103	0.065	0.081	0.049	0.739
9/16	0.5625	—	1.000	0.889	0.260	0.118	0.102	0.120	0.077	0.089	0.053	0.851
5/8	0.6250	—	1.125	1.002	0.298	0.133	0.116	0.137	0.088	0.097	0.058	0.962
3/4	0.7500	—	1.375	1.230	0.372	0.149	0.131	0.171	0.111	0.112	0.067	1.186

[1]Where specifying nominal size in decimals, zeros preceding decimal and in the fourth decimal place shall be omitted.
[2]Screws of these lengths and shorter shall have undercut heads as shown in Table 5.
[3]Tabulated values determined from formula for maximum H, Appendix V.
[4]No tolerance for gaging diameter is given. If the gaging diameter of the gage used differs from tabulated value, the protrusion will be affected accordingly and the proper protrusion values must be recalculated using the formulas shown in Appendix I.
*Not practical to gage.

TABLE 22   From The American Society of Mechanical Engineers—ANSI B18.6.3–1972–R1991

```
┌─────────────────────┐
│   CAP SCREWS        │
│                     │
│   R O U N D         │
│                     │
└─────────────────────┘
   Type of Head
```

AMERICAN NATIONAL STANDARD – SLOTTED HEAD CAP SCREWS,
SQUARE HEAD SET SCREWS, AND SLOTTED HEADLESS SET SCREWS

ASME/ANSI B18.6.2–1972 (R1993)

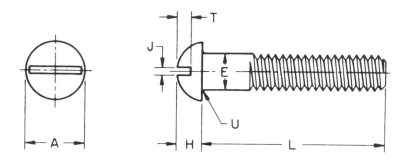

DIMENSIONS OF SLOTTED ROUND HEAD CAP SCREWS

Nominal Size[1] or Basic Screw Diameter		E Body Diameter		A Head Diameter		H Head Height		J Slot Width		T Slot Depth		U Fillet Radius	
		Max	Min	Max	Min	Max	Min	Max	Min	Max	Min	Max	Min
1/4	0.2500	0.2500	0.2450	0.437	0.418	0.191	0.175	0.075	0.064	0.117	0.097	0.031	0.016
5/16	0.3125	0.3125	0.3070	0.562	0.540	0.245	0.226	0.084	0.072	0.151	0.126	0.031	0.016
3/8	0.3750	0.3750	0.3690	0.625	0.603	0.273	0.252	0.094	0.081	0.168	0.138	0.031	0.016
7/16	0.4375	0.4375	0.4310	0.750	0.725	0.328	0.302	0.094	0.081	0.202	0.167	0.047	0.016
1/2	0.5000	0.5000	0.4930	0.812	0.786	0.354	0.327	0.106	0.091	0.218	0.178	0.047	0.016
9/16	0.5625	0.5625	0.5550	0.937	0.909	0.409	0.378	0.118	0.102	0.252	0.207	0.047	0.016
5/8	0.6250	0.6250	0.6170	1.000	0.970	0.437	0.405	0.133	0.116	0.270	0.220	0.062	0.031
3/4	0.7500	0.7500	0.7420	1.250	1.215	0.546	0.507	0.149	0.131	0.338	0.278	0.062	0.031

[1]Where specifying nominal size in decimals, zeros preceding decimal and in the fourth decimal place shall be omitted.

TABLE 23   From The American Society of Mechanical Engineers—ASME/ANSI B18.6.2–1972 (R1993)

AMERICAN NATIONAL STANDARD – SLOTTED HEAD CAP SCREWS,
SQUARE HEAD SET SCREWS, AND SLOTTED HEADLESS SET SCREWS   ASME/ANSI B18.6.2–1972 (R1993)

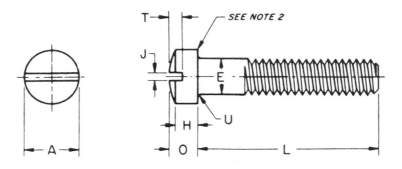

DIMENSIONS OF SLOTTED FILLISTER HEAD CAP SCREWS

Nominal Size[1] or Basic Screw Diameter		E Body Diameter		A Head Diameter		H Head Side Height		O Total Head Height		J Slot Width		T Slot Depth		U Fillet Radius	
		Max	Min	Max	Min	Max	Min	Max	Min	Max	Min	Max	Min	Max	Min
1/4	0.2500	0.2500	0.2450	0.375	0.363	0.172	0.157	0.216	0.194	0.075	0.064	0.097	0.077	0.031	0.016
5/16	0.3125	0.3125	0.3070	0.437	0.424	0.203	0.186	0.253	0.230	0.084	0.072	0.115	0.090	0.031	0.016
3/8	0.3750	0.3750	0.3690	0.562	0.547	0.250	0.229	0.314	0.284	0.094	0.081	0.142	0.112	0.031	0.016
7/16	0.4375	0.4375	0.4310	0.625	0.608	0.297	0.274	0.368	0.336	0.094	0.081	0.168	0.133	0.047	0.016
1/2	0.5000	0.5000	0.4930	0.750	0.731	0.328	0.301	0.413	0.376	0.106	0.091	0.193	0.153	0.047	0.016
9/16	0.5625	0.5625	0.5550	0.812	0.792	0.375	0.346	0.467	0.427	0.118	0.102	0.213	0.168	0.047	0.016
5/8	0.6250	0.6250	0.6170	0.875	0.853	0.422	0.391	0.521	0.478	0.133	0.116	0.239	0.189	0.062	0.031
3/4	0.7500	0.7500	0.7420	1.000	0.976	0.500	0.466	0.612	0.566	0.149	0.131	0.283	0.223	0.062	0.031
7/8	0.8750	0.8750	0.8660	1.125	1.098	0.594	0.556	0.720	0.668	0.167	0.147	0.334	0.264	0.062	0.031
1	1.0000	1.0000	0.9900	1.312	1.282	0.656	0.612	0.803	0.743	0.188	0.166	0.371	0.291	0.062	0.031

[1]Where specifying nominal size in decimals, zeros preceding decimal and in the fourth decimal place shall be omitted.
[2]A slight rounding of the edges at periphery of head shall be permissible provided the diameter of the bearing circle is equal to no less than 90 percent of the specified miminum head diameter.

TABLE 24   From The American Society of Mechanical Engineers—ASME/ANSI B18.6.2–1972 (R1993)

AMERICAN NATIONAL STANDARD
MACHINE SCREWS AND MACHINE SCREW NUTS

ASME/ANSI B18.6.3–1972 (R1992)

TYPE I RECESS
FLAT

Type of Head

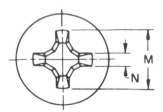

This type of recess has a large center opening, tapered wings, and blunt bottom, with all edges relieved or rounded

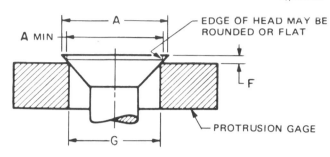

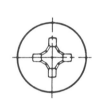

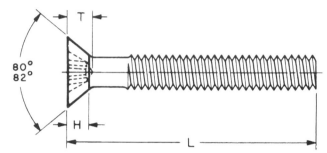

DIMENSIONS OF TYPE I CROSS RECESSED FLAT COUNTERSUNK HEAD MACHINE SCREWS

Nominal Size[1] or Basic Screw Diameter		L[2] These Lengths or Shorter are Undercut	A Head Diameter		H[3] Head Height	M Recess Diameter		T Recess Depth		N Recess Width	Driver Size	Recess Penetration Gaging Depth		F[4] Protrusion Above Gaging Diameter		G[4] Gaging Diameter
			Max, Edge Sharp	Min, Edge Rounded or Flat	Ref	Max	Min	Max	Min	Min		Max	Min	Max	Min	
0	0.0600	1/8	0.119	0.099	0.035	0.069	0.056	0.043	0.027	0.014	0	0.036	0.020	0.026	0.016	0.078
1	0.0730	1/8	0.146	0.123	0.043	0.077	0.064	0.051	0.035	0.015	0	0.044	0.028	0.028	0.016	0.101
2	0.0860	1/8	0.172	0.147	0.051	0.102	0.089	0.063	0.047	0.017	1	0.056	0.040	0.029	0.017	0.124
3	0.0990	1/8	0.199	0.171	0.059	0.107	0.094	0.068	0.052	0.018	1	0.061	0.045	0.031	0.018	0.148
4	0.1120	3/16	0.225	0.195	0.067	0.128	0.115	0.089	0.073	0.018	1	0.082	0.066	0.032	0.019	0.172
5	0.1250	3/16	0.252	0.220	0.075	0.154	0.141	0.086	0.063	0.027	2	0.075	0.052	0.034	0.020	0.196
6	0.1380	3/16	0.279	0.244	0.083	0.174	0.161	0.106	0.083	0.029	2	0.095	0.072	0.036	0.021	0.220
8	0.1640	1/4	0.332	0.292	0.100	0.189	0.176	0.121	0.098	0.030	2	0.110	0.087	0.039	0.023	0.267
10	0.1900	5/16	0.385	0.340	0.116	0.204	0.191	0.136	0.113	0.032	2	0.125	0.102	0.042	0.025	0.313
12	0.2160	3/8	0.438	0.389	0.132	0.268	0.255	0.156	0.133	0.035	3	0.139	0.116	0.045	0.027	0.362
1/4	0.2500	7/16	0.507	0.452	0.153	0.283	0.270	0.171	0.148	0.036	3	0.154	0.131	0.050	0.029	0.424
5/16	0.3125	1/2	0.635	0.568	0.191	0.365	0.352	0.216	0.194	0.061	4	0.196	0.174	0.057	0.034	0.539
3/8	0.3750	9/16	0.762	0.685	0.230	0.393	0.380	0.245	0.223	0.065	4	0.225	0.203	0.065	0.039	0.653
7/16	0.4375	5/8	0.812	0.723	0.223	0.409	0.396	0.261	0.239	0.068	4	0.241	0.219	0.073	0.044	0.690
1/2	0.5000	3/4	0.875	0.775	0.223	0.424	0.411	0.276	0.254	0.069	4	0.256	0.234	0.081	0.049	0.739
9/16	0.5625	—	1.000	0.889	0.260	0.454	0.431	0.300	0.278	0.073	4	0.280	0.258	0.089	0.053	0.851
5/8	0.6250	—	1.125	1.002	0.298	0.576	0.553	0.342	0.316	0.079	5	0.309	0.283	0.097	0.058	0.962
3/4	0.7500	—	1.375	1.230	0.372	0.640	0.617	0.406	0.380	0.087	5	0.373	0.347	0.112	0.067	1.186

[1]Where specifying nominal size in decimals, zeros preceding decimal and in the fourth decimal place shall be omitted.
[2]Screws of these lengths and shorter shall have undercut heads.
[3]Tabulated values determined from formula for maximum H, Appendix V, ANSI B18.6.3-1972.
[4]No tolerance for gaging diameter is given. If the gaging diameter of the gage used differs from tabulated value, the protrusion will be affected accordingly and the proper protrusion values must be recalculated using the formulas shown in Appendix I, ANSI B18.6.3-1972.

TABLE 25 From The American Society of Mechanical Engineers—ASME/ANSI B18.6.3–1972 (R1992)

AMERICAN NATIONAL STANDARD
MACHINE SCREWS AND MACHINE SCREW NUTS   ASME/ANSI B18.6.3–1972 (R1992)

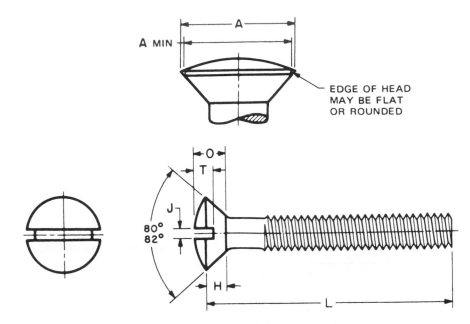

DIMENSIONS OF SLOTTED OVAL COUNTERSUNK HEAD MACHINE SCREWS

Nominal Size[1] or Basic Screw Diameter		L[2] These Lengths or Shorter are Undercut	A Head Diameter		H[3] Head Side Height	O Total Head Height		J Slot Width		T Slot Depth	
			Max, Edge Sharp	Min, Edge Rounded or Flat	Ref	Max	Min	Max	Min	Max	Min
00	0.0470	—	0.093	0.085	0.028	0.042	0.034	0.017	0.010	0.023	0.016
0	0.0600	1/8	0.119	0.099	0.035	0.056	0.041	0.023	0.016	0.030	0.025
1	0.0730	1/8	0.146	0.123	0.043	0.068	0.052	0.026	0.019	0.038	0.031
2	0.0860	1/8	0.172	0.147	0.051	0.080	0.063	0.031	0.023	0.045	0.037
3	0.0990	1/8	0.199	0.171	0.059	0.092	0.073	0.035	0.027	0.052	0.043
4	0.1120	3/16	0.225	0.195	0.067	0.104	0.084	0.039	0.031	0.059	0.049
5	0.1250	3/16	0.252	0.220	0.075	0.116	0.095	0.043	0.035	0.067	0.055
6	0.1380	3/16	0.279	0.244	0.083	0.128	0.105	0.048	0.039	0.074	0.060
8	0.1640	1/4	0.332	0.292	0.100	0.152	0.126	0.054	0.045	0.088	0.072
10	0.1900	5/16	0.385	0.340	0.116	0.176	0.148	0.060	0.050	0.103	0.084
12	0.2160	3/8	0.438	0.389	0.132	0.200	0.169	0.067	0.056	0.117	0.096
1/4	0.2500	7/16	0.507	0.452	0.153	0.232	0.197	0.075	0.064	0.136	0.112
5/16	0.3125	1/2	0.635	0.568	0.191	0.290	0.249	0.084	0.072	0.171	0.141
3/8	0.3750	9/16	0.762	0.685	0.230	0.347	0.300	0.094	0.081	0.206	0.170
7/16	0.4375	5/8	0.812	0.723	0.223	0.345	0.295	0.094	0.081	0.210	0.174
1/2	0.5000	3/4	0.875	0.775	0.223	0.354	0.299	0.106	0.091	0.216	0.176
9/16	0.5625	—	1.000	0.889	0.260	0.410	0.350	0.118	0.102	0.250	0.207
5/8	0.6250	—	1.125	1.002	0.298	0.467	0.399	0.133	0.116	0.285	0.235
3/4	0.7500	—	1.375	1.230	0.372	0.578	0.497	0.149	0.131	0.353	0.293

[1]Where specifying nominal size in decimals, zeros preceding decimal and in the fourth decimal place shall be omitted.
[2]Screws of these lengths and shorter shall have undercut heads.
[3]Tabulated values determined from formula for maximum H, Appendix V, ANSI B18.6.3-1972.

TABLE 26   From The American Society of Mechanical Engineers—ASME/ANSI B18.6.3–1972 (R1992)

SLOTTED

PAN

AMERICAN NATIONAL STANDARD
MACHINE SCREWS AND MACHINE SCREW NUTS

ASME/ANSI B18.6.3–1972 (R1993)

Type of Head

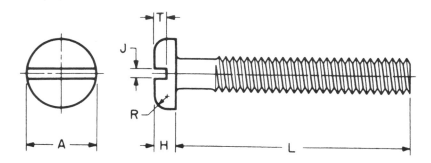

DIMENSIONS OF SLOTTED PAN HEAD MACHINE SCREWS

Nominal Size[1] or Basic Screw Diameter		A Head Diameter		H Head Height		R Head Radius	J Slot Width		T Slot Depth	
		Max	Min	Max	Min	Max	Max	Min	Max	Min
0000	0.0210	0.042	0.036	0.016	0.010	0.007	0.008	0.004	0.008	0.004
000	0.0340	0.066	0.060	0.023	0.017	0.010	0.012	0.008	0.012	0.008
00	0.0470	0.090	0.082	0.032	0.025	0.015	0.017	0.010	0.016	0.010
0	0.0600	0.116	0.104	0.039	0.031	0.020	0.023	0.016	0.022	0.014
1	0.0730	0.142	0.130	0.046	0.038	0.025	0.026	0.019	0.027	0.018
2	0.0860	0.167	0.155	0.053	0.045	0.035	0.031	0.023	0.031	0.022
3	0.0990	0.193	0.180	0.060	0.051	0.037	0.035	0.027	0.036	0.026
4	0.1120	0.219	0.205	0.068	0.058	0.042	0.039	0.031	0.040	0.030
5	0.1250	0.245	0.231	0.075	0.065	0.044	0.043	0.035	0.045	0.034
6	0.1380	0.270	0.256	0.082	0.072	0.046	0.048	0.039	0.050	0.037
8	0.1640	0.322	0.306	0.096	0.085	0.052	0.054	0.045	0.058	0.045
10	0.1900	0.373	0.357	0.110	0.099	0.061	0.060	0.050	0.068	0.053
12	0.2160	0.425	0.407	0.125	0.112	0.078	0.067	0.056	0.077	0.061
1/4	0.2500	0.492	0.473	0.144	0.130	0.087	0.075	0.064	0.087	0.070
5/16	0.3125	0.615	0.594	0.178	0.162	0.099	0.084	0.072	0.106	0.085
3/8	0.3750	0.740	0.716	0.212	0.195	0.143	0.094	0.081	0.124	0.100
7/16	0.4375	0.863	0.837	0.247	0.228	0.153	0.094	0.081	0.142	0.116
1/2	0.5000	0.987	0.958	0.281	0.260	0.175	0.106	0.091	0.161	0.131
9/16	0.5625	1.041	1.000	0.315	0.293	0.197	0.118	0.102	0.179	0.146
5/8	0.6250	1.172	1.125	0.350	0.325	0.219	0.133	0.116	0.197	0.162
3/4	0.7500	1.435	1.375	0.419	0.390	0.263	0.149	0.131	0.234	0.192

[1]Where specifying nominal size in decimals, zeros preceding decimal and in the fourth decimal place shall be omitted.

TABLE 27   From The American Society of Mechanical Engineers—ASME/ANSI B18.6.3–1972 (R1993)

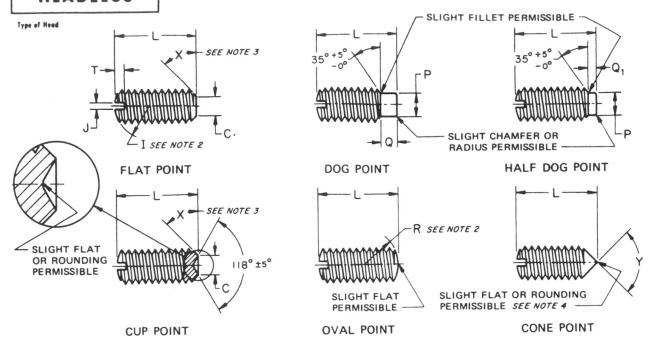

SET SCREWS
SLOTTED
HEADLESS

Type of Head

AMERICAN NATIONAL STANDARD—SLOTTED HEAD CAP SCREWS, SQUARE HEAD SET SCREWS, AND SLOTTED HEADLESS SET SCREWS   ASME/ANSI B18.6.2–1972 (R1993)

FLAT POINT   DOG POINT   HALF DOG POINT

CUP POINT   OVAL POINT   CONE POINT

DIMENSIONS OF SLOTTED HEADLESS SET SCREWS

Nominal Size[1] or Basic Screw Diameter		I[2] Crown Radius	J Slot Width		T Slot Depth		C Cup and Flat Point Diameters		P Dog Point Diameters		Q Point Length Dog		Q₁ Point Length Half Dog		R[2] Oval Point Radius	Y Cone Point Angle 90° ±2° For These Nominal Lengths or Longer; 118° ±2° For Shorter Screws
		Basic	Max	Min	Max	Min	Max	Min	Max	Min	Max	Min	Max	Min	Basic	
0	0.0600	0.060	0.014	0.010	0.020	0.016	0.033	0.027	0.040	0.037	0.032	0.028	0.017	0.013	0.045	5/64
1	0.0730	0.073	0.016	0.012	0.020	0.016	0.040	0.033	0.049	0.045	0.040	0.036	0.021	0.017	0.055	3/32
2	0.0860	0.086	0.018	0.014	0.025	0.019	0.047	0.039	0.057	0.053	0.046	0.042	0.024	0.020	0.064	7/64
3	0.0990	0.099	0.020	0.016	0.028	0.022	0.054	0.045	0.066	0.062	0.052	0.048	0.027	0.023	0.074	1/8
4	0.1120	0.112	0.024	0.018	0.031	0.025	0.061	0.051	0.075	0.070	0.058	0.054	0.030	0.026	0.084	5/32
5	0.1250	0.125	0.026	0.020	0.036	0.026	0.067	0.057	0.083	0.078	0.063	0.057	0.033	0.027	0.094	3/16
6	0.1380	0.138	0.028	0.022	0.040	0.030	0.074	0.064	0.092	0.087	0.073	0.067	0.038	0.032	0.104	3/16
8	0.1640	0.164	0.032	0.026	0.046	0.036	0.087	0.076	0.109	0.103	0.083	0.077	0.043	0.037	0.123	1/4
10	0.1900	0.190	0.035	0.029	0.053	0.043	0.102	0.088	0.127	0.120	0.095	0.085	0.050	0.040	0.142	1/4
12	0.2160	0.216	0.042	0.035	0.061	0.051	0.115	0.101	0.144	0.137	0.115	0.105	0.060	0.050	0.162	5/16
1/4	0.2500	0.250	0.049	0.041	0.068	0.058	0.132	0.118	0.156	0.149	0.130	0.120	0.068	0.058	0.188	5/16
5/16	0.3125	0.312	0.055	0.047	0.083	0.073	0.172	0.156	0.203	0.195	0.161	0.151	0.083	0.073	0.234	3/8
3/8	0.3750	0.375	0.068	0.060	0.099	0.089	0.212	0.194	0.250	0.241	0.193	0.183	0.099	0.089	0.281	7/16
7/16	0.4375	0.438	0.076	0.068	0.114	0.104	0.252	0.232	0.297	0.287	0.224	0.214	0.114	0.104	0.328	1/2
1/2	0.5000	0.500	0.086	0.076	0.130	0.120	0.291	0.270	0.344	0.334	0.255	0.245	0.130	0.120	0.375	9/16
9/16	0.5625	0.562	0.096	0.086	0.146	0.136	0.332	0.309	0.391	0.379	0.287	0.275	0.146	0.134	0.422	5/8
5/8	0.6250	0.625	0.107	0.097	0.161	0.151	0.371	0.347	0.469	0.456	0.321	0.305	0.164	0.148	0.469	3/4
3/4	0.7500	0.750	0.134	0.124	0.193	0.183	0.450	0.425	0.562	0.549	0.383	0.367	0.196	0.180	0.562	7/8

[1]Where specifying nominal size in decimals, zeros preceding decimal and in the fourth decimal place shall be omitted.
[2]Tolerance on radius for nominal sizes up to and including 5 (0.125 in.) shall be plus 0.015 in. and minus 0.000, and for larger sizes, plus 0.031 in. and minus 0.000. Slotted ends on screws may be flat at option of manufacturer.
[3]Point angle X shall be 45° plus 5°, minus 0°, for screws of nominal lengths equal to or longer than those listed in Column Y, and 30° minimum for screws of shorter nominal lengths.
[4]The extent of rounding or flat at apex of cone point shall not exceed an amount equivalent to 10 percent of the basic screw diameter.

TABLE 28   From The American Society of Mechanical Engineers—ASME/ANSI B18.6.2–1972 (R1993)

SET SCREWS

SQUARE

Type of Head

AMERICAN NATIONAL STANDARD—SLOTTED HEAD CAP SCREWS, SQUARE HEAD SET SCREWS, AND SLOTTED HEADLESS SET SCREWS   ASME/ANSI B18.6.2–1972 (R1993)

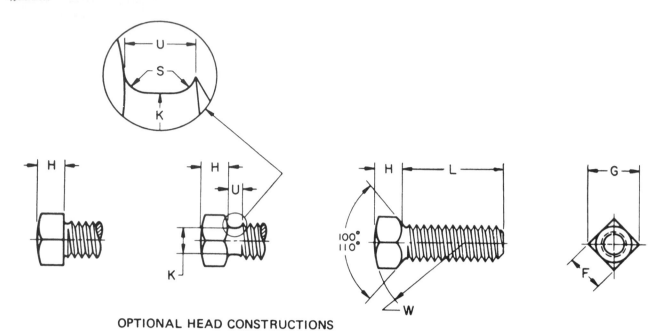

OPTIONAL HEAD CONSTRUCTIONS

DIMENSIONS OF SQUARE HEAD SET SCREWS

Nominal Size[1] or Basic Screw Diameter		F Width Across Flats		G Width Across Corners		H Head Height		K Neck Relief Diameter		S Neck Relief Fillet Radius	U Neck Relief Width	W Head Radius
		Max	Min	Max	Min	Max	Min	Max	Min	Max	Min	Min
10	0.1900	0.188	0.180	0.265	0.247	0.148	0.134	0.145	0.140	0.027	0.083	0.48
1/4	0.2500	0.250	0.241	0.354	0.331	0.196	0.178	0.185	0.170	0.032	0.100	0.62
5/16	0.3125	0.312	0.302	0.442	0.415	0.245	0.224	0.240	0.225	0.036	0.111	0.78
3/8	0.3750	0.375	0.362	0.530	0.497	0.293	0.270	0.294	0.279	0.041	0.125	0.94
7/16	0.4375	0.438	0.423	0.619	0.581	0.341	0.315	0.345	0.330	0.046	0.143	1.09
1/2	0.5000	0.500	0.484	0.707	0.665	0.389	0.361	0.400	0.385	0.050	0.154	1.25
9/16	0.5625	0.562	0.545	0.795	0.748	0.437	0.407	0.454	0.439	0.054	0.167	1.41
5/8	0.6250	0.625	0.606	0.884	0.833	0.485	0.452	0.507	0.492	0.059	0.182	1.56
3/4	0.7500	0.750	0.729	1.060	1.001	0.582	0.544	0.620	0.605	0.065	0.200	1.88
7/8	0.8750	0.875	0.852	1.237	1.170	0.678	0.635	0.731	0.716	0.072	0.222	2.19
1	1.0000	1.000	0.974	1.414	1.337	0.774	0.726	0.838	0.823	0.081	0.250	2.50
1 1/8	1.1250	1.125	1.096	1.591	1.505	0.870	0.817	0.939	0.914	0.092	0.283	2.81
1 1/4	1.2500	1.250	1.219	1.768	1.674	0.966	0.908	1.064	1.039	0.092	0.283	3.12
1 3/8	1.3750	1.375	1.342	1.945	1.843	1.063	1.000	1.159	1.134	0.109	0.333	3.44
1 1/2	1.5000	1.500	1.464	2.121	2.010	1.159	1.091	1.284	1.259	0.109	0.333	3.75

[1]Where specifying nominal size in decimals, zeros preceding decimal and in the fourth decimal place shall be omitted.

TABLE 29

AMERICAN NATIONAL STANDARD—SLOTTED HEAD CAP SCREWS,
SQUARE HEAD SET SCREWS, AND SLOTTED HEADLESS SET SCREWS   ASME/ANSI B18.6.2–1972 (R1993)

SET SCREWS

SQUARE

Type of Head

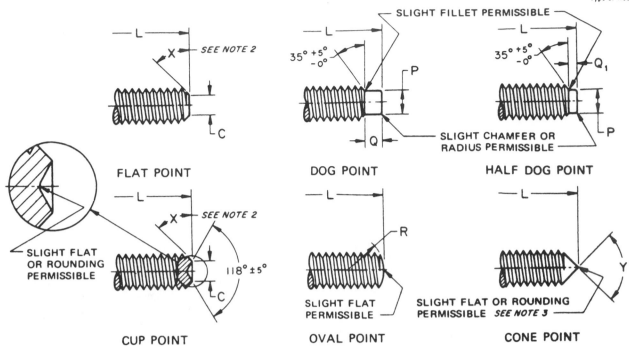

FLAT POINT   DOG POINT   HALF DOG POINT

CUP POINT   OVAL POINT   CONE POINT

DIMENSIONS OF SQUARE HEAD SET SCREWS (CONCLUDED)

Nominal Size[1] or Basic Screw Diameter		C		P		Q		Q₁		R	Y
		Cup and Flat Point Diameters		Dog and Half Dog Point Diameters		Point Length				Oval Point Radius +0.031 −0.000	Cone Point Angle 90° ±2° For These Nominal Lengths or Longer; 118° ±2° For Shorter Screws
						Dog		Half Dog			
		Max	Min	Max	Min	Max	Min	Max	Min		
10	0.1900	0.102	0.088	0.127	0.120	0.095	0.085	0.050	0.040	0.142	1/4
1/4	0.2500	0.132	0.118	0.156	0.149	0.130	0.120	0.068	0.058	0.188	5/16
5/16	0.3125	0.172	0.156	0.203	0.195	0.161	0.151	0.083	0.073	0.234	3/8
3/8	0.3750	0.212	0.194	0.250	0.241	0.193	0.183	0.099	0.089	0.281	7/16
7/16	0.4375	0.252	0.232	0.297	0.287	0.224	0.214	0.114	0.104	0.328	1/2
1/2	0.5000	0.291	0.270	0.344	0.334	0.255	0.245	0.130	0.120	0.375	9/16
9/16	0.5625	0.332	0.309	0.391	0.379	0.287	0.275	0.146	0.134	0.422	5/8
5/8	0.6250	0.371	0.347	0.469	0.456	0.321	0.305	0.164	0.148	0.469	3/4
3/4	0.7500	0.450	0.425	0.562	0.549	0.383	0.367	0.196	0.180	0.562	7/8
7/8	0.8750	0.530	0.502	0.656	0.642	0.446	0.430	0.227	0.211	0.656	1
1	1.0000	0.609	0.579	0.750	0.734	0.510	0.490	0.260	0.240	0.750	1 1/8
1 1/8	1.1250	0.689	0.655	0.844	0.826	0.572	0.552	0.291	0.271	0.844	1 1/4
1 1/4	1.2500	0.767	0.733	0.938	0.920	0.635	0.615	0.323	0.303	0.938	1 1/2
1 3/8	1.3750	0.848	0.808	1.031	1.011	0.698	0.678	0.354	0.334	1.031	1 5/8
1 1/2	1.5000	0.926	0.886	1.125	1.105	0.760	0.740	0.385	0.365	1.125	1 3/4

[1]Where specifying nominal size in decimals, zeros preceding decimal and in the fourth decimal place shall be omitted.
[2]Point angle X shall be 45° plus 5°, minus 0°, for screws of nominal lengths equal to or longer than those listed in Column Y, and 30° minimum for screws of shorter nominal lengths.
[3]The extent of rounding or flat at apex of cone point shall not exceed an amount equivalent to 10 percent of the basic screw diameter.

TABLE 29   *(Concluded)* From The American Society of Mechanical Engineers—ASME/ANSI B18.6.2–1972 (R1993)

AMERICAN STANDARD

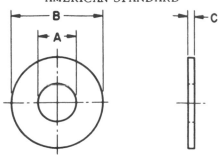

DIMENSIONS OF PREFERRED SIZES OF TYPE A PLAIN WASHERS **

Nominal Washer Size***			Inside Diameter A			Outside Diameter B			Thickness C		
			Basic	Tolerance		Basic	Tolerance		Basic	Max	Min
				Plus	Minus		Plus	Minus			
—	—		0.078	0.000	0.005	0.188	0.000	0.005	0.020	0.025	0.016
—	—		0.094	0.000	0.005	0.250	0.000	0.005	0.020	0.025	0.016
—	—		0.125	0.008	0.005	0.312	0.008	0.005	0.032	0.040	0.025
No. 6	0.138		0.156	0.008	0.005	0.375	0.015	0.005	0.049	0.065	0.036
No. 8	0.164		0.188	0.008	0.005	0.438	0.015	0.005	0.049	0.065	0.036
No. 10	0.190		0.219	0.008	0.005	0.500	0.015	0.005	0.049	0.065	0.036
$\frac{3}{16}$	0.188		0.250	0.015	0.005	0.562	0.015	0.005	0.049	0.065	0.036
No. 12	0.216		0.250	0.015	0.005	0.562	0.015	0.005	0.065	0.080	0.051
$\frac{1}{4}$	0.250	N	0.281	0.015	0.005	0.625	0.015	0.005	0.065	0.080	0.051
$\frac{1}{4}$	0.250	W	0.312	0.015	0.005	0.734*	0.015	0.007	0.065	0.080	0.051
$\frac{5}{16}$	0.312	N	0.344	0.015	0.005	0.688	0.015	0.007	0.065	0.080	0.051
$\frac{5}{16}$	0.312	W	0.375	0.015	0.005	0.875	0.030	0.007	0.083	0.104	0.064
$\frac{3}{8}$	0.375	N	0.406	0.015	0.005	0.812	0.015	0.007	0.065	0.080	0.051
$\frac{3}{8}$	0.375	W	0.438	0.015	0.005	1.000	0.030	0.007	0.083	0.104	0.064
$\frac{7}{16}$	0.438	N	0.469	0.015	0.005	0.922	0.015	0.007	0.065	0.080	0.051
$\frac{7}{16}$	0.438	W	0.500	0.015	0.005	1.250	0.030	0.007	0.083	0.104	0.064
$\frac{1}{2}$	0.500	N	0.531	0.015	0.005	1.062	0.030	0.007	0.095	0.121	0.074
$\frac{1}{2}$	0.500	W	0.562	0.015	0.005	1.375	0.030	0.007	0.109	0.132	0.086
$\frac{9}{16}$	0.562	N	0.594	0.015	0.005	1.156*	0.030	0.007	0.095	0.121	0.074
$\frac{9}{16}$	0.562	W	0.625	0.015	0.005	1.469*	0.030	0.007	0.109	0.132	0.086
$\frac{5}{8}$	0.625	N	0.656	0.030	0.007	1.312	0.030	0.007	0.095	0.121	0.074
$\frac{5}{8}$	0.625	W	0.688	0.030	0.007	1.750	0.030	0.007	0.134	0.160	0.108
$\frac{3}{4}$	0.750	N	0.812	0.030	0.007	1.469	0.030	0.007	0.134	0.160	0.108
$\frac{3}{4}$	0.750	W	0.812	0.030	0.007	2.000	0.030	0.007	0.148	0.177	0.122
$\frac{7}{8}$	0.875	N	0.938	0.030	0.007	1.750	0.030	0.007	0.134	0.160	0.108
$\frac{7}{8}$	0.875	W	0.938	0.030	0.007	2.250	0.030	0.007	0.165	0.192	0.136
1	1.000	N	1.062	0.030	0.007	2.000	0.030	0.007	0.134	0.160	0.108
1	1.000	W	1.062	0.030	0.007	2.500	0.030	0.007	0.165	0.192	0.136
$1\frac{1}{8}$	1.125	N	1.250	0.030	0.007	2.250	0.030	0.007	0.134	0.160	0.108
$1\frac{1}{8}$	1.125	W	1.250	0.030	0.007	2.750	0.030	0.007	0.165	0.192	0.136
$1\frac{1}{4}$	1.250	N	1.375	0.030	0.007	2.500	0.030	0.007	0.165	0.192	0.136
$1\frac{1}{4}$	1.250	W	1.375	0.030	0.007	3.000	0.030	0.007	0.165	0.192	0.136
$1\frac{3}{8}$	1.375	N	1.500	0.030	0.007	2.750	0.030	0.007	0.165	0.192	0.136
$1\frac{3}{8}$	1.375	W	1.500	0.045	0.010	3.250	0.045	0.010	0.180	0.213	0.153
$1\frac{1}{2}$	1.500	N	1.625	0.030	0.007	3.000	0.030	0.007	0.165	0.192	0.136
$1\frac{1}{2}$	1.500	W	1.625	0.045	0.010	3.500	0.045	0.010	0.180	0.213	0.153
$1\frac{5}{8}$	1.625		1.750	0.045	0.010	3.750	0.045	0.010	0.180	0.213	0.153
$1\frac{3}{4}$	1.750		1.875	0.045	0.010	4.000	0.045	0.010	0.180	0.213	0.153
$1\frac{7}{8}$	1.875		2.000	0.045	0.010	4.250	0.045	0.010	0.180	0.213	0.153
2	2.000		2.125	0.045	0.010	4.500	0.045	0.010	0.180	0.213	0.153
$2\frac{1}{4}$	2.250		2.375	0.045	0.010	4.750	0.045	0.010	0.220	0.248	0.193
$2\frac{1}{2}$	2.500		2.625	0.045	0.010	5.000	0.045	0.010	0.238	0.280	0.210
$2\frac{3}{4}$	2.750		2.875	0.065	0.010	5.250	0.065	0.010	0.259	0.310	0.228
3	3.000		3.125	0.065	0.010	5.500	0.065	0.010	0.284	0.327	0.249

*The 0.734 in., 1.156 in., and 1.469 in. outside diameters avoid washers which could be used in coin operated devices.
**Preferred sizes are for the most part from series previously designated "Standard Plate" and "SAE." Where common sizes existed in the two series, the SAE size is designated "N" (narrow) and the Standard Plate "W" (wide). These sizes as well as all other sizes of Type A Plain Washers are to be ordered by ID, OD, and thickness dimensions.
***Nominal washer sizes are intended for use with comparable nominal screw or bolt sizes.

TABLE 30   From The American Society of Mechanical Engineers—ASME/ANSI B18.22.1–1965 (R1990)

AMERICAN NATIONAL STANDARD
LOCK WASHERS

ASME/ANSI B18.21.1–1994

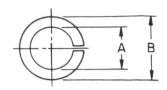

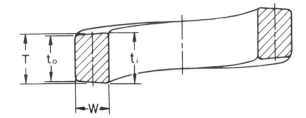

ENLARGED SECTION

DIMENSIONS OF REGULAR HELICAL SPRING LOCK WASHERS[1]

Nominal Washer Size		A Inside Diameter		B Outside Diameter	T Mean Section Thickness $\left(\dfrac{t_i + t_o}{2}\right)$	W Section Width
		Max	Min	Max[2]	Min	Min
No. 4	0.112	0.120	0.114	0.173	0.022	0.022
No. 5	0.125	0.133	0.127	0.202	0.030	0.030
No. 6	0.138	0.148	0.141	0.216	0.030	0.030
No. 8	0.164	0.174	0.167	0.267	0.047	0.042
No. 10	0.190	0.200	0.193	0.294	0.047	0.042
¼	0.250	0.262	0.254	0.365	0.078	0.047
⁵⁄₁₆	0.312	0.326	0.317	0.460	0.093	0.062
³⁄₈	0.375	0.390	0.380	0.553	0.125	0.076
⁷⁄₁₆	0.438	0.455	0.443	0.647	0.140	0.090
½	0.500	0.518	0.506	0.737	0.172	0.103
⅝	0.625	0.650	0.635	0.923	0.203	0.125
¾	0.750	0.775	0.760	1.111	0.218	0.154
⅞	0.875	0.905	0.887	1.296	0.234	0.182
1	1.000	1.042	1.017	1.483	0.250	0.208
1⅛	1.125	1.172	1.144	1.669	0.313	0.236
1¼	1.250	1.302	1.271	1.799	0.313	0.236
1⅜	1.375	1.432	1.398	2.041	0.375	0.292
1½	1.500	1.561	1.525	2.170	0.375	0.292
1¾	1.750	1.811	1.775	2.602	0.469	0.383
2	2.000	2.061	2.025	2.852	0.469	0.383
2¼	2.250	2.311	2.275	3.352	0.508	0.508
2½	2.500	2.561	2.525	3.602	0.508	0.508
2¾	2.750	2.811	2.775	4.102	0.633	0.633
3	3.000	3.061	3.025	4.352	0.633	0.633

[1]For use with 1960 Series Socket Head Cap Screws specified in American National Standard, ANSI B18.3.
[2]The maximum outside diameters specified allow for the commercial tolerances on cold-drawn wire.

TABLE 31   From The American Society of Mechanical Engineers—ASME/ANSI B18.21.1–1994

AMERICAN NATIONAL STANDARD
LOCK WASHERS

ASME/ANSI B18.21.1–1994

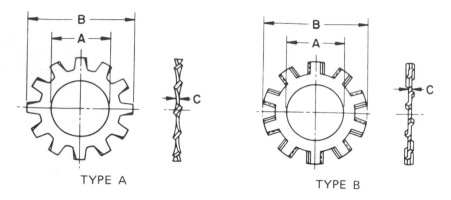

TYPE A   TYPE B

DIMENSIONS OF EXTERNAL TOOTH LOCK WASHERS

Nominal Washer Size		A		B		C	
		Inside Diameter		Outside Diameter		Thickness	
		Max	Min	Max	Min	Max	Min
No. 3	0.099	0.109	0.102	0.235	0.220	0.015	0.012
No. 4	0.112	0.123	0.115	0.260	0.245	0.019	0.015
No. 5	0.125	0.136	0.129	0.285	0.270	0.019	0.014
No. 6	0.138	0.150	0.141	0.320	0.305	0.022	0.016
No. 8	0.164	0.176	0.168	0.381	0.365	0.023	0.018
No. 10	0.190	0.204	0.195	0.410	0.395	0.025	0.020
No. 12	0.216	0.231	0.221	0.475	0.460	0.028	0.023
1/4	0.250	0.267	0.256	0.510	0.494	0.028	0.023
5/16	0.312	0.332	0.320	0.610	0.588	0.034	0.028
3/8	0.375	0.398	0.384	0.694	0.670	0.040	0.032
7/16	0.438	0.464	0.448	0.760	0.740	0.040	0.032
1/2	0.500	0.530	0.513	0.900	0.880	0.045	0.037
9/16	0.562	0.596	0.576	0.985	0.960	0.045	0.037
5/8	0.625	0.663	0.641	1.070	1.045	0.050	0.042
11/16	0.688	0.728	0.704	1.155	1.130	0.050	0.042
3/4	0.750	0.795	0.768	1.260	1.220	0.055	0.047
13/16	0.812	0.861	0.833	1.315	1.290	0.055	0.047
7/8	0.875	0.927	0.897	1.410	1.380	0.060	0.052
1	1.000	1.060	1.025	1.620	1.590	0.067	0.059

TABLE 32   From The American Society of Mechanical Engineers—ASME/ANSI B18.21.1–1994

AMERICAN NATIONAL STANDARD
LOCK WASHERS

ASME/ANSI B18.21.1–1994

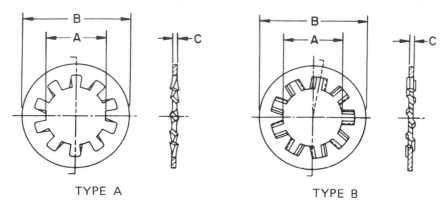

TYPE A   TYPE B

DIMENSIONS OF INTERNAL TOOTH LOCK WASHERS

Nominal Washer Size		A Inside Diameter		B Outside Diameter		C Thickness	
		Max	Min	Max	Min	Max	Min
No. 2	0.086	0.095	0.089	0.200	0.175	0.015	0.010
No. 3	0.099	0.109	0.102	0.232	0.215	0.019	0.012
No. 4	0.112	0.123	0.115	0.270	0.255	0.019	0.015
No. 5	0.125	0.136	0.129	0.280	0.245	0.021	0.017
No. 6	0.138	0.150	0.141	0.295	0.275	0.021	0.017
No. 8	0.164	0.176	0.168	0.340	0.325	0.023	0.018
No. 10	0.190	0.204	0.195	0.381	0.365	0.025	0.020
No. 12	0.216	0.231	0.221	0.410	0.394	0.025	0.020
¼	0.250	0.267	0.256	0.478	0.460	0.028	0.023
⁵⁄₁₆	0.312	0.332	0.320	0.610	0.594	0.034	0.028
³⁄₈	0.375	0.398	0.384	0.692	0.670	0.040	0.032
⁷⁄₁₆	0.438	0.464	0.448	0.789	0.740	0.040	0.032
½	0.500	0.530	0.512	0.900	0.867	0.045	0.037
⁹⁄₁₆	0.562	0.596	0.576	0.985	0.957	0.045	0.037
⅝	0.625	0.663	0.640	1.071	1.045	0.050	0.042
¹¹⁄₁₆	0.688	0.728	0.704	1.166	1.130	0.050	0.042
¾	0.750	0.795	0.769	1.245	1.220	0.055	0.047
¹³⁄₁₆	0.812	0.861	0.832	1.315	1.290	0.055	0.047
⅞	0.875	0.927	0.894	1.410	1.364	0.060	0.052
1	1.000	1.060	1.019	1.637	1.590	0.067	0.059
1⅛	1.125	1.192	1.144	1.830	1.799	0.067	0.059
1¼	1.250	1.325	1.275	1.975	1.921	0.067	0.059

TABLE 33   From The American Society of Mechanical Engineers—ASME/ANSI B18.21.1–1994

AMERICAN NATIONAL STANDARD SQUARE AND HEX NUTS

DIMENSIONS OF HEX NUTS AND HEX JAM NUTS

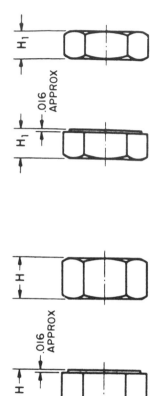

Nominal Size or Basic Major Dia. of Thread		F Width Across Flats			G Width Across Corners		H Thickness Hex Nuts			H₁ Thickness Hex Jam Nuts			Runout of Bearing Face, FIR Max		
													Hex Nuts Specified Proof Load		Jam Nuts All Strength Levels
		Basic	Max	Min	Max	Min	Basic	Max	Min	Basic	Max	Min	Up to 150,000 psi	150,000 psi and Greater	
1/4	0.2500	7/16	0.438	0.428	0.505	0.488	7/32	0.226	0.212	5/32	0.163	0.150	0.015	0.010	0.015
5/16	0.3125	1/2	0.500	0.489	0.577	0.557	17/64	0.273	0.258	3/16	0.195	0.180	0.016	0.011	0.016
3/8	0.3750	9/16	0.562	0.551	0.650	0.628	21/64	0.337	0.320	7/32	0.227	0.210	0.017	0.012	0.017
7/16	0.4375	11/16	0.688	0.675	0.794	0.768	3/8	0.385	0.365	1/4	0.260	0.240	0.018	0.013	0.018
1/2	0.5000	3/4	0.750	0.736	0.866	0.840	7/16	0.448	0.427	5/16	0.323	0.302	0.019	0.014	0.019
9/16	0.5625	7/8	0.875	0.861	1.010	0.982	31/64	0.496	0.473	5/16	0.324	0.301	0.020	0.015	0.020
5/8	0.6250	15/16	0.938	0.922	1.083	1.051	35/64	0.559	0.535	3/8	0.387	0.363	0.021	0.016	0.021
3/4	0.7500	1 1/8	1.125	1.088	1.299	1.240	41/64	0.665	0.617	27/64	0.446	0.398	0.023	0.018	0.023
7/8	0.8750	1 5/16	1.312	1.269	1.516	1.447	3/4	0.776	0.724	31/64	0.510	0.458	0.025	0.020	0.025
1	1.0000	1 1/2	1.500	1.450	1.732	1.653	55/64	0.887	0.831	35/64	0.575	0.519	0.027	0.022	0.027
1 1/8	1.1250	1 11/16	1.688	1.631	1.949	1.859	31/32	0.999	0.939	39/64	0.639	0.579	0.030	0.025	0.030
1 1/4	1.2500	1 7/8	1.875	1.812	2.165	2.066	1 1/16	1.094	1.030	23/32	0.751	0.687	0.033	0.028	0.033
1 3/8	1.3750	2 1/16	2.062	1.994	2.382	2.273	1 11/64	1.206	1.138	25/32	0.815	0.747	0.036	0.031	0.036
1 1/2	1.5000	2 1/4	2.250	2.175	2.598	2.480	1 9/32	1.317	1.245	27/32	0.880	0.808	0.039	0.034	0.039
See Notes	9	3			4									2	

TABLE 34   From The American Society of Mechanical Engineers—ASME/ANSI B18.2.2–1987 (R1993)

**AMERICAN NATIONAL STANDARD
SQUARE AND HEX NUTS**

ASME/ANSI B18.2.2–1987 (R1993)

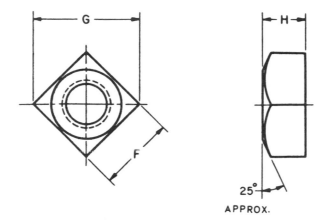

25°
APPROX.

DIMENSIONS OF SQUARE NUTS

Nominal Size or Basic Major Dia. of Thread		F Width Across Flats			G Width Across Corners		H Thickness		
		Basic	Max	Min	Max	Min	Basic	Max	Min
1/4	0.2500	7/16	0.438	0.425	0.619	0.584	7/32	0.235	0.203
5/16	0.3125	9/16	0.562	0.547	0.795	0.751	17/64	0.283	0.249
3/8	0.3750	5/8	0.625	0.606	0.884	0.832	21/64	0.346	0.310
7/16	0.4375	3/4	0.750	0.728	1.061	1.000	3/8	0.394	0.356
1/2	0.5000	13/16	0.812	0.788	1.149	1.082	7/16	0.458	0.418
5/8	0.6250	1	1.000	0.969	1.414	1.330	35/64	0.569	0.525
3/4	0.7500	1 1/8	1.125	1.088	1.591	1.494	21/32	0.680	0.632
7/8	0.8750	1 5/16	1.312	1.269	1.856	1.742	49/64	0.792	0.740
1	1.0000	1 1/2	1.500	1.450	2.121	1.991	7/8	0.903	0.847
1 1/8	1.1250	1 11/16	1.688	1.631	2.386	2.239	1	1.030	0.970
1 1/4	1.2500	1 7/8	1.875	1.812	2.652	2.489	1 3/32	1.126	1.062
1 3/8	1.3750	2 1/16	2.062	1.994	2.917	2.738	1 13/64	1.237	1.169
1 1/2	1.5000	2 1/4	2.250	2.175	3.182	2.986	1 5/16	1.348	1.276
See Notes	8	3							

TABLE 35   From The American Society of Mechanical Engineers—ASME/ANSI B18.2.2-1987 (R1993)

KEY AND KEYWAY SIZES

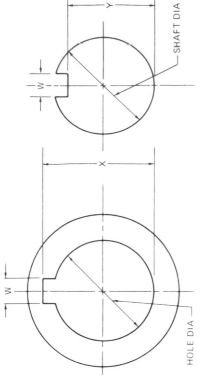

KEY SIZE = W x H
LENGTH (L) TO SUIT

Nom. Size	– DIA. – (Shaft)		'X' (Collar)		'Y' (Shaft)	
(Inch)	Inch	mm	Inch	mm	Inch	mm
1/2	.500	12.700	.560	14.224	.430	10.922
9/16	.562	14.290	.623	15.824	.493	12.522
5/8	.625	15.875	.709	18.008	.517	13.132
11/16	.688	17.470	.773	18.618	.581	14.757
3/4	.750	19.050	.837	21.259	.644	16.357
13/16	.812	20.640	.900	22.860	.708	17.983
7/8	.875	22.225	.964	24.485	.771	19.583
15/16	.938	23.820	1.051	26.695	.791	20.091
1	1.000	25.400	1.114	28.295	.850	21.818
1 1/16	1.062	26.985	1.178	29.921	.923	23.444
1 1/8	1.125	28.575	1.241	31.521	.986	25.044
1 3/16	1.188	30.165	1.304	33.121	1.049	26.644
1 1/4	1.250	31.750	1.367	34.722	1.112	28.244
1 5/16	1.312	33.340	1.455	36.957	1.137	28.879
1 3/8	1.375	34.923	1.518	38.557	1.201	30.505

From *Drafting for Trades and Industry—Mechanical and Electronic.* Nelson. Delmar Publishers Inc.

Shaft Nom. Size – DIA. –		Square (W = H)	Type	Square Key		Tolerance
From	To & Incl.			From	To & Incl.	
5/16 (8)	7/16 (11)	3/32 (2.38)	Bar Stock	—	3/4 (19.05)	+.000 – .002 (+.0000 – .0254)
7/16 (11)	9/16 (14)	1/8 (3.175)		3/4 (19.05)	1 1/2 (38.1)	+.000 – .003 (+.0000 – .0762)
9/16 (14)	7/8 (22)	3/16 (4.76)		1 1/2 (38.1)	2 1/2 (63.5)	+.000 – .004 (+.0000 – .1016)
7/8 (22)	1 1/4 (32)	1/4 (6.35)		2 1/2 (63.5)	3 1/2 (88.9)	+.000 – .006 (+.0000 – .1524)
1 1/4 (32)	1 3/8 (35)	5/16 (7.94)	Keystock	—	1 1/4 (31.75)	+.001 – .000 (+.0254 – .0000)
1 3/8 (35)	1 3/4 (44)	3/8 (9.53)		1 1/4 (31.75)	3 (76.2)	+.002 – .000 (+.0508 – .0000)
1 3/4 (44)	2 1/4 (57)	1/2 (12.7)		3 (76.2)	3 1/2 (88.9)	+.003 – .000 (+.0762 – .0000)
2 1/4 (57)	2 3/4 (70)	5/8 (15.88)				
2 3/4 (70)	3 1/4 (82)	3/4 (19.05)				
3 1/4 (82)	3 3/4 (95)	7/8 (22.23)				

(Figures in parenthesis = mm)

TABLE 36 From *Drafting for Trades and Industry—Mechanical and Electronic,* Nelson. Delmar Publishers Inc.

USA STANDARD

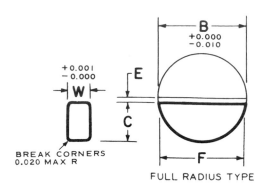

FULL RADIUS TYPE

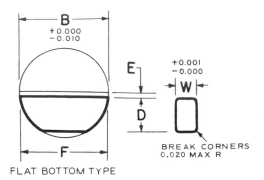

FLAT BOTTOM TYPE

WOODRUFF KEYS

Key No.	Nominal Key Size W × B	Actual Length F +0.000-0.010	Height of Key				Distance Below Center E
			C		D		
			Max	Min	Max	Min	
202	¹⁄₁₆ × ¼	0.248	0.109	0.104	0.109	0.104	¹⁄₆₄
202.5	¹⁄₁₆ × ⁵⁄₁₆	0.311	0.140	0.135	0.140	0.135	¹⁄₆₄
302.5	³⁄₃₂ × ⁵⁄₁₆	0.311	0.140	0.135	0.140	0.135	¹⁄₆₄
203	¹⁄₁₆ × ⅜	0.374	0.172	0.167	0.172	0.167	¹⁄₆₄
303	³⁄₃₂ × ⅜	0.374	0.172	0.167	0.172	0.167	¹⁄₆₄
403	⅛ × ⅜	0.374	0.172	0.167	0.172	0.167	¹⁄₆₄
204	¹⁄₁₆ × ½	0.491	0.203	0.198	0.194	0.188	³⁄₆₄
304	³⁄₃₂ × ½	0.491	0.203	0.198	0.194	0.188	³⁄₆₄
404	⅛ × ½	0.491	0.203	0.198	0.194	0.188	³⁄₆₄
305	³⁄₃₂ × ⅝	0.612	0.250	0.245	0.240	0.234	¹⁄₁₆
405	⅛ × ⅝	0.612	0.250	0.245	0.240	0.234	¹⁄₁₆
505	⁵⁄₃₂ × ⅝	0.612	0.250	0.245	0.240	0.234	¹⁄₁₆
605	³⁄₁₆ × ⅝	0.612	0.250	0.245	0.240	0.234	¹⁄₁₆
406	⅛ × ¾	0.740	0.313	0.308	0.303	0.297	¹⁄₁₆
506	⁵⁄₃₂ × ¾	0.740	0.313	0.308	0.303	0.297	¹⁄₁₆
606	³⁄₁₆ × ¾	0.740	0.313	0.308	0.303	0.297	¹⁄₁₆
806	¼ × ¾	0.740	0.313	0.308	0.303	0.297	¹⁄₁₆
507	⁵⁄₃₂ × ⅞	0.866	0.375	0.370	0.365	0.359	¹⁄₁₆
607	³⁄₁₆ × ⅞	0.866	0.375	0.370	0.365	0.359	¹⁄₁₆
707	⁷⁄₃₂ × ⅞	0.866	0.375	0.370	0.365	0.359	¹⁄₁₆
807	¼ × ⅞	0.866	0.375	0.370	0.365	0.359	¹⁄₁₆
608	³⁄₁₆ × 1	0.992	0.438	0.433	0.428	0.422	¹⁄₁₆
708	⁷⁄₃₂ × 1	0.992	0.438	0.433	0.428	0.422	¹⁄₁₆
808	¼ × 1	0.992	0.438	0.433	0.428	0.422	¹⁄₁₆
1008	⁵⁄₁₆ × 1	0.992	0.438	0.433	0.428	0.422	¹⁄₁₆
1208	⅜ × 1	0.992	0.438	0.433	0.428	0.422	¹⁄₁₆
609	³⁄₁₆ × 1⅛	1.114	0.484	0.479	0.475	0.469	⁵⁄₆₄
709	⁷⁄₃₂ × 1⅛	1.114	0.484	0.479	0.475	0.469	⁵⁄₆₄
809	¼ × 1⅛	1.114	0.484	0.479	0.475	0.469	⁵⁄₆₄
1009	⁵⁄₁₆ × 1⅛	1.114	0.484	0.479	0.475	0.469	⁵⁄₆₄

TABLE 37

WOODRUFF KEYS (CONCLUDED)

Key No.	Nominal Key Size W × B	Actual Length F +0.000-0.010	Height of Key				Distance Below Center E
			C		D		
			Max	Min	Max	Min	
610	³⁄₁₆ × 1¼	1.240	0.547	0.542	0.537	0.531	⁵⁄₆₄
710	⁷⁄₃₂ × 1¼	1.240	0.547	0.542	0.537	0.531	⁵⁄₆₄
810	¼ × 1¼	1.240	0.547	0.542	0.537	0.531	⁵⁄₆₄
1010	⁵⁄₁₆ × 1¼	1.240	0.547	0.542	0.537	0.531	⁵⁄₆₄
1210	³⁄₈ × 1¼	1.240	0.547	0.542	0.537	0.531	⁵⁄₆₄
811	¼ × 1³⁄₈	1.362	0.594	0.589	0.584	0.578	³⁄₃₂
1011	⁵⁄₁₆ × 1³⁄₈	1.362	0.594	0.589	0.584	0.578	³⁄₃₂
1211	³⁄₈ × 1³⁄₈	1.362	0.594	0.589	0.584	0.578	³⁄₃₂
812	¼ × 1½	1.484	0.641	0.636	0.631	0.625	⁷⁄₆₄
1012	⁵⁄₁₆ × 1½	1.484	0.641	0.636	0.631	0.625	⁷⁄₆₄
1212	³⁄₈ × 1½	1.484	0.641	0.636	0.631	0.625	⁷⁄₆₄

All dimensions given are in inches.

The key numbers indicate nominal key dimensions. The last two digits give the nominal diameter B in eighths of an inch and the digits preceding the last two give the nominal width W in thirty-seconds of an inch.

Example:

No. 204 indicates a key $\frac{2}{32} \times \frac{4}{8}$ or $\frac{1}{16} \times \frac{1}{2}$.

No. 808 indicates a key $\frac{8}{32} \times \frac{8}{8}$ or $\frac{1}{4} \times 1$.

No. 1212 indicates a key $\frac{12}{32} \times \frac{12}{8}$ or $\frac{3}{8} \times 1\frac{1}{2}$.

TABLE 37   *(Concluded)* From The American Society of Mechanical Engineers—ANSI B17.2–1967 (R1990)

WOODRUFF KEYS AND KEYSEATS

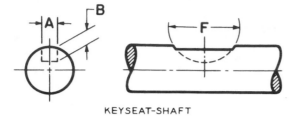

KEYSEAT-SHAFT

KEY ABOVE
SHAFT

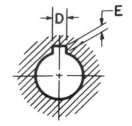

KEYSEAT-HUB

Key Number	Nominal Size Key	Keyseat — Shaft					Key Above Shaft	Keyseat — Hub	
		Width A*		Depth B	Diameter F		Height C	Width D	Depth E
		Min	Max	+0.005 −0.000	Min	Max	+0.005 −0.005	+0.002 −0.000	+0.005 −0.000
202	1/16 × 1/4	0.0615	0.0630	0.0728	0.250	0.268	0.0312	0.0635	0.0372
202.5	1/16 × 5/16	0.0615	0.0630	0.1038	0.312	0.330	0.0312	0.0635	0.0372
302.5	3/32 × 5/16	0.0928	0.0943	0.0882	0.312	0.330	0.0469	0.0948	0.0529
203	1/16 × 3/8	0.0615	0.0630	0.1358	0.375	0.393	0.0312	0.0635	0.0372
303	3/32 × 3/8	0.0928	0.0943	0.1202	0.375	0.393	0.0469	0.0948	0.0529
403	1/8 × 3/8	0.1240	0.1255	0.1045	0.375	0.393	0.0625	0.1260	0.0685
204	1/16 × 1/2	0.0615	0.0630	0.1668	0.500	0.518	0.0312	0.0635	0.0372
304	3/32 × 1/2	0.0928	0.0943	0.1511	0.500	0.518	0.0469	0.0948	0.0529
404	1/8 × 1/2	0.1240	0.1255	0.1355	0.500	0.518	0.0625	0.1260	0.0685
305	3/32 × 5/8	0.0928	0.0943	0.1981	0.625	0.643	0.0469	0.0948	0.0529
405	1/8 × 5/8	0.1240	0.1255	0.1825	0.625	0.643	0.0625	0.1260	0.0685
505	5/32 × 5/8	0.1553	0.1568	0.1669	0.625	0.643	0.0781	0.1573	0.0841
605	3/16 × 5/8	0.1863	0.1880	0.1513	0.625	0.643	0.0937	0.1885	0.0997
406	1/8 × 3/4	0.1240	0.1255	0.2455	0.750	0.768	0.0625	0.1260	0.0685
506	5/32 × 3/4	0.1553	0.1568	0.2299	0.750	0.768	0.0781	0.1573	0.0841
606	3/16 × 3/4	0.1863	0.1880	0.2143	0.750	0.768	0.0937	0.1885	0.0997
806	1/4 × 3/4	0.2487	0.2505	0.1830	0.750	0.768	0.1250	0.2510	0.1310
507	5/32 × 7/8	0.1553	0.1568	0.2919	0.875	0.895	0.0781	0.1573	0.0841
607	3/16 × 7/8	0.1863	0.1880	0.2763	0.875	0.895	0.0937	0.1885	0.0997
707	7/32 × 7/8	0.2175	0.2193	0.2607	0.875	0.895	0.1093	0.2198	0.1153
807	1/4 × 7/8	0.2487	0.2505	0.2450	0.875	0.895	0.1250	0.2510	0.1310
608	3/16 × 1	0.1863	0.1880	0.3393	1.000	1.020	0.0937	0.1885	0.0997
708	7/32 × 1	0.2175	0.2193	0.3237	1.000	1.020	0.1093	0.2198	0.1153
808	1/4 × 1	0.2487	0.2505	0.3080	1.000	1.020	0.1250	0.2510	0.1310
1008	5/16 × 1	0.3111	0.3130	0.2768	1.000	1.020	0.1562	0.3135	0.1622
1208	3/8 × 1	0.3735	0.3755	0.2455	1.000	1.020	0.1875	0.3760	0.1935
609	3/16 × 1 1/8	0.1863	0.1880	0.3853	1.125	1.145	0.0937	0.1885	0.0997
709	7/32 × 1 1/8	0.2175	0.2193	0.3697	1.125	1.145	0.1093	0.2198	0.1153
809	1/4 × 1 1/8	0.2487	0.2505	0.3540	1.125	1.145	0.1250	0.2510	0.1310
1009	5/16 × 1 1/8	0.3111	0.3130	0.3228	1.125	1.145	0.1562	0.3135	0.1622

TABLE 38   From The American Society of Mechanical Engineers—ANSI B17.2–1967 (R1990)

KEYS AND KEYSEATS

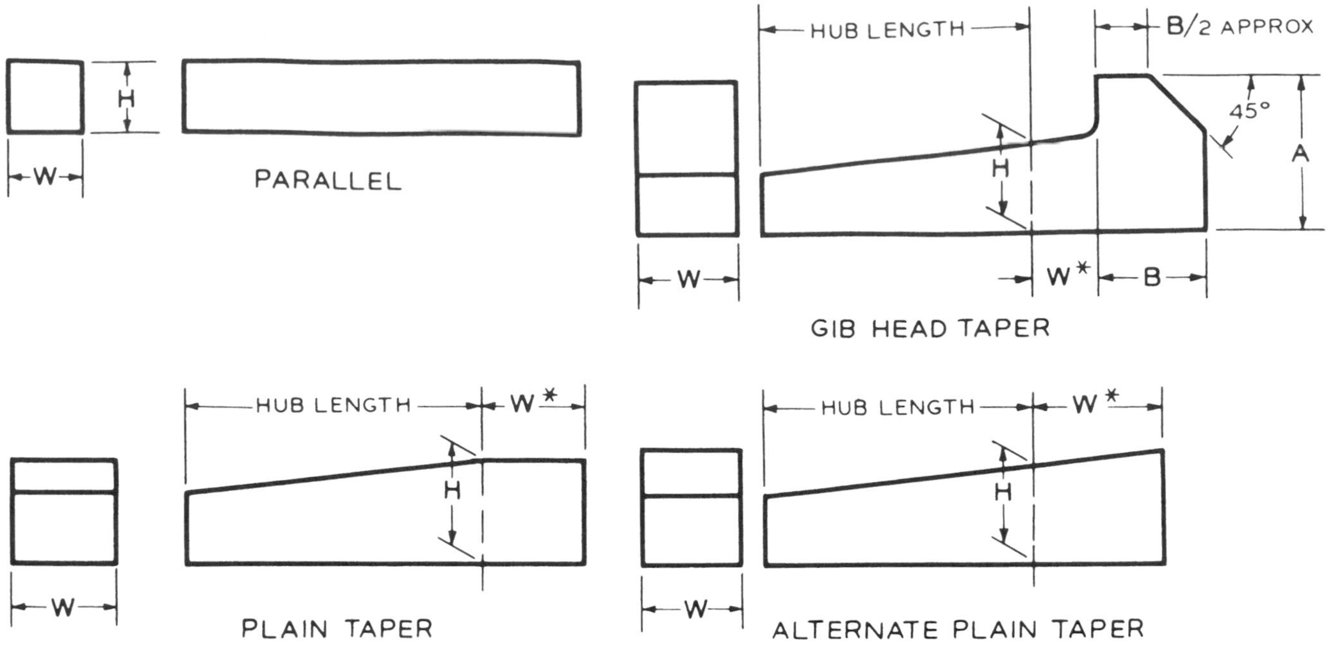

PARALLEL

GIB HEAD TAPER

PLAIN TAPER

ALTERNATE PLAIN TAPER

Plain and Gib Head Taper Keys Have a 1/8″ Taper in 12″

KEY DIMENSIONS AND TOLERANCES

KEY			NOMINAL KEY SIZE		TOLERANCE	
			Width, *W*		Width, *W*	Height, *H*
			Over	To (Incl)		
Parallel	Square	Bar Stock	—	3/4	+0.000 −0.002	+0.000 −0.002
			3/4	1-1/2	+0.000 −0.003	+0.000 −0.003
			1-1/2	2-1/2	+0.000 −0.004	+0.000 −0.004
			2-1/2	3-1/2	+0.000 −0.006	+0.000 −0.006
		Keystock	—	1-1/4	+0.001 −0.000	+0.001 −0.000
			1-1/4	3	+0.002 −0.000	+0.002 −0.000
			3	3-1/2	+0.003 −0.000	+0.003 −0.000
	Rectangular	Bar Stock	—	3/4	+0.000 −0.003	+0.000 −0.003
			3/4	1-1/2	+0.000 −0.004	+0.000 −0.004
			1-1/2	3	+0.000 −0.005	+0.000 −0.005
			3	4	+0.000 −0.006	+0.000 −0.006
			4	6	+0.000 −0.008	+0.000 −0.008
			6	7	+0.000 −0.013	+0.000 −0.013
		Keystock	—	1-1/4	+0.001 −0.000	+0.005 −0.005
			1-1/4	3	+0.002 −0.000	+0.005 −0.005
			3	7	+0.003 −0.000	+0.005 −0.005
Taper	Plain or Gib Head Square or Rectangular		—	1-1/4	+0.001 −0.000	+0.005 −0.000
			1-1/4	3	+0.002 −0.000	+0.005 −0.000
			3	7	+0.003 −0.000	+0.005 −0.000

*For locating position of dimension H. Tolerance does not apply.
See Table 41 for dimensions on gib heads.
All dimensions given in inches.

TABLE 39   From The American Society of Mechanical Engineers—ANSI B17.1–1967 (R1989)

WOODRUFF KEY SIZES FOR DIFFERENT SHAFT DIAMETERS

Shaft Diameter	5/16 to 3/8	7/16 to 1/2	9/16 to 3/4	13/16 to 15/16	1 to 1 3/16	1 1/4 to 1 7/16	1 1/2 to 1 3/4	1 13/16 to 2 1/8	2 3/16 to 2 1/2
Key Numbers	204	304 305	404 405 406	505 506 507	606 607 608 609	807 808 809	810 811 812	1011 1012	1211 1212

TABLE 40

USA STANDARD

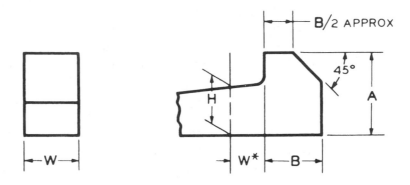

GIB HEAD NOMINAL DIMENSIONS

Nominal Key Size Width, W	SQUARE			RECTANGULAR		
	H	A	B	H	A	B
1/8	1/8	1/4	1/4	3/32	3/16	1/8
3/16	3/16	5/16	5/16	1/8	1/4	1/4
1/4	1/4	7/16	3/8	3/16	5/16	5/16
5/16	5/16	1/2	7/16	1/4	7/16	3/8
3/8	3/8	5/8	1/2	1/4	7/16	3/8
1/2	1/2	7/8	5/8	3/8	5/8	1/2
5/8	5/8	1	3/4	7/16	3/4	9/16
3/4	3/4	1-1/4	7/8	1/2	7/8	5/8
7/8	7/8	1-3/8	1	5/8	1	3/4
1	1	1-5/8	1-1/8	3/4	1-1/4	7/8
1-1/4	1-1/4	2	1-7/16	7/8	1-3/8	1
1-1/2	1-1/2	2-3/8	1-3/4	1	1-5/8	1-1/8
1-3/4	1-3/4	2-3/4	2	1-1/2	2-3/8	1-3/4
2	2	3-1/2	2-1/4	1-1/2	2-3/8	1-3/4
2-1/2	2-1/2	4	3	1-3/4	2-3/4	2
3	3	5	3-1/2	2	3-1/2	2-1/4
3-1/2	3-1/2	6	4	2-1/2	4	3

*For locating position of dimension *H*.

For larger sizes the following relationships are suggested as guides for establishing *A* and *B*.

$$A = 1.8\,H \qquad B = 1.2\,H$$

All dimensions given in inches.

TABLE 41 From The American Society of Mechanical Engineers—ANSI B17.1–1967 (R1989)

SHEET METAL AND WIRE GAGE DESIGNATION

GAGE NO.	AMERICAN OR BROWN & SHARPE'S A.W.G. OR B. & S.	UNITED STATES STANDARD	MANU-FACTURERS' STANDARD FOR SHEET STEEL	GAGE NO.
0000000	. . . .	.500	. . . .	0000000
000000	.5800	.469	. . . .	000000
00000	.5165	.438	. . . .	00000
0000	.4600	.406	. . . .	0000
000	.4096	.375	. . . .	000
00	.3648	.344	. . . .	00
0	.3249	.312	. . . .	0
1	.2893	.281	. . . .	1
2	.2576	.266	. . . .	2
3	.2294	.250	.2391	3
4	.2043	.234	.2242	4
5	.1819	.219	.2092	5
6	.1620	.203	.1943	6
7	.1443	.188	.1793	7
8	.1285	.172	.1644	8
9	.1144	.156	.1495	9
10	.1019	.141	.1345	10
11	.0907	.125	.1196	11
12	.0808	.109	.1046	12
13	.0720	.0938	.0897	13
14	.0642	.0781	.0747	14
15	.0571	.0703	.0673	15
16	.0508	.0625	.0598	16
17	.0453	.0562	.0538	17
18	.0403	.0500	.0478	18
19	.0359	.0438	.0418	19
20	.0320	.0375	.0359	20
21	.0285	.0344	.0329	21
22	.0253	.0312	.0299	22
23	.0226	.0281	.0269	23
24	.0201	.0250	.0239	24
25	.0179	.0219	.0209	25
26	.0159	.0188	.0179	26
27	.0142	.0172	.0164	27
28	.0126	.0156	.0149	28
29	.0113	.0141	.0135	29
30	.0100	.0125	.0120	30
31	.0089	.0109	.0105	31
32	.0080	.0102	.0097	32
33	.0071	.00938	.0090	33
34	.0063	.00859	.0082	34
35	.0056	.00781	.0075	35
36	.0050	.00703	.0067	36

TABLE 42

BEND ALLOWANCE FOR 90° BENDS (INCH)

Radii / Thickness	.031	.063	.094	.125	.156	.188	.219	.250	.281	.313	.344	.375	.438	.500	.531	.625
.013	.058	.108	.157	.205	.254	.304	.353	.402	.450	.501	.549	.598	.697	.794	.843	.991
.016	.060	.110	.159	.208	.256	.307	.355	.404	.453	.503	.552	.600	.699	.796	.845	.993
.020	.062	.113	.161	.210	.259	.309	.358	.406	.455	.505	.554	.603	.702	.799	.848	.995
.022	.064	.114	.163	.212	.260	.311	.359	.408	.457	.507	.556	.604	.703	.801	.849	.997
.025	.066	.116	.165	.214	.263	.313	.362	.410	.459	.509	.558	.607	.705	.803	.851	.999
.028	.068	.119	.167	.216	.265	.315	.364	.412	.461	.511	.560	.609	.708	.805	.854	1.001
.032	.071	.121	.170	.218	.267	.317	.366	.415	.463	.514	.562	.611	.710	.807	.856	1.004
.038	.075	.126	.174	.223	.272	.322	.371	.419	.468	.518	.567	.616	.715	.812	.861	1.008
.040	.077	.127	.176	.224	.273	.323	.372	.421	.469	.520	.568	.617	.716	.813	.862	1.010
.050		.134	.183	.232	.280	.331	.379	.428	.477	.527	.576	.624	.723	.821	.869	1.017
.064		.144	.192	.241	.290	.340	.389	.437	.486	.536	.585	.634	.732	.830	.878	1.026
.072			.198	.247	.296	.346	.394	.443	.492	.542	.591	.639	.738	.836	.885	1.032
.078			.202	.251	.300	.350	.399	.447	.496	.546	.595	.644	.743	.840	.889	1.036
.081			.204	.253	.302	.352	.401	.449	.498	.548	.598	.646	.745	.842	.891	1.038
.091			.212	.260	.309	.359	.408	.456	.505	.555	.604	.653	.752	.849	.898	1.045
.094			.214	.262	.311	.361	.410	.459	.507	.558	.606	.655	.754	.851	.900	1.048
.102				.268	.317	.367	.416	.464	.513	.563	.612	.661	.760	.857	.906	1.053
.109				.273	.321	.372	.420	.469	.518	.568	.617	.665	.764	.862	.910	1.058
.125				.284	.333	.383	.432	.480	.529	.579	.628	.677	.776	.873	.922	1.069
.156					.355	.405	.453	.502	.551	.601	.650	.698	.797	.895	.943	1.091
.188						.427	.476	.525	.573	.624	.672	.721	.820	.917	.966	1.114
.203								.535	.584	.634	.683	.731	.830	.928	.976	1.124
.218								.546	.594	.645	.693	.742	.841	.938	.987	1.135
.234								.557	.606	.656	.705	.753	.852	.950	.998	1.146
.250								.568	.617	.667	.716	.764	.863	.961	1.009	1.157

EXAMPLE: MATERIAL THICKNESS = 1/8", INSIDE RADII = 1/4" R. WHERE THESE CROSS = .480" BEND ALLOWANCE (B/A) PLUS TOTAL OF STRAIGHT LENGTHS = DEVELOPED LENGTH.

BEND ALLOWANCE FOR 90° BENDS (MILLIMETER)

Radii / Thickness	0.80	1.58	2.38	3.18	3.96	4.76	5.56	6.35	7.15	7.94	8.74	9.52	11.12	12.70	13.50	15.85
.330	1.46	2.74	3.98	5.22	6.45	7.73	8.96	10.20	11.44	12.71	13.95	15.19	17.70	20.18	21.42	25.16
.406	1.52	2.80	4.04	5.27	6.51	7.78	9.02	10.26	11.50	12.77	14.01	15.24	17.76	20.23	21.47	25.22
.508	1.59	2.87	4.11	5.34	6.58	7.86	9.09	10.33	11.57	12.84	14.08	15.32	17.83	20.30	21.54	25.29
.559	1.63	2.91	4.14	5.38	6.62	7.89	9.13	10.37	11.60	12.88	14.12	15.35	17.86	20.34	21.57	25.32
.635	1.68	2.96	4.20	5.43	6.67	7.95	9.18	10.42	11.66	12.93	14.17	15.40	17.92	20.39	21.63	25.38
.711	1.74	3.02	4.25	5.49	6.72	8.00	9.24	10.47	11.71	12.99	14.22	15.46	17.98	20.44	21.68	25.43
.813	1.81	3.08	4.32	5.56	6.79	8.07	9.30	10.54	11.78	13.05	14.29	15.53	18.04	20.52	21.75	25.50
.965	1.91	3.19	4.42	5.66	6.90	8.18	9.41	10.65	11.89	13.16	14.40	15.64	18.15	20.62	21.86	25.61
1.016	1.95	2.23	4.46	5.70	6.94	8.21	9.45	10.69	11.92	13.20	14.44	15.67	18.19	20.66	21.90	25.64
1.270		3.40	4.64	5.88	7.11	8.39	9.63	10.86	12.10	13.38	14.61	15.85	18.36	20.84	22.07	25.82
1.625		3.65	4.89	6.13	7.36	8.64	9.88	11.11	12.35	13.63	14.86	16.10	18.61	21.08	22.32	26.07
1.829			5.03	6.27	7.51	8.78	10.02	11.26	12.49	13.77	15.01	16.24	18.76	21.23	22.47	26.22
1.981			5.14	6.38	7.61	8.89	10.13	11.36	12.60	13.88	15.11	16.35	18.86	21.34	22.57	26.32
2.058			5.19	6.43	7.67	8.94	10.18	11.42	12.65	13.93	15.17	16.40	18.92	21.39	22.63	26.38
2.311			5.37	6.60	7.85	9.12	10.36	11.60	12.83	14.11	15.35	16.58	19.09	21.57	22.80	26.55
2.388			5.43	6.66	7.90	9.18	10.41	11.65	12.89	14.16	15.40	16.64	19.15	21.62	22.86	26.61
2.591				6.81	8.04	9.31	10.55	11.79	13.03	14.30	15.54	16.78	19.29	21.76	23.00	26.75
2.769				6.93	8.17	9.44	10.68	11.92	13.15	14.43	15.67	16.90	19.42	21.89	23.13	26.88
3.175				7.22	8.45	9.73	10.96	12.20	13.44	14.71	15.95	17.19	19.70	22.17	23.41	27.16
3.962					9.00	10.28	11.52	12.75	13.99	15.27	16.74	17.74	20.25	22.72	23.96	27.71
4.775					9.58	10.85	12.09	13.32	14.56	15.84	17.07	18.31	20.82	23.30	24.53	28.28
5.156						11.12	12.36	13.59	14.83	16.11	17.34	18.58	21.09	23.57	24.80	28.55
5.537								13.85	15.10	16.37	17.61	18.85	21.36	23.83	25.07	28.82
5.944								14.15	15.38	16.66	17.89	19.13	21.64	24.12	25.35	29.10
6.350								14.43	15.67	16.94	18.18	19.42	21.93	24.40	25.64	29.39

EXAMPLE: MATERIAL THICKNESS = 3.18 MM, INSIDE RADII = 6.35 MM R. WHERE THESE CROSS = 12.20 MM BEND ALLOWANCE (B/A) PLUS TOTAL OF STRAIGHT LENGTHS = DEVELOPED LENGTH.

TABLE 43   From *Drafting for Trades and Industry—Basic Skills,* Nelson. Delmar Publishers Inc.

BEND ALLOWANCE FOR EACH 1° OF BEND (INCH)

Radii Thickness	.031	.063	.094	.125	.156	.188	.219	.250	.281	.313	.344	.375	.438	.500	.531	.625
.013	.00064	.00120	.00174	.00228	.00282	.00338	.00392	.00446	.00500	.00556	.00610	.00664	.00774	.00883	.00937	.01101
.016	.00067	.00122	.00176	.00231	.00285	.00342	.00395	.00449	.00503	.00559	.00613	.00667	.00777	.00885	.00939	.01103
.020	.00069	.00125	.00179	.00233	.00287	.00343	.00397	.00452	.00506	.00561	.00616	.00670	.00780	.00888	.00942	.01106
.022	.00071	.00127	.00181	.00235	.00289	.00345	.00399	.00453	.00508	.00563	.00617	.00672	.00782	.00890	.00944	.01108
.025	.00074	.00129	.00184	.00238	.00292	.00348	.00402	.00456	.00510	.00566	.00610	.00674	.00784	.00892	.00946	.01110
.028	.00076	.00132	.00186	.00240	.00294	.00350	.00404	.00458	.00512	.00568	.00622	.00676	.00786	.00894	.00948	.01112
.032	.00079	.00134	.00189	.00243	.00297	.00353	.00407	.00461	.00515	.00571	.00625	.00679	.00789	.00897	.00951	.01115
.038	.00084	.00140	.00194	.00248	.00302	.00358	.00412	.00466	.00520	.00576	.00630	.00684	.00794	.00902	.00946	.01120
.040	.00085	.00141	.00195	.00249	.00303	.00359	.00413	.00468	.00522	.00577	.00632	.00686	.00796	.00904	.00958	.01122
.050		.00149	.00203	.00258	.00312	.00368	.00422	.00476	.00530	.00586	.00640	.00694	.00804	.00912	.00966	.01130
.064		.00160	.00214	.00268	.00322	.00378	.00432	.00486	.00540	.00596	.00650	.00704	.00814	.00922	.00976	.01140
.072			.00220	.00274	.00328	.00384	.00438	.00492	.00546	.00602	.00656	.00710	.00820	.00929	.00983	.01147
.078			.00225	.00279	.00333	.00389	.00443	.00497	.00551	.00607	.00661	.00715	.00825	.00933	.00987	.01152
.081			.00227	.00281	.00335	.00391	.00445	.00499	.00554	.00609	.00664	.00718	.00828	.00936	.00990	.01154
.091			.00235	.00289	.00343	.00399	.00453	.00507	.00561	.00617	.00671	.00725	.00835	.00944	.00998	.01162
.094			.00237	.00291	.00346	.00401	.00456	.00510	.00564	.00620	.00674	.00728	.00838	.00946	.00999	.01164
.102				.00298	.00352	.00408	.00462	.00516	.00570	.00626	.00680	.00734	.00844	.00952	.01006	.01170
.109				.00303	.00357	.00413	.00467	.00521	.00575	.00631	.00685	.00739	.00849	.00958	.01012	.01176
.125				.00316	.00370	.00426	.00480	.00534	.00588	.00644	.00698	.00752	.00862	.00970	.01024	.01188
.156					.00394	.00450	.00504	.00558	.00612	.00668	.00722	.00776	.00886	.00994	.01048	.01212
.188						.00475	.00529	.00583	.00637	.00693	.00747	.00802	.00911	.01019	.01073	.01237
.203								.00595	.00649	.00704	.00759	.00813	.00923	.01031	.01085	.01249
.218								.00606	.00660	.00716	.00770	.00824	.00934	.01042	.01097	.01261
.234								.00619	.00673	.00729	.00783	.00837	.00947	.01055	.01109	.01273
.250								.00631	.00685	.00741	.00795	.00849	.00959	.01068	.01122	.01286

EXAMPLE: MATERIAL THICKNESS = 1/8", INSIDE RADII = 1/4" R. WHERE THESE CROSS = .00534" IF THE INSIDE OF YOUR BEND IS 20° THE BEND ALLOWANCE (B/A) = .1068 (.00534 × 20°) BEND ALLOWANCE B/A PLUS TOTAL OF STRAIGHT LENGTHS = DEVELOPED LENGTH.

BEND ALLOWANCE FOR EACH 1° OF BEND (MILLIMETER)

Radii Thickness	0.80	1.58	2.38	3.18	3.96	4.76	5.56	6.35	7.15	7.94	8.74	9.52	11.12	12.70	13.50	15.85
.330	.0161	.0305	.0442	.0580	.0717	.0859	.0996	.1134	.1271	.1413	.1550	.1668	.1967	.2242	.2379	.2796
.406	.0169	.0311	.0448	.0586	.0723	.0865	.1002	.1140	.1277	.1419	.1556	.1694	.1973	.2248	.2385	.2802
.508	.0177	.0319	.0456	.0594	.0731	.0873	.1010	.1148	.1285	.1427	.1564	.1702	.1981	.2256	.2393	.2810
.559	.0181	.0323	.0460	.0598	.0735	.0877	.1014	.1152	.1289	.1431	.1568	.1706	.1985	.2260	.2397	.2814
.635	.0187	.0329	.0466	.0604	.0741	.0883	.1020	.1158	.1295	.1437	.1574	.1712	.1991	.2266	.2403	.2820
.711	.0193	.0335	.0472	.0610	.0747	.0889	.1026	.1164	.1301	.1443	.1580	.1718	.1997	.2272	.2409	.2826
.813	.0201	.0343	.0480	.0617	.0755	.0897	.1034	.1174	.1309	.1451	.1588	.1726	.2005	.2280	.2417	.2834
.965	.0213	.0355	.0492	.0629	.0767	.0909	.1046	.1183	.1321	.1463	.1600	.1737	.2017	.2291	.2429	.2845
1.016	.0217	.0358	.0496	.0633	.0771	.0913	.1050	.1187	.1325	.1467	.1604	.1741	.2021	.2295	.2433	.2849
1.270		.0378	.0516	.0653	.0790	.0932	.1070	.1207	.1345	.1486	.1624	.1761	.2040	.2315	.2453	.2869
1.625		.0406	.0543	.0681	.0818	.0960	.1097	.1235	.1372	.1514	.1652	.1789	.2066	.2343	.2480	.2897
1.829			.0559	.0697	.0834	.0976	.1113	.1251	.1388	.1530	.1667	.1805	.2084	.2359	.2496	.2913
1.981			.0571	.0709	.0846	.0988	.1125	.1263	.1400	.1542	.1679	.1817	.2096	.2371	.2508	.2925
2.058			.0577	.0715	.0852	.0994	.1131	.1269	.1406	.1548	.1685	.1823	.2102	.2377	.2514	.2931
2.311			.0597	.0734	.0872	.1014	.1151	.1288	.1426	.1568	.1705	.1842	.2122	.2396	.2534	.2950
2.388			.0603	.0740	.0878	.1020	.1157	.1294	.1432	.1574	.1711	.1848	.2128	.2402	.2540	.2956
2.591				.0756	.0894	.1035	.1173	.1310	.1448	.1589	.1727	.1864	.2143	.2418	.2556	.2972
2.769				.0770	.0907	.1049	.1187	.1324	.1461	.1603	.1741	.1878	.2157	.2432	.2570	.2986
3.175				.0802	.0939	.1081	.1218	.1356	.1493	.1635	.1772	.1910	.2189	.2464	.2601	.3018
3.962					.1001	.1142	.1280	.1417	.1555	.1696	.1834	.1971	.2250	.2525	.2663	.3079
4.775					.1064	.1206	.1343	.1481	.1618	.1760	.1897	.2035	.2314	.2589	.2726	.3143
5.156						.1235	.1373	.1510	.1648	.1789	.1927	.2064	.2344	.2618	.2756	.3172
5.537								.1540	.1677	.1819	.1957	.2094	.2373	.2648	.2785	.3202
5.944								.1572	.1709	.1851	.1988	.2126	.2405	.2680	.2817	.3234
6.350								.1603	.1741	.1883	.2020	.2157	.2437	.2711	.2849	.3265

EXAMPLE: MATERIAL THICKNESS = 3.175 MM, INSIDE RADII = 6.35 MM R. WHERE THESE CROSS = .1356 MM IF THE INSIDE OF YOUR BEND IS 20° THE BEND ALLOWANCE (B/A) = 2.712 MM (.1356 × 20°) BEND ALLOWANCE (B/A) PLUS TOTAL OF STRAIGHT LENGTHS = DEVELOPED LENGTH.

TABLE 44 From *Drafting for Trades and Industry—Basic Skills,* Nelson. Delmar Publishers Inc.

SPUR/PINION GEAR TOOTH PARTS
20 Degree Pressure Angles

Diametral Pitch	Circular Pitch	Circular Thickness	Addend.	Dedend.	Standard Fillet Radius
D.P.	C.P.	C.T.	A	D	
1	3.1416	1.5708	1.0000	1.2500	0.3000
2	1.5708	0.7854	0.5000	0.6250	0.1500
2.5	1.2566	0.6283	0.4000	0.5000	0.1200
3	1.0472	0.5236	0.3333	0.4167	0.1000
3.5	0.8976	0.4488	0.2857	0.3571	0.0857
4	0.7854	0.3927	0.2500	0.3125	0.0750
4.5	0.6981	0.3491	0.2222	0.2778	0.0667
5	0.6283	0.3142	0.2000	0.2500	0.0600
5.5	0.5712	0.2856	0.1818	0.2273	0.0545
6	0.5236	0.2618	0.1667	0.2083	0.0500
6.5	0.4833	0.2417	0.1538	0.1923	0.0462
7	0.4488	0.2244	0.1429	0.1786	0.0429
7.5	0.4189	0.2094	0.1333	0.1667	0.0400
8	0.3927	0.1963	0.1250	0.1563	0.0375
8.5	0.3696	0.1848	0.1176	0.1471	0.0353
9	0.3491	0.1745	0.1111	0.1389	0.0333
9.5	0.3307	0.1653	0.1053	0.1316	0.0316
10	0.3142	0.1571	0.1000	0.1250	0.0300
11	0.2856	0.1428	0.0909	0.1136	0.0273
12	0.2618	0.1309	0.0833	0.1042	0.0250
13	0.2417	0.1208	0.0769	0.0962	0.0231
14	0.2244	0.1122	0.0714	0.0893	0.0214
15	0.2094	0.1047	0.0667	0.0833	0.0200

TABLE 45

SURFACE TEXTURE ROUGHNESS

Surface Roughness Average Obtainable by Common Production Methods											
Roughness Height Rating in N Series of Roughness Grade, Microinches, μ in. AA											
Process	N12	N11	N10	N9	N8	N7	N6	N5	N4	N3	N2	N1
	2000	1000	500	250	125	63	32	16	8	4	2	1	.5

Process列:
Flame Cutting
Snagging
Sawing
Planing, Shaping

Drilling
Chemical Milling
Elect. Discharge Machining
Milling

Broaching
Reaming
Electron Beam
Laser
Electro-Chemical
Boring, Turning
Barrel Finishing

Electrolytic Grinding
Roller Burnishing
Grinding
Honing

Electro-Polishing
Polishing
Lapping
Superfinishing

Sand Casting
Hot Rolling
Forging
Perm. Mold Casting

Investment Casting
Extruding
Cold Rolling, Drawing
Die Casting

TYPICAL APPLICATION:
- Very rough surface. Equiv. to sand casting.
- Rough surface. Rarely used.
- Coarse finish. Equiv. to rolled surfaces & forgings.
- Medium finish. Commonly used. Reasonable appear.
- Good for close fits. Unsuitable for fast rotating members.
- Used on shafts & bearings with light loads & moderate speeds.
- Used on high speed shafts & bearings.
- Used on precision gauge & instrument work. Costly.
- Refined finish. Costly to produce.
- Super-finish. Costly. Seldom used.

The ranges shown above are typical of the processes listed.

Higher or lower values may be obtained under special conditions.

KEY: ▓ Average Application ▨ Less Frequent Application

From Interpreting Engineering Drawings. Jensen. Delmar Publishers Inc.

TABLE 46   From *Interpreting Engineering Drawings*, Jensen. Delmar Publishers Inc.

PROPERTIES, GRADE NUMBERS, AND USAGE OF STEEL ALLOYS

Class of Steel	*Grade Number	Properties	Uses
Carbon—Low 0.3% carbon	10xx	Tough—Less Strength	Rivet—Hooks—Chains – Shafts—Pressed Steel Products
Carbon—Medium 0.3% to 0.6% carbon	10xx	Tough & Strong	Gears—Shafts—Studs Various Machine Parts
Carbon—High 1.6% to 1.7%	10xx	Less Tough—Much Harder	Drills—Knives—Saws
Nickel	20xx	Tough & Strong	Axles—Connecting Rods—Crank Shafts
Nickel Chromium	30xx	Tough & Strong	Rings—Gears—Shafts – Piston Pins—Bolts – Studs—Screws
Molybdenum	40xx	Very Strong	Forgings—Shafts – Gears—Cams
Chromium	50xx	Hard W/Strength & Toughness	Ball Bearings—Roller Bearing—Springs – Gears—Shafts
Chromium Vanadium	60xx	Hard & Strong	Shafts—Axles—Gears – Dies—Punches—Drills
Chromium Nickel Stainless	60xx	Rust Resistance	Food Containers – Medical/Dental Surgical Instruments
Silicon—Manganese	90xx	Springiness	Large Springs

*The first two numbers indicate type of steel, the last two numbers indicate the approx. average carbon content—1010 steel indicates carbon steel w/approx. 0.10% carbon.

TABLE 47   From *Drafting for Trades and Industry—Mechanical and Electronic,* Nelson. Delmar Publishers Inc.

AMERICAN NATIONAL STANDARDS OF INTEREST TO DESIGNERS, ARCHITECTS, AND DRAFTERS

TITLE OF STANDARD

TABLE 48

G l o s s a r y

Abrasive Water Jet Machining (AWJM)   Similar to WJM except that an abrasive is added to the water jet after it exits the nozzle.

Labrado Abrasivo con Chorro de Agua   Similar al WJM, excepto que se le añade un abrasivo al chorro de agua luego que sale de la tobera.

Absolute System   A type of numerical control system in which all coordinates are located from a fixed or absolute zero point.

Sistema Absoluto   Un tipo de sistema de control numérico en el cual todas las coordenadas se localizan a base de un punto fijo o de cero absoluto.

Abstract   Part of the written report to the client, this includes one or two sentences about the content of the report to capture a potential reader's interest.

Abstracto   Parte de un informe escrito al cliente, éste incluye una o dos oraciones acerca del contenido del informe para capturar el interés de un posible lector.

Acme   Screw thread form.

Acme   Tipo de rosca de tornillo.

Active Coils   The total coils minus the two end coils in a compression spring.

Espirales Activas   El número total de espirales menos las dos espirales de los extremos en un resorte de compresión.

Actual Size   Actual measured size of a feature.

Tamaño Real   El tamaño real de un rasgo.

Acute Angle   An angle less than 90°.

Ángulo Agudo   Un ángulo menor de 90°.

Addendum   The radial distance from the pitch diameter to the top of the tooth. It is equal to 1/P.

Altura de Cabeza   La distancia radial desde el diámetro de paso hasta la altura del diente. Es igual a 1/P.

Adjacent Alignment Chart   Two related variables plotted along the same line, but on opposite sides.

Gráfica de Alineamiento Adyacente   Dos variables relacionadas que se grafican a lo largo de la misma línea, pero en lados opuestos.

Adjustable Curve   This instrument has a locking bow and is used to draw any radius from 6.75" to 200" (17 cm to 500 cm). It takes over where the ordinary compass leaves off and eliminates the beam compass.

Curva Ajustable   Este instrumento tiene un arco con seguro que se usa para dibujar cualquier radio desde 6.75″ hasta 200″ (17 cm a 500 cm). Se puede usar donde el compás ordinario deja de ser útil y elimina al compás de viga.

Aligning of Features   This occurs to correct the true projection concept. It is when the top hole is theoretically revolved to the cutting-plane line and projected to the sectional view.

Alineación de Rasgos   Esto ocurre para corregir el concepto de proyección verdadera. Es cuando un agujero teoréticamente se revuelve al plano cortante y se proyecta en la vista seccional.

Alignment Chart   A graphical representation of the relationship of two or more variables.

Gráfica de Alineamiento   Una representación gráfica de la relación entre dos o más variables.

Allen Screw   Special set screw or cap screw with hexagon socket in head.

Tornillo con Cabeza Hexagonal   Un tornillo de fijación o tornillo pasante especial con una cavidad hexagonal en la cabeza.

Allowance   Minimum clearance between mating parts. Specifically, the amount of room between two main parts, usually considered the tightest possible fit.

Concesión   La separación mínima entre piezas apareadas. Específicamente, la cantidad de espacio entre dos partes principales, generalmente considerado como el encaje más apretado posible.

Alloy   Two or more metals in combination, usually a fine metal with a baser metal.

Aleación   Dos o más metales en combinación, generalmente un metal fino con un metal base.

Alloy Steel   This is any steel that contains one or more alloying elements added to produce characteristics not available in unalloyed carbon steels.

Acero Aleado   Esto es cualquier acero que contiene uno o más elementos de aleación, que se le añaden para producir características que no están disponibles en los aceros con carbón sin alearse.

Alphanumeric   Refers to the totality of characters that are either alphabetic or numeric.

Alfanumérico   Se refiere a la totalidad de los caracteres que son alfabéticos o numéricos.

Alternate Section Lining   Combination of true projection and correct projection.

Revestimiento de Sección Alterna   Una combinación de una proyección real y una proyección correcta.

Alternatives   Convincing the client that a thorough investigation was done by the design group by including a brief summary of the alternatives that were considered but discarded and why.

Alternativas   Convencer al cliente que se hizo una investigación exhaustiva por el grupo de diseño por medio de un resumen breve de las alternativas que se consideraron pero que se desecharon y por qué.

Altitude (of a prism)   The perpendicular distance between its end polygons (or bases).

Altitud (de un prisma)   La distancia perpendicular entre sus dos polígonos extremos o bases.

Aluminum   A bright silvery metal that is very light. In addition to being light, it is an excellent conductor and able to resist corrosion. It is usually alloyed with copper to increase hardness and strength.

Aluminio   Un metal plateado brilloso que es muy liviano. Además de ser liviano, es un excelente conductor eléctrico y es capaz de resistir la corrosión. Generalmente se alea con cobre para aumentar su dureza y resistencia.

Ammonia   A colorless gas used in the development process of diazo and sepia prints.

Amoniaco   Un gas incoloro que se usa en el desarrollo de las reproducciones de diazo y sepia.

Analysis   Summarizing the major assumptions and calculations. Also, the study and evaluation of an object's characteristics.

Análisis   Resumir las suposiciones principales y los cálculos. También, el estudio y evaluación de las características de un objeto.

Angle Gear   Similar to a bevel gear except that the angles are at other than 90° to each other.

Engranaje Angular   Similar a un engranaje de bisel, excepto que los ángulos son otros diferentes de 90° uno con respecto al otro.

Angle Iron   A structural shape whose section is a right angle.

Hierro Angular   Una forma estructural cuya sección transversal es un ángulo recto.

Angle of Thread   The included angle between the sides of the thread.

Ángulo de Rosca   El ángulo incluido entre los lados de una rosca.

Angle Plate Jig   A modified form of a plate jig in which the surface to be machined is perpendicular to the locating surface.

Guía de Placa Angular   Una forma modificada de una guía de placa en la cual la superficie a labrarse está perpendicular a la superficie de localización.

Angularity   A feature control in which a given surface, axis, or center plane must form a specified angle other than 90° with a datum.

Angularidad   Un rasgo de control en el cual una superficie, eje o plano central dado, debe pasar por un ángulo diferente de 90° respecto a una referencia.

Annealing   The process of heating a part, soaking it until it is thoroughly heated, and slowly cooling it by turning off the furnace.

Recocer   El proceso de calentar una pieza hasta que esté calentada completamente, y luego enfriarla lentamente al apagar el horno.

Anodizing   The process of protecting aluminum by oxidizing in an acid bath using a direct current.

Anodizar   El proceso de proteger el aluminio oxidándolo en un baño ácido usando una corriente directa.

Application   A definable set of drafting tasks to be accomplished in a given drafting area; may be accomplished partly through manual procedures and partly through computerized procedures.

Aplicación   Un grupo definible de tareas de dibujo industrial que deben ser llevadas a cabo en un área de dibujo dada; pueden lograrse parcialmente por medio de procedimientos manuales y parcialmente por medio de procedimientos computarizados.

Application Engineering   This type of engineering takes previous designs into consideration when creating a new design. It is generally a team effort whose primary task is to modify and incorporate these existing concepts to meet the design requirements for the new system.

Ingeniería de Aplicación   Este tipo de ingeniería considera los diseños previos al crear diseños nuevos. Generalmente es un esfuerzo en equipo cuya tarea primaria es modificar e incorporar aquellos conceptos existentes para cumplir con los requisitos del nuevo sistema.

Appliqué A generic term used to describe a variety of shortcut products used in drafting.
Aplicado Un término genérico usado para describir una variedad de productos de atajos que se usan en el dibujo industrial.

Arbor Press A machine that performs internal broaching operations using a push-type, rather than a pull-type broach.
Prensa de Husillo Una máquina que lleva a cabo las operaciones de brochado internas usando un broche de tipo de empuje en vez uno de tipo de halar.

Arc A part of a circle.
Arco Una parte de un círculo.

Arc-weld To weld by electric arc; the work is usually the positive terminal.
Soldadura de Arco Soldar por medio de un arco eléctrico; la pieza de trabajo generalmente es el terminal positivo.

Arrowhead The most commonly used termination symbol for dimension and leader lines.
Flecha El símbolo de terminación más común para las líneas de dimensión y directrices.

Artificial Intelligence The ability of a machine to improve its own operation or the ability of a machine to perform functions that are normally associated with human intelligence such as learning, adapting, reasoning, self-correction, and automatic improvement.
Inteligencia Artificial La habilidad de una máquina de mejorar su propia operación o la habilidad de una máquina para llevar a cabo funciones que normalmente están asociadas con la inteligencia humana, tales como el aprendizaje, adaptación, razonamiento, auto corrección y mejoría automática.

Assembled Isometric Assembly Drawings This style shows how a single product looks when assembled.
Dibujo de Ensamblaje Isométrico Ensamblado Este estilo muestra cómo un solo producto se vería ensamblado.

Assembly Drawing Illustrates how a product is assembled when completed.
Dibujo de Ensamblaje Ilustra como un producto se ensambla cuando completado.

Assembly Section This section occurs when a sectional drawing is made up of two or more parts. It shows how the various parts go together.
Sección de Ensamblaje Esta sección ocurre cuando se crea un dibujo seccional de dos o más piezas. Muestra cómo se unen la varias piezas.

Attribute Listing A systematic guide for checking the completeness of technical detail drawings. It includes the questions: What is the function? What size is it? What materials are used? Can it be made—how and with what processes? What other factors need to be checked?

Listado de Atributos. Una guía sistemática para verificar la plenitud de detalles técnicos en los dibujos. Incluye las preguntas: ¿Cuál es su función? ¿De que tamaño es? ¿Qué materiales se usan? ¿Puede hacerse—cómo y con qué procesos? ¿Qué otros factores necesitan verificarse?

Attributes The characteristics of a part.
Atributos Las características de una pieza.

Automated Assembly Assembly by means of operations performed automatically by machines that are controlled and monitored by computers.
Ensamblaje Automatizado Ensamblaje por medio de operaciones que se hace automáticamente por máquinas que son controladas y monitoreadas por computadoras.

Automated Process Planning Using a computer in developing a process plan.
Planificación Automática de Proceso Usar una computadora al desarrollar un plan de proceso.

Automation The conversion of a process or system to automatic operation.
Automatización La conversión de un proceso o sistema a una operación automática.

Auxiliary Section An auxiliary view in section.
Sección Auxiliar Una vista auxiliar en sección.

Auxiliary View An additional view of an object, usually of a surface inclined to the principal surfaces of the object to provide a true size and shape view.
Vista Auxiliar Una vista adicional de un objeto, generalmente de una superficie inclinada a las superficies principales de un objeto para proveer una vista del tamaño y forma real.

Axes Plural of axis.
Ejes El plural de eje.

Axis A longitudinal center line of the thread. Also, an imaginary line around which parts rotate or are regularly arranged.
Eje Una línea de centro longitudinal de una rosca. También, una línea imaginaria alrededor de la cual las piezas rotan o están ordenadas regularmente.

Axonometric Projection Three-dimensional drawings.
Proyección Axonométrica Dibujos tridimensionales.

Axonometric Sketching This type of sketching may be one of three types—isometric, dimetric, or trimetric. All relate to proportional scales and angle positions of length, width, and depth.
Croquis Axonométrico Este tipo de croquis puede ser de uno de tres tipos—isométrico, dimétrico o trimétrico. Todos se relacionan con las escalas de proporción y la posición angular del largo, ancho y profundidad.

Azimuth Bearing In this method, the total angle is used going clockwise from due north.

Orientación de Azimut   En este método, el ángulo total se usa como si partiese a favor de las manecillas del reloj desde la dirección del norte.

Babbitt   A soft alloy for bearings, mostly of tin with small amounts of copper and antimony.

Babbit   Una aleación suave de cojinetes, hecha mayormente de estaño con pequeñas cantidades de cobre y antimonio.

Back Weld   A weld applied to the opposite side of a joint after the major weld has been applied.

Soldadura trasera   Una soldadura aplicada al lado posterior de la unión después de aplicar la soldadura principal.

Backlash   Lost motion between moving parts, such as threaded shaft and nut or the teeth of meshing gears.

Contragolpe   Perdida de movimiento entre partes móviles, tales como un eje de rosca y tuerca o los dientes de la malla de engranaje.

Balloon   Number within a circle, used to identify parts.

Globo   Número dentro de un círculo, utilizado para identificar las partes.

Bar Chart   Either horizontal or vertical, these charts most often show relative amounts. They often show the relative timing of activities in the form of a planning chart. Generated from a spreadsheet in a personal computer.

Grafica de barra   Estas graficas comúnmente demuestran cantidades relativas, ya sean horizontal o verticalmente. Por lo general demuestran el tiempo relativo de actividades en la forma de una grafica de planificación. Se generan en una hoja de calculo utilizando una computadora personal.

Base Circle   Used to lay out an offset follower. It is a circle with a radius equal to the distance from the center of the shaft to the center of the follower at its lowest position.

Círculo de Base   Utilizado para colocar un seguidor de desfase. Es un círculo con un radio igual a la distancia del centro del eje al centro del seguidor en su posición más baja.

Baseline   The bottom line of the displacement diagram.

Línea de Base   La línea inferior de un diagrama de desplazamiento.

Baseline Diagrams   This type of connection diagram shows all wire paths running into a single baseline. They are easier to read and follow than point-to-point diagrams, but they can be misleading because they do not show the electronic components in their proper position.

Diagramas de Línea de Base   Este tipo de diagrama de conexión demuestra todas las rutas de cable que atraviesan una sola línea de base. Son más fáciles de leer y de seguir que diagramas de punto a punto, pero pueden confundir porque no demuestran los componentes electrónicos en la posición apropiada.

Baseline Dimensioning   A system of dimensioning where as many features of a part as are functionally practical are located from a common set of datums.

Dimensionamiento de Línea de Base   Un sistema de dimensionamiento donde la mayoría de las características de una parte que sean funcionalmente prácticas están localizadas como un conjunto común de referencia.

Basic Dimension   A theoretically "perfect" dimension similar to a reference or nominal dimension. It is used to identify the exact location, size, orientation, or shape of a feature.

Dimensión Básica   Una dimensión teoréticamente perfecta similar a una referencia o una dimensión nominal. Esta es utilizada para identificar la localización exacta, tamaño, orientación o forma de una característica.

Basic Size   The size from which the limits of size are derived by the application of allowances and tolerances.

Tamaño Básico   El tamaño del cual los limites de tamaño se derivan por la aplicación de provisiones y tolerancias.

Batter   The slope of the grade is used. For each unit of rise (R), there are four horizontal (H) units.

Batter   La pendiente de un grado es utilizada. Para cada unidad de elevación (R), hay cuatro unidades horizontales (H).

Beam Diagrams   These are diagrams of beams of varying shapes, materials, and sizes. They consist mainly of free-body, shear, movement slope, and deflection diagrams.

Diagramas de Viga   Estos son diagramas de vigas de diferentes formas, materiales y tamaños. Estos consisten principalmente de diagramas de cuerpo-libre, esquilados, movimiento en pendiente y deflexión.

Bearing   A supporting member for a rotating shaft. Also, the bearing of a line is the direction of that line as it is drawn in the top view of a drawing.

Soporte   Un miembro de apoyo para un eje que rota. También el rumbo de una línea es la dirección de esa línea como ésta es dibujada en vista superior de un dibujo.

Orientación   La dirección de una línea según se dibuja en la vista superior de un dibujo.

Bend Allowance   The amount of sheet metal required to make a bend over a specific radius.

Provisión de Curvatura   La cantidad de plancha de metal requerida para doblar sobre un radio especifico.

Bend Allowance Charts   Two basic types of charts used to calculate bend allowances in both the English and metric systems. One is used for 90° bends, and the other is used for bends from 1° to 180°.

Tablas de Concesión de Curvatura   Son dos tipos básicos de tablas utilizadas para calcular la provisión de curvatura en el sistema métrico e inglés. Uno es utilizado para curvaturas de 90°, y el otro es utilizado para curvaturas entre 1° a 180°.

Bending An operation in which a metal part is simply bent to a desired angle.

Doblamiento Una operación en la cual una parte de metal es doblada a un ángulo deseado.

Between Symbol A symbolic means of indicating that the stated tolerance applies to a specified segment of a surface between designated points.

Entre Símbolo Un medio simbólico para indicar que la tolerancia establecida aplica a un segmento especifico de una superficie entre los puntos designados.

Bevel An inclined edge; not a right angle to joining surface.

Bisel Un borde inclinado, no en un ángulo recto a la superficie de unión.

Bevel Gear Cone-shaped in form with straight teeth that are on an angle to the axis of the shaft. It is used to transmit power and motion between intersecting shafts that are at 90° to each other.

Engranaje Biselado Conforme con dientes rectos que están en un ángulo con el eje. Este es utilizado para transmitir potencia y movimiento entre ejes que se intersecan que están a 90° el uno del otro.

Beveled Rings These have a 15° bevel on the groove-engaging edge. They are installed in grooves having a comparable bevel on the load-bearing wall.

Anillos Biselados Estos tiene 15° de bisel en el borde de ranura de encaje. Estos están instalados en las ranuras con un bisel comparable a una pared de carga.

Bilateral Tolerance When this tolerance is applied to a dimension, the tolerance applies in both directions, but is not necessarily evenly distributed.

Tolerancia Bilateral Cuando esta tolerancia es aplicada a una dimensión, la tolerancia se aplica en ambas direcciones, pero no es necesariamente distribuida equitativamente.

Bisect To divide into two equal parts.

Bisecar Dividir en dos partes iguales.

Blanking An operation that produces a part cut completely by the punch and die. Also, a stamping operation in which a press uses a die to cut blanks from flat sheets or strips of metal.

Cortar Una operación que produce una parte cortada completamente por el perforador y el troquel. Una operación de estampado en la cual una presa es utilizada como troquel para cortar blancos de planchas planas o tiras de metal.

Blind Hole A hole that does not go completely through an object.

Agujero ciego Un agujero que no atraviesa completamente el objeto.

Block Diagrams The most widely used types of diagrams in electronics engineering settings. There are three types: organizational charts, process flow charts, and functional block diagrams.

Diagrama de bloque El tipo de diagrama más utilizado en ingeniería electrónica. Hay tres tipos: tablas organizacionales, tablas de flujo de proceso y diagramas de bloque funcional.

Blueprint A copy of a drawing.

Fotocalco Una copia de un dibujo.

Boardroom Sketching This technique works well with larger groups than the conversational style. Since there are more people involved, a larger surface—for better viewing—is needed. It can be a chalkboard, but more likely will be an enamel, glass, or dry-erase surface.

Sala de Conferencia de Croquis Esta técnica trabaja bien con grupos más grandes que los de estilo conversacional. Dado a que hay más personas involucradas, una superficie grande es necesaria para mejor visualización. Ésta puede ser una pizarra, pero más probable será una superficie de esmalte, cristal o del borrar seco.

Bolt A threaded fastener that passes directly through parts to hold them together, and uses a nut to tighten or hold the parts together. The tip of the nut is flat.

Perno Un sujetador de rosca que pasa directamente a través de las partes que sujeta juntas y utiliza una tuerca para sujetar estas partes. La punta de la tuerca es plana.

Bolt Circle A circular center line on a drawing, containing the centers of holes about a common center.

Círculo del Perno Una línea circular central en el dibujo que contiene los centros de los agujeros alrededor de un centro común.

Bore To enlarge a hole with a boring bar or tool in a lathe, drill press, or boring mill.

Taladrar Agrandar un agujero con una barra taladrante o herramienta en un torno, presa taladrante o fresadora taladrante.

Boring Mill A machine tool normally used to machine large workpieces.

Fresadora Taladrante Una maquinaria normalmente utilizada para fresar piezas de trabajo grandes.

Boss A cylindrical projection on a casting or a forging.

Saliente Cilíndrico Una proyección cilíndrica en una fundición o en una pieza forjada.

Bow Compass One of two types of compass. This type uses a spring and an adjusting screw to set the compass to the desired radius.

Compás de arco Uno de dos tipos de compás. Este tipo utiliza un resorte y un tornillo ajustador para colocar el compás en un radio deseado.

Bowed Rings They differ from conventional types in that they are bowed around an axis perpendicular to the diameter bisecting the gap. Their construction permits the rings to function as springs as well as fasteners.

Anillos arqueados Estos difieren de los tipos convencionales en que están arqueados alrededor del eje perpendicular al diámetro que biseca la apertura. La construcción de los anillos le permite funcionar como resortes al igual que como ajustadores.

Box Jig Designed to be used for parts that require machining on several sides.

Guía de cajón Diseñada para ser utilizada en las partes que requieren trabajos en varios lados.

Brass An alloy of copper and zinc.

Latón Una aleación de cobre y zinc.

Brazing The process of joining metals together with a nonferrous filler rod. A temperature above 1000°F, but just below the melting point, is used to join the parts.

Soldadura de Latón El proceso de unir metales con una varilla de relleno no ferrosa. Se utiliza una temperatura sobre 1000°F, pero justo por debajo del punto de fusión, para unir estas partes.

Breakdown (bender) It is the station 2 of the drop forging process—it forms the parts into special contours.

Rotura (dobladora) Es la estación 2 del proceso de forjamiento de caída, ésta forma las partes en un contorno especial.

Broach A tool for removing metal by pulling or pushing it across the work; the most common use is producing irregular hole shapes such as squares, hexagons, ovals, or splines.

Brocha Una herramienta para remover metal cuando esta es halada o empujada a través de la pieza. La más utilizada comúnmente produce agujeros irregulares tales como cuadrados, hexágonos, óvalos o estrías.

Broaching Machine Used to modify the shape of a workpiece by pulling tools called broaches across or through the part.

Brochadora Utilizada para modificar la forma de una pieza halando herramientas llamadas brochas a través de la parte.

Broken-Out Section A portion removed from the drawing in order to make a particular feature easier to understand.

Sección Removida Una porción removida de un dibujo para hacer que una característica en particular sea fácil de entender.

Bronze An alloy of eight or nine parts of copper and one part of tin.

Bronce Una aleación de ocho o nueve partes de cobre y una parte de estaño.

Browser A client program that enables one to search through the information provided by a specific type of server.

Explorador Un programa de cliente que permite la búsqueda de información provista por un tipo especifico de servidor.

Buff To finish or polish on a buffing wheel composed of fabric with abrasive powders.

Lustrar Darle terminado o pulir en una rueda de pulido compuesta de tela con polvos abrasivos.

Burnish To finish or polish by pressure upon a smooth rolling or sliding tool.

Bruñir Darle terminado o pulir al presionar suavemente sobre una herramienta rodante o deslizante.

Burr The ragged edge or ridge left on metal after a cutting operation.

Borde No Terminado Borde sin acabar o saliente dejado en el metal luego de la operación de cortar.

Bushing A metal lining that acts as a bearing between rotating parts such as a shaft and pulley; also used on jigs to guide cutting tools.

Cojinete Revestimiento de metal que actúa como un cojinete entre las partes que rotan tales como el eje y la polea; también usado para guiar herramientas cortantes.

Butt Joint One of two kinds of basic rivet joints—the parts are butted and are held together by a cover plate or butt strap that is riveted to both parts.

Union de Tope Una de dos tipos de uniones básicas de remache—las partes se unen y se sujetan por una placa o banda que se remacha a ambas partes.

Cabinet Drawing Have the same range of receding axes as cavalier drawings, but the receding distances are drawn to half size.

Dibujo de Gabinete Tienen el mismo rango de ejes que se alejan tales como en los dibujos cavalier, pero las distancias que se alejan se dibujan a media escala.

CAD (Computer-Aided Drafting) The use of computers and peripheral devices to aid in the documentation of design projects.

CAD (Dibujo Industrial Asistido por Computadora) El uso de computadoras y aparatos periféricos para asistir en la documentación de los proyectos de diseño.

CAD System The hardware, software, and user in computer-aided drafting. It is usually referred to as a configuration of computer hardware. This is a misuse of the term. CAD hardware is just one of the previously mentioned three components needed to have a CAD system.

Sistema de CAD El equipo físico, software y el usuario en el dibujo asistido por computadora. Generalmente se conoce como la configuración del equipo físico de computadora. Esto es un uso inadecuado del término. El equipo físico para CAD es sólo uno de los tres componentes mencionados previamente necesarios para tener un sistema CAD.

CAE (Computer-Aided Engineering) Any use of a computer to assist in engineering tasks, analysis testing, calculating, design review, and experimentation.

Ingeniería Asistida por Computadora Cualquier uso de una computadora para asistir en las tareas de ingeniería, pruebas de análisis, cálculos, revisión de diseño y experimentación.

Calipers An instrument used for measuring diameters.

Calibrador Un instrumento usado para medir diámetros.

Callout A note on the drawing giving a dimension, specification, or machine process.

Nota Una nota en un dibujo que da una dimensión, especificación o proceso de labrado.

CAM (Computer-Aided Manufacturing) Refers to computer-controlled automation of the various manufacturing processes used in modern industry.

Manufactura Asistida por Computadora Se refiere a la automatización controlada por computadora de los varios procesos de manufactura usados en la industria moderna.

Cam A rotating shape for changing circular motion to reciprocating motion. Produces a simple means to obtain irregular or specified predictable, designed motion.

Leva Una figura que rota para cambiar el movimiento circular en un movimiento reciprocante. Produce una manera simple de obtener movimiento diseñado, irregular o predecible.

Cap Screw Usually used as a true screw, it passes through a clearance hole in one part and screws into another part.

Tornillo Pasante Generalmente se usa como un tornillo verdadero, pasa a través de un agujero pasante en una pieza y se atornilla en otra pieza.

Carburize To heat a low-carbon steel to approximately 2000°F in contact with material that adds carbon to the surface of the steel.

Carbonizar Calentar el acero bajo en carbón hasta aproximadamente 2000°F en contacto con material que le añade carbón a la superficie del acero.

Case Harden To harden the outer surface of a carburized steel by heating and then quenching.

Endurecer Endurecer la superficie exterior del acero carbonizado calentándolo y luego enfriándolo rápido.

Case Hardening The process of hardening the outside surface of a part to a preselected depth.

Endurecimiento Externo El proceso de endurecer la superficie exterior de una pieza hasta una profundidad predeterminada.

Cast Irons Metals in a family of cast ferrous metals.

Hierros de Fundición Metales en una familia de metales férreos de fundición.

Cast Steels Steels that are produced in ingot form by melting all component elements together and pouring the molten metal into a mold.

Aceros de Fundición Aceros que se producen en forma de lingotes derritiendo todos los elementos componentes juntos y vertiendo el metal derretido en un molde.

Casting A process of pouring molten metal into a mold that contains the desired shape in the form of a cavity.

Fundición de Molde Un proceso de verter metal derretido en un molde que contiene la forma deseada en la cavidad.

Cavalier Drawing A type of drawing in which a receding axis is drawn between 30° and 60° to the horizon, and the receding distances are drawn full size.

Dibujo Cavalier Un tipo de dibujo en el cual el eje que se aleja se dibuja entre 30° y 60° del horizonte y las distancias que se alejan se dibujan a escala completa.

CD-ROM Compact Disc, Read-Only Memory; a high-volume storage medium used to store drawing files electronically. Can be used to store large projects for permanent archiving.

CD-ROM Disco compacto, sólo para lecturas; un medio de almacenamiento de alta capacidad usado para almacenar archivos de dibujo electrónicamente. Puede usarse para almacenar grandes proyectos para archivo permanente.

Center Drill A special drill to produce bearing holes in the ends of a workpiece to be mounted between centers.

Taladro de Centro Un taladro especial para producir agujeros para cojinetes en los extremos de una pieza de trabajo montada entre centros.

Center Line A type of line used in drafting to indicate the center of an object; its main characteristic is that it is a broken line with one short dash in the center.

Línea de Centro Un tipo de línea que se usa en el dibujo industrial para indicar el centro de un objeto; su característica principal es que es una línea entrecortada con un guión corto en el centro.

Central Angle An angle formed by two radial lines from the center of the circle.

Ángulo Central Un ángulo formado por dos líneas radiales desde el centro del círculo.

Central Processing Unit (CPU) A unit of a computer that includes circuits controlling the interpretation and execution of instructions.

Unidad de Procesamiento Central Una unidad de una computadora que incluye circuitos que controlan la interpretación y ejecución de instrucciones.

Centrifugal Casting A process of pouring a measured amount of molten metal into a rotating mold.

Fundición Centrífuga El proceso de verter una cantidad medida de metal derretido en un molde que rota.

Ceramics Materials made by combining metallic and nonmetallic elements. Their main characteristics are that they have a high melting point, low conductivity, and are chemically and thermally stable. They are mainly used in electronics applications.

Cerámicas Materiales hechos combinando elementos metálicos y no metálicos. Sus características principales son que tienen un punto de fusión alto, conductividad eléctrica baja y son estables química y termalmente. Se usan mayormente para aplicaciones electrónicas.

Chain and Sprockets Used to transmit motion and power to shafts that are parallel to each other.

Cadena y Rueda Dentada Se usa para transmitir movimiento y poder a los ejes que están paralelos unos con otros.

Chain Dimensioning Successive dimensions that extend from one feature to another, rather than each originating at a datum.

Dimensionamiento de Cadena Dimensiones sucesivas que extienden de un rasgo al otro, en lugar de originarse cada uno de un conjunto común de referencia.

Chain Intermittent Weld A weld where each weld is applied directly opposite to each other on opposite sides of the joint.

Cadena de Soldadura Intermitente Una soldadura donde cada soldadura se aplica directamente opuesta una frente a la otra en lados opuestos de una unión.

Chamber of Commerce A city institution that lists all businesses within the area it covers, as well as the products these businesses produce.

Cámara de Comercio Una institución municipal que enumera todos los negocios adentro de su área, además que los productos que estos negocios producen.

Chamfer A beveled edge normally applied to cylindrical parts and a variety of different types of fasteners.

Chaflán Un borde biselado que normalmente se aplica a las partes cilíndricas y a una variedad de tipos de sujetadores.

Channel Jig A variation of the basic box jig (see also *Box Jig*).

Guía de Canal Una variación de la guía de cajón básica (vea también *Guía de Cajón*).

Character A letter, digit, or other symbol that is used as part of the organization, control, or representation of data.

Carácter Una letra, dígito u otro símbolo que se usa como parte de la organización, control o representación de datos.

Chase To cut threads with an external cutting tool.

Cincelar Cortar las roscas con una herramienta cortante externa.

Check Valves Used to restrict the flow of liquids through a pipe to one direction only.

Válvulas Reguladora Se usa para restringir el flujo de líquidos a través de una tubería hacia una dirección solamente.

Cheek In sand casting, this is a section that might be installed between the cope and the drag.

Mejilla En la fundición de molde en arena, esta es la sección que puede instalarse entre el armazón superior e inferior.

Chemical Milling A process that etches or shapes workpieces to close tolerances by chemically induced material removal.

Fresado Químico Un proceso que graba o forma las piezas de trabajo a tolerancias pequeñas por remoción de material químico.

Chill To harden the outer surface of cast iron by quick cooling, as in a metal mold.

Enfriar Endurecer la superficie externa del hierro fundido enfriándolo rápidamente, como en un molde de metal.

Chord Any straight line whose opposite ends terminate on the circumference of the circle.

Cuerda Cualquiera línea recta cuyos fines terminen en la circunferencia del círculo.

Chordal Addendum The distance from the top of the tooth to the point at which the chordal thickness is measured.

Altura de Cabeza Cordal La distancia desde la parte superior del diente al punto donde se mide el espesor cordal.

Chordal Thickness The length of the chord measured straight across at the pitch diameter. It is measured straight across tooth, not along the pitch diameter as is the circular thickness.

Espesor Cordal El largo de la cuerda medido directamente a través del diámetro de paso. Se mide directamente a través del diente, no a lo largo del diámetro de paso como es su espesor circular.

Chuck A device much like a vise, having movable jaws that grip and hold a rotating tool or workpiece.

Prensa de Sujeción Un aparato muy parecido a una prensa, que tiene mandíbulas movibles que aguantan una pieza o herramienta en rotación.

Circle A closed curve with all points at the same distance from the center point. Its major components are the diameter, the radius, and the circumference.

Círculo Una curva cerrada con todos los puntos a la misma distancia del punto central. Sus componentes principales son el diámetro, el radio y la circunferencia.

Circular Pitch The distance between corresponding points of adjacent teeth measured on the circumference of the pitch diameter. Also, the length of the arc along the pitch circle between the center of one gear tooth and the center of the next.

Largo de Paso Circular La distancia entre puntos correspondientes de dientes adyacentes medido en la circunferencia del diámetro de paso. También, el largo del arco a lo largo del círculo de paso, entre el centro de un diente de engranaje y el centro del próximo.

Circular Runout   A tolerance that identifies an infinite number of elemental tolerance zones on a feature when the feature is rotated 360° for each element.

Excentricidad Circular   Una tolerancia que identifica un número infinito de zonas de tolerancia elemental en un rasgo cuando el rasgo se rota 360° para cada elemento.

Circularity   Sometimes referred to as roundness, it is a feature control for a surface of revolution (cylinder, sphere, cone, etc.). It specifies that all points of a surface must be equidistant from the center line or axis of the object in question.

Circularidad   A veces se conoce como redondez, es un control de rasgo para una superficie de revolución (cilindro, esfera, cono, etc.). Especifica que todos los puntos en una superficie tienen que estar equidistantes de la línea de centro o eje del objeto en cuestión.

Circumference   The distance around the outer surface of a circle.

Circunferencia   La distancia alrededor de la superficie exterior de un objeto.

Civil Engineer's Scale   Also called a decimal-inch scale; measurements are read directly from the scale. Each graduation is equal to 1/10 of an inch if the number in the corner is 10.

Escala de Ingeniero Civil   También se conoce como la escala de pulgada decimal; las medidas se miden directamente de la escala. Cada gradación es equivalente a 1/10 de una pulgada si el número en la esquina es 10.

Clapper Box   This is where the cutting tool is mounted on the end of the ram in this unit.

Caja de Badajo   Esto es cuando la herramienta cortante se monta sobre el extremo del pistón en esta unidad.

Clearance Fit   A condition between mating parts in which there is always a clearance assembly. This is when the least material condition of the shaft is subtracted from the least material condition of the hole.

Encaje de Separación   Una condición entre partes apareadas en la cual siempre hay un ensamblaje de separación. Esto es cuando la condición de material mínimo de un eje se resta de la condición de material mínimo de un agujero.

Clearly Defined Ground Rules   In a team situation, these are rules that are thoroughly understood and subscribed to by all team members. They state clearly what behaviors are expected of team members, including what will be tolerated and what will not.

Normas Básicas Claramente Definidas   En una situación de equipo, éstas son las reglas que se entienden claramente y que son acogidas por todos los miembros del equipo. Declaran claramente qué comportamiento se espera de cada miembro del equipo, incluyendo qué se tolerará y qué no.

Clockwise   Rotation in the same direction as the hands of a clock.

A Favor de las Manecillas de Reloj   Rotación en el mismo sentido que las manecillas del reloj.

Closed Jigs   Jigs that enclose the part on more than one side, and are intended to machine the part on several sides without moving the part from the jig.

Guías Cerradas   Guías que encierran la pieza en uno o más lados, y su intención es poder labrar la pieza en varios lados sin mover la pieza de la guía.

Coil (turn)   One full turn or 360° of the wire about the corner axis.

Espiral   Un giro complete o 360° de un alambre alrededor del eje de esquina.

Coin   To form a part in one stamping operation.

Moneda   Formar una pieza en una operación de estampado.

Cold Rolled Steel   Bessemer steel containing .12% to .20% carbon that has been rolled while cold to produce a smooth, quite accurate stock.

Acero Rodado en Frío   Acero Bessemer que contiene .12% a .20% de carbón que se rueda mientras está frío para producir piezas en bruto que son lisas y bastante precisas.

Collar   A round flange or ring fitted on a shaft to prevent sliding.

Collar   Un reborde o anillo redondo que encaja en un eje para evitar deslizamiento.

Compass   An instrument for describing circles or transferring measurements. It consists of two pointed branches joined at the top by a pivot.

Compás   Un instrumento para describir círculos o transferir medidas. Consiste de dos ramas puntiagudas unidas en la parte superior por un pivote.

Composites   Materials created by combining two or more materials.

Material Compuesto   Materiales creados combinando dos o más materiales.

Compound   This occurs when more than one operation takes place on a part during a single stroke.

Combinado   Esto ocurre cuando ocurre más de una operación sobre una parte durante un solo tiro.

Compression Spring   Offers resistance to a compressive force or applies a pushing action. In its free state, the coils of the spring do not touch.

Resorte de Compresión   Ofrece resistencia a una fuerza de compresión o aplica una acción de empuje. En su estado libre, los espirales del resorte no se tocan.

Compressive Force   A force that "goes into" the joint.

Fuerza de Compresión   Una fuerza que actúa "hacia" la unión.

Computational Fluid Dynamics (CFD)   A means of using computers to mathematically simulate physical events. This involves the study of fluids and their characteristics under a variety of conditions.

Dinámica de Fluido Computacional   Una manera de usar computadoras para simular matemáticamente los eventos físicos. Esto consiste de estudiar los fluidos y sus características bajo una variedad de condiciones.

Computational Prototype   A three-dimensional computer-generated model that possesses the characteristics of a comparable physical object.

Prototipo Computacional   Un modelo tridimensional generado a computadora que posee las características de un objeto físico comparable.

Computer Graphics   A term that is sometimes misused a great deal. CG means using the computer and peripheral devices for producing any type of graphic image, technical or artistic.

Graficas Computarizadas   Un término que frecuentemente se usa mal. Graficas computarizadas significa el uso de computadoras y los aparatos periféricos para producir cualquier tipo de imagen gráfica, técnica o artística.

Concentric   Having a common center, as with circles or diameters.

Concéntrico   Que tiene el centro común, como en los círculos o diámetros.

Concentric Circle   Two or more circles with a common center point.

Círculo Concéntrico   Dos o más círculos con un punto central común.

Concentricity   A tolerance in which the axis of a feature must be coaxial to a specified datum regardless of the datum's and the feature's size.

Concentricidad   Una tolerancia en la cual el eje de un rasgo debe ser coaxial a una referencia independientemente de la referencia y del tamaño del rasgo.

Conceptual Design   Much use is made of known technical information from such sources as reference books, technical journals, and handbooks.

Diseño Conceptual   Hace mucho uso de la información técnica conocida de fuentes tales como libros de referencia, revistas técnicas y manuales.

Concurrency Charts   These provide designers with an additional graphical format with which to visualize equations with three or more variables, and to generate information.

Gráficas de Concurrencia   Estas le proveen a los diseñadores con un formato gráfico adicional para visualizar las ecuaciones de tres o más variables, y generar información.

Cone   A pyramid with a central axis and an infinite number of sides, which form a continuous curved lateral surface.

Cono   Una pirámide con un eje central y un número infinito de lados, los cuales forman una superficie lateral curva y continua.

Configuration   A group of machines, devices, parts, etc., that makes up a system.

Configuración   Un grupo de máquinas, aparatos, partes, etc. que forman un sistema.

Conic Section   A section cut by a plane passing through a cone.

Sección Cónica   Una sección cortada por un plano que pasa a través de un cono.

Connection Diagrams   Used to show how the components in an electronic system are connected.

Diagrama de Conexión   Se usa para mostrar cómo se conectan los componentes en un sistema electrónico.

Continuous Path   Machine controls that are used to control the movement of the tool throughout the cutting cycle.

Paso Continuo   Controles de máquina que se usan para controlar el movimiento de la herramienta a través del ciclo de corte.

Contour Interval   The difference in elevation between neighboring contour lines.

Intervalo de Contorno   La diferencia en elevación entre líneas de contorno adyacentes.

Contour Lines   Used on topographical maps to connect points of elevation. They are a constant elevation on the surface of the earth, above sea level.

Líneas de Contorno   Se usan en los mapas topográficos para conectar puntos de elevación. Muestran elevación constante en la superficie del terreno, por encima del nivel del mar.

Contract   A binding legal document.

Contrato   Un documento legal vinculante.

Conventional Break   An effective way of drawing long objects of constant cross section on a relatively small sheet of drawing paper.

Corte Convencional   Un método efectivo de dibujar los objetos largos de sección transversal constante en una hoja relativamente pequeña de papel.

Conversational Sketching   This occurs when two or more people are huddled around a drafting board or a cafeteria table. The person doing the talking combines several types of sketches in one drawing as the idea is talked through.

Croquis Conversacional   Esto ocurre cuando dos o más personas están reunidas alrededor de una mesa de dibujo o mesa de cafetería. La persona que habla combina distintos croquis en un dibujo a medida que se discute la idea.

Coordinate   An ordered set of data values, either absolute or relative, that specifies a location.

Coordenada   Un grupo ordenado de valores de datos, ya sean absolutos o relativos, que especifican una localización.

Coordinate Dimensioning   A type of rectangular datum dimensioning in which all dimensions are measured from two or three mutually perpendicular datum planes; all dimensions originate at a datum and include regular extension and dimension lines and arrowheads.

Dimensionamiento Coordinado   Un tipo de dimensionamiento de referencia rectangular en el cual las dimensiones se miden desde dos o tres planos de referencia mutuamente perpendiculares; todas las dimensiones se originan en una referencia e incluyen líneas de dimensión y de extensión regulares y flechas.

Cope   In sand casting, this is the top half of the flask.

Armazón Superior   En la fundición de molde en arena, ésta es la mitad superior del armazón.

Copper   A heavy metal that is salmon pink in its pure form. It is nonmagnetic and has excellent conductivity.

Cobre   Un metal pesado que es de color rosado como el salmón en su estado puro. No es magnético y tiene conductividad eléctrica excelente.

Core   To form a hollow portion in a casting by using a dry-sand core or a green-sand core in a mold.

Poner Núcleo   Formar una porción hueca en una fundición de molde usando un centro de arena seca o un centro de arena verde en el molde.

Core Prints   When sand cores are intended to be installed in a cavity, core prints must also be provided to locate and anchor the sand cores during the casting process. They are a simple extension of the pattern.

Impresos de Centro   Cuando se tienen que instalar los centros de arena en una cavidad, también deben proveerse impresos de centro para localizar y anclar los centros de arena durante el proceso de fundición. Son simples extensiones del patrón.

Costs   This item is part of the written report to the client. It details the basic direct costs that the client will have to lay out to have the design made.

Costos   Este artículo es parte del informe escrito al cliente. Detalla los costos básicos directos que el cliente tiene que pagar para hacer el diseño.

Counterbore   The enlargement of the end of a hole to a specified diameter and depth.

Avellanar   Engrandecer un extremo de un agujero a un diámetro y profundidad específica.

Counterbored Hole   Usually a plain through hole with an upper portion enlarged cylindrically to a specified diameter and depth.

Agujero Avellanado   Generalmente un agujero pasante simple con la porción superior engrandecida cilíndricamente a un diámetro y profundidad específica.

Countersink   To chamfer the end of a hole to receive a flathead screw.

Escariar   Biselar un extremo de un agujero para recibir un tornillo de cabeza plana.

Countersunk Hole   Usually a plain through hole with an upper portion enlarged conically, most often to 82°, to accommodate standard flathead machine screws.

Agujero Escariado   Generalmente un simple agujero pasante con la porción superior engrandecida cónicamente, frecuentemente a 82°, para acomodar la cabeza estándar de los tornillos de máquina.

CPU   Central Processing Unit (see also *Central Processing Unit*).

CPU   Unidad Central de Procesamiento (vea también *Unidad Central de Procesamiento*).

Creative Process   A process designed to produce a finished design. The elements include: accumulating the necessary information about the problem; "incubating" this information with other ideas that the brain retrieves from memory; illuminating its presence until a potential/probable solution is found; verifying its workability.

Proceso Creativo   Un proceso diseñado para producir un diseño terminado. Los elementos incluyen el acumular la información necesaria acerca del problema; "incubar" esta información con otras ideas que el cerebro tiene en memoria; iluminar su presencia hasta que se encuentre una solución potencial/probable; verificar su viabilidad.

Crest   The top point joining the sides of a thread.

Cresta   El punto superior que une los lados de una rosca.

Cursor   (1) On a CRT, a movable marker that is visible on the viewing surface and is used to indicate a position at which an action is to take place or the position on which the next device operation would normally be directed. (2) On digitizers, a movable reference, usually optical crosshairs used by an operator to indicate manually the position of a reference point or line where an action is to take place.

Cursor   (1) En un tubo de rayos catódicos, un marcador movible que se puede ver en la pantalla y que se usa para indicar la posición donde se llevará a cabo una acción o la posición adonde normalmente se dirigirá la próxima operación. (2) En los digitalizadores, una referencia movible, generalmente un "hilo de cabello" óptico usado por un operador para posicionar manualmente un punto o línea de referencia donde se llevará a cabo la acción.

Curve Problem   Differs slightly from the data problem in that the data points are to be part of a design development to generate curved lines.

Problema de Curva   Difiere un poco del problema de datos en que los puntos de datos serán parte del desarrollo de diseño para generar líneas curvas.

Cutoff Saw   Used to cut rough bar stock to lengths suitable for machining in other machine tools.

Sierra Tronzadora   Se usa para cortar en bruto las barras de materiales en largos apropiados para labrarse con otras herramientas de labrado.

Cuts and Fills   Leveling out a surface by excavating the earth of some of the higher points and using it to fill in the low points.

Cortar y Rellenar   Aplanar una superficie excavando el terreno en los puntos más altos y usándola para rellenar los puntos bajos.

Cutting Off   A cutting operation that is performed on a part to produce an edge other than a straight edge.

Cortar   Una operación de corte que se hace en una pieza para producir un borde diferente a un borde recto.

Cutting-Plane Line   Indicates that path an imaginary cutting plane follows to slice through an object.

Línea Plano Cortante   Indica el camino que sigue el plano cortante imaginario para cortar el objeto.

Cylindrical   A design where a shaft rotates exactly as it does in the radial design, but the action or direction of the follower differs greatly.

Cilíndrico   Un diseño en el cual el eje rota exactamente como en el diseño radial, pero la acción o dirección del seguidor varía grandemente.

Cylindrical Grinder   Used to precisely grind cylindrical or conical workpieces.

Moledora Cilíndrica   Se usa para moler piezas de trabajo cilíndricas o cónicas.

Cylindricity   A tolerance that simultaneously controls the surface of revolution for straightness, parallelism, and circularity of a feature, and is independent of any other features on a part. Also a feature control in which all elements of a surface of revolution form a cylinder. It gives the effect of circularity extended the entire length of the object, rather than just a specified cross section.

Cilindricidad   Una tolerancia que simultáneamente controla la superficie de revolución para rectitud, paralelismo y circularidad de un rasgo, y que es independiente de cualquier otro rasgo en la pieza. También un rasgo de control al en el cual todos los elementos de una superficie de revolución forman un cilindro. Proporciona el efecto de circularidad extendida a lo largo del largo completo de un objeto en lugar de sólo en la sección transversal.

Data   A representation of facts, concepts, or instructions in a formalized manner suitable for communication, interpretation, or processing by human or automatic means.

Datos   Una representación de los hechos, conceptos o instrucciones, en un formato adecuado para la comunicación, interpretación o procesamiento por medios automáticos o humanos.

Data Problem   Involves two changing variables, one independent and the other dependent.

Problema de Datos   Consiste de cambiar dos variables, una independiente y la otra dependiente.

Datum   Points, lines, planes, cylinders, and the like, assumed to be exact for purposes of computation, from which the location or geometric relationship (form) of features of a part may be established.

Referencia   Puntos, líneas, planos, cilindros y así por estilo, que se asume que son exactos para propósitos de computación, desde los cuales se puede establecer la localización o forma geométrica de los rasgos de una parte.

Datum Feature   The actual physical feature on a part used to establish a datum.

Rasgo de Referencia   El rasgo físico real de una pieza que se usa para establecer una referencia.

Datum Feature Simulator   A surface, the form of which is of such precise accuracy that it is used to simulate the datum.

Simulador de Rasgo de Referencia   Una superficie, cuya forma es de tanta precisión que se usa para simular la referencia.

Datum Feature Symbol   This symbol identifies datum features. It consists of a capital letter enclosed in a square frame. A leader line extends from the frame to the selected feature.

Símbolo de Rasgo de Referencia   Este símbolo identifica los rasgos de referencia. Consiste de una letra mayúscula en un cuadro cuadrado. Una directriz se extiende desde el cuadro hasta el rasgo seleccionado.

Datum Identification Symbol   A special rectangular box that contains the datum reference letter.

Símbolo de Identificación de Referencia   Un cuadro rectangular especial que contiene la letra de la referencia.

Datum Plane   A theoretically perfect plane from which measurements are made.

Plano de Referencia   Un plano teóricamente perfecto desde el cual se hacen las medidas.

Datum Reference Frame   A hypothetical three-dimensional frame that establishes the three axes of an X-Y-Z coordinate system into which the object being produced fits and from which measurements can be made.

Marco de Referencia   Un marco de referencia tridimensional e hipotético, que establece los tres ejes de un sistema de coordenadas X-Y-Z sobre el cual encaja el objeto en producción y desde el cual se hacen las medidas.

Datum Surface (feature)   A surface, the form of which is of such precise accuracy that it is used to simulate the datum.

Superficie de Referencia (rasgo)   Una superficie, cuya forma es de tal precisión que se usa para simular la referencia.

Datum Target Symbol   Identifies datum targets. It consists of a circle divided in half with a horizontal line.

Símbolo de Blanco de Referencia   Identifica los blancos de referencia. Consiste de un círculo dividido en dos con una línea horizontal.

Decimal-Inch Dimensioning Frequently used in the dimensioning of technical drawings. It is much less cumbersome than fractional and still used more than metric.

Dimensionamiento de Pulgada Decimal Se usa frecuentemente en el dimensionamiento de los dibujos técnicos. Es mucho menos incómodo que el dimensionamiento fraccional y se usa más que el métrico.

Decimal-Inch Scale Also called a civil engineer's scale (see *Civil Engineer's Scale*).

Escala de Pulgada Decimal También se conoce como la escala de ingeniero civil (vea *Escala de Ingeniero Civil*).

Decision Table A table that systematically compares solution alternatives with every other alternative as to how well each one satisfies the constraints and specifications from the definition of the problem.

Tabla de Decisiones Una tabla que sistemáticamente compara las alternativas de soluciones con otras alternativas acerca de cuán bien cada una satisface las restricciones y especificaciones de la definición del problema.

Dedendum The radial distance from the pitch diameter to the bottom of the tooth.

Altura de Cabeza Baja La distancia radial desde el diámetro de paso a la parte inferior del diente.

Delineation Pictorial representations; a chart, a diagram, or a sketch.

Delineación Representaciones pictóricas; una gráfica, un diagrama o un croquis.

Depth of Thread The distance between the crest and the root of the thread, as measured at a right angle to the axis.

Profundidad de Rosca La distancia entre la cresta y la raíz de la rosca, medida a ángulos rectos del eje.

Design File A file that includes freehand sketches, machine copies of technical information, copies of related drawings, summaries of telephone conversations, interviews, and vendor brochures.

Archivo de Diseño Un archivo que incluye dibujos a pulso, copias a máquina de información técnica, copias de dibujos relacionados y resúmenes de conversaciones telefónicas, entrevistas y folletos de vendedores.

Design Hints Simple ideas used to help create the finished product. They help in initial designing and become checklists as the project progresses.

Sugerencias de Diseño Ideas simples usadas para ayudar a crear el producto final. Ayudan en el diseño inicial y se convierten en listas de comprobación a medida que el proyecto progresa.

Design Procedure Acts as a general guideline for a team of designers, engineers, and drafters on the best way to accomplish the procedure sequentially.

Procedimiento de Diseño Sirve como una guía general para un equipo de diseñadores, ingenieros y delineantes acerca de la mejor manera de lograr el procedimiento secuencialmente.

Design Section This section illustrates the uses of the drawing and the related engineering practices. It includes: preparing assembly, detail, and layout drawings; following design procedures; making comprehensive parts lists; executing engineering revisions; handling invention agreements; keeping personal technical files; and preparing quotations.

Sección de Diseño Esta sección ilustra los usos del dibujo y las prácticas de ingeniería relacionas. Incluye: el preparar dibujos del ensamblaje, de detalles y del arreglo; seguir los procedimientos de diseño; hacer listas de partes comprensivas; ejecutar revisiones de ingeniería; manejar los acuerdos de invención; mantener archivos técnicos personales; y el preparar cotizaciones.

Design Size The size of a feature after an allowance for clearance has been applied and tolerances have been assigned.

Tamaño de Diseño El tamaño de un rasgo después de aplicarle una concesión para separación y asignar las tolerancias.

Detail Drawing A drawing of a single part that provides all the information necessary in the production of that part.

Dibujo Detallado Un dibujo de una sola pieza que provee toda la información necesaria en la producción de esa pieza.

Detail Notes Specific notes that pertain to one particular element or characteristic of a part.

Notas de Detalles Notas específicas relacionadas a un elemento o característica particular de una pieza.

Developed Length The total length of a pattern.

Largo de Desarrollo El largo total de un patrón.

Development Drawing of the surface of an object unfolded on a plane. Also, the pattern or template of a shape that is laid out in a single flat plane in preparation for the bending or folding of a material to a required shape.

Desarrollo Dibujo de la superficie de un objeto desdoblado en un plano. También, el patrón de una figura que se traza en un solo plano liso en preparación para el doblez de un material a una forma deseada.

Development Surface The exterior and/or interior of the sheet material used to form an object.

Superficie de Desarrollo El exterior y/o interior del material en hoja usado para formar un objeto.

Deviation The variance from a specified dimension or design requirement.

Desviación La variación de una dimensión o requisitos de dimensión especificada.

DFA (Design for Assembly) Procedure An iterative loop that occurs simultaneously with the DFM loop. The emphasis is on assembly of parts. The DFA loop first considers which parts should be assembled automatically, and next considers which parts need to be assembled manually.

Procedimiento de Diseño para Ensamblaje Un ciclo iterativo que ocurre simultáneamente con el ciclo DFM. El énfasis está en el ensamblaje de las piezas. El ciclo DFA considera primero cuáles partes deberían ser ensambladas manualmente, y después considera cuáles partes deberían ser ensambladas por mano.

DFD (Design for Disassembly) A design program with the goal of designing products that can be disassembled so that the greatest amount of reusable or recyclable materials, such as metal and plastic, can be recovered.

Diseño para Despiezado Un programa de diseño con el propósito de diseñar productos que pueden desensamblarse de manera que se pueda recuperar la cantidad más grande de materiales reciclables o reusables, tales como metales o plásticos.

DFM (Design for Manufacturability) A procedure in the form of an iterative loop that acts as a guide for a design team.

Diseño para Manufacturabilidad Un procedimiento en forma de un ciclo iterativo que actúa como una guía para el equipo de diseño.

DFR (Design for Recyclability) A design program with the goal of designing products made from materials that can be recycled for a variety of purposes.

Diseño para Reciclabilidad Un programa de diseño con el propósito de diseñar productos hechos de materiales que pueden reciclarse para una variedad de propósitos.

Diagram A figure or drawing that is marked out by lines; a chart or outline.

Diagrama Una figura o dibujo que está marcado por líneas; una gráfica o bosquejo.

Diameter The length of a straight line passing through the center of a circle and terminating at the circumference on each end.

Diametro El largo de una línea recta que pasa a través del centro de un círculo y que termina en la circunferencia en cada extremo.

Diametral Pitch A ratio equal to the number of teeth (N) on a gear per inch of pitch diameter (D).

Largo de Paso Diametral Una razón equivalente al número de dientes (N) en un engranaje por pulgada del diámetro de paso (D).

Diazo Material that is either a film or paper sensitized by means of azo dyes used for photocopying.

Diazo Material que es o una película o un papel hecho sensitivo por medio de tintes azo usados para fotocopias.

Die The term has a double meaning. It can be used to describe the entire assembled tool or the lower cutting part of the tool. Also a hardened metal piece shaped to cut or form a required shape in a sheet of metal by pressing it against a mating die.

Troquel El término tiene significado doble. Puede usarse para describir una entera herramienta ensamblada o la parte inferior de la herramienta que corta. También es un pedazo de metal endurecido formado para cortar la figura necesaria en una hoja de metal presionándolo contra otro troquel.

Die Casting A process in which molten metal is forced into metal dies under pressure, producing very accurate and smooth castings.

Fundición a Presión Un proceso en el cual el metal derretido se fuerza en discos metálicos bajo presión, produciendo piezas de fundición muy precisas y suaves.

Die Stamping The process of cutting or forming a piece of sheet metal with a die.

Estampado de Troquel El proceso de cortar y formar un pedazo de hojalata con un troquel.

Digitize To use numeric characters to express or represent data.

Digitalizar Usar caracteres numéricos para expresar o representar data.

Digitizer A device for converting positional information into digital signals; typically, a drawing or other graphic is placed on the measuring surface of the digitizer and traced by the operator using a cursor.

Digitalizador Un aparato para convertir la información posicional en señales digitales; típicamente, se coloca un dibujo u otra gráfica en la superficie de medida del digitalizador y se traza por operador usando un cursor.

Dimension Measurements given on a drawing, such as size and location.

Dimensión Medidas dadas en un dibujo, tales como tamaño y localización.

Dimension Line This line is drawn as a solid, thin line of approximately the same width as a hidden line. It should be broken for dimensions and have arrowheads for terminations. It is used to indicate graphically the linear distance being dimensioned.

Línea de Dimensión Esta línea se dibuja como una línea sólida y fina de aproximadamente el mismo espesor que una línea escondida. Debería ser entrecortada para dimensiones y debería tener flechas para las terminaciones. Se usa para indicar gráficamente la distancia lineal que se está dimensionando.

Dimensioning The process whereby size and location data for the subject of a technical drawing are provided.

Dimensionamiento El proceso en el cual se provee información de tamaño y localización para el sujeto de un dibujo técnico.

Dimetric Drawing Differs from isometric drawing in that only two angles are equal.

Dibujo Dimétrico   Difiere de los dibujos isométricos en que sólo dos ángulos son iguales.

Dip   The downward angle of the vein with respect to a horizontal plane at the site of the ore vein.

Depresión   El ángulo hacia abajo de una vena con respecto al plano horizontal en el lugar de la vena de mineral.

Direct Numerical Control (DNC)   The use of a common host computer for the distribution of part programs to remote machine tools.

Control Numérico Directo   El uso de una computadora común para la distribución de programas de parte para las herramientas de labrado remotas.

Direction   This is originally handed down from higher management. It outlines where the team should be going, what they are hoping to accomplish, and specific as well as general guidelines as to how to get there.

Dirección   Esto originalmente se envía desde la administración superior. Delinea adónde debería dirigirse un equipo, lo qué esperan lograr y guías específicas y generales de cómo llegar a ese punto.

Direction of Winding   The direction in which a spring is wound. It can be wound right or left hand.

Dirección de Enrollado   La dirección en la cual está enrollado un resorte. Puede estar enrollado a mano derecha o a mano izquierda.

Displacement Diagram   A designated layout of the required motion of a cam. It is laid out on a grid, and its length represents one complete revolution of the cam.

Diagrama de Desplazamiento   Una disposición designada que enseña la moción requisita de una leva. Se presenta sobre una rejilla, y su largo representa una revolución completa de la leva.

Display Monitor   An output device resembling a television that displays the information on which the CAD technician is working.

Monitor de Despliegue   Un aparato de salida parecido a un televisor que despliega la información en la cual el técnico de CAD está trabajando.

Double Weld   Similar to a multiweld, the same type of weld may be used on opposite sides of a single joint; for example, double V, double U, or double J welds. Used for strength and appearance.

Soldadura Doble   Similar a una soldadura múltiple, el mismo tipo de soldadura se puede usar en lados opuestos de una unión; por ejemplo, soldaduras V doble, U doble o J doble. Se usa para fortaleza y apariencia.

Double-Curved Surface   An object fully formed by curved lines with no straight lines.

Superficie de Doble Curva   Un objeto completamente formado por líneas curvas sin ninguna línea recta.

Double-Sided Board   A type of printed circuit board. These boards have components on one side, but the circuits are etched on both sides.

Tarjeta de Doble Lado   Un tipo de tarjeta de circuito impreso. Estas tarjetas tienen componentes en un lado, pero los circuitos están impresos en ambos lados.

Dowel   A cylindrical pin, commonly used to prevent sliding between two contacting flat surfaces.

Clavija   Un perno cilíndrico, comúnmente usado para evitar el deslizamiento entre dos superficies planas en contacto.

Draft   The slope or taper of the sides of a pattern, which permits it to be removed from the cope and drag without disturbing the cavity.

Calada   La pendiente o declive de los lados de un patrón, lo cual le permite que se remueva el armazón superior e inferior sin perturbar la cavidad.

Drafters   People who do drafting.

Delineante   Personas que hacen dibujos industriales.

Drafting   Creating technical drawings to support the design process.

Dibujo Industrial   El crear dibujos técnicos para apoyar al proceso de diseño.

Drag   In sand casting, this is the bottom half of the flask.

Armazón Inferior   En la fundición de molde en arena, ésta es la mitad inferior del armazón.

Draw   To temper steel.

Estirar   Templar el acero.

Drawing   A process of stretching and forming a metal sheet into a shape similar to a cup or top hat.

Estirado   El proceso de estirar y formar una hoja de metal en una forma similar a una taza o un sombrero de copa.

Drawing Surface   Whether a tabletop or drawing board, it must be flat, smooth, and large enough to accommodate the drawing and some drafting equipment.

Superficie de Dibujo   Ya sea la superficie de la mesa o tabla de dibujo, debe ser plana, lisa y lo suficientemente grande para acomodar el dibujo y el equipo de dibujo.

Drill   To cut a cylindrical hole with a drill; a blind hole does not go through the piece.

Taladrar   Cortar un agujero cilíndrico con una taladradora; un agujero ciego no atraviesa completamente la pieza.

Drill Bushings   Used to position and guide the cutting tools used in jigs. The three basic types are press-fit bushings, renewable bushings, and liner bushings.

Cojinetes de Taladro   Se usan para posicionar y guiar las herramientas de corte que se usan en guías. Los tres tipos básicos son los cojinetes de encaje a presión, los renovables y los de revestimiento.

Drill Press   A machine that is used mainly for producing holes.

Taladradora de Columna   Una máquina que se usa mayormente para producir agujeros.

Drop Forging   The process of forming a heated metal bar, or billet, into dies by using a drop hammer or with great pressure.

Forjado   El proceso de formar una barra de metal calentada, o lingote, en matrices usando un martillo de forja o con gran presión.

Drop-Bow Compass   This compass is used for circles with diameters of .3″ (0.08 cm) to .50″ (1.3 cm). It is rotated by holding the knurled head between the thumb and the second finger.

Compás de Arco Caído   Este compás se usa para círculos con diámetros de .3″ (0.08 cm) a .50″ (1.3 cm). Se rota aguantando la cabeza estriada entre el dedo pulgar y el segundo dedo.

Dry Transfer Sheet   Designed according to the same principles as transfer cards. The only major difference is that they are sheets—not cards.

Hoja de Transferencia en Seco   Diseñado de acuerdo a los mismos principios de las tarjetas de transferencia. La única diferencia mayor es que son hojas y no tarjetas.

Duplex Feature   A method of indexing the fixture to the machine in two parts. In use, the first part is machined while the second is loaded. The fixture is then rotated, and the first is unloaded while the second is worked on, while a third is loaded on the empty side.

Rasgo Doble   Un método de fijar un aditamento para labrar dos piezas. En uso, se labra la primera pieza mientras se carga la segunda. Luego se rota el aditamento, se descarga la primera pieza mientras se trabaja en la segunda y se carga una tercera pieza en el lado vacío.

Dwell   The period of time during which the follower does not move.

Tiempo de Permanencia   El periodo de tiempo durante el cual el seguidor no se mueve.

Eccentric   Not having the same center; off center.

Excéntrico   Que no tiene el mismo centro; fuera de centro.

ECO (Engineering Change Order)   An order for a change in a drawing to make improvements or correct an error. An ECO is issued by an Engineering Change Request (ECR) committee after all departments concerned have agreed to the change.

ECO (Orden de Cambio de Ingeniería)   Una orden para cambiar un dibujo para hacer mejoras o para corregir un error. Se emite una Orden de Cambio de Ingeniería por medio de un comité de Petición de Cambio de Ingeniería después que todos los departamentos involucrados están de acuerdo con el cambio.

ECR (Engineering Change Request)   A request to make a change in a drawing to make improvements or correct an error. The ECR explains what change is to be made, what it was before, why the change is to be made, whether it affects other parts, and who suggested the change. The request is made to an ECR committee, which issues an ECO after all departments concerned have agreed to the change.

ECR (Petición de Orden de Cambio)   Una petición para hacer un cambio en un dibujo para hacer mejoras o para corregir un error. La Petición de Cambio de Ingeniería explica qué cambios se harán, cómo era antes, por qué se harán los cambios, si afecta a otras piezas y quién sugirió el cambio. La petición se la hace ante un comité de Petición de Cambio de Ingeniería, el cual emite una Orden de Cambio de Ingeniería después que todos los departamentos involucrados están de acuerdo con el cambio.

Elastomers   Linear polymers that are able to stretch and return to their original shape and size (like an elastic band).

Elastómero   Polímeros lineales que pueden estirarse y que regresan a su forma y tamaño original (como una banda elástica).

Electrical Discharge Grinding (EDG)   A nontraditional machining process similar to EDM in that it employs rapidly recurring electrical discharges between an electrode and the workpiece in a dielectric fluid. The difference is that with EDG the electrode is a rotating wheel made of graphite or brass.

Amoladura por Descarga Eléctrica   Un proceso de labrado no tradicional similar al labrado por descarga eléctrica (EDM) en que utiliza descargas eléctricas rápidas y recurrentes entre un electrodo y la pieza de trabajo en un líquido dieléctrico. La diferencia es que con la amoladura por descarga eléctrica el electrodo es una rueda rotativa hecha de grafito o azófar.

Electrical Discharge Machining (EDM)   A nontraditional machining process that removes metal using a series of rapidly recurring electrical discharges between an electrically charged cutting tool and the workpiece submerged in a dielectric fluid.

Labrado por Descarga Eléctrica   Un proceso de labrado no tradicional que remueve metal usando una serie de descargas eléctricas rápidas y recurrentes entre una herramienta cortante cargada eléctricamente y la pieza de trabajo sumergida en un líquido dieléctrico.

Electrical Discharge Wire Cutting (EDWC)   A nontraditional machining process that is similar to the bandsawing process. Material removal occurs as the result of an electrical discharge as a wire electrode is drawn through the workpiece.

Corte con Alambre por Descarga Eléctrica   Un proceso de labrado no tradicional que es similar al proceso de corte con sierra de cinta. La remoción de material ocurre como resultado de una descarga eléctrica de un electrodo de alambre, a medida que se lo pasa a través de la pieza de trabajo.

Electrical Plan   This plan shows the house's electrical system superimposed on a floor plan without dimensions.

Plano Eléctrico   Este plano muestra el sistema eléctrico de la casa sobrepuesto en el plano de distribución sin dimensiones.

Electrochemical Discharge Grinding (ECDG)   A combination of two nontraditional machining processes: electrochemical grinding (ECG) and electrical discharge grinding (EDG). In ECDG, alternating current or pulsating direct current moves from a bonded graphite wheel to a positively charged workpiece.

Amoladura por Descarga Electroquímica   Una combinación de dos procesos de labrado no tradicionales: amoladura electroquímica y amoladura por descarga eléctrica. En la amoldadura por descarga electroquímica, una corriente alterna o corriente directa pulsante se mueve de una rueda de grafito a la pieza de trabajo cargada positivamente.

Electrochemical Grinding (ECG)   A nontraditional machining process that combines abrasion with an electrochemical reaction that occurs on the surface of the metal.

Amoladora Electroquímica   Un proceso de labrado no tradicional que combina la abrasión con una reacción electroquímica que ocurre en la superficie del metal.

Electrochemical Honing (ECH)   A nontraditional machining process similar to ECG. It combines electrochemical oxidation of surface metal and abrasion. The primary difference is that honing is a process of removing material inside a drilled hole.

Afinación Electroquímica   Un proceso de labrado no tradicional similar a la amoladura electroquímica. Combina la oxidación electroquímica de una superficie de metal y abrasión. La diferencia primaria es que la afinación es un proceso de remover material dentro de un agujero taladrado.

Electrochemical Machining (ECM)   The process, with many variations, of removing metal by combining electricity with a chemical to cause a reaction that is the opposite of metal plating.

Labrado Electroquímico   El proceso, con muchas variaciones, de remover metal combinando electricidad con un químico para causar una reacción que es lo opuesto del enchapado de metal.

Electrochemical Turning (ECT)   A variation of ECM. The primary difference is that the workpiece rotates.

Torneado Electroquímico   Una variación del labrado electroquímico. La diferencia principal es que la pieza de trabajo rota.

Electromechanical Machining (EMM)   A process that improves the performance of traditional machining processes. Metal removal is accomplished in the tradition-

al manner except that the workpiece is electrochemically polarized by applying a controlled voltage across the connection point between the workpiece and an electrolyte. This changes the surface of the workpiece, making it easier to machine.

Labrado Electromecánico   Un proceso que mejora el rendimiento de los procesos de labrado tradicional. La remoción de metal se logra de una manera tradicional, excepto que la pieza de trabajo se polariza electroquímicamente aplicándole un voltaje controlado a través de un punto de conexión entre la pieza de trabajo y un electrolito. Esto cambia la superficie de la pieza de trabajo, haciéndola más fácil de labrar.

Electron Beam Machining (EBM)   A thermal energy process. It uses thermal energy to remove metal.

Labrado con Rayo de Electrones   Un proceso de energía termal. Usa energía termal para remover metal.

Elevations   Orthographic views of the front, back, and sides of a structure. Most are exterior views; however, interior elevations of the kitchen, bathrooms, and utility areas are frequently included in residential plans.

Elevaciones   Vistas ortográficas del frente, la parte trasera y los lados de una estructura. La mayoría son vistas exteriores; sin embargo, elevaciones interiores de las cocinas, baños y áreas útiles se incluyen frecuentemente en los planos residenciales.

E-Mail   Short for electronic mail, or person-to-person written communication sent electronically.

E-mail   Abreviación para correo electrónico, o comunicación escrita de persona a persona enviada electrónicamente.

Empirical Equations   Equations that rely on experience of observation alone, often without due regard for system or theory. They can also be verified or disproved by observation or experiment.

Ecuaciones Empíricas   Ecuaciones que dependen de la experiencia de la observación solamente, frecuentemente sin consideración de sistema o teoría. Pueden verificarse o refutarse por observación o experimento.

Engineering Materials   In modern manufacturing settings this term refers to the raw materials.

Materiales de Ingeniería   En el ambiente de manufactura moderna, este término de refiere a los materiales brutos.

Epicycloid   If the span of a cycloid is convex, an epicycloid is drawn.

Epicicloide   Si la envergadura de un cicloide es convexa, se dibuja un epicicloide.

Equilateral Triangle   A triangle having three sides of equal length, as well as three equal interior angles.

Triángulo Equilátero   Un triángulo que tiene tres lados de largos iguales, al igual que tres ángulos interiores iguales.

Equilibrant   The equal and opposite reaction to a resultant if the vectors being added or subtracted are force vectors.

Equilibrante   La reacción igual y opuesta a una resultante si los vectores que se suman o restan son vectores de fuerza.

Equilibrium   The state of a solid body, either at rest or moving at a constant velocity, where the sum of the forces acting on the body in any direction is zero.

Equilibrio   El estado de un cuerpo sólido, ya sea en descanso o en movimiento a velocidad constante, donde la suma de las fuerzas que actúan sobre el cuerpo en cualquier dirección es cero.

Expanded Diameter   When metal is displaced to each side, forming a raised portion or flute extending along each side of the groove, the crest of the flute creates this diameter. The diameter is a few thousandths larger than the nominal diameter of the stock.

Diámetro Expandido   Cuando se despliega el metal a cada lado, formando una porción levantada o canaladura que se extiende a cada lado de la ranura, la cresta de la canaladura crea este diámetro. El diámetro es unas cuantas milésimas más grande que el diámetro nominal del material.

Exploded Isometric Drawing   This style shows all the components of an assembly as individual isometric drawings, including fasteners; however, the components are aligned in space the identical way they are to be assembled.

Dibujo Isométrico Despiezado   Este estilo muestra todos los componentes de un ensamblaje como dibujos isométricos individuales, incluyendo los sujetadores; sin embargo, los componentes se alinean en el espacio idénticamente como serán ensamblados.

Extension Lines   Solid, thin lines of approximately the same width as hidden lines.

Líneas de Extensión   Líneas sólidas y finas de aproximadamente el mismo grueso que las líneas escondidas.

Extension Spring   A spring that offers resistance to a pulling force. In its free state the coils are usually touching or very close.

Resorte de Extensión   Un resorte que ofrece resistencia a una fuerza de tirada. En su estado libre las espirales están tocándose o están muy cerca.

External Broaching   Typically used to produce some gear teeth, plier jaws, and other similar details.

Brochado Externo   Se usa típicamente para producir algunos dientes de engranajes, las mandíbulas de alicates y otros detalles similares.

External Client   A large variety of people and organizations, both in classrooms and in industry.

Cliente Externo   Una gran variedad de personas y organizaciones, tanto en los salones de clases y en la industria.

External Rings   For axial assembly, rings that are expanded with pliers so they may be slipped over the end of a shaft, stud, or similar part.

Anillos Externos   Para los ensamblajes axiales, los anillos que se expanden con alicates de manera que puedan colocarse alrededor de un extremo de un eje, un perno o una parte similar.

External Threads   Threads located on the outside of a part, such as those on a bolt.

Roscas Externas   Roscas que están localizadas en el exterior de la parte, tales como aquellos en los pernos.

Extruding   A process of forcing metal through a die of a desired form and cross section.

Formar por Extrusión   Un proceso de forzar metal a través de un troquel de la forma y corte transversal deseado.

Extrusion   Metal that has been shaped by forcing it in its hot or cold state through dies of the desired shape.

Extrusión   Metal que se forma forzándolo en su estado frío o caliente a través de troqueles de la forma deseada.

Fabrication   A term used to distinguish production operations from assembly operations.

Fabricación   Un término que se usa para distinguir las operaciones de producción de las operaciones de ensamblaje.

Face   To finish a surface at right angles, or nearly so, to the center line of rotation on a lathe.

Encarar   Acabar en un torno una superficie a un ángulo recto, o casi a un ángulo recto, de la línea de rotación de centro.

FAO   Finish All Over.

Acabar por Todos Lados   Acabar por Todos Lados.

Fastener   A mechanical device for holding two or more bodies in definite positions with respect to each other.

Sujetador   Un aparato mecánico para sujetar dos o más cuerpos en posiciones definidas, una con respecto a la otra.

Feature   A portion of a part, such as a diameter, hole, keyway, or flat surface.

Rasgo   Una porción de una parte, tal como un diámetro, agujero, muesca de posicionamiento o superficie plana.

Feature Control Symbol   A rectangular box in which all data referring to the subject feature control are placed. These include the symbol, datum references, the feature control tolerance, and modifiers.

Símbolo del Control de Rasgo   Una caja rectangular en que están colocados todos los datos que refieren al control del rasgo especificado. Puede incluir el símbolo, referencias del datum, la tolerancia del control de rasgo, y los modificadores.

Ferrous   A metal having iron as its base material.

Ferroso Un metal que tiene hierro como su material base.

Field Weld Any weld not made in the factory; that is, a weld that is to be made at a later date, perhaps at final assembly on-site.

Soldadura de Campo Cualquier soldadura que no se hace en la factoría; esto es, una soldadura que se hará en algún punto en el futuro, quizás en el lugar del ensamblaje final.

File To finish or smooth with a file.

Limar Acabar o suavizar con una lima.

Fillet An interior rounded intersection between two surfaces.

Filete Una intersección interior redondeada entre dos superficies.

Fillet Weld A weld made at a concave junction formed where two metal parts meet.

Soldadura de Filete Una soldadura que se hace en una unión cóncava donde se encuentran dos piezas de metal.

Fin A thin extrusion of metal at the intersection of dies or sand molds.

Aleta Una extrusión fina de metal en la intersección de los troqueles o moldes de arena.

Finish General finish requirements such as paint, chemical, or electroplating, rather than surface texture or roughness (see *Surface Texture*).

Terminado Requisitos generales de terminado tales como pintura, químico o galvanización, en lugar de una textura de superficie o aspereza (ver *Textura de Superficie*).

Finishing Impression The final station on a drop forge die. It is used to finish the part to the desired shape and size.

Impresión Final La estación final en un troquel de forja. Se usa para terminar la pieza a la forma y tamaño deseada.

Finite-Element Analysis (FEA) A virtual means of analyzing the structural response of objects. In this method, a model of a part is broken up into a finite number of polygonal elements called a mesh.

Análisis de Elementos Finitos Un método virtual de analizar la respuesta estructural de los objetos. En este método, un modelo de una pieza se divide en un número finito de elementos poligonales llamados una malla.

Finite-Element Model (FEM) A three-dimensional model of an object, with nodes containing specific information about the characteristics of the object at that point. The nodes are connected together to create an FEA mesh, which is then analyzed.

Modelo de Elementos Finitos Un modelo tridimensional de un objeto, con nodos que contienen información específica acerca de las características del objeto en ese punto. Los nodos están conectados para crear una malla de análisis de elementos finitos, que luego se analiza.

Fit The clearance or interference between two mating parts.

Encaje La separación o interferencia entre dos partes que encajan.

Fixed-Renewable Bushing Used for applications where the bushings do not need to be changed often, but where they are changed more often than press-fit bushings.

Cojinete Fijo Renovable Se usa para aplicaciones donde los cojinetes no necesitan cambiarse frecuentemente, pero que se cambian más frecuentemente que los cojinetes de encaje a presión.

Fixture A workholding tool that references the cutting tool to the workpiece by means of a set block.

Aditamento Una herramienta de trabajo que refiere la herramienta cortante a la pieza de trabajo por medio de un bloque.

Flange A relatively thin rim around a piece.

Reborde Un borde relativamente fino alrededor de una pieza.

Flange Weld Used to join thin metal parts together. Instead of welding to the surfaces of the parts to be joined, the weld is applied to the edges of thin material.

Soldadura de Reborde Se usa para unir piezas finas de metal. En lugar de soldar en las superficies a unirse, la soldadura se aplica a los bordes del material.

Flask In sand casting, this is a two-part box, or frame, used to contain the sand.

Armazón En los modelos de fundación de arena, ésta es la caja de dos partes, o armadura, que se usa para contener arena.

Flat Pattern A layout showing true dimensions of a part before bending; may be actual-size pattern on polyester film for shop use.

Patrón Plano Un trazado que muestra las dimensiones verdaderas de una pieza antes de doblarse; puede ser un patrón de tamaño real en una película de poliéster para usarse en el taller.

Flat Spring This spring is usually flat. Made of spring steel, it is designed to perform a special function. It is considered a power spring, and it resists a pressure.

Resorte Plano Este resorte generalmente es plano. Está hecho de acero de resorte, está diseñado para llevar a cabo una función específica. Se considera un resorte de potencia y resiste a la presión.

Flatbed Plotter A plotter that draws an image on a data medium such as paper or film mounted on a table.

Trazador Gráfico Plano Un trazador gráfico que dibuja una imagen en un medio tal como papel o película sobre una mesa.

Flatness A tolerance that controls the amount of variation that the perfect plane on a feature independent of any other features on the part. Also, a feature control of a surface that requires all elements of the surface to lie within two hypothetical parallel planes.

Llanura Una tolerancia que controla la cantidad de variación del plano perfecto en un rasgo independiente de otros rasgos en la pieza. También, un control de rasgo de una superficie que requiere que todos los elementos de la superficie estén dentro de dos planos paralelos hipotéticos.

Flat-Plane Surface A surface where all points can be interconnected to form straight lines without exception.

Superficie Plana Una superficie donde todos los puntos pueden conectarse para formar líneas rectas sin excepción.

Floor Plan A plan view of the structure in section with the ceiling and roof removed.

Plano de Distribución Una vista aérea de la estructura en sección con el techo removido.

Flowchart Students encounter these charts in computer programming courses. Other uses include flow of information in an organization, flow of materials in a manufacturing plant, and flow of energy uses in an agricultural facility.

Diagrama de Flujo Los estudiantes encuentran estas gráficas en los cursos de programas de computadoras. Otros usos incluyen el flujo de información en una organización, el flujo de materiales en una planta manufacturera y el flujo de uso de energía en una facilidad agricultural.

Flute Groove, as on twist drills, reamers, and taps.

Acanaladura Ranura, como en las barrenas, escariadores y los machos de rosca.

Fold Line The place where the imaginary perpendicular-orthographic viewing planes meet each other along the straight edge of intersection.

Línea de Doblez El lugar dónde los planos de vista imaginarios perpendiculares y ortográficos se encuentran a lo largo de los bordes rectos de intersección.

Foolproofing A method used to prevent parts from being loaded incorrectly. It involves a simple pin, which is normally enough to ensure proper loading of every part.

Hacer Infalible Un método usado para evitar que las partes sean cargadas incorrectamente. Consiste de un pasador simple, que normalmente es suficiente para asegurar que cada pieza esté cargada apropiadamente.

Force A directed action of one body on a second one that tends to change the state of motion, or rest, of the second body.

Fuerza Una acción dirigida de un cuerpo sobre un segundo cuerpo, que tiende a cambiar el estado de movimiento, o descanso, del segundo cuerpo.

Force Polygon These are drawn on force maps (same scale in all directions). Their major positive is their utility in science and technology courses. They are used in an optimization investigation.

Polígono de Fuerza Éstos se dibujan en mapas de fuerzas (misma escala en todas las direcciones). Su principal punto positivo es su utilidad en los cursos de ciencia y tecnología. Se usan en una investigación de optimización.

Foreshortening A phenomenon wherein the lengths of edges are shorter in the image than their actual lengths.

Escorzo Un fenómeno en el cual los largos de los bordes son más cortos en la imagen que sus largos actuales.

Forge To force metal while it is hot to take on a new shape by hammering or pressing.

Forjar Forzar el metal mientras está caliente para darle una nueva forma por medio de martilleo o prensado.

Form The form of a screw thread is actually its profile shape.

Forma La forma de una rosca de tornillo es en realidad su figura de perfil.

Forming An operation in which a part is bent or formed into a complex shape.

Moldeado Una operación en la cual una pieza se dobla o se forma en una figura compleja.

Form Tolerancing The permitted variation of a feature from the perfect form indicated on the drawing.

Tolerancia de Forma La variación permisible de un rasgo de su forma perfecta indicada en un dibujo.

Foundation Plan A plan view in section that shows the understructure of the building.

Plano de Cimiento Una vista aérea en sección que muestra la subestructura del edificio.

Fractional Dimensioning Generally not used on technical drawings, its primary use is on architectural or structural engineering drawings.

Dimensionamiento Fraccional Generalmente no se usa en los dibujos técnicos, su uso primario es en los dibujos arquitectónicos o estructurales.

Free Length The overall length of a spring when it is in its free state of unloaded condition.

Largo Libre El largo completo de un resorte cuando está en su estado libre de una condición sin carga.

Free-State Variation The concept that some parts cannot be expected to be contained.

Variación de Estado Libre El concepto de que no se puede esperar que algunas partes estén contenidas.

French Curve A thin plastic tool that comes in an assortment of curved surfaces. It is used to produce curved lines that cannot be made with a compass.

Curva Francesa Una herramienta plástica delgada que tiene una variedad de superficies curvas. Se usa para producir líneas curvas que no pueden hacerse con un compás.

Front View The orthographic view showing the front elevation.

Vista Frontal La vista ortográfica que muestra la elevación frontal.

Front View Auxiliary An auxiliary view projected from the front view.

Vista Frontal Auxiliar Una vista auxiliar proyectada desde la vista frontal.

Full Section A section of one of the regular multiviews that is sliced (cut) in two.

Sección Completa Una sección de una de las vistas múltiples regulares que fue cortada en dos.

Full Threads Threads that do not extend to the bottom of the hole.

Rosca Completa Las roscas que no se extienden hasta el fondo del agujero.

Full-Divided Scale A scale in which the units of measurements are subdivided throughout the length of the scale.

Escala Completa Dividida Una escala en la cual las unidades de medida están subdivididas a lo largo de la escala.

Fuller The first station on a drop forge die. This station is used to rough-form the part to fit the other cavities.

Batán La primera estación en un troquel de una forja. Esta estación se usa para formar la parte en bruto para que quepa en las otras cavidades.

Fusion Weld The intimate mating of molten metals producing the type of weld where a lot of heat is used to melt and fuse the joint.

Soldadura de Fusión La unión íntima de metales fundidos produciendo un tipo de soldadura en la cual se usa mucho calor para derretir y fundir los metales.

Gage A system of numbers used to specify the thickness of sheet metal.

Calibre Un sistema de números que se usan para especificar el espesor de una hoja de metal.

Galvanize To cover a surface with a thin layer of molten alloy, composed mainly of zinc.

Galvanizar Cubrir una superficie con una capa fina de una aleación derretida, compuesta mayormente por zinc.

Gasket A thin piece of rubber, metal, or some other material placed between surfaces to make a tight joint.

Junta Un pedazo fino de goma, metal u otro material que se coloca entre superficies para formar una unión fuerte.

Gate In sand casting, the opening in a sand mold at the bottom of the sprue through which the molten metal passes to enter the cavity or mold. Also, exit holes for the hot metal into the cavity of the mold.

Bebedero En la fundición en moldes de arena, la abertura en el molde de arena en la parte inferior de la entrada a través del cual pasa el metal derretido para entrar en la cavidad o molde. También, los orificios de salida para el metal caliente de la cavidad del molde.

Gate Valves Used to turn the flow of liquids through a pipe on or off without restricting the flow of the liquids through the valve or with as little restriction as possible.

Válvulas de Compuerta Se usan para encender o detener el flujo de metal derretido a través de una tubería sin restringir el flujo de los líquidos a través de la válvula o con tan poca restricción como sea posible.

Gear Tooth Caliper Used to check and measure an individual gear tooth.

Calibrador de Dientes de Engranaje Se usa para verificar y medir un diente de engranaje individual.

Gear Train Two or more gears used to achieve a designed RPM.

Tren de Engranajes Dos o más engranajes que se usan para obtener un número de revoluciones por minutos designado.

General Notes Broad items of information that have a job- or project-wide application rather than relating to one element or part.

Notas Generales Artículos amplios de información que tienen una aplicación a nivel del trabajo o de un proyecto en lugar de estar relacionadas con sólo un elemento o una parte.

General Oblique Drawing Similar to cabinet drawing except that the lines drawn along the direction of the receding distances may be at any scale the drafter chooses.

Dibujo Oblicuo General Similar a un dibujo de gabinete excepto que las líneas que se dibujan a lo largo de la dirección que se aleja pueden estar a cualquier escala escogida por el delineante.

Geometric Dimensioning and Tolerancing A means of dimensioning and tolerancing a drawing with respect to the actual function or relationship of part features that can be most economically produced; it includes positional and form dimensioning and tolerancing. It allows designers to set tolerance limits not just for the size of an object, but for all of the various critical characteristics of the part.

Dimensionamiento y Tolerancia Geométrica Un medio de dimensionamiento y de tolerar un dibujo con respecto a la función o relación actual de los rasgos de la pieza que pueden producirse más económicamente; incluye dimensionamiento y tolerancia de posición y de forma. Le permite a los diseñadores establecer límites de tolerancia no sólo para los tamaños de un objeto, pero para todas las características críticas de la parte.

Globe Valves Used to turn the flow of liquids through a pipe on and off.

Válvulas de Globo Se usan para permitir y detener el flujo de líquidos a través de una tubería.

Grade   The ratio of the vertical rise (R) to the horizontal run (H).

Pendiente   La razón entre la elevación vertical (R) y el tiro horizontal (H).

Graphic Communication   Using visual material to relate ideas.

Comunicación Gráfica   El uso de materiales visuales para comunicar ideas.

Graphical Differentiation   This technique, and the advantages to a designer, are similar to graphical integration, however, obtaining reasonably accurate results requires more skill. Used in the Pole and Ray method (see *Pole and Ray Method*).

Diferenciación Gráfica   Esta técnica, y sus ventajas al diseñados, es similar a la integración gráfica. Sin embargo, el obtener resultados razonablemente precisos requiere destreza. Se usa en el método de Polo y Rayo (vea *Método de Polo y Rayo*).

Graphical Integration   Provides designers with a visual, reasonably accurate approach to integration. This approach helps the designer "see" the results. Used in the Pole and Ray method (see *Pole and Ray Method*).

Integración Gráfica   Les provee a los diseñadores con un método visual y razonablemente preciso para integración. Este método ayuda a los diseñadores a "ver" los resultados. Se usa en el método de Polo y Rayo (vea *Método de Polo y Rayo*).

Graphical Method   A method of approximating areas by dividing them into convenient shapes, e.g., trapezoids and triangles, scaling appropriate lengths, and calculating the areas.

Método Gráfico   Un método de aproximar las áreas dividiéndolas en formas convenientes, por ejemplo, trapezoides y triángulos, ajustando los largos a la escala apropiada, y luego calculando las áreas.

Green Sand   A type of sand used in sand casting. The mold is prepared by ramming the green sand around a model of a part.

Arena Verde   Un tipo de arena que se usa en la fundición en moldes de arena. Los moldes se preparan compactando la arena verde alrededor de un modelo de una pieza.

Grind   To remove metal by means of an abrasive wheel, often made of carborundum; used where accuracy is required.

Moler   Remover material por medio de un material abrasivo, frecuentemente hecho de carborundo; se usa dónde se requiere precisión.

Groove Weld   Seven types of welds are considered to be groove welds: square, V, bevel groove, J, U, flare V, and flare bevel.

Soldadura de Ranura   Siete tipos de soldaduras se consideran soldaduras de ranuras: cuadrada, V, ranura bisel, J, U, V acampanado y bisel acampanado.

Grooved Fasteners   Used to solve metal-to-metal pinning needs with shear application.

Sujetadores con Ranura   Se usan para resolver las necesidades de sujetar metales a metales en aplicaciones con esquilado.

Grooved Studs   Widely used for fastening light metal or plastic parts to heavier members or assemblies.

Clavija Ranurada   Usado frecuentemente para sujetar partes de metal o plástico livianas a miembros de ensamblajes más pesados.

Group Technology   Grouping of parts into families based on similarities in design or production processes.

Tecnología de Grupo   Una agrupación de partes en familias basadas en similaridades de diseño o en el proceso de producción.

Gusset   A small plate used in reinforcing assemblies.

Placa de Refuerzo   Una pequeña placa que se usa para reforzar los ensamblajes.

Half Section   A type of section view in which the object is cut only halfway through and a quarter section is removed.

Media Sección   Un tipo de vista seccional en la cual el objeto se corta sólo hasta la mitad y se remueve un cuarto de sección.

Hard Copy   A copy of output in a visually readable form (e.g., printed reports, listings documents, summaries, and drawings).

Copia Impresa   Una copia de la salida en una forma leíble (por ejemplo, informes impresos, documentos de listados, resúmenes y dibujos).

Hard Disk Drive   A storage device used in computers; it may be contained within the computer console or in an external cabinet that is attached to the processing unit. Key concerns relating to hard disk drives are storage capacity (typically measured in megabytes and gigabytes of memory) and access time.

Unidad de Disco Duro   Un aparato de almacenamiento usado en las computadoras; puede estar contenido dentro de la consola de la computadora o en un gabinete externo que está conectado a la unidad de procesamiento. Las preocupaciones principales relacionadas a las unidades de disco duro son su capacidad de almacenamiento (típicamente medida en megabytes y gigabytes de memoria) y el tiempo de acceso.

Hardening   The process of heating a metal to a predetermined temperature, allowing the part to soak until thoroughly heated, and cooling it rapidly in a cooling material.

Endurecimiento   El proceso de calentar el metal a una temperatura predeterminada, permitiéndole que absorba el calor hasta estar calentado completamente, y luego enfriándolo rápidamente en un material de enfriamiento.

Hardness Test   Techniques used to measure the degree of hardness of heat-treated materials.

Prueba de Dureza   Técnicas usadas para medir el grado de dureza de materiales tratados con calor.

Hardware   The tools available to users of CAD for input, processing, and output.

Equipo Físico   Las herramientas disponibles a los usuarios de CAD para entrada, procesamiento y salida.

Harmonic Motion   A type of motion that is very smooth starting and stopping, but the speed is not uniform. It is good for fast speed.

Movimiento Harmónico   Un tipo de movimiento que es muy suave al comenzar y al detenerse, pero la velocidad no es uniforme. Es bueno para una velocidad rápida.

Heading   See *Bearing*.

Rumbo   Vea *Azimut*.

Heat Treatment   A series of processes used to alter or modify the existing properties in a metal to obtain a specific condition required for a workpiece. Also, to change the properties of metals by heating and then cooling.

Tratamiento de Calor   Una serie de procesos usados para alterar o modificar las propiedades existentes en un metal para obtener una condición especifica requerida para una pieza de trabajo. También, para cambiar las propiedades de los metales calentándolos y luego enfriándolos.

Heating/Ventilating/Air-Conditioning Plan   This plan shows the ductwork for the house's heating and cooling system superimposed on the floor plan without dimensions.

Plano de Calefacción/Ventilación/Aire Acondicionado   Este plano muestra los conductos para el sistema de calefacción y de enfriamiento superpuestos sobre el plano de distribución sin dimensiones.

Helical Gear   Operates on parallel shafts and transmits rotary motion in a manner identical to spur gears. The teeth have an involute shape but are not parallel to their rotational axis.

Engranaje Helicoidal   Opera en ejes paralelos y transmite el movimiento rotativo en una manera idéntica a los engranajes de dientes rectos. Los dientes tienen una forma intricada pero no son paralelos a su eje de rotación.

Helical Spring   Usually cyclindrical or conical in shape, there are three types: compression, extension, and torsion springs.

Resorte Helicoidal   Generalmente de forma cilíndrica o cónica, hay tres tipos: resortes de compresión, de extensión y de torsión.

Helix   A form of spiral used in screw threads, worm gears, and spiral stairways, among other things.

Hélice   Una forma de espiral que se usa en las roscas de tornillos, o engranaje de tornillo y las escaleras espirales, entre otras cosas.

Hemisphere   Hemispheres occur when a sphere is cut into two equal parts.

Hemisferio   Los hemisferios ocurren cuando una esfera se corta en dos partes iguales.

Herringbone Gear   A type of gear in which left-hand and right-hand teeth are combined.

Engranaje Doble Helicoidal   Un tipo de engranaje en el cual se combinan los dientes de vuelta derecha y de vuelta izquierda.

Hexagon   A polygon having six angles and six sides.

Hexágono   Un polígono que tiene seis ángulos y seis lados.

Hidden Lines   Lines not used in isometric drawings unless needed to show an important feature that otherwise would not be seen. Also feature outlines (for example, intersections of plane surfaces, real but invisible); used in each viewing plane where appropriate.

Líneas Escondidas   Líneas que no se usan en los dibujos isométricos a menos que sea necesario para mostrar rasgos importantes que no se verían de otra manera. También los bordes de rasgos (por ejemplo, las intersecciones de planos, reales pero invisibles); usados en cada plano de vista dónde sea apropiado.

High-Performance Alloys   A group of metals that have high strength and/or high corrosion resistance over a broad range of temperatures.

Aleación de Alto Rendimiento   Un grupo de metales que tienen alta resistencia y/o alta resistencia a la corrosión sobre un rango de temperaturas.

Highway Diagrams   This type of connection diagram collects wires that run along similar paths and combines them into groups. The components are located in the positions they will hold in the actual circuit.

Diagramas de Conexión   Este tipo de diagrama de conexión recolecta los cables que van a lo largo de caminos similares y los combina en grupos. Los componentes están ubicados en las posiciones que tendrían en el circuito actual.

Holography   This process produces images that simulate the effect of seeing the three-dimensionality of an object from multiple viewpoints.

Holografía   Este proceso produce imágenes que simulan el efecto de ver la tridimensionalidad de un objeto desde múltiples puntos de vista.

Home Page   The location on the Internet that is reached by using a given organization's Internet address. It is the Internet version of an organization's actual location.

Página de Portada   La localización en el Internet adonde se llega usando la dirección de Internet de una organización. Es la versión de Internet de la localización actual de una organización.

Hone A method of finishing a hole or other surface to a precise tolerance by using a spring-loaded abrasive block and rotary motion.

Afinar Un método de terminar un agujero u otra superficie a una tolerancia precisa usando un bloque accionado por resorte y un movimiento giratorio.

Horizontal Parallel to the horizon.

Horizontal Paralelo al horizonte.

Horizontal Boring Mills These machines perform a wide variety of different machining tasks normally associated with a milling machine but on a much larger scale.

Fresadoras de Taladrado Horizontal Estas máquinas llevan a cabo una variedad de diferentes tareas de labrado normalmente asociadas con una fresadora pero a una escala mucho más grande.

Human Factors Published tabulations of body dimensions of average males, females, and children.

Factores Humanos Tablas publicadas de dimensiones corporales de los varones, mujeres y niños promedios.

Hydrodynamic Machining (HDM) A nontraditional machining process that removes material using a high-velocity, high-pressure stream of liquid solution. It is used primarily for slitting and cutting nonmetallic materials.

Labrado Hidrodinámico Un proceso de labra no tradicional que remueve material usando un chorro de una solución líquida de alta velocidad y alta presión. Se usa principalmente para cortar en tiras y para cortar materiales no metálicos.

Icon A standardized graphical image designed to represent and communicate a computer application operation or function.

Icono Una imagen gráfica estandarizada diseñada para representar y comunicar la operación o función de una aplicación de computadora.

Inclined A line or plane at an angle to a horizontal line or plane.

Inclinado Una línea o plano a cualquier ángulo de una línea o plano horizontal.

Increment The length of a weld.

Incremento El largo de una soldadura.

Incremental System A system of numerically controlled machining that always refers to the preceding point when making the next movement; also known as continuous path or contouring method of NC machining.

Sistema Incremental Un sistema de labrado controlado numéricamente que siempre se refiere al punto anterior al hacer el próximo movimiento; también se conoce como el método de paso continuo o de contorno del labrado controlado numéricamente.

Indexing Jig Used primarily to machine parts that have machined details at intervals around the part.

Guía de Gradación Se usa principalmente para labrar partes que tienen detalles labrados a intervalos alrededor de una pieza.

Input The transfer of information into a computer or machine control unit (see *Output*).

Entrada La transferencia de información a una computadora o unidad de control de labrado (ver *Salida*).

Inside Diameter The inside diameter of a spring.

Diámetro Interior El diámetro interior de un resorte.

Intelligent Robot A robot that can make decisions based on data received from sensing devices.

Robot Inteligente Un robot que puede tomar decisiones basadas en los datos que recibe de los aparatos de sensores.

Interactive The capability of a computer application to permit an operator to change variables or test a design while the application is running. Virtual reality is an extreme example of an operator being able to interact with a computer application.

Interactiva La capacidad de una aplicación de computadora para permitirle a un operador cambiar variables o probar un diseño mientras la aplicación está ejecutándose. Realidad virtual es un ejemplo extremo de un operador capaz de interactuar con una aplicación de computadora.

Interchangeable Refers to a part made to limit dimensions so that it will fit any mating part similarly manufactured.

Intercambiable Referente a una pieza que fue hecha a dimensiones límites de manera que pueda encajar en cualquier pieza de acoplamiento manufacturada de manera similar.

Interface The means of interacting with a computer. Graphical user interfaces are becoming more prevalent and are designed to simplify applications by using graphical icons to guide the interaction with computer applications.

Interfaz El medio de interactuar con una computadora. Los interfaces de uso gráfico están convirtiéndose más comunes y son diseñados para simplificar las aplicaciones usando iconos para guiar la interacción con las aplicaciones de computadora.

Interference Fit This represents the smallest amount of interference.

Encaje de Interferencia Esto representa la cantidad más pequeña de interferencia.

Intermediate A translucent reproduction made on vellum, cloth, or film from an original drawing to serve in place of the original for making other prints.

Intermedio Una reproducción transparente hecha en vitela, tela o película desde un dibujo original para funcionar en lugar del original para hacer otras reproducciones.

Intermittent Weld   A series of short welds. When this type of weld is needed, an extra dimension is added to the welding symbol.

Soldadura Intermitente   Una serie de soldaduras cortas. Cuando se necesita este tipo de soldadura, se añade una dimensión al símbolo de soldadura.

Internal Clients   These clients can be instructors in classrooms and supervisory persons in industry.

Clientes Internos   Estos clientes pueden ser instructores en los salones de clases y supervisores en la industria.

Internal Rings   For axial assembly, these are compressed for insertion into a bore or housing.

Anillos Internos   Para un ensamblaje axial, éstos se comprimen para insertarlos en un agujero o un cárter.

Internal Thread   Threads located on the inside of a part, such as those on a nut.

Rosca Interna   Roscas ubicadas en el interior de una pieza, tales como las de una tuerca.

Internet   A worldwide network of loosely connected computer networks.

Internet   Una red mundial de redes de computadoras.

Invention Agreement   This form is signed by the drafter. It gives the company the right to any new invention designed while working with the company. It also (1) specifies that the employee will not reveal any of the company's discoveries or projects, and (2) is generally binding on the employee from six months to two years after leaving the company.

Acuerdo de Invención   Este formulario lo firma el delineante y le provee a la compañía el derecho de cualquier invención nueva diseñada por éste mientras trabaja con la compañía. También (1) especifica que el empleado no revelará ninguno de los descubrimientos o proyectos de la compañía, y (2) es generalmente obligatorio en el empleado desde seis meses hasta dos años después de irse de la compañía.

Investment Casting   A casting process in which a wax pattern is placed in a sand mold and the mold is fired to melt out the wax pattern, leaving a cavity into which molten metal is then poured.

Fundición en Molde de Cera   Un proceso de fundición en el cual se coloca un patrón de cera dentro de un molde de arena y se calienta el molde para derretir la cera, dejando una cavidad dentro de la cual se vierte el metal derretido.

Involute Curve   May be considered the path traced by the end of a string as it is unwound from a circle called a base circle. Also a spiral curve generated by a point on a chord as it unwinds from a circle or a polygon.

Curva Intricada   Puede considerarse como el camino trazado por el extremo de un hilo mientras se desenrolla de un círculo llamado un círculo base. También es una curva espiral generada por un punto en una cuerda a medida que se desenrolla de un círculo o un polígono.

Irregular Curve   The name for the type of curve that is drawn using a French curve.

Curva Irregular   El nombre dado al tipo de curva que se dibuja usando una curva francesa.

Isometric Assembly   These drawings appear in two popular formats: exploded and assembled (see *Exploded Isometric Drawing* and *Assembled Isometric Assembly Drawings*).

Ensamblaje Isométrico   Estos dibujos aparecen en dos formatos populares: despiezado y ensamblado (ver *Dibujo Isométrico Despiezado* y *Dibujo Isométrico Ensamblado*).

Isometric Drawing   The most used of the three common pictorials. They are easier to draw than perspective drawings, and they illustrate objects almost as well as perspective drawings. The drawings are so positioned that all three axes make equal angles with the picture plane, and measurements on all three axes are made to the same scale.

Dibujo Isométrico   El más usado de los tres tipos de dibujos pictóricos. Son más fáciles de dibujar que los dibujos de perspectiva e ilustran los objetos casi tan bien como los dibujos de perspectivas. Los dibujos se ubican de manera que los tres ejes forman ángulos iguales con el plano de dibujo, y las medidas en los tres ejes se hacen a la misma escala.

Isometric Projection   Causes dimensions along the three principal edges to be foreshortened to approximately 82% of their true length.

Proyección Isométrica   Causa que las dimensiones a lo largo de los tres ejes principales se acorten a aproximadamente un 82% de su largo verdadero.

Isometric Sketching   Sketching in which the length and width lines recede at 30° to the horizontal and the height lines are vertical.

Croquis Isométrico   Croquis en el cual las líneas de largo y ancho de las líneas retroceden a un ángulo de 30° de la horizontal y las líneas de alto son verticales.

Isopleth   The lines used to locate values in a nomogram.

Isolínea   Las líneas que se usan para localizar los valores en un nomograma.

Isosceles Triangle   A triangle with two sides of equal length and two equal interior angles.

Triángulo Isósceles   Un triángulo con dos lados de largos iguales y dos ángulos interiores iguales.

Iteration   A procedure in which repetition of a sequence of operations yields results successively closer to a desired result.

Iteración   Un procedimiento en el cual la repetición de una secuencia de operaciones provee resultados sucesivamente más cercanos al resultado deseado.

Jig   A workholding tool that guides the cutting tool through a hardened drill bushing during the cutting cycle.
Guía   Una herramienta que agarra piezas y guía la herramienta cortante a través de los cojinetes de taladro endurecidos durante el ciclo de corte.

Journal   Portion of a rotating shaft supported by a bearing.
Muñequilla   Porción de un eje que rota que está apoyada por un cojinete.

Just-in-Time Manufacturing (JIT)   An approach to manufacturing that seeks to completely eliminate waste.
Manufactura Justo a Tiempo   Un método de manufactura que busca eliminar el desperdicio por completo.

Kerf   Groove or cut made by a saw.
Corte   Ranura o corte hecho por una sierra.

Key   A small piece of metal sunk partly into both shaft and hub to prevent rotation of mating parts. It is a demountable part that provides a positive means of transferring torque between a shaft and a hub.
Lave   Un pedazo pequeño de metal parcialmente empotrado y en el eje y en el cubo para prevenir la rotación de partes que encajan. Es una parte removible que provee un medio positivo para transferir la fuerza de torsión entre un eje y un cubo.

Keyboard   The part of a CAD computer used for entering commands, text, annotation, and dimensions.
Teclado   La parte de una computadora de CAD que se usa para ingresar comandos, texto, anotaciones y dimensiones.

Keying   The process used for entering text on drawings, certain system commands, dimensions not entered automatically, and for logging on to the system.
Tecleo   El proceso usado para ingresar texto en los dibujos, ciertos comandos, dimensiones que no se ingresan automáticamente y para entrar al sistema.

Keyseat   An auxiliary-located rectangular groove machined into the shaft and/or hub to receive the key.
Asiento de Llave   Una ranura auxiliar rectangular que se labra en un eje y/o cubo para recibir la llave.

Keyslot   The slot machined in a shaft for Woodruff-type keys.
Ranura de Llave   Una ranura labrada en un eje para aceptar las llaves tipo Woodruff.

Keyway   A slot in a hub or portion surrounding a shaft to receive a key.
Muesca de Posicionamiento   Una ranura en un cubo o la porción que rodea a un eje para recibir una llave.

Kilo   Prefix meaning 1,000 units, as in kilometers (1,000 meters).
Kilo   Prefijo que significa 1,000 unidades, como en kilómetros (1,000 metros).

Knurl   A diamond-shaped or parallel pattern cut into cylindrical surfaces to improve gripping, the bonding between parts for permanent press fits and decoration.
Estrías   Un patrón en forma de diamante o paralelo que se corta en las superficies cilíndricas para mejorar su agarre, la adherencia de partes para encajes de presión permanentes y para decoración.

Laser Beam Machining (LBM)   A process that uses lasers for performing operations that are traditionally performed with cutting tools.
Labrado con Rayo Láser   Un proceso que usa láseres para llevar a cabo las operaciones que tradicionalmente se llevan a cabo con herramientas cortantes.

Lathe   One of the most commonly used and versatile machine tools in the shop. It can perform such operations as straight and taper turning, facing, straight and taper boring, drilling, reaming, tapping, threading, and knurling.
Torno   Una de las herramientas de labrado más frecuentemente usadas y más versátiles en el taller mecánico. El mismo puede llevar a cabo tales operaciones como el tornado, la cara, el taladrar derecho y gradualmente estrechado, el perforar, el escariar, el roscar y el estriar.

Layout   Lines drawn on material as guides for subsequent operations.
Trazado   Líneas que se dibujan en un material como guías para las operaciones subsiguientes.

Layout Development Drawing   A designer's freehand sketch redrawn to scale with instruments or a computer.
Dibujo de Desarrollo de Trazado   Un croquis hecho a mano libre por un diseñador dibujado nuevamente a escala con instrumentos o con una computadora.

Lead (noun. pronounced led)   A dark gray metal with the following characteristics: high density, low melting point, corrosion resistance, chemical stability, malleability, lubricity, and electrical properties.
Plomo   Un metal grisáceo oscuro con las siguientes características: densidad alta, punto de fundición bajo, resistencia a la corrosión, estabilidad química, maleabilidad, lubricidad y propiedades eléctricas.

Lead Time   The interval from the time the shop receives the drawing until production begins.
Tiempo de Acción   El intervalo de tiempo desde que el taller recibe el dibujo hasta que comienza la producción.

Leader Line   A thin line used to direct the information contained in notes or dimensions to the appropriate surface, feature, or point on a drawing.
Directriz   Una línea que se usa para dirigir la información contenida en notas o dimensiones a la superficie, rasgo o punto apropiado en un dibujo.

Least Material Condition (LMC)   A condition that exists when the least material is available for the features being dimensioned. It is the lower limit for an external feature and the upper limit for an internal feature.

Condición de Material Mínimo Una condición que existe cuando está disponible la cantidad mínima de material para los rasgos que se dimensionan. Es el límite inferior para un rasgo externo y el límite superior para un rasgo interno.

Letter of Transmittal A letter to a client, separate from but enclosed with a report, which states that the project has been completed, the report is enclosed, and that working with the client on future projects is anticipated.

Carta de Trámite Una carta a un cliente, aparte de pero que incluye un informe que afirma que el proyecto fue completado, que se incluye el informe, y que se anticipa trabajar con el cliente en proyectos futuros.

Lettering Machine A machine that outputs letters on a clear tape that is pressed onto the drawing surface.

Máquina de Letrado Una máquina que produce letras en una cinta transparente que se presiona sobre una superficie de dibujo.

Light Pen A handheld data-entry device used only with refresh displays. It consists of an optical lens and photocell, with associative circuitry mounted in a wand. Most light pens have a switch allowing the pen to be sensitive to light from the screen.

Pluma de Luz Un aparato de entrada de datos manual que se usa solamente con las pantallas refrescadas. Consiste de un lente óptico y una fotocelda, con el circuito asociado colocado en el mango. La mayoría de las plumas de luz tienen un interruptor que le permite a las plumas a ser sensibles a la luz de la pantalla.

Limit Dimensioning A tolerancing method that shows only the maximum and minimum dimensions that establish the limits.

Dimensionamiento de Límite Un método de tolerar que muestra sólo las dimensiones máximas y mínimas que establecen los límites.

Limits The maximum and minimum allowable sizes of a feature.

Límites Los tamaños máximos y mínimos permisibles para un rasgo.

Line Work A generic term given to all of the various techniques used in creating the graphic data on technical drawings.

Trabajo de Línea Un término genérico que se le da a todas las varias técnicas usadas en la creación de datos gráficos en los dibujos técnicos.

Lineless Diagrams A type of connection diagram that shows the electronic components with the connection points labeled, but it does not show wires. Instead of wires, the components are accompanied by a table that contains a designation, color, and destination code for each wire.

Diagramas sin Líneas Un tipo de diagrama de conexión que muestra los componentes electronicos con los puntos de conexión etiquetados, pero no muestra los cables. En lugar de cables, a los componentes les acompaña una tabla que contiene la designación, color y código de destino para cada cable.

Liner Bushings Used in conjunction with renewable bushings; they provide a hardened, wear-resistant mount for renewable bushings.

Cojinetes de Revestimiento Se usan en conjunto con los cojinetes renovables; proveen una montura endurecida, resistente al desgaste para los cojinetes renovables.

List Server A special program that forwards e-mail to all individuals on a list.

Servidor de Lista Un programa especial que envía mensajes de correo electrónico a todos los individuos en una lista.

Loaded Length The overall length of a spring with a special given or designed load applied to it.

Largo con Carga El largo completo de un resorte con una carga dada o de diseño aplicada sobre él.

Location Tolerance A tolerance that specifies the allowable variation from the perfect location.

Tolerancia de Localización Una tolerancia que especifica la variación permisible de una localización perfecta.

Log A record of activities completed and the time spent on each activity. It should take the who, where, what, when, and why style.

Registro Un registro de actividades completadas y el tiempo usado en cada actividad. Debería tener el estilo de quién, dónde, qué, cuándo y por qué.

Logarithm The exponent that indicates the power to which a number is raised to produce a given number.

Logaritmo El exponente que indica la potencia a la cual se eleva un número para producir un número dado.

Logic Diagrams Used in conjunction with the design of integrated circuits (ICs). They are most widely used with electronics that are based on the binary numbering concept.

Diagramas de Lógica Se una en conjunto con el diseño de circuitos integrados. Son más frecuentemente usados con electrónica que está basada en el concepto de numeración binaria.

Lowercase End Points Each line in space is identified by two of these.

Puntos de Extremos Minúsculos Cada línea en el espacio está identificada por dos de éstos.

Lowercase Letters Each point in space is called out by these.

Letras Minúsculas Cada punto en el espacio se llama por una de éstas.

Lug An irregular projection of metal, but not round as in the case of a boss; usually with a hole in it for a bolt or screw.
Saliente Una proyección de metal irregular, pero no redonda como en el caso de un saliente cilíndrico; generalmente con un agujero en él para aceptar un tornillo o un perno.

Machine Screw Used for screwing into thin materials. Most are threaded within a thread or two to the head. They sometimes incorporate a hex-head nut to fasten parts together.
Tornillo de Máquina Se usa para atornillar materiales finos. La mayoría tiene una rosca dentro de una rosca hasta la cabeza. A veces incorporan una tuerca hexagonal para unir las piezas.

Machining The process of removing metal with machine tools and cutters to achieve a desired form or feature.
Labrar El proceso de remover material con herramientas de labrado y cortantes para lograr una forma o un rasgo deseado.

Magnaflux A nondestructive inspection technique that uses a magnetic field and magnetic particles to locate internal flaws in ferrous metal parts.
Magnaflux Una técnica de inspección no destructiva que usa campos magnéticos y partículas magnéticas para localizar los defectos internos en las piezas de metales ferrosos.

Magnesium A soft, silvery white metal that is extremely light and machinable, but lacks the strength needed for structural uses. To increase its strength, it can be alloyed with various metals.
Magnesio Un metal blando, plateado blancuzco que es extremadamente liviano y fácil de labrar, pero carece de la fortaleza necesaria para usos estructurales. Para aumentar su fortaleza, se puede alear con varios metales.

Main Storage The general-purpose storage of a computer; usually main storage can be accessed directly by the operating registers.
Almacenamiento Principal El almacenamiento de aplicación general en una computadora; generalmente el almacenamiento principal lo pueden acceder directamente los registros operacionales.

Major Diameter The largest diameter of a screw thread, both external and internal.
Diámetro Mayor El diámetro más grande de una rosca de tornillo, tanto externo como interno.

Malleable Casting A casting that has been made less brittle and tougher by annealing.
Fundición Maleable Una fundición que se hizo menos quebradiza y fuerte por medio del recocido.

Manual of Steel Construction This manual contains the dimensions and other design and drafting-related information for structural steel shapes.
Manual de Construcción en Acero Este manual contiene las dimensiones y otra información relacionadas al dibujo industrial de las formas estructurales en acero.

Manufacturing The enterprise through which raw materials are converted into finished products.
Manufactura La empresa a través de la cual los materiales en bruto se convierten en productos acabados.

Manufacturing Resource Planning (MRPII) A concept that allows companies to systematically plan for the resources needed in manufacturing (i.e., materials, capacity of production systems, personnel, time, etc.).
Planificación de Recursos de Manufactura Un concepto que le permite a las compañías a planificar sistemáticamente para los recursos necesarios en la manufactura (esto es, materiales, capacidad de los sistemas de producción, personal, tiempo, etc.).

Maraging Steel A special class of high-strength steels in which nickel is the primary alloy.
Acero Marginizado Una clase especial de aceros de alta resistencia en los cuales el níquel es la aleación principal.

Mass Production Company An organization that produces large quantities of a product.
Compañía de Producción en Masa Una organización que produce grandes cantidades de un producto.

Mass Properties Analysis A process involving performing analysis computations concerning mass properties such as weight, volume, surface area, moment of inertia, and center of gravity.
Análisis de Propiedades de Masa Un proceso que consiste de hacer cálculos analíticos acerca de las propiedades de masa tales como peso, volumen, área de superficie, momento de inercia y centro de gravedad.

Material Handling The movement and handling of materials, either manually or through the use of powered equipment.
Manejo de Materiales El movimiento y manejo de materiales, ya sea manualmente o mediante el uso de equipo de potencia.

Maximum Material Condition (MMC) When a feature contains the maximum amount of material; that is, minimum hole diameter and maximum shaft diameter. It is the upper limit for an external feature and the lower limit for an internal feature.
Condición de Material Máximo Cuando un rasgo contiene la cantidad máxima de material, esto es, el diámetro de agujero mínimo y diámetro de eje máximo. Es el límite superior para un rasgo externo y el límite inferior para un rasgo interno.

Mean Diameter The theoretical diameter of the spring measured to the center of the wire diameter.

Diámetro Promedio El diámetro teórico de un resorte medido hasta el centro del diámetro del alambre.

Mechanical Method A method of approximating areas by using a planimeter.

Método Mecnico Un método de aproximar las áreas usando un planímetro.

Menu A list of options that are displayed either on the CRT or on a plastic or paper overlay.

Menú Una lista de opciones que se despliegan en el CRT o en una cubierta de plástico o papel.

Metes and Bounds A system used to identify the perimeters of any property. A metes and bounds land survey begins at a selected known point of origin called the point-of-beginning or POB and then describes property boundaries in relation to this by using distance measurements, compass bearings, and boundaries such as streets, fences, and rivers.

Medidas y Límites Un sistema usado para identificar los perímetros de cualquier propiedad. Cualquier estudio de planimetría de medidas y límites comienza con un punto conocido de origen llamado el punto de comienzo y luego describe las relaciones de límites en relación a éste usando medidas de distancia, orientaciones de compás y demarcaciones tales como calles, cercas y ríos.

Metric Dimensioning One of three dimensioning systems used on technical drawings.

Dimensionamiento Métrico Uno de tres sistemas de dimensionamiento que se usa en los dibujos técnicos.

Microcomputer A computer that is constructed using a microprocessor as a basic element of the CPU; all electronic components are arranged on one printed circuit board.

Microcomputadora Una computadora que se construyó usando un microprocesador como un elemento básico del CPU; todos los componentes electrónicos están puestos sobre una tarjeta de circuitos impresos.

Microprocessor The central processing unit of a microcomputer. The microprocessor is contained on a single integrated circuit chip.

Microprocesador La unidad de procesamiento central de una microcomputadora. El microprocesador está en un solo chip de circuito integrado.

Mill To remove material by means of a rotating cutter on a milling machine.

Fresar Remover material por medio de una cuchilla que rota en una fresadora.

Milli Prefix meaning one one-thousandth unit, as in millimeters (.001 meter).

Mili Prefijo que significa una milésima de unidad, como en milímetros (.001 metro).

Milling Machine Used to machine workpieces by feeding the part into a rotating cutter.

Fresadora Se usa para labrar las piezas de trabajo insertando la pieza en una cortadora giratoria.

Minor Diameter The smallest diameter of a screw thread, both external and internal.

Diámetro Menor El diámetro más pequeño de una rosca de un tornillo, tanto externa como interna.

Miter Gear Exactly the same as a bevel gear except that both mating gears have the same number of teeth. The shafts are at 90° to each other.

Engranaje de Inglete Exactamente igual al engranaje de bisel, excepto que los engranajes que se conectan tienen el mismo número de dientes. Los ejes están a 90° el uno del otro.

Modified Uniform Velocity It smoothes out the roughness slightly by adding a radius to the ends of the high and low points.

Velocidad Uniforme Modificada Suaviza la aspereza un poco añadiéndole un radio a los extremos de los puntos altos y bajos.

Modifier The application of MMC or RFS to alter the normally implied interpretation of a tolerance specification. There are symbols that can be attached to the standard geometric building block to alter their application or interpretation.

Modificador Una aplicación del MMC o RFS para alterar la interpretación normal implicada de una especificación de tolerancia. Los símbolos que se pueden adjuntar al bloque de construcción para alterar su aplicación o interpretación.

Mold The mass of sand or other material that forms the cavity into which molten metal is poured.

Molde La masa de arena u otro material que forma la cavidad en la cual se vierte el metal derretido.

Mouse A device that typically rolls on the desktop and directs the cursor on the display screen.

Ratón Un aparato que típicamente rueda sobre el escritorio y dirige el cursor en la pantalla del monitor.

Multilayered Boards These boards are made up of two or more single- and/or double-sided boards.

Tarjetas de Múltiples Capas Estas tarjetas están hechas de dos o más tarjetas de un solo lado, o de tarjetas de dos lados, o de ambas.

Multiple Reference Lines Used when there will be more than one operation on a particular joint. The first operation goes on the first line, the second on the second, and so on.

Líneas de Referencias Múltiples Se usan donde habrá más de una operación un una unión particular. La primera operación se indica en la primera línea, la segunda en la segunda y así sucesivamente.

Multiple Thread A type of screw threading in which two or more continuous ridges follow side-by-side. Used when speed or travel distance is an important design factor.

Rosca Múltiple Un tipo de rosca de tornillo en que dos o más resaltos continuos siguen uno al lado del otro. Usado cuando velocidad o distancia de viaje es un factor importante del diseño.

Multiweld A weld where more than one type of weld may be applied to a single joint—for example, a groove V weld on one side, and a backing weld on the other side. The weld is usually for strength or appearance.

Soldadura Múltiple Una soldadura en la cual más de un tipo de soldadura se puede aplicar a una sola unión—por ejemplo una soldadura de ranura V en un lado y una soldadura de espalda en el otro lado. Esta soldadura tiene usualmente el propósito de fuerza o apariencia.

Neck To cut a groove called a neck around a cylindrical piece.

Collarín Cortar una ranura llamada cuello alrededor una pieza cilíndrica.

Nodes Points of an object that contain specific information about the characteristics at that particular point.

Nodos Puntos en un objeto que contienen información específica acerca de las características de ese punto particular.

Nominal Dimension A dimension that represents a general classification size given to a part or product.

Dimensión Nominal Una dimensión que representa un tamaño de clasificación general que se le da a una pieza o producto.

Nomogram A graphical representation of a formula along straight or curved graduated lines that is used to find the unknown value of one variable in an equation, given the others. Uses alignment charts and concurrency charts to graphically represent equations.

Nomograma Una representación gráfica de una fórmula junto con curvas rectas o curvas graduadas que se usan para hallar el valor desconocido de una variable en una ecuación, dado los valores de las otras variables. Usa gráficas de alineación y gráficas de concurrencia para representar las ecuaciones gráficamente.

Nonferrous A description of metals not derived from an iron base or an iron alloy base; nonferrous metals include aluminum, magnesium, and copper, among others.

No Ferroso Una descripción de los metales que no se derivan de una base de hierro o una base de una aleación de hierro; los metales no ferrosos incluyen aluminio, magnesio y cobre, entre otros.

Nonisometric Line A line not parallel or perpendicular to any of the three principal edges in an isometric drawing.

Línea No Isométrica Una línea que no es paralela o perpendicular a ninguno de los tres bordes principales de un dibujo isométrico.

Nonroutine Designs Problems that are assigned to experienced engineers, designers, technologists, and drafters.

Dibujos No de Rutina Problemas que se asignan a los ingenieros, diseñadores, tecnólogos y delineantes con experiencia.

Nontraditional Machining Advanced machining processes that have been developed in response to the demands of modern manufacturing for more effective, more efficient ways to machine the many new materials that have been, and are still being, developed.

Labrado No Tradicional Procesos de labrado avanzados que se han desarrollado en respuesta a las demandas de la manufacturación moderna para maneras más efectivas y eficientes para labrar los muchos materiales que se han desarrollado o que aún se están desarrollando.

Normalize To heat steel above its critical temperature, and then cool it in air.

Normalizar Calentar el acero más allá de su temperatura crítica, y luego dejarlo enfriar en aire.

Normalizing The process used to remove the effects of machining or cold-working metals. The process in which ferrous alloys are heated and then cooled in still air to room temperatures to restore the uniform grain structure free of strains caused by the cold working or welding.

Normalizado Un proceso usado para remover los efectos del labrado o del trabajado en frío en los metales. El proceso en el cual las aleaciones férreas se calientan y luego se enfrían en aire quieto a temperatura ambiente para restaurar la estructura granular uniforme libre de la tensión causada por el trabajo en frío o de soldadura.

North/South Deviation A method of marking a drawing in which north is at the top of the page and south is at the bottom.

Desviación Norte/Sur Un método de marcar un dibujo en el cual el norte está en la parte superior de la página y el sur en la parte inferior de la página.

Notations Labeling that keeps track of all views and points in space and provides brief explanations/information on a drawing.

Notación Rótulos que llevan cuenta de todas las vistas y puntos en el espacio y proveen explicaciones breves o información en un dibujo.

Numerical Control A system of controlling a machine or tool by means of numeric codes that direct commands to control devices attached or built into the machine or tool.

Control Numérico Un sistema para controlar una máquina o herramienta por medio de códigos numéricos que dirigen los comandos que controlan aparatos que están conectados a o que están dentro de la máquina o herramienta.

Numerically Controlled Machine Tools Machine tools that are controlled by means of numeric codes that direct commands to control devices attached or built into the machine or tool.

Herramientas de Labrado Controladas Numéricamente Herramientas de labrado que se controlan por medio de códigos numéricos que dirigen los comandos que controlan aparatos que están conectados a o que están dentro de la máquina o herramienta.

Oblique Drawing A style of pictorial drawing in which one surface of the object, usually the front view, is drawn exactly as it would appear in a multiview drawing. One of its principal faces is parallel to the plane of projection and is projected in its true size and shape; the third set of edges is oblique to the plane of projection at some convenient angle.

Dibujo Oblicuo Un estilo de dibujo pictórico en el cual una superficie del objeto, generalmente la vista frontal, se dibuja exactamente como aparecería en un dibujo de vistas múltiples. Una de las caras principales está paralela al plano proyección y se proyecta en tamaño y forma real; el tercer grupo de bordes está oblicuo al plano de proyección en un ángulo conveniente.

Oblique Projection It shows the full size of one view.

Proyección Oblicua Muestra el tamaño completo de una vista.

Oblique Sketching A type of sketching involving a combination of a flat, orthographic front with depth lines receding at a selected angle (usually 45°).

Croquis Oblicuo Un tipo de croquis que consiste de una combinación de un frente plano y ortográfico con líneas de profundidad que van disminuyendo en un ángulo seleccionado (generalmente 45°).

Oblique Surfaces Surfaces that are not perpendicular to any of the three principal projection planes.

Superficies Oblicuas Superficies que no son perpendiculares a ninguno de los tres planos de proyección.

Obtuse Angle An angle larger than 90°.

Ángulo Obtuso Un ángulo mayor de 90°.

Obtuse Angle Triangle A triangle having an obtuse angle greater than 90° with no equal sides.

Triángulo de Ángulo Obtuso Un triángulo que tiene un ángulo obtuso mayor de 90° sin lados iguales.

Octagon A polygon having eight angles and eight sides.

Octágono Un polígono que tiene ocho ángulos y ocho lados.

Offset Measurements Used to locate one feature in relationship to another.

Medida de Compensación Se usa para localizar un rasgo en relación a otro.

Offset Section This section is accomplished by bending or offsetting the cutting-plane line. It is similar to a full section except that the cutting-plane line is not straight.

Sección de Compensación Esta sección se logra doblando o desplazando el plano cortante. Es similar a una sección completa, excepto que la línea del plano cortante no es recta.

Ogee Curve Used to join two parallel lines. It forms a gentle curve that reverses itself in a neat symmetrical geometric form.

Curva de "S" Se usa para unir dos líneas paralelas. Forma una curva suave que vira en si misma en una forma geométrica muy simétrica.

Open Jigs Jigs that cover only one side of a part and are used for relatively simple operations.

Guías Abiertos Guías que cubren sólo un lado de una pieza y se usan para operaciones relativamente simples.

Open-Carriage Typewriter Any brand of typewriter that has been especially designed to hold larger-than-normal media.

Máquina de Escribir de Transporte Abierto Cualquier marca de máquina de escribir que se ha diseñado para sostener hojas de papel más grandes de lo normal.

Open-Divided Scale It has its first unit of measurement subdivided, but the remaining units are open or free from subdivision.

Escala Abierta y Dividida Tiene la primera unidad de medida subdividida, pero las unidades restantes están abiertas o libres de subdivisiones.

Opitz System One of the most widely used classification and coding systems. It uses characters in 13 places to code the attributes of parts, and hence, to classify them.

Sistema Opitz Uno de los sistemas de clasificación y codificación más usado. Usa caracteres en 13 lugares para codificar los atributos de una pieza y por ende, clasificarlos.

Optimization The process of improving a process or system to its maximum functionality.

Optimización El proceso de mejorar un proceso o sistema a su funcionalidad máxima.

Ordinate The Y coordinate of a point (i.e., its vertical distance from the X axis measured parallel to the Y axis, or the vertical axis of a graph or chart).

Ordenada La coordenada Y de un punto (esto es, la distancia vertical desde el eje de X, medido paralelo al eje de Y o eje vertical de la gráfica).

Orthographic Projection A projection on a picture plane formed by perpendicular projectors from the object to the picture plane; third-angle projection is used in the United States, while first-angle projection is used in most countries outside the United States.

Proyección Ortográfica Una proyección en un plano de dibujo que está formado por proyecciones perpendiculares desde el objeto al plano de dibujo; en los Estados Unidos se usa la proyección de tercer ángulo, mientras que la proyección de primer ángulo se usa en la mayoría de los países fuera de Estados Unidos.

Orthographic Sketching This type of sketching deals with flat, graphic facsimiles of a subject showing no depth. The views selected depend on the nature of the subject and the judgment of the sketcher.

Croquis Ortográfico Este tipo de croquis trata con copias planas y gráficas de un sujeto que no muestra profundidad. Las vistas que se seleccionan dependen de la naturaleza del sujeto y el juicio de la persona que hace el croquis.

Output Data that have been processed from an internal storage to an external storage; opposite of input.

Salida Datos que se han procesado de un almacenamiento interno a un almacenamiento externo; lo opuesto a entrada.

Outside Diameter The measurement over the extreme outer edge of teeth. It is equal to the pitch diameter, plus two times the addendum.

Diámetro Externo La medida alrededor del extremo externo de los dientes. Es igual al diámetro de paso más dos veces la altura de cabeza.

Overlay Drafting A complete drafting process that uses advanced reproduction techniques to reduce the amount of time spent in preparing drafting documentation.

Dibujo Industrial por Superposición Un proceso de dibujo técnico completo que usa técnicas de reproducción avanzadas para reducir la cantidad de tiempo que se dedica en preparar la documentación de dibujo.

Parallel Having the same direction, such as two lines which, if extended indefinitely, would never meet.

Paralelo Que tiene la misma dirección, tales como dos líneas que, si se extienden infinitamente, nunca se encontrarían.

Parallel Alignment Chart The procedure for preparing a set of three parallel scales for a three-variable equation. The derivation aims at obtaining relationships from two of the scales so that the third can be located and graduated.

Gráfica de Alineación Paralela El procedimiento para preparar tres escalas paralelas para una ecuación de tres variables. La derivación aspira a obtener relaciones de dos de las escalas de manera que se pueda localizar y graduar la tercera.

Parallel Line Development Used for objects having parallel fold lines.

Desarrollo de Líneas Paralelas Se usa para objetos que tienen líneas de doblez paralelas.

Parallel Projection Is subdivided into the following three categories: orthographic, oblique, and axonometric projections.

Proyección Paralela Se subdivide en las siguientes tres categorías: proyección ortográfica, oblicua y axonométrica.

Parallel Station Lines A baseline that must always be 90° from the fold lines. Used for objects having parallel fold lines.

Líneas de Estación Paralelas Una línea base que siempre tiene que estar a 90° de las líneas de doblez. Se usa para los objetos que tienen líneas de doblez paralelas.

Parallelism A feature control that specifies that all points on a given surface, axis, line, or center plane must be equidistant from a datum.

Paralelismo Un control de rasgo que especifica que todos los puntos de una superficie dada, eje, línea o plano central, deben estar equidistantes de una referencia.

Parallelogram A quadrilateral whose opposite sides are parallel.

Paralelogramo Un cuadrilátero cuyos lados opuestos son paralelos.

Parallelogram Law Two vectors, not parallel, drawn to scale, having the same units, and acting at the same point, can be added or subtracted.

Ley del Paralelogramo Dos vectores, que no son paralelos, que están dibujados a escala, tienen las mismas unidades, y actúan sobre un mismo punto, pueden sumarse o restarse.

Parametric Design This allows drafters and designers to change dimensions in three-dimensional computer models and have all affected dimensions change accordingly.

Diseño Paramétrico Éste le permite al delineante y a los diseñadores cambiar las dimensiones de los modelos tridimensionales de computadora y que como consecuencia se cambien todas las dimensiones afectadas.

Part Individual component of an assembly.

Pieza Un componente individual de un ensamblaje.

Part Family A group of parts similar in design and/or the manufacturing process used to produce them.

Familia de Piezas Un grupo de piezas de diseños y/o procesos de manufactura similares.

Partial Auxiliary View A partial view makes it possible to eliminate one or more of the regular views which, in turn, saves drafting time and cost.

Vista Parcial Auxiliar Una vista parcial hace posible el eliminar una o más vistas regulares, lo que a su vez ahorra tiempo y gastos del proceso de dibujo industrial.

Parting Operation used to finish the edges of a part.

Partir Una operación que se usa para terminar los bordes de una pieza.

Parting Line This is the plane of division between the cope and drag.

Línea Divisoria Es el plano de división entre la caja superior y la caja inferior del molde.

Parts This is a listing, identified first by number, part name, and quantity needed. It also contains a cross reference to the corresponding balloon number.

Piezas Este es un listado de piezas, identificadas primero por número, nombre de la pieza y cantidad necesaria. También tiene una referencia al número de globo correspondiente.

Parts List A bill of material that itemizes the parts needed to assemble one complete full assembly of a product.

Lista de Piezas Una factura de materiales que desglosa las piezas necesarias para ensamblar un producto completo.

Patent Drawings A drawing that must accompany a patent application for a new idea or invention. Its purpose is to illustrate and then explain the invention's function. It must be mechanically correct and must adhere to strict patent drawing regulations.

Dibujos de Patentes Un dibujo que debe acompañar la aplicación para patente de una nueva idea o invención. Su propósito es ilustrar y explicar la función de la invención. Debe ser mecánicamente correcta y debe cumplir con los reglamentos estrictos para los dibujos de patentes.

Pattern A model, usually of wood, used in forming a mold for casting. In sand casting, this is a model of a part.

Patrón Un modelo, generalmente de madera, que se usa para formar el molde para la fundición. En la fundición en moldes de arena, éste es el modelo para una pieza.

Peen To hammer into shape with a ball peen hammer.

Martillar Formar mediante martillazos de un martillo de bola.

Pentagon A polygon having five angles and five sides.

Pentágono Un polígono que tiene cinco ángulos y cinco lados.

Perforating A punching operation performed on sheets to produce either a uniform hole pattern or a decorative form in the sheet.

Perforar Una operación de perforado que se hace en hojas para producir un agujero de patrón uniforme o una forma decorativa en una hoja.

Permanent Fasteners Used when parts will not be disassembled.

Sujetadores Permanentes Se usan cuando las partes no serán desensambladas.

Perpendicular A type of motion of the follower specific to the radial arm cam design. Also, lines or planes at a right angle to a given line or plane.

Perpendicular Un tipo de movimiento del seguidor que es específico al diseño de levas de brazo radial. También, líneas o planos que están a ángulos rectos de una línea o un plano dado.

Perpendicularity A feature control that specifies that all elements of a surface, axis, center plane, or line form a 90° angle with a datum.

Perpendicularidad Un control de rasgo que especifica que todos los elementos de una superficie, eje, plano central o línea forman un ángulo de 90° con una referencia.

Perspective Drawing A three-dimensional drawing of an object or scene that shows visible lines almost as a person's eyes would perceive them from one particular viewing location. Also a pictorial drawing in which receding lines converge at vanishing points on the horizon; the most natural of all pictorial drawings.

Dibujo de Perspectiva Un dibujo tridimensional de un objeto o escena que muestra las líneas visibles casi como los ojos de una persona las percibiría desde un punto de vista particular. También es un dibujo pictórico en el cual todas las líneas que se alejan convergen en puntos en el horizonte; el más natural de todos los dibujos pictóricos.

Perspective Projection A distortion phenomenon that occurs because the visual rays used to create the image in the viewer's eye are not parallel.

Proyección de Perspectiva Un fenómeno de distorsión que ocurre debido a que los rayos visuales que se usan para generar la imagen no son paralelos.

Perspective Sketching Involves creating a graphic facsimile of the subject in one-, two-, or three-point perspective using only a pencil.

Croquis de Perspectiva Consiste en crear una copia gráfica del sujeto en perspectiva de uno, dos y tres puntos, usando sólo un lápiz.

Phantom Line A type of line that resembles a center line except that it has two dashes.

Línea Fantasma Un tipo de línea que se parece a la línea de centro excepto que tiene dos guiones.

Photochemical Machining Also known as chemical blanking, this involves producing parts by chemical action. The chemical dissolves all of the metal except the desired part.

Labrado Fotoquímico También se conoce como troquelado químico, y consiste en producir piezas por acción química. El químico disuelve todo el metal excepto la parte deseada.

Photorealism A term used to describe images that are created in such a fashion that they might fool the viewer into thinking that they are photographs.

Fotorealismo Un término que se usa para describir las imágenes que se crean de tal manera que podrían engañar al espectador a pensar que están viendo fotografías.

Physical Prototype A real, physical object that can be handled, analyzed, and tested.

Prototipo Físico Un objeto real y físico que puede manipularse, analizarse y probarse.

Pickle To clean forgings or castings in dilute sulfuric acid.

Limpiar con Ácido Limpiar las piezas forjadas o de fundición en ácido sulfúrico diluido.

Pictorial Chart These charts employ varying sizes of figures. They illustrate relative amounts to the general public.

Gráfica Pictórica Estas gráficas utilizan varios tamaños de figuras. Ellas ilustran cantidades relativas al público en general.

Pictorial Drawing A drawing that depicts the object in a three-dimensional format; these drawings are typically isometric or perspective but may also be dimetric, trimetric, or oblique.

Dibujo Pictórico Un dibujo que muestra un objeto en tres dimensiones; estos dibujos típicamente son isométricos o de perspectiva pero también pueden ser dimétricos, trimétricos u oblicuos.

Pie Chart These charts look like circular pies and usually represent 100% of a budget. They are generated from a spreadsheet in a personal computer.

Gráfica Circular Estas gráficas parecen tortas circulares y generalmente representan 100% de un presupuesto. Se generan desde una hoja de cálculo en una computadora personal.

Piercing An operation similar to punching, except that no scrap is produced.

Agujerar Una operación similar al perforado, excepto que no produce material de desecho.

Piercing Point This is the exact location of the intersection of a surface and a line.

Punto Penetrante Ésta es la localización exacta de la intersección entre una superficie y una línea.

Pilot A piece that guides a tool or machine part.

Guía Una pieza que guía una herramienta o parte de una máquina.

Pilot Hole A small hole used to guide a cutting tool for making a larger hole.

Agujero Guía Un pequeño agujero que guía una herramienta cortante para hacer un agujero más grande.

Pinion Gear Exactly like the spur gear but usually smaller and with fewer teeth.

Engranaje de Piñón Exactamente como un engranaje de dientes rectos pero generalmente más pequeño y con menos dientes.

Pitch The distance from a point on a screw thread to a corresponding point on the next thread, as measured parallel to the axis. The center-to-center distance between increments.

Paso La distancia desde un punto en una rosca de un tornillo al punto correspondiente en la próxima rosca, medido paralelo al eje. La distancia centro a centro entre incrementos.

Pitch Circle An imaginary circle corresponding to the circumference of the friction gear from which the spur gear is derived.

Círculo de Paso Un círculo imaginario que corresponde a la circunferencia del engranaje de fricción del cual se deriva el engranaje de dientes rectos.

Pitch Diameter The diameter of an imaginary diameter centrally located between the major diameter and the minor diameter.

Diámetro de Paso El diámetro de un diámetro imaginario localizado centralmente entre el diámetro mayor y el diámetro menor.

Placing Drawings Shop drawings of concrete structural members that are poured on site; they show the placement of reinforcing bars in the on-site form. Shop drawings and placing drawings are the same; the difference is in where the work occurs, in a shop or on site.

Dibujos de Colocación Dibujos de taller de los miembros estructurales de concreto que se fundirán en el lugar de construcción; muestran la localización de las barras de refuerzo en el molde que está en el lugar. Los dibujos de taller y los dibujos de colocación son los mismos; la diferencia está en dónde se lleva a cabo el trabajo, en el taller o en el lugar de construcción.

Plane To remove material by means of a planer.

Acepillar Remover material por medio de una acepilladora.

Plane Surface If any two points anywhere on a surface are connected to form a straight line, and that line rests upon the surface, it is a plane surface.

Superficie Plana Si cualesquiera dos puntos en cualquier lugar en una superficie se conectan para formar una línea recta, y esa línea esta sobre la superficie, entonces es una superficie plana.

Planer Single-edged cutting tool used to perform a cutting operation. It is normally much larger than a shaper, and the work, rather than the cutter, moves on the planer. Seldom used today.

Acepilladora Una herramienta cortante de un solo filo que se usa para llevar a cabo una operación cortante. Normalmente es mucho más grande que una conformadora, y el trabajo en vez de la cortadora, se mueve en la acepilladora. Muy poco usado hoy día.

Planning Chart A method of visually displaying when phases and steps are scheduled to begin and end. Some steps on the chart (within a phase) may overlap.

Gráfica de Planificación Un método de mostrar visualmente cuándo las fases y pasos están programadas para comenzar y terminar. Algunos pasos en la gráfica (dentro de una fase) pueden solaparse.

Plasma A gas that has been heated to the point that it ionizes and becomes electrically conductive.

Plasma   Un gas que se ha calentado hasta el punto que se ioniza y conduce electricidad.

Plasma Arc Machining (PAM)   A process that makes use of a plasma arc to remove material.

Labrado con Arco de Plasma   Un proceso que usa un arco de plasma para remover material.

Plastics   Nonmetal materials that are made from either natural or synthetic resins. Their main characteristics are: thermal, mechanical, chemical and corrosion resistance, and fabricability; the ability to sustain high mechanical loads; and predictable performance.

Plásticos   Materiales no metálicos que están hechos de resinas naturales o sintéticas. Sus características principales son: resistencia termal, mecánica, química y a la corrosión, habilidad para fabricación; habilidad para soportar grandes cargas mecánicas; y funcionamiento predecible.

Plate   To coat a metal piece with another metal, such as chrome or nickel, by electrochemical methods.

Enchapar   Cubrir un pedazo de metal con otro metal, tal como cromo o níquel, por métodos electroquímicos.

Plate Jig   The most common form of jig. It consists of a simple plate that contains the required drill bushings, locators, and clamping elements.

Guía de Placa   La forma más común de guía. Consiste de una placa sencilla que contiene los cojinetes de taladro, localizadores y elementos de fijación necesarios.

Plot Plan   A plan view of the building site showing where the structure is located on the lot.

Plano del Terreno   Una vista plana del lugar de construcción que muestra dónde está ubicada la estructura en el lote de terreno.

Plotter   An output device that makes a drawing by converting digital data into text and graphics.

Trazador Gráfico   Un aparato de salida que crea un dibujo convirtiendo los datos digitales en texto y gráficos.

Plug Weld   A weld made through a round hole. The symbols are the same. (See also *Slot Weld*).

Soldadura de Tapón   Una soldadura que se hace a través de un agujero. Los símbolos son los mismos. (Vea también *Soldadura de Ranura*).

Point   An exact location in space or on a drawing surface.

Punto   Una localización exacta en el espacio o en la superficie de dibujo.

Point-to-Point   A variation in the machine (along with continuous-path machines). In this type, the machine operates on a series of preprogrammed coordinates to locate the position of the tool. When the tool finishes at one point, it automatically goes to the next point by the shortest route.

Punto a Punto   Una variación en la máquina (junto con las máquinas de camino continuo). En este tipo, la máquina opera con una serie de coordenadas preprogramadas para localizar la posición de la herramienta. Cuando la herramienta termina en un punto, automáticamente se mueve al próximo punto por el camino más corto.

Point-to-Point Diagram   A type of connection diagram showing the various terminal connection locations and the routing paths of every wire in an electronic system.

Diagrama de Punto a Punto   Un tipo de diagrama de conexión que muestra las localizaciones de varias conexiones terminales y las vías para cada cable en un sistema electrónico.

Polar Graph   A 360° map usually used to show the effect of a point source of energy (heat, light, electromagnetic, sound) at varying distances from the source.

Gráfica Polar   Un mapa de 360° que generalmente se usa para mostrar el efecto de una fuente puntual de energía (calor, luz, electromagnética, sonido) a diversas distancias desde la fuente.

Pole and Ray Method   A method of graphical integration and differentiation.

Método de Polo y Rayo   Un método gráfico de integración y diferenciación.

Poles   Two reference measuring positions on the surface of the sphere on opposite sides of its center.

Polos   Dos posiciones de referencia en la superficie de la esfera en lados opuestos de su centro.

Polish   To produce a highly finished or polished surface by friction, using a very fine abrasive.

Pulir   Producir una superficie bien acabada o pulida por fricción, usando un abrasivo fino.

Polygon   A plane geometric figure with three or more sides.

Polígono   Una figura geométrica con tres o más lados.

Polyhedron   A solid object bounded by plane surfaces. Each surface is called a face.

Poliedro   Un objeto sólido definido por superficies planas. Cada superficie se llama una cara.

Positional Tolerance   A tolerance that controls the position of a feature relative to the true dimension specified for the features, as related to a datum or datums.

Tolerancia de Posición   Una tolerancia que controla la posición de un rasgo relativo a la dimensión real especificada para los rasgos, según se relacionan con una referencia o referencias.

Powder Metallurgy   A process similar to casting in which metal particles, or powder, are blended and mixed. The powder is then forced into a die of the desired form under intense pressure. The resulting heat fuses the powder into a solid piece that can be machined.

Metalurgia de Polvo   Un proceso similar al proceso de moldes de fundición en el cual las partículas de metales o el polvo se mezclan. El polvo entonces se pasa a fuerza dentro de un troquel con la forma deseada bajo una presión intensa. El calor resultante funde el polvo en una pieza sólida que puede labrarse.

Powdered Metals Metal powders that are produced by atomization, electrolysis, chemical reduction, or oxide reduction. They can be mixed or sintered to produce alloys with an almost endless number of compositions. The properties of products made from powdered metals depend on which metals are mixed, in what amounts, and by what processes.

Metales en Polvo Polvos metálicos que se producen por atomización, electrólisis, reducción química o reducción de óxido. Las propiedades de los productos que se hacen de los metales en polvo dependen de cuáles metales se mezclan, en qué cantidades y por cuál proceso.

Power Saw A generic term for all types of electric or gas-operated saws. The types of power saws used most often in the machine shop are the contour band machine, or band saw, and the cutoff saw.

Sierra de Potencia Un término genérico para todo tipo de sierras eléctricas u operadas con gasolina. Los tipos de sierras de potencia que se usan con más frecuencia en un taller mecánico son la sierra de cinta de contorno o sierra de cinta y la sierra tronzadora.

Precision The quality or state of being precise or accurate; mechanical exactness.

Precisión La calidad o estado de ser preciso; exactitud mecánica.

Precision Grinder Grinders that are used for precision work. The principal types used are surface and cylindrical grinders.

Moledora de Precisión Moledoras que se usan para trabajo de precisión. Los tipos principales que se usan son las moledoras de superficie y cilíndricas.

Press Forging A single-step process in which the heated metal bar or billet is forced into a single die and is completed in a single press stroke.

Forjado a Presión Un proceso de un solo paso en el cual una barra de metal o un lingote caliente se presiona contra un troquel y se completa en un solo ciclo de presión.

Press-Fit Bushings A type of bushing used to position and guide the cutting tools used in jigs. They are pressed directly into the jig plate. They are used for short-run jigs or in applications where the bushings are not likely to be replaced frequently.

Cojinete de Encaje a Presión Un tipo de cojinete que se usa para posicionar y guiar las herramientas cortantes que se usan en las guías. Se encajan a presión directamente en la placa del guía. Se usan en las guías cortas o en las aplicaciones donde los cojinetes no serán reemplazados frecuentemente.

Primary Auxiliary View Views that have been projected perpendicular from an inclined surface and viewed looking directly at that surface.

Vista Primaria Auxiliar Vista que se proyecta perpendicularmente desde una superficie inclinada y se ve directamente en la superficie.

Principal Viewing Planes The top, front, and right-side planes; they are all perpendicular to each other.

Planos de Vistas Principales Los planos superior, frontal y del lado derecho; todos están perpendiculares unos a otros.

Printed Circuit Board A special laminated board upon which a predetermined pattern of printed circuits has been formed, and to which electronic components are mounted. There are three types: single-sided boards, double-sided boards, and multilayered boards.

Tarjeta de Circuito Impreso Una tarjeta especial laminada sobre la cual se ha formado un patrón predeterminado de circuitos impresos, y sobre la cual se montan componentes electrónicos. Hay tres tipos: tarjetas de un solo lado, tarjetas de dos lados y tarjetas de múltiples capas.

Printer A device that prints the output of a computer.

Impresora Un aparato que imprime el resultado de una computadora.

Prism A solid whose bases or ends are any congruent and parallel polygons, and whose sides are parallelograms.

Prisma Un sólido cuyas bases o extremos son cualesquiera polígonos paralelos y congruentes, y cuyos lados son paralelogramos.

Prismatic Pertaining to or like a prism.

Prismático Perteneciente a o parecido a un prisma.

Process Synchronization The state that has been achieved when all processes involved in producing a product are in balance and synchronized so that production flows smoothly and continuously.

Sincronización de Proceso El estado que se ha logrado cuando todos los procesos involucrados en producir un producto están balanceados y sincronizados de manera que su producción fluye suavemente y continuamente.

Processor The computer in a CAD system.

Procesador La computadora en un sistema de CAD.

Profile A feature control that specifies the amount of allowance variance of a surface or line elements on a surface.

Perfil Un control de rasgo que especifica la cantidad de variación permisible de una superficie o los elementos de línea en una superficie.

Profile Chart A chart that is used to illustrate the contours of a surface by plotting them on a graph. Used by material science texts to show the roughness of the surface of a metal by slicing the metal at a shallow angle and showing the cut surface in a magnified plot.

Gráfica de Perfil   Una gráfica que se usa para ilustrar los contornos de una superficie graficándolos en una gráfica. Se usa en los textos de ciencia de materiales para mostrar la aspereza de la superficie de un metal cortando el metal a un ángulo poco profundo y mostrando la superficie cortada en una gráfica magnificada.

Profile of a Line   A tolerance that controls the allowable variation of a surface from a basic profile or configuration. It establishes a tolerance for a given single element of a surface.

Perfil de una Línea   Una tolerancia que controla la variación permisible en una superficie del perfil o configuración básica. Establece la tolerancia para un elemento de una superficie.

Profile of a Surface   A tolerance that controls the allowable variation of a surface from a basic profile or configuration. It applies to the entire surface. The difference between the profile of a line and surface is similar to the difference between circularity and cylindricity.

Perfil de una Superficie   Una tolerancia que controla la variación permisible en una superficie del perfil o la configuración básica. Se aplica a toda la superficie. La diferencia entre el perfil de una línea y la superficie es similar a la diferencia entre la circularidad y la cilindricidad.

Program   A set of step-by-step instructions telling the computer to solve a problem with the information input to it or contained in memory.

Programa   Un grupo de instrucciones paso por paso que le dicen a una computadora que resuelva un problema con la información que le es dada o que está contenida en memoria.

Progress Report   A written report, directed to the client, to inform him/her of the things that have been done to complete the job and the things that still need to be done. They can be issued at agreed-upon intervals, depending on the length of the project.

Reportaje del Estado del Proyecto   Un reportaje escrito, dirigido al cliente, para informarle sobre las cosas que habían hecho para completar el trabajo, y las cosas que todavía necesitan completarse. Pueden publicarse a intervalos convenidos, dependiendo del largo del proyecto.

Progressive Die   A die in which several operations take place sequentially.

Troquel Progresivo   Un troquel en el cual se llevan a cabo varias operaciones en secuencia.

Projected Tolerance Zone   A modifier that allows a tolerance zone established by a locational tolerance to be extended a specified distance beyond a given surface.

Zona de Tolerancia Proyectada   Un modificador que permite que se establezca una zona de tolerancia por medio de una tolerancia que se extenderá una distancia más allá de una superficie dada.

Projection Weld   Similar to a spot weld and using the same welding symbol. This weld allows more penetration and, as a result, is a much better weld.

Soldadura de Proyección   Similar a una soldadura de punto y usa el mismo símbolo de soldadura. Esta soldadura permite más penetración y como resultado, es una soldadura mucho mejor.

Projectors   Imaginary lines of sight.

Proyecciones   Líneas de vista imaginarias.

Proposal   This is a combination of the selling price, specifications, delivery schedules, and the boilerplate.

Propuesta   Es la combinación del precio de venta, especificaciones, horarios de entregas y el texto modelo.

Prototype   An original model on which something is patterned.

Prototipo   Un modelo original según el cual se hace un patrón de algo.

Punch   To cut an opening of a desired shape with a rigid tool having the same shape.

Perforar   Cortar una abertura de una forma deseada con una herramienta rígida de la misma forma.

Punching   An operation that cuts a hole in a metal sheet or strip.

Perforado   Una operación que corta un orificio en una hoja o tira de metal.

Purchased Part   An item made elsewhere that is required to create a final product. Purchased parts are drawn on assembly drawings, usually in outline form, to show their function and relationship to the product.

Pieza Comprada   Un artículo que fue hecho en otro lugar y que se requiere para crear el producto final. Las piezas compradas se dibujan en dibujos de ensamblaje, generalmente en silueta, para mostrar su función y relación con el producto.

Pyramid   A polyhedron having a polygon as its base.

Pirámide   Un poliedro que tiene un polígono como su base.

Quadrant   A sector with a central angle of 90° and usually with one of the radial lines oriented horizontally.

Cuadrante   Un sector con un ángulo central de 90° y generalmente con una de las líneas radiales horizontales.

Quadrilateral   A plane figure bounded by four straight lines.

Cuadrilátero   Una figura plana bordeada por cuatro líneas rectas.

Quench   The rapid cooling of a machined part in a cooling material. To immerse a heated piece of metal in water or oil in order to harden it.

Enfriamiento Rápido   El enfriamiento rápido de una parte labrada en un material de enfriamiento. Sumergir un pedazo de metal calentado en agua o aceite para endurecerlo.

Quotation Department This department takes all factors, such as labor costs, material costs, manufacturing expenses, sales salaries and expenses, and general overhead costs, and uses them to arrive at the total cost to produce a product. It then adds the desired profit to the total cost to establish the selling price.

Departamento de Cotización Este departamento toma todos los factores, tales como los costos de mano de obra, costos de materiales, gastos de manufactura, gastos y salarios de ventas y los costos operativos generales, y los usa para llegar al costo total de producir un producto. Entonces le suma la ganancia deseada al costo total para establecer el precio de venta.

Rack A flat bar with gear teeth in a straight line to engage with teeth in a gear.

Cremallera Una barra plana con dientes de engrane alineados que encajan con los dientes de un engranaje.

Rack Gear A type of spur gear whose teeth are in a straight line or flat.

Engranaje de Cremallera Un tipo de engranaje de dientes rectos cuyos dientes están alineados o en un plano.

Radial Arm Design that changes rotary motion into either an up-and-down motion or a rocking action.

Brazo Radial Diseño que cambia el movimiento rotativo a un movimiento hacia arriba y hacia abajo o un movimiento de mecedora.

Radial Line Development A development in which the fold lines radiate from one point.

Desarrollo de Línea Radial Un desarrollo en el cual las líneas de doblez parten desde un punto.

Radius The distance from the center point of a circle to the outside curved surface. It is half the diameter.

Radio La distancia desde el punto central de un círculo a la superficie exterior curva. Es la mitad del diámetro.

RAM Random Access Memory.

RAM Memoria de Acceso Aleatorio.

Rapid Prototyping A prototyping process that builds the final product layer by layer by an electronic-laser process that uses data from 3-view drawings.

Creación Rápida de Prototipo Un proceso de creación de prototipo que construye el producto final capa por capa con un proceso electrónico con láser que usa datos de dibujos de 3 vistas.

Raster The coordinate grid dividing the display area of a graphics display.

Cuadrícula La cuadrícula coordinada que divide el área de una pantalla.

Real Time Response times in the interaction between humans and computers that are, for all intents and purposes, equal to real-world response times.

Tiempo Real Tiempos de respuesta en la interacción entre humanos y computadoras que son, para todo propósito práctico, iguales a los tiempos de respuesta del mundo real.

Rectangular System A method of surveying in which a large area of land is surveyed beginning from one basic reference point and divided into a grid of six-mile-square blocks called townships. Townships are further divided into 36 one-mile-square blocks called sections; these are subsequently divided into smaller and smaller rectangular sections of land.

Sistema Rectangular Un método de planimetría en el cual se mide un área grande de terreno, comenzando desde un punto básico de referencia y dividido en una cuadrícula de bloques de seis millas cuadradas llamados municipios. Los municipios se subdividen en 36 bloques de una milla cuadrada llamados secciones; éstos a su vez se subdividen en secciones de terreno cada vez más pequeñas.

Reference Dimension A nontoleranced dimension used for information purposes only, which does not govern production or inspection operations.

Dimensión de Referencia Una dimensión sin tolerar que se usa con propósitos informativos solamente y que no gobierna las operaciones de producción o inspección.

Reference Line A line that is always placed through the center of any symmetrical object.

Línea de Referencia Una línea que siempre se coloca a través del centro de cualquier objeto simétrico.

Refractory Metals Metals with an exceptional ability to withstand high temperatures; they also have a low coefficient of expansion, high modulus of elasticity, high resistance to thermal shock, high conductivity, and excellent hardness properties at high temperatures. Tungsten and molybdenum are two of the best known.

Metales Refractarios Metales con una habilidad excepcional para resistir altas temperaturas; también tienen un coeficiente de expansión termal bajo, un módulo de elasticidad alto, alta resistencia a shock termal, alta conductividad y propiedades excelentes de dureza a altas temperaturas. El tungsteno y el molibdeno son dos de los más conocidos.

Refresh Display A CRT device that requires the refreshing of its screen presentation at a high rate so that the image will not fade or flicker.

Pantalla Refrescada Un aparato de tubo de rayos catódicos (CRT) que requiere refrescar su presentación en pantalla a alta velocidad para que la imagen no se desvanezca o parpadee.

Regardless of Feature Size (RFS) The condition in which tolerance of position or form must be met irrespective of where the feature lies within its size tolerance. It tells machinists that the tolerance must be maintained regardless of the actual produced size of the object.

Independiente de Tamaño de Rasgo La condición en la cual la tolerancia de posición o forma debe de cumplirse independientemente de dónde caiga el rasgo dentro de la tolerancia de tamaño. Le dice a los labradores que las tolerancias deben mantenerse independiente del tamaño de producción del objeto.

Relief An offset of surfaces to provide clearance.

Relieve Un desbalance entre superficies para proveer separación.

Removed Section This section is very similar to a rotated section except that it is drawn removed or away from the regular views.

Sección Removida Esta sección es muy similar a la sección rotada excepto que se dibuja removida o separada de las vistas regulares.

Rendering Finishing a drawing to give it a realistic appearance; a representation.

Rendición Acabar un dibujo dándole una apariencia realista; una representación.

Renewable Bushing A type of drill bushing that is used in high-volume applications or where the bushings must be changed to suit different cutting tools used in the same hole.

Cojinete Renovable Un tipo de cojinete de taladro que se usa en las aplicaciones de alto volumen en el cual los cojinetes tienen que cambiarse para cumplir con las diferentes herramientas cortantes que se usan en el mismo agujero.

Repeatability The ability to redraw lines without creating a double image.

Reproducibilidad La habilidad de volver a dibujar líneas sin crear una imagen doble.

Resistance Welding The process of passing an electric current through the exact location where the parts are to be joined. This is usually done under pressure. Also, the process of welding metals by using the resistance of the metals to the flow of electricity to produce the heat for fusion of the metals.

Soldadura de Resistencia El proceso de pasar corriente eléctrica a través de una localización exacta dónde se unirán las partes. Esto generalmente se hace bajo presión. Además, el proceso de soldar los materiales usando la resistencia de los materiales al flujo de electricidad para producir calor para fundir los metales.

Resultant Addition of two vectors to obtain a single vector, which is equivalent to the effect of the two vectors.

Resultante La suma de dos vectores para obtener un solo vector, el cual es equivalente al efecto de los dos vectores.

Retaining Rings Precision-engineered fasteners. They provide removable shoulders for positioning or limiting the movement of parts in an assembly.

Anillos Retenedores Sujetadores diseñados con precisión. Éstos proveen bordes removibles para posicionar o limitar el movimientos de las partes de un ensamblaje.

Revolved Section Sometimes referred to as the rotated section, this is used to illustrate the cross section of ribs, webs, bars, arms, spokes, or other similar features of an object.

Sección Revuelta A veces llamada la sección girada, ésta es usada para ilustrar la sección transversal de costillas, entramados, barras, rayos o otros rasgos similares de un objeto.

Rib A relatively thin, flat member acting as a brace or support.

Costilla Un miembro plano y relativamente fino que sirve como refuerzo o apoyo.

Right Angle An angle of 90°.

Ángulo Recto Un ángulo de 90°.

Right Profile The right-side view of a drawing.

Perfil Derecho La vista del lado derecho de un dibujo.

Right Triangle A triangle having a right angle or an angle of 90°.

Triángulo Recto Un triángulo que tiene un ángulo recto o de 90°.

Right-Side View The orthographic view that is projected to the right of the front view; the right profile.

Vista Lateral La vista ortográfica que se proyecta al lado derecho de la vista frontal; el perfil derecho.

Ring Gear Similar to the spur, pinion, and rack gears, except that the teeth are internal.

Engranaje de Anillo Similar a un engranaje de dientes rectos, un piñón o un engranaje de cremallera, excepto que los dientes son internos.

Riser In sand casting, used to vent the mold and to allow gases to escape; also serves as a small reservoir for excess material so that the mold cavity will remain full as the metal begins to shrink during the cooling process.

Tubo de Subida En la fundición en moldes de arena, se usan como respiradero del molde para permitir el escape de los gases; también sirve como un pequeño estanque para el material en exceso de manera que la cavidad del molde permanezca llena a medida que el metal comienza a contraerse durante el proceso de enfriamiento.

Rivet (verb) To connect with rivets or to clench over the end of a pin by hammering.

Remachar Unir con remaches o remachar el extremo de un pasador martillándolo.

Rivets (noun) Permanent fasteners, usually used to hold sheet metal together.

Remache Sujetador permanente, generalmente se usa para sujetar hojalata.

Robot A reprogrammable multifunction manipulator designed to move material parts, tools, or specialized devices through variable programmed motions for the performance of a variety of tasks.

Robot Un manipulador multifuncional programable diseñado para mover partes de material, herramientas o aparatos especializados a través de movimientos variables programados para el funcionamiento de una variedad de tareas.

Rolling A process where the heated metal bar is formed by passing through rollers having the desired form impressed on their surfaces.

Rodar Un proceso en el cual se le da forma a una barra de metal caliente pasándola a través de rodillos que tienen la forma deseada en su superficie.

ROM (Read-Only Memory) Refers to computer chips that have programs built into them at the factory.

ROM (Memoria Sólo Lectura) Se refiere a un chip de computadora en el cual se almacenan los programas en la fábrica.

Roof Plan A plan view showing the location and spacing of rafters and/or trusses. It is superimposed on the floor plan without dimensions.

Plano de Techo Una vista de plano que muestra la localización y separación de las vigas y/o armaduras. Está sobrepuesta sobre el plano de distribución sin dimensiones.

Root The bottom point joining the sides of a thread.

Raíz El punto inferior que une los lados de una rosca.

Root Diameter The measurement over the extreme inner edges of teeth.

Diámetro de Raíz La medida sobre el extremo interior de los dientes.

Root Opening The allowed space between parts.

Abertura de Raíz El espacio permisible entre piezas.

Rotary Ultrasonic Machining (RUM) A nontraditional machining process whereby the grinding motion is rotary and diamond tools remove material. It is used in the development of prototypes.

Labrado Ultrasónico Rotativo Un proceso de labrado no tradicional en el cual el movimiento moledor es rotativo y las herramientas de diamante remueven el material. Se usa en el desarrollo de prototipos.

Rotate To revolve through an angle relative to an origin.

Rotar Girar por un ángulo relativo a un origen.

Round An exterior rounded intersection of two surfaces.

Redondo Una intersección exterior redondeada de dos superficies.

Routine Designs Relatively basic projects and problems used to introduce new members of an organization to the various people in the company and to the routes design information needs to flow through.

Diseños de Rutina Proyectos y problemas relativamente básicos que se usan para introducir a los nuevos miembros de la organización a las diferentes personas en la compañía y a las rutas por las cuales debe fluir la información de diseño.

Runout A feature control that limits the amount of deviation from perfect form that is allowed on surfaces or rotation through one full rotation of the object about its axis.

Excentricidad Un control de rasgo que limita la cantidad de desviación de una forma perfecta que es permisible en las superficies o en la rotación de un objeto alrededor de su eje a través de una rotación completa.

Sandblast To blow sand at high velocity with compressed air against castings or forgings to clean them.

Pulir con Chorro de Arena Soplar arena a alta velocidad con aire comprimido contra las piezas de moldes de fundición o forjadas para limpiarlas.

Sandwich Jig A variation of the plate jig in which the workpiece is held between two plates.

Guía de Sándwich Una variación de la guía de placa en la cual la pieza de trabajo se aguanta entre dos placas.

Scalars Magnitudes without directions, such as temperature and pressure.

Escalares Magnitudes sin direcciones, tales como temperaturas y presiones.

Scalene Triangle A triangle having no equal sides or angles.

Triangulo Escaleno Un triángulo que no tiene lados o ángulos iguales.

Schematic Chart A chart of a circuit or system, composed of a series of lines and symbols that show the flow of power or whatever the system carries through the circuit or system and its components.

Gráfica Esquemática Una gráfica de un circuito o un sistema, compuesta por una serie de líneas y símbolos que indica el flujo de potencia o lo que sea que el sistema carga a través del circuito o sistema y sus componentes.

Schematic Diagrams Used to show what components are contained on a circuit board and which components connect with which.

Diagramas Esquemáticos Se usan para mostrar qué componentes se contienen en una tarjeta de circuito y cuáles componentes se conectan entre sí.

Scientific Design A design process that uses concepts from physics, mechanics, materials science, chemistry, and mathematics.

Diseño Científico Un proceso de diseño que usa conceptos de física, mecánica, ciencia de materiales, química y matemática.

Scissors Drafting A technique used if any part of a set of drawings has been drawn previously; instead of redrawing this part, a slick is created and the details and other data needed from the slick are cut out and taped to a carrier sheet. A new slick is then created using the carrier sheet as the original.

Dibujo Industrial por Tijeras   Una técnica que se usa si alguna parte de un grupo de dibujos ya se ha dibujado; en lugar de volver a dibujar esta parte, se crea una reproducción en película, y los detalles y otros datos que se requieren de dicha reproducción se recortan y se adhieren a la hoja portadora. Entonces se crea una nueva reproducción en película usando la hoja portadora como original.

Scrape   To remove metal by scraping with a hand scraper, usually to fit a bearing.

Raspar   Remover metal raspándolo con raspador de mano, generalmente para que encaje en un cojinete.

Screw   A fastener that does not use a nut and is screwed directly into a part.

Tornillo   Un sujetador que no usa una tuerca y se atornilla directamente en una pieza.

Scriber Templates   Templates consisting of laminated strips with engraved grooves that are used to form letters. A tracing pin on the scriber moves in the grooves and guides the scriber in forming the letters.

Plantillas Inscribidas   Plantillas que consisten de tiras laminadas con ranuras grabadas que son usadas para formar letras. Un pasador de trazar en el puzón se mueve en las ranuras y guía el puzón en formar las letras.

Seam Weld   Like a spot weld, except that the weld is actually continuous from start to finish.

Soldadura Continua   Como una soldadura de punto, excepto que la soldadura es continua desde el principio hasta el fin.

Seams   The points at which two pieces of material are joined together by gluing, soldering, riveting, welding, and so on.

Junta   Los puntos dónde se unen dos pedazos de material por medio de pegamento, soldadura, remaches y así por el estilo.

Secondary Auxiliary View   A view that is projected directly from the primary auxiliary view; used when a primary auxiliary view is not enough to fully illustrate an object.

Vista Auxiliar Secundaria   Una vista que se proyecta directamente desde la vista primaria auxiliar; se usa cundo la vista primaria auxiliar no es suficiente para ilustrar un objeto completamente.

Section Lining   In welding, used to indicate where the weld is to be placed when a weld is not continuous and is needed in only a few areas. Otherwise, it shows where an object is sliced or cut by the cutting-plane line.

Línea de Sección   En soldadura, se usa para indicar dónde se debe colocar la soldadura cuando la soldadura no es continua y se necesita sólo en unas áreas. De lo con-

trario, muestra dónde se corta el objeto por la línea del plano cortante.

Sectional View   A view of an object obtained by the imaginary cutting away of the front portion of the object to show the interior detail.

Vista Seccional   Una vista de un objeto que se obtiene al hacer un corte imaginario y removiendo la porción frontal de un objeto para mostrar los detalles interiores.

Sector   The area of a circle lying between two radial lines and the circumference.

Sector   El área de un círculo que queda entre dos líneas radiales y la circunferencia.

Segment   The smaller portion of a circle separated by a chord.

Segmento   La porción más pequeña de un círculo separada por una cuerda.

Self-Locking Rings   A type of retaining ring; they do not require any grooves because they exert a frictional hold against axial displacement.

Anillos de Bloqueo Automático   Un tipo de anillo retenedor; no requieren ranuras porque ejercen un agarre de fricción contra el movimiento axial.

Selling Price   Prepared by the quotation department. The result of adding basic costs such as labor, engineering materials, salaries, sales costs, general overhead, and so on to come up with the cost of manufacturing the product, and then adding the desired profit to the cost.

Precio de Venta   Es preparado por el departamento de cotización. Es el resultado de sumar los costos básicos tales como mano de obra, materiales de ingeniería, salarios, gastos de venta, costos operativos y cosas por el estilo, para obtener el costo de manufacturar el producto y luego añadirle al costo la ganancia deseada.

Semicircle   Half of a circle.

Semicírculo   Mitad de un círculo.

Set Screw   A screw used to prevent motion between mating parts, such as the hub of a pulley on a shaft. It is screwed into and through one part so that it applies pressure against another part, thus preventing motion.

Tornillo de Fijación   Un tornillo para prevenir el movimiento entre piezas apareadas, tales como el centro de una polea en un eje. Se atornilla a través de una pieza de manera que aplica presión contra otra pieza, evitando de esta manera el movimiento.

Shape   To remove metal from a piece with a shaper.

Conformar   Remover metal de una pieza con una conformadora.

Shaper   A machine tool that uses a single-point cutting tool to perform the cutting operation. It is much smaller than the planer (see also *Planer*). The workpiece is mounted on a table or vise, and the cutter is driven by a reciprocating ram. Seldom used today.

Conformadora   Una maquina de labrado que usa una herramienta cortante de un solo punto para la operación cortante. Es mucho más pequeña que la acepilladora (ver también Acepilladora). La pieza de trabajo se monta sobre una mesa y el cortador lo maneja con un martinete reciprocante. Casi no se usa hoy día.

Shaving   An operation performed on a blanked part to produce an exact dimensional size or to square a blanked edge.

Rasurar   Una operación que se hace en una parte en una pieza bruta para producir un tamaño dimensional o para cuadrar un borde de una pieza bruta.

Shearing   A cutting action performed along a straight line. It uses two blades in sliding contact.

Cizallar   La acción cortante a lo largo de una línea recta. Usa dos cuchillas en contacto deslizante.

Shim   A thin piece of metal or other material used as a spacer in adjusting two parts.

Calce   Un pedazo fino de metal u otro material que se usa como espaciador para ajustar dos partes.

Shrink Rule   A standard steel rule that has the graduations marked for a specific amount of shrinkage.

Regla de Contracción   Una regla estándar de acero que tiene gradaciones marcadas para determinar la contracción.

Shrinkage   The percentage of size lost of a particular cast metal as it cools.

Contracción   El porcentaje del tamaño perdido en un metal fundido mientras se enfría.

Side View Auxiliary   An auxiliary view that is projected from the right-side view.

Vista Lateral Auxiliar   Una vista auxiliar que es proyectada desde la vista del lado derecho.

Simulation   The computer-generated emulation of a system or process.

Simulación   Una emulación generada a computadora de un sistema o proceso.

Single Thread   A screw thread type that is composed of one continuous ridge.

Rosca Sencilla   Un tipo de rosca de tornillo que consiste de un resalto contínuo.

Single-Curved Surface   A surface that can be unrolled to form a plane.

Superficie de una Curva   Una superficie que puede desenrollarse para formar un plano.

Single-Sided Boards   Printed circuit boards that have the electrical components mounted on one side and the connecting circuit etched on the other side.

Tarjetas de un solo Lado   Tarjetas de circuitos impresos que tienen componentes eléctricos montados en un lado y el circuito que los conecta grabado en el otro lado.

Size Tolerance   A tolerance where an ideal size is considered acceptable if it does not exceed certain parameters; for example, a size tolerance of ±.01″ for a piece that is ideally 4″ long means that it is acceptable if it is no longer than 4.01″ and no shorter than 3.99″.

Tolerancia de Tamaño   Una tolerancia en la cual el tamaño ideal se considera aceptable si no excede ciertos parámetros; por ejemplo, una tolerancia de tamaño de ±.01″ para una pieza que idealmente es de 4″ de largo significa que es aceptable si no es más larga de 4.01″ ni más corta de 3.99″.

Sketches   Rough drawings of what will be the final drawing.

Croquis   Dibujos rudos de lo qué será el dibujo final.

Slick   In drafting, a reproduction of a base sheet original drawing made on sensitized polyester film by fastening the two pieces of film together and exposing them to light. Used as a base sheet over which other types of plans may be overlaid while the base sheet original is being completed.

Reproducción en Película   En dibujo industrial, la reproducción de un dibujo original de hoja base que se hace sobre una película de poliéster sujetando los dos pedazos de película y exponiéndolos a la luz. Se usa como una hoja base sobre la cual se pueden sobreponer otros tipos de planos mientras la hoja base original se completa.

Slip-Renewable Bushings   A type of renewable drill bushing used where the bushing must be changed quickly. It is held in place by a lock screw but can be removed by turning it counterclockwise to the release position and lifting it out of the hole.

Cojinetes Deslizables Renovables   Un tipo de cojinete de taladro renovable que se usa dónde se debe cambiar un cojinete rápidamente. Se aguanta en su lugar por un tornillo de seguridad, pero puede removerse por girarlo de derecha a izquierda hasta la posición de libertad y después levantarlo desde su agujo.

Slitting   Performed on large sheets to produce thin stock strips. It may be performed by rollers or straight blades.

Cortar en Tiras   Se hace con hojas grandes para producir tiras finas de material. Puede hacerse por medio de rodillos o cuchillas rectas.

Slope   The angle, in degrees, that a line makes with a level plane.

Pendiente   El ángulo, en grados, que forma una línea con el plano de nivel.

Slot Weld   (see also *Plug Weld*) A weld made through an elongated hole. The symbols are the same.

Soldadura de Ranura   (ver también *Soldadura de Tapón*) Una soldadura hecha a través de un agujero alongado. Los símbolos son los mismos.

Software A set of programs, procedures, rules, and possibly associated documentation concerned with the operation of a CAD system.

Software Un grupo de programas, procedimientos, reglas y la posible documentación asociada con la operación de un sistema de CAD.

Solder To join with solder, a compound usually composed of lead and tin.

Estañar Unir con soldadura, un compuesto generalmente compuesto por plomo y estaño.

Solid Length The overall length of the spring when all coils are compressed together so they touch.

Largo Sólido El largo completo de un resorte cuando todas las espirales se comprimen de manera que estén tocándose.

Solid Modeling A type of modeling that involves actually constructing a mathematical, geometric model of a design part and displaying it on the monitor screen.

Modelado Sólido Un tipo de modelado que consiste en construir un modelo matemático y geométrico de una parte diseñada y mostrarlo en la pantalla del monitor.

Solution Description Describes the final solution, including the performance specifications and the design specifications.

Descripción de Solución Describe la solución final, incluyendo las especificaciones de funcionamiento y de diseño.

Specifications A detailed, precise presentation of something or a proposal for something. Usually divided into two main categories: performance and design. Performance specs describe what the design is supposed to accomplish; design specs describe how the product is made and the materials used.

Especificaciones Una presentación detallada y precisa de algo o una propuesta para algo. Generalmente se divide en dos categorías principales: funcionamiento y diseño. Las especificaciones de funcionamiento describen qué se supone que logre el diseño; las especificaciones de diseño describen cómo se hará el producto y qué materiales se usarán.

Sphere A closed surface, every point on which is equidistant from a common point or center.

Esfera Una superficie cerrada, cada punto de la cual está equidistante de un punto común o centro.

Spin To form a rotating piece of sheet metal into a desired shape by pressing it with a smooth tool against a rotating form.

Rotar Formar una pieza de hojalata que rota en la forma deseada prensándola con una herramienta lisa contra un molde que rota.

Spiral Bevel Gear Any bevel gear with curved teeth.

Engranaje Cónico en Espiral Cualquier engranaje cónico con los dientes curvos.

Spline A thin metal strip. Also a keyway, usually one of a series cut around a shaft or hole.

Lengüeta Una banda fina de metal. También se refiere a una muesca de posicionamiento, una de una serie de cortes alrededor de un eje o de un agujero.

Spot Welding A resistance-type weld that joins pieces of metal by welding separate spots rather than using a continuous weld.

Soldadura de Punto Un tipo de soldadura de resistencia que une pedazos de metal soldando puntos separados en lugar de formar una soldadura continua.

Spotface To produce a round spot or bearing surface around a hole, usually with a spotfacer; similar to a counterbore.

Avellanar Producir una superficie redonda o de apoyo alrededor de un agujero, generalmente con un avellanador.

Spotfaced Hole A plain through hole with a shallow counterbore deep enough to provide a flat surface for the head of a bolt or machine screw, or a nut for a threaded member.

Agujero Avellanado Un agujero pasante avellanado lo suficientemente para proveer una superficie plana a la cabeza de un tornillo o para la tuerca de un miembro con rosca.

Spring A mechanical device that is used to store and apply mechanical energy.

Resorte Un aparato mecánico que se usa para almacenar y aplicar energía mecánica.

Spring Pin A type of fastener; manufactured by cold forming strip metal in a progressive roll-forming operation. They are larger in diameter than the hole into which they are to be inserted and compress themselves as they are driven into the hole so that they exert radial forces around the entire circumference of the hole.

Pasador de Resorte Un tipo de sujetador, manufacturado por trabajo en frío en una operación de formación por rodillo. Son más grandes en diámetro que el agujero en el cual se van a insertar y se comprimen a medida que se meten a presión dentro del agujero de manera que ejercen fuerzas radiales a lo largo de toda la circunferencia del agujero.

Sprue Hole The hole in the top half of a mold flask through which the molten metal is poured to reach the gate or gates, which direct the metal to the cavity of the mold.

Orificio de Colada El orificio en la mitad superior de un molde a través del cual se vierte el metal derretido para que llegue a los canales que dirigen al metal a la cavidad del molde.

Spur Gear The most commonly used gear. It is cylindrical in form, with teeth that are cut straight across the face of the gear. All teeth are parallel to the axis of the shaft.

Engranaje de Dientes Rectos El tipo de engranaje más común. Es de forma cilíndrica, con dientes que se cortan rectos en la cara del engranaje. Todos los dientes están paralelos al eje.

Staggered Intermittent Weld A type of intermittent weld in which the welds are staggered directly opposite each other on opposite sides.

Soldadura Intermitente Alterna Un tipo de soldadura intermitente en la cual las soldaduras están alternadas directamente opuestas una de otra en lados opuestos.

Stainless Steels Steels that contain 10.5% or more chromium.

Aceros Inoxidables Aceros que contienen 10.5% o más de cromo.

Stamping A process of using dies to cut or form metal sheets or strips into a desired form.

Acuñado Un proceso de usar troqueles para cortar o formar hojas o tiras de metal en una forma deseada.

Station Lines These line segments must be assumed or chosen on the lateral surface.

Líneas de Estación Estos segmentos de línea tienen que asumirse o escogerse en la superficie lateral.

Statistical Process Control (SPC) A process that uses statistical control charts in monitoring production processes in an attempt to reduce the amount of scrap and rework.

Control de Proceso Estadístico Un proceso que usa tablas de control estadístico en el monitoreo de los procesos de producción en un intento de reducir la cantidad de desperdicio y de trabajo repetido.

Statistical Tolerancing Symbol A symbolic means of indicating that the stated tolerance is based on statistical process control (SPC).

Símbolo de Tolerancia Estadística Un medio simbólico de indicar que la tolerancia indicada está basada en un proceso de control estadístico.

Steps (in the Design Process) A set of recognizable tasks in the design process that must be done in a specific order.

Pasos (en el Proceso de Diseño) Un grupo de tareas reconocibles en el proceso de diseño que tienen que hacerse en un orden específico.

Stereolithography A process whereby solid models are created layer by layer by an electronic-laser process in a special liquid solution.

Estereolitografía Un proceso mediante el cual los modelos sólidos se crean capa por capa en un proceso electrónico por láser en una solución líquida especial.

Straight-Line Equation The equation $y = mx + b$ (for example) specifies that a dependent variable y is a function of an independent variable x, m is the slope of the straight line, and b is the intercept along the vertical axis.

Ecuación de Línea Recta La ecuación $y = mx + b$ (por ejemplo) especifica que la variable dependiente y es una función de la variable independiente x, m es la pendiente de la línea recta y b es el intercepto a lo largo del eje vertical.

Straightness A tolerance that controls the allowable variation of a surface or an axis from a theoretically perfect plane or line. It can be used to control surface elements, an axis, or a center plane.

Rectitud Una tolerancia que controla la variación permisible de una superficie o un eje de un plano o línea perfecta. Puede usarse para controlar los elementos de superficies, un eje o un plano central.

Stress Relieving Heating a metal part to a suitable temperature and holding that temperature for a determined time, then gradually cooling it in air; this treatment reduces the internal stresses induced by casting, quenching, machining, cold working, or welding.

Alivio de Tensión Calentar una pieza metálica a una temperatura adecuada y mantenerla a esa temperatura por un tiempo determinado y luego enfriarla gradualmente; este tratamiento reduce las tensiones inducidas por los procesos de fundición, enfriado, labrado, trabajado en frío o de soldadura.

Stretchout A flat pattern development for use in laying out, cutting, and folding lines on flat stock, such as paper or sheet metal, to be formed into an object.

Patrón Extendido Un patrón de desarrollo plano que se usa para establecer las líneas de corte y de dobleces en material plano, tales como papel u hojalata, para formarlos en un objeto.

Strike A compass direction of a horizontal line lying in the plane of an ore vein.

Dirección de Descubrimiento La dirección de compás de una línea horizontal que está en el plano de una vena de mineral.

Structure Anything constructed of parts.

Estructura Cualquier cosa que está formada de partes.

Stud A fastener that is a steel rod with threads at both ends. It is screwed into a blind hole and holds other parts together by a nut on its free end.

Perno Un sujetador que es una barra de metal con roscas en ambos extremos. Se atornilla en un agujero ciego y sujeta otras piezas por medio de una tuerca en su extremo libre.

Stylus A penlike device that provides input or output of coordinate data, usually for the purpose of indicating where the next entered character will be displayed.

Estilete Un aparato parecido a un bolígrafo que provee data de coordenadas de entrada y salida, generalmente con el propósito de indicar dónde se desplegará el próximo carácter.

Subassembly Drawing A drawing of two or more parts permanently fastened together; very similar to an assembly drawing.

Dibujo de Subensamblaje   Un dibujo de dos o más partes que están permanentemente unidas; muy similar al dibujo de ensamblaje.

Surface Grinder   A grinder mainly used to produce flat, angular, or special contours on flat workpieces.
Moledora de Superficie   Una moledora que se usa principalmente para producir contornos planos, angulares o especiales en las piezas de trabajos planas.

Surface Texture   The roughness, waviness, lay, and flaws of a surface.
Textura de Superficie   La aspereza, ondulación y fallas de una superficie.

Surface Weld   This weld is done when a surface must have material added to it or built up.
Soldadura de Superficie   Esta soldadura se hace cuando hay que añadirle material a una superficie o ésta tiene que acrecentarse.

Swage   To hammer metal into shape while it is held over a swage or die that fits into a hole in the swage block or anvil.
Estampar   Martillar el metal para darle forma mientras se aguanta sobre un estampador o un troquel que encaja en un agujero en el bloque de estampar o yunque.

Sweat   To fasten metal together by the use of solder between the pieces and the application of heat and pressure.
Soldadura Sudada   Unir metal por medio del uso de soldadura entre los pedazos y la aplicación de calor y presión.

Symbol   A letter, character, or schematic design representing a unit or component.
Símbolo   Una letra, carácter o diseño esquemático que representa una unidad o componente.

Tabular Dimension   A type of rectangular datum dimensioning in which dimensions from mutually perpendicular datum planes are listed in a table on the drawing instead of on the pictorial portion.
Dimensión Tabular   Un tipo de dimensionamiento de referencia rectangular en el cual las dimensiones de planos de referencia mutuamente perpendiculares se enumeran en una tabla en el dibujo en lugar de en la parte pictórica del dibujo.

Tangent Plane   This concept uses a modifying symbol with an orientation tolerance to modify the intended control of the surface.
Plano Tangente   Este concepto usa un símbolo modificado con una tolerancia de orientación para modificar el control intencionado de una superficie.

Tap   A tool used to produce internal threads by hand or machine.
Macho de Rosca   Herramienta que se usa para producir las roscas internas a mano o por máquina.

Taper Pin   A small, tapered pin for fastening, usually to prevent a collar or hub from rotating on a shaft.
Pasador Cónico   Un pasador pequeño y cónico que se usa para sujetar, generalmente para evitar que un collar o cubo rote sobre un eje.

Taper Reamer   A reaming tool that produces accurately tapered holes, as for a taper pin.
Escariador Cónico   Una herramienta para escariar que produce agujeros cónicos precisos, como para un pasador cónico.

Tapered Hole   A hole that must first be drilled .015″ smaller in diameter than the smallest diameter of the tapered hole, and then reamed.
Agujero Cónico   Un agujero que tiene que taladrarse .015″ más pequeño que el diámetro más pequeño del agujero cónico y luego tiene que escariarse.

Tempering   The process of heating a hardened workpiece to a temperature below the hardening temperature, allowing the part to soak for a specific time period, and cooling.
Templar   El proceso de calentar una pieza de trabajo endurecida a una temperatura por debajo de su temperatura de endurecimiento, permitiéndole a la pieza que se remoje por un tiempo específico y se enfríe.

Template   A guide or pattern used to mark out the work.
Plantilla   Una guía o patrón que se usa para marcar el trabajo.

Temporary Fasteners   Used when the parts will be disassembled at some future time.
Sujetadores Provisionales   Se usan cuando las piezas se desarmarán en algún momento futuro.

Tensile Force   A force that "goes away from" the joint.
Fuerza en Tensión   La fuerza que "es hacia fuera" de una unión.

Tensile Strength   The maximum load (pull) a piece supports without breaking or failing.
Resistencia a la Tensión   La carga máxima (en tensión) que una pieza aguanta sin romperse o fallar.

Tesselation   Flat shading, where any curved surfaces on a model are broken up into facets.
Mosaico   Sombrado plano donde cualquier superficie curva de un modelo se divide en facetas.

Thermal Analysis   A means of evaluating the thermal characteristics of a part or product.
Análisis Termal   Una manera de evaluar las características termales de una pieza o producto.

Thickness Gage   A standard gage used for determining what type of metal to use on any given project. It deals with the thickness—rather than the weight—of various metals used in construction. In current usage, it replaced the older but more confusing weight gage system.

Calibre de Espesor Un calibre estándar que se usa para determinar qué tipo metal usar en cualquier proyecto. Utiliza el espesor—en lugar del peso—de varios metales que se usan en construcción. En su uso actual, reemplazó al sistema más viejo y más confuso del calibre de peso.

Thinwall Section This section is any very thin object that is drawn in section. It should be filled in solid black.

Sección de Pared Fina Esta sección es cualquier objeto muy fino que se dibuja en secciones. Debería ser rellenado de negro sólido.

Through Hole As its name implies, this hole goes completely through an object.

Agujero Pasante Como su nombre implica, es un agujero que pasa completamente a través de un objeto.

Time Interval The time it takes a cam to move the follower to the designed height.

Intervalo de Tiempo El tiempo que le toma a una leva mover al palpador a la altura designada.

Time-Dependent Variables Variable characteristics that change over time; time-dependent variables are frequently best viewed as animations.

Variables Dependientes del Tiempo Características variables que cambian con el tiempo; las variables dependientes del tiempo se ven mejor como animaciones.

Timing Diagram A diagram used to visualize graphically and study the interaction of cams on the same shaft by placing the different displacement diagrams in a column with the starting and ending points in line.

Diagrama de Sincronización Un diagrama que se usa para visualizar gráficamente y estudiar la interacción de las levas en el mismo eje colocando los diferentes diagramas de desplazamiento en una columna con los puntos de comienzo y de fin en línea.

Timing Graph A graph that aids designers in coordinating electrical and electronic components in a system driven by clock pulses.

Gráfica de Sincronización Una gráfica que le ayuda a los diseñadores a coordinar los componentes eléctricos y electrónicos en un sistema controlado por pulsos de reloj.

Tin A silvery white metal that is soft and ductile. It is too weak to be used in its pure form, so it is alloyed with other metals such as antimony, copper, lead, zinc, or silver. It is most commonly used to coat steel containers used to preserve food.

Estaño Un metal plateado blancuzco que es blando y dúctil. Es muy débil para usarse en su estado puro, así que se alea con otros metales tales como antimonio, cobre, plomo, zinc o plata. Comúnmente se usa para recubrir los envases de acero que se usan para preservar alimentos.

Title Block A section on the drawing sheet whose primary function is to provide information not normally given on the drawing itself.

Bloque Titular Una sección en una hoja de dibujo cuya función principal es proveer información que normalmente no se provee en el dibujo mismo.

Tolerance Regarding locators, acceptable deviation limits determined by the specific size of the part to be machined. As a general rule, the tolerance of jigs and fixtures should be approximately 30% to 50% of the part tolerance.

Tolerancia Referente a los localizadores, son los límites aceptables de desviación determinados por un tamaño específico de la pieza a ser labrada. Como regla general, la tolerancia de las guías y aditamentos debería ser aproximadamente 30% a 50% de la tolerancia de la pieza.

Tolerancing Setting acceptable limits of deviation.

Tolerar Establecer los límites aceptables de desviación.

Tool Forces Those forces generated by the cutting tool during the machining cycle.

Fuerzas de Corte Aquellas fuerzas que son generadas por la herramienta de corte durante el ciclo de labrado.

Tooling Standard and/or special tools for producing a particular part, including jigs, fixtures, gauges, cutting tools, and so on.

Herramientas Herramientas estándares y/o especiales para producir una pieza particular, incluyendo guías, aditamentos, instrumentos de medición, herramientas cortantes, y cosas por el estilo.

Top Auxiliary View An auxiliary view projected from the top view.

Vista Superior Auxiliar Una vista auxiliar que es proyectada desde la vista superior.

Top View A viewing plane that is usually above the object and perpendicular to the other two; the plan view.

Vista Superior Un plano de vista que generalmente está encima el objeto y perpendicular a los otros dos planos; vista aérea.

Torque The rotational or twisting force in a turning shaft.

Fuerza de Torsión La fuerza rotacional o de torsión en un eje que rota.

Torsion Spring Offers resistance to a torque or twisting action. In its free state, the coils are usually either touching or very close.

Resorte de Torsión Ofrece resistencia a una fuerza de torsión o a una acción de torcer. En su estado libre, las espirales del resorte se están tocando o están muy cerca.

Total Runout A tolerance that simultaneously controls a surface related to a datum in all directions.

Excentricidad Total Una tolerancia que simultáneamente controla una superficie en relación a una referencia en todas las direcciones.

Trade-off A compromise, usually between costs and materials.

Concesión Un compromiso, generalmente entre costos y materiales.

Trammel An instrument consisting of a straightedge with two adjustable fixed points for drawing curves and ellipses.

Elipsógrafo Un instrumento que consiste de una regla con dos puntos fijos ajustables y se usa para dibujar curvas y elipses.

Transfer Cards Cards containing lettering or any frequently used type of graphic data. They are designed to fit against a straightedge for ease of alignment, and the symbols are transferred to the drawing by rubbing them with a blunt point.

Tarjetas de Transferencia Tarjetas que contienen letras o cualquier tipo de datos gráficos usados frecuentemente. Están diseñadas para encajar contra una regla para fácil alineación, y los símbolos se transfieren al dibujo frotando las tarjetas con una punta roma.

Transition Fit A condition in which two parts are toleranced so that either a clearance or an interference can result when the parts are assembled.

Encaje de Transición Una condición en la cual dos piezas se han tolerado de manera que resulte en una separación o una interferencia cuando se ensamblen las piezas.

Translucent A quality of material that passes light but diffuses it so that objects are not identifiable.

Translúcido La cualidad de un material que permite que la luz pase pero que la difunde de manera que los objetes no son identificables.

Triangle A closed plane figure with three straight sides.

Triángulo Una figura de plano cerrado con tres lados rectos.

Triangulation A method of dividing a surface into a number of triangles and then transferring each triangle's true size and shape to the development.

Triangulación Un método de dividir una superficie en un número de triángulos y luego transferir el tamaño real y forma de cada triangulo al desarrollo.

Triangulation Development The development process of breaking up an object into a series of triangular plane surfaces.

Desarrollo de Triangulación El proceso de desarrollo de romper un objeto en una serie de superficies planas triangulares.

Trilinear Graphs These graphs illustrate how differing percentages of three elements of a mixture behave or should behave.

Gráficas Trilineales Estas gráficas ilustran cómo se comportan o deberían comportarse diferentes porcentajes de tres elementos en una mezcla.

Trimetric Drawing A pictorial drawing technique that uses axes that are rotated so that each axis is drawn at a different angle, and each axis uses a different scale.

Dibujo Trimétrico Una técnica de dibujo pictórico en el que se rotan los ejes de manera que cada eje se dibuja a un ángulo diferente y a una escala diferente.

Trimming An operation performed on formed or drawn parts to remove the ragged edge of the blank.

Desorillar Una operación que se hace en las piezas formadas para remover el borde irregular de una pieza bruta.

True Distance The true actual straight distance between any two points.

Distancia Real La distancia recta real entre cualesquiera dos puntos.

True Length True length of a line is the actual straight distance between its two ends.

Largo Real El largo real de una línea es la distancia recta real entre sus dos extremos.

True Position The basic or theoretically exact position of a feature.

Posición Real La posición básica o teórica exacta de un rasgo.

True Size True actual size of a surface or object.

Tamaño Real El tamaño real de una superficie u objeto.

True Slope This slope can only be seen in the view that has a level line and where it is drawn at a true length.

Pendiente Real Esta pendiente sólo puede verse en una vista que tiene una línea nivelada y que se dibujó al largo real.

Truncated Having the apex, vertex, or end cut off by a plane.

Truncado Que tiene un ápice, vértice o un extremo cortado por un plano.

Trusses Frames that are usually two-dimensional and are made of structural shapes, such as angles, tubes, rounds, and channels.

Armadura Armazones que generalmente son de dos dimensiones y que están compuestas por formas estructurales como ángulos, tubos, círculos y canales.

T-Square A piece of equipment used with a drafting board. It is used to provide horizontal lines. Used mainly by the beginning drafter.

T Cuadrada Un pedazo de equipo que se usa con una mesa de dibujo industrial. Se usa para proveer líneas horizontales. Lo usan mayormente los delineantes principiantes.

Turn To produce on a lathe a cylindrical surface parallel to the center line.

Tornear Producir en un torno una superficie cilíndrica paralela a la línea de centro.

Two-Force Members Members that have force at each end. The forces are equal in magnitude, in opposite directions, and collinear.

Miembros de Dos Fuerzas Miembros que tienen una fuerza en cada extremo. Las fuerzas son iguales en magnitud, en direcciones opuestas y son colineles.

Typical This term, when associated with any dimension or feature, means the dimension or feature applies to the locations that appear to be identical in size and shape.

Típico Este término, al asociarlo con algún rasgo o dimensión, significa que el rasgo o dimensión se aplica a los lugares que parecen ser idénticos en tamaño y forma.

Ultrasonic Machining (USM) A nontraditional machining process that removes material using an abrasive slurry driven by a special tool vibrating at a high frequency along a longitudinal axis.

Labrado Ultrasónico a Maquina Un proceso no tradicional de labrar a máquina que remueve material usando una lechada abrasiva que la guía una herramienta especial que vibra a altas frecuencias a lo largo del eje longitudinal.

Ultrasonically Assisted Machining (UAM) A nontraditional machining process that involves assisting such traditional machining processes as drilling and turning by coupling the tool with an external source of vibrating energy.

Labrado a Maquina Asistido por Ultrasonido Un proceso no tradicional de labrar a máquina que consiste en asistir un proceso de labrado tradicional, tal como el taladrado o el torneado, acoplando la herramienta a una fuente externa de energía vibratoria.

Uniform Having the same form or characteristic; unvarying.

Uniforme Que tiene la misma forma o característica; invariante.

Uniform Acceleration The smoothest motion of all cams, and its speed is constant through the cam travel. The acceleration curve is actually a parabolic curve; the first half of the curve is exactly the reverse of the second half. This form is best for high speed.

Aceleración Uniforme El movimiento más suave de todos las levas, y su velocidad es constante a través del viaje de la leva. La curva de aceleración es en realidad una curva parabólica; la primera mitad de la curva es exactamente el reverso de la segunda mitad. Esta forma es la mejor para velocidad alta.

Uniform Resource Locator (URL) An address on the Internet. For example, the publisher's address is http://www.delmar.com.

Localizador Uniforme de Recursos Una dirección en el Internet. Por ejemplo, la dirección de la publicadora es http://www.delmar.com.

Uniform Velocity Regarding a cam, a type of motion. That is, it rises and falls at a constant speed, but the start and stop are very abrupt and rough.

Velocidad Uniforme Referente a una leva, un tipo de movimiento. Esto es, sube y baja a una velocidad constante, pero el comenzar y el detener son muy abruptos y ásperos.

Unilateral Tolerance When applied to a dimension, the tolerance applies in one direction (for example, the object may be larger but not smaller, or it may be smaller but not larger).

Tolerancia Unilateral Cuando se aplica a una dimensión, la tolerancia aplica en una dirección (por ejemplo, el objeto puede ser más grande pero no más pequeño, o puede ser más pequeño pero no más grande).

Unlimited Plane A term meaning that there is a limitless extension beyond the indicated boundaries.

Plano Ilimitado Un término que significa que hay una extensión ilimitada más allá de los bordes indicados.

Uppercase Letters A lettering system where all the letters are uppercase. It is used on the viewing plane.

Letras Mayúsculas Un sistema de letras en el cual todas las letras son mayúsculas. Se usa en el plano de vista.

Upset To form a head or enlarged end on a bar by pressure or by hammering between dies.

Recalcar Formar una cabeza o un extremo engrandecido en una barra por medio de presión o martillando entre troqueles.

Upsetting A process used to enlarge selected sections of a metal part.

Recalcadura Un proceso usado para engrandecer secciones selectas de una parte metálica.

Vector A graphical symbol that represents direction by means of an arrowhead at one end of a straight line. A directed line segment which, in computer graphics, is always defined by its two end points.

Vector Un símbolo gráfico que representa dirección por medio de una flecha en un extremo de una línea recta. Un segmento de línea con dirección que, en gráficos de computadoras, siempre se define por sus dos puntos extremos.

Vendor Catalogs Catalogs published by vendors of materials that list materials and their typical uses to aid in their selection.

Catálogos de Vendedores Catálogos publicados por los vendedores de materiales que listan materiales y sus usos típicos para ayudar en su selección.

Verification A technique used in preparing notes on drawings; the notes are first written down on a print of the drawing, and then another drafter checks the notes to ensure that they are accurate and not open to multiple interpretations.

Verificación Una técnica que se usa para preparar notas en los dibujos; las notas se escriben primero en una impresión del dibujo y luego otro redactor verifica las notas para asegurar que son precisas y no están abiertas a interpretaciones múltiples.

Vernier Scale   A small, movable scale, attached to a larger fixed scale, for obtaining fractional subdivisions of the fixed scale.
Escala Vernier   Una escala pequeña y movible que está pegada a una escala más grande, y se usa para obtener subdivisiones fraccionales de la escala fija.

Vertex   The highest point of something; the top; the summit. Plural: vertices.
Vértice   La posición más alta de algo; la cimbre; la cumbre. Plural: vértices.

Vertical   Perpendicular to the horizon.
Vertical   Perpendicular al horizonte.

Vertical Boring Mill   A machine tool commonly used to turn, bore, and face large parts in much the same way as a lathe.
Fresa de Taladrado Vertical   Una herramienta para labrar que se usa comúnmente para tornar, taladrar y encarar partes grandes de manera similar a un torno.

Vertical Turret Lathe   A variation of the vertical boring mill. Rather than using a single tool, this machine uses a turret arrangement to mount and position the cutting tools.
Torno de Torre Vertical   Una variación de la fresa de taladrado vertical. En lugar de usar una herramienta sola, esta máquina usa un arreglo de torre para montar y posicionar las herramientas cortantes.

Virtual Condition   A condition of a feature in which all tolerances of location, size, and form are considered. Also, where straightness applied to the axis or center line of a feature creates a boundary condition.
Condición Virtual   Una condición de algún rasgo en la cual se consideran todas las tolerancias de localización, tamaño y forma. También, dónde la rectitud aplicada al eje o la línea central de un rasgo crea una condición de frontera.

Virtual Reality   A technology designed to simulate either a physical or hypothetical environment. It is intended to involve all the natural senses.
Realidad Virtual   Una tecnología diseñada para simular un ambiente físico o hipotético. Su intención es involucrar a todos los sentidos.

Visible Line   One of twelve basic types of lines, it is thin and dark.
Línea Visible   Una de doce tipos básicos de líneas. Es fina y oscura.

Visualization   A pictorial representation of data for the purpose of more clearly and concisely communicating information.
Visualización   Una representación pictórica de los datos con el propósito de comunicar información más clara y concisamente.

Vorticity   The stage of a fluid in vortical (rotational) motion; the whirling of a fluid (like blood in the heart).
Vorticidad   El estado de un fluido en movimiento voraginoso (rotacional); el arremolinar de un fluido (como la sangre en el corazón).

Warped Surface   A surface that cannot be developed because it is neither a plane surface nor a curved surface.
Superficie Combada   Una superficie que no puede desarrollarse porque no es una superficie plana ni es una superficie curva.

Waste   Used here in the broadest sense and encompasses rejected parts, wasted personnel time, wasted machine time, excessive inventory, and any other situation that wastes time, material, or any other manufacturing resource.
Desperdicio   Aquí se usa en el sentido más amplio y comprende las partes rechazadas, el tiempo perdido del personal, el tiempo perdido de las máquinas, el inventario en exceso y cualquier otra situación que desperdicia tiempo, material o cualquier otro recurso manufacturero.

Water Jet Machining (WJM)   A nontraditional machining process used with nonmetal workpieces. It uses a thin, pressurized jet of water against the workpiece.
Labrado con Chorro de Agua   Un proceso no tradicional de labrado que se usa con piezas que no son de metal. Usa un chorro fino de agua presurizada contra la pieza de trabajo.

Web   A thin, flat part joining larger parts; also known as a rib.
Entremado   Una pieza plana y fina que une piezas más grandes; también se conoce como una costilla.

Weight Gage   An obsolete system for determining metal usage. It was based on the weight of wrought iron. It has been replaced by the thickness gage.
Calibre de Peso   Un sistema obsoleto para determinar el uso de metal. Estaba basado en el peso del hierro forjado. Se ha reemplazado por el calibre de espesor.

Welded Tool   The bodies offer a faster lead time, but must be machined frequently to remove any distortion caused by the heat of welding.
Herramienta Soldada   Las piezas ofrecen un tiempo de procesamiento más rápido, pero tienen que labrarse a máquina frecuentemente para remover cualquier distorsión causada por el calor de la soldadura.

Welding   The process of holding two pieces of metal together, permanently, by means of pressure or fusion-welding processes.
Soldadura   El proceso de unir dos pedazos de metal, permanentemente, por medio de procesos de soldadura a presión o por fusión.

Welding Symbol   A symbol used to convey welding instructions on a drawing. It consists of a reference line, a leader line and arrow, and, if needed, a tail.

Símbolo de Soldadura   Un símbolo que se utiliza para comunicar las instrucciones de soldadura en un dibujo. Consiste de una línea de referencia, una línea principal y una flecha y, si es necesario, una cola.

Whiteprinter   A machine that reproduces a drawing through a chemical process. An original drawing is placed on a sheet of coated whiteprint paper and exposed to light, and the exposed sheet is passed through ammonia vapor for developing. This causes the portions of the sheet that were shielded from the light by the lines on the original to turn blue or black.

Impresora de Blanco   Máquina que reproduce los dibujos por medio de un proceso químico. El dibujo original se coloca sobre una hoja de papel de impresión en blanco y se expone a la luz, y la hoja expuesta se pasa por vapor de amoníaco para revelarse. Esto causa que las partes de la hoja que fueron bloqueadas de la luz por las líneas en el dibujo original se tornen azul o negras.

Whole Depth   The total depth of a tooth space.

Profundidad Total   La profundidad total de un espacio de engranaje.

Wire Size   The diameter of the wire used to make up the spring.

Tamaño de alambre   El diámetro del alambre usado para formar un resorte.

Woodruff Key   A semicircular flat key.

Llave Woodruff   Una llave plana y semicircular.

Working Circle   Considered a distance equal to the distance from the center of the camshaft to the highest point on the cam.

Círculo de Trabajo   Se considera como una distancia igual a la distancia desde el centro del eje de levas al punto más alto de la leva.

Working Depth   The distance that a tooth projects into the mating space. It is equal to gear addendum (a) plus pinion addendum (a).

Profundidad de Trabajo   La distancia que se proyecta un diente dentro de su pieza de engrane. Es igual a la altura de cabeza del engranaje (a) más la altura de cabeza del piñón (a).

Working Drawings   A set of drawings that provides details for the production of each part, and information for the correct assembly of the finished product.

Dibujos de Trabajo   Grupo de dibujos que provee detalles para la producción de cada pieza e información para el ensamblaje correcto del producto final.

Workpiece   A part in any stage of manufacturing before it becomes a finished part.

Pieza de Trabajo   Un parte en cualquier paso de la manufactura antes de convertirse en una pieza acabada.

Workstation   The assigned location where a worker performs his or her job (i.e., the keyboard and the system display).

Estación de Trabajo   El lugar asignado dónde un trabajador hace su trabajo (esto es, el teclado y la pantalla del sistema).

Worm Gear   Used to transmit power and motion at a 90° angle between nonintersecting shafts. It is normally used as a speed reducer.

Engranaje de Tornillo   Se usa para transmitir potencia y movimiento a un ángulo de 90° entre los ejes que no se cruzan. Normalmente se usa como un reductor de velocidad.

Written Notes   Brief, carefully worded statements placed on drawings to convey information not covered or not adequately explained using graphics.

Notas Escritas   Declaraciones cortas, redactadas cuidadosamente que se colocan en los dibujos para comunicar información que no está cubierta o que no se explica adecuadamente usando gráficos.

Wrought Iron   Iron of low carbon content; useful because of its toughness, ductility, and malleability.

Hierro Forjado   Hierro con poco contenido de carbón; es útil debido a su dureza, ductilidad y maleabilidad.

Yards   Volume in cubic yards; used in construction.

Yardas   Volumen en yardas cúbicas; utilizado en construcción.

Zinc   A bluish-white metal that is widely used in manufacturing. It is produced in castings and is widely used for coating other materials. Its advantages include ease of casting, strength, ductility, and the ability to hold close tolerances.

Zinc   Un metal azul-blancuzco que se usa extensamente en la manufactura. Se produce en moldes fundidos y se usa extensamente para recubrir otros materiales. Sus ventajas incluyen la facilidad de fundición, fortaleza, ductilidad y su habilidad para mantener tolerancias pequeñas.

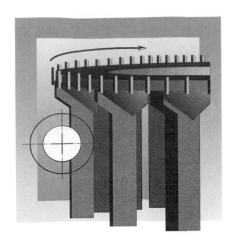

Index